TREATISE ON WATER SCIENCE

TREATISE ON WATER SCIENCE

EDITOR-IN-CHIEF
Peter Wilderer
Technische Universitaet Muenchen, Institute for Advanced Study, Munich, Germany

VOLUME 2
THE SCIENCE OF HYDROLOGY

VOLUME EDITOR
STEFAN UHLENBROOK
Department of Water Engineering, UNESCO-IHE, Delft, The Netherlands

AMSTERDAM • BOSTON • HEIDELBERG • LONDON • NEW YORK • OXFORD
PARIS • SAN DIEGO • SAN FRANCISCO • SINGAPORE • SYDNEY • TOKYO

Elsevier
Radarweg 29, PO Box 211, 1000 AE Amsterdam, The Netherlands
The Boulevard, Langford Lane, Kidlington, Oxford OX5 1GB, UK
30 Corporate Drive, Suite 400, Burlington MA 01803, USA

Copyright © 2011 Elsevier B.V. All rights reserved

No part of this publication may be reproduced, stored in a retrieval system or transmitted in any form or by any means electronic, mechanical, photocopying, recording or otherwise without the prior written permission of the publisher

Permissions may be sought directly from Elsevier's Science & Technology Rights Department in Oxford, UK: phone (+44) (0) 1865 843830; fax (+44) (0) 1865 853333; email: permissions@elsevier.com. Alternatively you can submit your request online by visiting the Elsevier web site at (http://elsevier.com/locate/permissions), and selecting *Obtaining permission to use Elsevier material*

Notice
No responsibility is assumed by the publisher for any injury and/or damage to persons or property as a matter of products liability, negligence or otherwise, or from any use or operation of any methods, products, instructions or ideas contained in the material herein, Because of rapid advances in the medical sciences, in particular, independent verification of diagnoses and drug dosages should be made

British Library Cataloguing in Publication Data
A catalogue record for this book is available from the British Library

Library of Congress Catalog Number: 2010940213

ISBN: 978-0-444-53193-3

For information on all Elsevier publications
visit our website at books.elsevier.com

Printed and bound in Italy

07 08 09 10 10 9 8 7 6 5 4 3 2 1

Working together to grow libraries in developing countries
www.elsevier.com | www.bookaid.org | www.sabre.org
ELSEVIER   BOOK AID International   Sabre Foundation

Editorial: Fiona Geraghty
Production: Gareth Steed
Edward Taylor

CONTENTS

Editorial Board		vii
Contributors to Volume 2		ix
Contents of All Volumes		xi
The Importance of Water Science in a World of Rapid Change: A Preface to the *Treatise on Water Science*		xv
Preface – The Science of Hydrology	S Uhlenbrook	1
2.01	Global Hydrology T Oki	3
2.02	Precipitation D Koutsoyiannis and A Langousis	27
2.03	Evaporation in the Global Hydrological Cycle AJ Dolman and JH Gash	79
2.04	Interception AMJ Gerrits and HHG Savenije	89
2.05	Infiltration and Unsaturated Zone JW Hopmans	103
2.06	Mechanics of Groundwater Flow M Bakker and EI Anderson	115
2.07	The Hydrodynamics and Morphodynamics of Rivers N Wright and A Crosato	135
2.08	Lakes and Reservoirs D Uhlmann, L Paul, M Hupfer, and R Fischer	157
2.09	Tracer Hydrology C Leibundgut and J Seibert	215
2.10	Hydrology and Ecology of River Systems A Gurnell and G Petts	237
2.11	Hydrology and Biogeochemistry Linkages NE Peters, JK Böhlke, PD Brooks, TP Burt, MN Gooseff, DP Hamilton, PJ Mulholland, NT Roulet, and Turner	271
2.12	Catchment Erosion, Sediment Delivery, and Sediment Quality DE Walling, SN Wilkinson, and AJ Horowitz	305
2.13	Field-Based Observation of Hydrological Processes M Weiler	339
2.14	Observation of Hydrological Processes Using Remote Sensing Z Su, RA Roebeling, J Schulz, I Holleman, V Levizzani, WJ Timmermans, H Rott, N Mognard-Campbell, R de Jeu, W Wagner, M Rodell, MS Salama, GN Parodi, and L Wang	351
2.15	Hydrogeophysics SS Hubbard and N Linde	401
2.16	Hydrological Modeling DP Solomatine and T Wagener	435
2.17	Uncertainty of Hydrological Predictions A Montanari	459
2.18	Statistical Hydrology S Grimaldi, S-C Kao, A Castellarin, S-M Papalexiou, A Viglione, F Laio, H Aksoy, and A Gedikli	479
2.19	Scaling and Regionalization in Hydrology G Blöschl	519
2.20	Stream–Groundwater Interactions KE Bencala	537

EDITORIAL BOARD

Editor-in-Chief
Peter Wilderer
Technische Universitaet Muenchen, Institute for Advanced Study, Munich, Germany

Editors

Peter Rogers
Harvard School for Engineering and Applied Sciences, Cambridge, MA, USA

Stefan Uhlenbrook
Department of Water Engineering, UNESCO-IHE, Delft, The Netherlands

Fritz Frimmel
Karlsruhe Institute of Technology, Karlsruhe, Germany

Keisuke Hanaki
The University of Tokyo, Tokyo, Japan

Tom Vereijken
European Water Partnership, Grontmij, The Netherlands

CONTRIBUTORS TO VOLUME 2

H Aksoy
Istanbul Technical University, Istanbul, Turkey

EI Anderson
WHPA, Bloomington, IN, USA

M Bakker
Delft University of Technology, Delft, The Netherlands

KE Bencala
US Geological Survey, Menlo Park, CA, USA

G Blöschl
Vienna University of Technology, Vienna, Austria

JK Böhlke
US Geological Survey, Reston, VA, USA

PD Brooks
University of Arizona, Tucson, AZ, USA

TP Burt
Durham University, Durham, UK

A Castellarin
Università degli Studi di Bologna, Bologna, Italy

A Crosato
UNESCO-IHE, Delft, The Netherlands

AJ Dolman
VU University Amsterdam, Amsterdam, The Netherlands

R Fischer
Consulting Engineers for Water and Soil Limited, Possendorf, Germany

JH Gash
VU University Amsterdam, Amsterdam, The Netherlands

A Gedikli
Istanbul Technical University, Istanbul, Turkey

AMJ Gerrits
Delft University of Technology, Delft, The Netherlands

MN Gooseff
Pennsylvania State University, University Park, PA, USA

S Grimaldi
Università degli Studi della Tuscia, Viterbo, Italy

A Gurnell
Queen Mary, University of London, London, UK

DP Hamilton
University of Waikato, Hamilton, New Zealand

I Holleman
Royal Netherlands Meteorological Institute, De Bilt, The Netherlands

JW Hopmans
University of California, Davis, CA, USA

AJ Horowitz
US Geological Survey, Atlanta, GA, USA

SS Hubbard
Lawrence Berkeley National Laboratory, Berkeley, CA, USA

M Hupfer
Leibniz-Institute of Freshwater Ecology and Inland Fisheries, Berlin, Germany

R de Jeu
VU University Amsterdam, Amsterdam, The Netherlands

S-C Kao
Oak Ridge National Laboratory, Oak Ridge, TN, USA

D Koutsoyiannis
National Technical University of Athens, Athens, Greece

F Laio
Politecnico di Torino, Torino, Italy

A Langousis
Massachusetts Institute of Technology (MIT), Cambridge, MA, USA

C Leibundgut
University of Freiburg, Freiburg, Germany

V Levizzani
ISAC-CNR, Bologna, Italy

N Linde
University of Lausanne, Lausanne, Switzerland

N Mognard-Campbell
OMP/LEGOS, Toulouse, France

A Montanari
University of Bologna, Bologna, Italy

PJ Mulholland
Oak Ridge National Laboratory, Oak Ridge, TN, USA

T Oki
The University of Tokyo, Tokyo, Japan

S-M Papalexiou
National Technical University of Athens, Zographou, Greece

GN Parodi
University of Twente, Enschede, The Netherlands

L Paul
University of Technology, Dresden, Germany

NE Peters
US Geological Survey, Atlanta, GA, USA

G Petts
University of Westminster, London, UK

M Rodell
NASA/GSFC, Greenbelt, MD, USA

RA Roebeling
Royal Netherlands Meteorological Institute, De Bilt, The Netherlands

H Rott
University of Innsbruck, Innsbruck, Austria

NT Roulet
McGill University, Montreal, QC, Canada

MS Salama
University of Twente, Enschede, The Netherlands

HHG Savenije
Delft University of Technology, Delft, The Netherlands

J Schulz
Deutscher Wetterdienst, Offenbach, Germany

J Seibert
University of Zurich, Zurich, Switzerland

DP Solomatine
UNESCO-IHE Institute for Water Education and Delft University of Technology, Delft, The Netherlands

Z Su
University of Twente, Enschede, The Netherlands

WJ Timmermans
University of Twente, Enschede, The Netherlands

JV Turner
CSIRO Land and Water, Wembley, WA, Australia

S Uhlenbrook
Department of Water Engineering, DA Delft, The Netherlands

D Uhlmann
University of Technology, Dresden, Germany

A Viglione
Technische Universität Wien, Vienna, Austria

T Wagener
The Pennsylvania State University, University Park, PA, USA

W Wagner
Vienna University of Technology, Vienna, Austria

DE Walling
University of Exeter, Exeter, UK

L Wang
University of Twente, Enschede, The Netherlands

M Weiler
Albert-Ludwigs University of Freiburg, Freiburg, Germany

SN Wilkinson
CSIRO Land and Water, Townsville, QLD, Australia

N Wright
University of Leeds, Leeds, UK

CONTENTS OF ALL VOLUMES

Volume 1 Management of Water Resources

Preface – Management of Water Resources

1.01 Integrated Water Resources Management

1.02 Governing Water: Institutions, Property Rights, and Sustainability

1.03 Managing Aquatic Ecosystems

1.04 Water as an Economic Good: Old and New Concepts and Implications for Analysis and Implementation

1.05 Providing Clean Water: Evidence from Randomized Evaluations

1.06 Pricing Water and Sanitation Services

1.07 Groundwater Management

1.08 Managing Agricultural Water

1.09 Implementation of Ambiguous Water-Quality Policies

1.10 Predicting Future Demands for Water

1.11 Risk Assessment, Risk Management, and Communication: Methods for Climate Variability and Change

Volume 2 The Science of Hydrology

Preface – The Science of Hydrology

2.01 Global Hydrology

2.02 Precipitation

2.03 Evaporation in the Global Hydrological Cycle

2.04 Interception

2.05 Infiltration and Unsaturated Zone

2.06 Mechanics of Groundwater Flow

2.07 The Hydrodynamics and Morphodynamics of Rivers

2.08 Lakes and Reservoirs

2.09 Tracer Hydrology

2.10 Hydrology and Ecology of River Systems

2.11 Hydrology and Biogeochemistry Linkages

2.12 Catchment Erosion, Sediment Delivery, and Sediment Quality

2.13 Field-Based Observation of Hydrological Processes

2.14 Observation of Hydrological Processes Using Remote Sensing

2.15 Hydrogeophysics

2.16 Hydrological Modeling

2.17   Uncertainty of Hydrological Predictions

2.18   Statistical Hydrology

2.19   Scaling and Regionalization in Hydrology

2.20   Stream–Groundwater Interactions

Volume 3   Aquatic Chemistry and Biology

Preface – Aquatic Chemistry and Biology

3.01   Sum Parameters: Potential and Limitations

3.02   Trace Metal(loid)s (As, Cd, Cu, Hg, Pb, PGE, Sb, and Zn) and Their Species

3.03   Sources, Risks, and Mitigation of Radioactivity in Water

3.04   Emerging Contaminants

3.05   Natural Colloids and Manufactured Nanoparticles in Aquatic and Terrestrial Systems

3.06   Sampling and Conservation

3.07   Measurement Quality in Water Analysis

3.08   Identification of Microorganisms Using the Ribosomal RNA Approach and Fluorescence *In Situ* Hybridization

3.09   Bioassays for Estrogenic and Androgenic Effects of Water Constituents

3.10   Online Monitoring Sensors

3.11   Standardized Methods for Water-Quality Assessment

3.12   Waterborne Parasitic Diseases: Hydrology, Regional Development, and Control

3.13   Bioremediation: Plasmid-Mediated Bioaugmentation of Microbial Communities – Experience from Laboratory-Scale Bioreactors

3.14   Drinking Water Toxicology in Its Regulatory Framework

3.15   Characterization Tools for Differentiating Natural Organic Matter from Effluent Organic Matter

3.16   Chemical Basis for Water Technology

Volume 4   Water-Quality Engineering

Preface – Water-Quality Engineering

4.01   Water and Wastewater Management Technologies in the Ancient Greek and Roman Civilizations

4.02   Membrane Filtration in Water and Wastewater Treatment

4.03   Wastewater Reclamation and Reuse System

4.04   Seawater Use and Desalination Technology

4.05   Abstraction of Atmospheric Humidity

4.06   Safe Sanitation in Low Economic Development Areas

4.07   Source Separation and Decentralization

4.08   Modeling of Biological Systems

4.09 Urban Nonpoint Source Pollution Focusing on Micropollutants and Pathogens

4.10 Constructed Wetlands and Waste Stabilization Ponds

4.11 Membrane Technology for Water: Microfiltration, Ultrafiltration, Nanofiltration, and Reverse Osmosis

4.12 Wastewater as a Source of Energy, Nutrients, and Service Water

4.13 Advanced Oxidation Processes

4.14 Biological Nutrient Removal

4.15 Biofilms in Water and Wastewater Treatment

4.16 Membrane Biological Reactors

4.17 Anaerobic Processes

4.18 Microbial Fuel Cells

4.19 Water in the Pulp and Paper Industry

4.20 Water in the Textile Industry

4.21 Water Availability and Its Use in Agriculture

Index

THE IMPORTANCE OF WATER SCIENCE IN A WORLD OF RAPID CHANGE: A PREFACE TO THE *TREATISE ON WATER SCIENCE*

The world in which we live is currently undergoing rapid changes, triggered by outstanding advances in natural sciences, medicine, and technology. As a result, the human population grows to levels never known before. Innovative communication and transportation means permit globalization of economy and urban lifestyle. Cities and city life exert an unprecedented pull. More than half of the world's population already live in urban settings – the tendency is rising.

Cities meet the expectations of immigrants, citizens, and businesses only when served by an appropriate infrastructure. Unfortunately, in many parts of the world cities grow faster than the required infrastructure can be planned, financed, and installed. In many cases, installation of water distribution networks and sewer systems, waterworks, and wastewater treatment plants is often lagging far behind schedule – be it because of the lack of financial resources or because higher priority is given to other infrastructural projects, roads, and highways, for instance.

At a larger scale, the water demand of agriculture and industry is growing overproportionally with respect to population size as people shift preference to products requiring particularly high volumes of water during the growth season or during the fabrication process, respectively. Two examples underline this statement – the shift toward meat consumption and the preference of clothing made of cotton fibers. The consumers are often unaware of the water required to raise cattle, swine, and poultry, and to keep cotton fields productive particularly when such fields are located in arid regions as is the case in Uzbekistan, for instance.

Although the water demand is increasing, worldwide, the capacity of local water resources is not. It is even decreasing in very many areas of the world, resulting from pollution of water bodies and soil, from over-abstraction of water, and from effects caused by climate change. Water deficits in municipal, industrial, and agricultural settings are the result.

In many cases, urban and agricultural areas developed in regions where *ab initio* freshwater is scarce. Drought situations caused by global warming and climate change amplify the deficit between water demand and water availability. Overabstraction of groundwater to meet the local water demand is a common but unsustainable solution to the problem of water shortage. In areas close to the ocean, over-abstraction causes seawater intrusion and subsequent increase of the salinity of groundwater. Rising sea level caused by melting of shelf ice intensifies the intrusion of seawater not only in aquifers but in estuaries as well. In addition, deterioration of ground- and surface water is caused by excess usage of fertilizers and pesticides, and by uncontrolled dumping of solid and liquid wastes onto land. Aggravation of water deficits in municipal, industrial, and agricultural environments is the result.

In the nineteenth and twentieth centuries, health problems and eutrophication caused by pollution of surface- and groundwater were recognized and solved by legal frameworks and enforcement of regulations, and by investing large amounts of money in the development and implementation of infrastructural concepts and technologies. In high-income countries, design engineers and operators of water distribution and sewer systems, water works, and wastewater treatment plants are well trained, nowadays – a major prerequisite of proper functioning of technical installations. In the medium- and low-income countries, however, responsible management of water resources and effective operation and maintenance of water technology are often foreign words.

In the twenty-first century, we are confronted with a comparably much larger and much more complex problem of water management compared to the years past. A new approach to water management and water technology is required in response to the rapid increase at the demand side, and rapid loss of capacity and quality at the supply side. A paradigm shift appears to be urgently necessary.

The old paradigm was the answer to the conditions prevailing in the highly industrialized and water-rich regions of the world. Over the past decades, considerable time was available to develop, implement, and upgrade measures capable of solving the specific local and regional problems. This, however, is not the situation we have to deal with today and in the years to come. In future, we have to support people with effective and robust water and wastewater services even if the capacity of the local water resources is critically short. To avoid evolvement of economic and societal instabilities, we are obliged to develop techniques and management concepts which can be implemented in virtually no time. We have to serve people, industry, and agriculture alike while keeping the function of aquatic and terrestrial ecosystems preserved. We need methods which are adjustable to the changing climatic boundary conditions. We need well-educated water professionals in academia, water services, and water authorities who understand the local environmental, economic, and societal framework conditions, to draw appropriate decisions and take responsible action. We need methods which are financially affordable. These methods are to be safe with respect to public health. Moreover, they must guarantee ecosystems to exert their generic life-supporting function.

The task to solve the complex issue of water-related problems caused by urbanization and lifestyle changes is challenging because of the speed of change at both the demand and the supply side, and also because of the limitations at the financial side. Business as usual is not a tolerable approach.

In the course of a shifting paradigm, we should realize that sectoral approaches (as they were usually taken in the past) are to be overcome. We need to understand that the water quantity and quality issue are inextricably linked to the issue of energy and food supply, and with the issue of land management as well. What we need is a holistic approach. Measures are to be taken which permit solution of the energy, water, and food crisis in conjunction with measures which enable restoration of the self-regulating capacity of terrestrial and aquatic ecosystems in harmony with the human demand for land.

Scientists and engineers are called to take up the task of problem solving as a challenge and as a chance. Solutions have to be found on the basis of the existing portfolio of knowledge and experience, but open minded with respect to the very local conditions in rapid transition. The *Treatise of Water Science* is to be considered as a platform on which innovative research and development may proceed. It summarizes the contemporary state of knowledge in the field of water science and technology and paves the way toward a new horizon. Serving humanity with safe water while keeping the self-regulating capacity of the aquatic ecosystems intact – this has to be our common goal.

Peter Wilderer

Preface – The Science of Hydrology

S Uhlenbrook, Department of Water Engineering, DA Delft, The Netherlands

© 2011 Elsevier B.V. All rights reserved.

The world is changing and it seems that the speed of changes is accelerating. In the overall introduction of the *Treatise on Water Sciences*, the editor-in-chief Peter Wilderer (The Importance of Water Science in a World of Rapid Change: A Preface to the Treatise on Water Science) discusses the prevailing changes, its drivers, and possible impacts on different water disciplines. A major challenge is that all changes and their various impacts are interacting with each other, although how and to what extent is often poorly understood. For scientists and practitioners, this makes the problem identification and the development of sustainable solutions for water problems a very difficult task. Therefore, it is very timely to summarize the contemporary state of the knowledge in the different fields of water sciences and technology, and to provide a platform for innovative research and development. I am pleased to conclude that this volume on hydrology is an important piece of the complex puzzle.

What is hydrology? The International Association of Hydrological Sciences (IAHS) in collaboration with UNESCO defined hydrology as the "science that deals with the water of the earth, their occurrence, circulation and distribution, their chemical and physical properties, and their reaction with their environment, including their relation to living beings." In addition, it states that hydrology is the "science that deals with the processes governing the depletion and replenishment of the water resources of the land areas of the earth, and various phases of the hydrological cycle." This is indeed a very wide definition. Many aspects of the chemical properties and interactions with the environment are part of Volume 3 of this treatise. Topics that are directly related to the management of the water resources are part of (Preface – Management of Water Resources) of this treatise. However, this volume (The Science of Hydrology) deals with all major components of the water cycle and key water-quality aspects. It also discusses the linkages to closely related disciplines.

The aims of the science of hydrology were well summarized by the Dutch Foresight Committee on Hydrological Science (KNAW, 2005) as follows:

1. to understand the mechanisms and underlying processes of the hydrological cycle and its interactions with the lithosphere, atmosphere, and biosphere;
2. to enhance our knowledge of interactions between the hydrosphere and atmosphere, the hydrosphere and lithosphere, and the hydrosphere and biosphere, thereby increasing our understanding of the role that water plays in the Earth system;
3. to quantify human impact on the past, present, and future conditions of hydrological systems; and
4. to develop strategies for sustainable use and protection of water resources, hydrological systems, and the associated environmental conditions.

The science of hydrology is special, as it holds a place, on the one hand, in the field of Earth System Sciences, where it is directly linked to earth science disciplines, such as atmospheric sciences, geomorphology, geology, soil sciences, geobiology, and ecology. On the other hand, hydrology is an applied science and, as such, a part of engineering. This makes the discipline highly relevant to the management and development of the water resources and the prediction and mitigation of water-related natural hazards (floods, droughts, landslides, etc.) to finally support life, civilization, and sustainable development. These complementary aspects of hydrology (Earth System Sciences and the basis for water management/engineering) make it an exciting and very relevant discipline. It is quite a dynamic discipline given the significant developments of the past decades; many of them are reviewed in this volume.

The volume starts with a comprehensive overview of global hydrology and the spatio-temporal variability of hydrological fluxes and water resources on a large scale. It continues with several chapters on the main variables of the water balance, such as precipitation, evaporation and interception, and stream discharge; then it goes on to discuss the storage components of groundwater, soil water, lakes, and reservoirs. Unfortunately, a chapter on snow and ice, the globally largest and regionally/locally often very important water storage component, was withdrawn at a late stage and could not be replaced in time.

The volume continues with several chapters discussing the state of the art and the possible future developments of observation methods for ground-based techniques (i.e., field-based methods, tracer techniques, and hydrogeophysics) and remote-sensing techniques. Key data analysis and modeling techniques as well as theoretical considerations are reviewed in four, mainly theoretical, chapters on scaling and regionalization, statistical methods, hydrological modeling, and uncertainty estimation techniques. The linkages between hydrology and aquatic ecology and biogeochemistry are discussed in two comprehensive chapters. Two chapters are related to the processes and issues of erosion and sedimentation as well as surface water–groundwater interactions.

The inclusion of all these topics results in a sizable volume with 20 chapters, exceeding 500 pages. However, several hydrology-related topics are not or could be only partly covered (e.g., urban hydrology, snow and ice, coastal hydrological systems, landscape evolution, and hydrogeomorphology). Perhaps this can be seen as an invitation to redo the exercise in a few years from now, and to review the latest developments in this dynamic field and strive for more completeness.

References

KNAW (2005) *Turning the Water Wheel Inside Out. Foresight Study on Hydrological Science in the Netherlands.* Amsterdam: Royal Academy of Arts and Sciences.

2.01 Global Hydrology

T Oki, The University of Tokyo, Tokyo, Japan

© 2011 Elsevier B.V. All rights reserved.

2.01.1	**Introduction**	3
2.01.1.1	The Earth System and Water	3
2.01.1.2	Water Reserves, Fluxes, and Residence Time	4
2.01.2	**Global Water Cycle**	5
2.01.2.1	Existence of Water on Earth	5
2.01.2.2	Water Cycle on the Earth	7
2.01.3	**Global Water-Balance Requirements**	9
2.01.3.1	Water Balance at Land Surface	9
2.01.3.2	Water Balance in the Atmosphere	11
2.01.3.3	Combined Atmosphere–River Basin Water Balance	11
2.01.3.3.1	Estimation of large-scale evapotranspiration	12
2.01.3.3.2	Estimation of total water storage in a river basin	12
2.01.3.3.3	Estimation of zonally averaged net transport of freshwater	12
2.01.3.4	Bottom Line of Global Water Balance	12
2.01.4	**Global Water Balance**	13
2.01.4.1	Uncertainties in Global Water-Balance Estimates	14
2.01.4.2	Water Balance and Climate	15
2.01.4.3	Annual Water Balance in Climatic Regions	16
2.01.5	**Challenges in the Global Hydrology and Research Gaps**	17
2.01.5.1	Macroscale Hydrological Modeling	17
2.01.5.2	Global Changes and Global Hydrology	19
2.01.5.3	Global Trace of Water Cycles	20
2.01.5.4	Interactions of Global- and Local-Scale Hydrology	22
2.01.5.5	Research Opportunities in Global Hydrology	22
2.01.5.6	Research Gaps in Global Hydrology	23
Acknowledgments		24
References		24

2.01.1 Introduction

2.01.1.1 The Earth System and Water

The Earth system is unique in that water exists in all three phases, that is, water vapor, liquid water, and solid ice, when compared with the forms of water on other planets. The transport of water vapor is regarded as energy transport because of the large amount of latent-heat exchange that occurs during its phase change to liquid water (approximately 2.5×10^6 J kg^{-1}); therefore, the water cycle is closely linked to the energy cycle. Even though the energy cycle on the Earth is an open system driven by solar radiation, the amount of water on the Earth does not change during shorter than geological timescales (Oki, 1999), and the water cycle itself is a closed system.

On a global scale, hydrologic cycles are associated with atmospheric circulation, which is driven by the unequal heating of the Earth's surface and atmosphere in latitude (Peixoto and Oort, 1992).

Annual mean absorbed solar energy at the top of the atmosphere is highest near the equator with approximately 300 W m^{-2}, and decreases rapidly at higher latitudes, and is approximately 60 W m^{-2} at the Arctic and Antarctic regions. Emitted terrestrial radiative energy from the Earth at the top of the atmosphere is approximately 250 W m^{-2} for the areas between 20° N and 20° S, gradually decreases at higher latitudes, and is approximately 175 W m^{-2} at the Arctic region and 150 W m^{-2} at the Antarctic region. As a consequence, the net annual energy balance is positive (absorbing) over tropical and subtropical regions between 30° N and 30° S, and negative in higher latitudes (Dingman, 2002).

Without atmospheric and oceanic circulations on the Earth, temperature differences on the Earth would have been more drastic. Temperatures in the equatorial zone would have been much higher such that the outgoing terrestrial radiation balances the absorbed solar energy, and the temperatures in the polar regions would have been much lower as well. Both the atmosphere and the ocean carry much energy from the equatorial regions toward the polar regions. In the case of atmosphere, the energy transport consists of sensible heat and latent-heat fluxes (Masuda, 1988). The global water circulation includes the latent-heat transport in which water vapor plays an active role in the atmospheric circulation. Water vapor is not a passive component of the atmosphere system;

2.01.1.2 Water Reserves, Fluxes, and Residence Time

The total volume of water on the Earth is estimated as approximately $1.4 \times 10^{18}\,\text{m}^3$, and it corresponds to a mass of $1.4 \times 10^{21}\,\text{kg}$ (**Figure 1**, revised from Oki and Kanae (2006)). Compared with the total mass of the Earth ($5.974 \times 10^{24}\,\text{kg}$), the mass of water constitutes only 0.02% of the planet, but it is critical for the survival of life on the Earth, and the Earth is called the Blue Planet and the Living Planet.

There are various forms of water on the Earth's surface. Approximately 70% of its surface is covered with salty water, the oceans. Some of the remaining areas (continents) are covered by freshwater (lakes and rivers), solid water (ice and snow), and vegetation (which implies the existence of water). Even though the water content of the atmosphere is comparatively small (approximately 0.3% by mass and 0.5% by volume of the atmosphere), approximately 60% of the area of the Earth is always covered by cloud (Rossow *et al.*, 1993). The Earth's surface is dominated by the various phases of water.

Water on the Earth is stored in various reserves, and various water flows transport water from one to another. Water flow (mass or volume) per unit time is also called water flux. The mean residence time in each reserve can be simply estimated from total storage volume in the reserve and the mean flux rate to and from the reserve:

$$T_\text{m} = V/F \qquad (1)$$

where T_m, V, and F are mean residence time, total storage, and the mean flux rate, respectively. We can also represent the distribution of flux rate of water flow that comes in and goes out from the storage Chapman, 1972). The last column of **Table 1** (simplified from the table in Korzun (1978)) presents global values of the mean residence time of water. Evidently, the water cycle on the Earth is a stiff differential system with variability on many timescales, from a few weeks to thousands of years.

The mean residence time is also important when considering water-quality deterioration and restoration, since it can be an index of how much water is turned over. Apparently, river water or surface water is more vulnerable to pollution than groundwater; however, any measure to increase water-quality recovery tends to be more efficient for river water than groundwater, and, as suggested from **Table 1**, the mean

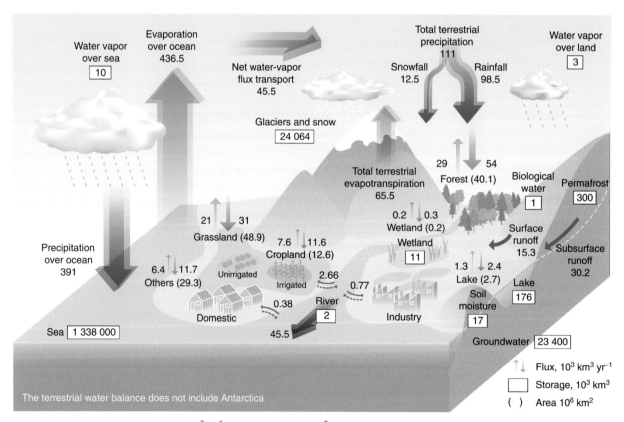

Figure 1 Global hydrological fluxes (1000 km³ yr⁻¹) and storages (1000 km³) with natural and anthropogenic cycles are synthesized from various sources. Big vertical arrows show total annual precipitation and evapotranspiration over land and ocean (1000 km³ yr⁻¹), which include annual precipitation and evapotranspiration in major landscapes (1000 km³ yr⁻¹) presented by small vertical arrows; parentheses indicate area (million km²). The direct groundwater discharge, which is estimated to be about 10% of the total river discharge globally, is included in river discharge. The values of area sizes for cropland and others are corrected from original ones. From Oki T, Nishimura T, and Dirmeyer P (1999) Assessment of annual runoff from land surface models using total runoff integrating pathways (TRIP). *Journal of the Meteorological Society of Japan* 77: 235–255 and Ok T and Kanae S (2006) Global hydrological cycles and world water resources. *Science* 313(5790): 1068–1072.

Table 1 World water reserves[a]

Form of water	Covering area (km²)	Total volume (km³)	Mean depth (m)	Share (%)	Mean residence time
World ocean	361 300 000	1 338 000 000	3700	96.539	2500 years
Glaciers and permanent snow cover	16 227 500	24 064 100	1463	1.736	1600 years
Ground water	134 800 000	23 400 000	174	1.688	1400 years
Ground ice in zones of permafrost strata	21 000 000	300 000	14	0.0216	10 000 years
Water in lakes	2 058 700	176 400	85.7	0.0127	17 years
Soil moisture	82 000 000	16 500	0.2	0.0012	1 year
Atmospheric water	510 000 000	12 900	0.025	0.0009	8 days
Marsh water	2 682 600	11 470	4.28	0.0008	5 years
Water in rivers	148 800 000	2120	0.014	0.0002	16 days
Biological water	510 000 000	1120	0.002	0.0001	A few hours
Artificial reservoirs		8000			72 days
Total water reserves	510 000 000	1 385 984 610	2718	100.00	

[a]Simplified from Table 9 of "World water balance and water resources of the earth" by UNESCO Korzun, 1978. The last column, mean residence time, is from Table 34 of the report.

residence time of river water is shorter than that of groundwater. Since the major interests of hydrologists have been the assessment of volume, inflow, outflow, and the chemical and isotopic composition of water, the estimation of mean residence time of a certain domain has been one of the major targets of hydrology. It should be recalled that the residence time estimated with isotope tracers often differs from the hydrological residence time derived from Equation (1) (Uhlenbrook et al., 2002, 2004). This is due to the fact that in the subsurface system, the diffusive exchange processes between mobile and immobile parts make the residence time usually much longer. This process is particularly important for hydrochemical processes.

2.01.2 Global Water Cycle

2.01.2.1 Existence of Water on Earth

Table 1 and Figure 1 denote the quantity of water stored in each of the reserves on the Earth. Most of the storage values given in Table 1 are taken from Korzun (1978), except for water vapor in the atmosphere which is calculated from atmospheric data (Oki et al., 1995). The various reserves of water on the Earth are discussed in the following:

The proportion of water in the ocean is large (96.5%). Even though in classical hydrology, ocean processes are traditionally excluded, the global hydrological cycle is never closed without including them. The ocean circulations carry large amounts of energy and water. The surface ocean currents are driven by surface wind stress, and the atmosphere itself is sensitive to the sea-surface temperature. Temperature and salinity determine the density of ocean water, and both factors contribute to the overturning and deep-ocean general circulation

Other major reserves are solid waters on the continent (glaciers and permanent snow cover) and groundwater. Glaciers are the accumulation of ice originated from the atmosphere, and they generally move slowly on land over a long period. Glaciers form a discriminative U-shaped valley over land, and remain moraine when they retreat. If a glacier flows into an ocean, its terminated end often forms an iceberg. Glaciers react in comparatively longer timescales against climatic change, and they also induce isostatic responses of continental-scale upheavals or subsidence in even longer timescales. Even though it is predicted that the thermal expansion of oceanic water dominates the anticipated sea-level rise due to global warming, glaciers over land are also a major concern as the cause for sea-level rise associated with global warming.

Groundwater is the subsurface water occupying the saturated zone. It contributes to runoff in the low-flow regime between storm events. Deep groundwater may also reflect the long-term climatological situation. Groundwater in Table 1 includes both gravitational and capillary water, but groundwater in the Antarctica (roughly estimated as 2×10^6 km³) is excluded. Gravitational water is the water in the unsaturated zone (vadose zone) which moves under the influence of gravity. Capillary water is water found in the soil above the water table by capillary diffusion, a phenomenon associated with the surface tension of water in soils acting as porous media. In terms of groundwater recharge, Döll and Fiedler (2008) estimated the global groundwater recharge flux to be 12 666 km³ yr^{-1} and approximately 1000 mm yr^{-1} in the Amazon region. They assumed the recharge flux is a fraction of the total runoff. Koirala (2010) estimated groundwater recharge flux by coupling a land-surface scheme, namely Minimal Advanced Treatments of Surface Interaction and RunOff (MATSIRO; see Takata et al., 2003), with a macro-scale groundwater representation Yeh et al., 2005). The global distribution of model-simulated groundwater recharge is illustrated in Figure 2(a). Total groundwater recharge flux is estimated as 31 789 km³ yr^{-1} and the value is close to the flux of subsurface runoff in Figure 1 (30 200 km³ yr^{-1}).

Soil moisture is the water that is held above the groundwater table. It influences the energy balance at the land surface, such that a lack of available moisture suppresses evapotranspiration (which consists of soil evaporation, plant transpiration, and interception loss), and changes surface albedo. Soil moisture also alters the fraction of precipitation partitioned into direct runoff and infiltration. The precipitation water becoming direct runoff cannot be evaporated from the same place, while the water infiltrated into soil may be taken up by hydraulic suction and evaporated back into the atmosphere. The global distribution of model-estimated mean

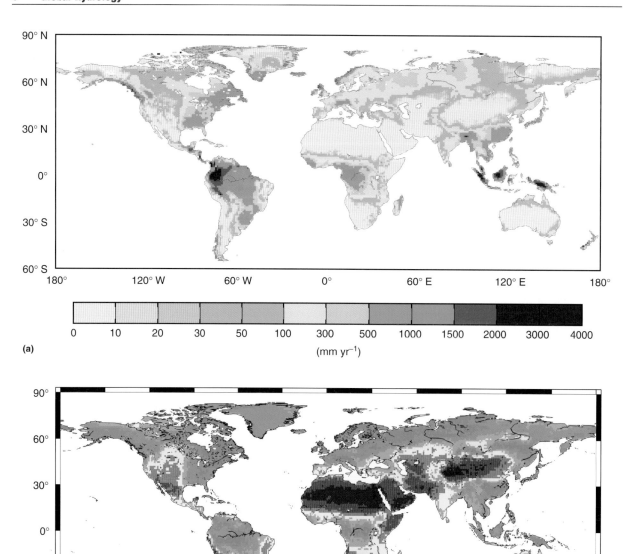

Figure 2 (a) Global map of long-term mean net groundwater recharge (mm yr^{-1}) estimated by an LSM coupled with macroscale groundwater model (Yeh and Eltahir, 2005; S. Koirala, 2010); (b) global distribution of annual mean soil wetness index estimated by 13 LSMs averaged for 1986–95 through the second phase of the Global Soil Wetness Project; (c) same as (b) but for annual precipitation (mm yr^{-1}) used in the GSWP2 based on observation; (d) same as (b) but for estimated annual runoff (mm yr^{-1}) by LSMs; (e) same as (b) but for mean river discharge (10^6 m^3 yr^{-1}); (f) annual vapor-flux convergence (mm year^{-1}) for 1989–92 (Oki et al., 1995). (g) Annual mean evapotranspiration (mm yr^{-1}) estimated as a residual of (f) and precipitation corresponding to the period; (h) same as (b) but for annual mean evapotranspiration (mm yr^{-1}) by LSMs for 1986–95. Data of (a) from Takata K, Emori S, and Watanabe T (2003) Development of minimal advanced treatments of surface interaction and runoff. *Global and Planetary Change* 38: 209–222 and Koirala S (2010) Explicit Representation of Groundwater Process in a Global-Scale Land Surface Model to Improve Hydrological Predictions. PhD Thesis, The University of Tokyo; (b) from Dirmeyer PA, Gao XA, Zhao M, Guo ZC, Oki T, and Hanasaki N (2006) GSWP-2 multimodel anlysis and implications for our perception of the land surface. *Bulletin of the American Meteorological Society* 87: 1381–1397; and (f) from Oki T, Musiake K, Matsuyama H, and Masuda K (1995) Global atmospheric water balance and runoff from large river basins. *Hydrological Processes* 9: 655–678.

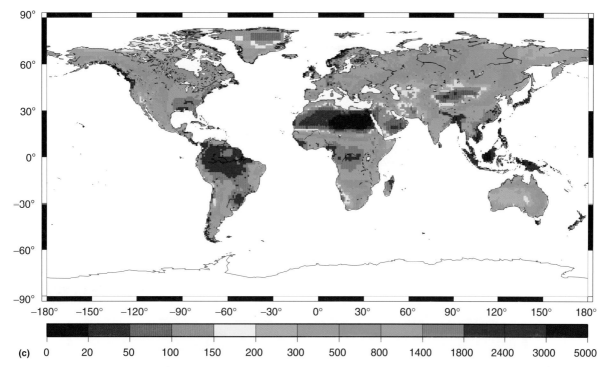

Figure 2 Continued.

soil wetness index is shown in **Figure 2(b)**. Generally, the distribution is correlated with precipitation distribution (**Figure 2(c)**) as well as with runoff distribution (**Figure 2(d)**), but the global distribution of river discharge (**Figure 2(e)**) accumulates total runoff generated in the upper watershed, and the shape of the river channels can be seen in the distribution.

The atmosphere carries water vapor, which influences the heat budget via latent-heat-exchange processes. Condensation of water vapor releases latent heat, which warms the atmosphere and affects the atmospheric general circulation. Liquid water (droplets, clouds, etc.) in the atmosphere is another result of condensation. Clouds significantly change the radiation in the atmosphere and at the Earth's surface. However, the volume of liquid (and solid) water contained in the atmosphere is relatively small, as most of the water in the atmosphere exists as water vapor. Water vapor is also the major absorber in the atmosphere of both short-wave and long-wave radiation. Precipitable water is the total water vapor in the atmospheric column integrated from land surface to the top of the atmosphere. Vertically integrated water-vapor flux convergence is a useful tool to diagnose global water balance (see **Figure 2(f)** for its global distribution).

The amount of water stored in rivers (**Figure 2(e)**) is rather tiny compared to other reserves at any time; however, the recycling speed, which can be estimated as the inverse of the mean residence time (Equation (1)), of river water (river discharge) is relatively high, and it is important because most societal applications ultimately depend on river water as a renewable and sustainable resource.

The amount of water stored transiently in a soil layer, in the atmosphere, and in the river channels is relatively minute, and the time spent through these subsystems is relatively short. However, they play a dominant role in the global hydrological cycle.

2.01.2.2 Water Cycle on the Earth

The water cycle plays many important roles in the climate system, and **Figure 1** schematically illustrates various flow paths of water in the global hydrologic system (Oki and Kanae, 2006). Precipitation is calculated from global estimates based on observations from the forcing data of the Global Soil Wetness Project (GSWP2, the second phase the project, see the discussion in Section 2.01.4) over land, and data from Climate Modeling Analysis and Prediction (CMAP; Xie and Arkin, 1996) over ocean. Land-surface fluxes, such as evapotranspiration and surface and subsurface runoff, are the estimated results from GSWP2. Differentiation of precipitation between snow and rain over land is also either estimated by land-surface models that participated in the GSWP2, or given by individual forcing determined by temperature. Values on human water withdrawals for irrigation, industry, and households are taken from Shiklomanov (1997). Water-vapor transport and its convergence are estimated using the European Centre for Medium-Range Weather Forecast (ECMWF) objective analyses, obtained as the 4-year mean from 1989 to 1992 (Oki et al., 1995). The roles of these water fluxes in the global hydrologic system are now briefly reviewed:

- Precipitation is the water flux from atmosphere to land or ocean surface. It drives the hydrological cycle over the land surface, also changes surface salinity (and temperature) over the ocean, and affects its thermohaline circulation. Rainfall

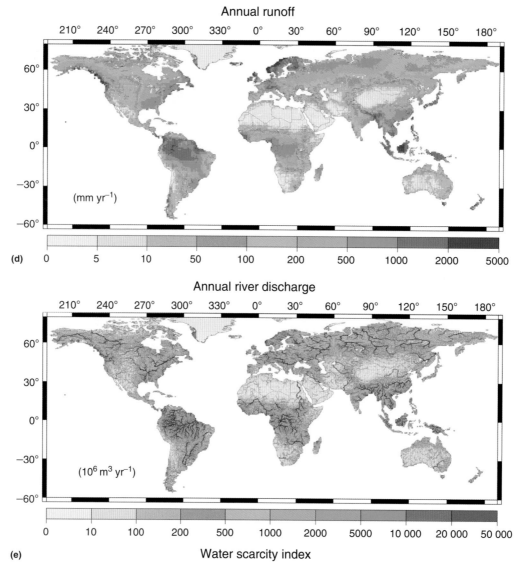

Figure 2 Continued.

refers to the liquid phase of precipitation. A part of it is intercepted by canopy over vegetated areas, and the remaining part reaches the Earth's surface as through-fall. The highly variable, intermittent, and concentrated behavior of precipitation in time and space domain compared to other major hydrological fluxes mentioned below makes the observation of this quantity and the aggregation of the process complex and difficult. Global distribution of precipitation is presented in **Figure 2(c)**. Currently, satellite-based estimates merged with *in situ* observational data have been produced and revealed to the public (e.g., Kubota *et al.*, 2007).

- Snow has special characteristics compared with rainfall. Snow may be accumulated and the surface temperature will not rise above 0 °C until the completion of snowmelt. The albedo of snow is quite high (as high as clouds). Consequently, the existence of snow changes the surface energy and water budget enormously. A snow surface typically reduces the aerodynamic roughness, and therefore may also have a dynamical effect on the atmospheric circulation and hydrologic cycle.

- Evaporation is the return flow of water from the surface to the atmosphere and the latent-heat flux from the surface. The amount of evaporation is determined by both atmospheric and hydrological conditions. From the atmospheric point of view, the partition of incoming solar energy to the surface between latent and sensible heat flux is important. Wetness at the surface influences this partition significantly because the ratio of actual evapotranspiration to the potential evaporation is reduced due to drying stress. The stress is sometimes formulated as a resistance under which evaporation is classified as hydrology driven (soil controlled). If the land surface is wet enough compared to available energy for evaporation, the condition is classified as radiation driven (atmosphere controlled).

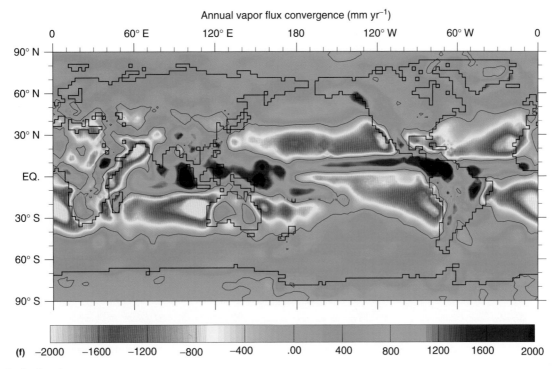

Figure 2 Continued.

- Transpiration is the evaporation of water through stomata of leaves. It has two special characteristics different from evaporation from soil surfaces. One is that the resistance of stomata is related not only to the dryness of soil moisture but also to the physiological conditions of vegetation, through the opening and closing of stomata. Another feature is that roots of plants can uptake water from deeper soil layers and transpirate the water, compared to the case of evaporation from bare soil without plants. Vegetation also modifies the surface energy and water balance by altering surface albedo and by intercepting some precipitation and evaporating this rainwater. The global distribution of total evapotranspiration is shown in **Figure 2(g)**, which is estimated using the atmospheric water-balance computation (Equation (7) in Section 2.01.3.3.1), and in **Figure 2(h)**, estimated by land-surface models (in Section 2.01.4.1).
- Runoff returns water from the land to the ocean, which may otherwise be transported in vapor phase by evaporation and atmospheric advection. The runoff into the ocean is also important for the freshwater balance and the salinity of the ocean. Rivers carry not only water mass but also sediment, chemicals, and various nutritional matters from continents to seas. Without rivers, global hydrologic cycles on the Earth are not closed. Runoff at hillslope scale is a nonlinear and complex process. Surface runoff can be generated when the intensity of rainfall or snowmelt exceeds the infiltration rate of the soil, or when precipitation falls on the saturated land surface. Saturation at land surface can be formed mostly by topographic-concentration mechanism along the hillslopes. Infiltrated water in the upper part of the hillslope flows down the slope and discharges at the bottom of the hillslope. Due to the highly variable heterogeneity of topography, soil properties (such as hydraulic conductivity and porosity) and precipitation, basic equations such as Richard's equation, which are fairly valid at a point scale or hillslope scale, cannot be directly applied in the macroscale because of the nonlinearity involved. The global distributions of runoff and river discharge are illustrated in **Figures 2(d)** and **2(e)**.

The global water cycle integrates these components, which consist of the state variables (precipitable water, soil moisture, etc.) and the fluxes (precipitation, evaporation, etc.).

2.01.3 Global Water-Balance Requirements

The conservation law of water mass in any arbitrary control volume indicates the water balance. In this section, the water balance over land, for an atmospheric column, and its combination is presented (Oki, 1999). Some applications of these water-balance equations for estimating some of the water-balance components are introduced as well.

2.01.3.1 Water Balance at Land Surface

In the field of hydrology, river basins have commonly been selected for study, and water balance has been estimated using ground observations, such as precipitation, runoff, and storage in lakes and/or groundwater. The water balance over the land is described as

$$\partial S/\partial t = P - E - R_o - R_u \qquad (2)$$

where S represents the water storage within the area, t is the time, $(\partial S/\partial t)$ is the change of total water storage with time,

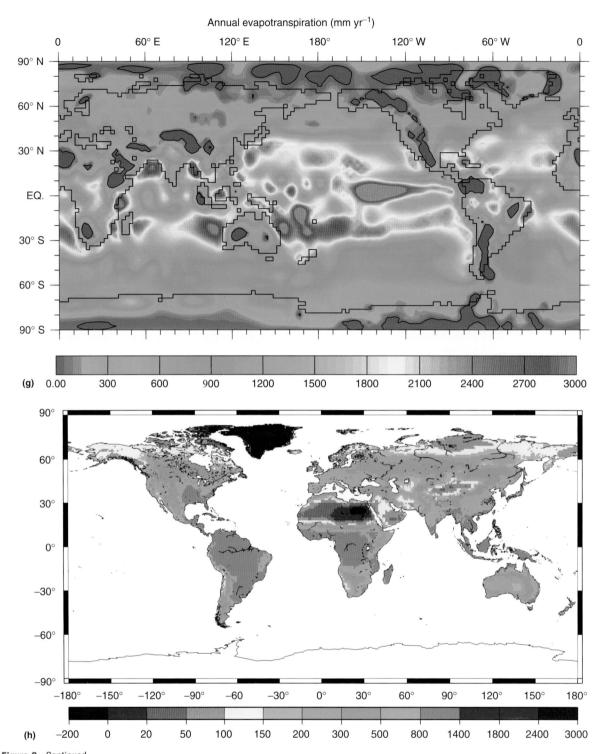

Figure 2 Continued.

P is the precipitation, E is the evapotranspiration, R_o is the surface runoff, and R_u is the groundwater movement (all fluxes above are given in the unit volume per time step). S includes snow accumulation in addition to soil moisture, groundwater, and surface-water storage including retention water within the control volume. The control volume is defined by the area of interest over the land with its bottom generally at the impermeable bedrock. These terms are shown in **Figure 3(a)**. Equation (2) implies that water storage over land is increased by precipitation, and decreased by evapotranspiration, surface runoff, and groundwater movement.

If the considered area of water balance is set within an arbitrary boundary, R_o represents the net outflow of water from this area (i.e., the total outflow minus total inflow from surrounding areas). Although, in general, it is not easy to estimate groundwater movement R_u, the net flux per unit area within a large area is expected to be comparatively small. If all groundwater movement is considered to be that observed at the gauging point of a river ($R_u = 0$), then Equation (2) becomes

$$\partial S / \partial t = P - E - R_o \qquad (3)$$

This assumption is generally valid at the outlet of a catchment. In most cases, surface runoff R_o becomes river discharge through the transport of river-channel network. The river discharge is an integrated quantity over the whole catchment and can be observed at a downstream point in contrast to other fluxes, such as P and E, which have to be spatially measured.

2.01.3.2 Water Balance in the Atmosphere

Atmospheric water-vapor flux convergence contains water-balance information that can complement the traditional hydrological elements such as precipitation, evapotranspiration, and discharge. The basic concepts as well as the application of atmospheric data to estimate terrestrial water balance were first presented by Starr and Peixöto (1958). The atmospheric water balance for a column of atmosphere from the bottom at land surface to the top of the atmosphere is described by

$$\partial W / \partial t = Q + (E - P) \qquad (4)$$

where W represents the precipitable water (i.e., column storage of water vapor) and Q is the convergence of water-vapor flux in the atmosphere (all fluxes given in the unit volume per time step). Since the atmospheric water content in both solid and liquid phases are generally small, only the water vapor is considered in Equation (4). The balance is schematically illustrated in **Figure 3(b)**, which describes that the water storage in an atmospheric column is increased by the lateral convergence of water vapor and evapotranspiration through the bottom of the column (i.e., land surface), and decreases by the precipitation falling out from the bottom of the atmosphere column to the land.

2.01.3.3 Combined Atmosphere–River Basin Water Balance

Since there are common terms in Equations (3) and (4), they can be combined as

$$\partial W / \partial t + Q = (P - E) = \partial S / \partial t + R_o \qquad (5)$$

Figure 3(c) illustrates the balance in Equation (5), and shows that the difference of precipitation and evapotranspiration is equal to the sum of the decrease of atmospheric water-vapor storage and lateral (horizontal) convergence, and also to the sum of the increase of water storage over the land and runoff. Theoretically, Equation (5) can be applied for any control volume of land area combined with the atmosphere above, even though the practical applicability depends on the accuracy and availability of atmospheric and hydrologic information.

The following further assumptions are often employed in annual water-balance computations:

- Annual change of atmospheric water-vapor storage is negligible ($(\partial W / \partial t) = 0$).
- Annual change of water storage at the land is negligible ($(\partial S / \partial t) = 0$).

With these assumptions, Equation (5) simplifies into

$$Q = (P - E) = R_o \qquad (6)$$

If a river basin is considered as the water-balance region, R_o is simply the discharge from the basin. The simplified Equation (6) demands that the water-vapor convergence, that is, precipitation minus evaporation and net runoff should balance over the annual period when the temporal change of all storage terms can be neglected.

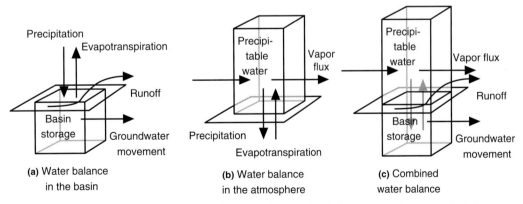

Figure 3 (a) Terrestrial water balance, (b) atmospheric water balance, and (c) combined atmosphere–land surface water balance. (a), (b), and (c) correspond to Equations (2), (4), and (5), respectively.

2.01.3.3.1 Estimation of large-scale evapotranspiration

Generally, it is not an easy task to obtain large-scale evapotranspiration E based on observations except for the annual timescale in which E can be estimated as the residual of P and R_o. However, the combined water balance can help estimate E at a shorter timescale, for example, monthly. Note that Equation (4) can be rewritten as

$$E = \partial W / \partial t - Q + P \qquad (7)$$

which can be applicable over a period shorter than a year, unlike the assumption in Equation (6). If atmospheric and precipitation data are available over a short timescale such as a month or a day, evapotranspiration can be estimated at the corresponding timescales; however, it is also subject to severe limitations imposed by the data accuracy. The region over which the evapotranspiration is estimated is not limited to a river basin; rather, it depends on the scale and the associated accuracy of the available atmospheric and precipitation data.

The global distributions of precipitation P, integrated water-vapor convergence Q, and evapotranspiration E estimated by Equation (7) are presented in **Figures 2(c), 2(f)**, and **2(g)**, respectively. The zonal mean precipitation P, integrated water-vapor convergence Q, and evapotranspiration E estimated by Equation (7) are presented in **Figures 4(a)–4(c)**. As can be seen from the zonal mean precipitation along the midlatitude, storm tracks over the North Pacific and Atlantic oceans are stronger in December–January–February (DJF) than in June–July–August (JJA). The Intertropical Convergence Zone (ITCZ) is enhanced in JJA, when the southeastern part of the Asian continent is covered by the southwest monsoon rainfall. The distribution of E is less dependent on the latitude and has smaller seasonal changes compared to P (see also Trenberth and Guillemot, 1998). The mean evapotranspiration in tropical areas is approximately 4 mm d^{-1}.

2.01.3.3.2 Estimation of total water storage in a river basin

Total terrestrial water storage, as the sum of surface water (such as river water, snow water, and water in lakes), soil moisture, and groundwater, is generally difficult to estimate on the global scale. Combining Equations (3) and (4) yields

$$\partial S / \partial t = \partial W / \partial t + Q - R_o \qquad (8)$$

which indicates that the change of water storage in the control volume over the land can in principle be estimated from the atmospheric and runoff data. Although an initial value is required to obtain the absolute value of storage, the atmospheric water balance can be useful in estimating the seasonal change of total water storages in large river basins.

2.01.3.3.3 Estimation of zonally averaged net transport of freshwater

The meridional (north–south direction) distribution of the zonally averaged annual energy transports by the atmosphere and the ocean has been evaluated, even though there are quantitative problems in estimating such values (Trenberth and Solomon, 1994). However, the corresponding distribution of water transport has not often been studied, although the cycles of energy and water are closely related. Wijffels et al. (1992) used values of atmospheric water-vapor convergence Q from Bryan and Oort (1984) and discharge data from Baumgartner and Reichel (1975) to estimate the freshwater transport by oceans and atmosphere, but their results seem to have large uncertainties and they did not present the freshwater transport by rivers.

The annual freshwater transport in the meridional (north–south) direction can be estimated from Q and river discharge with geographical information such as the location of river mouths and basin boundaries (Oki et al., 1995). The estimated result is shown in **Figure 5**, in which it shows that in the case of oceans net transport is the residual of northward and southward freshwater flux by all ocean currents globally, and it cannot be compared directly with individual ocean currents such as the Kuroshio and the Gulf Stream. Transports by the atmosphere and by the ocean have almost the same absolute values for most of the latitudes, but with a different sign. The transport by rivers is about 10% of these other fluxes globally (there may be an underestimation because average Q tends to be smaller than average river discharge observed at the global land surface). The negative (southward) peak by rivers at 30°S is mainly due to the Parana River in South America, and the peaks at the equator and 10°N are due to rivers in South America, such as the Magdalena and Orinoco. Large Russian rivers, such as the Ob, Yenisey, and Lena, carry freshwater toward the north between 50 and 70°N.

These results indicate that the hydrological processes over land play non-negligible roles in the climate system, not only by the exchange of energy and water at the land surface, but also through the transport of freshwater by rivers which affects water balance of the oceans and forms a part of the hydrological circulation on the Earth among the atmosphere, land, and oceans.

2.01.3.4 Bottom Line of Global Water Balance

Water balance on the global scale with consideration of land and ocean areas separately can be expressed as

$$P_l - E_l - \partial S_l / \partial t = R = -(P_o - E_o) + \partial S_o / \partial t \qquad (9)$$

where P, E, S, and R represent precipitation, evapotranspiration, total water storage, and continental runoff, respectively, with the subscript l indicating values for land and o for ocean.

For the steady state, the temporal changes of S_l and S_o can be neglected and Equation (9) becomes

$$P_l - E_l = R = -(P_o - E_o) \qquad (10)$$

which indicates that continental runoff can be estimated as a residual of total evapotranspiration (E_o) and precipitation (P_o) over the ocean. It could be an effective method to estimate continental runoff since a macroscale estimation of precipitation and evaporation is relatively easier over ocean than over the land. If precise estimates of the long-term trend of global mean precipitation and evapotranspiration over both land and ocean are available, there is a potential to infer the trend of the water stored over the land or ocean as suggested by Equation (9).

From Equation (10)

$$P_e = E_e \qquad (11)$$

can be derived which states that precipitation all over the Earth $P_e = P_l + P_o$ and evapotranspiration all over the Earth $E_e = P_l + P_o$ should be identical under the conditions when the temporal changes of water storage over land and ocean are negligible.

2.01.4 Global Water Balance

The values quoted in **Table 1** and **Figure 1** are estimated based on various observations with some assumptions in order to obtain global perspectives. These values are sometimes different in other references probably because the source of observed data, methodology of estimation, and assumptions are different. In some cases, global water balances are estimated using empirical relationship of evapotranspiration to precipitation in each latitude belt (Baumgartner and Reichel, 1975).

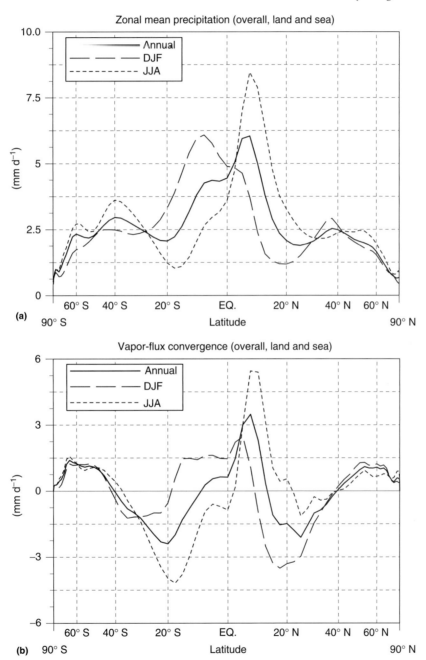

Figure 4 (a) Meridional distribution of precipitation (P) for mean over land and sea; (b) same as (a) but for vapor-flux convergence (Q); (c) same as (a) but for evapotranspiration (E) calculated as a residual of P and Q. Data of (a) from Xie P and Arkin PA (1996) Analyses of global monthly precipitation using gauge observations, satellite estimates, and numerical model predictions. *Journal of Climate* 9: 840–858 and (b) from Oki T, Musiake K, Matsuyama H, and Masuda K (1995) Global atmospheric water balance and runoff from large river basins. *Hydrological Processes* 9: 655–678.

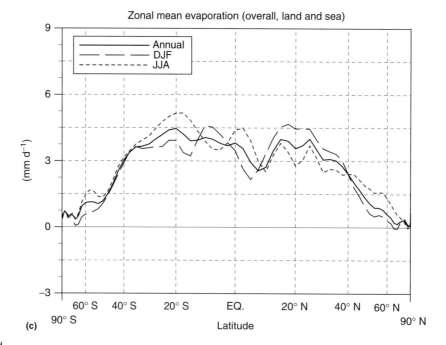

Figure 4 Continued.

Under an international research project, the land-surface models (LSMs) were used to estimate global water and energy balances for 1986–95 in order to obtain global distribution of surface soil moisture, which is not easy to obtain but relevant for understanding the land–atmosphere interactions (Dirmeyer *et al.*, 2006). The project was called the second phase of GSWP2 and its goal was to produce state-of-the-art global data sets of land-surface fluxes, state variables, and related hydrologic quantities.

2.01.4.1 Uncertainties in Global Water-Balance Estimates

In GSWP2, meteorological forcing data are hybrid products of the National Center for Environmental Prediction (NCEP)/ Department of Energy (DOE) reanalysis data and observational data based on *in-situ* and satellite monitoring, provided at a 3-hourly time step for a period of 13.5 years from July 1982 to December 1995. The first 3.5 years' data are used for spin up. The land-surface parameters are specified from the Earth Resources Observation and Science Data Center (EDC) for the land-cover data and the International Geosphere–Biosphere Programme Data Information System (IGBP-DIS) for the soil data. Both land-surface parameters and meteorological forcing are at 1° resolution for all land grids excluding Antarctica.

Figure 6(a) illustrates the model-derived global water balance over the global land excluding ice, glacier, and lake. The numeric in the box corresponds to the 10-year mean annual value of eight LSMs participated in the GSWP2 project (Oki *et al.*, 2005). All the simulations were performed using identical forcing data given by $1° \times 1°$ longitudinal and latitudinal grid boxes, and typical time steps of the calculations are 5 min to 3 h. The vertical ranges shown above and below the boxes indicate the maximum and minimum values in the interannual variation of mean annual value among the eight LSMs. The horizontal ranges shown left and right to the boxes indicate the maximum and minimum values of the intermodel variation of the 10-year mean value of eight LSMs. Generally, intermodel variation exceeds interannual variation, which suggests that the uncertainty associated with model selection is larger than the sampling error of estimating global water balance. In the case of rainfall, intermodel variation is small because identical precipitation forcing was given to LSMs so that the differences among LSM estimates were merely caused by the rain/snow judgment made by each modeling group.

The advantage of using models to estimate global water balance is the capability to have more detailed insights than using observations. For example, snow over the land excluding ice and glacier areas is approximately 10% of total precipitation, and the ratio of surface runoff and subsurface runoff is approximately 2:3 in **Figure 6(a)**, but the latter is approximately 1:2 in **Figure 1**. This is because the model results used for calculating average value are different for **Figures 1** and **6(a)**. In some LSMs, neither surface nor subsurface runoff process is considered, which is the reason why the minimum values are zero. Even though at present it is difficult to assess the validity of these breakdowns, due to the lack of observations on the partitions between snow and rain or between surface and subsurface runoff, such model-based estimates will stimulate scientific interest to collect and compile global information on these important hydrological quantities in the future.

Further, evapotranspiration was estimated separately by bare-soil evaporation (E_s), evaporation from intercepted water on leaves (E_i), evaporation from open water (E_w), and transpiration from vegetation (E_t), as shown in **Figure 6(b)**, even though the intermodel variations are quite large partially because some LSMs do not consider all of these components of evapotranspiration. Even though the values in **Figure 6(b)** are not definitive, it is interesting to see that bare-soil evaporation

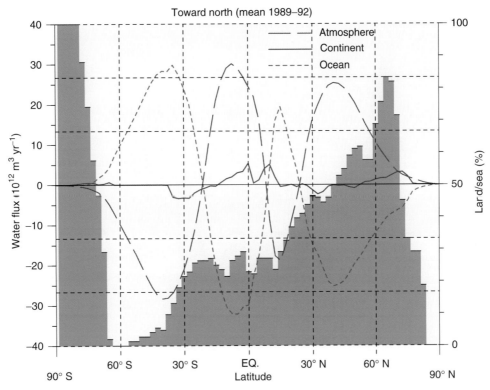

Figure 5 The annual freshwater transport in the meridional (north–south) direction by atmosphere, ocean, and rivers (land). Water-vapor flux transport of $20 \times 10^{12} \, m^3 \, yr^{-1}$ corresponds to approximately 1.6×10^{15} W of latent-heat transport. Shaded bars behind the lines indicate the fraction of land at each latitudinal belt.

and transpiration from vegetation are closely comparative, and interception loss is approximately 10% of the total evapotranspiration. It would be interesting if these estimates can be revised and validated by certain observation-based measures, and intermodel discrepancies can therefore be reduced.

2.01.4.2 Water Balance and Climate

Based on observed precipitation (Xie and Arkin, 1996) and river-discharge records archived at the Global Runoff Data Centre, annual water balance for most river basins worldwide was estimated (Oki et al., 1999), and this is presented in **Figure 7**. The ordinate in **Figure 7** is the residual of long-term mean annual precipitation and runoff, and this annual loss should correspond to long-term mean annual evapotranspiration. Different symbols are used for plotting: red stars indicate water balance of the river basins where gauging stations are located between 20° S and 20° N. The plus symbols indicate the river basins where gauging stations are located between 20° and 40° in both hemispheres, and the blue circles are 40° or higher. The line connects the mean precipitation and annual loss for each 5° latitudinal belt. As seen, even though the scatter is large, approximately 70% of precipitation is evapotranspirated in high-latitude river basins. On the other hand, mean evapotranspiration in tropical river basins is approximately $1000 \, mm \, yr^{-1}$ with less dependency on annual precipitation. Such analyses on the relationship between P and E have long been used for estimation of global water balance, for example, in Baumgartner and Reichel (1975). It is also clear from **Figure 7** that river basins with annual precipitation less than $800 \, mm \, yr^{-1}$ have marginal amounts of river runoff since most precipitation is used for evapotranspiration. In these river basins, evapotranspiration is mainly controlled by the availability of water (water controlled), and this is in contrast to tropical river basins with precipitation higher than $1000 \, mm \, yr^{-1}$ where annual evapotranspiration is limited by the available energy (energy- or radiation controlled).

Budyko (1974) proposed an equation

$$E/P = [\zeta(\tanh 1/\zeta)(1 - \cosh \zeta + \sinh \zeta)]^{(1/2)} \quad (12)$$

where E and P are annual evapotranspiration and precipitation respectively, and the Budyko's dryness index is defined as

$$\zeta = R_n/lP \quad (13)$$

where R_n and l are net radiation and the coefficient of latent heat, respectively. This equation is derived by considering that the E/P should be asymptotic to 1.0 for dry regions (large ζ) since E should be less than P, and E/P should be asymptotic to R_n/lP for wet regions (small ζ) since E should be less than R_n/l.

Budyko's equation (12) is conceptual, but it can provide a realistic water balance as shown in **Figure 8**. Mean water balance averaged for each $0.2\zeta(=R_n/lP)$ bin estimated by an LSM corresponds fairly well with the curve according to the Budyko's equation (12), even though large scatters are found in the plots of each 1° longitudinal and latitudinal grid box. Yang et al.

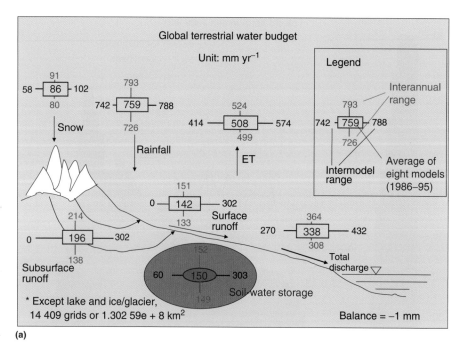

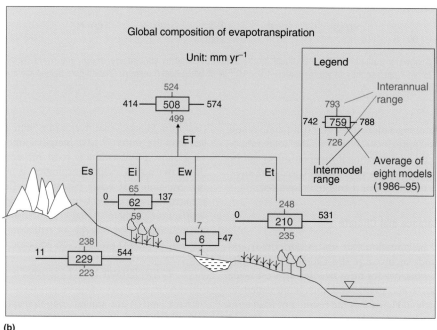

Figure 6 (a) Global terrestrial water balance averaged for 1986–95 estimated by eight LSMs in boxes. Interannual variation range (vertical) for 1986–95 and intermodel discrepancies (horizontal) among eight models are presented for the annual mean estimates; (b) same as (a) but for global composition of evapotranspiration.

(2009) analyzed annual water balance in 99 river basins in China and concluded that this scatter can at least partially be explained by the vegetation cover in the river basin.

2.01.4.3 Annual Water Balance in Climatic Regions

Annual water balances estimated by GSWP2 were analyzed and each $1° \times 1°$ grid box was classified into one of the following six climatic regions according to the Budyko's dryness index ζ and annual precipitation:

- arid region: $\zeta > 4.0$;
- semiarid region: $4.0 > \zeta > = 2.0$;
- semi-humid region: $2.0 > \zeta > = 1.2$;
- humid region: $1.2 > \zeta > = 0.7$;
- tropical humid region: $0.7 > = \zeta$ and annual precipitation larger than 2000 mm yr^{-1}; and

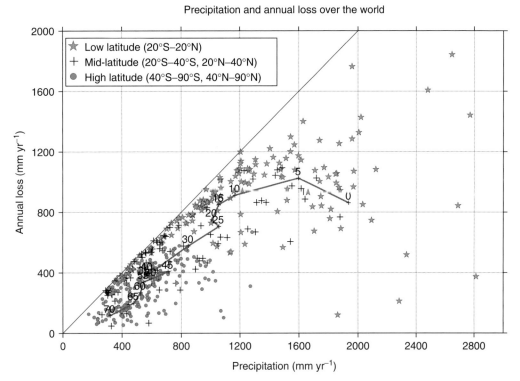

Figure 7 Annual water balance in major river basins. Annual loss is estimated as a residual of annual precipitation and observed runoff in catchments of 250 gauging stations of river discharge. From Oki T, Nishimura T, and Dirmeyer P (1999) Assessment of annual runoff from land surface models using total runoff integrating pathways (TRIP). *Journal of the Meteorological Society of Japan* 77: 235–255.

- very humid region: $0.7 > = \zeta$ and annual precipitation less than $2000\,\mathrm{mm\,yr^{-1}}$.

The six classified regions are illustrated in **Figure 9** along with the ice-covered region. The differentiation is difficult between tropical humid region and very humid region only by using the Budyko's dryness index ζ. Therefore, annual precipitation is considered in the classification. It is interesting to see that ζ is similar in both tropical and high-latitude regions in addition to the Asian monsoon region. Perhaps, it is also necessary to consider the seasonal change of major water-balance terms for better differentiation of these regions.

Long-term mean water balance for each climatic region classified by ζ is presented in **Figure 10(a)**. Separation of mean annual precipitation into evapotranspiration, surface runoff, and subsurface runoff is also illustrated. The sum of these corresponds to annual precipitation, and the ratio of annual evapotranspiration to precipitation is close to the mean ζ of each region. Slightly negative evapotranspiration in the ice region indicates net sublimation in the region, that is, the land surface obtains energy from the atmosphere through sublimation.

Evapotranspiration is also divided into four components: bare-soil evaporation, transpiration, evaporation from intercepted water, and evaporation from open water, as presented in **Figure 10(b)**. Note that not all the LSMs that participated in GSWP2 have considered all of these four components, and some of these values could be underestimated. Transpiration and evaporation from intercepted water are proportional to the vegetation biomass in each region, and it is interesting to note that the magnitude of bare-soil evaporation is relatively uniform globally than other components, except for arid, very humid, and ice regions.

2.01.5 Challenges in the Global Hydrology and Research Gaps

2.01.5.1 Macroscale Hydrological Modeling

The development of macroscale hydrological models was a serious topic of discussion among Japanese scientists researching land–atmosphere interaction studies in the early 1990s when Global Energy and Water Experiment (GEWEX) Asian Monsoon Experiment (GAME) was under preparation. Two approaches were identified: one to expand a conventional microscale rainfall-runoff hydrological model into a macroscale model, which can run on the continental scale with a detailed energy balance and vegetation representation, and the other to enhance hydrological processes in LSMs and couple them with horizontal water-flow processes, particularly with river flow.

A river-routing scheme was hence developed with a global-flow direction map, and named as the total runoff integrating pathways (TRIPs) (Oki and Sud, 1998). Such a river-routing scheme can be coupled with any LSM, and can also be used as a post-processor integrating the runoff estimated by LSMs into

18 Global Hydrology

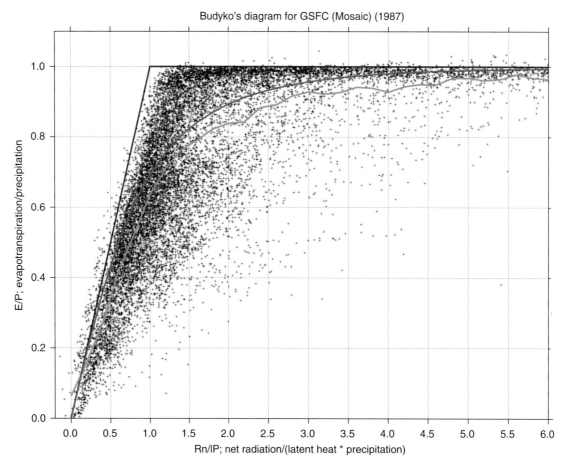

Figure 8 Annual energy and water balance of 1° longitude and latitudinal grids calculated by an LSM for 1987 (Koster *et al.*, 1999). Plots indicate the energy and water balances in each grid box and green line presents the mean E/P for each 0.2 $\zeta = R_n/IP$ bin. The red curve indicates Equation (12). From Budyko MI (1974) *Climate and Life*, Miller DH (trans.). San Diego, CA: Academic Press.

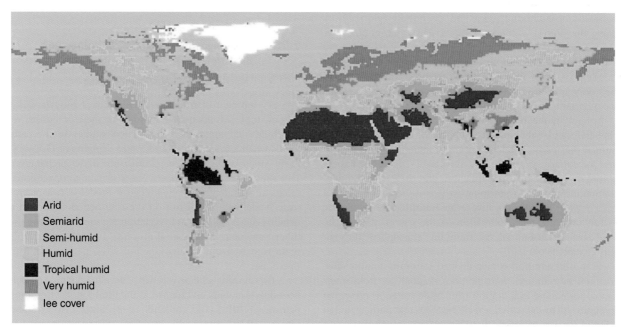

Figure 9 Classification of energy and water-balance regime using Budyko's dryness index $\zeta = R_n/IP$.

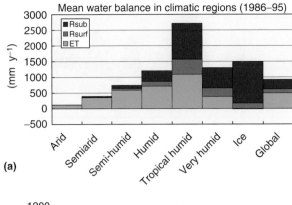

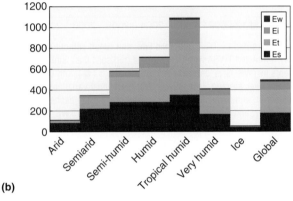

Figure 10 Components of (a) the surface hydrologic balance, and (b) total evapotranspiration estimated under the second phase of the Global Soil Wetness project. From Dirmeyer PA, Gao XA, Zhao M, Guo ZC, Oki T, and Hanasaki N (2006) GSWP-2 multimodel anlysis and implications for our perception of the land surface. *Bulletin of the American Meteorological Society* 87: 1381–1397.

river discharge (Oki *et al.*, 1999). The first version of TRIP adopted a primitive fixed-velocity scheme (Miller *et al.*, 1994), while the variable-velocity version was also developed (Ngo-Duc *et al.*, 2007). TRIP was coupled in some general circulation model (GCM) projections used in the Intergovernmental Panel on Climate Change (IPCC) Assessment Report 4 (AR4) to identify the impact of climate change on hydrological cycles (Falloon and Betts, 2006), and there have been some studies of future assessment on the world water resources and global flood disasters utilizing the TRIP model, as well (Oki and Kanae, 2006; Hirabayashi and Kanae, 2009). Further, Kim *et al.* (2009) underscored the importance of river component in terrestrial water storage (TWS) variation over global river basins. To reduce simulation uncertainty, ensemble simulations were performed with multiple precipitation data, and a localized Bayesian model averaging technique was applied to TRIP simulation.

Figure 11 shows that river storage not only explains different portions of total TWS variations, but also plays different roles in different climatic regions. It is the most dominant water-storage component in wet basins (e.g., the Amazon) in terms of amplitude, and it acts as a buffer which smoothes the seasonal variation of total TWS especially in snow-dominated basins (e.g., the Amur). It signifies that model simulation of TWS may not be able to properly reproduce the amplitude and seasonal pattern of observed TWS variation by Gravity Recovery and Climate Experiment (GRACE; see Tapley *et al.*, 2004) without an appropriate representation of river-storage component. The dominant role of river storage was already indicated in a pilot study which compared total TWS changes estimated by the atmospheric water-balance method and a GCM simulation coupled with TRIP in the Amazon river basin (Oki *et al.*, 1996). However, the message was not undoubtedly convincing until recent years when satellite-observed GRACE data became available. Using a geodesy approach, Han *et al.* (2009) employed a fixed-velocity version of TRIP in the Amazon river basin and its vicinity, and compared the model simulations to the residual of GRACE raw measurements derived from removing all the gravity-influencing factors except for the horizontally moving water. They demonstrated that the optimal flow velocity of TRIP in the Amazon varies between rising and falling water levels.

2.01.5.2 Global Changes and Global Hydrology

Macroscale hydrological models have also been developed in response to the societal expectations for solving current and future world water issues. There has been a concrete demand for the information on how much water resources are available now and what kinds of changes are projected for the future. Conventionally, available freshwater resources are commonly defined as annual runoff estimated by historical river-discharge data or water-balance calculation (Lvovitch, 1973; Baumgartner and Reichel, 1975; Korzun, 1978). Such an approach has been used to provide valuable information on the annual freshwater resources in many countries. Atmospheric water balance using the water-vapor flux-convergence data could be alternatively used to estimate global distribution of runoff owing to the advent of atmospheric reanalysis and data-assimilation system (Oki *et al.*, 1995).

Simple analytical water-balance models have been widely used to estimate global-scale available freshwater resources in the world since the beginning of this century (Alcamo *et al.*, 2000; Vörösmarty *et al.*, 2000; Döll *et al.*, 2003; Rockström *et al.*, 2009). Later, LSMs were used to simulate global water cycles (Oki *et al.*, 2001; Dirmeyer *et al.*, 2006). Changes of hydrological cycles during the twenty-first century associated with anticipated climate change are projected (Milly *et al.*, 2005; Nohara *et al.*, 2006), and their impacts on the demands and supplies of global water resources are estimated assuming future climatic and social-change scenarios (Arnell, 2004; Alcamo *et al.*, 2007; Shen *et al.*, 2008). Some of those estimations were calibrated by multiplying an empirical factor for the river basins, where observed river-discharge data are available. However, recent model simulations with advanced climate forcing data can estimate global runoff distribution with adequate accuracy without the need of calibration (Hanasaki *et al.*, 2008a). Changes in extreme river discharge are also of interest now (Hirabayashi *et al.*, 2008).

Several recently developed macroscale hydrological models for water-resource assessment also include a reservoir-operation scheme (Haddeland *et al.*, 2006; Hanasaki *et al.*, 2006) in order to simulate the real hydrological cycles, which are significantly influenced by anthropogenic activities and modified from natural hydrological cycles even on the global

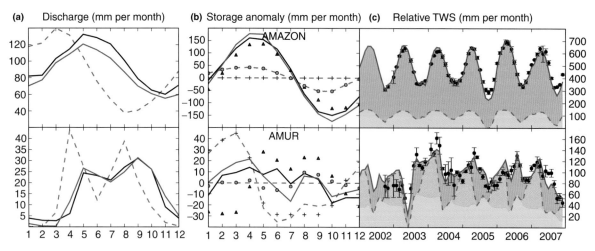

Figure 11 (a) Seasonal variations of gauged discharge (black solid line), discharge routed by TRIP (red solid line) and runoff without routing (gray dashed line). (b) Seasonal variations of GRACE observed TWSA (black solid line), simulated TWSA with river storage (red solid line), simulated TWSA without river storage (gray dashed line), and the major water storage components in TWS. Gray crosses (+), green circles (●), and blue triangles (▲) represent the individual storage component of snow water, soil moisture, and river storage, respectively. (c) Interannual variations of relative TWS: the GRACE observation (black dot), simulation with river storage (red solid line), and simulation without river storage (gray dashed line). Each area shaded by blue, gray, and green indicates the portion of river storage, snow water, and soil moisture in the simulated relative TWS, respectively. From Kim H, Yeh P, Oki T, and Kanae S (2009) The role of river storage in the seasonal variation of terrestrial water storage over global river basins. *Geophysical Research Letters* 36: L17402 (doi:10.1029/2009GL039006).

scale in the Anthropocene (Crutzen, 2002). An integrated water-resources model is further coupled with a crop-growth submodel, which can simulate the timing and quantity of irrigation requirement, and a submodel, which can estimate environmental flow requirement (Hanasaki et al., 2008a). Such an approach is able to assess the balances of water demand and supply on a daily timescale, and a gap in the subannual distribution of water availability and water use can be detected in the Sahel, the Asian monsoon region, and southern Africa, where conventional water-scarcity indices such as the ratio of annual water withdrawal to water availability and available annual water resources per capita (Falkenmark and Rockström, 2004) cannot properly detect the stringent balance between demand and supply (Hanasaki et al., 2008b).

2.01.5.3 Global Trace of Water Cycles

Numerical models can be associated with a scheme tracing the origin and flow path as if tracing the isotopic ratio of water (Yoshimura et al., 2004; Fekete et al., 2006). Such a flow-tracing function of water in the integrated water-resources model (Hanasaki et al., 2008a) considering the sources of water withdrawal from stream flow, medium-size reservoirs, and nonrenewable groundwater, in addition to precipitation to croplands, enabled the assessment of the origin of water producing major crops (Hanasaki et al., 2010). **Figure 12(a)** illustrates the ratio of blue water to total evapotranspiration during cropping period in irrigated croplands. Here, the blue water is defined as that part of evapotranspiration originating from irrigation, whereas the green water is from precipitation (see Falkenmark and Rockström, 2004). **Figure 12(a)** shows a distinctive geographical distribution in the dependence on blue water. In addition, the ratios of the source of blue water for stream flow including the influence of large reservoirs, medium-size reservoirs, and nonrenewable and nonlocal blue water are shown in **Figures 12(b)–12(d)**. Areas highly dependent on nonrenewable and nonlocal blue water were detected in Pakistan, Bangladesh, western part of India, north and western parts of China, some regions in the Arabian Peninsula, and the western part of the United States through Mexico. Cumulative nonrenewable and nonlocal blue-water withdrawals estimated by the model correspond fairly well with the country statistics of total groundwater withdrawals (Hanasaki, 2009, personal communication), and such an integrated model has the ability to quantify the global virtual water flow (Allan, 1998; Oki and Kanae, 2004) or water footprint Hoekstra and Chapagain, 2007) through major crop consumption (Hanasaki et al., 2010).

It is apparent that these achievements illustrate how the framework of global off-line simulation of LSMs, coupled with lateral river-flow model and/or anthropogenic activities, driven by the best-available meteorological forcing data, such as precipitation and downward radiation, is relevant for estimating global energy and water cycles, validating the estimates and sometimes the quality of forcing data with independent observations, and improving the models themselves. There are attempts to utilize this framework for assessing the impacts of climate change on future hydrological cycles which would demand adaptation measures in water-resources management, flood management, and food production. For such purposes, it is necessary to develop reliable forcing data for the future, based on GCM projections probably with bias corrections and spatial and temporal downscaling, as well as developing best estimates for the future boundary conditions for hydrological simulations such as vegetation type and land use/land cover.

Figure 13 summarizes the concept of how the forcing data have a large impact on the accuracy of the output from theories, equations, and numerical models. Certainly, the spatial

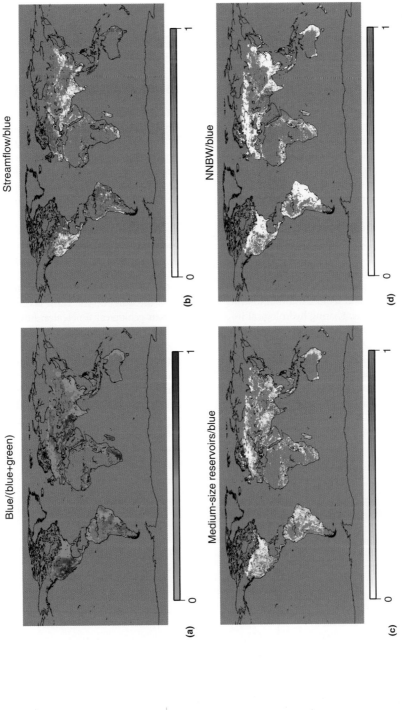

Figure 12 (a) The ratio of blue water to the total evapotranspiration during a cropping period from irrigated cropland (the total of green and blue water). The ratios of (b) streamflow, (c) medium-size reservoirs, and (d) nonrenewable and nonlocal blue-water withdrawals to blue water. From Hanasaki N, Inuzuka T, Kanae S, and Oki T (2010) An estimation of global virtual water flow and sources of water withdrawal for major crops and livestock products using a global hydrological model. *Journal of Hydrology* 384: 232–244.

and temporal boundary conditions as well as the field information characterizing the target region, such as land cover and land use, are critically important to obtain reasonable estimates. Therefore, it is recommended to examine the quality of forcing data and boundary conditions along with revising the core theory, principle equations, and model code, or tuning model parameters, for hydrological modeling. It is particularly important for global-scale studies since uncertainties in the forcing data and field information are relatively large compared to those at the local scale. It should also be recalled that the applicability and accuracy of the model are highly dependent on the specific temporal and spatial scales.

2.01.5.4 Interactions of Global- and Local-Scale Hydrology

As described in Oki et al. (2006), water, as one of the major components of the global climate system, is one of the major cross-cutting axes in the Earth system science. The water cycle transports various materials, such as sediments and nutrients, from land to the oceans. Water resources are closely related to energy, industry, and agricultural production. Of course, water is indispensable for life and supports health. Water issues are related to poverty, and providing access to safe drinking water is one of the key necessities for sustainable development. In the past, water issues remained local issues; however, due to the increase in international trade and mutual interdependence among countries, water issues now often need to be dealt with on the global scale, and require information on global hydrology for their solutions. Sharing hydrological information relating to the transboundary river basins and shared aquifers will help reduce conflict between relevant countries, and quantitative estimates of recharge amounts or potentially available water resources will assist in implementing sustainable water use.

Global hydrology is not merely concerned with global monitoring, modeling, and world water-resources assessment. Owing to recent advancements in global earth-observation technology and macroscale modeling capacity, global hydrology can now provide basic information on the regional hydrological cycle which may support the decision-making process in the integrated water-resources management.

It should also be examined to what extent such a framework of off-line simulation of LSMs can be applied to finer spatial and temporal scales, such as 1-km grid spacing and hourly time interval. For such research efforts, observational data from regional studies can provide significant information, and efforts to integrate data sets from various regional studies should be promoted.

2.01.5.5 Research Opportunities in Global Hydrology

There are still challenging scientific issues to be resolved in global hydrology. For instance, separation of rain and snow on the global scale, as illustrated in **Figure 1**, is of interest. However, this is based on numerical-model estimates and it is quite uncertain about the accuracy of the numbers or even the ratio between rain and snow (here, approximately 8:1). The situation is similar for the ratio between surface and sub-surface runoff (here, approximately 1:2). It is reasonable to infer direct groundwater discharge from land to oceans to be a residual between total runoff estimated by numerical models subtracted by observation-based total discharge; however, the accuracy of such an estimate is yet to be determined. As described in Section 2.01.1, it is believed that total amount of the water on the Earth is conserved on a timescale shorter than geological timescales; however, do we have any reliable observational evidence for it? Our knowledge seems to be incomplete on the water exchange between Earth's surface and mantle, although a recent report suggested that the lower mantle may store 5 times more water than the ocean (Murakami et al., 2002).

The direct groundwater discharge to the ocean, estimated to be about 10% of total river discharge globally (Church, 1996),

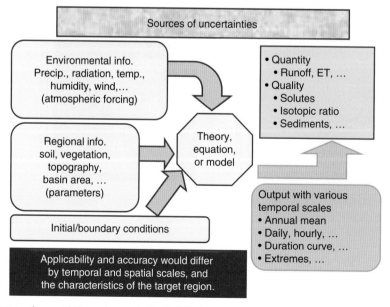

Figure 13 Sources and causes of uncertainties and errors in the hydrological simulations and estimations.

is included in the river discharge plotted in **Figure 1**. According to the model estimates by Koirala (2010), groundwater recharge in the grid boxes of $1° \times 1°$ in longitude and latitude near the coastal line is totally $5890 \, km^3 \, yr^{-1}$, and it is approximately 13% of the annual total runoff from the continents to oceans of $45\,500 \, km^3 \, yr^{-1}$. Even though the total amount of groundwater recharge depends on the grid size defining the coastal region, is it just a coincidence or is it that groundwater recharge in the coastal areas provides a good proxy of direct groundwater runoff from continents to oceans?

Scientific interest in global hydrology has increased since the 1980s as public awareness of global environmental issues and process interactions, such as El Niño events and anthropogenic climate change, has risen. Further work is required for the detection and attribution of the present-day hydrological changes – in particular, changes in the intra-seasonal water availability and in the frequency and magnitude of extreme events. Uncertainties in the future projections of hydrological quantities and water qualities should be reduced and quantitative accuracy should be enhanced, particularly for runoff regime changes. It is further anticipated that feedbacks of mitigation and adaptation measures for the concerned climate change on water sectors will be assessed for proper policymaking.

In addition to anthropogenic climate change, it is of interest in global hydrology to assess the impact of land-use change, such as deforestation and urbanization, human activities, such as reservoir construction and water withdrawals for irrigation, industry, and domestic water uses, and emission of air pollutants which would have been suppressing weak rainfall and modulate precipitation occurrence weekly.

2.01.5.6 Research Gaps in Global Hydrology

Hydro-meteorological monitoring networks need to be maintained and further expanded to enable the analysis of hydro-climatic trends at the local level and the improvement in the accuracy of predictions, forecasts, and early warnings. As clearly illustrated in **Figure 14** (from Oki et al., 1999), global hydrological simulations are relatively poor in areas with little *in-situ* observations. Basic observational networks on the ground are critically indispensable for proper monitoring and modeling of global hydrology; however, it is also required to utilize remotely sensed information in order to fill the gaps of *in-situ* observations. One of the current trends in the utilization of remote sensing technique is the so-called data

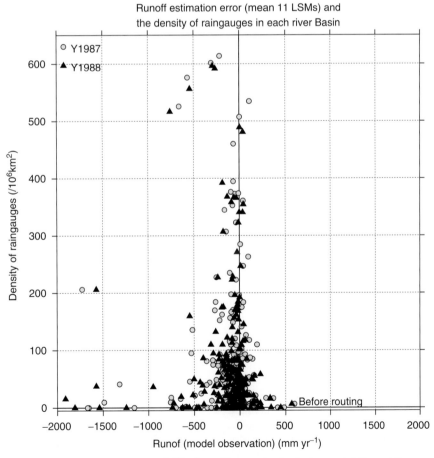

Figure 14 Comparisons between the density of rain gauge ($/10^6 \, km^2$) used in preparing the forcing precipitation and the mean bias error ($mm \, yr^{-1}$) of 11 LSMs for 150 major river basins in the world in 1987 and 1988. From Oki T, Nishimura T, and Dirmeyer P (1999) Assessment of annual runoff from land surface models using total runoff integrating pathways (TRIP). *Journal of the Meteorological Society of Japan* 77: 235–255.

assimilation, which optimally merges observations (not limited to remote sensing) with numerical-model estimates.

Reliable observational data are essentially necessary not only as the forcing data for global hydrological modeling, but also for the validation of model estimates. River discharge and soil-moisture data are critically important for global hydrological studies. However, contributions from the operational agencies in the world are not yet well established and need to be enhanced.

Isotopic ratio of rainwater is collected and the data set is available through the Global Network of Isotopes in Precipitation (GNIP) by the International Atomic Energy Agency (IAEA). These observational data can be used to investigate the routes and mechanisms of how the evaporated waters from the ocean surface are transported and precipitated at particular locations, which can be estimated by water-vapor transfer models with the consideration of isotopic processes (Yoshimura et al., 2008). The information on the isotopic ratio of river waters is not well organized and cannot be used easily on the global scale.

Current global hydrological modeling has not yet integrated most of the latest achievements in process understanding and regional- or local-scale modeling studies. Global simulation of solutes and sediments are emerging. Both natural and anthropogenic sources should be considered, as for nutrients, and probably such models should be coupled with agricultural models which simulate crop growth. Moreover, it is rather difficult that river ice jams can be simulated properly by current hydrologic models. For both problems of water quality and ice jams, a proper simulation of water temperature in rivers and lakes are requisite. Moreover, the representation of groundwater has been rather simple in global hydrological modeling.

Some of the above issues have not been emphasized in the current global hydrological modeling due to their relatively minor impact on the climatic feedbacks from the land surface to the atmosphere. It is meaningful to recall that global hydrology has been developed in cooperation with global climate modeling; however, it is time to develop global hydrological models primarily for responding to the demands of understanding the hydrological cycle on the land surface and for supporting better water-resources management. From this point of view, integrated hydrological and water-resource models, which consider natural and anthropogenic water cycles and are coupled with crop models and reservoir operation models in order to provide a more realistic impact assessment and support the design of practical adaptation measures, should be developed and implemented.

Acknowledgments

This study was funded by Grants-in-Aid for Scientific Research from the Japan Society for the Promotion of Science (19106008 >), the Global Environment Research Fund of the Ministry of Environment (S-5), and the Innovation Program of Climate Change Projection for 21st Century of the Ministry of Education, Culture, Sports, Science, and Technology of Japan.

References

Alcamo J, Floerke M, and Maerker M (2007) Future long-term changes in global water resources driven by socio-economic and climatic changes. *Hydrological Sciences Journal* 52(2): 247–275.

Alcamo J, Henrichs T, and Rösch T (2000) World water in 2025 – global modeling and scenario analysis for the World Commission on water for the 21st century. *Kassel World Water Series, Technical Report No. 2.* Kassel, Germany: Centre for Environmental Systems Research, University of Kassel.

Allan JA (1998) Virtual water: A strategic resource. Global solution to regional deficits. *Groundwater* 36(4): 545–546.

Arnell NW (2004) Climate change and global water resources: SRES scenarios and socio-economic scenarios. *Global Environmental Change* 14: 31–52.

Baumgartner A and Reichel E (1975) *The World Water Balance: Mean Annual Global, Continental and Maritime Precipitation, Evaporation and Runoff.* 179pp. Munich: Ordenbourg.

Bryan F and Oort A (1984) Seasonal variation of the global water balance based on aerological data. *Journal of Geophysical Research* 89: 11717–11730.

Budyko MI (1974) *Climate and Life.* Miller DH (trans.). San Diego, CA: Academic Press.

Chapman TG (1972) Estimating the frequency distribution of hydrologic residence time. In: World Water Balance, vol. 1, pp. 136–152. IASH-UNESCO-WMO.

Church TM (1996) An underground route for the water cycle. *Nature* 380: 579–580.

Crutzen PJ (2002) Geology of mankind-the Anthropocene. *Nature* 415(23). doi:10.1038/415023a

Dingman SL (2002) *Physical Hydrology*, 2nd edn. Upper Saddle River, NJ: Prentice-Hall. 646pp

Dirmeyer PA, Gao XA, Zhao M, Guo ZC, Oki T, and Hanasaki N (2006) GSWP-2 multimodel anlysis and implications for our perception of the land surface. *Bulletin of the American Meteorological Society* 87: 1381–1397.

Döll P and Fiedler K (2008) Global-scale modeling of groundwater recharge. *Hydrology and Earth System Sciences* 12: 863–865.

Döll P, Kaspar F, and Lehner B (2003) A global hydrological model for deriving water availability indicators: Model tuning and validation. *Journal of Hydrology* 270(1): 105–134.

Falkenmark M and Rockström J (2004) *Balancing Water for Humans and Nature.* 247pp. London: Earthscan.

Falloon PD and Betts RA (2006) The impact of climate change on global river flow in HadGEM1 simulations. *Atmospheric Science Letters* 7: 62–68.

Fekete BM, Gibson JA, Aggarwal P, and Vörösmarty CJ (2006) Application of isotope tracers in continental scale hydrological modeling. *Journal of Hydrology* 330: 444–456.

Haddeland I, Lettenmaier DP, and Skaugen T (2006) Effects of irrigation on the water and energy balances of the Colorado and Mekong river basins. *Journal of Hydrology* 324: 210–223.

Han S-C, Kim H, Yeo I-Y, et al. (2009) Dynamics of surface water storage in the Amazon inferred from measurements of inter-satellite distance change. *Geophysical Research Letters* 36: L09403 (doi:10.1029/2009GL037910)

Hanasaki N, Inuzuka T, Kanae S, and Oki T (2010) An estimation of global virtual water flow and sources of water withdrawal for major crops and livestock products using a global hydrological model. *Journal of Hydrology* 384: 232–244.

Hanasaki N, Kanae S, and Oki T (2006) A reservoir operation scheme for global river routing models. *Journal of Hydrology* 327: 22–41.

Hanasaki N, Kanae S, Oki T, et al. (2008a) An integrated model for the assessment of global water resources – part 1: Model description and input meteorological forcing. *Hydrology and Earth System Sciences* 12: 1007–1025.

Hanasaki N, Kanae S, Oki T, et al. (2008b) An integrated model for the assessment of global water resources – part 2: Applications and assessments. *Hydrology and Earth System Sciences* 12: 1027–1037.

Hirabayashi Y and Kanae S (2009) First estimate of the future global population at risk of flooding. *Hydrological Research Letters* 3: 6–9.

Hirabayashi Y, Kanae S, Emori S, Oki T, and Kimoto M (2008) Global projections of changing risks of floods and droughts in a changing climate. *Hydrological Sciences Journal* 53(4): 754–772.

Hoekstra AY and Chapagain AK (2007) Water footprints of nations: Water use by people as a function of their consumption pattern. *Water Resource Management* 21: 35–48.

Kim H, Yeh P, Oki T, and Kanae S (2009) The role of river storage in the seasonal variation of terrestrial water storage over global river basins. *Geophysical Research Letters* 36: L17402 (doi:10.1029/2009GL039006)

Koirala S (2010) Explicit Representation of Groundwater Process in a Global-Scale Land Surface Model to Improve Hydrological Predictions. PhD Thesis, Graduate School of Engineering, The University of Tokyo, Tokyo, Japan.

Korzun VI (1978) *World Water Balance and Water Resources of the Earth – Studies and Reports in Hydrology*, 25, Paris: UNESCO.

Koster RD, Oki T, and Suarez MJ (1999) Assessing success in the offline validation of land surface models. *Journal of the Meteorological Society of Japan* 77: 257–263.

Kubota T, Shige S, Hashizume H, et al. (2007) Global precipitation map using satellite-borne microwave radiometers by the GSMaP project. *IEEE Transactions on Geoscience and Remote Sensing* 45(7): 2259–2275.

Lvovitch MI (1973) The global water balance. *Transactions – American Geophysical Union* 54: 28–42.

Masuda K (1988) Meridional heat transport by the atmosphere and the ocean: Analysis of FGGE data. *Tellus* 40A: 285–302.

Miller JR, Russell GL, and Caliri G (1994) Continental-scale river flow in climate models. *Journal of Climate* 7: 914–928.

Milly PCD, Dunne KA, and Vecchia AV (2005) Global pattern of trends in streamflow and water availability in a changing climate. *Nature* 483: 347–350.

Murakami M, Hirose K, Yurimoto H, Nakashima S, and Takafuji N (2002) Water in Earth's lower mantle. *Science* 295(5561): 1885–1887.

Ngo-Duc T, Oki T, and Kanae S (2007) A variable streamflow velocity method for global river routing model: Model description and preliminary results. *Hydrology and Earth System Sciences Discussions* 4: 4389–4414.

Nohara D, Kitoh A, Hosaka M, and Oki T (2006) Impact of climate change on river discharge projected by multi-model ensemble. *Journal of Hydrometeorology* 7: 1076–1089.

Oki T (1999) The global water cycle. In: Browning K and Gurney R (eds.) *Global Energy and Water Cycles*, pp. 10–27. Cambridge: Cambridge University Press.

Oki T, Agata Y, Kanae S, Saruhashi T, Yang DW, and Musiake K (2001) Global assessment of current water resources using total runoff integrating pathways. *Hydrological Sciences Journal* 46: 983–995.

Oki T, Hanasaki N, Shen Y, Kanae S, Masuda K, and Dirmeyer PA (2005) Global water balance estimated by land surface models participated in the GSWP2. In: *The 19th Conference on Hydrology, AMS Annual Meeting*. San Diego, CA, USA, January 2005.

Oki T and Kanae S (2004) Virtual water trade and world water resources. *Water Science and Technology* 49(7): 203–209.

Oki T and Kanae S (2006) Global hydrological cycles and world water resources. *Science* 313(5790): 1068–1072.

Oki T, Kanae S, and Musiake K (1996) River routing in the global water cycle, *GEWEX News*, WCRP, International GEWEX Project Office, 6: 4–5.

Oki T, Musiake K, Matsuyama H, and Masuda K (1995) Global atmospheric water balance and runoff from large river basins. *Hydrological Processes* 9: 655–678.

Oki T, Nishimura T, and Dirmeyer P (1999) Assessment of annual runoff from land surface models using total runoff integrating pathways (TRIP). *Journal of the Meteorological Society of Japan* 77: 235–255.

Oki T and Sud YC (1998) Design of total runoff integrating pathways (TRIP) – a global river channel network. *Earth Interactions* 2: 1–37.

Oki T., Valeo C., and Heal K. (eds.) (2006) *Hydrology 2020*, ISSN 0144-7815. Wallingford: IAHS.

Peixöto JP and Oort AH (1992) *Physics of Climate*. 520pp. New York: American Institute of Physics.

Rockström J, Falkenmark M, Karlberg L, Hoff H, Rost S, and Gerten D (2009) Future water availability for global food production: The potential of green water for increasing resilience to global change. *Water Resources Research* 45: W00A12 (doi:10.1029/2007WR006767)

Rossow WB, Walker AW, and Garder LC (1993) Comparison of ISCCP and other cloud amounts. *Journal of Climate* 6: 2394–2418.

Shen Y, Oki T, Utsumi N, Kanae S, and Hanasaki N (2008) Projection of future world water resources under SRES scenarios: Water withdrawal. *Hydrological Sciences Journal* 53(1): 11–33.

Shiklomanov IA (ed.) (1997) Assessment of water resources and water availability in the world. *Background Report for the Comprehensive Assessment of the Freshwater Resources of the World*. Geneva: WMO/SEI.

Starr VP and Peixöto J (1958) On the global balance of water vapor and the hydrology of deserts. *Tellus* 10: 189–194.

Takata K, Emori S, and Watanabe T (2003) Development of minimal advanced treatments of surface interaction and runoff. *Global and Planetary Change* 38: 209–222.

Tapley BD, Bettadpur S, Ries JC, Thompson PF, and Watkins MM (2004) GRACE measurements of mass variability in the Earth system. *Science* 305(5683): 503–505.

Trenberth KE and Guillemot CJ (1998) Evaluation of the atmospheric moisture and hydrological cycle in the NCEP/NCAR reanalyses. *Climate Dynamics* 14: 213–231.

Trenberth KE and Solomon A (1994) The global heat balance: Heat transports in the atmosphere and ocean. *Climate Dynamics* 10: 107–134.

Uhlenbrook S, Frey M, Leibundgut Ch, and Maloszewski P (2002) Hydrograph separations in a mesoscale mountainous basin at event and seasonal timescales. *Water Resources Research* 38(1096). doi:10.1029/2001WR000398

Uhlenbrook S, Roser S, and Tilch N (2004) Development of a distributed, but conceptual catchment model to represent hydrological processes adequately at the meso scale. *Journal of Hydrology* 291: 278–296.

Vörösmarty CJ, Green P, Salisbury J, and Lammers RB (2000) Global water resources: Vulnerability from climate change and population growth. *Science* 289: 284–288.

Wijffels SE, Schmitt RW, Bryden HL, and Stigebrandt A (1992) Transport of freshwater by the oceans. *Journal of Physical Oceanography* 22: 155–162.

Xie P and Arkin PA (1996) Analyses of global monthly precipitation using gauge observations, satellite estimates, and numerical model predictions. *Journal of Climate* 9: 840–858.

Yang D, Shao W, Yeh PJ-F, Yang H, Kanae S, and Oki T (2009) Impact of vegetation coverage on regional water balance in the nonhumid regions of China. *Water Resources Research* 45: W00A14.

Yeh PJ-F and Eltahir EAB (2005) Representation of water table dynamics in a land surface scheme. Part I: Model development. *Journal of Climate* 18: 1861–1880.

Yoshimura K, Kanamitsu M, Noone D, and Oki T (2008) Historical isotope simulation using reanalysis atmospheric data. *Journal of Geophysical Research* 113: doi:1029.10/2008JD010074

Yoshimura K, Oki T, Ohte N, and Kanae S (2004) Colored moisture analysis estimates of variations in 1998 Asian monsoon water sources. *Journal of the Meteorological Society of Japan* 82: 1315–1329.

2.02 Precipitation

D Koutsoyiannis, National Technical University of Athens, Athens, Greece
A Langousis, Massachusetts Institute of Technology (MIT), Cambridge, MA, USA

© 2011 Elsevier B.V. All rights reserved.

2.02.1	**Introduction**	27
2.02.1.1	The Entrancement of Precipitation	28
2.02.1.2	Forms of Precipitation	29
2.02.1.3	Precipitation Metrics	29
2.02.1.4	The Enormous Variability of Precipitation	30
2.02.1.5	Probability and Stochastic Processes as Tools for Understanding and Modeling Precipitation	32
2.02.1.5.1	Basic concepts of probability	35
2.02.1.5.2	Stochastic processes	36
2.02.1.5.3	Stationarity	37
2.02.1.5.4	Ergodicity	38
2.02.1.5.5	Some characteristic stochastic properties of precipitation	38
2.02.2	**Physical and Meteorological Framework**	42
2.02.2.1	Basics of Moist Air Thermodynamics	43
2.02.2.2	Formation and Growth of Precipitation Particles	44
2.02.2.3	Properties of Precipitation Particles	45
2.02.2.3.1	Terminal velocity	45
2.02.2.3.2	Size distribution	45
2.02.2.4	Clouds and Precipitation Types	46
2.02.2.4.1	Cumulus cloud systems	46
2.02.2.4.2	Stratus cloud systems	47
2.02.2.5	Precipitation-Generating Weather Systems	47
2.02.2.5.1	Fronts	47
2.02.2.5.2	Mechanical lifting and orographic precipitation	48
2.02.2.5.3	Extratropical cyclones	48
2.02.2.5.4	Isolated extratropical convective storms	50
2.02.2.5.5	Extratropical squall lines and rainbands	50
2.02.2.5.6	Monsoons	51
2.02.2.5.7	Tropical cyclones	51
2.02.3	**Precipitation Observation and Measurement**	53
2.02.3.1	Point Measurement of Precipitation	53
2.02.3.1.1	Measuring devices	53
2.02.3.1.2	Typical processing of rain gauge data	54
2.02.3.1.3	Interpolation and integration of rainfall fields	56
2.02.3.2	Radar Estimates of Precipitation	58
2.02.3.2.1	Basics of radar observation and measurement	58
2.02.3.2.2	Radar observation of distributed targets and the estimation of precipitation	59
2.02.3.3	Spaceborne Estimates of Precipitation	60
2.02.3.3.1	The IR signature of cloud tops	60
2.02.3.3.2	The visible reflectivity of clouds	61
2.02.3.3.3	The microwave signature of precipitation	61
2.02.4	**Precipitation modeling**	62
2.02.4.1	Rainfall Occurrence	62
2.02.4.2	Rainfall Quantity	64
2.02.4.3	Space–Time Models	65
2.02.4.4	Rainfall Disaggregation and Downscaling	66
2.02.4.5	Multifractal Models	67
2.02.5	**Precipitation and Engineering Design**	68
2.02.5.1	Probabilistic versus Deterministic Design Tools	68
2.02.5.2	Extreme Rainfall Distribution	70
2.02.5.3	Ombrian Relationships	71
References		72

2.02.1 Introduction

2.02.1.1 The Entrancement of Precipitation

Precipitation and its related phenomena, such as cloud formation and movement, thunders, and rainbow, are spectacular (Figure 1) due to their huge diversity and complexity. This complexity makes them difficult to comprehend, model, and predict. Hence, it is understandable that ancient civilizations explained these phenomena in a hyperphysical manner, assuming that deities were responsible for their creation. For example, in Greek mythology, some of the phenomena were deified (e.g., Iris is the name of a goddess as well as of the rainbow), whereas the most impressive among them, thunders in particular, were attributed to the action of the King of the Gods, Zeus (Jupiter in Roman mythology; similar deities are Indra in Hinduism, Thor in Norse, etc.). Demystification of these processes and formation of the physical concept of the hydrological cycle was closely related to the birth of science, by the turn of the seventh century BC. While the hydrological cycle was founded as a concept in the sixth century BC by Anaximander, Anaximenes, and Xenophanes, and was later advanced by Aristotle (Koutsoyiannis *et al.*, 2007), certain aspects related to precipitation can be understood only within the frame of modern science. The fact that a solid or liquid hydrometeor resists gravity and remains suspended in the atmosphere in a cloud is counterintuitive, and needs advanced knowledge of physics, fluid dynamics, and statistical thermodynamics to be understood and modeled.

The complexity of the processes involved in precipitation and their enormous sensitivity to the initial conditions (where tiny initial differences produce great differences in the final phenomena), retain, to this day, some of the ancient mythical and magical magnificence of the societal perception of precipitation. People still believe in hyperphysical interventions in matters concerning precipitation. As put by Poincaré (1908), father of the notion of chaos:

> Why do the rains, the tempests themselves seem to us to come by chance, so that many persons find it quite natural to pray for rain or shine, when they would think it ridiculous to pray for an eclipse?

Figure 1 Precipitation and related phenomena (from upper-left to lower-right): Monsoon rainfall (Pune, India, September 2009; photo by D Koutsoyiannis); snowy mountainous landscape (Mesounta, Greece, December 2008; from http://www.mesounta.gr/mesounta/ist_eik1/07_xion_03.htm); thunder (Athens, Greece, November 2005; from the photo gallery of Kostas Mafounis); rainbow (Mystras, Greece, April 2008, from laspistasteria.wordpress.com/2008/04/08/rainbow-3/).

Amazingly, however, and at the very same time, there is little disbelief in some climate modelers' prophecies (or outputs of global circulation models (GCMs)) of the precipitation regimes over the globe in the next 100 years or more. This indicates an interesting conflict between perceptions of precipitation – that it is so unstable, uncertain, and unpredictable that prayers are needed to invoke precipitation, and that for some scientists, the future evolution of precipitation on Earth is still predictable in the long term. The latter belief concerns not only the general public, but also the scientific community. For example, a Google Scholar search with either of the keywords 'precipitation' or 'rainfall', plus the keywords 'climate change' and 'GCM', locates 21 700 publications (as of August 2009), of which about 200 have been cited 100 times or more. This huge list of results appears despite the fact that climate modelers themselves admit to the performance of their models being low, as far as precipitation is concerned (Randall et al., 2007). An independent study by Koutsoyiannis et al. (2008), which compares model results for the twentieth century with historical time series, has shown that the models are not credible at local scales and do not provide any basis for assessment of future conditions. These findings demonstrate that, even today, the perception of precipitation, not only by the general public, but even by scientists specialized in the study of precipitation, meteorologists, climatologists, and hydrologists, continues to be contradictory, problematic, and, in some sense, mysterious.

2.02.1.2 Forms of Precipitation

Precipitation occurs in a number of forms, either liquid or solid, or even mixed (sleet). Liquid precipitation includes rainfall and drizzle, where the former is the most common and most significant, and the latter is characterized by much smaller drop sizes and lighter intensity. Dew is another liquid form, formed by condensation of water vapor (mostly at night) on cold surfaces (e.g., on tree leaves).

Most important among the solid forms of precipitation are snow and hail. At high latitudes or at high altitudes, snow is the predominant form of precipitation. Snowfall may occur when the temperature is low and snow accumulates on the ground until the temperature rises sufficiently for it to melt. On the other hand, hail may fall in relatively high temperature and usually melts rapidly. While hailstones are amorphous and usually large (one to several centimeters in diameter), snowflakes are symmetrical and visually appealing with a tremendous variety of shapes, so that no two snowflakes are the same.

Occult precipitation is induced when clouds or fog is formed in forested areas, and it includes liquid (fog drip) and solid (rime) forms. Fog drip occurs when water droplets are deposited on vegetative surfaces, and the water drips to the ground. Rime is formed when supercooled air masses encounter exposed objects, such as trees, that provide nucleation sites (see Sections 2.02.2.1 and 2.02.2.2) for formation and buildup of ice, much of which may fall to the ground in solid or liquid form. In some places (e.g., in humid forested areas), precipitation of this type may reach significant amounts; for example, rime constitutes about 30% of the annual precipitation in a Douglas fir forest in Oregon (Harr, 1982; Dingman, 1994) and about 30% of total precipitation in fir-forested mountainous areas of Greece (Baloutsos et al., 2005), and it is the sole precipitation type on the rainless coast of Peru (Lull, 1964; Dingman, 1994).

2.02.1.3 Precipitation Metrics

The principal metric of precipitation is the rainfall depth h (commonly expressed in millimeters) that falls at a specified point in a specified period of time t; this can be easily perceived and measured by a bucket exposed to precipitation. A derivative quantity is the precipitation intensity

$$i := \frac{dh}{dt} \quad (1)$$

with units of length divided by time (typically $mm\,h^{-1}$, $mm\,d^{-1}$, and $mm\,yr^{-1}$). Since it cannot be measured directly (at an instantaneous time basis), it is typically approximated as

$$i = \frac{\Delta h}{\Delta t} \quad (2)$$

where Δh is the change of the depth in a finite time interval Δt. The intensity derived from Equation (2) is a time-averaged value – but at a point basis. Spatial averaging at various scales is always very useful as can be seen in Section 2.02.1.4. This averaging needs precipitation measurement at several points, followed by appropriate numerical integration methods (see Section 2.02.3.1). While traditional precipitation-measurement networks are sparse, thus making the estimation of areal precipitation uncertain, in recent decades, new measurement techniques have been developed implementing radar and satellite technologies (Sections 2.02.3.2 and 2.02.3.3). These provide a detailed description of the spatial distribution of precipitation, thus enabling a more accurate estimation. The latter techniques inherently involve the study of other metrics of precipitation such as the distribution of the size, velocity, and kinetic energy of the precipitation particles, and the so-called radar reflectivity (Section 2.02.3.2).

Furthermore, the quantitative description of the processes related to the fall, accumulation, and melting of snow involves a number of additional metrics, such as the snowfall depth (new snow falling), the snowcover depth or snowpack depth (the depth of snow accumulated at a certain point at a particular time), the snow density ρ_s, and the water equivalent of snowfall or of snowpack, defined as $h = h'\rho_s/\rho_w$ where h' is the snowfall or the snowcover depth, and $\rho_w = 1000\,kg\,m^{-3}$ is the liquid water density. Typical values of ρ_s for snowfall range between 0.07 and $0.15\rho_w$ (e.g., Dingman, 1994) but a commonly used value is $\rho_s = 0.1\rho_w = 100\,kg\,m^{-3}$. For this value, a snowfall depth of say, 10 cm, corresponds to a precipitation water equivalent of 10 mm. The density of snowpack is generally larger than $0.1\rho_w$ (because of compaction due to gravity and other mechanisms) and depends on the elapsed time and snowpack depth. After a few days, it is about $0.2\rho_w$ whereas after some months it may become about $0.4\rho_w$.

2.02.1.4 The Enormous Variability of Precipitation

The different phases and forms of precipitation, and the different shapes of precipitation particles (drops, flakes, and hailstones) are just a first indicator of the great diversity of the precipitation phenomena. At a macroscopic level and in quantitative terms, this diversity is expressed by the enormous variability of the precipitation process, in space and time, at all spatial and temporal scales. Intermittency is one of the aspects of variability, but even in areas or time periods in which precipitation is nonzero, the precipitation depth or average intensity is highly variable.

Figure 2 shows the spatial variability of precipitation over the globe in $mm\,d^{-1}$ at the climatic scale (average for the 30-year period 1979–2008) and at an annual scale (average for the year 2006), based mostly on satellite data (see **Figure 2** caption and Section 2.02.3.3). While the average precipitation rate over the globe and over the specified 30-year period is $2.67\,mm\,d^{-1}$ or $977\,mm\,yr^{-1}$, we observe huge differences in different areas of the globe. In some areas, mostly in tropical seas and in equatorial areas of South America and Indonesia, this rate exceeds $10\,mm\,d^{-1}$ or $3.65\,m\,yr^{-1}$. On the other hand, in large areas in the subtropics, where climate is dominated by semi-permanent anticyclones, precipitation is lower than $1\,mm\,d^{-1}$ or $365\,mm\,yr^{-1}$. Significant portions of these areas in Africa, Australia, and America are deserts, where the average precipitation is much lower than $1\,mm\,d^{-1}$. In addition, in polar regions, where the available atmospheric moisture content is very low due to low temperature (see Section 2.02.2.1 and **Figure 14**), the amounts of precipitation are very small or even zero. For example, it is believed that certain dry valleys in the interior of Antarctica have not received any precipitation during the last 2 million years (Uijlenhoet, 2008).

Figure 3 depicts the zonal precipitation profile and shows that the climatic precipitation rate at an annual basis is highest at a latitude of $5°\,N$, reaching almost $2000\,mm\,yr^{-1}$ and has a second peak of about $1500\,mm\,yr^{-1}$ at $5°\,S$. Around the Tropics of Cancer and Capricorn, at $23.4°\,N$ and S, respectively, the rainfall rate displays troughs of about $600\,mm\,yr^{-1}$, whereas at mid-latitudes, between $35°$ and $60°$ both N and S, rainfall increases again and remains fairly constant, close to the global average of $977\,mm\,yr^{-1}$. Then, toward the poles, it decreases to about $100\,mm$ in Antarctica and slightly more, to $150\,mm\,yr^{-1}$, in the Arctic. **Figure 3** also shows monthly climatic profiles for the months of January and July. It can be seen that the rainfall conditions for the 2 months are quite different, with the largest differences appearing at about $15°\,N$ and S and the smallest at about $30°\,N$ and S. Below $30°$ in the Northern Hemisphere, as well as above the Arctic Circle ($66.6°$), rainfall is higher during summer (July) than during winter (January), but at mid-latitudes, this relationship is reversed. Similar conditions are met in the Southern Hemisphere (where January and July are summer and winter months, respectively).

In both **Figures 2** and **3**, apart from climatic averages, the specific values for a certain year, namely 2006, are also shown. We observe that there are differences in the climatic values, manifesting temporal variability over the different years. This variability seems to be lower in comparison to the spatial variability over the globe, as well as to the seasonal variability reflected in the profiles of different months. However, while the spatial variability over the globe and the seasonal variability are well comprehended and roughly explainable in terms of basic physical and astronomical knowledge (i.e., solar radiation, relationship of temperature and atmospheric moisture content, and motion of Earth), in other words, they are regular, the interannual variability is irregular, and difficult or even impossible to predict.

Such irregular variability appears at finer timescales as well as at finer spatial scales. In fact, as easily understood from elementary statistics, as the spatial and/or temporal scale becomes finer, the variability increases. **Figure 4** demonstrates how the variability of the spatial distribution of rainfall at a monthly temporal scale (January 2006) increases when the spatial scale decreases from $2.5° \times 2.5°$ (upper panel) to $0.25° \times 0.25°$ (lower panel). Clearly, the areas of equal rainfall amount (including areas of negligible rainfall, i.e., $<1\,mm\,d^{-1} \approx 0.04\,mm\,h^{-1}$), which are smooth in the upper panel become rough and erratic in the lower panel. Moreover, the maximum observed rainfall is $21\,mm\,d^{-1}$ (monthly $651\,mm$) in the upper panel and $1.2\,mm\,h^{-1}$ (monthly $893\,mm$) in the lower panel.

Figure 5 demonstrates the increasing variability with the decreasing timescale. Specifically, it depicts how the image of the rainfall distribution changes at a daily scale (9 January 2006) and at a sub-daily scale, at 3-hourly intervals of the same day. The differences between **Figure 5** and **Figure 4** are prominent. Especially at the 3-hourly scale, a vast part of the globe receives no rainfall, and the part that receives rainfall, it is irregularly distributed, yet not showing a totally random pattern. The maximum observed rate during this 3-hourly interval is $22\,mm\,h^{-1}$, about 18 times higher than the maximum rate at the monthly scale shown in **Figure 4**. The lowest panels of **Figure 5** provide a zoom-in over the area lying between $9°\,N–5°\,S$ and $78–92°\,E$, which is located in the Indian Ocean, south-east of Sri Lanka. This area received a large amount of rainfall on this particular day, with a rate that is nonuniform in space and time.

Figures 6–8 focus on the temporal variability of precipitation. **Figure 6** depicts the monthly and annual variation of the average precipitation over the globe. We can see that at both scales, the variability is remarkable. Thus, the annual precipitation in the last 30 years has varied between 957 and 996 mm and obviously much higher variation should have occurred in the past – but data of this type covering the entire globe do not exist for earlier periods. However, we can get an idea of earlier variation using rain gauge data (see Section 2.02.3.1) at certain locations.

Perhaps, the oldest systematic observations of rainfall quantity in the world were made in Korea, in the fifteenth century. Rainfall records for the city of Seoul ($37.57°\,N$, $126.97°\,E$, and $85\,m$) exist for the period since 1770, and are considered to be reliable (Arakawa, 1956; Wang et al., 2006, 2007). The recorded annual rainfall in Seoul is plotted in **Figure 7** along with the running climatic averages at 10-year and 30-year timescales. The data are now available at a monthly scale from the climatic database of the Dutch Royal Netherlands Meteorological Institute (KNMI), while the monthly data for 1770–1907 appear also in Arakawa (1956).

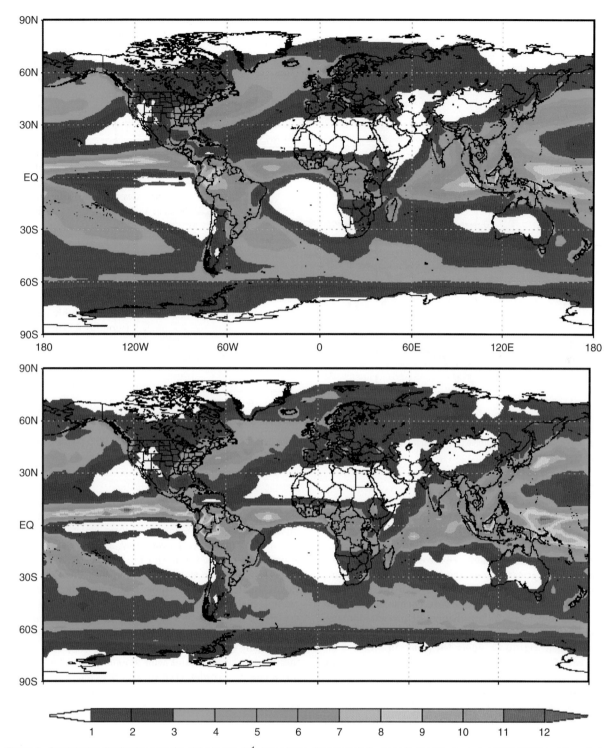

Figure 2 Precipitation distribution over the globe in mm d^{-1}, (upper) at a climatic scale (average for the 30-year period 1979–2008) and (lower) at an annual scale (average for year 2006). Data and image generation due to the Global Precipitation Climatology Project (GPCP) made available by NASA at http://disc2.nascom.nasa.gov/Giovanni/tovas/rain.GPCP.2.shtml; resolution 2.5° × 2.5°.

Comparisons show that the two time series are generally consistent, but not identical. The more modern data series has a few missing values, which generally correspond to high values of the older version (and it has been common practice in hydrometeorological data processing to delete very high values or outliers, which are regarded suspect, see Section 2.02.3.1.2). In the time series plotted in **Figure 7**, these gaps have been filled in using the values of the older time series, and a few other missing values have been filled in with the average of the four nearest monthly values of the same month

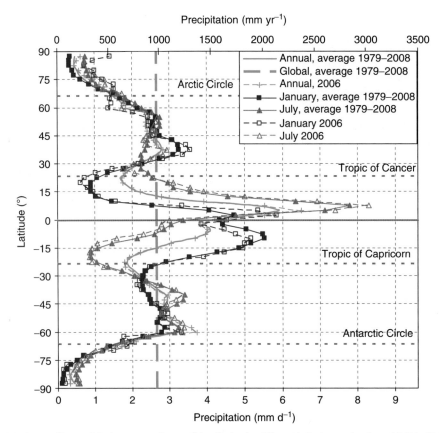

Figure 3 Zonal precipitation profile: precipitation averaged over all longitudes for different latitudes varying from 90°N to 90°S (−90°N); data from GPCP.

(see justification in Section 2.02.3.1.3). The plot shows that during the 238 years of record, the annual rainfall varied between 634 and 3057 mm and the climatic 30-year average varied between 1139 and 1775 mm. These figures indicate a huge variability: the maximum observed annual rainfall is almost 5 times greater than the minimum and the maximum 30-year climatic rainfall is 55% higher than the minimum. Such observed changes underscore the ever-changing character of climate, and render future changes of precipitation predicted by climate modelers (which typically vary within 10–20%; compare Fig. 10.12, upper left panel, in Meehl *et al.*, 2007, with **Figure 2** herein) to be unrealistically low and too unsafe to support planning.

Figure 7 also includes a plot of another long time series, for Charleston City, USA (32.79° N, 79.94° W, and 3 m); the record begins in 1835. This time series is also available at the KNMI database, and a few missing monthly values have been filled in by the average of the four nearest monthly values of the same month. Here, the annual rainfall varied between 602 and 1992 mm (3.3 times higher than minimum) and the climatic 30-year average varied between 1135 and 1425 mm (25% higher than minimum).

Finally, **Figure 8** depicts the time series of a storm measured at unusually high temporal resolution, that is, 10 s. This storm, with duration 96 790 s or about 27 h starting at 1990-02-12T17:03:39, is one of several storms that were measured at the University of Iowa using devices that support high sampling rates (Georgakakos *et al.*, 1994). **Figure 8** also includes plots at 5-min and hourly timescales. The minimum intensity was virtually zero at all three scales, whereas the maximum rainfall intensity was 118.7, 38.9, and 18.1 mm h^{-1} at timescales of 10 s, 5 min, and 1 h, respectively. As the mean intensity during the storm is 3.89 mm h^{-1}, these maximum values are 30, 10, and 4.6 times higher than the mean. This example highlights the spectacular variability of rainfall, particularly at fine timescales (see also Uijlenhoet and Sempere-Torres, 2006). As the total rainfall amount of this storm event only slightly exceeds 100 mm, it could be thought of as a rather modest event. Storms with amounts much higher than this are often recorded even in semi-dry climates and, obviously, the variability of rainfall intensity during such storms is even higher.

2.02.1.5 Probability and Stochastic Processes as Tools for Understanding and Modeling Precipitation

The high variability and the rough and irregular patterns in observed fields and time series are much more prominent in precipitation than in other meteorological variables such as atmospheric pressure or temperature. High variability implies high uncertainty and, unavoidably, this affects predictability in deterministic terms. Considering weather prediction as an example, it is well known that the forecasts of atmospheric pressure and temperature are much more reliable than those

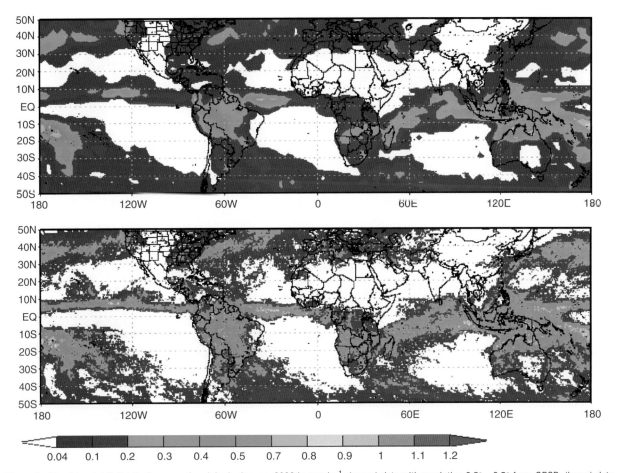

Figure 4 Monthly rainfall distribution over the globe in January 2006 in mm h^{-1}: (upper) data with resolution 2.5° × 2.5° from GPCP; (lower) data with resolution 0.25° × 0.25° from the Tropical Rainfall Measuring Mission (TRMM) and Other Rainfall Estimate (3B42 V6) archive, made available by NASA at http://disc2.nascom.nasa.gov/Giovanni/tovas/TRMM_V6.3B42.shtml.

of precipitation. Numerical weather prediction (NWP) uses current weather conditions as an input to mathematical models of the atmosphere, which solve the flow (Navier–Stokes) equations, the thermodynamic energy equation, the state equation of gases, and the equation for conservation of water vapor, over a grid covering the entire atmosphere. The processes related to cloud formation and precipitation (see Section 2.02.2.2) are less accurately represented in these models. While the continuous improvement of NWP models resulted in a considerable reduction of forecast errors on pressure and temperature, the improvement in the so-called quantitative precipitation forecast (QPF) has been slower (Olson et al., 1995). Further, although the advances in computing infrastructure permitted the increase in model resolution that leads generally to an improvement of precipitation forecasts, recently, many authors have highlighted the limitations of such an approach (e.g., Mass et al., 2002; Lagouvardos et al., 2003; Kotroni and Lagouvardos, 2004). The major advancement in QPF in the last decades was the abandonment of the pure deterministic approach, which seeks a unique prediction, and the adoption of a more probabilistic approach to precipitation forecast, based on earlier ideas of Epstein (1969) and Leith (1974). In this approach, known as ensemble forecasting, the same model produces many forecasts. To produce different forecasts, perturbations are introduced, for example, in the initial conditions, and, because of the nonlinear dynamics with sensitive dependence on the initial conditions (e.g., Lorenz, 1963), these perturbations are magnified in time, thus giving very different precipitation amounts in a lead time of 1 or more days. The different model outputs can then be treated in a probabilistic manner, thus assigning probabilities to rainfall occurrence as well as to the exceedance of a specified rainfall threshold. In this manner, although the model uses deterministic dynamics, the entire framework is of the Monte Carlo or stochastic type.

This method is satisfactory for a time horizon of forecast of a few days. In hydrology, this time horizon is relevant in real-time flood forecasting. However, in hydrological design, horizons as long as 50 or 100 years (the lifetimes of engineering constructions) are typically used. For such long horizons, the use of deterministic dynamics and of the related laborious models would not be of any help. However, a probabilistic approach is still meaningful – in fact the only effective approach – and, in this case, it can be formulated irrespective of the dynamics. Rather, the probabilistic approach should be based, in this case, on historical records of precipitation, such as those displayed in **Figures 7** and **8**. The reasoning behind neglecting the deterministic dynamics is

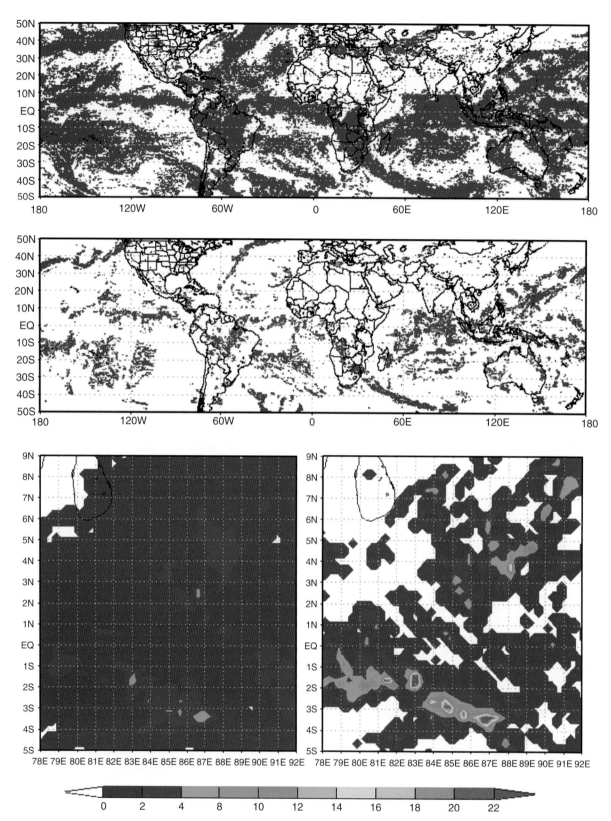

Figure 5 Spatial rainfall distribution at daily and sub-daily scale: (upper) daily rainfall over the zone between 50°N and 50°S on 9 January 2006; (middle) 3-hourly rainfall at 09:00 on the same day; (lower left) zoom-in of the upper panel for daily rainfall in the Indian Ocean south-east of Sri Lanka (shown in figure); (lower right) zoom-in of the middle panel for 3-hourly rainfall for the same area. Data in mm h^{-1} with resolution 0.25° × 0.25° from the TRMM 3B42 V6 archive.

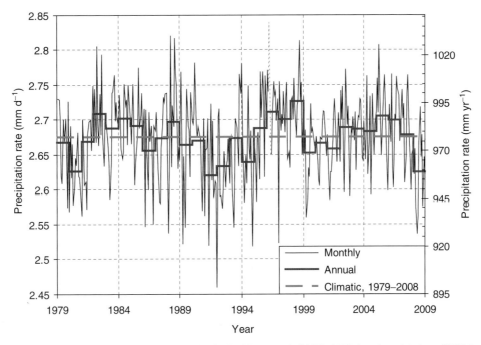

Figure 6 Evolution of the globally averaged monthly precipitation in the 30-year period 1979–2008, based on data from GPCP (see caption of Figure 2).

that, beyond a certain time horizon (which in precipitation is of the order of several days), even the simplest nonlinear systems tend to a statistical equilibrium state. In this state, the probability distribution of the system properties, conditioned on the initial state, is practically equal to the marginal (i.e., unconditional) probability distribution of the same properties (Koutsoyiannis, 2009). This equilibrium, which is different from the typical thermodynamic equilibrium, corresponds to the maximization of the entropy of the vector of random variables defining the system state.

2.02.1.5.1 Basic concepts of probability

Probability is thus not only a mathematical tool to model precipitation uncertainty, but also a concept for understanding the behavior of precipitation. Probabilistic thinking provides insights into phenomena and their mathematical descriptions, which may not be achievable in deterministic terms. It should be recalled that, according to the Kolmogorov (1933) system, probability is a normalized measure, that is, a function P that maps sets (areas where unknown quantities lie) to real numbers (in the interval [0, 1]). Furthermore, a random variable $\underline{x}$ is a single-valued function of the set of all elementary events (so that to each event, it maps a real number) and is associated with a probability distribution function. The latter is defined as

$$F_{\underline{x}}(x) := P\{\underline{x} \leq x\} \qquad (3)$$

where x is any real number, which should be distinguished from the random variable $\underline{x}$. (Distinction of random variables from their values is usually done by denoting them with upper case and lower case letters, respectively. This convention has several problems – e.g., the Latin x and the Greek χ, if put in upper case, are the same symbol X – other texts do not distinguish the two at all, thus creating another type of ambiguity. Here, we follow a different convention, in which random variables are underscored and their values are not.) $F_{\underline{x}}(x)$ is a nondecreasing function of x with the obvious properties $F_{\underline{x}}(-\infty) = 0$ and $F_{\underline{x}}(+\infty) = 1$. For continuous random variables (as is, for instance, the representation of a nonzero rainfall depth), the probability that a random variable $\underline{x}$ would take any particular value x is $P\{\underline{x} = x\} = 0$. Thus, the question of whether one particular value (say $x_1 = 10$ mm, assuming that $\underline{x}$ denotes daily rainfall at a location) is more probable than another value (say $x_2 = 10$ m, which intuitively seems extremely improbable) cannot be answered in terms of the probability function P, as all particular values have probability equal to zero. The derivative of F, that is,

$$f_{\underline{x}}(x) := \frac{dF_{\underline{x}}(x)}{dx} \qquad (4)$$

termed the probability density function, can provide this answer, as the quantity $f_{\underline{x}}(x) \, dx$ is the probability that rainfall will lie in an interval of length dx around x. Apparently, then the ratio $f_{\underline{x}}(x_1)/f_{\underline{x}}(x_2)$ equals the ratio of the probabilities at points x_1 and x_2.

These rather simple notions allow quantification of uncertainty and enable the production of different type of predictions, which offer a concrete foundation of rational decisions for the design and management of water-resources projects. This quantification is sometimes (mostly in Bayesian statistics) referred to as 'probabilization of uncertainty' that is meant to be the axiomatic reduction from the notion of unknown to the notion of a random variable (Robert, 2007).

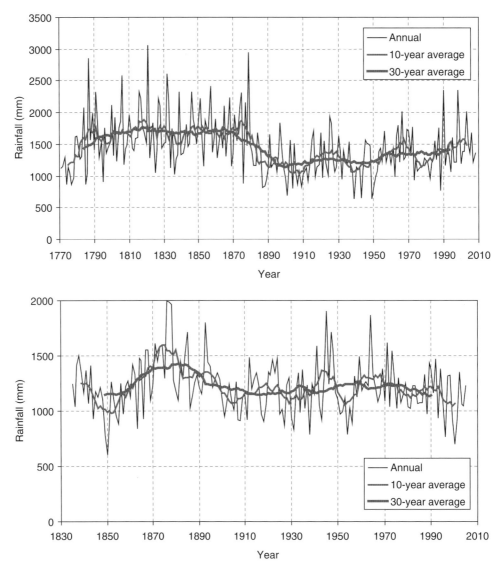

Figure 7 Annual precipitation time series in two of the stations with the longest records worldwide: (upper) Seoul, Korea; (lower) Charleston City, USA. Data from the database of the Dutch Royal Netherlands Meteorological Institute (KNMI; http://climexp.knmi.nl) and additional information as shown in text.

2.02.1.5.2 Stochastic processes

In the study of rainfall variation in time, the notion and the theory of stochastic processes provide the necessary theoretical framework. A stochastic process is defined as an arbitrarily (usually infinitely) large family of random variables $\underline{x}(t)$ (Papoulis, 1991). In most hydrological applications, time is discretized using an appropriate time step δ; for integer i, the average of the continuous time process $\underline{x}(t)$ from $t = (i-1)\delta$ to $t = i\ \delta$, is usually denoted $\underline{x}_i$ and forms a discrete time stochastic process. The index set of the stochastic process (i.e., the set from which the index t or i takes its values) can also be a vector space, rather than the real line or the set of integers. This is the case, for instance, when we assign a random variable (e.g., rainfall depth) to each geographical location (a two-dimensional (2D) vector space) or to each location and time instance (a 3D vector space). Stochastic processes with a multidimensional index set are also known as random fields.

A realization $x(t)$ (or x_i) of a stochastic process $\underline{x}(t)$ (or $\underline{x}_i$), which is a regular (numerical) function of the time t (or a numerical sequence in time i), is known as a sample function. Typically, a realization is observed at countable time instances (and not in continuous time, even if the process is of continuous-time type). This sequence of observations is also referred to as a time series. Clearly then, a time series is a sequence of numbers, whereas a stochastic process is a family of random variables. (Unfortunately, a large body of literature does not make this distinction and confuses stochastic processes with time series.)

The distribution and the density functions of the random variable $\underline{x}_i$, that is,

$$F_i(x) := P\{\underline{x}_i \leq x\}, \quad f_i(x) := \frac{\mathrm{d}F_i(x)}{\mathrm{d}x} \quad (5)$$

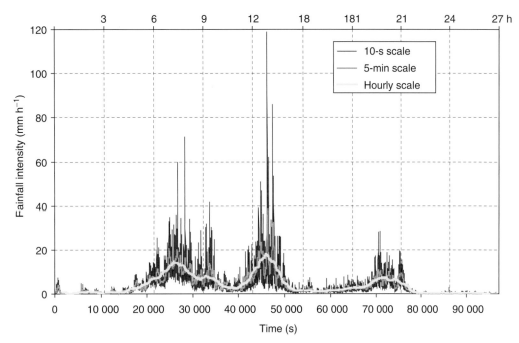

Figure 8 Time series of a storm in Iowa, USA, measured at the University of Iowa with temporal resolution of 10 s; time zero corresponds to 1990-02-12T17:03:39. From Georgakakos KP, Carsteanu AA, Sturdevant PL, and Cramer JA (1994) Observation and analysis of Midwestern rain rates. *Journal of Applied Meteorology* 33: 1433–1444.

are called, respectively 'first-order distribution function' and 'first-order density function' of the process. Likewise, the second-order distribution function is $F_{i_1 i_2}(x_1, x_2) = P\{\underline{x}_{i_1} \leq x_1, \underline{x}_{i_2} \leq x_2\}$ and this can be generalized to define the nth-order distribution function. It should be recalled that the expected value of a function g of one, two, or more random variables is the integral of g multiplied by the density f, that is,

$$E[g(\underline{x}_i)] := \int_{-\infty}^{\infty} g(x) f_i(x) \, dx$$

$$E[g(\underline{x}_{i_1}, \underline{x}_{i_2})] = \int_{-\infty}^{\infty} \int_{-\infty}^{\infty} g(x_1, x_2) f_{i_1 i_2}(x_1, x_2) \, dx_1 \, dx_2 \tag{6}$$

The use of square brackets in $E[\ ]$ and the random variables $\underline{x}_i$ rather than their values x signifies the fact that the expected value is not a function of the real number x; rather, it depends solely on the distribution function associated with the random variable $\underline{x}_i$. Of particular interest are the cases where $g(\underline{x}_i) = \underline{x}_i$, where $E[\underline{x}_i] =: \mu_i$ is the mean value of $\underline{x}_i$, and $g(\underline{x}_{i_1}, \underline{x}_{i_2}) = (\underline{x}_{i_1} - \mu_{i_1})(\underline{x}_{i_2} - \mu_{i_2})$, where $E[(\underline{x}_{i_1} - \mu_{i_1})(\underline{x}_{i_2} - \mu_{i_2})] =: C_{i_1 i_2}$ is the process autocovariance, that is, the covariance of the random variables $\underline{x}_{i_1}$ and $\underline{x}_{i_2}$. The process variance (the variance of the variable $\underline{x}_i$), is a special case of the latter, that is, $\text{Var}[\underline{x}_i] = C_{ii}$, whereas the standard deviation is the square root of the latter, that is, $\sigma_i := \sqrt{C_{ii}}$. Consequently, the process autocorrelation (the correlation coefficient of the random variables $\underline{x}_{i_1}$ and $\underline{x}_{i_2}$) is $\rho_{i_1 i_2} := C_{i_1 i_2}/(\sigma_{i_1} \sigma_{i_2})$.

2.02.1.5.3 Stationarity

As implied by the above notation, in the general setting, the statistics of a stochastic process, such as the mean and autocovariance, depend on time i and thus vary with time. However, the case where these statistical properties remain constant in time is most interesting. A process with this property is termed a 'stationary' process. More precisely, a process is called 'strict-sense' stationary, if all its statistical properties are invariant with a shift in the time origin. That is, the distribution function of any order of $\underline{x}_{i+j}$ is identical to that of $\underline{x}_i$. A process is called 'wide-sense stationary' if its mean is constant and its autocovariance depends only on time differences (lags), that is,

$$E[X_i] = \mu, \quad E[(X_{i+j} - \mu)(X_i - \mu)] = C_j \tag{7}$$

Evidently, the standard deviation is constant too, that is, $\sigma_i = \sigma$, and the autocorrelation is a function of the time lag only, that is, $\rho_{i+j,\, i} = \rho_j$. A strict-sense stationary process is also wide-sense stationary, but the reverse is not true.

A process that is not stationary is called nonstationary. In a nonstationary process, one or more statistical properties depend on time. A typical case of a nonstationary process is the cumulative rainfall depth whose mean obviously increases with time. For instance, let us assume that the instantaneous rainfall intensity $\underline{i}(t)$ at a geographical location and period of the year is a stationary process, with a mean μ. Let us further denote by $\underline{h}(t)$, the rainfall depth collected in a large container (a cumulative rain gauge) at time t, and assume that at the time origin, $t = 0$, the container is empty. Clearly $E[\underline{h}(t)] = \mu t$. Thus $\underline{h}(t)$ is a nonstationary process.

It should be stressed that stationarity and nonstationarity are properties of a stochastic process, not of a sample function or time series. There is some confusion in the literature about this, as there are several studies that refer to a time series as stationary or nonstationary. As a general rule, to characterize a process as nonstationary, it suffices to show that some statistical property is a deterministic function of time (as in the above example of the cumulative rainfall), but this cannot be directly inferred merely from a time series. To understand this, let us consider the time series of annual rainfall in Seoul, plotted in the upper panel of **Figure 7**. Misled by the changing regime of precipitation at the climatic scale, as manifest in the plot of the 30-year average, it would be tempting to note (1) an increasing trend in the period 1770–90; (2) a constant climate with high precipitation during 1790–1870; (3) a decreasing trend between 1870 and 1900; and (4) a constant climate with low precipitation thereafter. It is then a matter of applying a fitting algorithm to determine, say, a broken-line type of function to the time series, which would be called a deterministic function of time. The conclusion would then be that the time series is nonstationary. However, this is a wrong *ex-post* argument, which interprets the long-term variability of the processes as a deterministic function. Had the function been indeed deterministic, it would also apply to future times, which obviously is not the case. Comparison with the previous example (cumulative rainfall), where the deterministic function $E[\underline{x}(t)] = \mu t$ was obtained by theoretical reasoning (deduction) rather than by inspection of the data, demonstrates the real basis of nonstationarity. Koutsoyiannis (2006b) has provided a more detailed study of this issue.

Stochastic processes describing periodic phenomena, such as those affected by the annual cycle of the Earth, are clearly nonstationary. For instance, the daily rainfall at a mid-latitude location cannot be regarded as a stationary process. Rather, a special type of a nonstationary process, whose properties depend on time in a periodical manner (are periodic functions of time), should be used. Such processes are called 'cyclostationary' processes.

2.02.1.5.4 Ergodicity

The concept of ergodicity (from the Greek words *ergon*, work; and *odos*, path) is central to the problem of determining the distribution function of a process from a single sample function (time series). A stationary stochastic process is ergodic if any statistical property can be determined from a sample function. Given that, in practice, the statistical properties are determined as time averages of time series, the above statement can be formulated alternatively – a stationary stochastic process is ergodic if time averages equal ensemble averages (i.e., expected values). For example, a stationary stochastic process is mean ergodic if

$$E[\underline{x}_i] := \lim_{N \to \infty} \frac{1}{N} \sum_{i=1}^{N} \underline{x}_i \qquad (8)$$

The left-hand side in the above equation represents the ensemble average, whereas the right-hand side represents the time average, for the limiting case of infinite time. While the left-hand side is a parameter, rather than a random variable, the right-hand side is a random variable (as a sum of random variables). Equating a parameter with a random variable implies that the random variable has zero variance. This is precisely the condition that makes a process ergodic, a condition that does not hold true for every stochastic process.

2.02.1.5.5 Some characteristic stochastic properties of precipitation

It has been widely accepted that rainfall exhibits some autocorrelation (or time dependence) if the timescale of study is daily or sub-daily, but this dependence vanishes at larger timescales, such as monthly or yearly. Thus, for timescales monthly and above, rainfall data series have been traditionally treated as independent samples. Mathematically, such a perception corresponds to a Markovian dependence at fine timescales, in which the autocorrelation decreases rapidly with time lag in an exponential manner, that is,

$$\rho_j = \rho^j \qquad (9)$$

where $\rho := \rho_1$. Then for a large lag j, or for a large scale of aggregation and even for the smallest lag (one), the autocorrelation is virtually zero (e.g., Koutsoyiannis, 2002). If $\underline{x}_i$ denotes the stochastic process at an initial timescale, which is designated as scale 1, then the averaged process at an aggregated timescale $k = 2, 3, \ldots,$ is

$$\underline{x}_i^{(k)} := \frac{\underline{x}_{(i-1)k+1} + \ldots + \underline{x}_i}{k} \qquad (10)$$

(with $\underline{x}_i^{(1)} \equiv \underline{x}_i$). Let $\sigma^{(k)}$ be the standard deviation at scale k. In processes $\underline{x}_i$ independent of time, $\sigma^{(k)}$ decreases with scale according to the well-known classical statistical law of inverse square-root, that is,

$$\sigma^{(k)} = \frac{\sigma}{\sqrt{k}} \qquad (11)$$

However, this law hardly holds in geophysical time series, including rainfall time series, whatever the scale is. This can be verified based on the examples presented in Section 2.02.1.4. A more plausible law is expressed by the elementary scaling (power-law) property

$$\sigma^{(k)} = \frac{\sigma}{k^{1-H}} \qquad (12)$$

where H is the so-called Hurst exponent, named after Hurst (1951) who first studied this type of behavior in geophysical time series. Earlier, Kolmogorov (1940), when studying turbulence, had proposed a mathematical model to describe this behavior. This behavior has been known by several names, including the Hurst phenomenon, long-term persistence, and long-range dependence, and a simple stochastic model that reproduces it is known as a simple scaling stochastic model or fractional Gaussian noise (due to Mandelbrot and van Ness, 1968). Here, the behavior is referred to as the Hurst–Kolmogorov (HK) behavior or HK (stochastic) dynamics and the model, as the HK model.

This behavior implies that the autocorrelation decreases slowly, that is, according to a power-type function, with lag j:

$$\rho_i^{(k)} = \rho_j = (1/2)[(|j+1|)^{2H} + (|j-1|)^{2H}] - |j|^{2H} \approx H(2H-1)j^{2H-2} \quad (13)$$

so that independence virtually never holds, unless $H = 0.5$, a value which reinstates classical statistics including the law in Equation (11). Most often, natural processes including rainfall are positively correlated and H varies in the range (0.5, 1).

The above framework is rather simple and allows easy exploration of data to detect whether they indicate consistence with classical statistics or with the HK behavior. A simple exploration tool is a double logarithmic plot of the estimates of standard deviation $\sigma^{(k)}$ versus scale k, which is known as a 'climacogram'. (Climacogram < Greek Κλιμακόγραμμα < (climax (κλιμαξ) = scale) + (gramma (γράμμα) = written).) In such a plot, the classical law and the HK law are manifested by a linear arrangement of points with slopes -0.5 and $H - 1$, respectively. We must bear in mind, however, that a consequence of the HK law in Equation (12) is that the classical estimator of the variance

$$\underline{s}^2 = \frac{1}{n-1}\sum_{i=1}^n (\underline{x}_i - \bar{\underline{x}})^2 \quad (14)$$

where n is the sample size, $\bar{\underline{x}} \equiv \underline{x}_1^{(n)}$ is the estimator of the mean, and $\underline{s}$ the estimator of standard deviation, implies negative bias if there is temporal dependence. The bias becomes very high for HK processes with H approaching 1. Apparently then, $\underline{s}$ could be a highly biased estimator of σ; an approximately unbiased estimator is (Koutsoyiannis, 2003a; Koutsoyiannis and Montanari, 2007):

$$\tilde{\underline{s}} := \sqrt{\frac{n'}{n'-1}}\underline{s} \quad (15)$$

where n' is the equivalent (or effective) sample size, that is, the sample size that in the framework of classical statistics would lead to the same uncertainty (in the estimation of μ by $\bar{x} \equiv x_1^{(n)}$) as an HK series yields with sample size n. For an HK process, n' is related to n by

$$n' = n^{2(1-H)} \quad (16)$$

It can be seen that n' can be very small even for high n if H is high, and thus the correcting factor $\sqrt{n'/(n'-1)}$ in Equation (15) can be very large (see Koutsoyiannis and Montanari, 2007).

Returning to the time series of globally averaged monthly precipitation in the 30-year period 1979–2008, which has been discussed earlier and is displayed in **Figure 6**, we may now study its statistical properties for several timescales. As the precipitation amounts are averaged over the entire globe, the effect of seasonality is diminished and the time series can be modeled using a stationary process rather than a cyclostationary one. **Figure 9** depicts the climacogram, that is, a logarithmic plot of standard deviation versus scale. Empirical estimates of standard deviations have been calculated using both the classical estimator in Equation (14) and the HK estimator in Equation (15). Theoretical curves resulting from the classical statistical model (assuming independence), the Markovian model, and the HK model have also been plotted. For the Markovian model, the lag one autocorrelation coefficient, estimated from the monthly data, is $\rho = 0.256$ and for the HK model, the estimate of the Hurst coefficient is $H = 0.70$. This can be obtained readily from the slope of the straight line fitted to the group of empirical points in **Figure 9**, which should be $H - 1$. Here, a slightly modified algorithm from Koutsoyiannis (2003a) has been used for the estimation of H. Overall, **Figure 9** clearly demonstrates that the empirical

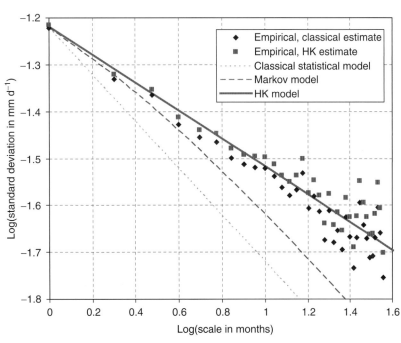

Figure 9 Climacogram of the time series of globally averaged monthly precipitation in the 30-year period 1979–2008 shown in **Figure 6**. The estimate of the Hurst coefficient for the HK model is $H = 0.70$.

points are inconsistent with the classical and Markovian models and justify an assumption of HK behavior.

Similar plots have been constructed, and are shown in **Figure 10**, for the annual precipitation time series from Seoul, Korea, and Charleston City, USA, displayed in **Figure 7**. Again, the empirical evidence from data precludes the applicability of the classical statistical model and favors the HK statistics. An additional plot for the 10-s precipitation time series in Iowa, USA, displayed in **Figure 8**, is depicted in **Figure 11**. Here, the Hurst coefficient is very high, $H = 0.96$. The difference between the empirical points based on classical statistics on the one hand and the HK statistics on the other hand is quite distinctive. Apparently, the classical model is completely inappropriate for the rainfall process.

The HK stochastic processes can be readily extended in a 2D setting (or even a multidimensional one). The 2D version of Equation (12) is

$$\sigma^{(k)} = \frac{\sigma}{k^{2-2H}} \quad (17)$$

This can be obtained by substituting k^2 for k in Equation (12). Equations (15) and (16) still hold, provided that n is the

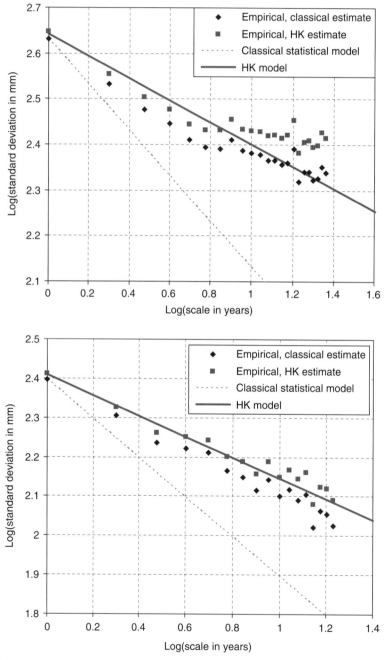

Figure 10 Climacogram of the annual precipitation time series at: (upper) Seoul, Korea and (lower) Charleston City, USA, which are shown in **Figure 7**; the estimated Hurst coefficients are 0.76 and 0.74, respectively.

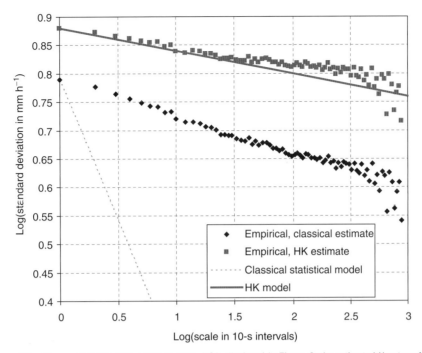

Figure 11 Climacogram of the 10-s precipitation time series in Iowa, USA, displayed in **Figure 8**; the estimated Hurst coefficient is 0.96.

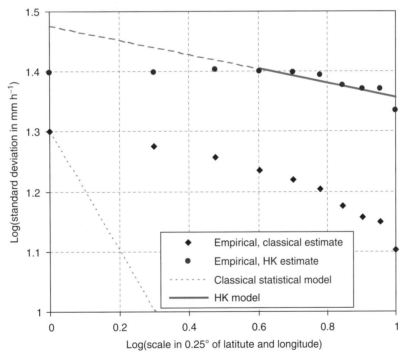

Figure 12 Climacogram of the spatial daily rainfall over the area 9° N–5° S and 78–92° E (Indian Ocean south-east of Sri Lanka) on 9 January 2006, as shown in the lower-left panel of **Figure 5**; the estimated Hurst coefficient is 0.94.

number of points, which is inversely proportional to k^2. Figure 12 demonstrates this behavior by means of a climacogram for the spatial daily rainfall over the area 9° N–5° S and 78–92° E (Indian ocean south-east of Sri Lanka) on 9 January 2006, displayed in the lower-left panel of **Figure 5**. Here, the estimated Hurst coefficient is again very high, $H = 0.94$. As in all previous cases, the classical model is completely inappropriate, while the HK model seems reasonable for scales ⩾4, which correspond to a resolution of 1° × 1° and beyond.

Thus, the evidence presented using several examples of different spatial and temporal scales indicates that HK dynamics is consistent with the nature of rainfall. These dynamics appear as scaling behavior, either in time or in space, which is either full, applicable to the entire range of scales, or asymptotic, applicable to large scales. Both these scaling behaviors are manifested as power laws of standard deviation versus temporal or spatial scale and of autocorrelation versus lag. There exists another type of scaling behavior in precipitation, the scaling in state, which is sometimes confused with the other two scaling behaviors, but is fundamentally different. Scaling in state is a property of the marginal distribution function of rainfall (it has no relation to the dependence structure of the process unlike other types of scaling) and is expressed by power laws of the tails of (1) the probability density function $f(x)$, (2) the survival function (or exceedance probability) $F^*(x) := P\{\underline{x} > x\} = 1 - F(x)$, and (3) the return period $T = \delta/F^*(x)$ where δ is the length of the timescale examined. These scaling properties are expressed as

$$x \propto T^\kappa, \quad F^*(x) \propto x^{-1/\kappa}, \quad f(x) \propto x^{-1-1/\kappa} \quad (18)$$

and are equivalent to each other. All these are asymptotic, that is, they hold only for large values of x or, in other words, for the distribution tails. Such tails are known by several names, such as long, heavy, strong, power-type, overexponential, algebraic, or Pareto tails. The latter name comes from the Pareto distribution, which in its simplest form is given in Equation (18), although its generalized form is applicable to rainfall (see Section 2.02.5.2). As this is an asymptotic behavior, long records are needed to observe it. **Figure 13** shows a logarithmic plot of the empirical distribution (expressed in terms of return period T) of a large data set of daily rainfall. This data set was created using records of 168 stations worldwide, each of which contained data of 100 years or more (Koutsoyiannis, 2004b). For each station with n years of record, n annual maximum values of daily rainfall were extracted. These values were standardized by their mean and merged into one sample of length 17 922 station-years. From the theoretical distributions, also plotted in **Figure 13**, it is observed that the Pareto distribution (whose right tail appears as a straight line in the logarithmic plot; see Section 2.02.5.2) with $\kappa = 0.15$ provides the best fit, thus confirming the applicability of asymptotic scaling in state and the inappropriateness of the exponential-type tail. This has severe consequences, particularly in hydrological design, as distributions with exponential tails have been most common in hydrological practice, whereas it is apparent that the power-type tails are more consistent with reality. As shown in **Figure 13**, the difference between the two types can be substantial.

Koutsoyiannis (2005a, 2005b) produced the aforesaid different types of scaling from the principle of maximum entropy. As entropy is a measure of uncertainty, the applicability of the principle of maximum entropy and its consistence with observed natural behaviors characterizing the precipitation process underscores the dominance of uncertainty in precipitation.

2.02.2 Physical and Meteorological Framework

Atmospheric air is a heterogeneous mixture of gases, also containing suspended particles in liquid and solid phase. The most abundant gases are nitrogen (N_2) and oxygen (O_2) that account for about 78% and 21%, respectively, by volume of the atmospheric permanent gases, followed by argon (Ar) and traces of other noble gases. Their concentrations are almost constant worldwide and up to an altitude of about 90 km. Water vapor (H_2O) appears in relatively low concentrations, which are highly variable. However, water vapor is very important for energy exchange on Earth (it accounts for 65% of the radiative transfer of energy in the atmosphere; Hemond

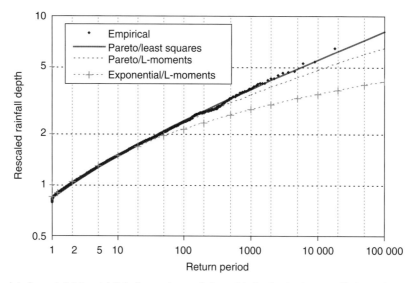

Figure 13 Logarithmic plot of rescaled daily rainfall depth vs. return period: empirical estimates from a unified sample over threshold, formed using rainfall data from 168 stations worldwide (17 922 station-years). The unified sample was rescaled by the mean of each station, and fitted using a Pareto and an exponential distribution model. Adapted from Koutsoyiannis D (2004b) Statistics of extremes and estimation of extreme rainfall: 2. Empirical investigation of long rainfall records. *Hydrological Sciences Journal* 49(4): 591–610.

and Fechner-Levy, 2000), as well as for mass transfer processes in the hydrological cycle. Under certain conditions (i.e., pressure and temperature), water vapor can transform into droplets or ice crystals with subsequent release of latent heat (see Sections 2.02.2.1 and 2.02.2.2). More generally, the water-vapor content of atmospheric air affects its density, and it is of central importance in atmospheric thermodynamics (Section 2.02.2.1). The varying content and importance of water vapor in precipitation processes and thermodynamics has led to the study of atmospheric air as a mixture of two (ideal) gases: dry air and water vapor. This mixture is usually referred to as 'moist air' and has thermodynamic properties determined by its constituents (e.g., Rogers and Yau, 1996; Cotton and Anthes, 1989).

The particles of solid and liquid material suspended in air are called 'aerosols'. Common examples of aerosols are water droplets and ice crystals (called 'hydrometeors'), smoke, sea salt (NaCl), dust, and pollen. The size distribution of solid aerosols depends strongly on their location. For example, the size spectrum of aerosols over land is narrow with high concentrations of small particles (e.g., kaolinite, dust, and pollen), whereas the size spectrum of aerosols over sea is wider with small concentrations of larger particles (e.g., sea salt; Ryan et al., 1972). Existence of aerosols in the atmosphere is of major importance, since a select group of aerosols called 'hydroscopic nuclei', is crucial for the nucleation of liquid water and initiation of rain (e.g., Brock, 1972, and Section 2.02.2.2).

When moist air is cooled (i.e., below its dew point; see Section 2.02.2.1), an amount of water vapor condenses and a cloud forms, but precipitation may or may not occur. Initiation of rain requires the formation of hydrometeors (i.e., water droplets and ice crystals) of precipitable size (e.g., Gunn and Kinzer, 1949; Twomey, 1964, 1966; Brock, 1972). Formation and growth of these particles are governed by processes that take place at scales comparable to their size (micrometers to millimeters). The latter processes form the core of cloud microphysics, whereas large-scale processes, related to thermodynamics of moist air and motion of air masses, form the core of cloud dynamics. Importantly, precipitation is the combined effect of both large- and microscale processes, and both processes are equally important and necessary for precipitation to occur.

2.02.2.1 Basics of Moist Air Thermodynamics

In a parcel of moist air at temperature T with volume V and mass $M = M_d + M_v$ with the two components denoting mass of dry air and water vapor, respectively, the density is $\rho = M/V$ and the concentration of water vapor, known as specific humidity, is $q := M_v/M$. The quantity $r = M_v/M_d := q/(1-q)$ is usually referred to as the mixing ratio. The total pressure of the moist air in the parcel, p (the atmospheric pressure), equals the sum of the partial pressures of dry air p_d and that of water vapor e (i.e., $p = p_d + e$). Specific humidity and vapor pressure are interrelated through

$$q = \frac{\varepsilon e}{p - (1-\varepsilon)e} \quad (19)$$

where $\varepsilon = 0.622$ is the ratio of molar masses of water vapor and dry air. Air cannot hold an arbitrarily high quantity of vapor. Rather, there is an upper limit of the vapor pressure e^*, called the saturation vapor pressure, which depends on the temperature T and is given by the Clausius–Clapeyron equation. A useful approximation to this equation is

$$e^*(T) = 6.11 \exp\left(\frac{17.67T}{T + 237.3}\right) \quad (20)$$

where e^* is in hPa and T is in °C. Consequently, from Equations (19) and (20) we can calculate the saturation specific humidity q^*, which is a function of T, and expresses the water vapor holding capacity of air. As shown in **Figure 14**, this capacity changes drastically, almost exponentially, with temperature, so that a change of temperature from -40 to $40\,°C$ increases this capacity by 2.5 orders of magnitude.

The ratio of the actual to saturation vapor pressure, that is, $e/e^* =: U$, called the relative humidity, is normally lesser than 1. When an air parcel cools, while e remains constant, e^*

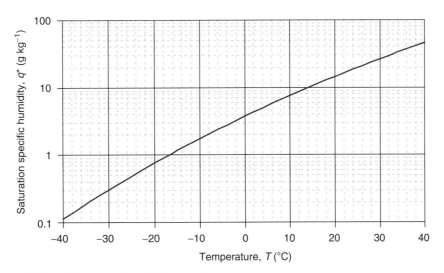

Figure 14 Saturation specific humidity as a function of air temperature.

decreases and hence U increases, up to the saturation value 1 or 100%. The temperature T_d at which saturation occurs is called the dew point temperature and is calculated using Equation (20) by setting $e^*(T_d) = e$. Therefore, cooling of the air parcel below the dew point temperature results in condensation, or transformation of the excess water vapor into liquid water in the form of droplets. During this change of phase, the relative humidity remains 100%. Condensation releases heat at a fairly constant rate ($L \approx 2.5$ MJ kg^{-1}); this rate equals that of evaporation of water at a constant temperature and is thus called latent heat.

For an air parcel to ascend and expand spontaneously, so that condensation and cloud formation can occur, the ambient (atmospheric) temperature gradient $\gamma := -dT/dz$, where z denotes altitude, also known as lapse rate, must be high (otherwise, an uplifted air parcel will sink again). While the parcel ascends and expands adiabatically (i.e., in a way that no heat transfer takes place between the air parcel and its ambient air), its own lapse rate is $\gamma_d = 9.8\,°C\,km^{-1}$ if the expansion is dry adiabatic (i.e., if it takes place without condensation of water vapor) and somewhat smaller, γ^*, if the expansion is moist adiabatic (i.e., if the temperature has fallen below dew point, so that some of the water vapor in the parcel condenses to liquid form). The gradient γ^* is not constant but varies with temperature T and air pressure p so that $\gamma^* = 4\,°C\,km^{-1}$ for $T = 25\,°C$ and $p = 1000$ hPa, whereas $\gamma^* = 9\,°C\,km^{-1}$ for $T = -25\,°C$ and $p = 1000$ hPa; an average value is $\gamma^* = 6.5\,°C\,km^{-1}$ (Koutsoyiannis, 2000b; see also Wallace and Hobbs, 1977). When the ambient lapse rate γ is smaller than γ^*, the atmosphere is stable, and no spontaneous lift occurs and no clouds are formed. When $\gamma > \gamma_d$, the atmosphere is unstable and favors air lift and formation of clouds. The case $\gamma^* < \gamma < \gamma_d$ is known as conditional instability and it serves as an important mechanism for mesoscale precipitation processes (see Sections 2.02.2.4 and 2.02.2.5).

2.02.2.2 Formation and Growth of Precipitation Particles

The Clausius–Clapeyron equation describes the equilibrium condition of a thermodynamic system consisting of bulk water and vapor. A state out of the equilibrium, in which $e > e^*$ ($U > 1$) is possible, but is thermodynamically unstable, and is called supersaturation. Detailed study of the transition of water vapor to liquid or ice at or above saturation is associated with certain free-energy barriers. An example of such an energy barrier is the dynamic energy associated with the surface tension, σ, of a water droplet. For a spherical droplet, σ is proportional to the pressure of water within the droplet p and inversely proportional to its radius r (i.e., $\sigma = p/2r$). This means that a high vapor pressure is needed for a very small droplet to be maintained and not evaporate. In essence, the free-energy barrier of surface tension makes droplet formation solely by condensation of water vapor (a process usually referred to as 'homogeneous nucleation'), almost impossible in nature. However, if the surface-tension barrier is bypassed, common supersaturations of the order of 1–2% (i.e., $U = 1.01$–1.02) are sufficient for water vapor to diffuse toward the surface of the droplet. The rate of diffusional growth is proportional to the supersaturation $U - 1$ of the ambient air, and inversely proportional to the radius r of the droplet: that is, $dr/dt \propto (U - 1)/r$ (Mason, 1971; Rogers and Yau, 1996).

While homogeneous nucleation requires large supersaturations, formation of droplets is drastically facilitated by particulated matter of the size of micrometers or lower, the aerosols, some of which, called condensation nuclei, are hydrophilic and serve as centers for droplet condensation (Brock, 1972; Slinn, 1975; Hobbs et al., 1985). This process is usually referred to as 'heterogeneous nucleation' and it is almost exclusively the process that governs water vapor condensation in the atmosphere (Houze, 1993).

When the temperature in the cloud drops below the freezing point, water droplets are said to be supercooled, and they may or may not freeze. For pure water droplets, homogeneous freezing does not occur until the temperature drops below $-40\,°C$ (Rogers and Yau, 1996). However, the presence of certain condensation nuclei, called ice nuclei, may allow freezing of water droplets at temperatures a few degrees below $0\,°C$. These nuclei are particles of the size of micrometers, or lower, which form strong bonds with water and closely match the crystallic structure of ice. Different particles serve as condensation nuclei at different subfreezing temperatures. For example, silver iodide (AgI) serves as an ice nucleator at $-4\,°C$ and kaolinite at $-9\,°C$ (e.g., Houghton, 1985).

Evidently, a cloud is an assembly of tiny droplets with usually met concentrations of several hundreds per cubic centimeter, and radii of several micrometers. This structure is very stable and the only dominant process is vapor diffusion, which accounts for the size-growth evolution of the whole droplet population (Telford and Chai, 1980; Telford and Wagner, 1981). Precipitation develops when the cloud population becomes unstable and some droplets grow faster relative to others.

In general, two main mechanisms account for the cloud microstructure becoming unstable. The first mechanism is the collision and coalescence (i.e., sticking) of larger (and faster-moving) collector drops with smaller (and slower-moving) collected droplets. This mechanism is particularly important for precipitation development in warm clouds (i.e., at temperatures in excess of $0\,°C$; see, e.g., Houze, 1993) and, for a long time, it has formed an active research area in cloud and precipitation physics (e.g., Langmuir, 1948; Bowen, 1950; Telford, 1955; Scott, 1968, 1972; Long, 1971; Drake, 1972a, 1972b; Gillespie, 1972, 1975; Robertson, 1974; Berry and Reinhardt, 1974a, 1974b; Vohl et al., 1999; Pinsky et al., 1999, 2000; Pinsky and Khain, 2004; and review in Testik and Barros (2007)). Its significance for precipitation processes depends considerably on the droplet-size spectra, with larger effectiveness for wider spectra with small concentrations of larger particles (Berry and Reinhardt, 1974a, 1974b).

The second mechanism is related to interaction between water droplets and ice crystals, and is limited to clouds with tops that extend to subfreezing temperatures (i.e., cold clouds). In particular, when an ice crystal develops in the presence of a large number of supercooled droplets, the situation becomes immediately unstable and the ice crystal grows due to diffusion of water vapor from the droplets toward the crystal. This is due to the fact that the equilibrium vapor pressure over ice is less than that over water at the same subfreezing temperature. Thus, the ice crystal grows by diffusion of water vapor and the supercooled droplets evaporate to compensate for this. The transfer rate of water vapor depends on the difference between the equilibrium vapor pressure of water and ice, a quantity that becomes sufficiently large at about $-15\,°C$ (Uijlenhoet, 2008). The latter process is called the Bergeron–Findeisen mechanism, named after the scientists who first studied it (Bergeron, 1935; Findeisen, 1938).

Once the ice crystals have grown by vapor diffusion to sizes sufficiently large for gravitational settling to dominate, they start falling and colliding with their ambient droplets and ice crystals, a process usually referred to as 'accretional growth'. In the first case (i.e., when ice crystals collide with droplets), graupel or hail may form, whereas in the second case, snowflakes are likely to form.

As the frozen particles fall, it is possible to enter layers with temperatures higher than $0\,°C$ and start melting. If the particles have relatively small terminal velocities (or equivalently small size; see Section 2.02.2.3), they may reach the ground as raindrops indistinguishable from those formed by coalescence. Alternatively, in cold weather, or when large hailstones are formed, the precipitation particles may reach the ground unmelted.

Additional discussion on the mechanisms of formation and growth of precipitation particles, and the potential human intervention on the mechanisms by technological means are discussed in **Chapter 4.05 Abstraction of Atmospheric Humidity**.

2.02.2.3 Properties of Precipitation Particles

2.02.2.3.1 Terminal velocity

The terminal velocity $U_X(D)$ of a precipitable particle of type $X = R$ (rain), H (hail), S (snow), and effective diameter D is the maximum velocity this particle may develop under gravitational settling relative to its ambient air. In theory, $U_X(D)$ can be obtained by balancing the weight of the particle with the sum of the static and dynamic buoyancy (i.e., drag forces) on the particle. For a rigid spherical raindrop, one obtains $U_R(D) \propto \sqrt{D}$ (e.g., Rogers and Yau, 1996).

Theoretical calculation of $U_X(D)$ becomes more complicated when the dynamical characteristics of the falling particles depend on their linear size D and the ambient temperature T. For example, droplets with diameters D smaller than about 0.35 mm are approximately spherical, drops with diameters in the range 0.35–1 mm tend to deform by the aerodynamic shear receiving a more elliptical shape, whereas larger drops frequently break down into smaller droplets due to excessive elongation or surface vibrations (e.g., Testik and Barros, 2007; Uijlenhoet, 2008). Moreover, the crystallic structure, shape, size, and, hence, the aerodynamic properties of snowflakes depend on the ambient temperature T (Fletcher, 1962; Locatelli and Hobbs, 1974; Houghton, 1985; Rogers and Yau, 1996).

In the absence of exact theoretical solutions for the terminal velocity $U_X(D)$ of precipitation particles under complex atmospheric conditions, several empirical formulae have been developed (e.g., Gunn and Kinzer, 1949; Liu and Orville, 1969; Wisner et al., 1972; Locatelli and Hobbs, 1974; Atlas and Ulbrich, 1977; Lin et al., 1983). According to Liu and Orville (1969), who performed a least squares analysis of Gunn and Kinzer's (1949) data, the terminal velocity of raindrops of diameter D can be approximated by a power-law type relationship:

$$U_R(D) = a\,D^b \qquad (21)$$

where $a = 2115\,\mathrm{cm^{1-b}\,s^{-1}}$ and $b = 0.8$ are empirical constants. For raindrops with diameters in the range $0.5 \leqslant D \leqslant 5$ mm, Atlas and Ulbrich (1977; see also Uijlenhoet, 2008) suggest the use of Equation (21) with parameters $a = 1767\,\mathrm{cm^{1-b}\,s^{-1}}$ and $b = 0.67$.

For hail, Wisner et al. (1972) suggest

$$U_H(D) = D^{1/2}\left(\frac{4g\rho_H}{3C_D\rho}\right)^{1/2} \qquad (22)$$

where $g = 9.81\,\mathrm{m\,s^{-2}}$ is the acceleration of gravity, $\rho \approx 1.2\,\mathrm{kg\,m^{-3}}$ is the density of air, $\rho_H = 800\text{–}900\,\mathrm{kg\,m^{-3}}$ is the density of the hailstone, and $C_D = 0.6$ is a drag coefficient.

For graupel-like snow of hexagonal type, Locatelli and Hobbs (1974) suggest:

$$U_S(D) = c\,D^d \qquad (23)$$

where $c = 153\,\mathrm{cm^{1-d}\,s^{-1}}$ and $d = 0.25$ are empirical constants that, in general, depend on the shape of the snowflakes (e.g., Stoelinga et al., 2005).

$U_X(D)$ relationships other than power laws have also been suggested (e.g., Beard (1976) and review by Testik and Barros (2007)). However, the power-law form in Equations (21)–(23) is the only functional form that is consistent with the power-law relations between the radar reflectivity factor Z (see Section 2.02.3.2) and the rainfall intensity i (Uijlenhoet, 1999, 2008).

2.02.2.3.2 Size distribution

A commonly used parametrization for the size distributions of precipitation particles is that introduced by Marshall and Palmer (1948). According to this parametrization, precipitation particles have exponential size distributions of the type

$$n_X(D) = n_{0X}\exp(-b_X D),\quad X = R, H, S \qquad (24)$$

where the subscript X denotes the type of the particle: rain (R), hail (H), or snow (S); D is the effective diameter of the particle; b_X is a distribution scale parameter with units of (length^{-1}) (see below); and n_{0X} is an intercept parameter that depends on the type of the particle with units of (length^{-4}): that is, number of particles per unit diameter and per unit volume of air (see below).

To determine the parameters n_{0R} and b_R in Equation (24) for rainfall, Marshall and Palmer (1948) used observations from summer storms in Canada. The study reported a constant value of the intercept parameter $n_{0R} = 8 \times 10^{-2}\,\mathrm{cm^{-4}}$, whereas the scale parameter b_R was found to vary with the rainfall intensity i at ground level as: $b_R = 41\,i^{-0.21}\,\mathrm{cm^{-1}}$, where i is in millimeters per hour. Clearly, the mean raindrop size $1/b_R$ increases with increasing rainfall intensity i.

Gunn and Marshall (1958) used snowfall observations from Canada to determine the parameters n_{0S} and b_S for snow. The study concluded that both n_{0S} and b_S depend on the precipitation rate as:

$$n_{0S} = 0.038\, i^{-0.87}\, \text{cm}^{-4}, \quad b_S = 25.5\, i^{-0.48}\, \text{cm}^{-1} \quad (25)$$

where i is the water equivalent (in millimeters per hour) of the accumulated snow at ground level. Similar to the mean raindrop size, the mean snowflake size $1/b_S$ increases with increasing i. A modification to the distribution model of Gunn and Marshall (1958) has been proposed by Houze et al. (1979) and Ryan (1996). According to these authors, the intercept parameter for snow, n_{0S}, is better approximated as a decreasing function of the temperature T of the ambient air. The latter is responsible for the properties and structures of ice crystals (see Section 2.02.2.2).

Federer and Waldvogel (1975) used observations from a multicell hailstorm in Switzerland to determine the parameters n_{0H} and b_H for hail. The study showed pronounced variability of the intercept parameter $n_{0H} = 15 \times 10^{-6}$ to $5.2 \times 10^{-4}\,\text{cm}^{-4}$, moderate variability of the scale parameter $b_H = 3.3$–$6.4\,\text{cm}^{-1}$, and concluded showing an exponential mean size distribution for hailstones with constant parameters: $n_{0H} \approx 1.2 \times 10^{-4}\,\text{cm}^{-4}$ and $b_H \approx 4.2\,\text{cm}^{-1}$.

Alternative models, where the size distributions of precipitation particles are taken to be either gamma or lognormal, have also been suggested (e.g., Ulbrich, 1983; Feingold and Levin, 1986; Joss and Waldvogel, 1990). However, the exponential distribution model introduced by Marshall and Palmer (1948) has been empirically validated by a number of studies (see, e.g., Kessler, 1969; Federer and Waldvogel, 1975; Joss and Gori, 1978; Houze et al., 1979; Ryan, 1996; Ulbrich and Atlas, 1998; Hong et al., 2004), and has found the widest application by being used in the cloud-resolving schemes of many state-of-the-art NWP models (e.g., Cotton et al., 1994; Grell et al., 1995; Reisner et al., 1998; Thompson et al., 2004; Skamarock et al., 2005).

A more general formulation for the size distribution of precipitation particles, which includes the exponential model of Marshall and Palmer (1948), and the gamma and lognormal models as special cases, was suggested by Sempere-Torres et al. (1994, 1998). According to their formulation, the size distribution of precipitation particles can be parametrized as

$$n_X(D) \propto i^f g(D_X/i^z), \quad X = R, H, S \quad (26)$$

where f and z are constant exponents, i is the precipitation rate, and $g(x)$ is a scalar function with parameter vector $\boldsymbol{a}$. For a certain form of g, the functional dependence of the parameters f, z, and $\boldsymbol{a}$ is obtained by satisfying the equation for the theoretical precipitation rate originating from particles with size distribution $n_X(D)$ (Sempere-Torres et al., 1994, 1998; Uijlenhoet, 2008):

$$i = \frac{\pi}{6} \int_0^\infty n_X(D) U_X(D) D^3 dD, \quad X = R, H, S \quad (27)$$

where $U_X(D)$ is given by Equations (21)–(23). Note, however, that the units of n_X depend on those used for D and i and, of course, the functional form of $g(x)$.

2.02.2.4 Clouds and Precipitation Types

Clouds owe their existence to the process of condensation, which occurs in response to several dynamical processes associated with motions of air masses, such as orographic or frontal lifting (see Section 2.02.2.5), convection, and mixing. At the same time, clouds and the resulting precipitation influence the dynamical and thermodynamical processes in the atmosphere. For example, clouds affect air motions through physical processes, such as the redistribution of atmospheric water and water vapor, the release of latent heat by condensation, and the modulation of the transfer of solar and infrared (IR) radiation in the atmosphere.

A cloud system is formed by a number of recognizable isolated cloud elements that are identifiable by their shape and size (e.g., Scorer and Wexler, 1963; Austin and Houze, 1972; Orlanski, 1975). On the lowest extreme, cloud systems with a scale of about 1 km or less are classified as microscale systems. On the highest extreme, atmospheric phenomena of linear extent of 1000 km and upward are classified in the synoptic scale and include the cloud systems associated with baroclinic instabilities, and extratropical cyclones (i.e., low-pressure centers). In between these two extreme scales, atmospheric phenomena with linear extent between a few kilometers and several hundred kilometers are the so-called mesoscale phenomena. These phenomena are more likely associated with atmospheric instabilities, as well as frontal and topographic lifting. Mesoscale phenomena include many types of clouds and cloud systems that are usually classified into two main categories: stratiform and convective (cumulus) cloud systems. In general, stratiform cloud systems have the shape of a flat appearing layer and produce widespread precipitation associated with large-scale ascent, produced by frontal or topographic lifting, or large-scale horizontal convergence. By contrast, convective cloud systems have large vertical development, produce localized showery precipitation, and are associated with cumulus-scale convection in unstable air. Next, we focus on the structure of these systems and the forms of precipitation they produce.

2.02.2.4.1 Cumulus cloud systems

Cumulus clouds are formed by small thermals (upward-moving air parcels heated by contact to the warm ground) where condensation occurs and they grow to extend vertically throughout the troposphere. Their vertical extent is controlled by the depth of the unstable layer, while their horizontal extent is comparable to their vertical extent. A typical linear dimension of a cumulus cloud is 3–10 km, with updraft velocities of a few meters per second (Rogers and Yau, 1996).

Observations performed by Byers and Braham (1949; see also Weisman and Klemp, 1986) revealed that convective storms are formed by a number of cells, each one of which passes through a characteristic cycle of stages (**Figure 15**). The cumulus stage of a cell is characterized by an updraft throughout most of the cell. At this stage, which lasts approximately 10–20 min, the cell develops and expands

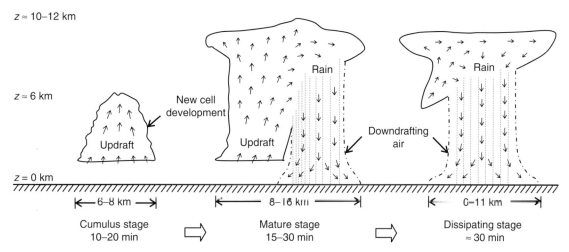

Figure 15 Stages of development of convective cells. Adapted from Weisman ML and Klemp JB (1986) Characteristics of isolated convective storms. In: Ray PS (ed.) *Mesoscale Meteorology and Forecasting*, ch. 15, pp. 331–358. Boston, MA: American Meteorological Society.

vertically while the air becomes saturated and hydrometeors grow due to vapor condensation and turbulent coalescence (see Section 2.02.2.2).

Some ice and water particles grow large enough to fall relative to the ambient updraft and initiate a downdraft within the cell. The downdraft is initially in saturated condition, but as it moves toward the lower troposphere and mixes with subsaturated air, evaporational cooling occurs, which introduces negative buoyancy and accelerates the downdraft. This is the start of the mature stage of the cell, which lasts for approximately 15–30 min. The air of the downdraft reaches the ground, as a cold core, and changes the surface wind pattern. This change may initiate a new thermal at a neighboring location, which might grow to a new cell. The downdraft interferes with the updraft at the lower levels of the cloud and finally cuts off the updraft from its source region. At this point, the cell enters its dissipating stage. At this stage, which lasts for about 30 min, the updraft decays and consequently, the precipitation source is eliminated.

2.02.2.4.2 Stratus cloud systems

Stratus clouds are associated with mesoscale, or even synoptic, vertical air motions that arise from large-scale horizontal convergence and frontal or orographic lifting of moist air masses. The ascending motion of air is weak (i.e., a few tens of centimeters per second) relative to cumulus convection, but it extends over large areas and durations to produce widespread rain or snow.

The lifetime of a stratus formation is of the order of days, and its size may extend over hundreds of kilometers horizontally. The ascended air masses, having the form of a flat appearing layer, remain convectively stable even after they are lifted to higher altitudes. Since atmospheric turbulence is not intense, initiation of rain is mainly dominated by the ice particle growth due to vapor deposition (the Bergeron–Findeisen mechanism; Section 2.02.2.2), when the ascended air masses are thick enough to reach subfreezing temperatures. In general, thin stratus clouds are usually nonprecipitating, whereas thick stratus clouds (i.e., 1–2 km vertical extent) are capable of producing substantial widespread rain or snow.

Although the classification of cloud systems in stratiform and convective is useful for observation purposes, it cannot be considered sharp (Harrold and Austin, 1974). Observations from radars or rain gauges show that widespread precipitation has a fine-scale structure with intense precipitation regions confined to elements with size of a few kilometers, while rainfall features of convective origin (e.g., cells) can grow and/or cluster over a large region producing continuous precipitation similar to that of stratiform formations.

In general, convective rainfall patterns are nonuniform and are associated with locally intense rainfall regions ranging in size from 3 to 10 km. The latter evolve rapidly in time and are separated by areas free of precipitation. By contrast, stratiform patterns are associated with less-pronounced small-scale structures and a wider overall extent that persists in time.

2.02.2.5 Precipitation-Generating Weather Systems

2.02.2.5.1 Fronts

Atmospheric circulation is formed by advecting air masses with fairly uniform characteristics. Depending on their source of origin, different air masses may have different temperatures and moisture contents. For example, continental air masses are drier and their temperatures vary in a wider range relative to maritime air masses. The interface of two opposing air masses with different temperatures and moisture contents is usually referred to as a front. Along this interface, the warmer and lighter air rises above the colder and denser air. The vertical lifting causes the warmer air to cool adiabatically, the water vapor to condense, and, hence, precipitation to form.

A cold front occurs when advancing cold air wedges itself under warmer air and lifts it (**Figure 16(a)**), whereas a warm front develops when faster-moving warm air overrides a colder and denser air mass (**Figure 16(b)**; Koutsoyiannis and Xanthopoulos, 1999). An occluded front forms when warm air is trapped between two colder and denser air masses. An example of an occluded front is shown in **Figure 16(c)**, where a cold front catches up a slower-moving warm front.

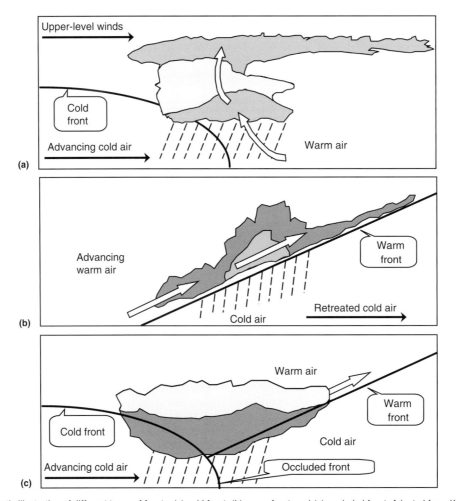

Figure 16 Schematic illustration of different types of fronts: (a) cold front, (b) warm front, and (c) occluded front. Adapted from Koutsoyiannis D and Xanthopoulos T (1999) *Engineering Hydrology*, 3rd edn., p. 418. Athens: National Technical University of Athens (in Greek).

Fronts may extend over hundreds of kilometers in the horizontal direction and are associated with vertical wind speeds of the order of a few tens of centimeters per second. This range of values is in accordance with vertical motions caused by the horizontal wind convergence of synoptic-scale low-level flow. Hence, frontal precipitation is mostly stratiform with widespread rain or snow over large areas and durations. Note, however, that embedded within the areas of frontal precipitation there are mesoscale regions that exhibit cellular activity.

2.02.2.5.2 Mechanical lifting and orographic precipitation

Orographic precipitation occurs when horizontally moving warm and humid air meets a barrier such as a mountain range. In this case, the barrier causes uplift of the incoming air. As the moist air moves upslope, it cools adiabatically, water vapor condenses to liquid water or ice (depending on the altitude where the dew point temperature occurs), and precipitation is likely to form (e.g., Smith, 1993; Hemond and Fechner-Levy, 2000). In general, orographic precipitation (unless combined with other mechanisms such as cyclonic activity and fronts) is narrow banded since it occurs in association with water-vapor condensation by mechanical lifting, a process that becomes effective at a certain elevation along the topography. After surpassing the top of the mountain range, on the lee side, the air moves downward and this causes adiabatic warming, which tends to dissipate the clouds and stop the precipitation, thus producing a rain shadow.

2.02.2.5.3 Extratropical cyclones

Extratropical cyclones are synoptic scale low-pressure systems that occur in the middle latitudes (i.e., pole-ward of about 30° latitude) and have length scales of the order of 500–2500 km (e.g., Hakim, 2003). They usually form when two air masses with different temperatures and moisture contents that flow in parallel, or are stationary, become coupled by a preexisting upper-level disturbance (usually a low-pressure center) near their interface.

An example is the formation of extratropical cyclones along the interface of mid-latitude westerlies (i.e., winds that flow from West to East; e.g., Lutgens and Tarbuck, 1992), with the equator-ward-moving polar, and thus colder, air masses (i.e., polar easterlies). As shown in **Figure 17**, which refers to the Northern Hemisphere, the motion of both warm and cold air masses is caused by pressure gradients and their direction is south–north and north–south, respectively (Koutsoyiannis

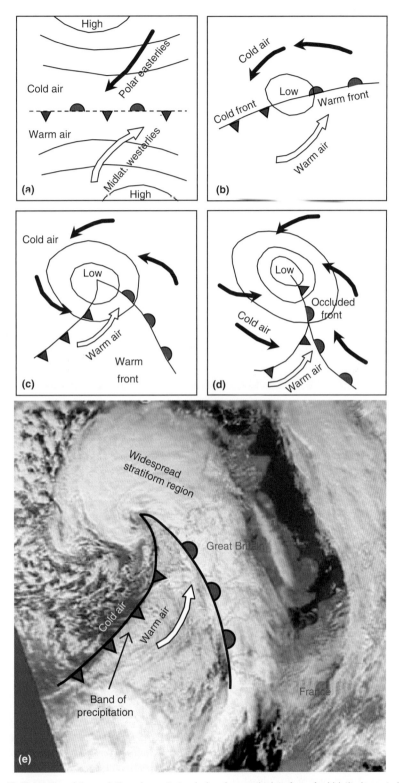

Figure 17 (a)–(d) Schematic illustration of the evolution of an extratropical cyclone at the interface of mid-latitude westerlies and the equator-ward-moving polar easterlies; and (e) extra-tropical cyclone over the British Isles on 17 January 2009: motion of air masses, fronts, and characteristic precipitation regions. (a)–(d), Adapted from Koutsoyiannis D and Xanthopoulos T (1999) *Engineering Hydrology*, 3rd edn., p. 418. Athens: National Technical University of Athens (in Greek); and (e) from http://www.ncdc.noaa.gov/sotc/index.php?report = hazards&year = 2009&month = jan).

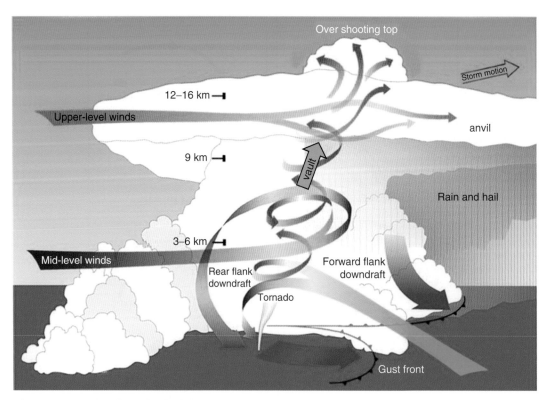

Figure 18 Schematic illustration of the wind circulation in a super-cell storm. Adapted from http://www.nssl.noaa.gov/primer/tornado/.

and Xanthopoulos, 1999). However, these directions are diverted to the right (in the Northern Hemisphere) by Coriolis forces. The initial disturbance formed by the shear along the interface of the two air masses (**Figure 17(a)**) grows as the warmer and lighter air rises above the colder air and starts rotating in an emerging spiral called the cyclone (**Figure 17(b)**). As the cyclone evolves, the cold front approaches the slower-moving warm front (**Figure 17(c)**) and then catches up with it forming an occluded front (**Figure 17(d)**). Finally, mixing between the two air masses causes the fronts to lose their identities and the cyclone to dissipate. The adiabatic cooling of the warm and moist air results in a widespread region of stratiform precipitation that propagates with the upper-level flow far beyond the fronts (**Figure 17(e)**).

2.02.2.5.4 Isolated extratropical convective storms

A short-lived single-cell is the simplest storm of convective origin. Single cells have horizontal cross sections of the order of 10–100 km^2 and move with the mean environmental flow over the lowest 5–7 km of the troposphere. The stages of development of a single-cell storm were discussed in Section 2.02.2.4. The multicell storm is a cluster of short-lived single cells with cold outflows (i.e., downdrafts) that combine to form a large gust front (Weisman and Klemp, 1986). The convergence along the leading edge of the front triggers new updraft development and subsequent formation of new cells. Because of the new cell development, multicell storms may last several days and span over large areas with linear extents of hundreds of kilometers.

The super-cell storm is the most intense of all isolated convective storms. It has a lifetime of several hours, it exhibits large vertical development, and produces strong winds, heavy rainfall, or hail, and long-lived tornadoes, that is, intense vortices with diameters of the order of 100–500 m (e.g., Browning and Ludlam, 1962; Rotunno, 1986; Weisman and Klemp, 1986; Bluestein, 2003), where the updrafts and downdrafts are displaced horizontally and interact mutually to sustain a long-lived circulation (**Figure 18**). The updraft enters at low levels and ascends in a region called the vault, which might penetrate into the stratosphere. Super-cell storms usually evolve from multicell formations when the magnitude of the vertical wind shear, defined as the difference between the density-weighted mean wind over the lowest 6 km and a representative surface layer wind (e.g., 500 m mean wind), suffices to produce a long-lived rotating updraft that mutually interacts with the downdraft (e.g., Weisman and Klemp, 1982, 1984, 1986).

2.02.2.5.5 Extratropical squall lines and rainbands

Intense rainfall events are usually organized in lines (i.e., squall lines) and bands (i.e., rainbands) with characteristic scales of hundreds of kilometers. According to Hane (1986), rainbands are sufficiently elongated rainfall areas that are nonconvective or weakly convective, and squall lines include all linear convective structures stronger than rainbands. These large-scale features are considered to be manifestations of the large mesoscale horizontal circulation, in association with spatial fluctuations of the surface temperature and moisture content of atmospheric air masses.

The conditions for squall line formation are (1) a convectively unstable near-surface environment (i.e., moist and warm near-surface air with relatively cold air aloft) to maintain the development of convective cells, (2) a layer of dry air directly above the near-surface moist air to enhance development of an intense and wide cold downflow by evaporative cooling (i.e., the dry middle-level air causes precipitation particles to evaporate and a negatively buoyant cold front to form), and (3) a triggering mechanism for release of the convective instability (e.g., frontal or orographic lifting). Once the squall line has formed, it feeds itself through convergence along the cold gust front. This convergence produces strong ascent and forms new cells ahead of the storm.

Rainbands in extratropical regions occur primarily in association with well-organized extratropical cyclones (Hane, 1986). In this case, precipitation is maintained by the ascent resulting from the warm advection of the advancing cyclone, with subsequent formation of a widespread region of stratiform precipitation (Section 2.02.2.5.3). Extratropical rainbands can also be formed in synoptic-scale environments other than those associated with cyclonic circulation. An example is the environment associated with the development of symmetric instabilities (e.g., Bennetts and Sharp, 1982; Seltzer et al., 1985).

2.02.2.5.6 Monsoons

The term monsoon generally applies to climates that exhibit long, distinct, and remarkably regular rainy and dry periods associated with the spatial distribution of solar heating during summer and winter. According to a definition proposed by Ramage (1971), a monsoon climate is characterized by (1) prevailing wind directions that shift by at least 120° between January and July, (2) prevailing wind direction that persists at least 40% of the time in January and July, (3) mean wind speeds that exceed $3\,m\,s^{-1}$ in either January or July, and (4) fewer than one cyclone–anticyclone alternation every 2 years in either January or July in a 5° latitude–longitude rectangle. In essence, Ramage's (1971) criteria exclude most extratropical regions with prevailing synoptic-scale cyclonic and anticyclonic circulations and, in addition, require the mean wind direction to be driven and sustained exclusively by the seasonally varying temperature contrast between continental and oceanic masses. Under these constraints, only India, South-Eastern Asia, Northern Australia, and West and central Africa have monsoon climates (Slingo, 2003). For example, in India, about 80% of the mean annual rainfall accumulation (about 2 m) occurs during the months of June, July, and August (Smith, 1993).

The main driving mechanism for monsoons is the temperature contrast between continental and oceanic masses due to the seasonal cycle of solar heating. More precisely, the lower thermal inertia of continental masses relative to oceans causes the former to heat up more rapidly during spring and summer by the solar radiation. This results in a sharp temperature gradient, which causes a humid flow of oceanic near-surface air to move toward the land (something similar to a massive sea breeze). As it reaches the land, the humid air warms up and rises, water vapor condenses to liquid water, and rain falls. A similar process occurs during winter, when the continental air masses cool up more rapidly than the surrounding ocean water, with subsequent formation of a cold and dry massive low-level flux toward the ocean.

An important factor that determines the intensity of monsoon rainfall is the geographical orientation of continents and oceans relative to the equator (Slingo, 2003). For example, the north–south orientation of the South-Eastern Asian and Northern Australian monsoon system allows the dry outflow from the winter continent to warm up and load moisture from the ocean, flow across the equator toward the summer hemisphere, and, eventually, feed the monsoon rains over the summer continent. This is also the reason why the largest rainfall accumulations for durations larger than 24 h are associated with the Asian–Australian monsoon system (Smith, 1993).

2.02.2.5.7 Tropical cyclones

Tropical cyclones form a particular class of synoptic-scale low-pressure rotating systems that develop over tropical or subtropical waters (Anthes, 1982; Landsea, 2000). These systems have linear extent of the order of 300–500 km and are characterized by well-organized convection and cyclonic (counterclockwise in the Northern Hemisphere) surface wind circulation around a relatively calm low-pressure region, called the eye of the storm (**Figures 19** and **20**). Tropical cyclones with sustained wind speeds in the range $17-32\,m\,s^{-1}$ are called tropical storms whereas stronger tropical cyclones are usually referred to as hurricanes (i.e., when observed in the North Atlantic Ocean, in the Northeast Pacific Ocean east of the dateline, and in the South Pacific Ocean east of 160° E) or typhoons (i.e., when observed in the Northwest Pacific Ocean west of the dateline). Note, however, that extreme rainfall accumulations for durations of the order of a day, or higher, are usually produced by moderate or even low-intensity tropical cyclones (Langousis and Veneziano, 2009b). An example is the tropical storm Allison in 2001, which looped over the Houston area causing rainfall accumulations in excess of 850 mm. According to the US National Oceanic and Atmospheric Administration (NOAA; Stewart, 2002), Allison (2001) ranks as the costliest and deadliest tropical storm in the history of the US with 41 people killed, 27 of who were drowned, and more than $6.4 billion (2007 USD) in damages.

The genesis and development of tropical cyclones require the following conditions to be maintained (e.g., Gray, 1968, 1979): (1) warm ocean waters (surface temperature $T > 27\,°C$); (2) a conditionally unstable atmosphere where the air temperature decreases fast with height; (3) a relatively moist mid-troposphere to allow the development of widespread thunderstorm activity; (4) a minimum distance of about 500 km from the equator in order for the Coriolis force to be sufficiently large to maintain cyclonic circulation; (5) a near-surface disturbance with sufficient vorticity and low-level convergence to trigger and maintain cyclonic motion; and (6) low magnitude of vertical wind shear (less than $10\,m\,s^{-1}$), defined as the difference between the 200- and 850-hPa horizontal wind velocities in the annular region between 200 and 800 km from the tropical cyclone center (Chen et al., 2006). The latter condition is important for the maintenance of the deep convection around the center of the cyclone.

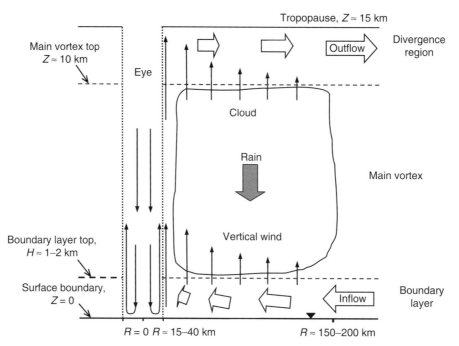

Figure 19 Schematic representation of the structure of a mature hurricane.

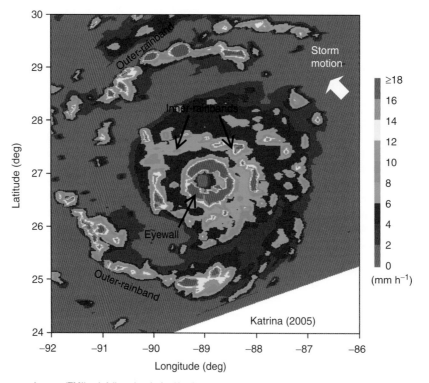

Figure 20 TRMM microwave imager (TMI) rainfall retrievals for Hurricane Katrina on 28 August (2005) at 21:00 UTC (frame 44373): different types of rainbands and their locations relative to the center of the storm.

At a first approximation, a tropical cyclone can be seen as a heat engine fueled by the buoyant motion of warm and saturated (hence convectively highly unstable) air masses that lie directly above the warm tropical and subtropical ocean waters (e.g., Emanuel, 1986, 1989; Renno and Ingersoll, 1996; Marks, 2003). By contrast, extratropical cyclones obtain their energy from the horizontal temperature gradients in the atmosphere (Section 2.02.2.5.3).

During its mature stage, a tropical cyclone includes four distinct flow regions (Yanai, 1964; Smith, 1968; Frank, 1977; Willoughby, 1990; Smith, 2000), as depicted in **Figure 19**:

1. Away from the surface boundary (in the altitude range from 2–3 km to about 10 km), frictional stresses are negligible and the horizontal winds are in approximate gradient balance (e.g., La Seur and Hawkins, 1963; Hawkins and Rubsam, 1968; Holland, 1980; Willoughby, 1990, 1991; Vickery et al., 2000). In this region, usually referred to as the main vortex, the radial inflow is negligible, whereas the tangential flow is maintained by the balance between the inward-directed pressure gradient force and the sum of the outward-directed centrifugal and Coriolis forces.
2. Within the boundary layer (in the altitude range below 1–2 km), frictional stresses decelerate the tangential flow, reduce the magnitude of the Coriolis and centrifugal forces, and result in an inward net force that drives low-level convergence. Calculations performed by Smith (1968, 2003), Kepert (2001), Kepert and Wang (2001), and Langousis et al. (2008) show that the radial inflow in the boundary layer turns upward before it reaches the tropical cyclone center causing vertical fluxes of moisture. Langousis and Veneziano (2009a) showed that these fluxes can be used to obtain accurate estimates for the large-scale mean rainfall intensity field in tropical cyclones as a function of the tropical cyclone characteristics.
3. At altitudes in excess of about 10 km, the curved isobars, which are responsible for the tropical cyclone formation and maintenance, start to flatten. As a consequence, the inward-directed pressure gradient force that maintains the cyclonic circulation decreases with increasing height leading to an outward-directed net force that drives high-level divergence.
4. Finally, there is a core flow region, called the eye of the tropical cyclone, with diameters of the order of 15–40 km. This region is free of cloud with light tangential winds and a downflow close to the axis.

The condensation of water vapor caused by the ascending motion of humid near-surface air leads to the formation of cloud systems. These systems, which are usually precipitating, are organized around the cyclone center into long quasi-circular formations usually referred to as rainbands. Despite variations of rainband characteristics from one storm to another, and during the evolution of a single storm (e.g., Miller, 1958; Barnes et al., 1983; Marks, 1985; Molinari et al., 1999), a number of studies (Willoughby et al., 1984; Powell, 1990; Molinari et al., 1994, 1999, among others) have shown that rainbands, depending on their location relative to the storm center, share similar structural characteristics and can be organized into three distinct classes: eyewall, inner-rainbands, and outer-rainbands (**Figure 20**):

1. The eyewall is a well-developed convective band that surrounds the eye of the tropical cyclone. This band has a width of approximately 10–15 km with upward-directed quasi-steady velocities in the range of 0.5–3 m s^{-1} or more, with the larger values being associated with more intense systems. The quasi-steady updrafts mostly reflect the radial convergence of horizontal fluxes, which become maximum close to the eye of the tropical cyclone (Smith, 1968; Shapiro, 1983; Kepert, 2001). The eyewall almost always has the highest cloud tops (Jorgensen, 1984a), contains the largest annular mean rainfall intensity (Marks, 1985; Houze et al., 1992), and exhibits weak cellular structure as evidenced by radar observations (e.g., Jorgensen, 1984b; Marks, 1985).
2. The inner-rainbands (Molinari et al., 1994, 1999) are a group of spiral bands located outside the eyewall at radial distances smaller than approximately 120 km, and are also referred to as a stationary band complex (Willoughby et al., 1984). This group moves slowly, if at all, and maintains a rather fixed position relative to the vortex. Rainfall inside the inner-rainband region is mostly stratiform, with active convection covering 5–10% of the total rainfall area and contributing 40–50% of the total rainfall volume (e.g., Marks, 1985; Marks and Houze, 1987; Marks et al., 1992).
3. Outer-rainbands typically occur at radial distances larger than approximately 150 km from the tropical cyclone center (e.g., Powell, 1990; Molinari et al., 1994). They develop by the increased convergence at the boundary of the vortex envelope, where the convectively unstable environmental air flows around the storm and gives rise to formation of convective cells (e.g., Beer and Giannini, 1980; Ooyama, 1982; Molinari et al., 1994). Consequently, outer-rainbands have more cellular structure than inner ones, which develop in a less-unstable atmosphere.

2.02.3 Precipitation Observation and Measurement

2.02.3.1 Point Measurement of Precipitation

2.02.3.1.1 Measuring devices

The measurement of precipitation at a point is as easy as placing a bucket at the point of observation and periodically measuring the quantity of water it collects. The collected volume divided by the area of the opening is the precipitation depth. Due to this simplicity, such gauges have been used systematically since many centuries, and must have been discovered independently in different times, perhaps even in the antiquity, and in different places in the world, such as in ancient Greece and ancient India (Kosambi, 2005). However, their records have not survived; so the oldest available records now are those in Seoul, Korea, already presented in Section 2.02.1.3 and **Figure 7** (upper), which go back to 1770, even though measurements must have been taken in much earlier periods since 1441 (Arakawa, 1956).

The traditional device for rainfall measurement, known as rain gauge or pluviometer, is still in use today and, in fact, remains the most accurate device also providing the calibration basis for new measurement devices and techniques. It is a simple cylinder whose opening has an area (e.g., 200–500 cm^2 according to World Meteorological Organization (1983)) larger than (e.g., 10-fold) the cross section of the cylinder, which allows a greater sensitivity of the reading of the rainfall depth in a millimetric ruler attached to the cylinder. In another type of instrument, known as cumulative gauge, which is placed in inaccessible areas, the diameter of the cylinder may be larger than that of the opening, to enable the storage of a large volume of precipitation between the times of two visits to the place.

In an autographic (or recording) rain gauge, also known as a pluviograph, the water depth in the cylinder is recorded with the help of a mechanism involving a floating device. Another type of recording gauge, known as a tipping-bucket gauge, introduces the rainwater to one of a pair of vessels with a known small capacity (typically equivalent to 0.2 mm of rainfall) that is balanced on a fulcrum; when one vessel is filled, it tips and empties, while the time of this event is recorded, and the other vessel is brought into position for filling. In traditional autographic devices, these recordings are done on a paper tape attached to a revolving cylinder driven by a clockwork motor that is manually wound. In modern instruments, this device is often replaced by electronic systems, which provide digital recordings on a data logger and/or a computer connected by a cable or radio link.

A rain gauge does not include all precipitation forms, snow in particular, except in light snowfalls when the temperature is not very low and the snow melts quickly. Generally, accurate measurement of snow precipitation (the water equivalent) needs specific instruments, equipped with a heating device to cause melting of snow. If such an instrument is not available, the snow precipitation is estimated as 1/10 of the snowfall depth (see justification in Section 2.02.1.3).

2.02.3.1.2 Typical processing of rain gauge data

Measurement of precipitation in rain gauges is followed by several consistency checks to locate measurement errors and inconsistencies. Errors are caused due to numerous reasons, including human lapses and instrument faults, which may be systematic in case of inappropriate maintenance. Inconsistencies are caused by changes of installed instruments, changes in the environmental conditions (e.g., growing of a tree or building of a house near the rain gauge), or movement of the gauge to a new location. When errors are detected, corrections of the measurements are attempted.

The standard meteorological practices include checks of outliers (a measured value is rejected if it is out of preset limits), internal consistency (checks are made whether different variables, e.g., precipitation and incoming solar radiation are compatible with each other), temporal consistency (the consistency of consecutive measurements is checked), and spatial consistency (the consistency of simultaneous measurements in neighboring stations is checked). Such checks are done in the timescale of measurement (e.g., daily for pluviometers or hourly for pluviographs) but systematic errors can only be located at aggregated (e.g., annual) timescales.

The most popular method applied at an aggregated timescale for consistency check and correction of inconsistent precipitation data is that of the double-mass curve, which is illustrated in **Figure 21**. The method has a rather weak statistical background and is rather empirical and graphical (but there is a more statistically sound version in the method by Worsley (1983)). The double-mass curve is a plot of the successive cumulative annual precipitation Σy_i at the gauge that is checked versus the successive cumulative annual precipitation Σx_i for the same period of a control gauge (or the average of several gauges in the same region). If the stations are close to each other and lie in a climatically homogeneous region, the annual values should correlate to each other. *A fortiori*, if the two series are consistent with each other, the cumulated values Σy_i and Σx_i are expected to follow a proportionality relationship. A departure from this proportionality can be interpreted as a systematic error or inconsistency, which should be corrected. Such a departure is usually reflected in a change in the slope of the trend of the plotted points. The aggregation of annual values x_i and y_i to calculate Σx_i and Σy_i is typically done from the latest to the oldest year. **Figure 21** (upper) shows the double-mass curve for 50 pairs of values representing annual precipitation at two points, whose cross-correlation (between x_i and y_i) is 0.82. The newest 25 points form a slope of $m = 0.70$, whereas the oldest 25 form a much greater slope, $m' = 0.95$. Assuming that the newest points are the correct ones (with the optimistic outlook that things are better now than they were some years before), we can correct the older 25 annual y_i by multiplying them with the ratio of slopes, $\lambda = m/m' := 0.737$. A second double-mass curve, constructed from the corrected measurements, that is, from $y'_i := y_i$ for $i \leqslant 25$ (the newest years) and $y'_i := \lambda y_i$ for $i > 25$ (the oldest years) is also shown in **Figure 21** (upper).

In fact, the data values used in **Figure 21** are not real rainfall data but rather are generated from a stochastic model (Koutsoyiannis, 2000a, 2002) so that both stations have equal mean and standard deviation (1000 and 250 mm, respectively), be correlated to each other (with correlation coefficient 0.71) and, most importantly, exhibit HK behavior (with $H = 0.75$, compatible with the values found in the real-world examples of Section 2.02.1.5). Hence, evidently, all values are correct, consistent, and homogeneous, because they were produced by the same model assuming no change in its parameters. Thus, the example illustrates that the method can be dangerous, as it can modify measurements, seemingly inconsistent, which however are correct. While this risk inheres even in time-independent series, it is largely magnified in the presence of HK behavior. **Figure 21** (lower) provides a normal probability plot of the departure of the ratio λ from unity (where the horizontal axis z is the standard normal distribution quantile and the distributions were calculated by the Monte Carlo method) for two cases: assuming independence in time and assuming HK behavior with $H = 0.75$ as in the above example. The plots clearly show that, for the same probability, the departure of λ from unity in the HK case is twice as high as in the classical independence case. For the HK case, departures of ± 0.25 from unity appear to be quite normal for 25-year trends and even more so for finer timescales, that is, ± 0.35 to ± 0.40 for 10-year to 5-year consecutive trends (not shown in **Figure 21**). Note that the method is typically applied even for corrections of as short as 5-year trends (Dingman, 1994), and so its application most probably results in distortion rather than correction of rainfall records.

Apparently, the correction of the series using the double-mass curve method removes these trends that appear in one of the two time series. Removal of trends results in reduction of the estimated Hurst coefficient or even elimination of the exhibited HK behavior (Koutsoyiannis, 2003a, 2006b). Thus, if we hypothesize that the HK behavior is common in precipitation, application of methods such as the double-mass curves may have a net effect of distortion of correct data, based on a vicious circle logic: (1) we assume time independence of

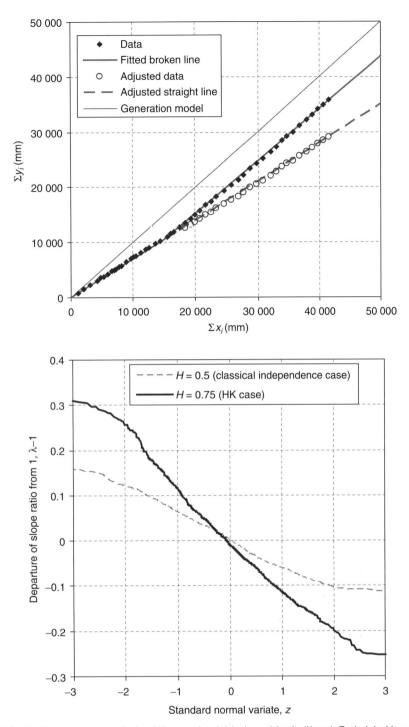

Figure 21 Illustration of the double-mass curve method and the associated risks in applying it. (Upper): Typical double-mass curve for 50 pairs of points, where the first 25 (newest) and the last 25 (oldest) form slopes $m = 0.7$ and $m' = 0.95$, respectively; the adjusted points with $\lambda = m/m' = 0.737$ are also shown. (Lower): Comparison of probability distributions of the departure of the ratio λ from unity for series independent in time or with HK behavior with $H = 0.75$; the distributions were calculated using the Monte Carlo method based on synthetic series with a total size of 1000.

the rainfall process; (2) we interpret manifestation of dependence (the HK behavior in particular) as incorrectness of data; (3) we modify the data so as to remove the influence of dependence; and (d) we obtain a series that is much closer to our faulty assumption of independence. The widespread use of the double-mass curve method in routine processing of precipitation time series may thus have caused enormous distortion of the real history of precipitation at numerous stations worldwide, in addition to masking HK behavior.

The above discourse aims to issue a warning against unjustified use of consistency check and correction methods that could eliminate the extreme values (see, e.g., the note about the Seoul station in Section 2.02.1.3) and the long-term variability implied by the HK behavior; the effect of both these mistreatments of data is a serious underestimation of design precipitation and flow in engineering constructions and management decisions. As a general advice for their correct application, we can stress that all methods of this type should never be applied blindly. An inspection of local conditions (environment of the rain gauge station and practices followed by the observer) as well as of the station's archive history is necessary before any action is taken toward altering the data. Unless information on local conditions and archive history justify that inconsistencies or errors exist, corrections of data should be avoided.

2.02.3.1.3 Interpolation and integration of rainfall fields

The interpolation problem, that is, the estimation of an unmeasured precipitation amount y from related precipitation quantities x_i ($i = 1, \ldots, n$) is encountered very often in routine hydrologic tasks, such as the infilling of missing values of recorded precipitation at a station or the estimation of precipitation at an ungauged location. The integration problem refers to the estimation of an average quantity y over a specified area (or time period) based on measurements x_i ($i = 1, \ldots, n$) of the same quantity and the same time period at different points (or respectively, at different time periods at the same point). The literature provides a huge diversity of methods, most of which, however, could be reduced to a linear statistical relationship applicable to both the interpolation and the integration problem:

$$\underline{y} = w_1 \underline{x}_1 + \ldots + w_n \underline{x}_n + \underline{e} \qquad (28)$$

where w_i denotes a numerical coefficient (weighting factor) and $\underline{e}$ denotes the estimation error. The same could be written in vector form:

$$\underline{y} = \mathbf{w}^T \underline{\mathbf{x}} + \underline{e} \qquad (29)$$

with $\mathbf{w} := [w_1, \ldots, w_n]^T$ and $\underline{\mathbf{x}} := [\underline{x}_1, \cdots, \underline{x}_n]^T$, and the superscript T denotes the transpose of a vector (or a matrix). The notation in Equations (28) and (29) suggests that $\underline{\mathbf{x}}$, $\underline{y}$, and $\underline{e}$ are treated as random variables, even though this may not be necessary in some of the existing methods. All interpolation techniques provide a means for estimating the numerical coefficients w_i, either conceptually or statistically, whereas the statistical methods provide, in addition, information about the error. Most commonly, the latter information includes the expected value $\mu_e := E[\underline{e}]$ and its standard deviation $\sigma_e := (\text{Var}[\underline{e}])^{1/2}$. A statistical estimation in which $E[\underline{e}] = 0$ is called unbiased, and one in which the mean square error $\text{MSE} := E[\underline{e}^2] = \sigma_e^2 + \mu_e^2$, is the smallest possible, is called best; if both these happen, the estimation is called best linear unbiased estimation (BLUE). While the BLUE solution is in principle quite simple (see below), the estimation of its weighting factors is not always straightforward. Hence, several simplified statistical methods as well as empirical conceptual methods are in common use. Another reason that explains why such a diversity of methods has emerged is the different type of objects that each of the elements of the vector $\underline{\mathbf{x}}$ represents. For instance, in temporal interpolation, these elements can be observed values at times before and after the time of interpolation. In spatial interpolation these can be simultaneously observed values for stations lying in the neighborhood of the point of interpolation. Simultaneous temporal and spatial interpolation, although unusual, may be very useful. For example, an optimal way to infill a missing value in a time series at a specific time would be to include in $\underline{\mathbf{x}}$ measurements taken in neighboring gauges at this specific time, as well as measurements taken at the point of interest at preceding and subsequent times.

Let us first examine the different methods in which the estimation of y is based on a single observation $\underline{x} \equiv \underline{x}_i$ at one neighboring (in space or time) point only. Here is a list of options, in which the following notation has been used: $\mu_x := E[\underline{x}]$ and $\mu_y := E[\underline{y}]$ are the expected values of $\underline{x}$ and $\underline{y}$, respectively; $\sigma_x^2 := E[(\underline{x} - \mu_x)^2]$ and $\sigma_y^2 := E[(\underline{y} - \mu_y)^2]$ are the variances of $\underline{x}$ and $\underline{y}$, respectively; $\sigma_{xy} := E[(\underline{x} - \mu_x)(\underline{y} - \mu_y)]$ is the covariance of $\underline{x}$ and $\underline{y}$; and $\rho_{xy} := \sigma_{xy}/(\sigma_x \sigma_y)$ is the correlation of $\underline{x}$ and $\underline{y}$.

1. *Equality*: $y = x$. The single point of observation considered in this naive type of interpolation is the station i nearest to the interpolation point, with $\underline{x} \equiv \underline{x}_i$. As discussed below, this simple interpolation forms the background of the Thiessen method of spatial integration. It is generally biased, with bias $\mu_e = \mu_y - \mu_x$ and its MSE is $\sigma_y^2 + \sigma_x^2 - 2\sigma_{xy} + \mu_e^2$. However, if the precipitation field is stationary (so that the means and variances at all points are equal to global parameters μ and σ^2, respectively), it becomes unbiased, with $\text{MSE} = 2\sigma^2(1 - \rho_{xy})$. Evidently, for $\rho_{xy} < 0.5$, the method results in $\text{MSE} > \sigma^2$ and therefore there is no meaning in adopting it for low correlation coefficients (an estimate $x = \mu$ would be more effective).

2. *Normal ratio*: $y = w\,x$ with $w = \mu_y/\mu_x$. This is a better alternative to the equality case, but it requires a sample of measurements to be available for y in order to estimate the average μ_y. This estimation is unbiased ($\mu_e = 0$) but not best ($\text{MSE} = \sigma_y^2 + \sigma_x^2\, \mu_y^2/\mu_x^2 - 2\,\sigma_{xy}\,\mu_y/\mu_x$).

3. *Homogenous linear regression*: $y = wx$ with $w = E[\underline{y}\,\underline{x}]/E[\underline{x}^2] = (\sigma_{xy} + \mu_x\mu_y)/(\sigma_x^2 + \mu_x^2)$. This is a biased estimation ($\mu_e = \mu_y - w\,\mu_x$) albeit best ($\text{MSE} = \sigma_y^2 + (\mu_y^2\sigma_x^2 - 2\mu_x\mu_y\sigma_{xy} - \sigma_{xy}^2)/(\mu_x^2 + \sigma_x^2)$).

4. *Linear regression*: $y = w\,x + b$ with $w = \text{Cov}[\underline{y}\,\underline{x}]/\text{Var}[\underline{x}] = \sigma_{xy}/\sigma_x^2$ and $b = \mu_y - w\,\mu_x$. This can be derived from Equation (28) by adding an auxiliary variable whose values are always 1 (i.e., $y = w\,x + b\cdot 1$). It has the properties of being both unbiased and best, with $\text{MSE} = \sigma_y^2(1 - \rho_{xy}^2)$. However, it has the deficiency of potentially resulting in negative values, if $b < 0$, or of excluding values between 0 and b if $b > 0$. Another drawback emerges when many values of y are estimated in an attempt to extend a record of y based on a longer record of x. In this case, the resulting extended record has negatively biased variance, because the method does not preserve variance. To remedy this, a random error e should be added (using the probability distribution of e), which however is not determined in a unique manner and makes the method no longer best.

5. *Organic correlation*: $y = wx + b$ with $w = \text{sign}[\rho_{xy}]\,\sigma_y/\sigma_x$ and $b = \mu_y - w\,\mu_x$. This preserves both mean (i.e., it is unbiased) and variance, but it is not best ($\text{MSE} = 2\sigma_y^2(1 - |\rho_{xy}|)$.

Evidently, for $|\rho_{xy}|<0.5$, the method results in MSE $> \sigma_y^2$ and therefore adopting it is pointless for low correlation coefficients. Similar to the standard linear regression, the organic correlation retains the deficiency of producing negative values or excluding some positive values.

Coming to the interpolation based on multiple x_i, in the simplest case, all weights w_i are assumed equal for all i, that is, $w_i = 1/n$ so that y is none other than the average of x_i (the arithmetic mean method). This simple version is used very often to fill in sparse missing values of rain gauge records. The quantities x_i could be simultaneous measurements at neighboring points (say, within a radius of 100 km), or at neighboring times, or both. Here, neighboring times should not necessarily be interpreted in the literal meaning, but with an emphasis on similarity of states. For example, a missing value of monthly precipitation in April 2000 could be estimated by, say, the average of the precipitation of the April months of 1998, 1999, 2001, and 2002. In another version, the average of all April months with available data are used, but a local average (as we have already discussed in Section 2.02.1.4) is preferable over an overall average, assuming that precipitation behaves like an HK process rather than a purely random one; this is similar to taking the average of points within a certain distance rather than a global average in spatial interpolation. This is not only intuitive but it can have a theoretical justification (D. Koutsoyiannis, personal notes), according to which for an HK process with $H = 0.7$, a local average based on 3 time steps before and 3 after the interpolation time is optimal (produces lowest MSE); the optimal number of points becomes $2+2$ and $1+1$ for $H = 0.75$ and $H \geqslant 0.8$, respectively.

This simple method does not impose any requirement for calculation of statistical quantities for its application. Another method of this type, which takes account of the geographical locations and, in particular, the distances d_i between the interpolated stations, is the method of inverse distance weighting (IDW). In each of the basis stations, it assigns weights as

$$w_i = \frac{d_i^{-b}}{\sum_{j=1}^{k} d_j^{-b}} \quad (30)$$

where the constant b is typically assumed to be 2.

Among methods whose application requires statistical quantities to be known, the simplest is a direct extension of the normal ratio method, in which $w_i = (1/n)(\mu_y/\mu_{x_i})$. The BLUE method itself belongs to this type. Initially, we can observe that a simple but biased solution for w in Equation (29) can be easily obtained as

$$w = C^{-1}\eta, \quad \mu_e = \mu_y - w^T \mu_x,$$
$$\sigma_e^2 = \sigma_y^2 - \eta^T C^{-1} \eta = \sigma_y^2 - w^T \eta \quad (31)$$

where $\eta := \mathrm{Cov}[y, \underline{x}]$ is the vector whose elements are the covariances of y with $\underline{x}$ (see Section 2.02.1.5) and $C := \mathrm{Cov}[\underline{x},\underline{x}]$ is the positive definite symmetric matrix whose elements are the covariances of the vector $\underline{x}$ with itself. One way to make it unbiased is to add an auxiliary variable $\underline{x}_{n+1}$ whose values are always 1. This is the multivariate extension of the typical linear regression described in point 4 of the previous list, and thus it retains the deficiency of potentially producing negative values or excluding some positive values. A better way to make it unbiased is to add a constraint $\mu_y = w^T \mu_x$ (the bivariate analog of this is the equality case, described in point 1 of the list). In the latter case, the MSE becomes

$$\mathrm{MSE} = \sigma_e^2 = \sigma_y^2 + \mu_y^2 + w^T(C + \mu_x \mu_x^T)w$$
$$- 2w^T(\eta + \mu_y \mu_x) \quad (32)$$

Minimization of the MSE with the above constraint using a Lagrange multiplier -2λ results in the system of equations

$$Cw + \mu_x \lambda = \eta, \quad \mu_x^T w = \mu_y \quad (33)$$

whose solution for the $n+1$ unknowns $w_1, \ldots, w_n, \lambda$ is

$$w' = C'^{-1} \eta' \quad (34)$$

where

$$w' := \begin{bmatrix} w \\ \lambda \end{bmatrix}, C' := \begin{bmatrix} C & \mu_x \\ \mu_x^T & 0 \end{bmatrix}, \eta' := \begin{bmatrix} \eta \\ \mu_y \end{bmatrix} \quad (35)$$

The value of the error is then calculated as

$$\mathrm{MSE} = \sigma_e^2 = \sigma_y^2 + w^T C w - 2w^T \eta \quad (36)$$

As seen in Equations (31) and (34), the application of the method requires a number of covariances to be estimated (specifically, this number is $(n^2 + 3n)/2$, given that C is symmetric). Not only does this restrict the method's application to points where measurements exist, in order to estimate the covariances, but, when n is large, it is infeasible to reliably estimate so many parameters from data and to derive a positive definite C. The viable alternative is to assume a parametric stochastic model for the precipitation field. In the simplest case, the field could be assumed stationary and isotropic, where $\mu_{x_i} = \mu_y = \mu, \sigma_{x_i} = \sigma_y = \sigma$, and the covariance among any two points i, j is a function f of the geographical distance d_{ij} between these points, that is, $\sigma_{ij} := \mathrm{Cov}[\underline{x}_i, \underline{x}_j] = f(d_{ij})$. In this case, Equation (35) simplifies to

$$C' := \begin{bmatrix} C & 1 \\ 1^T & 0 \end{bmatrix}, w' := \begin{bmatrix} w \\ \lambda' \end{bmatrix}, \eta' := \begin{bmatrix} \eta \\ 1 \end{bmatrix} \quad (37)$$

where $\lambda' = \lambda \mu$ and 1 is a vector with all its elements equal to 1. The last solution is widely known as 'kriging' (although kriging is sometimes formulated not in terms of covariance as in here, but in terms of the so-called semivariogram, a notion that is not appropriate for processes with HK behavior). We can observe from Equation (37) that the solution is now independent of μ, as is also the error, which is still calculated from Equation (36). It only depends on the covariance function $f(d)$. A function $f(d)$ compatible with the HK behavior of precipitation, as discussed in Section 2.02.1.5, is of the form

$$f(d) = \min(c, \alpha\, d^{4H-4}) \quad (38)$$

where H is the Hurst coefficient and $c \gg 0$ and α are parameters; in particular, c violates theoretical consistency but has been introduced to avoid problems related to the infinite covariance for distance tending to zero.

It can be observed that if the point of interpolation coincides with any one of the basis points i, then η is identical to one of the columns of C and η' is identical to one of the columns of C'. Thus, given the symmetry of C and C', from Equation (31) or (34), we obtain that w is a unit vector, that is, all elements are zero except one, which will be equal to 1. This shows the consistency of the method, that is, its property to reproduce the measurements at gauged points with zero error.

All of the above methods that can interpolate at an arbitrary point (rather than only at a gauged one) provide a basis for numerical integration to find the average precipitation over a specific area A. Eventually, these methods result again in Equation (28) or (29), where now y is the areal average precipitation. In particular, in the arithmetic mean and the normal ratio methods, because they do not make any assumption about the position of the point to which interpolation refers, the estimate y is an interpolation at any point and a spatial average as well. The equality method works as follows: the geographical area of interest is divided into polygons, the so-called Thiessen polygons, each of which contains the points nearest to each of the stations. All points belonging to a specified polygon are regarded to have received a precipitation amount equal to that of the station corresponding to this polygon. Thus, in the integration, we use either Equation (28) or (29), where all gauged x_i in the area are considered with weights $w_i = A_i/A$, whereas A_i and A are the areas of the polygon corresponding to x_i. The remaining methods (IDW and BLUE) can be explicitly put in the form of Equation (29), but this is rather tedious if done analytically. A simpler alternative is to make interpolations to many points, for example, on a dense square grid. In turn, the gridded interpolations could be used for integration using equal weights for all grid points (i.e., arithmetic mean).

2.02.3.2 Radar Estimates of Precipitation

Radio detection and ranging (radar) was developed at the beginning of World War II as a remote-sensing technique to measure the range and bearing of distant objects (such as ships and airplanes) by means of radio echoes (e.g., Battan, 1973). Since the early 1970s, radar techniques have also been used for the identification (i.e., shape, size, motion, and thermodynamic properties) of precipitation particles. The latter are weather-related distributed targets, which in contrast to ships and airplanes, have characteristics that evolve in time and depend on the atmospheric conditions.

Because of their ability to provide estimates of areal precipitation quickly (i.e., at time intervals of about 5–15 min), at high resolutions (i.e., down to spatial scales of about 1 km) and over wide areas (i.e., with an effective range of about 200–400 km), radars have found wide application in atmospheric research, weather observation, and forecasting (e.g., Atlas et al., 1984; Doviak and Zrnic, 1993; Uijlenhoet, 1999, 2008; Bringi and Chandrasekar, 2001; Krajewski and Smith, 2002; Testik and Barros, 2007). An example is the next generation weather radar (NEXRAD) network with 159 operational weather surveillance radar 88 Doppler (WSR-88D) units (as of February 2009), deployed throughout the continental United States and at selected locations overseas. According to NOAA's weather service (US National Oceanic and Atmospheric Administration, 2009), since its establishment in 1988, the NEXRAD project has provided significant improvements in severe weather and flash flood warnings, air traffic control, and management of natural resources.

2.02.3.2.1 Basics of radar observation and measurement

A typical weather radar has three main components (Battan, 1973): (1) the transmitter, which generates short pulses of energy in the microwave frequency portion of the electromagnetic spectrum, (2) the antenna, which focuses the transmitted energy into a narrow beam, and (3) the receiver, which receives the backscattered radiation from distant targets that intercept the transmitted pulses.

Some important parameters, and their range of values, that characterize the radar equipment are (Rogers and Yau, 1996): (1) the instantaneous power of the pulse $P_t \approx 10\text{--}10^3$ kW (also referred to as peak power), (2) the duration of the pulse $\tau \approx 0.1\text{--}5\,\mu s$, (3) the frequency of the signal $\nu \approx 3\text{--}30$ GHz, (4) the pulse repetition frequency (PRF) $f_r \approx 200\text{--}2000$ Hz, defined as the reciprocal of the time interval t_{max} that separates two distinct pulses (i.e., $t_{max} = f_r^{-1} \approx 0.5 - 5$ ms), and (5) the beamwidth of the antenna $\theta \approx 1°$, defined as the angular separation between points where the power of the transmitted signal is reduced to half of its maximum value (or equivalently 3 dB below the maximum). The latter is attained at the beam axis.

The wavelength λ of the signal is defined as the distance between two sequential crests (or troughs) of the electromagnetic wave and it is related to its frequency as

$$\lambda \nu = c \quad (39)$$

where $c = 3 \times 10^8$ m s^{-1} is the velocity of light in vacuum. It follows from Equation (39) that typical frequencies $\nu = 3\text{--}30$ GHz correspond to wavelengths λ between 10 and 1 cm, but most weather radars operate at wavelengths $\lambda = 3\text{--}10$ cm (X-, C-, and S-band; see, e.g., Uijlenhoet and Berne, 2008). Shorter wavelengths are more effectively attenuated by atmospheric hydrometeors and precipitation particles (hence the transmitted signal has a small effective range), whereas for longer wavelengths the backscattered radiation from the precipitation particles does not have sufficient power to be detected by the receiver without noise induced by ground targets (e.g., Uijlenhoet, 2008).

When conducting radar observations and measurements, the direction of the target is obtained from the azimuth and elevation of the antenna when the returning echo is received. The range r of the target is calculated from the relation

$$r = c\,t/2 \quad (40)$$

where t is the time interval between the transmission of the pulse and the reception of the echo. If the target is moving, the radial velocity u_r of the target (i.e., in the radar-pointing direction) can be obtained from the frequency shift $\Delta \nu$ of the received relative to the transmitted signal. The frequency shift is caused by the Doppler effect and it is related to u_r as:

$$\Delta v = -2u_r/\lambda \quad (41)$$

with positive Δv being associated with targets that move toward the radar.

If t (the time interval between transmission and reception) is larger than t_{max} (the reciprocal of the pulse repetition frequency, f_r), the echo from the target will reach the receiver after a new pulse has been transmitted. Hence, targets that return enough energy to be detected by the receiver (see below) and are located at distances $r > r_{max} = c/(2f_r)$, will appear unrealistically close to the antenna. Thus, r_{max} is the maximum range within which targets are indicated correctly on the radar screen and it is usually referred to as the unambiguous range (Battan, 1973; Rogers and Yau, 1996).

The visibility of a target by the radar depends on whether the returning signal has sufficient power P_r to be detected by the receiver. As an example, we consider a point target (i.e., a target with linear dimension smaller than about 10% of λ) with cross section A_t located at distance r from the radar. We suppose that the radar transmits pulses with peak power P_t that propagate isotropically in space (i.e., in a 3D sphere). It follows from simple geometric considerations that the power P_i intercepted by the target is

$$P_i = \frac{P_t A_t}{4\pi r^2} \quad (42)$$

where $4\pi r^2$ is the surface area of a sphere with radius r. If the transmitted signal is focused in a narrow beam by the antenna (as is commonly the case), Equation (42) becomes

$$P_i = G\frac{P_t A_t}{4\pi r^2} \quad (43)$$

where $G = (4\pi A_e)/\lambda^2$ is a dimensionless constant called the antenna axial gain that depends on the characteristics (i.e., the wavelength λ) of the signal and the aperture A_e of the antenna.

Assuming that the target scatters the intercepted signal isotropically in space, the power P_r that reaches the radar is

$$P_r = \frac{P_i A_e}{4\pi r^2} = G\frac{P_t A_t A_e}{(4\pi r^2)^2} = P_t A_t \frac{\lambda^2 G^2}{(4\pi)^3 r^4} \quad (44)$$

If the power P_r is large enough to be detected by the receiver without unwanted echoes (e.g., noise from ground targets), the target is visible to the radar and it is indicated on the radar screen.

For nonisotropic scatterers, the cross section of the target A_t should be replaced by the backscatter cross section σ of the target. For spherical particles with diameter $D < \lambda/10$, usually referred to as Rayleigh scatterers, σ can be calculated from the relation (Battan, 1973)

$$\sigma = \frac{\pi^5 |K|^2 D^6}{\lambda^4} \quad (45)$$

where $|K|$ is the amplitude of the complex refraction index ($|K|^2 \approx 0.93$ for liquid water and 0.21 for ice), which characterizes the absorptive and refractive properties of the spherical scatterer. Due to the much higher value of $|K|^2$ for liquid water relative to ice (about 4.5 times higher), the melting layer of ice particles in precipitation-generating weather systems appears on the radar screen as a bright band of high reflectivity.

2.02.3.2.2 Radar observation of distributed targets and the estimation of precipitation

For a typical weather radar that operates in the C-band portion of the electromagnetic spectrum ($\lambda = 3.75-7.5$ cm), raindrops and snowflakes (i.e., particles with effective diameters $D < 5-6$ mm) can be approximated as Rayleigh spherical scatterers with backscatter cross section σ given by Equation (45). However, there are reasons why atmospheric hydrometeors should not be treated as isolated point targets. One reason is that the pulse transmitted by the radar illuminates simultaneously, numerous precipitation particles that are included in a certain volume of air V, referred to as the resolution volume of the radar. Hence, the returned signal contains spatially averaged information from the whole population of raindrops and snowflakes in V.

For parabolic antennas, where the beam pattern is approximately the same in all directions, an accurate estimate of V can be obtained by assuming that the resolution volume is a cylinder with effective height equal to half of the pulse length $l = c\tau$ and diameter $d_V = r\theta$, that is, the separation distance between points where the power of the transmitted signal is reduced to half of its maximum value. This gives

$$V = \pi\left(\frac{r\theta}{2}\right)^2\frac{c\tau}{2} \quad (46)$$

where θ is in radians. Equation (46) assumes that all energy in the radar transmitted pulse is contained within the half-power beamwidth; assuming a Gaussian shape of the beam pattern, the denominator of (46) (and, likewise, that of (49) below) should be multiplied by a factor $2 \ln 2$ (Probert-Jones, 1962). Another reason why raindrops and snowflakes cannot be treated as isolated point targets is that their turbulent motion that causes the power P_r of the returned signal to fluctuate in time. To this extent, an accurate approximation of the time-averaged power $\bar{P}_r$ (over a sufficiently long interval of about 10^{-2} s), which accounts also for multiple backscattering cross sections, is given by (Rogers and Yau, 1996)

$$\bar{P}_r = P_t\frac{\lambda^2 G^2}{(4\pi)^3 r^4}\sum_V \sigma \quad (47)$$

where r is the time-averaged range of the resolution volume V, and the summation is taken over all σ in V. For Rayleigh scatterers, Equations (45) and (47) are combined to give

$$\bar{P}_r = P_t\frac{\pi^2 G^2 |K|^2}{64 r^4 \lambda^2}\sum_V D^6 \quad (48)$$

Assuming homogeneity of the population of hydrometeors in V, Equation (48) can be written as

$$\bar{P}_r = P_t \frac{\pi^2 G^2 |K|^2}{64 r^4 \lambda^2} V \int_0^\infty n(D) D^6 dD$$
$$= P_t \frac{\pi^3 G^2 |K|^2 \theta^2 c\tau}{512 r^2 \lambda^2} Z = C \frac{|K|^2 Z}{r^2} \quad (49)$$

where $n(D)$ is the size distribution of precipitation particles in V (i.e., number of particles per unit diameter and per unit volume of air), C is the so-called radar constant that depends solely on the characteristics of the system under consideration, and

$$Z := \int_0^\infty n(D) D^6 dD \quad (50)$$

is the reflectivity factor with units (length^3) that depend solely on the size distribution of the precipitation particles. For the Marshall and Palmer (1948) parametrization described by Equation (24), Equation (50) takes the form

$$Z = 720 \, n_0 b^{-7} \quad (51)$$

where n_0 and b are the intercept and scale parameters of the exponential size distribution. For the expressions given in Section 2.02.2.3.2, for rain and snow we obtain

$$\begin{align} \text{(a)} \;\; & Z = 296 \, i^{1.47} \; (\text{rain}) \\ \text{(b)} \;\; & Z = 3902 \, i^{2.49} \; (\text{snow}) \end{align} \quad (52)$$

where Z has units of mm^6 m^{-3} and i is the rainfall intensity (or the water equivalent of the accumulated snow at ground level) in millimeters per hour.

For rain, Equation (52a) is very close to the empirical $Z-i$ relationships (usually referred to as $Z-R$ relationships, where $R \equiv i$ denotes the rainfall intensity) found in the literature (e.g., Marshall et al., 1955; Battan, 1973; Uijlenhoet, 1999, 2001, 2008), whereas for snow there is more variability and Equation (52b) should be seen only as an approximation. When combined, Equations (40), (49), and (52) allow conversion of radar measurements (i.e., $\bar{P}_r$, t and r) to precipitation intensity i.

2.02.3.3 Spaceborne Estimates of Precipitation

The history of observation of Earth from space started on 4 October 1957, when the Soviet Union successfully launched *Sputnik-I*, the first artificial satellite. *Sputnik-I* provided information on the density of the highest layers of the atmosphere and on the radio-signal distribution in the ionosphere. The first launch was immediately followed by the launch of *Sputnik-II* by the Soviet Union on 3 November 1957 and the launches of *Explorer-I* (1 February 1958), *Vanguard-I* (17 March 1958), *Vanguard-II* (17 February 1959), and *TIROS-I* (1 April 1960) by the United States of America. The success of *TIROS-I* in surveying atmospheric conditions (in particular, the cloud coverage of Earth) opened a new era for meteorological research and development using spaceborne observations.

Since the 1970s, meteorological satellites have become essential in studying the development and evolution of weather-related phenomena over the 71% of the Earth's surface covered by sea, where other types of measurements are unavailable. For example, the Tropical Rainfall Measuring Mission (TRMM; Simpson et al., 1988; Kummerow et al., 1998), which started in November 1997 by the National Aeronautics and Space Administration (NASA) of the United States and the National Space Development Agency (NASDA) of Japan, has provided vast amounts of rainfall and energy estimates in tropical and subtropical regions and advanced the understanding and modeling of extreme rainfall events caused by tropical cyclones (e.g., Lonfat et al., 2004, 2007; Chen et al., 2006, 2007; Langousis and Veneziano, 2009a, 2009b). TRMM data have also been used to improve the accuracy of high-resolution weather forecasts produced by limited-area models (e.g., Lagouvardos and Kotroni, 2005) and to investigate the relationship between lighting activity, microwave brightness temperatures (see below), and spaceborne radar-reflectivity profiles (Katsanos et al., 2007).

We can distinguish two types of sensing by satellites, passive and active. Passive sensing is based on measuring the radiative intensity emitted or reflected by particles in the atmosphere, such as cloud droplets and hydrometeors of precipitable size. Active sensing is conducted using radar equipment carried by the satellite. Next, we discuss some basic principles of passive remote sensing in the visible (V, $\lambda \approx 0.39-0.77\,\mu m$), IR (wavelengths $\lambda \approx 0.77\,\mu m - 0.1\,mm$), and microwave (MW, $\lambda \approx 0.1\,mm - 10\,cm$) portions (channels) of the electromagnetic spectrum. The basic principles of operation of active sensors are similar to those of radars, reviewed in Section 2.02.3.2. For a more detailed review on the principles and techniques of remote sensing, the reader is referred to Barrett and Martin (1981), Elachi (1987), Stephens (1994), and Kidder and Vonder Haar (1995).

2.02.3.3.1 The IR signature of cloud tops

The high absorptivity of cloud droplets in the IR spectral range causes clouds to appear opaque in the IR channel. Hence, the IR radiation received by the satellite's radiometer originates mostly from the cloud tops, which can be approximated with sufficient accuracy as black bodies, that is, as objects that absorb all incident radiation and emit it at a rate that depends solely on their temperature. In this case, we can use Stefan–Boltzman's law of radiation (e.g., Barrett and Martin, 1981) to calculate the temperature T_b of the cloud tops from the intensity J of the received IR radiation:

$$T_b = (J/\sigma_{SB})^{1/4} \quad (53)$$

where $\sigma_{SB} = 5.7 \times 10^{-8}\,W\,m^{-2}\,K^{-4}$ is the Stefan–Boltzman constant and T_b is in kelvins. T_b is usually referred to as brightness temperature (e.g., Smith, 1993) and, for a given atmospheric lapse rate γ (see Section 2.02.2.1), it can be used to calculate cloud top heights.

Evidently, lower brightness temperatures T_b correspond to clouds with higher tops and larger probabilities of rain.

Hence, we can develop regression equations to relate brightness temperatures to observed surface rainfall rates (e.g., Griffith et al., 1978; Stout et al., 1979; Arkin, 1979; Richards and Arkin, 1981; Arkin and Meisner, 1987; Adler and Negri, 1988). Two important limitations apply (Richards and Arkin, 1981; Liu, 2003): (1) due to the statistical character of the regressed quantities, the accuracy of the rainfall-retrieval algorithm increases with increasing scale of spatial or temporal averaging, and (2) the parameters of the regression depend on the climatology of the region and, therefore, cannot be used at regions with different climatic characteristics.

An example of surface rainfall estimation from IR images is the temperature threshold method developed by Arkin (1979), Richards and Arkin (1981) and Arkin and Meisner (1987). Arkin (1979) used IR imagery from the Synchronous Meteorological Satellite-1 (SMS-1) and radar data from Global Atmosphere Research Program (GARP) Atlantic Tropical Experiment (GATE) to investigate the correlation between radar-estimated precipitation rates and the fraction of areas with brightness temperature T_b below a certain threshold T_{min}. The study found a maximum correlation (around 0.85) for a brightness temperature threshold $T_{min} \approx 235\,K$ ($-38\,°C$). Richards and Arkin (1981) showed that a linear relationship is sufficient to describe the dependence between spatially averaged surface rainfall and the fraction of areas with $T_b < 235\,K$, with error variance that increases with decreasing scale of spatial averaging. Based on these results, Arkin and Meisner (1987) suggested the use of the Geostationary Operational Environmental Satellite (GOES) Precipitation Index (GPI) to calculate spatial rainfall averages in the tropics:

$$GPI = 3(mm/h)F_c\,H \qquad (54)$$

where GPI is the spatially averaged rainfall accumulation in a grid box of $2.5°$ latitude $\times\,2.5°$ longitude, F_c is the mean fraction (a dimensionless quantity between 0 and 1) of the grid box covered by brightness temperatures $T_b < 235\,K$, and H is the length of the observation period in hours.

The temperature threshold method of Arkin (1979), Richards and Arkin (1981), and Arkin and Meisner (1987) produces accurate estimates of the spatially averaged rainfall in the tropical belt ($30°S$ to $30°N$), at grid scales larger than $2.5° (\approx 275\,km)$ (Arkin and Meisner, 1987) and for averaging durations greater than about a month (Ba and Nicholson, 1998). The error increases significantly as we move to mid-latitudes, especially during cold seasons (e.g., Liu, 2003). Extensions of the method include the use of the upper tropospheric humidity (UTH) in the vicinity of convective clouds as an additional predictive variable (Turpeinen et al., 1987), and the combination of IR and visible imagery (i.e., bi-spectral methods; see below) to exclude nonprecipitating clouds with high tops.

2.02.3.3.2 The visible reflectivity of clouds

The signature of Earth in the visible (V) channel is due to the reflection of the sunlight by clouds and, when the sky is clear, the surface features. Consequently, visible imagery is available only during daylight hours. Due to its shorter wavelength, visible radiation can penetrate deeper into clouds than the infrared portion of the electromagnetic spectrum, but similar to the IR channel, it still represents the upper portion of clouds and serves as an indirect signature of surface rainfall. However, visible reflectivity can complement the IR brightness temperatures to allow better classification of clouds and qualitative assessment of the probability of precipitation. This is the basis of the well-known bi-spectral methods (e.g., Lovejoy and Austin, 1979; Bellon et al., 1980; Tsonis and Isaac, 1985; Tsonis, 1987; O'Sullivan et al., 1990; Cheng et al., 1993; Cheng and Brown, 1995; King et al., 1995; Liu, 2003).

The visible reflectivity of clouds increases fast with the increasing liquid water path, that is, the vertically integrated liquid water in the atmospheric column. Hence, we can use IR brightness temperatures to calculate the altitude of the cloud tops and visible reflectivities to obtain a qualitative estimate of the vertically averaged liquid water of the cloud, which is indicative of the rainfall potential. For example, low brightness temperatures (i.e., cold cloud tops) and high visible reflectivities (i.e., thick clouds) indicate cumulonimbus formations with high probability of precipitation (see Section 2.02.2.4.1), warm cloud tops and high visible reflectivities indicate stratiform rainfall (see Section 2.02.2.4.2), whereas cold cloud tops and low visible reflectivies indicate cirrus clouds, which are usually nonprecipitating.

An example of bi-spectral methods is the RAINSAT technique developed by Lovejoy and Austin (1979) and Bellon et al. (1980). This technique uses visible reflectivities to reduce the number of false alarms obtained from the IR channel and more accurately estimate surface rainfall rates. The RAINSAT method was developed using GOES infrared and visible imagery and radar data from tropical (i.e., GATE) and mid-latitude (i.e., McGill weather radar, Quebec, Canada) locations as ground truth. The method was optimized by Cheng et al. (1993) and Cheng and Brown (1995) for the area of the UK, using IR and visible imagery from the European geostationary satellites *Meteosat-2*, *Meteosat-3*, and *Meteosat-4* and rainfall retrievals from nine weather radars located in the United Kingdom and Ireland. A similar cloud classification technique has been proposed by Tsonis and Isaac (1985) and Tsonis (1987). This technique is based on cluster analysis of pixels with different brightness temperatures and visible reflectivities and has been developed using *GOES* satellite data and rainfall retrievals from the Woodbridge weather radar in Ontario, Canada.

2.02.3.3.3 The microwave signature of precipitation

Contrary to the IR and visible spectral ranges, microwave radiation can effectively penetrate through cloud and rain layers and provide the signature of the integrated contribution of precipitation particles in the atmospheric column. Hence, brightness temperatures obtained from the MW channel are better linked to surface rainfall rates than the visible reflectivities and IR brightness temperatures.

The type and size of the precipitation particles detected by the microwave radiometer depends on the frequency of the upwelling radiation. Above 80 GHz (i.e., wavelengths $\lambda < 3.75\,mm$), ice crystals scatter the upwelling MW radiation and fade the signature of raindrops. Hence, above 80 GHz, the radiometer senses only ice, where lower brightness temperatures are associated with more scattering, larger ice particles, and higher precipitation intensities at ground level.

Below about 20 GHz (i.e., $\lambda > 1.5$ cm), the radiative intensity of raindrops dominates the microwave signature of hydrometeors in the atmospheric column, whereas ice particles are virtually transparent. Thus, below 20 GHz, the microwave radiometer detects the vertically integrated signature of rainwater, where higher brightness temperatures are associated with more intense rainfall at ground level. Low-frequency microwave imagery is especially useful when calculating surface rainfall rates over oceans, where the almost constant sea surface temperature and emissivity allow translation of the spatial and temporal variations of brightness temperatures to variations of sea-level rainfall rates (e.g., Liu, 2003). The same is not true over land, where the surface features cause the ground temperature and emissivity to vary significantly in space and time. Another limitation of low-frequency microwave images is the saturation of the microwave channel at high rainfall rates, which causes negative biases of the obtained rainfall intensity (e.g., Liu, 2003; Viltard et al., 2006).

Between 20 GHz and 80 GHz, scattering and emission by raindrops and ice particles occur simultaneously and the microwave radiation undergoes multiple transformations. Hence, the microwave radiometer detects different rain paths at different microwave frequency ranges.

Combining brightness temperatures from different MW channels to more accurately assess surface rainfall rates is an open research problem and it has driven the development of many rainfall-estimation algorithms (Grody, 1991; Spencer et al., 1989; Alishouse et al., 1990; Berg and Chase, 1992; Hinton et al., 1992; Liu and Curry, 1992, 1993; Ferriday and Avery, 1994; Petty, 1994a, 1994b, 2001a, 2001b; Kummerow and Giglio, 1994a, 1994b; Ferraro and Marks, 1995; Kummerow et al., 1996, 2001; Berg et al., 1998; Aonashi and Liu, 2000; Levizzani et al., 2002). For a review of microwave methods of estimation over ocean and land, and their advantages and limitations, the reader is referred to Wilheit et al. (1994), Petty (1995), and Kidd et al. (1998) respectively.

2.02.4 Precipitation modeling

As already clarified in Section 2.02.1.5, modeling of precipitation is not possible without using any type of a stochastic approach. Even the deterministic numerical weather forecast models, which determine the state and motion in the atmosphere by solving differential equations, to model precipitation, use parametrization schemes. These schemes, instead of describing the detailed dynamics of the precipitation process, establish and use equations of statistical type to quantify the output of the dynamical system. In addition, as mentioned in Section 2.02.1.5, the modern framework for predicting precipitation particularly as input to hydrological models (the ensemble forecasting), is of the Monte Carlo or stochastic type. The description of these stochastic techniques belongs to the sphere of weather forecasting and is not within the scope of this chapter. In more engineering-oriented applications, precipitation is typically modeled as an autonomous process, without particular reference to the atmospheric dynamics. Next, we outline some of the most widespread modeling practices for precipitation but without details and mathematical formulations, which the interested reader can find in the listed references.

2.02.4.1 Rainfall Occurrence

From the early stages of the analysis of precipitation intermittency, it was recognized that rainfall occurrences are not purely random. In other words, rainfall occurrence cannot be modeled (effectively) as a Bernoulli process in discrete time or, equivalently, as a Poisson process in continuous time. It should be recalled that in a Bernoulli process, an event (rainfall/wet state) occurs with a probability p (and does not occur with probability $1 - p$) constant in time, and each event is independent of all preceding and subsequent events. In a Poisson process, the times of occurrence of events (i.e., the starting times of rainfalls) are random points in time. In this process, the time differences between consecutive occurrences are independent identically distributed (IID) with exponential distribution.

Both discrete time and continuous time representations of the rainfall occurrence process, which in fact are closely related (e.g., Foufoula-Georgiou and Lettenmaier, 1986; Small and Morgan, 1986), have been investigated. The most typical tool of the category of discrete time representations is the Markov chain model (Gabriel and Neumann, 1962; Feyerherm and Bark, 1964; Hershfield, 1970; Todorovic and Woolhiser, 1975; Haan et al., 1976; Chin, 1977; Katz, 1977a, 1977b; Kottegoda and Horder, 1980; Roldan and Woolhiser, 1982). In this model, any time interval (e.g., day) can be in one of two states, dry or wet, and it is assumed that the state in a time interval depends on the state in the previous interval.

It was observed, however, that Markov chain models yield unsatisfactory results for rainfall occurrences, especially for dry intervals (De Bruin, 1980). Moreover, the interannual variance of monthly (or seasonal) total precipitation is greater than that predicted by Markov chain models, an effect usually referred to as overdispersion (Katz and Parlange, 1998). Extended versions of the binary state Markov chains using a higher number of past states may improve performance. Additional states in such model versions have been defined based on a combination of states of two consecutive periods (Hutchinson, 1990) or on accounting for the rainfall depth of each interval (Haan et al., 1976). A more effective enhancement is to use transition probabilities taking into account more than one previous interval, which leads to stochastic binary chains of order higher than one (Pegram, 1980; Katz and Parlange, 1998; Clarke, 1998). In more recent developments, to account for a long number of previous time intervals and simultaneously avoid an extremely high number of transition probabilities, it was proposed that, instead of the sequence of individual states of these intervals, one could use conditional probabilities based on aggregation of states of previous intervals (Sharma and O'Neill, 2002). Similarly, one could use a discrete wetness index based on the number of previous wet intervals (Harrold et al., 2003). An extension of the Markov chain approach to multiple sites has been studied by Pegram and Seed (1998).

In a more recent study, Koutsoyiannis (2006a) used the principle of maximum entropy, interpreted as maximum uncertainty, to explain the observed dependence properties of the rainfall-occurrence process, including the overdispersion or clustering behavior and persistence. He quantified intermittency by the probability $p^{(1)}$ that a time interval of length 1 h is dry, and dependence by the probability that two consecutive intervals are dry, that is by $p^{(2)}$, where in general $p^{(k)}$ denotes the probability that an interval of length k is dry. Using these two probabilities and a multiscale entropy-maximization framework, he was able to determine any conditional or unconditional probability of any sequence of dry and wet intervals at any timescale. Thus, he described the rainfall occurrence process including its dependence structure at all scales using only two parameters. The dependence structure appeared to be non-Markovian, yet not over-exponential. Application of this theoretical framework to the rainfall data set of Athens indicated good agreement of theoretical predictions and empirical data at the entire range of scales for which probabilities dry and wet can be estimated (from 1 h to several months). An illustration is given in **Figure 22**.

In the continuous time representation of the rainfall occurrence process, the dominant tools are the cluster-based point processes (Waymire and Gupta, 1981a, 1981b, 1981c). These are essentially based on the prototype of the spatial distribution of galaxies devised by Neyman and Scott (1952) to describe their property of clustering relative to the Poisson process. With reference to storms, if they were regarded as instantaneous pulses positioned at random points in time, the logarithm of probability that the interarrival time exceeds a value x, or the log survival function, would be proportional to x. However, empirical evidence suggests that the log survival function is a nonlinear concave function of x, which indicates a tendency for clustering of rainfall events relative to the Poisson model (Foufoula-Georgiou and Lettenmaier, 1986). This clustering has been modeled by a cascade of two Poisson processes, corresponding to two characteristic timescales of arrivals of storms and storm cells.

The Neyman–Scott process with instantaneous pulses was the first one applied to rainfall occurrence (Kavvas and Delleur, 1981; Rodriguez-Iturbe et al., 1984), later succeeded by the Neyman–Scott rectangular pulses and the very similar Bartlett–Lewis rectangular pulse models (Rodriguez-Iturbe et al., 1987). The Bartlett–Lewis rectangular pulse model, which is the most typical and successfully applied model of this type, assumes that rainfall occurs in the form of storms of certain durations and that each storm is a cluster of random cells. The general assumptions of the rainfall occurrence process are:

1. Storm origins t_i occur according to a Poisson process with rate λ.
2. Origins t_{ij} of cells of each storm i arrive according to a Poisson process with rate β.
3. Arrivals of each storm i terminate after a time v_i exponentially distributed with parameter γ.
4. Each cell has a duration w_{ij} exponentially distributed with parameter η.

In the original version of the model, all model parameters are assumed constant. In a modified version, the parameter η is randomly varied from storm to storm with a gamma distribution with shape parameter α and scale parameter v.

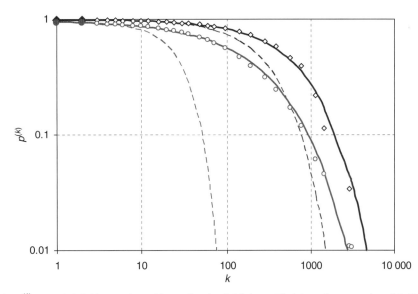

Figure 22 Probability dry $p^{(k)}$ vs. scale k (in h), as estimated from a hourly rainfall data set in Athens, Greece, and predicted by the maximum entropy model in Koutsoyiannis (2006a) for the entire year (circles and red full line) and the dry season (June–September; diamonds and blue full line). The model was fitted using two data points in each case (marked in full in the plot), that is, the probability dry for 1 h, $p \equiv p^{(1)}$, and 2 h, $p^{(2)}$, which are respectively 0.9440 and 0.9335 for the entire year and 0.9888 and 0.9860 for the dry season. The final model is expressed as $p^{(k)} = p^{[1+(\xi^{-1/\eta}-1)(k-1)]^{\eta}}$, where the parameters are respectively $\eta = 0.63$ and $\xi = 0.816$ for the entire year and $\eta = 0.83$ and $\xi = 0.801$ for the dry season. For comparison, lines resulting from the Markov chain model are also plotted (dashed lines). From Koutsoyiannis D (2006a) An entropic-stochastic representation of rainfall intermittency: The origin of clustering and persistence. *Water Resources Research* 42(1): W01401 (doi:10.1029/2005WR004175).

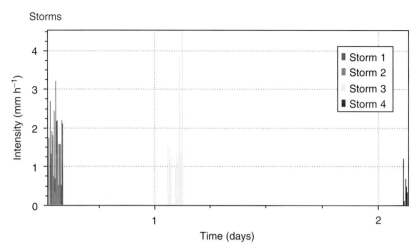

Figure 23 Simulated realization of a series of four storms from the Bartlett–Lewis rectangular pulse model (modified version with randomly varying η) occurring within three days (notice the overlap of storms 1 and 2, which is allowed by the model), implemented by the Hyetos software (see Section 2.02.4.4). The model parameters are $\lambda = 0.94\,d^{-1}$, $\kappa = \beta/\eta = 1.06$, $\varphi = \gamma/\eta = 0.059$, $\alpha = 2.70$, $\nu = 0.0068\,d^{-1}$, and $\mu_x = \sigma_x = 24.3\,\text{mm}\,d^{-1}$.

Subsequently, parameters β and γ also vary so that the ratios $\kappa := \beta/\eta$ and $\varphi := \gamma/\eta$ are constant.

A major problem of these models was their inability to reproduce the probability of zero rainfall at multiple timescales (Velghe et al., 1994). In this respect, Foufoula-Georgiou and Guttorp (1986) noted that the Neyman–Scott model parameters are scale dependent and thus cannot be attributed a physical meaning. To ameliorate this, modifications of both the Neyman–Scott model (Entekhabi et al., 1989) and the Bartlett–Lewis model (Rodriguez-Iturbe et al., 1988; Onof and Wheater, 1993, 1994) were proposed. These are in fact based on the randomization of the mean interarrival time of one of the two Poisson processes. Evaluation and comparison of several cluster-based rectangular pulse models for rainfall were done by Velghe et al. (1994) and Verhoest et al. (1997), whereas a comprehensive review of Poisson-cluster models has been provided by Onof et al. (2000). An extension of the concept introducing a third Poisson process was proposed by Cowpertwait et al. (2007).

2.02.4.2 Rainfall Quantity

In the discrete time representations of rainfall occurrence, the rainfall quantity in each wet interval is modeled separately from the occurrence process, usually based on statistical analysis of the observed record. In the point-process representations, the storms and cells are abstract quantities that do not fully correspond to real-world objects. Therefore, they cannot be identified in the recorded time series. An assumption is typically made that each cell has a uniform intensity x_{ij} with a specified distribution, and based on all assumptions, the statistical characteristics of the rainfall process at one or more timescales are derived analytically (Rodriguez-Iturbe et al., 1987, 1988). These statistical characteristics are compared to the empirically derived statistics, and, by minimizing the departures of the two, the model parameters are estimated. The distribution of the uniform intensity x_{ij} is typically assumed to be exponential with parameter $1/\mu_x$. Alternatively, one can choose a two-parameter gamma distribution with mean μ_x and standard deviation σ_x. In this manner, the point-process models describe the entire rainfall process, including occurrence and quantity. A demonstration of the model is shown in **Figure 23**. However, in some cases (e.g., Gyasi-Agyei and Willgoose, 1997), point processes have been used to simulate merely rainfall occurrences and then have been combined with other models that simulate rainfall depths. Other modeling approaches for the rainfall process (including its intermittency) are reviewed in Srikanthan and McMahon (2001).

With their typical assumptions, including those of the exponential or gamma distribution for rain-cell amount, the point-process models, despite providing satisfactory representation of the process at a specific timescale or a small range of timescales, cannot really perform satisfactorily over a wide range of scales and also lead to exponential distribution tails, whereas it has been recently recognized that the tails must be of power type (see Sections 2.02.1.5 and 2.02.5.2). Generally, the distribution function of rainfall varies among different timescales. At very fine scales, the density is J-shaped, that is, with a mode at zero, and perhaps with density tending to infinity as the rainfall depth or the intensity tends to zero. At coarse timescales such as monthly (for wet months) and annual, the distribution becomes bell-shaped and tends to become normal as the scale increases. However, its tail always departs from the exponential tail of the normal distribution. In fact, for theoretical reasons, if at the right tail, the survival function is a power function of the rainfall depth or intensity x, with exponent $1/\kappa$, that is, $F^*(x) \propto x^{-1/\kappa}$ (see Equation (18)), then it will be of the same type and will have precisely the same exponent $1/\kappa$ at any timescale (the proof is omitted). This behavior of the tail is perhaps the only invariant distributional property across all scales, whereas the shape of the body of the distribution varies significantly across different scales. However, even this variation must have a simple and unique explanation, which is the principle of maximum entropy. Specifically, Koutsoyiannis (2005a) has shown that all diverse shapes of the distribution across different scales can be derived from the principle of maximum entropy constrained on known mean and variance.

Papalexiou and Koutsoyiannis (2008) proposed a single distribution (a power-transformed beta prime distribution, also known as generalized beta of second kind; see also Koutsoyiannis, 2005a) with four parameters, which provides good fits for rainfall intensity at timescales from hourly to annual. Only one of the four parameters (corresponding to the exponent of the tail) is invariant across scales. If the range of scales of interest is smaller, then specific special cases of this distribution can be used as good approximations. For example, the three-parameter Burr type VII distribution, which has the advantage of providing a closed form of the quantile function, can be used effectively for timescales from a few minutes to a couple of months (Papalexiou and Koutsoyiannis, 2009).

2.02.4.3 Space–Time Models

Space–time modeling of precipitation is one of the most demanding tasks of stochastic modeling in hydrology and geophysics. Rainfall intermittency should be modeled in both space and time, along with the motion of rainfall fields, the rainfall quantity, and its temporal and spatial structure. One of the relatively simple solutions has been provided by the extension of point-process models used for the rainfall process at a single site. This extension introduces a description of rainfall cells in space, in addition to that in time, and a motion of the cells. As an example, we summarize here the Gaussian displacement spatial–temporal rainfall model (GDSTM; Northrop, 1996, 1998). This model, is a spatial analog of a point-process model having a temporal structure similar to that of the Bartlett-Lewis rectangular pulse model described above and a spatial structure known as the Gaussian displacement structure, introduced by Cox and Isham (1988).

Similar to its single-site analog, GDSTM assumes that rainfall is realized as a sequence of storms, each consisting of a number of cells. Both storms and cells are characterized by their centers, durations, and areal extents (see sketch in **Figure 24**) and, in addition, cells have certain uniform rainfall intensity. Specifically, the following assumptions characterize storms and cells.

Storm centers arrive according to a homogeneous Poisson process of rate λ in 2D space (denoted by x, y) and time (denoted by t) and move with a uniform velocity (V_x, V_y). Each storm has a finite duration $\underline{L}$ (assumed exponentially distributed with parameter $\beta = 1/\mu_L$) and an infinite areal extent, represented by an elliptical geometry with eccentricity ϵ and orientation θ, and incorporates a certain number of rainfall cells. However, a storm can be assigned a finite storm area, the area that contains a certain percentage of rainfall cells. The storm area varies randomly and in each storm, it is determined in terms of the realization of a random variable $\underline{w}$, which determines uniquely (for the specific storm) a set of parameters σ_x^2, σ_y^2, and ρ that determine the displacement of cell centers from the storm center. Specifically, $\underline{w}$ is Gamma-distributed with shape and scale parameters determined in terms of the eccentricity ε and the mean storm area μ_s. At the same time, the parameter ρ is determined in terms of the eccentricity ε and the storm orientation, θ. Following the generation of $\underline{w}$, the parameters σ_x^2 and σ_y^2 are determined

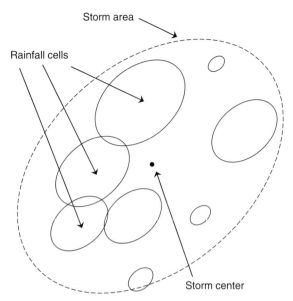

Figure 24 Sketch of the spatial structure of the Gaussian displacement spatial–temporal rainfall model.

in terms of the eccentricity ε, the storm orientation θ, and the value of $\underline{w}$.

Each rainfall cell is assigned a center $(\underline{x}_c, \underline{y}_c, \underline{t}_c)$. The time origin $\underline{t}_c$ follows a Poisson process starting at the time ordinate of the storm origin $\underline{t}_0$ (with the first cell being located at this point) and ending at $\underline{t}_0 + \underline{L}$. The expected number of cells within that time interval is $\mu_c = 1 + \beta/\gamma$, where γ is the cell-generation Poisson process parameter. The spatial displacements from the storm center are random variables jointly normally distributed with zero means, variances σ_x^2 and σ_y^2, and correlation ρ. Given these parameters, the displacement $\Delta \underline{x}$ of each cell is generated as a normal variate $(0, \sigma_x)$ and the displacement $\Delta \underline{y}$ as a normal variate $(\mu_{y|x}, \sigma_{y|x})$. Furthermore, each cell has a finite duration $\underline{D}$ (assumed exponentially distributed with parameter $1/\mu_D$) and an elliptical area with major axis $\underline{a}$, forming an angle θ with the x-axis (west–east), and minor axis $\underline{b} = \sqrt{1 - \varepsilon^2}\underline{a}$. It is assumed that $\underline{a}$ is a random variable gamma distributed with shape and scale parameters depending on the mean storm area μ_A and the eccentricity ε, respectively. Finally, each cell has an intensity $\underline{x}$ independent of any other variable, exponentially distributed with parameter $1/\mu_x$.

The model is defined in terms of 11 independent parameters, namely: (1) the rate of storm arrival (number of storms per area per time), λ; (2) the mean cell duration, μ_D; (3) the mean storm duration, μ_L; (4) the mean cell area, μ_A; (5) the mean storm area, μ_s; (6) the mean number of cells per storm, μ_c; (7) the mean cell intensity, μ_x; (8 and 9) the components of the cell and storm velocity in the x direction (east), V_x, and in the y direction (north), V_y; (10) the cell and storm eccentricity, ε; and (11) the cell and storm orientation, θ.

Similar to its single-point analog, the entities of the spatial point-process model are abstract. To make the model outputs comparable to reality, integration from continuous time over a specific timescale and/or spatial scale is needed, from which the first- and second-order rainfall statistics are calculated. The latter serve as the basis for parameter estimation using either rain gauge or radar data. Due to model complexity, the calculation of the statistics can be done only numerically; hence, the entire model application (and the parameter estimation in particular, which needs numerical optimization, e.g., using the generalized reduced-gradient method) is laborious.

2.02.4.4 Rainfall Disaggregation and Downscaling

Both disaggregation and downscaling refer to the generation of a precipitation field at a specific temporal and/or spatial scale given a known precipitation field (measured or simulated) at a certain larger temporal and/or spatial scale (lower resolution). Disaggregation and downscaling are very useful procedures and have several applications, such as in the following cases:

1. Global-scale weather-prediction models provide rainfall forecasts at a low resolution, for example, grid size of 50 km. Hydrologic models require the description of the precipitation field at a much higher resolution, with grid size of the order of 1 km.
2. Satellite precipitation estimates are available at a spatial scale ⩾0.25° (latitude and longitude), or about 28 km at the equator, and a temporal scale of 3 h. Again, hydrologic applications require higher resolutions.
3. The majority of historical point rainfall records come from daily rain gauges, which have often been operational for several decades. The number of rain gauges providing hourly or sub-hourly resolution data is smaller by about an order of magnitude. However, hydrologic applications, especially flood studies, usually need hourly or even sub-hourly data.
4. In complex problems of stochastic generation of precipitation time series or precipitation fields, it is difficult to reproduce simultaneously, the long-term and the short-term stochastic structure of precipitation using a single model. A better approach is to couple several models, starting from a large-scale model to represent the long-term behavior. The outputs of the latter are then disaggregated into finer scales. Note, however, that in a recent study Langousis and Koutsoyiannis (2006) developed a stochastic framework capable of reproducing simultaneously the long-term and the short-term stochastic structure of hydrological processes, avoiding the use of disaggregation.

While disaggregation and downscaling are similar in nature, they also have a difference that distinguishes them. Downscaling aims at solely producing a precipitation field $\underline{y}$ with the required statistics at the scale of interest, being statistically consistent with the given field $\underline{x}$ at the finer scale. Disaggregation demands full and precise consistency, which introduces an equality constraint in the problem of the form

$$\underline{C}\,\underline{y} = \underline{x} \qquad (55)$$

where $\underline{C}$ is a matrix of coefficients. For example, assuming that $\underline{x}$ is an annual amount of precipitation at a station and $\underline{y}$ is the vector consisting of the 12 monthly precipitation values at the same station, $\underline{C}$ will be a row vector with all its elements equal to 1, so that Equation (55) represents the requirement that the sum of all monthly precipitation amounts must equal the annual amount.

Task 1 could be accomplished by running a second meteorological model at the limited area of interest. Such models, known as limited-area models, can have much higher resolution than global models. The description of this type of downscaling, known as dynamical downscaling, because it is based on the atmospheric dynamics, is not within the scope of this chapter. In contrast, a stochastic procedure need not refer to the dynamics, and is generic and appropriate for both downscaling and disaggregation and for all above tasks 1–4. This generic procedure resembles the interpolation procedure described in Section 2.02.3.1.3, but there are two important differences. First, it is necessary to include the error terms in the generation procedure (recall that in interpolation, which is a point estimation, knowing only the mean and variance of the error was sufficient). Second, the generated values $\underline{y}$ at the different points should be statistically consistent to each other. This precludes the separate application of an algorithm at each point of interest and demands simultaneous generation at all points. In turn, this demands that the error terms in different points should be correlated to each other. All these requirements could be summarized in the linear generation scheme

$$\underline{y} = A\,\underline{x} + B\,\underline{v} \qquad (56)$$

where A and B are matrixes of coefficients and $\underline{v}$ is a vector of independent random variables, so that the term $B\underline{v} =: \underline{e}$ corresponds to the error term in interpolation (cf. Equation (29)). In disaggregation, Equation (56) should be considered simultaneously with Equation (55).

For Gaussian random fields without intermittency, the application of Equations (55) and (56) is rather trivial. However, the intermittency of the rainfall processes and the much-skewed distributions at fine timescales are severe obstacles for rainfall disaggregation. To overcome such obstacles, several researchers have developed a plethora of rather *ad hoc* disaggregation models (see review by Koutsoyiannis (2003b)). However, the application of the above theoretically consistent scheme is still possible, if combined with a stochastic model, accounting for intermittency (e.g., a Bartlett–Lewis model), and if an appropriate strategy is used to implement Equation (55). Such a strategy includes recursive application of Equation (56) until the error in Equation (55) becomes relatively low, and is followed by correction of the error of the accepted final iteration by appropriate adjusting procedures, which should not alter the covariance structure of the precipitation field. The general strategy of stochastic disaggregation is described in Koutsoyiannis (2001) and two implementations for temporal rainfall disaggregation at a fine (hourly) scale at a single site and at multiple sites are described in Koutsoyiannis and Onof (2001) and Koutsoyiannis et al. (2003), respectively. The models described in the latter two papers, named Hyetos and MuDRain, respectively, are available online, and have been used in several applications worldwide. Typical results of the two models are shown in **Figures** 25 and 26, respectively.

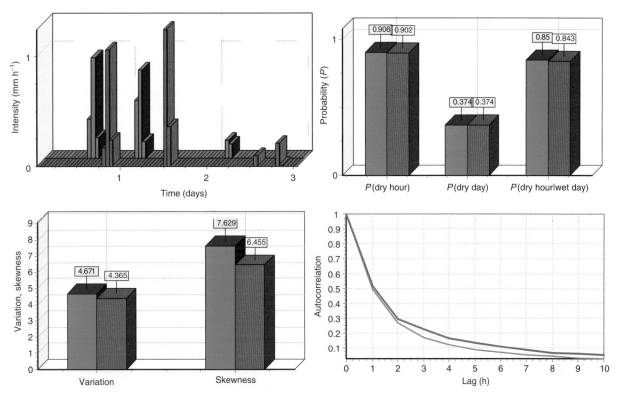

Figure 25 Typical screens produced by the Hyetos software during disaggregation of daily to hourly rainfall data, where plots in green and red refer to disaggregated and original data respectively. Upper-left panel shows typical hyetographs, where the green (disaggregated) plot is the result of the storms shown in **Figure 23** converted to a hyetograph at an hourly scale. Notice that while daily totals match, the temporal distribution of rainfall differs in the disaggregated and original hyetographs. However, in the statistical sense, the disaggregated series resembles the original, as shown in the other panels comparing statistics of disaggregated and original series.

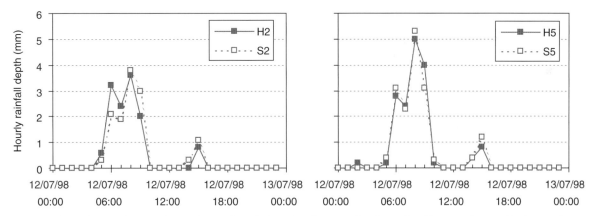

Figure 26 While, as shown in **Figure 25** (upper-left panel), in single-variate disaggregation, the produced hyetographs resemble the actual ones only in a statistical sense, multivariate disaggregation reproduces the actual shapes of hyetographs provided that fine-scale (e.g., hourly) data exist in at least one of the stations. The two panels show a comparison of historical (marked H) and simulated (by the MuDRain disaggregation model; marked S) hyetographs on a day with relatively high rainfall (~ 16 mm) at two rain gauges (2 and 5) in the Brue catchment located in South-Western England. From Koutsoyiannis D, Onof C, and Wheater HS (2003) Multivariate rainfall disaggregation at a fine timescale. *Water Resources Research* 39(7): 1173 (doi:10.1029/2002WR001600).

2.02.4.5 Multifractal Models

Rainfall models of multifractal type have for a long time been known to accurately reproduce several statistical properties of actual rainfall fields in finite but practically important ranges of scales: typically from below 1 h to several days in time and from below 10 km to more than 100 km in space (Schertzer and Lovejoy, 1987, 1989; Tessier *et al.*, 1993; Fraedrich and Larnder, 1993; Olsson, 1995; Lovejoy and Schertzer, 1995; Over and Gupta, 1996; Carvalho *et al.*, 2002; Nykanen and

Harris, 2003; Kundu and Bell, 2003; Deidda *et al.*, 2004, 2006; Gebremichael and Krajewski, 2004; Calenda *et al.*, 2005; Gebremichael *et al.*, 2006; Veneziano and Langousis, 2005; García-Marín *et al.*, 2007; Langousis and Veneziano, 2007). These properties include the scaling of the moments of different orders (Schertzer and Lovejoy, 1987; Menabde *et al.*, 1997; Deidda *et al.*, 1999; Deidda, 2000), the power law behavior of spatial and temporal spectral densities (Olsson, 1995; Tessier *et al.*, 1996; Deidda *et al.*, 2004, 2006), the alteration of wet and dry intervals (Over and Gupta, 1996; Schmitt *et al.*, 1998; Olsson, 1998; Güntner *et al.*, 2001; Langousis and Veneziano, 2007), and the distribution of extremes (Hubert *et al.*, 1998; Veneziano and Furcolo, 2002; Veneziano and Langousis, 2005; Langousis and Veneziano, 2007; Langousis *et al.*, 2007; Veneziano *et al.*, 2009). Significant deviations of rainfall from multifractal scale invariance have also been pointed out. These deviations include breaks in the power-law behavior of the spectral density (Fraedrich and Larnder, 1993; Olsson, 1995; Menabde *et al.*, 1997), lack of scaling of the non-rainy intervals in time series (Schmitt *et al.*, 1998), differences in scaling during the intense and moderate phases of rainstorms (Venugopal *et al.*, 2006), the power deficit at high frequencies relative to multifractal models (Perica and Foufoula-Georgiou, 1996a, 1996b; Menabde *et al.*, 1997; Menabde and Sivapalan, 2000), and more complex deviations as described in Veneziano *et al.* (2006a).

Next, we review some basic properties of stationary multifractal processes and discuss a simple procedure to construct discrete multifractal fields based on the concept of multiplicative cascades. For a detailed review on the generation of multifractal processes and their applications in hydrological modeling and forecasting, the reader is referred to Veneziano and Langousis (2010).

Let $\underline{i}^{(d)}(t)$ be the average rainfall intensity averaged over timescale d at time t. The stochastic process $\underline{i}^{(d)}(t)$ is said to be stationary multifractal if, for any timescale d, its statistics remain unchanged when the time axis is contracted by a factor $r > 1$ and the intensity is multiplied by a random variable $\underline{a}_r$, that is,

$$\underline{i}^{(d/r)}(t) \stackrel{d}{=} \underline{a}_r \, \underline{i}^{(d)}(t) \quad (57)$$

where $\stackrel{d}{=}$ denotes equality in (any finite-dimensional) distribution. The notation implies that the distribution of $\underline{a}_r$ depends only on r and not on time t or the intensity $\underline{i}^{(d)}$. Obviously, the mean of $\underline{a}_r$ is 1 and furthermore $\underline{a}_r$ is assumed to be stochastically independent of $\underline{i}^{(d)}$ at the higher scale d. The distribution of $\underline{a}_r$ characterizes the scaling properties as well as many other characteristics of the rainfall process including the marginal distribution, intermittency, distribution of extremes, etc. Equation (57) need not apply for arbitrarily large timescales but rather applies up to a maximum scale $d = d_{\max}$. In rainfall, $d_{\max}$ seems to be of the order of several days and it is representative of the mean interarrival time of rainfall events (Langousis and Veneziano, 2007; Langousis *et al.*, 2007; Veneziano *et al.*, 2007). We note for comparison that the related equation in the simple scaling (HK) representation of Section 2.02.1.5. is $(\underline{i}^{(d/r)} - \mu) \stackrel{d}{=} r^{1-H}(\underline{i}^{(d)} - \mu)$ or $\underline{i}^{(d/r)} \stackrel{d}{=} \mu(1 - r^{1-H}) + r^{1-H} \underline{i}^{(d)}$, so that, when the HK process has zero mean, it can be viewed as a special case of the multifractal process in which the random variable $\underline{a}_r$ is replaced by a deterministic power function of resolution r.

A property of stationary multifractal processes, which has been used to verify multifractality, is that the spectral density $s(\omega)$ behaves like ω^{-b} where ω is the frequency, and $b < 1$ is a constant (e.g., Fraedrich and Larnder, 1993; Olsson, 1995; Deidda *et al.*, 2004; Hsu *et al.*, 2006). More comprehensive checks of multifractality involve the dependence of statistical moments of different orders on scale. In particular, under perfect multifractality $E[(\underline{i}^{(d)})^q] \propto E[(\underline{a}_r)^q] \propto d^{-K(q)} \propto r^{K(q)}$, where $K(q)$ is a convex function, usually referred to as moment-scaling function (Gupta and Waymire, 1990; Veneziano, 1999). All concepts and methods are readily extended to space–time rainfall (Veneziano *et al.*, 2006b).

A simple procedure to construct discrete stationary multifractal fields is based on iterative application of Equation (57) starting from a large timescale $d \leqslant d_{\max}$ and gradually decreasing the timescale (i.e., at resolutions $r \propto m^n$, where $m > 1$ and $n \geqslant 1$ are integers). The contraction by the same factor $r = m$ at each step simplifies generation, since only the distribution of $\underline{a}_r \equiv \underline{a}_m$ is needed. This forms the concept of so-called isotropic discrete multiplicative cascade. Its construction in the D-dimensional cube S_D starts at level 0 with a single tile $\Omega_{0_1} \equiv S_D$ with constant unit intensity inside Ω_{0_1}. At level $n = 1, 2, \ldots$ (or equivalently at resolutions $r = m^D$, m^{2D}, ...) each tile at the previous level $n - 1$ is partitioned into m^D tiles where $m > 1$ is the integer multiplicity of the cascade. The intensity inside each cascade tile Ω_{n_i} ($i = 1, \ldots, m^{nD}$) is obtained by multiplying that of the parent tile at level $n - 1$ by an independent copy $\underline{y}_i$ of a unit-mean random variable $\underline{y}$, called the generator of the cascade. Clearly, for $r = m^{nD}$, $\underline{a}_r = \underline{y}_1 \underline{y}_2 \cdots \underline{y}_n$. For illustration, **Figure 27** shows a simulated realization of a 2D binary (i.e., $m = 2$) discrete multiplicative cascade developed to level $n = 8$.

2.02.5 Precipitation and Engineering Design

2.02.5.1 Probabilistic versus Deterministic Design Tools

The design and management of flood protection works and measures require reliable estimation of flood probability and risk. A solid empirical basis for this estimation can be offered by flow-observation records with an appropriate length, sufficient to include a sample of representative floods. In practice, however, flow measurements are never enough to support flood modeling. The obvious alternative is the use of hydrologic models with rainfall input data to generate streamflow. Notably, even when flow records exist, rainfall probability still has a major role in hydrological practice; for instance, in major hydraulic structures, the design floods are estimated from appropriately synthesized design storms (e.g., US Department of the Interior, Bureau of Reclamation, 1977, 1987; Sutcliffe, 1978). The need to use rainfall data as the basis of hydrologic design becomes even more evident in the study of engineering structures and urban water-management systems that modify the natural environment, so that past flood records, even if they exist, are no longer representative of the future modified system.

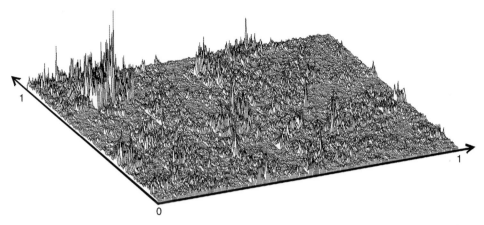

Figure 27 Simulated realization of a 2D stationary multifractal field. The random variable $\underline{y}$ is taken to be lognormal with unit mean value and log-variance $(\sigma_{\ln y})^2 = 0.2\ln(2)$.

Hydrologic design does not necessarily require full modeling of the rainfall process, of the type discussed in Section 2.02.4. Usually, in design studies, the focus is on extreme rainfall, which, notably, may not be represented well in such models, which are better for the average behavior of rainfall. However, historically, the perception of intense rainfall and the methodologies devised to model it have suffered from several fallacies spanning from philosophical to practical issues, which we describe next to cast a warning against their acceptance and use.

The first fallacy is of a rather philosophical type. As discussed in Section 2.02.1.5, the modeling of the rainfall process in pure deterministic terms has been proven to be problematic. However, deterministic thinking in science is strong enough, so that after the failure in providing full descriptions, it was headed to determining physical bounds to precipitation in an attempt to design risk-free constructions or practices. The resulting concept of probable maximum precipitation (PMP), that is, an upper bound of precipitation that is physically feasible (World Meteorological Organization, 1986), is perhaps one of the biggest failures in hydrology. Using elementary logic, we easily understand that even the terminology is self-contradictory, and thus not scientific. Namely, the word probable contradicts the existence of a deterministic limit.

Several methods to determine PMP exist in literature and are described in World Meteorological Organization (1986). However, examination in depth of each of the specific methods separately will reveal that they are all affected by logical inconsistencies. While they are all based on the assumption of the existence of a deterministic upper limit, they determine this limit statistically. This is obvious in the so-called statistical approach by Hershfield (1961, 1965), who used 95 000 station-years of annual maximum daily rainfall belonging to 2645 stations, standardized each record, and found the maximum over the 95 000 standardized values. Naturally, one of the 95 000 standardized values would be the greatest of all others, but this is not a deterministic limit to call PMP (Koutsoyiannis, 1999). If one examined 95 000 additional measurements, one might have found an even higher value. Thus, the logical problem here is the incorrect interpretation that an observed maximum in precipitation is a physical upper limit.

The situation is perhaps even worse with the so-called moisture maximization approach of PMP estimation (World Meteorological Organization, 1986), which seemingly is more physically (hydrometeorologically) based than the statistical approach of Hershfield. In fact, however, it suffers twice by the incorrect interpretation that an observed maximum is a physical upper limit. It uses a record of observed dew point temperatures to determine an upper limit, which is the maximum observed value. Then it uses this limit for the so-called maximization of an observed sample of storms, and asserts the largest value among them as PMP. Clearly, this is a questionable statistical approach, because (1) it does not assign any probability to the value determined and (2) it is based only on one observed value (known in statistics as the highest-order statistic), rather than on the whole sample, and thus it is enormously sensitive to one particular observation of the entire sample (Papalexiou and Koutsoyiannis, 2006; Koutsoyiannis, 2007). Thus, not only does the determination of PMP use a statistical approach (rather than deterministic physics), but it uses bad statistics. The arbitrary assumptions of the approach extend beyond the confusion of maximum observed quantities with physical limits. For example, the logic of moisture maximization at a particular location is unsupported given that a large storm at this location depends on the convergence of atmospheric moisture from much greater areas.

Rational thinking and fundamental philosophical and scientific principles can help identify and dispel such fallacies. In particular, the Aristotelian notions of *potentia* (Greek, dynamis) and of potential infinite (Greek, *apeiron*; Aristotle, Physics, 3.7, 206b16) that "exists in no other way, but ... potentially or by reduction" (and is different from mathematical complete infinite) would help us to avoid the PMP concept. In fact, this does not need a great deal of philosophical penetration. The same thing is more practically expressed as "conceptually, we can always imagine that a few more molecules of water could fall beyond any specified limit" (Dingman, 1994). Yet, the linkage to the Aristotelian notions of *potentia* and potential infinity may make us more sensitive in seeing the logical inconsistencies (see also Koutsoyiannis, 2007).

According to Popper (1982) the extension of the Aristotelian idea of *potentia* in modern terms is the notion of probability. Indeed, probability provides a different way to perceive the intense rainfall and flood and to assign to each value a certain probability of exceedance (see next session) avoiding the delusion of an upper bound of precipitation and the fooling of decision makers that they can build risk-free constructions. In this respect, the criticism of the PMP and the probable maximum flood (PMF) involves logical, technical, philosophical, and ethical issues (e.g., Benson, 1973).

One typical argument against the use of probabilistic approaches, in favor of PMP, which is very old yet popular even today, has been stated by Horton (1931; from Klemes, 2000), "It is, however, important to recognize the nature of the physical processes involved and their limitations in connection with the use of statistical methods. ... Rock Creek cannot produce a Mississippi River flood any more than a barnyard fowl can lay an ostrich egg." However, this argument reveals an incorrect perception of probability and statistics. In a probability theoretic context, there is not a logical inconsistency. Assuming, for example, that the annual peak flood of the Mississippi river (x_M) is on the average (μ_M), a million times larger than the average (μ_C) flood of a certain small creek (x_C), and assuming that both x_M and x_C have a lognormal distribution with standard deviation $\sigma_{\ln x}$ of logarithms of about 0.3 (which is roughly equal to the coefficient of variation of the annual flood peaks, assumed equal in the two streams), one can readily find that the probability that the flood in the creek x_C in some year exceeds the mean annual flood μ_M of Mississippi is $\Phi^*(\zeta) := 1 - \Phi_G(\zeta)$ where Φ_G is the standard normal distribution function and $\zeta = \ln(\mu_M/\mu_C)/\sigma_{\ln x}$ or $\zeta = \ln(10^6)/0.3 = 46$. For large ζ, the approximation $\ln \Phi^*(\zeta) = -(1/2)[\ln(2\pi\zeta^2) + \zeta^2]$ holds (e.g., Abramowitz and Stegun, 1965); hence $\ln \Phi^*(\zeta) = -1062.75$, so that the probability of exceedance is $\Phi^*(\zeta) = 10^{-462}$. That is, according to the probabilistic approach, the return period of the event that the small creek flood matches or exceeds the mean annual flood of the Mississippi is 10^{462} years. Assuming that the age of the universe is of the order of 10^{10} years, one would wait, on the average, 10^{452} times the age of the universe to see this event happen – if one foolishly hoped that the creek, the Mississippi, and the Earth would exist for such a long time. Evidently, such a low probability could be regarded as synonymous to impossibility, which shows that the probabilistic approach does not regard the floods of Mississippi equivalent to those of a small creek (see also an example about the age of a person by Feller (1950)).

2.02.5.2 Extreme Rainfall Distribution

Having been exempted from the concept of an upper limit to precipitation and having adopted a probabilistic approach, the real problem is how the rainfall intensity grows as the probability of exceedance decreases. Clearly, as the probability of exceedance tends to zero, the intensity tends to infinity. There exists a mathematically proven lower limit to the rate of this growth, which is represented by an exponential decay of the probability of exceedance with intensity. The alternative is a power-low decay and, as already mentioned in Section 2.02.1.5, the two options may lead to substantial differences in design quantities for high return periods. In this respect, the most important questions, which have not received definite answers yet, are again related to the notion of infinity.

Accordingly, the distribution tails are important to know in engineering design. However, the study of the tails is difficult and uncertain because the tails refer to infrequent events that require very long records to appear. Traditionally, rainfall records are analyzed in two ways. The most frequent is to choose the highest of all recorded precipitation intensities (for a given averaging timescale) at each year and form a statistical sample with size equal to the number of years of the record. The other is to form a sample with all recorded intensities over a certain threshold irrespective of the year they occurred. Usually, the threshold is chosen high enough, so that the sample size is again equal to the number of years of the record. This however is not necessary: it can well be set equal to zero, so that all recorded intensities are included in the sample. However, the threshold simplifies the study and helps focus the attention on the distribution tail.

If $\underline{x}_1, \underline{x}_2, ..., \underline{x}_n$ are random variables representing the recorded average intensities within a year at nonoverlapping time periods equal to a chosen timescale d, then the maximum among them $\underline{y} := \max(\underline{x}_1, \underline{x}_2, ..., \underline{x}_n)$ has a distribution function $H_n(y)$ fully dependent on the joint distribution function of $\underline{x}_i$. Assuming that $\underline{x}_i$ are IID with common distribution function $F(x)$, then $H_n(x) = [F(x)]^n$. If n is not constant, but rather can be regarded as a realization of a random variable (corresponding to the fact that the number of rainfall events is not the same in each year) with Poisson distribution with mean ν, then the distribution function H becomes (e.g., Todorovic and Zelenhasic, 1970; Rossi et al., 1984)

$$H(x) = \exp\{-\nu[1 - F(x)]\} \tag{58}$$

In particular, if the threshold has been chosen with the above rule (to make the sample size equal to the number of years of the record) then obviously $\nu = 1$. Equation (58) expresses in a satisfactory approximation, the relationship between the above two methodologies and the respective distributions F and H. The two options discussed above are then represented as follows:

1. *Exponential tail*.

$$F(x) = 1 - \exp(-x/\lambda + \psi),$$
$$H(x) = \exp[-\exp(-x/\lambda + \psi)], \quad x \geq \lambda\psi \tag{59}$$

where $\lambda > 0$ and $\psi > 0$ are parameters, so that $\lambda\psi$ represents the specified threshold. Here F is the exponential distribution and H is the Gumbel distribution, also known as extreme value type I (EV1) distribution.

2. *Power tail*.

$$F(x) = 1 - \left[1 + \kappa\left(\frac{x}{\lambda} - \psi\right)\right]^{-1/\kappa},$$
$$H(x) = \exp\left\{-\left[1 + \kappa\left(\frac{x}{\lambda} - \psi\right)\right]^{1/\kappa}\right\}, \quad x \geq \lambda\psi \tag{60}$$

where $\lambda > 0$, $\psi > 0$ and $\kappa > 0$ are parameters, and $\lambda\psi$ represents the specified threshold. Here F is the generalized Pareto distribution (a generalized form of Equation (18)) and H is the generalized extreme value (GEV) distribution.

In the case $\kappa > 0$ considered here, GEV is also called the extreme value type II (EV2) distribution. The case $\kappa < 0$ is mathematically possible and is called the extreme value type III (EV3) distribution. However, this is inappropriate for rainfall as it puts an upper bound ($\lambda\psi$) for x, which is inconsistent. The case $\kappa = 0$, corresponds precisely to the exponential tail (exponential and Gumbel distributions).

For years, the exponential tail and the Gumbel distribution have been the prevailing models for rainfall extremes, despite the fact that they yield unsafe (the smallest possible) design rainfall values. Recently, however, their appropriateness for rainfall has been questioned. Koutsoyiannis (2004a, 2005a, 2007) discussed several theoretical reasons that favor the power/EV2 over the exponential/EV1 case. As already mentioned (Section 2.02.1.5.5), Koutsoyiannis (2004b, 2005a) compiled an ensemble of annual maximum daily rainfall series from 169 stations in the Northern Hemisphere (28 from Europe and 141 from the USA) roughly belonging to six major climatic zones and all having lengths from 100–154 years. The analysis provides sufficient support for the general applicability of the EV2 distribution model worldwide. Furthermore, the ensemble of all samples was analyzed in combination and it was found that several dimensionless statistics are virtually constant worldwide, except for an error that can be attributed to a pure statistical sampling effect. This enabled the formation of a compound series of annual maxima, after standardization by the mean, for all stations (see **Figure 13**, which shows the distribution of a compound sample over threshold of all stations, except one in which only annual maxima existed). The findings support the estimation of a unique κ for all stations, which was found to be 0.15.

Additional empirical evidence with the same conclusions is provided by the Hershfield's (1961) data set, which was the basis of the formulation of Hershfield's PMP method. Koutsoyiannis (1999) showed that this data set does not support the hypothesis of an upper bound in precipitation, that is, PMP. Rather, it is consistent with the EV2 distribution with $\kappa = 0.13$, while the value $\kappa = 0.15$ can be acceptable for that data set too (Koutsoyiannis, 2004b). This enhances the trust that an EV2 distribution with $\kappa = 0.15$ can be regarded as a generalized model appropriate for mid-latitude areas of the Northern Hemisphere.

In a recent study, Veneziano et al. (2009) used multifractal analysis to show that the annual rainfall maximum for timescale d can be approximated by a GEV distribution and that typical values of κ lie in the range 0.09–0.15 with the larger values being associated with more arid climates. This range of values agrees well with the findings of Koutsoyiannis (1999, 2004b, 2005a). Similar results were provided by Chaouche (2001) and Chaouche et al. (2002). Chaouche (2001) explored a database of 200 rainfall series of various time steps (month, day, hour, and minute) from the five continents, each including more than 100 years of data. Using multifractal analyses, it was found that (1) an EV2/Pareto type law describes the rainfall amounts for large return periods; (2) the exponent of this law is scale invariant over scales greater than an hour (as stated in Section 2.02.4.2, it cannot be otherwise because this is dictated by theoretical reasons); and (3) this exponent is almost space invariant. Other studies have also expressed skepticism for the appropriateness of the Gumbel distribution for the case of rainfall extremes and suggested hyper-exponential tail behavior. Coles et al. (2003) and Coles and Pericchi (2003) concluded that inference based on the Gumbel model for annual maxima may result in unrealistically high return periods for certain observed events and suggested a number of modifications to standard methods, among which is the replacement of the Gumbel model with the GEV model. Mora et al. (2005) confirmed that rainfall in Marseille (a rain gauge included in the study by Koutsoyiannis (2004b)) shows hyper-exponential tail behavior. They also provided two regional studies in the Languedoc-Roussillon region (south of France) with 15 and 23 gauges, for which they found that a similar distribution with hyper-exponential tail could be fitted. This finding, when compared to previous estimations, leads to a significant increase in the depth of rare rainfall. On the same lines, Bacro and Chaouche (2006) showed that the distribution of extreme daily rainfall at Marseille is not in the Gumbel-law domain. Sisson et al. (2006) highlighted the fact that standard Gumbel analyses routinely assign near-zero probability to subsequently observed disasters, and that for San Juan, Puerto Rico, standard 100-year predicted rainfall estimates may be routinely underestimated by a factor of two. Schaefer et al. (2006) using the methodology by Hosking and Wallis (1997) for regional precipitation-frequency analysis and spatial mapping for 24-h and 2-h durations for the Washington State, USA, found that the distribution of rainfall maxima in this State generally follows the EV2 distribution type.

2.02.5.3 Ombrian Relationships

One of the major tools in hydrologic design is the ombrian relationship, more widely known by the misnomer rainfall intensity-duration-frequency (IDF) curve. An ombrian relationship (from the Greek ombros, rainfall) is a mathematical relationship estimating the average rainfall intensity i over a given timescale d (sometimes incorrectly referred to as duration) for a given return period T (also commonly referred to as frequency, although frequency is generally understood as reciprocal to period). Several forms of ombrian relationships are found in the literature, most of which have been empirically derived and validated by the long use in hydrologic practice. Attempts to give them a theoretical basis have often used inappropriate assumptions and resulted in oversimplified relationships that are not good for engineering studies.

In fact, an ombrian relationship is none other than a family of distribution functions of rainfall intensity for multiple timescales. This is because, the return period is tied to the distribution function, that is, $T = \delta/[1 - F(x)]$, where δ is the mean interarrival time of an event that is represented by the variable $\underline{x}$, typically 1 year. Thus, a distribution function such as one of those described in Section 2.02.4.2, is at the same time an ombrian relationship. This has been made clear in Koutsoyiannis et al. (1998) who showed that the empirical considerations usually involved in the construction of ombrian curves are not necessary at all, and create difficulties and confusion.

However, the direct use in engineering design of a fully consistent multiscale distribution function may be too complicated. Simplifications are possible to provide satisfactory approximations, given that only the distribution tail is of interest and that the range of scales of interest in engineering studies is relatively narrow. Such simplifications, which were tested recently and were found to be reasonable (Papalexiou and Koutsoyiannis, 2009) are

1. The separability assumption, according to which the influences of return period and timescale are separable (Koutsoyiannis et al., 1998), that is,

$$i(d, T) = \frac{a(T)}{b(d)} \quad (61)$$

where $a(T)$ and $b(d)$ are mathematical expressions to be determined.

2. The use of the Pareto distribution for the rainfall intensity over some threshold at any timescale, as discussed in Section 2.02.5.2; this readily provides a simple expression for $a(T)$.

3. The expression of $b(d)$ in the simple form

$$b(d) = (1 + d/\theta)^\eta \quad (62)$$

where $\theta > 0$ and $\eta > 0$ are parameters. A justification of this relationship, which is a satisfactory approximation for timescales up to a few days, can be found in Koutsoyiannis (2006a).

Based on assumptions 1–3, we easily deduce that the final form of the ombrian relationship is

$$i(d, T) = \lambda' \frac{(T/\delta)^\kappa - \psi'}{(1 + d/\theta)^\eta} \quad (63)$$

where $\psi' > 0$, $\lambda' > 0$ and $\kappa > 0$ are parameters. In particular, as discussed in Section 2.02.5.2, κ is the tail-determining parameter and unless a long record exists, which could support a different value, it should be assumed $\kappa = 0.15$. Equation (63) is dimensionally consistent, if θ has units of time (as well as δ), λ' has units of intensity, and κ and ψ are dimensionless. The numerator of Equation (63) differs from a pure power law that has been commonly used in engineering practice, as well as in some multifractal analyses. Consistent parameter-estimation techniques for ombrian relationships have been discussed in Koutsoyiannis et al. (1998) as well as in **Chapter 2.18 Statistical Hydrology**.

References

Abramowitz M and Stegun IA (1965) *Handbook of Mathematical Functions*. New York: Dover.

Adler RF and Negri AJ (1988) A satellite infrared technique to estimate tropical convective and stratiform rainfall. *Journal of Applied Meteorology* 27: 30–51.

Alishouse JC, Snyder SA, Vongsathorn J, and Ferraro RR (1990) Determination of oceanic total precipitable water from the SSM/I. *IEEE Transactions on Geoscience and Remote Sensing* 28(5): 811–816.

Anthes RA (1982) *Tropical Cyclones: Their Evolution, Structure, and Effects*. Meteorological Monographs Number 41. Boston, MA: American Meteorological Society.

Aonashi K and Liu G (2000) Passive microwave precipitation retrievals using TMI during the Baiu period of 1998. Part I: Algorithm Description and Validation. *Journal of Applied Meteorology* 39: 2024–2037.

Arakawa H (1956) On the secular variation of annual totals of rainfall at Seoul from 1770 to 1944. *Theoretical and Applied Climatology* 7(2): 205–211.

Arkin PA (1979) The relationship between fractional coverage of high cloud and rainfall accumulations during GATE over the B-scale array. *Monthly Weather Review* 107: 1382–1387.

Arkin PA and Meisner BN (1987) The relationship between large-scale convective rainfall and cold cloud over the Western Hemisphere during 1982–84. *Monthly Weather Review* 115: 51–74.

Atlas D and Ulbrich CW (1977) Path and area integrated rainfall measurement by microwave attenuation in the 1–3 cm band. *Journal of Applied Meteorology* 16: 1322–1331.

Atlas D, Ulbrich CW, and Meneghini R (1984) The multiparameter remote measurement of rainfall. *Radio Science* 19: 3–21.

Austin PM and Houze RA (1972) Analysis of the structure of precipitation patterns in New England. *Journal of Applied Meteorology* 11: 926–935.

Ba MB and Nicholson SE (1998) Analysis of convective activity and its relationship to the rainfall over the rift valley lakes of East Africa during 1983–90 using the Meteosat infrared channel. *Journal of Applied Meteorology* 37: 1250–1264.

Bacro J-N and Chaouche A (2006) Incertitude d'estimation des pluies extrêmes du pourtour méditerranéen: Illustration par les données de Marseille. *Hydrological Sciences Journal* 51(3): 389–405.

Baloutsos G, Bourletsikas A, and Kaoukis K (2005) Study and investigation of the fog precipitation characteristics in the fir forest of Agios Nikolaos, Evritania, Greece, in Greek. *Geotechnica Epistemonica Themata* 2(16): 34–45.

Barnes GM, Zipser EJ, Jorgensen D, and Marks FD (1983) Mesoscale and convective structure of hurricane rainband. *Journal of the Atmospheric Sciences* 40: 2125–2137.

Barrett EC and Martin DW (1981) *The Use of Satellite Data in Rainfall Monitoring*. London: Academic Press.

Battan LJ (1973) *Radar Observations of the Atmosphere*. Chicago, IL: The University of Chicago Press.

Beard KV (1976) Terminal velocity and shape of cloud and precipitation drops aloft. *Journal of the Atmospheric Sciences* 33: 851–864.

Beer T and Giannini L (1980) Tropical cyclone cloud bands. *Journal of the Atmospheric Sciences* 37: 1511–1520.

Bellon A, Lovejoy S, and Austin G (1980) Combining satellite and radar data for the short-range forecasting of precipitation. *Monthly Weather Review* 108: 1554–1566.

Bennetts DA and Sharp JC (1982) The relevance of conditional symmetric instability to the prediction of mesoscale frontal rainbands. *Quarterly Journal of the Royal Meteorological Society* 108: 595–602.

Benson MA (1973) Thoughts on the design of design floods *Floods and Droughts, Proceedings of the 2nd International Symposium in Hydrology*, pp. 27–33. Fort Collins, CO: Water Resources Publications.

Berg W and Chase R (1992) Determination of mean rainfall from the special sensor microwave/imager (SSM/I) using a mixed lognormal distribution. *Journal of Atmospheric and Oceanic Technology* 9: 129–141.

Berg W, Olson W, Ferraro R, Goodman SJ, and LaFontaine FJ (1998) An assessment of the first- and second-generation navy operational precipitation retrieval algorithms. *Journal of the Atmospheric Sciences* 55: 1558–1575.

Bergeron T (1935) On the physics of cloud and precipitation. In: *Proceedings of the Fifth Assembly of the International Union of Geodesy and Geophysics (IUGG)*, pp. 156–178. Lisbon, Portugal.

Berry EX and Reinhardt RL (1974a) An analysis of cloud drop growth by collection: Part I. Double distributions. *Journal of the Atmospheric Sciences* 31: 1814–1824.

Berry EX and Reinhardt RL (1974b) An analysis of cloud drop growth by collection: Part II. Single initial distributions. *Journal of the Atmospheric Sciences* 31: 1825–1831.

Bluestein HB (2003) Tornadoes. In: Holton JR, Curry JA, and Pyle JA (eds.) *Encyclopedia of Atmospheric Sciences*, pp. 2290–2297. London: Academic Press.

Bowen EG (1950) The formation of rain by coalescence. *Australian Journal of Scientific Research* A3: 192–213.

Bringi VN and Chandrasekar V (2001) *Polarimetric Doppler Weather Radar*. New York, NY: Cambridge University Press.

Brock JR (1972) Condensational growth of atmospheric aerosols. In: Hidy GM (ed.) *Aerosols and Atmospheric Chemistry*, pp. 149–153. New York, NY: Academic Press.

Browning KA and Ludlam FH (1962) Airflow in convective storms. *Quarterly Journal of the Royal Meteorological Society* 88: 117–135.

Byers HR and Braham RR (1949) The thunderstorm. *Final Report of the Thunderstorm Project, US Deptartment of Commerce*. Washington, DC: US Government Printing Office.

Calenda G, Gorgucci E, Napolitano F, Novella A, and Volpi E (2005) Multifractal analysis of radar rainfall fields over the area of Rome. *Advances in Geosciences* 2: 293–299.

Carvalho LMV, Lavallée D, and Jones C (2002) Multifractal properties of evolving convective systems over tropical South America. *Geophysical Research Letters* 29: doi: 10.1029/2001GL014276.

Chaouche K (2001) Approche multifractale de la modélisation stochastique en hydrologie. Thèse, Ecole Nationale du Génie Rural, des Eaux et des Forêts, Centre de Paris, France. http://www.engref.fr/thesechaouche.htm (accessed March 2010).

Chaouche K, Hubert P, and Lang G (2002) Graphical characterisation of probability distribution tails. *Stochastic Environmental Research and Risk Assessment* 16(5): 342–357.

Chen SS, Lonfat M, Knaff JA, and Marks FD Jr. (2006) Effects of vertical wind shear and storm motion on tropical cyclone rainfall asymmetries deduced from TRMM. *Monthly Weather Review* 134: 3190–3208.

Chen SS, Price JF, Zhao W, Donelan M, and Walsh EJ (2007) The CBLAST-hurricane program and the next-generation fully coupled atmosphere–wave–ocean models for hurricane research and prediction. *Bulletin of the American Meteorological Society* 88: 311–317.

Cheng M and Brown R (1995) Delineation of precipitation areas by correlation of Meteosat visible and infrared data with radar data. *Monthly Weather Review* 123: 2743–2757.

Cheng M, Brown R, and Collier C (1993) Delineation of precipitation areas using Meteosat infrared and visible data in the region of the United Kingdom. *Journal of Applied Meteorology* 32: 884–898.

Chin EH (1977) Modelling daily precipitation occurrence process with Markov chains. *Water Resources Research* 13(6): 956–959.

Clarke RT (1998) *Stochastic Processes for Water Scientists*. New York, NY: Wiley.

Coles S and Pericchi L (2003) Anticipating catastrophes through extreme value modelling. *Applied Statistics* 52: 405–416.

Coles S, Pericchi LR, and Sisson S (2003) A fully probabilistic approach to extreme rainfall modeling. *Journal of Hydrology* 273(1–4): 35–50.

Cotton WR and Anthes RA (1989) *Storm and Cloud Dynamics*, International Geophysics Series, vol. 44. San Diego, CA: Academic Press.

Cotton WR, Thompson G, and Mielke PW (1994) Realtime mesoscale prediction on workstations. *Bulletin of the American Meteorological Society* 75: 349–362.

Cowpertwait P, Isham V, and Onof C (2007) Point process models of rainfall: Developments for fine-scale structure. *Proceedings of the Royal Society A* 463(2086): 2569–2587.

Cox D and Isham V (1988) A simple spatial–temporal model of rainfall. *Proceedings of the Royal Society A* 415: 317–328.

De Bruin HAR (1980) A stochastic description of wet and dry spells in terms of an effective number of days. *Journal of Hydrology* 45: 91–99.

Deidda R (2000) Rainfall downscaling in a space–time multifractal framework. *Water Resources Research* 36(7): 1779–1794.

Deidda R, Badas MG, and Piga E (2004) Space–time scaling in high-intensity tropical ocean global atmosphere coupled ocean–atmosphere response experiment (TOGA-COARE) storms. *Water Resources Research* 40: W02506.1 (doi:10.1029/2003WR002574).

Deidda R, Benzi R, and Siccardi F (1999) Multifractal modeling of anomalous scaling laws in rainfall. *Water Resources Research* 35(6): 1853–1867.

Deidda R, Grazia-Badas M, and Piga E (2006) Space–time multifractality of remotely sensed rainfall fields. *Journal of Hydrology* 322: 2–13 (doi:10.1016/j.jhydrol.2005.02.036).

Dingman SL (1994) *Physical Hydrology*. Englewood Cliffs, NJ: Prentice Hall.

Doviak R and Zrnic D (1993) *Doppler Radar and Weather Observations*. p. 562. New York, NY: Elsevier.

Drake RL (1972a) A general mathematical survey of the coagulation equation. In: Hidy GM and Brock JR (eds.) *Topics in Current Aerosol Research (Part 2)*, pp. 201–384. Oxford: Pergamon.

Drake RL (1972b) The scalar transport equation of coalescence theory: Moments and kernels. *Journal of the Atmospheric Sciences* 29: 537–547.

Elachi C (1987) *Introduction to the Physics and Techniques of Remote Sensing*. Wiley Series in Remote Sensing. Hoboken, NJ: Wiley.

Emanuel KA (1986) An air–sea interaction theory for tropical cyclones. Part I: Steady state maintenance. *Journal of the Atmospheric Sciences* 43: 585–604.

Emanuel KA (1989) Polar lows as artic hurricanes. *Tellus* 41A: 1–17.

Entekhabi D, Rodriguez-Iturbe I, and Eagleson PS (1989) Probabilistic representation of the temporal rainfall process by the modified Neyman–Scott rectangular pulses model: Parameter estimation and validation. *Water Resources Research* 25(2): 295–302.

Epstein ES (1969) Stochastic dynamic prediction. *Tellus* 21: 739–757.

Federer B and Waldvogel A (1975) Hail and raindrop size distributions from a Swiss multicell storm. *Journal of Applied Meteorology* 14: 91–97.

Feingold G and Levin Z (1986) The lognormal fit to raindrop spectra from frontal convective clouds in Israel. *Journal of Climate and Applied Meteorology* 25: 1346–1363.

Feller W (1950) *An Introduction to Probability Theory and Its Applications*. New York, NY: Wiley.

Ferraro RR and Marks GF (1995) The development of SSM/I rain-rate retrieval algorithms using ground-based radar measurements. *Journal of Atmospheric and Oceanic Technology* 12: 755–770.

Ferriday JG and Avery SK (1994) Passive microwave remote sensing of rainfall with SSM/I: Algorithm development and implementation. *Journal of Applied Meteorology* 33: 1587–1596.

Feyerherm AM and Bark LD (1964) Statistical methods for persistent precipitation patterns. *Journal of Applied Meteorology* 4: 320–328.

Findeisen W (1938) Die Kolloid Meteorologischen Vorgang der Niederschlags Bildung. *Meteorologische Zeitschrift* 55: 121–133.

Fletcher NH (1962) *The Physics of Rainclouds*. Cambridge: Cambridge University Press.

Foufoula-Georgiou E and Guttorp P (1986) Compatibility of continuous rainfall occurrence models with discrete rainfall observations. *Water Resources Research* 22(8): 1316–1322.

Foufoula-Georgiou E and Lettenmaier DP (1986) Continuous-time versus discrete-time point process models for rainfall occurrence series. *Water Resources Research* 22(4): 531–542.

Fraedrich K and Larnder C (1993) Scaling regimes of composite rainfall time series. *Tellus* 45A: 289–298.

Frank WM (1977) The structure and energetics of the tropical cyclone: I. Storm structure. *Monthly Weather Review* 105: 1119–1135.

Gabriel KR and Neumann J (1962) A Markov chain model for daily rainfall occurrences at Tel Aviv. *Quarterly Journal of the Royal Meteorological Society* 88: 90–95.

García-Marín AP, Jiménez-Hornero FJ, and Ayuso JL (2007) Applying multifractality and the self-organized criticality theory to describe the temporal rainfall regimes in Andalusia (southern Spain). *Hydrological Processes* (doi:10.1002/hyp.6603).

Gebremichael M and Krajewski WF (2004) Assessment of the statistical characterization of small-scale rainfall variability from radar: Analysis of TRMM ground validation datasets. *Journal of Applied Meteorology* 43(8): 1180–1199.

Gebremichael M, Over TM, and Krajewski WF (2006) Comparison of the scaling properties of rainfall derived from space- and surface-based radars. *Journal of Hydrometeorology* 7: 1277–1294.

Georgakakos KP, Carsteanu AA, Sturdevant PL, and Cramer JA (1994) Observation and analysis of Midwestern rain rates. *Journal of Applied Meteorology* 33: 1433–1444.

Gillespie DT (1972) The stochastic coalescence model for cloud droplet growth. *Journal of the Atmospheric Sciences* 29: 1496–1510.

Gillespie DT (1975) Three models for the coalescence growth of cloud drops. *Journal of the Atmospheric Sciences* 32: 600–607.

Gray WM (1968) Global view of the origins of tropical disturbances and storms. *Monthly Weather Review* 96: 669–700.

Gray WM (1979) Hurricanes: Their formation, structure and likely role in the tropical circulation. In: Shaw DB (ed.) *Meteorology Over the Tropical Oceans*. London: Royal Meteorological Society.

Grell GA, Dudhia J, and Stauffer DR (1995) A description of the fifth-generation Pennsylvania State/NCAR mesoscale model (MM5), NCAR Tech. Note TN-398 + STR122 pp. (Available from UCAR Communications, P.O. Box 3000, Boulder, CO 80307.)

Griffith CG, Woodley WL, Grube PG, Martin DW, Stout J, and Sikdar DN (1978) Rain estimation from geosynchronous satellite imagery – visible and infrared studies. *Monthly Weather Review* 106: 1153–1171.

Grody NC (1991) Classification of snow cover and precipitation using the special sensor microwave imager. *Journal of Geophysical Research* 96(D4): 7423–7435.

Gunn KLS and Marshall JS (1958) The distribution with size of aggregate snowflakes. *Journal of Meteorology* 15: 452–461.

Gunn R and Kinzer GD (1949) The terminal velocity of fall for water drops in stagnant air. *Journal of Meteorology* 6: 243–248.

Güntner A, Olsson J, Calver A, and Gannon B (2001) Cascade-based disaggregation of continuous rainfall time series: The influence of climate. *Hydrology and Earth System Sciences* 5: 145–164.

Gupta VK and Waymire E (1990) Multiscaling properties of spatial rainfall and river flow distributions. *Journal of Geophysical Research* 95: 1999–2009.

Gyasi-Agyei Y and Willgoose GR (1997) A hybrid model for point rainfall modelling. *Water Resources Research* 33(7): 1699–1706.

Haan CT, Allen DM, and Street JO (1976) A Markov chain model of daily rainfall. *Water Resources Research* 12(3): 443–449.

Hakim GJ (2003) Cyclogenesis. In: Holton JR, Curry JA, and Pyle JA (eds.) *Encyclopedia of Atmospheric Sciences*, pp. 589–594. London: Academic Press.

Hane CE (1986) Extratropical squall lines and rainbands. In: Ray PS (ed.) *Mesoscale Meteorology and Forecasting*, ch. 1, pp. 359–389. Boston: American Meteorological Society.

Harr RD (1982) Fog drip in the Bull Run municipal watershed, Oregon. *Water Resources Bulletin* 18: 785–789.

Harrold TI, Sharma A, and Sheather SJ (2003) A nonparametric model for stochastic generation of daily rainfall occurrence. *Water Resources Research* 39(10): 1300 (doi: 10.1029/2003WR002182).

Harrold TW and Austin PM (1974) The structure of precipitation systems – a review. *Journal de Recherches Atmospheriques* 8: 41–57.

Hawkins HF and Rubsam DT (1968) Hurricane Hilda, 1964. II. Structure and budgets of the hurricane on October 1, 1964. *Monthly Weather Review* 96: 617–636.

Hemond HF and Fechner-Levy EJ (2000) *Chemical Fate and Transport in the Environment*, 2nd edn. London: Academic Press.

Hershfield DM (1961) Estimating the probable maximum precipitation. *ASCE: Journal of the Hydraulics Division* 87(HY5): 99–106.

Hershfield DM (1965) Method for estimating probable maximum precipitation. *Journal of the American Water Works Association* 57: 965–972.

Hershfield DM (1970) A comparison of conditional and unconditional probabilities for wet- and dry-day sequences. *Journal of Applied Meteorology* 9(5): 825–827.

Hinton BB, Olson WS, Martin DW, and Auvine B (1992) A passive microwave algorithm for tropical oceanic rainfall. *Journal of Applied Meteorology* 31: 1379–1395.

Hobbs PV, Bowdle DA, and Radke LF (1985) Particles in the lower troposphere over the high plains of the United States. Part I: Size distributions, elemental compositions and morphologies. *Journal of Climate and Applied Meteorology* 24: 1344–1356.

Holland GJ (1980) An analytic model of the wind and pressure profiles in hurricanes. *Monthly Weather Review* 108: 1212–1218.

Hong S-Y, Dudhia J, and Chen S-H (2004) A revised approach to ice microphysical processes for the bulk parameterization of clouds and precipitation. *Monthly Weather Review* 132: 103–120.

Horton RE (1931) The field, scope and status of the science of hydrology. In: *Trans. Am. Geophys. Union, 12th Annu. Mtg.*, National Research Council, pp. 189–202. Washington, DC.

Hosking JRM and Wallis JR (1997) *Regional Frequency Analysis – an Approach Based on L-Moments*. New York, NY: Cambridge University Press.

Houghton HG (1985) *Physical Meteorology*. Cambridge, MA: MIT Press.

Houze RA (1993) *Cloud Dynamics*. San Diego, CA: Academic Press.

Houze RA, Hobbs PV, Herzegh PH, and Parsons DB (1979) Size distributions of precipitation particles in frontal clouds. *Journal of the Atmospheric Sciences* 36: 156–162.

Houze RA, Marks FD Jr., and Black RA (1992) Dual-aircraft investigation of the inner core of hurricane Norbert. Part II: Mesoscale distribution of ice particles. *Journal of the Atmospheric Sciences* 49: 943–962.

Hsu HM, Moncrieff MW, Tung WW, and Liu C (2006) Multiscale temporal variability of warm-season precipitation over North America: Statistical analysis of radar measurements. *Journal of the Atmospheric Sciences* 63: 2355–2368.

Hubert P, Bendjoudi H, Schertzer D, and Lovejoy S (1998) A Multifractal explanation for rainfall intensity–duration–frequency curves. In: Llasat C, Versace P, and Ferrari E (eds.) *Heavy Rains and Flash Floods*, pp. 21–28, Cosenza, Italy: Natl. Res. Counc., Group for Prev. from Hydrol. Disasters.

Hurst HE (1951) Long term storage capacities of reservoirs. *Transactions of the American Society of Civil Engineers* 116: 776–808.

Hutchinson MF (1990) A point rainfall model based on a three-state continuous Markov occurrence process. *Journal of Hydrology* 114: 125–148.

Jorgensen DP (1984a) Mesoscale and convective-scale characterisitcs of mature hurricanes. Part I: General observations by research aircraft. *Journal of the Atmospheric Sciences* 41(8): 1268–1286.

Jorgensen DP (1984b) Mesoscale and convective-scale characterisitcs of mature hurricanes. Part II: Inner core structure of hurricane Allen (1980). *Journal of the Atmospheric Sciences* 41(8): 1287–1311.

Joss J and Gori E (1978) Shapes of raindrop size distributions. *Journal of Applied Meteorology* 17: 1054–1061.

Joss J and Waldvogel A (1990) Precipitation measurement and hydrology. In: Atlas D (ed.) *Radar in Meteorology, Battan Memorial and 40th Anniversary Radar Meteorology Conference*, American Meteorological Society, pp. 577–597. Chicago, IL: The University of Chicago Press.

Katsanos D, Lagouvardos K, Kotroni V, and Argiriou A (2007) Relationship of lightning activity with microwave brightness temperatures and spaceborne radar reflectivity profiles in the Central and Eastern Mediterranean. *Journal of Applied Meteorology* 46: 1901–1912.

Katz RW (1977a) An application of chain-dependent processes to meteorology. *Journal of Applied Probability* 14: 598–603.

Katz RW (1977b) Precipitation as a chain-dependent process. *Journal of Applied Meteorology* 16: 671–676.

Katz RW and Parlange MB (1998) Overdispersion phenomenon in stochastic modeling of precipitation. *Journal of Climate* 11: 591–601.

Kavvas ML and Delleur JW (1981) A stochastic cluster model of daily rainfall sequences. *Water Resources Research* 17(4): 1151–1160.

Kepert J (2001) The dynamics of boundary layer jets within the tropical cyclone core. Part I: Linear theory. *Journal of the Atmospheric Sciences* 58: 2469–2484.

Kepert J and Wang Y (2001) The dynamics of boundary layer jets within the tropical cyclone core. Part II: Nonlinear enhancement. *Journal of the Atmospheric Sciences* 58: 2485–2501.

Kessler E (1969) *On the Distribution and Continuity of Water Substance in Atmospheric Circulation*. Meteorological Monographs 10, no. 32, 84pp. Boston, MA: American Meteorological Society.

Kidd C, Kniveton D, and Barrett EC (1998) The advantages and disadvantages of statistically derived–empirically calibrated passive microwave algorithms for rainfall estimation. *Journal of the Atmospheric Sciences* 55: 1576–1582.

Kidder SQ and Vonder Haar TH (1995) *Satellite Meteorology*. London: Academic Press.

King PW, Hogg WD, and Arkin PA (1995) The role of visible data in improving satellite rain-rate estimates. *Journal of Applied Meteorology* 34: 1608–1621.

Klemes V (2000) Tall tales about tails of hydrological distributions. *Journal of Hydrologic Engineering* 5(3): 227–239.

Kolmogorov AN (1933) Grundbegrijfe der Wahrscheinlichkeitsrechnung *Ergebnisse der Math.* (2), Berlin; 2nd English edn. (1956): *Foundations of the Theory of Probability*, 84pp. New York, NY: Chelsea Publishing Company.

Kolmogorov AN (1940) Wienersche Spiralen und einige andere interessante Kurven in Hilbertschen Raum. *Doklady Akademii Nauk SSSR* 26: 115–118.

Kosambi DD (2005) *The Culture and Civilization of Ancient India in Historical Outline*. Delhi: Vikas Publishing House.

Kotroni V and Lagouvardos K (2004) Evaluation of MM5 high-resolution real-time forecasts over the urban area of Athens, Greece. *Journal of Applied Meteorology* 43: 1666–1678.

Kottegoda NT and Horder MA (1980) Daily flow model based on rainfall occurrences using pulses and a transfer function. *Journal of Hydrology* 47: 215–234.

Koutsoyiannis D (1999) A probabilistic view of Hershfield's method for estimating probable maximum precipitation. *Water Resources Research* 35(4): 1313–1322.

Koutsoyiannis D (2000a) A generalized mathematical framework for stochastic simulation and forecast of hydrologic time series. *Water Resources Research* 36(6): 1519–1533.

Koutsoyiannis D (2000b) *Lecture Notes on Hydrometeorology – Part 1*, 2nd edn., 157pp. Athens: National Technical University of Athens. http://www.itia.ntua.gr/en/docinfo/116 (accessed March 2010).

Koutsoyiannis D (2001) Coupling stochastic models of different time scales. *Water Resources Research* 37(2): 379–392.

Koutsoyiannis D (2002) The Hurst phenomenon and fractional Gaussian noise made easy. *Hydrological Sciences Journal* 47(4): 573–595.

Koutsoyiannis D (2003a) Climate change, the Hurst phenomenon, and hydrological statistics. *Hydrological Sciences Journal* 48(1): 3–24.

Koutsoyiannis D (2003b) Rainfall disaggregation methods: Theory and applications. In: Piccolo D and Ubertini L (eds.) *Proceedings, Workshop on Statistical and Mathematical Methods for Hydrological Analysis*, pp. 1–23. Rome: Università di Roma "La Sapienza".

Koutsoyiannis D (2004a) Statistics of extremes and estimation of extreme rainfall: 1. Theoretical investigation. *Hydrological Sciences Journal* 49(4): 575–590.

Koutsoyiannis D (2004b) Statistics of extremes and estimation of extreme rainfall: 2. Empirical investigation of long rainfall records. *Hydrological Sciences Journal* 49(4): 591–610.

Koutsoyiannis D (2005a) Uncertainty, entropy, scaling and hydrological stochastics: 1. Marginal distributional properties of hydrological processes and state scaling. *Hydrological Sciences Journal* 50(3): 381–404.

Koutsoyiannis D (2005b) Uncertainty, entropy, scaling and hydrological stochastics: 2. Time dependence of hydrological processes and time scaling. *Hydrological Sciences Journal* 50(3): 405–426.

Koutsoyiannis D (2006a) An entropic-stochastic representation of rainfall intermittency: The origin of clustering and persistence. *Water Resources Research* 42(1): W01401 (doi:10.1029/2005WR004175).

Koutsoyiannis D (2006b) Nonstationarity versus scaling in hydrology. *Journal of Hydrology* 324: 239–254.

Koutsoyiannis D (2007) A critical review of probability of extreme rainfall: Principles and models. In: Ashley R, Garvin S, Pasche E, Vassilopoulos A, and Zevenbergen C (eds.) *Advances in Urban Flood Management*, pp. 139–166. London: Taylor and Francis.

Koutsoyiannis D (2010) A random walk on water. *Hydrology and Earth System Sciences* 14: 585–601.

Koutsoyiannis D, Efstratiadis A, Mamassis N, and Christofides A (2008) On the credibility of climate predictions. *Hydrological Sciences Journal* 53(4): 671–684.

Koutsoyiannis D, Kozonis D, and Manetas A (1998) A mathematical framework for studying rain-fall intensity–duration–frequency relationships. *Journal of Hydrology* 206(1–2): 118–135.

Koutsoyiannis D, Mamassis N, and Tegos A (2007) Logical and illogical exegeses of hydrometeorological phenomena in ancient Greece. *Water Science and Technology: Water Supply* 7(1): 13–22.

Koutsoyiannis D and Montanari A (2007) Statistical analysis of hydroclimatic time series: Uncertainty and insights. *Water Resources Research* 43(5): W05429 (doi: 10.1029/2006WR005592).

Koutsoyiannis D and Onof C (2001) Rainfall disaggregation using adjusting procedures on a Poisson cluster model. *Journal of Hydrology* 246: 109–122.

Koutsoyiannis D, Onof C, and Wheater HS (2003) Multivariate rainfall disaggregation at a fine timescale. *Water Resources Research* 39(7): 1173 (doi:10.1029/2002WR001600).

Koutsoyiannis D and Xanthopoulos T (1999) *Engineering Hydrology*, 3rd edn., p. 418. Athens: National Technical University of Athens (in Greek).

Krajewski WF and Smith JA (2002) Radar hydrology: Rainfall estimation. *Advances in Water Resources* 25(8–12): 1387–1394 (doi: 10.1016/S0309-1708(02)00062-3).

Kummerow C and Giglio L (1994a) A passive microwave technique for estimating rainfall and vertical structure information from space. Part I: Algorithm description. *Journal of Applied Meteorology* 33: 3–18.

Kummerow C and Giglio L (1994b) A passive microwave technique for estimating rainfall and vertical structure information from space. Part II: Applications to SSM/I data. *Journal of Applied Meteorology* 33: 19–34.

Kummerow C, Barnes W, Kozu T, Shiue J, and Simpson J (1998) The tropical rainfall measuring mission (TRMM) sensor package. *Journal of Atmospheric and Oceanic Technology* 15: 809–817.

Kummerow C, Hong Y, Olson WS, et al. (2001) The evolution of the Goddard profiling algorithm (GPROF) for rainfall estimation from passive microwave sensors. *Journal of Applied Meteorology* 40: 1801–1820.

Kummerow CD, Olson WS, and Giglio L (1996) A simplified scheme for obtaining precipitation and vertical hydrometeor profiles from passive microwave sensors. *IEEE Transactions on Geoscience and Remote Sensing* 34: 1213–1232.

Kundu PK and Bell TL (2003) A Stochastic model of space–time variability of mesoscale rainfall: Statistics of spatial averages. *Water Resources Research* 39(12): 1328 (doi:10.1029/2002WR001802).

Lagouvardos K and Kotroni V (2005) Improvement of high resolution weather forecasts through humidity adjustment, based on satellite data. *Quarterly Journal of the Royal Meteorological Society* 131: 2695–2712.

Lagouvardos K, Kotroni V, Koussis A, Feidas C, Buzzi A, and Malguzzi P (2003) The meteorological model BOLAM at the National Observatory of Athens: Assessment of two-year operational use. *Journal of Applied Meteorology* 42: 1667–1678.

Landsea CW (2000) Climate variability of tropical cyclones. In: Pielke R Jr. and Pielke R Sr. (eds.) *Storms*, vol. 1, New York, NY: Routledge.

Langmuir I (1948) The production of rain by a chain reaction in cumulus clouds at temperatures above freezing. *Journal of Meteorology* 5: 175–192.

Langousis A and Koutsoyiannis D (2006) A stochastic methodology for generation of seasonal time series reproducing over-year scaling behavior. *Journal of Hydrology* 322(1–4): 138–154.

Langousis A and Veneziano D (2007) Intensity–duration–frequency curves from scaling representations of rainfall. *Water Resources Research* 43(2): W02422 (doi: 10.1029/2006WR005245).

Langousis A and Veneziano D (2009a) Theoretical model of rainfall in tropical cyclones for the assessment of long-term risk. *Journal of Geophysical Research* 114: D02106 (doi:10.1029/2008JD010080).

Langousis A and Veneziano D (2009b) Long term rainfall risk from tropical cyclones in coastal areas. *Water Resources Research* 45: W11430 (doi:10.1029/2008WR007624).

Langousis A, Veneziano D, Furcolo P, and Lepore C (2007) Multifractal rainfall extremes: Theoretical analysis and practical estimation. *Chaos Solitons and Fractals* 39(3): 1182–1194 (doi:10.1016/j.chaos.2007.06.004).

Langousis A, Veneziano D, and Chen S (2008) Boundary layer model for moving tropical cyclones. In: Elsner J and Jagger TH (eds.) *Hurricanes and Climate Change*, p. 265. New York, NY: Springer.

La Seur NE and Hawkins HF (1963) An analysis of hurricane Cleo (1958) based on data from research reconnaissance aircraft. *Monthly Weather Review* 91: 694–709.

Leith CE (1974) Theoretical skill of Monte Carlo forecasts. *Monthly Weather Review* 102: 409–418.

Levizzani V, Amorati R, and Meneguzzo F (2002) A review of satellite-based rainfall estimation methods, MUSIC – multiple sensor precipitation measurements, integration, calibration and flood forecasting, deliverable 6.1, Italy.

Lin YL, Farley RD, and Orville HD (1983) Bulk parameterization of the snow field in a cloud model. *Journal of Climate and Applied Meteorology* 22: 1065–1092.

Liu G (2003) Precipitation. In: Holton JR, Curry JA, and Pyle JA (eds.) *Encyclopedia of Atmospheric Sciences*, pp. 1972–1979. London: Academic Press.

Liu G and Curry JA (1992) Retrieval of precipitation from satellite microwave measurement using both emission and scattering. *Journal of Geophysical Research* 97: 9959–9974.

Liu G and Curry JA (1993) Determination of characteristic features of cloud liquid water from satellite microwave measurements. *Journal of Geophysical Research* 98(D3): 5069–5092.

Liu JY and Orville HD (1969) Numerical modeling of precipitation and cloud shadow effects on mountain-induced cumuli. *Journal of the Atmospheric Sciences* 26: 1283–1298.

Locatelli JD and Hobbs PV (1974) Fall speeds and masses of solid precipitation particles. *Journal of Geophysical Research* 79: 2185–2197.

Lonfat M, Marks FD Jr., and Chen SS (2004) Precipitation distribution in tropical cyclones using the tropical rainfall measuring mission (TRMM) microwave imager: A global perspective. *Monthly Weather Review* 132: 1645–1660.

Lonfat M, Rogers R, Marchok T, and Marks FD Jr. (2007) A parametric model for predicting hurricane rainfall. *Monthly Weather Review* 135: 3086–3097.

Long AB (1971) Validity of the finite-difference droplet equation. *Journal of the Atmospheric Sciences* 28: 210–218.

Lorenz EN (1963) Deterministic nonperiodic flow. *Journal of the Atmospheric Sciences* 20: 130–141.

Lovejoy S and Austin GL (1979) The delineation of rain areas from visible and IR satellite data for GATE and mid-latitudes. *Atmosphere-Ocean* 17(1): 77–92.

Lovejoy S and Schertzer D (1995) Multifractals and rain. In: Kundzewicz ZW (ed.) *New Uncertainty Concepts in Hydrology and Hydrological Modelling*, pp. 61–103. Cambridge: Cambridge University Press.

Lull HW (1964) Ecological and silvicultural aspects. In: Chow VT (ed.) *Handbook of Applied Hydrology*, sec. 6. New York, NY: McGraw-Hill.

Lutgens FK and Tarbuck EJ (1992) *The Atmosphere: An Introduction to Meteorology*, 5th edn.. Englewood Cliffs, NJ: Prentice Hall.

Mandelbrot BB and van Ness JW (1968) Fractional Brownian motion, fractional noises and applications. *SIAM Review* 10: 422–437.

Marks FD (1985) Evolution of the structure of precipitation in hurricane Allen (1980). *Monthly Weather Review* 113: 909–930.

Marks FD (2003) Hurricanes. In: Holton JR, Curry JA, and Pyle JA (eds.) *Encyclopedia of Atmospheric Sciences*, pp. 942–966. London: Academic Press.

Marks FD and Houze RA Jr. (1987) Inner-core structure of hurricane Alicia from airborne Doppler radar observations. *Journal of the Atmospheric Sciences* 44: 1296–1317.

Marks FD, Houze RA Jr., and Gamache JF (1992) Dual-aircraft investigation of the inner core of hurricane Norbert. Part I: Kinematic structure. *Journal of the Atmospheric Sciences* 49: 919–942.

Marshall JS, Hitschfeld W, and Gunn KLS (1955) Advances in radar weather. *Advances in Geophysics* 2: 1–56.

Marshall JS and Palmer WM (1948) The distribution of raindrops with size. *Journal of Meteorology* 5: 165–166.

Mason BJ (1971) *The Physics of Clouds*. Oxford: Clarendon.

Mass CF, Ovens D, Westrick K, and Colle B (2002) Does increasing horizontal resolution produce more skillful forecasts? *Bulletin of the American Meteorological Society* 83: 407–430.

Meehl GA, Stocker TF, Collins WD, et al. (2007) Global climate projections. In: Solomon S, Qin D, Manning M, et al. (eds.) *Climate Change 2007: The Physical Science Basis. Contribution of Working Group I to the Fourth Assessment Report of the Intergovernmental Panel on Climate Change*, pp. 747–845. Cambridge: Cambridge University Press.

Menabde M, Harris D, Seed A, Austin G, and Stow D (1997) Multiscaling properties of rainfall and bounded random cascades. *Water Resources Research* 33(12): 2823–2830.

Menabde M and Sivapalan M (2000) Modelling of rainfall time series and extremes using bounded random cascades and levy-stable distributions. *Water Resources Research* 36(11): 3293–3300.

Miller BL (1958) Rainfall rates in Florida hurricanes. *Monthly Weather Review* 86(7): 258–264.

Molinari J, Moore PK, and Idone VP (1999) Convective structure of hurricanes as revealed by lightning locations. *Monthly Weather Review* 127: 520–534.

Molinari J, Moore PK, Idone VP, Henderson RW, and Saljoughy AB (1994) Cloud-to-ground lightning in hurricane Andrew. *Journal of Geophysical Research* 99: 16/665–16/676.

Mora RD, Bouvier C, Neppel L, and Niel H (2005) Approche régionale pour l'estimation des distributions ponctuelles des pluies journalières dans le Languedoc-Roussillon (France). *Hydrological Sciences Journal* 50(1): 17–29.

National Oceanic and Atmospheric Administration (2009) About the Radar Operations Center. http://www.roc.noaa.gov/WSR88D/About.aspx (accessed March 2010).

Neyman J and Scott EL (1952) A theory of the spatial distribution of galaxies. *Astrophysical Journal* 116: 144–163.

Northrop P (1996) Modelling and Statistical Analysis of Spatial–Temporal Rainfall Fields. PhD Thesis, Department of Statistical Science, University College, London.

Northrop PJ (1998) A clustered spatial–temporal model of rainfall. *Proceedings of the Royal Society A* 454: 1875–1888.

Nykanen DK and Harris D (2003) Orographic influences on the multiscale statistical properties of precipitation. *Journal of Geophysical Research* 108(D8): CIP6.1–CIP6.13 (doi:10.1029/2001JD001518).

Olson DA, Junker NW, and Korty B (1995) Evaluation of 33 years of quantitative precipitation forecasting at the NMC. *Weather and Forecasting* 10: 498–511.

Olsson J (1995) Limits and characteristics of the multifractal behavior of a high-resolution rainfall time series. *Nonlinear Processes in Geophysics* 2: 23–29.

Olsson J (1998) Evaluation of a scaling cascade model for temporal rainfall disaggregation. *Hydrology and Earth System Sciences* 2: 19–30.

Onof C, Chandler RE, Kakou A, Northrop P, Wheater HS, and Isham V (2000) Rainfall modelling using Poisson-cluster processes: A review of developments. *Stochastic Environmental Research and Risk Assessment* 14: 384–411.

Onof C and Wheater HS (1993) Modelling of British rainfall using a random parameter Bartlett–Lewis rectangular pulse model. *Journal of Hydrology* 149: 67–95.

Onof C and Wheater HS (1994) Improvements to the modeling of British rainfall using a modified random parameter Bartlett–Lewis rectangular pulses model. *Journal of Hydrology* 157: 177–195.

Ooyama KV (1982) Conceptual evolution of theory and modeling of the tropical cyclone. *Journal of the Meteorological Society of Japan* 60: 369–379.

Orlanski I (1975) A rational subdivision of scales for atmospheric processes. *Bulletin of the American Meteorological Society* 56(5): 527–530.

O'Sullivan F, Wash CH, Stewart M, and Motell CE (1990) Rain estimation from infrared and visible GOES satellite data. *Journal of Applied Meteorology* 29: 209–223.

Over TM and Gupta VK (1996) A space–time theory of mesoscale rainfal using random cascades. *Journal of Geophysical Research* 101(D21): 26319–26331.

Papalexiou SM and Koutsoyiannis D (2006) A probabilistic approach to the concept of probable maximum precipitation. *Advances in Geosciences* 7: 51–54.

Papalexiou SM and Koutsoyiannis D (2008) Probabilistic description of rainfall intensity at multiple time scales. IHP 2008 Capri Symposium: "The Role of Hydrology in Water Resources Management", UNESCO, International Association of Hydrological Sciences. Capri, Italy. http://www.itia.ntua.gr/en/docinfo/884 (accessed March 2010).

Papalexiou SM and Koutsoyiannis D (2009) Ombrian curves: From theoretical consistency to engineering practice. In: *8th IAHS Scientific Assembly/37th IAH Congress*. Hyderabad, India, 2009. http://www.itia.ntua.gr/en/docinfo/926 (accessed March 2010).

Papoulis A (1991) *Probability, Random Variables, and Stochastic Processes*, 3rd edn. New York, NY: McGraw-Hill.

Pegram GGS (1980) An auto-regressive model for multi-lag Markov chains. *Journal of Applied Probability* 17: 350–362.

Pegram GGS and Seed AW (1998) The feasibility of stochastically modelling the spatial and temporal distribution of rainfields. *WRC Report No. 550/1/98 to the Water Research Commission*, Pretoria, South Africa.

Perica S and Foufoula-Georgiou E (1996a) Linkage of scaling and thermodynamic parameters of rainfall: Results from midlatitude mesoscale convective systems. *Journal of Geophysical Research* 101(D3): 7431–7448.

Perica S and Foufoula-Georgiou E (1996b) Model for multiscale disaggregation of spatial rainfall based on coupling meteorological and scaling descriptions. *Journal of Geophysical Research* 101(D21): 26347–26361.

Petty GW (1994a) Physical retrievals of over-ocean rain rate from multichannel microwave imaging. Part I: Theoretical characteristics of normalized polarization and scattering indices. *Meteorology and Atmospheric Physics* 54: 79–99.

Petty GW (1994b) Physical retrievals of over-ocean rain rate from multichannel microwave imaging. Part II: Algorithm implementation. *Meteorology and Atmospheric Physics* 54: 101–121.

Petty GW (1995) The status of satellite-based rainfall estimation over land. *Remote Sensing of Environment* 51: 125–137.

Petty GW (2001a) Physical and microwave radiative properties of precipitating clouds. Part I: Principal component analysis of observed multichannel microwave radiances in tropical stratiform rainfall. *Journal of Applied Meteorology* 40: 2105–2114.

Petty GW (2001b) Physical and microwave radiative properties of precipitating clouds. Part II: A parametric 1D rain-cloud model for use in microwave radiative transfer simulations. *Journal of Applied Meteorology* 40: 2115–2129.

Pinsky M and Khain A (2004) Collisions of small drops in a turbulent flow. Part II: Effects of flow accelerations. *Journal of the Atmospheric Sciences* 61: 1926–1939.

Pinsky M, Khain A, and Shapiro M (1999) Collisions of small drops in a turbulent flow. Part I: Collision efficiency. Problem formulation and preliminary results. *American Meteorological Society* 56: 2585–2600.

Pinsky M, Khain A, and Shapiro M (2000) Stochastic effects of cloud droplet hydrodynamic interaction in a turbulent flow. *Atmospheric Research* 53: 131–169.

Poincaré H (1908) *Science et Méthode*. Reproduced in Poincaré H (1956) Chance. In: Newman JR (ed.) *The World of Mathematics*. New York, NY: Simon and Schuster.

Popper K (1982) *Quantum Physics and the Schism in Physics*. London: Unwin Hyman.

Powell MD (1990) Boundary layer structure and dynamics in outer hurricane rainbands. Part I: Mesoscale rainfall and kinematic structure. *Monthly Weather Review* 118: 891–917.

Probert-Jones JR (1962) The radar equation in meteorology. *Quarterly Journal of the Royal Meteorological Society* 88: 485–495.

Ramage C (1971) *Monsoon Meteorology*, International Geophysics Series, vol. 15. San Diego, CA: Academic Press.

Randall DA, Wood RA, Bony S, et al. (2007) Climate models and their evaluation. In: Solomon S, Qin D, Manning M, et al. (eds.) *Climate Change 2007: The Physical Science Basis*, Contribution of Working Group I to the Fourth Assessment Report of the IPCC, pp. 589–662. Cambridge, UK and New York, NY: Cambridge University Press.

Reisner J, Rasmussen RJ, and Bruintjes RT (1998) Explicit forecasting of supercooled liquid water in winter storms using the MM5 mesoscale model. *Quarterly Journal of the Royal Meteorological Society* 124B: 1071–1107.

Renno N and Ingersoll A (1996) Natural convection as a heat engine: A theory for CAPE. *Journal of the Atmospheric Sciences* 53(4): 572–585.

Richards F and Arkin P (1981) On the relationship between satellite-observed cloud cover and precipitation. *Monthly Weather Review* 109: 1081–1093.

Robert CP (2007) *The Bayesian Choice, From Decision-Theoretic Foundations to Computational Implementation*. New York, NY: Springer.

Robertson D (1974) Monte Carlo simulations of drop growth by accretion. *Journal of the Atmospheric Sciences* 31: 1344–1350.

Rodriguez-Iturbe I, Cox DR, and Isham V (1987) Some models for rainfall based on stochastic point processes. *Proceedings of the Royal Society A* 410: 269–298.

Rodriguez-Iturbe I, Cox DR, and Isham V (1988) A point process model for rainfall: Further developments. *Proceedings of the Royal Society A* 417: 283–298.

Rodriguez-Iturbe I, Gupta VK, and Waymire E (1984) Scale considerations in the modeling of temporal rainfall. *Water Resources Research* 20(11): 1611–1619.

Rogers RR and Yau MK (1996) *A Short Course in Cloud Physics*, 3rd edn. Woburn, MA: Butterworth-Heinemann.

Roldan J and Woolhiser DA (1982) Stochastic daily precipitation models: 1. A comparison of occurrence processes. *Water Resources Research* 18(5): 1451–1459.

Rossi F, Fiorentino M, and Versace P (1984) Two-component extreme value distribution for flood frequency analysis. *Water Resources Research* 20(7): 847–856.

Rotunno R (1986) Tornadoes and tornadogenesis. In: Ray PS (ed.) *Mesoscale Meteorology and Forecasting*, ch. 18, pp. 414–436. Boston, MA: The American Meteorological Society.

Ryan BF (1996) On the global variation of precipitating layer clouds. *Bulletin of the American Meteorological Society* 77: 53–70.

Ryan BT, Blau HH, Thuna PC, Cohen ML, and Roberts GD (1972) Cloud microstructure as determined by optical cloud particle spectrometer. *Journal of Applied Meteorology* 11: 149–156.

Schaefer MG, Barker BL, Taylor GH, and Wallis JR (2006) Regional precipitation-frequency analysis and spatial mapping for 24-hour and 2-hour durations for Wahington State. *Geophysical Research Abstracts* 8: 10899.

Schertzer D and Lovejoy S (1987) Physical modeling and analysis of rain and clouds by anisotropic scaling of multiplicative processes. *Journal of Geophysical Research* 92: 9693–9714.

Schertzer D and Lovejoy S (1989) Generalized scale invariance and multiplicative processes in the atmosphere. *Pure and Applied Geophysics* 130: 57–81.

Schmitt F, Vannitsem S, and Barbosa A (1998) Modeling of rainfall time series using two-state renewal processes and multifractals. *Journal of Geophysical Research* 103(D18): 23181–23193.

Scorer RS and Wexler H (1963) *A Colour Guide to Clouds*. Oxford: Pergamon.

Scott WT (1968) On the connection between the Telford and kinetic equation approaches to droplet coalescence theory. *Journal of the Atmospheric Sciences* 25: 871–873.

Scott WT (1972) Comments on "Validity of the finite-difference droplet collection equation". *Journal of the Atmospheric Sciences* 29: 593–594.

Seltzer MA, Passarelli RE, and Emanuel KA (1985) The possible role of symmetric instability in the formation of precipitation bands. *Journal of the Atmospheric Sciences* 42: 2207–2219.

Sempere-Torres D, Porrà JM, and Creutin JD (1994) A general formulation for raindrop size distribution. *Journal of Applied Meteorology* 33: 1494–1502.

Sempere-Torres D, Porrà JM, and Creutin JD (1998) Experimental evidence of a general description for raindrop size distribution properties. *Journal of Geophysical Research* 103(D): 1785–1797.

Shapiro LJ (1983) The asymmetric boundary layer flow under a translating hurricane. *Journal of the Atmospheric Sciences* 40: 1984–1998.

Sharma A and O'Neill R (2002) A nonparametric approach for representing interannual dependence in monthly streamflow sequences. *Water Resources Research* 38(7): 1100 (doi: 10.1029/2001WR000953).

Simpson J, Adler RF, and North GR (1988) Proposed tropical rainfall measuring mission (TRMM) satellite. *Bulletin of the American Meteorological Society* 69: 278–295.

Sisson SA, Pericchi LR, and Coles SG (2006) A case for a reassessment of the risks of extreme hydrological hazards in the Caribbean. *Stochastic Environmental Research and Risk Assessment* 20: 296–306.

Skamarock W, Klemp J, Dudhia J, et al. (2005) *A Description of the Advanced Research WRF Version 2*. Mesoscale and Microscale Meteorology Division, National Center for Atmospheric Research, Boulder, CO, USA.

Slingo J (2003) Monsoon. In: Holton JR, Curry JA, and Pyle JA (eds.) *Encyclopedia of Atmospheric Sciences*, pp. 1365–1370. London: Academic Press.

Slinn WGN (1975) Atmospheric aerosol particles in surface-level air. *Atmospheric Environment* 9: 763–764.

Small MJ and Morgan DJ (1986) The relationship between a continuous-time renewal model and a discrete Markov chain model of precipitation occurrence. *Water Resources Research* 22(10): 1422–1430.

Smith JA (1993) Precipitation In: Maidment DA (ed.) *Handbook of Applied Hydrology*, ch. 3, pp. 3.1–3.47. New York, NY: McGraw-Hill.

Smith RK (1968) The surface boundary layer of a hurricane. *Tellus* 20: 473–484.

Smith RK (2000) The role of cumulus convection in hurricanes and its representation in hurricane models. *Reviews of Geophysics* 38: 465–489.

Smith RK (2003) A simple model of hurricane boundary layer. *Quarterly Journal of the Royal Meteorological Society* 129: 1007–1027.

Spencer RW, Goodman HM, and Hood RE (1989) Precipitation retrieval over land and ocean with the SSM/I: Identification and characteristics of the scattering signal. *Journal of Atmospheric and Oceanic Technology* 6: 254–273.

Srikanthan R and McMahon TA (2001) Stochastic generation of annual, monthly and daily climate data: A review. *Hydrology and Earth System Sciences* 5(4): 653–670.

Stephens GL (1994) *Remote Sensing of the Lower Atmosphere: An Introduction*. New York, NY: Oxford University Press.

Stewart SR (2002) Tropical cyclone report. *Tropical Storm Allison, 5–17 June 2001*, US National Oceanic and Atmospheric Administration. http://www.nhc.noaa.gov/2001allison.html (accessed March 2010).

Stoelinga MT, Woods CP, Locatelli JD, and Hobbs PV (2005) On the representation of snow in bulk microphysical parameterization schemes. *WRF/MM5 Users' Workshop*, June 2005.

Stout JS, Martin DW, and Sikdar DN (1979) Estimating GATE rainfall with geosynchronous satellite images. *Monthly Weather Review* 107: 585–598.

Sutcliffe JV (1978) *Methods of Flood Estimation, A Guide to Flood Studies Report*, Report 49. Wallingford: Institute of Hydrology.

Telford JW (1955) A new aspect of coalescence theory. *Journal of Meteorology* 12: 436–444.

Telford JW and Chai SK (1980) A new aspect of condensation theory. *Pure and Applied Geophysics* 118: 720–742.

Telford JW and Wagner PB (1981) Observations of condensation growth determined by entity type mixing. *Pure and Applied Geophysics* 119: 934–965.

Tessier Y, Lovejoy S, Hubert P, Schertzer D, and Pecknold S (1996) Multifractal analysis and modeling of rainfall and river flows and scaling, causal transfer functions. *Journal of Geophysical Research* 101(D21): 26427–26440.

Tessier Y, Lovejoy S, and Schertzer D (1993) Universal multifractals in rain and clouds: Theory and observations. *Journal of Applied Meteorology* 32: 223–250.

Testik FY and Barros AP (2007) Toward elucidating the microstructure of warm rainfall: A survey. *Reviews of Geophysics* 45: RG2003 (doi:10.1029/2005RG000182).

Thompson G, Rasmussen RM, and Manning K (2004) Explicit forecasts of winter precipitation using an improved bulk microphysics scheme. Part I: Description and sensitivity analysis. *Monthly Weather Review* 132: 519–542.

Todorovic P and Woolhiser DA (1975) A stochastic model of n-day precipitation. *Journal of Applied Meteorology* 14: 17–24.

Todorovic P and Zelenhasic E (1970) A stochastic model for flood analysis. *Water Resources Research* 6(6): 1641–1648.

Tsonis A (1987) Determining rainfall intensity and type from GOES imagery in the midlatitudes. *Pure and Applied Geophysics* 21: 29–36.

Tsonis A and Isaac G (1985) On a new approach for instantaneous rain area delineation in the midlatitudes using GOES data. *Journal of Applied Meteorology* 24: 1208–1218.

Turpeinen OM, Abidi A, and Belhouane W (1987) Determination of rainfall with the ESOC precipitation index. *Monthly Weather Review* 115: 2699–2706.

Twomey S (1964) Statistical effects in the evolution of a distribution of cloud droplets by coalescence. *Journal of the Atmospheric Sciences* 21: 553–557.

Twomey S (1966) Computation of rain formation by coalescence. *Journal of the Atmospheric Sciences* 23: 405–411.

Uijlenhoet R (1999) Parameterization of Rainfall Microstructure for Radar Meteorology and Hydrology. Doctoral Dissertation, Wageningen University, The Netherlands.

Uijlenhoet R (2001) Raindrop size distributions and radar reflectivity –rain rate relationships for radar hydrology. *Hydrology and Earth System Sciences* 5: 615–627.

Uijlenhoet R (2008) Precipitation physics and rainfall observation. In: Bierkens MFP, Dolman AJ, and Troch PA (eds.) *Climate and the Hydrological Cycle*, IAHS Special Publication 8, pp. 59–97. Wallingford: IAHS.

Uijlenhoet R and Berne A (2008) Stochastic simulation experiment to assess radar rainfall retrieval uncertainties associated with attenuation and its correction. *Hydrology and Earth System Sciences* 12: 587–601.

Uijlenhoet R and Sempere-Torres D (2006) Measurement and parameterization of rainfall microstructure. *Journal of Hydrology* 328(1–2): 1–7 (ISSN 0022-1694; doi: 10.1016/j.jhydrol.2005.11.038).

Ulbrich C (1983) Natural variations in the analytical form of the raindrop size distribution. *Journal of Climate and Applied Meteorology* 22(10): 1764–1775.

Ulbrich CW and Atlas D (1998) Rainfall microphysics and radar properties: Analysis methods for drop size spectra. *Journal of Applied Meteorology* 37: 912–923.

US Department of the Interior, Bureau of Reclamation (1977) *Design of Arch Dams*. Denver, CO: US Government Printing Office.

US Department of the Interior, Bureau of Reclamation (1987) *Design of Small Dams*, 3rd edn.. Denver, CO: US Government Printing Office.

Velghe T, Troch PA, De Troch FP, and Van de Velde J (1994) Evaluation of cluster-based rectangular pulses point process models for rainfall. *Water Resources Research* 30(10): 2847–2857.

Veneziano D (1999) Basic properties and characterization of stochastically self-similar processes in R^D. *Fractals* 7(1): 59–78.

Veneziano D and Furcolo P (2002) Multifractality of rainfall and intensity–duration–frequency curves. *Water Resources Research* 38(12): 1306–1317.

Veneziano D, Furcolo P, and Iacobellis V (2006a) Imperfect scaling of time and space–time rainfall. *Journal of Hydrology* 322(1–4): 105–119.

Veneziano D and Langousis A (2005) The areal reduction factor a multifractal analysis. *Water Resources Research* 41: W07008 (doi:10.1029/2004WR003765).

Veneziano D and Langousis A (2010) Scaling and fractals in hydrology. In: Sivakumar B (ed.) *Advances in Data-Based Approaches for Hydrologic Modeling and Forecasting*, ch. 4, p. 145. Singapore: World Scientific.

Veneziano D, Langousis A, and Furcolo P (2006b) Multifractality and rainfall extremes: A review. *Water Resources Research* 42: W06D15 (doi:10.1029/2005WR004716).

Veneziano D, Langousis A, and Lepore C (2009) New asymptotic and pre-asymptotic results on rainfall maxima from multifractal theory. *Water Resources Research* 45: W11421 (doi:10.1029/2009WR008257).

Veneziano D, Lepore C, Langousis A, and Furcolo P (2007) Marginal methods of intensity–duration–frequency estimation in scaling and nonscaling rainfall. *Water Resources Research* 43: W10418 (doi:10.1029/2007WR006040).

Venugopal V, Roux SG, Foufoula-Georgiou E, and Arneodo A (2006) Revisiting multifractality of high resolution temporal rainfall using a wavelet-based formalism. *Water Resources Research* 42: W06D14 (doi:10.1029/2005WR004489).

Verhoest N, Troch PA, and De Troch FP (1997) On the applicability of Bertlett–Lewis rectangular pulses models in the modeling of design storms at a point. *Journal of Hydrology* 202: 108–120.

Vickery PJ, Skerlj PF, and Twisdale LA (2000) Simulation of hurricane risk in the US using empirical track model. *Journal of Structural Engineering* 126(10): 1222–1237.

Viltard N, Burlaud C, and Kummerow C (2006) Rain retrieval from TMI brightness temperature measurements using a TRMM PR-based data base. *Journal of Applied Meteorology and Climatology* 45: 455–466.

Vohl O, Mitra S, Wurzler S, and Pruppacher H (1999) A wind tunnel study on the effects of turbulence on the growth of cloud drops by collision and coalescence. *Journal of the Atmospheric Sciences* 56: 4088–4099.

Wallace JW and Hobbs PV (1977) *Atmospheric Science: An Introductory Survey*. San Diego, CA: Academic Press.

Wang B, Ding Q, and Jhun JG (2006) Trends in Seoul (1778–2004) summer precipitation. *Geophysical Research Letters* 33(15): L15803.

Wang B, Jhun JG, and Moon BK (2007) Variability and singularity of Seoul, South Korea, rainy season (1778–2004). *Journal of Climate* 20(11): 2572–2580.

Waymire E and Gupta VK (1981a) The mathematical structure of rainfall representations: 1. A review of the stochastic rainfall models. *Water Resources Research* 17(5): 1261–1272.

Waymire E and Gupta VK (1981b) The mathematical structure of rainfall representations: 2. A review of the theory of point processes. *Water Resources Research* 17(5): 1273–1285.

Waymire E and Gupta VK (1981c) The mathematical structure of rainfall representations: 3. Some applications of the point process theory to rainfall processes. *Water Resources Research* 17(5): 1287–1294.

Weisman ML and Klemp JB (1982) The dependence of numerically simulated convective storms on vertical wind shear and buoyancy. *Monthly Weather Review* 110: 504–520.

Weisman ML and Klemp JB (1984) The structure and classification of numerically simulated convective storms in directionally varying wind shears. *Monthly Weather Review* 112: 2479–2498.

Weisman ML, and Klemp JB (1986) Characteristics of isolated convective storms. In: Ray PS (ed.) *Mesoscale Meteorology and Forecasting*, ch. 15, pp. 331–358. Boston, MA: American Meteorological Society.

Wilheit T, Adler R, Avery S, *et al.* (1994) Algorithms for the retrieval of rainfall from passive microwave measurements. *Remote Sensing Reviews* 11: 163–194.

Willoughby HE (1990) Gradient balance in tropical cyclones. *Journal of the Atmospheric Sciences* 47: 265–274.

Willoughby HE (1991) Reply. *Journal of the Atmospheric Sciences* 48: 1209–1212.

Willoughby HE, Marks FD Jr., and Feinberg RJ (1984) Stationary and moving convective bands in hurricanes. *Journal of the Atmospheric Sciences* 41(22): 3189–3211.

Wisner C, Orville HD, and Myers C (1972) A numerical model of hail-bearing cloud. *Journal of the Atmospheric Sciences* 29: 1160–1181.

World Meteorological Organization (1983) *Guide to Meteorological Instruments and Methods of Observation*. publication 8. 5th edn. Geneva: World Meteorological Organization.

World Meteorological Organization (1986) Manual for estimation of probable maximum precipitation. *Operational Hydrology Report 1*, 2nd edn., publication 332. Geneva: World Meteorological Organization.

Worsley KJ (1983) Testing a two-phase multiple regression. *Technometrics* 25(1): 35–41.

Yanai M (1964) Formation of tropical cyclones. *Reviews of Geophysics* 2: 367.

Relevant Websites

http://www.itia.ntua.gr
 Itia National Technical University of Athens, Greece.
http://laspistasteria.wordpress.com
 Mud to the Stars, Reports of Nature.
http://disc2.nascom.nasa.gov
 National Aeronautics and Space Administration.
http://www.ncdc.noaa.gov
 National Climatic Data Center, National Oceanic and Atmospheric Administration.
http://www.nssl.noaa.gov
 National Severe Storms Laboratory, National Oceanic and Atmospheric Administration.
http://climexp.knmi.nl
 Royal Dutch Meteorological Institute.

2.03 Evaporation in the Global Hydrological Cycle

AJ Dolman and JH Gash, VU University Amsterdam, Amsterdam, The Netherlands

© 2011 Elsevier B.V. All rights reserved.

2.03.1	Introduction	79
2.03.2	General Theory of Evaporation	80
2.03.2.1	Vegetated Surfaces	81
2.03.2.2	Bare Soil	82
2.03.2.3	Open Water and Lakes	82
2.03.3	Regional and Equilibrium Evaporation	83
2.03.4	Trends and Variability in Global Evaporation	83
2.03.5	Summary and Conclusions	85
References		86

2.03.1 Introduction

Evaporation is the transfer of moisture from a particular surface to the overlying atmosphere. The physical process of evaporation consists of the exchange of water molecules between a free water surface and the air. The surface can be any among the following nonexhaustive list: a lake, the inside of plant leaves, the water surface adhering to a soil conglomerate, or the surface of a soil or canopy during or just after rain. The evaporation rate is expressed as the quantity of water evaporated per unit area per unit time from a (water) surface under existing atmospheric conditions. This chapter describes the progress in understanding evaporation at the local scale, both from an observational and a conceptual perspective. It then moves on to the global scale, through a discussion of regional scale feedbacks. The main conclusion is that while we have gained considerable understanding in local scale evaporation, the data sets required to study the impact of evaporation on the global water cycle are still lacking.

Evaporation at the Earth's surface is constrained both by the energy available to convert liquid water into vapor and by the capability of the surrounding air to transfer moisture away from the saturated surface. At the surface, these constraints are best expressed through the energy balance equation and the transfer equations of latent and sensible heat:

$$R_g(1-\zeta) + L_\downarrow - L_\uparrow = \lambda E + H + G + \delta S \quad (1)$$

$$\lambda E = -\rho \lambda K_v \frac{\partial q}{\partial z} \quad (2)$$

$$H = -\rho c_p K_h \frac{\partial T}{\partial z} \quad (3)$$

$$\beta = \frac{H}{\lambda E} \quad (4)$$

with R_g the incoming short-wave radiation (W m^{-2}), ζ the short-wave albedo, $L_\downarrow$ and $L_\uparrow$ the incoming and outgoing long-wave radiation (W m^{-2}), λE the evaporation (or latent heat flux, W m^{-2}) (with λ the latent heat of vaporization and E the mass flux), H the sensible heat flux (W m^{-2}), G the soil heat flux (W m^{-2}), δS the change in heat storage in the biomass and atmosphere below a reference height above the surface (W m^{-2}), K_v and K_h the transfer coefficients for water vapor and heat, respectively, $\partial q/\partial z$ the vertical gradient in specific humidity (g kg^{-1}), and similarly $\partial T/\partial z$ is the vertical gradient in temperature (K), ρ is the density of air (kg m^{-3}), and c_p the specific heat of air (J kg^{-1} K^{-1}). β is the Bowen ratio and a useful indicator of the dryness of the surface because it shows how the available energy is partitioned into sensible and latent heat. Factors influencing the rate of evaporation can easily be determined from the above equations: radiation as the limit of available energy, the gradient in moisture, or the specific humidity deficit and factors that influence the transfer coefficients such as wind speed and roughness of the surface, and, crucially, when dealing with evaporation from leaves (transpiration), water availability (soil moisture).

Thus, despite the complexity of the interaction between a partially wet surface and the atmosphere, it may be possible to simplify this interaction and to approximate evaporation by considering only two factors, available energy (radiation) and water availability. Available energy determines the maximum evaporation possible for the given climatic conditions and unlimited water availability, that is, potential evaporation (see also Allen *et al.*, 1998). Water availability can be characterized by the amount of precipitation. Under dry conditions, potential evaporation exceeds precipitation, and the actual evaporation from an area will approach the amount of precipitation received. Conversely, under wet conditions, water availability exceeds potential evaporation and actual evaporation will approach asymptotically potential evaporation. This relationship was first used by Budyko (1974) to set a constraint on global evaporation. Later, Milly and Dunne (2002) suggested that the long-term water balance of a catchment is determined by the interaction of supply (precipitation) and demand (potential evaporation), mediated by soil moisture storage. At longer timescales, the change in soil moisture can be neglected. **Figure 1** shows results of Zhang *et al.* (2001), from a simple model based on the above considerations for grass and forested catchments, using catchment discharge and precipitation data, that allow evaporation to be calculated as the residual of these two. These results suggest that at the scale of catchments such a simple approach works remarkably well, with a linear (1:1) relation between rainfall and evaporation for catchments that receive up to 500 mm yr^{-1}, and an

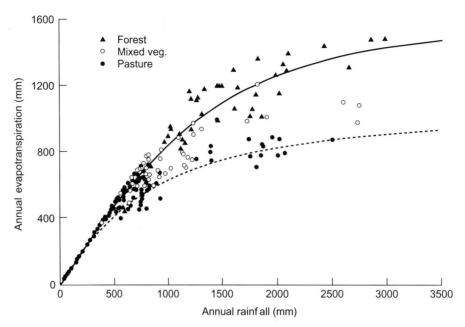

Figure 1 Relationship between annual precipitation and evaporation. The dashed line represents the best fit of a theoretical model (Zhang et al., 2001) for grass-only catchments, the solid line the best fit for forests. From Zhang et al. (2001). (Note that we use the word evapotranspiration here in this graph consistent with its origin. This rather loosely defined term refers to the total of (wet canopy and bare soil) evaporation and transpiration. Throughout the text, we prefer to use the word evaporation as the physical term denoting the process of transforming liquid water into its vapor form.)

asymptotic relation (driven by available energy and not by water availability) for catchments receiving more than 500 mm yr^{-1}. The difference between forest and grassland is a result of the high rate of evaporation of intercepted rainfall from forest; interception loss creates a considerable additional evaporative loss for forest (see also **Chapter 2.04 Interception**). Total evaporation from forest saturates at about 1400 mm yr^{-1}, while for grasslands this value is around 900 mm yr^{-1}.

Direct observations of evaporation have been made since the mid-1990s using the eddy-covariance technique in a global network, Fluxnet (see Aubinet et al., 2001). While there are other techniques based on scintillometry available, these have not been used extensively in networks such as Fluxnet. Evaporation measured by micrometeorological techniques usually refers to dry-canopy evaporation only (transpiration) and contains little information on evaporation during wet-canopy conditions (interception). Consequently, total evaporation values such as those given by Law et al. (2002) should be treated with caution, when interception losses are not explicitly treated. With this caveat in mind, the annual evaporation from Fluxnet data for coniferous forests is 397 ($\pm$31) mm yr^{-1}; for mixed evergreen and deciduous forest, 386 ($\pm$18) mm yr^{-1}; for deciduous broadleaf, 512 ($\pm$69) mm yr^{-1}; for grassland, 494 ($\pm$104) mm yr^{-1}; and for crops, 666 ($\pm$67) mm yr^{-1}. Grassland and crops have higher dry-canopy evaporation than forest, because they are less strongly coupled to the atmosphere and thus show less stomatal control (e.g., Shuttleworth and Calder, 1979). This also gives forest the possibility to survive occasional drought that would kill off annual grassland species.

The Fluxnet data can be used to identify the main controls on evaporation. Wilson et al. (2002) showed how the partitioning of energy, as expressed by the ratio of sensible heat to latent heat, the Bowen ratio (β), can be used to classify different vegetation types in climate space. **Figure 2** shows the position of the individual sites with respect to the magnitude of the latent heat (evaporation) and sensible heat fluxes. In contrast to the annual rates discussed earlier, these data refer to the growing season only. Also shown are lines of constant available energy (the sum of latent and sensible heat) and lines of constant β. Moving from the lower left corner of the diagram to the upper right, the available energy increases, whereas moving from the lower right to the upper left, the value of β increases. Low evaporation rates are found in Sitka spruce, tundra, and boreal forest, and high evaporation rates in deciduous forests and agriculture. Most of the forests have high β's, implying that much of the energy received is transmitted back to the atmosphere as sensible heat. Deciduous forests tend to have lower β's with higher evaporation rates than coniferous forests. The average β at tundra sites appears to be close to 1.

2.03.2 General Theory of Evaporation

Penman (1948) was among the first to achieve the crucial combination of the energy balance equation (1) with the transfer Equations (2) and (3) to derive an expression for actual evaporation from vegetation well supplied with water. Although the vertical gradients in Equations (2) and (3) could be derived from the differences between air and surface values, measurements of surface temperature are difficult and not made routinely. Penman overcame this problem by introducing the slope of the saturated vapor pressure versus temperature curve, Δ, approximated as a linear function and

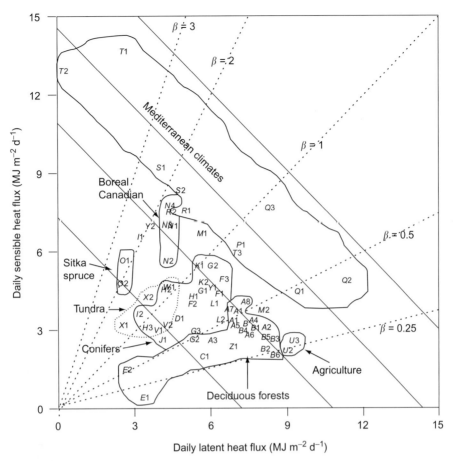

Figure 2 The daily cumulative sensible heat flux vs. the daily cumulative latent heat flux between days 165 and 235 for the Fluxnet sites analyzed by Wilson *et al.* (2002). The letter and number codes refer to the sites as given by Wilson *et al.* (2002). Also shown are lines of constant Bowen ratio (dashed lines) and lines of constant total turbulent energy fluxes (solid diagonal lines). Enclosed areas denote subjective delineations between different vegetation types and climates. From Wilson *et al.* (2002).

evaluated at air temperature:

$$\Delta = \frac{e^*(T_s) - e^*(T_a)}{T_s - T_a} \quad (5)$$

where $e^*(T_s)$ is the saturated vapor pressure at surface temperature T_s (K), and $e^*(T_a)$ is the saturated vapor pressure at air temperature T_a. The evaporation, E, in mm d^{-1} is given by

$$E = \frac{\Delta Q + \gamma E_a}{\Delta + \gamma} \quad (6a)$$

where Q is water equivalent of the net radiation (from the left-hand side of Equation (1)), γ the psychrometric constant and

$$E_a = f(U)(e^*(T_a) - e_a) \quad (6b)$$

E_a is the aerodynamic or demand term for evaporation, with $f(U)$ a wind function, and $(e^*(T_a) - e_a)$ the vapor pressure deficit.

The original Penman equation contains an empirical wind function that replaces the transfer coefficients in Equations (2) and (3). This wind function was difficult to generalize, and subsequently Thom and Oliver (1977) provided a more physical basis including considerations of aerodynamic transfer over rough surfaces by introducing an explicit aerodynamic resistance. Penman applied his equation to bare soil evaporation, vegetated or cropped surfaces, and open water bodies. The equation is still widely used to calculate evaporation from well-watered, short vegetation.

2.03.2.1 Vegetated Surfaces

Monteith (1965) (see also Gash and Shuttleworth, 2007) introduced the control of vegetation on evaporation by including a canopy scale resistance r_s. This resistance represents the restriction on the transfer of water from the collective saturated surfaces inside the plant stomatal cavities to the air outside the leaves. The resulting equation, now carrying both an aerodynamic (to replace the Penman wind function) and a canopy resistance, is known as the Penman–Monteith (PM) equation and is arguably still the most elegant yet advanced resistance model of evaporation used in hydrological practice today (Shuttleworth, 1993). For a mathematically precise and exact definition and brief historical overview of its development, the reader is referred to Raupach (2001). Most commonly, the PM equation is expressed in specific humidity units (rather than vapor pressure deficit as in Equation (6)) with evaporation expressed in energy units (W m^{-2}) and reads as

$$\lambda E = \frac{\Delta(Q_* - G) + \frac{\rho c_p (q^*(T_a) - q_a)}{r_a}}{\Delta + \frac{c_p}{\lambda}\left(1 + \frac{r_s}{r_a}\right)} \quad (7)$$

with Δ the slope of the saturated specific humidity versus temperature curve and Q_* the net radiation (W m^{-2}). $q^*(T_a)$ is the saturated specific humidity of the air (g kg^{-1}) at the reference level and q_a the specific humidity at the same level. Two important variables appear in this equation that replace the transfer coefficients of the transfer equations of heat and moisture: the aerodynamic and surface resistance, r_a and r_s (s m^{-1}). (Only in the case of a full canopy cover, can the surface resistance be equal to the canopy resistance. For the derivation of the big-leaf version of the PM model as presented here, this is not important. When the canopy cover is not full, and there is bare ground directly in contact with the overlying atmosphere, the surface resistance is not equal to the canopy resistance and approaches the reciprocal sum of the canopy and an assumed soil resistance.) The PM equation assumes that evaporation and sensible heat originate from the same source in the canopy. The main advantage of the PM equation is that the meteorological driving variables, wind speed, specific humidity deficit, and temperature are required only at a single level above the surface, removing the need for the notoriously difficult observation of surface values. The main obstacle for the practical application of this equation is the estimation of the values for aerodynamic and surface resistance. When there is unlimited supply of water, the PM equation can be used to calculate potential evaporation. It can be shown then to collapse to the Penman equation with an aerodynamic resistance rather than a wind function. The PM equation is now the preferred method of estimating crop water requirements as reference crop evaporation (Allen et al., 1998). When the canopy is wet, the surface resistance equals zero and the PM equation can be used to calculate evaporation of intercepted rainfall (see also **Chapter 2.04 Interception**). Although there is a wide range of empirical evaporation equations of which some are particular cases of the PM equation (see Brutsaert, 1982; Shuttleworth, 1993), the use of the PM equation is widespread because of its clear physical interpretation.

2.03.2.2 Bare Soil

Evaporation from bare soil can be a significant component of the water balance, particularly in semi-arid environments (Wallace and Holwill, 1997). Soil evaporation can be described as a two-stage process. The first stage occurs when the available soil moisture is sufficient to meet the atmospheric demand. This occurs immediately after rainfall or irrigation events. Soil evaporation under these conditions equals potential evaporation. Typically, this stage lasts 1–2 days, although in some cases when evaporative demand is low and the soil contains a high amount of clay, this stage may last for up to 5 days. In the second stage, the amount of soil moisture has dropped and soil evaporation is no longer only restricted by evaporative demand but also by availability of moisture. In these conditions, the change of soil moisture with time can be described as a desorption process with evaporation proportional to the square root of the time since the start of the process:

$$\lambda E_s = \frac{1}{2}\alpha(t - t_0)^{-1/2} \quad (8)$$

with λE_s the soil evaporation, t the time and t_0 the time since the start of second stage drying, D_s is a desorptivity (in units of W m^{-2} d$^{-1/2}$ when evaporation is expressed as a heat flux, or in mm d$^{-1/2}$ when evaporation is expressed as water flux). The desorptivity is assumed constant for a particular soil type. It varies from a value of 2.1 for sandy loam with gravel, to a value of 5 mm d^{-1} for a clay loam soil (Kustas et al., 2002). Although the two-stage process describes soil evaporation at diurnal timescales, extension of the theory to (sub) hourly timescales is straightforward (Brutsaert and Chen, 1996; Porté-Agel et al., 2000). The determination of the desorptivity coefficient can be problematic, as can the identification of the switch from stage 1 to 2 (Kustas et al., 2002).

The observed dependence of soil evaporation on available soil moisture suggests the feasibility of a resistance approach that incorporates a dependence of soil surface resistance on soil moisture. Mahfouf and Noilhan (1991) review several such formulations. These approaches can be divided into so-called α- and β_s-approaches. In the α-approach, the saturated humidity in the soil pore space is adjusted by a factor α that may be related to soil matrix potential and takes into account that, averaged over a certain depth, the evaporation takes place from a nonsaturated surface. In the β_s-approach, the humidity in the pore space at the evaporation front is assumed to be saturated, and β_s is the ratio of an aerodynamic resistance to the sum of the aerodynamic and soil surface resistance.

2.03.2.3 Open Water and Lakes

The key process controlling evaporation from large lakes (or reservoirs) is the absorption of solar energy. This energy is not absorbed at the surface; because water is a semitransparent medium, the solar radiation is absorbed over depth. The rate of change of absorption with depth depends on the turbidity, but may be significant down to several meters below the surface. Solar energy heats the water up during the spring and summer, and this energy is not available for evaporation; but energy released as the water cools in autumn and winter is available and enhances the evaporation. This creates a phase lag between lake evaporation and the annual radiation cycle. While in the tropics this lag will be small, at high latitudes the phase lag may be as much as 5 or 6 months (Blanken et al., 2000) with the rate of change in storage, dS/dt, being the dominant source of energy for evaporation. The energy available for evaporation is given by

$$A = Q_* + \frac{dS}{dt} + A_q \quad (9)$$

where A_q is the rate of net energy advection due to inflow and outflow of water. Net radiation, Q_*, is given by Equation (1) with long-wave radiation emitted by the lake as

$$L_\uparrow = \varepsilon\sigma T_w^4 \qquad (10)$$

ε is emissivity, σ the Stefan–Boltzmann constant, and T_w the surface temperature of the water (K). Q_* over the lake can be estimated from measurements over land, but must take account of the different albedo and different surface temperature of the lake. Like the ocean, lake albedo varies strongly with solar elevation (see Finch and Gash, 2002; Finch and Hall, 2005; Payne, 1972).

The Penman (1948) equation (Equation (6)) is often used to estimate lake evaporation, as it appears to remove the need for surface temperature measurement; however, this is not the case as water temperature is still needed to calculate the emitted long-wave radiation. Nevertheless, it should give good results if used with the available energy calculated from Equation (9) and measurements made over the lake (see Linacre, 1993). Working in the tropics where the annual cycle in water temperature is small and energy storage can be neglected, Sene et al. (1991) found good agreement between daily estimates made with the Penman equation and eddy-covariance measurements of evaporation.

To overcome the lack of water temperature measurements, Finch and Gash (2002) applied a simple numerical, finite difference scheme to calculate a running balance of lake energy storage. A new value of the water temperature required to force energy closure was calculated at each time step. For a well-mixed lake of known depth, the evaporation could then be calculated from land-based, daily meteorological observations of sunshine hours, relative humidity, wind run, and average air temperature. The model gave good agreement with mass-balance measurements of the water loss from a reservoir with no inflow or outflow.

2.03.3 Regional and Equilibrium Evaporation

Priestley and Taylor (1972) showed that the Bowen ratio would approach a constant value defined by $\beta = s/\lambda$ when air moves over a moist surface, and gradients of temperature and specific humidity with height are small or become saturated with respect to moisture. Combining this insight with the PM equation, and setting the second term above the nominator to zero as well as the surface or canopy resistance ($r_s = 0$) as would be appropriate for a moist surface, yields the equilibrium evaporation (see also Brutsaert, 1982; Raupach, 2001):

$$\lambda E = \frac{\Delta}{\Delta + \frac{c_p}{\lambda}} A \qquad (11)$$

Equilibrium evaporation as defined by these authors refers to the lower limit of evaporation from a moist surface where the specific humidity deficit of the second term in the nominator of the PM equation has become zero as a result of contact of air with a moist surface over a very long fetch.

It can easily be shown that the second term in the nominator of the PM equation (Equation (7)) now represents the departure from this equilibrium. Priestley and Taylor (1972) represented this departure from equilibrium evaporation by the parameter α. α can be shown to be unity only when the specific humidity deficit in the PM equation is zero, in other words, when advection is negligible. That this is hardly ever the case proves the fact that most empirical values of α for short crops are of the order 1.2–1.3 (e.g., Brutsaert, 1982). For tall crops, Shuttleworth and Calder (1979) showed convincingly that the equilibrium approach is not appropriate because the physiological control of the forest transpiration reduces α below a value of 1 in dry canopy conditions, while in wet canopy conditions large-scale advection and negative sensible heat fluxes (Stewart, 1977) may form an additional supply of energy and force α to be well above the value for short crops. This emphasizes the important point that for tall crops in particular, it is important to estimate dry and wet canopy evaporation separately (see also **Figure 1**).

Thus, at larger scale, atmospheric conditions can override the surface control by exerting a strong feedback on evaporation through the humidity and temperature of the atmospheric boundary layer. A number of concepts have been derived that use the feedback power of the atmosphere to estimate regional-scale evaporation (e.g., Bouchet, 1963; Morton, 1983). Although McNaughton and Jarvis (1991) show that at larger scale the feedback of the atmospheric boundary layer dampens the effects of surface controls – and thus makes precise estimation of the surface resistance less important – the physical basis of the Bouchet and Morton schemes remains doubtful (see de Bruin, 1983; McNaughton and Spriggs, 1989). Nevertheless, de Bruin showed that the feedback of the increasing atmospheric boundary-layer humidity during the day causes the regional surface conductance to vary less than if the feedback were neglected. The relatively large confidence of hydrologists in using potential evaporation formulas probably finds its physical explanation in this feedback.

2.03.4 Trends and Variability in Global Evaporation

Globally, evaporation from the land surface to the atmosphere amounts to $71 \times 10^3 \, km^3 \, yr^{-1}$ (Baumgartner and Reichel, 1975). It is the key return flow in the hydrological cycle from the surface on which the precipitation falls, back to the atmosphere. Evaporation from the oceans is a far larger component of the hydrological cycle at an estimated $428 \times 10^3 \, km^3 \, yr^{-1}$ (Baumgartner and Reichel, 1975). A recent multi-model ensemble of 11 state-of-the-art land surface models (Dirmeyer et al., 2006) estimated annual evaporation over a range of 58×10^3 to $85 \times 10^3 \, km^3$, indicating the uncertainty in our ability to model evaporation from land. Oki and Kanae (2006) estimated total terrestrial evaporation at $66 \times 10^3 \, km^3 \, yr^{-1}$. Thus, although it is an important component of the hydrological cycle, the exact magnitude and variability, both spatially and temporally, of evaporation from land remains highly uncertain.

Figure 3 shows one of the few available estimates of the latitudinal distribution of evaporation in mm yr^{-1} for both ocean and land. This estimate (Baumgartner and Reichel, 1975) is based on the balance between precipitation and runoff on land and a variety of other methods (Peixoto and Oort, 1996). Note the relatively large contribution of Southern

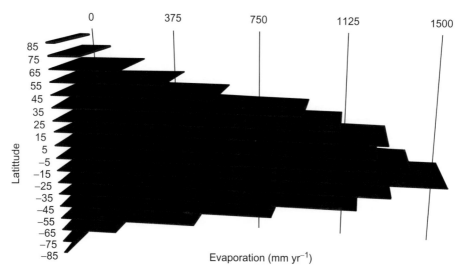

Figure 3 Global evaporation according to Baumgartner and Reichel (1975) for different 10° latitude bands.

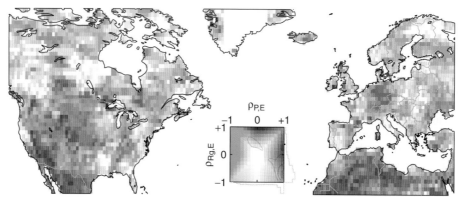

Figure 4 Multi-model analysis of controls on yearly evaporation. Correlation between yearly evaporation and global radiation ($r_{Rg,E}$), and precipitation ($r_{P,E}$), for the period 1986–95. Each color corresponds to a unique combination of $r_{Rg,E}$ and $r_{P,E}$. The gray lines (legend) show the global frequency distribution (Teuling et al., 2009).

Hemisphere latitude bands 20–40° S compared to their northern equivalents. This difference is largely due to the greater area of land in the Northern Hemisphere, which evaporates at a significantly lower rate than the ocean. The largest evaporative flux is found in the humid tropics, mainly as a result of large amounts of precipitation and high solar radiation. Compared to the latitudinal distribution of precipitation, the evaporation is more smoothly distributed with a general tendency of decreasing evaporation when moving poleward.

Decreases in radiation, global dimming, have caused a debate about an observed decline in pan measured evaporation (e.g., Roderick and Farquhar, 2002) that would be contrary to expectations for a warming climate. Peterson et al. (1995), using data from a network of pan evaporimeters in the US and the former Soviet Union, found a decrease in pan evaporation between 1950 and 1990. However, Fu et al. (2009) in a more comprehensive analysis suggested that, although many observations across the world indicate a general trend of pan evaporation decreasing over the last 50 years, this trend is not universal. A decrease in evaporation presents a paradox, as with global warming one would expect an increase due to the larger water holding capacity of the atmosphere through the Clausius–Clapeyron equation. An increase in evaporation would also match an increase in precipitation, although this is regionally very variable. Roderick and Farquhar (2002) explained the paradox by relating the decrease in evaporation to a decrease in solar radiation (global dimming). However, the dimming trend has recently reversed into a brightening trend and thus cannot singularly be held responsible for the decrease in pan evaporation.

Teuling et al. (2009) presented an analysis of the major controls on evaporation using an ensemble of land surface models forced off line with meteorological data. They investigate the control of two key drivers on evaporation, incoming solar radiation and soil moisture. Figure 4 shows that Europe, North Africa, and North America are characterized by two evaporation regimes: a humid regime with high correlation with radiation expressed through the correlation coefficient $r_{Rg,E}$, but low correlation with precipitation, $P(r_{P,E})$; and a more arid regime with high $r_{P,E}$, but low $r_{Rg,E}$. Because radiation and precipitation tend to be negatively correlated,

Teuling et al. (2009) concluded that yearly variations in evaporation reflect either variations in R_g or P, but not in both. Central Europe is among the regions with the highest $r_{Rg,E}$ correlation, while in more arid regions such as the US Midwest and the Sahara, evaporation correlates only with precipitation. Using data from direct observations of evaporation (Fluxnet; see Baldocchi et al., 2001), Teuling et al. were able to reproduce and validate these modeled patterns quite well.

Teuling et al. (2009) made a strong argument for a regional approach to explain some of the evaporation trends. Based on the different sensitivities of the various drivers of evaporation (see **Figure 4**) and the conclusion of other work that a dimming trend has been reversed, they concluded that scenarios of both decreasing actual evaporation with decreasing pan evaporation in regions with ample supply of water (e.g., central Europe), and of increasing evaporation with decreasing pan evaporation (e.g., the US Midwest) are consistent. Using basin-scale discharge data and precipitation they found that evaporation decreased over Europe during the dimming period and increased later, consistent with the high sensitivity of European evaporation to radiation rather than precipitation. During the dimming period, the positive trend in runoff is induced by reduced evaporation, rather than increased precipitation. After 1983, evaporation derived as the residual of precipitation minus runoff, increased in all central European basins during the brightening phase. These results suggest that evaporation trends follow radiation trends in central Europe. In contrast, in the US Midwest the upward trends in evaporation derived as a catchment residual before 1983 are followed by decreasing trends. These may be explained by trends in precipitation combined with high correlation between solar radiation and evaporation as inferred from **Figure 2**.

Next to the analysis of Teuling et al. (2009) that attribute changes in evaporation to either radiation or water availability (precipitation), recent studies of changes in pan evaporation in Australia attribute most of the reduction in pan evaporation to reduced wind speed (Roderick et al., 2007; Rayner, 2007; McVicar et al., 2008). The cause of such a reduction in regional wind speed is not certain, although wind speed reductions have been widely reported in mid-latitudes in both hemispheres (see also Shuttleworth, 2009).

2.03.5 Summary and Conclusions

Evaporation is an important, but regionally a still poorly quantified term in the global water balance. At global scale its determination as a residual of the continental scale water balance hinges on adequate estimation of precipitation and river discharge. Estimates obtained by direct bottom-up modeling vary considerably. This is in some contrast to our understanding of the basic physics of evaporation, that is well known, as for instance is shown by the Penman and PM equations. Application of these equations for use at local scales, for instance irrigation practice, is widespread (e.g., Allen et al., 1998), and to a large extent very successful.

The importance of evaporation in the global hydrological cycle critically encompasses two related aspects: its direct role as a term in the water budget, and its potential to impact weather and climate processes, by changing aspects of the surface energy balance and boundary layer. Both these roles depend on the balance between the controlling forces of evaporation, surface moisture and available energy. Where only water availability is limiting and radiation plentiful, such as in the (semi-)arid tropics, evaporation may add moisture to the overlying air that can then tip the balance to produce precipitation. These semi-arid areas have been found (e.g., Koster et al., 2002) to be sensitive to land surface precipitation feedbacks. In these feedbacks the role of evaporation is critical. On the other hand, in areas where both precipitation and radiation are not limiting, large-scale evaporation appears constrained by the available energy. Under these conditions, equations such as the Priestly–Taylor equation that do not explicitly take account of water availability on average perform well.

Thus, to be able to predict the effect of, for instance, land-use change on evaporation (and resulting catchment discharge), one would need to determine first which process, if any, is limiting. In cases where water availability is limiting, atmospheric feedbacks may also become important and simple bottom-up estimates of the change in evaporation may be wrong. In cases where neither water nor energy is limiting, to first order a bottom-up estimate based on energy constraints would be appropriate. Changes in the pattern of large-scale moisture recycling, such as found in the Amazon (e.g., Meesters et al., 2009), may however be important to estimate changes in the resulting precipitation climate.

The debate whether and where evaporation is increasing or decreasing is fundamental to our understanding of the role of evaporation in the global water cycle and climate. Improved, high-quality data sets are needed to provide benchmarks for climate models and to increase our process understanding. Currently, such data sets unfortunately do not exist. It is worth noting that the role of evaporation in climate is not related only to its direct effects on the hydrological cycle. Through the influence evaporation exerts on the partitioning of the energy balance, the effects on climate are also seen in surface temperatures. High evaporation keeps surfaces cool; low evaporation makes them hot. Seneviratne et al. (2006) and Fischer et al. (2007) used regional and global climate modeling to investigate the role of land surface atmosphere feedbacks on temperature in a changing climate. They concluded that soil moisture through its effect in reducing evaporation has a major impact on the variability and mean and maxima of surface temperatures in Europe. Fischer et al. (2007) concluded that land–atmosphere interactions over drought regions account for typically 50–80% of the number of hot days in a Northern Hemisphere summer. This is mainly due to local effects through the limitation of evaporation (and increased sensible heat flux) due to drought conditions. Drought conditions may also have remote effects on areas around or outside the actual drought region, through changes in atmospheric circulation and advection of air masses. These mechanisms can enhance an existing anticyclonic circulation over, or slightly downstream of, a drought anomaly.

Evaporation plays a key role in the global water cycle and hence in the global climate system. There have, however, been very few attempts to produce robust estimates of evaporation based on a global approach. Yet such data are urgently needed to validate and constrain current climate models. Thus, despite that at the practical level, considerable advances have been made in our ability to estimate and observe evaporation at local level, our understanding of evaporation in the global climate system (e.g., Kleidon and Schymanski, 2008) still shows significant gaps. The challenge for the next decade in evaporation research is to fill these gaps.

References

Allen RG, Pereira LS, Raes D, and Smith M (1998) *Crop Evapotranspiration – Guidelines for Computing Crop Water Requirements*, p. 300. Rome: Irrigation and Drainage, FAO.

Aubinet M, Chermanne B, Vandenhaute M, Longdoz B, Yernaux M, and Laitat E (2001) Long term carbon dioxide exchange above a mixed forest in the Belgian Ardennes. *Agricultural and Forest Meteorology* 108: 293–315.

Baldocchi D, Falge E, Gu L, et al. (2001) FLUXNET: A new tool to study the temporal and spatial variability of ecosystem-scale carbon dioxide, water vapor, and energy flux densities. *Bulletin of the American Meteorological Society* 82: 2415–2434.

Baumgartner A and Reichel E (1975) *Die Weltwasserbilanz*, 179pp. München: Oldenbourg Verlag.

Blanken PD, Rouse WR, Culf AD, et al. (2000) Eddy covariance measurements of evaporation from Great Slave Lake, Northwest Territories, Canada. *Water Resources Research* 36: 1069–1077.

Bouchet RJ (1963) Evapotranspiration réelle et potentielle, signification climatique, Int. Assoc. Sci. Hydrol., Proc. Berkeley, Calif. Symp., Publ. 62: 134–142.

Brutsaert W (1982) *Evaporation into the Atmosphere*. Dordrecht: Kluwer.

Brutsaert W and Chen D (1996) Diurnal variation of surface fluxes during thorough drying (or severe drought) of natural prairie. *Water Resources Research* 32: 2013–2019.

Budyko MI (1974) *Climate and Life*, 508p. New York: Academic Press.

De Bruin HAR (1983) A model of the Priestley–Taylor parameter, α. *Journal of the Applied Meteorology* 22: 572–578.

Dirmeyer PA, Gao X, Zhao M, Guo Z, Oki T, and Hanasaki N (2006) GSWP-2: Multimodel analysis and implications for our perception of the land surface. *Bulletin of the American Meteorological Society* 87: 1381–1397.

Finch JW and Gash JHC (2002) Application of a simple finite difference model for estimating evaporation from open water. *Journal of Hydrology* 255: 253–259.

Finch JW and Hall RL (2005) Evaporation from lakes. In: Anderson MG (ed.) *Encyclopedia of Hydrological Sciences*, pp. 635–646. Chichester: Wiley.

Fischer EM, Seneviratne S, Luethi M, and Schaer C (2007) Contribution of land–atmosphere coupling to recent European summer heat waves. *Geophysical Research Letters* 34: L06707 (doi:10.1029/2006GL029068).

Fu G, Charles SP, and Yu J (2009) A critical overview of pan evaporation trends over the last 50 years. *Climatic Change* 97(1–2): 193–214.

Gash JHC and Shuttleworth WJ (2007) *Evaporation*. Wallingford: IAHS Press.

Gedney N, Cox PM, Betts RA, Boucher O, Huntingford C, and Stott PA (2006) Detection of a direct carbon dioxide effect in continental river runoff records. *Nature* 439: 835–838.

Kleidon A and Schymanski S (2008) Thermodynamics and optimality of the water budget on land: A review. *Geophysical Research Letters* 35: L20404.

Koster RD, Dirmeyer PA, Guo Z, et al. (2004) Regions of strong coupling between soil moisture and precipitation. *Science* 305: 1138–1140.

Kustas WP, Prueger JH, and Hipps LE (2002) Impact of using different time-averaged inputs for estimating sensible heat flux of riparian vegetation using radiometric surface temperature. *Journal of Applied Meteorology* 41: 319–332.

Law BE, Falge E, Gu L, et al. (2002) Environmental controls over carbon dioxide and water vapor exchange of terrestrial vegetation. *Agricultural and Forest Meteorology* 113: 97–120.

Linacre ET (1993) Data-sparse estimation of lake evaporation, using a simplified Penman equation. *Agricultural and Forest Meteorology* 64: 237–256.

Mahfouf JF and Noilhan J (1991) Comparative study of various formulations of evaporation from bare soil using *in situ* data. *Journal of Applied Meteorology* 30: 1354–1365.

McNaughton KG and Jarvis PG (1991) Effects of spatial scale on stomatal control of transpiration. *Agricultural and Forest Meteorology* 54: 279–302.

McNaughton KG and Spriggs TW (1989) An evaluation of the Priestley–Taylor equation. In: Black TA, Spittlehouse DL, Novak MD, and Price DT (eds.) *Estimation of Areal Evaporation*, IAHS Publication No. 177, pp. 89–104. Wallingford: IAHS Press.

McVicar TR, Van Niel TG, Li LT, et al. (2008) Wind speed climatology and trends for Australia, 1975–2006: Capturing the stilling phenomenon and comparison with near-surface reanalysis output. *Geophysical Research Letters* 35: L20403 (doi:10.1029/2008GL035627)

Meesters AGCA, Dolman AJ, and Bruijnzeel LA (2009) Comment on "Biotic pump of atmospheric moisture as driver of the hydrological cycle on land" by AM Makarieva and VG Gorshkov, Hydrol. Earth Syst. Sci., 11, 1013–1033, 2007. *Hydrology and Earth System Sciences* 13: 1299–1305.

Milly PCD and Dunne KA (2002) Macroscale water fluxes, 2, Water and energy supply control of their interannual variability. *Water Resources Research* 38(10): 1206 (doi:10.1029/2001WR000760)

Monteith JL (1965) Evaporation and environment. In: *The State and Movement of Water in Living Organisms. Proceedings of the 19th Symposium Society for Experimental Biology*, pp. 205–234. Swansea: Cambridge University Press.

Morton FI (1983) Operational estimates of areal evapotranspiration and their significance to the science and practice of hydrology. *Journal of Hydrology* 66: 1–76.

Oki T and Kanae S (2006) Global hydrological cycles and world water resources. *Science* 313: 1068–1072.

Payne RE (1972) Albedo of the sea surface. *Journal of the Atmospheric Sciences* 29: 959–970.

Peixoto JP and Oort AH (1996) The climatology of relative humidity in the atmosphere. *Journal of Climate* 9: 3443–3463.

Penman HL (1948) Natural evaporation from open water, bare soil and grass. *Proceedings of the Royal Society of London, Series A: Mathematical and Physical Sciences* 193: 120–145.

Peterson TC, Golubev VS, and Groisman PY (1995) Evaporation losing its strength. *Nature* 377: 687–688.

Piao S, Friedlingstein P, Ciais P, De Noblet-Ducoudre N, Labat D, and Zaehle S (2007) Changes in climate and land use have a larger direct impact than rising CO_2 on global river runoff trends. *Proceedings of the National Academy of Sciences of the United States of America* 104: 15242–15247.

Porté-Agel F, Parlange MB, Cahill AT, and Gruber A (2000) Mixture of time scales in evaporation: Desorption and self-similarity of energy fluxes. *Agronomy Journal* 92: 832–836.

Priestley CHB and Taylor RJ (1972) On the assessment of surface heat flux and evaporation using large-scale parameters. *Monthly Weather Review* 100: 81–92.

Raupach MR (2001) Combination theory and equilibrium evaporation. *Quarterly Journal of the Royal Meteorological Society* 127: 1149–1181.

Rayner DP (2007) Wind run changes are the dominant factor affecting pan evaporation trends in Australia. *Journal of Climate* 20: 3379–3394.

Roderick ML and Farquhar GD (2002) The cause of decreased pan evaporation over the past 50 years. *Science* 298: 1410–1411.

Roderick ML, Rotstayn LD, Farquhar GD, and Hobbins MT (2007) On the attribution of changing pan evaporation. *Geophysical Research Letters* 34: L17403 (doi:10.1029/2007GL031166).

Sene KJ, Gash JHC, and McNeil DD (1991) Evaporation from a tropical lake: Comparison of theory with direct measurements. *Journal of Hydrology* 127: 193–217.

Seneviratne SI, Luethi D, Litschi M, and Schaer C (2006) Land–atmosphere coupling and climate change in Europe. *Nature* 443, doi:10.1038/nature05095.

Shuttleworth WJ (1993) Evaporation. In: Maidment DR (ed.) *Handbook of Hydrology*, pp. 4.1–4.53. New York: McGraw-Hill.

Shuttleworth WJ (2009) On the theory relating changes in area-average and pan evaporation. *Quarterly Journal of the Royal Meteorological Society* 135: 1230–1247.

Shuttleworth WJ and Calder IR (1979) Has the Priestley–Taylor equation any relevance to forest evaporation? *Journal of Applied Meteorology* 18: 639–646.

Stewart JB (1977) Evaporation from the wet canopy of a pine forest. *Water Resources Research* 13: 915–921.

Sweers HE (1976) A nomogram to estimate the heat-exchange coefficient at the air–water interface as a function of wind speed and temperature: A critical survey of some literature. *Journal of Hydrology* 30: 375–401.

Teuling AJ, Hirschi M, Ohmura A, et al. (2009) A regional perspective on trends in continental evaporation. *Geophysical Research Letters* 36(2): L02404.

Thom AS and Oliver HR (1977) On Penman's equation for estimating regional evaporation. *Quarterly Journal of the Royal Meteorological Society* 103: 345–357.

Valentini R, Mateucci G, Dolman AJ, et al. (2000) Respiration as the main determinant of carbon balance in European forests. *Nature* 404: 861–865.

van der Molen MK, Dolman AJ, Waterloo MJ, and Bruijnzeel LA (2006) Climate is affected more by maritime than by continental land use change: A multiple scale analysis. *Global and Planetary Change* 54: 128–149.

Wallace JS and Holwill CJ (1997) Soil evaporation from tiger-bush in south-west Niger. *Journal of Hydrology* 188–189: 426–442.

Willett KM, Gillett NP, Jones PD, and Thorne PW (2007) Attribution of observed surface humidity changes to human influence. *Nature* 449, doi:10.1038/nature06207.

Wilson KB, Baldocchi DD, Aubinet M, et al. (2002) Energy partitioning between latent and sensible heat flux during the warm season at FLUXNET sites. *Water Resources Research* 38(12): 1294 (doi:10.1029/2001WR000989).

Zhang L, Dawes WR, and Walker GR (2001) Response of mean annual evapotranspiration to vegetation changes at catchment scale. *Water Resources Research* 37: 701–708.

2.04 Interception

AMJ Gerrits and HHG Savenije, Delft University of Technology, Delft, The Netherlands

© 2011 Elsevier B.V. All rights reserved.

2.04.1	Introduction	89
2.04.2	Importance of Interception	90
2.04.3	Types of Interception	91
2.04.3.1	Canopy Interception	91
2.04.3.2	Forest Floor Interception	91
2.04.3.3	Fog Interception	92
2.04.3.4	Snow Interception	92
2.04.3.5	Urban Interception	93
2.04.4	Methods to Measure Interception	93
2.04.4.1	Canopy	93
2.04.4.2	Forest Floor	94
2.04.5	Interception Models	95
2.04.5.1	Conceptual Rutter Model	95
2.04.5.2	Analytical Gash Model	96
2.04.5.3	Stochastic Interception Models	96
2.04.5.3.1	Poisson distribution	96
2.04.5.3.2	Markov chains	97
2.04.5.3.3	Gamma probability density function and transfer functions	98
2.04.6	Consequences of Underestimating Interception for Hydrological Modeling and Water Resource Assessment	99
2.04.7	Outlook	99
References		99

Nomenclature

b	constant in Rutter (1971) model (L^{-1})		p_t	trunk fraction coefficient (−)
B_r	Bowen ratio (−)		P	precipitation ($L\,T^{-1}$)
c	canopy coverage (−)		P_g	gross precipitation ($L\,T^{-1}$)
c_p	specific heat ($L\,M\,T^{-3}\,K^{-1}$)		P'_g	gross precipitation necessary for canopy saturation ($L\,T^{-1}$)
D	drainage rate from the canopy ($L\,T^{-1}$)		P''_g	gross precipitation necessary for trunk saturation ($L\,T^{-1}$)
D_i	interception threshold ($L\,T^{-1}$)			
E	actual evaporation ($L\,T^{-1}$)		q	maximum amount of rain drops on element (−); specific humidity ($M\,T^{-1}L^{-1}$)
E_i	interception evaporation ($L\,T^{-1}$)			
$E_{i,c}$	interception evaporation from canopy ($L\,T^{-1}$)			
			r	amount of rain drops on element (−)
$E^l_{i,c}$	evaporation from leaves (without trunk) ($L\,T^{-1}$)		S_c	storage of canopy (L)
			S^l_c	storage of leaves (without trunk) (L)
$E^t_{i,c}$	evaporation from the trunk ($L\,T^{-1}$)		S^t_c	storage of trunk (L)
$E_{i,f}$	interception evaporation from forest floor ($L\,T^{-1}$)		S_f	storage of forest floor (L)
			S_i	interception storage (L)
E_l	evaporation from lower basin ($L\,T^{-1}$)		S_l	storage of the lower basin (L)
E_p	potential evaporation ($L\,T^{-1}$)		S_u	storage of the upper basin (L)
H	sensible heat flux ($M\,T^{-3}$)		T_f	throughfall ($L\,T^{-1}$)
I	interception process ($L\,T^{-1}$)		T_s	stemflow ($L\,T^{-1}$)
L	number of elemental surface areas per unit ground (L^{-2})		v	mean volume of raindrops (L^3)
m	mean number of raindrops striking an element (−)		z	height (L)
			β	scaling factor ($L\,T^{-1}$)
n	mean number of drops retained per element (−)		ϵ	constant in Rutter (1971) model (−)
			λ	latent heat of vaporization coefficient ($L^2\,T^{-2}$)
$n_{r,d}$	number of rain days per month (−)			
n_m	days within a month (=30.5) (−)		ρ	density of water ($M\,L^{-3}$)
p	throughfall coefficient (−)		θ	potential temperature (K)

2.04.1 Introduction

When it rains the entire surface becomes wet: trees, shrubs, grass, forest floor, footpaths, etc. Also in urban areas, roads and roofs become wet, sometimes forming pools of stagnant water. After rainfall has ceased these surfaces soon become dry again. This process is called 'interception'. It is the part of the rainfall that is captured by surface storage (i.e., vegetation, roofs, etc.) before it can run off or infiltrate into the soil. The intercepted water generally evaporates during the event and shortly after the rainfall ceased, so that it can repeat its function during the next rainfall event.

In the literature, interception is defined in different ways: sometimes as a stock, sometimes as a flux, or, more appropriately, as the entire interception process (Savenije, 2005). If only interception storage (S_i [L]) is considered, interception is defined as the amount of rainfall which is temporarily stored on the Earth's surface. Actually, this is the interception capacity or water-holding capacity. If interception is defined as a flux, then it is the intercepted water which evaporated over a certain time [L T^{-1}] during and after the event. When interception is considered as a process (I [L T^{-1}]), it is defined as the part of the rainfall flux which is intercepted on the wetted surface after which it is fed back to the atmosphere. The interception process equals the sum of the change of interception storage (S_i) and the evaporation from this stock (E_i):

$$I = \frac{dS_i}{dt} + E_i \qquad (1)$$

The timescale of the interception process is in the order of 1 day. After 1 day, it is fair to assume for most climates that the first term on the right-hand side in Equation (1) approaches zero, and $I = E_i$. Of course, in the case of snow under cold climates, this may take longer.

How much of the precipitation is intercepted depends on several factors, which can be divided into three groups:

- *Vegetation characteristics*. Large vegetation types, such as trees, have a high aerodynamic roughness, causing high potential evaporation rates. Grasses, crops, or bushes, on the other hand, have a much lower roughness and thus do not have as high potential evaporation rates. The storage capacity also depends on the vegetation type. The shape of the leaves, the thickness, the density (leaf area index), and the configuration of the branches determine how much water can be stored. For example, the capacity of a coniferous or a deciduous tree is different (e.g., Rutter et al., 1975; Baird and Wilby, 1999; Bryant et al., 2005; Toba and Ohta, 2005). Although intuitively one might think that a deciduous tree can hold more water in its bucket-like leaves, a coniferous tree can hold much more water by adhesion. Furthermore, it is also important to take the seasonality into account. Deciduous trees lose their leaves in the dormant season, causing a large reduction in the canopy storage capacity. Vegetation also determines the amount of understorey growth and forest floor. The forest floor of different vegetation types can have significantly different interception behavior (e.g., a thick needle layer or a thin leaf litter layer).
- *Rainfall characteristics*. Rainfall has a large influence on the interception process. The rainfall frequency is a major determining factor. It makes a big difference if rainfall occurs as one continuous storm or as a sequence of several small events with dry spells in between. Even if the total rainfall depth is the same, the last scenario intercepts much more rainwater, because between the events the storage can be (partly) emptied by evaporation and thus more storage is available. Second, the rainfall intensity is important, although there is no consensus in literature. Horton (1919) and Wang et al. (2007) concluded that the interception capacity is lower at higher intensity because high rainfall intensities cause splashing and shaking of leaves. On the other hand, Aston (1979) and Keim et al. (2006b) noted the opposite: high rainfall intensities coincide with high storage capacities, due to dynamic storage.
- *Evaporative demand*. If the potential evaporation (i.e., open water evaporation) is high, the intercepted water can evaporate more easily during and after the event. Wind plays an important role in removing moisture from the surface providing a higher vapor deficit, particularly in the canopy. Moreover, the roughness of the vegetation increases the evaporative power, by causing turbulence which makes it easier to take up the intercepted water. However, wind can also reduce the amount of interception by reducing the storage. Horton (1919), Klaassen et al. (1996), and Hörmann et al. (1996) noted that with increasing wind speed the measured storage capacity is less, due to the fact that the wind shakes the rainwater off the leaves.

Of the above three factors, the rainfall characteristics are most dominant for evaporation from interception. Although both the storage capacity (mainly vegetation characteristic) and the available energy form a constraint to the evaporation flux per event, the number of events is a more important factor. This is confirmed by the sensitivity analysis of Gerrits et al. (2009c).

2.04.2 Importance of Interception

Although most surfaces can store only a few millimeters of rainfall, which is often not much in comparison to other stocks in the water balance, interception is generally a significant process. The impact becomes evident at longer timescales. Although interception storage is generally small, the number of times that the storage is filled and depleted can be so large that the interception flux is generally of the same order of magnitude as the transpiration flux. In addition, the interception process smooths the rain intensities, causing more gradual infiltration. Interception redistributes the rainfall as well. Some parts of a field receive less water due to interception, whereas other parts receive more due to funneling of the vegetation (e.g., Germer et al., 2006; Gerrits et al., 2009b). Subsequently, this has an influence on the soil moisture patterns, and this is again important for flood generation (Roberts and Klingeman, 1970).

Besides the hydrological effects, there are influences on the nutrient cycle of a forest, and on agricultural applications.

For example, interception affects the efficiency of insecticides and fertilizers (Aston, 1979). Besides, fire retardants are more effective if they are stored by vegetation. Finally, interception may reduce soil erosion by preventing rain drops to directly hit and erode the soil layer (Walsh and Voigt, 1977), although in the case of canopy interception the opposite can be true due to the formation of larger rain drops with a higher impact on the forest floor.

2.04.3 Types of Interception

As already stated in Section 2.04.1, it is possible to define an infinite number of interception types. In principle, every surface that can store water can be considered as an interception type. In this chapter, we focus on the major types, mainly occurring in a natural environment plus some special mechanisms. However, more often than not, it is a combination of mechanisms. For example, in a forest, it is likely that a part of the rainfall is intercepted by the canopy of a tree, while the remaining part can be intercepted by epiphytes on the branches and/or bark, and, finally, the understorey and forest floor intercept the throughfall before infiltration starts.

2.04.3.1 Canopy Interception

Canopy interception is the rainwater that is stored on the leaves and branches of a tree which is subsequently evaporated. This interception can be calculated by measuring rainfall above the trees or measured in an open area nearby (gross rainfall P_g) and subtracting the throughfall (T_f) and stemflow (T_s) (**Figure 1**):

$$E_{i,c} + \frac{dS_c}{dt} = P_g - T_f - T_s \qquad (2)$$

Many research studies have been carried out on canopy interception. In **Table 1** an overview is given. We can see in the table and also in tables in Kittredge (1948), Zinke (1967), and Breuer et al. (2003) that there is a large difference in the canopy interception by deciduous and coniferous trees (e.g., Kittredge, 1948; Bryant et al., 2005; Toba and Ohta, 2005). Not only because deciduous trees lose their leaves, but also because the leaf area of coniferous trees is much larger than of deciduous trees; coniferous trees can store much more water. Furthermore, leaves may swing over when they become too heavy, causing a (sudden) decrease of the storage capacity. However, Herbst et al. (2008) found counterintuitive results, where higher evaporation rates were found in deciduous trees in winter caused by rougher aerodynamics of the bare canopy and deeper penetration of the wind.

In most cases, the storage of water on the branches is small; however, in some environments, the branches can be overgrown by epiphytes. Pypker et al. (2006) showed that in a Douglas fir forest the canopy water storage can potentially be increased by >1.3 mm and Hölscher et al. (2004) found that epiphytes can account for 50% of the storage capacity. However, this large increase in storage capacity is not necessarily resulting in high interception values (storage + evaporation), because the water uptake and release by the epiphytes is delayed. It takes a while to saturate the epiphytes, and already before saturation, runoff generation can take place. Successively, after wetting, the drying of the epiphytes takes much longer than drying of the canopy, causing less storage to be available.

Another special type of canopy interception is interception by agricultural crops. In essence, there is no difference between crops and other vegetation types. They both can store water up to a certain threshold and then drain water to the floor as throughfall. However, whereas vegetation has a gradual seasonal pattern (summer vs. winter), crops have a phenological growth cycle (seeding to harvesting) which is therefore more abrupt. Hence, when modeling crop interception the appropriate description of the variation in the storage capacity is important.

2.04.3.2 Forest Floor Interception

Forest floor interception is the part of the throughfall that is temporarily stored in the top layer of the forest floor and successively evaporated within a few hours or days during and after the rainfall event. The forest floor can consist of short vegetation (like grasses, mosses, bushes, and creeping vegetation), litter as described by Hoover and Lunt (1952) as the litter and fermentation (L and F) layer (i.e., leaves, twigs, and small branches), or bare soil. Although the latter seems to have an overlap with soil evaporation, we distinguish them by the fact that soil evaporation refers to the water that is stored in the root zone (Groen and Savenije, 2006).

Figure 1 Two major interception types in the natural environment.

Table 1 Canopy interception values in literature, with $S_{c,max}$ the water storage capacity and $E_{i,c}$ the interception evaporation as percentage of gross precipitation

Source	Specie	Location	$S_{c,max}$ (mm)	$E_{i,c}$ (%)
Rutter et al. (1975)	Corsian pine (*Pinus nigra*)	United Kingdom	1.05	35
	Douglas fir (*Pseudotsuga menziesii*)	United Kingdom	1.2	39
	Norway spruce (*Picea abies*)	United Kingdom	1.5	48
	Hornbeam (*Carpinus betulus*)	United Kingdom	1.0 (leafy) 0.65 (leafless)	36
	Oak (*Quercus robur*)	United Kingdom	0.875 (leafy) 0.275 (leafless)	18
Gash and Morton (1978)	Scots pine (*Pinus sylvestris*)	United Kingdom	0.8	
Gash et al. (1980)	Sitka spruce (*Picea sitchensis*)	United Kingdom	0.75–1.2	27–32
	Scots pine (*Pinus sylvestris*)	United Kingdom	1.02	42
Rowe (1983)	Beech (*Nothofagus*)	New Zealand	1.5 (leafy) 1.2 (leafless)	35 (leafy), 22 (leafless)
Bruijnzeel and Wiersum (1987)	Acacia auriculiformis	Indonesia	0.5–0.6	11–18
Viville et al. (1993)	Norway spruce (*Picea abies*)	France		34.2
Hörmann et al. (1996)	Beech (*Asperulo-fagetum*)	Germany	1.28 (leafy) 0.84 (leafless)	18
Valente et al. (1997)	Pinus pinaster	Portugal	0.41	10.8
	Eucalyptus globulus	Portugal	0.21	17.1
Navar et al. (1999)	Tamaulipan thornscrub	Mexico		18.9
Bryant et al. (2005)	Loblolly (*Pinus taeda*) & shortleaf pine (*Pinus echinata*)	USA (GA)	1.97	22.3
	Longleaf pine (*Pinus palustris*)	USA (GA)	1.70	17.6
	Scrub oak (*Quercus berberidifolia*)	USA (GA)	1.40	17.4
	White oak (*Quercus alba*) & shortleaf pine (*Pinus echinata*) & loblolly pine (*Pinus palustris*)	USA (GA)	1.58	18.6
	Hardwood	USA (GA)	0.98	17.7
Toba and Ohta (2005)	Larc (*Larix cajanderi*)	Siberia		29
	Red pine (*Pinus sylvester*)	Siberia		36
	Red pine (*Pinus densiflora*)	Japan		13–17
	Sawtooth oak (*Quercus acutissima*)	Japan		24
	Oak (*Quercus serrata*)	Japan		18
Cuartas et al. (2007)	Rain forest	Brazil	1.0	13–22

See also tables in Kittredge (1948), Zinke (1967), and Breuer et al. (2003).

In Table 2 some results are presented of previous work on forest floor interception.

2.04.3.3 Fog Interception

A special type of interception is fog interception or cloud interception. Vegetation can intercept not only rain, but also moisture (in the form of small water droplets) from the air. Fog can occur due to different processes. Bruijnzeel et al. (2005) distinguished nine types: radiation fog, sea fog, stream fog, advection fog, ice fog, coastal fog, valley fog, urban fog, and mountain fog.

Fog interception is mainly important in tropical montane environments (table in Bruijnzeel (2005): 6–53% of rainfall), and can also play a significant role in semi-arid regions near the coast (e.g., Hursh and Pereira, 1953; Hutley et al., 1997; Hildebrandt et al., 2007). In both environments, the main problem with fog interception studies is to measure precipitation and throughfall (Equation (2)), which is especially important because fog deposition can be twice as high as normal rainfall. Since conventional rain gauges are not suitable to measure fog deposition, special fog collectors have been developed with often wire meshes to intercept the moisture. These instruments suffer from various limitations. An overview of fog collectors can be found in Bruijnzeel et al. (2005).

2.04.3.4 Snow Interception

Snowfall is also intercepted by trees. Especially, coniferous trees can store so much snow, that they collapse under its weight. As an example, Storck et al. (2002) found in a Douglas-fir-dominated forest that up to 60% of the snowfall was intercepted, equaling 40 mm of snow water equivalent (swe).

The storage of snow on the canopy is different from rain. For rainfall interception the storage capacity is mainly a function of the leaf surface area, whereas for snow interception the branch strength and canopy shape are more important (Ward and Trimble, 2004). Furthermore, the snow storage is also dependent on the temperature. If snow falls with temperatures close to freezing point, the cohesion of snow is higher causing more snow to be accumulated on the canopy (Ward and Trimble, 2004).

Another difference between rainfall interception and snow interception is the way in which interception storage is depleted. Rainfall interception is a real threshold process,

Table 2 Forest floor interception values in literature, with the water storage capacity $S_{f,max}$ and the interception evaporation $E_{i,f}$ as percentage of net precipitation (i.e., throughfall)

Source	Forest floor type	Location	$S_{f,max}$ (mm)	$E_{i,f}$ (%)
Haynes (1940)	Kentucky bluegrass (*Poa pratensis*)	?		56[a]
Kittredge (1948)	Californian grass (*Avena, Stipa, Lolium, Bromus*)	USA (CA)		26[a]
Beard (1956)	*Themeda* and *Cymbopogon*	South Africa		13[a]
Helvey (1964)	Poplar	USA (NC)		34
Brechtel (1969)	Scot's pine	USA (NY)		21
	Norway spruce	USA (NY)		16
	Beech	USA (NY)		16
	Oak	USA (NY)		11
Pathak *et al.* (1985)	*Shorea robusta* and *Mallotus philippensis*	India		11.8
	Pinus roxburghii and *Quercus glauca*	India		7.8
	Pinus roxburghii	India		9.6
	Quercus leucotrichophora and *Pinus roxburghii*	India		10.6
	Quercus floribunda and *Quercus leucotrichophora*	India		11.0
	Quercus lanuginosa and *Quercus floribunda*	India		11.3
Clark (1940) in Thurow *et al.* (1987)	Blue stem *Andropogon gerardi* Vitman	USA (TX)		57–84
Walsh and Voigt (1977)	Pine (*Pinus sylvestris*)	United Kingdom	0.6–1.7	
	Beech (*Fagus sylvaticus*)	United Kingdom	0.9–2.8	
Pitman (1989)	Bracken litter (*Pteridium aquiliunum*)	United Kingdom	1.67	
Miller *et al.* (1990)	Norway spruce	Scotland		18[a]
	Sitka spruce	Scotland		16[a]
Thamm and Widmoser (1995)	Beech (*Asperulo-Fagetum*)	Germany	2.5–3.0	12–28
Putuhena and Cordery (1996)	*Pinus radiata*	Australia	2.78	
	Eucalyptus	Australia	1.70	
Schaap and Bouten (1997)	Douglas fir	Netherlands		0.23 mm d^{-1}
Li *et al.* (2000)	Peble mulch (5–9 cm)	China	0.281	11.5[a]
	Peble mulch (2–6 cm)	China	0.526	17.4[a]
Sato *et al.* (2004)	*Cryptomeria japonica*	Japan	0.27–1.72	
	Lithocarpus edulis	Japan	0.67–3.05	
Guevara-Escobar *et al.* (2007)	Grass (*Aristida divaricata*)	Mexico	2.5	
	Woodchips (*Pinus*)	Mexico	8	
	Poplar leaves (*Populus nigra*)	Mexico	2.3	

[a] % of gross precipitation instead of net precipitation.

whereby throughfall starts when the storage capacity is exceeded. The storage capacity is then emptied by evaporation. Snow, on the other hand, can only be removed from the canopy by three ways: sublimation, mechanical removal (sliding leading to mass release), and melt water drip (Miller, 1966).

2.04.3.5 Urban Interception

Most hydrological studies focus on natural environments and not on urbanized areas, which is also the case for interception studies. However, recently, with the increasing interest for alternative sources of water for nonpotable domestic use (so-called 'gray water'), water balance studies on (interception) evaporation in urban areas increased (Grimmond and Oke, 1991; Ragab *et al.*, 2003; Gash *et al.*, 2008; Nakayoshi *et al.*, 2009).

The difference between urban and rural interception is not only that the typical storage capacities of buildings, roads, etc., are unknown, but also that the entire energy balance is different in a city. Oke (1982) discovered the so-called 'Urban Heat Island', that is, higher temperatures in urban areas compared to the surrounding rural areas. The Urban Heat Island is mainly caused by the (relatively warm) buildings that block the cold night sky. Furthermore, the thermal properties of a city are different: concrete and asphalt have much higher heat capacities than forests and also the surface radiative properties differ (e.g., albedo and emissivity). The lack of vegetation in urban areas, which reduces cooling by transpiration, also causes a difference in the energy balance.

2.04.4 Methods to Measure Interception
2.04.4.1 Canopy

There exist already many methods to measure canopy interception. The most-often used method is by measuring rainfall above the canopy and subtract throughfall and stemflow (e.g., Helvey and Patric, 1965). However, the problem with this method is that the canopy is not homogeneous, which causes it to be difficult to obtain representative throughfall data. Using multiple rain gauges under the canopy (Helvey and Patric, 1965; Keim *et al.*, 2005; Gerrits *et al.*, 2009b) reduces this problem. Sometimes the collectors are moved to achieve a better representation of throughfall (e.g., Lloyd and Marques, 1988; Tobón-Marin *et al.*, 2000; Manfroi *et al.*, 2006; Ziegler et al., 2009). Another method to avoid the problem with the spatial distribution of the canopy was introduced by Calder and Rosier (1976) and applied by, for example, Shuttleworth et al. (1984), Calder et al. (1986), and Calder (1990). They covered the forest floor with plastic sheets and collected the throughfall. The disadvantage of this method is that for long periods irrigation is required, because otherwise, in the end, the trees will dry out and may even die due to water shortage. The method by Hancock and Crowther (1979) avoided these problems, by making use of the cantilever effect of branches. If leaves on a branch hold water, it becomes more heavy and will bend. By measuring the displacement, it is possible to determine the amount of intercepted water. Huang et al. (2005) refined this method by making use of strain gauges. However, the disadvantages of these methods are that only information about one single branch is obtained and it is quite laborious to measure an entire tree. Edwards (1986), Fritschen and Kinerson (1973), and Storck et al. (2002) made use of weighing lysimeters with trees. Although interception of a whole tree is measured with this method, the big disadvantage of this method is that it is expensive and destructive. Friesen et al. (2008) developed a nondestructive method to measure canopy interception of a whole tree. With mechanical displacement sensors, Friesen et al. (2008) measured the stem compression due to interception water, which is an integration of the whole canopy. However, although this method looks promising, it is still under development.

A totally different way of measuring canopy interception of a forest plot is to make use of ray attenuation. Calder and Wright (1986) used the attenuation of gamma rays. They transmitted from a tower gamma-rays through the canopy at different heights and measured the gamma-ray density at a receiving tower. The ratio between transmitted and received gamma-ray density during dry conditions is successively compared to this ratio during a rainfall event. This gives an estimate of the amount of water stored on the canopy over time. Although the method gives interception estimated of an entire forest, the method becomes inaccurate under windy conditions. Furthermore, safety standards inhibits unattended use of this method. Bouten et al. (1991) overcame this problem by making use of microwave attenuation. It appears to be a suitable method to measure canopy wetness, although it is an expensive method.

Evaporation can also be measured by flux measurements. By measuring temperature (θ) and specific humidity (q) at several heights (z) above the canopy, one can calculate the Bowen ratio (B_r), which is the sensible heat flux, H, divided by the latent heat flux (λE):

$$B_r = \frac{H}{\rho \lambda E} = \frac{c_p \delta\theta/\delta z}{\lambda \delta q/\delta z} \quad (3)$$

Combined with the energy balance, evaporation can be calculated (Gash and Stewart, 1975). The main difficulty with the Bowen ratio method is to measure the humidity gradient more accurately (Stewart, 1977). Another method is the eddy covariance technique, where the net upward or downward flux is determined by fast-response three-dimentional (3D) wind speed measurements combined with a concentration measurement. This concentration can be humidity, temperature, or CO_2 concentrations (Amiro, 2009).

2.04.4.2 Forest Floor

In the literature, little can be found on forest floor interception, although some researchers have tried to quantify the interception amounts. Generally, these methods can be divided into two categories (Helvey and Patric, 1965):

1. lab methods, whereby field samples are taken to the lab and successively the wetting and drying curves are determined by measuring the moisture content and
2. field methods, whereby the forest floor is captured into trays or where sheets are placed underneath the forest floor.

An example of the first category is that of Helvey (1964), who performed a drainage experiment on the forest floor after it was saturated. During drainage, the samples were covered, and after drainage had stopped (24 h), the samples were taken to the lab, where the samples were weighed and successively dried until a constant weight was reached. By knowing the oven dry weight of the litter per unit area and the drying curve, the evaporation from interception could be calculated. In this way, they found that about 3% of the annual rainfall evaporated from the litter. Similar work was done by Bernard (1963), Walsh and Voigt (1977), and Sato et al. (2004). However, what they all measured was not the flux, but the storage capacity.

Another example of lab experiments was carried out by Putuhena and Cordery (1996). First, field measurements were carried out to determine the spatial variation of the different forest floor types. Second, storage capacities of the different forest floor types were measured in the lab using a rainfall simulator. Finally, the lab experiments were extrapolated to the mapping step. In this way, Putuhena and Cordery (1996) found average storage capacities of 2.8 mm for pine and 1.7 mm for eucalyptus forest floors. Moreover, Guevara-Escobar et al. (2007) made use of a rainfall simulator.

Examples of the second category have been, for example, carried out by Pathak et al. (1985), who measured the weight of a sample tray before and after a rainfall event. They found litter interception values of 8–12% of the net precipitation. In addition, here, they measured the storage capacity, rather than the flux. Schaap and Bouten (1997) measured the interception flux by the use of a lysimeter and found that 0.23 mm d^{-1} evaporated from a dense Douglas fir stand in early spring and summer. Also, Brechtel (1969) and Thamm and Widmoser (1995) made use of lysimeters. Brechtel (1969) manually measured the infiltrated water and Thamm and Widmoser (1995) developed an automatic and more sophisticated method, whereby the suction under the forest floor is controlled by a tensiometer. Gerrits et al. (2007) developed a method whereby both the forest floor interception and the infiltrated water are continuously weighed in suspended trays with strain gauges. In **Figure 2** the schematic setup is shown.

Measurements with sheets were done, for example, by Li et al. (2000), who found that pebble mulch intercepts 17% of the gross precipitation. Miller et al. (1990) found comparable results (16–18%) for a mature coniferous plantation in Scotland.

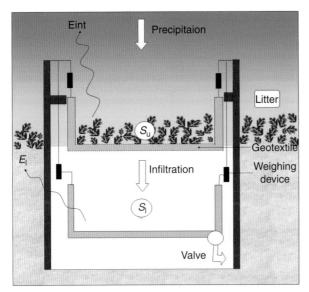

Figure 2 Forest floor interception device by Gerrits AMJ, Savenije HHG, Hoffmann L, and Pfister L (2007) New technique to measure forest floor interception – an application in a beech forest in Luxembourg. *Hydrology and Earth System Sciences* 11: 695–701.

2.04.5 Interception Models

In literature, several models have been developed to simulate forest interception. Almost all of these models are concentrated on canopy interception, sometimes including stem interception (**Table 3**). In principle, these models can be expanded to include forest floor or any surface interception as well.

The most often used interception models are the conceptual model of Rutter *et al.* (1971) (Section 2.04.5.1) and the analytical model of Gash (1979) (Section 2.04.5.2) or revisions of these models. Furthermore, there exist some stochastic models, which will be described in Section 2.04.5.3. In **Table 3** an overview and summary of the models are given. A more detailed overview and comparison can be found in Muzylo *et al.* (2009).

2.04.5.1 Conceptual Rutter Model

The conceptual framework of the original Rutter model is depicted in **Figure 3**. As can be seen the rainfall is divided into three parts:

1. free throughfall, that is, throughfall which did not touch the canopy at all (pP_g),
2. trunk input ($p_t P_g$), and
3. canopy input (($1 - p - p_t)P_g$).

The rain that falls on the canopy can drain to the ground (i.e., canopy drainage, D), or evaporate ($E_{i,c}^l$), or it can be stored on the canopy (S_c^l):

$$(1 - p - p_t) \int P_g dt = \int D dt + \int E_{i,c}^l dt + \int dS_c^l \quad (4)$$

The rain that falls on the trunk can evaporate from the trunk ($E_{i,c}^t$), or drain in the form of stemflow (T_s), or it can be stored on the trunk (S_c^t):

$$p_t \int P_g dt = \int T_s dt + \int E_{i,c}^t dt + \int dS_c^t \quad (5)$$

with $E_{i,c} = E_{i,c}^l + E_{i,c}^t$ and $S_c = S_c^l + S_c^t$ for the total canopy interception.

The evaporation from the wet canopy is calculated with the Penman equation (Penman, 1948). Because the canopy is not always completely wet ($S_c^l < S_{c,max}^l$), the actual evaporation rate can be calculated by the fraction of the potential evaporation: $E_p S_c^l / S_{c,max}^l$. The same concept is applied for the trunks. However, for the determination of the potential evaporation of the trunks, the potential evaporation of the canopy is multiplied with and extra constant ϵ.

Stemflow is modeled as a threshold process, whereby no stemflow is generated when $S_c^t < S_{c,max}^t$, and when the threshold is exceeded stemflow equals the difference between S_c^t and $S_{c,max}^t$.

Canopy drainage is modeled in a similar way; however, when the threshold $S_{c,max}^l$ is exceeded, drainage is defined as

$$D = D_s \exp\left[b(S_c^l - S_{c,max}^l)\right] \quad (6)$$

with D_s being the rate of drainage when the canopy is saturated and b [L^{-1}] an empirical coefficient.

Valente *et al.* (1997) revised the original Rutter model, to model interception in a more realistic way for sparse canopies. The main drawbacks of the original model were the partitioning of free throughfall and canopy input, and the conceptual error that evaporation from interception can theoretically be higher than potential evaporation (Valente *et al.*, 1997). Therefore, they divided the conceptual model into two areas: a covered area (c) and an uncovered area ($1 - c$). Second, in the revised Rutter model, only water can reach the trunk after it has flowed through the canopy as a part of the canopy drainage. Water which is not drained by the trunk is directly dripping to the ground. The final change was made that evaporation from the saturated canopy is not equal to the potential evaporation, but is reduced by a factor $1 - \epsilon$ ($0 < \epsilon < 1$). The remaining energy (($\epsilon)E_p$) is then available for evaporating water from the saturated trunk (**Figure 4**).

Table 3 Characteristics of interception models

Main author	Model type	Interception element: canopy	stem	forest floor	Timescale
Rutter	Conceptual	x	x		≤ hourly
Gash	Analytical	x	x		event
C alder	Stochastic	x			≤ hourly
De Groen	Concept./stoch.	x	x	x	monthly
Keim	Concept./stoch.	x			6-hourly

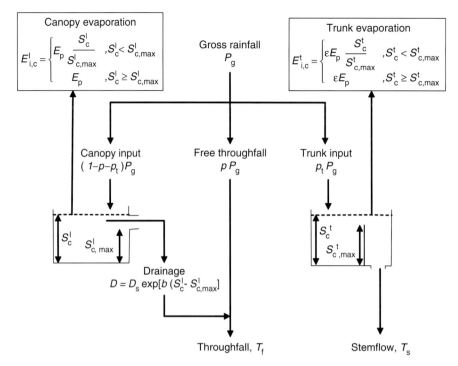

Figure 3 Conceptual framework of the Rutter model. Modified from Valente F, David JS, and Gash JHC (1997) Modelling interception loss for two sparse eucalypt and pine forest in central Portugal using reformulated Rutter and Gash analytical models. *Journal of Hydrology* 190: 141–162.

2.04.5.2 Analytical Gash Model

The original Gash model is conceptually the same as the Rutter model (see Section 2.04.5.1); however, it does not require meteorological data of high temporal resolution (hourly) and requires less computation time. The main assumption of the Gash model is that it is possible to represent the real rainfall pattern by different discrete rainfall events, each consisting of three phases:

1. wetting phase,
2. saturation phase, and
3. drying phase (long enough to dry the entire canopy).

Similar to the Rutter model, rainfall is divided into canopy input $(1-p-p_t)$, free throughfall (p), and trunk input (p_t).

The Gash model makes a distinction between storms which are not large enough to saturate the canopy ($P_g < P'_g$: m storms) and storms which are large enough to saturate the canopy ($P_g \geq P'_g$: n storms). The amount of gross rainfall necessary to saturate the canopy is P'_g (see Table 4). Interception evaporation is then calculated for the canopy and the trunk.

Although the original Gash model appears to work fine for several types of forests, it contains some weaknesses for modeling sparse forests, similar to the Rutter model. Hence, Gash et al. (1995) revised their existing model according to the revised Rutter model (Rutter et al., 1975). An overview of the formulas of the revised Gash model can be found in Table 4.

2.04.5.3 Stochastic Interception Models

2.04.5.3.1 Poisson distribution

Calder (1986) developed a stochastic interception model, where he assumes that a tree consists of several elemental areas, which all have the same probability to be struck by raindrops. The Poisson probability of an element to be struck by r drops equals

$$P_r = \frac{m^r}{r!} e^{-m} \qquad (7)$$

with m the mean number of raindrops striking an element per storm.

If an element can hold q raindrops, the mean number of drops per element (n) can be expressed as

$$n = \sum_{r=0}^{q} r \cdot P_r + q \cdot P(r>q) \qquad (8)$$

$$= q + \sum_{r=0}^{q} P_r \cdot (r-q) \qquad (9)$$

with $P(r>q)$ the probability of elements being struck by more than q drops and is equal to $1 - \sum_{r=0}^{q} P_r$. To upscale from elemental area to canopy area, the number of elemental surface areas per unit ground (L) is required and the mean volume of raindrops (v):

$$S_c = nvL \qquad (10)$$

$$S_{c,\max} = qvL \qquad (11)$$

$$P_g = mvL \qquad (12)$$

Evaporation is then obtained by (with $dS_c dE_{i,c} = -1$)

$$\frac{dn}{dE_{i,c}} = \frac{dS_c}{dE_{i,c}} \times \frac{dn}{dS_c} = \frac{-1}{vL} \qquad (13)$$

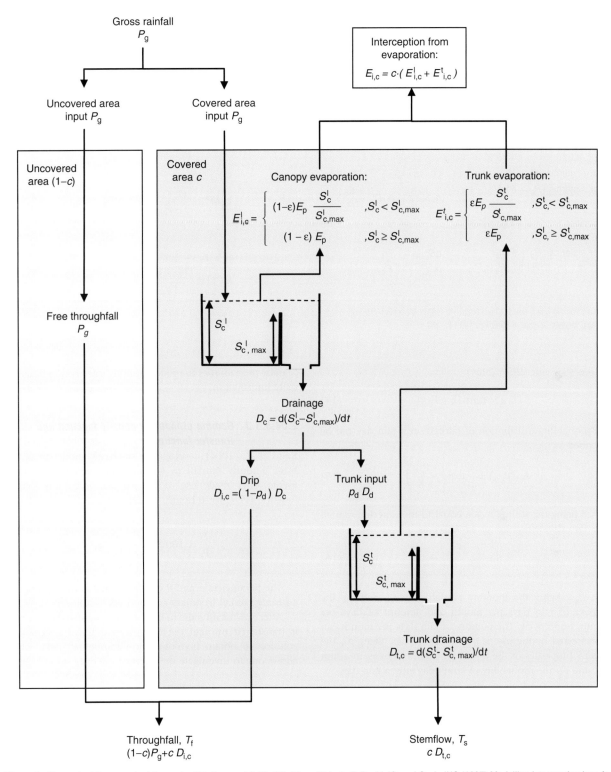

Figure 4 Conceptual framework of the revised Rutter model. Modified from Valente F, David JS, and Gash JHC (1997) Modelling interception loss for two sparse eucalypt and pine forest in central Portugal using reformulated Rutter and Gash analytical models. *Journal of Hydrology* 190: 141–162.

The Calder model is very simple and describes the threshold behavior of interception very well; however, it is difficult to upscale from drop size scale to forest size scale. This hinders the applicability of the model.

2.04.5.3.2 Markov chains

Groen and Savenije (2006) developed a monthly interception model based on a daily interception model and the daily rainfall characteristics. They assumed interception on a daily

Table 4 Components of interception of the original Gash (1979) model and the revised Gash et al. (1995) model for sparse canopies

	Original Gash (1979)	Revised (sparse canopy) Gash et al. (1995)
Amount of gross rainfall necessary to saturate the canopy (P'_g) and trunk (P''_g)	$P'_g = -\dfrac{\bar{P}_g S_{c,max}}{\bar{E}_p} \ln\left[1 - \dfrac{\bar{E}_p}{(1-p-p_t)\bar{P}_g}\right]$ $P''_g = S^t_{c,max}/p_t$	$P'_g = -\dfrac{\bar{P}_g}{(1-\epsilon)\bar{E}_p} \dfrac{S_{c,max}}{c} \ln\left[1 - \dfrac{(1-\epsilon)\bar{E}_p}{\bar{P}_g}\right]$ $P''_g = \dfrac{\bar{P}_g}{\bar{P}_g - (1-\epsilon)\bar{E}_p} \dfrac{S^t_{c,max}}{p_t c} + P'_g$
Evaporation from canopy interception ($E^t_{i,c}$):		
1. for m storms ($P_g < P'_g$)	$(1-p-p_t)\sum_{j=1}^m P_{g,j}$	$c\sum_{j=1}^m P_{g,j}$
2. for n storms ($P_g \geq P'_g$)	$n(1-p-p_t)P'_g + \dfrac{\bar{E}_p}{\bar{P}_g}\sum_{j=1}^n (P_{g,j}-P'_g)$	$c\left[nP'_g + \dfrac{(1-\epsilon)\bar{E}_p}{\bar{P}_g}\sum_{j=1}^n (P_{g,j}-P'_g)\right]$
Evaporation from trunk interception ($E^t_{i,c}$):		
1. for q storms ($P_g \geq P''_g$)	qS^t_c	qS^t_c
2. for $m+n-q$ storms ($P_g < P''_g$)	$p_t \sum_{j=1}^{m+n-q} P_{g,j}$	$p_t c\left[1 - \dfrac{(1-\epsilon)\bar{E}_p}{\bar{P}_g}\sum_{j=1}^n (P_{g,j}-P'_g)\right]$

Modified from Valente F, David JS, and Gash JHC (1997). Modelling interception loss for two sparse eucalypt and pine forest in central Portugal using reformulated Rutter and Gash analytical models. *Journal of Hydrology* 190:141–162.

scale as (Savenije, 1997, 2004)

$$E_{i,d} = \min(D_{i,d}, P_{g,d}) \quad (14)$$

The probability distribution of rainfall on a rain day can be described as

$$f_{i,d}(P_{g,d}) = \frac{1}{\beta}\exp\left(-\frac{P_{g,d}}{\beta}\right) \quad (15)$$

with β being the scaling factor, equal to the expected rainfall on a rain day, which can be expressed as

$$\beta = \frac{P_{g,m}}{E(n_{r,d}/n_m)} \quad (16)$$

with $P_{g,m}$ being the monthly rainfall and $n_{r,d}$ and n_m the number of rain days per month and amount of days per month, respectively. The number of rain days per month can be expressed by the use of Markov properties. Being p_{01} the Markov properties of the transition from a dry day to a rain day, and p_{11} the probability of a rain day after a rain day:

$$n_{r,d} = n_m \frac{p_{01}}{1-p_{11}+p_{01}} \quad (17)$$

Multiplying Equations (14) and (15) and successively integrating results in monthly evaporation from interception:

$$E_{i,m} = E(n_{r,d}|n_m)\int_0^\infty E_{i,d} \cdot f_{i,d}(P_d)dP_d \quad (18)$$

$$= P_m\left(1 - \exp\left(\frac{-D_{i,d}}{\beta}\right)\right) \quad (19)$$

Hence, the model of Groen and Savenije (2006) is a parsimonious model with only one measurable parameter, and Markov probabilities to model monthly interception based on daily information.

2.04.5.3.3 Gamma probability density function and transfer functions

Keim et al. (2004) developed a stochastic model to obtain from 6-hourly rainfall to 6-hourly throughfall for extreme events. They made use of the gamma probability density function (PDF; for 6 hourly rainfall):

$$\frac{T_f}{P_g} \times 100\% = \frac{P_g^{\alpha-1} e^{-P_g/\theta}}{\Gamma(\alpha)\theta^\alpha} \quad (20)$$

The parameters α and θ can be estimated by dividing the 6-hourly rainfall in ranges and find the best-fit sets.

After downscaling the rainfall and throughfall data, rainfall is transferred through the canopy by a linear system convolution to obtain high-resolution throughfall data, which allows one to investigate the effect of intensity smoothing:

$$T_f(t) = \int_0^t P(t)g(t-\tau)dt \quad (21)$$

with the transfer function $g(t-\tau)$. Keim et al. (2004) found that the transfer function can be best described with the exponential distribution:

$$g(t) = \alpha\, e^{-\alpha t} \quad (22)$$

By coupling the stochastic model with the the intensity smoothing transfer function, effects of forest canopies on extreme rainfall events can be investigated.

2.04.6 Consequences of Underestimating Interception for Hydrological Modeling and Water Resource Assessment

Hydrologists often consider precipitation as the start of the hydrological cycle. After a rainfall event, the first separation point in the cycle is on the Earth surface. Part of the rainwater is intercepted by the vegetation or ground surface and the remainder infiltrates into the unsaturated zone or runs off. The part of the rainfall that is intercepted successively evaporates from the temporary storage.

This first separation point in the hydrological cycle is not always considered a significant process. This is partly due to the technical difficulties that are inherent to interception measurements (Lundberg et al., 1997; Llorens and Gallart, 2000), but it is also generally considered a minor flux, although previous studies tell us that interception can amount to 10–50% of the precipitation depending on the vegetation type (Klaassen et al., 1998). Even then, these studies mostly refer to canopy interception only. If forest floor interception is taken into account as well, the percentage is substantially higher.

Furthermore, it is often stated that interception is particularly not important for the generation of floods. This is not true. Interception strongly influences the antecedent soil moisture conditions, which are very important for the generation of floods (Roberts and Klingeman, 1970). Still, interception is regularly (partly) disregarded in hydrological models, or taken as a fixed percentage of the precipitation. As a result, after model calibration, interception is generally compensated by other processes such as transpiration, soil evaporation, or even recharge (Savenije, 2004).

Zhang and Savenije (2005) showed that the hydrograph at the outlet of the Geer basin in Belgium improved significantly when interception was included in a rainfall–runoff model using the representative elementary watershed (REW) approach. Both the Nash–Sutcliffe efficiency and the percentage bias improved. They also showed that, in calibration, the soil moisture storage capacity compensated for the neglect of the interception process.

Keim et al. (2006a) investigated the effects of (canopy) interception. They looked at the influence on the subsurface stormflow generation and concluded that interception caused a delay in the onset of subsurface stormflow, lowered and delayed stormflow peaks, and decreased total flow and the runoff ratio. They also found that simply reducing the rainfall by a constant factor did not result in a satisfactory peak flow response.

Fenicia et al. (2008) looked at the change in the movement of the Pareto front when stepwise new processes were included in a variable model structure. They concluded that when interception was included and especially, when spatially distributed interception was included, the Pareto front moved significantly to the origin. Hence, their conclusion was that interception is an important process and should therefore be included in hydrological models.

2.04.7 Outlook

More than 2000 articles have been published on interception studies (source: Scopus and ISI Web of Knowledge^{SM}) and still new articles are being published. Most of these articles focus on canopy interception and describe in detail the process for different tree species in different climates, resulting in long reference tables as, for example, presented in **Table 1** and by Breuer et al. (2003). Although this information is of high value for modeling purposes, it would have been more logical if these tables had also been available for the other types of interception, such as described in Section 2.04.3. Especially, since generally more than one mechanism occur, these mechanisms interact (see, e.g., between canopy and forest floor interception (Gerrits et al., 2009a). It would really be a way forward, if a broader scope was systematically considered in interception studies.

Although a more balanced database on interception values will help, it is not the complete solution for hydrological modeling. Often, experimental results are site and time specific. Therefore, it is difficult to upscale literature values on interception for catchment modeling. This problem may be solved by considering the energy balance. If we would know how the available energy is partitioned over the different fluxes and compartments, we would be able to determine interception evaporation as well. However, this would require intensive field experiments where both the energy fluxes and the evaporation processes are measured simultaneously. Remote sensing could provide the necessary spatial and temporal information on energy partitioning. Through a combination of methods, interception could be more adequately incorporated in hydrological models.

References

Amiro B (2009) Measuring boreal forest evapotranspiration using the energy balance residual. *Journal of Hydrology* 366(1–4): 112–118.

Aston A (1979) Rainfall interception by eight small trees. *Journal of Hydrology* 42(3–4): 383–396.

Baird AJ and Wilby RL (eds.) (1999) *Eco-Hydrology-Plants and Water in Terrestrial and Aquatic Environments*. London: Routledge.

Beard JS (1956) Results of the mountain home rainfall interception and infiltration project on black wattle, 1953–1954. *Journal of South African Forestry* 27: 72–85.

Bernard JM (1963) Forest floor moisture capacity of the New Jersey pine barrens. *Ecology* 44(3): 574–576.

Bouten W, Swart PJF, and De Water E (1991) Microwave transmission, a new tool in forest hydrological research. *Journal of Hydrology* 124(1–2): 119–130.

Brechtel HM (1969) Wald und Abfluss-Methoden zur Erforschung der Bedeutung des Waldes fur das Wasserdargebot. *Deutsche Gewasserkundliche Mitteilungen* 8: 24–31.

Breuer L, Eckhardt K, and Frede H-G (2003) Plant parameter values for models in temperate climates. *Ecological Modelling* 169(2–3): 237–293.

Bruijnzeel LA (2005) Tropical montane cloud forest: A unique hydrological case. In: Bonell M and Bruijnzeel LA (eds.) *Forests, Water and People in the Humid Tropics*, pp. 462–483. Cambridge: Cambridge University Press.

Bruijnzeel LA, Eugster W, and Burkard R (2005) Fog as a hydrologic input. In: Anderson MG (ed.) *Encyclopedia of Hydrological Sciences*, pp. 559–582. Chichester: Wiley.

Bruijnzeel LA and Wiersum KF (1987) Rainfall interception by a young Acacia Auriculiformis (a. cunn) plantation forest in West Java, Indonesia: Application of Gash's analytical model. *Hydrological Processes* 1: 309–319.

Bryant ML, Bhat S, and Jacobs JM (2005) Measurements and modeling of throughfall variability for five forest communities in the southeastern US. *Journal of Hydrology* 312: 95–108.

Calder IR (1986) A stochastic model of rainfall interception. *Journal of Hydrology* 89: 65–71.

Calder IR (1990) *Evaporation in the Uplands*. Chichester: Wiley.

Calder IR and Rosier PTW (1976) The design of large plastic-sheet net-rainfall gauges. *Journal of Hydrology* 30(4): 403–405.

Calder IR and Wright IR (1986) Gamma ray attenuation studies of interception from Sitka Spruce: Some evidence for an additional transport mechanism. *Water Resources Research* 22: 409–417.

Calder IR, Wright IR, and Murdiyarso D (1986) A study of evaporation from tropical rain forest–West Java. *Journal of Hydrology* 89: 13–31.

Clark OR (1940) Interception of rainfall by prairie grasses, weeds and certain crop plants. *Ecological Monographs* 10: 243–277.

Cuartas LA, Tomasella J, Nobre AD, Hodnett MG, Waterloo MJ, and Mnera JC (2007) Interception water-partitioning dynamics for a pristine rainforest in Central Amazonia: Marked differences between normal and dry years. *Agricultural and Forest Meteorology* 145(1–2): 69–83.

de Groen MM and Savenije HHG (2006) A monthly interception equation based on the statistical characteristics of daily rainfall. *Water Resources Research* 42: W12417.

Edwards WRN (1986) Precision weighing lysimetry for trees, using a simplified tared-balance design. *Tree Physiology* 1: 127–144.

Fenicia F, Savenije HHG, Matgen P, and Pfister L (2008) Understanding catchment behavior through stepwise model concept improvement. *Water Resources Research* 44: 1–13.

Friesen J, Beek C van, Selker J, Savenije HHG, and Giesen N van de (2008) Tree rainfall interception measured by stem compression. *Water Resources Research* 44: W00D15.

Fritschen LJ, Cox L, and Kinerson R (1973) A 28-meter Douglas-fir in a weighing lysimeter. *Forest Science* 19: 256–261.

Gash JHC (1979) An analytical model of rainfall interception by forests. *Quarterly Journal of the Royal Meteorological Society* 105: 43–55.

Gash JHC, Lloyd CR, and Lauchaud G (1995) Estimation sparse forest rainfall interception with an analytical model. *Journal of Hydrology* 170: 79–86.

Gash JHC and Morton AJ (1978) An application of the rutter model to the estimation of the interception loss from Thetford Forest. *Journal of Hydrology* 38(1–2): 49–58.

Gash JHC, Rosier PTW, and Ragab R (2008) A note on estimating urban roof runoff with a forest evaporation model. *Hydrological Processes* 22(8): 1230–1233.

Gash JHC and Stewart JB (1975) The average surface resistance of a pine forest derived from Bowen ratio measurements. *Boundary-Layer Meteorology* 8: 453–464.

Gash JHC, Wright IR, and Lloyd CR (1980) Comparative estimates of interception loss from three coniferous forests in Great Britain. *Journal of Hydrology* 48(1–2): 89–105.

Germer S, Elsenbeer H, and Moraes JM (2006) Throughfall and temporal trends of rainfall redistribution in an open tropical rainforest, South-Western Amazonia (Rondonia, Brazil). *Hydrology and Earth System Sciences* 10: 383–393.

Gerrits AMJ, Pfister L, and Savenije HHG (2009a) Spatial and temporal variability of canopy and forest floor interception in a beech forest. *Hydrological Processes*, doi: 10.1002/hyp. 7712, published online, 7 june 2010.

Gerrits AMJ, Savenije HHG, Hoffmann L, and Pfister L (2007) New technique to measure forest floor interception–an application in a beech forest in Luxembourg. *Hydrology and Earth System Sciences* 11: 695–701.

Gerrits AMJ, Savenije HHG, and Pfister L (2009b) Canopy and forest floor interception and transpiration measurements in a mountainous beech forest in Luxembourg. *IAHS Redbook* 326: 18–24.

Gerrits AMJ, Savenije HHG, Veling EJM, and Pfister L (2009c) Analytical derivation of the Budyko curve based on rainfall characteristics and a simple evaporation model. *Water Resources Research* 45: W04403.

Grimmond CSB and Oke TR (1991) An evapotranspiration-interception model for urban areas. *Water Resources Research* 27: 1739–1755.

Guevara-Escobar A, Gonzalez-Sosa E, Ramos-Salinas M, and Hernandez-Delgado GD (2007) Experimental analysis of drainage and water storage of litter layers. *Hydrology and Earth System Sciences* 11(5): 1703–1716.

Hancock NH and Crowther JM (1979) A technique for the direct measurement of water storage on a forest canopy. *Journal of Hydrology* 41: 105–122.

Haynes JL (1940) Ground rainfall under vegetation canopy of crops. *Journal of the American Society of Agronomy* 32: 176–184.

Helvey JD (1964) Rainfall interception by hardwood forest litter in the southern Appalachians *U.S. Forest Service Research Paper SE*, vol. 8, pp. 1–8. Asherille, NC: Department of Agriculture, Forest Science, Southeastern Forest Experiment station.

Helvey JD and Patric JH (1965) Canopy and litter interception of rainfall by Hardwoods of Eastern United States. *Water Resources Research* 1(2): 193–206.

Herbst M, Rosier PT, McNeil DD, Harding RJ, and Gowing DJ (2008) Seasonal variability of interception evaporation from the canopy of a mixed deciduous forest. *Agricultural and Forest Meteorology* 148(11): 1655–1667.

Hildebrandt A, Al Aufi M, Amerjeed M, Shammas M, and Eltahir EAB (2007) Ecohydrology of a seasonal cloud forest in Dhofar: 1. Field experiment. *Water Resources Research* 43: W10411.

Hölscher D, Köhler L, Dijk AIJM van, and Bruijnzeel LAS (2004) The importance of epiphytes to total rainfall interception by a tropical montane rain forest in Costa Rica. *Journal of Hydrology* 292(1–4): 308–322.

Hoover MD and Lunt HA (1952) A key for the classification of forest humus types. *Soil Science Society Proceedings* 16: 368–371.

Hörmann G, Branding A, Clemen T, Herbst M, Hinrichs A, and Thamm F (1996) Calculation and simulation of wind controlled canopy interception of a beech forest in Northern Germany. *Agricultural and Forest Meteorology* 79(3): 131–148.

Horton RE (1919) Rainfall interception. *Monthly Weather Review* 47(9): 603–623.

Huang YS, Chen SS, and Lin TP (2005) Continuous monitoring of water loading of trees and canopy rainfall interception using the strain gauge method. *Journal of Hydrology* 311: 1–7.

Hursh CR and Pereira HC (1953) Field moisture balance in the Shimba Hills, Kenya. *East African Agricultural Journal* 18: 139–148.

Hutley LB, Doley D, Yates DJ, and Boonsaner A (1997) Water balance of an Australian subtropical rainforest at altitude: The ecological and physiological significance of intercepted cloud and fog. *Australian Journal of Botany* 45: 311–329.

Keim R, Skaugset A, Link T, and Iroum A (2004) A stochastic model of throughfall for extreme events. *Hydrology and Earth System Sciences* 8(1): 23–34.

Keim RF, Meerveld HJT van, and McDonnell JJ (2006a) A virtual experiment on the effects of evaporation and intensity smoothing by canopy interception on subsurface stormflow generation. *Journal of Hydrology* 327: 352–364.

Keim RF, Skaugset AE, and Weiler M (2005) Temporal persistence of spatial patterns in throughfall. *Journal of Hydrology* 314: 263–274.

Keim RF, Skaugset AE, and Weiler M (2006b) Storage of water on vegetation under simulated rainfall of varying intensity. *Advances in Water Resources* 29: 974–986.

Kittredge J (ed.) (1948) *Forest Influences*. New York: McGraw-Hill.

Klaassen W, Bosveld F, and de Water E (1998) Water storage and evaporation as constituents of rainfall interception. *Journal of Hydrology* 212–213: 36–50.

Klaassen W, Lankreijer HJM, and Veen AWL (1996) Rainfall interception near a forest edge. *Journal of Hydrology* 185(1–4): 349–361.

Li XY, Gong JD, Gao QZ, and Wei XH (2000) Rainfall interception loss by pebble mulch in the semi arid region of China. *Journal of Hydrology* 228: 165–173.

Llorens P and Gallart F (2000) A simplified method for forest water storage capacity measurement. *Journal of Hydrology* 240: 131–144.

Lloyd CR and Marques ADO (1988) Spatial variability of throughfall and stemflow measurements in amazonian rainforest. *Agricultural and Forest Meteorology* 42(1): 63–73.

Lundberg A, Eriksson M, Halldin S, Kellner E, and Seibert J (1997) New approach to the measurement of interception evaporation. *Journal of Atmospheric and Oceanic Technology* 14: 1023–1035.

Manfroi OJ, Kuraji K, Suzuki M, et al. (2006) Comparison of conventionally observed interception evaporation in a 100-m^2 subplot with that estimated in a 4-ha area of the same Bornean Lowland tropical forest. *Journal of Hydrology* 329(1–2): 329–349.

Miller HD (1966) Transport of intercepted snow from trees during snowstorms *US Forest Service–Research Paper*, vol. 33, pp. 1–30. Berkeley, CA: US department of Agriculture, Forest Service, Pacific Southwest Forest & Range Experiment Station.

Miller JD, Anderson HA, Ferrier RC, and Walker TAB (1990) Comparison of the hydrological budgets and detailed hydrological responses in two forested catchments. *Forestry* 63(3): 251–269.

Muzylo A, Llorens P, Valente F, Keizer J, Domingo F, and Gash J (2009) A review of rainfall interception modelling. *Journal of Hydrology* 370(1–4): 191–206.

Nakayoshi M, Moriwaki R, Kawai T, and Kanda M (2009) Experimental study on rainfall interception over an outdoor urban-scale model. *Water Resources Research* 45: W04415.

Navar J, Charles F, and Jurado E (1999) Spatial variations of interception loss components by Tamaulipan thornscrub in Northeastern Mexico. *Forest Ecology and Management* 24: 231–239.

Oke TR (1982) The energetic basis of the urban heat island. *Quarterly Journal of the Royal Meteorological Society* 108(455): 1–24.

Pathak PC, Pandey AN, and Singh JS (1985) Apportionment of rainfall in central Himalayan forests (India). *Journal of Hydrology* 76: 319–332.

Penman HL (1948) Natural evaporation from open water, bare soil and grass. *Proceedings of the Royal Society of London* 193: 120–146.

Pitman JI (1989) Rainfall interception by bracken litter–relationship between biomass, storage and drainage rate. *Journal of Hydrology* 111: 281–291.

Putuhena W and Cordery I (1996) Estimation of interception capacity of the forest floor. *Journal of Hydrology* 180: 283–299.

Pypker TG, Unsworth MH, and Bond BJ (2006) The role of epiphytes in rainfall interception by forests in the Pacific Northwest. I. Laboratory measurements of water storage. *Canadian Journal of Forest Research* 36: 808–818.

Ragab R, Bromley J, Rosier P, Cooper JD, and Gash JHC (2003) Experimental study of water fluxes in a residential area: 1. Rainfall, roof runoff and evaporation: The effect of slope and aspect. *Hydrological Processes* 17(12): 2409–2422.

Roberts MC and Klingeman PC (1970) The influence of landform and precipitation parameters on flood hydrograph. *Journal of Hydrology* 11: 393–411.

Rowe L (1983) Rainfall interception by an evergreen beech forest, Nelson, New Zealand. *Journal of Hydrology* 66(1–4): 143–158.

Rutter AJ, Kershaw KA, Robins PC, and Morton AJ (1971) A predictive model of rainfall interception in forests. I. Derivation of the model and comparison with observations in a plantation of Corsican pine. *Agricultural Meteorology* 9: 367–384.

Rutter AJ, Morton AJ, and Robins PC (1975) A predictive model of rainfall interception in forests. II. Generalization of the model and comparison with observations in some coniferous and hardwood stands. *Journal of Applied Ecology* 12: 307–300.

Sato Y, Kumagai T, Kume A, Otsuki K, and Ogawa S (2004) Experimental analysis of moisture dynamics of litter layers – the effect of rainfall conditions and leaf shapes. *Hydrological Processes* 18: 3007–3018.

Savenije HHG (1997) Determination of evaporation from a catchment water balance at a monthly time scale. *Hydrology and Earth System Sciences* 1: 93–100.

Savenije HHG (2004) The importance of interception and why we should delete the term evapotranspiration from our vocabulary. *Hydrological Processes* 18: 1507–1511.

Savenije HHG (2005) Interception. In: Lehr JH and Keeley J (eds.) *Water Encyclopedia: Surface and Agricultural Water*. Hoboken, NJ: Wiley Publishers.

Schaap MG and Bouten W (1997) Forest floor evaporation in a dense Douglas fir stand. *Journal of Hydrology* 193: 97–113.

Shuttleworth WJ, Gash JHC, Lloyd CR, Moore CJ, Roberts JM, *et al.* (1984) Eddy correlation measurements of energy partition for Amazonian forest. *Quarterly Journal of the Royal Meteorological Society* 110: 1143–1162.

Stewart JB (1977) Evaporation from the wet canopy of a pine forest. *Water Resources Research* 13(6): 915–921.

Storck P, Lettenmaier DP, and Bolton SM (2002) Measurement of snow interception and canopy effects on snow accumulation and melt in a mountainous maritime climate, Oregon, United States. *Water Resources Research* 38: 1223.

Thamm F and Widmoser P (1995) Zur hydrologischen Bedeutung der organischen Auflage im Wald: Untersuchungsmethoden und erste Ergebnisse. *Zeitschrift fur Pflanzenernahrung und Bodenkunde* 158: 287–292.

Thurow TL, Blackburn WH, Warren SD, and Taylor CA Jr. (1987) Rainfall interception by midgrass, shortgrass, and live oak mottes. *Journal of Range Management* 40(5): 455–460.

Toba T and Ohta T (2005) An observational study of the factors that influence interception loss in boreal and temperate forests. *Journal of Hydrology* 313: 208–220.

Tobón-Marin C, Bouten IW, and Dekker S (2000) Gross rainfall and its partitioning into throughfall, stemflow and evaportation of intercepted water in four forest ecosystems in Western Amazonia. *Journal of Hydrology* 237: 40–57.

Valente F, David JS, and Gash JHC (1997) Modelling interception loss for two sparse eucalypt and pine forest in central Portugal using reformulated Rutter and Gash analytical models. *Journal of Hydrology* 190: 141–162.

Viville D, Biron P, Granier A, Dambrine E, and Probst A (1993) Interception in a mountainous declining spruce stand in the Strengbach catchment (Vosges, France). *Journal of Hydrology* 144: 273–282.

Walsh RPD and Voigt PJ (1977) Vegetation litter: An underestimated variable in hydrology and geomorphology. *Journal of Biogeography* 4: 253–274.

Wang A, Diao Y, Pei T, Jin C, and Zhu J (2007) A semi-theoretical model of canopy rainfall interception for a broad-leaved tree. *Hydrological Processes* 21(18): 2458–2463.

Ward AD and Trimble SW (2004) *Environmental hydrology*, 2nd edn. Boca Raton, FL: CRC Press.

Zhang GP and Savenije HHG (2005) Rainfall-runoff modelling in a catchment with a complex groundwater flow system: Application of the Representative Elementary Watershed (REW) approach. *Hydrology and Earth System Sciences* 9: 243–261.

Ziegler AD, Giambelluca TW, Nullet MA, *et al.* (2009) Throughfall in an evergreen-dominated forest stand in Northern Thailand: Comparison of mobile and stationary methods. *Agricultural and Forest Meteorology* 149(2): 373–384.

Zinke PJ (1967) Forest interception studies in the United States. In: Sopper WE and Lull HW (eds.) *Forest Hydrology*, pp. 137–161. Oxford: Pergamon.

2.05 Infiltration and Unsaturated Zone

JW Hopmans, University of California, Davis, CA, USA

© 2011 Elsevier B.V. All rights reserved.

2.05.1	Introduction	103
2.05.2	**Soil Properties and Unsaturated Water Flow**	103
2.05.2.1	Soil Water Retention	104
2.05.2.2	Unsaturated Hydraulic Conductivity	105
2.05.2.3	Modeling of Unsaturated Water Flow and Transport	106
2.05.2.4	Infiltration Processes	107
2.05.3	**Infiltration Equations**	109
2.05.3.1	Philip Infiltration Equation	109
2.05.3.2	Parlange et al. Model	109
2.05.3.3	Swartzendruber Model	110
2.05.3.4	Empirical Infiltration Equations	110
2.05.4	**Measurements**	110
2.05.4.1	Infiltration	110
2.05.4.2	Unsaturated Water Flow	111
2.05.5	**Scaling and Spatial Variability Considerations**	112
2.05.6	**Summary and Conclusions**	113
Acknowledgments		113
References		113

2.05.1 Introduction

As soils make up the upper part of the unsaturated zone, they are subjected to fluctuations in water and chemical content by infiltration and leaching, water uptake by plant roots, and evaporation from the soil surface. It is the most dynamic region of the subsurface, as changes occur at increasingly smaller time and spatial scales when moving from the groundwater toward the soil surface. Environmental scientists are becoming increasingly aware that soils make up a critically important component of the earth's biosphere, because of their food production and ecological functions, and the soil's important role in controlling water quality. For example, prevention or remediation of soil and groundwater contamination starts with proper management of the unsaturated zone.

Water entry into the soil by infiltration is among the most important soil hydrological processes, as it controls the partitioning between runoff and soil water storage. Runoff water determines surface water quantity and quality, whereas infiltrated water determines plant available water, evapotranspiration, groundwater recharge, and groundwater quality. Also through exfiltration, infiltrated water affects water quality in waterways and associated riparian zones. Despite its relevance and our reliable physical understanding of infiltration, we have generally many difficulties predicting infiltration at any scale. Mostly, this is so because the infiltration rate is a time-varying parameter of which its magnitude is largely controlled by spatially variable soil properties, in both vertical and horizontal directions of a hydrologic basin. Moreover, infiltration rate and runoff are affected by vegetation cover, as it protects the soil surface from the energy impacts of falling raindrops or intercepting rainfall, serving as temporary water storage. The kinetic energy of rainfall causes soil degradation, leading to soil surface sealing and decreasing infiltration.

Historically, solutions to infiltration problems have been presented by way of analytical solutions or empirically. Analytical solutions provide values of infiltration rate or cumulative infiltration as a function of time, making simplifying assumptions of soil depth variations of water content, before and during infiltration. Instead, we now often use powerful computers to conduct numerical simulations of unsaturated water flow to solve for water content and water fluxes throughout the unsaturated soil domain in a single vertical direction or in multiple spatial dimensions, allowing complex initial and boundary conditions. However, although the modeling of multidimensional unsaturated water flow is extremely useful for many vadose zone applications, it does not necessarily improve the soil surface infiltration rate prediction, in light of the large uncertainty of the soil physical properties and initial and boundary conditions that control infiltration. In contrast, empirical infiltration models serve primarily to fit model parameters to measured infiltration, but have limited power as a predictive tool.

2.05.2 Soil Properties and Unsaturated Water Flow

The soil consists of a complex arrangement of mostly connected solid, liquid, and gaseous phases, with the spatial distribution and geometrical arrangement of each phase, and the partitioning of solutes between phases, controlled by physical, chemical, and biological processes. The unsaturated zone is bounded by the soil surface and merges with the groundwater in the capillary fringe. Water in the unsaturated soil matrix is held by capillary and adsorptive forces. Water is a primary factor leading to soil formation from the weathering of parent material such as rock or transported deposits, with additional factors of climate, vegetation, topography, and parent material determining soil physical properties.

Defining the soil's dry bulk density by ρ_b (ML^{-3}), soil porosity, ε ($L^3 L^{-3}$), is defined by

$$\varepsilon = 1 - \frac{\rho_b}{\rho_s} \qquad (1)$$

with ρ_s being the soil's particle density (ML^{-3}). Equation (1) shows that soil porosity has lower values as bulk soil density is increased such as by compaction.

Unsaturated water flow is largely controlled by the physical arrangement of soil particles in relation to the water and air phases within the soil's pore space, as determined by pore-size distribution and water-filled porosity or volumetric water content, θ (L^3 water/L^3 bulk soil). The volumetric water content θ expresses the volume of water present per unit bulk soil as

$$\theta = \frac{w\rho_b}{\rho_w} \qquad (2)$$

where w is defined as the mass water content (M of water/M dry soil) and we take $\rho_w = 1000$ kg m^{-3}. Alternatively, the soil water content can be described by the degree of saturation S (–) and the equivalent depth of stored water D_e (L), or

$$S = \frac{\theta}{\varepsilon} \quad \text{and} \quad D_e = \theta D_{soil} \qquad (3)$$

so that θ can also be defined by the equivalent depth of water per unit depth of bulk soil, D_{soil} (L). The volumetric water content ranges between 0.0 (dry soil) and the saturated water content, θ_s, which is equal to the porosity if the soil were completely saturated. The degree of saturation varies between 0.0 (completely dry) and 1.0 (all pores completely water-filled). When considering water flow, the porosity term is replaced by the saturated water content, θ_s, and both terms in Equation (3) are corrected by subtracting the so-called residual water content, θ_r (soil water content for which water is considered immobile), so that the effective saturation, S_e, is defined as

$$S_e = \frac{\theta - \theta_r}{\theta_s - \theta_r} \qquad (4)$$

In addition to the traditional thermogravimetric method to determine soil water content, many other measurement techniques are available, including neutron thermalization, electrical conductivity, dielectric, and heat pulse methods. A recent review on soil moisture measurement methods was presented by Robinson *et al.* (2008), focusing on measurement constraints between the many available methods across spatial scales.

In soils, the driving force for water to flow is the gradient in total water potential. The total potential of bulk soil water can be written as the sum of all possible component potentials, so that the total water potential (ψ_t) is equal to the sum of osmotic, matric, gravitational, and hydrostatic pressure potential. Whereas in physical chemistry the chemical potential of water is usually defined on a molar or mass basis, soil water potential is usually expressed with respect to a unit volume of water, thereby attaining units of pressure (Pa); or per unit weight of water, leading to soil water potential expressed by the equivalent height of a column of water (L). The resulting pressure head equivalent of the combined adsorptive and capillary forces in soils is defined as the matric pressure head, h. When expressed relative to the reference potential of free water, the water potential in unsaturated soils is negative (the soil water potential is less than the water potential of water at atmospheric pressure). Hence, the matric potential decreases or is more negative as the soil water content decreases. In using head units for water potential, the total water potential (H) is defined as the sum of matric potential (h), gravitational potential (z), hydrostatic pressure potential (p), and osmotic potential (π). For most hydrological applications, the contribution of the osmotic potential can be ignored, so that for unsaturated water flow ($p = 0$) the total soil water potential can be written as

$$H = h + z \qquad (5)$$

The measurement of the soil water matric potential *in situ* is difficult and is usually done by tensiometers in the range of matric head values larger (less negative) than -6.0 m. A tensiometer consists of a porous cup, usually ceramic, connected to a water-filled tube (Young and Sisson, 2002). The suction forces of the unsaturated soil draw water from the tensiometer into the soil until the water pressure inside the cup (at pressure smaller than atmospheric pressure) is equal to the pressure equivalent of the soil water matric potential just outside the cup. The water pressure in the tensiometer is usually measured by a vacuum gauge or pressure transducer. Other devices that are used to indirectly measure the soil water matric potential include buried porous units (Scanlon *et al.*, 2002), for which either the electrical resistance or the thermal conductivity is measured *in situ*, after coming into hydraulic equilibrium with the surrounding soil (h in sensor and soil are equal). Although widely used, these types of sensors require laboratory calibration, before field installation.

2.05.2.1 Soil Water Retention

The soil water retention function determines the relation between the volume of water retained by the soil, expressed by θ, and the governing soil matric, or suction forces (Dane and Hopmans, 2002). These suction forces are typically expressed by the soil water matric head (strictly negative) or soil suction (strictly positive). These suction forces increase as the size of the water-filled pores decreases, as may occur by drainage, water uptake by plant roots, or soil evaporation. Also known as the soil water release or soil water characteristic function, this soil hydraulic property describes the increase of θ and the size of the water-filled pores with an increase in matric potential, as occurs by infiltration. Since the matric forces are controlled by pore-size distribution, specific surface area, and type of physico-chemical interactions at the solid–liquid interfaces, the soil water retention curve is very soil specific and highly nonlinear. It provides an estimate of the soil's capacity to hold water after irrigation and free drainage (field capacity), minimum soil water content available to the plant (wilting point), and root zone water availability for plants.

The soil water retention curve exhibits hysteresis, that is, the θ value is different for wetting (infiltration) and drying (drainage).

By way of the unique relationship between soil water matric head and the radius of curvature of the air–water interface in the soil pores, and using the analogy between capillary tubes and the irregular pores in porous media, a relationship can be derived between soil water matric head (h) and effective pore radius, r_e, or

$$\rho g h = \frac{2\sigma \cos \alpha}{r_e} \quad (6)$$

where σ and α are defined as the surface tension and wetting angle of wetting fluid with soil particle surface (typically values for σ and α are 0.072 N m^{-1} and 0°, respectively), ρ is the density of water, and g is the acceleration due to gravity (9.8 m s^{-2}). Because of capillary equation, the effective pore-size distribution can be determined from the soil water retention curve in the region where matric forces dominate. Laboratory and field techniques to measure the soil water retention curve, and functional models to fit the measured soil water retention data, such as the van Genuchten (1980) and Brooks and Corey (1964) models, are described by Kosugi et al. (2002). Alternatively, knowledge of the particle size distribution may provide information on the shape of the soil water retention curve, as presented by Nasta et al. (2009). An example of measured and fitted soil water retention data for two different soils is presented in **Figure 1** (Tuli and Hopmans, 2004).

2.05.2.2 Unsaturated Hydraulic Conductivity

The relation between the soil's unsaturated hydraulic conductivity, K, and volumetric water content, θ, is the second essential fundamental soil hydraulic property needed to describe unsaturated soil water flow. K is a function of the water and soil matrix properties, and controls water infiltration and drainage rates, and is strongly affected by water content and possibly by hysteresis. It is defined by the Darcy–Buckingham equation, which relates the soil water flux density to the total driving force for flow, with K being the proportionality factor. Except for special circumstances, the total driving force for water flow in soils is determined by the sum of the matric and gravitational forces, expressed by the total water potential head gradient, $\Delta H/L$ (L L^{-1}), where ΔH denotes the change in total water potential head over the distance L. For vertical flow, the application of Darcy's law yields the magnitude of water flux from

$$q = -K(\theta)\left(\frac{\mathrm{d}h}{\mathrm{d}z} + 1\right) \quad (7)$$

where q is the Darcy water flux density (L^3 water L^{-2} soil surface T^{-1}) and z defines the vertical position ($z > 0$, upwards, L). A soil system is usually defined by the bulk soil, without consideration of the size and geometry of the individual flow channels or pores. Therefore, the hydraulic conductivity (K) describes the ability of the bulk soil to transmit water, and is expressed by volume of water flowing per unit area of bulk soil per unit time (L T^{-1}).

Functional models for unsaturated hydraulic conductivity are based on pore-size distribution, pore geometry, and connectivity, and require integration of soil water retention functions to obtain analytical expressions for the unsaturated hydraulic conductivity. The resulting expressions relate the relative hydraulic conductivity, K_r, defined as the ratio of the unsaturated hydraulic conductivity, K, and the saturated hydraulic conductivity, K_s, to the effective saturation, S_e, and can be written in the following generalized form (Kosugi et al., 2002):

$$K_r(S_e) = S_e^l \left[\frac{\int_0^{S_e} |h|^{-\eta} \mathrm{d}S_e}{\int_0^{S} |h|^{-\eta} \mathrm{d}S_e}\right]^\gamma \quad (8)$$

where l and η are parameters related to the tortuosity and connectivity of the soil pores, and the value of the parameter γ is determined by the method of evaluating the effective pore radii. For values of $l = 0.5$, $\eta = 1.0$, and $\gamma = 2.0$, Equation (8) reduces to the so-called Mualem (1976) model, that is routinely combined with the van Genuchten (1980) soil water retention model to yield a closed-form expression for the unsaturated hydraulic conductivity function. The moisture dependency is highly nonlinear, with a change in K of five or more orders of magnitude across field-representative changes in unsaturated soil water content. Methods to measure the saturation dependency of the hydraulic conductivity are involved and time consuming. A variety of methods are described in Dane and Topp (2002) and Dirksen (2001). Measurement errors are generally large due to (1) the difficulty of flow measurements in the low water content range and (2) the dominant effect of large pores (macropores), cracks, and fissures in the high water content range. An example of the unsaturated hydraulic conductivity for water, relative to its

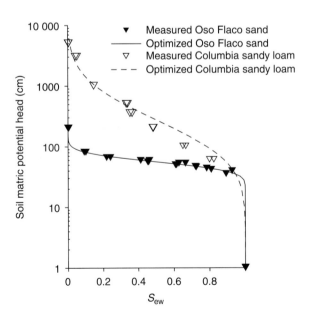

Figure 1 Measured (symbols) and fitted (lines) soil water retention data. From Tuli AM and JW Hopmans (2004) Effect of degree of saturation on transport coefficients in disturbed soils. *European Journal of Soil Science* 55: 147–164.

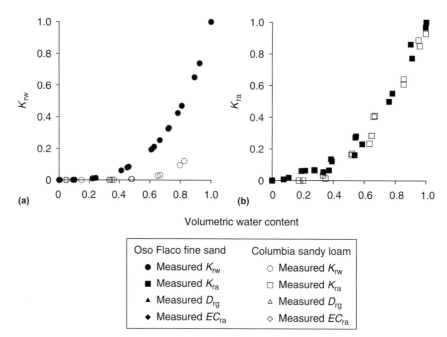

Figure 2 (a) Measured relative hydraulic conductivity for water (K_{rw}) and (b) air conductivity (K_{ra}) as a function of degree of water (S_{ew}) and air (S_{ea}) saturation. From Tuli AM and JW Hopmans (2004) Effect of degree of saturation on transport coefficients in disturbed soils. *European Journal of Soil Science* 55: 147–164.

saturated values (K_{rw}), is presented in Figure 2(a), for the same two soils as in **Figure 1**. We note that θ_s in the vadose zone is typically about 85% of the porosity, so that a saturated soil (e.g., as the result of ponded infiltration) is really a satiated soil due to entrapped air, with a saturated hydraulic conductivity that is significantly smaller than the true K_s.

The unsaturated hydraulic conductivity is related to the intrinsic soil permeability, k (L^2), by

$$K = \frac{\rho g k}{\mu} \qquad (9)$$

where μ denotes the dynamic viscosity of water (F T L^{-2}). The usage of permeability instead of conductivity allows application of the flow equation to liquids other than water with different density and viscosity values. In addition to unsaturated hydraulic conductivity, **Figure 2** also includes data for the saturation dependency of the relative air conductivity (K_{ra}), as might be important for water infiltration in soil, when the soil gas phase is trapped and increasing in pressure, so that water infiltration is partly controlled by soil air permeability (Latifi *et al.*, 1994).

2.05.2.3 Modeling of Unsaturated Water Flow and Transport

Numerous studies have been published addressing different issues in the numerical modeling of unsaturated water flow using the Richards' equation. In short, the dynamic water flow equation is a combination of the Darcy expression and a mass balance formulation. Using various solution algorithms, the soil region of interest is discretized in finite-size elements, i, that can be one, two, or three dimensional, to solve for temporal changes in h, θ, or water flux, q, for each element or voxel i at any time t.

Most multidimensional soil water flow models use a finite-element, Picard time-iterative numerical scheme (Šimunek *et al.*, 2008) to solve the Richards equation. For isotropic conditions and one-dimensional vertical flow, the general water flow equation simplifies to

$$\frac{\partial \theta}{\partial t} = \frac{\partial}{\partial z}\left[K(h)\left(\frac{\partial h}{\partial z} + 1\right)\right] - S(z,t) \qquad (10)$$

where S (L^{3}L^{-3}T^{-1}) is the sink term, accounting for root water uptake. Boundary and initial conditions must be included to allow for specified soil water potentials or fluxes at all boundaries of the soil domain. Richards' equation is a highly nonlinear partial differential equation, and is therefore extremely difficult to solve numerically because of the largely nonlinear dependencies of both water content and unsaturated hydraulic conductivity on the soil water matric head. Both the soil water retention and unsaturated hydraulic conductivity relationships must be known *a priori* to solve the unsaturated water flow equation. Specifically, it will need the slope of the soil water retention curve, or water capacity $C(h)$, defined as $C(h) = \mathrm{d}\theta/\mathrm{d}h$.

As dissolved solutes move through the soils with the water, various physical, chemical, and biological soil properties control their fate. In addition to diffusion and dispersion, fate and transport of chemicals in the subsurface are influenced by sorption to the solid phase and biological transformations. Both diffusion and dispersion of the transported chemical are a function of pore-size distribution and water content. Mechanical or hydrodynamic dispersion is the result of water mixing within and between pores as a result of variations in pore water velocity. Increasing dispersivity values cause greater spreading of the chemical, thereby decreasing peak concentration. Sorbed chemicals move through the vadose zone slower than noninteracting chemicals, and the degree of sorption will largely depend on mineral type, specific surface area of the solid phase, and organic matter fraction. In addition, biogeochemical processes and radioactive decay affect contaminant concentration, such as by cation exchange, mineral precipitation and dissolution, complexation, oxidation–reduction reactions, and by microbial biodegradation and transformations. However, all these mechanisms depend on soil environmental conditions, such as temperature, pH, water saturation, and redox status, and their soil spatial variations. The solute transport equation is generally referred to as the convection–dispersion equation (CDE), and includes the relevant transport mechanisms to simulate and predict temporal changes in soil solute concentration within the simulation domain (Šimunek et al., 2008).

2.05.2.4 Infiltration Processes

For one-dimensional infiltration, the infiltration rate (L T^{-1}), $i(t)$, can be defined by Equation (7) at the soil surface (subscript surf), or

$$i(t) = -K(\theta)\left(\frac{\partial h}{\partial z} + 1\right)_{surf} \quad (11a)$$

Cumulative infiltration $I(t)$, expressed as volume of water per unit soil surface area (L), is defined by

$$I(t) = \int_0^t i(t) dt \quad (11b)$$

Analytical solutions of infiltration generally assume that the wetted soil profile is homogeneous in texture with uniform initial water content. They also make distinction between ponded ($h > 0$ or p) and nonponded soil surface (unsaturated, $h < 0$) infiltration. The infiltration capacity of the soil is defined by $i_c(t)$, the maximum rate at which a soil can absorb water for ponded soil surface conditions. Its maximal value is at time zero, and decreases with time to its minimum value approaching the soil's saturated hydraulic conductivity, K_s, as the total water potential gradient decreases, and tends to unity, with the downward moving wetting front. As defined by Equation (11b), the soil's cumulative infiltration capacity, $I_c(t)$, is defined by the area under the capacity curve. It represents the maximum amount of water that the soil can absorb at any time. Typically, at the onset of infiltration ($t = 0$), the rainfall rate, $r(t)$, will be lower than $i_c(t)$, so that the infiltration rate is equal to the rainfall rate (i.e., $r(t) < i_c(t)$ for $h_{surf} < 0$). If at any point in time, the rainfall rate becomes larger than the infiltration capacity, ponding will occur ($h_{surf} > 0$), resulting in runoff. The time at which ponding occurs is defined as t_p (time to ponding). Thus, the actual infiltration rate will depend on the rainfall rate and its temporal changes. This makes prediction of infiltration and runoff much more difficult for realistic time-variable rainfall patterns.

Therefore, infiltration rate prediction is often described as a function of the cumulative infiltration, I, or $i(I)$, independent of the time domain, and with $i(I)$ curves that are independent of rainfall rate (Skaggs, 1982). An example of such a time-invariant approach is the IDA or infiltrability-depth approximation (Smith et al., 2002). The main IDA assumption is that time periods between small rainfall events are sufficiently small so that soil water redistribution and evaporation between events do not affect infiltration rate. IDA implies that the infiltration rate at any given time depends only on the cumulative infiltration volume, regardless of the previous rainfall history. Following this approach, t_p is defined as the time during a storm event when I becomes equal to $I_c(t_p)$, or

$$R = \int_{t=0}^{t_p} r(t) dt = I_c(t_p)$$

whereas $i(t) = r(t)$ for $t < t_p$. The time invariance of $i(I)$ holds true also when a layered/sealed soil profile is considered (Mualem and Assouline, 1989).

For illustration purposes, we present a hypothetical storm event with time-varying $r(t)$ in **Figure 3** (from Hopmans et al., 2007) in combination with an assumed soil-specific infiltration capacity curve, $i_c(t)$. At what time will ponding occur? It will not be at $t = 7$, when $r(t)$ exceeds i_c for the first time. In order to approximate t_p, we plot both I_c and R for the storm in **Figure 4(a)**, as a function of time and determine t_p as the time at which both curves intersect ($t_p = 13$, for $R = I_c = 110$), since at that time, the cumulative infiltration of the storm is identical to the soil's infiltration capacity. The final corresponding $i(I)$ for this soil and storm event is presented in **Figure 4(b)**, showing that the soil infiltration rate is equal to $r(t)$ until $I = R(t) = I_c(t_p) = 110$, after which the infiltration rate is soil-controlled and determined by $I_c(t)$. More accurate approximations to the time-invariant approach can be found in Sivapalan and Milly (1989) and Brutsaert (2005), using the time compression or time condensation approximation that more accurately estimates infiltration prior to surface ponding.

In addition to whether the soil is ponded or not, solutions of infiltration distinguish between cases with and without gravity effects, as different analytical solutions apply. As Equation (11a) shows, infiltration rate $i(t)$ is determined by both the soil water matric potential gradient, dh/dz, and gravity. However, at the early stage of infiltration into a relatively dry soil, infiltration rate is dominated by the matric potential gradient so that the gravity effects on infiltration can

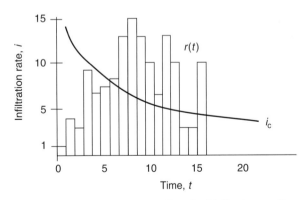

Figure 3 Hypothetical rainfall event, $r(t)$, and soil infiltration capacity, $i_c(t)$. The rainfall event starts at $t = 0$. From Hopmans JW, Assouline S, and Parlange J-Y (2007) Soil infiltration. In: Delleur JW (ed.) *The Handbook of Groundwater Engineering*, pp. 7.1–7.18. Boca Raton, FL: CRC Press.

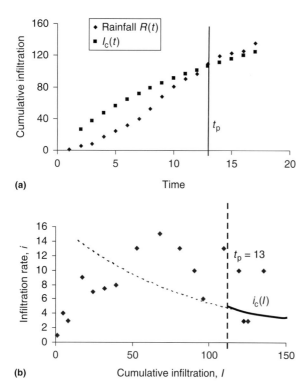

Figure 4 (a) Cumulative infiltration corresponding with infiltration capacity, $I_c(t)$, and cumulative rainfall, $R(t)$ and (b) actual infiltration rate vs. I. Ponding starts only after $t_p = 13$, or $I = 110$. From Hopmans et al. (2007).

horizontal infiltration, I is a linear function of $t^{1/2}$, with S being defined as the slope of this line. Hence, for saturated soil conditions where $\theta_1 = \theta_s$, the infiltration capacity is computed from $i_c(t) = \frac{1}{2}St^{-1/2}$. Incidentally, this also leads to $I_c = S^2/2i_c$.

A relatively simple analytical solution without and with gravity effects was suggested by Green and Ampt (1911) for a ponded soil surface, with $\theta_{surf} = \theta_1$. The assumptions are that the wetting front can be approximated as a step function with a constant effective water potential, h_f, at the wetting front, a wetting zone hydraulic conductivity of $K(\theta_1) = K_1 = K_s$, and a constant soil water profile of $\Delta\theta = \theta_1 - \theta_0$. Using this so-called delta-function assumption of a $D(\theta)$ with a Dirac-delta function form, both solutions for horizontal and vertical infiltration can be relatively easily obtained (Jury et al., 1991; Haverkamp et al., 2007). Assuming that K_0 at the initial water content, θ_0, is negligible, the Green and Ampt (GA) solution of vertical infiltration for ponded conditions is ($h = h_{surf} > 0$):

$$I = I_c = K_1 t + (h_{surf} - h_f)\Delta\theta \ln\left(1 + \frac{I}{(h_{surf} - h_f)\Delta\theta}\right) \quad (14)$$

which can be solved iteratively for I. This simple, yet physically based, solution appears to work best for dry coarse-textured soils. A theoretical expression for the wetting front potential head, h_f, was defined by Mein and Farrell (1974), to yield that $h_f = \int_0^{h_0} K_r(h) dh$, where the relative conductivity $K_r = K(h)/K_s$. The so-called S-form of the GA equation can be obtained by comparing the gravity-free solution of GA with the Boltzmann solution, to yield $S_0^2 = -2K_1 h_f \Delta\theta$:

$$S_1^2(\theta_1) = 2K_1\Delta\theta(h_{surf} - h_f) = S_0^2 + 2K_1 h_{surf}\Delta\theta \quad (15a)$$

so that

$$I = K_1 t + h_{surf}\Delta\theta + \frac{S_0^2}{2K_1}\ln\left(1 + \frac{I}{h_{surf}\Delta\theta + S_0^2/2K_1}\right) \quad (15b)$$

In reality, the wetting front is not a step function, but will consist of a time-dependent transition zone where water content changes from θ_1 to θ_0. The shape of this transition zone will be a function of time and is controlled by soil type. The step function assumption is better for uniform coarse-textured soils that have a Dirac-like $D(\theta)$, for which there is a sharp decline in K with a decrease in water content near saturation. The wetting front is generally much more diffuse for finer-textured soils that have a wide pore-size distribution.

By now, it must be clear that infiltration and its temporal changes are a function of many different soil factors. In addition to rainfall intensity and duration and the soil physical factors, such as soil water retention and hydraulic conductivity, infiltration is controlled by the initial water content, surface sealing and crusting, soil layering, and the ionic composition of the infiltrated water (Kutilek and Nielsen, 1994; Assouline, 2004). For example, Vandervaere et al. (1998) applied the GA model to sealed soil profiles, by assuming that the wetting front potential decreases suddenly as it leaves the seal and enters the soil. This results in a discontinuous drop in the infiltration rate. Many relatively simple infiltration equations have been proposed and are successfully used to

be ignored. Gravity becomes important in the later stages of infiltration, when the wetting front has moved further down. For gravity-free drainage, a simple analytical solution can be found, after transforming Equation (11a) into a θ-based form by defining the diffusivity $D(\theta) = K(\theta)dh/d\theta$, so that

$$i(t) = -D(\theta)\left(\frac{\partial\theta}{\partial z}\right)_{surf} \quad (12)$$

Using the Boltzmann transformation for a constant head boundary condition (Bruce and Klute, 1956), and defining the scaling variable $\varphi = z/t^{1/2}$, combination of Equations (10) without gravity and sink term and (12) resulted in a unique solution of θ as a function of φ, from which the wetting profile can be computed for any time t (Kirkham and Powers, 1972). Defining θ_1 and θ_0 as the surface water content during infiltration and the initial uniform profile water content, respectively, cumulative infiltration, I, is computed from

$$I = \int_{\theta_0}^{\theta_1} z \, d\theta = t^{1/2}\int_{\theta_0}^{\theta_1} \varphi \, d\theta \quad (13a)$$

and results in the simple infiltration equation $I = St^{1/2}$, where the sorptivity S (L T$^{-1/2}$) is defined as

$$S(\theta_1) = \int_{\theta_0}^{\theta_1} \varphi \, d\theta \quad (13b)$$

Equation (13a) states that for gravity-free infiltration during the early times of vertical infiltration, and at all times for characterize infiltration. This has been achieved despite that these equations apply for homogeneous soils only, in theory.

2.05.3 Infiltration Equations

In addition to the solutions in Section 2.05.2, other physically based analytical solutions have been presented, using different assumptions allowing for a closed-form solution. These can potentially be used to predict infiltration from known soil hydraulic properties of homogeneous soils. However, in practice, this is difficult as soil physical characteristics near the soil surface are time dependent because of soil structural changes and their high spatial variability. Alternatively, various empirical infiltration models have been proposed that are very useful for describing measured infiltration data. A parameter sensitivity analysis of many of the presented infiltration models, analyzing the effects of measurement error, was given by Clausnitzer et al. (1998). This section presents the most frequently used infiltration models in both categories.

2.05.3.1 Philip Infiltration Equation

Philip (1957a) presented an analytical infinite-series solution to the water-content-based form of Richards' equation for the case of vertical infiltration:

$$\frac{\partial \theta}{\partial t} = \frac{\partial}{\partial z}\left(D(\theta)\frac{\partial \theta}{\partial z} + K(\theta)\right) \quad (16)$$

For the boundary condition of $h_{surf} = 0$ and $\theta_1 = \theta_s$, the Philip (1957a) solution converged to the true solution for small and intermediate times, but failed for large times. In this case, an alternative solution was presented (Philip, 1957b). With additional assumptions regarding the physical nature of soil water properties, Philip (1987) proposed joining solutions that are applicable for all times. Philip (1957c) introduced a truncation of the small-time series solution that is a simple two-parameter model equation (PH model):

$$I_c = At + St^{1/2} \quad (17a)$$

which should be accurate for all but very large t, and suitable for applied hydrological studies. The sorptivity S depends on several soil physical properties, including initial water content θ_0, and the hydraulic conductivity and soil water retention functions. S is equal to the expression defined in Equation (13b). Philip (1969) showed that A may take values between $0.38K_s$ and $0.66K_s$. The physical interpretation of A is not straightforward; however, for long times when gravity is dominant and $h_{surf} = 0$, one would expect A to be equal to K_s.

Differentiation of Equation (17a) yields the infiltration rate, or

$$i_c = 1/2St^{-0.5} + A \quad (17b)$$

Using (17b) to express t as a function of i_c and substituting in Equation (17a) yields $I(i)$, or

$$I = \frac{S^2(i - A/2)}{2(i - A)^2} \quad (17c)$$

For positive pressure heads (h_{surf}), the correction of Equation (15a) to S can be applied. In many cases, values of S and A are obtained from curve fitting. We note that for gravity-free flow, the pH solution without the gravity term corresponds with the Boltzmann solution for horizontal flow in Equation (13).

2.05.3.2 Parlange et al. Model

Parlange et al. (1982) proposed the following universal model (Parlange et al., model, PA model):

$$t = \frac{S^2}{2K_1^2(1-\delta)}\left[\frac{2K_1}{S^2}I - \ln\frac{\exp\left(\frac{2\delta K_1 I}{S^2}\right) + \delta - 1}{\delta}\right] \quad (18a)$$

assuming that K_0 is small so that the ΔK in Parlange et al. (1982) is equal to K_1. The value of the parameter δ can be chosen to approach various closed-form solutions. For example, Equation (18a) reduces to the GA solution for δ equal to zero. Its value is a function of $K(\theta)$, and is defined by (Parlange et al., 1985):

$$\delta = \frac{1}{\theta_s - \theta_0}\int_{\theta_0}^{\theta_s}\frac{K_s - K(\theta)}{K_s}d\theta \quad (18b)$$

An approximate value of $\delta = 0.85$ was suggested by Parlange et al. (1982) for a range of soil types. After taking the time derivative of I, the following $i(I)$-relationship can be derived (Espinoza, 1999):

$$i = K_1 + \delta K_1\left(1 - \exp\left(\frac{2I\delta K_1}{S^2}\right)\right)^{-1} \quad (18c)$$

Because Equation (18) is based on integration of the water-content-based form of Richards' equation, its theoretical scope is limited to nonponded conditions. A generalization of Equation (18) to include ponded conditions without affecting the value of S was introduced by Parlange et al. (1985). Haverkamp et al. (1990) presented a modification of their model to include upward water flow by capillary rise. The resulting infiltration model contained six physical parameters, in addition to the interpolation parameter δ (Haverkamp et al., 1990). Both the PA and the Haverkamp et al. (1990) model require an iterative procedure to predict $I(t)$. Barry et al. (1995) presented an explicit approximation to the Haverkamp et al. (1990) model, retaining all six physical parameters (BA model):

$$I = K_1 t + \frac{S^2 + 2K_1 h_{surf}\Delta\theta}{2\Delta K}$$
$$\times \left[t^* + 1 - \gamma - \exp\left(\frac{-6(2t^*)^{0.5}}{6 + (2t^*)^{0.5}} - \frac{2t^*}{3}\right)\right.$$
$$+ \frac{\gamma}{1+t^*}\left\{\exp\left(-\frac{2t^*}{3}\right)[1 - (1-\gamma)^8 t^{*2.5}]\right.$$
$$\left.\left. + (2\gamma + t^*)\ln\left(1 + \frac{t^*}{\gamma}\right)\right\}\right] \quad (19a)$$

where

$$t^* = \frac{2t(\Delta K)^2}{S^2 + 2K_1 h_{\text{surf}}\Delta\theta}, \quad \gamma = \frac{2K_1(h_{\text{surf}} + h_a)\Delta\theta}{S^2 + 2K_1 h_{\text{surf}}\Delta\theta} \quad (19b)$$

and h_a denotes the absolute value of the soil water pressure head at which the air phase becomes discontinuous upon wetting. By defining

$$B_1 = (h_{\text{surf}} + h_a)\Delta\theta \quad \text{and} \quad B_2 = \frac{2}{S^2 + 2K_1 h_{\text{surf}}\Delta\theta} \quad (19c)$$

Equation (19a) can be expressed by only four fitting parameters K_0, K_1, B_1, and B_2. The Clausnitzer *et al.* (1998) study concluded that both the PA and BA models described infiltration equally well; however, the BA model, while most advanced, was not as well suited to serve as a fitting model due to nonuniqueness problems caused by the larger number of fitting parameters.

2.05.3.3 Swartzendruber Model

Swartzendruber (1987) proposed an alternative series solution that is applicable and exact for all infiltration times, and also allows for surface ponding. Its starting point is similar to the GA approach; however, its derivation does not require a step function for the wetted soil profile. Its simplified form is a three-parameter infiltration equation (SW model):

$$I = K_1 t + \frac{S}{A_0}\Big(1 - \exp(-A_0 t^{1/2})\Big) \quad (20)$$

where A_0 is a fitting parameter of which its value depends on the surface water content, θ_1. As $A_0 \to 0$, it reduces to a form of the Philip (1957b) model with K_1 as the coefficient of the linear term, and for which dI/dt approaches K_1 as $t \to \infty$. As for the GA model, the S-term can be corrected using Equation (15a) to account for ponded conditions.

2.05.3.4 Empirical Infiltration Equations

For most of these types of infiltration equations, the fitting parameters do not have a physical meaning and are evaluated by fitting to experimental data only. However, in many cases, the specific form of the infiltration equation is physically intuitive. For example, the empirical infiltration equation by Horton (1940) is one the most widely used empirical infiltration equations. It considers infiltration as a natural exhaustion process, during which infiltration rate decreases exponentially with time from a finite initial value, $i_c|_{t=0} = (\alpha_1 + \alpha_2)$, to a final value, $\alpha_1 = K_1$. Accordingly, cumulative infiltration I (L) is predicted as a function of time t (HO model):

$$I = \alpha_1 t + \frac{\alpha_2}{\alpha_3}[1 - \exp(-\alpha_3 t)] \quad (21)$$

with the soil parameter $\alpha_3 > 0$, representing the decay of infiltration rate with time. In Equation (21), α_1 can be associated with the hydraulic conductivity (LT^{-1}) of the wetted soil portion, K_1, for $t \to \infty$.

Another simple empirical infiltration equation is the Kostiakov (1932) model (KO):

$$i = at^{-b} \quad (22)$$

Clearly, this equation will not fit infiltration data at long times, as it predicts zero infiltration rate as $t \to \infty$. The value of a should be equal to the infiltration rate at $t = 1$, and $0 < b < 1$.

Mezencev (1948) proposed another infiltration model, and modified the KO model by including a linear term with a coefficient β_1, so that $\beta_1 \to K_1$ for $t \to \infty$ provided $0 < \beta_3 < 1$ and $\beta_2 > 0$ (ME model):

$$I = \beta_1 t + \frac{\beta_2}{1 - \beta_3}t(1 - \beta_3) \quad (23)$$

Other models include the Soil Conservation Service (1972) method and the Holtan solution (Kutilek and Nielsen, 1994; Espinoza, 1999).

2.05.4 Measurements

2.05.4.1 Infiltration

Infiltration measurements can serve various purposes. In addition to characterizing infiltration, for example, to compare infiltration between different soil types, or to quantify macropore flow, it is often measured to estimate the relevant soil hydraulic parameters from the fitting of the infiltration data to a specific physically based infiltration model. This is generally known as inverse modeling. Infiltration is generally measured using one of three different methods: a sprinkler method, a ring infiltration method, or a permeameter method. The sprinkler method is mostly applied to determine time of ponding for different water application rates, whereas the ring infiltrometer method is used when the infiltration capacity is needed. The permeameter method provides a way to measure infiltration across a small range of h-values ≤ 0. A general review of all three methods was recently presented by Smettem and Smith (Smith *et al.*, 2002), whereas a comparison of different infiltration devices using seven criteria was presented by Clothier (2001).

Rainfall sprinklers or rainfall simulators are also sprinkler infiltrometers, but they are typically used to study runoff and soil erosion (e.g., Morin *et al.*, 1967). They mimic the rainfall characteristics (e.g., kinetic energy) of natural storms, specifically the rainfall rate, rainfall droplet size distribution, and drop velocity. Most of these devices measure infiltration by subtracting runoff from applied water. Using a range of water application rates, infiltration measurements can be used to determine the $i(I)$ curve for a specific soil type, with specific soil hydraulic properties such as K_s or S. Various design parameters for many developed rainfall simulators, specifically nozzle systems, were presented by Peterson and Bubenzer (1986). A portable and inexpensive simulator for infiltration measurements along hillslopes was developed by Battany and Grismer (2000). This low-pressure system used a hypodermic syringe needle system to form uniform droplets at rainfall intensities ranging from 20 to 90 mm h^{-1}.

Ring infiltrometers have historically been used to characterize soil infiltration by determining the infiltration capacity, i_c. A ring is carefully inserted in the soil so that water can be ponded over a known area. Since a constant head is required, a constant water level is maintained either by manually adding water and using a measuring stick to maintain a constant depth of ponded water, by using a Mariotte system, or by a valve connected to a float that closes at a predetermined water level. Measurements are usually continued until the infiltration rate is essentially constant. Water seepage around the infiltrometer is prevented by compaction of the soil around and outside of the infiltrometer. Multidimensional water flow under the ring is minimized by pushing the ring deeper into the soil, or by including an outer buffer ring. In the latter case, the soil between the two concentric rings is ponded at the same depth as the inner ring, to minimize lateral flow directed radially outward. The deviation from the assumed one-dimensionality depends on ring insertion depth, ring diameter, measurement time and soil properties such as its hydraulic conductivity, and the presence of restricting soil layers. A sensitivity analysis on diverging flow of infiltrometers was presented by Bouwer (1986) and Wu et al. (1997).

Permeameters are generally smaller than infiltrometers and allow easy control of the soil water pressure head at the soil surface. Generally, multidimensionality of flow must be taken into account, using Wooding's (1968) equation for steady flow (Q_∞, L^3 T^{-1}) from a shallow, circular surface pond of free water, or

$$Q_\infty = K_s \left(\pi r_0^2 + \frac{4 r_0}{\alpha} \right) \quad (24a)$$

The first and second terms in parentheses denote the gravitational and capillary components of infiltration and α denotes the parameter in Gardner's (1958) unsaturated hydraulic conductivity function:

$$K(h) = K_s \exp(\alpha h) \quad (24b)$$

In this model of the so-called Gardner soil, the macroscopic capillary length, λ_c, is equivalent to $1/\alpha$. The basic analysis for most permeameter methods relies on Wooding's solution. An extensive review of the use of permeameters was presented by Clothier (2001), including the tension infiltrometers and disk permeameters, by which the soil water pressure at water entry is controlled by a bubble tower. Their use is relatively simple, and based on analytical solutions of steady-state water flow. The permeameter method is economical in water use and portable. The soil hydraulic properties (S and K), in an inverse way, can be inferred from measurements using (1) both short- and long-time observations, (2) disks with various radii, or (3) using multiple water pressure heads. Transient solutions of infiltration may be preferable, as it allows analysis of shorter infiltration times, so that the method is faster and likely will better satisfy the homogeneous soil assumption. Differences between one- and three-dimensional solutions for transient infiltration were analyzed by Haverkamp et al. (1994), Vandervaere et al. (2000), and Smith et al. (2002) from multidimensional numerical modeling analysis. These effects were reported to be small if gravity effects were included.

Nowadays, permeameters are most often applied to estimate the soil's hydraulic characteristics in an inverse way, by fitting infiltration data to analytical solutions. In many cases, auxiliary water content or matric potential data are required to yield unique solutions.

2.05.4.2 Unsaturated Water Flow

Whereas infiltration measures are typically conducted along the soil surface only, measurement of unsaturated water flow requires installation of instruments and sensors below ground, thereby largely complicating measurement procedures and analysis. The simplest expression for unsaturated water flow estimation is the Darcy equation (7), but still requires the measurement of soil water content (θ) or soil water matric potential (h) at various soil depths, and knowledge of the unsaturated hydraulic function, $K(\theta)$, as expressed by Equation (8). Installation of soil moisture or potential sensors requires extreme care, because of issues of soil disturbance, inadequate soil sensor contact, and inherent soil heterogeneities. In addition, it is not always straightforward to determine installation depth of sensors, as it will depend on *a priori* knowledge of soil horizon differentiation. Inherently problematic is the fact that no soil water flux meters are available to accurately measure the unsaturated soil water flux q in Equation (7). A review by Gee et al. (2003) provides possible direct and indirect methods, but none of them are adequate because of problems with divergence of water flow near the flux measurement device. Recently, the heat pulse probe was developed (Kamai et al., 2008) for indirect measurement of soil water flux, but is limited to fluxes of 6 mm d^{-1} or higher. Finally, very few routine measurements are available to determine the $K(\theta)$ relationship. In fact, the lack of the unsaturated conductivity information is the most limiting factor of *in situ* application of the Darcy equation. Most promising is the application of inverse modeling for parameter estimation of the soil hydraulic functions, using both laboratory and field techniques (Hopmans et al., 2002b), which can be used in conjunction with *in situ* water content and soil water potential measurements to estimate temporal changes in depth distribution of soil water flux.

Selected steady-state solutions are provided in Jury et al. (1991), but are only of limited use for real field conditions since soil water content and matric potential values change continuously. Most realistically, one must apply the transient unsaturated water flow (Equation (10)) that arises from combination of the Darcy equation with mass conservation. However, its solution also requires *a priori* knowledge of the soil water capacity, C, as determined from the slope of the soil water retention curve, and time measurements of θ and h, at the various soil horizon interfaces and at the boundaries of the soil domain of interest, including at the soil profile bottom. Although certainly possible, relatively few of such field experiments are conducted routinely because they are time consuming and wrought with complications. However, in combination with inverse modeling, such field experiments can provide a wealth of information, including plant root water uptake dynamics, plant transpiration, and drainage rates (Vrugt et al., 2001). Therefore, large lysimeters with selected water content and soil water potential measurements may be very useful.

2.05.5 Scaling and Spatial Variability Considerations

Soil hydrologists need to apply locally measured soil physical data to characterize flow and transport processes at large-scale heterogeneous vadose zones. For example, prediction of soil water dynamics, such as infiltration at the field scale, is usually derived from the measurement of soil hydraulic properties from laboratory cores, as collected from a limited number of sampling sites across large spatial extents. Soil parameters obtained from these small-scale measurements are subsequently included in numerical models with a grid or element size many times larger, with the numerical results extrapolated to predict large-scale flow and transport behavior. Because of the typical nonlinearity of soil physical properties, their use across spatial scales is inherently problematic. Specifically, the averaging of processes determined from discrete small-scale samples may not describe the true soil behavior involving larger spatial structures. Moreover, the dominant physical flow processes may vary between spatial scales. Considering that soil physical, chemical, and biological measurements are typically conducted for small measurement volumes and that the natural variability of soils is enormous, the main question asked is how small-scale measurements can provide information about large-scale flow and transport behavior. In their treatise of scale issues of vadose zone modeling, Hopmans et al. (2002a) offer a conceptual solution, considering the control of small-scale processes on larger-scale flow behavior. Hence, vadose zone properties are nonunique and scale dependent, resulting in effective properties that vary across spatial scales and merely serve as calibration parameters in simulation models. Therefore, their accurate prediction in heterogeneous materials can only be accomplished using scale-appropriate measurements, including those that measure at the landscape scale.

In addition, infiltration measurements are typically conducted at measurement scales in the range of 0.2–1.0 m. This is relevant for irrigation purposes, especially for micro-irrigation applications. Yet, infiltration information is often needed for much larger spatial scales, at the pedon scale, hillslope scale, and watershed scale. Very little work has been done relating infiltration process to measurement or support scale. Exceptions are the studies by Sisson and Wierenga (1981) and Haws et al. (2004), who measured steady-state infiltration at three spatial scales, ranging from 5 to 127-cm-diameter infiltrometer rings. Their results showed that much of the larger-scale infiltration occurs through smaller-scale regions, and that the spatial variability of infiltration decreased as the measurement scale increased. Thus, in general, we find that the process of infiltration might vary with spatial scale, and that larger spatial scales are required to estimate representative infiltration characteristics across a typical landscape.

Many field studies have dealt with the significant areal heterogeneity of soil hydraulic properties, and particularly that of the saturated hydraulic conductivity, K_s (Nielsen et al., 1973). The heterogeneity in K_s is recognized to have a major effect on unsaturated flow, leading to significant variation in local infiltration. In general, accounting for areal heterogeneity leads to shorter ponding times and to a more gradual decrease of the infiltration flux with time (Smith and Hebbert, 1979; Sivapalan and Wood, 1986). To characterize spatial variable infiltration rates, Sharma et al. (1980) measured infiltration with a double-ring infiltrometer at 26 sites in a 9.6-ha watershed. The infiltration data were fitted to the PH infiltration Equation (17a), and fitting parameters S and A were scaled to express their spatial variability and to describe the ensemble-average or composite infiltration curve of the watershed. A simpler but similar scaling technique for infiltration data was presented by Hopmans (1989), who measured transient infiltration at 50 sites along a 100-m transect. Data were fitted to both the PH and a modified KO model that includes an additional constant c as a second term in Equation (23). This paper showed that spatial variability of infiltration can be easily described by the probability density function of a single scaling parameter, to be used for applications in Monte Carlo simulation of watershed hydrology, as suggested for the first time by Peck et al. (1977). For application at the field scale, the so-called one-point method was presented by Shepard et al. (1993) to estimate furrow-average infiltration parameters of PH Equation (17a), across a furrow-irrigated agricultural field. They used the volume-balance principle from furrow advance time across the field, water inflow rate, and flow area measurements.

For modeling surface hydrology, by subtracting the infiltration rate, $i(t)$, from the rainfall rate, $r(t)$, it is possible to estimate spatial and temporal distributions of rainfall excess or runoff. The influence of spatial heterogeneity in rainfall and soil variability on runoff production was studied by Sivapalan and Wood (1986) from an analytical solution of infiltration and making use of the IDA approximation. Statistical characteristics of ponding time and infiltration rate were presented for two cases, one with a spatially variable soil with a lognormal K_s distribution and uniform rainfall, and the other for a homogeneous soil with spatially variable rainfall. Among the various results, this study concluded that the ensemble infiltration approach is biased for spatially variable soils. Their results also showed that the cumulative distribution of ponding times or proportion of ponded area is an excellent way of analyzing mean areal infiltration. Moreover, the spatial correction of infiltration rate is time dependent and varies depending on the correlation lengths of rainfall and soil K_s. This study neglected the effects of surface water run-on, as caused by accumulated water upstream, running on to neighboring areas, thereby contributing locally to infiltration. A quantitative analysis of soil variability effects on watershed hydraulic response that included surface water interactions, such as run-on, was presented by Smith and Hebbert (1979), through analysis of the effects of deterministic changes of infiltration properties in the direction of surface water flow, using a kinematic watershed model. In a subsequent study by Woolhiser et al. (1996), it was clearly demonstrated that runoff hydrographs along a hillslope are significantly affected by spatial trends in the soil's saturated hydraulic conductivity. We expect that important new information can be collected by linking this interactive modeling approach with remote sensing and geographical information system (GIS) tools. A detailed analysis and review of the control of spatially variable hydrologic properties on overland flow are presented by Govindaraju *et al.* (2007).

Yet another concern regarding nonideal infiltration, causing spatially variable infiltration at small spatial scales, comes from the presence of water-repellent or hydrophobic soils. Since the 1980s much new research and findings have been presented, improving the understanding of the underlying physical processes and its relevance to soil water flow and water infiltration (DeBano, 2000; Wang *et al.*, 2000). Infiltration may be controlled by soil surface crust-forming dynamics, which is another complex phenomenon dominated by a wide variety of factors involving soil properties, rainfall characteristics, and local water flow conditions. Two types of rainfall-induced soil seals can be identified: (1) structural seals that are directly related to rainfall through the impact of raindrops and sudden wetting and (2) depositional or sedimentary seals that are indirectly related to rainfall as it results from the settling of fine particles carried in suspension by runoff in soil depressions. A recent review on concepts and modeling of rainfall-induced soil surface sealing was presented by Assouline (2004).

2.05.6 Summary and Conclusions

Although important and seemingly simple, infiltration is a complicated process that is a function of many different soil properties, rainfall, land use, and vegetation characteristics. In addition to rainfall intensity and duration as well as the soil physical factors, such as soil water retention and hydraulic conductivity, infiltration is controlled by the initial water content, surface sealing and crusting, hydrophobicity, soil layering, and the ionic composition of the infiltrated water. Many relatively simple infiltration equations have been proposed historically, and are successfully used to characterize infiltration. Other physically based analytical solutions have been presented that can potentially be used to predict infiltration. However, in practice, this is difficult as soil physical characteristics near the soil surface show naturally high soil spatial variability and are often time dependent because of soil structural changes. Alternatively, infiltration is often measured to estimate the relevant soil hydraulic parameters from the fitting of the infiltration data to a specific infiltration model by inverse modeling, such as by using permeameters.

Whereas most infiltration measurement techniques and infiltration models apply to relatively small spatial scales, infiltration information is often needed at the watershed and hillslope scales. Yet, it has been shown that much of the larger-scale infiltration occurs through smaller-scale regions, for example, because infiltration is largely controlled by spatial variations of the soil's physical characteristics at the land surface, vegetation cover, and topography. In general, we expect that the process of infiltration varies with spatial scale, and that measurements at larger spatial scales are needed to estimate representative infiltration characteristics across hillslope and larger spatial scales. For that purpose, improved solutions to infiltration across scales from the field to basin scale are needed, such as may become available using rapidly developing techniques including remote sensing, GIS, and new measurement devices.

Acknowledgments

This chapter is partly based on the paper by Hopmans *et al.* (2007), and includes edited sections of that paper. The author acknowledges the significant input received by Drs. J.-Y Parlange and S. Assouline in writing the 2007 paper.

References

Assouline S (2004) Rainfall-induced soil surface sealing: A critical review of observations, conceptual models and solutions. *Vadose Zone Journal* 3: 570–591.

Barry DA, Parlange J-Y, Haverkamp R, and Ross PJ (1995) Infiltration under ponded conditions: 4. An explicit predictive infiltration formula. *Soil Science* 160: 8–17.

Battany MC and Grismer ME (2000) Development of a portable field rainfall simulator for use in hillside vineyard runoff and erosion studies. *Hydrological Processes* 14: 1119–1129.

Bouwer H (1986). Intake rate: Cylinder infiltrometer, In: Klute A, (ed.) *Methods of Soil Analysis, Part 1*. Number 9 in the Series Agronomy, pp. 825–844. Madison, WI: American Society of Agronomy.

Brooks RH and Corey AT (1964) *Hydraulic Properties of Porous Media*, Hydrology Paper No. 3. Fort Collins, CO: Colorado State University.

Bruce RR and Klute A (1956) The measurement of soil moisture diffusivity. *Soil Science Society American Proceedings* 20: 458–462.

Brutsaert W (2005) *Hydrology – An Introduction*. New York, NY: Cambridge University Press.

Clausnitzer V, Hopmans JW, and Starr JL (1998) Parameter uncertainty analysis of common infiltration models. *Soil Science Society of America Journal* 62: 1477–1487.

Clothier BE (2001) Infiltration. In: Smith KA and Mullins CE (eds.) *Soil and Environmental Analysis, Physical Methods*, 2nd edn., Revised and Expanded, pp. 239–280. New York: Dekker.

Dane JH and Hopmans JW (2002) Soil water retention and storage – introduction. In: Dane JH and Topp GC (eds.) *Methods of Soil Analysis. Part 4. Physical Methods*, pp. 671–674. Madison, WI: Soil Science Society of America.

Dane JH and Topp GC (eds.) (2002) *Methods of Soil Analysis. Part 4. Physical Methods*, vol. 5, Madison, WI: Soil Science Society of America.

DeBano LF (2000) Water repellency in soils: A historical overview. *Journal of Hydrology* 231: 4–32.

Dirksen C (2001) Hydraulic conductivity. In: Smith KA and Mullins CE (eds.) *Soil and Environmental Analysis*, pp. 141–238. New York: Dekker.

Espinoza RD (1999) Infiltration. In: Delleur JW (ed.) *The Handbook of Groundwater Engineering*, pp. 7.1–7.18. Boca Raton, FL: CRC Press.

Gardner WR (1958) Some steady state solutions of unsaturated moisture flow equations with application to evaporation from a water table. *Soil Science* 85: 228–232.

Gee GW, Zhang F, and Ward AL (2003) A modified vadose zone fluxmeter with solution collection capability. *Vadose Zone Journal* 2: 627–632.

Govindaraju RS, Nahar N, Corradini C, and Morbidelli R (2007) Infiltration and run-on under spatially-variable hydrologic properties. In: Delleur JW (ed.) *The Handbook of Groundwater Engineering*, pp. 8.1–8.15. Boca Raton, FL: CRC Press.

Green WA and Ampt GA (1911) Studies on soils physics: 1. The flow of air and water through soils. *Journal of Agricultural Science* 4: 1–24.

Haverkamp R, Debionne S, Viallet P, Angulo-Jaramillo R, and de Condappa D (2007) Soil properties and moisture movement in the unsaturated zone. In: Delleur JW (ed.) *The Handbook of Groundwater Engineering*, pp. 6.1–6.59. Boca Raton, FL: CRC Press.

Haverkamp R, Parlange J-Y, Starr JL, Schmitz G, and Fuentes C (1990) Infiltration under ponded conditions: 3. A predictive equation based on physical parameters. *Soil Science* 149: 292–300.

Haverkamp R, Ross PJ, Smettem KRJ, and Parlange J-Y (1994) Three-dimensional analysis of infiltration from the disc infiltrometer. 2. Physically based infiltration equation. *Water Resources Research* 30: 2931–2935.

Haws NW, Boast CW, Rao PSC, Kladivko EJ, and Franzmeier DP (2004) Spatial variability and measurement scale of infiltration rate on an agricultural landscape. *Soil Science Society of America Journal* 68: 1818–1826.

Hopmans JW (1989) Stochastic description of field-measured infiltration data. *Transactions of the American Society of Agricultural Engineers* 32: 1987–1993.

Hopmans JW, Assouline S, and Parlange J-Y (2007) Soil infiltration. In: Delleur JW (ed.) *The Handbook of Groundwater Engineering*, pp. 7.1–7.18. Boca Raton, FL: CRC Press.

Hopmans JW, Nielsen DR, and Bristow KL (2002a) How useful are small-scale soil hydraulic property measurements for large-scale vadose zone modeling. In: Smiles D, Raats PAC, and Warrick A (eds.) *Heat and Mass Transfer in the Natural Environment, the Philip Volume*. Geophysical Monograph Series No. 129, pp. 247–258. Washington, DC: American Geophysical Union.

Hopmans JW, Šimunek J, Romano N, and Durner W (2002b) Inverse methods. In: Dane JH and Topp GC (eds.) *Methods of Soil Analysis. Part 4. Physical Methods*, pp. 963–1008. Madison, WI: Soil Science Society of America.

Horton RE (1940) An approach towards a physical interpretation of infiltration capacity. *Soil Science Society American Proceedings* 5: 399–417.

Jury WA, Gardner WR, and Gardner WH (1991) *Soil Physics*. New York: Wiley.

Kamai T, Tuli A, Kluitenberg GJ, and Hopmans JW (2008) Soil water flux density measurements near 1 cm/day using an improved heat pulse probe. *Water Resources Research* 44: doi: 10.1029/2008WR007036.

Kirkham D and Powers WL (1972) *Advanced Soil Physics*. New York: Wiley.

Kostiakov AN (1932) On the dynamics of the coefficient of water percolation in soils and on the necessity of studying it from a dynamic point of view for purposes of amelioration. In: *Transactions of the Sixth Commission of the International Society of Soil Science A*, pp. 17–21.

Kosugi K, Hopmans JW, and Dane JH (2002) Water retention and storage – parametric models. In: Dane JH and Topp GC (eds.) *Methods of Soil Analysis. Part 4. Physical Methods*, pp. 739–758. Madison, WI: Soil Science Society of America.

Kutilek M and Nielsen DR (1994) *Soil Hydrology. GeoEcology Textbook*. Cremlingen-Destedt. Germany: Catena Verlag.

Latifi H, Prasad SN, and Helweg OJ (1994) Air entrapment and water infiltration in two-layered soil column. *Journal of Irrigation and Drainage Engineering* 120: 871–891.

Mein RG and Farrell DA (1974) Determination of wetting front suction in the Green–Ampt equation. *Soil Science Society of America Proceedings* 38: 872–876.

Mezencev VJ (1948) Theory of formation of the surface runoff (Russian). *Meteorologia i Gidrologia* 3: 33–40.

Morin J, Goldberg D, and Seginer I (1967) A rainfall simulator with a rotating disc. *Transactions of the American Society of Agricultural Engineers* 10: 74–77.

Mualem Y (1976) A new model for predicting the hydraulic conductivity of unsaturated porous media. *Water Resources Research* 12: 513–522.

Mualem Y and Assouline S (1989) Modeling soil seal as a non-uniform Layer. *Water Resources Research* 25: 2101–2108.

Nasta P, Kamai T, Chirico GB, Hopmans JW, and Romano N (2009) Scaling soil water retention functions using particle-size distribution. *Journal of Hydrology* 374: 223–234.

Nielsen DR, Biggar JB, and Ehr KT (1973) Spatial variability of field measured soil water properties. *Hilgardia* 42: 215–260.

Parlange J-Y, Haverkamp R, and Touma J (1985) Infiltration under ponded conditions: 1. Optimal analytical solution and comparison with experimental observations. *Soil Science* 139: 305–311.

Parlange J-Y, Lisle I, Braddock RD, and Smith RE (1982) The three-parameter infiltration equation. *Soil Science* 133: 337–341.

Peck AJ, Luxmoore RJ, and Stolzy JL (1977) Effects of spatial variability of soil hydraulic properties in water budget modeling. *Water Resources Research* 13: 348–354.

Peterson AE and Bubenzer GD (1986). Intake rate: Sprinkler infiltrometer, In: Klute A, (ed.) *Methods of Soil Analysis, Part 1*. Number 9 in the series Agronomy, pp. 45–870. Madison, WI: American Society of Agronomy.

Philip JR (1957a) The theory of infiltration: 1. The infiltration equation and its solution. *Soil Science* 83: 345–357.

Philip JR (1957b) The theory of infiltration: 2. The profile at infinity. *Soil Science* 83: 435–448.

Philip JR (1957c) The theory of infiltration: 4. Sorptivity and algebraic infiltration equations. *Soil Science* 84: 257–264.

Philip JR (1969) Theory of infiltration. In: Chow VT (ed.) *Advances in Hydroscience*, vol. 5, pp. 215–296. New York, NY: Academic Press.

Philip JR (1987) The infiltration joining problem. *Water Resources Research* 12: 2239–2245.

Robinson DA, Campbell CS, Hopmans JW, *et al.* (2008) Soil moisture measurement for ecological and hydrological watershed-scale observatories: A review. *Vadose Zone Journal* 7: 358–389.

Scanlon BR, Andraski BJ, and Bilskie J (2002) Miscellaneous methods for measuring matric or water potential. In: Dane JH and Topp GC (eds.) *Methods of Soil Analysis. Part 4. Physical Methods*, pp. 643–670. Madison, WI: Soil Science Society of America.

Sharma ML, Gander GA, and Hunt CG (1980) Spatial variability of infiltration in a watershed. *Journal of Hydrology* 45: 101–122.

Shepard JS, Wallender WW, and Hopmans JW (1993) One-point method for estimating furrow infiltration. *Transactions of American Society of Agricultural Engineers* 36: 395–404.

Šimunek J, Van Genuchten MTh, and Sejna M (2008) Development and applications of the HYDRUS and STANMOD software packages and related codes. *Vadose Zone Journal* 7: 587–600.

Sisson JB and Wierenga PJ (1981) Spatial variability of steady-state infiltration rates as a stochastic process. *Soil Science Society of America Journal* 45: 699–704.

Sivapalan M and Milly PCD (1989) On the relationship between the time condensation approximation and the flux-concentration relation. *Journal of Hydrology* 105: 357–367.

Sivapalan M and Wood EF (1986) Spatial heterogeneity and scale in the infiltration response of catchments. In: Gupta VK, Rodriguez-Iturbe I, and Wood EF (eds.) *Scale Problems in Hydrology*, pp. 81–106. Hingham, MA: Reidel.

Skaggs RW (1982) Infiltration, In: Haan CT, Johnson HP, and Brakensiek DL, (eds.) *Hydrologic Modeling of Small Watersheds*, ASAE Monograph No. 5, 121–166. St. Joseph, MI: ASAE.

Smith RE and Hebbert RHB (1979) A Monte-Carlo analysis of the hydrologic effects of spatial variability of infiltration. *Water Resources Research* 15: 419–429.

Smith RE, Smettem KRJ, Broadbridge P, and Woolhiser DA (2002) *Infiltration Theory for Hydrologic Applications*. Water Resources Monograph 15, Washington, DC: American Geophysical Union.

Soil Conservation Service (1972) Estimation of direct runoff from storm rainfall *National Engineering Handbook, Section 4: Hydrology*, pp. 10.1–10.24. Washington, DC: USDA.

Swartzendruber D (1987) A quasi-solution of Richards' equation for the downward infiltration of water into soil. *Water Resources Research* 23: 809–817.

Tuli AM and Hopmans JW (2004) Effect of degree of saturation on transport coefficients in disturbed soils. *European Journal of Soil Science* 55: 147–164.

Vandervaere J-P, Vauclin M, and Elrick DE (2000) Transient flow from tension infiltrometers: I. The two-parameter equation. *Soil Science Society of America Journal* 64: 1263–1272.

Vandervaere J-P, Vauclin M, Haverkamp R, Peugeot C, Thony J-L, and Gilfedder M (1998) Prediction of crust-induced surface runoff with disc infiltrometer data. *Soil Science* 163: 9–21.

Van Genuchten MTh (1980) A closed-form equation for predicting the hydraulic conductivity of unsaturated soils. *Soil Science Society of America Journal* 44: 892–898.

Vrugt JA, Hopmans JW, and Šimunek J (2001) Calibration of a two-dimensional root water uptake model. *Soil Science Society of America Journal* 65: 1027–1037.

Wang Z, Wu QJ, Wu L, Ritsema CJ, Dekker LW, and Feyen J (2000) Effects of soil water repellency on infiltration rate and flow instability. *Journal of Hydrology* 231: 265–276.

Wooding RA (1968) Steady infiltration from a shallow circular pond. *Water Resources Research* 4: 1259–1273.

Woolhiser DA, Smith RE, and Giraldez J-V (1996) Effects of spatial variability of saturated hydraulic conductivity on Hortonian overland flow. *Water Resources Research* 32: 671–678.

Wu L, Pan L, Robertson MJ, and Shouse PJ (1997) Numerical evaluation of ring-infiltrometers under various soil conditions. *Soil Science* 162: 771–777.

Young MH and Sisson JB (2002) Tensiometry. In: Dane JH and Topp GC (eds.) *Methods of Soil Analysis, Part 4: Physical Methods*, pp. 575–606. Madison, WI: Soil Science Society of America.

Relevant Websites

http://www.decagon.com
  Decagon Devices, Mini-Disk Infiltrometer.
http://hopmans.lawr.ucdavis.edu
  Jan W. Hopmans, Vadose Zone Hydrology.
http://www.pc-progress.com
  PC-Progress: Engineering Software Developer; HYDRUS 2D/3 D for Windows, Version 1.xx.
http://ag.arizona.edu/sssa-s1
  SSSA Soil Physics Division S-1.
http://en.wikipedia.org
  Wikipedia, Infiltration (Hydrology).

2.06 Mechanics of Groundwater Flow

M Bakker, Delft University of Technology, Delft, The Netherlands
El Anderson, WHPA, Bloomington, IN, USA

© 2011 Elsevier B.V. All rights reserved.

2.06.1	Introduction	115
2.06.2	Brief History	116
2.06.3	Hydraulic Head	116
2.06.4	Darcy's Law	116
2.06.5	Steady Conservation of Mass	118
2.06.6	Flow Types	118
2.06.6.1	Spatial Dimension	118
2.06.6.2	Time Dependence	118
2.06.6.3	Geologic Setting	118
2.06.7	The Dupuit Approximation	119
2.06.8	Potential Flow and the Discharge Vector	120
2.06.9	One-Dimensional Flow	120
2.06.9.1	Confined Flow between Two Rivers	121
2.06.9.2	Combined Flow between Two Rivers	121
2.06.9.3	Unconfined Flow in a River Valley	121
2.06.10	One-Dimensional Radial Flow	122
2.06.10.1	Flow to a Well at the Center of a Circular Island without Recharge	122
2.06.10.2	Recharge on a Circular Island	122
2.06.10.3	Well at the Center of a Circular Island with Recharge	122
2.06.11	The Principle of Superposition	123
2.06.11.1	A Well in Uniform Flow	123
2.06.11.2	The Method of Images	124
2.06.11.3	Flow to a Pumping Well in an Alluvial Valley	125
2.06.12	The Stream Function and the Complex Potential	126
2.06.12.1	Evaluation of the Capture Zone Envelope Using the Complex Potential	128
2.06.13	Transient Flow	128
2.06.13.1	One-Dimensional Periodic Flow	129
2.06.13.2	Transient Wells	130
2.06.13.3	Convolution	130
2.06.14	Computer Models	132
2.06.15	Discussion	133
Acknowledgments		133
References		133

2.06.1 Introduction

Groundwater is the most important resource of freshwater on earth. It moves very slowly through the top part of the earth's crust from areas of recharge (often originating from rainfall) to discharge in springs, wells, rivers, lakes, and oceans. The baseflow of rivers, the flow between rainfall or snowmelt events, is caused predominantly by inflow from groundwater. In many parts of the world, groundwater is the only source for drinking water or irrigation. Groundwater resources are threatened by over-exploitation and contamination. Major problems include a rapid decline of the groundwater table caused by pumping of groundwater for irrigated agriculture, salinization of groundwater resources due to heavy pumping in coastal areas, and contamination of groundwater by leakage of toxic chemicals. Accurate tools are needed to predict whether the current and proposed uses of groundwater resources are sustainable and safe.

The field of groundwater flow, also called hydrogeology, is large and only the basic physical principles of groundwater flow through porous media are discussed in this chapter. Detailed textbooks include Verruijt (1970), Bear (1972), Strack (1989), and Fitts (2002). This chapter focuses on groundwater flow through porous materials such as sand, silt, or clay. Significant amounts of groundwater may flow through fractured rock formations. The concepts outlined in this chapter apply to such formations when the fractured rock may be represented by an equivalent porous medium.

Compared to other areas of hydrology, the governing equations for groundwater flow are relatively well known. Exact solutions can be obtained for many important flow systems. These exact solutions provide important insights into the flow of groundwater and groundwater interactions with the accessible environment. This chapter begins with a description of the governing equations. Flow principles are explained through discussion of a set of steady and transient flow problems. This chapter concludes with a brief discussion of available modeling tools for solving more complicated problems.

2.06.2 Brief History

The foundation of the quantitative description of groundwater flow was laid by Henry Darcy and Jules Dupuit. Darcy (1856) peformed column experiments that led to what is now called Darcy's law for groundwater flow. Jules Dupuit was a classmate of Darcy and his replacement as director of Water and Bridges in Paris in the 1850s. Dupuit (1863) recognized that in many cases the vertical variation of the horizontal components of flow may be neglected, in essence reducing the mathematical description of groundwater flow by one dimension; one of his examples was a formula for flow to a well. At the end of the nineteenth century, Forchheimer (1886) combined Darcy's law with the continuity equation to show that steady groundwater flow through piecewise homogeneous aquifers is governed by Laplace's equation. This opened the door to many existing solutions that were derived for other problems governed by the same equation. Equations that describe transient groundwater flow take into account that the aquifer can store water. Phreatic storage, storage through movement of the groundwater table, was included by Boussinesq (1904). The process of elastic storage was conceptualized by Meinzer (1928), Theis (1935), and Jacob (1940), and led to the definition of storativity.

2.06.3 Hydraulic Head

The mechanical energy per unit weight in an incompressible fluid is given at a point by the following sum:

$$\frac{p}{\gamma} + Z + \frac{V^2}{2g} \tag{1}$$

where p [ML^{-1}T^{-2}] is the pressure, γ [MT^{-2}L^{-2}] is the specific weight of the fluid, Z [L] is the elevation above a fixed datum, V [LT^{-1}] is the speed of the fluid, and g [LT^{-2}] is the acceleration due to gravity. The first term, referred to as pressure head, reflects the pressure energy of the fluid. The second term reflects the potential energy of the system and is called the elevation head. The third term, known as the velocity head, reflects the kinetic energy of the fluid.

In groundwater applications, typical fluid speeds within the porous medium are so small that the velocity head is negligible. The combination of pressure head and elevation head is known as the hydraulic head, also called piezometric head or simply head. The dimension of head h is length [L]:

$$h = \frac{p}{\gamma} + Z \tag{2}$$

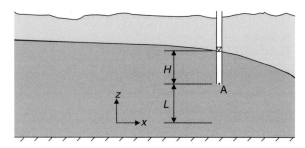

Figure 1 Measurement of the hydraulic head at a point in an aquifer using a piezometer. Saturated zone is darker gray and is bounded on top by the groundwater table.

The hydraulic head provides a good estimate of the available energy per unit weight at a point in a groundwater flow field, and is fairly easy to measure. **Figure 1** shows a piezometer set into the saturated portion of an aquifer. A piezometer is a hollow tube, open to the aquifer only at the bottom (point A in the figure).

The hydraulic head at point A is the sum of the pressure head and elevation head. The fluid that rises into the piezometer is hydrostatic and the pressure at point A is

$$p_A = \gamma H \tag{3}$$

or the pressure head at point A is H. The elevation head at point A, measured with respect to the datum shown in the figure, is L. The hydraulic head at point A is given by $L + H$ and is equal to the height above the datum to which water rises in the piezometer.

The hydraulic head may be measured at a point in a groundwater flow field once a piezometer has been placed. Hydraulic head data at surface water features such as lakes and streams, which form the natural boundaries of many aquifers, are often available as time series. Head data are the most abundant information a groundwater engineer has at his disposal. The importance of the hydraulic head in groundwater calculations will become clear in the remainder of this chapter.

2.06.4 Darcy's Law

In 1856, Henry Darcy performed experiments from which he concluded that the flow of groundwater is proportional to the head gradient (Darcy, 1856). The general setup of the experiment is simple. A cylinder is filled with aquifer material. The ends of the cylinder are attached to two reservoirs with different levels (**Figure 2**). Water flows through the aquifer material from the higher reservoir to the lower reservoir. In this fashion, Darcy showed that the discharge Q [L^{3}T^{-1}] through the soil column is proportional to the head difference $h_1 - h_2$ [L] and the cross-sectional area A [L^2] of the column, and inversely proportional to the length of the soil column L [L]:

$$Q = kA\frac{h_1 - h_2}{L} \tag{4}$$

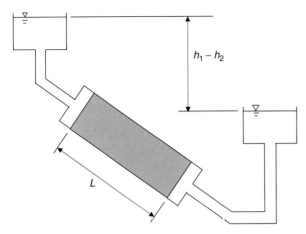

Figure 2 The experiment of Darcy. Darcy used a vertical column, but the flow is independent of the angle of the soil column.

The proportionality constant is k [L/T] and is called the hydraulic conductivity. Equation (4) is adequate to describe flow through a soil column, but not flow through an aquifer. Flow through an aquifer is expressed in terms of the specific discharge vector $\vec{q}$ [LT^{-1}], the discharge per unit area of aquifer normal to the direction of flow:

$$\vec{q} = -k\nabla h \quad (5)$$

The components of $\vec{q}$ in the Cartesian x, y, and z directions may be written as

$$q_x = -k\frac{\partial h}{\partial x}, \quad q_y = -k\frac{\partial h}{\partial y}, \quad q_z = -k\frac{\partial h}{\partial z} \quad (6)$$

Equation (5) is known as Darcy's law, although it is an empirical formula relating the head gradient to the specific discharge vector. Note the equivalence between Darcy's law and other physical laws such as Fourier's law for heat flux, Ohm's law for current density, and Fick's law for diffusive flux.

The hydraulic conductivity of aquifers may be anisotropic. Sedimentary aquifers often consist of a sequence of thin layers of slightly coarser or slightly finer material. In such aquifers, the average vertical hydraulic conductivity is smaller than the average horizontal hydraulic conductivity. Hence, the hydraulic conductivity is anisotropic and is written as a tensor **K** so that Darcy's law becomes

$$\vec{q} = -\mathbf{K}\nabla h \quad (7)$$

When the principal directions of the hydraulic conductivity tensor coincide with the horizontal and vertical directions of sedimentary aquifers, the Cartesian components of Darcy's law become

$$q_x = -k_h\frac{\partial h}{\partial x}, \quad q_y = -k_h\frac{\partial h}{\partial y}, \quad q_z = -k_v\frac{\partial h}{\partial z} \quad (8)$$

where k_h is the horizontal hydraulic conductivity and k_v the vertical hydraulic conductivity.

The hydraulic conductivity is a function of the fluid that flows through the aquifer and of the shapes, sizes, and interconnectedness of the pores of the aquifer. For example, the hydraulic conductivity is smaller for oil than for water when flowing through the same aquifer, as oil is more viscous than water. Similarly, warmer water results in a larger hydraulic conductivity than colder water, as the viscosity of water decreases with temperature. The hydraulic conductivity may be written as

$$k = \frac{\kappa \rho g}{\mu} \quad (9)$$

where κ [L^2] is called the intrinsic permeability of the aquifer and is a characteristic of the pore-size distribution and tortuosity of the porous medium, ρ [ML^{-3}] is the density of the fluid, g [LT^{-2}] is the acceleration of gravity, and μ [MLT^{-1}] is the dynamic viscosity. In this way, the property of the porous material (κ) is separated from the properties of the fluid (ρ and μ). Variations of ρ and μ may play a role in coastal aquifers because of changes in salinity, or in cases where the temperature of the groundwater varies significantly, for example, in river bank filtration projects or systems for aquifer thermal energy storage.

The hydraulic conductivity of an aquifer may be measured with a Darcy experiment. This requires, however, that undisturbed and representative samples are taken from an aquifer. As this is a difficult, if not impossible, task, field measurements are more likely to give accurate results. Representative values for the hydraulic conductivity are given in Table 1 for water flowing through different aquifer materials. The hydraulic conductivity generally varies spatially throughout an aquifer. The heterogeneity is rarely known well. The head and flow in the aquifer may often be simulated accurately by treating the aquifer properties as piecewise homogeneous. Travel times are, however, strongly affected by heterogeneity of the aquifer (e.g., Moore and Doherty, 2006).

The specific discharge vector has the same units as a velocity and is sometimes called the Darcy flux or the Darcy velocity. It is important to note, however, that a water particle that flows through the aquifer does not flow with an average velocity equal to the specific discharge. The specific discharge is the discharge through a unit area of aquifer. Only part of this unit area consists of pores while the larger part consists of solid particles. The ratio of the volume of pores to the volume of aquifer is called the porosity n. Water can only flow through the pores, so that the average velocity vector $\vec{v}$ may be obtained

Table 1 Representative values of hydraulic conductivity for various aquifer materials

Material	k (m d^{-1})
Clay	<0.0001
Sandy clays	0.0001–0.001
Peat	0.0001–0.01
Silt	0.001–0.01
Very fine sands	0.1–1
Fine sands	1–10
Coarse sands	10–100
Sands with gravel	100–1000
Gravels	>1000

Modified from Verruijt A (1970) *Theory of Groundwater Flow*. New York: MacMillan.

from the specific discharge as

$$\vec{v} = \vec{q}/n \tag{10}$$

This is called an average velocity. The velocity through a larger pore, or through the center of a pore is likely to be larger than the velocity through a smaller pore or along the edge of a soil particle. The velocity of groundwater is generally very small. The head may drop 1 or 2 m every 1000 m. A gradient of 0.002 in a sand with a hydraulic conductivity of $10\,\mathrm{m\,d}^{-1}$ and a porosity of 0.2 gives an average velocity of only $0.1\,\mathrm{m\,d}^{-1}$. Larger velocities occur in very specific cases only, such as near pumping wells.

2.06.5 Steady Conservation of Mass

Darcy's law provides three scalar equations for the four unknowns: q_x, q_y, q_z, and h. The fourth equation for solving the system is obtained from conservation of mass.

A derivation of the differential statement of conservation of mass for a flowing fluid is given in any standard fluid mechanics text (i.e., Munson *et al.*, 2002). The result states that the divergence of the mass flow rate and the rate of accumulation of fluid mass are in balance at every point in the flow field:

$$\nabla \cdot (\rho\vec{v}) + \frac{\partial \rho}{\partial t} = 0 \tag{11}$$

where $\vec{v}$ is the fluild velocity and ρ the fluid density. By analogy, a statement for conservation of mass made for groundwater flowing through a porous material of porosity n is

$$\nabla \cdot (\rho\vec{q}) + \frac{\partial(\rho n)}{\partial t} = 0 \tag{12}$$

The fluid density is multiplied by the porosity in the second term as the fluid mass occurs only in the pore spaces. If the porous media is rigid ($\partial n/\partial t = 0$) and the fluid density is constant in time, or if the flow is steady, (12) reduces to

$$\nabla \cdot (\rho\vec{q}) = 0 \tag{13}$$

If, in addition, the fluid density is constant in space the simplest form of conservation of mass emerges,

$$\nabla \cdot \vec{q} = 0 \tag{14}$$

Equation (14) is also known as the continuity of flow equation; when the density is constant, conservation of mass is equivalent to continuity of flow. Conservation of mass (14) may be combined with Dary's law (5) to obtain a single differential equation governing three-dimensional groundwater flow through a homogeneous aquifer:

$$\nabla^2 h = 0 \tag{15}$$

This result was first obtained by Forcheimer (1886).

2.06.6 Flow Types

Groundwater flow may be classified according to the spatial dimensions of the flow field, the dependence of the flow on time, and the aquifer setting in which the flow occurs. The focus in this chapter is on one- and two-dimensional, steady and transient flow in single aquifer systems with isotropic and homogeneous properties. These flow types and others are described in the following.

2.06.6.1 Spatial Dimension

Flow in an aquifer may be one, two, or three dimensional depending on the boundary conditions associated with the flow. Most aquifers are relatively thin in comparison to their areal extent. In these settings, which are referred to as shallow aquifers, one- and two-dimensional analyses are often adequate. In shallow aquifers the vertical variations in the hydraulic head are negligibly small when compared to horizontal variations in head. For problems where three-dimensional flow is important, near local features such as partially penetrating or horizontal wells, or near partially penetrating streams, the effects of concentrated vertical flow can be incorporated approximately into two-dimensional models.

2.06.6.2 Time Dependence

Groundwater flow is either steady or transient. In steady flow, there are no changes in flow or hydraulic head in time. Analyses of steady flow are used to reflect long-term, average conditions in an aquifer, for example, the dewatering of an aquifer for a large construction project, or delineation of wellhead protection areas for municipal water supply wells. Transient flow occurs when aquifer boundary conditions change in time, for example, changing aquifer recharge, changing river levels, and varying pumping rates of wells. A specific application of a transient flow analysis is the evaluation of aquifer properties by field tests, such as pumping tests, when it is not practical to run the test until steady conditions are reached.

2.06.6.3 Geologic Setting

The geologic setting of an aquifer may be used to further define the flow type as confined, unconfined, combined, or multiaquifer flow. The subsoil may be divided in more permeable and less permeable layers. The permeable layers may transmit significant amounts of water. They are called aquifers and can be used as the source for drinking water or irrigation. The less permeable layers transmit little or no water and cannot be used for water supply; they are commonly called aquicludes, confining layers, aquitards, or leaky layers.

An aquifer is confined when it is bounded on the top and bottom by impermeable layers, or layers with significantly lower permeability than the aquifer. In contrast, an unconfined aquifer is not bounded on top by an impermeable layer. Flow in an aquifer is called confined when the head in the aquifer is above the impermeable top of the aquifer (Figure 3(a)).

For unconfined flow, the saturated part of the aquifer is bounded on top by the groundwater table, also called the

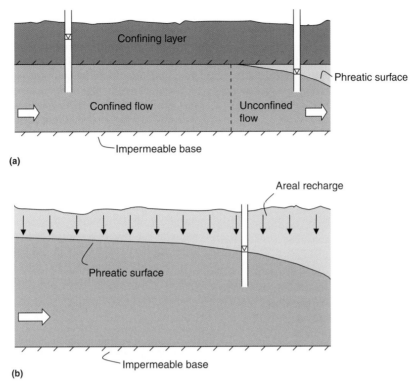

Figure 3 Definition of aquifer types and flow types: (a) combined confined and unconfined flow and (b) unconfined flow with recharge.

phreatic surface (**Figure 3(b)**). The concept of a groundwater table seems simple: when one digs a deep enough hole, it will fill up with water to the level of the groundwater table. Upon closer examination, the concept is less clear, however. When digging down, the soil gets wetter and wetter until the groundwater table is reached. In the section from the surface to the groundwater table, the pores of the soil are filled with both water and air; this section is called the unsaturated zone and is described in detail in **Chapter 2.05 Infiltration and Unsaturated Zone**. The saturated zone starts at the phreatic surface, which is defined as the depth where the pressure in the water is equal to atmospheric. The phreatic surface is curved when there is flow in the aquifer. The surface goes down in the direction of flow; thus, the velocity of a water particle always has a downward component; in most cases, this component is relatively small.

In unconfined flow, the saturated thickness varies with the elevation of the water table. Flow in a confined aquifer becomes unconfined when the head falls below the impermeable top of the aquifer. In a confined aquifer, the flow may consist of both regions where the head is above the confining layer and regions where the head is below the confining layer. This is referred to as combined confined and unconfined flow, or simply combined flow. Combined flow in a confined aquifer is illustrated in **Figure 3(a)**.

Often, aquifers are stratified with alternating layers of relatively permeable material separated by layers of less permeable materials. The flow in these systems may move from one aquifer through a leaky layer to another aquifer, and is referred to as multiaquifer flow.

2.06.7 The Dupuit Approximation

The basic idea behind the Dupuit approximation (also called the Dupuit–Forchheimer approximation) is to approximate groundwater flow in an aquifer as two-dimensional flow in a horizontal plane. The approximation allows many problems to be solved in simple form that otherwise could not be solved. Conditions of the Dupuit approximation are commonly stated as (e.g., Bear, 1972):

1. the flow is horizontal ($q_z = 0$);
2. the hydraulic head is constant in the vertical ($h = h(x,y)$); and
3. the hydraulic gradient is equal to the slope of the water table.

There are various interpretations of the physical meaning of the Dupuit approximation. Bear (1972) shows that the head predicted with a Dupuit model in a single aquifer represents the average head over the depth of the aquifer. Polubarinova-Kochina (1962) shows that Dupuit models are exact for anisotropic aquifers with infinite vertical hydraulic conductivity. This idea was explored further by Kirkham (1967).

Strack (1984) showed that conditions 2 and 3 listed above are consequences of neglecting the resistance to vertical flow in an aquifer, and that condition 1 is unnecessary. Strack's interpretation allows for the calculation of nonzero vertical flow components (q_z) and three-dimensional pathlines in two-dimensional Dupuit models of single- and multiaquifer flow. Also, this interpretation clearly identifies where errors may be introduced by making the Dupuit approximation. Strack's interpretation is adopted here and the Dupuit approximation is defined as neglecting the resistance to vertical flow in an aquifer. As stated, the major advantage of the Dupuit approximation is a two-dimensional head field ($h = h(x,y)$) with two-dimensional horizontal flow components ($q_x = q_x(x,y)$, and $q_y = q_y(x,y)$), while the vertical flow remains a function of all three coordinates ($q_z = q_z(x,y,z)$).

2.06.8 Potential Flow and the Discharge Vector

The discharge vector $\vec{Q}$ [L²/T] is defined as the depth-integrated specific discharge vector. The x-component of the discharge vector is obtained as

$$Q_x = \int_{Z_b}^{Z_t} q_x(x,y,z)\mathrm{d}z \quad (16)$$

where Z_b and Z_t are the bottom and top elevations of the saturated portion of the aquifer, respectively. Upon making the Dupuit approximation in a shallow aquifer (16) becomes

$$Q_x = q_x(x,y)\int_{Z_b}^{Z_t} \mathrm{d}z = q_x(Z_t - Z_b) \quad (17)$$

The term within parentheses is the saturated thickness of the aquifer. For confined flow, the saturated thickness equals the aquifer thickness H. For unconfined flow, it is equal to $h - Z_b$. In this chapter, the datum for h is chosen at the bottom of the aquifer ($Z_b = 0$), so that the saturated aquifer thickness is h. Substituting Darcy's law into (17) and applying the appropriate saturated thicknesses gives

$$Q_x = \begin{cases} q_x H = -kH\dfrac{\partial h}{\partial x}, & \text{confined flow} \\ q_x h = -kh\dfrac{\partial h}{\partial x}, & \text{unconfined flow} \end{cases} \quad (18)$$

The y-component of the discharge vector is obtained in a similar manner. Equation (18) suggests the existence of a discharge potential, Φ [L³/T], from which the discharge vector may be calculated:

$$\vec{Q} = -\nabla\Phi \quad (19)$$

The following function satisfies both (19) and (18) and is the discharge potential (e.g., Strack, 1989):

$$\Phi = \begin{cases} kHh - \tfrac{1}{2}kH^2, & \text{confined flow} \\ \tfrac{1}{2}kh^2, & \text{unconfined flow} \end{cases} \quad (20)$$

When $h \geq H$, flow is confined, otherwise (or in the absence of a confining layer) it is unconfined.

Equation (20) represents a single potential for combined flow; the potential is continuous across the interface where flow changes from confined to unconfined and the head in the aquifer is equal to the aquifer thickness ($h = H$). For confined flow, the product kH in (20) is referred to as the transmissivity T [L²/T] of the aquifer.

Groundwater flow may be written as potential flow when the base of the aquifer is horizontal and the aquifer properties are piecewise constant. Writing groundwater flow as potential flow simplifies the formulation of confined and unconfined flow and allows the use of the many potential flow solutions that exist in other fields. The definition of potential flow is that the flow is equal to the gradient of the potential. For groundwater flow, the definition is modified by adding a minus sign (19). The discharge potential has no useful physical meaning, but is merely a convenient quantity in mathematical modeling.

Using the discharge potential, Darcy's law has been rewritten in terms of the discharge vector (19). The differential statement of conservation of mass may also be written in terms of the discharge vector. This may be done either by writing a flow balance on an elementary volume of aquifer (e.g., Strack, 1989), or by integrating the continuity equation over the depth of the aquifer (e.g., Bear, 1972). The result is

$$\nabla \cdot \vec{Q} = N \quad (21)$$

where N [L/T] is the steady areal recharge rate, or the rate at which water infiltrates through the unsaturated zone into the saturated portion of the aquifer. If the aquifer is confined, or there is no recharge to the aquifer, $N = 0$. Combining (19) and (21) results in Poisson's equation

$$\nabla^2\Phi = -N \quad (22)$$

where the Laplacian is now understood to mean differentiation in the horizontal plane only. Confined and unconfined flow are handled in the same way in terms of the discharge vector. Boundary conditions are written in terms of Φ using (20), in terms of components of $\vec{Q}$, or a combination of Φ and $\vec{Q}$. The resulting boundary-value problem is solved for Φ and the results translated to heads, using the inverse of (20).

2.06.9 One-Dimensional Flow

The governing differential equation for steady one-dimensional, confined, unconfined, or combined flow in a shallow aquifer is (see (22))

$$\frac{\mathrm{d}^2\Phi}{\mathrm{d}x^2} = -N \quad (23)$$

where Φ is related to hydraulic head, h, by (20). When the recharge rate N is constant, the general solution to this differential equation is

$$\Phi = -\tfrac{1}{2}Nx^2 + Ax + B \quad (24)$$

where A and B are constants that must be evaluated from boundary conditions. If the flow is confined, or there is no areal recharge, $N = 0$. Three examples of one-dimensional flow that demonstrate various boundary conditions to evaluate the constants are presented in the following.

2.06.9.1 Confined Flow between Two Rivers

The simplest case of confined flow is one-dimensional flow between two fixed-head boundaries, for example, two fully penetrating rivers as illustrated in Figure 4(a). When the aquifer is homogeneous, the solution shows that the head varies linearly between the head on the left and the head on the right.

The head at the river to the left is equal to h_L ($h_L \geq H$), and the head at the river to the right is h_R ($h_R \geq H$). The boundary conditions must be written in terms of the discharge potential. From (20),

$$\Phi_L = \Phi(x=0) = kHh_L - \tfrac{1}{2}kH^2$$
$$\Phi_R = \Phi(x=L) = kHh_R - \tfrac{1}{2}kH^2 \qquad (25)$$

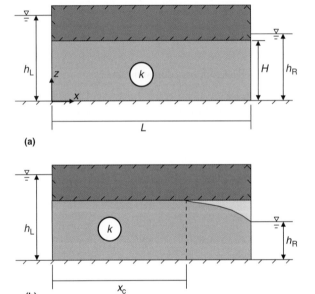

(a)

(b)

Figure 4 One-dimensional flow between two rivers: (a) confined flow and (b) combined confined and unconfined flow.

Application of the boundary conditions to the general solution with $N=0$ results in the following discharge potential:

$$\Phi = -\left(\frac{\Phi_L - \Phi_R}{L}\right)x + \Phi_L \qquad (26)$$

The discharge vector is obtained by differentiating (19):

$$Q_x = -\frac{d\Phi}{dx} = \frac{\Phi_L - \Phi_R}{L} \qquad (27)$$

where Q_x is constant throughout the aquifer. Alternatively, the discharge potential may be written as

$$\Phi = -Q_x x + \Phi_L \qquad (28)$$

2.06.9.2 Combined Flow between Two Rivers

If the head at the river on the right is below the aquifer confining unit ($h_R < H$), the flow in the aquifer will be combined flow. In this case, Φ_R is computed as (see (20))

$$\Phi_R = \tfrac{1}{2}kh_R^2 \qquad (29)$$

The head at the left remains above the confining unit, and therefore Φ_L is defined as before (25).

The solution, written in terms of the discharge potential, (26), is still valid, as well as the expression for Q_x (27). The location where the flow changes from confined to unconfined flow, as shown in Figure 4(b), is found by setting the discharge potential equal to $kH^2/2$ and solving for the x-coordinate

$$x_c = \frac{\Phi_L - kH^2/2}{Q_x} \qquad (30)$$

2.06.9.3 Unconfined Flow in a River Valley

Consider unconfined flow in a buried bedrock valley to a river of constant head, illustrated in Figure 5. There is no confining unit on the aquifer. The discharge potential is related to head by (20). The right aquifer boundary ($x=L$) is the impermeable valley wall and the left boundary ($x=0$) is the river. The

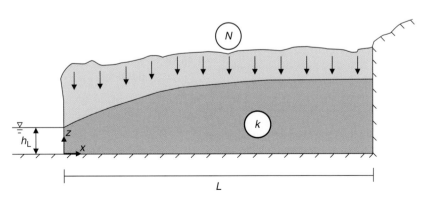

Figure 5 One-dimensional unconfined flow in a river valley with recharge.

corresponding boundary conditions are

$$Q_x(x = L) = 0 \tag{31}$$

$$\Phi(x = 0) = \tfrac{1}{2}kh_L^2 = \Phi_L \tag{32}$$

Application of the boundary conditions (31) and (32) to the general solution (24) results in

$$\Phi = -Nx\left(\frac{x}{2} - L\right) + \Phi_L \tag{33}$$

2.06.10 One-Dimensional Radial Flow

Radial flow in an aquifer is another case of one-dimensional flow. The governing equation for one-dimensional, radial, potential flow, written in radial coordinates r, is

$$\nabla^2 \Phi = \frac{d^2\Phi}{dr^2} + \frac{1}{r}\frac{d\Phi}{dr} = -N \tag{34}$$

The general solution to this differential equation is

$$\Phi = -\tfrac{1}{4}Nr^2 + A\ln r + B \tag{35}$$

where A and B are constants to be evaluated from boundary conditions. The discharge vector is obtained, as before, as minus the gradient of the discharge potential. In polar coordinates, the radial component of the discharge vector is $Q_r = -d\Phi/dr$.

Radial flow problems and solutions are important in groundwater engineering because they represent the local flow field around pumping wells. Solutions to three example problems of radial flow are provided below.

2.06.10.1 Flow to a Well at the Center of a Circular Island without Recharge

The discharge potential that is the solution to this problem is referred to as Φ_1. The following boundary conditions fix the head at the perimeter of the island ($r = R$) and the perimeter of the well ($r = r_w$), respectively:

$$\Phi_1(r = R) = \Phi_R \tag{36}$$

$$\Phi_1(r = r_w) = \Phi_w \tag{37}$$

Application of the boundary conditions to the general solution, using $N = 0$, results in the following discharge potential:

$$\Phi_1 = \frac{\Phi_R - \Phi_w}{\ln(R/r_w)}\ln(r/R) + \Phi_R \tag{38}$$

The discharge rate of the well, Q [L^3/T], may be obtained by evaluating the discharge vector $Q_r(r = r_w)$ and multiplying by the perimeter of the well:

$$Q = -2\pi r_w Q_r(r = r_w) = 2\pi\frac{\Phi_R - \Phi_w}{\ln(R/r_w)} \tag{39}$$

Equation (39) is useful to compute the discharge for a desired head at the well. In practice, it is more common to know the discharge of a well. If the discharge of the well is known, the solution may be written as (combine (38) and (39))

$$\Phi_1 = \frac{Q}{2\pi}\ln(r/R) + \Phi_R \tag{40}$$

Equation (40) is known as the Thiem equation. The radial component of the discharge vector for a steady well is

$$Q_r = -\frac{d\Phi}{dr} = -\frac{Q}{2\pi r} \tag{41}$$

In many two-dimensional problems or problems with recharge, the condition at the well is approximated as

$$Q = \lim_{r_w \to 0} -2\pi r_w Q_r(r = r_w) \tag{42}$$

Condition (42) produces accurate results as the radius of the well is often much smaller than the horizontal scale of the groundwater problem being considered. However, only in simple radial flow cases are conditions (42) and (39) equivalent.

2.06.10.2 Recharge on a Circular Island

The solution for recharge on a circular island is referred to as Φ_2. Once again, the head is fixed at the perimeter of the island

$$\Phi_2(r = R) = \Phi_R \tag{43}$$

By considering symmetry, the second boundary condition may be written as

$$Q_r(r = 0) = 0 \tag{44}$$

Application of the boundary conditions (43) and (44) to the general solution (35) results in the following discharge potential:

$$\Phi_2 = -\tfrac{1}{4}N(r^2 - R^2) + \Phi_R \tag{45}$$

Note that, by continuity of flow, the total groundwater discharge at the perimeter of the island is

$$Q_r(r = R)2\pi R = N\pi R^2 \tag{46}$$

which may also be derived by taking the derivative of (45).

2.06.10.3 Well at the Center of a Circular Island with Recharge

This problem contains both the features of the first two examples: recharge and a pumping well. Here the boundary conditions are specified as

$$\Phi_3(r = R) = \Phi_R \tag{47}$$

$$Q = \lim_{r_w \to 0} -2\pi r_w Q_r(r = r_w) \tag{48}$$

Application of the boundary conditions to the general solution yields the following discharge potential:

$$\Phi_3 = -\frac{1}{4}N(r^2 - R^2) + \frac{Q}{2\pi}\ln(r/R) + \Phi_R \qquad (49)$$

The head as a function of radial distance from the well is shown in **Figure 6**. The solid line represents the case for which the well pumps half the total areal recharge entering the aquifer, and the dashed line represents the case for which the well pumps exactly all the recharge on the island. Note that for the latter case, the flow at the perimeter of the island is zero and thus the phreatic surface is horizontal there; furthermore, the drawdown at the well is much larger than for the former case. It is emphasized that it is easy to create a case for which the well cannot pump all the infiltrated water. Theoretically, the maximum discharge is reached when the water level at the well is at the bottom of the aquifer; the practical limit is much less, of course. When the specified discharge in formula (49) is not possible, the potential at the well will be negative, and thus a head cannot be computed with the inverse of formula (20) for unconfined flow.

2.06.11 The Principle of Superposition

Comparison of the solutions to the three example problems above reveals that the third solution is the sum of the first two solutions with the additive constant modified. Addition of multiple solutions to obtain another solution is an example of the principle of superposition, which is applicable to all linear differential equations, including the equations of Laplace and Poisson.

The sum of the potentials of the first two problems in the previous section satisfies the differential equation of the third problem:

$$\nabla^2\Phi_3 = \nabla^2(\Phi_1 + \Phi_2) = \nabla^2\Phi_1 + \nabla^2\Phi_2 = 0 - N$$
$$= -N \qquad (50)$$

Similarly, the boundary condition at the well (48) is satisfied by the sum of the two potentials. Finally, the value of the sum of the two potentials at $r = R$ is a constant

$$\Phi_3(r = R) = \Phi_1(r = R) + \Phi_2(r = R) = 2\Phi_R \qquad (51)$$

This is not the value specified in the boundary condition (47), but this is easily corrected by modification of the additive constant.

In this example of superposition, two radial solutions are added such that the resulting solution is also radial, one-dimensional flow. In general, however, superposition of radial flow solutions results in two-dimensional flow. As an example, consider two pumping wells of strengths Q_1 and Q_2 at locations (x_1, y_1) and (x_2, y_2) in an infinite aquifer. By superposition, the discharge potential is

$$\Phi = \frac{Q_1}{2\pi}\ln r_1 + \frac{Q_2}{2\pi}\ln r_2 + A \qquad (52)$$

where $r_1 = \sqrt{(x-x_1)^2 + (y-y_1)^2}$, $r_2 = \sqrt{(x-x_2)^2 + (y-y_2)^2}$, and A is a constant that may be determined from a reference point of known potential. The resulting contours of head are presented in **Figure 7** with dotted lines and indicate that the superposition of the two one-dimensional solutions results in a truly two-dimensional flow field. The solid lines in **Figure 7** are streamlines.

2.06.11.1 A Well in Uniform Flow

The case of confined flow to a well in an otherwise uniform flow field is another example of the principle of superposition. The potential for a uniform flow field with a head gradient G in the positive x-direction is given by

$$\Phi = -TGx \qquad (53)$$

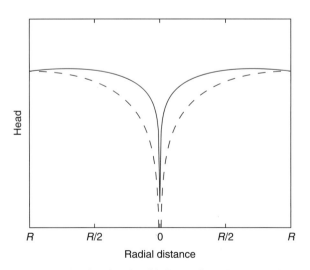

Figure 6 Head as function of radial distance for well at center of circular island with recharge: half the recharge is pumped by the well (solid), and all the recharge is pumped by the well (dashed).

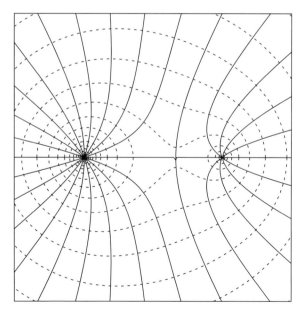

Figure 7 Flow net for two wells with Q_1 (left) larger than Q_2: head contours (dotted) and streamlines (solid).

such that $Q_x = TG$, where T is the transmissivity. The potential for a well in uniform flow is obtained through superposition of the potential for uniform flow (53) and a steady well located at the origin (40) plus an arbitrary constant Φ_0:

$$\Phi = -TGx + \frac{Q}{2\pi}\ln(r) + \Phi_0 \qquad (54)$$

An example of head contours obtained from this solution is shown in **Figure 8**. The heavy line in **Figure 8** is part of two streamlines that separate the groundwater flowing to the well from the groundwater that flows past the well. This dividing streamline forms the capture zone envelope of the well.

It is important to protect drinking water wells from contamination and many countries have guidelines for the delineation and protection of capture zones for water supply wells. Guidelines commonly require protection of the zone of groundwater around the well that will be captured by the well within a certain period of time, for example, 5 years or 20 years. These capture zones and capture zones for other time periods all lie within the capture zone envelope. The dashed lines in **Figure 8** represent the 5- and 20- year capture zones for this case. In most cases, capture zones are actually three-dimensional parts of the aquifer, but they are commonly approximated as two-dimensional zones on a map.

The width W of the capture zone envelope far upstream of the well in **Figure 8** may be computed from continuity as

$$W = Q/(GT) \qquad (55)$$

At the well ($x = 0$), the width of the capture zone is reduced to $W/2$. Special attention is paid to the point on the capture zone envelope farthest downstream of the well. This is a stagnation point, as the discharge vector is, theoretically, zero there. At the stagnation point, the effect of the well is exactly balanced by the hydraulic gradient of the uniform flow. The capture zone boundaries for large times approach the stagnation point, but only the boundary of the capture zone envelope passes through it.

For this simple problem, the capture zones for any time period may be evaluated analytically (Bear and Jacobs, 1965). It is more common, however, to evaluate the capture zone boundaries by particle tracking methods (e.g., Strack, 1989; Bakker and Strack, 1996).

2.06.11.2 The Method of Images

The method of images is an application of the superposition principle. Wells or other singularities are placed outside of the problem domain using symmetry to satisfy conditions specified along a boundary. For example, if the two wells in (52) have the same discharge ($Q_1 = Q_2 = Q$), and are placed symmetrically about the y-axis ($x_2 = -x_1$, $y_1 = y_2 = 0$), the discharge potential becomes

$$\Phi = \frac{Q}{2\pi}\ln(r_1 r_2) + A \qquad (56)$$

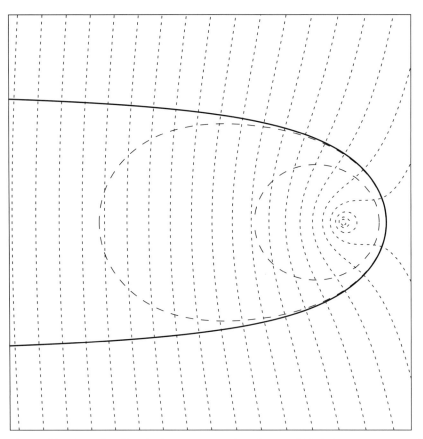

Figure 8 Head contours for a well in uniform flow (dotted). Capture zone envelope (heavy solid line), 5-year capture zone (small dashed contour), 20-year capture zone (large dashed contour).

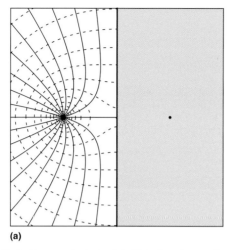 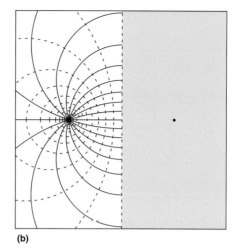

Figure 9 The method of images. Equipotentials (dotted) and streamlines (solid) for a well pumping near (a) an impermeable boundary and (b) a boundary of constant potential. The dots to the right of the flow field indicate the locations of the image wells.

Investigation of the behavior of this solution along the line passing midway between the wells ($x=0$) shows that the x-component of the discharge vector is zero. This potential is the solution to the problem of a well pumping next to an infinitely long impermeable boundary in a semi-infinite aquifer. As the problem domain lies to the left of the impermeable line, the well operating at ($+x_1$, 0) is referred to as the image well. Contours of the discharge potential are shown in **Figure 9(a)**.

Another solution is obtained when the image well at ($+x_1$, 0) is given the opposite discharge of the pumping well:

$$\Phi = \frac{Q}{2\pi} \ln \frac{r_1}{r_2} + A \quad (57)$$

Investigation of the behavior of this discharge potential shows that the potential is constant and equal to A along the line $x=0$. This discharge potential is the solution to the problem of a well pumping near a large lake or fully penetrating stream of constant potential A whose boundary lies along $x=0$. Again, the image wells lie outside the problem domain. Contours of the discharge potential are shown in **Figure 9(b)**.

Superposition and the method of images are two of the primary tools available to hydrologists and engineers for developing analytical solutions to steady and transient groundwater flow problems. Many analytical solutions to problems with wells and equipotential and/or impermeable boundaries may be obtained by the method of images. The method is also applicable to heterogeneity boundaries (Maxwell, 1873; Muskat, 1933) and leaky (Cauchy-type) boundaries (Keller, 1953; Anderson, 2000). The solution to a more complex and practical problem of groundwater flow is developed below.

2.06.11.3 Flow to a Pumping Well in an Alluvial Valley

The problem of groundwater flowing in an alluvial aquifer in a bedrock valley is considered. The aquifer is unconfined and receives areal recharge at a rate N; the governing differential equation is Poisson's equation (22). The aquifer is bounded below by impermeable bedrock, and to the right by the bedrock wall of the buried valley as illustrated in **Figure 10**. The condition specified at the valley wall is

$$Q_x(x=L) = 0 \quad (58)$$

To the left the aquifer is bounded by a flowing river; the sloping head at the river is approximated with the condition

$$\Phi(0, y) = Ay + \Phi_0 \quad (59)$$

where Φ_0 is the potential of the river at $x=0$ and A is approximately the slope of the water surface of the river. The condition at the pumping well is

$$Q = \lim_{r \to 0} -2\pi r Q_r \quad (60)$$

where $r^2 = (x - x_w)^2 + (y - y_w)^2$, and ($x_w, y_w$) are the coordinates of the pumping well.

A solution is obtained by considering three simpler problems, each representing a particular feature of the whole problem, and applying superposition to the results. First, the effects of the recharge are considered. The problem of one-dimensional flow along the x-axis from the valley wall ($x=L$) to a boundary of zero constant potential at the river ($\Phi_1(0, y) = 0$) was solved previously. Substitution of 0 for Φ_L in (33) gives

$$\Phi_1 = -Nx\left(\frac{x}{2} - L\right) \quad (61)$$

Second, the effects of the well are considered. Laplace's equation is solved subject to (58), (60), and $\Phi_2(0, y) = 0$. The solution to this problem is obtained by repetitive use of the method of images about the two aquifer boundaries, using the elementary solutions (57) and (56), which results in an

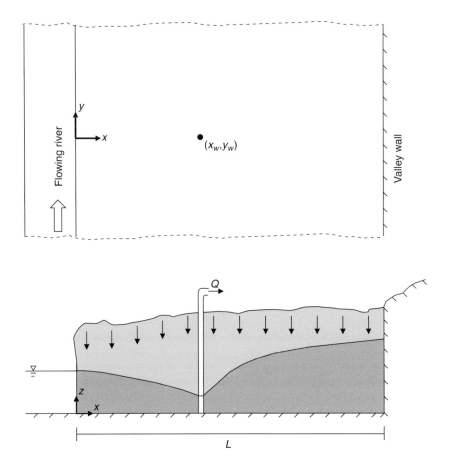

Figure 10 Definition sketch: flow to a well in an alluvial valley.

infinite sum of image wells:

$$\Phi_2 = \frac{Q}{4\pi}\ln\frac{(x-x_w)^2+(y-y_w)^2}{(x+x_w)^2+(y-y_w)^2}$$
$$+\frac{Q}{4\pi}\sum_{n=1}^{\infty}(-1)^n\ln\left[\left\{\frac{(x-x_w-2nL)^2+(y-y_w)^2}{(x+x_w-2nL)^2+(y-y_w)^2}\right\}\right.$$
$$\left.\left\{\frac{(x-x_w+2nL)^2+(y-y_w)^2}{(x+x_w+2nL)^2+(y-y_w)^2}\right\}\right] \quad (62)$$

Third, the effect of the sloping stream is included. This solution satisfies Laplace's equation and represents one-dimensional flow along the y-axis:

$$\Phi_3 = A y + \Phi_0 \quad (63)$$

As this solution produces no flow in the x-direction, condition (58) is satisfied. By comparison of (63) and (59), it is seen that this boundary condition is also satisfied.

The full solution is the sum of the three potentials (61), (62), and (63):

$$\Phi = \Phi_1 + \Phi_2 + \Phi_3 \quad (64)$$

A careful check of the solution shows that the correct differential equation is satisfied (Poisson's equation), and that the boundary conditions (58) through (60) are satisfied exactly. Finally, as the flow is unconfined, the discharge potential is related to the head through equation (20). A three-dimensional depiction of the groundwater table is shown for this example in **Figure 11**. The distance between the river and valley wall is $L=500$ m, and the distance between the well and the river is 100 m. The bottom of the aquifer is at $z=0$ m. In addition, contours are shown on the bottom of the figure.

2.06.12 The Stream Function and the Complex Potential

Head contours have been defined previously as curves of constant head which allow us to visualize the variation of mechanical energy within a flow field. In homogeneous aquifers, head contours are equal to potential contours, called equipotentials. Streamlines allow for the visualization of the average paths of groundwater flow. Streamlines are defined as lines that are everywhere tangent to the discharge vector. Using this definition, a differential equation may be written for a

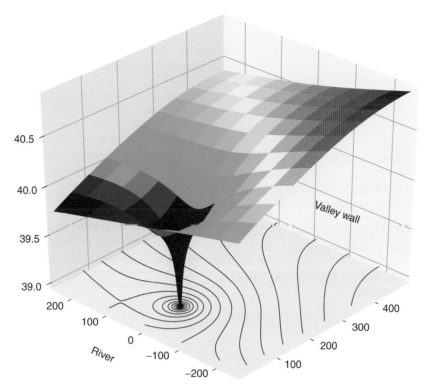

Figure 11 A three-dimensional depiction of the groundwater table for a well pumping in an alluvial valley as shown in **Figure 10**. Contours of the head are shown on the bottom of the figure.

streamline

$$\frac{Q_y}{Q_x} = \frac{\cos\alpha}{\sin\alpha} = \frac{dy/ds}{dx/ds} \quad (65)$$

where s is the position along the streamline and α is the angle between the s and x axes. Substituting in Darcy's law the equation becomes

$$-\frac{\partial \Phi}{\partial y}\frac{dx}{ds} + \frac{\partial \Phi}{\partial x}\frac{dy}{ds} = 0 \quad (66)$$

By Darcy's law, the discharge vector is everywhere normal to equipotentials. Examples are provided in **Figure 9** which shows streamlines and equipotentials for flow to a well near an impermeable boundary and near an equipotential boundary. Note that the two sets of lines cross everywhere at right angles.

Equipotentials and streamlines may both be drawn for problems of steady groundwater flow governed by the equations of Laplace and Poisson. However, in the case of Laplace's equation, a special function exists – the stream function, or $\Psi(x, y)$ – whose contours represent streamlines. The stream function exists for a steady, two-dimensional, divergence-free groundwater flow. As the value of the stream function does not change along a streamline,

$$\frac{d\Psi}{ds} = \frac{\partial \Psi}{\partial x}\frac{dx}{ds} + \frac{\partial \Psi}{\partial y}\frac{dy}{ds} = 0 \quad (67)$$

By comparing (66) with (67), the following relationships are obtained between derivatives of the discharge potential and the stream function:

$$\frac{\partial \Phi}{\partial x} = \frac{\partial \Psi}{\partial y} \quad (68)$$

$$\frac{\partial \Phi}{\partial y} = -\frac{\partial \Psi}{\partial x} \quad (69)$$

Equations (68) and (69) are known as the Cauchy–Riemann equations (e.g., Strack, 1989). It may also be shown that the stream function is single-valued, and harmonic ($\nabla^2\Psi = 0$). These properties of the stream function indicate that it is the harmonic conjugate of the discharge potential. Given a discharge potential that satisfies Laplace's equation, the corresponding stream function may be evaluated from (68) and (69).

The properties of the discharge potential and the stream function suggest the use of complex variables to solve groundwater flow problems governed by Laplace's equation. There are many texts on complex variables including their use in solving groundwater flow problems (e.g., Polubarinova-Kochina, 1962; Verruijt, 1970; Bear, 1972; Strack, 1989). The topic is only briefly discussed here.

For groundwater problems governed by Laplace's equation, a complex potential Ω exists which is an analytic function of the complex coordinate $z = x + iy$. The real part of the complex potential is the discharge potential and the imaginary part is the stream function:

$$\Omega(z) = \Phi(x,y) + i\Psi(x,y) \quad (70)$$

The negative derivative of the complex potential is the complex discharge

$$W(z) = -\frac{d\Omega}{dz} = Q_x - iQ_y \quad (71)$$

Introduction of complex variables allows for the use of more sophisticated tools, including conformal mapping, to solve many groundwater flow problems. In particular, using complex variables allows for the simultaneous solution of the discharge potential and the stream function. An example demonstrating the utility of the stream function and the complex potential is presented in the following.

2.06.12.1 Evaluation of the Capture Zone Envelope Using the Complex Potential

The complex potential for a well in an otherwise uniform flow field is

$$\Omega = -TGz + \frac{Q}{2\pi}\ln z + \Phi_0 \quad (72)$$

Separation into real and imaginary parts shows that the discharge potential is the same as obtained previously (54):

$$\Phi = -TGx + \frac{Q}{2\pi}\ln r + \Phi_0 \quad (73)$$

$$\Psi = -TGy + \frac{Q}{2\pi}\theta \quad (74)$$

where (r, θ) are polar coordinates. The location of the stagnation point, z_s, is evaluated as

$$W(z = z_s) = 0 \quad (75)$$

which gives

$$z_s = \frac{Q}{2\pi TG} \quad (76)$$

The value of the complex potential at the stagnation point is

$$\Omega(z = z_s) = -\frac{Q}{2\pi}\left[1 - \ln\left(\frac{Q}{2\pi TG}\right)\right] + \Phi_0 \quad (77)$$

which is a purely real number. Therefore, the value of the stream function at the stagnation point is zero. The contour $\Psi = 0 = \Psi_s$ defines the capture zone envelope:

$$-TGy + \frac{Q}{2\pi}\theta = 0 \quad (78)$$

The equation for the capture zone envelope is obtained in polar coordinates using $y = r \sin\theta$ and solving (78) for r:

$$r = \frac{Q}{2\pi}\frac{\theta}{TG\sin\theta} \quad (79)$$

2.06.13 Transient Flow

In the previous sections, steady-state flow was treated: the head was only a function of the spatial coordinates. In reality, the head is often also a function of time. When the head increases, more water is stored in the aquifer, and when the head decreases, less water is stored in the aquifer. For steady flow, continuity of flow states that the divergence of the discharge vector (21) is equal to the areal recharge rate N. When groundwater flow is transient, the divergence of the discharge vector is equal to the areal recharge plus the decrease in storage of water in the aquifer. The physics of the storage process is different for unconfined aquifers than for confined aquifers, but with suitable approximations, both lead to the same governing differential equation. The derivation of the governing equation for transient flow from the general statement of conservation of mass includes many approximations which are not discussed here. Rigorous derivations stating all necessary approximations are provided by Verruijt (1969) and Brutsaert (2005).

First, consider a column of an unconfined aquifer with constant surface area A. When the head in the column is increased by an amount dh (i.e., the phreatic surface is raised dh), the volume of water in the column increases by an amount

$$dV = SdhA \quad (80)$$

were S [-] is the storativity of the unconfined aquifer. When the aquifer material above the phreatic surface is dry, the storativity of the unconfined aquifer is equal to the porosity. In practice, the storativity is always smaller than the porosity, as there is water present in the pores above the phreatic surface. The storativity of an unconfined aquifer is also called the specific yield.

Next, consider a column of a confined aquifer with constant surface area A. When the head is now increased by dh, the volume of water still increases by an amount dV (80), but the storage coefficient is much smaller. Additional water can only be stored in the column through compression of the water and expansion of the aquifer. For most unconsolidated aquifers, the ability of the aquifer to expand is significantly larger than the ability of the water to compress, so that the compression of the water may be neglected. The storage coefficient of a confined aquifer is a function of the aquifer thickness: an aquifer of the same material but twice the thickness has a storage coefficient that is twice as large. The storage coefficient of a confined aquifer may be written as

$$S = S_s H \quad (81)$$

where S_s [L^{-1}] is the specific storage of the aquifer. Typical values for the specific storage of sand are between $S_s = 10^{-3}$ m^{-1} and $S_s = 10^{-5}$ m^{-1}.

Inclusion of the storage term in the divergence of the discharge vector (21) gives

$$\nabla \cdot \vec{Q} = -S\frac{\partial h}{\partial t} + N \quad (82)$$

where the areal recharge N may now vary with time. Using the potential for confined flow, this equation may be converted to

$$\nabla^2 \Phi = \frac{1}{D} \frac{\partial \Phi}{\partial t} - N \qquad (83)$$

where the aquifer diffusivity D is defined as

$$D = T/S \qquad (84)$$

and T is the transmissivity. The governing differential equation reduces to the diffusion equation when the areal recharge equals zero:

$$\nabla^2 \Phi = \frac{1}{D} \frac{\partial \Phi}{\partial t} \qquad (85)$$

The diffusion equation governs the transient behavior of many other physical processes.

Using the potential for unconfined flow, the continuity equation (82) may be written as

$$\nabla^2 \Phi = \frac{S}{kh} \frac{\partial \Phi}{\partial t} - N \qquad (86)$$

This nonlinear differential equation for transient unconfined flow is called the Boussinesq equation (Boussinesq, 1904). A common way to linearize the equation is to replace the head h in front of the time derivative on the right-hand side by an average head $\bar{h}$ (Strack, 1989), so that the diffusivity of an unconfined aquifer becomes $D = S/(k\bar{h})$. Note that after linearization, unconfined flow is also described by the diffusion equation (in absence of areal infiltration). Another way to linearize the differential equation for transient unconfined flow is to use the differential equation for transient confined flow, to approximate the transmissivity by $T \approx k\bar{h}$ and to use the storage coefficient for unconfined flow. The latter approach is used in this chapter.

The solution of combined transient confined and transient unconfined flow is not as easy as it was for steady flow, because the storage coefficients differ between confined and unconfined flow. Exact solutions for transient groundwater flow are, not surprisingly, more difficult to obtain than those for steady flow. Common mathematical approaches include separation of variables, Fourier series, and Laplace or other transforms (e.g., Bruggeman, 1999). In this chapter solutions are presented, without derivation, for one-dimensional flow. These solutions are valid for both confined and for unconfined flow as long as the linearization of the differential equation for unconfined flow is reasonable.

2.06.13.1 One-Dimensional Periodic Flow

Consider one-dimensional transient flow where the boundary condition varies periodically through time. The aquifer is semi-infinite and is bounded by open water at $x = 0$; there is no areal infiltration and no flow at infinity. The water table at the boundary varies sinusoidally:

$$h(0, t) = h_0 + A\cos(2\pi t/\tau) \qquad (87)$$

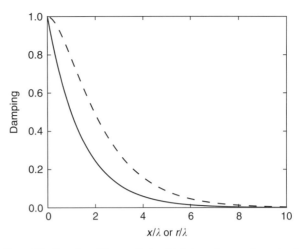

Figure 12 Damping of the amplitude with distance for one-dimensional periodic flow. Damping of the head when head at boundary varies as (87) (solid) and damping of the radial flow when discharge of a well varies as (90) (dashed).

where A is the amplitude of the fluctuation and τ is the time period of the fluctuation. The sinusoidal fluctuation in the surface water (87) may be caused, for example, by tides, by the periodic operation of hydroelectric dams, or by seasonal fluctuations of the surface water level. Solutions to problems of periodic flow may be obtained by separtation of variables. The solution to this problem is

$$\Phi = Th_0 + AT\Re \, \exp(-x\sqrt{\mathrm{i}}/\lambda + 2\pi\mathrm{i}t/\tau) \qquad (88)$$

where $\Re$ stands for taking the real part of the complex function, and λ is a characteristic length defined as

$$\lambda = \sqrt{\tau D/2\pi} \qquad (89)$$

The amplitude A dampens away from the open water as $\exp[-x/(\lambda\sqrt{2})]$, and is shown in **Figure 12**. At a distance of 3λ, the amplitude has damped to less than 5% of the amplitude at $x = 0$, and at a distance of 6λ, the amplitude has damped to less than 0.25% of the amplitude at $x = 0$. This result may be used as a rule of thumb to assess whether fluctuations in surface water levels need to be taken into account when considering the head and flow in an aquifer. If the area of interest is farther away from a surface water body than 6λ, periodic fluctuations of the surface water level with a period of τ may be neglected. Note that λ is a function of the period τ: the longer the period τ, the larger the characteristic length λ. Fluctuations with different periods and amplitudes may be superimposed in time. An arbitrary fluctuation of the water level may be approximated by a Fourier series.

A similar analysis may be carried out for a well with an average discharge of Q_0 and a sinusoidal discharge with an amplitude of Q_0:

$$Q(t) = Q_0 + Q_0\cos(2\pi t/\tau) \qquad (90)$$

At a certain distance from the well, the sinusoidal fluctuation of the discharge is unnoticeable and it seems that the well pumps with a steady discharge Q_0. This distance depends again on the characteristic length λ (89). The discharge vector for a well with a constant discharge Q_0 is given by Equation (41). The relative difference between the radial flow caused by the well with sinusoidal discharge (90) and the flow caused by a well with constant discharge Q_0 is 4.6% at a distance of 6λ, reducing to 0.3% at 10λ (see **Figure 12**). Hence, a well with a periodic discharge (90) varying between 0 and $2Q_0$ may be represented by a well with steady discharge Q_0 beyond a distance of 10λ from the well.

2.06.13.2 Transient Wells

In Equation (40) the solution was presented for steady flow to a well with discharge Q. Here, the transient equivalent is discussed. At time $t = t_0$ the head in the aquifer is constant and equal to h_0 everywhere and a well starts pumping with discharge Q. The head h_0, and thus the corresponding potential Φ_0, at infinity remains constant throughout time:

$$\Phi(\infty, t) = \Phi_0 \quad (91)$$

This problem may be solved as a similarity solution or by Laplace transforms. The potential as a function of time and the radial distance from the well is

$$\Phi = \Phi_0 + \frac{Q}{4\pi} E_1\left(\frac{Sr^2}{4T(t-t_0)}\right), \quad t \geq t_0 \quad (92)$$

where E_1 is the exponential integral defined as

$$E_1(u) = \int_u^\infty \frac{\exp(-s)}{s} ds \quad (93)$$

Solution (92) is known as the Theis solution (Theis, 1935). The head is a function of only one dimensionless parameter, u

$$u = \frac{Sr^2}{4T(t-t_0)} \quad (94)$$

Hence, if a certain drawdown $h_0 - h(r_1, t_1)$ is reached at a distance r_1 at time t_1, the same drawdown is reached at a distance $2r_1$ at time $4t_1$. A common approximation for E_1 is the series

$$E_1(u) = -\gamma - \ln u - \sum_{n=1}^{\infty} \frac{(-u)^n}{n(n!)} \quad (95)$$

where $\gamma = 0.5772\ldots$ is Euler's constant. The infinite series in (95) converges quickly (when $u < 1$), so that in practice only a small number of terms needs to be used.

One might expect that if the well is pumped for a long-enough period of time, the head will approach a steady-state position. This is not the case: the Theis solution (92) does not approach the Thiem solution (40) for large time. For the Thiem solution, the head approaches infinity when r approaches infinity, because the source of water for the Thiem solution lies at infinity. The Theis solution approaches h_0 when r approaches infinity according to (91), and all the pumped water comes from storage. In reality, there is always a water source closer than infinity, and if that source is included in the solution, the transient solution will approach a steady solution for large time. For example, consider a well at $(-x_1, y_1)$ near a large lake with a constant potential A along $y = 0$; the steady solution was obtained with the method of images and is given in Equation (57). A transient solution may also be obtained with the method of images as

$$\Phi = \frac{Q}{4\pi}\left[E_1\left(\frac{Sr_1^2}{4T(t-t_0)}\right) - E_1\left(\frac{Sr_2^2}{4T(t-t_0)}\right)\right] + A \quad (96)$$

When time approaches infinity, u approaches zero, and E_1 may be represented with the first two terms of (95). Substitution of these terms for E_1 in (96) leads to the steady solution (57).

Even though the head of the Theis solution by itself does not approach the steady-state head of the Thiem solution, the discharge vector does approach the steady-solution. The radial flow Q_r of the Theis solution may be obtained through differentiation of (92) to give

$$Q_r = -\frac{Q}{2\pi r} \exp(-u) \quad (97)$$

It is seen from this equation that when time approaches infinity, and u approaches zero, Q_r approaches the steady discharge vector (41). The consequence is that head gradients in the Theis solution approach the steady head gradients obtained with the Thiem solution, even though the head values themselves do not.

The Theis solution is very useful to determine aquifer parameters from a pumping test. During a pumping test, a well is turned on and the drawdown is measured in a nearby observation well. The Theis solution may be fit to observed head data to determine the transmissivity T and the storage coefficient S in the neighborhood of the well.

Tansient solutions may be superimposed in time as well as in space. For example, consider a well with a discharge Q operating from $t = t_0$ to $t = t_1$ and with zero discharge after t_1. For the period $t > t_1$, the potential may be represented by two Theis wells, one with a discharge Q starting at $t = t_0$ and one with a discharge $-Q$ starting at $t = t_1$:

$$\Phi = \frac{Q}{4\pi}\left[E_1\left(\frac{Sr^2}{4T(t-t_0)}\right) - E_1\left(\frac{Sr^2}{4T(t-t_1)}\right)\right], \quad t \geq t_1 \quad (98)$$

This is called a pulse solution, where the pulse lasts from t_0 until t_1.

2.06.13.3 Convolution

In the last example of the previous section, a solution was presented for a well that pumped with a discharge Q from t_0 to t_1. When the pumping period is 1 day, this solution may be used to compute the head variation caused by a well for which daily discharge records are available. This requires the repeated superposition through time of solution (98), called convolution. This solution approach is an example of a standard technique to solve differential equations. More formally, the approach is based on the determination of the solution for a unit impulse, in this case a discharge of unit volume over a short period, theoretically an infinitely short period. The

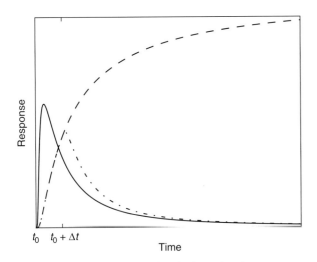

Figure 13 Examples at one specific point for an impulse response (solid), step response (dashed), and pulse response (dash-dotted) for a well near a long straight river; the pulse response is identical to the step response for the period of the pulse Δt.

response due to a unit impulse is called the impulse response function (e.g., **Figure 13**). For a well, the impulse response function θ of the potential is

$$\theta(r,t) = \frac{1}{4\pi} \frac{\exp(-u)}{t} \quad (99)$$

The potential for a time-varying discharge $Q(t)$ is obtained with the convolution integral (Duhamel's principle):

$$\Phi(r,t) = \int_{-\infty}^{t} Q(t')\theta(r,t-t')dt' \quad (100)$$

The Theis equation (92) may be obtained with the convolution integral by specifying the discharge as $Q(t') = Q$ for $t' \geq t_0$ in (99), and using that

$$\frac{dt'}{t-t'} = \frac{du}{u(r,t-t')} \quad (101)$$

The Theis equation is an example of a step response (e.g., **Figure 13**): at time t_0, the discharge changes from 0 to Q. In general, the unit step response Θ is obtained from the impulse response through integration

$$\Theta(x,y,t,t_0) = \int_{t_0}^{t} \theta(x,y,t-t')dt' \quad (102)$$

where the step occurs at time t_0. For practical application, the convolution integral is often written as a sum of pulse response functions. An example of a pulse response was given by the last example in the previous section, where a well was pumped at a constant discharge for a finite period. In general, the pulse response F_j for a pulse of length Δt starting at $t = t_j$ is defined as

$$F_j(x,y,t) = \Theta(x,y,t,t_j) - \Theta(x,y,t,t_{j+1}), \quad t \geq t_{j+1} \quad (103)$$

where $t_j = j\Delta t$. An example of a pulse response is given in **Figure 13**. Consider the case for which the applied stress is known over periods of equal length Δt, t_n is defined as $t_n = n\Delta t$, and Q_n is the stress from $t = t_n$ until $t = t_{n+1}$. The potential at time t_n may be computed with the convolution sum:

$$\Phi(x,y,t_n) = \sum_{j=0}^{n-1} Q_j F_j(x,y,t_n), \quad n \geq 1 \quad (104)$$

The convolution approach assumes that the system is linear. Nonlinear behavior may occur, for example, in the summer time when pumping is at its peak and there is little rainfall. During such periods, ditches or streams may go dry, which means that the hydrological system and thus the impulse response function change. In such cases it is not possible to simulate the head variation with a straightforward convolution. In practice, when a system is sufficiently linear, the convolution approach works very well.

The pulse response is different for different stresses (areal recharge, pumping, lake-level changes) and needs to take into account all nearby boundary conditions. Once the pulse response for a stress is known at a point, the head variation may be simulated using the convolution approach. Consider again the problem of unconfined flow in a buried bedrock valley to a river of constant head, as illustrated in **Figure 5**. The solution for a constant recharge rate is given in Equation (33). Instead of a constant recharge rate, the recharge now varies daily as shown in **Figure 14(a)** for a period of 7 years. Note that the recharge is negative for days without rainfall due to evaporation. The step response for this problem may be obtained with the Laplace transform technique and is given in Bruggeman (1999, Eq. 133.16). Alternatively, the pulse response may be obtained with a computer model; computer models are discussed in the next section. Convolution of the recharge with the step response gives the head variation. The head variation at the valley wall ($x = L$) and near the river at $x = 0.1L$ are shown in **Figure 14(b)**. Note that the total head variation at the valley wall (~ 1.8 m) is much larger than near the head boundary (~ 0.4 m). The head variation at the valley wall has a long memory: the head value depends on the recharge that fell almost 2 years ago. In other words, the pulse response approaches zero after approximately 2 years. The head variation is not shown for the first 2 years in **Figure 14(b)** as it would require recharge information prior to the record shown in **Figure 14(a)**.

When heads are measured, they always show the effect of recharge, as shown in **Figure 14**. Most head measurements also show the effect of barometric variations and earth tides. The latter are often undesirable and need to be removed; a computer program to remove these variations is called BETCO, which is available for download from the Internet. Time series of head observations always show the effect of the different time-varying stresses that act on the groundwater system. A stochastic approach called 'time series analysis' may

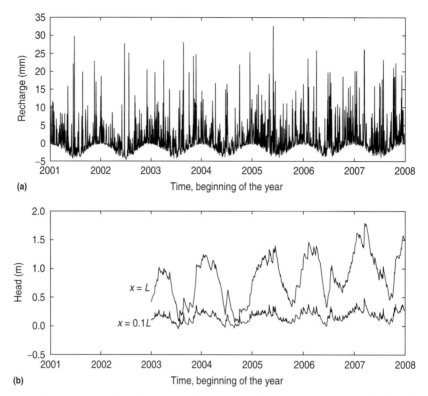

Figure 14 Daily recharge rate (a) and head variation (b) at valley wall ($x = L$, large fluctuation), and near specified-head boundary ($x = 0.1L$, small fluctuation) for the system shown in **Figure 10**.

be used to unravel the series and compute the head variations due to the individual stresses. The traditional method for time series analysis is the Box–Jenkins method (Box and Jenkins, 1970). Recently, the PIRFICT method was developed for time series analysis of hydrological data (Von Asmuth et al., 2008). The PIRFICT method uses predefined, parameterized shapes for the impulse response functions and allows for irregular time series, and time series with missing data.

2.06.14 Computer Models

The relatively simple solutions presented in this chapter may be used to solve real problems. They may be used for first estimates, to verify more complicated models, and to gain insight in the flow problem. In many cases, however, the setting is more complicated than, for example, a well near a long, straight lake boundary. To obtain solutions for more complicated problems, general solution approaches for the governing differential equations are implemented in computer programs. These computer programs may be applied to simulate groundwater flow in domains with more complicated boundary shapes, with a variety of boundary conditions, as well as flow in aquifers that are not homogeneous. The resulting computer models remain an approximation of reality and the modeler must decide what details to put into the model based on the purpose of the model. Most existing computer programs for modeling groundwater flow are based on one of three methods to solve the mathematical problem: the analytic element method, the finite difference method, or the finite element method. Characteristics of these three methods are discussed here briefly and some references to free software are given.

The analytic element method is based on the superposition of analytic functions (Strack, 1989; Haitjema, 1995; Fitts, 2002). In this respect, it is an extension of many of the solutions presented in this chapter. Each analytic function represents a hydrogeologic feature in the aquifer, such as a well, the section of a stream, or the boundary between two geologic formations. Each analytic element has at least one free parameter. The free parameter of an element may be specified or a condition may be specified so that the computer program can compute the value of the free parameter. For example, the free parameter of a well is its discharge. The discharge may be specified, or the modeler may specify the desired head (or drawdown) at the well, and the computer program computes the corresponding discharge (often simultaneously with the other free parameters in the model). An advantage of the analytic element method is that the head and flow can be computed analytically at any point in the aquifer. The model extends, theoretically, to infinity, but has no practical significance beyond the area where sufficient analytic elements are defined to simulate the flow. Many analytic elements exist, including wells, stream segments with and without leaky bottoms, boundaries between zones with different aquifer properties, leaky and impermeable walls, areal recharge, and lakes.

Most currently available analytic element programs are restricted to steady flow in piece-wise homogeneous aquifer systems. Analytic element approaches have been developed for transient flow through piecewise homogeneous aquifers (e.g., Bakker, 2004; Kuhlman and Neuman, 2009) and for flow through aquifers with continuously varying properties (e.g., Craig, 2009). Analytic elements are ideally suited for implementation in an object-oriented computer code; a simple design is presented in Bakker and Kelson (2009). Several analytic element programs are available for modeling steady flow. Single aquifer codes include WhAEM, which contains a graphical interface, and Split. An approach for steady multi-aquifer flow was developed by Bakker and Strack (2003) and is implemented in the program TimML. A graphical user interface for both Split and TimML is VisualAEM. Commercial programs are available as well, but are not discussed here.

The most popular computer program for modeling groundwater flow is MODFLOW (Harbaugh et al., 2000), which is based on the finite difference method and is available by the download form the internet. MODFLOW model domains are discretized in a grid of rectangles, called cells. Heads are computed at cell centers and flows are computed between cell centers. Hydrogeologic features need to be simplified to fit the chosen grid. The governing differential equation is solved approximately by replacing the derivatives in the differential equation by difference equations. Transient solutions are obtained by stepping through time. The cell size and time step need to be chosen sufficiently small to obtain an accurate solution. A condition must be specified along the entire model boundary. This is in contrast to analytic element models, which do not have a formal boundary. The model boundary should be chosen, where possible, along real hydrogeologic boundaries, such as rivers or impermeable rock outcrops. This is never possible everywhere, however, especially in deeper aquifers. Hence, along parts of the boundary artificial boundary conditions need to be specified based on the modeler's expert knowledge. The finite difference method is relatively easy to implement in computer codes and allows for continuously varying aquifer properties. Many specialized packages exist for MODFLOW to model a variety of features (e.g., wells, lakes, and drains) and flow types (e.g., unsaturated flow). Seawater intrusion may be simulated with the SWI package or SEAWAT. Creation of input files for MODFLOW is cumbersome. Many powerful graphical interfaces are available commercially. A free graphical interface is version 5 of PMWIN. Python scripts to run MODFLOW are available by the download from the Internet.

The finite element method also requires a discretization of the model area, although the common choice is a discretization in triangles. Heads are computed at the corners of the triangles (called nodes) and flows are computed between them. The mathematics behind the finite element method is based on a minimization principle, which is more complex than the finite difference method. Grids of triangles are more flexible than grids of rectangles as it is much easier to represent shapes of hydrogeologic features with triangles, and small triangles can be used where needed. Sophisticated grid builders are available to construct complicated grids of triangles. Other practical advantages and disadvantages of the finite element method are similar to the finite difference method. It is more complicated to implement the finite element method in a computer code than the finite difference method. A commonly used free finite element code is SUTRA, which can also be used to model unsaturated flow and variable density flow.

2.06.15 Discussion

The basic principles of groundwater flow were discussed in this chapter. Essential principles, such as Darcy's law, the Dupuit approximation, and the derivation of Laplace's equation, originate from the nineteenth century. Many analytic solutions were developed in the twentieth century, while numerical methods and computer models became important in the last few decades. The size of numerical models has grown over the years with the available computational power. Evaluation in other fields shows that the size of modeling grids follows Moore's law and doubles approximately every 1.5 years, as does the computational power (Voller and Porté Agel, 2003). Analytic solutions play an important role in the evaluation of numerical model results. Analytic formulas may be used to assess whether the results of a numerical model are approximately correct. Such a comparison often shows that application of a simple analytic model provides a very reasonable solution.

A number of discussions have been published on the future of hydrogeology in the first decade of the twenty-first century (e.g., Voss, 2005; Miller and Gray, 2008). Progress is being made in the modeling of heterogeneous domains using advanced calibration tools and assessment of the predictive ability of groundwater models (e.g., Doherty, 2008) and stochastic modeling (e.g., Zhang and Zhang, 2004). Another active area of research is the linkage of groundwater models with unsaturated zone models, surface water models, and ultimately atmospheric models. Such linkages run into serious issues of differences in temporal and spatial scales that have yet to be resolved satisfactorily. As in other areas of hydrology, some groundwater models try to include details because they exist, not because they matter (Haitjema, 1995). The inability to capture the full complexity of systems and processes makes the search for accurate simplifications a continuing endeavor.

Acknowledgments

We gratefully acknowledge Tanja Euser for creating **Figures 1–5 and 32.10**.

References

Anderson EI (2000) The method of images for leaky boundaries. *Advances in Water Resources* 23: 461–474.
Bakker M (2004) Transient analytic elements for periodic Dupuit–Forchheimer flow. *Advances in Water Resources* 27(1): 3–12.
Bakker M and Kelson VA (2009) Writing analytic element programs in Python. *Ground Water* 47(6): 828–834.
Bakker M and Strack ODL (1996) Capture zone delineation in two-dimensional groundwater flow models. *Water Resources Research* 32(5): 1309–1315.
Bakker M and Strack ODL (2003) Analytic elements for multiaquifer flow. *Journal of Hydrology* 271(1–4): 119–129.
Bear J (1972) *Dynamics of Fluids in Porous Media*. New York: Dover Publications.
Bear J and Jacobs M (1965) On the movement of water bodies injected into aquifers. *Journal of Hydrology* 3: 37–57.
Boussinesq J (1904) Recherches théoriques sur le coulement des nappes d'eau infiltrées dans le sol. *Journal de Mathématiques Pures et Appliquées* 10: 5–78.
Box GEP and Jenkins GM (1970) *Time Series Analysis, Forecasting and Control*. San Francisco, CA: Holden-Day.

Bruggeman GA (1999) Analytical Solutions of Geohydrological Problems, Developments in Water Science, 46, Amsterdam: Elsevier.

Brutsaert W (2005) *Hydrology – An Introduction*. New York: Cambridge University Press.

Craig JR (2009) Analytic elements for flow in harmonically heterogeneous aquifers. *Water Resources Research* 45: W06422 (doi:10.1029/2009WR007800).

Darcy H (1856) *Les Fourntaines Publiques de la Vlle de Dijon*. Paris: Dalmont.

Doherty J (2008) *Manual and Addendum for PEST: Model Independent Parameter Estimation*. Brisbane, Australia: Watermark Numerical Computing.

Dupuit J (1863) *Études Théoriques et Practiques sur le Mouvement des Eaux dans les Canaux Decouverts et á Travers les Terrains Perméables*, 2nd ed. Dunod: Paris.

Fitts CR (2002) *Groundwater Science*. New York: Academic Press.

Forchheimer P (1886) Ueber die Ergiebigkeit von Brunnenanlagen und Sickerschlitzen. *Z. Architekt. Ing. Ver. Hannover* 32: 539–563.

Haitjema HM (1995) *Analytic Element Modeling of Groundwater Flow*. San Diego, CA: Academic Press.

Harbaugh AW, Banta ER, Hill MC, and McDonald MG (2000) MODFLOW-2000, the US Geological Survey modular ground-water model-user guide to modularization concepts and the ground-water flow process. USGS Open-File Report 00–92.

Jacob CE (1940) The flow of water in an elastic artesian aquifer. *Transactions, American Geophysical Union* 21: 574–586.

Keller JB (1953) The scope of the image method. *Communications on Pure and Applied Mathematics* VI: 505–512.

Kirkham D (1967) Explanation of paradoxes in Dupuit–Forchheimer seepage theory. *Water Resources Research* 3(2): 609–622.

Kuhlman KL and Neuman SP (2009) Laplace-transform analytic-element method for transient porous-media flow. *Journal of Engineering Mathematics* 64(2). 113–13.

Maxwell JC (1873) *A Treatise on Electricity and Magnetism*, vol. 1, Oxford: Clarendon Press.

Meinzer OE (1928) Compressibility and elasticity of artesian aquifers. *Economic Geology* 23: 263–291.

Miller CT and Gray WG (2008) Hydrogeological research, education, and practice: A path to future contributions. *Journal of Hydrologic Engineering* 13(7): 7–12.

Moore C and Doherty J (2006) The cost of uniqueness in groundwater model calibration. *Advances in Water Resources* 29: 605623.

Munson BR, Young DF, and Okiishi TH (2002) *Fundamentals of Fluid Mechanics*, 4th edn. New York: John Wiley and Sons.

Muskat M (1933) Potential distribution about an electrode on the surface of the Earth. *Physics* 4(4): 129–147.

Polubarinova-Kochina PY (1962) *Theory of Groundwater Movement*. De Wiest JMR. (trans.). Princeton, NJ: Princeton University Press.

Strack ODL (1984) Three-dimensional streamlines in Dupuit-Forchheimer models. *Water Resources Research* 20(7): 812–822.

Strack ODL (1989) *Groundwater Mechanics*. Englewood Cliffs, NJ: Prentice Hall.

Theis CV (1935) The relation between the lowering of the piezometric surface and the rate and duration of discharge of a well using ground-water storage. *Transactions, American Geophysical Union* 16: 519524.

Verruijt A (1969) Elastic stage of aquifers. In: de Wiest RJM (ed.) *Flow Through Porous Media*, pp. 331–376. New York: Academic press.

Verruijt A (1970) *Theory of Groundwater Flow*. New York: MacMillan.

Voller VR and Porté Agel (2003)) Moore's law and numerical modeling. *Journal of Computational Physics* 172(2): 698–703.

Von Asmuth JR, Maas K, Bakker M, and Petersen J (2008) Modeling time series of ground water head fluctuations subjected to multiple stresses. *Ground Water* 46(1): 30–40.

Voss CI (2005). The future of hydrogeology. *Hydrogeology Journal* 13(1): 1–6.

Zhang Y-K and Zhang D (2004) Forum: The state of stochastic hydrology. *Journal of Stochastic Environmental Research and Risk Assessment* 18(4): 265.

Relevant Websites

http://www.civil.uwaterloo.ca
 Civil and Environmental Engineering, University of Waterloo; James R. Craig, Visual AEM.

http://www.epa.gov
 EPA: United States Environmental Protection Agency; Ecosystems Research Division, WLAEM2000.

http://code.google.com
 Google.Wigaem; timml; flopy.

http://www.groundwater.buffalo.edu
 Groundwater Research Group, UB Groundwater Group Software.

http://www.hydrology.uga.edu
 Hydrology@University of Georgia; BETCO: Barometric and Earth tide Correction.

http://bakkerhydro.org
 Mark Bakker, SWI package.

http://www.pmwin.net
 PMWiN.NET by Wen-Hsing Chiang, PMWIN Version 5.3.

http://water.usgs.gov
 USGS: U.S. Geological Survey. SEAWAT; SUTRA Version 2.1; MODFLOW-2000 version 1.18.01.

2.07 The Hydrodynamics and Morphodynamics of Rivers

N Wright, University of Leeds, Leeds, UK
A Crosato, UNESCO-IHE, Delft, The Netherlands

© 2011 Elsevier B.V. All rights reserved.

2.07.1	Early History of Hydrodynamics and Morphodynamics in Rivers and Channels	135
2.07.2	State of the Art in Hydrodynamics and Morphodynamics	138
2.07.2.1	Fluid Flow	138
2.07.2.1.1	Mass	138
2.07.2.1.2	Momentum	138
2.07.2.1.3	Energy	139
2.07.2.2	Numerical Solution	139
2.07.2.2.1	Boundary conditions	139
2.07.2.3	Depth and Process Scales	139
2.07.2.4	Cross-Section Scale	140
2.07.2.5	River Reach Scale	141
2.07.2.6	Spatial Scales in River Morphodynamics	142
2.07.2.7	Geomorphological Forms in Alluvial River Beds	143
2.07.2.7.1	Ripples and dunes	145
2.07.2.7.2	Bars	146
2.07.2.8	River Planimetric Changes	147
2.07.2.9	Bed Resistance and Vegetation	148
2.07.2.10	Discussion of Current Research and Future Directions	152
2.07.2.10.1	Incremental changes	152
2.07.2.10.2	Step changes	152
References		152

2.07.1 Early History of Hydrodynamics and Morphodynamics in Rivers and Channels

The study of flow in open channels and their shape is inextricably linked to the study of fluid dynamics more generally, and hydrodynamics can perhaps be best defined as the application of the theory of fluid dynamics to flows in open channels. Early work on the general properties of fluids was carried out by the ancient Greeks. They studied many fluid phenomena, and the work of Archimedes on hydrostatics is well known. However, it was the Romans who demonstrated a more practical knowledge of fluid flow and open-channel flow in particular. They constructed advanced water-supply systems including aqueducts and water wheels. Archaeological evidence confirms their use of sophisticated siphon systems that required advanced techniques to seal the pipes in order to maintain the necessary pressures and this is likely to have required an understanding of pressure and fluid potential energy. Unfortunately, there is no documentary evidence of the knowledge that they had, as it was a practical skill.

In Islamic civilizations around the ninth century, engineers and physicists studied fluid flow and made use of hydraulics through water wheels in order to process grain and carry out other mechanical tasks. They also engineered channels for irrigation and developed the systems of qanats for irrigation. Chinese engineers also harnessed energy by using water wheels to power furnaces.

Despite its widespread use and study the theory of open channel flow did not advance, and by the beginning of the nineteenth century the study of flow in pipes was probably more advanced, particularly in its mathematical description. This reflects the intrinsic difficulty of open-channel flow that is often not fully appreciated by a cursory examination. Under more detailed examination, it becomes clear that we do not know *a priori* what the depth will be in a channel as opposed to full pipe flow where the cross-sectional area is known: that is, the relationship between depth (m), discharge ($m^3 s^{-1}$), and cross-sectional geometry cannot be expressed in a simple formula. In essence, this is the fundamental question to be answered by both theoreticians and practitioners. The situation is further complicated by the high variation in bed and bank material. Due to this complexity, early studies were empirical.

The first step to a more mathematics- and physics-based approach had been taken by Leonardo da Vinci (1452–1519). His book entitled *Del moto e misura dell'acqua* (*Water Motion and Measurement*), written in around 1500 and published in 1649 after his death, is a treatise of nine individual books, of which the first four deal with open-channel flows (Graf, 1984). In this book, da Vinci made an early attempt to formulate the law of continuity linking the water flow to channel width, depth, slope, and roughness. Nevertheless, the founder of river hydraulics has been traditionally viewed as Benedetto Castelli (1577–1644), a pupil of Galileo Galilei, who wrote the book entitled *Della misura delle acque correnti* (*Measurement of Water Flows*) (1628), in which he explained the law of continuity in more precise terms. It is perhaps worth noting that Castelli was also engaged by the Pope as a consultant on the management of rivers in the Papal States, reflecting the combination of theoretical and practical approaches.

Sir Isaac Newton (1642–1727) discussed fluid statics and dynamics at length in his *Principia Mathematica* (1687) (Anderson, 2005). He proposed his law of viscosity stating that shear stress was proportion to the velocity gradient with the constant of proportionality being the viscosity. Newton's work informed later studies and Prandtl used the shear stress relationship to create an analogy for turbulent flow. In the eighteenth century, work on the fundamental mathematical description of fluid mechanics was advanced by Daniel Bernoulli (1700–82), Jen le Rond d'Alembert (1717–83), and Leonhard Euler (1707–83). The latter used momentum and mass conservation to derive the Euler equations for fluid flow and these were not surpassed until the Navier–Stokes equations were derived with their treatment of viscous shear stress. These were derived independently by Claude-Louis Navier in 1822 and George Stokes in 1845 (Anderson, 2005).

The Navier–Stokes equations were of a general nature. In terms of open-channel flow, it was realized that the key parameters were discharge ($m^3\ s^{-1}$), depth (m), cross-section geometry, longitudinal bed slope, and the nature of the bed and banks. The cross-section geometry is clearly infinitely variable and difficult to encapsulate in a formula and so key geometric properties were chosen to represent it. These are wet area (A (m^2)) and wetted perimeter (P (m)), and these are often used to derive the hydraulic radius, R ($=A/P$ (m)). Based on this theory, Chézy (1717–98) developed his theory of open-channel flow as balance of the frictional and gravitational force. He proposed the formula

$$V = C\sqrt{RS} \qquad (1)$$

where C is the Chézy coefficient ($m^{1/2}\ s^{-1}$), R the hydraulic radius (m), and S the longitudinal bed slope ($m\ m^{-1}$). Although C is often assumed to be constant for a given channel, it has dimensions and does vary with the water depth.

Later Manning proposed an alternative formula based on his measurements and this has been widely adopted in the English-speaking world:

$$\bar{V} = \frac{1}{n} R^{2/3} S^{1/2} \qquad (2)$$

where n is Manning's coefficient. Again, this is dimensional ($m^{-1/3}\ s^{-1}$) and varies with water depth.

The formulations by Chézy and Manning are valid for flows that are steady state and uniform. These assumptions clearly do not apply in many cases, particularly in natural rivers. In a treatise published in 1828, Bélanger put forward an equation for a backwater in steady, one-dimensional (1-D) gradually varied flows, that is, flows with constant discharge, but gradually varied depth (Chanson, 2009). This equation can be used to qualitatively assess the flow profile in a section of a river and further allows for the analysis of the profile across a series of different reaches with different characteristics (Chanson, 1999). It still uses Chézy or Manning to calculate a friction slope, but it must be borne in mind that this takes these equations beyond their validity. A full solution of the backwater equation is not possible with a closed or continuous solution, but it is possible to use discrete, stepping methods to calculate solutions as a set of points moving away from a control section. This is one of the early examples of numerical solution. Bélanger used the direct step method to calculate the longitudinal distance taken for a given depth change, and other methods such as the standard step method, Euler method, and predictor–corrector methods have subsequently been developed. Bélanger also recognized the importance of the Froude number, which is the ratio of momentum to gravitational effects in an open channel and which governs whether information can flow upstream, in a similar way to its analogy, the Mach number, in compressible gas dynamics. Bélanger also identified that there were singular points in the solution of the backwater equations where the flow was critical and where the Froude number has the value of 1.

The ability to calculate gradually varied flow allowed for the calculation of water profiles between control points and critical points, but it is not applicable at the control points themselves. These control points include structures such as weirs, sluices, and bridges which were increasingly being used in the nineteenth century as a result of the industrial development in Europe. Bélanger paid much attention to the phenomenon of the hydraulic jump. This is observed when the water flow changes from a shallow, fast flow with a Froude number greater than 1 to a flow that is deep and slow with a Froude number less than 1. This transition cannot occur smoothly and is therefore highly turbulent and complex. Bélanger used the momentum concept to derive an equation relating the depths upstream and downstream of the jump (the conjugate depths). After a first attempt, he presented his complete theory in 1841 (Chanson, 2009) and the equation bearing his name is still in use today. Bélanger also went on to examine other control structures such as the broad-crested weir. This formed the basis of the study of rapidly varied flows using the concept of specific energy to obtain insight into the phenomena.

Further progression in 1-D open-channel flow led to the development of the full shallow water equations by Barré de Saint Venant (1871) but these are discussed in the next section in view of their continued widespread use in modern river modeling software.

The next major development of relevance to open-channel flow came in the more general field of boundary layer theory. The boundary where the main flow in a channel meets the bed and banks is of crucial importance particularly in steady flows where there is a balance between gravity and the friction generated at the interface. The contribution of Ludwig Prandtl (1875–1953) to fluid dynamics was significant and comprehensive (Anderson, 2005), but the most significant contribution was to identify the concept of the boundary layer. He postulated that the flow at a surface was zero and that the effect of friction was experienced in a narrow layer adjacent to the surface: away from this boundary layer, the flow was inviscid and could be studied with simpler techniques such as those of Euler. Prandtl then used his theory to derive equations for the velocity profile and consequent shear stresses in the boundary layer. These concepts are particularly relevant to open-channel flows as they demonstrate that the friction effects are confined to a narrow region adjacent to the bed and banks; they also provide a theoretical framework for studying these. Nikuradse used these concepts to study the effect of roughness in pipes and this led to his seminal work that produced the concept of sand grain roughness in pipes. He used the latter to derive friction factors for pipes and much of this theory was later transferred to the study of resistance due to friction in open channels.

In the above, we can see that there has been a move from empiricism to a more physical and mathematical basis for the equations used in open-channel flow. However, a completely nonempirical formulation is still not available and is arguably impossible to achieve. This distinction should always be borne in mind and it is vital to remember that although we can find accurate solutions to the equations, these solutions represent models of reality and whoever is conducting the analysis must also use their knowledge and judgment in drawing conclusions.

So far, this brief history has focused on hydrodynamics, but in addition to the movement of water, an understanding of rivers needs a sound understanding of the movement of sediment and changes in the shape and location of the river channel. The balance between entrainment and deposition of sediment by water flow is the fundamental process governing the geomorphological changes of alluvial rivers at all spatial and temporal scales. The water flow over a mobile bed generates spatial and temporal variations of the sediment transport capacity, causing either net entrainment or net deposition of sediment. Subtractions and additions of sediment are the cause of local bed level changes that in turn alter the original flow field. The discipline of river morphodynamics deals with the interaction between water flow and sediment, which is controlled by the bed shape evolution. Morphodynamic studies use the fundamental techniques of fluid mechanics and applied mathematics to describe these changes and to treat related problems, such as local scour formation, bank erosion, river incision, and river planimetric changes (Parker's e-book).

River morphodynamics became a science with Leonardo da Vinci, who annotated and sketched several morphodynamic phenomena (Manuscript I, 1497), such as bed erosion and deposit formation generated by flow disturbances due to obstacles, channel constrictions, and river bends. Leonardo reported two possible experiments, one on bed excavation by water flow and another on near-bank scour (Marinoni, 1987; Macagno, 1989).

Initiation of sediment motion was first described by Albert Brahams (1692–1758), who wrote the two-part book *Anfangsgründe der Deich und Wasserbaukunst* (*Principles of Dike and Hydraulic Engineering*) between 1754 and 1757. Brahams suggested that initiation of sediment motion takes place if the near-bed velocity is proportional to the submerged bed material weight to the one-sixth power, using an empirically based proportionality coefficient. Later Shields (1936) proposed a general relationship for initiation of sediment motion based on the analysis of data gathered in numerous experiments. He provided an implicit relation between shear velocity, u_* (m s^{-1}), and critical shear stress, τ_c (Pa), at the point of initiation of motion. His relationship is still the one most used for issues dealing with sediment transport.

Although sediment transport is the basic process leading to geomorphological changes in rivers, it is the balance between the volume of sediment entrained by the water flow and the volume of deposited sediment that governs the shape of river beds. Pierre Louis George Du Buat (1734–1809), in his *Principes d'hydraulique* (Du Buat, 1779), realized the importance of bed material for the river cross-sectional shape and conducted experiments to study the cross-section formation in channels excavated in different soil materials ranging from clay to cobbles. However, the first attempt to treat a morphodynamic problem in quantitative terms was made only about one century and a half later by the Austrian Exner (1925), who is consequently considered the founder of morphodynamics. Exner was interested in describing the formation of dunes in river beds, for which he derived one of the existing versions of the conservation laws of bed sediment that are now known as Exner equations. His equation, however, does not describe dune generation, but the evolution of existing dunes:

$$(1-p)\frac{\partial z_b}{\partial t} = -\frac{\partial q_s}{\partial x} \qquad (3)$$

where p is the soil porosity (–); z_b the bed level (positive upward) (m); t the time (s); q_s the sediment transport rate per unit of channel width (m^2 s^{-1}); and x the longitudinal direction (m).

By substituting the sediment transport rate, q_s, with a monotonic function of flow velocity in Equation (3), the obtained relation reads

$$\frac{\partial z_b}{\partial t} = -\left[\frac{dq_s}{du}\right]\frac{\partial u}{\partial x} \quad \text{with} \quad q_s = q_s(u) \qquad (4)$$

where u is the flow velocity (m s^{-1}).

The amount of transported sediment q_s increases when the velocity increases, which means that the term

$$\frac{dq_s}{du} \qquad (5)$$

in Equation (4) is always positive. The result is that erosion occurs in areas of accelerating flow, whereas sedimentation occurs in areas of decelerating flow. This could explain why dunes move downstream. Exner had assumed sediment transport capacity to be simply proportional to the flow velocity, whereas in reality sediment transport capacity is related to the flow velocity to the power three or more (Graf, 1971).

The combination of Exner's relation (Equation (3)) to a relation for sediment transport and to the continuity and momentum equations for water flow leads to a fully integrated 1-D morphodynamic model. Several models of this type have been developed after Exner and it is not easy to establish who was the first to do this. Already in 1947, van Bendegom developed a mathematical model describing the geomorphological changes of curved channels in two dimensions (2-D). The model consisted in coupling the 2-D (depth-averaged) momentum and continuity equations for shallow water with the sediment balance equation (Exner's equation in two dimensions) and a relation describing the sediment transport capacity of the flow. He corrected the sediment transport direction to take into account the effects of spiral flow and channel bed slope. van Bendegom carried out the first simulation of 2-D morphological changes of a river bend with fixed banks by hand, since computers were not available then. Bank erosion was finally introduced in 1-D morphodynamic models in the 1980s (Ikeda *et al.*, 1981) and in 2-D models about 10 years later (Mosselman, 1992).

Only in recent decades it has been realized that river morphology may be strongly influenced by the presence of aquatic plants and animals, as well as by floodplain vegetation (Tsujimoto, 1999). For a long period, vegetation in open channels was only considered as an additional static flow resistance factor to bed roughness, although already at the end of the nineteenth century some pioneer concepts suggested links between the river geomorphology and plants (Davis, 1899).

Over the past few decades the move from empiricism to a more theoretical description of hydrodynamics and morphodynamics has been followed by a move from the expression of theory in equations to computer-based methods. Initially, the latter involved numerical solution of the theoretical equations, but more recently it has been developed with machine-learning techniques for extracting information from measured data which can be seen as a return to empiricism but with vast computing resources compared with past centuries.

2.07.2 State of the Art in Hydrodynamics and Morphodynamics

Rivers convey water and sediment through the catchment to the sea. Moving water and sediment are subjected to forces such as gravity, friction, viscosity, turbulence, and momentum. In order to quantify the system we consider physical variables, such as velocity, depth, discharge, sediment concentration, and channel shape. Hydrodynamics and morphodynamics seek to relate these variables to the forces using the concepts of momentum and energy.

2.07.2.1 Fluid Flow

The concept of scale, both spatial and temporal, is vital to any study of hydrodynamics or morphodynamics and so in the discussions below we consider the following spatial scales:

- *Reach scale (entire river reach)*. A river reach is a large part of the river, which can reasonably be considered as uniform. River reach studies focus on the longitudinal variations of flow field, water depth, and other variables, such as sediment concentration. Often, one value of the variable per river cross section is enough.
- *Cross-section scale (main channel cross section)*. This is the spatial scale of studies for which the transverse variations of flow field, water depth, roughness, etc., are relevant. In this case it is often sufficient to derive the depth-averaged value of the variable and its variation in transverse direction.
- *Depth scale (water depth)*. This is the spatial scale of those studies for which the vertical variations of flow field are relevant.

- *Process scale (local)*. This is the spatial scale at which processes, such as sediment entrainment, deposition, and turbulence, occur.

Whatever scale is being considered, the fundamental principles used in fluid dynamics are conservation of mass, momentum (Newton's second law), and energy. These may need to be simplified according to the scale under consideration, the data available, and the level of detail required in the analysis, but they cannot be violated.

2.07.2.1.1 Mass

Conservation of mass is based on the fact that mass can be neither created nor destroyed; therefore, within a general control volume the accumulation of mass is equivalent to the difference between the input and the output. For a definitive derivation the reader is referred to Batchelor (1967) and for a more accessible derivation to Versteeg and Malalasekera (2007). Expressed in partial differential form, conservation of mass is governed by

$$\frac{\partial}{\partial t}(\rho) + \frac{\partial}{\partial x}(\rho \cdot u) + \frac{\partial}{\partial y}(\rho \cdot v) + \frac{\partial}{\partial z}(\rho \cdot w) = 0 \qquad (6)$$

where ρ is the water density (kg m^{-3}); x the longitudinal distance (m); y the transversal distance (m); z the vertical distance (m); t the time (s); u the flow velocity component in longitudinal direction (m s^{-1}); v the flow velocity component in transversal direction (m s^{-1}); and w the flow velocity component in vertical direction (m s^{-1}).

Equation (6) states that the change in density ρ with respect to time within a volume element plus the change in mass flow ($\rho \times u$) in *x*-direction plus the change in mass flow ($\rho \times v$) in *y*-direction plus the change in mass flow ($\rho \times w$) in *z*-direction is equal to zero.

In comparison, the equation for the conservation of mass in integral form for an arbitrary volume is

$$\frac{\partial}{\partial t}\iiint_V \rho \cdot dV + \iint_S \rho \cdot \mathbf{u} \cdot dS = 0 \qquad (7)$$

where the change in density ρ with respect to time within the control volume plus the change in mass flow $\rho \times \mathbf{u}$ over the surface S of the control volume is zero.

More compactly, the equation in divergent form is

$$\frac{\partial}{\partial t}(\rho) + \nabla \cdot (\rho \cdot \mathbf{u}) = 0 \qquad (8)$$

with the velocity vector $\mathbf{u} = u \times i + v \times j + w \times k$ in the three directions *i*, *j*, *k* in space.

2.07.2.1.2 Momentum

Newton's second law states that the rate of change of momentum of a body is equal to the force applied. In the case of a fluid, this principle is applied to the general control volume and the net momentum flux (inflow less outflow) is equated to the forces. The forces considered depend on the situation under consideration, but the main ones are gravity, shear stress, and pressure. Again the reader is referred to other texts for detailed derivation (Batchelor, 1967; Versteeg and Malalasekera, 2007).

$$\rho \frac{Du}{Dt} = -\frac{\partial p}{\partial x} + \frac{\partial}{\partial x}\left[2\mu\frac{\partial u}{\partial x} + \lambda \operatorname{div} \underline{u}\right] + \frac{\partial}{\partial y}\mu\left[\frac{\partial u}{\partial y} + \frac{\partial v}{\partial x}\right]$$
$$+ \frac{\partial}{\partial z}\left[\mu\left(\frac{\partial u}{\partial z} + \frac{\partial w}{\partial x}\right)\right] + F_x \quad (9a)$$

$$\rho \frac{Dv}{Dt} = -\frac{\partial p}{\partial y} + \frac{\partial}{\partial x}\left[\mu\left(\frac{\partial u}{\partial y} + \frac{\partial v}{\partial x}\right)\right] + \frac{\partial}{\partial y}\left[2\mu\frac{\partial v}{\partial y} + \lambda \operatorname{div} \underline{u}\right]$$
$$+ \frac{\partial}{\partial z}\left[\mu\left(\frac{\partial v}{\partial z} + \frac{\partial w}{\partial y}\right)\right] + F_y \quad (9b)$$

$$\rho \frac{Dw}{Dt} = -\frac{\partial p}{\partial z} + \frac{\partial}{\partial x}\left[\mu\left(\frac{\partial u}{\partial z} + \frac{\partial w}{\partial x}\right)\right] + \frac{\partial}{\partial y}\left[\mu\left(\frac{\partial v}{\partial z} + \frac{\partial w}{\partial y}\right)\right]$$
$$+ \frac{\partial}{\partial z}\left[2\mu\frac{\partial w}{\partial z} + \lambda \operatorname{div} \underline{u}\right] + F_z \quad (9c)$$

where u, v, and w are the components of velocity in the x, y, and z directions respectively; ρ the density; p the pressure; μ the dynamic viscosity; λ the second viscosity; and F_x, F_y, and F_z are the components of body force.

Using the divergent form again gives the Navier–Stokes equations as

$$\rho \frac{Du}{Dt} = -\frac{\partial p}{\partial x} + \nabla \cdot (\mu \nabla u) + F_x \quad (10)$$

2.07.2.1.3 Energy
Conservation of energy comes from the first law of thermodynamics

$$\frac{dE}{dt} = \dot{W} + \dot{Q} \quad (11)$$

which states that the change in the total energy E in the volume element equals the power $\dot{W}$ plus the heat flux $\dot{Q}$ in the volume element. Its application is dependent on the exact situation in which it is applied, and given the large variation in situations it will not be considered in detail here.

2.07.2.2 Numerical Solution

It is possible to solve Equations (6)–(11) analytically in a few, simplified cases, and pioneers such as Prandtl were able to obtain significant insight through doing this. However, the full equations are not amenable to closed solutions and only with the advent of digital computing it has become possible to obtain solutions, albeit approximated ones. To derive a form that is suitable for computer solution, the continuous partial derivatives are converted to difference equations for discrete, point values. There are many ways of doing this and specific cases are discussed below in the relevant context. However, numerical techniques for partial differential equations fall into three main categories: finite differences, finite volumes, and finite elements.

The initial task, as mentioned above, is to convert the differential equations, which have continuously defined functions as solutions, to a set of algebraic equations that connect values at various discrete points that can be manipulated by a computer. This process is called discretization. Various methods are used for this and the main three are finite difference, finite element, and finite volume. More details can be found elsewhere (Wright, 2005).

2.07.2.2.1 Boundary conditions
Whether seeking an analytical or numerical solution, it is necessary to specify boundary conditions for any problem. In open-channel flow, these are specific and tend to be different from those encountered in other fields. In most cases the flow in a reach of river or channel is controlled by a specified discharge at the upstream and downstream boundary, a condition that specifies the depth. The latter includes a fixed depth, a time-varying depth, a critical flow condition, or a depth-discharge relationship.

2.07.2.3 Depth and Process Scales

Viewed at a local scale, the flow is complex and 3-D. It has a predominant downstream flow direction, but the flow can be separated into a boundary layer, where the effects of the boundary and its nature are predominantly felt, and the free stream flow. Within the latter, there are relatively low gradients as the speed of the water increases toward the free surface. The maximum speed is achieved just below the free surface and there is a slight reduction at the surface due to the effects of air resistance and the attenuation of turbulence toward the surface.

At channel bends, a particular flow structure is observed. The water higher in the column travels faster than that at a lower position and therefore does not change its direction in as short a distance. This leads to an increase in the water surface elevation at the outer, concave bank, which in turn drives fluid down and along the bed toward the inner, convex bank. In this way, we observe a super-elevation at the outer bend and a secondary circulation. Further counter-rotating circulations may be induced by the main secondary circulation if the bend is sharp (Blanckaert, 2002). The particular configuration of the flow inside river bends should be taken into account for the modeling of sediment transport and river morphodynamics.

The complete description of fluid flow, based on the continuum hypothesis which ignores the molecular nature of a fluid, is given by the Navier–Stokes equations described above. For a laminar flow, these equations can be discretized to give a highly accurate representation of the real fluid flow. However, laminar flow rarely occurs in open-channel flows so we must address one of the fundamental phenomena of fluid dynamics: turbulence. As the Reynolds number (Reynolds number is defined by $Re = \rho u L/\mu$, where ρ is the density, u the velocity, L the representative length scale, and μ the viscosity) of a flow increases, random motions are generated that are not suppressed by viscous forces as in laminar flows. The resulting turbulence consists of a hierarchy of eddies of differing sizes. They form an energy cascade which extracts energy from the mean flow into large eddies and in turn smaller eddies extract energy from these which are ultimately dissipated via viscous forces.

In straight prismatic channels, secondary circulations are present just as in curved ones, but at a much smaller magnitude. Although the main flow is in the downstream direction with no deviation, the effect of the walls on turbulence causes secondary circulations of the order of 1–2% of the main flow (Beaman *et al.*, 2007).

Turbulence is perhaps the most important remaining challenge for fluid dynamics generally. In theory, it is possible to predict all the eddy structures from the large ones down to the smallest. This is known as direct numerical simulation (DNS). However, for practical flows this requires computing power that is not available at present and may not be available for many years. A first level of approximation can be made through the use of large eddy simulations (LESs). These use a length scale to differentiate between larger and smaller eddies. The larger eddies are predicted directly through the use of an appropriately fine grid that allows them to be resolved. The smaller eddies are not directly predicted, but are accounted for through what is known as a subgrid scale model (Smagorinsky, 1963). This methodology can be justified physically through the argument that large eddies account for most of the effect on the mean flow and are highly anisotropic whereas the smaller eddies are less important and mostly isotropic. Care is needed in applying these methods as an inappropriate filter or grid size and low accuracy spatio-temporal discretization can produce spurious results. If this is not done, LES is not much more than an inaccurate laminar flow simulation. Although less computationally demanding than DNS, LES still requires fine grids and consequently significant computing resources that still mean it is not a viable, practical solution.

In view of the demands of DNS and LES, most turbulence modeling still relies on the concept of Reynolds averaging where the turbulent fluctuations are averaged out and included as additional modeled terms in the Navier–Stokes equations. The most popular option is the k–ε model, which is usually the default option in Computational Fluid Dynamics (CFD) software, where k represents the kinetic energy in the turbulent fluctuations and ε represents the rate of dissipation of k. Interested readers are referred to CFD texts (Versteeg and Malalasekera, 2007) for further details.

Given the complexities and computational demands of 3-D modeling in rivers, it has largely remained a research tool. Notable work has been done by Rastogi and Rodi (1978), Olsen and Stokseth (1995), Hodskinson and Ferguson (1998), and Morvan *et al.* (2002), and a more comprehensive review is given by Wright (2001).

2.07.2.4 Cross-Section Scale

The fully 3-D equations while being a complete representation are computationally expensive to solve and in many situations unnecessarily complex. It is therefore necessary to simplify them and this is often done in the case of open-channel flow. The assumption is made that the flow situation being considered is shallow, that is to say, the lateral length scale is much greater than the vertical one (note: in this regard the Pacific Ocean is shallow in that it is much wider than it is deep!). Once we have assumed shallow water, we can further assume that streamlines are parallel and that there is no acceleration in the vertical leading to the vertical momentum equation being replaced by an equation for hydrostatic pressure. In turn, once we have assumed that there is no vertical velocity, we can depth-integrate the two horizontal velocities, resulting in three equations: one for conservation of mass and two for momentum in the horizontal. These equations can be derived rigorously by either considering the physical situation or applying the assumptions to the Navier–Stokes equations.

These 2-D equations are less time consuming to solve than the Navier–Stokes equations and there is a significant body of research devoted to this. This has culminated in a number of computer codes that are available both commercially and as research codes. These can be classified into those based on the finite difference, finite element, or finite volume methodology. In the present context one significant difference is relevant. The finite element method minimizes the error in the solution to the underlying mathematical equations in a global sense while finite volume minimizes it in a local sense. This means that a finite volume method will always conserve mass at each time step and throughout a simulation. The finite element and finite difference methods will only have true mass conservation once the grid is refined to a level where further refinement makes no further change to the solution.

A number of codes based on the finite difference method have been developed and used in practice. Details of each can be found on the developers' websites. Examples are ISIS2 D (Halcrow), MIKE21 (DHI), TUFLOW (WBM), and Sobek & Delft3d (Deltares).

Codes using the finite element method are less common in river applications, but have been popular for flows in estuaries and coastal areas where the geometries can be complex. Examples are TELEMAC-2 D (EDF) (Bates, 1996), SMS produced by Brigham Young University based on codes from the USACE such as RMA2 D (King, 1978), and CCHE2 D produced by NCCHE, University of Mississippi (Wang *et al.*, 1989).

Codes using the finite volume method have been developed more recently as their strength in mass conservation and their ability to correctly model transitions have been realized. The latter is based on the use of Godunov-based methods (Sleigh *et al.*, 1998; Alcrudo and Garcia-Navarro, 1993; Bradford and Sanders, 2002) or on the use of total variation diminishing (TVD) schemes (Garcia-Navarro and Saviron, 1992). In recent decades, there has been significant development of unstructured finite volume codes (Anastasiou and Chan, 1997; Sleigh *et al.*, 1998; Olsen, 2000). These can be considered as a combination of finite element and finite volume approaches. They use the same unstructured grids as finite element and solve the mathematical equations in a finite volume manner that ensures conservation. In this way, they ensure physical realism and ease of application.

The issue of wetting and drying is a perennially difficult one for 2-D models (Bates and Horritt, 2005). As water levels drop, areas of the domain may become dry and the calculation procedure must remove these from the computation in a way that does not compromise mass conservation or computational stability. Most available codes can deal with this phenomenon, but they all compromise between accuracy and stability. This issue must be carefully examined in results from any 2-D simulation where wetting and drying are significant. There is active research in this area with a number of recent contributions that may well improve matters (Liang, 2008; Lee and Wright, 2009).

In assuming a depth-averaged velocity, 2-D models neglect vertical accelerations and make no prediction of vertical velocities. This, in turn, means that they do not predict or model the effects of the secondary circulations described above. The neglect of secondary circulations can lead to inappropriate model predictions for velocity and depth and in turn this can cause inaccuracies in morphological studies where the secondary circulations are a significant contribution to bed/bank erosion. There are a number of amendments to 2-D models to take an account of this phenomenon. The simplest calculates a measure of helical flow from an analysis of the velocity and acceleration vector at a point. This, in turn, is used to calculate a vertical velocity profile and vertical velocities. This approach is adopted in different forms in MIKE21C (DHI 1998), CCHE3D (NCCHE, University of Mississippi; Kodama, 1996), and CH3D (USACE; Engel et al., 1995) among others.

A more accurate but computationally expensive method is the layered model (TELEMAC-3D, EDF; Delft3D, Deltares; TRIVAST; Falconer and Lin, 1997). This establishes a number of vertical layers and solves equations for the horizontal velocities in each layer. Subsequently, equations are solved for a vertical velocity based on an analysis of the interactions between each layer and the water depth is calculated appropriately. This is mainly suitable for wide bodies of water with significant vertical variations of velocity, temperature, salinity, or other variables in the vertical such as estuaries, lakes, and coastal zones. Nex and Samuels (1999) applied TELEMAC-3D to the River Severn. They reported some success and qualitative agreement with measurements. A further development of this technique is to include the treatment of nonhydrostatic pressure variations (Stansby and Zhou, 1998; Casulli and Stelling, 1998).

A 2-D model of a river and its floodplains require information about the channel bed topography and the terrain heights of the surrounding floodplain. In the past this required a mixture of time-consuming measurements and interpolation from published, paper-based maps. A significant advance over the past 10–15 years has been the use of remotely sensed data, which offer both increased accuracy and density of data along with reduced collection times. This comes at some expense, but the cost continues to come down. Current techniques such as light detection and ranging (LiDAR) can provide data every 25 cm at accuracies down to 10 cm. More experimental techniques can also be used to measure through the water surface to give detailed and accurate bed topography. Besides providing accurate data for model construction, remote sensing can also provide data on flood extents for use in validation. These procedures are now in regular use in commercial work and continue to be an area of active research. More details can be found in the literature (Horritt et al., 2001; Wright et al., 2008). Remotely sensed data need to be used with careful consideration of accuracy and the level of detail required in specific areas. For example, in modeling the interaction of a main channel with a floodplain it is necessary to have accurate data along the embankments of the main channel, and commonly used LiDAR data can miss these features through the use of a regular rectangular grid.

In this case, the LiDAR may need to be supplemented by other techniques such as Global Positioning System (GPS) (Wright et al., 2008).

2.07.2.5 River Reach Scale

When considering long river reaches even a 2-D model can become cumbersome. In such cases, the length of the river is of several orders of magnitude greater than the width. It is therefore assumed that lateral variations in velocity and free surface height can be neglected and that the flow direction is entirely along stream. Under these assumptions the equations first formulated by Jean-Claude Barré de Saint Venant apply and these have formed the basis for the most widely used commercial river modeling packages. Each of these conceptualizes the river as a series of cross sections. At each the velocity is assumed perpendicular to the cross section. The resistance due to the bed and banks is based on one of the steady-state formulations for normal flow such as Mannings, Chézy, or Colebrook-White (Chanson, 1999).

Early numerical methods for solving this system of equations were pioneered by Abbott and Ionescu (1967) and Preissmann (1961). Both of these methods are essentially parabolic in nature, while the equations are hyperbolic. In view of this more recent methods have drawn on the body of research from compressible gas dynamics which has a similar set of equations. This has produced algorithms that are more robust and able to correctly represent transitions (Garcia-Navarro et al., 1999; Crossley et al., 2003), but which are not so straightforward to implement particularly with regard to the incorporation of hydraulic structures such as weirs and sluices.

Another recent development that is proving popular in some countries is the linking of 1-D and 2-D models. The former offers efficiency and lower data requirements while the latter can give better results on floodplains. A number of techniques have been proposed for linking these models (Dhonda and Stelling, 2003; Wright et al., 2008), but which one is the most reliable or successful is not yet clear. In fact, there is evidence to suggest that there are considerable differences among the different formulations and even among the different users of the same software package (Kharat, 2009).

Although the 1-D approach is based on an analysis of the situation at a cross section, it can be applied to rivers of significant lengths up to hundreds if not thousands of kilometers. Further through the incorporation of junction equations relating flows and depths at confluences and difluences, it can be used to model complex networks of rivers and channels.

Over the past three decades, several commercial packages have been developed based on the 1-D shallow water equations (InfoWorks, ISIS, MIKE11, and Sobek, among others). In the US, the USACE Hydrologic Engineering Center has also developed the HEC-RAS software that is freely available. These software packages combine the basic numerical solution with sophisticated tools for data input and graphical output. They are designed to make use of remotely sensed data and to provide 2-D and 3-D output in both steady and animated formats.

2.07.2.6 Spatial Scales in River Morphodynamics

River morphodynamics deals with the shape and, in a wider sense, composition of the river bed. The shape of alluvial rivers is made up by the combination of many geomorphological forms, which can be recognized at specific spatial scales, from small ripples to large bars and meanders. The development of geomorphological forms is related to the balance between entrainment and deposition of sediment over different control volumes and times. In modeling, every factor influencing sediment motion has to be taken into account, but in different ways depending on the spatial and temporal scale of the study (Schumm and Lichty, 1965; Phillips, 1995). In particular, processes that operate at smaller scales are parametrized to take into account their effects at larger spatial and temporal scales. Processes that operate at larger scales may be represented as boundary conditions for the studies focusing on smaller scales.

At the largest spatial scale, the one of the entire river basin or single sub-basins, we can recognize the entire river network. Typical river basin-scale issues involve soil erosion, reservoir or lake sedimentation, as well as solid and water discharge formation. Basin-scale studies are characterized by the description of the entire river drainage network or large parts of it, such as the delta or a sub-basin. Geographic information systems, 0-D and 1-D morphodynamic models, as well as 1-D or 2-D runon–runoff models, are the typical tools used. The river basin scale is not further treated here, since its issues generally fall under the other related disciplines of hydrology and physical geography.

Lowering the observation point and zooming in on the river system, different river reaches, each one characterized by planform style and sinuosity, are highlighted. A single river reach is characterized by one value of the water discharge, but changing with time. Depending on the reach characteristics, the typical temporal variations range from hours to days for the discharge; from years to several tens of years for the longitudinal bed slope. A river reach in morphodynamic equilibrium is characterized by a longitudinal bed slope that can be considered constant at a chosen temporal scale (de Vries, 1975). Reach-scale issues mainly deal with the assessment of the environmental impact of human interventions, such as river training, and with the natural river evolution on the long term. For this, morphodynamic studies need to determine bed aggradation and degradation, along the river reach, changes in sinuosity and planform style. The typical tools are 0-D reach-averaged formula (e.g., Chézy, 1776; Lane, 1955), describing the water flow at reach-scale morphodynamic equilibrium, as well as 1-D cross-sectionally averaged models. Commercial 1-D codes updating the riverbed elevation are: MIKE11 (DHI) and SOBEK-RE (Deltares).

By further zooming in on the river, the attention moves to the river corridor, or river belt, the area including main river channel and floodplains. Specific morphological features recognizable at this spatial scale are scroll bars inside river bends (**Figure 1**), a sign of past bend grow. Corridor-scale studies mainly deal with flood risk, river rehabilitation projects, as well as river planimetric changes. The typical tools are 2-D, depth-averaged, or a combination of 1-D (cross-sectionally averaged) and 2-D (depth-averaged) morphodynamic models. These models often have to include formulations for bank retreat and advance and for the effects of (partly) submerged vegetation on water levels, sediment transport, and deposition. Commercial codes developed for the study of the river morphological changes at this and smaller spatial scales are (among others): MIKE21 (DHI), Delft3D (Deltares), and SOBEK-1D-2D (Deltares). Examples of free 2-D codes are: FaSTMECH (Geomorphology and Sediment Transport

Figure 1 Aerial view of a tributary of the Ob River (Russia). Scroll bars on floodplains and point bars inside river bends are clearly visible. Courtesy of Saskia van Vuren.

Laboratory of USGS) and RIC-Nays (Hokkaido University). These two models adopt the user interface IRIC, developed in the Geomorphology and Sediment Transport Laboratory of USGS (USA).

Central and multiple bars, either migrating or static, are the characteristic geomorphological features to be studied at the cross-section scale (Figure 2). Typical engineering issues are river navigation and the design of hydraulic works, such as trains of groynes, bridges, and offtakes. Typical tools are 2-D, depth-averaged, models, formulated for curved flow (van Bendegom, 1947), often including bank retreat and advance (Mosselman, 1998). Modeling often regards bar formation, bar migration, and channel widening and narrowing as the natural development or as the effects of human interventions.

If the observation point moves from a point above the river to a point inside the river channel, the vertical contour of the river cross section becomes visible (Figure 3). Water-depth variations in transverse direction, due to the presence of local deposits and scours, as well as water-depth variations in longitudinal direction, due to the presence of dunes, are the major morphological features observable at this spatial scale. Typical depth-scale studies deal with scour formation around structures, bank erosion, bank accretion, as well as dune development and migration. Typical tools are either 3-D or 2-D and 1-D vertical morphodynamic models, often focusing on local bed level changes or on vertical variations, of, for instance, salinity, suspended solid concentration, soil stratification, and bank slope.

The smallest spatial scale that is relevant for the river morphodynamics is called the process scale. This is the scale of fundamental studies describing processes such as sediment entrainment and deposition, for which phenomenon such as turbulence plays a major role. The typical geomorphological forms to be studied at this small spatial scale are ripples (Figures 4–6). The typical tools are detailed morphodynamics models in one, two, and three dimensions.

In morphodynamics, temporal and spatial scales are strongly linked. Phenomena with small spatial scales also have small temporal scales, and phenomena with large spatial scales have large temporal scales (de Vriend, 1991, 1998; Blöschl and Sivapalan, 1995). The linkage between spatial and temporal scales is formed by sediment transport. For the development or migration of a small bedform, only a small amount of sediment needs to be displaced, whereas large amounts of sediment are needed for the development of large geomorphological forms, such as bars.

Phenomena interact dynamically when they occur more or less on the same scale. Small-scale phenomena, such as ripples, appear as noise in the interactions with phenomena on larger scales, such as bar migration, but they can produce residual effects, such as changes of bed roughness (Figure 5). Their effect on larger scales can be accounted for by parametrization procedures (upscaling). Phenomena operating on much larger spatial and temporal scales can be treated as slowly varying or constant conditions. They define scenarios, described in terms of boundary conditions, when studying their effects on much smaller scales. Thus, basin-scale studies are essential for the generation of the input (boundary conditions) for the morphodynamic studies on smaller spatial scales.

2.07.2.7 Geomorphological Forms in Alluvial River Beds

Geomorphological forms in rivers can be caused by the presence of geological forcing, human interventions, and man-made structures, but they also arise as a natural instability of the interface between the flowing water and

Figure 2 Multiple bars in the braided Hii River (Japan). Courtesy of Takashi Hosoda.

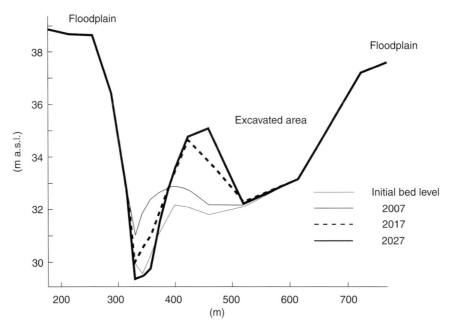

Figure 3 River Meuse (the Netherlands): temporal bed level changes during the period 2007–27. On the vertical, the bed elevation in meters above sea level (Villada Arroyave and Crosato, 2010).

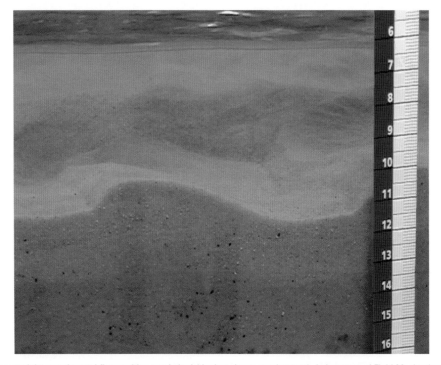

Figure 4 Ripples in a straight experimental flume with a sandy bed (the bar shows centimeters). Laboratory of Fluid Mechanics of Delft University of Technology.

sediment. In analogy with the interaction between air moving above water (wind), the instability of the water–sediment interface produces waves of different sizes, which can coexist and interact with each other.

Ripples are the smallest ones, originating from the instability of the viscous sublayer near the river bed (**Figure 4**).

Dunes are the main source of hydraulic resistance of a river and hence a key factor in raising water levels during floods (**Figure 7**). They are also the first parts of the river bed that need to be dredged to improve navigation. Dune formation and propagation is so intimately linked to sediment transport, that the latter cannot be modeled properly without accounting for dunes (ASCE Task Committee on Flow and Transport over Dunes, 2002). Bars are the largest waves in the river bed; they can be scaled with the channel cross section (**Figure 2**).

Figure 5 The presence of 3-D ripples acts as noise for the study of alternate bars in this laboratory experiment carried out at the Laboratory for Fluid Mechanics of Delft University of Technology.

2.07.2.7.1 Ripples and dunes

For increasing Froude numbers the river bed is first plane and then covered by ripples and dunes. The flow regime close to the critical Froude number ($Fr \approx 1$) is again characterized by plane bed. If the Froude number increases further (supercritical flow), antidunes begin to form with upstream breaking waves over the crest (Simons and Richardson, 1961). Southard and Boguchwal (1990) provided the most extensive bedform phase diagrams showing the possible occurrence of ripples, dunes, antidunes, or plane bed under different sediment size and flow conditions.

Bedforms may have either a 2-D or a 3-D pattern. 2-D ripples and dunes have fairly regular spacing, heights, and lengths. Their crest lines tend to be straight or slightly sinuous, and are oriented perpendicular to the mean flow lines (**Figure 6**). In contrast, 3-D features have irregular spacing, heights, and lengths with highly sinuous or discontinuous crest lines (Ashley, 1990), as in **Figures 4** and **5**.

In general, ripples scale with the sediment diameter while dunes scale with the water depth (Bridge, 2003), but there is no clear distinction between ripples and dunes for limited water depths, as for instance, in flume experiments. Extensive data compilations by Allen (1968) and Flemming (1988) demonstrated that there is a break in the continuum of observed bedforms discriminating ripples from dunes. For instance, ripples are only present for fine sediment with $D<1$ mm. However, there are no generally valid techniques to divide ripple from dune regimes and some authors choose to make no distinction at all.

The first theoretical study of dune instability was carried out by Kennedy (1969). Spectacular progress in knowledge of dune dynamics is linked to the increasing sophistication of numerical modeling (Nelson et al., 1993). Recent models produce detailed simulations of the instantaneous structure of flow over a dune-covered bed. Giri and Shimizu (2006)

Figure 6 2-D ripples in the Het Swin Estuary (the Netherlands).

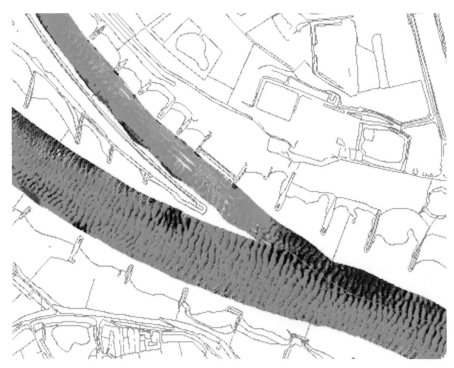

Figure 7 Dunes in the Waal River and Pannerdense Canal (the Netherlands) on 4 November 1998. Flow from right to left. Courtesy of Rijkswaterstaat. Upstream of bifurcation: discharge 9600 m³ s⁻¹, water depth 10.7 m, mean grain size 3.3 mm, flow velocity 2.1 m s⁻¹, dune height 1.0 m, and dune length 22 m. Analysis by Wilbers, Department of Physical Geography, Utrecht University, Utrecht, The Netherlands.

developed a 2-D model for the prediction of dunes under unsteady flow regime. Nabi *et al.* (2009) provided the first detailed 3-D model of dune formation.

2.07.2.7.2 Bars

Bars are shallow parts of river bed topographies that become visible at low flows and can be either migrating or steady. Bars occur in more or less regular, periodic patterns as a result of interactions between flowing water and sediment. In a river channel one or more parallel rows of bars may be present; the number of rows present is called bar mode m. Alternate bars have mode m equal to 1 (**Figure 8**); multiple bars have mode larger than 2. Migrating bars as well as steady bars (Crosato and Desta, 2009) develop spontaneously as a result of morphodynamic instability and for this they are often referred to as free bars. Confined sediment deposits caused by local changes of the channel geometry, such as point bars inside river bends (**Figure 1**), should therefore be distinguished from free bars. The stability analyses performed by, among others, Hansen (1967), Callander (1969), and Engelund (1970) define the conditions that govern the development of bars in alluvial river channels. The width-to-depth ratio of the river channel is the dominating parameter for free bar formation: the larger the width-to-depth ratio, the larger the bar mode. This means that multiple bars form at larger width-to-depth ratios than alternate bars. Moreover, no bars can form for width-to-depth ratios that are smaller than a certain critical value. Parker (1976) and Fredsøe (1978) related the presence or absence of free bars to the channel planform, that is, meandering or braided. By persistently enhancing opposite bank erosion, steady alternate bars (**Figure 8**, left) are seen as a

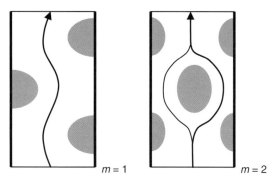

Figure 8 Left: alternate bars ($m=1$). Right: central bars ($m=2$).

key ingredient for the evolution of straight water courses into meandering water courses (Olesen, 1984). Multiple bars are a characteristic of braided rivers.

The linear theory by Seminara and Tubino (1989) defines marginal stability curves separating the conditions in which a certain number of bars per cross section grows from the conditions in which the same bar mode decays. The river is supposed to select the bar mode with the fastest growth rate, which is a function of the width-to-depth ratio, the Shields parameter, the sediment grain size, and the particle Reynolds number. A single physics-based formula was recently derived by Crosato and Mosselman (2009) from a stability analysis. The formula allows one to compute directly the mode of free bars that develop in an alluvial channel, but it is limited to rivers having width-to-depth ratio smaller than 100. By assuming that meandering rivers are characterized by the presence of alternate bars and braided rivers by multiple bars, the same formula can also be used to determine the type of planform that can be expected to develop after widening or narrowing of a river channel.

2.07.2.8 River Planimetric Changes

The study of the river planimetric changes requires the assessment of both bank erosion and bank accretion rates and for braided-anabranched rivers also to the assessment of the stability of channel bifurcations.

Meandering rivers have single-thread channels with high sinuosity and almost constant width (**Figure 1**). They could be regarded as a particular type of braided rivers (Murray and Paola, 1994), those having bar mode equal to 1. River meandering is governed by the interaction between bank accretion, bank erosion, and alluvial bed changes (**Figure 9**). Bank erosion causes channel widening and enhances opposite bank accretion. Conversely, bank accretion causes river narrowing and enhances opposite bank erosion. The two processes of bank erosion and accretion do not occur contemporarily, and for this reason the river width is subject to continuous fluctuations. However, generally a stable time-averaged width is achieved in the long term. Understanding the process of bank accretion and width formation is therefore a fundamental prerequisite for the modeling of meandering river processes and, more in general, for the modeling of the river morphology.

All existing meander migration models (Ikeda et al., 1981; Johannesson and Parker, 1989; Crosato, 1989; Sun et al., 1996; Zolezzi, 1999; Abad and Garcia, 2005; Coulthard and van de Wiel, 2006) assume the rate of bank retreat to be the same as the rate of opposite bank advance. This means that the lateral migration rate of the river channel can be assumed to be equal to the retreat rate of the eroding bank. This is in turn assumed to be proportional to the near-bank flow velocity excess with respect to the normal flow condition, following the approach by Ikeda et al. (1981). Some meander migration models take also into account the effects of the near-bank water depth excess on the bank retreat rate (e.g., Crosato, 1990). The proportionality coefficients in the channel migration formula are supposed to weigh the bank erosion rates.

These coefficients should be a function of the bank characteristics only, but are in fact bulk parameters incorporating the effects of opposite bank advance and some numerical features (Crosato, 2007).

Existing theories on river meandering focus on the assessment of bank retreat rates without defining the conditions for the opposite bank to advance with the same speed. However, it is just the balance between the rate of bank advance and the rate of opposite bank retreat that makes the difference between braiding and meandering (**Figure 10**). A meandering river requires that, in the long term, the bank retreat rate is counterbalanced by the bank advance rate at the other side. If bank retreat exceeds bank advance, the river widens and, by forming central bars or by cutting through the point bar, assumes a multi-thread (braided) pattern. If bank advance exceeds bank retreat, the river narrows and silts up.

So far, most research has focused on the processes of bank erosion (e.g., Partheniades, 1962, 1965; Krone, 1962; Thorne, 1988, 1990; Osman and Thorne, 1988; Darby and Thorne, 1996; Rinaldi and Casagli, 1999; Dapporto et al., 2003; Rinaldi et al., 2004) and bed development, whereas the equally important bank accretion has received little attention

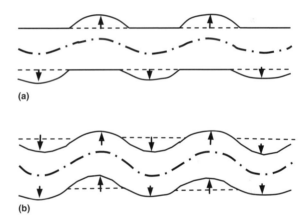

Figure 10 A sinuous water flow is not sufficient for meandering. (a) straight river planform with bank retreat, but without bank advance. (b) meandering river planform in which bank advance counterbalances bank retreat.

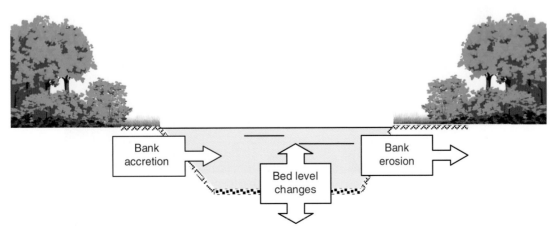

Figure 9 Morphological processes shaping the river cross section.

(Parker, 1978a, 1978b; Tsujimoto, 1999; Mosselman et al., 2000). As a result, there are no comprehensive physics-based river width predictors.

Bank accretion is governed by the dynamic interaction between riparian vegetation, flow distribution, frequency as well as intensity of low and high flow stages, local sedimentation, soil strengthening and by the interaction between opposite (eroding and accreting) banks. Bank erosion and accretion strongly depend on climate (Crosato, 2008). Climate changes can therefore alter the river cross section and the river pattern. Present knowledge on river morphological processes is insufficient to fully assess these effects.

A number of existing 2-D and 3-D morphological models, such as Delft3D, treat bank accretion as near-bank bed aggradation and bank erosion as near-bank bed degradation. These models are suitable for the prediction of width changes of channels without vegetation and with mildly sloping banks, but fail to predict the morphodynamics of meandering rivers, which are characterized by cohesive banks and riparian vegetation. A few 2-D morphological models simulate bank erosion, but not bank accretion. One example is the model RIPA, which was developed at Delft University of Technology by Mosselman (1992) and further extended by the University of Southampton (Darby et al., 2002).

In large valleys or near the sea, the river can split into several channels. In anabranched rivers each anabranch is a distinct, rather permanent, channel with bank lines (**Figure 11**). The river bed is mainly constituted by loose sediment, such as sand and gravel, whereas silt prevails at the inner parts of bends and in general where the water is calm. Anabranches are commonly formed within deposits of fine material. Vegetation and soil cohesiveness stabilize the river banks and the islands separating the anabranches, so that the planimetric changes are slow if compared to the river bed changes. Studying the morphological changes of this type of rivers requires the assessment of the stability of bifurcations (Wang et al., 1995; Kleinhans et al., 2008). Experimental and theoretical research started almost 100 years ago (Bulle, 1926; Riad, 1961) and are still going on (Bolla Pittaluga et al., 2003; De Heer and Mosselman, 2004; Ten Brinke, 2005; Bertoldi et al., 2005; Kleinhans et al., 2008). The major difficulty rises in the assessment of sediment distribution between the two branches of the bifurcating channel, which is a function of water discharge distribution, sediment characteristics, channel curvature at the bifurcation point, and presence of bars.

2.07.2.9 Bed Resistance and Vegetation

In any study of a river, whether it is experimental, full-scale measurement or numerical in 1-D, 2-D, or 3-D, the irregular geometry of the boundary (bed and banks) cannot be directly represented. Even with full-scale measurements, the boundary cannot be accurately mapped at the scale of the bed material. In all cases, a conceptual representation of the effect of the boundary on the flow is used to account for momentum and energy dissipation. There is much misunderstanding of the nature of these resistance or roughness laws and inconsistencies in their application (Morvan et al., 2008). Clearly, given the importance of boundary resistance determining flow and depth it is necessary to have a clear understanding of the various methods and any limitations on their applicability.

If it were possible to solve the Navier–Stokes equations on a grid that was fine enough to resolve the smallest scale of turbulence (the Kolmogorov scale), then there would be no need for a turbulence model or a model of the effect of the boundary. However, this is not yet generally possible and even in 3-D solutions there is a need to simplify the equations. In 3-D a turbulence model is used as well as the resistance model, but in 2-D and 1-D models both these phenomena tend to be included in a resistance term. This in turn can lead to uncertainty in the definition and a lack of rigor in its application. This uncertainty together with lack of rigor

Figure 11 Anabranched planform: the Amazon River near Iquitos, Peru. Courtesy of Erik Mosselman.

increases as the dimensionality decreases. Another consequence of the difference between 3-D, 2-D, and 1-D modeling is that the value of a parameter such as roughness height will vary between each dimensionality even if the physical situation under consideration is identical due to the fact that the resistance model incorporates different physical phenomena in each case. This can be seen in **Figure 12** from Morvan *et al.* (2009).

The results from Manning's equation differ from those of the 3-D model as k_s is varied indicating that the two are quite different. In fact, the Manning's equation results are more sensitive to changes in the roughness which is due to the 3-D model representing phenomena such as turbulence and secondary circulations directly rather than in the resistance parametrization.

In view of its significance there has been much work in this area over the last century and the reader is referred to Davies and White (1925), Ackers (1958), ASCE (1963), Rouse (1965), Yen (1991, 2002), and Dawson and Fisher (2004). Specific types of roughness are considered by Sayre and Albertson (1963) and ESDU (1979). Reynolds and Schlicting have written useful textbooks on the wider subject (Reynolds, 1974; Schlichting *et al.*, 2004). A good review of the topic in the context of modeling is given in the paper by Morvan *et al.* (2008).

Early work on roughness was performed in pipes by Nikuradase, mentioned above as building on the work of Prandtl. The Darcy–Weisbach equations for pipe flow uses a friction factor that is based on geometry (diameter in the case of pipes), mean velocity, and surface characteristics based on a relative roughness defined by a quantity known as the Nikuradse equivalent sand grain roughness nondimensionalized by the diameter of the pipe. In 1-D open-channel studies the geometric parameter of diameter is replaced by hydraulic radius (area divided by wetted perimeter) which leads to discontinuities when the flow moves onto flood plains as the wetted perimeter increases abruptly while the cross-sectional area does not. It is clear that using the theory from pipe flow in open channels raises difficult issues with complex cross sections and using the hydraulic radius to capture geometrical effects is problematic. In practice, many people use hydraulic radius and then adjust the value of the Nikuradse roughness or equivalent to ensure that the frictional head loss per unit channel length matches the bed slope. This demonstrates that the roughness parameter is often related to energy loss in the model as much as any physical measurement of the nature of the surface. In fact, the parameter is a function of local bed geometry, flow regime, cross-section geometry, and turbulence. Given the wide range of effects, it is clear that the parameterization depends on the model used for the overall fluid flow.

It is worth considering in a little more detail the nature of the forces acting on the fluid due to presence of the bed. Morvan *et al.* described these as

- skin drag (e.g., roughness due to surface texture, grain roughness);
- form drag (e.g., roughness due to surface geometry, bedforms, dunes, separation, etc.); and
- shape drag (e.g., roughness due to overall channel shape, meanders, bends, etc.).

Skin and form drag can be considered to occur on a plane, but shape drag is due to larger-scale 3-D patterns. Again, it is clear that the way each of these is represented depends on the sort of model used. A resistance parameter such as Manning's coefficient n or Chézy used at a reach scale is based on the concept of bed resistance, although in practice it is also calibrated to account for shape drag.

In many representations, roughness is characterized by a roughness height. It is often not appreciated that although this quantity has the units of length, it is not a measure of the height of the roughness elements. It is rather a parameter in an analytical model of flow at the wall (i.e., in 3-D):

$$\frac{u_\tau}{u_*} = \frac{1}{\kappa} \ln(E(k_s^+) \cdot y^+) \qquad (12)$$

where $E(k_s^+)$ is a function of the nondimensional roughness height, $k_s^+ = k_s u_*/\nu$, in which k_s is the roughness height, κ the von Karman's constant usually taken equal to 0.41, and ν is the kinematic viscosity.

It seems attractive to base our estimates for roughness heights on work such as Nikuradse's on relatively zsmooth experimental channels. This has led to formulations such as $k_s = 3.5 \times D_{84}$ or $k_s = 6.8 \times D_{50}$, where D_{XX} stands for the grain diameter for which $xx\%$ of the particles are finer, reported in Clifford *et al.* (1992). The latter paper makes interesting reading and shows that the grain–roughness relationship is inadequate. This is because there are several momentum loss mechanisms in these flows and they are not represented by such a simple equation. A further complication is that in some 3-D simulations values of the roughness height are derived from these formulas that are in fact greater than the size of the grid perpendicular to the wall. This could suggest that the grid resolves flows at a scale less than the size

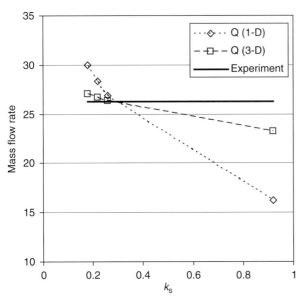

Figure 12 Variation of the mass flow rate in 1-D model and 3-D model for a trapezoidal channel compared against the measured value (Morvan *et al.*, 2009).

of the roughness which contradicts the fact that the roughness features have been removed to give a smooth planar surface.

The above discussion has focused mainly on 3-D models, but the situation when we consider 2-D and 1-D models is even less clear. Continuing the approach of considering surface roughness as the parameter governing resistance, various formulas have been proposed to connect the roughness height with a parameter such as Manning's n for 1-D models:

HR Wallingford tables (Ackers, 1958):

$$k_s(\text{mm}) = (n/0.038)^6 \qquad (13)$$

Massey (Massey 1995):

$$k_s(\text{SI}) = 14.86R/\exp_{10}\left(\frac{0.0564R^{1/6}}{n}\right) \qquad (14)$$

Chow (1959):

$$k_s(\text{SI}) = 12.20R/\exp_{10}\left(\frac{0.0457R^{1/6}}{n}\right) \qquad (15)$$

Strickler (1923):

$$k_s(\text{ft}) = (n/0.0342)^6 \qquad (16)$$

These differ not only in the numerical values used, but also in the functional form. They also give large ranges for roughness height for small variations in Manning's n. This indicates the uncertainties in this process, which have led authors to seek better means of characterizing the geometry and surface characteristics in order to approximate resistance.

In some cases, particularly with large cross section covering a main channel and floodplains, there are zones with quite different resistances within the cross section. In such cases divided channel method (DCM) can be used where the cross section is divided into panels, and a conveyance is calculated in each one before being combined into a composite value (Knight, 2005). This has been shown to be successful and is incorporated in most commercial software. All these methods assume quasi-straight river reaches, and do not include lateral momentum transfer effects. Thus, they cannot predict accurately either the water level in compound river channels or the proportion of flow between the main channel and floodplains. More recent developments include the effect of flow structure, through the adoption of improved methods (Knight, 2005). These may be grouped under the headings: the DCM, the coherence method (COHM), the Shiono and Knight method (SKM), and the lateral division method (LDM). Several authors have presented examples of these methods applied to fluvial problems (Knight et al., 1989; Knight, 2005). The SKM, for example, uses three parameters rather than just the one used by approaches such as Manning's or Chézy. In fully 2-D shallow water models the flow is considered in separate vertical water columns and the variables are depth and two perpendicular velocities or discharges per unit width. In this case, the resistance is applied only to the surface at the base of the water column (the bed) and the roughness height will be different even from a 1-D model and for the same bed material.

It is clear from this discussion that parametrizing resistance in open-channel flows is not straightforward and needs knowledge and experience from numerical and physical modeling. A number of conclusions can be drawn (based on those in Morvan et al. (2009)):

- roughness varies between models, which represent different dimensions and therefore reach-scale roughness is a different concept from local roughness;
- using roughness to represent features other than sand-grain roughness lessens the validity of the underlying theory and is questionable;
- models of roughness in 1-D hydraulic models are valid and will continue to be useful when based on sound analysis and calibrated appropriately; and
- 1-D modelers should focus more on estimating conveyance than establishing one sole value of Manning's n or Chézy's C for a channel.

This shows that the representation of resistance in real rivers is a complex task. It could therefore lead to the conclusions that hydraulic modeling is fraught with difficulty and that it is of little benefit. This is not the case and when used with care they are extremely useful (Knight et al., 2009).

If the representation of the resistance due to the nonuniform surface of the bed and banks presents a significant challenge to modelers, the representation of the effects of vegetation is perhaps an even greater one. Further, the need to represent vegetation is becoming greater with the design of more natural channels and the need to model inundation flows across vegetated floodplains. Besides being nonuniform, vegetation experiences changes in its resistance as it deforms as the velocity of the water increases.

The effects of vegetation on river processes are many, complex, and difficult to quantify (Fisher and Dawson, 2003; Rinaldi and Darby, 2005; Gurnell et al., 2006). The ability of vegetation to stabilize river banks (Ott, 2000) partly depends upon scale, with both size of vegetation relative to the watercourse and absolute size of vegetation being important (Abernethy and Rutherfurd, 1998). Vegetation stabilization is most effective along small watercourses. On relatively large rivers, fluvial processes tend to dominate (Thorne, 1982; Pizzuto, 1984; Nanson and Hickin, 1986). The effect of vegetation on the conveyance of a channel depends on a number of factors such as density, type, height, and distribution of plants and their development stage (Allmendiger et al., 2005; Dijkstra, 2003).

At the local scale, single plants act as roughness elements. Isolated trees and relative small clusters of plants increase turbulence around them leading to local scour, just as bridge piers do. Dense vegetation, instead, reduces the flow velocity between and above plants and sediment transport, enhancing local siltation. In this way, riparian vegetation increases the development of natural levees during floods as well as bank accretion. Rooted plants reduce local soil erosion by binding the soil with the roots (**Figure 13**) and by covering it. In this way, riparian vegetation decreases bank erosion. Heavy trees, however, can enhance gravitational bank failure by increasing the load on the bank (Ott, 2000). Finally, vegetation causes local accumulation of organic material (falling leaves,

Figure 13 Roots protecting the river bank against erosion. Geul River (The Netherlands). Courtesy of Eva Miguel.

branches, and dead plants), which further reinforces the soil cohesion and strength (Baptist, 2005; Baptist and De Jong, 2005; Baptist et al., 2005).

At the cross-section scale vegetation affects the river morphodynamics by acting on (Crosato, 2008) (1) river bed degradation/aggradation, (2) bank erosion, and (3) bank accretion by:

- Deflecting the water flow. Aquatic and riparian vegetation increase the local hydraulic roughness and for this reason, the flow concentrates where vegetation is absent (Tsujimoto, 1999; Pirim et al., 2000; Rodrigues et al., 2006). This lowers the flow velocity within the plants, where sedimentation increases, and causes bed degradation in the nonvegetated area of the channel, where the flow velocity becomes higher. By deflecting the flow toward the opposite bank, riparian vegetation enhances opposite bank erosion (Dijkstra, 2003).
- Protecting the vegetated parts of the riverbed and bank against erosion (**Figure 13**).
- Accelerating the vertical growth of accreting banks and bars.
- Raising water levels. By increasing the hydraulic roughness, aquatic vegetation increases the water levels.

At the river-reach scale vegetation affects the water levels as well as the river planform formation (e.g., Murray and Paola, 2003; Jang and Shimizu, 2007; Samir Saleh and Crosato, 2008; Crosato and Samir Saleh, 2010). Murray and Paola studied the effects of soil strengthening by floodplain vegetation on the river planform, whereas Jang and Shimizu and Samir Saleh and Crosato studied the effects of increased hydraulic roughness. All works demonstrated that vegetation decreases the degree of braiding of river systems and might even transform a braiding into a meandering system.

Early studies considered the effects of vegetation on flow qualitatively (Powell, 1978; Dawson and Robinson, 1984) and demonstrated that the effects of vegetation varied over the seasons and that the relationship between resistance and vegetation varied greatly with depth. Later, semiquantitive relationships (Stephens et al., 1963; Shih and Rahi, 1982; Pitlo, 1982) were studied and demonstrated that if Manning's n is used to represent the resistance in a vegetated channel, values of up to 20 times the nonvegetated value can be found, but that such changes were more pronounced in smaller channels.

These semiquantitative approaches of increasing the amount of numerical resistance by changing the resistance parameter are still widely used by many practitioners. This is, however, based on the flawed concept resistance due to vegetation, whether emergent or submerged, stems from a boundary layer phenomenon while it is actually a mixing layer phenomenon (Ghisalberti and Nepf, 2002). This implies that the resistance from vegetation depends on depth and can therefore never be fully accounted for by a resistance parameter that is based on a surface representation rather than extending through the water column. These limitations have led to the proposal of more quantitative methods and a number of these were given by Fisher and Dawson (Table 1).

The work in **Table 1** and that of others (Larsen et al., 1990; Bakry, 1992; Salama and Bakry, 1992; Watson, 1997) indicate that while there may be a relationship between resistance and vegetation, it is complex and there is, as yet, no ideal equation for this relationship. The limitations of this approach have led a number of authors to propose more sophisticated representations based on analyzing the drag coefficient of vegetation. Most work (Wu et al., 1999; Fischer-Antze et al., 2001; Ghisalberti and Nepf, 2002, 2004; Wilson et al., 2003) has focused on laboratory channels which is vital to reduce the uncertainties in full-scale cases and to allow for well-founded fundamental conclusions to be drawn. However, work that has been carried out on real rivers is scarce (Stoesser et al., 2003; Nicholas and McLelland, 2004), which has had little or no measured data for comparison. Stoesser et al. (2003) applied a 3-D model for vegetative resistance on the Restrhein and Nicholas and McLelland (2004) used a 3-D model on the floodplains of a natural river.

The drag coefficient is often based on that for a nonflexible cylinder, but this is clearly not the case with vegetation. More recent work has studied the effect of flexibility (Kouwen, 1988; Querner, 1994; Rahmeyer et al., 1996; Fathi-Maghadam and Kouwen, 1997). Further fundamental understanding has been advanced by Japanese researchers and are reviewed by Hasegawa et al. (1999).

The reduction-factor approach outlined in Baptist (2005) and Baptist et al. (2007) quantifies the hydraulic effect that vegetation can exert on the flow by considering the distribution of shear stress within the water column rather than

Table 1 Different methods to derive the Manning's roughness coefficient of vegetated channels (Fisher and Dawson, 2003)

Authors	VR^a range ($m^2 s^{-1}$)	Discharge ($m^3 s^{-1}$)	Area^b (m^2)	Equation^{c,d}
Marshall and Westlake (1990)	0.24–1.3	0.2	1	$n = 0.1 + 0.153\frac{K_{va}}{VR}$
Pepper (1970)	0.58–8.46	2.4		$n = 0.06 + 0.17\frac{K_{va}}{VR}$
Wessex Scientific Environmental Unit (1987)	0.24–1.3	15	43	$n = 0.032 + 0.027\frac{K_{va}}{Vd}$
Wessex Scientific Environmental Unit (1987)	0.15–1.1	15	43	$n = 0.041 + 0.022\frac{K_{va}}{Vd}$
Wessex Scientific Environmental Unit (1987)	0.15–1.1	15	43	$n = 0.029 + 0.022\frac{K_{va}}{Vd}$
Larsen et al. (1990)	0.025–0.15	0.1	0.7	$n = 0.057 + 0.0036\frac{K_{va}}{VR}$
HR Wallingford (1992)	0.04–0.11	4	3.5	$n = 0.035 + 0.0239\frac{K_{va}}{VR}$

^{a}VR, product of the flow velocity V (m s^{-1}) and the hydraulic radius R (m).
^{b}A, channel cross-sectional area (m^2).
$^c K_{va}$, vegetation coverage coefficient.
$^d d$, water depth (m).

considering the forces on individual vegetation stands. In order to include this approach in 2-D and 3-D models, an equivalent value of Chézy's roughness coefficient is calculated based on characteristics of the vegetation such as drag and density. Unlike the standard approach, this value changes with vegetation density and depth as the simulation progresses.

As observed by Baptist (2005), other 3-D models for the resistance due to vegetation have been developed. The models mentioned earlier by Stoesser et al. (2003) and Nicholas and McLelland (2004) did not add any further source terms to the turbulence model, because they were not certain that this would improve the simulation results. Baptist's model includes the effects of vegetation in the turbulence closure. This has been shown by Uittenbogaard (2003) to fit laboratory measurements of mean flow, eddy viscosity, Reynolds stress, and turbulence intensity well.

2.07.2.10 Discussion of Current Research and Future Directions

Any discussion of future directions quickly becomes dated and in view of this the authors restrict themselves to outlining the areas where new developments are anticipated or required. As a precursor the overall context for river studies should be mentioned and a significant challenge that is already being addressed is how to position river science and engineering within the overall framework of modern river management which entails full recognition of environmental, societal, and economic issues.

Overall the major issue in rivers, as in all studies of the natural environment, is how to account for physical features and phenomena that are not directly incorporated into the models (whether conceptual or numerical). In rivers this means, amongst others, bed resistance, vegetation, turbulence, each of which is a significant challenge in its own right. It is perhaps best to consider future directions as progressing by either increments or step changes.

2.07.2.10.1 Incremental changes
Incremental changes are as follows:

- improvements in the estimation of the parameters for bed resistance and better end-user tools that acknowledge uncertainty and encourage a rigorous approach to calibration;
- improvements in our understanding of flow through vegetation and the ways in which this can be parameterized; and
- increased understanding of which models to use in which circumstance which should take account of spatial and temporal scales, uncertainty, and levels of acceptable risk; this includes more knowledge of the role of reduced complexity modeling (Hunter et al., 2007).

2.07.2.10.2 Step changes
Step changes are as follows:

- new methods of representing resistance parameterization based on improved encapsulation of knowledge from experimental and full-scale measurement;
- development of fundamental understanding and models for bank accretion to bring this to the level of current work on bank erosion;
- development of new paradigms to explicitly acknowledge all sources of uncertainty in modeling; and
- development of a scientific basis for an understanding of the generation, movement, and impact of floating debris.

References

Abad JD and Garcia MH (2005) Hydrodynamics in Kinoshita-generated meandering bends: Importance for river planform evolution. In: Parker G and García MH (eds.) *River, Coastal and Estuarine Morphodynamics: RCEM 2005*, pp. 761–771. London: Taylor and Francis (ISBN 0 415 39270 5).

Abbott MB and Ionescu F (1967) On the numerical computation of nearly horizontal flows. *Journal of Hydraulic Research* 5: 97–117.

Abernethy B and Rutherfurd ID (1998) Where along a river's length will vegetation most effectively stabilise stream banks? *Geomorphology* 23: 55–75.

Ackers P (1958) Hydraulics research paper: Resistance of fluids flowing in channels and ducts. *HMSO* 1: 1–39.

Alcrudo F and Garcia-Navarro P (1993) A high-resolution Godunov-type scheme in finite volumes for the 2D shallow-water equations. *International Journal for Numerical Methods in Fluids* 16(6): 489–505.

Allen JRL (1968) The nature and origin of bedform hierarchies. *Sedimentology* 10: 161–182.

Allmendiger NE, Pizzuto JE, Potter N Jr., Johnson TE, and Hession WC (2005) The influence of riparian vegetation on stream width, eastern Pennsylvania. *USA GSA Bulletin* 117(1/2): 229–243 (doi: 10.1130/B25447.1).

Anastasiou K and Chan CT (1997) Solution of the 2D shallow water equations using the finite volume method on unstructured triangular meshes. *International Journal for Numerical Methods in Fluids* 24: 1225–1245.

Anderson JD (2005) Ludwig Prandtl's boundary layer. *Physics Today* 58(12): 42–48.

ASCE (1963) Friction factors in open channels [Task Force on Friction Factors in Open Channels]. *Journal of the Hydraulics Division*, Proc. ASCE, Vol 89, HY2, March, 97-143 (Discussion in *Journal of the Hydraulics Division*, Vol 89, July, Sept. & Nov., 1963 and closure in Vol. 90, HY4 July 1964).

ASCE Task Committee on Flow and Transport over Dunes (2002) Flow and transport over dunes. *Journal of Hydraulic Engineering* 128: 726–728.

Ashley GM (1990) Classification of large-scale subaqueous bedforms: A new look at an old problem. *Journal of Sedimentary Petrology* 60: 160–172.

Bakry MF (1992) Effect of submerged weeds on the design: Procedure of earthen Egyptian canals. *Irrigation and Drainage Systems* 6: 179–188.

Baptist MJ (2005) Modelling floodplain biogeomorphology. PhD thesis, Delft University of Technology, Delft, The Netherlands, ISBN 90-407-2582-9 (http://repository.tudelft.nl/view/ir/uuid%3Ab2739720-e2f6-40e2-b55f-1560f434cbee/).

Baptist MJ and De Jong JF (2005) Modelling the influence of vegetation on the morphology of the Allier, France. In: Harby A, et al. (eds.) *Proceedings of Final COST 626*, Silkeborg, Denmark, pp. 15–22. 19–20 May 2005.

Baptist MJ, van den Bosch LV, Dijkstra JT, and Kapinga S (2005) Modelling the effects of vegetation on flow and morphology in rivers. *Large Rivers* 15(1–), HYPERLINK "http://journalseek.net/cgi-bin/journalseek/journalsearch.cgi?field=issn&query=0365-284X" Archiv für Hydrobiologie–Supplement 155(1): 339–357.

Baptist MJ, Babovic V, Uthurburu JR, Keijzer M, Uittenbogaard RE, Mynett A, and Verwey A (2007) On inducing equations for vegetation resistance. *Journal of Hydraulic Research* 45(4): 435–450.

Batchelor G (1967) *An Introduction to Fluid Dynamics*. Cambridge: Cambridge University Press.

Bates PD, Anderson MG, Price DA, Hardy RJ, and Smith CN (1996) Analysis and development of hydraulic models for floodplain flows. In: Anderson M, Walling D, and Bates P (eds.) *Floodplain Processes*. Chichester: Wiley.

Bates PD and Horritt MS (2005) Modelling wetting and drying processes in hydraulic models. In: *Computational Fluid Dynamics: Applications in Environmental Hydraulics*, ch. 6, pp. 121–146. Chichester: Wiley (doi:10.1002/0470015195).

Barré de Saint Venant AJC (1871) Théorie du mouvement non permanent des eaux avec application aux crues des rivières et à l'introduction des marées dans leur lits. *Compte rendu des seances de l'Academie des Sciences* 73: 147–154 and 237–240 (in French).

Beaman F, Morvan HP, and Wright NG (2007) *Estimating Parameters for Conveyance in 1D Models of Open Channel Flow from Large Eddy Simulation*. IAHR Congress, Venice, Italy.

Bertoldi W, Pasetto A, Zanoni L, and Tubino M (2005) Experimental observations on channel bifurcations evolving to an equilibrium state. In: Parker G and Garcia MH (eds.) *Proceedings of the 4th Symposium on River, Coastal and Estuarine Morphodynamics (RCEM)*, pp. 409–420. MN, USA: ASCE.

Blanckaert K (2002) Flow and Turbulence in Sharp Open-Channel Bends. PhD Thesis, EPFL, Lausanne, Switzerland.

Blöschl G and Sivapalan M (1995) Scale issues in hydrological modelling: A review. *Hydrological Processes: An International Journal* 9: Also in: Kalma JD and Sivapalan M (eds.) *Advances in Hydrological Processes, Scale Issues in Hydrological Modelling*, John Wiley & Sons, 1995, ISBN 0-471-95847-6.

Boguchwal LA and Southard JB (1990) Bed configurations in steady unidirectional water flows. Part 1: Scale model study using fine sands. *Journal of Sedimentary Research* 60: 649–657.

Bolla Pittaluga M, Repetto R, and Tubino M (2003) Channel bifurcation in braided rivers: Equilibrium configurations and stability. *Water Resources Research* 39(3): 1046.

Bradford SF and Sanders BF (2002) Finite-volume model for shallow-water flooding of arbitrary topography. *Journal of Hydraulic Engineering* 128(3): 289–298.

Brahams A (1754). *Anfangs-Gründe der Deich und Wasserbaukunst Teil 1 und 2, Unveränderter Nachdruck der Ausgabe Aurich. Tapper, 1767 u. 1773* (in German) Leer: Marschenrat, Schuster, 1989.

Bridge JS (2003) *Rivers and Floodplains: Forms, Processes and Sedimentary Record*, 491pp. Bodmin: Blackwell.

Bulle H (1926) *Untersuchungen über die Geschiebeableitung bei der Spaltung von Wasserläufen (Investigations on the Sediment Diversion at the Division of Channels)* (in German). Berlin: VDI Verlag.

Callander RA (1969) Instability and river channels. *Journal of Fluid Mechanics* 36(3): 465–480.

Casulli V and Stelling GS (1998) Numerical simulation of 3D quasi-hydrostatic free-surface flows. *Journal of Hydraulic Engineering* 124(7): 678–686.

Chanson H (1999) *The Hydraulics of Open Channel Flow*. Oxford: Elsevier.

Chanson H (2009) Jean-Baptiste Bélanger, hydraulic engineer, researcher and academic. In: Ettema R (ed.) *The 33rd IAHR Congress: Water Engineering for a Sustainable Environment*. Vancouver, BC, Canada. Vancouver: IAHR.

Chézy A (1776) Formule pour trouver la vitesse constant que doit avoir l'eau dans une rigole ou un canal dont la pente est donnée. Dossier 847 (MS 1915) of the manuscript collection of the Ecole des Ponts et Chaussées. Reproduced as Appendix 4, pp. 247–251 of Mouret (1921).

Chow VT (1959) *Open Channel Flow*. New York: McGraw-Hill.

Clifford NJ, Robert A, and Richards KS (1992) Estimation of flow resistance in gravel-bedded rivers: A physical explanation of the multiplier of roughness length. *Earth Surface Processes and Landforms* 17: 111–126.

Coulthard TJ and Van de Wiel MJ (2006) A cellular model of river meandering. *Earth Surface Processes and Landforms* 31: 123–132 (doi:10.1002/esp.1315).

Crosato A (1989) Meander migration prediction. *Excerpta* 4: 169–198.

Crosato A (1990) *Simulation of Meandering River Processes*. Communications on Hydraulic and Geotechnical Engineering, Delft University of Technology, Report No. 90-3, ISSN 0169-6548.

Crosato A (2007) Effects of smoothing and regridding in numerical meander migration models. *Water Resources Research* 43(1): W01401 (doi:10.1029/2006WR005087).

Crosato A (2008) Analysis and Modelling of River Meandering. PhD Thesis, Delft University of Technology.

Crosato A and Desta FB (2009) Intrinsic steady alternate bars in alluvial channels. Part 1: Experimental observations and numerical tests. In: Vionnet CA, García MH, Latrubesse EM, and Perillo GME (eds.) *Proceedings of the River, Coastal and Estuarine Morphodynamics: RCEM 2009*, vol. 2, pp. 759–765. London: Taylor and Francis.

Crosato A and Mosselman E (2009) Simple physics-based predictor for the number of river bars and the transition between meandering and braiding. *Water Resources Research* 45: W03424 (doi:10.1029/2008WR007242).

Crosato A and Samir Saleh M (2010) Numerical study on the effects of floodplain vegetation on reach-scale river morphodynamics. *Earth Surface Processes and Landforms*. Wiley-InterScience (in press).

Crossley AJ, Wright NG, and Whitlow CD (2003) Local time stepping for modeling open channel flows. *Journal of Hydraulic Engineering* 129(6): 455–462.

Dapporto S, Rinaldi M, Casagli N, and Vannocci P (2003) Mechanisms of riverbank failure along the Arno river, central Italy. *Earth Surface Processes and Landforms* 28: 1303–1323.

Darby SE, Alabyan AM, and Van De Wiel MJ (2002) Numerical simulation of bank erosion and channel migration in meandering rivers. *Water Resources Research* 38(9): 1163–1174.

Darby SE and Thorne CR (1996) Stability analysis for steep, eroding, cohesive riverbanks. *Journal of Hydraulic Engineering* 122(8): 443–454.

Davis WM (1899) The geographical cycle. *Geographical Journal* 14: 481–504.

Davies SJ and White CM (1925) A review of flow in pipes and channels. Reprinted by the Offices of Engineering, London, 1–16.

Dawson H and Fisher KR (2004) Roughness Review & Roughness Advisor, UK Defra/Environment Agency Flood and Coastal Defence R&D Programme.

Dawson FH and Robinson WN (1984). Submerged macrophytes and the hydraulic roughness of a lowland chalk stream. *Verhandlungen der Internationalen Vereinungen fur Theroetische und Angwandte Limnologie*, pp. 1944–1948.

De Heer AFM and Mosselman E (2004) Flow structure and bedload distribution at alluvial diversions. In: Greco M, Carravetta A, and Morte RD (eds.) *Riverflow 2004: Proceedings of the Second International Conference on Fluvial Hydraulics*, pp. 802–807. Naples, Italy, 23–25 June 2004. London: Taylor and Francis.

De Vriend HJ (1991) Mathematical modelling and large-scale coastal behaviour. Part 1: Physical processes. *Journal of Hydraulic Research* 29(6): 727–740.

De Vriend HJ (1998) Large-scale coastal morphological predictions: A matter of upscaling? In: *Proceedings of the 3rd Conference on Hydroscience and -Engineering*. Brandenburg University of Technology at Cottbus Cottbus/Berlin, Germany, 31 August–3 September, 1998 (on CDROM).

De Vries M (1975) A morphological time scale for rivers. In: *Proceedings of the 16th Congress of the IAHR*, vol. 2, paper B3, pp. 17–23. São Paulo, Brazil.

Dhondia JF and Stelling GS (2002) Application of one dimensional–two dimensional integrated hydraulic model for flood simulation and damage assessment. *Hydroinformatics 2002*. Cardiff: IWA Publishing.

Dijkstra JT (2003) The influence of vegetation on scroll bar development. MSc Thesis, Delft University of Technology, Delft, the Netherlands (http://repository.tudelft.nl/search/ir/?q=dijkstra&faculty=&department=&type=&year=).

Du Buat PLG (1779) *Principes d'hydraulique*. Paris: L'imprimerie de monsieur.

Engel J, Hotchkiss R, and Hall B (1995) Three dimensional sediment transport modelling using CH3D computer model *Proceedings of the First International Water Resources Engineering Conference*. New York: ASCE.

Engelund F (1970) Instability of erodible beds. *Journal of Fluid Mechanics* 42(3): 225–244.

Engelund F and Fredsøe J (1982) Sediment ripples and dunes. *Annual Review of Fluid Mechanics* 14: 13–37.

ESDU (1979) Losses caused by friction in straight pipes with systematic roughness elements, Engineering Sciences Data Unit (ESDU), pp. 1–40. London, September.

Exner FM (1925) *Uber die Wechselwirkung zwischen Wasser und Geschiebe in Flussen*. Sitzungber Akad. Wiss Wien, Part IIa, Bd. 134, pp. 165–180 (in German).

Falconer RA and Lin B (1997) Three-dimensional modelling of water quality in the humber estuary. *Water Research, IAWQ* 31(5).

Fathi-Maghadam M and Kouwen N (1997) Nonrigid, nonsubmerged, vegetative roughness on floodplains. *Journal of Hydraulic Engineering* 123(1): 51–57.

Fischer-Antze T, Stösser T, Bates PD, and Olsen NRB (2001) 3D numerical modelling of open-channel flow with submerged vegetation. *Journal of Hydraulic Research* 39(3): 303–310.

Fisher K and Dawson H (2003) Reducing uncertainty in river flood conveyance: Roughness review. Project W5A-057, DEFRA/Environment Agency Flood and Coastal Defence R and D Programme. http://www.river-conveyance.net/ces/documents/RoughnessReviewFinal_July07.pdf (accessed May 2010).

Flemming BW (1988) Zur Klassifikation subaquatischer, strömungstransversaler Transportkörper. *Bochumergeologische und geotechnische Arbeiten* 29: 44–47 (in German).

Fredsøe J (1974) On the development of dunes in erodible channels. *Journal of Fluid Mechanics* 64: 1–16.

Fredsøe J (1978) Meandering and braiding of rivers. *Journal of Fluid Mechanics* 84(4): 609–624.

Fredsøe J (1982) Shape and dimensions of stationary dunes in rivers. *Journal of Hydraulic Engineering* 108(HY8): 932–946.

Garcia-Navarro P, Fras A, and Villanueva I (1999) Dam-break flow simulation: Some results for one-dimensional models of real cases. *Journal of Hydrology* 216(3–4): 227–247.

Garcia-Navarro P and Saviron JM (1992) McCormack's method for the numerical-simulation of one-dimensional discontinuous unsteady open channel flow – reply. *Journal of Hydraulic Research* 30(6): 862–863.

Ghisalberti M and Nepf HM (2002) Mixing layers and coherent structures in vegetated aquatic flows. *Journal of Geophysical Research-Oceans* 107(C2).

Ghisalberti M and Nepf HM (2004) The limited growth of vegetated shear layers. *Water Resources Research* 40(7).

Ghisalberti M and Nepf H (2006) The structure of the shear layer in flows over rigid and flexible canopies. *Environmental Fluid Mechanics* 6(3): 277–301.

Giri S (2008) Computational modelling of bed form evolution using detailed hydrodynamics: A brief review on current developments. In: van Os AG and Erdbrink CD (eds.) *Proceedings of NCR-Days 2008*, NCR-Publications 33-2008, pp. 74–75 (ISSN 1568-234X). Minneapolis, MN: NCR.

Giri S and Shimizu Y (2006) Numerical computation of sand dune migration with free surface flow. *Water Resources Research* 42: W10422 (doi:10.1029/2005WR004588).

Gurnell AM, van Oosterhout MP, de Vlieger B, and Goodson JM (2006) Reach-scale impacts of aquatic plant growth on physical habitat. *River Research and Applications* 22: 667–680.

Graf WH (1971) *Hydraulics of Sediment Transport*. 513pp. New York: McGraw-Hill.

Graf WH (1984) *Hydraulics of Sediment Transport. Part One: A Short History of Sediment Transport*. 521pp. (ISBN 0-918334-56-X). Highlands Ranch, CO: Water Resources Publication.

Hansen E (1967) On the formation of meanders as a stability problem. *Progress Report 13*, 9pp. Lyngby: Coastal Engineering Laboratory, Technical University of Denmark, Basic Research.

Hasegawa K, Asai S, Kanetaka S, and Baba H (1999) Flow properties of a deep open experimental channel with a dense vegetation bank. *Journal of Hydroscience and Hydraulic Engineering* 17(2): 59–70.

Hodskinson A and Ferguson R (1998) Numerical modelling of separated flow in river bends: Model testing and experimental investigation of geometric controls on the extent of flow separation at the concave bank. *Hydrological Processes* 12: 1323–1338.

Horritt MS, Mason D, and Luckman AJ (2001) Flood boundary delineation from synthetic aperture radar imagery using a statistical active contour model. *International Journal of Remote Sensing* 22(13): 2489–2507.

Hunter NM, Bates PD, Horrritt MS, and Wilson MD (2007) Simple spatially-distributed models for predicting flood inundation: A review. *Geomorphology* 90: 208–225.

Ikeda S, Parker G, and Sawai K (1981) Bend theory of river meanders. Part 1: Linear development. *Journal of Fluid Mechanics* 112: 363–377.

Jang C-L and Shimizu Y (2007) Vegetation effects on the morphological behavior of alluvial channels. *Journal of Hydraulic Research* 45(6): 763–772.

Johannesson H and Parker G (1989) Velocity redistribution in meandering rivers. *Journal of Hydraulic Engineering* 115(8): 1019–1039.

Kennedy JF (1969) The formation of sediment ripples, dunes and antidunes. *Annual Review of Fluid Mechanics* 1: 147–168.

Kharat DB (2009) Practical Aspects of Integrated 1D–2D Flood Modelling of Urban Floodplains using LiDAR Topography Data. PhD Thesis, School of the Built Environment, Heriot Watt University, Edinburgh.

King IP and Norton WR (1978) Recent application of RMA's finite element models for two dimensional hydrodynamics and water quality *Second International Conference on Finite Elements in Water Resources*. London: Pentech Press.

Kleinhans MG, Jagers HRA, Mosselman E, and Sloff CJ (2008) Bifurcation dynamics and avulsion duration in meandering rivers by one-dimensional and three-dimensional models. *Water Resources Research* 44: W08454 (doi:10.1029/2007WR005912).

Knight DW (2005) River flood hydraulics: Theoretical issues and stage-discharge relationships. In: Knight DW and Shamseldin AY (eds.) *River Basin Modelling for Flood Risk Mitigation*, ch. 17, pp. 301–334. Leyden: Balkema.

Knight DW, McGahey C, Lamb R, and Samuels P (2009) *Practical Channel Hydraulics: Roughness, Conveyance and Afflux*. Taylor and Francis.

Knight DW, Shiono K, and Pirt J (1989) Prediction of depth mean velocity and discharge in natural rivers with overbank flow. In: Falconer RA, Goodwin P, and Matthew RGS (eds.) *International Conference on Hydraulic and Environmental Modelling of Coastal, Estuarine and River Waters*. University of Bradford. Aldershot: Gower Technical Press.

Kodama T, Wang S, and Kawaharam M (1996) Model verification on 3D tidal current analysis in Tokyo Bay. *IInternational Journal for Numerical Methods in Fluids* 22: 43–66.

Kouwen N (1988) Field estimation of the biomechanical properties of grass. *Journal of Hydraulic Research* 26(5): 559–568.

Krone RB (1962) Flume studies of the transport of sediment in estuarial shoaling processes. Final Report, Hydraulic Engineering Laboratory and Sanitary Engineering Research Laboratory, Berkeley, CA, prepared for US Army Engineer District, San Francisco, CA, under US Army Contract No. DA-04-203 CIVENG-59-2.

Krone RB (1963) *A Study of Rheological Properties of Estuarial Sediments*. Technical Bulletin No. 7, Committee on Tidal Hydraulics, Corps of Engineers, US Army; prepared by US Army Engineer Waterways Experiment Station, Vicksburg, MS.

Lane EW (1955) The importance of fluvial morphology in hydraulic engineering. In: *Proceedings of the American Society of Civil Engineers*, vol. 81, paper 745, 17pp. San Diego, Reston, VA: ASCE.

Larsen T, Frier J, and Vestergaard K (1990) Discharge/stage relations in vegetated Danish streams. In: White WR (ed.) *Proceedings of the International Conference on River Flood Hydraulics*, paper F1. 17–20 September 1990. Wallingford: Wiley.

Lee S-H and Wright NG (2010) Simple and efficient solution of the shallow water equations with source terms. *International Journal for Numerical Methods in Fluids* 63(3): 313–340.

Liang QH and Borthwick AGL (2008) Adaptive quadtree simulation of shallow flows with wet-dry fronts over complex topography. *Computers and Fluids* 38(2): 221–234.

Macagno E (1989) *Leonardian Fluid Mechanics in the Manuscript I*. IIHR Monograph, No. 111. Iowa City, IA: The University of Iowa.

Marinoni A (1987) *Il Manoscritto I*. Florence: Giunti-Barbera (in Italian).

Marshall EJP and Westlake DF (1990) Water velocities around water plants in chalk streams. *Folia Geobotanica* 25: 279–289.

Massey BS (1995) *Mechanics of Fluids*. Chapman and Hall.

Morvan HP, Knight DW, Wright NG, Tang XN, and Crossley AJ (2008) The concept of roughness in fluvial hydraulics and its formulation in 1-D, 2-D & 3-D numerical simulation models. *Journal of Hydraulic Research* 46(2): 191–208.

Morvan H, Pender G, Wright NG, and Ervine DA (2002) Three-dimensional hydrodynamics of meandering compound channels. *Journal of Hydraulic Engineering* 128(7): 674–682.

Mosselman E (1992) Mathematical Modelling of Morphological Processes in Rivers with Erodible Cohesive Banks. PhD Thesis, Communications on Hydraulic and Geotechnical Engineering, No. 92-3, Delft University of Technology, ISSN 0169-6548.

Mosselman E (1998) Morphological modelling of rivers with erodible banks. *Hydrological Processes* 12: 1357–1370.

Mosselman E, Shishikura T, and Klaassen GJ (2000) Effect of bank stabilisation on bend scour in anabranches of braided rivers. *Physics and Chemistry of the Earth, Part B* 25(7–8): 699–704.

Murray AB and Paola C (1994) A cellular model of braided rivers. *Nature* 371: 54–57.

Murray AB and Paola C (2003) Modelling the effects of vegetation on channel pattern in bedload rivers. *Earth Surface Processes and Landforms* 28: 131–143 (doi: 10.1002/esp.428).

Nabi M, De Vriend HJ, Mosselman E, Sloff CJ, and Shimizu Y (2009) Simulation of subaqueous dunes using detailed hydrodynamics. In: Vionnet CA, García MH, Latrubesse EM, and Perillo GME (eds.) *Proceedings of the River, Coastal and Estuarine Morphodynamics: RCEM 2009*, pp. 967–974. London: Taylor and Francis.

Nanson GC and Hickin EJ (1986) A statistical analysis of bank erosion and channel migration in Western Canada. *Bulletin of the Geological Society of America* 97(4): 497–504.

Nelson JM, McLean SR, and Wolfe SR (1993) Mean flow and turbulence fields over two-dimensional bedforms. *Water Resources Research* 29: 3935–3953.

Nex A and Samuels P (1999) The use of 3 D CFD models in river flood. defence, Report No. SR542, HR Wallingford.

Nicholas AP and McLelland SJ (2004) Computational fluid dynamics modelling of three-dimensional processes on natural river floodplains. *Journal of Hydraulic Research* 42(2): 131–143.

Olesen KW (1984) Alternate bars in and meandering of alluvial rivers. In: Elliott CM (ed.) *River Meandering, Proceedings of the Conference Rivers '83*, pp. 873–884. New Orleans, LA, USA, 24–26 October 1983. New York: ASCE.

Olsen NRB (2000) Unstructured hexahedral 3D grids for CFD modelling in fluvial geomorphology. Fourth International Conf. Hydroinformatics 2000, Iowa, USA.

Olsen NRB and Stokseth S (1995) 3-Dimensional numerical modeling of water-flow in a river with large bed roughness. *Journal of Hydraulic Research* 33(4): 571–581.

Osman AM and Thorne CR (1988) Riverbank stability analysis I: Theory. *Journal of Hydraulic Engineering* 114(2): 134–150.

Ott RA (2000) Factors affecting stream bank and river bank stability, with an emphasis on vegetation influences. Prepared for the Region III Forest Practices Riparian Management Committee, Tanana Chiefs Conference, Inc. Forestry Program, Fairbanks, Alaska.

Parker G (1976) On the cause and characteristic scales of meandering and braiding in rivers. *Journal of Fluid Mechanics* 76(3): 457–479.

Parker G (1978a) Self-formed straight rivers with equilibrium banks and mobile bed. Part 1: The sand–silt river. *Journal of Fluid Mechanics* 89: 109–125.

Parker G (1978b) Self-formed straight rivers with equilibrium banks and mobile bed. Part 2: The gravel river. *Journal of Fluid Mechanics* 89: 127–146.

Partheniades E (1962) A Study of Erosion and Deposition of Cohesive Soils in Salt Water. PhD Thesis, University of California, Berkeley, CA, USA.

Partheniades E (1965) Erosion and deposition of cohesive soils. *Journal of the Hydraulic Division* 91(HY1): 105–139.

Pepper AT (1970) Investigation of Vegetation and Bend Flow Retardation of a Stretch of the River Ousel. BSc Dissertation, National College of Agricultural Engineering, Silsoe.

Phillips JD (1995) Biogeomorphology and landscape evolution: The problem of scale. *Geomorphology* 13: 337–347.

Pitlo RH (1982) Flow resistance of aquatic vegetation. In: *Proceedings of the 6th EWRS Symposium on Aquatic Weeds*. European Weed Research Society.

Pizzuto JE (1984) Bank erodibility of shallow sandbed streams. *Earth Surface Processes and Landforms* 9: 113–124.

Powell KEC (1978) Weed growth – a factor in channel roughness. In: Herschy RW (ed.) *Hydrometry, Principles and Practice*, pp. 327–352. Chichester: Wiley.

Preissmann A (1961) Propagation des intumescences dans les canaux et rivieres. In: *Proceedings of the First Congress of the French Association for Computation*, pp. 433–442. Grenoble, France.

Pirim T, Bennet SJ, and Barkdoll BD (2000) Restoration of degraded stream corridors using vegetation: An experimental study. United States Department of Agriculture, Channel & Watershed Processes Research Unit, National Sedimentation Laboratory, Research Report No 14.

Querner EP (1994) Aquatic weed control within an integrated water management framework. In: White WR (ed.) *Report 67 of the Agricultural Research Department*. Wageningen: Agricultural Research Department.

Rahmeyer W, Werth D, and Freeman GE (1996) Flow resistance due to vegetation in compound channels and floodplains. In: *Semi-Annual Conference on Multiple Solutions to Floodplain Management*. Utah, November 1996.

Rastogi A and Rodi W (1978) Predictions of heat and mass transfer in open channels. *Journal of Hydraulic Division* 104(HY3): 397–420.

Reynolds AJ (1974) *Turbulent Flows in Engineering*. London: Wiley.

Riad K (1961) Analytical and Experimental Study of Bed Load Distribution at Alluvial Diversions. Doctoral Thesis, Delft University of Technology, Delft, The Netherlands.

Rinaldi M and Casagli N (1999) Stability of streambanks formed in partially saturated soils and effects of negative pore water pressure: The Sieve River (Italy). *Geomorphology* 26(4): 253–277.

Rinaldi M, Casagli N, Dapporto S, and Gargini A (2004) Monitoring and modelling of pore water pressure changes and riverbank stability during flow events. *Earth Surface Processes and Landforms* 29: 237–254.

Rinaldi M and Darby SE (2005) Advances in modelling river bank erosion processes. In: Proceedings of the 6th International Gravel Bed Rivers VI, Lienz, Austria, 5–9 September, 2005.

Rodrigues S, Bréhéret J-G, Macaire J-J, Moatar F, Nistoran D, and Jugé P (2006) Flow and sediment dynamics in the vegetated secondary channels of an anabranching river: The Loire River (France). *Sedimentary Geology* 186: 89–109.

Rouse H (1965) Critical analysis of open-channel hydraulics. *Journal of Hydraulic Engineering* 91: 1–25.

Salama MM and Bakry MF (1992) Design of earthen, vegetated open channels. *Water Resources Management* 6: 149–159.

Sayre WW and Albertson ML (1963) Roughness spacing in rigid open channels. *Transactions of American Society of Civil Engineers, ASCE* 128(1): 343–427.

Schlichting H, Gersten K, Krause E, and Oertel HJ (2004) *Boundary-Layer Theory*. Springer

Samir Saleh M and Crosato A (2008) Effects of riparian and floodplain vegetation on river patterns and flow dynamics. In: Gumiero G, Rinaldi M, and Fokkens B (eds.) *Proceedings of 4th ECRR International Conference on River Restoration*, Italy, Venice S. Servolo Island, 16–21 June 2008, ECRR-CIRF Publication. Printed by Industrie Grafiche Vicentine S.r.l., pp. 807–814.

Schumm SA and Lichty RW (1965) Time, space and casuality in geomorphology. *American Journal of Science* 263: 110–119.

Seminara G and Tubino M (1989) Alternate bar and meandering: Free, forced and mixed interactions. In: Ikeda S and Parker G (eds.) *River Meandering*, Water Resources Monograph, vol. 12, pp. 267–320 (ISBN 0-87590-316-9). Washington, DC: AGU.

Shields AF (1936) *Application of Similarity Principles and Turbulence Research to Bed-Load Movement*. Hydrodynamics Laboratory Publication No. 167, Ott WP and Van Uchelen JC (trans.), US Department of Agriculture, Soil Conservation Service, Cooperative Laboratory, California Institute of Technology, Pasadena, CA.

Shih SF and Rahi GS (1982) Seasonal variations of Manning's roughness coefficient in a subtropical marsh. *American Society of Agricultural Engineers* 25: 116–119.

Simons DB and Richardson EV (1961) Forms of bed roughness in alluvial channels. *Journal of Hydraulics Division* 87(1): 87–105.

Sleigh PA, Gaskell PH, Berzins M, and Wright NG (1998) An unstructured finite-volume algorithm for predicting flow in rivers and estuaries. *Computers and Fluids* 27(4): 479–508.

Smagorinsky J (1963) General circulation experiments with primitive equations. Part 1: Basic experiments. *Monthly Weather Review* 91: 99–164.

Southard JB and Boguchwal LA (1990) Bed configuration in steady unidirectional water flows. Part 2: Synthesis of flume data. *Journal of Sedimentary Research* 60(5): 658–679.

Stansby PK and Zhou JG (1998) Shallow-water flow solver with nonhydrostatic pressure: 2D vertical plane problems. *International Journal for Numerical Methods in Fluids* 28: 541–563.

Stephens JC, Blackburn RD, Seamon DE, and Weldon LW (1963) Flow retardance by channel weed and their control. *Journal of the Irrigation and Drainage Division* 89(IR2): 31–56.

Stoesser T, Wilson CAME, Bates PD, and Dittrich A (2003) Application of a 3D numerical model to a river with vegetated floodplains. *Journal of Hydroinformatics* 5(2): 99–112.

Strickler A (1923) Beitrage auf Frage der eschwindigkeitsformel under der Rauhiskeitszahlen frustrome, kanale under gescholossene Leitungen. Mitteilungen des eidgenossischen Amtes fur Wasserwirtschaft 16.

Sun T, Meakin P, and Jøssang T (1996) A simulation model for meandering rivers. *Water Resources Research* 32(9): 2937–2954.

Ten Brinke WBM (2005) The Dutch Rhine: A restrained river. Uitgeverij Veen Magazines, Diemen, The Netherlands, ISBN 9076988 919.

Thorne CR (1982) Processes and mechanisms of river bank erosion. In: Hey RD, Bathurst JC and Thorne CR (eds.) *Gravel-Bed Rivers*, pp. 227–259. Chichester: Wiley.

Thorne CR (1988) Riverbank stability analysis. II: Applications. *Journal of Hydraulic Engineering* 114(2): 151–172.

Thorne CR (1990) Effects of vegetation on riverbank erosion and stability. In: Thornes JB (ed.) *Vegetation and Erosion*, pp. 125–144. Chichester: Wiley.

Tsujimoto T (1999) Fluvial processes in streams with vegetation. *Journal of Hydraulic Research* 37(6): 789–803.

Uittenbogaard R (2003) Points of view and perspectives of horizontal large-eddy simulation at Delft, CERI, Sapporo, http://www.wldelft.nl/rnd/publ/docs/Ui_CE_2003.pdf.

van Bendegom L (1947) Enige beschouwingen over riviermorfologie en rivierverbetering. De Ingenieur B. Bouw- en Waterbouwkunde 1 59(4): 1–11 (in Dutch). (*Some considerations on river morphology and river improvement.* English translation, Natural Resources Council Canada, 1963, Technical Translation No. 1054.)

Versteeg HK and Malalasekera W (2007) *Introduction to Computational Fluid Dynamics: The Finite Volume Method*, 503pp. Harlow: Pearson.

Villada Arroyave JA and Crosato A (2010) Effects of river floodplain lowering and vegetation cover. In: *Proceedings of the Institution of Civil Engineers, Water Management*, vol. 163, pp. 1–11 (doi:10.1680/wama2010.163.1.1).

Wallingford (1992) The hydraulic roughness of vegetated channels. *Report No. SR 305*, March 1992. HR Wallingford.

Wang SSY, Alonso VV, Brebbia CA, Gray WG, and Pinder GF (1989) Finite elements in water resources. *Third International Conference, Finite Elements in Water Resources*, Mississippi, USA.

Wang ZB, Fokkink RJ, De Vries M, and Langerak A (1995) Stability of river bifurcations in 1D morphodynamic models. *Journal of Hydraulic Research* 33(6): 739–750.

Watson D (1987) Hydraulic effects of aquatic weeds in UK rivers. *Regulated Rivers: Research and Management* 1: 211–227.

Wessex Scientific Environmental Unit (1987) The Effect of Aquatic Macrophytes on the Hydraulic Roughness of a Lowland Chalk River.

Wright NG (2001) Conveyance implications for 2D and 3D modelling. Scoping Study for Reducing Uncertainty in River Flood Conveyance. Environment Agency (UK).

Wright NG (2005) Introduction to numerical methods for fluid flow. In: Bates P, Ferguson R, and Lane SN (eds.) *Computational Fluid Dynamics: Applications in Environmental Hydraulics*. Chichester: Wiley.

Wright NG, Villanueva I, Bates PD, *et al.* (2008) A case study of the use of remotely-sensed data for modelling flood inundation on the River Severn, UK. *Journal of Hydraulic Engineering* 134(5): 533–540.

Wu FC, Shen HW, and Chou YJ (1999) Variation of roughness coefficients for unsubmerged and submerged vegetation. *Journal of Hydraulic Engineering* 125(9): 934–942 (doi:10.1061/(ASCE)0733-9429(1999)).

Yen BC (1991) *Channel Flow Resistance: Centennial of Manning's Formula*. Colorado, USA: Water Resources Publications.

Yen BC (2002) Open channel flow resistance. *Journal of Hydraulic Engineering* 128(1): 20–39.

Zienkiewicz OZ and Cheung YK (1965) Finite elements in the solution of field problems. *Engineer* 507–510.

Zolezzi G (1999) River Meandering Morphodynamics. PhD Thesis, 180pp. Department of Environmental Engineering, University of Genoa.

Relevant Websites

http://delftsoftware.wldelft.nl
 Deltares; Delft Hydraulics Software: SOBEK and Delft3D.
http://www.halcrow.com
 Halcrow; ISIS Software.
http://www.hec.usace.army.mil
 Hydrologic Engineering Center; HEC-RAS Software.
http://www.mikebydhi.com
 MIKE by DHI.
http://www.river-conveyance.net
 Reducing Uncertainty in Estimation of Flood Levels; Conveyance and Afflux Estimation System (CES/AES).
http://wwwbrr.cr.usgs.gov
 US Geological Survey Central Region Research; Geomorphology and Sediment Transport Laboratory of USGS.
http://vtchl.uiuc.edu
 Ven Te Chow Hydrosystems Laboratory; Gary Parker's e-book.
http://www.wallingfordsoftware.com
 Wallingford Software; InfoWorks Software.

2.08 Lakes and Reservoirs

D Uhlmann and L Paul, University of Technology, Dresden, Germany
M Hupfer, Leibniz-Institute of Freshwater Ecology and Inland Fisheries, Berlin, Germany
R Fischer, Consulting Engineers for Water and Soil Limited, Possendorf, Germany

© 2011 Elsevier B.V. All rights reserved.

2.08.1	**Morphometry, Hydrodynamics, Chemistry, and Biology of Lakes**	157
2.08.1.1	Origin and Development of Lakes	157
2.08.1.1.1	Origin of lakes	157
2.08.1.1.2	Lake development with a large contribution of photosynthesis	159
2.08.1.1.3	Development of reservoirs	159
2.08.1.2	Structure and Functioning of Drainage Basins of Natural and Man-Made Lakes	160
2.08.1.3	Lake and Reservoir Morphometry	160
2.08.1.4	Influx and Vertical Distribution of Solar Energy	164
2.08.1.4.1	Underwater light conditions	164
2.08.1.4.2	Heat budget and thermal structure	167
2.08.1.5	Water Movement	170
2.08.1.6	Basic Chemistry	175
2.08.1.6.1	Systematics of lakes with respect to water quality	175
2.08.1.6.2	Ionic balance	177
2.08.1.6.3	Inorganic compounds and buffer properties	177
2.08.1.6.4	Sequence of microbially mediated redox processes	178
2.08.1.6.5	Iron, manganese, and sulfur compounds	180
2.08.1.6.6	Nutrients (nitrogen and phosphorus) and trace substances	182
2.08.1.6.7	Organic carbon – humic compounds	185
2.08.1.7	Biotic Structure	187
2.08.1.8	Photosynthesis: Generation and Consumption of Dissolved Oxygen	189
2.08.1.9	Oxygen Stratification: Circulation/Quality Types of Lakes	193
2.08.2	**Fundamental Properties of Reservoirs**	197
2.08.2.1	Functions of Reservoirs	197
2.08.2.2	Characteristic Differences between Natural Lakes and Reservoirs	199
2.08.2.3	Environmental Impacts of Reservoirs	203
2.08.3	**Management, Protection, and Rehabilitation of Lakes and Reservoirs**	204
2.08.3.1	Main Water-Quality Problems	204
2.08.3.2	General Management Strategies	205
2.08.3.3	Measures for Eutrophication Control	205
2.08.3.3.1	External measures	205
2.08.3.3.2	Internal measures	206
2.08.4	**Current Knowledge Gaps and Future Research Needs**	209
2.08.4.1	Lakes and Reservoirs as Constituents of Their Catchment Areas	209
2.08.4.2	Responses of Lakes and Reservoirs to Climate Change	210
2.08.4.3	Biodiversity and Its Role in the Functioning of Lake and Reservoir Ecosystems	210
2.08.4.4	Integrated Management of Lakes and Reservoirs	211
References		211

2.08.1 Morphometry, Hydrodynamics, Chemistry, and Biology of Lakes

2.08.1.1 Origin and Development of Lakes

2.08.1.1.1 Origin of lakes

Lakes are hollows which are filled, at least partially, with water. The nature of the physico-chemical and biological events taking place in a lake is related to its shape and size, as well as to the characteristics of the drainage basin. These characteristics are in turn largely determined by the mode of origin of the lake. Thus, ascertaining the mode of origin of a lake is useful in understanding some of the general characteristics of the lake and drawing comparisons with other lakes of similar origins.

In general, the forces forming a lake are

1. catastrophic, or sudden in geological terms;
2. regional in nature, often giving rise to several similar lakes forming a lake district; and
3. caused by erosion (of the outlet) and sedimentation of the basin so that lakes become temporary features of the landscape.

From a geological point of view, not only reservoirs but also lakes are short-lived (with a few exceptions such as Lake Baikal and Lake Tanganyika, the first of which is more than 56 million years old).

Lakes are formed in two ways: (1) by the filling with water into a natural depression or (2) by impoundment behind a natural dam.

A depression may be formed by a geological deformation (tectonic or volcanic), by elutriation, by ice or water erosion, or by the slow melting of dead ice blocks in the subsoil. Lake pans may also have been deflated by wind action.

Natural dams may be formed by glacial moraines, by landslides, or by biogenic $CaCO_3$ deposition.

Most of the existing lakes have been formed by glacial activity during the last ice age. The following statements on the genesis of lakes are essentially based upon Wetzel (2001), Hutchinson (1957), and Keller (1962):

1. Lakes formed by glacial ice movement are more in number than lakes formed by other processes. With the retreat of the large Pleistocene glaciers, an immense number of lakes were formed, creating a large variety of glacially formed lake types.

    Glacial ice-scour lakes occur in extended rock areas where loosened rock materials have been removed by glaciers. Examples are the upland peneplains of Scandinavia, the United Kingdom, and the great Canadian Shield. Due to scouring of preexisting valleys to great depths, the Great Bear Lake and the Great Slave Lake were formed. The most impressive examples of lakes on the North American continent produced by ice erosion of rocks, however, are the Laurentian Great Lakes.

2. In glaciated valleys, chains of smaller lakes could also have evolved by scouring.

    Morainal damming of pre-glacial valleys created many lakes in the Northern Hemisphere. Dams of moraine material may, in high mountains, occasionally attain a height of more than 120 m (Keller, 1962). Cirque lakes are frequently arranged along a valley in stairways. Many lakes also emerged from cavities in tillite, a metamorphosed old bedrock. The closely related Kettle lakes originated from the melting of ice blocks which had previously been buried for up to several hundred years, in moraine material.

3. In the flat regions of Siberia and in northern America, millions of small, shallow cryogenic lakes, which evolved from the thawing of local permafrost soils, were formed.

    Biogenic formation of $CaCO_3$ dams occurs when due to biological (photosynthesis) or mechanical (very turbulent flow) removal of (carbonate-balanced) CO_2 from Ca$(HCO_3)_2$-rich water (cf. Section 2.08.1.6.3), calcareous crusts may be generated with a growth of 1 cm yr^{-1} or more. In this way, barriers which are able to impound a river can be formed. In front of these dams, there are usually waterfalls and behind them there are lakes (**Figure 1**). One of the organisms which may cause this growth of high-calc-sinter dams is the filamentous cyanobacterium, *Schizothrix*.

4. The lake basins of tectonic lakes are depressions which have been formed by movements of comparatively deep portions of the Earth's crust. Most of these lakes are a result of faulting with single-fault displacements, or exist in downfaulted troughs (**Figure 2**; Wetzel, 2001).

    The latter type is called a 'graben'. Well-known examples are Lake Baikal in Siberia and Lake Tanganyika in equatorial Africa. These lakes have maximum depths of 1620 m and 1435 m, respectively. Both lakes contain a large number of plant and animal species which are endemic, that is, they occur only in these particular water bodies. Both lakes were already in existence in the Mesozoic period. Another

Figure 1 Some of the Plitvice Lakes, Croatia, the dams of which have been formed by biogenic precipitation of $CaCO_3$. The brown color is caused by Fe^{3+}. Courtesy of Dr. Anita Belanovic.

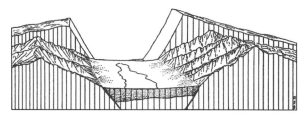

Figure 2 Diagram of a tectonic lake basin: a depressed fault-block between two upheaved fault-blocks. In the foreground, situation after a considerable period of erosion and deposition. From Wetzel RG (2001) *Limnology. Lake and River Ecosystems*, 3rd edn. San Diego, CA: Academic Press.

well-known example of a graben lake is Lake Tahoe (California/Nevada).

From a moderate uplifting of the seabed connected to tectonic movements, the Caspian Sea and the Aral Sea in Western Asia were formed in the Miocene period. Upwarping of the Earth's crust also resulted in the formation of other large lakes such as Lake Okeechobee, Florida, and Lake Victoria, Central Africa.

5. Volcanic lakes are formed when depressions that may be formed due to volcanic activity are undrained, and usually are filled with water. The basins and their drainage areas often have a basaltic nature and thus a low concentration of dissolved solids, inclusive of nutrients. Volcanic crater lakes are often circular and are called 'maar lakes' if they have small diameters (up to 2000 m) and 'calderas' when of a larger size. A well-known example of a maar lake is Crater Lake, Oregon, with a depth of 608 m.

Lava streams may flow into a preexisting river valley and form a dam wall. Behind this dam, a lake may be created.

6. Lake origins are also formed by landslides when large quantities of unconsolidated material suddenly move into the floors of valley streams to create dams and lakes (Figure 3; Wetzel, 2001). Such landslides occur frequently in glaciated mountains. The landslides are usually brought about by abnormal events such as excessive rain acting on unstable slopes, or by earthquake activity. Disastrous floods may be caused downstream if such dams break.

7. Solution lakes have been created by the dissolution of carbonate, also of sulfate or other soluble rock. They are mostly connected, similar to many other lake types, with the groundwater.

8. Among several lake types formed by river activity, flood-plain lakes are the best known. Oxbow lakes were created from truncated meanders.

9. Deflation lakes result from the erosive effect of wind, mostly in arid areas. They are often ephemeral. The fine structure of sediments/soils in very large dry depressions in North and South Africa, in many cases, reflect climatic changes over the last millennia.

Figure 3 Lake formed by a large landslide into a steep-sided stream-eroded canyon. From Wetzel RG (2001) *Limnology. Lake and River Ecosystems*, 3rd edn. San Diego, CA: Academic Press.

2.08.1.1.2 Lake development with a large contribution of photosynthesis

As soon as in a shallow lake the higher emergent vegetation (see Figure 33; Uhlmann, 1979) starts to predominate, the accumulation of biomass residues, mostly, cellulose and lignin mud, considerably increases. The thickness of the organic sediment layer may be a multiple of the water depth in the senescent stage of a lake (see Figure 41; Kusnezow, 1959). The accelerated sedimentation and resulting shallowness favor both the further spreading of emergent vegetation and the increase in water losses by evapotranspiration. This often results in the final disappearance of the water body.

The biomass residues may also originate from floating mats of *Sphagnum* moss. These peat-forming mosses initially colonize the outer margins of the water body.

The drainage patterns of lakes in the flood plains of tropical rivers can be largely altered by massive growths, not only of emergent, but also of floating-leaved vegetation. This can probably also go along with increased sediment accumulation.

In the early stages of development of clear-water lakes formed by glaciation, the biotic productivity is limited by the lack of nutrients, the long winter period, and a high removal rate of dissolved organic materials by photolysis. During the long intermediate stage of lake ontogeny, phytoplankton production governs sediment accumulation, with a small contribution of macrophyte biomass. Organic sedimentation here is largely balanced by microbial decomposition. In the later and terminal stages, the proportion of emergent and wetland vegetation and thus the biogenic silting-up rapidly increases, due to the great amount of lignified and cellulosic residues which now accumulate. This also applies to lakes which are shallow initially.

There are also other terminal stages in the development of lakes in temperate climates, for example, various types of persistent mire ecosystems and shallow lakes which are durable due to a permanent high inflow and level of groundwater.

2.08.1.1.3 Development of reservoirs

Compared with natural lakes, reservoirs are extremely short-lived. They are designed for a lifetime of at least 50 years, but this is, in many cases, only realistic if the deposited silt is collected in pre-reservoirs and mechanically removed at intervals of several decades. Without such countermeasures, a reservoir may be completely filled with sediment within a few years in areas with very heavy erosion. On the other hand, reservoirs with an estimated lifetime of several centuries also exist (Nilsson, 2009).

In the first phase of impoundment, the plankton benefit from the release of nutrients, due to the degradation of the submerged terrestrial vegetation. This may even increase the fish yields. The second phase is characterized by an oligotrophication if the inflowing water is poor in nutrients, or by a eutrophication if the water is fertilized by domestic animals or agricultural effluent. The final phase is an advanced deposition of silt whereby conditions for biogenic siltation are substantially improved. In tropical reservoirs, luxuriant growths of floating-leaved plants become possible under conditions of comparatively small water-level fluctuations.

2.08.1.2 Structure and Functioning of Drainage Basins of Natural and Man-Made Lakes

The drainage basin clearly regulates the characteristics of lakes and reservoirs, which include soil, ionic composition, slope, and, in combination with the climate, vegetation cover (Wetzel, 2001). Soil and vegetation not only influence the runoff, but also the composition and quantity of organic matter that enters the tributaries.

The area of the drainage basin as related to the area of the water body is normally < 10:1 in the case of lakes, but it is > 10:1 for reservoir catchments. The drainage basin of a reservoir is usually large enough to fill the man-made lake within a period of 1 year or less. The influence of the structure and the uses of the drainage basin upon the water quality are therefore quite substantial. Reservoirs are also subjected much more to the stochastics in the relationship between atmospheric precipitation and runoff, than is the case in Pleistocene lakes, which are largely supplied by groundwater.

Lakes are often close to the center of their drainage basin, whereas reservoirs are located at the margin. This is predetermined by the morphometry of the territory.

Densely forested drainage basins provide a good water quality in reservoirs because of their anti-erosion function. Furthermore, a forest cover promotes the infiltration of rainwater as a pre-treatment step in the context of drinking-water supply. The potable water supply of big cities such as New York, Tokyo, Beijing, Rio de Janeiro, and Los Angeles is completely, or largely, based upon water from forested drainage basins (K. H. Feger, personal communication).

From agricultural areas, mainly soil particles, nitrogen and phosphorus from fertilizers, as well as herbicides and pesticides are lost by surface runoff, due to heavy rains. Pastures may not only be sources of N- and P-compounds, but also release cysts of parasitic protists from manure depositions.

The nitrate concentration in a reservoir or lake is often an indicator of the state of the environment. In the Saidenbach Reservoir (Germany), the increase in agricultural production, and particularly the application of liquid manure to the catchment, has caused a threefold increase of nitrate concentration in 17 years (rise from $10\,\text{mg}\,\text{l}^{-1}$ in 1962 to $30\,\text{mg}\,\text{l}^{-1}$ in 1979, W. Horn in Uhlmann and Horn, 2001).

Densely populated drainage basins are generally not compatible with the safe operation of drinking-water reservoirs. The introduction of purified domestic effluent (with advanced treatment) into drinking-water reservoirs is extremely problematic not only due to the loads of potentially harmful microorganisms, but also due to unacceptable N and P concentrations in storm-water outlets, subsequent to heavy rains. The allowable P concentrations for purified effluent in flowing waters are normally set at a level which does not affect the ecosystem. However, the very low P concentrations of around $10\,\mu\text{g}\,\text{l}^{-1}$, which are required for drinking-water reservoirs, may not be achieved downstream of wastewater-treatment plants even if these are operated in full accordance with internationally accepted regulations.

In temperate climates, drinking-water reservoirs are situated mostly in hilly areas with igneous rock as the mineral subsoil. Consequently, they often have water of low hardness or they can even be weakly acidic. This facilitates the binding of phosphate to Al- and Fe^{3+}-complexes (also in the colloidal-size class) and favors an oligotrophic state of the water body.

In the past decades, many sites in Europe and North America have become subject to an acidification of soil and water due to atmospheric depositions. Liming of the forests in the drainage basins has been used as a counteractive measure, but it simultaneously increases the trophic state (i.e., phytoplankton production) as is well known from fishponds.

In former decades, bogs in the catchments of drinking-water reservoirs were often drained. If clearing of the drainage ditches is not done on a regular basis, the resulting waterlogging leads not only to an increased leaching of (coniferous) soils, but also to the growth of bogs. This may result in an increased concentration of humic substances/dissolved organic carbon in reservoir waters (Sudbrack et al., 2005). Thus, the costs for water treatment may largely increase.

Sometimes reservoir systems are interconnected. Downstream water bodies generally receive better quality water which is improved due to the retention in the upstream water bodies, and they serve as pre-reservoirs for water treatment. Thus, the concentration of imported suspended solids is generally much lower downstream. In many cases, the trophic state and phytoplankton production (Sections 2.08.1.7 and 34.1.8) likewise decrease. The quality may also be improved by introducing water from reservoirs which are situated in a bypass.

2.08.1.3 Lake and Reservoir Morphometry

The morphology of lakes, their size and shape, is often related to their origin and age. Lake morphometry is the quantification of characteristic morphological dimensions whose fundamental limnological importance was emphasized by Kalff (2002):

> Regardless of how lakes are formed, their surface shape, surface area, underwater form, depth and the irregularity of their shorelines have a major impact on turbulence, lake stratification, sedimentation and resuspension, and the extent of littoral-zone wetlands that determine lake functioning.

The determination of morphometric measures requires a bathymetric map of the lake with a scaled outline of the shoreline and submerged contour lines in several depths below the surface. In the past, the depth development of a lake had to be determined by lowering a plumb line from a boat or the frozen surface at many stations. Nowadays, precise bathymetric maps of water bodies are created using digital sonars coupled with a global positioning system (GPS)-receiver and data evaluation using geographic information system GIS-software (**Figure 4**; Sytsma et al., 2004).

The interaction of a lake with the atmosphere (e.g., radiative energy balance, gas exchange, and direct matter import by precipitation) and the impact of driving meteorological forces (particularly the effect of wind on mixing and stratification, surface waves, and water movements) depend primarily on its surface area A_0 and form described by the maximum length l_{max} and width b_{max} measured perpendicular to l_{max}. The maximum length is the distance between the two most distant points of the lake surface. It is often measured as a straight line that may cross islands or promontories. Sometimes, it is

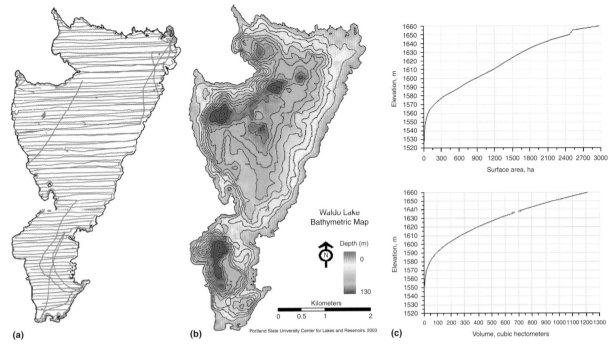

Figure 4 Example of a bathymetric study at the Waldo Lake (Oregon, USA): (a) data-collection cruise paths; (b) bathymetric map interpolated from data collected; (c) resulting hypsographic curve A_z (above) and volume–depth distribution V_z (below). From Sytsma M, Rueter J, Petersen R, et al. (2004) Waldo Lake Research in 2003. Center for Lakes and Reservoirs, Department of Environmental Sciences and Resources, Department of Civil and Environmental Engineering, Portland State University, Portland, Oregon 97201–0751. http://www.clr.pdx.edu/docs/2003report.pdf (accessed April 2010).

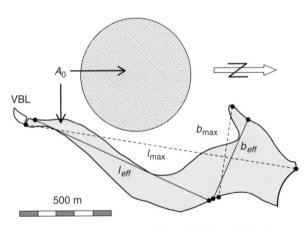

Figure 5 Morphometric characteristics of the Neunzehnhain II reservoir (Germany, 50° 42.6′ N, 13° 09.13′ E; VBL: pre-dam). Reservoir and circle have identical areas ($A_0 = 28.9$ ha). The shoreline development D_L of the reservoir is 2.0 and that of the circle is 1.0.

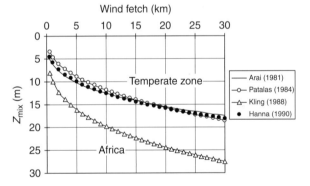

Figure 6 Empirical relationships between wind fetch and mixing depth z_{mix} established for lakes in Japan, Europe and North America, and in tropical Africa. Data for lakes in Japan from Arai T (1981) Climatic and geomorphological influences on lake temperature. *Verhandlungen Internationale Vereinigung für Theoretische und Angewandte Limnologie* 21: 130–134; for lakes in Europe and North America from Patalas K (1984) Mid-summer mixing depths of lakes of different latitudes. *Verhandlungen Internationale Vereinigung für Theoretische und Angewandte Limnologie* 22: 97–102 and Hanna M (1990) Evaluation of models predicting mixing depth. *Canadian Journal of Fisheries and Aquatic Sciences* 47: 940–947; and for lakes in tropical Africa from Kling GW (1988) Comparative transparency, depth of mixing and stability of stratification in lakes of Cameroon, West Africa. *Limnology and Oceanography* 33: 27–40.

measured along the thalweg of the lake. In contrast with b_{max}, the mean width b_{mean} is given as the quotient of A_0 and l_{max}. However, most important are the effective length l_{eff} and width b_{eff} (perpendicular to l_{eff}) which represent the longest distances from shore to shore, not interrupted by land (**Figure 5**). The effective lake axis d_{eff} is the average of l_{eff} and b_{eff}.

Surface dimensions are used to quantify the wind effect on a lake. The parameter l_{eff} is usually called the maximum 'effective wind fetch' F (Håkanson, 1981; Kling, 1988; Hanna, 1990). Other authors define F as $\sqrt{A_0}$ (Arai, 1981) or d_{eff} (Patalas, 1984). No matter how F is defined, it was frequently found to correlate significantly with mixing depth z_{mix} (**Figure 6**; Arai, 1981; Patalas, 1984; Kling, 1988; Hanna, 1990). Furthermore, the maximum height of surface waves and thus, their erosive impact on the shores, sediment resuspension, and transport into deeper regions of the lake are related to F (see Section 2.08.1.5).

The length of the shoreline L_0 and the dimensionless shoreline development D_L with

$$D_L = \frac{L_0}{2\sqrt{\pi A_0}} \quad (1)$$

characterize the land–water and littoral–pelagial interactions of a lake. D_L relates L_0 to the circumference of a circle with an area identical to the lake's surface A_0 (**Figure 5**). Thus, the minimum of D_L is 1, and the more the lake's surface differs from a circular shape, the higher is the D_L, indicating stronger linkage of the lake to the drainage basin and more extended shallow littoral zones. Many reservoirs exhibit high D_L due to their dendritic surface shape.

The depth distribution of a lake is described by the hypsographic curve. The areas A_z are dependent on depths z. A_z is calculated from determinations of the areas enclosed by k contour lines in different depths, from surface down to the maximum depth z_{max}, drawn on a bathymetric map. The volume $V_i - V_{i+1}$ of the layer between neighboring contour lines at depths z_i and z_{i+1} can be estimated as follows:

$$V_i - V_{i+1} \approx \tfrac{1}{3}(A_i + A_{i+1} + \sqrt{A_i A_{i+1}})(z_{i+1} - z_i) \quad (2)$$

with $0 \leq i \leq k-1$, $z_k = z_{max}$, $A_k = 0$, and $V_k = 0$. The sum of the volumes of the layers below z_i is the volume V_i and consequently, the sum of the volumes of all layers is the total volume V_0 of the lake. Finally, the volume–depth development V_z for $0 \leq z \leq z_{max}$ can be constructed. It is clear that the accuracy of the curves A_z and V_z increases with the increasing number k of contour lines.

Maximum depth z_{max} and average depth $\bar{z}$ with

$$\bar{z} = \frac{V_0}{A_0} \quad (3)$$

are very important parameters influencing the vertical distribution and zonation of, for example, temperature, underwater light conditions, nutrients, oxygen, primary productivity, sediment resuspension, and many others. Thienemann (1927) has already stated that shallow lakes generally tend toward a higher eutrophy than deep lakes. He defined the boundary between eutrophy and oligotrophy at $\bar{z} = 18$ m for German lakes. Kalff (2002) named $\bar{z}$ probably the most useful single morphometric feature available.

The shallowness of a lake can be characterized by its relative depth z_{rel} (%), which is the ratio between z_{max} and the diameter of a circle with area A_0:

$$z_{rel} = 50 z_{max} \sqrt{\frac{\pi}{A_0}} \quad (4)$$

Characteristic values of z_{rel} lie between 1% (large and shallow lakes) and about 4% (deep lakes). Calderas, maars, fjords, or solution basins may have $z_{rel} > 10\%$. The record $z_{rel} = 374\%$ is held by the Hawaiian volcanic crater lake, Kauhako ($A_0 = 0.35$ ha, $z_{max} = 250$ m) (Cole, 1994).

The extension of the littoral zone, the potential development of submerged macrophytes, the near-shore sediment transport and quality (water and organic content, particle size; e.g., Håkanson and Boulion, 2002), and the colonization of littoral sediments with benthic organisms, depend on the slope s (%) of the shore. The slope s_i between two contour lines at depths z_i and z_{i+1} is calculated as follows (lengths and depths in m, areas in m^2; see **Figure 7**)

$$s_i = 100 \frac{(L_i + L_{i+1})(z_{i+1} - z_i)}{2(A_i - A_{i+1})} \quad (5)$$

The average basin slope $\bar{s}$ is

$$\bar{s} = 50 \frac{z_{max}}{n A_0} \sum_{i=0}^{k-1} (L_i + L_{i+1}) \quad (6)$$

Lakes of identical A_0 and z_{max} may have different volumes due to different volume–depth distributions. In order to classify lake types, the index D_V called 'volume development' was defined. D_V is the ratio between the lake's real volume

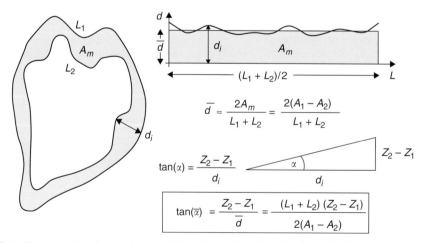

Figure 7 Schematic illustration on how to determine the slope between two contour lines L_1 and L_2 at depths z_1 and z_2 (L_1, L_2, z_1, and z_2 in m, areas A_1 and A_2 enclosed by L_1 and L_2 given in m^2, $A_m = A_1 - A_2$, d_i – distance between L_1 and L_2 at a certain position, $\bar{d}$ – average distance).

$V_0 = A_0 \bar{z}$ (Equation (3)) and the volume $\frac{1}{3}A_0 z_{max}$ of an inverted cone with base area equal to the lake's surface A_0 and height coincident with the lake's z_{max}:

$$D_V = \frac{3A_0\bar{z}}{A_0 z_{max}} = 3\frac{\bar{z}}{z_{max}} \quad (7)$$

A total of 202 out of 243 lakes evaluated by Carpenter (1983) came under the range 1 (V-shaped or cone) $\leq D_V \leq 2$ (U-shaped or ellipsoid). The U-shaped basins of many old natural lakes are the result of lake aging, that is, the deposition and focusing of large quantities of allochthonous and autochthonous sediments in the deepest parts, over a long time. Relatively young reservoirs, however, often have V-shaped basins.

Morphometric models were developed that describe the geometric shape of lakes as quadric surfaces and sinusoids based on D_V. Junge (1966) introduced the transformation

$$n = 2D_V - 3 \quad \text{(Junge's shape index)} \quad (8)$$

and derived the following formulas for the relative area A_x and volume V_x using the normalized depth $x = z/z_{max}$ (z downward positive):

$$A_x = \frac{A_z}{A_0} = 1 - nx^2 - (1-n)x \quad (9)$$

$$V_x = \frac{V_z}{V_0} = 1 - \frac{6x - 3(1-n)x^2 - 2nx^3}{3+n} \quad (10)$$

Junge (1966) found satisfactory agreement between measured and calculated volume–depth distributions for most lakes and ponds considered in the survey. Significant deviations are characteristic of lakes with singular deep pits ($D_V < 1$) or large flat-bottom areas ($D_V > 2$).

Based on Junge's model, lake types can be classified by principal geometric characteristics (**Table 1**; Junge, 1966). In order to elucidate the most important differences between the basic geometric lake types, circular basins with identical A_0 and z_{max} are assumed.

The areas and volumes of the upper (epilimnetic) water layers decrease much faster in the cone than in the ellipsoid (**Figure 8**; Junge, 1966). This has many very important consequences:

- The epilimnion of V-shaped basins is shallower. Thus, the impact of sediment-related processes, such as the extension of the littoral area and its colonization with submerged macrophytes, as well as its role as a habitat for fishes, the influence of benthic organisms, sediment resuspension, and nutrient remobilization at the sediment–water interface on the epilimnetic matter turnover, is potentially highest in V-shaped water bodies.

- The area of the threshold between epilimnion and hypolimnion is smallest in V-shaped and largest in U-shaped basins. Therefore, the probability that particles and algae settle on the epilimnetic sediment area is highest in V-shaped lakes. The dilution of nutrients released from epilimnetic sediments is greatest in those lakes. Both features were found to influence the primary productivity of lakes (Fee, 1979).

- The ratio r_V between the epilimnion and hypolimnion volumes is highest and increases faster in the cone and thus, the hypolimnetic oxygen balance is more critical in V-shaped basins. Thienemann (1927) had already postulated that lakes with $r_V > 1$ tend toward a eutrophic state while those with $r_V < 1$ are more oligotrophic. Sedimentation from epilimnion into hypolimnion is an important loss factor for phytoplankton, primarily for fast-settling diatoms. The probability of algae settling into the hypolimnion is much higher in U-shaped than in V-shaped lakes.

- The average epilimnetic light intensity is higher in V-shaped systems, due to the lower volume in the deeper zones compared with U-shaped ones. Light limitation of phytoplankton growth may be more significant in U-shaped water bodies.

- The transport of sediments into deeper regions of the lake, the so-called sediment focusing, depends on the average basin slope (Blais and Kalff, 1995), which is highest in U-shaped lakes.

- The total volume V_0 of the ellipsoid is twice the volume of the cone and $4/3 V_0$ of the paraboloid (**Table 1**). Consequently, the cone has a much lower heat capacity. Furthermore, its thermal stability is much lower due to the low depth of the gravity center, if identical vertical temperature distributions are assumed. Both facts may influence the timing of the periods of full turnover and stratification and the beginning of ice covering.

Table 1 Parameters describing the shape of basic geometric lake types

		Basic geometric lake type		
Parameter	Symbol	Cone	Paraboloid	Ellipsoid
Volume development	D_V	1	3/2	2
Junge's shape index	n	-1	0	1
Normalized area	A_x	$(1-x)^2$	$1-x$	$1-x^2$
Normalized volume	V_x	$(1-x)^3$	$(1-x)^2$	$1-x(3-x^2)/2$
Depth of gravity center of the completely mixed lake	x_{gc}	1/4	1/3	3/8
Total volume (m³)	V_0	$A_0 z_{max}/3$	$A_0 z_{max}/2$	$2 A_0 z_{max}/3$

From the morphometry model of Junge CO (1966) Depth distributions for quadric surfaces and other configurations. In: Hrbacek J (ed.) *Hydrobiological Studies*, Academia Publishing House of the Czechoslovak Academy of Sciences, pp. 257–265. Prague.

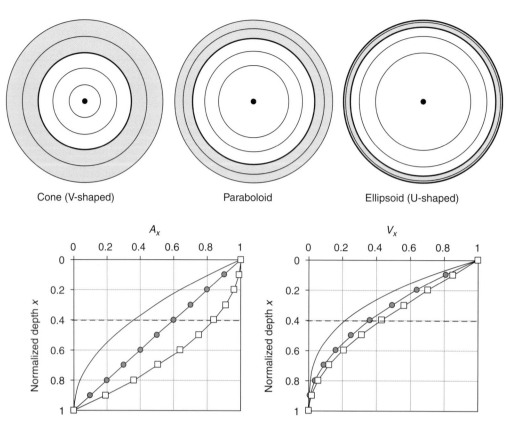

Figure 8 (Top) Bathymetric maps of idealized (identical circular surface and maximum depth) lake types based on the morphometry model. Contour lines are drawn for the normalized depths $x = 0, 0.2, 0.4, \ldots, 1$. Attention should be paid to the different portions of the gray (epilimnion) and white (hypolimnion) areas (assumed a mixing depth of $x_{mix} = 0.4$). (Bottom) Normalized hypsographic curves $A_x = f(x)$ (left) and volume–depth distributions $V_x = f(x)$ (right) for idealized lake types (——— cone, —●— paraboloid, and —□— Ellipsoid). The dashed lines mark the assumed mixing depth of $x_{mix} = 0.4$. Bathymetric maps based on the morphometry model of Junge CO (1966) Depth distributions for quadric surfaces and other configurations. In: Hrbacek J (ed.) *Hydrobiological Studies*, Academia Publishing House of the Czechoslovak Academy of Sciences, pp. 257–265. Prague.

- V-shaped lakes have a shorter theoretical residence time $\bar{t} = V_0/Q_a$ (a), where Q_a (m³ a⁻¹) is the mean annual discharge. The dilution of inflowing water-carrying nutrients, suspended matter, and other substances is low in V-shaped lakes, but much higher in U-shaped ones. Hence, the resistance of U-shaped basins against changing external loading is greater. The delay of an aggravation of the trophic state, in the case of increasing nutrient imports, is longer in those lakes. V-shaped water bodies may respond faster to reduced external loading.

It can be concluded that not only the size, but also the shape of lakes considerably influences their physical, chemical, and biological structure and functioning.

2.08.1.4 Influx and Vertical Distribution of Solar Energy

Solar radiation is the Earth's most important natural energy source and has a prominent ecological role. The global solar irradiance I_G (W m⁻²) is the solar radiation measured on a horizontal plane at the Earth's surface and spans wavelengths λ between about 200 and 3000 nm. The spectrum is divided into the ranges of ultraviolet (UV) radiation ($\lambda < 380$ nm; may be harmful to organisms), visible light (380 nm $\leq \lambda \leq$ 750 nm), and infrared radiation ($\lambda > 750$ nm; thermal radiation). I_G is the sum of direct sun radiation and diffuse sky radiation (measured at full cloud cover, radiation reflected from clouds, water and dust particles, and other aerosols suspended in the atmosphere). It depends on latitude (**Figure 9**; Stras'kraba, 1980), altitude (thickness of the atmosphere, higher I_G at higher altitudes), and penetrability of the atmosphere (higher I_G in dry regions compared with wet regions). If measured values of the local global radiation are not available, they can be approximately calculated from daily integrals of the radiation reaching the Earth's surface on totally cloudless days, observations of sunshine duration and day length (Straškraba, 1980).

2.08.1.4.1 Underwater light conditions

The ratio of the irradiance reflected at surfaces to the incident flux is called 'albedo' r (in parts of 1). The reflection at water surfaces is mostly mirror like (specular reflection), with respect to the surface normal (angle of incidence α equals angle of reflection). Albedo r of direct sun radiation is a function of α and, therefore, the daily average varies geographically. However, r decreases quickly with decreasing α ($r < 0.13$ if $\alpha < 70°$, and $r < 0.03$ if $\alpha < 45°$). The reflection of diffuse sky radiation is lower at high α and higher at low α, than that of direct sun radiation. Surface waves reduce r at $\alpha > 70°$. For Central

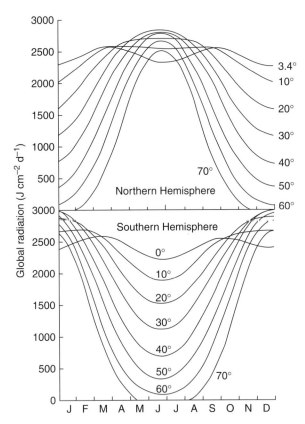

Figure 9 Annual variations of daily integrals of the radiation reaching the Earth's surface on totally cloudless days calculated for selected latitudes (atmospheric transmission factor 0.6). Modified from Straškraba M (1980) The effects of physical variables on freshwater production: Analysis based on models. In: le Cren ED and McConnell RH (eds.) *The Functioning of Freshwater Ecosystems*, IBP 22, pp. 13–84. Cambridge University Press.

Europe, a daily average of $r \approx 0.1$ can be assumed for water surfaces. Fresh snow reflects about 80–90%, old snow about 40–70%, and ice c. 25–35% of the irradiation.

Radiation entering the water surface changes its direction due to the higher density and lower velocity of propagation in water than in air. This phenomenon is called refraction. The angle of refraction β is lower than α. The opposite applies to radiation, which is scattered from particles in the water back into the air. The radiation is angled away from the surface normal. If the angle of incidence of the backscattered radiation is greater than 49°, it is completely reflected at the water–air interface.

The photosynthetically active radiation (PAR) spanning 400–700 nm, within the range of visible light, is potentially a growth-limiting factor with regard to plants. Considering the intensity I_{0+} of PAR above the water surface, the approximation $I_{0+} \approx 0.46 I_G$ is widely accepted and with an average daily albedo $r = 0.1$, the intensity I_{0-} of PAR just below the water surface can be expressed as

$$I_{0-} = 0.414 I_G \qquad (11)$$

Divers realize that colored objects become much paler in greater water depths. This phenomenon is the consequence of the changing spectral composition of light, with increasing water depth (**Figure 10**; Vollenweider, 1961; Uhlmann and Horn, 2001), due to the wavelength-specific transmission

$$T_{\lambda z} = 100 \, \frac{I_{\lambda z}}{I_{\lambda 0-}} (\%) \qquad (12)$$

with the wavelength λ and the light intensities $I_{\lambda 0-}$ and $I_{\lambda z}$ just below the surface and at depth z, respectively. While the transmission of pure water is high in the blue–green part of the spectrum ($\lambda = 450$–500 nm), the range of the most penetrating light component shifts toward green–yellow ($\lambda = 530$–580 nm) or even yellow–red ($\lambda = 580$–650 nm), depending on the concentration of suspended particles and dissolved substances (e.g., humic acids). Thus, the visual images of lakes change correspondingly: clear lakes appear blue–green, and eutrophicated lakes or those influenced by colored dissolved organic substances look greenish or brownish, due to the predominant backscattering of the respective range of the light spectrum.

Although some phytoplankton species were found to react specifically to changing light quality (chromatic adaptation), the decrease of absolute light intensity with increasing water depth is far more important for photosynthesis. The light attenuation is described by the Lambert–Beer's law:

$$I_{\lambda z} = I_{\lambda 0-} \exp(-k_\lambda z) \qquad (13)$$

The attenuation coefficient k_λ (m^{-1}) is a measure for the combined effect of absorption (i.e., transformation of radiation energy into heat or biochemical energy) and scattering (i.e., change of propagation direction caused by particles or water-density inhomogeneities) on the intensity of the light of the wavelength λ. Although Equation (13) is, strictly speaking, only valid for parallel monochromatic light beams, it can also be applied to relatively narrow spectral bands such as PAR. The average PAR attenuation coefficient k_{PAR} is derived from the spectral attenuation coefficients k_λ

$$k_{PAR} = \frac{1}{300} \int_{400}^{700} k_\lambda \, d\lambda \approx \frac{1}{3}(k_{450} + k_{550} + k_{650}) \qquad (14)$$

or, nowadays, directly calculated from underwater light measurements using spherical quantum sensors whose spectral response is adapted to the PAR range. The value k_{PAR} is the sum of k_W (pure water), k_S (dissolved or colloidal matter), k_P (phytoplankton), and k_D (nonliving particles)

$$k_{PAR} = k_W + k_S + k_P + k_D = k_W + \varepsilon_S C_S + \varepsilon_P C_P + \varepsilon_D C_D \qquad (15)$$

where the ε_i (l m^{-1} mg^{-1}) are the substance-specific attenuation coefficients, and the C_i (mg l^{-1}) the substance concentrations, respectively. The mean extinction of the light flux directed downward is a suitable index for evaluating its spectral distribution (Vollenweider, 1961; see **Figure 10**). The property of substances to preferably absorb and reflect light in specific wavelength ranges is utilized in the remote sensing of water-quality criteria (e.g., the distribution of chlorophyll and water temperature) by air or satellite-borne reflectance measurements.

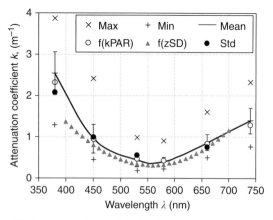 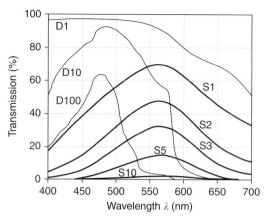

Figure 10 (a) Mean (±standard deviation) and range of variation (min, max) of spectral attenuation coefficients (measured bi-weekly during the ice-free seasons from 1975 to 1985) as well as approximations derived from the average attenuation coefficient of PAR (f (kPAR)) and the average Secchi disk transparency (f (zSD)) of the Saidenbach Reservoir (Germany) compared with the spectral standard distribution (Std); (b) average spectral underwater light transmission in the Saidenbach Reservoir (S1: 1m, S2: 2m, ..., S10: 10m) and of distilled water (thickness of water layer D1: 1m, D10: 10m, D100: 100m). (a) Std from Vollenweider RA (1961) Photometric studies in inland waters: I. Relations existing in the spectral extinction of light in water. *Memorie dell'Istituto Italiano di Idrobiologia* 13: 87–113. (b) Curves D1, D10, and D100 from Uhlmann D and Horn W (2001) *Hydrobiologie der Binnengewässer. Ein Grundriss für Ingenieure und Naturwissenschaftler.* Stuttgart: Eugen Ulmer; and data of the Saidenbach Reservoir from Paul L (1989) Interrelationships between optical parameters. *Acta Hydrophysica, Berlin* 33: 41–63.

The transparency or Secchi depth z_{SD} provides a clear impression of the optical properties of standing waters. It is easily measured using a Secchi disk, a white disk, usually 25 cm in diameter, which is lowered on the shady side of a boat down to the depth of its visual disappearance. Relationships such as $k_{PAR} z_{SD} = a$ have been frequently published, since the early work of Poole and Atkins (1929). The value of a must be considered as water-body specific and, is consequently found to be widely scattering between about 1.1 and 4.6. This is quite understandable, because of the fact that the Secchi disk visually disappears, not only because of light attenuation but primarily because the contrast between disk and background becomes imperceptible. This is strongly influenced by both the concentration and size of the suspended particles in the water column above the disk. Thus, the percentage of transmission at z_{SD} must be higher in a turbid lake. Nevertheless, the Secchi depth can be used as a predictor for spectral underwater light distribution. This was shown by Paul (1989), who found highly significant correlations of the type

$$k_i = a_i + \frac{b_i}{z_{SD}} \quad (16)$$

where i stands for both λ and PAR, respectively (**Figure 10**). The value a_i can be considered as a first-order approximation for the lake-specific attenuation coefficients of water without suspended particles.

The light requirements of phototrophic organisms are quite different and, moreover, depend on their physiological state. Water plants adapted to low-light intensities are very sensitive to small enhancements of illumination or, on the other hand, their photosynthetic response is inhibited if the light intensity increases sharply. Thus, it is impossible to define a general minimum-compensation light intensity I_{comp} (W m^{-2}) that is necessary to compensate for the oxygen consumption at night (respiration), with the respective photosynthetic oxygen production in the daytime. Therefore, the underwater light situation is characterized by the depth z_{eu} of the euphotic zone, which is the water layer expanding from surface down to the depth, where the PAR intensity I_{zeu} is 1% of I_0:

$$z_{eu} = \frac{4.6}{k_{PAR}} \quad (17)$$

In the euphotic zone (phototrophic layer), the light intensity is considered to be sufficiently high to allow photosynthesis. In the layers below z_{eu} (tropholytic zone), respiration exceeds production and the phytoplankton development is light limited. It has to be borne in mind that z_{eu} is only an approximate guiding principle. It is related to full daylight and sufficient day length. The light intensity I_{zeu} in medium or high latitudes is much lower in winter than in summer and thus, z_{eu} may substantially overestimate the depth of the compensation point in winter, and vice versa in summer.

Phytoplankton cells are vertically transported by wind-driven and/or convective currents throughout the mixed layer bounded by the mixing depth z_{mix} and are exposed to an average light intensity

$$\bar{I}_{zmix} = \frac{I_{0-}}{z_{mix}} \int_0^{z_{mix}} \exp(-k_{PAR}z) dz$$

$$= \frac{I_{0-}}{k_{PAR} z_{mix}} (1 - \exp(-k_{PAR} z_{mix})) \quad (18)$$

or, if $k_{PAR} z_{mix} > 3$ and Equations (11) and (17) are considered

$$\bar{I}_{zmix} \approx \frac{I_{0-}}{k_{PAR} z_{mix}} = \frac{z_{eu} I_{0-}}{4.6 z_{mix}} \approx 0.09 I_G \frac{z_{eu}}{z_{mix}} \quad (19)$$

Hence, if the ratio between mixing depth z_{mix} and euphotic depth z_{eu} exceeds a critical value, phytoplankton development is severely restricted, due to a very low average light intensity in the mixed water column. This principle is utilized in the artificial destratification of lakes, to control mass development of nuisance algae (see Section 2.08.3.3).

2.08.1.4.2 Heat budget and thermal structure

The heat budget of a lake is determined by seasonal variations of

- the absorption of short-wave solar radiation,
- the net exchange of long-wave radiation between lake surface and atmosphere,
- the conductive exchange of heat at the water surface depending on the temperature difference between water and air,
- the evaporative heat loss,
- the heat import and export by inflow and outflow (only important in lakes or reservoirs with short retention times or in lakes significantly fed by the meltwater of glaciers), and
- the heat exchange at the lake bottom (only important in instances of geothermal activities).

Heat is vertically transported by convective and advective currents and turbulent eddy-diffusion. Currents generated by wind have a particular impact, and heat conduction is of minor importance. Therefore, the thermal structure of a lake greatly depends on the size and shape of the surface and its exposure to wind.

The amount of heat stored by a lake, the heat content θ (kJ), is calculated as follows:

$$\theta = c_p \rho \int_0^{z_{max}} A_z \vartheta_z \mathrm{d}z \approx 4186 * V_0 \bar{\vartheta} \quad (20)$$

with area A_z (m^2) and temperature ϑ_z (°C) at depth z (m), maximum depth z_{max}, specific heat of water $c_p = 1$ kcal kg^{-1} K^{-1} = 4.1855 kJ kg^{-1} K^{-1}, density of water ρ (kg m^{-3}), lake volume V_0 (m^3), and average temperature $\bar{\vartheta}$ (°C). The annual heat budget, the quantity of heat gained during warming and released during autumnal cooling, is the difference between the annual maximum and minimum heat content. Halbfaß (1921) described the immense magnitude of a lake's annual heat turnover, comparing it with the length of a wagon train loaded with an equivalent amount of coal for heating. For the relatively small Klingenberg reservoir (near Dresden, Germany; $V_0 \approx 16 * 10^6$ m^3 and $z_{max} \approx 33$ m), he calculated the annual heat budget equivalent to the combustion heat of a coal train, 17 km in length. Thus, large lakes may significantly influence local climate.

The vertical temperature distribution represents density stratification. Generally, the water density ρ depends on salinity s (‰), pressure P (bar), and temperature ϑ. A set of formulas to precisely calculate $\rho(\vartheta,s,P)$, was provided by Chen and Millero (1986). However, the influence of salinity (normally $s < 1$ ‰) and pressure P with $P \approx 0.1z$ is negligible in freshwater and ρ is primarily determined by ϑ (**Figure 11**, see also **Figure 12**; Chen and Millero, 1986). The unique quality of water, the decrease of density at temperatures <4 °C (anomaly of water), and the fact that the density of ice only is

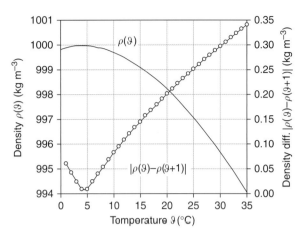

Figure 11 Density of water $\rho(\vartheta)$ and density difference $\Delta\rho = |\rho(\vartheta) - \rho(\vartheta+1)|$ vs. temperature ϑ at sea level and normal air pressure.

about 92% of the density of water at 0 °C, ensures the survival of aquatic organisms even in cold winters. Ice develops and floats at the lake surface and the water temperature in deep layers is not much lower than 4 °C. Therefore, aquatic organisms need to resist a much smaller annual temperature variation than terrestrial organisms.

The density difference $\Delta\rho$ increases considerably with increasing temperature. For instance, $\Delta\rho$ between water temperatures of 19 °C and 4 °C is almost identical with $\Delta\rho$ between 28 °C and 22 °C. Thus, relatively small temperature differences in tropical lakes may represent stronger density gradients than in temperate lakes. At the beginning of the warmer seasons, the heat gain in surface water is higher than wind-driven currents that can distribute vertically, and a temperature stratification is established in deep or even shallower, but wind-sheltered lakes, that usually consists of three characteristic layers (**Figure 12**):

- Epilimnion – the upper warm, less dense, and turbulently mixed layer of almost homogenous temperature and density.
- Metalimnion – the intermediate stratum with strongly decreasing temperature and increasing density.
- Hypolimnion – the cold, more dense, and relatively quiescent bottom layer with low temperature and density gradients.

The static stability of temperature stratification can be characterized by density gradients, for example, by the buoyancy or Brunt–Väisälä frequency N (s^{-1}) with

$$N^2 = \frac{g}{\rho}\frac{\mathrm{d}\rho}{\mathrm{d}z} \approx \frac{2 * 9.81 * (\rho_{z+\Delta z} - \rho_z)}{\rho_{z+\Delta z} + \rho_z} \quad (21)$$

(g (m s^{-2}) the local acceleration due to gravity and $\Delta z = 1$ m), or by the relative thermal resistance to mixing R (Wetzel and Likens, 1991) with

$$R = \frac{\rho_{z+\Delta z} - \rho_z}{\rho(4\,°C) - \rho(5\,°C)} \approx \frac{10^3}{8}(\rho_{z+\Delta z} - \rho_z) \quad (22)$$

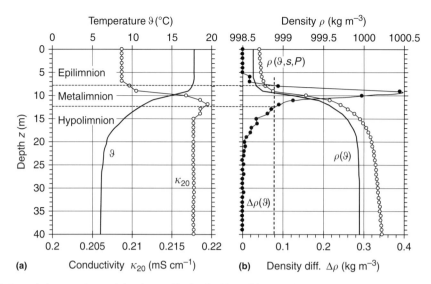

Figure 12 Characteristic vertical temperature and density stratification for lakes of the transient zone (about 40°–60° N or S) in summer. (a) Vertical profiles of water temperature ϑ and conductivity κ_{20} (mS cm^{-2}, related to $\vartheta = 20$ °C) measured in the Saidenbach Reservoir on 10 July 2007. (b) Respective distributions of water density $\rho(\vartheta)$ as a function of ϑ alone, density $\rho(\vartheta,s,P)$ depending on ϑ, salinity $s \approx 0.5\kappa_{20}$, and pressure $P \approx 0.1z$, and density difference $\Delta\rho(\vartheta) = \rho(\vartheta_{z+1}) - \rho(\vartheta_z)$. The maximum deviation of $\rho(\vartheta)$ from $\rho(\vartheta,s,P)$ is less than 0.04%, which clearly shows the primary impact of temperature on density in freshwaters. (b) Curve $\rho(\vartheta,s,P)$ from Chen C-TA and Millero FJ (1986) Precise thermodynamic properties of natural waters covering only the limnological range. *Limnology and Oceanography* 31: 657–662.

Schmidt (1915) defined thermal stability S_0 (Nm m^{-2}) as the required energy per square meter of the lake surface, to completely mix the stratified water body, without change in its heat content

$$S_0 = \frac{g}{A_0} \int_0^{z_{max}} A_z(z_{fc} - z)(\rho_{fc} - \rho_z) dz$$
$$= \frac{g}{A_0} \int_0^{z_{max}} A_z \rho_z(z - z_{fc}) dz = \frac{gM}{A_0}(z_{st} - z_{fc}) \quad (23)$$

with area A_z and density ρ_z at depth z, acceleration due to gravity $g = 9.81$ m s^{-2}, surface area A_0, depth z_{fc} of the gravity center, and density ρ_{fc} during the full circulation period. Thus, $S_0 A_0$ corresponds to the work (to be accomplished by the wind) that is required to lift the total mass M of the lake by the distance $z_{st} - z_{fc}$, which is the difference between the depths z_{st} and z_{fc} of the gravity center of the stratified, and the completely circulating lake. In the case of a stable stratification (lighter, less dense above heavier water layers with higher density), it is $z_{st} > z_{fc}$ and $S_0 > 0$. The stability characterizes the degree of separation of the hypolimnetic water layers from the epilimnetic ones. Deep lakes have a higher S_0 than shallow water bodies. The difference $z_{st} - z_{fc}$ usually amounts to only a few millimeters. However, the energy needed for mixing a stably stratified lake is huge, due to the enormous mass to be raised. For instance, the mass of the relatively small Klingenberg reservoir, mentioned above, is about 1600 times the mass of the Eiffel tower in Paris. Thermal stability is an important parameter that has to be considered in planning artificial lake destratification, as a measure to prevent hypolimnetic oxygen depletion and/or the mass development of noxious phytoplankton by light limitation, resulting from a too high ratio of z_{mix} and z_{eu}.

The metalimnion is usually defined as the stratum where the temperature gradient exceeds a certain limit (e.g., 1 K m^{-1}). However, considering lakes at different latitudes with quite different temperature ranges, the upper and the lower threshold of the metalimnion should be related to density gradients. In temperate zones, the beginning of a stable summer stratification is often observed when the surface temperature exceeds 10 °C. Accordingly, the depth of the 10 °C-isotherm is a good predictor for the threshold between meta- and hypolimnion. Consequently, the metalimnion could more generally be defined as the layer with density gradients greater than 0.08 kg m^{-4} (see **Figure 12**). This limit corresponds approximately to the water-density difference between 9 °C and 10 °C and, thus, to $N^2 \approx 0.0008$ s^{-2} (Equation (21)) or to $R \approx 10$ (Equation (22)). The level of the maximum density gradient is called 'thermocline'.

Talling (1971) determined the mixing depth z_{mix} of the upper mixed layer as the depth with a temperature 0.5 K below the temperature at a depth of 2 m. In this manner, superficial thermal gradients, which may develop during transitional calm weather periods, are largely excluded. Referring to this principle, but transferred to density gradients, z_{mix} can be defined as the depth with a density 0.08 kg m^{-3} higher than that at the depth of 2 m. Thus, for a given temperature $\vartheta_2 > 7.2$ °C in $z = 2$ m, the temperature ϑ_{zmix} (°C) at the depth z_{mix} amounts to

$$\vartheta_{zmix} = \vartheta_2 - 0.28 - 3031200 \exp(-2\vartheta_2) - 5.2 \exp(-0.2\vartheta_2) \quad (24)$$

Latitude and altitude determine a lake's seasonal temperature range and mixing scheme, depending on the regional air temperature range (**Figure 13**; Stras 'kraba 1980). For latitudes up to about 40° N or S, the bottom temperature ϑ_B of deep lakes corresponds approximately to the minimum annual water temperature and decreases from very high values in the tropics, to 4 °C. At higher latitudes, ϑ_B remains constant at the temperature of the density maximum. The resulting mixing type depends on the absolute temperature range and the seasonal surface temperature variation. Increasing distance from the equator and increasing altitude have the same effect (**Figure 14**; Hutchinson and Lo' ffler, 1956). The following thermal lake types are distinguished depending on the principal mixing behavior:

1. *Amictic lakes*. They are permanently frozen lakes at high latitudes and/or altitudes that never overturn; and
2. *Holomictic lakes*. They refer to lakes that mix at least once per year, further specified as:

   - *Oligomictic lakes*. They are mostly very deep tropical lakes, with high heat capacities, that are rarely and irregularly mixed (usually under extreme weather situations, e.g., tropical storms).
   - *Polymictic lakes*. They refer to shallower lakes with low vertical-density gradients that mix frequently, sometimes daily.
   - *Monomictic lakes*. They are lakes with one mixing period, either in winter at water temperatures $\geq 4\,°C$ (warm monomictic subtropical lakes, or large and deep lakes, in the temperate zone with high heat capacity, that do not freeze), or in summer at water temperatures $\leq 4\,°C$ (cold monomictic lakes at high latitudes, where ice cover melts only in summer).
   - *Dimictic lakes*. They are sufficiently deep lakes of the temperate zone that circulate in spring and autumn and are ice covered in winter.
   - *Meromictic lakes*. They refer to partially mixed lakes with a deep-water layer enriched by dissolved salts, or are sufficiently wind sheltered, small but deep lakes.

A sequence of typical vertical temperature profiles of a dimictic water body (Lake Stechlin, Germany) is shown in **Figure 15**.

An inverse temperature distribution (colder above warmer water) is observed during the winter stagnation when the lake is ice covered (e.g., 15 February 2006 in **Figure 15**). Vertical mixing is strongly reduced due to the cutoff of wind action by the ice. Some convective mixing just below the ice is possible on sunny days, if the ice is clear and irradiance heats the uppermost water layers.

After the disappearance of the ice cover, complete mixing of the entire water body is likely. As long as the water temperature is lower than 4 °C, warming at the surface provokes an increase in the density and mixing is induced, even in dead calm. After the water has reached the temperature of maximum density (profile from 7 April 2006 in **Figure 15**), further heating produces less-dense water at the surface and mixing requires sufficient wind energy. Thus, mixing becomes more and more episodic and depends on the actual weather

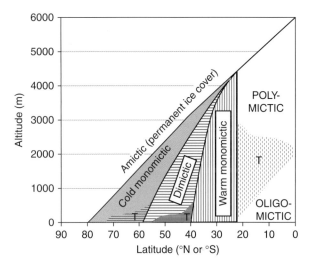

Figure 14 Scheme of the distribution of thermal lake types depending on latitude and altitude. Modified from Hutchinson GE and Löffler H (1956) The thermal classification of lakes. *Proceedings of the National Academy of Sciences of the United States of America* 42: 84–86. T, transitional regions.

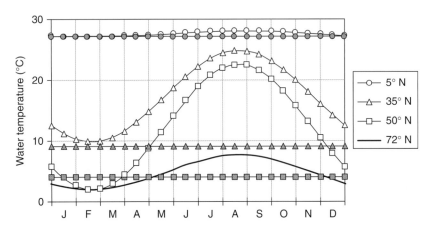

Figure 13 Trends of seasonal variations of surface (open symbols) and corresponding bottom temperatures (filled symbols) of medium-sized lakes at low elevations calculated for selected northern latitudes from empirical equations provided by Straškraba M (1980) The effects of physical variables on freshwater production: Analysis based on models. In: le Cren ED and McConnell RH (eds.) *The Functioning of Freshwater Ecosystems*, IBP 22, pp. 13–84. Cambridge University Press.

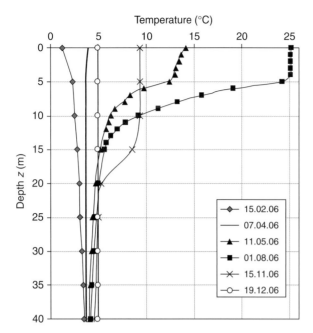

Figure 15 Sequence of temperature profiles characterizing the seasonal change between mixing and stratification typical for a dimictic lake (upper 40 m of Lake Stechlin, 13° 02′ E, 53° 09′ N, Germany; maximum depth of the lake is 69 m). Courtesy of Dr. P. Kasprzak, IGB Berlin.

situation. Inconsistent and relatively cold weather (typical April-weather) may prolong the period of spring full circulation and foster the warming of the whole water column (increase in temperature of the deep-water layers to more than 5 °C). Conversely, warm and calm weather immediately after the temperature homogeneity at 4 °C, may quickly form density gradients at the surface, which even strong winds cannot equalize any further. Thus, the spring full circulation period is short and the deep-water layers remain relatively cold (as was observed on 11 May 2006, **Figure 15**).

Once a stable thermocline is established and the summer stagnation has started, further increasing air and, consequently, surface water temperatures, strengthen the temperature and density differences (1 August 2006 in **Figure 15**) and, thus, the thermal, hydrodynamic, chemical, and biological decoupling between the illuminated, warm, flushed, wind-mixed epilimnion and the usually dark, cold, quiescent hypolimnion takes place. The thermal structure fundamentally influences the temporal development and spatial distribution of biological and chemical food-web components. The water column is subdivided into two reaction spaces with completely different physical, chemical, and biological properties. Therefore, Ruttner (1962) characterized thermics as the pivotal point of lake limnology.

After midsummer, irradiation and air-temperature decline and successive cooling and mixing increase the depth of the epilimnion, and the metalimnion slowly propagates downward. The metalimnetic density gradients decrease, and heat and matter exchange by eddy-diffusion, between epi- and hypolimnion, increases. For instance, the maximum density gradients of the temperature profile from 15 August 2006 in Lake Stechlin (**Figure 15**) were much smaller than 0.08 kg m^{-4} and thus, by definition, the stratification could no longer be considered as stable. Eventually, the stratification disappeared (19 December 2006) and the lake went into the phase of autumn full circulation.

Further cooling favors convective overturn, until the water temperature reaches 4 °C. From then on, the ice cover on an entire lake may be established in a single, calm, and frosty night and winter stagnation will be initiated. If this happens early, the deep-water temperature remains relatively high ($\sim$4 °C) all through the winter. Paradoxically, the temperature of the water column may decrease much more in a mild winter with a late ice-up.

Climate change is expected to significantly influence seasonal temperature development, the duration of the mixing and stagnation periods, the solubility of gases, and the exchange of heat and matter between water and sediment (Blenckner et al., 2002). Milder winters result in later freeze-up and earlier ice break-up and, in extreme cases, ice cover and winter stagnation do not even develop. Thus, formerly dimictic lakes may become monomictic. Recent model simulations predict opposite effects of climate change in some regions of the temperate latitudes, for example, in Northern Atlantic regions (Hansen et al., 2004). Decreasing temperatures are forecast, due to changes in the thermohaline circulation of the ocean. Therefore, climate change will influence the duration of summer stagnation, the epilimnetic temperatures and density gradients, and, thus, the hypolimnetic oxygen budget of stagnant water bodies. It will also likely affect the phytoplankton species composition, succession, and abundance.

2.08.1.5 Water Movement

Unlike rivers, lakes are identified as stagnant or standing water bodies. However, natural waters are never completely quiescent. Horizontal and vertical water movements of quite different spatial and temporal scales transport dissolved and particulate materials and heat. They influence the gas exchange with the atmosphere and affect the basin morphology, due to erosion and deposition of sediments. Therefore, knowledge about the hydrodynamic structure is important for the understanding of the matter turnover of lakes.

Water flows in lakes and reservoirs are largely turbulent, that is, chaotic, swirling, multidirectional, and disordered. Unidirectional and smooth laminar flows can only be observed at very low flow velocities, for example, in thin boundary layers between water and sediment in deep, stratified lakes or in the metalimnion of wind-sheltered, small basins during calm weather.

Wind, solar radiation, and in- and outflows are the most important forces generating water movement. In large lakes, air-pressure differences along the surface, the Coriolis force, resulting from the Earth's rotation, and the gravitational attraction of the sun and moon, may also cause or influence water movement.

The spatial and temporal variations in wind force are of the greatest importance for the formation of nonperiodic currents. Wind acts at the water surface and, thus, the size and shape of the lake and its orientation to the prevailing wind direction are decisive factors. The wind exposure of a lake is described as a wind fetch, defined as the unobstructed distance that wind can travel over water in a constant direction. The kinetic energy of the wind is proportional to u^3, where u (m s^{-1}) is the wind velocity, normally measured 10 m above the surface. The velocity of wind-driven currents is about $0.02u$ and is independent of the height of surface waves. In the open water of large and deep lakes, the Coriolis force causes a deflection of the wind drift to the prevailing wind direction of about 45° to the right in the Northern hemisphere, and to the left in the Southern hemisphere, respectively. This deflection increases with water depth and therefore the currents in the deepest water layers may flow opposite to the wind direction. This phenomenon is called the 'Ekman spiral'. In smaller lakes, the water feels the shore and the bottom, and boundary effects influence the flow-field. Currents parallel to the shores prevail. A downwind drift of water masses, unavoidably causes the leeward drift of a corresponding amount of water and, consequently, large-scale horizontal and, for example, in the mixed epilimnion, vertical circular motions are formed (**Figure 16**; Hutter K (1983)). Such gyres may produce inhomogeneous (patchy) distributions of chemical or biological constituents (e.g., patchiness of phytoplankton or water-quality parameters). The circulation patterns are strongly influenced by lake-basin irregularities (e.g., islands and bays).

Attentive observers may occasionally notice streaks of foam or debris (windrows) at uniform distances from one another, at the surface of lakes on windy days, that are deflected 5°–15° to the right of the wind direction (in the Northern Hemisphere). This appearance is an indication of the Langmuir circulation, a wind-driven helical circulation system, rotating clockwise and counterclockwise alternatively, that is initiated at wind speeds of more than about 3 m s^{-1} (**Figure 17**). Air bubbles, produced by breaking waves and floating materials, flow from the upwelling range (divergence zone) to the range of the downwelling motion (convergence zone) and concentrate at the surface. The distances between the windrows increase with increasing wind speed. The diameter of the vortices is about half of the distance between the streaks, but never larger than the depth of the epilimnion. Langmuir cells may significantly affect the development of the phytoplankton, which are passively transported vertically, through the underwater light field, within short time intervals (Vincent, 1980). Patchiness of zooplankton may also be caused by Langmuir circulation (Malone and McQueen, 1983).

Propagating surface-gravity waves imply a horizontal transport of water. However, wind waves only cause surface water particles to move in circular orbits, with almost no drift of water. Wind waves are characterized by their height H from trough to crest, length L from crest to crest, and period of oscillation (**Figure 18**). Wave formations and their dimensions, depend on wind speed, wind fetch F, wind duration, and water depth. In the open water, the maximum height H_{max} of the waves can be estimated from fetch F as $H_{max} = 0.105\sqrt{F}$. H_{max} is identical to the diameter of the orbital motion at the surface. The diameters D_z of the orbits shrink with increasing depth z and no vertical displacement of water parcels, attributed to surface waves is found below $z \approx 0.5L$ (**Figure 18**). Waves approaching the shore regions or in shallow lakes with $z_{max} < 0.5L$ feel the bottom, and the shape of the orbital motions close to the bottom becomes

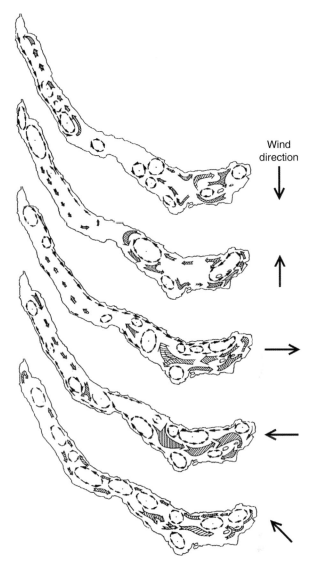

Figure 16 Qualitative distribution of the mean steady state transport in Lake Zürich for spatially uniform constant winds blowing from N, S, W, E, and SE. Modified from Hutter K (1983) Strömungsdynamische Untersuchungen im Zürich- und im Luganersee – Ein Vergleich von Feldmessungen mit Resultaten theoretischer Modelle. *Schweizerische Zeitschrift fur Hydrologie* 45: 101–144.

more and more elliptical. Lightweight particles are resuspended and washed downward into the deeper regions of the basin. With further decreasing water depth, the waves at the surface become higher and steeper, the wavelengths shorter, and their erosive impact increases. Finally, if $z < 0.05L$, the waves break and strong erosion of the shore may be observed. Thus, wind waves strongly affect the development of the shorelines and the littoral zones of lakes.

Strong wind, persistently blowing from a constant direction, pushes the upper warm water masses of a stratified lake to the downwind side and generates a tilt of the whole surface and thus, an unstable position. Due to the restoring force of gravity, the water flows back and, due to inertia, a swinging, oscillating motion of the surface is caused, which is called a surface or 'external seiche'. Periods of external seiches are

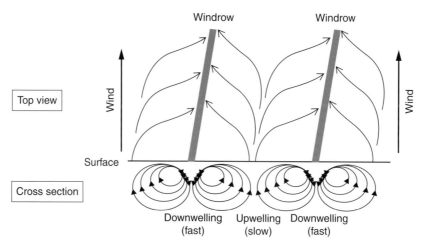

Figure 17 Schematic representation of Langmuir circulation cells. Air bubbles and debris are flowing from the divergence (upwelling) zone to the convergence (downwelling) zone and create streaks (windrows) of almost constant distance at the surface that are nearly parallel to the wind direction.

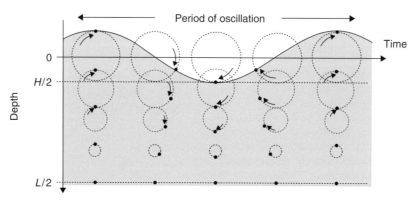

Figure 18 Circular motion of water particles in five layers from surface down to the depth $L/2$ at five moments during one period of oscillation of a wind wave. L, wave length in cm; H, wave height in cm.

rather short – seconds in small lakes and minutes or a few hours in large basins. The amplitudes vary between a few centimeters in small water bodies and about 2 m in large lakes. However, internal seiches, that is, the periodic up- and downwelling of water layers of different density and depth, forming standing waves of much larger amplitudes and longer periods of oscillation, are more important in terms of matter transport and impact on phytoplankton development. This phenomenon is not visible to observers at the surface, but becomes evident from considerable periodic temperature variations in the depths of the metalimnion and below. The example shown in **Figure 19** indicates superimposed, one-nodal, internal seiches of different periods of oscillation (about 24 h and 8 h) in a reservoir. The temperature variations at the West and East stations almost mirror each other, while they are comparably low at the central station, which is apparently close to the position of the wave node of the oscillation. Large horizontal, but low vertical water movements are observed in the nodal areas. The opposite applies to the crest regions, where up- and downwelling prevails. However, these vertical movements generate highly turbulent currents along the sloped bottom and create so-called internal surges, similar to breaking surface waves (Mortimer and Horn, 1982). The temperature stratification at the sediment surface periodically varies from stable (warm water over colder sediment; situation 1 in **Figure 19**) to unstable (colder water over warmer sediment that fosters the release of interstitial water; situation 2 in **Figure 19**). The velocities of the vertical movements are usually in the range of several millimeters per second, while those of the horizontal current components may be up to a hundred times higher. The vertical displacements of the layers from their stable positions depend on the size of the lake basin, density gradient, depth of and vertical distance to the thermocline, and can be higher than 10 m. Sudden changes of wind direction or periodic (e.g., diurnal) fluctuations of wind speed may cause phase shifts of the oscillations in different depths, resulting in the interference of waves with several lake-specific periods of oscillation (**Figure 20**).

Internal seiches may be observed almost permanently in stratified lakes during the summer stagnation (**Figure 21**). Even in calm weather, the oscillations continue for a long time (days or even weeks) with, however, decreasing amplitudes after their excitement has ceased. Wave structures rotating around large lake basins, may occur under the influence of the Coriolis force and wind-driven, horizontal large-scale circulations and, finally, highly complex current patterns result.

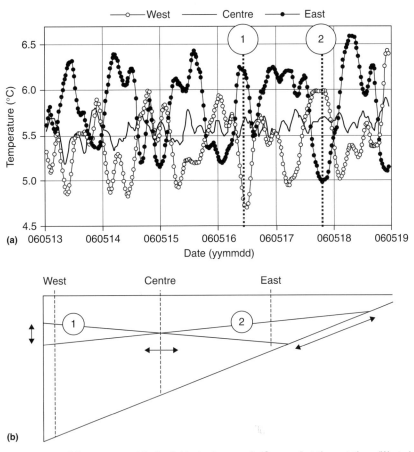

Figure 19 (a) Periodic temperature variations measured in the Saidenbach reservoir (Germany) at three stations (West about 50 m in front of the dam, Center ~1 km, and East about 2 km apart from the West station) in time intervals of 30 min in May 2006. (b) Diagram of the position of a water interface at the two moments marked in the graph above. Arrows qualitatively indicate prevailing water movements.

Internal seiches have a great impact on the turbulent vertical exchange of heat and the transport of materials. They resuspend sediment particles, accelerate their dislocation into the deepest regions of the lake, and enhance the release of dissolved substances from the sediment. The periodical transport of phytoplankton cells, throughout the vertical light field in the crest regions of the internal waves, substantially increases the photosynthesis rate in water layers at the base of the euphotic zone (Paul, 1987).

Discharge-related currents, especially floods, may generate basin-wide water movements, particularly in lakes with short retention times. As the import of nutrients, allochthonous particulate matter and other substances by the tributaries, is most important to the materials budget, knowledge about the seasonal variability of the depth of inflow and the propagation of the inflowing water is crucial for the understanding of the trophic situation and the availability of nutrients in the euphotic zone. In reservoirs, the balance between inflow and outflow from different depths, decisively determines the development of the fill-level and the volume of the hypolimnion during summer stratification. The withdrawal of water from the deep-water layers of reservoirs causes currents in the hypolimnion, which is more or less quiescent in natural lakes. The entrainment depth to which the inflowing water plunges characteristically varies seasonally, depending on the temperature (density) distribution in the lake and the temperature (density) of the tributaries (**Figure 22**; Carmack et al., 1979). Surface inflow is observed when the density of the river water is lower than that of the lake; underflow occurs in the reverse instance. Interflows are typical in situations in which the river density is between that of the lake's surface and the bottom of the lake. Hydraulic short-circuiting, that is, the longitudinal distribution of river water from the mouth of the tributary to the dam in a relatively thin metalimnetic layer, within a very short time (a few hours), has frequently been observed in reservoirs. Such events are critical in the case of drinking-water reservoirs, because harmful substances (e.g., turbidity and microbial pollution) in high concentrations may contaminate the raw water (Clasen and Bernhardt, 1983). Intrusion far below the depth of the respective lake temperature can be observed, if the density of the inflowing water is considerably enhanced due to very high flood-induced turbidity, caused by suspended mineral particles. Such turbidity currents may import oxygen into the hypolimnion of seldom fully circulating (e.g., deep pre-alpine) lakes (Lambert et al., 1984; DeCesare et al., 2006).

Turbidity currents are a special form of density currents. Density currents are, in general, water movement, caused by density differences. They can also result from water-temperature differences, as a consequence of differential heating or

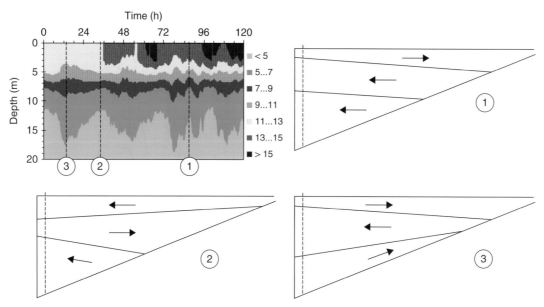

Figure 20 Vertical and temporal temperature variability caused by superimposed internal waves of different periods of oscillation and phase shifts in the upper 20 m water layer of the West station c. 50 m in front of the dam ($z_{max} = 45$ m) of the Saidenbach reservoir observed from 18 May 2005 0 a.m. to 22 May 2005 12 p.m. (top left). For the marked points in time, wave modes and principal water movements are schematically shown.

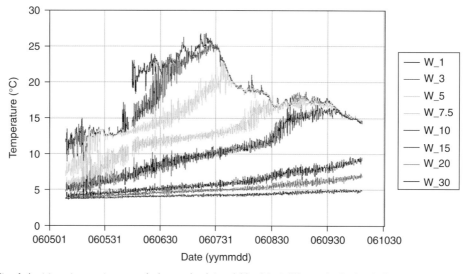

Figure 21 Results of short-term temperature records (measuring interval 30 min) at different depths (m, indicated by the numbers given in the legend) at station West (~ 50 m in front of the dam) of the Saidenbach reservoir (Germany) during the summer stratification in 2006 (dates are given in the yymmdd-format). The permanent temperature fluctuations at depths below the epilimnion show the ubiquitary nature of internal seiches. Those at the surface may also result from short-term changes of irradiation and air temperature.

cooling in lake segments. For instance, shallow bays on the margins of lakes, may heat up during the day and cool down at night, more rapidly than the open water. The resulting density differences generate convective exchange of water and of dissolved materials between littoral and pelagic zones (Wells and Sherman, 2001). Strongly increased conductivity (salinity), for example, due to thaw salt from roads in winter, may also generate density currents and vertical temperature inversions. Convective currents are generally upwelling movements of less dense, lighter water (e.g., plumes of heated waste water), or downwelling movements of denser, heavier water (resulting from surface cooling in summer or warming in spring when the deeper water layers have temperatures lower than $4\,°C$).

As mentioned above, water movement in lakes is mostly turbulent. Turbulence results from friction between water layers moving with different velocities (Baumert et al., 2005). Shear forces produce vortices and, if they collapse, they dissipate the energy of motion and cause mixing of water. The spatial dimensions of these vortices, that is, the intensity of eddy-diffusion, decrease with increasing density gradient and/or reduction of velocity differences between adjacent water

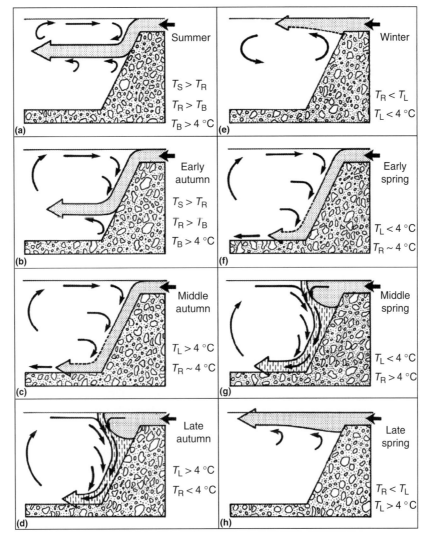

Figure 22 Schematic representations of seasonal riverine circulation patterns of the Kamloops Lake, British Columbia. Stippled areas denote river water; dashed areas denote lake and river water mixtures involved in cabbeling process (mixing of water of identical density but slightly different temperature and salinity). From Carmack EC, Gray CBJ, Pharo CH, and Daley RJ (1979) Importance of lake–river interaction on seasonal patterns in the general circulation of Kamloops Lake, British Columbia. *Limnology and Oceanography* 24: 634–644. T_R, river temperature; T_L, lake water temperature; T_S, lake surface temperature; and T_B lake bottom temperature.

strata. Turbulence transports heat, dissolved substances, and gases vertically between the epilimnion and the hypolimnion. This can be observed by small-scale vertical temperature inversions in temporally and spatially highly resolved, thermal microstructure measurements (Wüest *et al.*, 2000).

2.08.1.6 Basic Chemistry

2.08.1.6.1 Systematics of lakes with respect to water quality

There are approximately 8 million natural lakes with surface areas of >1 ha, on the Earth (Ryanzhin *et al.*, 2001). The majority of them are freshwater lakes, which are of vital importance for humankind, animals, and plants. A global model based on the Pareto distribution, shows that the global extent of natural lakes ≥ 0.1 ha is about 304 million lakes (Downing *et al.*, 2006).

Some of the different types of lakes classified on the basis of their water quality are as follows:

1. *Soft- and hard-water lakes.* As many lakes are connected with the groundwater, their water chemistry is influenced by the geological substrate of the watershed. Hard-water lakes dominate when the catchment is rich in calcium. Soft-water lakes are characterized by the low content of the hardness components, calcium and magnesium. Lakes with a calcium deficit are normally fed by rainwater without soil contact, or exist in Ca-deficient, sandy, outwash plains.

In calcareous, oligotrophic (i.e., clear water) lakes that are mainly fed by groundwater, such as Lake Stechlin, Northern Germany, submerged plants use HCO_3^-/CO_2 as a C source for photosynthesis

$$CO_2 * H_2O \Leftrightarrow HCO_3^- + H^+ \quad (25)$$

$$HCO_3^- \Leftrightarrow CO_3^{2-} + H^+ \quad (26)$$

Suspensions or deposits of hardly soluble $CaCO_3$ (biogenic decalcification) may be formed by CO_2 uptake and H^+ consumption (pH increase)

$$Ca^{2+} + CO_3^{2-} \Leftrightarrow CaCO_3(s) \quad (27)$$

Another example of hard-water lakes are the acidic hard-water lakes. Here, $CaCO_3$ formation does not occur and sulfate is the dominating anion. These lakes are, in many cases, impacted or man made (mining lakes). Acid mine drainage (AMD), a result of the mining and milling of sulfur-bearing coal and ores, plays a dominant role in surface-water chemistry and pollution, in many areas of the world. Oxidation of disulfide minerals (e.g., pyrite or marcasite (FeS_2)) occurs mostly from the reactions of tailing and mining wastes with oxygen and water in underground workings, tailings, open pits, and waste rock dumps. This produces acidic water, rich in metals, commonly referred to as AMD. The most noticeable environmental change is the pollution of flowing water with severe impacts on aquatic life.

The following reactions result from acidified runoff on regulated mine sites. Representative species of bacteria engaged in these processes are also mentioned

1.1. Iron disulfide (pyrite and marcasite) is oxidized to sulfate by oxygen with a very high energy yield for bacteria (*Thiobacillus ferrooxydans* and *Thiobacillus thiooxydans*)

$$2FeS_2(s) + 7O_2 + 2H_2O$$
$$\Rightarrow 2Fe^{2+} + 4SO_4^{2-} + 4H^+ \quad (28)$$

1.2. Ferrous iron is oxidized to ferric iron

$$14Fe^{2+} + 3.5O_2 + 14H^+ \Rightarrow 14Fe^{3+} + 7H_2O$$
$$\text{(slow under acidic conditions)} \quad (29)$$

1.3. Sulfur/sulfide is oxidized with ferric ions to sulfate

$$FeS_2(s) + 14Fe^{3+} + 8H_2O$$
$$\Rightarrow 15Fe^{2+} + 2SO_4^{2-} + 16H^+ \quad (30)$$

1.4. Ferric iron is hydrolyzed to ferric hydroxide

$$Fe^{3+} + 3H_2O \Rightarrow Fe(OH)_3(s) + 3H^+ \quad (31)$$

In the presence of nitrate, the FeS_2 oxidation proceeds by autotrophic denitrification (*Thiobacillus denitrificans*)

$$5FeS_2(s) + 14NO_3^- + 4H^+$$
$$\Rightarrow 5Fe^{2+} + 10SO_4^{2-} + 7N_2 + 2H_2O \quad (32)$$

whereas Fe(II) is further oxidized to Fe(III) by species such as *Gallionella ferruginea*

$$10Fe^{2+} + 2N_3^- + 14H_2O$$
$$\Rightarrow 10FeOOH(s) + N_2 + 18H^+ \quad (33)$$

Reactions (28), (30), (31), and (33) lead to an enormous production of acid.

Sulfur concentrations in many surface waters have increased greatly as a result of acid mine run off and SO_2 emissions.

2. *Saline lakes*. In semiarid and arid climates, lakes may show a high concentration of dissolved solids, due to the surplus of evapotranspiration above the runoff into the lakes. In relation to their main ingredients, salt lakes may be subdivided into soda, chloride, and sulfate lakes.

The majority of saline lakes, in terms of area, are chloride lakes. Historically, they are remnants of isolated seawater bodies in continental locations.

3. *Soda lakes*. Lakes with a very high alkalinity level, mainly due to soda (Na_2CO_3), occur in southeast Europe and are common in the East African rift valley. The water in soda lakes becomes alkaline with a pH of approximately 10, due to the alkalis, carbonate, and bicarbonate. The water tastes bitter and feels oily. These salts accumulate in lakes without discharge, if the subsoil consists of carbonate or volcanic rock, and whose water budget is characterized by high evaporation rates. Therefore, soda lakes are usually found in semi-deserts and steppe areas. Mono Lake (California) contains about 280 million tons of dissolved solids and, depending on its seasonally fluctuating water level, is 2–3 times more salty than the ocean. It is also rich in borate and potassium. Periodic eruptions of volcanic ash have also considerably contributed to Lake Mono's chemical mix. Soda lakes are often rich in biomass, provided they are not too deep. Due to the high pH values and salt concentration, alkaliphilic/alkalitolerant and simultaneously, halophilic organisms, are characteristic. The limited biodiversity essentially comprises specialized bacteria (among others: cyanobacteria such as *Spirulina* and *Archaea*) and algae. They may appear in great abundance and reduce the Secchi disk transparency to a few centimeters. Soda lakes thus rank among the most productive ecosystems. Special protophytes (flagellates) are characterized by accessory-colored pigments (carotenoids, phycobiline, and rhodopsin). They are responsible for the conspicuous coloration of numerous soda lakes. Many sodium carbonate lakes are utilized for the production of natural soda.

4. *Bog lakes*. These lakes are normally poor in electrolytes. Bog lakes are found in all geographic latitudes of the humid climate zones, from the wetlands in the hills to the plains, to the marshes adjacent to large rivers. They are among the aquatic systems with high species diversity. In bogs and bog lakes, the production of organic C compounds is greater than microbial mineralization. The slow and incomplete decomposition of vegetation residues, under continuous water surplus from rainfall or soil water, is accompanied by a high oxygen deficit, resulting in peat deposition and siltation (see Section 2.08.1.6). Dystrophic lakes are poor in nutrients and calcium as well as phytoplankton, and are mostly strongly acidic and rich in dissolved humic materials. They are clear, but mostly brownish. Their watershed is often small; therefore, it is not remarkable that some species typical of bog lakes are also found in acidic mine lakes.

5. *Crater lakes or volcanic lakes.* Crater lakes covering active (fumarolic) volcanic vents are sometimes termed volcanic lakes – a cap of meteoric water over the vent of an active volcano. The chemistry of the water may be dominated by high-temperature volcanic gas components or by a lower temperature fluid that has interacted extensively with volcanic rock. Precipitation of minerals such as gypsum ($CaSO_4 * H_2O$) and silica (SiO_2) can determine the concentration of Ca and Si (Kusakabe, 1994).

The water of these lakes may be extremely acidic (e.g., pH ∼ 0.3). Lakes located in dormant or extinct volcanoes tend to contain freshwater, and the water clarity in such lakes may be exceptionally high due to the lack of inflowing streams and sediments. Crater lakes form as incoming precipitation fills the depression.

The lake deepens until equilibrium is reached between water inflow, losses due to evaporation, subsurface drainage, and possibly also surface outflow, if the lake fills the crater up to the lowest point of its rim. Surface outflow can erode the deposits damming the lake, lowering its level. If the dam erodes rapidly, a breakout flood can be produced.

2.08.1.6.2 Ionic balance

Apart from living organisms, lakes contain a wide array of ions, molecules, and complexes from the weathering of soils and bedrock in the watershed, the atmosphere, and the sediments. Therefore, the chemical composition of a lake is fundamentally a function of its climate, which affects its hydrology and its basin geology. An ion balance based on equivalents for typical freshwater is presented in **Figure 23**.

These ions (bicarbonate, sulfate, chloride, calcium, magnesium, sodium, and potassium) are usually present in concentrations of $mg\,l^{-1}$ (ppm), whereas other ions such as phosphate, nitrate, ammonium, and heavy metals are present at $\mu g\,l^{-1}$ (ppb) levels.

2.08.1.6.3 Inorganic compounds and buffer properties

Main inorganic compounds and buffering properties present in lakes and reservoirs are as follows:

1. *Alkalines.* The alkali ions Na^+ and K^+ are mostly discharged in the K- and Na-feldspar weathering processes and represent an important part of the ion balance of water

$$\underset{\text{(albite)}}{NaAlSi_3O_8(s)} + 5.5H_2O \Rightarrow Na^+ + OH^-$$
$$+ 2H_4SiO_4 + \tfrac{1}{2}\underset{\text{(kaolinite)}}{Al_2Si_2O_5(OH)_4(s)} \quad (34)$$

$$\underset{\text{(albite)}}{NaAlSi_3O_8(s)} + 4.5H_2O + CO_2 * H_2O$$
$$\Rightarrow Na^+ + HCO_3^- + 2H_4SiO_4 + \tfrac{1}{2}Al_2Si_2O_5(OH)_4 \quad (35)$$

$$3\underset{\text{(K-feldspar)}}{KAlSi_3O_8} + 2CO_2 * H_2O + 12H_2O \Rightarrow 2K^+$$
$$+ 2HCO_3^- + 6H_4SiO_4 + \underset{\text{(mica-illite)}}{KAl_3Si_3O_{10}(OH)_2} \quad (36)$$

The weathering of aluminosilicates is accompanied by a release of cations and of silic acid. As a result of these reactions, alkalinity is released. Minerals of the kaolinite group are the main metabolites of feldspar weathering.

2. *Alkaline earths.* The hardness components Ca^{2+} and Mg^{2+} are quantitatively the most important cations in freshwater lakes. Calcium usually enters the water as either calcium carbonate ($CaCO_3$), or calcium sulfate ($CaSO_4$):

$$\underset{\text{(calcite)}}{CaCO_3} + H_2O \Rightarrow Ca^{2+} + HCO_3^- + OH^- \quad (37)$$

$$CaCO_3 + CO_2 {}^*H_2O \Rightarrow Ca^{2+} + 2HCO_3^- \quad (38)$$

$$\underset{\text{(anhydrite)}}{CaSO_4} \Rightarrow Ca^{2+} + SO_4^{2-} \quad (39)$$

The predominant source of magnesium is dolomite:

$$\underset{\text{(dolomite)}}{CaMg(CO_3)_2} + 2H_2O \Rightarrow Ca^{2+} + Mg^{2+}$$
$$+ 2HCO_3^- + 2OH^- \quad (40)$$

3. *Carbonate species* ($CO_2/HCO_3^-/CO_3^{2-}$). The reactive inorganic forms of environmental carbon are carbon dioxide ($CO_2 * H_2O$), carbonic acid (H_2CO_3), bicarbonate (HCO_3^-) and carbonate (CO_3^{2-}). Carbon dioxide plays a fundamental role in determining the pH in lakes. An important element in acid–base chemistry is the bicarbonate ion, HCO_3^-, which may act as either an acid or a base. Aqueous

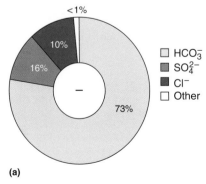
(a)

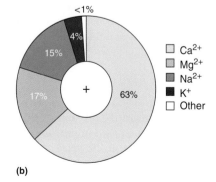
(b)

Figure 23 Major anions (a) and cations (b) in freshwater systems.

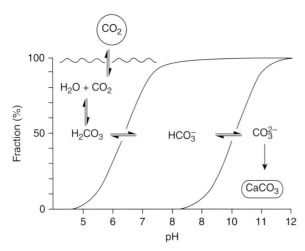

Figure 24 States of inorganic carbon depending on water pH (CO$_2$-system). From Uhlmann D and Horn W (2001) *Hydrobiologie der Binnengewässer. Ein Grundriss für Ingenieure und Naturwissenschaftler.* Stuttgart: Eugen Ulmer.

CO$_2$ solutions react acidically by forming carbonic acid. Carbonic acid can subsequently dissociate in two steps to release protons:

$$CO_2 + H_2O \Leftrightarrow H_2CO_3 \quad pK = -2.8 \quad (41)$$

$$H_2CO_3 \Leftrightarrow HCO_3^- + H^+ \quad pK_{s1} = 6.35 \quad (42)$$

$$HCO_3^- \Leftrightarrow CO_3^{2-} + H^+ \quad pK_{s2} = 10.33 \quad (43)$$

The pH dependence of the CO$_2$/HCO$_3^-$/CO$_3^{2-}$-system is shown in **Figure 24** (Uhlmann and Horn, 2001).

4. *Buffer intensity, base-neutralizing capacity, acid-neutralizing capacity.* Most lake waters are pH-buffered, because of the existence of the carbonate buffer system.

Buffer solutions are necessary to maintain the optimal pH for the enzyme systems of plants and animals. The buffer intensity is a measure of the ability of water to compensate for the addition of strong acid or base without appreciable pH change.

2.08.1.6.4 Sequence of microbially mediated redox processes

These processes become noticeable at the sediment/water interface and from there may extend into the hypolimnion. Many chemical changes that take place during the early diagenesis of sediments, depend on the redox environment in the interstitial water. The redox environment is determined by the degree to which organic compounds are preserved or undergo microbial decomposition.

Important redox processes are based on

- the surplus of organic matter being oxidized by oxygen or by an oxidized category of nitrogen (NO$_3^-$), iron (Fe(III)), sulfur (SO$_4^{2-}$), or by organic carbon itself in methane fermentation (the reduction reactions in the upper half of **Figure 25**; Stumm and Morgan, 1981), or
- reduced species of iron (Fe(II)), nitrogen (NH$_4^+$, NO$_2^-$), sulfur (HS$^-$, S, and S$_2$O$_3^{2-}$), or methane being oxidized by oxygen (the oxidation reactions can be observed in the bottom half of **Figure 25**).

The two types of reactions do not occur in the same place or under the same redox conditions, but may very well occur at different depths in the same stratified lake. For example, it is likely that nitrate (NO$_3^-$) might be reduced in the hypolimnion of a lake with a clinograde oxygen curve, producing ammonia (NH$_3$). Should the ammonia subsequently be transported into the oxygenic epilimnion, the ammonia will be oxidized back to nitrate. In general, the first type of reaction (upper half of **Figure 25**) will take place under more reducing conditions, and the second type (lower half of **Figure 25**) will occur under more oxidizing conditions and in the presence of oxygen.

The reactions occur as a hierarchy. In general, nitrate and ferric iron will be reduced before any reduction of sulfate occurs. Similarly, sulfate reduction will proceed to nearly total consumption of sulfate, before methane fermentation occurs. Thus, the pH and redox circumstances in which each of the reactions may be expected to occur can be anticipated, based on the hierarchy of reactions.

In detail, the following oxidation processes concerning organic matter, take place:

1. Aerobic microorganisms use oxygen (O$_2$) as the electron acceptor for the microbial degradation of organic compounds (aerobic respiration)

$$\underset{\text{(organic matter)}}{[CH_2O]} + O_2 \Rightarrow CO_2 + H_2O \quad (44)$$

Bacteriological oxygen demand (BOD) refers to the amount of oxygen needed by aerobic organisms in a given water volume.

2. In case of O$_2$ depletion, nitrate (NO$_3^-$) is the next electron acceptor to be used by microorganisms for the oxidation of organic carbon. Nitrate is then converted to gaseous nitrogen, under anoxic conditions, by denitrifying bacteria (denitrification). The reduction product is a nitrogen-containing gas mixture, usually molecular nitrogen (see Equation (45)), but nitrogen oxides may also be produced:

$$5[CH_2O] + 4NO_3^- + 4H^+$$
$$\Rightarrow 2N_2(g) + 5CO_2 + 7H_2O \quad (45)$$

3. Before Fe(III) reduction, the reduction of Mn(IV)-compounds (in most cases as MnO$_2$) will take place, which will also be released through the interstitial water of the sediment into the deeper hypolimnion (Davison, 1993):

$$[CH_2O] + 2MnO_2(s) + 4H^+$$
$$\Rightarrow 2Mn^{2+} + CO_2 + 3H_2O \quad (46)$$

4. Fe(III) (e.g., FeOOH) in the sediment or in the hypolimnion is reduced to Fe(II) by chemo-autotrophic, facultative/obligatory anaerobic microorganisms of the genera *Geobacter* sp., *Geovibrio*, and *Shewanella* sp. (Jones et al., 1983):

$$[CH_2O] + 4FeOOH(s) + 8H^+$$
$$\Rightarrow 4Fe^{2+} + CO_2 + 7H_2O \quad (47)$$

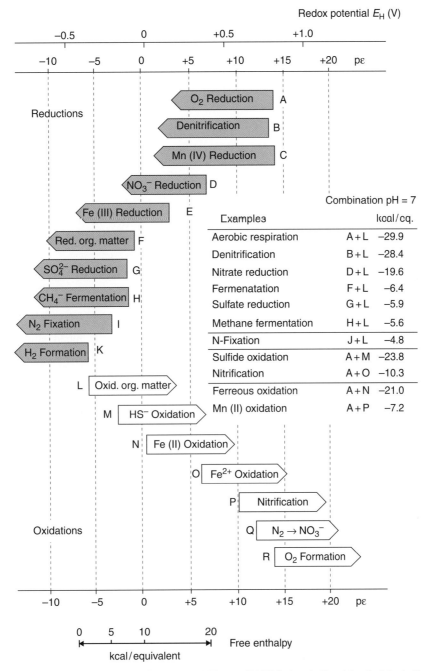

Figure 25 Microbially mediated redox processes. From Stumm W and Morgan JJ (1981) *Aquatic Chemistry: An Introduction Emphasizing Chemical Equilibria in Natural Waters*, 2nd edn., 780pp. New York: Wiley.

As FeOOH occurs in the sediment, the Fe(II) formed will first appear in the interstitial water, with the highest concentrations in the top layer, from where it diffuses into the overlying water.

The presence of NO_3^- suppresses the FeOOH-reducing activity and will therefore keep the P-adsorbing capacity of the sediment intact. Davison (1992) showed that Fe(II) does not form complexes with organic compounds, but may fix phosphate as vivianite $(Fe_3(PO_4)_2 \cdot 8H_2O)$.

5. The next electron acceptor to be discussed is sulfate (SO_4^{2-}). It will be reduced to H_2S, mostly by obligatory anaerobic microorganisms of the genera *Desulfovibrio* or *Desulfotomaculum*:

$$2[CH_2O] + SO_4^{2-} + H^+ \Rightarrow 2CO_2 + HS^- + H_2O \quad (48)$$

6. In the complete absence of oxygen and other electron acceptors (anaerobic conditions), microorganisms

decompose organic material to CO_2 and CH_4 in various steps (methane fermentation):

6.1. Hydrolysis. – extracellular splitting of macromolecules:

- fats $\Rightarrow$ fatty acids, glycerine by lipolytic bacteria (*Bacillus, Alcaligenes,* and *Pseudomonas*);
- proteins $\Rightarrow$ amino acids by proteolytic bacteria (*Peptococcus, Staphylococcus,* and *Clostridiun*);
- cellulose $\Rightarrow$ glucose, acetic acid, alcohol, H_2, and CO_2 by anaerobic cellulolytic bacteria. There are many species of cellulolytic bacteria and fungi;
- starch $\Rightarrow$ glucose by amylolytic bacteria (*Micrococcus* and *Clostridium*).

6.2. Anaerobic fermentation – performed in sequencing steps:

- *Acidogenic step.*
  $C_6H_{12}O_6 \Rightarrow C_2H_5COO^- + CH_3COO^- + CO_2 + H_2 + 2H^+$
  $\Delta_R G = -286\, kJ\, mol^{-1}$
- *Acetogenic step.*
  $C_2H_5COO^- + 2H_2O \Rightarrow CH_3COO^- + CO_2 + 3H_2$
  $\Delta_R G = -81.9\, kJ\, mol^{-1}$
- *Methanogenic step.*
  Acetate decarboxylization/methane formation
  $CH_3COOH \Rightarrow CH_4 + CO_2$
  $\Delta_R G = -56\, kJ\, mol^{-1}$
  $CO_2 + 4H_2 \Rightarrow CH_4 + 2H_2O$
  $\Delta_R G = -139\, kJ\, mol^{-1}.$

In summation

$$2(nCH_2O) \Rightarrow nCO_2(g) + nCH_4(g) \qquad (49)$$

The methane-producing bacteria consist of four major genera: *Methanbacterium, Methanobacillus, Methanococcus,* and *Methanosarcina.*

2.08.1.6.5 Iron, manganese, and sulfur compounds

The following list explains the presence of compounds of iron, manganese, and sulfur:

1. *Iron.* The primary oxidation states of iron in water are Fe(II) and Fe(III). In most aerobic surface water, Fe(III) predominates and is nearly insoluble at neutral pH values:

$$Fe^{3+} + 3OH^- \Rightarrow Fe(OH)_3(s) \qquad (50)$$

Fe(II), on the other hand, is soluble and dominates under anaerobic conditions.

The oxidation–reduction cycle controls the fate of iron in most lakes, but it varies seasonally, particularly in lakes that develop an anoxic hypolimnion during the stagnation period. Oxygen concentration at the water–sediment interface often approaches zero. This causes the reduction of Fe(III) to soluble Fe(II), which is then transported upward into the water column. The oxygenated water produces re-oxidation into the insoluble Fe(III), which settles at the bottom to repeat the cycle.

An oxidation–reduction cycle also controls the fate of manganese; however, Fe(II) is oxidized to particulate Fe(III) much more rapidly than the corresponding species of Mn. In addition, Fe(III) is reduced at a lower redox potential than Mn(IV).

Many elements, including P (as phosphate), are scavenged by iron through adsorption onto particles when Fe(III) is formed as part of the cycle. In the presence of phosphate, a basic iron phosphate ($Fe_2(OH)_3PO_4$) is formed with a Fe:P = 2:1 stoichiometry:

$$2Fe(OH)_2^+ + H_2PO_4^- + OH^-$$
$$\Rightarrow Fe_2(OH)_3PO_4(s) + 2H_2O \qquad (51)$$

Another possible reaction occurs when phosphate is adsorbed on to the hydrolyzed sediment surface (Me = Al, Fe, and Mn):

$$= Me - OH + H_2PO_4^-$$
$$\Leftrightarrow\; = Me - H_2PO_4 + OH^- \; (\text{adsorption}) \qquad (52)$$

Fe(III) reduction takes place as explained next. In surface sediments, iron-bound phosphate is solubilized as follows:

$$Fe_2(OH)_3PO_4(s) + \tfrac{1}{2}[CH_2O] + 3H^+$$
$$\Leftrightarrow 2Fe^{2+} + H_2PO_4^- + \tfrac{1}{2}CO_2 + \tfrac{5}{2}H_2O \qquad (53)$$

Iron is predominately associated with sulfides. Sulfides are an important sink for trace metals in reduced sediments. The proportion of heavy metals bound by Fe- and Mn-hydrous oxides is highly variable and depends on water depth and redox conditions.

As iron plays such an important role in the fate of trace metals and nutrients, a breach in the iron redox cycle may also lead to the mobilization of toxic elements into the environment, for example, redistribution of sulfidic sediments during dredging.

2. *Manganese.* Total Mn concentration in freshwater is extremely variable, ranging from 0.002 to $>4\, mg\, l^{-1}$, whereas particulate Mn accounts for >90%, and often >95% of the total waterborne residue. Mn is of little direct toxicological significance, but may limit the growth of algae.

Mn exists in the oxidation states from +2 to +7, mostly as Mn(II)-(manganous) and Mn(IV)-(manganic).

The oxidation–reduction cycle is important in controlling the fate of Mn in most lake waters. Oxygen concentration at the sediment/water interface often approaches zero. This causes the reduction of Mn(IV) to soluble Mn(II), which is then transported upward into the water column. The oxygenated water causes re-oxidation to insoluble Mn(IV) which settles at the bottom to repeat the cycle. The Mn(II) oxidation is autocatalytic and may be represented as follows (Stumm and Morgan, 1981):

$$Mn^{2+} + \tfrac{1}{2}O_2 + H_2O \Rightarrow MnO_2(s) + 2H^+ \qquad (54)$$

$$Mn^{2+} + MnO_2(s) \Rightarrow Mn^{2+}MnO_2(s) \qquad (55)$$

$$Mn^{2+} \quad MnO_2(s) + \tfrac{1}{2}O_2 + H_2O$$
$$\Rightarrow 2MnO_2(s) + 2H^+ \qquad (56)$$

At circum-neutral pH, oxidation leads to considerable sorption of Mn^{2+} to MnO_2. The rate of Mn oxidation increases through the presence of manganese-oxidizing bacteria. Mn(IV) is reduced to dissolved Mn(II) at higher redox potentials than Fe(III) oxides. Oxidation of Mn(II) to Mn(IV) proceeds much more slowly than the oxidation of Fe(II) to Fe(III). Mn(II) and Mn(IV) follow essentially the same cycle as iron in lakes, where O_2 may be in short supply during one or more seasons.

3. *Sulfur compounds.* Anthropogenic emission to the atmosphere dominates the S cycle in many parts of the world. In fact, 80% of the global SO_2 emissions and >45% of the total river-borne sulfate-sulfur comes from man-made sources. In addition, sulfur occurs in surface waters as a result of natural weathering and due to emission from volcanoes, sea-salt aerosols, forest fires, and microbial decomposition of organic material. The main components produced by microbial decomposition are dimethyl disulfide (($CH_3)_2S_2$), hydrogen sulfide (H_2S), carbon disulfide (CS_2), dimethyl sulfide (($CH_3)_2S$), and methane thiol (CH_3SH).

The dominant S species, under normal pH and E_h conditions in lake waters, are sulfate (SO_4^{2-}), sulfide (H_2S, HS^-), and elemental sulfur (**Figure 26**; Zehnder and Zinder, 1980).

S(+VI) and S(-II) are the dominant stable oxidation states, but under reducing conditions thionates, thiosulfates, polysulfides, and sulfites may also be present.

Sulfur has an environmental significance because it

3.1. forms complexes with many toxic agents, organic materials, and hydrogen in many surface waters and
3.2. is the primary agent of acidification in many lakes and reservoirs (see Section 2.08.3.1).

The S cycle is shown in **Figure 27**. Sulfide is often present as dissolved anion HS^-, especially in hot springs. It is a common product of microbial processes in wetlands and eutrophic lakes. There are two important sources of H_2S in the environment:

1. the anaerobic decomposition of organic matter containing sulfur and
2. the reduction of sulfates and sulfites to sulfide.

Both mechanisms require reducing, anaerobic conditions and are strongly accelerated by the presence of sulfur-reducing bacteria. Some possibilities include

2.1. *Microbial sulfate and sulfur reduction.* The sulfur-reducing bacteria are strictly anaerobic. *Desulfovibrio*, *Desulfotomaculum*, *Desulfomonas*, and *Desulfolobus* reduce SO_4^{2-} to HS^-:

$$4CH_3COCOOH + SO_4^{2-} + 2H^+$$
$$\Rightarrow H_2S + 4CH_3COO^- + 4H^+ + 4CO_2 \qquad (57)$$

whereas e-donors are lactate, acetate, ethanol, malate, formiate, and fatty acids. Dissimilatoric S reducers use fatty acids, lactate, benzoate, and succinate to reduce

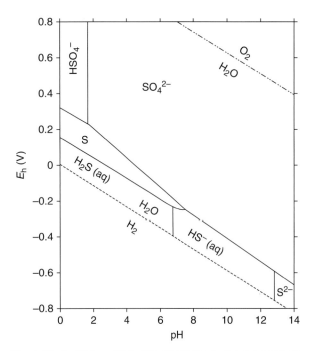

Figure 26 Equilibrium distribution of total dissolved sulfur species in the presence of iron. From Zehnder AJB and Zinder SH (1980) The sulfur cycle. In: Hutzinger O (ed.) *The Handbook of Environmental Chemistry*, vol. 1A, pp. 105–145. Berlin: Springer.

S to H_2S. H_2S may have two stages of dissociation under reducing conditions depending on pH:

$$H_2S(aq) + H_2O \Leftrightarrow HS^- + H_3O^+ \quad pK_{s1} = 7.02 \qquad (58)$$

$$HS^- + H_2O \Leftrightarrow S^{2-} + H_3O^+ \quad pK_{s2} = 13.9 \qquad (59)$$

2.2. *Aerobic sulfide oxidation.* It is catalyzed by bacteria such as *Achromatium oxaliforum* (sometimes with calcium carbonate crystals) and *Beggiatoa*. These oxidize HS^- to SO_4^{2-}

$$HS^- + 2O_2 \Rightarrow SO_4^{2-} + H^+ \qquad (60)$$

The genera *Thioploca*, *Thiothrix*, and *Lamprocystis* oxidize HS^- with nitrate anoxically (chemolithotrophic). The C-source is CO_2. The microorganisms form carpets at the bottom. *Thiothrix* can store sulfur in its cells just as *Beggiatoa* sp. does.

The obligate aerobic, gram-negative *Acidithiobacillus* uses the e-donors sulfide, sulfur, or thiosulfate for the chemolithoautotrophic metabolism, whereas adenosine triphosphate (ATP) is obtained from the respiration chain:

$$H_2S + \tfrac{1}{2}O_2 \Rightarrow S + H_2O \quad \Delta G_R = -209 \text{ kJ mol}^{-1} \qquad (61)$$

$$H_2S + 2O_2 \Rightarrow 2H^+ + SO_4^{2-}$$
$$\Delta G_R = -798 \text{ kJ mol}^{-1} \qquad (62)$$

$$S + \tfrac{3}{2}O_2 + H_2O \Rightarrow 2H^+ + SO_4^{2-}$$
$$\Delta G_R = -587 \text{ kJ mol}^{-1} \qquad (63)$$

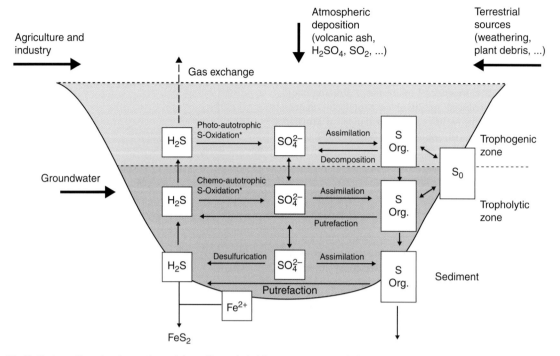

Figure 27 Vertical zonation of main reactions of the sulfur cycle in lakes and reservoirs. S Org, sulfur in cells of organisms (e.g., amino acids and gluthione); S_0, elementary sulfur; *photo-autotropic and chemoautotrophic sulfurication in aerobic/anaerobic interfaces.

$$S_2O_3^{2-} + H_2O + 2O_2 \Rightarrow 2SO_4^{2-} + 2H^+$$
$$\Delta G_R = -818 \text{ kJ mol}^{-1}. \quad (64)$$

Acidithiobacillus ferrooxydans can oxidize sulfides and ferrous ions.

3. *Anaerobic phototrophic sulfide oxidation.* This is performed by green sulfur bacteria (*Chlorobium*) that can release elementary sulfur into the water, and by red (purple) sulfur bacteria (*Chromatium*) that can oxidize sulfur to sulfate. Both can store sulfur in their cells

$$2H_2S + CO_2 \stackrel{h*v}{\Rightarrow} [CH_2O] + H_2O + 2S \quad (65)$$

Further reactions of S to thiosulfate ($S_2O_3^{2-}$) and/or sulfate SO_4^{2-} are possible.

Chemical and microbial sulfide oxidation produces large amounts of H^+, as illustrated by the following equation:

$$HS^- + \tfrac{3}{2}O_2 \Rightarrow SO_3^{2-} + H^+ \quad (66)$$

Sulfur has following effects on the chemistry of surface water:

- increase in the production of H^+,
- change in redox potential through several reactions, including:

$$HS^- \Leftrightarrow S^0 + H^+ + 2e \quad (67)$$

- decrease in alkalinity through the oxidation of H_2S, and
- mobilization of metals and phosphate from sediments, due to changes in pH and Eh (also noted for natural and man-made derived radionuclides).

All these effects have been noted in response to atmospheric deposition of sulfur and discharge of acid mine runoff.

2.08.1.6.6 Nutrients (nitrogen and phosphorus) and trace substances

In temperate climate regions, essential nutrients for the growth of aquatic organisms, such as bioavailable phosphorus and nitrogen, typically increase in spring due to snowmelt runoff and due to the mixing of accumulated nutrients into the water column, from the bottom, during spring turnover. In less-productive systems, significant amounts of nitrogen compounds may be deposited during rainfall or snowfall events (wet deposition), and during the less-obvious deposition of aerosols and dust particles (dry deposition). Nitrogen and phosphorus, in dry fallout and wet precipitation, may also originate from fertilizers in agricultural areas. The two principal nutrients P and N exist in the anionic form (NO_3^-/NO_2^- and $HPO_4^{2-}/H_2PO_4^-$) and are not subject to retention by cation-exchange processes. However, the ammonium cation (NH_4^+), which is formed mainly in the sediment, is fixed by cation exchange. The nitrate anion, unlike phosphate, does not form insoluble compounds with metals and is therefore readily leached from the soil. It dissolves easily into surface water and groundwater.

Phosphorus tends to be a minor element in natural waters because most inorganic P compounds have low solubility. Dissolved P concentration is generally in the range of 0.01–0.1 mg l^{-1} and rarely exceeds 0.2 mg l^{-1}. The dissolved

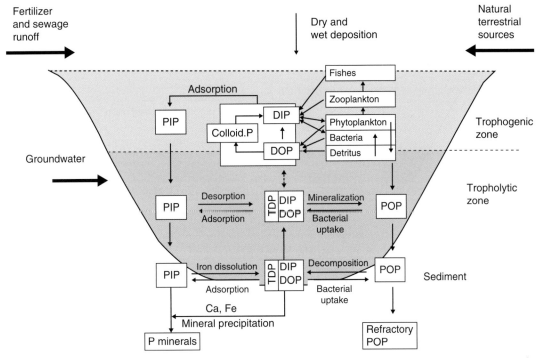

Figure 28 Main processes of the phosphorus cycle in lakes and reservoirs. DIP, dissolved inorganic P; DOP, dissolved organic P; TDP, total dissolved P; PIP, particulate inorganic P; POP, particulate organic P.

phosphorus in lakes is most often the principal growth-limiting nutrient for the development of phytoplankton, planktonic bacteria, and aquatic plants. The critical level of inorganic phosphorus for algal-growth blooms can be as low as $0.01–0.005\,mg\,l^{-1}$ in summer, but is more frequently around $0.05\,mg\,l^{-1}$. Phytoplankton is able to use phosphorus only from ortho-phosphate ($PO_4^{3-}/HPO_4^{2-}/H_2PO_4^-$), and not, for example, the polyphosphate form for growth. The concentration of dissolved polyphosphates in lakes is normally negligible.

The excess of phosphorus in domestic wastewater (synthetic detergents and human waste, including about $1.5\,g\,d^{-1}$ per person in urine), effluent from agriculture (liquid manure, mineral fertilizers, and several insecticides), and in corrosion-control agents in water supply and industrial cooling water systems, is frequently the main cause of algal blooms and other symptoms of lake eutrophication (see overview in **Figure 28**).

In lakes, phosphorus is cycled between organically and inorganically bound forms. Phosphate anions are largely immobilized, both in the soil and in oxidized sediment layers, by the formation of insoluble Fe, Ca, and Al phosphates, or by adsorption to soil/sediment particles.

The two major steps of the phosphorus cycle, the conversion of organic P into inorganic P and back to organic P, are both microbially mediated. The conversion of insoluble P forms, such as $Ca_3(PO_4)_2$ into soluble HPO_4^{2-} is also carried out by microorganisms (**Figure 28**). Organic P in the tissue of dead plants and animals is also converted, microbially, to ortho-phosphate.

Phosphorus mobility increases under anoxic conditions, because solid ferric iron, into which phosphate is strongly adsorbed, is reduced to soluble ferrous iron, thereby releasing adsorbed phosphate:

$$Fe(OH)_3\cdots HPO_4^{2-} + e^- \leftrightarrow Fe^{2+} + HPO_4^{2-} + 3OH^- \quad (68)$$

Aluminum and iron phosphates precipitate in acidic sediments. In calcite-rich sediments, Ca-phosphate is deposited. The immobilization of phosphorus in sediments is therefore controlled by properties such as pH, redox potential, texture, cation-exchange capacity, the amount of Ca, Al, and Fe-oxides present, and uptake by microorganisms and rooted plants.

Emergent aquatic plants often obtain large quantities of phosphorus from the sediment and can release large amounts into the water. When the ortho-phosphate concentration in the water is low, which is usually the case, phytoplankton excretes extracellular enzymes, alkaline phosphatases, which decompose the organic P (e.g., phosphate esters) that is excreted by higher aquatic plants. The released ortho-phosphate is then available for the phytoplankton.

Phosphate (in contrast to nitrate) is strongly adsorbed to soil particles and does not move freely within the groundwater in the majority of cases. High inputs of total phosphorus are due to particle erosion from steep slopes with easily erodible soils.

In deep stratified lakes, transport of mobilized phosphate into the upper water layers is limited and the availability of dissolved phosphate in late winter may predetermine the phytoplankton growth in spring and summer. Direct P supply of the euphotic zone from the sediment is important in the littoral and shallow lakes during the whole season.

Nitrogen compounds that are of the greatest interest to water-quality management are those that are biologically available as nutrients, or are toxic to humans or aquatic life.

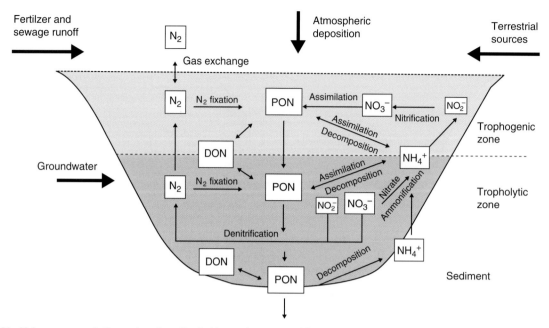

Figure 29 Main processes of nitrogen transformation in lakes and reservoirs. PON, particulate organic N; DON, dissolved organic N.

Atmospheric nitrogen (N_2) is the primary source of all nitrogen species, but it is not directly available to plants, because it is normally not bioavailable. The conversion of atmospheric nitrogen into other chemical forms is called nitrogen fixation and is accomplished by a few types of heterotrophic bacteria. Some autotrophs, namely several cyanobacteria, also have the ability to fix nitrogen from the air in heterocysts (special thick-walled cells that assure strict anaerobic conditions) by means of the enzyme nitrogenase. The N cycle is illustrated in **Figure 29**.

When nitrogen is cycled, it undergoes a series of oxidation–reduction reactions that convert it from N-containing organic molecules, such as proteins, to ammonia (NH_3). This is called 'desamination/ammonification'

$$R-NH_2 + H_2O \Rightarrow NH_3 + R-OH \quad (69)$$

followed by the acid–base-reaction

$$NH_3 + H_2O \Rightarrow NH_4^+ + OH^- \quad (70)$$

In oligotrophic lakes, high oxygen concentrations permit metabolism of ammonia to nitrate, resulting in low levels of nitrite and ammonia and comparatively high levels of nitrate in the hypolimnion (nitrification). Nitrification is the oxidation of nitrogen (III) to nitrogen (V). Two steps are indicated:

1. Ammonia oxidation by *Nitrosomonas* (obligatory autotrophic, CO_2 as C source), *Nitrosococcus*, *Nitrosospira*, and *Nitrosolobus*:

$$2NH_4^+(aq) + 3O_2 \Rightarrow 2NO_2^-(aq) + 2H_2O + 4H^+ \quad \Delta_R G^\circ = -272 \text{ kJ mol}^{-1}. \quad (71)$$

2. Nitrite oxidation by *Nitrobacter* (facultatively autotrophic, CO_2 and organic C as C source), *Nitrococcus*, and *Nitrospira*:

$$2NO_2^- + O_2 \Rightarrow 2NO_3^- \quad \Delta_R G = -5.2 \text{ kJ mol}^{-1} \quad (72)$$

Considering the formation of bacterial biomass, it follows that:

$$NH_4^+ + 0.1325(CO_2^*H_2O) + 1.8275O_2 \\ + 0.00125HPO_4^{2-} \Rightarrow 0.00125C_{106}H_{263}O_{110}N_{16}P \\ + 0.98NO_3^- + 1.9775H^+ + 0.98H_2O \quad (73)$$

Denitrification is the biological reduction of nitrate to nitrogen (by-products/intermediates NO_2^-, NH_4^+, NO, and N_2O) under anoxic conditions ($O_2 < 1 \text{ mg l}^{-1}$) in the pH range 7–8. It is typically restricted to the sediment and/or occurs in a fully or almost deoxygenated hypolimnion.

The nitrite concentration in lakes is usually very low:

$$NO_3^- \overset{I}{\Rightarrow} NO_2^- \overset{II}{\Rightarrow} NO \overset{III}{\Rightarrow} N_2O \overset{IV}{\Rightarrow} N_2 \quad (74)$$

with the enzymes: I = nitrate reductase – enzyme associated with the respiration chain; II = nitrite reductase; III = NO reductase; and IV = N_2O reductase.

Denitrification is performed by heterotrophic bacteria, such as *Paracoccus denitrificans*, and by autotrophic denitrifiers (e.g., *Thiobacillus denitrificans*).

Another reaction pathway is the reduction of NO_3^- to NH_4^+ (nitrate ammonification):

$$NO_3^- + 2H^+ + 8(H) \Rightarrow NH_4^+ + 3H_2O \quad (75)$$

Nitrate ammonification is carried out by several groups of microorganisms, especially of the Enterbacteriaceae (*Escherichia coli* and *Enterobacter aerogenes*). These are all facultative anaerobes which are able to work under anaerobic conditions. This process does not lead to the liberation of molecular nitrogen.

In eutrophic lakes, anoxia results in increased levels of ammonia with increasing sediment depth, and also in the hypolimnion.

2.08.1.6.7 Organic carbon – humic compounds

Organic matter in lakes, seas, and oceans originates predominantly from photosynthesis. The fate of organic matter from plants, animals, and microbes is extremely complex because many organisms and many compounds are involved. The detritus is composed of a broad spectrum of substrates. Structural polymers of plant cell walls are most abundant. These include cellulose, hemicelluloses, pectins, and lignin.

The biodegradation of organic matter in the aquatic system by microorganisms occurs by way of a number of stepwise, microbially catalyzed processes (e.g., oxidation, reduction, dehalogenation, and others). Biodegradation is the natural way of recycling waste, or breaking down organic matter into nutrients that can be used by other organisms. Degradation means decay and the bio prefix means that the decay is carried out by a huge assortment of bacteria, fungi, insects, worms, and other organisms.

Primary biotic substances and decomposition products are summarized in **Table 2** (Sigg and Stumm, 1991).

Next, we discuss the carbon cycle. Many organic compounds (see **Figure 30**; Thurman, 1995) and their decomposition products interact with both suspended material and sediments in water bodies. Colloids can also play a significant role in the transport of organic pollutants in surface waters. Settling of suspended material, containing adsorbed organic matter, carries organic compounds into the sediment. Some organics are transported into the sediments by particulate remains of organisms, or by fecal pellets from zooplankton. Suspended particulate matter affects the mobility of organic compounds adsorbed to particles. Furthermore, this organic matter undergoes biodegradation and chemical degradation, by different pathways and at different rates, when compared with organic matter in solution. The most common types of sediments considered for their organic binding abilities are clays, organic (humic) substances, and complexes of clays and humic materials. Both clays and humic substances act as cation exchangers. Therefore, these materials adsorb cationic, organic compounds through ion exchange. Since most sediments lack strong anion-exchange sites, negatively charged organics do not adhere strongly. Thus, these compounds are relatively mobile and biodegradable in water, despite the presence of solids.

Microorganisms are strongly involved in the carbon cycle, mediating crucial biochemical reactions. Photosynthetically, active organisms are the predominant C-fixing components in water, as they assimilate CO_2. The pH of the water is raised enabling precipitation of $CaCO_3$ and $CaMg(CO_3)_2$.

Humic substances are refractory, irregularly built-on substances of high molecular weight (structure proposal, see **Figure 31**; Stottmeister, 2008) that occur in soil and water. These complex compounds are formed by microbial

Table 2 Primary biotic substances and decomposition products

Primary biotic substances	Decomposition products	Intermediates and end products that appear in natural freshwater systems
Proteins	Polypeptides ⇒ Amino acids ⇒ RCOOH RCH$_2$OHCOOH RCH$_2$OH RCH$_3$ RCH$_2$NH$_2$	NH_4^+, CO_2, HS^-, CH_4, peptides, amino acids, urea, phenols, indols, fatty acids, mercaptans
Lipids Fat Waxes Oils Hydrocarbons	Fatty acids + Glycerin ⇒ RCH$_2$OH, RCOOH, RCH$_3$, RH	Aliphatic acids, acetic -, lactic-, citric-, glycolic-, maleic-, stearic-, oleic acids, carbohydrate, hydrocarbons
Hydrocarbons Cellulose Starch Hemicellulose Lignin	Monosaccharide ⇒ Hexogene Oligosaccharide Pentogene Chitin Glucosamine	Glucose, laevulose, galactose, arabinose, ribose, xylose
Porphyrin and plant pigments Chlorophyll Hemin Carotene Xanthophylls	Chlorine ⇒ Pheophytin ⇒ Hydrocarbons	Phytan, pristan, carotinoide, isoprenoid, alcohol, ketone, porphyrin
Polynucleotides	Nucleotides ⇒ Purines and pyrimidine bases	
Complex substances, formed from intermediate products	Phenols, quinoides, amino acids and decomposition products of hydrocarbons	Melanin, humic substances, humic-, fulvic acids, tannines

From Sigg L and Stumm W (1991) *Aquatische Chemie*. Stuttgart: B.G. Teubner Verlag.

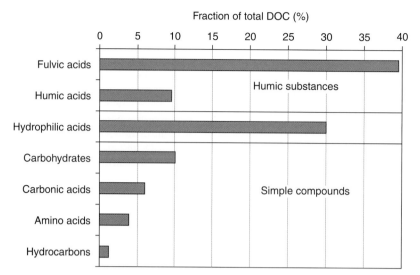

Figure 30 Distribution of dissolved organic carbon in typically freshwater with DOC = 5 mg l^{-1}. From Thurman EM (1995) *Organic Geochemistry of Natural Waters*, 497pp. Boston, MA: Martinus Nijhoff/Dr. W. Junk Publishers.

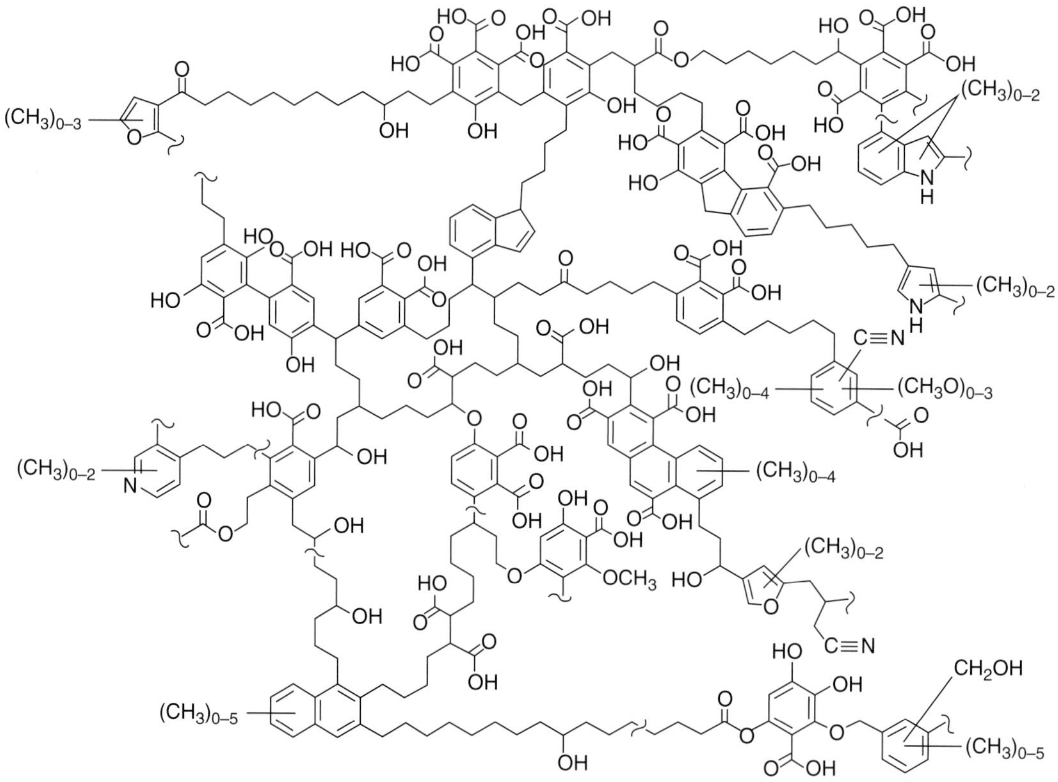

Figure 31 Structural proposal of a humic molecule. After Stottmeister U (2008) *Altlastensanierung mit Huminstoffsystemen. Prinzipien der Natur in der Umwelttechnologie*. Chemie in unserer Zeit, vol. 42, pp. 24–41. Weinheim: Wiley-VCH Verlag GmbH.

decomposition and/or partial new syntheses of vegetable and animal material. Humic acids, a fraction of the humic substances, exist at neutral pH values, in dissolved form. However, under strongly acidic conditions, they are insoluble. The fraction of humic acids that are soluble at pH = 1 are termed fulvic acids. They have smaller, average molar masses than humic acids. Humic and fulvic acids both have a cyclic aromatic basic structure (Figure 31), but fulvic acids are more soluble in water because they have more number of hydrophilic functional groups than humic acids.

From soil leaching, terrestrial detritus input, shallow groundwater, and overland flow, organic compounds (e.g., humics) can generally be discharged into lakes. The greater the area of the drainage basin, relative to the surface of the lake, the higher the input of humic acids from the soil, as against extraction and dilution by rain. Humic acids undergo slow degradation by sunlight and biodegradation. As the retention time of water in the lake increases, the color of the lake reduces because the humic acids are exposed to sunlight and the attack by microorganisms continues for a longer period. When the slope of a watershed is steep, surface runoff moves rapidly to the lake, allowing less time for contact with the soil during which humic acids may be picked up. Steep watersheds also tend to have less organic content in their soils and therefore yield less humic acid. Relatively flat watersheds allow rainwater to penetrate the soil for longer periods. They also have larger accumulations of organic material in the soil and larger areas of wetland, which contribute color to runoff. Calcium is known to precipitate humic acids. Lakes with high Ca^{2+} concentrations and/or with $CaCO_3$ as sedimentary rock are lighter in color than those in hard, igneous rock catchments because of the precipitation of humic acids.

2.08.1.7 Biotic Structure

The biotic structure of a lake is governed by vertical gradients in light intensity and temperature. In temperate climates, thermal gradients occur in the water column, right down to the lake bottom. The organisms which are characteristically found in these layers are described later in the chapter.

The livelihood of the biota in lakes and reservoirs is provided mainly by photosynthesis. The photosynthetic organisms, that is, cyanobacteria, algae, and macrophytes, are called 'primary producers'.

Their biomass is eaten by animals, that is, by primary consumers. These consumers comprise a very broad range of organisms, from aquatic protozoa to many species of freshwater fish. There are usually differing trophic levels of consumers, with the carnivores occupying the highest level (**Figure 32**; Ligvoet and Witte, 1991). There are also secondary and tertiary consumers.

The biological communities of the water body, that is, in the pelagic zone, are closely linked to benthic communities, which are localized at the bottom, or in the sediment. This also applies to nutrients in the water column, which are linked to nutrients in the sediment through sedimentation–resuspension and diffusion.

Suspended nonliving particles with a high content of organic materials (e.g., cellulose) and associated bacteria, called 'detritus', are an important source of biochemical energy for microorganisms and animals, especially those with a filtering apparatus such as several types of zooplankton. Very little of the organic material produced by photosynthesis is utilized directly in the consumer food chain (Wetzel, 2001). Instead, it enters the detritus pool. After deposition as organic bottom sediment, the detritus is processed by bacteria, fungi, and bottom-dwelling invertebrate animals such as oligochaete worms, mussels, and, in particular, insect larvae. The quantity of this particulate organic material is huge, but its (mainly microbial) metabolism is slower than that of biomass because not enough oxygen is available in the sediment.

The principal controlling factor in the supply of biochemical energy is solar radiation and its penetration depth. For the photosynthetic production of organic matter in a lake ecosystem, not only the open water (pelagic zone) with its

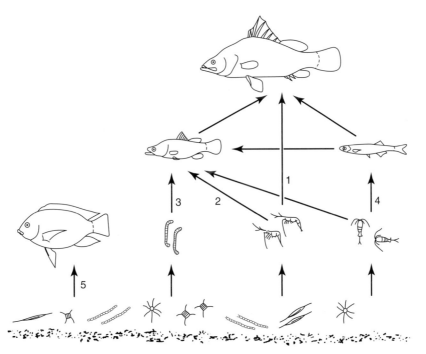

Figure 32 The current food web of Lake Victoria. The main food chains are: 1. via *Caridina* (freshwater prawn) to *Lates* (Nile perch, top predator); 2. to juvenile *Lates*, 3. via benthonic insect larvae to juvenile *Lates*, 4. via zooplankton to *Rastrineobola* (Silver Cyprinid) and *Lates*, 5. from micro-algae to *Oreochromis* (Tilapia). From Ligvoet W and Witte F (1991) Perturbation through predator introduction: Effects on the food web and fish yields in Lake Victoria (East Africa). In: Ravera O (ed.) *Terrestrial and Aquatic Ecosystems. Perturbation and Recovery*, pp. 263–268. New York: Ellis Horwood.

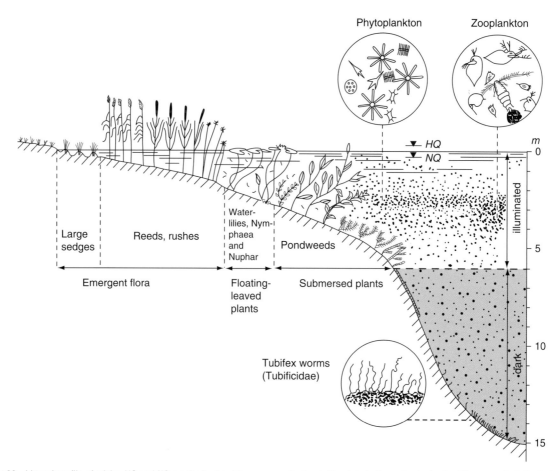

Figure 33 Littoral profile of a lake. HQ and NQ are the limits of the mean water level. The dots in the water body should represent phytoplankton and zooplankton, respectively. From Uhlmann D (1979) *Hydrobiology. A Text for Engineers and Scientists*. Chichester: Wiley.

phytoplankton, but also the riparian and the (illuminated) bottom zones containing macrophytes and algae, are responsible (**Figure 33**).

Among the aquatic macrophytes, there are emergent plants such as reeds, submerged vegetation such as milfoil, and floating-leaved species such as water lilies. Some of the algae are filamentous; others are unicellular or grow in colonies. These sessile, phototrophic organisms are called the 'phytobenthos'. The riparian area is called the 'littoral zone'. The phytobenthos also includes microscopic algae, which grow on solid surfaces (living and nonliving) together with bacteria, as biofilms (Aufwuchs). These biofilms cohere by extracellular polymeric substances (EPSs) which consist mainly of polysaccharides and proteins. Biofilms in which microscopic algae dominate are also called 'periphyton'.

Invertebrate animals living in free water are called zooplankton (small crustaceans, some of which are effective grazers of phytoplankton and bacteria, and rotifers). Many of these animals are filter feeders and some of them are able to feed on cells with diameters smaller than 1 μm. In the free water, there are, in addition, rapidly swimming water insects (larvae) and fish. Fish larvae may also rank among the zooplankton. Several species of fish (such as some species of whitefish, *Coregonus*) and fish larvae eat mainly zooplankton, while others feed on smaller fish or on zoobenthos, that is, invertebrate animals living at the bottom of the lake. Several types of zoobenthos are mobile, for example, insect larvae, snails, worms, and protozoa. Others are sessile or slow moving, for example, sponges, different types of worms, certain insect larvae, and mussels. Only a small number of animal species in lakes are omnivorous.

We now discuss the seasonal variations in lakes. In many lakes and reservoirs of the temperate zones, rapid growth of phytoplankton dominates early in spring, because the nutrient supply (from the winter) is still high, and the activity of filtering zooplankton is, on account of the low temperatures, still far below the maximum values found in summer. Therefore, during this season, the vertical flux of nutrients from the lower water layer, rich in dissolved N and P, is still not restricted by the thermal stratification of the water body. At the height of the summer season, mass growth of particular phytoplankton species (water blooms of cyanobacteria) may become a nuisance, because of toxic metabolites or of odors. This, however, is mostly to be expected if the nutrient load from outside is too high. Solar radiation not only controls the photosynthesis of organic materials, but also controls the generation of molecular oxygen. In many lakes, this process is even more relevant to oxygenation than the diffuse introduction of oxygen from the atmosphere. The light-penetration depth separates an upper water layer (euphotic zone, normally located within the epilimnion), and this layer is warmed in the summer season, from a lower water layer (aphotic zone) in the hypolimnion, which is dark and remains cold. The latter is called the 'profundal zone' as far as it concerns the lake bottom.

The phototrophic biota is inseparably connected to the dynamics of several chemical elements, in particular inorganic C, N, and P. In the drainage basin, the duration of contact of water with soil and associated microorganisms influences the content of dissolved salts, all the above-mentioned nutrients, and organic materials such as humic acids (Wetzel, 2001).

A lake ecosystem consists of the lake and its entire drainage basin. Lake ecosystems require a continual input of organic matter, produced mainly by photosynthesis (in part outside the water body). This organic matter is used by animals, who are the consumers, or degraded by microorganisms. Biochemical energy and inorganic nutrients must continually be replenished, basically by import from the drainage basin. A usually small proportion of the produced biomass sometimes becomes unavailable due to sedimentation.

The average, elementary composition of biomass is approximately in accordance with the formula

$$C_{106}H_{263}O_{110}N_{16}P \qquad (76)$$

Thus, the organic substance of the organisms comprises, as related to the dry mass, about 40% carbon.

The nitrogen content is usually comparatively constant, because of the irreplaceable amino and nucleic acids, whereas phosphate may be stored by many microorganisms and algae as polyphosphate, and is thus highly variable.

The biotic communities in lakes and reservoirs are self-sustaining. This is also true for single components such as the phytoplankton. Nearly all species are free floating and find the best conditions for propagation in a level not too deep (because of the lack of light energy), but also not at the surface because of potential damage by UV radiation. Nearly all species aim at neutral buoyancy and avoid losses by sinking toward the lake bottom, through storage of substances low in weight, such as oil droplets in the case of diatoms. The zooplankton, on the other hand, is actively motile and prefers depths with abundant food, but with protection from predation by fish.

The entity of fluxes of energy/organic materials, which are necessary to supply/maintain the structural and functional stability of a lake ecosystem, or of its compartments, is called the food web (**Figure 32**). There are unidirectional transport pathways of biochemical energy through the ecosystem from the primary producers to the top carnivores. At each trophic level, up to 90% of the collected energy is lost through respiration and defecation, and thus released as heat energy or as waste material. The sum of energy losses increases with the number of trophic levels within a food web. This also implies that, in principle, for fisheries/human nutrition, the use of plant-eating fish is more economical than the production of fish biomass on the uppermost trophic level, because the number of trophic levels between producers and consumers is less.

The flux of biochemical energy within the lake ecosystem may be controlled not only bottom-up, for example, by increasing or decreasing the supply of nutrients to the phototrophic organisms, but also, alternatively, top-down, for example, by a higher or lower grazing pressure from herbivorous consumers, or by fish species feeding on phytoplankton. Thus, mass growth of zooplankton may cause clear water, whereas a very dense stock of zooplankton-eating fish may sustain a high turbidity, induced by phytoplankton. The bulk of the biochemical energy flow is usually carried by a comparatively low number of dominant species. Most of the many thousands of different species, in a large lake ecosystem, probably play an ancillary role in terms of biochemical energy flux. The names and the portion of most species of bacteria and fungi in the community metabolism of lakes and reservoirs, is still unknown. Their metabolic rate is often higher than that of algae, higher plants, and animals.

Shallow lakes, rich in nutrients, may exhibit alternative equilibria (Uhlmann 1980; Scheffer et al., 1993): (1) status 1 – if there is abundant submerged vegetation, the water is clear and light can penetrate to the lake bottom. The vegetation also operates as a screen, which eliminates particles from water moved by wind currents. This is a self-preserving mechanism; (2) status 2 – the alternative is characterized by dense and long-lasting growth of phytoplankton, so that the submerged vegetation cannot become relevant. The shift from phytoplankton turbidity to clear water, is mostly caused by the disappearance of zooplankton-eating fish, for example, by ice and snow cover, which may cause oxygen deficiency and winterkill of fish.

The overall species diversity is specifically high in aquatic systems which are not subject to anthropogenic impacts such as acidification, eutrophication, or organic pollution. Thus, the biotic diversity is (with some notable exceptions) a measure of organization and integrity within the lake ecosystems (Wetzel, 2001).

2.08.1.8 Photosynthesis: Generation and Consumption of Dissolved Oxygen

Apart from temperature, oxygen is the most significant parameter in water bodies, because it is essential for the metabolism of the bulk of the biota. The dynamics of oxygen distribution are basic to the understanding of the growth and distribution of aquatic organisms.

The dissolved oxygen in water bodies originates from two sources:

1. entry from the atmosphere and
2. photosynthesis.

Air contains approximately 21% oxygen (by volume) and the remainder is mainly nitrogen. The amount of molecular oxygen, which can be dissolved in a water body in equilibrium with the atmosphere, is low and increases markedly at low temperatures (**Figure 34**).

Thus, at the end of the vernal circulation, the O_2 concentration is 13.1 mg l^{-1} in lakes of the temperate climate belt at sea level, but substantially lower in tropical, lowland lakes, with their higher temperatures. With increased salinity, this concentration becomes even lower. The higher the barometric (atmospheric) pressure, the higher the oxygen saturation concentration.

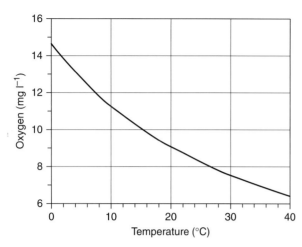

Figure 34 Relationship between water temperature and O_2 concentration (at saturation) under normal air pressure for pure water.

Photosynthetically active organisms normally produce oxygen, as well as organic materials, from inorganic substances, using sunlight as the sole energy source. In the process, they reduce carbon dioxide and in part cleave the water molecule:

$$\underset{\text{carbon dioxide}}{6CO_2} + \underset{\text{water}}{12H_2O} + \underset{\text{light energy}}{2.872 \text{ kJ}} \Rightarrow$$
$$\underset{\text{glucose}}{C_6H_{12}O_6} + \underset{\text{water}}{6H_2O} + \underset{\text{oxygen}}{6O_2} \quad (77)$$

The synthesis of algal biomass may be (simplified) described as follows:

$$106CO_2 + 16NO_3^- + HPO_4^{2-} + 122H_2O + 18H^+ + \Leftrightarrow$$
$$C_{106}H_{263}O_{110}N_{16}P + 138O_2 \quad (78)$$

[H] denotes biochemically bound hydrogen. From this equation, it is evident how important nitrogen and phosphorus are as sources for the synthesis of biomass.

When no free carbon dioxide is available, phototrophic organisms precipitate carbonate in accordance with Equation (79):

$$Ca^{2+} + 2HCO_3^- \Leftrightarrow CaCO_3 + H_2O + CO_2 \quad (79)$$

Thus, over long periods of time, large depositions of limestone may evolve by biogenic decalcification.

With photosynthetic activity, deep lakes and reservoirs are vertically subdivided into two levels: the upper, fully illuminated level which is called 'euphotic', that is, with generation of O_2 and of organic material, and a lower one, called 'aphotic', that is, very dark with consumption of both O_2 and organic reserves and formation of CO_2. In clear water bodies, photosynthesis is inhibited by UV radiation immediately below the surface; thus, the light intensity is optimal at depth z_{opt}, which may be measured using a series of vertically oriented light and dark bottles, filled with ambient water and phytoplankton.

The maximum possible photosynthesis is regulated by the upper limit of the quantum yield. It amounts to 30–40 g m^{-2} d^{-1} C in some tropical water bodies with a very high density of phototrophic biomass (Baumert and Uhlmann, 1983). In many clearwater lakes with a very low concentration of nutrients, not even the annual primary production attains such a high level. As many species of phytoplankton are able to grow at low temperatures, the photosynthetic production may reach high supersaturation, even under a cover of clear ice.

Instead of solar energy, several species of bacteria (and only bacteria) are capable of utilizing the energy obtained by the oxidation of energy-rich inorganic compounds, such as sulfide (oxidation to sulfate) and ammonia (oxidation to nitrite/nitrate). This ability is known as chemosynthesis (as opposed to photosynthesis).

As an intermediate reservoir of biochemical energy, obtained by photosynthesis or chemosynthesis, all organisms use an organic phosphate, ATP (**Figure 35**; Uhlmann and Horn, 2001) as bioenergetic currency. Moreover, many species of microorganisms are able to store energy-rich polyphosphates.

For the process of photosynthesis, only visible sunlight can be utilized, which represents about 46% of the total irradiation. From glucose (cf. Equation (77)), many organic materials and biomass (for cell propagation) are produced inside the cells of the photosynthetically or chemosynthetically active organisms. According to their energy source, autotrophic organisms, which depend on solar energy or on the oxidation energy of inorganic compounds such as H_2S, can be distinguished from heterotrophic organisms, which need an organic source of biochemical energy.

All organisms require biochemical energy (ATP) for the maintenance of life functions and for growth/reproduction. Energy is released mainly from stored carbohydrates, by biochemical combustion with oxygen:

$$C_6H_{12}O_6 + 6O_2 \Rightarrow 6CO_2 + 6H_2O + 2.872 \text{ kJ} \quad (80)$$

This equation is the reverse of Equation (77) and an expression of respiration, another principal metabolic process. Nearly all organisms require oxygen for energy generation. O_2 is also needed by bacteria and fungi for the decomposition of organic material. In polluted water bodies, microbial decomposition may require a high amount of dissolved oxygen, sometimes up to total depletion.

If in a water body, the concentration of nutrients (beside CO_2, mainly N and P compounds) for the growth of phototrophic biomass is very high, the concentration of produced molecular oxygen may substantially exceed the O_2-saturation level, which results from the equilibrium between atmospheric and aquatic oxygen (**Figure 36**; Uhlmann and Horn, 2001). On the other hand, during the night, when respiration is not counterbalanced by photosynthesis, the oxygen concentration may, in extreme cases, decrease to zero.

For a natural water body, this situation is completely undesirable. Thus, a basic principle of water-quality management is to prevent too-high diurnal and seasonal amplitudes in oxygen concentration, caused by photosynthesis and respiration, as far as possible. There should only be a low supersaturation level, if any, and a decline down to about 3 mg l^{-1} dissolved oxygen in the water at the most. In a lake, this concentration level has to be maintained in the hypolimnion up to the autumnal circulation, if salmonid fish/whitefish are

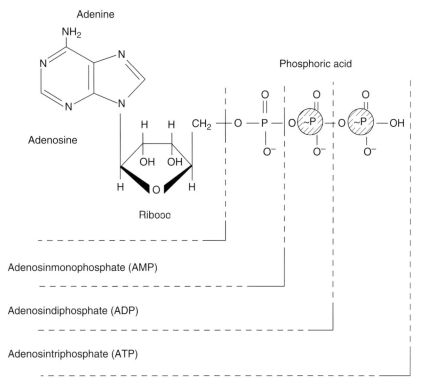

Figure 35 Structure of adenosine triphosphate, ATP, the principal biochemical energy store. From Uhlmann D and Horn W (2001) *Hydrobiologie der Binnengewässer. Ein Grundriss für Ingenieure und Naturwissenschaftler.* Stuttgart: Eugen Ulmer.

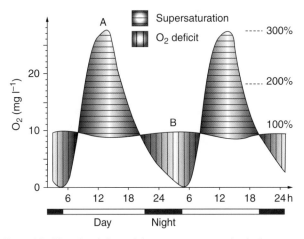

Figure 36 Diurnal variations of the oxygen concentration in the upper water layer of a wastewater-treatment lagoon, very rich in phytoplankton. Diagrammatically presented after empirical data of I. Röske. From Uhlmann D and Horn W (2001) *Hydrobiologie der Binnengewässer. Ein Grundriss für Ingenieure und Naturwissenschaftler.* Stuttgart: Eugen Ulmer.

to survive. In populated catchments, such a water-quality level needs to be maintained, principally by the advanced treatment of wastewater and by the control of nutrient fluxes from fertilized agricultural areas.

In the upper water layers of lakes, O_2-depleting processes are not usually noticeable, because these are fully compensated for by photosynthesis and by atmospheric aeration. Thus, the oxygen concentration here does not deviate much from the saturation level. Generally, a comparison of the rate of oxygen consumption, in relation to the photosynthesis rate, permits an approximate evaluation of the metabolism in a lake or a reservoir.

Diffusion of oxygen in water is a very slow process and its vertical transport requires external energy for turbulent mixing. However, oxygen produced by photosynthesis at depths greater than 1–4 m can remain dissolved there, even if the turbulent mixing of the water body is very low.

In lakes and reservoirs of the temperate climate belt, atmospheric oxygen is introduced immediately after the disappearance of the ice cover in early spring. At 4 °C, the O_2 concentration is near 100% saturation, that is, at an altitude around sea level $\leq 13\,\mathrm{mg\,l^{-1}}$.

As shown in **Figure 37** (Uhlmann and Horn, 2001), thermal stratification of the water body is usually associated with oxygen stratification. However, at very low nutrient levels, the phytoplankton production remains low. Such a lake is described as oligotrophic. Accordingly, oxygen saturation in the upper layers barely exceeds 100% and the O_2 deficit in the deeper strata remains low. At the end of the summer stagnation, the oxygen concentration in the hypolimnion is still approximately as high as in the epilimnion, despite the much lower temperature. In most of the lakes in populated areas, however, the amount of organic matter (plankton biomass) that arrives in the hypolimnion by sinking from the upper, illuminated (i.e., euphotic) water layer is high. Therefore, the oxygen concentration in the hypolimnion is progressively reduced. This results in an O_2 deficit. Lakes with increased nutrient content and increased production are called 'eutrophic'.

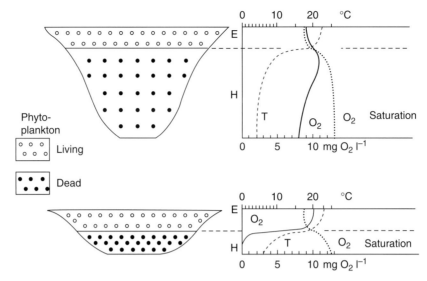

Figure 37 The vertical distribution of dissolved oxygen in thermally stratified lakes. It is presumed that the phytoplankton cells, after sinking down into the dark hypolimnion, die and are microbially decomposed due to O_2 consumption. Situation at the end of the summer stagnation period, O_2, actual concentration; T, water temperature. In the top portion of the figure, the lake has a large, and in the lower one a comparatively small, initial O_2 store. Based on Thienemann, from Uhlmann D and Horn W (2001) *Hydrobiologie der Binnengewässer. Ein Grundriss für Ingenieure und Naturwissenschaftler.* Stuttgart: Eugen Ulmer.

The loss of oxygen in the hypolimnion and, in particular, at the sediment–water interface, primarily results from microbial decomposition of plankton residues and other organic particles. Oxygen consumption by bacterial respiration is intensive in all water layers, but in the epilimnion is offset by turbulent mixing and photosynthesis. It is most intensive at the sediment/water interface, which may rapidly become anaerobic.

The smaller the hypolimnion volume is, compared with the volume of the epilimnion, the higher is the probability that the oxygen pool is not large enough to prevent O_2 depletion down to a very low, or even a zero level (**Figure 37**).

According to the O_2 stratification, increased hypolimnic concentrations of redox-sensitive components may result, which are also associated with the decreased oxygen concentration. This relates to NO_3^-/NH_4^+, Mn^{2+}/Mn^{4+}, Fe^{3+}/Fe^{2+}, and SO_4^{2-}/H_2S. These first develop within the sediment, but then diffuse into the water.

Most species of fish cannot survive, even at low temperatures, at oxygen concentrations of less than $2\,mg\,l^{-1}$. The development of the O_2 deficit in the hypolimnion from the beginning of the stratification period quantifies the metabolic relationship between the illuminated, that is, trophogenic (euphotic) zone, and the underlying tropholytic (aphotic) layer, which is dark.

In tropical or subtropical zones, the hypolimnion temperature will never be as low as $4\,°C$, and the microbial degradation processes are thus substantially enhanced. If the bottom of a future reservoir in the humid tropics is not cleared, the submerged terrestrial vegetation will give rise to rapid oxygen depletion and microbial sulfate reduction. Thus, the personnel in the powerhouse of a newly constructed reservoir in humid Latin America had to wear gas masks for several years, because of the hydrogen sulfide that escaped from the hypolimnetic water used (to obtain a sufficiently high power head) for the operation of the turbines.

Figure 38 (Kusnezow, 1959) depicts the H_2S accumulation in an eastern European lake.

If dissolved humic materials are present, chemical oxidation due to photochemical reactions caused by UV radiation near the lake surface may be relevant. The biochemical, largely inert, humic acid molecules (**Figure 31**) are split into smaller fragments which are accessible to microbial degradation and may thus introduce biochemical energy into the food web.

With the initiation of the autumnal circulation, oxygen-rich epilimnetic water disperses deeper and deeper into the hypolimnion. When circulation is complete, the oxygen concentration is high, right down to the bottom. An ice cover prevents O_2 exchange with the atmosphere. An ice cover is usually also covered by snow and thus the photosynthetic oxygen production is low, or zero, due to the strongly reduced underwater light intensity.

In the warm season, the oxygen concentration in the metalimnion may be higher than above (metalimnetic O_2 maximum), if there is still enough light for photosynthesis and a lower temperature (i.e., higher O_2 solubility) than in the overlying epilimnion (**Figure 39**; Uhlmann and Horn, 2001).

In the metalimnion, the vertical diffusion is lower than both in epi- and hypolimnion. Conversely, absence of light and decreased vertical turbulence may cause a metalimnetic O_2 minimum. This can, *inter alia*, be caused by the microbial oxidation of methane ascending from the bottom sediment into an upper water layer (under conditions of stratification), the temperature of which is higher than that below.

According to the respective O_2 stratification, gradients of redox-sensitive components of nitrogen, manganese, iron (**Figure 40**; Uhlmann and Horn, 2001), and sulfur (SO_4^{3-}/H_2S) also may develop here.

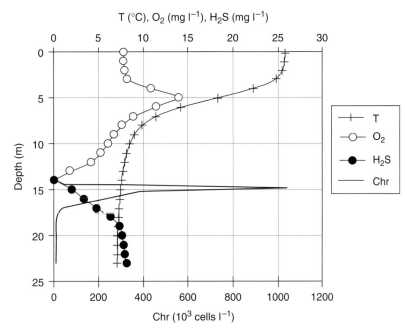

Figure 38 The vertical distribution of temperature (*T*), oxygen (O_2), hydrogen sulfide (H_2S), and the bacterium *Chromatium* (Chr) at the end of July in Lake Belowod, Russia. Note also the metalimnetic O_2 maximum. Redrawn from Kusnezow SI (1959) *Die Rolle der Mikroorganismen im Stoffkreislauf der Seen*, 301pp. Berlin: Deutscher Verlag der Wissenschaften.

This outflow of interstitial water also concerns nonreducing substances with an increased density, such as $CaHCO_3$ in hard-water lakes. Such a layer with increased $CaHCO_3$ concentration may then act as a physical barrier against the vertical (turbulent) flow of dissolved materials within the hypolimnion.

The volume loading of the hypolimnion with oxidizable matter, that is, mainly oxygen-depleting residues of phytoplankton, is high if the hypolimnion volume is small and vice versa (**Figure 39**). Unfortunately, many lakes in lowland areas and also many reservoirs have small hypolimnion volumes and are therefore often very sensitive to increased phytoplankton production caused by human impacts (eutrophication by sewage-borne nutrients or by runoff from fertilized agricultural areas). The organic material deposited in the bottom sediment is largely unavailable for rapid biotic conversion, due to the lack of oxygen. Exceptions are the microbial utilization of nitrate and sulfate as electron acceptors, and the anaerobic digestion of organic materials, with methane (CH_4) as an organic, but volatile end product. Over long periods, mud deposition may raise the bed of a lake to such an extent (**Figure 41**) that light can penetrate everywhere, right down to the bottom.

Thus, macrophytes may spread over the whole area if very dense phytoplankton blooms are absent. Consequently, an already shallow lake may be converted into a reed swamp. In periods of high transpiration of emergent plants, a drop in water level may be caused. While some lakes progress through this sequence, it is not always the rule. Other types of wetlands, in succession stages, may similarly be formed.

In many cases, the biogenic fine structure of the sediment layers (with pollen grains and spores) reflects former climatic conditions.

2.08.1.9 Oxygen Stratification: Circulation/Quality Types of Lakes

The combination of thermal stratification and biological activity causes characteristic patterns in water chemistry. Typical seasonal changes in dissolved oxygen and temperature are shown in **Figure 42** (Wetzel, 1975).

In spring and autumn, both oligotrophic and eutrophic lakes tend to have uniform, well-mixed conditions throughout the water column. During summer and winter stratification, the conditions diverge.

The O_2 concentration in the epilimnion remains high throughout the summer. However, conditions in the hypolimnion vary with trophic status. In eutrophic (i.e., phytoplankton-rich) lakes, hypolimnetic O_2 concentration declines during the summer because it is cut off from all sources of oxygen, while organisms continue to consume oxygen. The bottom layer of the lake and even the entire hypolimnion may eventually become anoxic. Epilimnetic oxygen concentrations vary on a daily basis in eutrophic lakes. Fluctuations between O_2 supersaturation and deficit are typical.

In oligotrophic lakes with thermal stratification in summer, the oxygen content of the epilimnion decreases as the water temperature increases. The O_2 content of the hypolimnion is initially higher than that of the epilimnion, because the saturated colder water contains more oxygen (from spring turnover).

This oxygen distribution is known as an 'orthograde oxygen profile' (**Figure 42**; Wetzel, 1975). In oligotrophic lakes, low phytoplankton biomass favors deeper light penetration and less decomposition. The O_2 concentration may therefore increase with depth below the thermocline, where colder water with higher O_2 concentration occurs (oxygen is more

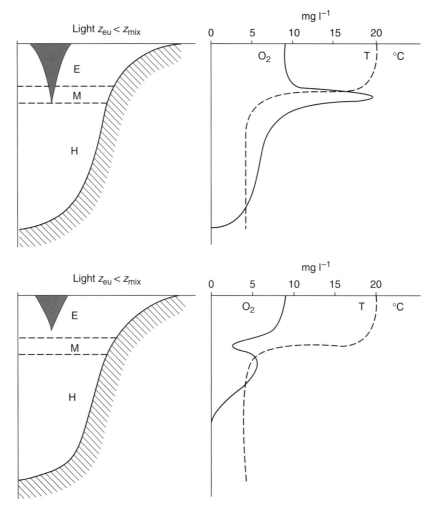

Figure 39 Differing relationships between light-penetration depth z_{eu} and epilimnion depth z_{mix}. If $z_{eu} \geq z_{mix}$, planktonic photosynthesis is possible in the metalimnion (sometimes even in the hypolimnion). This results in a metalimnetic O_2-maximum. If $z_{eu} \leq z_{mix}$, the result is a metalimnetic O_2-minimum. From Uhlmann D and Horn W (2001) *Hydrobiologie der Binnengewässer. Ein Grundriss für Ingenieure und Naturwissenschaftler.* Stuttgart: Eugen Ulmer.

soluble in colder water). In eutrophic lakes, on the other hand, the oxygen profile is clinograde because of the sharp decline of O_2 concentration in the hypolimnion (**Figure 42**), after only a few weeks of summer stratification. Later, the hypolimnion is often anaerobic.

These differences between eutrophic and oligotrophic lakes tend to disappear with autumn turnover.

In winter, oligotrophic lakes generally have uniform O_2 conditions along the vertical axis. Ice-covered eutrophic lakes however, may develop a winter stratification of dissolved oxygen. If there is only limited, or no snow cover to block sunlight, phytoplankton and several macrophytes may continue to photosynthesize, resulting in an increase in O_2 content just below the ice. However, microorganisms continue to decompose organic material in the water column and in the sediment. No oxygen inflow from the air occurs because of the ice cover, and if snow covers the ice, it becomes too dark for photosynthesis. This condition can cause fish mortality during the winter (winterkill). When the dissolved oxygen level drops below $1\,mg\,l^{-}\;O_2$, biochemical processes at the sediment–water interface accelerate the release of phosphate and ammonium from the sediment as a basis for increased phytoplankton growth.

The dynamics of oxygen distribution in inland waters are essential to the dynamics of the biota, because many animal species cannot survive/propagate at low ($\leq 3\,mg\,l^{-1}$) O_2 concentrations.

For lakes and reservoirs in the temperate climate regions, the following seasonality in temperature, with direct impacts on the oxygen balance, is representative:

1. *Spring.* Complete overturn subsequent to the disappearance of the ice cover. Atmospheric oxygen is absorbed and evenly distributed throughout the water body. Its concentration corresponds with the saturation level. If the lake is not too deep, and the average light intensity is high enough, a phytoplankton spring bloom may even result in remarkable oxygen supersaturation within the entire water column.

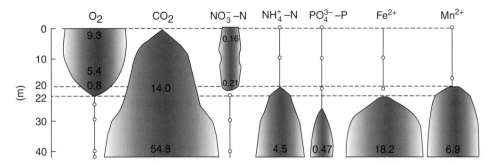

Figure 40 Chemical stratification in an Austrian lake due to hypolimnetic O_2 depletion. The numbers in the graphs correspond to the cubic roots of the concentrations (mg l^{-1}). From Ruttner From Uhlmann D and Horn W (2001) *Hydrobiologie der Binnengewässer. Ein Grundriss für Ingenieure und Naturwissenschaftler*. Stuttgart: Eugen Ulmer.

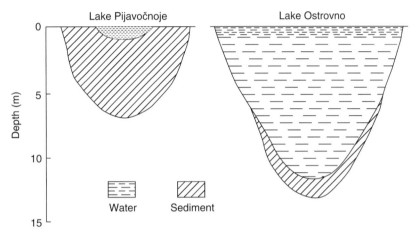

Figure 41 Profiles of two lakes in different stages of silting-up (biogenic sediment accumulation). From Kusnezow SI (1959) *Die Rolle der Mikroorganismen im Stoffkreislauf der Seen*, 301pp. Berlin: Deutscher Verlag der Wissenschaften.

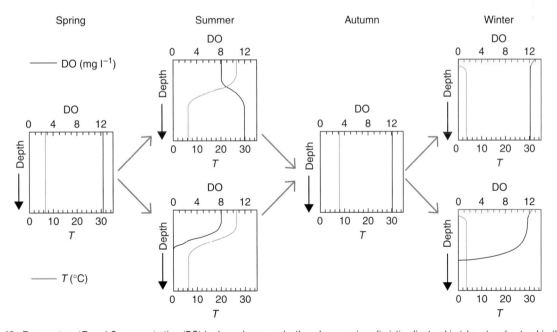

Figure 42 Temperature (T) and O_2 concentration (DO) in dependence on depth and season in a dimictic oligotrophic (above) and eutrophic (below) lake. Modified after Wetzel RG (1975) *Limnology*. Philadephia, PA: WB Saunders.

2. *Summer.* Heating of the upper water layers with formation of the epi-, meta-, and hypolimnion. In eutrophic lakes, oxygen at saturation level is present only in the epilimnion, but a noticeable deficit exists in the hypolimnion. The hypolimnetic concentration further decreases toward the end of the summer.
3. *Autumn.* Convective mixing of the epilimnion due to the cooling down of the overlying air. This turbulence goes far into the depth. At the end of the season, a homothermal state is achieved with 4C and near O_2 saturation.
4. *Winter.* During calm weather and frost, the water surface may freeze. The upper water layers are now not only colder, but also lighter than the bulk of the water body. Such an inverse thermal stratification is maintained as long as the ice cover exists. On the other hand, if the ice is clear, a thin surface layer of water may even attain a temperature of up to 10 °C.

The O_2 concentration below a cover of clear ice corresponds, at least, with the saturation level, because respiration is usually low in winter and because photosynthetic oxygenation is possible. In shallow lakes however, below snow-covered ice, the accumulation of sediment-borne methane and/or hydrogen sulfide may cause oxygen depletion and fish kills. (Below clear ice, photosynthetic O_2 production may be high enough to counteract depletion.) Lakes with two overturn periods per year are called 'dimictic' and they undergo at least two phases of atmospheric oxygenation.

In lakes, the average water temperature correlates with the mean air temperature. The hypolimnion temperature shows an inverse, nearly linear relationship with the altitude above sea level (**Figure 43**; Uhlmann and Horn, 2001).

The seasonal change between overturn and stratification periods is governed by the seasonality in air temperature. As is evident from **Figure 44** (Uhlmann and Horn, 2001), this has drastic consequences for oxygenation.

In the humid tropics, the variations in air temperature are often so slight and wind action so low that overturn in oligomictic lakes does not occur and oxygen depletion is high. This is all the more valid as hypolimnion temperatures are also high and thus accelerate rapid microbial degradation. The formation of H_2S by microbial sulfate reduction is not uncommon.

If, on the other hand, the wind is violent and nocturnal cooling likewise strong, a lake may be subjected to a frequent overturn, sometimes at diurnal intervals. Such a lake is called 'polymictic' and is situated either in an arid/semiarid climate or at high elevations in the tropic belt. A diurnal, deep overturn introduces so much atmospheric oxygen that no marked deficit can evolve.

As in tropical/subtropical lakes in the lowlands, no ice cover can form, the stratification period is interrupted by a single long-lasting circulation period in the cool season. Such a lake is 'warm monomictic'. The dissolved oxygen here may diminish after a prolonged stratification period.

Conversely, lakes at very high altitudes, or in the Arctic, often have an ice cover for more than 9 months of the year. An overturn occurs only in the summer. As long as ice still remains, the water temperature is not higher than 4 °C. This type

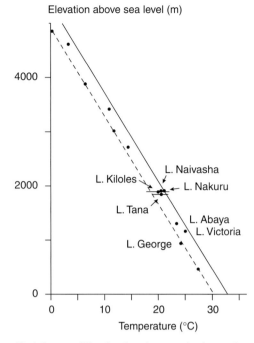

Figure 43 Influence of the elevation above sea level upon the hypolimnetic temperature of East African lakes. The names of some well-investigated lakes are also mentioned. From Talling and Lemoalle, modified from Uhlmann D and Horn W (2001) *Hydrobiologie der Binnengewässer. Ein Grundriss für Ingenieure und Naturwissenschaftler.* Stuttgart: Eugen Ulmer.

of lake is called 'cold monomictic'. Such lakes are usually well oxygenated.

In reservoirs, the spatial and temporal variability in oxygen is even higher than in lakes (Cole and Hannan, 1990).

We now discuss the trophic level of lakes and reservoirs. Lakes are generally classified according to their potential primary production, based on nutrient supply:

1. Oligotrophic water bodies have a low chlorophyll concentration ($<4\,\mathrm{mg\,m^{-3}}$), which is based upon a low nutrient supply (up to $2.5\,\mathrm{mg\,m^{-3}}$ P). The water is very clear, usually with a low Ca concentration and is circum-neutral to low acidic. The bedrock is mostly siliceous. The water body is deep and exhibits only a low O_2 hypolimnetic deficit. The light penetration depth often exceeds 20 m.
2. Mesotrophic lakes are, according to climatic conditions and the geological substrate, the most frequent type of non-shallow water bodies in Eurasia and Northern America. They are clear, have a P content $<12\,\mathrm{mg\,m^{-3}}$, and a light penetration depth of approximately 8 m. If their bedrock is poor in Ca, the pH is slightly <7.0. There are also mesotrophic lakes with higher alkalinity levels, that are well buffered against acidification. The submerged vegetation in which Characeae often dominate, may be covered by $CaCO_3$ crusts. This is often the case if inflows from groundwater are relevant. The planktonic chlorophyll *a* concentration is about 4–$7\,\mathrm{mg\,m^{-3}}$ and the Secchi transparency is more than 2 m. The oxygen deficit in the hypolimnion at the end of the summer stagnation is still

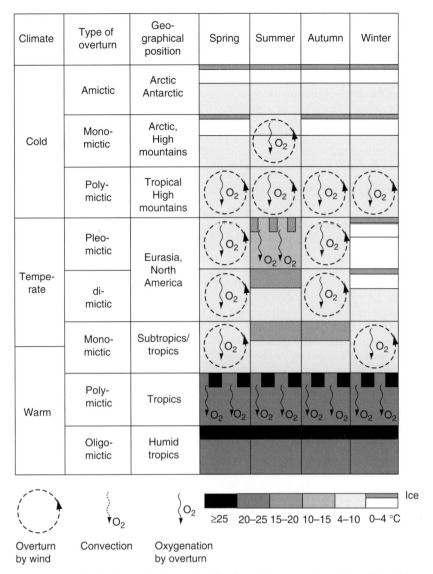

Figure 44 Influence of different geographic locations on the thermal stratification and overturn periods in lakes. From Uhlmann D and Horn W (2001) *Hydrobiologie der Binnengewässer. Ein Grundriss für Ingenieure und Naturwissenschaftler.* Stuttgart: Eugen Ulmer.

modest. If mesotrophic lakes are poor in Ca, they are sensitive to an increased P load, that is, to eutrophication.

3. Eutrophic water bodies are mostly rich in Ca and subject to a P concentration of <30 mg m^{-3}. They have a Chl a concentration of up to 1 mg l^{-1} and high phytoplankton production. Light penetration depth is often not more than 2 m. Although oxygen concentration in the epilimnion may achieve supersaturation, it may go down to zero in the hypolimnion at the end of the summer stagnation.

Polytrophic/hypertrophic water bodies may have a total P concentration of more than 100 mg m^{-3}. They are characterized by long-lasting phytoplankton growth in the warm seasons, which is often caused by Cyanobacteria. These may produce liver-toxic metabolites. The water bodies are subject to heavy variations in O$_2$ concentration with supersaturation during the day and depletion during the night (**Figure 36**), or long-lasting depletion in the hypolimnion, due to the high phytoplankton concentration. The quality status of such water bodies in temperate climates is exclusively man made (mostly by the introduction of untreated domestic effluent).

2.08.2 Fundamental Properties of Reservoirs

2.08.2.1 Functions of Reservoirs

A reservoir is an artificial lake in which water is impounded. The storage in a reservoir increases the availability of water for various purposes. Reservoirs also provide the most effective control of floods, if they are properly designed (enough retention capacity) and operated. The degree of regulation expresses the proportion of the mean annual discharge of a river, which can be stored in a reservoir. Most reservoirs have a degree of regulation below 100%, but Lake Volta, Ghana, holds the world record of 428%. This implies that more than 4 years of average discharge could be stored in the reservoir without releasing any water downstream (Nilsson, 2009).

Reservoirs are usually constructed in areas in which the groundwater resources are not sufficient to satisfy the demand for drinking or irrigation water, or where the discharge of a river and the morphology of the site favor the installation of a man-made lake to supply power. Reservoirs increase the hydraulic head. They provide 19% of the world's total electricity supply.

Often, the sites of reservoir construction are in areas where natural lakes seldom occur. The most common purpose of reservoirs is for irrigation.

There are nearly 50 000 dams in the world with heights above 15 m, which store more than 15% of the global runoff (Nilsson, 2009). There are innumerable smaller dams.

Reservoirs were already built about 5000 years ago in China, Mesopotamia, and Egypt. In view of the anticipated global warming, reservoirs are expected to become even more important in the future for water management during dry periods. In this context, even the storage and final purification of (preferably treated) domestic effluent in reservoirs, for the irrigation of crops, become increasingly important in dry climates, in an anticipated dramatic depletion of groundwater resources. In this way, not only will wastewater effluent be reused, but in addition, nitrogen and phosphorus compounds, which otherwise generate excessive growths of photosynthetically active organisms in water bodies (eutrophication), may serve as fertilizers for crop production and contribute to an increased production of food.

With the construction of a reservoir, a single, high-priority use is usually crucial, but most reservoirs are subject to multiple uses. It is impossible to operate each function at its optimum level. Some of the uses are conflicting, such as hydropower and irrigation, or drinking-water supply and commercial fisheries. Secondary uses of drinking-water reservoirs in central Europe are for flood protection, power generation, and recreation (which is not necessarily water bound).

In dry climates, the supply of irrigation water is often the principal function of reservoirs. For example, the agricultural productivity of Egypt doubled after the completion of the Aswan Dam (Lake Nasser/Nubia). Fisheries in reservoirs are very important as they provide fish, which are a source of protein in warm regions. However, the drawdown of the water level may lead to the temporary loss of important habitats for propagation and shelter of certain fish species (unlike the situation in natural lakes). Drawdown also has an adverse effect on water-based recreation and the development of macrophytes.

In temperate and in warm, semi-humid climates, the primary function of many of the large dams is power generation. For example, more than 70% of the energy production in Brazil is provided by dams (Tundisi and Straškraba, 1999). Due to the construction of such dams, the residents are provided with a steady source of electricity. Many hydroelectric power stations in Europe and North America generate peak current, or compensate for variations in power consumption by means of pumped-storage systems.

Most reservoirs, including many drinking-water reservoirs, have a considerable flood-retention capacity. Nearly all reservoirs operate as hydraulic buffer systems and thus bring about, not only equalization of the flow, but also of the concentrations of dissolved materials, for example, of nitrate, in the case of water supply. Other water uses, such as power generation and navigation, are also dependent on a more equalized flow.

For several reservoirs, low-flow augmentation of rivers is a principal function. For this reason, maximum water-level fluctuations of 125 m and 140 m, respectively, are legislated for two reservoirs in Norway (Nilsson, 2009).

For the different uses of reservoirs, there are different standards required with regard to water quality. The highest demands have to be met in the case of a drinking-water supply.

Many reservoirs are situated in attractive, scenic, hilly areas and may thus become appealing for tourism and recreational uses such as sport fishing, bathing, diving, rowing, sailing, surfing, hiking, rafting, excursions on motorboats, and camping. Several reservoirs in Northwest Germany have an occupancy rate of four or more sailing boats and surfers per hectare (LAWA, 2001).

With regard to drinking-water supply, there are different views on the clearing of vegetation and soil beneath an envisaged reservoir. In several countries, clearing is ordered to remove sources of low redox potential, which might induce the release of dissolved manganese and iron from the sediment. In subtropical and tropical conditions, hypolimnion temperatures may exceed 15 or even 20 °C. At such a high level, microbial degradation processes are accelerated, which facilitate not only oxygen depletion, but also microbial sulfate reduction, with the accumulation of hydrogen sulfide in the hypolimnion. For this very reason, in the Brokopondo reservoir, Suriname, turbines were damaged by the microbial generation of sulfuric acid, when atmospheric oxygen entered the turbines.

In lakes having a high priority for fisheries, oxygen depletion and H_2S generation are unwanted, but the losses of dissolved nitrogen and phosphorus compounds by the hypolimnetic outflow are likewise undesirable. These substances are mainly released by the microbial breakdown of vegetation at the bottom of the water body and by leaching from the soil. This so-called 'trophic upsurge' is absolutely undesirable in drinking-water reservoirs, but in the case of reservoirs such as Lake Kariba (Zambia), fishermen were disappointed because the plankton and fish productivity stabilized at a very low (oligotrophic) level.

A reservoir normally operates as a biochemical reactor. In the longitudinal direction, the concentration of microbially degradable substances, introduced by the inflow, is reduced. The same applies to the reduction in the number of hygienically relevant microorganisms. In the high-temperature range, a water residence time of at least 20 days is considered to be appropriate in reducing the number of potentially pathogenic microorganisms by 99% (Thornton et al., 1996). In this regard, the residence time of the microbial cells in the surface layer exposed to a high intensity of UV radiation, and in some cases, grazing by zooplankton, seem to be among the most relevant mechanisms for pathogen elimination.

In former decades, reservoirs were used for the biological polishing of wastewater effluent, often in Europe and North

Figure 45 Retention of highly turbid inflowing water (from right) after a summer flood, by a submerged flexible curtain in the Saidenbach Reservoir (Germany). Courtesy of LTV, Saxony.

America. Nowadays, many reservoirs are still used as advanced wastewater-treatment plants in nearly all the other areas of the world (with exceptions such as Australia and Japan). Worldwide, reservoirs are used for the storage and quality amelioration of river water, as a source of raw water for drinking-water supply. This applies to, among others, the densely populated areas around the lower courses of large rivers such as the Rhine and the Meuse.

When river water with a very high content of dissolved phosphorus and nitrogen compounds is impounded, phytoplankton growth may become excessive; therefore, water quality in terms of particulate and dissolved organic carbon may substantially deteriorate. Thus, costly measures may become necessary to control phytoplankton growth. The situation is even worse in warm climates, where equipment for nutrient elimination from wastewater effluent is often lacking and where the growth of cyanobacteria is favored, which produces not only unsightly water blooms, but also hepatotoxic metabolites.

With further progress in wastewater treatment, reclaimed domestic effluent may even become one of the sources of raw water for drinking-water reservoirs in semiarid climates. There are also reservoirs that serve to improve navigation on rivers.

One of the unintended functions of multipurpose reservoirs is the retention of sediments. Sedimentation reduces the lifetime of the water body, sometimes to an intolerable level. An effective countermeasure, in addition to sediment-bypassing and sediment-flushing, is the construction of pre-reservoirs (Paul and Pütz, 2008), which may more easily be mechanically de-silted than the bottom of the main dam. The application of a floating underflow baffle, for the retention of turbidity currents near the surface, is a low-cost solution when compared with a concrete structure. Such a plastic curtain may, however, be operated in combination with an already-existing underwater wall (Paul *et al.*, 1998) (**Figure 45**).

2.08.2.2 Characteristic Differences between Natural Lakes and Reservoirs

Reservoirs are frequently regarded as man-made lakes. However, there are remarkable morphological and hydraulic differences between natural lakes and reservoirs that are primarily related to their origin, use, and management practices. Reservoirs are generally not older than 100 years and are created by damming a river, while natural lakes are usually much older and were formed by natural geologic processes. The ratio between length and width, the shoreline length and the shoreline development (see Section 2.08.1.3) of dendritic river-type reservoirs, as a rule, is much larger than that of lakes with comparable surface areas.

The retention time of reservoirs depends on their main usage. Although almost all reservoirs today are multifunctional, those that are primarily used for hydropower generation and/or flood control have mostly short residence times of a few days, up to some weeks. Drinking-water dams largely exhibit theoretical retention times of about 1 yr. The flushing of natural lakes, however, is much lower (retention times of many years, frequently even of several decades). The consequence of the larger shoreline development and the much shorter theoretical retention time of reservoirs is their stronger coupling to the drainage basins. Thus, reservoirs respond faster and are more sensitive to changes in the tributary water quality, caused by alterations of the catchment usage. Extended droughts or floods have stronger and more immediate impacts on reservoirs. The quality of the sediment is also affected. While in lakes the sediments are primarily of an autochthonous (internally produced) organic nature, the distribution of allochthonous (imported) mineral material is higher in reservoirs. Siltation as a consequence of erosion and, hence, faster aging, is a serious problem with reservoirs in catchment areas with small areas of forested regions. The import of suspended particles and nutrients in reservoirs can be reduced by pre-impoundments, located upstream of the tributaries' mouths (Paul, 2003; DWA, 2005; Paul and Pütz, 2008).

As a result of aging (concentration of sediments in the deepest parts of a lake basin), the majority of natural lakes exhibit U-shaped basins, unlike the comparatively young, predominantly V-shaped reservoirs in mountainous regions. The deepest areas of lakes are more or less in their centers, while those of reservoirs are, normally, not far from the dam walls (**Figure 46**; Stras'kraba and Gnauck, 1983). Hence, reservoirs are characterized by longitudinally differing morphometric structures:

- strongly flushed riverine, shallow and narrow regions below the tributaries' mouths;
- transitional region of moderate depth; and
- lesser drained lacustrine and deep basin near the dam.

This hydro-morphological structuring and the high flushing of reservoirs cause considerably stronger longitudinal differences in the physical, chemical, and biological characteristics of the water quality (**Figure 47**; UNEP, 2000) than in natural lakes, which are predominantly vertically structured (**Figure 33**). The shape of comparable shallow-bounded reservoirs situated in the lowlands and filled with water pumped from an adjacent river is more like a natural lake.

Thienemann (1913) emphasized that the principal difference between a man-made reservoir and a natural lake lies in the flow-through conditions. While a normal lake exhibits

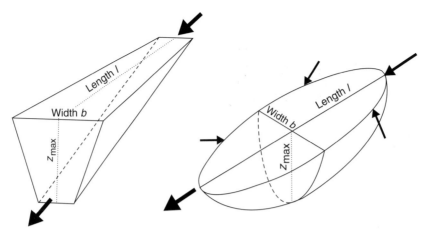

Figure 46 Diagrammatic comparison of principal basin-shape characteristics of reservoirs (left) and lakes (right). Modified from Straškraba M and Gnauck A (1983) *Aquatische Ökosysteme – Modellierung und Simulation*, 279pp. Fischer Jena.

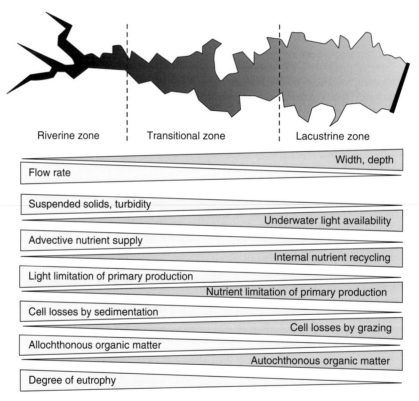

Figure 47 Diagram of the longitudinal zonation of the hydro-morphology and water quality in large reservoirs. From Kimmel and Groeger modified from UNEP (2000) Lakes and reservoirs – similarities, differences and importance. *IETC Short Report 1* http://www.unep.or.jp/ietc/Publications/Short_Series/LakeReservoirs-1 (accessed April 2010).

surface outflow, most of the water of a reservoir is withdrawn from deep-water layers and overflow seldom occurs. Consequently, Thienemann characterized three main points that distinguish lakes from reservoirs:

1. *The absence of a shallow shore-bank in reservoirs*, usually formed by long-term wave action at a more or less constant water level in natural lakes.

2. *The fast growth of vegetation on dry shores during periods of water level drawdown*. Thienemann considers the emerging vegetation as positive for the productivity of reservoirs. The organic substances produced during dry periods can be utilized by zoobenthos when the shores are flooded again.

3. *The influence of the deep-water release on the thermal structure of reservoirs*. Thienemann describes the farreaching consequences of the accelerated downward displacement of

the metalimnion, resulting from the permanent withdrawal of cold hypolimnion water during the summer stratification.

The main purpose of most reservoirs is the storage of water during seasons with high discharge and the release of water during dry periods. Thus, considerable seasonal ups and downs of the surface level are normal in reservoirs. A reduced fill level generally has potentially negative impacts on the water quality of reservoirs. All hydrographic parameters affecting the trophic state deteriorate: (1) average and maximum depth, retention time, and ratio between lake area A_0 and catchment area, decrease; (2) ratio between epilimnion volume V_{epi} and hypolimnion volume V_{hypo}, and nutrient load per unit surface area, increase. The impact of extreme nutrient-, suspended matter-, and microbial loads resulting from floods is much higher at reduced fill level, due to diminished dilution in the reservoir.

During the drawdown of the water level, not only fine substrate but also coarse sediments are successively washed into deeper regions of the basin and the slopes of the shores remain steep. Therefore, extended belts of macrophytes and the rich fauna of the warmed, highly productive littoral regions of lakes are almost completely missing and the pelagic-matter turnover obtains higher importance in reservoirs. Recent investigations (Kahl et al., 2008) have shown the strong influence of water-level fluctuations on the reproduction of fish species spawning in the shallow littoral.

With regard to drinking-water reservoirs, in particular, the resuspension of sediments during the drawdown of the water level, may cause substantial turbidity that complicates the raw water treatment in waterworks. Conversely, the terrestrial vegetation growing quickly on the dry shores reduces sediment resuspension when the fill level rises again. However, the flooded terrestrial plants have mostly negative effects on the water quality. Their microbial degradation may cause oxygen depletion, remobilization of nutrients, and the development of substances producing odor and flavor (Scharf, 2002). Furthermore, they represent an excellent habitat for planktivorous fish feeding on zooplankton.

As with lakes, the development of the thermocline in reservoirs at the beginning of the summer, depends primarily on the morphological and meteorological conditions (above all, wind exposure) and occurs annually in a relatively narrow time span. At that time, a high level of filling of a reservoir guarantees the formation of a large hypolimnion volume with high oxygen content. The lower the ratio V_{epi}/V_{hypo} at the onset of the summer stagnation, the lower the risk of critical oxygen deficits in the deep-water layers, at the end of the summer.

During the summer stratification, the thermal structure of the epilimnion, both in lakes and reservoirs, is primarily meteorologically determined. The depth propagation of the metalimnion in lakes is controlled by heat transport and mixing processes. In deep reservoirs with stable stratification, however, the increase of the ratio V_{epi}/V_{hypo} and the temperature distribution in the hypolimnion depends mainly on the balance of water inflow and outflow. The inlet temperatures are in the range of metalimnetic or even epilimnetic temperatures in summer. Thus, inflow from the inlets into the hypolimnion can usually be excluded. Therefore, the downward movement of the threshold between meta- and hypolimnion, characterized in stagnant water bodies of the temperate zone by the depth of the 10 °C contour line can be estimated with sufficient accuracy only from the reduction of the hypolimnion volume caused by the deep-water release (**Figure 48**). The depth of a reservoir's epilimnion (i.e., mixing depth z_{mix}), increases much faster than in a natural lake, which in turn affects the summer phytoplankton species composition and abundance. The significance of the underwater light intensity as a potential growth-limiting factor may be higher in dams during the summer stagnation. Phytoplankton circulates vertically through the deeper epilimnion, as it receives lower average light intensity. The enlargement of z_{mix} is also associated with resuspension and transport of sediments into the deeper parts of the dam, due to the higher turbulence in the epilimnion.

The hypolimnetic water temperatures in summer and, thus, the temperatures of the entire water body at the beginning of the autumn circulation, are higher in reservoirs than in natural lakes of corresponding depth. If the quantity of water withdrawn from a reservoir's hypolimnion during the summer exceeds its volume at the beginning of the summer stagnation, the autumn full circulation is initiated much earlier than in a lake of comparable size and depth (as was the case in the Saidenbach reservoir in 1987, see **Figure 48**). The hypolimnetic warming and the continuous increase of the ratio V_{epi}/V_{hypo} are critical, because the higher specific loading of the smaller hypolimnion with suspended organic material settling down from the epilimnion, and the higher temperatures, accelerate the oxygen consumption in the bottom waters. Conversely, the temperature of the deep-water layers in winter is lower in reservoirs than in lakes.

The facts described above seem to suggest that the release of water from the hypolimnion of reservoirs is generally negative. However, other factors should be considered. First, there is no alternative, because economic drinking-water production from reservoirs during the summer is only possible from the cold and comparatively clear hypolimnetic water. Second, partial renewal of the deepest water layers is highly recommendable. The export of oxygen-free water enriched with nutrients, dissolved manganese, and other substances, remobilized from the sediment, has an oligotrophication effect and cause a continuous oxygenation of the deep water, due to the downward displacement of water from above with higher oxygen concentrations (**Figure 49**).

Therefore, this principle is utilized in the artificial deep-water withdrawal from lakes as a restoration measure (Section 2.08.3.3).

Lakes and reservoirs influence the downstream river in quite different ways. Except in lakes of the arid or semiarid zones, the hydrological coupling between the tributaries and the downstream river is, although damped, intact. Reservoirs, however, noticeably isolate the upper river reaches from the lower ones. The seasonal discharge variation of the downstream river is totally different from that of the tributaries. The often-missing, or changed, temporal sequence of floods and droughts completely alters the processes of river-bed formation and ecological structuring below a reservoir. During summer, the effluent of a lake is warm and exhibits high

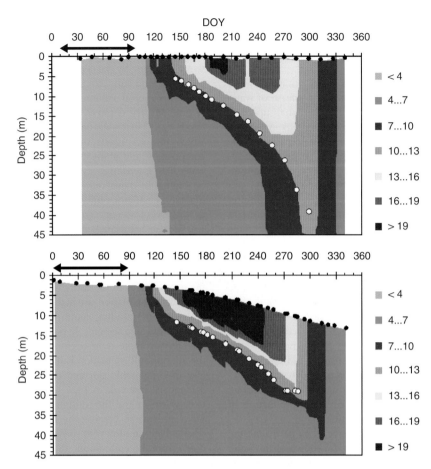

Figure 48 Development of the water temperature (ranges given in °C) of the Saidenbach Reservoir vs. depth and time (DOY – day of the year), over 2 years with different hypolimnetic raw water release (1987: ~109 000 m³ d⁻¹, 2003: ~62 500 m³ d⁻¹). 1987 was a wet year and 2003, an extremely dry year. The black dots mark the sampling dates and the surface-level development, white dots the course of the 10 °C contour line calculated from only the volume balance of the hypolimnion, and the double-arrows the duration of ice covering.

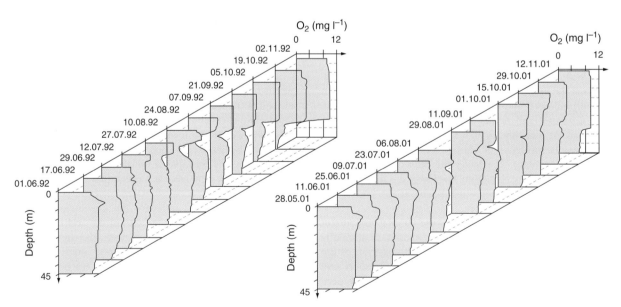

Figure 49 Influence of different raw water-intake levels on the development of the oxygen concentration above the bottom of the Saidenbach reservoir. The raw water was released from an intake at a depth of 30 m in 1992 and thus, anaerobic conditions resulted in the deep layers of the reservoir at the end of the summer stagnation (left). The oxygen concentration above the bottom remained higher than 5 mg l⁻¹ until the end of the summer stratification in 2001, when the raw water was withdrawn via the bottom outlet of the reservoir (right).

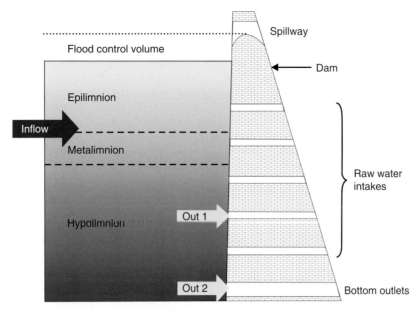

Figure 50 Characteristic vertical segmentation of a drinking-water reservoir's volume during the summer stagnation and assembly of outlet structures for the regulation of outflows. Out 1: raw water to the water-treatment plant, Out 2: discharge into the downstream river.

concentrations of oxygen, but is low in dissolved nutrients. Mineral turbidity is low as well, but the export of phytoplankton may be high. The water released from the hypolimnion of a dam, however, is cold, contains almost no particles, and may have low oxygen, but high dissolved nutrient, iron, and manganese concentrations. Even substances potentially harmful to organisms, such as nitrite, ammonia, and redox-sensitive dissolved heavy metals (e.g., arsenic) may be released. In winter, the outflow of a dam is warmer, compared with a lake. Thus, the structures of the biotic community beneath a lake or reservoir may be quite different. Finally and importantly, most reservoirs are impassable barriers for ascending organisms and may cause genetic degradation in upstream rivers.

As mentioned earlier, the discharge management of reservoirs, including the proportioning of the amounts of water released through outlets in different depths at a certain time of the year, has a strong impact on the materials turnover and water quality in the reservoir. This is particularly important in drinking-water reservoirs that are also used for flood control (**Figure 50**). In general, raw water supply for drinking-water production is optimal if the reservoir is entirely filled and allowed to overflow in cases of strong inflows. Under these conditions, the largest possible hypolimnion volume is formed at the beginning of the summer stagnation and its decrease is low during summer. However, flood control works best if the flood-control storage is large and kept permanently empty, except in cases of disastrous, uncontrollable discharge, which is therefore in direct conflict with the drinking-water supply. Flood control implies that overflow is avoided and, consequently, the water inflow has to be released through outlets below the surface, after the water surface has reached the full supply level (lower boundary of the flood control storage). In many (especially older) dams, the raw water intakes cannot be used for discharge into the downstream river or their capacity is too small. Thus, there is no alternative for the release of discharge beyond the demand for raw water through the bottom outlets (see **Figure 50**). Consequently, turbid, nutrient-enriched, warm inflowing water is stored in the reservoir and valuable clear, cold hypolimnetic water has to be released into the underlying stream. Hence, the risk of serious problems with raw water treatment on the basis of a too-small or prematurely depleted hypolimnion with all the consequences described above is high in rainy summers, especially in a reservoir with a theoretical retention time of much less than a year. It is, in a certain way, paradoxical that not dry, but wet summers, are especially critical for drinking-water reservoirs.

2.08.2.3 Environmental Impacts of Reservoirs

A reservoir imposes serious impacts on the natural environment. Even more adverse is the cumulative effect of multiple dams built along the length of a river.

These impacts concern not only the downstream reaches of the flowing water, but also the sites which once were fluviatile or terrestrial, inclusive of the biota. Due to the construction of reservoirs, large areas of wetland, with a high biotic diversity, have been lost.

In a temperate climate region, a storage reservoir may reverse the hydrological conditions during summer: early lowwater and a later flood.

Dams impede the migration of river fish species into the upstream or the downstream reaches. From this ecological fragmentation (Nilsson, 2009) of the river continuum, many species have lost their propagation sites and have become extinct. In the Columbia River (Washington State), which has been blocked by a series of dams, migrating salmon populations, along its total length, have dropped by as much as 95%. Along the Yangtse River, many species of fish have become extinct, due to the construction of the Three Gorges Dam, which otherwise protects the lives of hundreds of thousands of people in the densely populated downstream plains of the river.

To ensure public water supplies in semiarid climates, dams are, in many instances, the only alternative.

In many cases, the economic benefits of dams are so high that all environmental impacts have been neglected. Turbines, which produce more than 10% of the electric energy in the USA, and even more in Brazil, cause very high mortality in river fish.

There are many cases in which the environmental impacts of dam construction are higher than the economic benefits. For this reason, the operation licenses for several dams in North America have not been renewed.

The number of people who have been forced to migrate because of the construction of dams is estimated to be 40–80 million (Nilsson, 2009).

Among the most important impacts of reservoirs in warm climates is the excessive growth of submerged or floating-leaved vegetation. This, in turn, provides a favorable habitat for water snails, which host trematode worms, causing Schistosomiasis (bilharzia), a very serious disease of the inner organs (bladder and liver). The larval vectors may rapidly penetrate the human skin after a brief contact with the reservoir water. About 200 million people are affected worldwide.

In sheltered bays, the floating-leaved vegetation hosts myriads of larvae of biting midges.

Schistosomiasis and other waterborne diseases, such as malaria, substantially reduce the capability of a person to do physical work. Thus, the construction of reservoirs in warm climates has, in many cases, had dramatic socioeconomical impacts. One of many examples is the area surrounding the Volta reservoir, Ghana (Thornton et al., 1996).

Intensive public discussion on the impacts of very large dams underpinned the foundation of the World Commission on Dams in 1997. This institution, mandated by the World Conservation Union and the World Bank, evaluates existing reservoirs, develops guidelines for the design, construction, and sustainable operation of dams, and even requests a potential decommissioning, if deemed necessary.

2.08.3 Management, Protection, and Rehabilitation of Lakes and Reservoirs

2.08.3.1 Main Water-Quality Problems

Lakes and reservoirs are exposed to numerous natural and anthropogenic stress factors. The impact of human activities on the aquatic environment has increased during the past centuries, due to an exponentially increasing population and industrialization. Intensified agriculture, increasing industrial and sewage discharge, and the direct use of water resources, for example, for drinking-water supply, energy production, or fisheries, has resulted in the degradation of many aquatic ecosystems and a dramatic loss of biodiversity. Multiple pressures are caused mainly by eutrophication, acidification, salinization, and contamination by hazardous substances, which are all accompanied by human-induced climate change and by the invasion of neobiota (Hupfer and Kleeberg, 2007). Eutrophication is the most common problem in lakes and reservoirs. It is defined as the increasing intensity of primary production (trophy) due to enhanced availability and uptake of nutrients. About 30–40% of lakes and reservoirs worldwide are affected by unnaturally high nutrient concentrations. The current changes in the trophic state are termed cultural eutrophication, which can clearly be separated from the natural eutrophication that occurs during the aging of a lake over thousands of years. Obvious indications of eutrophication are high turbidity caused by algal blooms, dense macrophyte growths, food-web changes, mass development of toxin-producing cyanobacteria (blue-green algae), reduced species diversity, oxygen depletion, enrichment of reductive substances (e.g., toxic hydrogen sulfide), fish kills, and odors. Consequently, eutrophication has a strong influence on anthropogenic water uses such as drinking-water supply, fisheries, and recreation. Eutrophication is one of the major global environmental problems, because of its importance in health and food production in densely populated areas (Smith, 2003). In most lakes and reservoirs, phosphorus (P) is the minimum factor controlling the trophic state. The P input originates from point sources (e.g., wastewater-treatment plants and industrial wastewater) or nonpoint sources (e.g., erosion, atmospheric deposition, surface runoff, and groundwater). Recent studies suggest that nitrogen (N) is apparently of greater importance as a limiting nutrient, than previously assumed (Sterner, 2008). Acidification appears, after eutrophication, as the most important threat and stress factor for lakes and reservoirs. Acidification is the additional input of acids into water, decreasing the pH value and usually eliminating the carbonate buffer system. The decrease in pH may cause an extreme loss of biodiversity, as many invertebrates, fish, and other vertebrates cannot survive or reproduce in acidic environments. Biota is also influenced by indirect consequences of acidification, such as the increased release of potentially toxic metal ions (in particular aluminum, copper, cadmium, zinc, and lead) from soils and sediments. Acid deposition has changed the natural water chemistry and, thus, the biological structure in 50 000–100 000 lakes and watercourses in Europe and North America. This adverse situation prohibits the use of the impacted water for irrigation, fishery, and aquaculture as well as its use as drinking water. Many lakes and reservoirs are atmospherically acidified due to emissions of sulfur and nitrogen oxides mainly from the burning of fossil fuels or from agriculture, as well as from natural sources (e.g., volcanic eruptions and emissions from soils and wetlands due to oxidation of sulfur-containing minerals). Declining groundwater levels, enhanced nitrate concentrations in the groundwater, the artificial drainage of wetlands, or long-lasting droughts in soils due to global warming and mining activities can lead to the oxidation of reduced sulfur minerals (pyrite and marcasite), whereby the acid input into surface waters is geogenically increased (see Section 2.08.1.6). Salinization also leads to marked changes in biotic communities, since freshwater organisms usually have only a limited tolerance for enhanced Cl^- concentrations. Salinization is caused by changes in the hydrological regime, which may be due to enhanced evaporation or discharge of salt-rich water from mining, oil production, and agriculture (irrigation). Hazardous substances comprise a variety of organic and inorganic substances, such as toxic metals, pesticides, organic surfactants, pharmaceuticals, and mineral oils. Metals and organic compounds, for example, polychlorinated biphenyls (PCBs), are accumulated in sediments or in food chains and can build up toxic concentrations.

Finally, aquatic ecosystems are very vulnerable to climate change. Climate influences the physical, chemical, and biological structure of temperate lakes and reservoirs due to fluctuating water levels, increase of water temperature, shorter ice-cover periods, invasion of nonindigenous animal and plant species from tropical and subtropical regions, and structural changes in catchment areas (e.g., soil erosion and land use), including boundaries between terrestrial and aquatic systems. Most scenarios show that the risk of eutrophication problems, such as critical oxygen conditions and mass development of toxic blue–green algae, will increase due to a warmer climate.

2.08.3.2 General Management Strategies

The different kinds and causes of water-quality problems have resulted in the development of numerous strategies to restore the functioning of degraded aquatic ecosystems. Surface waters are closely connected to their terrestrial and atmospheric environment and therefore, preventive protection of lakes and reservoirs begins in the catchment areas. External measures are aimed at the reduction of pollutant sources in the catchment area as primary reason for water-quality problems. These inputs become restricted when the terrestrial matter cycles are mostly closed. These measures include the construction of sewage treatment plants for the purification of municipal and industrial wastewater, extensification measures in agriculture, the recycling of industrial waste, and the establishment of buffer systems such as ponds, pre-dams, or constructed wetlands. Critical load models are helpful for achieving qualitative aims of aquatic ecosystem management (e.g., Vollenweider, 1976). In this way, it is possible to assess a necessary limit of nutrients and harmful substances not only as concentrations at the location of emissions, but also with respect to their effects, depending on the characteristics of the lakes or reservoirs. This implies that potential control strategies may include the optimization of the structure and processes within the waters, so that negative symptoms of excessive nutrient loadings can be minimized (Benndorf, 2008). Ecological engineering, also called ecotechnology, involves several ecological approaches for optimizing the ecosystem structure or to support specific ecosystem functions in a way such that the management targets are being maximally supported. Integrative concepts of lake and reservoir management consider the control of multiple external factors as well as internal measures.

As eutrophication is still the major water-quality problem with far-reaching ecological and economical consequences, it is not surprising that a broad spectrum of lake and reservoir rehabilitation/restoration methods has been developed to combat eutrophication and its symptoms. The following section describes management options in more detail using eutrophied lakes as an example.

2.08.3.3 Measures for Eutrophication Control

Measures for eutrophication control are focused on decreasing phosphorus availability in water and minimizing the symptoms of eutrophication. Depending on the location, water managers can select both external and internal measures.

2.08.3.3.1 External measures

The reduction of excessive nutrient loading as the cause of eutrophication is the key to sustainable effects. Phosphorus load can be reduced by (1) decreasing P emissions, (2) increasing the P-retention capacity in the catchment, or (3) purification of inlet water immediately before it enters the lake. Decreasing P emissions includes measures such as building treatment plants for municipal and industrial effluent, extensifying agricultural land use (e.g., reducing fertilizers and lowering stocking densities), and using phosphate-free detergents. Diversion of wastewater outside the watershed has been often used when the reduction of nutrients in wastewater-treatment systems is not sufficient. Alternatively, the nutrient reduction can be achieved by so-called ring canalization, collecting sewage and storm water for treatment in a central plant downstream of the protected lake. Thereby, the storm water with its high P load is kept away from the lake in the case of combined sewage systems. Land-management procedures, generally known as 'best management practice', are the primary methods for protecting surface waters from nonpoint sources. Increasing P-retention capacity in the landscape is a strategy that takes advantage of the ability of structural landscape elements to retain P by reestablishing effective former sinks (e.g., re-wetting of fens) or by constructing equivalent systems (constructed wetlands). Purification of inlet water includes buffer systems between highly productive agricultural areas and the water, such as pre-dams, macrophyte belts, and ponds. A P-elimination plant (PEP) at the main inflow of a lake is the costlier technical alternative.

Despite external load reduction, lakes and reservoirs have often not shown the expected improvement of water quality in an acceptable time. The resistance is explained by the following main delay mechanisms (Hupfer and Hilt, 2008):

1. Long hydraulic water-retention time prevents a fast response to decreased P loading.
2. The biological response of a lake to changes in the nutrient level is nonlinear. One example is the top-down control of phytoplankton growth by fish. High abundances of planktivorous fish established during the eutrophic period prevent both the appearance of large herbivorous zooplankton and thus a higher grazing pressure on phytoplankton (**Figure 51**; Hupfer and Hilt, 2008). A second example is the occurrence of two alternative stable states in shallow lakes (Scheffer et al., 1993). As a consequence of top-down and bottom-up processes, the threshold P concentration for a shift from clear (macrophyte-dominated) to turbid (plankton-dominated) state (eutrophication) is higher than the P concentration necessary to shift the system backward from turbid to clear-water stage (re-oligotrophication).
3. The P concentrations remain high due to release of P from the pool accumulated in the sediment during high external P loading. Examples in the literature show that this process can continue over a time span of more than 10 years before the surplus P pool is released or permanently buried

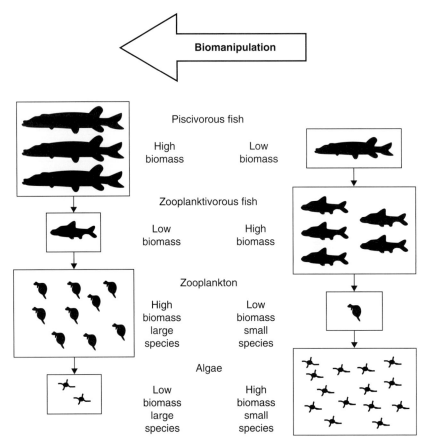

Figure 51 Schematic view of top-down control of phytoplankton abundance in eutrophic lake. Effect of manipulation of zooplanktivorous fish biomass (left), compared with an unmanipulated food web. From Hupfer M and Hilt S (2008) Lake restoration. In: Jorgensen SE and Fath BD (eds.) *Encyclopedia of Ecology*, pp. 2080–2093. Oxford: Elsevier.

(Søndergaard et al., 2005). If the problem of eutrophication is not solved by reducing external nutrient input alone, additional internal measures can shorten the adaptation time to reach the desired water quality. This can support the regime shift according to management targets. Additionally, some internal measures are able to compensate for a too-high residual external loading.

2.08.3.3.2 Internal measures

Physical, chemical, and biological measures aim at nutrient control or at changes of biological structure to reinforce the recovery. **Table 3** summarizes the main internal methods for controlling excessive phytoplankton growth due to eutrophication, as well as for reducing other undesired symptoms of a high trophic state.

The lake P concentration can be influenced by increasing P export and by increased P retention in the sediment. The addition of chemicals is aimed at supplying new sorption sites for phosphate, leading to P removal from the water and a subsequent sedimentation to the bottom. High doses of chemicals not only remove P from the water, but also increase the P-binding capacity in the sediment, so that P release from sediments is decreased for a longer period. For inactivation, salts of aluminum, iron, and calcium have been widely used. A newly developed product is Phoslock® – a bentonite (95%) artificially enriched with lanthanum (5%). Depending on the lake size and chemicals, the treatment is realized by piping, by distribution on the ice cover, by airplane or boats, and by aeration devices. In-lake P inactivation was successfully performed in many stratified and nonstratified lakes in North America and Europe, but only short-term effects were observed in cases of continued external loading that quickly replaced the eliminated P.

Hypolimnetic withdrawal increases the P export, since nutrient-rich hypolimnetic water instead of P-poor epilimnetic water is removed from the lake or reservoir. Coincidentally, the retention time in the hypolimnion is shortened and the risk for enrichment of reduced substances is reduced. Typically, a withdrawal pipe is installed near the deepest point of the lake with an outlet below the water level so that it acts as a siphon. In reservoirs, hypolimnetic withdrawal can be achieved with a variable deep-water outlet in the dam. To prevent hydrological imbalances, the hypolimnetic withdrawal of P-rich water can be combined with an external P-elimination plant or constructed wetland with returning of the water to the lake or reservoir after treatment. The P export is also increased by flushing of the lake water with water low in nutrients, or by destratification (increase of P at water surface and in outflow).

Permanent or intermittent destratification destroys or prevents density gradients within the water column, thus

Table 3 Physical, chemical, and biological measures for decreasing the primary production and for abatement of negative symptoms of eutrophication in lakes and reservoirs

	Principle	Main control variable	Further effects
Chemical methods			
Al-/Fe-salts	In-lake P inactivation by metal hydroxides, addition to water surface or into the hypolimnion	$P_{Lake}\downarrow$	Mobile sedimentary $P\downarrow$
Phoslock®	Addition of lanthanum containing bentonite	$P_{Lake}\downarrow$	Mobile sedimentary $P\downarrow$
Nitrate addition	Application of nitrate into the deep water or sediment	Redox potential $\uparrow$, stabilizing of iron-bound phosphorus at the sediment surface	Mineralization of organic matter $\uparrow$ Reduced substances $\downarrow$ P release $\downarrow$
External P elimination	Treatment of P-rich (hypolimnetic) water by Al or Fe salts outside the lake	$P_{Lake}\downarrow$	Hypolimnetic oxygen $\uparrow$ Reduced substances $\downarrow$
Physical methods			
Aeration/oxygenation	Introducing of air or molecular oxygen into the deep water	Hypolimnetic oxygen $\uparrow$	Reduced substances $\downarrow$ P release $\downarrow$
Hypolimnetic withdrawal	Natural discharge of P-poor surface waters is (partly) replaced by P-rich deep water	P export $\uparrow$	Hypolimnetic oxygen $\uparrow$ Reduced substances $\downarrow$
Destratification	Temporary or intermittent increasing of mixing layer by aerators, jet-stream pumps, or by inducing a thermal convection flux	Light for algae growth $\downarrow$	P export $\uparrow$ Zooplankton $\uparrow$ CO_2 inhibition of blue–green algae
Sediment dredging	Partial or complete removal of surface sediments	Water depth $\uparrow$	Reductive potential $\downarrow$ P release $\downarrow$ Toxic substances $\downarrow$
Dilution	Transfer of P-poor water from outside the catchment	Water residence time $\downarrow$	P export $\uparrow$
Biological methods			
Food-web manipulation	Increasing of grazing pressure on phytoplankton by (1) removal of zooplanktivorous and benthivorous fish or (2) protection or stocking of piscivorous fishes	Algae biomass $\downarrow$	$P_{Lake}\downarrow$
Macrophyte removal	Mechanical harvesting methods, water level drawdown, sediment covers, and surface shading	Macrophytes $\downarrow$	Phytoplankton $\uparrow$ Prevention of siltation
Macrophyte transplantation	Planting or seeding of submerged plants	Macrophytes $\uparrow$	Zooplankton $\uparrow$ Allelopathic depression of phytoplankton Sedimentation $\uparrow$ Sediment resuspension $\uparrow$

increasing the oxygen supply of deeper water layers. Increasing of mixing layers can lead to light-limitation of algae growth. Additionally, cyanobacteria are outcompeted by green algae or diatoms when the carbon dioxide concentration in the euphotic zone is increased (Shapiro, 1984). The destratification is often realized by introduction of compressed air, jet-stream pumps, or by inducing a convection flux by introducing warmer surface water into the hypolimnion.

Aeration and oxygenation (introduction of oxygen as air or liquid oxygen) are applied to prevent acute oxygen depletion. Thus, the enrichment of dissolved iron, manganese, ammonium, hydrogen sulfide, and free carbonic acid in the hypolimnetic water is prevented, and internal P loading may be reduced. Additionally, oxygen availability improves the conditions for cold-water fish and invertebrates in the hypolimnion. Analogously to molecular oxygen, nitrate is used for *in situ* oxidation of reduced substances and of biodegradable organic matter. The advantage of nitrate instead of oxygen is that much higher oxidation equivalents can be added, because nitrate is more soluble than molecular oxygen. Thus, nitrate penetrates deeper into the sediment than oxygen. On the other hand, oxygen is introduced from the atmosphere during mixing, whereas nitrate must be applied repeatedly. Nitrate is added as solution of calcium nitrate directly into the deep water or by injection into the sediment. In some cases, nitrate is added in granular form. Nitrate and oxygen enhance P retention in sediments only under special geochemical circumstances (mobility or surplus of iron). In many cases, oxidation has not shown expected results (Liboriussen et al., 2009).

The dredging of sediments is the partial or complete removal of sediment layers rich in nutrients and organic matter. Dredging aims at (1) deepening of lakes, (2) elimination of toxic substances, and (3) reducing the P release rate. Additionally, dredging serves to sustain several technical functions of lakes and reservoirs (pre-dams, flood protection, and shipping lanes). In some cases, dredging did not successfully lower eutrophication by P control because the mobile P pool (temporary P) in the sediment was small, or the internal P cycle was controlled by the newly settled P (Annadotter et al., 1999). Similar reasons can minimize the effects of sediment capping for eutrophication control (Hupfer et al., 2000). During this measure, an artificial barrier is inserted between sediment and water. The barrier minimizes the transport of nutrients and other harmful substances from the sediment (or groundwater) into the lake. The material (e.g., foil of polyethylene, clay, calcite, sand, and zeolites) for capping can act as physical or chemical barriers. Another use of sediment capping is the prevention of excessive growth of rooted macrophytes.

Biomanipulation influences the biological structure within a lake/reservoir to improve the water quality. The main applications for lake restoration include (1) food-web manipulation and (2) macrophyte biomass control. Food-web manipulations in lakes are man-made alterations of the lake biota and their habitats to facilitate reduction of algal (phytoplankton) biomass. In most cases, food-web manipulation refers to the reduction of zooplanktivorous fish-biomass that leads to a reduction of phytoplankton (**Figure 51**) and clearer water. Food webs are regulated either by resources (bottom-up) or by predation (top-down). A reduction of the biomass of zooplanktivorous and benthivorous fish can be achieved by stocking with piscivorous fish, such as pike (*Esox lucius* L.), pike-perch (*Sander lucioperca* L.), or perch (*Perca fluviatilis* L.). A direct reduction can also be achieved by poisoning, removal by conventional fishery techniques, or a temporary drainage of the lake. Stocking of piscivorous fish has often been less successful than fish removal. A marked reduction of the biomass of zooplanktivorous fish such as roach (*Rutilus rutilus* L.) is often followed by an increase in the abundance and size of zooplankton (predominantly *Daphnia* species). This increases the grazing pressure on phytoplankton and potentially leads to the top-down control of phytoplankton biomass, in which case, the water becomes clear and extreme values of oxygen and pH are avoided. The success of food-web manipulation may also be triggered by bottom-up forces. Benthivorous fish, such as bream (*Abramis brama* L.) or common carp (*Cyprinus carpio* L.), exert bottom-up effects on water quality as they increase sediment resuspension, water turbidity, and internal nutrient loading. The removal of benthivorous fish may therefore also strongly determine the success of a food-web manipulation. Top-down control of phytoplankton biomass was found to occur in shallow lakes and in deep lakes of slightly eutrophic or mesotrophic state. It is unlikely to be effective in eutrophic or hypertrophic deep lakes. The substantially higher success rates of food-web manipulations in shallow lakes can be attributed to positive feedback mechanisms, triggered by the recovery of submerged macrophytes. Experience has shown that lake water quality can only be improved by food-web manipulation if annual loading is lower than 0.5–1.0 g of total P per square meter of lake surface (see Benndorf, 2008), or if the in-lake P concentration is lower than $50 \mu g \, l^{-1}$ in shallow lakes (see Jeppesen et al., 2009).

Macrophyte biomass control includes measures to restore aquatic plant communities in order to take advantage of the beneficial aspects of plants in lakes, as well as measures to control excessive growth that results in conflict with certain lake uses, or to eradicate nonindigenous species. Submerged macrophytes are of crucial importance in shallow lakes, due to the vegetation-turbidity feedback. They stabilize the clear, vegetation-dominated state due to the reduction of nutrient and light availability to phytoplankton, enhancement of top-down control of algae by providing refuge for zooplankton, suppression of phytoplankton by the excretion of allelopathic substances, facilitation of phytoplankton sedimentation, and prevention of sediment resuspension. In shallow lakes, the successful establishment of submerged macrophytes is therefore a prerequisite for the long-term success of other rehabilitation measures such as food-web manipulations. Submerged vegetation will develop naturally in most cases when light and sediment conditions in the lake are favorable, for example, after the application of other internal measures to reduce phytoplankton development. Artificial support by planting or seeding of submerged plants might be useful if viable propagules are lacking in the sediment and no macrophyte stands are present in the vicinity of the lake, if a rehabilitation method applied only decreased turbidity for a period too short for natural re-colonization, if submerged macrophytes are immediately needed for the successful development of introduced pike, or if the promotion of specific (e.g., low growing) macrophyte species in particular areas of the lake is required to enable recreational use (Hilt et al., 2006).

Excessive macrophyte growth can be a result of eutrophication, or of increasing water transparency after the application of rehabilitation measures, as well as of invasion of neophytes. Measures against macrophyte growth are only needed when macrophytes hinder certain lake uses (e.g., recreation and navigation). Methods to control or eradicate aquatic macrophytes include water-level drawdown for a period sufficient to kill the plants and their reproductive structures, mechanical harvesting methods, sediment covering and surface shading, aquatic herbicides, and biological controls, such as phytophagous insects, plant-feeding fish, and plant pathogens. In shallow lakes, macrophyte-control measures should be applied with caution, due to the risk of a return to the turbid, phytoplankton-dominated state.

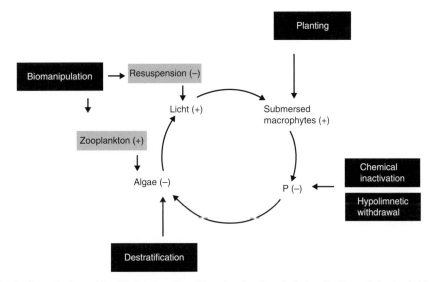

Figure 52 Positive feedback mechanisms (simplified) induced by different restoration strategies. Positive spiral extended from Hansson LA and Brönmark C (2009) Biomanipulation of aquatic ecosystems. In: Likens G (ed.) *Encyclopedia of Inland Waters*, pp. 242–248. Elsevier.

Figure 52 (Hansson and Brönmark, 2009) provides an overview of processes triggered by biomanipulation and other internal measures, which can lead to self-stabilization of the system. The positive spiral of reduced algal biomass, improved light conditions, macrophyte growth, and P lowering can be enforced by technical measures at several starting points.

Experiences with long-term efficiency of internal measures to control eutrophication are summarized in several publications (e.g., Cooke *et al.*, 2005; Søndergaard *et al.*, 2007; Gulati *et al.*, 2008).

2.08.4 Current Knowledge Gaps and Future Research Needs

Worldwide, lakes and reservoirs are of tremendous importance for diverse human usage. One of the urgent global problems is the supply of freshwater of adequate quality, for use as drinking water and for food production. The demands and impacts on freshwater ecosystems not only influence the human population directly, but also increasingly threaten their natural environment, to varying spatial scales. For example, exported pollutants degrade coastal waters (Behrendt and Dannowski, 2005; Feistel *et al.*, 2008) and modifications of freshwater ecosystems may even alter the carbon cycle on a global scale (Tranvik *et al.*, 2009). Lakes and reservoirs are also impacted upon by far-off matter transport via atmosphere, surface water, and groundwater. Long-term changes in climate are expected to exacerbate usage problems and modify the linkages between terrestrial environment and freshwater systems, in a complex way. The impacts on freshwater ecosystems have drastically decreased their biodiversity (Dudgeon *et al.*, 2006; Weijters *et al.*, 2009), although the role of biodiversity for resilience and functioning of lakes and reservoirs and for adjacent systems is not well known. The complexity of mechanisms determining the response, cannot be adequately predicted with the existent understanding of processes and the available modeling tools. Quantification of changes and better understanding of cause–effect relationships are necessary for new concepts of protection and adaptive management. The challenges for scientists in the field of lake and reservoir research are manifold and enormous. Therefore, the following sections are restricted to selected topics, which are under intense discussion.

2.08.4.1 Lakes and Reservoirs as Constituents of Their Catchment Areas

Lakes and reservoirs are linked with terrestrial environments and connect different types of surface waters in complex landscapes. Surface waters are depressions in watersheds, integrating different influences from the surrounding area. The accumulation of organic matter and other substances in lakes and reservoirs can intensify the elemental cycles. Therefore, lakes and reservoirs can be considered as hot spots for biogeochemical processes, with disproportionately large effects on mass flows in the landscape. On a smaller scale, boundary layers, for example, interfaces between groundwater and surface water, the sediment and water interface, and interfaces between separate water bodies, are places with high turnover rates. Ecological boundaries are characterized by steep gradients, which stimulate intensive reactions that determine their sink or source function (Cadenasso *et al.*, 2003). The biogeochemical reaction rates and their temporal dynamics are influenced mainly by the hydrological regimes and transport processes. The understanding of these processes is essential to estimate regional or global budgets of nutrients, carbon, and greenhouse gases namely CO_2, CH_4, and N_2O (Tranvik *et al.*, 2009). The emission of these greenhouse gases is considered the most important impact of global warming. It is not yet possible to rate the contribution of lakes and reservoirs to the global greenhouse gas balance. It is assumed that particularly CH_4 and possibly N_2O emissions from inland waters, primarily in the tropics, might have a significant impact. Furthermore, increasing temperatures may stimulate greenhouse-gas production, not only in wetlands, but also in lakes and reservoirs. The greenhouse-gas production could also be fostered by an enhanced input of carbon. Observations of increasing concentrations of dissolved humic acids have recently been reported from many regions worldwide. The reasons for this phenomenon are not yet definitely established. It is probably a consequence of an intensified microbial turnover of organic soil substances, due to higher temperatures and increasing soil pH, resulting from rapidly declining anthropogenic acidification in large regions in Europe and North America. From investigations into the drainage basins of drinking-water reservoirs of the German federal state, Saxony, it was concluded that the re-wetting of formerly drained, swampy, coniferous forest sites (revitalization of fens) may have considerably enforced the leaching of dissolved humic substances and caused higher costs for water treatment (Sudbrack et al., 2005). On the other hand, the re-wetting of fens as management options to use the natural P retention in the catchment, is of great importance to decrease diffusive P loading into surface waters.

Research in the following decades should be aimed at (1) the determination of precise rates of carbon and nutrient cycling and storage in lakes and reservoirs, and to quantify the regional and global budget of nutrients, carbon, and greenhouse gases and (2) the quantification of substance fluxes in the watershed to the lake/reservoir due to land-use changes and climate variations. These approaches need (1) a better implementation of single studies to a global level, (2) interdisciplinary science and effective collaboration among aquatic and terrestrial scientists, (3) more process studies concerning the interplay between physical, biological, and chemical processes in highly reactive boundary layers, and (4) the development of models integrating landscape, climate change, and aquatic systems stimulating the generation of hypotheses and new empirical studies.

2.08.4.2 Responses of Lakes and Reservoirs to Climate Change

Worldwide temperature increases of 1–5 K within the next 50, years caused by climate change, have been predicted (IPCC, 2007). Responses of lakes and reservoirs to climate change are estimated to include (1) warming of waters with impacts on duration of ice covering, on timing and intensity of mixing and stratification, on the gas exchange between atmosphere and water as well as the solubility of gases in water (e.g., oxygen saturation), and on heat and matter exchange at the sediment–water interface; (2) alterations of flow through quantity and dynamics, import of suspended matter and dissolved substances such as nutrients and dissolved organic matter and water-residence time; (3) water-level fluctuations with implications on sediment transport, resuspension, and accumulation, matter exchange between sediment and water, hypolimnetic oxygen depletion, gas-exchange rate, and drying and re-wetting of littoral areas; and (4) changes of irradiation and underwater light intensity and possibly increasing effects of UV radiation on organisms (climate-driven effects; see e.g., Blenckner et al., 2002; Keller et al., 2008; Wantzen et al., 2008; Boulding et al. 2008; Livingstone and Adrian, 2009; Vincent, 2009).

However, the effects of global-change phenomena on a specific water body may differ considerably, depending on geographic location (latitude and altitude) and the morphology of surface waters, and can also be masked by many other impacts. The local effects of global change are difficult to predict and remain largely uncertain. Little is currently known about the changes in the amount and timing of seasonal runoff, derived from rainfall and snowmelt and the frequency of wind-induced disturbances on the vertical structure in the water bodies.

Further research is needed (1) to improve understanding of the sensitivity of different types of lakes/reservoirs pertaining to changes of climate drivers, (2) on effects of climate change on biodiversity, and (3) on improvements of precise weather forecasts as precondition for reservoir management, the optimization of flood control and water-quality requirements, and the development of reservoir-safety plans and risk assessment.

To fill these gaps, the following actions are necessary: (1) exploration of long-term data sets (e.g., long-time series of phytoplankton succession) and their integration into ecological sub-models for the water-quality management of standing waters; (2) combination of climate models with lake/reservoir models; (3) merging of paleolimnological records with hydrological and hydrochemical modeling, with changes in land uses, and application of improved methods such as DNA techniques to recognize changes in biota and isotope methods; (4) improvement of models for hydrological forecasting, especially of discharge fluctuations in the tributaries; and (5) collaborative research among climatologists, hydrologists, and limnologists to understand critical transitional events.

2.08.4.3 Biodiversity and Its Role in the Functioning of Lake and Reservoir Ecosystems

Stability and service functions of lakes and reservoirs are closely linked with the conservation of biological diversity. Although freshwater ecosystems cover only 0.8% of the Earth's surface, it is estimated that more than 10% of all animals and 35% of all vertebrates live (at least during one growth stage) in inland waters. Related to area, inland waters have a 10-times higher biodiversity than other ecosystems. Therefore, inland waters are biodiversity hotspots in the landscape with huge importance for distribution and speciation. One explanation of this concentration of diversity is the high potential for isolation of inland waters. Compared to species in marine and terrestrial ecosystems, freshwater species had 4–6-times higher extinction rates in the recent past. Anthropogenic changes of biotopes (e.g., construction of dams, pollution, draining of wetlands, and climate change) and invasion of exotic species are the main causes of losses in biodiversity. The complexity of current changes is, however, not fully understood. Biodiversity plays a crucial role in the resilience (elasticity) and dynamics of aquatic ecosystems. The role of biodiversity and the role of single species for different ecological functions are the subject of an increasing number of theoretical and empirical studies. The species composition of bacteria and fungi is still largely unknown, although these microorganisms are responsible for the flow and cycling of elements in reservoirs and lakes. The identification of the aquatic microbial species, which in most cases are noncultivable, became possible in the last decade by the broad availability of molecular probes, which specifically bind to the DNA of taxonomic units and species.

Thus, future research should be focused on (1) the monitoring and documentation of biodiversity, (2) the improved understanding of ecosystem functioning with respect to evolutionary processes, (3) the consequences of biodiversity on biogeochemical cycles, (4) the relationships between different aspects of biodiversity and anthropogenic disturbances, and (5) the effects of invading species on the structure, function, and achievements of the aquatic ecosystems. Further tasks include the development of a global database for freshwater systems, the quantification and assessment of biodiversity for ecosystem services, including the identification of keystone species, and the development of a preventive management of biodiversity.

2.08.4.4 Integrated Management of Lakes and Reservoirs

Measures for protection and remediation of damaged aquatic ecosystems are mainly focused on the control of external loading of nutrients, toxic substances, organic matter, and other pollutants. Currently, the reduction of the external loading is oriented to prescriptive concentration limits for the ambient water or to the minimization of emissions according to the best available technology (e.g., the four purification steps in wastewater-treatment plants). Critical load concepts such as the Vollenweider model are helpful for achieving qualitative aims of lake and reservoir management (Vollenweider, 1976). These models enable the user to predict qualitative states depending on the characteristics of lakes and reservoirs. Consequently, the water quality is not only the result of external pollution, but also a function of the physical, chemical, and biological structures of lakes and reservoirs. This means that a potential control strategy may include changes of internal structures within the reservoirs/lakes. Optimized ecosystem structures may tolerate higher emissions and mitigate negative consequences of pollution, which means that integrated control strategies, including both remediation measures in the catchments (reduction of loading) and lake-internal measures (optimization of ecosystem structure), are most promising and effective (Benndorf, 2008). In the last few decades, numerous techniques for lake and reservoir management were developed to optimize the structure of lake/reservoir ecosystems and their catchments, to achieve specific targets. The application of ecological approaches is referred to as ecological engineering or ecotechnology mainly used to control eutrophication and acidification. Many case studies have demonstrated that internal measures are successful when they are combined with a sufficient external load reduction. On the other hand, actual effects of internal measures have often not achieved the expected results. Although limnologists and hydrologists have focused for a long time on the causes and consequences of eutrophication, it remains a burning problem (primarily in tropical regions) and requires further research. A great deal is known about the role of macronutrients, particularly those that are consumed in large quantities, especially carbon, phosphorus, and nitrogen, their sources, import and cycling, as well as their passage through the food web. Far less, however, is known about the impact of micronutrients, which are required in relatively small quantities, such as iron, manganese, some trace elements, and vitamins. Furthermore, the influence of the huge variety of xenobiotics (e.g., antibiotics, drogues, herbicides, pesticides, insecticides, and fungicides) and their degradation products on aquatic ecosystems and (upon re-entering the human food chain) on human health, is currently largely unknown. The imperative for an integrative management, which includes social, ecological, and economic aspects, can be illustrated through reservoir management: The drinking-water supply from reservoirs is of great and growing importance in many countries. It is of decisive social and economic interest to have a strict control on the costs for production, processing, and distribution of drinking water. Competing uses of drinking-water reservoir (e.g., flood control, irrigation, delivery of water to the downstream river, hydropower production, and if necessary, tourist and leisure usage), as well as ecological aspects have to be taken into account. Therefore, a far-sighted and scientifically substantiated and integrated dam-management approach, considering all processes and factors influencing water quantity and quality in a complex manner, becomes increasingly important. Reservoir management must include ecological aspects of the entire river system from the origins of the tributaries to the downstream reaches. Medium-term and long-term concepts have to incorporate forecasts of development of water resources and the drinking-water consumption, as well as changing water quality in the reservoirs and downstream, resulting from possible structural alterations in the catchment area, as well as local climate trends.

The following future tasks have been identified: (1) improvement of the theoretical basis for ecotechnological principles for lakes/reservoirs and their catchment areas through a better understanding of the complexity of influenced processes and functions; (2) development of a consistent concept for the implementation of internal measures (definition of target concentration for initiating in-lake measures, time sequence of external and internal measures, determination of the long-term stability of restoration methods, and cost–benefit calculations); (3) improvement of decision support systems (DSSs) with implementation of predictive models; (4) development of an adaptive management to mitigate changes in climate and land usage; and (5) definition of critical threshold concentrations of pollutants in different climate regions, for moving toward the good ecological state of surface waters.

References

Annadotter H, Cronberg G, Aagren R, Lundstedt B, Nilsson P-A, and Ströbeck S (1999) Multiple techniques for lake restoration. *Hydrobiologia* 395/396: 77–85.

Arai T (1981) Climatic and geomorphological influences on lake temperature. *Verhandlungen Internationale Vereinigung für Theoretische und Angewandte Limnologie* 21: 130–134.

Baumert H and Uhlmann D (1983) Theory of the upper limit to phytoplankton production per unit area in natural waters. *International Review of Hydrobiology* 68: 753–783.

Baumert HZ, Simpson J, and Sündermann J (eds.) (2005) *Marine Turbulence – Theories, Observations and Models*, 630pp. (ISBN: 0521837898). New York: Cambridge University Press.

Behrendt H and Dannowski R (2005) *Nutrients and Heavy Metals in the Odra River System: Emissions from Point and Diffuse Sources, Their Loads, and Scenario Calculations on Possible Changes*. Berlin: Weißensee-Verlag.

Benndorf J (2008) Ecotechnology and emission control. Alternative or mutually promoting strategies in water resources management? *International Review of Hydrobiology* 93: 466–478.

Blais JM and Kalff J (1995) The influence of lake morphometry on sediment focusing. *Limnology and Oceanography* 40: 582–588.

Blenckner T, Omstedt A, and Rummukainen M (2002) A Swedish case study of contemporary and possible future consequences of climate change on lake function. *Aquatic Science* 64: 171–184.

Boulding AN, Rees GN, Baldwin DS, Suter PJ, and Watson GO (2008) Changes in sediment microbial community structure within a large water-storage reservoir during an extreme drawdown event. *Marine and Freshwater Research* 59: 890–896.

Cadenasso ML, Picket STA, Weathers KC, and Jones CG (2003) A framework for a theory of ecological boundaries. *Bioscience* 53: 750–758.

Carmack EC, Gray CBJ, Pharo CH, and Daley RJ (1979) Importance of lake–river interaction on seasonal patterns in the general circulation of Kamloops Lake, British Columbia. *Limnology and Oceanography* 24: 634–644.

Carpenter S (1983) Lake geometry: Implications for production and sediment accretion rates. *Journal of Theoretical Biology* 105: 273–286.

Chen C-TA and Millero FJ (1986) Precise thermodynamic properties of natural waters covering only the limnological range. *Limnology and Oceanography* 31: 657–662.

Clasen J and Bernhardt H (1983) In-situ-Trübungsmessungen in der Wahnbachtalsperre. *Gwf – Wasser/Abwasser* 124: 575–581.

Cole GA (1994) *Textbook of Limnology*, 4th edn. xii + 412pp. Prospect Heights, IL: Waveland Press. (ISBN 0-88133-800-1)

Cole TM and Hannan HH (1990) Dissolved oxygen dynamics. In: Thornton KW, Kimmel BL, and Payne FE (eds.) *Reservoir Limnology. Ecological Perspectives*, pp. 71–107. New York: Wiley.

Cooke GD, Welch EB, Peterson S, and Nichols SA (2005) *Restoration and Management of Lakes and Reservoirs*, 591pp. Boca Raton, FL: CRC Press.

Davison W (1992) Iron particles in freshwater. In: Buffle J and Van Leeuwen HP (eds.) *Environmental Particles*, vol. 1, pp. 317–352. Boca Raton, FL: Lewis.

Davison W (1993) Iron and manganese in lakes. *Earth-Science Reviews* 34: 119–163.

DeCesare G, Boillat J-L, and Schleiss AJ (2006) Circulation in stratified lakes due to flood-induced turbidity currents. *Journal of Environmental Engineering* 132: 1508–1517.

Downing JA, Prairie YT, Cole JJ, et al. (2006) The global abundance and size distribution of lakes, ponds, and impoundments. *Limnology and Oceanography* 51: 2388–2397.

Dudgeon D, Arthington AH, Gessner MO, et al. (2006) Freshwater biodiversity: Importance, threats, status and conservation challenges. *Biological Reviews* 81: 163–182.

DWA (2005) Merkblatt DWA-M 605: Wirkung, Bemessung und Betrieb von Vorsperren zur Verminderung von Stoffeinträgen in Talsperren. Deutsche Vereinigung für Wasserwirtschaft, Abwasser und Abfall e.V., Hennef.

Fee EJ (1979) A relation between lake morphometry and primary productivity and its use in interpreting whole-lake eutrophication experiments. *Limnology and Oceanography* 24: 401–416.

Feistel R, Nausch G, and Wasmund N (2008) *State and Evolution of the Baltic Sea, 1952–2005: A Detailed 50-Year Survey of Meteorology and Climate, Physics, Chemistry, Biology, and Marine Environment*, 703pp. Hoboken, NJ: Wiley-Interscience.

Gulati RD, Pires LMD, and Van Donk E (2008) Lake restoration studies: Failures, bottlenecks and prospects of new ecotechnological measures. *Limnologica* 38: 233–247.

Håkanson L (1981) *A Manual of Lake Morphometry*. ix + 78pp. (ISBN 3-540-10480-1). Springer

Håkanson L and Boulion VV (2002) Empirical and dynamical models to predict the cover, biomass and production of macrophytes in lakes. *Ecological Modelling* 151: 213–243.

Halbfaß W (1921) Die Thermik der Klingenberger Talsperre in Sachsen. *Internationale Revue der gesamten Hydrobiologie und Hydrographie* 9(1/2): 166–188.

Hanna M (1990) Evaluation of models predicting mixing depth. *Canadian Journal of Fisheries and Aquatic Sciences* 47: 940–947.

Hansen B, Østerhus S, Quadfasel D, and Turrell W (2004) Already the day after tomorrow? *Science* 305(5686): 953–954.

Hansson LA and Brönmark C (2009) Biomanipulation of aquatic ecosystems. In: Likens G (ed.) *Encyclopedia of Inland Waters*, pp. 242–248. Elsevier.

Hilt S, Gross E, Hupfer M, et al. (2006) Restoration of submerged vegetation in eutrophied shallow lakes – a guideline and state of the art in Germany. *Limnologica* 36: 155–171.

Hupfer M and Hilt S (2008) Lake restoration. In: Jorgensen SE and Fath BD (eds.) *Encyclopedia of Ecology*, pp. 2080–2093. Oxford: Elsevier.

Hupfer M and Kleeberg A (2007) State and pollution of freshwater ecosystems – warning signals of a changing environment. In: Lozan JL, Graßl H, Hupfer P, and Schönwiese C-D (eds.) *Climate Change: Enough Water for All? Wissenschaftliche Fakten in Zusammenarbeit Mit GEO*, pp. 126–132. Hamburg.

Hupfer M, Pöthig R, Brüggemann R, and Geller W (2000) Mechanical resuspension of autochthonous calcite (Seekreide) failed to control internal phosphorus cycle in a eutrophic lake. *Water Research* 34: 859–867.

Hutchinson GE (1957) *A Treatise on Limnology. I. Geography, Physics, and Chemistry*, 1015pp. New York: Wiley.

Hutchinson GE and Löffler H (1956) The thermal classification of lakes. *Proceedings of the National Academy of Sciences of the United States of America* 42: 84–86.

Hutter K (1983) Strömungsdynamische Untersuchungen im Zürich- und im Luganersee – Ein Vergleich von Feldmessungen mit Resultaten theoretischer Modelle. *Schweizerische Zeitschrift für Hydrologie* 45: 101–144.

IPCC (2007) Intergovernmental panel on climate change. *Fourth Assessment Report*. http://www.ipcc.ch (accessed April 2010).

Jeppesen E, Sondergaard M, Jensen HS, and Ventälä AM (2009) Lake and reservoir management. In: Likens G (ed.) *Encyclopedia of Inland Waters*, pp. 295–309. Elsevier.

Jones JG, Gardener S, and Simon BM (1983) Bacterial reduction to ferric ions in a stratified eutrophic lake. *Journal of General Microbiology* 129: 131–139.

Junge CO (1966) Depth distributions for quadric surfaces and other configurations. In: Hrbacek J (ed.) *Hydrobiological Studies*, pp. 257–265. Prague: Academia Publishing House of the Czechoslovak Academy of Sciences.

Kahl U, Hülsmann S, Radke RJ, and Benndorf J (2008) The impact of water level fluctuations on the year class strength of roach: Implications for fish stock management. *Limnologica* 38: 258–268.

Kalff J (2002) *Limnology*, xii + 592pp. Upper Saddle River, NJ: Prentice Hall. (ISBN 0-13-033775-7).

Keller R (1962) *Gewässer und Wasserhaushalt des Festlandes*. Leipzig: B. G. Teubner Verlagsgesellschaft.

Keller W, Patterson AM, Somers KM, Dillon PJ, Heneberry J, and Ford A (2008) Relationships between dissolved organic carbon concentrations, weather, and acidification in small boreal shield lakes. *Canadian Journal of Fisheries and Aquatic Sciences* 65: 786–795.

Kling GW (1988) Comparative transparency, depth of mixing and stability of stratification in lakes of Cameroon, West Africa. *Limnology and Oceanography* 33: 27–40.

Kusakabe M (ed.) (1994) Special Issue: Geochemistry of Crater Lakes. *Geochemical Journal* 28: 137–306.

Kusnezow SI (1959) *Die Rolle der Mikroorganismen im Stoffkreislauf der Seen*, 301pp. Berlin: Deutscher Verlag der Wissenschaften.

Lambert A, Kelts K, and Zimmermann U (1984) Trübeströme in Seen: Sauerstoffeintrag durch grundnah eingeschichtetes Flußwasser. *Schweizerische Zeitschrift für Hydrologie* 46: 41–50.

LAWA (2001) *Gewässerbewertung – Stehende Gewässer. Vorläufige Richtlinie für die Trophieklassifikation von Talsperren*, 36pp. Berlin: Länderarbeitsgemeinschaft Wasser (LAWA), Kulturbuch-Verlag. (ISBN: 3-88961-237-7).

Liboriussen L, Sondergaard M, Jeppesen E, et al. (2009) Effect of hypolimnetic oxygenation on water quality: Results from five Danish lakes. *Hydrobiologia* 625: 157–172.

Ligvoet W and Witte F (1991) Perturbation through predator introduction: Effects on the food web and fish yields in Lake Victoria (East Africa). In: Ravera O (ed.) *Terrestrial and Aquatic Ecosystems. Perturbation and Recovery*, pp. 263–268. New York: Ellis Horwood.

Livingstone DM and Adrian R (2009) Modeling the duration of intermittent ice cover on a lake for climate-change studies. *Limnology and Oceanography* 54: 1709–1722.

Malone BJ and McQueen DJ (1983) Horizontal patchiness in zooplankton populations in two Ontario Kettle lakes. *Hydrobiologia* 99: 101–124.

Mortimer CH and Horn W (1982) Internal wave dynamics and their implications for plankton biology in the Lake of Zurich. *Vierteljahresschrift Naturforschende Gesellschaft Zürich* 127: 299–318.

Nilsson C (2009) Reservoirs. In: Likens G (ed.) *Encyclopedia of Inland Waters*, pp. 625–633. Elsevier.

Patalas K (1984) Mid-summer mixing depths of lakes of different latitudes. *Verhandlungen Internationale Vereinigung für Theoretische und Angewandte Limnologie* 22: 97–102.

Paul L (1987) Influence of Seiche-generated light field fluctuations on phytoplankton growth. *Internationale Revue der gesamten Hydrobiologie und Hydrographie* 72: 269–281.

Paul L (1989) Interrelationships between optical parameters. *Acta Hydrophysica, Berlin* 33: 41–63.

Paul L (2003) Nutrient elimination in pre-dams – results of long-term studies. *Hydrobiologia* 504: 289–295.

Paul L and Pütz K (2008) Suspended matter elimination in a pre-dam with discharge dependent storage level regulation. *Limnologica* 38: 388–399.

Paul L, Schrüter K, and Labahn J (1998) Phosphorus elimination by longitudinal subdivision of reservoirs and lakes. *Water Science and Technology* 37(2): 235–243.

Poole HJ and Atkins WRG (1929) Photoelectric measurements of submarine illumination throughout the year. *Journal of the Marine Biological Association of the United Kingdom* 16: 297–324.

Ruttner F (1962) *Grundriß der Limnologie*. 3 Auflage 332pp. Berlin: Walter De Gruyter.

Ryanzhin SV, Straškraba M, Geller W (2001) Developing WORLDLAKE Database and GIS for limnological studies. In: *Proceedings of the 9th International Conference on the Conservation and Management of Lakes*, pp. 25–28, Session 5, Otsu, Japan, 10–15 November 2001. Otsu: ILEC Publications.

Scharf W (2002) Refilling, ageing and water quality management of Brucher Reservoir. *Lakes and Reservoirs: Research and Management* 7: 13–23.

Scheffer M, Hosper SH, Meijer ML, Moss B, and Jeppesen E (1993) Alternative equilibria in shallow lakes. *Trends in Ecology and Evolution* 8: 275–279.

Schmidt W (1915) Über den Energie-Gehalt der Seen. Mit Beispielen vom Lunzer Untersee nach Messungen mit einem einfachen Temperaturlot. *Internationale Revue der gesamten Hydrobiologie und Hydrographie* 6(supplement): 1–25.

Shapiro J (1984) Blue-green dominance in lakes. The role and management significance of pH and CO_2. *Internationale Revue der gesamten Hydrobiologie und Hydrographie* 69: 765–780.

Sigg L and Stumm W (1991) *Aquatische Chemie*, 388pp. Stuttgart: B.G. Teubner Verlag.

Smith VH (2003) Eutrophication of freshwater and marine ecosystems: A global problem. *Environmental Science and Pollution Research* 10: 126–139.

Søndergaard M, Jensen JP, and Jeppesen E (2005) Seasonal response of nutrients to reduced phosphorus loading in 12 Danish lakes. *Freshwater Biology* 50: 1605–1615.

Søndergaard M, Jeppesen E, Lauridsen TL, et al. (2007) Lake restoration: Successes, failures and long-term effects. *Journal of Applied Ecology* 44: 1095–1105.

Sterner RW (2008) On the phosphorus limitation paradigm for lakes. *Internationale Revue der gesamten Hydrobiologie und Hydrographie* 93: 433–445.

Stottmeister U (2008) Altlastensanierung mit Huminstoffsystemen. Prinzipien der Natur in der Umwelttechnologie. *Chemie in unserer Zeit*, vol. 42, pp. 24–41. Weinheim: Wiley-VCH Verlag GmbH.

Straškraba M (1980) The effects of physical variables on freshwater production: Analysis based on models. In: le Cren ED and McConnell RH (eds.) *The Functioning of Freshwater Ecosystems*, IBP 22, pp. 13–84. Cambridge University Press.

Straškraba M and Gnauck A (1983) *Aquatische Ôkosysteme – Modellierung und Simulation*. 279pp. Fischer Jena.

Stumm W and Morgan JJ (1981) *Aquatic Chemistry: An Introduction Emphasizing Chemical Equilibria in Natural Waters*, 2nd edn, 780pp. New York: Wiley.

Sudbrack R, Freier K, Grunewald K, Scheithauer J, Schmidt W, and Wolf C (2005) Verstärkte Huminstoffeinträge in Trinkwassertalsperren im Erzgebirge (Freistaat Sachsen). *GWF Wasser Abwasser* 146: 847–851.

Sytsma M, Rueter J, Petersen R, et al. (2004) *Waldo Lake Research in 2003*. Center for Lakes and Reservoirs, Department of Environmental Sciences and Resources, Department of Civil and Environmental Engineering, Portland State University, Portland, Oregon 97201–0751. http://www.clr.pdx.edu/docs/2003report.pdf (accessed April 2010).

Talling JF (1971) The underwater light climate as a controlling factor in the production ecology of freshwater phytoplankton. *Mitteilungen Internationale Vereinigung für Theoretische und Angewandte Limnologie* 19: 214–243.

Thienemann A (1913) Die Besiedlung der Talsperren. *Die Naturwissenschaften* 1(48): 1163–1167.

Thienemann A (1927) Der Bau des Seebeckens in seiner Bedeutung für den Ablauf des Lebens im See. *Verhandlungen der kaiserlich-königlichen. Zoologisch-Botanischen Gesellschaft in Wien* 77: 87–91.

Thornton J, Steel A, and Rast W (1996) Reservoirs. In: Chapman D (ed.) *Water Quality Assessments – A Guide to the Use of Biota, Sediments and Water in Environmental Monitoring*, 2nd edn.. UNESCO/WHO/UNEP.

Thurman EM (1995) *Organic Geochemistry of Natural Waters*. 497pp. Boston, MA: Martinus Nijhoff/Dr. W. Junk Publishers.

Tranvik LJ, Downing JA, Cotner JB, et al. (2009) Lakes and reservoirs as regulators of carbon cycling and climate. *Limnology and Oceanography* 54: 2298–2314.

Tundisi JG and Straškraba M (eds.) (1999) *Theoretical Reservoir Ecology and Its Applications*, Brazilian Academy of Sciences, International Institute of Ecology, 585pp. Leiden: Backhuys Publishers.

Uhlmann D (1979) *Hydrobiology. A Text for Engineers and Scientists*, 313pp. Chichester: Wiley.

Uhlmann D (1980) Stability and multiple steady states of hypereutrophic ecosystems. In: Mur L and Barica J (eds.) *Hypertrophic Ecosystems. Developments in Hydrobiology 2*, pp. 235–247. The Hague: Junk.

Uhlmann D and Horn W (2001) *Hydrobiologie der Binnengewässer. Ein Grundriss für Ingenieure und Naturwissenschaftler*, 528pp. Stuttgart: Eugen Ulmer.

UNEP (2000) Lakes and reservoirs – similarities, differences and importance. *IETC Short Report 1* http://www.unep.or.jp/ietc/Publications/Short_Series/LakeReservoirs-1 (accessed April 2010).

Vincent WF (1980) Mechanisms of rapid photosynthetic adaptation in natural phytoplankton communities. II. Changes in photochemical capacity as measured by DCMU-induced chlorophyll fluorescence. *Journal of Phycology* 16: 568–577.

Vincent WF (2009) Effects of climate change on lakes. In: Likens G (ed.) *Encyclopedia of Inland Waters*, pp. 55–60. Elsevier.

Vollenweider RA (1961) Photometric studies in inland waters: I. Relations existing in the spectral extinction of light in water. *Memorie dell'Istituto Italiano di Idrobiologia* 13: 87–113.

Vollenweider RA (1976) Advances in defining critical loading levels for phosphorus in lake eutrophication. *Memorie dell'Istituto Italiano di Idrobiologia* 33: 53–83.

Wantzen KM, Rothhaupt KO, Mörtl M, Cantonati MG, Tóth L, and Fischer P (2008) Ecological effects of water-level fluctuations in lakes: An urgent issue. *Hydrobiologia* 613: 1–4.

Weijters MJ, Janse JH, Alkemade R, and Verhoeven JTA (2009) Quantifying the effect of catchment land use and water nutrient concentrations on freshwater river and stream biodiversity. *Aquatic Conservation: Marine and Freshwater Ecosystems* 19: 104–112.

Wells MG and Sherman B (2001) Stratification produced by surface cooling in lakes with significant shallow regions. *Limnology and Oceanography* 46: 1747–1759.

Wetzel RG (1975) *Limnology*. Philadephia, PA: WB Saunders.

Wetzel RG (2001) *Limnology. Lake and River Ecosystems*, 3rd edn, 1006pp. San Diego, CA: Academic Press.

Wetzel RG and Likens GE (1991) *Limnological Analyses*, 2nd edn, 391pp. New York: Springer.

Wüest A, Piepke G, and van Senden DC (2000) Turbulent kinetic energy balance as a tool for estimating vertical diffusivity in wind-forced stratified waters. *Limnology and Oceanography* 45: 1388–1400.

Zehnder AJB and Zinder SH (1980) The sulfur cycle. In: Hutzinger O (ed.) *The Handbook of Environmental Chemistry*, vol. 1A, pp. 105–145. Berlin: Springer.

2.09 Tracer Hydrology

C Leibundgut, University of Freiburg, Freiburg, Germany
J Seibert, University of Zurich, Zurich, Switzerland

© 2011 Elsevier B.V. All rights reserved.

2.09.1	**Introduction**	215
2.09.1.1	Patterns of Development	215
2.09.1.2	Questions that Tracer Hydrology Helps to Answer	216
2.09.2	**Principal Conception and Approaches of Tracer Hydrology**	217
2.09.2.1	Hydrological System Approach	217
2.09.2.2	Mathematical Models	218
2.09.2.3	Design of Tracer Hydrology Studies	219
2.09.3	**Fundamentals of Environmental and Artificial Tracers**	220
2.09.3.1	Different Types of Tracers	220
2.09.3.2	Environmental Tracers	221
2.09.3.2.1	Isotope tracers ^{18}O and ^{2}H	221
2.09.3.2.2	Other environmental tracers	223
2.09.3.3	Artificial Tracers	223
2.09.3.3.1	Characteristics of artificial tracers	223
2.09.3.3.2	Fluorescent tracers	223
2.09.3.3.3	Nonfluorescent artificial tracers	225
2.09.4	**Tracer Hydrology Applications**	226
2.09.4.1	Hydrograph Separation	227
2.09.4.2	Catchment Transit Time Estimation	227
2.09.4.3	Analysis of Sources of Nitrogen in Streams	228
2.09.4.4	Advection–Dispersion Modeling	229
2.09.4.5	Discharge Measurement	230
2.09.4.6	Chloride-Based Groundwater-Recharge Estimation	232
2.09.4.7	Tracer Experiment in a Porous Aquifer	232
2.09.5	**Concluding Remarks**	233
2.09.5.1	Guidance on Further Reading	233
2.09.5.2	Reflections and Future Research	234
References		235

2.09.1 Introduction

Tracers are substances which can be detected in the water at very low concentrations and allow following, or tracing, the flow of water. The ability to trace the flow of water is crucial for understanding the complex processes in hydrological systems. This understanding is important in many respects such as predicting water quality, which is often controlled by different water sources, flow pathways, and transit times, or the impacts of climate or land-use/land-cover changes on catchment's hydrological response. Tracer hydrology helps to address questions such as how runoff is generated, which flow pathways the water takes, how long the water is in the catchment, where runoff comes from, or where water from a pollution source will flow. Tracer methods are of special importance in soil- and groundwater systems, because of the limited availability of other observation methods for subsurface waters. In this chapter, the use of tracer methods in hydrology is presented. In addition to the description of different tracers and methods, tracer hydrology is also presented as an advanced way to holistically study the characteristics of hydrological systems.

We begin with describing the development and the concept of tracer hydrology and a discussion of the hydrological questions which can be treated by tracer techniques. The different tracers used in hydrology and the methods used to interpret tracer experiments are then described. Finally, selected tracer studies are reviewed as examples of how tracer hydrology has contributed to the advancement of hydrology. This chapter presents a short and informative overview of the current state of tracer hydrology. We are aware that there are many aspects of tracer hydrology, which cannot be addressed within this short chapter. For more detailed information several textbooks are available and guidance on further reading is provided in Section 2.09.5.1. This chapter largely builds on the recent textbook by Leibundgut et al. (2009). Further information on many issues, which are briefly discussed in this chapter, can be found in the textbook even if there is not always an explicit reference to the textbook.

2.09.1.1 Patterns of Development

First tracer experiments were conducted about 150 years ago (Käss, 1998). It followed a slow but fascinating development of tracer hydrology, before, in the 1960s, tracer techniques began to be developed and used more widely in hydrology. This development was possible in particular because of the progress made in measurement techniques and by the digitalization of data processing. Almost simultaneously, the computer era began and opened up new possibilities for environmental modeling. During this fascinating phase of development of natural sciences came the evolution of holistic approaches. In the case of tracer hydrology, this implied an increased focus on using tracers, often in combination with other investigation methods, to characterize hydrological system behavior.

Most of the fundamental principles of tracer hydrology have been developed during this phase. Environmental isotopes have provided major inputs to the study of hydrological processes such as runoff generation, runoff component separation, transit times, recharge, and groundwater flow, and are still central for developing perceptual understanding and conceptual models of hydrological processes. The application of artificial tracers moved from being a measurement technique to a powerful tool for the development, parametrization, and validation of models for solute transport in ground- and surface waters.

Besides many other factors, three fortunate milestones marked this development. A powerful framework for realizing numerous ideas was created through the founding and activities of the Association of Tracer Hydrology (ATH). The ATH promoted the use of tracer techniques in Europe between the late 1950s and the end of the twentieth century in many ways. The second milestone was the establishment of the Isotope Hydrology Laboratory at the International Atomic Energy Agency (IAEA) in Vienna in 1961. It propelled the rapid development of isotope techniques, beginning with environmental tritium, as a research tool for investigating the hydrological cycle worldwide. The XXth General Assembly of the International Union of Geodesy and Geophysics in 1991 in Vienna can be considered to be the third milestone. The International Commission on Tracers (ICT), within the International Association of Hydrological Sciences (IAHS), was established. Its aim was, among others, to bring together experimental hydrologists with modelers for the integrated investigation of the hydrological system. This event is significant since at the that time there was a much more uncritical belief in the potential of (pure) modeling in hydrology than today and, thus, the establishment of a clearly experimentally oriented commission within the IAHS was not without criticism. The following years showed an increasing integration of the tracer methods into hydrological research and applied hydrology by the international community, which validated this structural development. Experimental hydrology, in particular the strongly emerging catchment hydrology, increasingly used tracer methods in order to assess hydrological processes and system functions. In particular, the calibration and validation of mathematical models were based increasingly on tracer hydrological research.

2.09.1.2 Questions that Tracer Hydrology Helps to Answer

The objective of tracer hydrology is the investigation of water in all its various phases, behaviors, and characteristics within the different media and substrates represented in the water cycle. The use of tracers in hydrology, therefore, defines the scientific field that aims at understanding the hydrological system by making use of environmental and artificial tracers and modeling. Tracing of water provides unique methods for a direct insight into the dynamics of surface and subsurface water bodies.

The large degree of complexity often found in hydrological systems is one of the main reasons why tracer techniques are needed both in hydrological research and in applied hydrology to characterize these systems. Tracers provide empirical data of real and often unexpected flow patterns, whereas models provide tools for flow and transport predictions. There is a fruitful connection between empirical tracer-based observations of flow and transport processes and the theoretical formulation of these processes, which has resulted in beneficial combinations and co-developments of both approaches to characterize hydrological systems. How does water flow through a hillslope or a glacier? How much runoff in rivers during an event originates directly from the event rainstorm? How much water is stored in an aquifer? Where and when water entered an aquifer as recharge? Phase changes, such as evaporation, condensation, and sublimation, can be identified and quantified using tracers. Tracers provide information on the origin of pollution and, thus, assist in deciding on the appropriate remediation approaches. Finally, tracer techniques are useful tools in understanding and quantifying transport processes. Thus, tracer techniques are applicable in all general fields of hydrology and experience can be gained in all the components of the water cycle. The possibilities of tracer techniques are vast, as described above. Tracer hydrology provides an integrative system approach for hydrological studies and other water-related research. The system approach is based on the determination of a function characterizing the system based on time series of known (measured) inputs to a system and known (measured) outputs. A more extensive review of publications in the field of tracer hydrology can be found in Leibundgut *et al.* (2009).

There are two basic groups of tracers: (1) artificial tracers are actively brought into the hydrological system, so we refer to their application and (2) environmental tracers are defined as specific components of the water cycle, thus we discuss their utilization.

Artificial tracers are defined as substances which are added intentionally to hydrological systems in planned experiments. The scales of application of artificial tracers are limited in both time and space. In general, artificial tracers are used in systems, which have a residence time of less than 1 year. On the other hand, artificial tracers allow labeling specific parts of a hydrological system. Typical fields of applications of artificial tracers include

- the detection of hydrological connections,
- flow paths and flow directions in catchments and aquifers,
- delineation of catchments and aquifers (qualitative),
- determination of flow velocities and further aquifer flow parameters based on the tracer breakthrough curves (TBCs),
- hydrodynamic dispersion,
- runoff separation,
- residence time,
- infiltration and runoff generation processes,

Table 1 Relevant hydrological tracers

Environmental tracers	Artificial tracers	
Stable isotopes	*Solute tracers*	*Activatable radionuclides*
Deuterium (^{2}H)	*Fluorescent dyes*	*Dissolved gas tracers*
Oxygen-18 (^{18}O)	Naphthionate	Helium
Carbon-13 (^{13}C)	Pyranine	Neon
Nitrogen-15 (^{15}N)	Uranine	Stable isotopes of krypton
Sulfur-34 (^{34}S)	Eosine	Sulfur hexafluoride (SF$_6$)
	Rhodamines	
Radioactive isotopes	*Nonfluorescent dyes*	
Tritium (^{3}H) (and helium-3 (^{3}He))	e.g., Brilliant blue	
Carbon-14 (^{14}C)		
Argon-39 (^{39}Ar)	*Salts*	
Krypton-85 (^{85}Kr)	Sodium/potassium chloride	
Radon-222 (^{222}Rn)	Sodium/potassium bromide	*Particulate tracers*
Radium-226 (^{226}Ra)	Lithium chloride	Lycopodium spores
Silicium-32 (^{32}Si)	Potassium iodide	Bacteria
Chlorine-36 (^{36}Cl)	Sodium borate (borax)	Viruses
		Phages
Noble gases	*Fluorobenzoic acids*	DNA
Anthropogenic trace gases	*Deuterated Water* (^{2}H)	Synthetic microspheres
Chlorofluorocarbons (CFCs)	*Radionuclides*	Phytoplankton
Sulfur hexafluoride (SF$_6$)	e.g., Tritium (^{3}H)	
Geochemical compounds	Chrome-51 (^{51}Cr)	
e.g., silicate, chloride, DOC, heavy metals	Bromide-82 (^{82}Br)	
	Indium-131 (^{131}I)	
Physio-chemical parameters		
e.g., Electrical conductivity, temperature		

- labeling of unsaturated zone water movement,
- convection–diffusion processes in surface water,
- simulation of contaminant transport, and
- discharge measurement applying dilution methods.

A major characteristic of environmental tracers is that the input, or the injection, of the tracer to the hydrological system is provided by nature (Kendall and McDonnell, 1998). This can be both an advantage and a disadvantage. The major advantage is that catchment-wide injections are possible. Therefore, environmental tracers can be used at different scales such as catchment scale studies and even at the global scale. On the other hand, it is usually impossible to trace the contributions from single locations with environmental tracers. Another advantage of environmental tracers is that they have been injected for a long period of time. Therefore, these tracers allow systems with long transit times to be characterized. Substances such as sulfur hexafluoride (SF$_6$) or cesium (anthropogenic tracers) brought to the hydrological cycle by accidents or as pollution may in some way be classified as artificial tracers but are often used in similar ways as environmental tracers (Table 1).

Environmental tracers have been used in studies on all components of the water cycle. Thus, the utilization of environmental tracers provides methods for the investigation of some major components of the hydrological systems such as precipitation processes and origin assignment, open water evaporation, transpiration and stem flow, soil water dynamics, groundwater flow and recharge studies, subsurface flow mechanisms, and runoff components. Major applications of environmental tracers include

- hydrological process studies: direct or indirect recharge mechanisms, identification of runoff components, and subsurface flow mechanisms;
- origin of water and water constituents: for instance, the discrimination of summer or winter recharge, the assignment of recharge altitude or the detection of origin of nitrate or dissolved inorganic carbon;
- determination of residence times: age dating or analysis of the amplitude of the variation of stable isotopes of water;
- quantitative determination of flow components: estimation of evaporation from open water surface, hydrograph separation; and
- paleohydrological studies.

2.09.2 Principal Conception and Approaches of Tracer Hydrology

2.09.2.1 Hydrological System Approach

Artificial and environmental tracer experiments provide an improved understanding of hydrological systems such as catchments, streams, and aquifers. Environmental and artificial tracer methods have both been developed into important tools for various aspects of hydrology. These methods are used to estimate water resources to both reconstruct past hydrological conditions and investigate runoff generation processes at scales from plots to catchments. Without tracer methods many hydrological processes would not be possible to observe. In other words, tracer hydrology provides tools for understanding and characterizing complex flow processes through soils, on surfaces, in channels, through and along hillslopes, in aquifers, or in artificial systems. Environmental and artificial tracer approaches have both advantages and limitations; both approaches might, thus, also complement each other in experimental studies. When trying to understand hydrological systems, hydrometric data, such as runoff or groundwater levels, are often not sufficient on their own, but tracer information provides important information on system characteristics needed to derive a perceptual model of the investigated system. Tracer hydrology aims to develop, test, and validate such representations of hydrological systems that are in sound agreement with the available data by making use of environmental and artificial tracer experiments and modeling.

Describing hydrological systems such as catchments, hillslopes, soil columns, or stream segments is generally based on

1. measured or otherwise known inputs (water fluxes, constituent loads, and energy) as a function of time and space;
2. a characteristic functioning of the system, which can be described by a set of equations representing, for instance, the flow in surface water bodies, in the subsurface, or at the soil–vegetation–atmosphere interface; and
3. a known or measured output of the investigated variables, again as a function of space and time (**Figure 1**).

In the case of tracer hydrology approaches, the input is usually the concentration or load of a particular tracer, for instance, in groundwater recharge or precipitation (for environmental tracers) or the injected mass for artificial tracers. The output from a system is usually the tracer load leaving the system, being controlled by runoff volume or well yield and the tracer concentrations. Interpreting the information contained in the input–output relationship of specific variables of a system is, thus, a key in tracer hydrology. Transfer functions between input and output are identified from tracer data and can be used for system characterization and, subsequently, predictions of system behavior. Mathematical modeling is an important tool for the interpretation of both environmental tracers and artificial tracer experiments.

The application of the convergence approach in tracer hydrology can be used to derive concepts of hydrological systems (Leibundgut, 1987; Leibundgut et al., 2009). The convergence approach utilizes the fact of flow path converge toward a spring or a branch of a river. At the spring, the information represented by the tracers and the hydrograph respectively can be deciphered and combined by adequate techniques. These concepts and the corresponding mathematical models can be varying in complexity, ranging from simple to more complex model structures, representing the principal functioning of the investigated hydrological system. Tracer experiments and mathematical modeling can be performed in an iterative way, where predictions derived from a system model can be used for an improved design of further experiments (e.g., better planning of the observation network).

Using several independent methods and techniques simultaneously when investigating a hydrological question is generally beneficial. In tracer hydrology, such an approach is common, for instance, when using different tracer techniques (environmental and artificial tracers) in combination with other hydrological methods (such as hydrometric or geophysical methods). The combination of different tracers, if possible, ensures that any specific limitation of a single tracer does not bias the characterization of the hydrological system. Often, different artificial tracers are combined and there are also many studies where different environmental tracers are used. The combination of environmental and artificial tracers, however, might often be the most promising approach. Moreover, it is always valuable to combine or even integrate tracer methods with other experimental hydrological investigation methods (e.g., hydrometry, geophysics, hydrochemistry, and remote sensing).

The advantage of such an integrated approach is the added value of the combination of results obtained by different, often independent, methods. If different methods provide consistent results, more general conclusions can be drawn. Uncertainties of individual methods can be reduced if different methods point in the same direction. Disagreements between different methods, on the other hand, can be even more valuable as they provide impetus for an improved experimental design or research aiming at resolving the contradiction.

2.09.2.2 Mathematical Models

For the interpretation of a tracer experiment, it is in most cases necessary to use some kind of a mathematical model. Such a model has to be based on adequate concepts of tracer transport and tracer behavior in the system. The conclusions which might be derived from a tracer hydrology study might largely depend on the model used for interpretation as already discussed by Eriksson (1958). Therefore, it is important to be clear about the underpinning assumptions.

A perceptual model is a qualitative description of a system and its most important characteristics, such as geometry, hydraulic connections, parameters, and initial and boundary conditions related to the intended use of the model. In practice, the perceptual model demonstrates the principal idea of

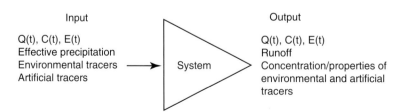

Figure 1 Hydrological system approach adapted to tracer hydrology by the convergence approach. From Leibundgut Ch, Malozewski P, and Külls Ch (2009) *Tracers in Hydrology*. Chichester: Wiley.

water circulation in the system. In tracer hydrology, the term 'conceptual model' has also been used for this qualitative description, but here the term 'perceptual model' is used to avoid confusion with the use of the terms in catchment runoff modeling (Beven, 2001).

Based on a perceptual model, a mathematical model can be derived. Here, the hydrological, physical, and/or hydrochemical system is represented by mathematical functions, including model parameters. Typically, features such as geometry, hydraulic connections, and initial and boundary conditions need to be specified in some way. Mathematical models in tracer hydrology typically describe the storage and movement of both water and tracer mass. In certain cases, analytical solutions of the mathematical formulation exist for given boundary conditions. Depending on the system being described, the mathematical model can more or less be based on physical concepts such as Richard's equation or on a system approach such as transit-time distribution functions.

Calibration and validation are important steps in any model application. During calibration, model parameters are varied until a good fit between model simulations and experimental observations is obtained. Sometimes, model structures are also varied during calibration. Trial-and-error calibration refers to the procedure where model parameters are changed by hand based on, often subjective, decisions by the model user. The advantage is that prior knowledge can be considered; the disadvantages are the subjectivity and the large time demand. Automatic optimization procedures based on some objective functions are often used as a more objective and time-efficient way to calibrate model parameters. This approach, on the other hand, has been criticized as advanced, but still simplistic curve fitting. The calibration of a mathematical model to experimental data can also be interpreted as inverse modeling of parameter values for a certain system. Depending on the experimental data and the model used, it might not be possible to find a unique solution through optimization due to the problem of equifinality (Beven, 2001).

Tracer methods are often used to describe poorly known systems. Therefore, mathematical models should be as simple as possible to allow determination of parameters through calibration. The aim of validation is to increase confidence that a model is a suitable representation of the system for which it is applied. One way of validation is comparing the calibrated model parameters with independent measurements of the parameters (e.g., filter velocity). Applying the model to make predictions using data, which is independent from those used in calibration, is another way of model validation. While this independent data are often data of the same type but from a different time period, different types of data might also be used. In catchment modeling, for instance, models are usually calibrated using runoff data, but tracers might be used to falsify or confirm certain model structures (Seibert and McDonnell, 2002; Seibert et al. 2003; Uhlenbrook and Sieber, 2005). In addition, regional circulation models might be tested based on isotope data (Sturm et al., 2005). It must be noted, however, that the simulation of isotopes requires model extensions. Therefore, parameter uncertainty often is not reduced despite additional validation information. However, including tracer data in the model testing will lead to internally more consistent models. The role of isotopes in the validation of global and regional atmospheric circulation models and their coupling with hydrological models as well as the response of ecosystems to climate change has not been fully investigated yet.

For modeling of environmental tracers, usually, the injection of tracer, which occurs naturally over an area and continues over a longer time period, has to be considered. This injection can be either by precipitation or by solution of minerals from earth substrate. Generally, system approaches such as the convolution integral for transit time analyses are useful in these cases. The tracer injection for artificial tracer, on the other hand, is typically concentrated to a single point or line. In this case, mathematical models based on dispersion theory are generally used. Analytical solutions for advection–dispersion processes in all dimensions and different boundary conditions are available and further described in Section 2.09.4.4. For heterogeneous systems and complex boundary conditions, transport equations usually have to be solved numerically.

Hydrological systems, such as an aquifer or a catchment, can be considered as systems, for which the characteristic behavior can be evaluated if both input and output time series of water and tracer concentrations are known. Usually, some forms of mean or effective parameters are used to describe the system in such approaches. Examples of parameters include volumes of water, transit times, and flow rates through the system. The tracer transport through the system, that is, between input and output, can then be described by a lumped-parameter approach (cf. **Figure 1**). The transit-time distribution function, or its first moment, the mean transit-time (MTT), is an especially important system characteristic, which is often derived for investigated systems (Maloszewski and Zuber, 1982). When stagnant water can be neglected, the MTT of water in the system can be further used to estimate the volume of water stored in a system. In more complex systems with mobile and stagnant water (e.g., fissured aquifers), it is important to distinguish between the MTT of tracer and the MTT of water, as these will usually not be the same in these cases. The transit time of a tracer in such systems is controlled by both the flow of the mobile water component and the diffusive exchange of tracer between mobile and stagnant water (see also Section 2.09.4.4).

2.09.2.3 Design of Tracer Hydrology Studies

Tracer studies and experiments are always carried out as part of a hydrological or a water resource issue. The success of an integrated tracer study as well as an artificial tracer experiment very much depends on careful planning. **Figure 2** structures the process of planning and executing an experiment using a flowchart. A clear idea of the aims of the experiment and of the influencing factors is fundamental. A major tracer study program (master plan) must be established in advance, in order to minimize problems and inconsistencies. For instance, it is important to estimate an appropriate mass for tracer injections or to plan sampling at a proper temporal (and/or spatial) resolution. Guidelines for carrying out practical experiments are provided by Leibundgut et al. (2009). When using environmental tracers it is important to ensure that it is possible to clearly distinguish different sources or flow pathways by

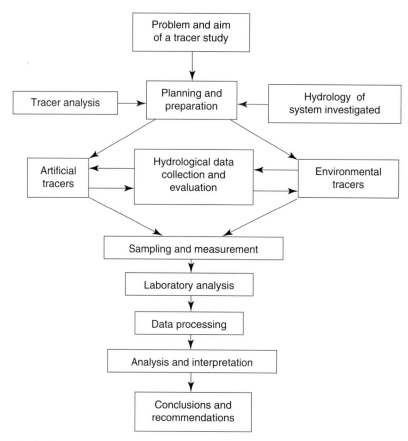

Figure 2 Flowchart illustrating the planning and execution of a tracer study and a tracer experiment. From Leibundgut Ch, Malozewski P, and Külls Ch (2009) *Tracers in Hydrology*. Chichester: Wiley.

their tracer fingerprint. This means that differences must be larger than the analytical accuracy and the natural variation within certain sources or pathways.

Practical aspects of tracer applications are further discussed in Section 2.09.4, together with the discussion of specific types of tracer studies, such as hydrograph separation or discharge measurement.

2.09.3 Fundamentals of Environmental and Artificial Tracers

2.09.3.1 Different Types of Tracers

There are basically two groups of hydrological tracers. On the one hand, there are tracers which are added in defined amounts to hydrological systems for planned experiments. On the other hand, there are the environmental tracers which are naturally occurring in hydrological systems. Here, concentration differences, which exist for some reason between different parts of the investigated hydrological system, are used to trace water fluxes. Isotopes and geochemical compounds are examples of this latter group of tracers, which are usually called environmental or natural tracers. Examples of tracers which can be artificially added to hydrological systems (so called artificial tracers) are fluorescent or salt tracers. In both groups, there are many different possible tracers, which all have their advantages and disadvantages and are more or less suitable for particular applications (Table 1). Pollutants, or the so-called pollution tracers, can be seen as a group of tracers in between environmental and artificial tracers. While they are not naturally occurring, they are not added to the hydrological system in well-defined ways either.

Environmental tracers are defined as properties or constituents of water that are occurring naturally and have not been added within an intended experiment, providing qualitative or quantitative information about a hydrological system. Naturally occurring stable and radioactive isotopes are the most commonly used natural tracers, especially the isotopes found in the water molecule (^{18}O and ^{2}H; see Section 2.09.3.2.1). Some of the environmental tracers result from anthropogenic releases to the atmosphere or to the hydrological cycle. Constituents such as krypton-85, chlorofluorocarbons (CFCs), and SF_6, for instance, have been released to the atmosphere as a result of industrial activities, and tritium (^{3}H) has been released due to atomic bomb tests. While their injections were not planned and definitely not aimed at the purpose of providing age dating methods to hydrologists, these natural tracers can be used in similar ways to naturally occurring constituents for tracer studies. Different pollution tracers, such as nitrate, organic pollutants, or remnants of past mining activities, can also provide information about hydrological processes.

Environmental tracers allow major components of the hydrological cycle to be studied. These tracers have, for instance, been used in studies on precipitation processes, open water evaporation, transpiration and stem flow, soil water dynamics, groundwater recharge, subsurface flow mechanisms, and runoff generation. Natural tracers enter hydrological systems by diffuse and continuous processes via precipitation, the outflow of certain reservoirs or the solution from minerals. The fact that the input function or the injection of tracer to the hydrological system is provided by nature is a major advantage of natural tracers. This enables investigations on a large scale with respect to both space and time, and natural tracers are, thus, particularly useful for integrated approaches at the catchment scale and water balance studies. Environmental tracers can also be used for very long timescales, including paleohydrological studies such as for the analysis of rainfall origin or recharge in the Holocene or for relict groundwater formation in arid areas, if the past input of an environmental tracer can be reconstructed from data or from physical principles. In other words, environmental tracers allow information from experiments to be obtained which have been started long before a particular research project actually started. A disadvantage is that often the input function is difficult to define. In addition, the input and output signals might be relatively weak.

Artificial tracers are added to a hydrological system in planned experiments by well-defined injections. The boundary conditions for such experiments are generally much easier to define than in the case for natural tracers. The input signal can also be much more pronounced and it is possible to label a specific component of the investigated system such as an inflow to a lake. On the other hand, the scales in time and space for application are limited, and it might only be possible to gain insights into a part of the system during the time of the experiment. For example, artificial tracers cannot usually be used for systems having transit times larger than 1 year for practical reasons. Besides, catchment-wide injections are usually not possible.

2.09.3.2 Environmental Tracers

2.09.3.2.1 Isotope tracers ^{18}O and ^{2}H

Isotopes are different types of atoms of a certain element having a different atomic mass due to a different number of neutrons. Some isotopes are stable, whereas others are radioactive. The stable isotopes oxygen-18 (^{18}O) and deuterium (^{2}H) are the most commonly used environmental tracers. Since both isotopes are naturally occurring as part of the water molecule, they provide almost perfect tracers for the flow of water. The relative occurrence of these isotopes compared to ^{16}O and ^{1}H can be expressed as abundance ratio.

Usually, the isotopic content of a water sample is expressed by δ-values. These are computed based on the isotope fractions in a certain sample (R_{sample}) compared to the ratio of a standard ($R_{standard}$). Generally, the internationally accepted standard, Vienna Standard Mean Ocean Water (VSMOW), is used. In this water, the average ratio of ^{18}O compared to ^{16}O is $2.005\ 2 \cdot 10^{-3}$ and the ratio for ^{2}H compared to ^{1}H is $1.557\ 5 \cdot 10^{-4}$:

$$\delta = \frac{R_{sample} - R_{standard}}{R_{standard}} \times 1000(‰) \quad (1)$$

The values of $\delta^{18}O$ or $\delta^{2}H$ are expressed as ‰ difference from the standard being used. Positive δ-values indicate an increased concentration of ^{18}O or ^{2}H compared to the standard being used, whereas negative δ-values denote a decrease of heavier isotopes in the sample. When VSMOW is used as standard (as is usually done), ocean water by definition has a $\delta^{18}O$ value close to 0‰. In most parts of the hydrological cycle, ^{18}O occurs less frequently than in ocean water, which means that δ-values are usually negative with most negative values found in ice samples from cold, arctic regions (about −50 to −25‰). While δ-values do not directly express concentrations, they can be used as concentrations in most cases. The δ-notation can be confusing because there are different ways to compare the δ of two water samples: lower or higher, more or less negative, lighter or heavier, and depleted or enriched (**Table 2**). Lower, that is, more negative δ-values imply a lower concentration of the heavier isotopes and, thus, lighter water (e.g., water from glaciers). A sample with higher, that is, less negative δ-values, on the other hand, can also be described as heavier water as it contains more ^{18}O isotopes (e.g., ocean water or rainwater during the summer season). Sometimes a sample is also described to be enriched or depleted; here, it is important to be clear about what isotope is enriched or depleted in the sample relative to another.

Fractionation processes. These processes cause changes of the isotopic composition through an exchange of isotopes due to physical or chemical processes. In the case of $^{18}O/^{16}O$ and $^{2}H/^{1}H$, such processes by which the isotopic concentrations are changed are mainly related to phase changes of the water. Until equilibrium is reached, the exchange continues. Note that an equilibrium state does not imply that the phases have the same isotope concentrations, but that the concentrations no longer change over time. Evaporation is of special importance for the fractionation of $^{18}O/^{16}O$ and $^{2}H/^{1}H$. Since the light isotopic species of water ($^{1}H_2^{16}O$) has a higher vapor pressure than heavier species ($^{1}H_2^{18}O$ or $^{1}H^{2}H^{16}O$), the latter are less likely to be evaporated than the light water species (**Figure 3**). In other words, when water is evaporating, the evaporated water will have lower δ-values (more negative) than the remaining water. Consequently, the effect can be used to determine evaporation rates of lakes (Gibson et al., 1996). Larger increases in the δ-values (i.e., values becoming less negative or even positive) of the lake water over time correspond to higher evaporation rates.

Variation in precipitation. As a result of the fractionation processes, clear geographic and seasonal patterns of isotopic precipitation composition can be observed (Rozanski et al., 1992). These patterns form the basis for many hydrological studies using isotopes. The two major factors controlling the

Table 2 Different expressions used to compare the isotopic composition of two samples

$\delta^{18}O = -8‰$	$\delta^{18}O = -15‰$
Higher δ-value	Lower δ-value
Less negative δ-value	More negative δ-value
Higher concentration of ^{18}O	Lower concentration of ^{18}O
Enriched in ^{18}O	Depleted in ^{18}O

isotopic composition of precipitation are temperature, which controls fractionation, and the amount of the original water vapor, which has not become precipitation earlier. As a result of these two factors, the following effects can be observed:

- *Latitude effect.* The concentration of heavier isotopes is lower at higher latitudes. The explanation is that a larger part of the vapor has already been precipitated and that fractionation during phase transitions between water and vapor is more pronounced at low temperatures.
- *Elevation effect.* Precipitation is increasingly depleted in heavier isotopes at higher altitudes. This is the combined result of increased fractionation due to lower temperatures and moisture depletion by adiabatic cooling during orographic precipitation. Equilibrium fractionation increases with lower temperatures, making fractionation more efficient at higher altitudes. Repeated rainout during the uplift of air masses further causes a decreasing concentration of heavier isotopes.
- *Continental effect.* Precipitation becomes more depleted in heavier isotopes with increasing distance from the source of the water vapor, which is usually the ocean. On its way over a continent cloud water in the concentration of heavier isotopes will decrease with each rainfall event along the trajectory.
- *Seasonal effect.* Typical seasonal variations of stable isotope compositions are observed for many regions. This is a result of different sources and trajectories of the air masses reaching the region and varying fractionation processes in the source area of atmospheric moisture as well as along its way to the region.
- *Amount effect.* Rainwater during small events usually has higher concentrations of heavier isotopes than rainwater collected from large rainfall events. This effect is caused by evaporation of rain on its way toward the ground and is mainly observed for light rains or early rains during an event. Note that this effect does not apply to snowfall.

The above effects can be seen in long-term average data. For individual precipitation events, there can be a large variability both between and within events due to air masses of different origin and history.

There is usually a clear relation between $\delta^{18}O$ and δ^2H in precipitation (**Figure 4(a)**). At a global scale, $\delta^{18}O$ and δ^2H in precipitation is characterized by the equation $\delta^2H = 8\delta^{18}O + 10$ (Craig, 1961). The equation is also called the global meteoric water line (GMWL). Based on the data of the IAEA global network of isotopes in precipitation (GNIP), a revised version of the GMWL with a slope of 8.17 ± 0.07‰ and an intercept of 11.27 ± 0.65‰ has been proposed (Rozanski *et al.*, 1992). Deviations from this global correlation exist at a regional scale and, for several regions, specific meteoric water lines have been determined. Deviations from the GMWL are especially significant in coastal areas, on islands, and in tropical mountainous regions with typical slopes and intercepts. Some regional meteoric water lines should be interpreted with care as local meteoric water lines might be influenced by deficiencies of the sampling network or less cautious procedures than those used by IAEA. When water is affected by evaporation from free water surfaces, the increase in heavier isotopes in the remaining water differs for ^{18}O and ^2H. As a result of this, the slope of the $\delta^{18}O$–δ^2H relationship differs from the slope of the meteoric water line in the $\delta^{18}O$-δ^2H diagram with typical evaporation slopes varying between 4 and 5.5 (**Figure 4(b)**). This allows determining, for instance, whether a lake contributes to groundwater recharge.

Sampling and measuring. The accuracy of stable isotope measurements depends on sampling procedures and the analytical technique used. The most important issue when

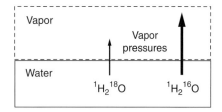

Figure 3 Differences in the vapor pressure for the two isotope species $^1H_2^{18}O$ and $^1H_2^{16}O$ during equilibrium exchange with water vapor. From Leibundgut Ch, Malozewski P, and Külls Ch (2009) *Tracers in Hydrology*. Chichester: Wiley.

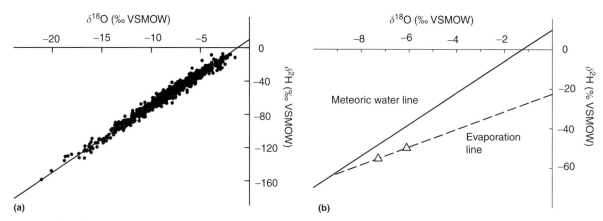

Figure 4 $\delta^{18}O$–δ^2H diagrams/global meteoric water line (GMWL). (a) GMWL and isotopic composition of samples (1998–2008) from two precipitation stations in the Black Forest, Southern Germany (Katzensteig and Schweizerhof). (b) GMWL and samples of surface water (triangles) influenced by evaporation. Modified from Leibundgut Ch, Malozewski P, and Külls Ch (2009) *Tracers in Hydrology*. Chichester: Wiley.

taking water samples for isotope measurements is to avoid evaporation because of the associated artificial fractionation processes. Using double inlet mass spectrometry, the analytical error amounts to about ±0.1‰ for $\delta^{18}O$ and ±1.0 to 1.5‰ for δ^2H. As an alternative, isotope concentrations can be measured using tunable diode laser spectroscopy, which is an innovative technique for stable isotope measurements and has gained importance recently. This new technique is used by an increasing number of research groups and allows isotopes to be analyzed at considerably lower costs with a precision of about 0.3‰ for $\delta^{18}O$ and of about 1.0‰ for δ^2H. Another advantage is the smaller sample volume needed for the analysis. To analyze both $\delta^{18}O$ and δ^2H by traditional mass spectrometry normally requires a minimum sample volume of about 15 ml, while 2 ml are sufficient when using tunable diode laser spectroscopy. This represents important progress, for instance, for hydrological isotope studies investigating sap flow and soil water. However, for practical reasons and to minimize sources of uncertainties, it is usually recommendable to take for hydrological purposes samples not <50 ml in the field.

2.09.3.2.2 Other environmental tracers

There are several other environmental tracers which can provide useful information. First, there are additional isotopes. The $^{13}C/^{12}C$ ratio can be used to trace sources of carbon dioxide and the $^{15}N/^{14}N$ ratio can be used to trace sources of nitrogen. Radioactive isotopes, such as tritium or ^{14}C, have been used for age dating of runoff and groundwater. Geochemical compounds, such as silicate, chloride, and dissolved organic carbon (DOC), can also be used as tracer. However, here it has to be considered that these tracers might not be conservative on their way through a hydrological system. Temperature and electrical conductivity provide simple-to-measure tracers, the latter being a lumped measure of ions in the water.

Some tracers, such as krypton-85, CFCs, and SF_6, have been released to the atmosphere as a result of industrial activities. They can nevertheless be considered as environmental tracers because they can now be utilized in a similar way to naturally occurring tracers. In particular, they are applied over large areas in nonplanned experiments, a characteristic they have in common with tracers such as ^{18}O or ^2H. They are in particular used for estimating the residence of water in groundwater.

2.09.3.3 Artificial Tracers

2.09.3.3.1 Characteristics of artificial tracers

Classically, four main groups of suitable artificial tracers are distinguished based on their chemical appearance: fluorescent, salt, radioactive, and particulate tracers. In recent years, additional substances, such as dissolved gases, nonfluorescent dyes, fluorobenzoic acids (FBAs), or deuterium, were also applied successfully (and increasingly) in hydrological tracer studies. Fluorescent tracers, however, are still the most important and most applied tracers (cf. Table 1).

For the primary purpose of tracing the flow of water, the substance needs to be conservative in the aquatic environment under the conditions of the hydrological system. A tracer is considered conservative if it is physico-biochemically stable

Table 3 Required properties of artificial tracers in general and in the example of fluorescent tracers in particular

Properties to be considered	Requirements of an ideal (conservative) tracer
1. Solubility in water	High
2. Fluorescence intensity	High
3. Detection limit	Low
4. *pH dependency*	Low
5. *Temperature dependency*	Low
6. *Photolytic stability*	High
7. *Sorption processes*	Negligible
8. *Chemical and biological stability*	High
9. Toxicity and related environmental effects	None or minimal
10. Costs and other practical aspects	Low or moderate

The properties in italic are the main characteristics of conservative tracers. From Leibundgut Ch, De Carvalho-Dill A, Maloszewski P, Müller I, and Schneider J (1992) Investigation of solute transport in the porous aquifer of the test site Wilerwald (Switzerland). *Steirische Beiträge zur Hydrogeologie* 43: 229–250.

(nonreactive in natural water) and not sorptive. Particularly in classical quantitative tracer investigations, it is essential to use conservative tracers to allow correct determination of hydrodynamic system properties and flow and transport parameters. Besides conservative behavior, further characteristics are needed to make a substance an appropriate tracer in hydrological field experiments from a more practical point of view. These properties are listed in Table 3. If these requirements are met completely, we refer to the tracer as ideal, albeit only nearly ideal artificial tracers occur in reality.

In principle, when choosing an artificial tracer for application in (field) experiments, the guideline followed is that an ideal tracer represents the water flow, but nonideal tracers can also be useful for special applications. In any case, a sound knowledge of the characteristics of the tracer substances and the respective measurement techniques is required in order to perform experiments successfully both in the field and in the laboratory. In the following, the principles of (ideal) tracer properties are discussed using fluorescent tracers as an example, but are valid principally for all other groups of artificial tracers. Thereafter, a short summary of the other artificial tracer groups is given.

2.09.3.3.2 Fluorescent tracers

Fluorescence is a luminescence that occurs where energy is emitted as visible light after being absorbed as electromagnetic radiation of a different wavelength. The substances used for tracing purposes are situated within the small range of visible light between the higher ultraviolet and the infrared wavelengths (~350–750 nm). They are characterized by specific fluorescence spectra corresponding to their wavelengths of excitation and emission. The emission takes just as long as the activation is driven by an excitation energy source, causing only transient fluorescence effects compared to longer-lived phosphorescence. The intensity of fluorescent emission follows a linear dependency involving the intensity of incident light and the tracer concentration over a wide range. At very high concentrations, the fluorescence intensity is reduced, because a self-shadowing effect of the molecules, called 'quenching', occurs. However, as the expression tracer implies, only very low concentrations are used as far as possible.

The spectral fluorometer, which features a light source that excites the sample and a detector that measures the emitted light spectrum, is the most common laboratory device for fluorescent tracer analysis today. In the future, advanced multicoupled analytical techniques, such as high-performance thin-layer chromatography with automated multiple development (HPTL/AMD), nano-chip liquid chromatography/quadruple time-of-flight mass spectrometry (Nano-Chip-LC/QTOF-MS), and laser spectroscopy, might become important instruments. On-site measurements to monitor the tracer breakthroughs are often desirable. For fluorescent tracers, different types of *in situ* devices are available (fiberoptic-, flow-through-, pocket-, and Xe-flashlight fluorometers). A completely different sampling approach is placing active charcoal bags (probe, fluocapteur) in the system under investigation, with the subsequent tracer extraction in the lab, which provides only the cumulative tracer amount. The technique is applied to ensure that no tracer passes unobserved at remote sampling points or where site access is difficult (e.g., in karst caves or glaciers) during long-term experiments or at places where no tracer propagation is expected.

All fluorescent tracers in use are organic dyes. Relevant characteristics of fluorescent tracers are listed in **Table 4**. In the following, the different characteristics are briefly summarized and, as already mentioned, are exemplarily valid for the other artificial tracers.

The detection limit plays a key role for a substance's applicability as artificial tracer because it is strongly linked to rather practical criteria of experiment realization, which include questions about required tracer mass, possibility of adequate execution of tracer injection, costs, and, not least, the degree of interference to the aquatic environment of an experiment. The level of detection limits exhibited by the whole group of fluorescents is quite appropriate for hydrological tracing purposes. The detection limit depends, on the one hand, on the fluorescence intensity to which it is positively correlated, and, on the other hand, on the background fluorescence of the sample. Relative fluorescence intensities of some commonly applied fluorescent tracers are shown in **Table 4**. Uranine plays a dominant role among the fluorescent tracers, due to its much higher fluorescence intensity.

The solubility of tracers in water is a crucial requirement of tracers used to investigate water flows in the hydrological cycle, as the tracer should be as close to the characteristics of water as possible. Pyranine, uranine, eosine, and naphthionate are characterized by good-to-very-good solubility. By contrast, the solubility of the rhodamines is considerably lower. The temperature dependency of the fluorescent tracers is usually unproblematic.

A changing pH value has the potential to change the electrical charge of the molecule from negative, through neutral to a positive value, and vice versa, which again affects the excitation and emission spectra of the fluorescent tracers. If the medium reaches a certain degree of acidity, the compounds partially lose their fluorescence. The fluorescent tracers' sensitivities to pH vary, in the pH range of natural waters uranine and pyranine react most sensitively. This needs to be considered especially for *in situ* measurements. Problems occur principally in peat-bog and swamp regions, and generally in acid soils and crystalline geological settings, whereas in laboratory analysis of samples, the problem can be managed quite simply as the pH dependency of dye tracers is reversible by means of adequate buffering prior to fluorometric analysis.

The exposure to light, called 'photolytic dependency', has an irreversible effect on fluorescence. The different sensitivities of tracers to light should be considered when planning the use of fluorescent tracers in experiments. In principle, the two standard tracers, uranine and eosine, are clearly not suitable for surface water experiments during daylight, except for experiments of short duration.

The sorption behavior is the most important criterion relevant to the use of artificial tracers generally, and fluorescent tracers in particular, because sorptivity is strongly linked to the question of conservativity. Sorption is a highly complex but crucial process in the performance of experiments in the unsaturated and saturated zones. Depending on their molecular makeup, fluorescent tracers exhibit widely contrasting

Table 4 Summary of the relevant characteristics of the fluorescent tracers

Tracer	Ex/Em (nm)	Rel. fluorescence yield (%)	Detection limit (mg m^{-3})	Toxicity (–)	Solubility (20 °C) (g l^{-1})	Light sensitivity (–)	Absorption behavior (–)
Naphthionate	325/420	18	0.2	Harmless	240	High	Very good
Pyranine	455/510	18	0.06	Harmless	350	High	Good
Uranine	491/516	100	0.001	Harmless	300	High	Very good
Eosine	515/540	11.4	0.01	Harmless	300	Very high	Good
Amidorhodamine G	530/555	32	0.005	Sufficient	3	Low	Sufficient
Rhodamine B	555/575	9.5	0.02	Toxic	3–20	Low	Insufficient
Rhodamine WT	561/586	10	0.02	Toxic	3–20	Very low	Insufficient
Sulforhodamine B	564/583	7	0.03	Sufficient	10 (10 °C)	Low	Insufficient

Ex/Em: excitation and emission wavelength respectively. Relative fluorescence yield with respect to the fluorescence intensity of uranine, which is among all fluorescent tracers the most sensitive one. From Leibundgut Ch, De Carvalho-Dill A, Maloszewski P, Müller I, and Schneider J (1992) Investigation of solute transport in the porous aquifer of the test site Wilerwald (Switzerland). *Steirische Beiträge zur Hydrogeologie* 43: 229–250.

reactions upon contact with different substrates. Cationic tracers usually interact more strongly with the substrates than anionic ones; however, both groups usually react in a way that is referred to as reversible sorption. As was mentioned above, changing pH values have the potential to change the electrical charge of the molecule. Thus, the sorption affinity of fluorescent tracers also depends on the pH. The two parameters, K_d [$L^3 M^{-1}$] (distribution coefficient) and R_d [-] (retardation coefficient), characterizing sorption processes are commonly used in tracer studies. K_d represents the mobility of the tracers in a certain medium, describing the thermodynamic equilibrium of the tracer between substrate and solution. Describing the partitioning between liquids and solids using K_d is only valid if the involved reactions are fast and reversible, and if the isotherm is linear. Commonly K_d is determined by means of batch experiments (Equation (2)) and depends on the ionic composition of the exchanger and the solution used:

$$K_d = \frac{V}{m} \frac{c_i - c_s}{c_s} \quad (2)$$

where V is the volume of the solution [L^3], m the mass of the dry substrate [M], c_i the initial tracer concentration [$M L^{-3}$], and c_s is the dissolved tracer concentration (equilibrium solution) [$M L^{-3}$].

R_d [-] describes the retardation of average tracer transport velocity v_t [$L T^{-1}$] compared to the average flow velocity of water v [$L T^{-1}$] (Equation (3)):

$$R_d = \frac{v}{v_t} \quad (3)$$

R_d can also be expressed as a function of K_d, porosity n [-], and the bulk mass density ρ_b [$M L^{-3}$] (Equation (4)):

$$R_d = 1 + \frac{\rho_b}{n} K_d \quad (4)$$

Both parameters assume a linear sorption isotherm with instantaneous equilibrium, which is applicable only to a limited extent under the natural conditions of experiments since most sorption and desorption processes follow kinetic reactions in reality. Direct transfer of reference values of K_d and R_d obtained from laboratory experiments to field conditions is not possible; nevertheless, they provide a useful guideline for the estimation of the sorption loss of a given tracer and, consequently, the estimation of the tracer injection mass required. Sorption tests have been performed and distribution and retardation coefficients can be found in Leibundgut et al. (2009).

The chemical and biological stability of fluorescent tracers may readily decompose as a consequence of oxidation and other chemical changes. However, oxidative processes affect the dye tracers, as they are organic compounds, to different degrees. In addition, microbial degradation of fluorescent tracers is known to occur both under natural conditions during experiments and in stored samples. High salinity may decrease fluorescence intensity but generally to a much lesser degree than either pH effects or exposure to light. Further processes that may cause problems, especially in the case of long-term experiments, were reported, but there are examples indicating fluorescents to remain rather stable in unpolluted aquifers. However, it is strongly recommended to use brown glass bottles for sampling and to execute analysis as quickly as possible to minimize these problems. Dealing with less pure water, one must be aware of the potential effects of degradation or chemical reactions on the experimental results.

Toxicity and related environmental effects have to be considered. Each injection of artificial tracer in a hydrological system is in a sense a contamination of the water body in question. Therefore, it is a strong requirement to users to apply tracer masses as low as possible. Carefully planned and correctly prepared tracer experiments generally involve only minimal quantities of fluorescent tracer substances. Thus, the contamination is usually tolerable. Summarizing the results of studies related to human and eco-toxicological aspects, it can be stated that uranine is harmless and eosine, pyranine, and naphthionate appear to be harmless. It is suspected that the rhodamine group as a whole is toxic, except amidorhodamine G and sulforhodamine B, which are less problematic. When preparing a tracer experiment, it is important that national regulations pertaining to tracer experiments are consulted.

Due to their relatively easy handling, the high sensitivity of the analysis, the low detection limit, and, consequently, the small quantity of tracer mass needed in field experiments fluorescent tracers are popular, in general, among tracer hydrologists. They are also attractive because of the linearity of the calibration curve in the measuring scale, and comparable low toxicity levels. For a better evaluation in the evaluation of tracer substances, a summary assessment of the relevant properties of the commonly used fluorescent tracers is provided in **Table 4**. For further information the reader is referred to Leibundgut et al. (2009), where a comprehensive presentation and discussion of the fundamental methodological basics concerning the hydrological application of fluorescent tracers are given.

2.09.3.3.3 Nonfluorescent artificial tracers

Salts are inorganic compounds which break up into cations and anions when dissolved in water. Various salts have been used as tracers such as sodium chloride (NaCl), potassium chloride (KCl), sodium bromide (NaBr), lithium chloride (LiCl), borax ($Na_2B_4O_7$), and potassium iodide (KI). The advantages of salt tracers are their common availability, relative low costs, simple handling, and the potential for continuous measurements. The disadvantages of salt tracers include high natural background concentrations in many cases, relatively high detection limits, and issues related to sorption and ion exchange. The first two points often result in the need for a large mass injection, which in turn might cause transport problems and environmental concerns. While many salt tracer experiments have been performed in the past, these disadvantages limit the usability of salts as tracers. Salt tracers are mainly useful in small-scale experiments such as soil column tests or investigations in small surface water bodies. It should be noted that sodium chloride is widely used as a tracer for discharge measurements using the dilution method (see Section 2.09.4.5). Of the several anions tested as hydrological tracers, bromide (Br^-) might be the most suitable. Bromide is often used for tracer experiments in the vadose zone, and as a reference tracer for comparison purposes (e.g., Onodera and Kobayashi, 1995; Sambale *et al.*, 2000; Parsons *et al.*, 2004; Einsiedl, 2005; Hangen *et al.*, 2005). The advantages of bromide compared to other anions are lower background concentrations in natural waters (~ 300 times lower than Cl^-), low toxicity, and low sorption to soil particles.

Drifting particles, such as spores, phytoplankton, bacteria, viruses, phages, DNA tracers, and microspheres, constitute another group of artificial tracers. Their most characteristic feature is that they are not in solution. Obviously, they do not correspond to the properties of a conservative (ideal) tracer. They particularly have considerable potential as tracers for special applications, as in the investigation of the filtration capacity of the unsaturated zone and aquifers. They are being used to study and investigate the flow behavior of microorganisms and particles in saturated, unsaturated, and surface water systems with respect to the infiltration of contaminants in sewage or irrigation water, and for all applications with hygiene implications, and impacts on water supply installations and water protection zones (Sabir *et al.*, 1999; Auckenthaler *et al.*, 2002; Zvikelsky and Weisbrod, 2006).

The potential health and environmental risks posed by radioactive tracers mean that the use of human applied radioactive substances is very limited nowadays. The particular suitability of radioactive tracers is due to the very high sensitivity and the possibility of a selective detection, the disappearance of the tracer from the system due to decay, and the ability to follow the flow path of water and tracer using a Geiger counter. One of the most important methods when using radioactive tracers are the single-well technique and groundwater recharge estimation (Moser and Rauert, 1980; IAEA, 1983).

A variety of FBAs have been used as tracers, especially for vadose zone hydrology because of the lack of background concentrations and the relatively good mobility properties. Their use in hydrological applications has received considerable attention over the past 20 years. They appear to be useful tracers that behave under most conditions found in soils and aquifers nearly conservative. FBA tracers are applied in studies of both porous and fractured aquifers as well as in water flow and solute transport in the unsaturated zone. Most FBAs are quite expensive. The greatest potential use of FBAs would appear to be in multitracer tests due to the wide variety of available isomers in this tracer family displaying comparable characteristics (Dahan and Ronen, 2001; Hu and Moran, 2005).

The stable isotopes ^{18}O and ^{2}H are used as environmental tracers, but the isotopes can also be used as artificial tracers when water with a specified isotope composition (usually with enriched concentrations of ^{18}O or ^{2}H) is injected into a system. In laboratory tests, lysimeter and groundwater studies, and for investigations of solute transport in the unsaturated zone and water flow in the soil–vegetation system, deuterated water has been used successfully (Garcia Gutiérrez *et al.*, 1997; Himmelsbach *et al.*, 1998; Becker and Coplen, 2001; Stamm *et al.*, 2002; Mali *et al.*, 2007). An advanced application of deuterated water is the estimation of tree transpiration and the investigation of plant water uptake in the xylem flow (Calder, 1992). Deuterium can be purchased in concentrated form (i.e., water with a high portion of $^{2}H_2O$ molecules), but only applications involving relatively small volumes of water are feasible because of the high costs. Despite this restriction, deuterium is an attractive artificial tracer, particularly for investigations in the unsaturated zone and plant water transport (Königer *et al.*, 2010). Due to even higher costs, ^{18}O is hardly used as artificial tracer.

The use of dissolved gas tracers as environmental tracers in paleohydrological studies was already proposed 50 years ago. The widespread use as tracers was inhibited by technical difficulties related to the injection, sampling, and analysis of the gases. Many of these problems have been overcome in recent times and, since the 1990s, their application as artificial tracers in hydrology has increased. Usually, applied gas tracers include helium (He), neon (Ne), stable isotopes of krypton (Kr), and SF_6 (Wilson and McKay, 1996; Solomon *et al.*, 1998; Divine *et al.*, 2003). The advantages are inert and nontoxic behavior in hydrological systems as well as low background concentrations compared to the concentrations of the injected dissolved gas. This means that dissolved gas tracers can be used for tracer experiments, including large volumes of water. The volatile nature of these tracers makes them obviously different from other tracers and demands specific considerations for tracer injection, sampling, and analysis to prevent unwanted degassing. The volatile nature of gas tracers can, however, also be used as advantage when degassing is used on purpose. This approach has been suggested, for instance, to delineate unsaturated zones or to detect pools and residual zones of nonaqueous phase liquids (NAPLs) in the subsurface using dissolved gas as a partitioning tracer.

Nonfluorescent dye tracers applied in staining techniques have attracted remarkable interest as a tool to demonstrate and study the occurrence of preferential flow in soils (Flury and Wai, 2003; see also **Chapter 2.13 Field-Based Observation of Hydrological Processes**). The visualization staining experiments in vadose zone hydrology rely on the otherwise problematic sorption effect of the tracer, ensuring its distinct visibility along its flow pathways. This is in contrast to the classical idea of tracer applications, as the tracer remaining along the flow pathway is of interest rather than the tracer outflow.

2.09.4 Tracer Hydrology Applications

Tracer methods allow numerous hydrological questions to be studied. In this section, some of these applications are discussed. The applications used as examples here help to answer the following questions: Where does storm runoff come from (Section 2.09.4.1)? How long does water travel in a catchment from rainfall to runoff (Section 2.09.4.2)? What are the main sources for nitrogen in streams (Section 2.09.4.3)? How can the transport in hydrological systems be described mathematically (Section 2.09.4.4)? How much water is flowing in a stream (Section 2.09.4.5)? How can groundwater recharge be estimated based on chloride concentrations (Section 2.09.4.6)? Also, how can flow in a complex porous aquifer be described (Section 2.09.4.7)? While the tracer approaches listed here include some of the most important applications of tracer hydrological methods, this section is of course by no means a complete review of all approaches by which tracer methods make important contributions to hydrology (see also Chapter 2.13 Field-Based Observation of Hydrological Processes and Chapter 2.20 Stream–Groundwater Interactions).

2.09.4.1 Hydrograph Separation

One of the most significant contributions of tracer methods to hydrological science is the result of isotopic hydrograph separation studies. The idea is that one can estimate the contribution of groundwater to stream runoff during events based on the mixing ratio of precipitation and groundwater in the stream. Starting with the influential works by Pinder and Jones (1969) and Sklash and Farvolden (1979), many studies have used this approach to estimate the amount of old water in stream flow events based on the isotopic composition of the different components. This old water is usually interpreted as groundwater already being stored in the catchment before the event compared to precipitation as new water. Isotopic hydrograph separation has also been used for studying snowmelt runoff (Rodhe, 1981, 1987). Results, especially in humid catchments, have repeatedly shown that this old water is a surprisingly large portion of the total event runoff for both rainfall and snowmelt events (Genereux and Hooper, 1998). These results dramatically changed the view on runoff generation from theories involving significant amounts of overland flow to theories where groundwater plays a more prominent role.

The tracer approach to hydrograph separation requires the concentration of isotopes or other tracers in the different sources of stream water (e.g., precipitation and groundwater) to be significantly different. Hydrograph separation is then the computation of the relative contributions of these sources based on end-member-mixing analysis (EMMA; Hooper et al., 1990). The basic idea is that different flow components can be separated based on mass balance calculations for both water and tracer(s) based on known end-member compositions. Details on the method are given by Genereux and Hooper (1998) and Buttle (2005). In principle, end-member analysis can be used to differentiate any number of flow components by using the same number of end-members. In practice, however, hydrograph separation is mainly applied for two or three components, as it is difficult to find more suitable, distinctly independent tracers and also error propagation becomes an increasingly problematic issue when more end-members are used.

In its simplest version, the end-member analysis is used to distinguish two components which are often called new and old water or event and pre-event water. Old, or pre-event, water is the water that is already in the catchment before the event starts, whereas new, or event, water is the water that enters the catchment as rainfall or snowmelt. Combining the mass balance equations for water (Equation (5)) and tracer (Equation (6)), the fraction of old water, X [–], in the event stream flow can then be computed (Equation (7)). Q $[L^3 T^{-1}]$ is the flow of water, C $[M L^{-3}]$ is the concentration of the tracer and the subscripts T, P, and E refer to total, pre-event, and event, respectively. While the δ-values are, strictly speaking, no concentrations, the δ-values can be generally used directly as concentrations in these calculations:

$$Q_T = Q_P + Q_E \quad (5)$$

$$c_T Q_T = c_P Q_P + c_E Q_E \quad (6)$$

$$X = \frac{Q_P}{Q_T} = \frac{C_T - C_E}{C_P - C_E} \quad (7)$$

The isotope concentrations of both event water and stream flow are measured during the event, whereas the isotopic composition of the pre-event water is usually assumed to equal the composition of stream flow before the event. This seemingly simple method of hydrograph separation is complicated by the fact that the isotopic composition of both event and pre-event water might vary in time and space.

The problem of defining input concentrations is of special concern when performing hydrograph separations for snowmelt events, because the isotopic composition of the event (=melt) water may, in this case, vary considerably with time. Laudon et al. (2002) suggested a method to consider these temporal variations of the event water concentration. Their so-called runCE method considers the timing and amount of melt water entering the soil water reservoir and the portion of previously melted water which already left the catchment (i.e., runoff of event water). By assuming full mixing of the soil reservoir, the event-water isotopic composition is computed based on the cumulative snowmelt (including rain water) from the snow lysimeters and the cumulative volume of melt water that has drained from the snow pack but has not yet left the catchment as stream flow during the event. In theory, using ^{18}O and ^2H for hydrograph separation should provide the same results. Lyon et al. (2009), however, demonstrate that results might differ for the two isotopes and attribute this to spatial and temporal isotopic variability.

Most often, hydrograph separations have been performed in single catchments. There is great value in comparative studies with simultaneous measurements for hydrograph separation in nearby catchments as demonstrated for spring-flood hydrograph separations for 15 nested catchments in Northern Sweden (Laudon et al., 2007). There was a good correlation between the portion of new water (i.e., snowmelt) in the streams and wetland percentage in the catchments, which was interpreted as an effect of surficial runoff on top of frozen wetlands. Only by using several catchments could these differences in runoff generation mechanisms be inferred.

2.09.4.2 Catchment Transit Time Estimation

Water spends different amounts of time in a catchment from entering the catchment as precipitation until leaving the catchment as runoff due to a variety of possible flow pathways. Transit time distributions can be used to describe this variation and the mean of these distributions, the MTT, is an important characteristic of a catchment (McGuire and McDonnell, 2006). In the literature, often, the terms transit (or travel) time and residence time are used with no distinction. However, in most systems, there is a difference between the time water parcels need to travel from input to output and the time a water parcel has been in the system. If the simple case of a piston flow system is considered (i.e., each water parcel travels the same flow path and there is no dispersion), the MTT will be twice the mean residence time (McGuire and McDonnell, 2006).

The transit time distribution $g(t)$ can be thought of as the tracer concentration response in runoff at the outlet to an instantaneous, conservative tracer addition over the entire catchment area (for simplicity, a zero background concentration is assumed; Maloszewski and Zuber, 1982; McGuire and McDonnell, 2006). The transit time distribution can be described by

$$g(t) = \frac{c(t)}{\int_0^\infty c(t)\,dt} = \frac{c(t)Q}{M} \qquad (8)$$

$c(t)$ [M L^{-3}] is the tracer concentration caused by an injection of a certain mass of tracer (M [M]) at $t=0$ and Q [L^{3}T^{-1}] is the discharge. The transit time distribution $g(t)$ describes how much of the tracer at a given time is leaving the catchment. The MTT of the tracer (τ_M) is then the average arrival time of the tracer at the catchment outlet (Equation (9)):

$$\tau_M = \frac{\int_0^\infty t\, c(t)\,dt}{\int_0^\infty c(t)\,dt} = \int_0^\infty t\, g(t)\,dt \qquad (9)$$

Commonly used transit time distribution functions, g(t), include the piston flow model (no mixing, Equation (10)), the exponential model (complete mixing, Equation (11)), the combined exponential piston flow model (Equation (12)), or the advection–dispersion model (Equation (13)) (see also Section 2.09.4.4):

$$g(t) = d(t - \tau_M) \qquad (10)$$

where $d(t)$ denotes the Dirac delta function:

$$g(t) = \frac{1}{\tau_M} \exp\left(\frac{t}{\tau_M}\right) \qquad (11)$$

$$g(t) = \frac{\eta}{\tau_M} \exp\left(-\frac{\eta t}{\tau_M} + \eta - 1\right) \quad \text{for } \tau > (\eta - 1)\tau_M/\eta \qquad (12)$$
$$g(t) = 0 \quad \text{for } \tau \leq (\eta - 1)\,\tau_M/\eta$$

$$g(t) = \frac{1}{t\sqrt{4\pi P_D t/\tau_M}} \exp\left(-\frac{(1 - t/\tau_M)^2}{4\,P_D\,t/\tau_M}\right) \qquad (13)$$

P_D [-] is the dispersion parameter and η [-] is the ratio of the total water volume in the system to the volume of that part of the water characterized by the exponential transit time distribution.

The roof project at Gårdsjön, Sweden, provided a special opportunity for transit time estimations (Rodhe et al., 1996). In this catchment scale experiment, a transparent plastic roof was constructed below the tree crowns to intercept the acid deposition. Beneath the roof, water input to the catchment was simulated by an irrigation system using water from a nearby lake for irrigation. The isotopic composition of the irrigated water differed clearly from that of the natural precipitation due to fractionation during lake evaporation. This resulted in a step change of input water isotope concentrations, which could be used to estimate the transit time distribution directly.

In most cases, however, there is no instantaneous input (or change of input) of a tracer to the catchment, but transit time distributions are derived from time series of the input and output of a conservative tracer such as ^{18}O. In this case, the transit time distribution $g(t)$ can be derived by inverse modeling using the convolution integral approach. This means that the concentration at the outlet at time t, $c_{out}(t)$, can be expressed as a function of $g(t)$ and the input concentrations, c_{in}, during the preceding period (Equation (14)):

$$c_{out}(t) = \int_0^\infty g(\tau)\, c_{in}(t - \tau)\,d\tau \qquad (14)$$

In Equation (14), steady state is assumed, that is, flow does not change over time. This can be relaxed if t and τ are expressed as flow time instead of clock time. Equation (14) also assumes that the flow pathways do not change significantly over time and that the transit time distribution, thus, is time-invariant. In reality, $g(\tau)$ might vary with time, because, for instance, different, faster flow pathways are activated during wetter conditions. These issues make transit time estimates challenging (McDonnell et al., 2010).

As for the hydrograph, MTT estimates are of special interest when they can be compared for different catchments. McGuire et al. (2005) computed MTT estimates for seven catchments in Oregon and found that these values could be related to a topography-based travel time estimate. Tetzlaff et al. (2009) extended this approach to 55 catchments in five different regions and found that the relations found in Oregon hold in some regions, whereas soil controls on transit times caused reversed relations between the topography-based travel time estimate and MTT in other regions.

2.09.4.3 Analysis of Sources of Nitrogen in Streams

Tracer methods can also be used to trace the sources of nutrients and pollutants and, thus, support sustainable watershed management by providing information about origins and flow paths of nutrients and pollutants. The basic idea is similar to hydrograph separation, namely that different sources can be identified by their specific chemical composition. The different composition of water chemistry can thereby be used as a fingerprint to identify different sources of pollutions. This approach can be mathematically formalized as EMMA. In the case of nitrogen, the isotope ^{15}N can be used to identify sources of nitrogen in the stream flow, especially when combined with ^{18}O measurements, since different sources of nitrogen are characterized by different isotopic compositions (**Figure 5** Kendall, 1998). Several recent studies, mainly in North America, have highlighted the potential of using the combination of the two isotopes to determine the primary sources, flows, and fates of nitrogen at the landscape scale. While there is a great potential of using isotopes for the identification of sources, several studies also point out as-yet unsolved problems with these approaches (Bedard-Haughn

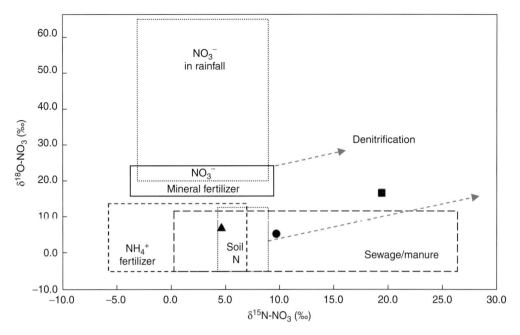

Figure 5 Results of nitrate isotope analysis and isotopic composition of various sources of nitrate. From Leibundgut Ch, Malozewski P, and Külls Ch (2009) *Tracers in Hydrology*. Chichester: Wiley.

et al., 2003). Fractionation processes of NO_3^-, for instance, might alter the $\delta^{15}N$ source signature. Wide ranges of $\delta^{18}O$ values for different nitrogen sources can make a quantitative source apportionment difficult. Furthermore, there are issues related to the sample collection of nitrate and the analysis of nitrogen and oxygen isotope ratios, as discussed by Silva et al. (2000). Despite all these difficulties, the combination of N and O isotope information may provide a valuable tracer approach for distinguishing among potential sources of nitrate. Battaglin et al. (2001), for instance, concluded based on isotope data, that "in-stream N assimilation and not denitrification accounts for most of the N loss in the lower Mississippi River during the spring and early summer months."

2.09.4.4 Advection–Dispersion Modeling

The advection–dispersion equations (ADEs), also called transport equations, are often used to describe the solute transport in systems such as aquifers or stream reaches. Advection refers to the transport of dissolved substances with the bulk water flow, whereas dispersion is caused by mechanisms such as velocity variations, mixing of different flow pathways, and molecular diffusion. Together with tracer experiments, ADE can be used to infer system parameters by inverse modeling. Obviously, it is important that the mathematical description of tracer transport and tracer behavior in the studied system is adequate and, in many situations, the simple ADE might not be appropriate because of processes such as chemical reactions along the flow pathways or an exchange of mobile and immobile water.

In a porous media, saturated flow can be described by a three-dimensional (3D) ADE when there is a steady, Darcian flow, no exchange with mobile water, and nonreactive solutes (ideal tracers). In most cases, the solution of such an ADE can only be found by applying numerical techniques. In cases where the flow lines in the medium can be assumed to be always parallel to the x-axis and where the tracer is already well mixed through the whole vertical distance of a homogeneous system, for example, by injection into an aquifer through a fully penetrating well, the 3D ADE can be simplified to the 2D form (Equation (15)), which describes tracer transport in the horizontal plane along the flow direction:

$$D_L \frac{\partial^2 C}{\partial x^2} + D_T \frac{\partial^2 C}{\partial y^2} - v \frac{\partial C}{\partial x} = \frac{\partial C}{\partial t} \quad (15)$$

D_L [$L^2 T^{-1}$] and D_T [$L^2 T^{-1}$] are the longitudinal and transverse dispersion coefficients, v [$L T^{-1}$] is the mean water velocity, t [T] is time, and x and y [L] are the Cartesian coordinates in a horizontal plane.

Equation (15) can only be solved numerically in most cases. One special case for which there is an analytical solution is an input which can be mathematically described by the Dirac function. This means that an instantaneous injection of the tracer can be assumed, as is adequate for many practical experiments. Equation (16) provides the analytical solution of the 2D ADE for the case of a homogenous aquifer where a tracer of mass M [M] is instantaneously injected over the entire depth at the origin of the coordinate system (i.e., $x = y = 0$) at time $t = 0$ (Lenda and Zuber, 1970):

$$C(x, y, t) = \frac{M}{nH} \frac{x}{4\pi v t^2 \sqrt{D_L D_T}} \exp\left[-\frac{(x-vt)^2}{4 D_L t} - \frac{y^2}{4 D_T t}\right] \quad (16)$$

where H [L] is the mean thickness of the aquifer and n [-] is the effective porosity. The velocity and dispersion coefficients (v, D_L, D_T) can then be estimated based on a tracer experiment where observation wells are situated perpendicular to the flow direction.

Transverse dispersion is usually much smaller than the longitudinal dispersion. In some cases, transverse dispersion can be neglected; examples are transport through soil columns when the tracer is injected throughout the whole cross section of the column perpendicular to the flow direction, or transport in streams, when the tracer is injected throughout the whole cross section of the stream. If the x-axis is defined to be parallel to the flow direction, the transport equation can then be reduced to its 1D form (Equation (17)), which can be solved analytically for an instantaneous injection (Equation (18); Lenda and Zuber, 1970; Kreft and Zuber, 1978):

$$D_L \frac{\partial^2 C}{\partial x^2} - v \frac{\partial C}{\partial x} = \frac{\partial C}{\partial t} \quad (17)$$

$$C(x,t) = \frac{M}{Q} \frac{x}{\sqrt{4\pi D_L t^3}} \exp\left[-\frac{(x-vt)^2}{4 D_L t}\right] \quad (18)$$

Q [$L^3 T^{-1}$] is the volumetric flow of water (flow rate through the column, river discharge or pumping rate in a combined pumping-tracer experiment). The two unknown parameters (v, D_L) can be calculated from the observed concentration (tracer breakthrough) curve.

Both the MTT of water, $t_0 = x/v$ [T], and the dispersion parameter, $P_D = D_L/(vx)$ [–] can be derived from tracer experiments. The latter is a measure of system heterogeneity and affects the shape of the theoretical tracer concentration curve. With increasing P_D values, the time to peak concentrations decreases.

The solution of the inverse problem (i.e., the estimation of the transport parameters) can be obtained in an automatic fitting procedure that combines the least squares method with Taylor series approximation of both solutions (Maloszewski, 1981). The assumption is that the inverse problem (best fit) is solved when the values of parameters (t_0 and P_D) are chosen so that the sum of squared differences between the theoretical and observed concentrations is minimal. The parameters can also be obtained by manually fitting the appropriate theoretical solutions to experimental concentrations using a trial-and-error procedure.

Often, parameter estimation is also carried out using approximate solutions of the ADE combined with the method of moments or the cumulative curve method. However, the usefulness of these methods in practice is limited because their application is possible only under very restricted conditions. Experimental TBCs are observed which deviate from the idealized tracer concentration curve described theoretically by the dispersion equation, for instance, TBCs with multiple peaks or featuring a strong tailing effect. To address the underlying processes, advanced models for tracer experiments in complex systems, such as multiflow and double-porosity systems, have been developed.

In cases where mass transfer between mobile and immobile zones or states by physical or chemical processes is of importance, the late tailing of TBCs might be heavily influenced by these processes. Haggerty et al. (2004) analyzed results from 316 tracer experiments compiled from literature and found evidence of multiple timescales of mass transfer in aquifers and soils. They also concluded that mass transfer parameters were related to experimental duration, which implies that single-rate mass transfer model predictions over longer (or shorter) timescales than the duration of the experiment should be interpreted with care. In streams, there are two types of transient storage, in-channel dead zones or out-of-channel hyporheic zones. While these storages are associated with different biogeochemical processes, it is difficult to distinguish between these two types of transient storage based on stream tracer experiments. The two storages might also be characterized by different residence time distributions (Gooseff et al., 2005). Similar to aquifers, the late tailing in TBCs is also in streams heavily influenced by the mass transfer between mobile and immobile water (transient storage), which implies that both results are very sensitive to the method used to derive mass transfer parameters and to the length scale of the experiment (Wörman and Wachniew, 2007).

2.09.4.5 Discharge Measurement

The tracer dilution method is widely used to determine discharge in small catchments, in headwaters, and especially in poorly accessible regions. It is the only adequate technique to measure discharge in strongly turbulent rivers and springs with bouldery cross sections where other techniques are not suitable.

The approach of discharge measurement using artificial tracers (tracer dilution gauging) is based on the principle that the dilution of a known mass of tracer injected into the flow system is in proportion to discharge, if complete mixing of the tracer is ensured. After complete mixing of the injected artificial tracer into the surface water stream, the concentration is measured downstream. The dilution is in proportion to the discharge. An indispensable prerequisite when applying the dilution method is a uniform, complete (3D) mixing of the tracer over the entire body of water at the observed cross section of the watercourse. Downstream of the full mixing point a measurement is possible at any point of the cross section. The distance needed to achieve uniform mixing can differ considerably depending on the channel roughness. The estimation of an appropriate distance requires some experience. In general, increasing roughness results in increased mixing and, thus, shorter distances are needed in streams with a high roughness. Sampling for background concentrations upstream of the injection site is required.

For the slug injection method, a known mass of tracer (M [M]) as concentrated solution is poured in bulk into the system as a (quasi) instant impulse (Dirac impulse). The tracer pulse spreads due to vertical, lateral, and longitudinal dispersion as well as turbulent mixing. The longitudinal dispersion causes the typical TBC with a relatively steep increase from the background concentration to the peak, and a gradual decline (tailing) back to the background (**Figure 6**). Measurements and/or sampling are needed at the measuring cross section during the entire passage of the tracer cloud downstream. Assuming 100% recovery of tracer at the sampling point, provided there are no losses of the tracer mass, and the TBC is measured until it reaches the background level (c_b), the discharge (Q) can then be calculated as follows:

$$Q = \frac{M}{\int_0^\infty (c(t) - c_b)\, dt} \quad (19)$$

M [g] is the injected tracer mass, $c(t)$ [g l^{-1}] is the measured concentration at time t, and c_b [g l^{-1}] is the background concentration.

For the constant rate injection method, the tracer solution is injected into the system at a constant rate over an extended period of time. The flux rate of the injected solution and the tracer concentration in the solution are required to be constant over time. The resulting TBC measured downstream typically rises from the background concentration to a constant value, also called plateau concentration (**Figure 7**). At this point, a sample is taken to determine this constant plateau value at the end of the mixing section. In order to obtain the constant plateau concentration, the duration of the injection has to be sufficiently long. The discharge can then be calculated based on the tracer solution inflow rate, q_{in} [l s^{-1}], the tracer solution concentration, c_{in} [g l^{-1}], the measured sustained plateau concentration, c_p [g l^{-1}], and the background concentration, c_b [g l^{-1}] (Equation (20)):

$$Q = q_{in} \frac{c_{in} - c_b}{c_p - c_b} \quad (20)$$

Both the slug injection and the constant injection approaches have their advantages as well as problematic issues. A main advantage of the slug injection is that the injection is experimentally rather simple and does not require any special instruments. On the other hand, continuous registration of the TBC is required. The measurement is also directly sensitive to any error in the value used for the injected tracer mass, M. Errors might be caused by incomplete dissolution of salt when preparing the injection solution. The constant rate injection approach requires accurate concentration measurements only at two points in time, but the injection requires special arrangements to ensure a constant rate of tracer injection over a long period. Although any good soluble chemical is a potential tracer for the dilution method, only the salt tracers (NaCl) and fluorescent tracers are routinely used for operational purposes.

When choosing a tracer for the dilution discharge measurement, the following requirements should be considered. A tracer which does not undergo photolytic decay is absolutely required, since the solution to be measured is more or less completely exposed to daylight due to the turbulences in rivers. Since all fluorescent tracers are at least slightly affected by the photolytic decay, the tracer experiment should be short (usually unproblematic) or scheduled for a night test using this type of tracer.

Furthermore, sorption onto streambed materials should be negligibly low for the tracer used. Sorption leads to retention of the tracer cloud, thus the measurement is incorrect. Finally, pH independence, low detection limit, and toxicological harmlessness are required.

Among others, the fluorescent tracer amidorhodamine G (sulforhodamine G) is the most applied fluorescent tracer in dilution gauging. However, due to the strong sorption effects of the rhodamines group, uranine is increasingly used. It may be used especially in tests conducted at night or for short tests in order to avoid photolytic decay. Background concentrations of fluorescent tracers in most rivers are negligible. Hence, only small amounts of tracer are required even for higher discharge. This is a major advantage compared to salt tracers.

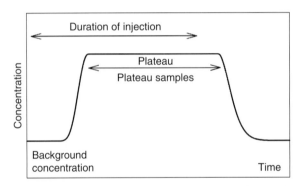

Figure 6 Schematic presentation of the tracer breakthrough curve after slug injection. At the downstream sampling point, after the mixing length, the tracer breakthrough curve is recorded. From Leibundgut Ch, Malozewski P, and Külls Ch (2009) *Tracers in Hydrology*. Chichester: Wiley.

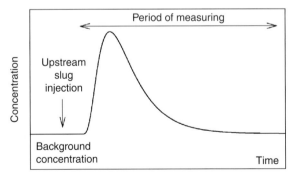

Figure 7 Schematic presentation of the tracer breakthrough after constant rate injection (cf. via Mariotte bottle). At the downstream sampling point, the tracer concentration rises to a plateau value. From Leibundgut Ch, Malozewski P, and Külls Ch (2009) *Tracers in Hydrology*. Chichester: Wiley.

For salt tracers, the measurement is based on the relation between salt concentration and electrolytic conductivity, as there is a linear correlation between these quantities. The natural electrolytic background conductivity, due to the mineralization of the stream water, has to be considered. Calibration of the measuring instruments with stream water is essential to estimate the specific correlation between salt concentration and the electrolytic conductivity of the water to be measured. Site-specific calibration is needed because the specific chemical composition of the stream water may influence the relation between salt concentration and electrolytic conductivity. For slug injection measurements, the amount of salts required are usually relatively high in order to achieve a peak concentration which has an electrolytic conductivity of more than 100 µS cm^{-1} above the background concentration. As a general rule, about 1 kg of salt (NaCl) should be used per

100 l s^{-1} estimated discharge, depending on the background concentration. With fluorescent tracers, the required injection mass is much smaller because of their lower detection limits (see Section 2.09.3.3.2). Using electrolytic conductivity as proxy, the continuous measurement of salt concentrations has become very convenient, especially with the advancement of hand-held field loggers for conductivity measurements. Overall, the slug injection method is usually much easier to use than the constant rate injection method. More detailed instructions for salt dilution measurements can also be found in Leibundgut *et al.* (2009) and in a series of papers by Moore (2004a, 2004b, 2004c, 2005).

Today, fluorescent tracer concentrations can also be easily measured continuously in the stream. Robust portable fluorescent fiber optical fluorometers for *in situ* measurement are useful and reliable. These instruments operate using a light conductor sensor. Through the light conductor, the excitation signal enters the river water in the same way that the emission signal arrives at the measuring instrument. The evaluated quantity is the intensity of the emission signal. As there is a linear correlation between fluorescence and the dye concentration, the fluorescence is used as a measure for the dye concentrations.

When using fluorescent tracers measured by *in situ* fluorometers, calibration of the instruments with the river water to be measured is required. The background concentration, as well as the whole calibration curve, needs to be specified correctly for any water to be measured. There is a general linearity of the dependency between tracer concentration and fluorescent intensity of the water to be measured. Stream water is filled into a measuring cup, and the tracer concentration is gradually increased while the fluorescence intensity is measured.

2.09.4.6 Chloride-Based Groundwater-Recharge Estimation

Chloride concentration measurements can be used for estimating recharge fluxes in the saturated or unsaturated zone in arid and semi-arid regions using the chloride mass-balance (CMB) method (Eriksson and Khunakasem, 1969; Allison and Hughes, 1978; Bazuhaira and Wood, 1996). The method is based on the mass balance of chloride considering long-term input (precipitation) and groundwater (or soil water) concentrations. Due to evaporation, the chloride concentration in the soil increases and the relationship between chloride concentrations in the precipitation and in the soil water or groundwater is thus a measure of evaporation and recharge. The recharge R [mm a^{-1}] can be estimated based on the precipitation amount P [mm a^{-1}], the chloride concentration in the groundwater (or soil water), $c_{Cl,G}$ [mg l^{-1}], and the precipitation-weighted mean chloride concentration in the input, $c_{Cl,P}$ [mg l^{-1}] (Equation (21)):

$$R = \frac{P\, c_{Cl,P}}{c_{Cl,G}} \qquad (21)$$

It is important to note that the CMB only considers vertical flow processes and is limited in its applicability by a set of assumptions which have to be fulfilled. These assumptions include only direct recharge can take place (i.e., recharge resulting from direct infiltration of precipitation), no additional internal source or lateral inflow of chloride exists, chloride concentration in groundwater or soil water experiences no increase or decrease by dissolution, plant uptake or other sink terms, and precipitation amount and chloride concentration in precipitation are not correlated. If these assumptions are fulfilled, in general, the method provides more accurate estimates with increasing evaporation rates.

2.09.4.7 Tracer Experiment in a Porous Aquifer

A study performed at a test site situated in Quaternary sediments in the Swiss Central Plateau is presented here as an example of tracer experiments in porous aquifers (Leibundgut *et al.*, 1992). Postglacial Holocene sandy gravel aquifers in the proximity of the Alps are generally known to be strongly heterogeneous. This heterogeneity was also confirmed for the test site through geophysical prospecting and tracer tests. However, in the range of the tracer tests, the aquifer was assumed to be relatively homogenous with a fairly regular thickness of approximately 15 m and a hydraulic gradient uniformly decreasing in the longitudinal axe. In this section, the depth to the groundwater table was found to be <5 m which provided good accessibility. The water table contour lines indicated a convergent flow pattern (**Figure 8**). The chemical composition of the groundwater indicated no problems caused by potential interactions between the groundwater and the applied tracers. According to preliminary hydrogeological and geophysical tests, the aquifer was found to consist of three distinct layers with different hydraulic conductivities. This multilayered structure of the aquifer required a multilevel sampling within the three layers, which was realized using a well sampling system with packers and sampling at depths of 5, 8, and 11 m.

Uranine (1 kg) and naphtionate (20 kg) dissolved in water (90 l) were injected evenly as mixed tracer solution over the whole depth of the aquifer over a period of 1 h (Dirac impulse). The observation wells were sampled for a period of nearly 400 hs using automatic sampling devices. The topmost layer (5 m) and the bottom layer (11 m) were sampled in 4-hr intervals and the middle layer (8 m) in 1-h intervals. Here, only data from one sampling site (D5) located 100 m from the injection well are presented (see **Figure 8**). TBCs were derived from these observations for uranine and naphtionate, respectively.

The evaluation of the breakthrough curves of the three layers (or flow domains) took place under the assumption of constant transport parameters. The resulting dispersivities were far larger than values observed for similar formations and, thus, a reinterpretation of the data was performed. The superposition of three tracer concentration curves shows a multipeak shape (**Figure 9**). This can be interpreted as a result of flow through parallel layers. The following mean values of aquifer parameters have been assumed for modeling: mean thickness of 10 m and average saturated porosity of 12–17%. The modeling was performed using the transport model described in Section 2.09.4.4 (Maloszewski and Zuber, 1993) and, in particular, the modeling in multilayered systems (Maloszewski *et al.*, 2006).

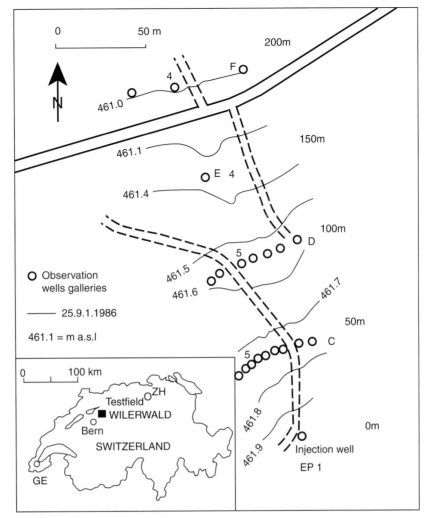

Figure 8 Test site with location of wells and water table contours. From Leibundgut Ch, De Carvalho-Dill A, Maloszewski P, Müller I, and Schneider J (1992) Investigation of solute transport in the porous aquifer of the test site Wilerwald (Switzerland). *Steirische Beiträge zur Hydrogeologie* 43: 229–250.

The transport parameters found for the D5 sampling site, for all of three layers and both applied tracers, allowed to reproduce the measured breakthrough curve based on modeled partial curves for the three layers (**Figure 9**). The partial curves are calculated from the fitted concentration-time curve (breakthrough curve). Both the obtained transport parameters and the aquifer parameters are listed in **Table 5**. The aquifer parameters were calculated based on the modeling of a naphtionate breakthrough curve, so should be considered as an approximation of the real parameters. The results show a strong heterogeneity of the aquifer in both the vertical and the horizontal direction. In the original paper, the complex issue of a tracer test in porous multilayered aquifer and its modeling is discussed in further detail (Leibundgut *et al.*, 1992).

2.09.5 Concluding Remarks

2.09.5.1 Guidance on Further Reading

The following standard textbooks are suitable as further information on tracer methods.

Tracers in Hydrology (Leibundgut *et al.*, 2009) provides a comprehensive, up-to-date presentation of the use of tracer methods in hydrology including modeling and the presentation of selected case studies illustrating the theory.

The book *Isotope Tracers in Catchment Hydrology* (Kendall and McDonnell, 1998) consists of 22 chapters treating the use of isotope tracers in catchment hydrology from different perspectives. This book is a valuable source of information on fundamentals of catchment hydrology, principles of isotope geochemistry, and the isotope variability in the hydrological cycle. Many case studies using isotope tracer methods are described and illustrate the opportunities for using isotope techniques for a wide range of investigations.

Environmental Isotopes in the Hydrological Cycle (Mook, 2000), published by IHP and IAEA, is a series of six volumes which give a comprehensive review of theoretical concepts and practical applications in isotope tracer hydrology. The first volume provides a general overview of theory and methods followed by three volumes dealing with atmospheric water, surface water, as well as water in the saturated and unsaturated zone. The fifth volume focuses on human impact on groundwater and the final volume on modeling.

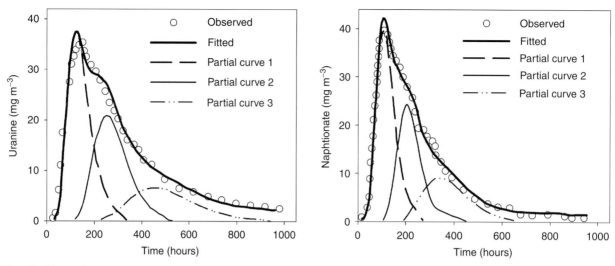

Figure 9 Observed tracer concentration and calculated (best fit) curves (in well D 5) for uranine (right) and naphtionate (left). From Leibundgut Ch, De Carvalho-Dill A, Maloszewski P, Müller I, and Schneider J (1992) Investigation of solute transport in the porous aquifer of the test site Wilerwald (Switzerland). *Steirische Beiträge zur Hydrogeologie* 43: 229–250.

Table 5 Parameters obtained as a result of modeling of uranine and naphtionate concentration curves (in well D 5); where t_{oj} are mean transit time, v_j mean water velocity, α_{Lj} longitudinal dispersivity each for the respective layer j, R_j relative contribution of partial curve j to the total concentration curve and obtained from those the hydrogeological parameters: t_o and k are mean values for all layers of the mean transit time and the hydraulic conductivity respectively, k_j is the hydraulic conductivity, and h_j the thickness of single layers

Curve/layer j	Uranine			Naphtionate							
	t_{oj} (days)	v_j (m d⁻¹)	α_{Lj} (m)	t_{oj} (days)	v_j (m d⁻¹)	α_{Lj} (m)	R_j (–)	t_o (days)	k (m s⁻¹)	k_j (m s⁻¹)	h_j (m s⁻¹)
1	6.5	15.4	7.7	5.4	18.4	6.8	0.45	9.2	4.7×10^{-3}	8.0×10^{-3}	2.7
2	12.1	8.3	3.7	9.3	10.8	2.6	0.32			4.8×10^{-3}	3.2
3	22.3	4.5	4.0	16.3	6.1	3.4	0.23			2.7×10^{-3}	4.1

From Leibundgut Ch, De Carvalho-Dill A, Maloszewski P, Müller I, and Schneider J (1992) Investigation of solute transport in the porous aquifer of the test site Wilerwald (Switzerland). *Steirische Beiträge zur Hydrogeologie* 43: 229–250.

Isotopes in the Water Cycle: Past, Present and Future of a Developing Science (Aggarwal et al., 2005) presents the history of isotope hydrology, state-of-the-art applications, and new developments. The applications described in this book include quantification of groundwater resources, water balance studies, and investigations of past and present global environmental and climate changes.

The books by Clark and Fritz (1997) and Mazor (2004) focus on the use of environmental tracers in groundwater systems, the first in particular on hydrogeology and environmental tracers, the latter in particular on environmental tracers, noble gases, and chemistry. The book *Tracing Technique in Geohydrology* by Käss (1998) describes the various tracing techniques.

Specific aspects of tracer hydrology have also been discussed in review papers on topics such as transit times (McGuire and McDonnell, 2006) or dye tracers in the vadose zone (Flury and Wai, 2003).

2.09.5.2 Reflections and Future Research

Tracer hydrology will continue to play an important role for the advancement in understanding of hydrological systems. Developments of analytical techniques provide new opportunities. The reduction of costs per sample allows more samples to be analyzed. Decreasing the required sample volume allows samples to be analyzed from sources where only limited amounts of water can be sampled. Reduced detection limits allow tracer signals at higher dilutions to be measured. New measurement techniques also increased the interest in using temperature recently, because fiber optic distributed temperature sensors allow measurements in high spatial and temporal resolution (Selker et al., 2006). Finally, there are new natural and artificial substances that could be used as tracers such as DNA (Smith et al., 2008) and diatoms (Pfister et al., 2009). The integration of tracer data into the development and testing of hydrological and environmental models is still not fully explored. Tracer approaches can provide important additional information for the evaluation of models. On the other hand, how to use this information is often not straightforward and additional model parameters or routines might be needed to allow comparison of model simulations with experimental tracer data. How to best integrate experimental approaches, mathematical methods for their interpretation, and hydrological models, such as catchment and groundwater models or soil–water–atmosphere transfer

(SVAT) schemes, will continue to be important research questions in tracer hydrology.

References

Aggarwal PK, Gat JR, and Froehlich KF (eds.) (2005) *Isotopes in the Water Cycle: Past, Present and Future of a Developing Science*. Berlin: Springer.

Allison GB and Hughes MW (1978) The use of environmental chloride and tritium to estimate total local recharge to an unconfined aquifer. *Australian Journal of Soil Research* 16: 181–195.

Auckenthaler A, Raso G, and Huggenberger P (2002) Particle transport in a karst aquifer: Natural and artificial tracer experiments with bacteria, bacteriophages and microspheres. *Water Science and Technology* 46(3): 131–138.

Battaglin WA, Kendall C, Chang CCY, Silva SR, and Campbell DH (2001) Chemical and isotopic evidence of nitrogen transformation in the Mississippi River, 1997–98. *Hydrological Processes* 15(7): 1285–1300.

Bazuhaira AS and Wood WW (1996) Chloride mass-balance method for estimating ground water recharge in arid areas: Examples from western Saudi Arabia. *Journal of Hydrology* 186: 153–159.

Becker MW and Coplen TB (2001) Technical note. Use of deuterated water as a conservative artificial groundwater tracer. *Hydrogeology Journal* 9: 512–516.

Bedard-Haughn A, van Groenigen JW, and van Kessel C (2003) Tracing ^{15}N through landscapes: Potential uses and precautions. *Journal of Hydrology* 272: 175–190.

Beven KJ (2001) *Rainfall-Runoff Modelling: The Primer*. New York: Wiley.

Buttle J (2005) Isotope hydrograph separation of runoff sources. In: Anderson MG and McDonnell JJ (eds.) *Encyclopaedia of Hydrological Sciences*, ch. 116, pp. 1763–1774. Chichester: Wiley.

Calder IR (1992) Deuterium tracing for the estimation of transpiration from trees. Part 2: Estimation of transpiration rates and transpiration parameters using a time-averaged deuterium tracing method. *Journal of Hydrology* 130: 27–35.

Clark I and Fritz P (1997) *Environmental Isotopes in Hydrogeology*. Boca Raton, FL: CRC Press.

Craig H (1961) Isotopic variations in meteoric waters. *Science* 133: 1702–1703.

Dahan O and Ronen Z (2001) Analytical procedure for simultaneous use of seven fluorobenzoates in multitracer tests. *Ground Water* 39(3): 366–370.

Divine CE, Sanford WE, and McCray JE (2003) Helium and neon groundwater tracers to measure residual DNAPL: Laboratory investigation. *Vadose Zone Journal* 2: 382–388.

Einsiedl F (2005) Flow system dynamics and water storage of a fissured-porous karst aquifer characterized by artificial and environmental tracers. *Journal of Hydrology* 312: 312–321.

Eriksson E (1958) The possible use of tritium for estimating groundwater storage. *Tellus* 10: 472–478.

Eriksson E and Khunakasem V (1969) Chloride concentration in groundwater, recharge rate and rate of deposition of chloride in the Israel Coastal Plain. *Journal of Hydrology* 7(2): 178–197.

Flury M and Wai NN (2003) Dyes as tracers for vadose zone hydrology. *Reviews of Geophysics* 41(1): 2.1–2.37.

Garcia Gutiérrez MG, Guimerà J, Illera de Llano A, Benitez AH, Humm J, and Saltino M (1997) Tracer test at El Berrocal site. *Journal of Contaminant Hydrology* 26: 179–188.

Genereux DP and Hooper RP (1998) Oxygen and hydrogen isotopes in rainfall-runoff studies. In: Kendall C and McDonnell JJ (eds.) *Isotope Tracers in Catchment Hydrology*, pp. 319–346. Amsterdam: Elsevier.

Gibson JJ, Edwards TWD, and Prowse TD (1996) Development and validation of an isotopic method for estimating lake evaporation. *Hydrological Processes* 10(10): 1369–1382.

Gooseff MN, LaNier J, Haggerty R, and Kokkeler K (2005) Determining in-channel (dead zone) transient storage by comparing solute transport in a bedrock channel-alluvial channel sequence, Oregon. *Water Resources Research* 41(6): W06014. 1–7

Haggerty R, Harvey CF, von Schwerin CF, and Meigs LC (2004) What controls the apparent timescale of solute mass transfer in aquifers and soils? A comparison of experimental results. *Water Resources Research* 40(1): W01510.

Hangen E, Gerke HH, Schaaf W, and Hüttl RF (2005) Assessment of preferential flow processes in a forest-reclaimed lignitic mine soil by multicell sampling of drainage water and three tracers. *Journal of Hydrology* 303: 16–37.

Himmelsbach T, Hötzl P, and Maloszewski P (1998) Solute transport processes in a highly permeable fault zone of Lindau fractured rock test site (Germany). *Ground Water* 36(5): 792–800.

Hooper RP, Christophersen N, and Peters NE (1990) Modelling stream water chemistry as a mixture of soil water end-members – an application to the Panola Mountain catchment, Georgia, USA. *Journal of Hydrology* 116: 321–343.

Hu Q and Moran JE (2005) Simultaneous analyses and applications of multiple fluorobenzoate and halide tracers in hydrologic studies. *Hydrological Processes* 19: 2671–2687.

IAEA (ed.) (1983) *Guidebook on Nuclear Techniques in Hydrology*. Vienna: IAEA.

Käss W (1998) *Tracing Technique in Geohydrology*. Rotterdam: Balkema.

Kendall C (1998) Tracing nitrogen sources and cycling in catchments. In: Kendall C and McDonnell JJ (eds.) *Isotope Tracers in Catchment Hydrology*, pp. 519–576. Amsterdam: Elsevier.

Kendall C and McDonnell JJ (eds.) (1998) *Isotope Tracers in Catchment Hydrology*. Amsterdam: Elsevier.

Königer P, Leibundgut C, Link TJ, and Marshall JD (2010) Stable isotopes applied as water tracers in column and field studies. *Organic Geochemistry* 41(1): 31–40 (doi:10.1016/j.orggeochem.2009.07.006).

Kreft A and Zuber A (1978) On the physical meaning of the dispersion equation and its solutions for different initial and boundary conditions. *Chemical Engineering Science* 33: 1471–1480.

Laudon H, Hemond HF, Krouse R, and Bishop KH (2002) Oxygen 18 fractionation during snow melt: Implications for spring flood hydrograph separation. *Water Resources Research* 38(11): 1258 (doi:10.1029/2002WR00510).

Laudon H, Sjöblom V, Buffam I, Seibert J, and Mörth M (2007) The role of catchment scale and landscape characteristics for runoff generation of boreal streams. *Journal of Hydrology* 344: 198–209.

Leibundgut Ch (1987) Hydroökologische Untersuchungen in einem alpinen Einzugsgebiet – Grindelwald. *Schlussbericht zum Schweizerischen MAB-Programm Nr. 30*. Berne: FOEN (Federal Office for the Environment).

Leibundgut Ch, De Carvalho-Dill A, Maloszewski P, Müller I, and Schneider J (1992) Investigation of solute transport in the porous aquifer of the test site Wilerwald (Switzerland). *Steirische Beiträge zur Hydrogeologie* 43: 229–250.

Leibundgut Ch, Malozewski P, and Külls Ch (2009) *Tracers in Hydrology*. Chichester: Wiley.

Lenda A and Zuber A (1970) Tracer dispersion in groundwater experiments *Isotope Hydrology 1970*, pp. 619–641. Vienna: IAEA.

Lyon SW, Desilets SLE, and Troch PA (2009) A tale of two isotopes: Differences in hydrograph separation for a runoff event when using δD versus $\delta^{18}O$. *Hydrological Processes* 23(14): 2095–2101.

Mali N, Urbanc J, and Leis A (2007) Tracing of water movement through the unsaturated zone of a coarse gravel aquifer by means of dye and deuterated water. *Environmental Geology* 51(8): 1401–1412.

Maloszewski P (1981) Computerprogramm für die Berechnung der Dispersion und der effektiven Porosität in geschichteten porösen Medien. *GSF-Bericht* R 269 Munich-Neuherberg: GSF (Gesellschaft für Strahlen- und Umweltforschung), 33p.

Maloszewski P, Wachniewski P, and Czuprynski P (2006) Study of hydraulic parameters in heterogeneous gravel beds: Constructed wetland in Nowa Słupia (Poland). *Journal of Hydrology* 331: 630–642.

Maloszewski P and Zuber A (1982) Determining the turnover time of groundwater systems with the aid of environmental tracers. 1. Models and their applicability. *Journal of Hydrology* 57: 207–231.

Maloszewski P and Zuber A (1993) Principles and practice of calibration and validation of mathematical models for the interpretation of environmental tracer data in aquifers. *Advances in Water Resources* 16: 173–190.

Mazor E (2004) *Chemical and Isotopic Groundwater Hydrology: The Applied Approach*, 3rd (revised and expanded) edn. New York: Dekker.

McDonnell JJ, McGuire K, Aggarwal P, Beven KJ, Biondi D, Destouni G, Dunn S, James A, Kirchner J, Kraft P, Lyon S, Maloszewski P, Newman B, Pfister L, Rinaldo A, Rodhe A, Sayama T, Seibert J, Solomon K, Soulsby C, Stewart M, Tetzlaff D, Tobin C, Troch P, Weiler M, Western A, Wörman A, and Wrede S (2010) How old is streamwater? Open questions in catchment transit time conceptualization, modelling and analysis. *Hydrological Processes* 24(12): 1745–1754 (doi: 10.1002/hyp.7796).

McGuire KJ and McDonnell JJ (2006) A review and evaluation of catchment transit time modeling. *Journal of Hydrology* 330: 543–563.

McGuire KJ, McDonnell JJ, Weiler M, *et al.* (2005) The role of topography on catchment-scale water residence time. *Water Resources Research* 41(5): W05002 (doi:10.1029/2004WR003657).

Mook WG (2000) *Environmental Isotopes in the Hydrological Cycle Principles and Applications*, 6 vols., Paris; Vienna: UNESCO; IAEA.

Moore RD (2005) Introduction to salt dilution gauging for streamflow measurement. Part III: Slug injection using salt in solution. *Streamline Watershed Management Bulletin* 8(2): 1–6.

Moore RD (2004a) Introduction to salt dilution gauging for streamflow measurement: Part I. *Streamline Watershed Management Bulletin* 7(4): 20–23.

Moore RD (2004b) Introduction to salt dilution gauging for streamflow measurement. Part II: Constant-rate injection. *Streamline Watershed Management Bulletin* 8(1): 11–15.

Moore RD (2004c) Construction of a Mariotte bottle for constant-rate tracer injection into small streams. *Streamline Watershed Management Bulletin* 8(1): 15–16.

Moser H and Rauert W (eds.) (1980) *Isotopenmethoden in der Hydrologie. Lehrbuch der Hydrogeologie 8*. Berlin/Stuttgart: Gebrüder Borntraeger.

Onodera S and Kobayashi M (1995) Evaluation of seasonal variation in bypass flow and matrix flow in a forest soil layer using bromide ion. *IAHS Publication* 229: 99–108.

Parsons DF, Hayashi M, and van der Kamp G (2004) Infiltration and solute transport under a seasonal wetland: Bromide tracer experiments in Saskatoon, Canada. *Hydrological Processes* 18: 2011–2027.

Pfister L, McDonnell JJ, Wrede S, et al. (2009) The rivers are alive: On the potential for diatoms as a tracer of water source and hydrological connectivity. *Hydrological Processes* 23: 2841–2845.

Pinder GF and Jones JF (1969) Determination of the groundwater component of peak discharge from the chemistry of total runoff. *Water Resources Research* 5: 438–445.

Rodhe A (1987) The Origin of Streamwater Traced by Oxygen-18. PhD Thesis, Uppsala University, UNGI Report Series A. no. 41.

Rodhe A (1981) Spring flood melt water or groundwater. *Nordic Hydrology* 12(1): 21–30.

Rodhe A, Nyberg L, and Bishop K (1996) Transit times for water in a small till catchment from a step shift in the oxygen-18 content of the water input. *Water Resources Research* 32: 3497–3511.

Rozanski K, Araguasaraguas L, and Gonfiantini R (1992) Relation between long-term trends of O-18 isotope composition of precipitation and climate. *Science* 258(5084): 981–985.

Sabir IH, Orgersen J, Haldorson S, and Aleström P (1999) DNA tracers with information capacity and high detection sensitivity tested in groundwater studies. *Hydrogeology Journal* 7: 264–272.

Sambale Ch, Peschke G, Uhlenbrook S, Leibundgut Ch, and Markert B (2000) Simulation of vertical water flow and bromide transport under temporarily very dry conditions. *IAHS Publication* 262: 339–345.

Seibert J and McDonnell JJ (2002) On the dialog between experimentalist and modeler in catchment hydrology: Use of soft data for multicriteria model calibration. *Water Resources Research* 38(11): 1241 (doi:10.1029/2001WR000978).

Seibert J, Rodhe A, and Bishop K (2003) Simulating interactions between saturated and unsaturated storage in a conceptual runoff model. *Hydrological Processes* 17(2): 379–390.

Selker JS, Thévenaz L, Huwald H, et al. (2006) Distributed fiber-optic temperature sensing for hydrologic systems. *Water Resources Research* 42: W12202 (doi:10.1029/2006WR005326).

Silva SR, Kendall C, Wilkison DH, Ziegler AC, Chang CCY, and Avanzino RJ (2000) A new method for collection of nitrate from fresh water and the analysis of nitrogen and oxygen isotope ratios. *Journal of Hydrology* 228: 22–36.

Sklash M and Farvolden R (1979) The role of groundwater in storm runoff. *Journal of Hydrology* 43: 45–65.

Smith J, Gao B, Funabashi H, et al. (2008) Pore-scale quantification of colloid transport in saturated porous media. *Environmental Science and Technology* 42(2): 517–523 (doi: 10.1021/es070736x).

Solomon DK, Cook PG, and Sanford WE (1998) Dissolved gases in subsurface hydrology. In: Kendall C and McDonnell JJ (eds.) *Isotope Tracers in Catchment Hydrology*, pp. 291–318. Amsterdam: Elsevier.

Stamm Ch, Sermet R, Leuenberger L, et al. (2002) Multiple tracing of fast solute transport in a drained grassland soil. *Geoderma* 109: 245–268.

Sturm K, Hoffmann G, Langmann B, and Stichler W (2005) Simulation of delta O-18 in precipitation by the regional circulation model REMOiso. *Hydrological Processes* 19(17): 3425–3444.

Tetzlaff D, Seibert J, McGuire KJ, et al. (2009) How does landscape structure influence catchment transit times across different geomorphic provinces? *Hydrological Processes* 23(6): 945–953.

Uhlenbrook S and Sieber A (2005) On the value of experimental data to reduce the prediction uncertainty of a process-oriented catchment model. *Environmental Modelling and Software* 20(1): 19–32.

Wilson RD and McKay DM (1996) SF_6 as conservative tracer in saturated media with high intragranular porosity or high organic carbon content. *Ground Water* 34(2): 241–249.

Wörman A and Wachniew P (2007) Reach scale and evaluation methods as limitations for transient storage properties in streams and rivers. *Water Resources Research* 43(10): W10405. 1–13.

Zvikelsky O and Weisbrod N (2006) Impact of particle size on colloid transport in discrete fractures. *Water Resources Research* 42(12): W12S08 (doi:10.1029/2006WR004873).

2.10 Hydrology and Ecology of River Systems

A Gurnell, Queen Mary, University of London, London, UK
G Petts, University of Westminster, London, UK

© 2011 Elsevier B.V. All rights reserved.

2.10.1	Introduction	237
2.10.2	**Key Hydrological Characteristics of River Networks**	239
2.10.2.1	Flow Indices and Regimes	239
2.10.2.2	Hydrological Connectivity	241
2.10.3	**River-Corridor Dynamics**	243
2.10.3.1	River Regimes and River Styles	243
2.10.3.2	Changing River Styles	245
2.10.3.3	The River-Corridor Habitat Mosaic	247
2.10.3.4	Distribution and Connectivity of Physical Habitats	248
2.10.3.5	Habitat Dynamics and the Role of Plants as Ecosystem Engineers	249
2.10.4	**Aquatic Ecosystems**	252
2.10.4.1	Instream Flows and Flow Regimes	252
2.10.4.1.1	The significance of multidimensional variations in flow characteristics	252
2.10.4.1.2	Adaptations of biota to running water	254
2.10.4.2	Ecohydraulics and Mesohabitats	255
2.10.4.2.1	Hydraulic stream ecology and the mesohabitat template	255
2.10.4.2.2	Habitat-suitability criteria	257
2.10.4.2.3	Models of biological responses to changing flows	258
2.10.5	**Managing River Flows to Protect Riverine Ecosystems**	259
References		263

2.10.1 Introduction

Scientists have long explored relationships between hydrological processes and river ecosystems and many notable concepts, frameworks, and hypotheses have been proposed and empirically tested as the field has advanced over the last 50 years. These advances in scientific understanding (summarized in **Table 1**) demonstrate the fundamental and complex controls that hydrological processes impose on aquatic and riparian ecosystems. In many cases, these hydrological controls are direct, for example, specific properties of river flows are related to the dynamics and life cycles of many organisms (e.g., the flood pulse concept (FPC), Junk et al., 1989). In other cases, the controls are indirect. Of particular significance here is the influence of hydrological processes on river-channel forms and dynamics that in turn determine the mosaic of habitat patches available for organisms (e.g., the impact of flood disturbance on the succession of riparian plant communities (Décamps and Tabacchi, 1994), patch dynamics, and the shifting-habitat mosaic within river systems (Pringle et al., 1988; Stanford et al., 2005; Latterell et al., 2006). Most recently, research has started to explore how some organisms found in aquatic and riparian environments, particularly certain plant species, not only respond to but also control hydrological and geomorphological processes (Corenblit et al., 2007), supporting interactions that increase both the physical- and bio-complexity of river landscapes (e.g., Gurnell et al., 2005).

In this chapter, we explore the ways in which flows, directly and indirectly, drive riverine ecosystems. We investigate three main themes, which progressively move focus from the catchment (river-flow dynamics) to the river corridor (river-corridor dynamics and the riparian zone) and then to the river (in-river dynamics and the aquatic ecosystem). In Section 2.10.2, we explore the key spatial and temporal properties of river flows that have been shown to have significance for river-corridor (riparian and aquatic) ecosystems. In Section 2.10.3, we consider the impact of hydrological processes on river corridors, ranging from their impact on gross corridor morphology to their influence on the character and turnover of the mosaic of physical habitats contained within the corridor, placing particular emphasis on the riparian zone and the style of river that it encloses. In Section 2.10.4, we focus on the importance of hydrological processes for aquatic ecosystems and organisms, first, emphasizing the importance of the river's flow regime and then on the finer spatial scale of mesohabitats and hydraulic stream ecology. The chapter concludes by considering how river flows can be managed to protect riverine ecosystems.

In providing an overview of relationships between hydrology and ecology within river corridors, there is inevitable overlap with the themes of other chapters in this volume. We deliberately exclude discussion of biogeochemical processes, since these are the theme of Chapter 38. However, we include some topics that are fundamental to our theme, while being the focus of other chapters. Flow hydraulics (**Chapter 2.07 The Hydrodynamics and Morphodynamics of Rivers**), erosion and deposition of sediment (**Chapter 2.12 Catchment Erosion, Sediment Delivery, and Sediment Quality** and **Chapter 2.20 Stream–Groundwater Interactions**), and

Table 1 Some river ecosystem concepts, frameworks, and hypotheses that explicitly incorporate hydrological processes

Concept/framework/hypothesis	Dimensions emphasized	Source
Stream order within river networks	Longitudinal	Horton (1945)
Hydraulic geometry approach: linking channel form to discharge	Longitudinal	Leopold and Maddock (1953)
Zonation of fish communities along rivers	Longitudinal	Huet (1959)
Concept of different river-flow regimes	Temporal	Pardé (1955)
Slope-discharge control of river-channel patterns	Longitudinal	Leopold and Wolman (1957), Lane (1957)
The hyphoreic zone	Vertical	Orghidan (1959)
Dynamic equilibrium	Temporal	Hack (1960)
Zonation of macroinvertebrates and fish along rivers	Longitudinal	Illies and Botosaneanu (1963)
Fluvial process-form dynamics	Four dimensions	Leopold et al. (1964)
The ecology of running waters	Longitudinal	Hynes (1970, 1975)
Drainage basin form and process	Four dimensions	Gregory and Walling (1973)
Nutrient recycling	Longitudinal	Webster et al. (1975)
Instream flow needs	Longitudinal	Orsborn and Allman (eds., 1976)
Structure and process-form dynamics of the fluvial system	Four dimensions	Schumm (1977)
River continuum concept	Longitudinal	Vannote et al. (1980)
Resource spiraling	Longitudinal	Newbold et al. (1981, 1982), Elwood et al. (1983)
Ecosystem perspective of riparian zones	Four dimensions	Gregory et al. (1991)
Groundwater and stream ecology	Vertical	Hynes (1983)
Serial discontinuity concept	Longitudinal	Ward and Stanford (1983a)
Tributaries modify the river continuum concept	Longitudinal	Bruns et al. (1984)
Hydraulic stream ecology	Longitudinal	Statzner and Higler (1986), Statzner et al. (1988)
Nested-hierarchical framework for stream habitat classification	Four dimensions	Frissell et al. (1986)
River–floodplain connectivity	Lateral	Amoros and Roux (1988)
Patch dynamics in lotic systems	Temporal	Pringle et al. (1988)
Role of disturbance in stream ecology	Temporal	Resh et al. (1988)
Flood pulse concept	Temporal and lateral	Junk et al. (1989)
Four-dimensional nature of lotic systems	Four dimensions	Ward (1989)
Aquatic–terrestrial ecotone	Lateral	Naiman and Décamps (1990)
Hyporheic corridor concept	Vertical	Stanford and Ward (1993)
Flood-disturbance regime and succession of riparian plant communities	Temporal	Décamps and Tabacchi (1994)
Hydraulic food-chain models	Temporal	Power et al. (1995)
Fluvial hydrosystem approach	Four dimensions	Petts and Amoros (1996)
Indicators of hydrologic alteration	Temporal	Richter et al. (1996)
The natural-flow regime	Temporal	Poff et al. (1997)
Process domains and the river continuum	Longitudinal	Montgomery (1999)
Flow pulse concept	Temporal and lateral	Tockner et al. (2000)
Geomorphic thresholds in riverine landscapes	Temporal, longitudinal, lateral	Church (2002)
Flow-sediment–biota relations	Four dimensions	Osmundson et al. (2002)
Processes and downstream linkages of headwater streams	Longitudinal	Gomi et al. (2002)
Ecological effects of drought perturbation	Temporal	Lake (2003)
Network dynamics hypothesis	Longitudinal	Benda et al. (2004)
Effective discharge for ecological processes	Temporal	Doyle et al. (2005)
River styles framework	Longitudinal, lateral	Brierley and Fryirs (2005)
The riverine ecosystem synthesis	Four dimensions	Thorp et al. (2006)
Fish environmental guilds		Welcomme et al. (2006)
Vegetation as a driver of physical- and bio-complexity in fluvial corridors	Four dimensions	Gurnell et al. (2005), Corenblit et al. (2007)
Hydrologic spirals	Longitudinal and vertical	Poole et al. (2008)
Ecological limits of hydrologic alteration	Temporal	Poff et al. (2009)

interactions between surface and groundwater all contribute to the definition of river-corridor habitats. These processes also drive the geomorphological features (landforms and sedimentary structures) of river corridors that are the foundation of their habitat mosaic. We integrate aspects of all these themes to provide a robust context for our discussions of the relationships between hydrology and the river corridor.

2.10.2 Key Hydrological Characteristics of River Networks

2.10.2.1 Flow Indices and Regimes

A fundamental premise is that the characteristic community of species that comprise the riverine ecosystem is adapted to the natural-flow regime (Naiman *et al.*, 2002). Despite incomplete understanding of precisely how hydrological processes support river ecosystems, it is accepted that the ecological integrity of river systems depends upon their dynamic character (Poff *et al.*, 1997). As a result, a great deal of research has been devoted to (1) extracting key parameters of river flows that appear to be of ecological importance (e.g., Olden and Poff, 2003; Doyle *et al.*, 2005); (2) identifying direct human modifications of river flows (e.g., Vörösmarty *et al.*, 1997; Vörösmarty and Sahagian, 2000) and indirect human-induced hydrological changes attributable to catchment land-use change (e.g., Allan, 2004; Gurnell *et al.*, 2007) as well as climate change (e.g., Huntington, 2006); and (3) understanding how to manipulate impacted river flows to reinstate those crucial elements of the flow regime that sustain river ecosystems (e.g., Petts, 2009; Poff *et al.*, 2009).

A fundamental but simple method of describing flow conditions at a site is to construct a flow-duration curve, which graphically displays the proportion of time that any particular river flow is exceeded at a site. Flow-duration curves provide a summary of both the central tendency and dispersal of flows and so support comparison of flow characteristics across space and time as well as permit extraction of summary flow indices such as flow percentiles (e.g., Patel, 2007). Their simplicity and high information content have made flow-duration curves an important tool in water resource and river-flow management. Various flow percentiles, particularly the 95th percentile, have been associated with the maintenance of biological and chemical quality in surface waters (e.g., Dakova *et al.*, 2000), and have been used to prescribe flows for regulated systems, such as the minimum acceptable flow (MAF) that has been required for river systems in England and Wales since the 1963 Water Resources Act (Petts, 1996). Similar proposals based on specific proportions of median (50th percentile) or mean annual flow have been adopted in many parts of the world (Arthington *et al.*, 2006). Moreover, as pressures on water resources have increased over the last 50 years, the flow-duration curve has been used to underpin operational rules for delivering ecologically acceptable flow regimes that balance the requirement for water abstractions with the hydrological needs of the river ecosystem (e.g., Petts *et al.*, 1999).

A more sophisticated approach to characterizing river flows, which incorporates not only flow magnitude but also timing, is to define the (mean) annual flow regime. This describes the typical annual sequence of flows based on average monthly or weekly discharges over a period of years (typically 20 years of records). Pardé (1955) was the first to recognize that strong spatial variations in the annual flow regime existed, whereas the ecological importance of the annual flow regime was encapsulated in the flood pulse concept of Junk *et al.* (1989). Haines *et al.* (1988) undertook a global analysis of flow regimes, defining 15 classes from a cluster analysis of 32 000 station years of data from 969 stations (**Figure 1(a)**). A more recent global analysis of monthly streamflows from 1345 sites by Dettinger and Diaz (2000) generated similar results, recognizing 10 broad classes of annual regime (**Figure 1(b)**) and showing wide regional contrasts in the timing of monthly flow extremes and the amplitude of variations in flow between months (i.e., regime seasonality) as well as variability in regime between years. At the regional level, Harris *et al.* (2000) applied principal components analysis (PCA) to 25 years of monthly river flow and air-temperature data from four sites in the UK to classify subregional shifts in both temperature and river-flow-regime magnitude and timing (**Table 2**). Since both the temperature regime (Caissie, 2006) and flow regime (Junk and Wantzen, 2004) are recognized as major controls on river ecology, such joint analyses may prove a profitable line of research to further advance understanding of river-ecosystem dynamics. For example, the flow and temperature regime classification approach described by Harris *et al.* (2000) was combined with analysis of 6 years of macroinvertebrate sample data for a groundwater-fed river in South East England by Wood *et al.* (2001). The analysis revealed a significant difference in macroinvertebrate community abundance in relation to flow-regime class and its component timing and magnitude classes across all scales of analysis (entire river, upstream and downstream sectors, and habitat type), and the influence of air-temperature regime, used as a surrogate for water temperature, varied significantly between riffle sites.

Characterization of flow regime as a context for ecological studies has more frequently been based on a series of indices that represent magnitude, frequency, duration, and timing of flow properties rather than an integration of the entire annual regime. Richter *et al.* (1996) devised a method for calculating indicators of hydrologic alteration to support aquatic ecosystem management. This method compares ecologically relevant hydrological indices extracted from pre- and post-impact (or reference and impacted system) data series, so that contrasts in the properties of the index frequency distributions can be used to define key hydrological changes within impacted systems. Olden and Poff (2003) investigated 171 published hydrological indices to identify a reduced index set capable of explaining the major part of the statistical variation encompassed in the full set. By subjecting a data set containing estimates of all of the indices for 420 US gauging sites to PCA, they illustrated considerable redundancy between many of the indices. They interpreted the results of their analysis to aid researchers' choice of a subset of indices suited to the hydroclimatic region and ecological question being addressed. More recently, Monk *et al.* (2007) presented a similar PCA-based analysis of hydroecological data for 83 rivers across England and Wales.

Three recent studies illustrate how index-based characterizations of flow regimes differ according to geographical variations in controlling factors such as climate, land cover, and catchment characteristics. Poff *et al.* (2006) analyzed 10 indices describing properties of peak flows, low flows, flow duration, and variability for 158 catchments from four hydroregions of the United States. They showed how regional flow regimes are modified by land cover and the presence of dams. Snelder *et al.* (2005) demonstrated the effectiveness with which an *a priori*, map-based, classification of New Zealand river systems and river sectors (the River Environment Classification (REC)) discriminated between sites with

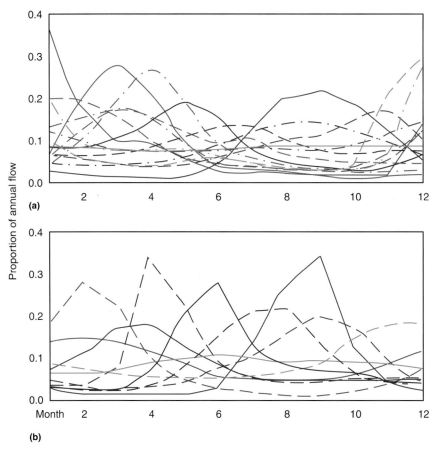

Figure 1 Classifications of global river flow regimes by (a) Haines AT, Finlayson BL, and McMahon TA (1988) The global classification of flow regimes. *Applied Geography* 8: 255–272 and (b) Dettinger MD and Diaz HF (2000) Global characteristics of stream flow seasonality and variability. *Journal of Hydrometeorology* 1: 289–310.

different flow regimes. The REC is based on a hierarchy of climate and topographic factors that are assumed to influence the physical and biological characteristics of rivers. Following Richter *et al.* (1996) and Poff *et al.* (1997), Snelder *et al.* (2005) selected flow variables that characterized five ecologically relevant properties of the intra-annual flow regime: magnitude, frequency, duration, and timing of high and low flows and measures of rate of change of flow. Finally, McMahon and Finlayson (2003) derived a variety of hydrological indices to explore the low-flow hydrology of Australia. They investigated styles, trends, and cycles of low-flow sequences within reference rivers located in different Australian flow-regime zones and considered their implications for indigenous aquatic biota.

Given the acknowledged importance of flow regimes for river ecosystems, it is important to consider the degree to which river-flow regimes may respond to climate change, with or without direct human impacts on land use and regulation of river flows. While broad overviews of projected changes to global water resources can be found in Arthurton *et al.* (2007) and Kundzewicz *et al.* (2007), the underlying science specific to adjustments in river-flow regimes is briefly outlined here. Allen and Ingram (2002) explained the scientific challenges involved in making predictions of future adjustments in the hydrological cycle, but Huntington (2006) reviewed the hydroclimatological evidence and suggested that, despite substantial uncertainties, the global water cycle has been intensifying during much of the twentieth century. These observations give support to predictions of future increases in precipitation and evaporation under global warming, with hydrological feedbacks involving more atmospheric heat trapping associated with increased atmospheric water vapor; changes in cloud properties and extent, which impact on surface warming; and changes in snow and ice melt that influence albedo and thus surface reflection – absorption of radiation (Huntington, 2006). Arora and Boer (2001) modeled the impacts of such increases in global temperature, precipitation, and evapotranspiration on the annual flow regime for rivers in different parts of the world. In particular, they predicted a decrease in the amplitude and earlier high- and low-flow periods in mid- and high-latitude rivers as a result of a decrease in the proportion of precipitation falling as snow and the earlier annual occurrence of snowmelt. They also predicted a less-marked reduction in the amplitude of the flow regime for low-latitude rivers. Arnell (2003) drew similar conclusions regarding flow-regime changes across the globe and also suggested that inter-annual variability in runoff is likely to increase in most catchments. Increasing runoff variability was also emphasized by Milly *et al.* (2002), who suggested a notable increase in the occurrence of great floods (i.e.,

Table 2 The 14 tenets of the river-ecosystem synthesis

Tenet set/number	Description
A	Factors influencing species distributions/the composition of the species pool
1	Species distributions are associated primarily with the distribution of small to large spatial patches formed by hydrogeomorphological forces and modified by climate and vegetation
2	Distributions of species and ecotypes and community diversity from headwaters to river mouth primarily reflect the nature of the functional-process zone rather than the position along the the river network
3	Species diversity is maximum at ecological nodes/transitions between hydrogeomorphological patches or areas of marked habitat convergence/divergence within functional process zones
4	Throughout river networks, species diversity and density vary significantly with flow velocity and, in large rivers, are positively correlated with hydrological retention, except where other abiotic environmental conditions (e.g., oxygen, temperature, and substrate type) restrict taxa
B	Factors controlling species diversity and abundance in the context of the assemblage of species potentially present
5	The most important environmental feature regulating community composition is the hierarchical habitat template, primarily determined by interactions between landforms and flow characteristics
6	Deterministic and stochastic factors contribute significantly to community regulation. Their relative importance is scale- and habitat-dependent, although stochastic factors are more important overall
7	A quasi-equilibrium is maintained by a dynamic patch mosaic
8	(a) Classical (facilitative) succession is primarily limited to terrestrial elements of river landscapes (riparian and floodplain habitats including on islands) and occurs in response to hydrogeomorphological processes; (b) the relative importance of simple seasonal species replacement vs. true, non-facilitative succession (e.g., a blend of inhibition and tolerance succession) within wetted portions of the riverscape varies directly with stream size and inversely with hydrological variability
C	Processes at the ecosystem and landscape levels
9	(a) Annually, through an algal-grazer food web pathway, autochthonous autotrophy provides for most metazoan productivity across the river network, but allochthonous organic matter may be more important for some species, in some seasons and in shallow, heavily canopied headwaters; (b) A collateral, weakly linked, decomposer food pathway (the microbialviral loop) is primarily responsible (sometimes with algal respiration) for a river's heterotrophic state (P/R < 1)
10	Algal production is the primary source of organic energy fueling aquatic metazoan food webs in the floodplains of most river systems during over-bank floods especially in rivers with seasonal, warm-weather floods
11	Average current velocity and nutrient spiral length are positively correlated with river discharge; both decrease in functional process zones with extensive lateral components
12	Naturally dynamic hydrological patterns are necessary to maintain the evolved biocomplexity in river networks
13	The frequency of flood-linked life-history characteristics increases directly with seasonal predictability of floods and their concurrence with periods of maximum system primary productivity.
14	Biocomplexity generally peaks at intermediate levels of connectivity between the main channel and lateral aquatic habitats of the river landscape, but the relationship varies among the types of connectivity, evolutionary adaptation of taxa to flowing water, and functional processes examined

Adapted from Thorpe et al. (2003).

discharges exceeding 100-year levels from basins larger than 200 000 km^2).

Overall, river-flow regimes are highly sensitive to changes in climate, particularly where temperature increases are experienced in snow-fed catchments (e.g., Krasovskaia, 1996; Krasovskaia and Saelthun, 1997). Such changes, even in the absence of direct human manipulations of catchments and river systems, are likely to have profound ecological consequences. However, across many areas of the globe, changes are superimposed on extensive human modifications of flow regimes. For example, Vörösmarty et al. (1997) estimated that in the mid-1980s, the maximum water storage behind 746 of the world's largest dams (generally over 15 m high with maximum storage capacities ⩾ 0.5 km^3) was equivalent to 20% of the global mean annual runoff. Crucially, for downstream flow regimes as well as the physical and chemical properties of river water, the median water-residence time behind these impoundments was 0.40 years.

2.10.2.2 Hydrological Connectivity

A second fundamental principle is that the connectivity between habitats along a river and between a river and its floodplain is essential to the viability of populations of many riverine species (Bunn and Arthington, 2002). Pringle (2001) defined hydrological connectivity in a river-ecosystem context as "water-mediated transfer of matter, energy, or organisms within or between elements of the hydrologic cycle" (p. 981). A brief summary of the concept can be found in Pringle (2003). The idea of hydrological connectivity is built on concepts of transfers across the three spatial dimensions of fluvial systems (longitudinal, lateral, and vertical) and along the fourth dimension – time (Stanford and Ward, 1988; Ward, 1989). Specifically, the hydrological dynamics of a river's natural-flow regime (Section 2.10.2.1) lead to spatially and temporally complex inundation dynamics; surface and subsurface flow pathways; organism movement pathways; and

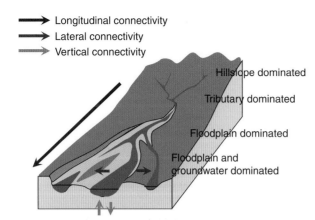

Figure 2 Schematic representation of longitudinal, lateral, and vertical connectivity along river systems.

mobilization, transport, and deposition of matter throughout the three-dimensional (3-D) corridors defined by the river network (**Figure 2**).

The ecological significance of longitudinal (upstream to downstream) connectivity was encapsulated in the river continuum concept (RCC, Vannote et al., 1980). The RCC considers inputs, transport, and processing of organic matter down river systems from headwater to lower reaches (the river continuum). High allochthonous inputs of organic matter from riparian vegetation are conceptualized to dominate narrow, shaded headwater streams. These allochthonous inputs are gradually replaced by autochthonous production as the river widens and receives more light in its middle reaches. However, autochthonous production may decrease toward lower river reaches as water depth and turbidity increase. These downstream changes in the nature and quantity of both locally produced organic matter and the products of organic-matter processing from upstream are associated with progressive downstream shifts in the species composition of macroinvertebrate communities. Ward and Stanford (1983a) extended the RCC to regulated rivers in their serial discontinuity concept (SDC), which considered the impacts of impoundments located in the headwaters, middle, and lower reaches of river systems on ecosystem character and processes. In parallel with the development of the RCC and SDC, the concept of nutrient spiraling (Newbold et al., 1981, 1982; Elwood et al., 1983) described how nutrient cycles are stretched into spirals by downstream transport processes within stream ecosystems.

The crucial role of geomorphology in supporting physically distinct segments along river systems with distinct assemblages of physical habitats, rather than a smooth downstream continuum, is embedded in the sector level of the nested, hierarchical approach to river characterization and classification proposed by Frissell et al. (1986) (**Figure 3**). The definition of physically distinct river stretches and the role of fluvial processes in supporting such distinct stretches and defining thresholds between them in alluvial systems (Church, 2002) are examples of the way in which river systems can be conceptualized as forming a discontinuum (Poole, 2002) or a series of beads on a string (Ward et al., 2002, **Figure 4(a)**), where geomorphologically different stretches exhibit different responses to flow disturbances (Resh et al., 1988) and different dynamic, mosaics of habitats (Pringle et al., 1988; Townsend, 1989). Indeed, Montgomery (1999) introduced the geomorphologically based process domain concept as an alternative to the RCC to encompass how "spatial variability in geomorphological processes governs temporal patterns of disturbances that influence ecosystem structure and dynamics" (p. 397) (**Figure 4(b)**).

Following the proposal of the RCC, the importance of lateral (river to floodplain) connectivity was soon recognized, particularly along large floodplain rivers (Welcomme, 1979; Sedell et al., 1989). This formed the core of the FPC (Junk et al., 1989), which emphasized the crucial importance of the connection and disconnection of the floodplain from the river during seasonal flood events. The FPC highlights the importance of inputs of dissolved and suspended sediments from the river into the floodplain, which contribute to nutrient cycles, production, and decomposition processes on the floodplain, and also the role of flood disturbances in resetting floodplain community development. While the river provides a refuge and dispersal route for aquatic organisms during lower flows, much of the primary and secondary production of the river system occurs within the floodplain. More recently, building on the role of the relatively predictable (usually annual) flood pulse, emphasis has been placed on the ecological importance of all river-flow events, from high-frequency, within-channel pulses to low-frequency, high-magnitude floods. The role of this spectrum of events was formalized in the FPC (Tockner et al., 2000), and its importance has been demonstrated empirically in relation to connecting and disconnecting water bodies (Arscott et al., 2002; Malard et al., 2006), and sustaining a shifting mosaic of habitats (Gurnell et al., 2005; Stanford et al., 2005) and organisms (Arscott et al., 2003) within complex, near-natural river landscapes.

In the vertical dimension, the hyporheic zone (the zone of surface water–groundwater interactions below the river bed) was first recognized as a distinct biotope by Orghidan (1959). Subsequent research has succeeded in demonstrating the remarkable spatial extent and temporal dynamics of the hyporheic corridor (Stanford and Ward, 1993), its high vertical connectivity with surface water and groundwater bodies (e.g., Ward et al., 1999), the importance of geomorphological controls for its structure and functioning (Poole et al., 2006), as well as its role in supporting diverse subsurface and surface aquatic faunal communities (e.g., Danielopol, 1989; Malard et al., 2003) often at considerable distances from surface river channels (e.g., Stanford and Ward, 1988). Brunke and Gonser (1997) provided an exhaustive review of the ecological significance of exchange processes between rivers and groundwater across the hyporheic corridor. More recently, Poole et al. (2008) focused on surface–groundwater connections to develop the concept of hydrologic spirals, which describes streams "as a collection of hierarchically organized, individual flow paths that spiral across ecotones within streams and knit together stream ecosystems." This concept provides a very dynamic perspective on longitudinal, lateral, and vertical connectivity that incorporates biogeochemical as well as hydrological spiraling. It links tightly with the concepts of patch mosaics and their dynamics, longitudinal (dis)continua, and the interactions between hydrology and geomorphology

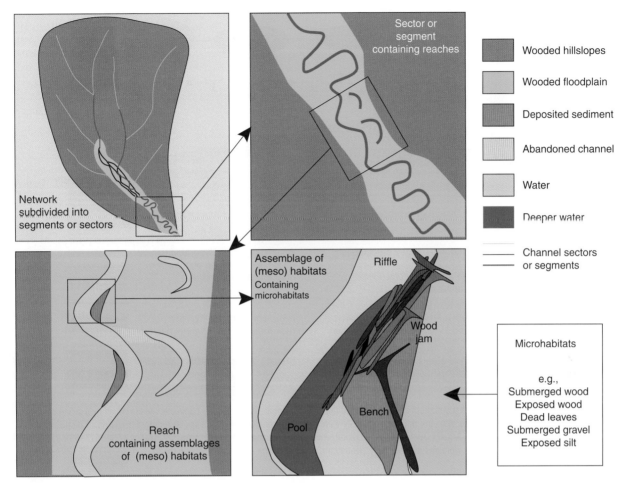

Figure 3 The nested, hierarchical structure of river systems. Adapted from Frissell CA, Liss WJ, Warren CE, and Hurley MD (1986) A hierarchical framework for stream habitat classification: Viewing streams in a watershed context. *Environmental Management* 10: 199–214.

that are increasingly being recognized as generating hierarchies of landscape filters onto which species map according to their traits.

2.10.3 River-Corridor Dynamics

2.10.3.1 River Regimes and River Styles

The character of river corridors depends upon the flow regime and the longitudinal, lateral, and vertical fluxes of water, sediment, and organic matter that the flow regime drives. If maintained in their natural state, these fluxes support heterogeneous, biodiverse landscapes (e.g., Nilsson and Svedmark, 2002) and give rise to a wide spectrum of dynamic river planform styles (e.g., Gurnell et al., 2009) that provide a diversity of physical habitats. Interactions between the flow regime and the form of river channels and their floodplains are the central theme of fluvial geomorphology, and understanding of these interactions has advanced rapidly over the last 50 years (e.g., see overviews by Leopold et al. (1963), Schumm (1977), Knighton (1998), and Brierley and Fryirs (2005)). The FPC (Junk et al., 1989) was the first ecosystem-focused concept to link the flow regime with river–floodplain connectivity, encapsulating both the core multidimensional processes that produce the landforms and dynamism of river landscapes and also the adaptations of aquatic and riparian organisms to these dynamics. Recent ecologically based reviews have highlighted this crucial role of flow–form–organism interactions. For example, Bunn and Arthington (2002) noted that flow largely determines physical habitat in streams and that physical habitat is a significant control on biotic composition, as the first of their four key principles linking hydrology and aquatic biodiversity. Changes in climate and the direct and indirect activities of humans have led to major changes in flow regimes and thus fluxes of water and sediment, which have resulted in dramatic changes to river-corridor characteristics worldwide (e.g., Tockner and Stanford, 2002; Petts and Gurnell, 2005).

The potential morphological complexity that fluvial processes can confer on river corridors has been revealed through major advances in scientific understanding over the last 50 years. In reviewing progress in the classification and scientific understanding of river planform styles over this period, Gurnell et al. (2009) identified four main research phases:

- The earliest phase yielded simple classifications that discriminated between straight, meandering, and braided planform patterns and defined threshold conditions of

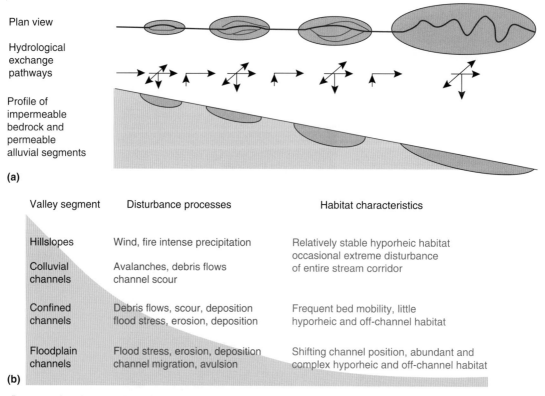

Figure 4 Representation of the river discontinuum as (a) a series of three-dimensional, hydrologically connected beads linked by a string of confined reaches and (b) a sequence of process domains reflecting spatially variable geomorphological responses to temporal patterns of disturbances. (a) Adapted from Ward JV, Tockner K, Arscott DB, and Claret C (2002) Riverine landscape diversity. *Freshwater Biology* 47(4): 517–539. (b) Adapted from Montgomery DR (1999) Process domains and the river continuum. *Journal of the American Water Resources Association* 35: 397–410.

channel forming (e.g., bankfull, mean, or median annual flood) discharge (Q) and slope (S) at which rivers might switch from one form to another (e.g., Lane, 1957; Leopold and Wolman, 1957).

- A second phase, demonstrated that different Q–S threshold conditions existed in different river environments and that bed-sediment caliber could be added to Q and S to improve the identification of thresholds or more gradual transitions between different river planform styles (e.g., Osterkamp, 1978; Begin, 1981; Bray 1982; Carson, 1984a, 1984b; Ferguson, 1987; van den Berg, 1995; Bledsoe and Watson, 2001).
- A third phase recognized an increasing number of channel planform styles, including a transitional wandering style between braided and single-thread planforms (Desloges and Church, 1989), and also the importance of sediment supply as well as sediment caliber in influencing river planform styles and blurring the transitions between them. Schumm (1985) described three groups of alluvial river according to their dominant caliber and mode of sediment transport (bedload, mixed load, and suspended load). He noted that suspended load channels tended to be fairly narrow with a predominantly single thread, stable planform, whereas bedload channels were wider and tended to have more unstable, multithread planforms. Mixed-load channels possessed intermediate characteristics between the other two groups. Building on Schumm's (1985) and Mollard's (1973) work on river planform styles, Church (1992) defined sequences of river planform styles along gradients in stream power (a combination of Q and S), sediment caliber, and sediment supply, whereas Nanson and Knighton (1996) derived a detailed subdivision of multithread (anabranching) river styles discriminated by specific stream power, bed, and bank-material caliber. There was also, within this third phase, increasing recognition that river planform styles produced distinct river corridor and floodplain styles. In particular, Nanson and Croke (1992) identified 15 floodplain styles associated with particular styles of river planform and reflecting gradients in specific stream power, sediment caliber, and valley confinement.
- The fourth phase had its origins in the work of Nanson and Knighton (1996) who explicitly mentioned the role of living and/or dead vegetation in association with virtually all of the anabranching styles that they identified. At the same time, Gurnell (1995) also reviewed research on vegetation along river corridors in relation to different styles of river channel and floodplain. It is increasingly evident that plants not only form a crucial contribution to the biodiversity of river corridors but that there are also many different ways in which vegetation exerts a direct influence on river planforms. For example, within very low energy anastomosing systems of the upper Narew River, Poland, Gradzinski *et al.* (2003) stressed the overwhelming impact

of vegetation, which produces an erosion-resistant peat layer that stabilizes channel banks, stimulates aggradation of channel beds, and induces avulsions to maintain the multithread network. Within the low-energy, arid environment of the Northern Plains, Central Australia, channels adopt a multithread form separated by long sinuous ridges and islands that are developed, maintained, and reinforced by the action of trees on flow patterns, sediment transport, retention, and reinforcement (Tooth and Nanson, 1999, 2000). Huang and Nanson (2006) suggested that this development of multiple vegetated ridges can be an important mechanism for achieving stability in anabranching systems, particularly where the flow regime and sediment supply are highly irregular. The ridges reduce channel width allowing an increase in flow and sediment-transport efficiency without any adjustment in channel slope.

Gurnell *et al.* (2001, 2005) and Gurnell and Petts (2006), working on the Tagliamento River, Italy, established the importance of vegetation for channel planform style in high-energy multithread systems. A model of island development within braided reaches of this system explains their transition from bar-braided to island-braided styles. Island formation depends upon supply and rapid sprouting of large numbers of uprooted trees on gravel bar surfaces and the ability of these sprouting deposited trees to trap, grow through, and reinforce fine sediment. In this way, tree-reinforced patches aggrade and coalesce to form mature, elevated, vegetated islands and extensions to the wooded floodplain in periods between large destructive floods. Other researchers have made similar observations of the impact of vegetation on multithread systems in both field and flume studies (Gran and Paola, 2001; Tal *et al.*, 2004; Coulthard, 2005). Recently, Gurnell *et al.* (2009) reviewed evidence for direct and indirect impacts of vegetation on European multithread river-channel styles. Indirect influences operate at the catchment scale, where changing vegetation cover influences the drainage basin hydrological cycle and, as a consequence, both runoff and sediment delivery to the river network. These changes then combine with direct influences of vegetation at the channel margins (e.g., Zanoni *et al.*, 2008), including changes in the rate of vegetation growth with fluctuations in moisture supply from the alluvial aquifer (e.g., Gurnell and Petts, 2006). In summary, the flood magnitude and frequency required to maintain a particular channel planform style increase with the rate of riparian vegetation growth particularly tree growth, and the vegetation-growth stage. Vegetation-growth performance in turn reflects local climate and the subsurface hydrological regime. Strong rates of riparian vegetation (particularly tree) colonization and growth can shift the threshold conditions at which planform change can occur, despite no change in the flow and sediment regimes (Gurnell *et al.*, 2009, **Figure 5**).

2.10.3.2 Changing River Styles

Human impacts on flow and sediment regimes are manifest in changing river styles, and are further moderated by changes in vegetation colonization, growth, and management. For example, Petts and Gurnell (2005) reviewed scientific progress in understanding the impact of dam construction and associated flow regulation on downstream river-channel characteristics. Reductions in flood disturbance combined with increasing moisture availability attributable to raised baseflows provide improved vegetation survival and growth, particularly where dams are constructed in semiarid areas (e.g., Williams and Wolman, 1984). As a result of the moist, less-disturbed environment downstream of dams, vegetation encroaches across river margins and bar surfaces, inducing channel narrowing (Johnson, 1997, 2000; Merritt and Cooper, 2000). This process is usually underpinned by construction of marginal bench-type landforms through sediment trapping in the encroaching vegetation (e.g., Sherrard and Erskine, 1991). Petts and Gurnell (2005) conceptualized the spectrum of channel responses to reduced discharge and sediment delivery in river channels downstream of dams (**Figure 6**). The impact of a reduced sediment load on downstream river channels (**Figure 6(a)**) is mediated by the resistance of the channel margins to erosion. Erosion resistance results from the caliber and cohesiveness of bed and bank sediment plus any additional protective cover and cohesion provided by vegetation. These responses are usually confined to a relatively short reach below a dam, although significant responses may be expressed in downstream reaches whose susceptibility to erosion is high. Accommodation (no observable channel change) may occur in reaches where the regulated flows are not large enough to be competent to erode and transport sediment. Elsewhere, channel-bed incision and bank erosion will progress, unless limited by bed armoring, or the exposure of resistant banks, or until the channel slope is reduced sufficiently to reduce stream energy below competent levels. In reaches receiving sediment from upstream erosion or tributary inputs, bed incision and channel narrowing, enhanced by riparian vegetation encroachment, may occur simultaneously. Moreover, desynchronization of sediment delivery from upstream reaches and tributaries can create highly unstable phases of scour and fill. Channel narrowing following the removal of flood flows (**Figure 6(b)**) can also occur in an unstable manner and at different rates but it is accelerated by riparian vegetation encroachment onto sediment deposits. However, where sediment sources are limited or vegetation establishment and growth is slow, flows are accommodated within the existing channel form. In the vicinity of unregulated tributary confluences (**Figure 6(c)**), bed aggradation, lateral berm construction, and channel migration can all occur and extend downstream, with vegetation encroachment again reinforcing the development of depositional landforms.

Illustrations of the impact of human manipulation of vegetation on channel planform can be drawn from case studies where vegetation biomass has been reduced or removed. Brooks and Brierley (2002) and Brooks *et al.* (2003) demonstrated the massive influence of riparian vegetation and large wood removal from alluvial rivers in southeast Australia, by reconstructing changes in the heavily impacted Cann River over 150 years of European settlement in comparison with the relatively unimpacted Thurra River. They showed that their study reach on the Cann River has experienced a 3.6-fold increase in channel depth, 2.4-fold increase in channel slope, 7-fold increase in channel capacity, and 150-fold increase in its lateral migration rate, since European settlement. In extreme cases, reduction or removal of riparian vegetation can

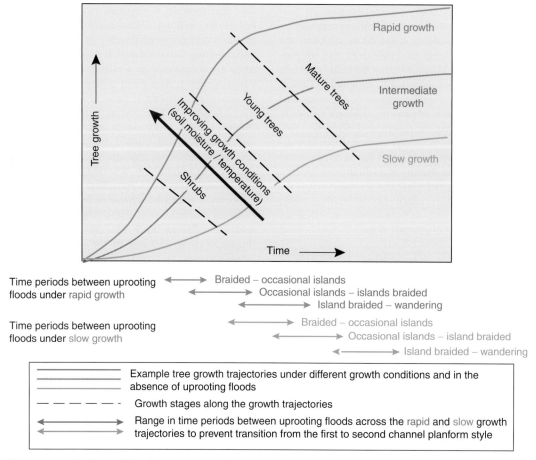

Figure 5 Conceptual model of associations between tree-growth performance and flood magnitude/frequency in relation to the maintenance of different channel styles. Three trajectories of riparian tree growth (green lines) pass through tree-growth stages (black dashed lines) at different rates according to different growing conditions (blue arrow). The impact of these different trajectories on transitions between channel styles vary with different growing conditions (the growth trajectories) and the maximum time period between floods capable of uprooting trees across the three trajectories (horizontal green arrows below the main graph), which prevent transitions between the given river planform styles (green text associated with the horizontal green arrows). Sediment supply is assumed to be sufficient to support the different channel styles. Adapted from Gurnell AM, Surian N, and Zanoni L (2009) Multi-thread river channels: A perspective on changing European alpine river systems. *Aquatic Sciences* (71(3): 253–265.).

lead to floodplain unraveling and transition from single thread to braided channel styles, as was observed by Griffin and Smith (2004) on the Plum Creek, Colorado, USA, in response to the coincidence of overgrazing of floodplain shrubs with a high-energy flow event. Conversely, Smith (2004) attributes the maintenance of a single thread channel following a 300-year flood on the Clark Fork of the Columbia River, USA, to the presence of a high riparian shrub biomass.

Changes in flow regime can allow alien species to invade (Bunn and Arthington, 2002). Within riparian zones, alien invaders can result in significant changes to vegetation cover and biomass, leading to major adjustments in channel planform style. The invasion of many semiarid systems in the USA by Tamarix species (e.g., Lite and Stromberg, 2005; Birken and Cooper, 2006; Stromberg et al., 2007) is particularly well documented. These species have proved to be particularly invasive where natural perennial flow regimes that support native willow and poplar species have been replaced by intermittent flow regimes with changed peak-flow timing, depressed alluvial water tables, and often a more-saline water quality (Glenn and Nagler, 2005). In some areas, invasion by Tamarix species has produced a dense cover of shrubs across the river corridor, confining channel widths and simplifying channel patterns (e.g., Graf, 1978).

In summary, river styles and their dynamics reflect the interaction of three key ingredients: (1) hydrological processes (catchment hydrological cycle, river-flow regime including flow extremes, and alluvial groundwater dynamics); (2) sediment supply, transport, and caliber; and (3) vegetation performance. Rivers of different style support different types and patterns of physical habitats and different rates of habitat turnover. They are also maintained by, as well as act as, controls on hydrological connectivity. Thus, river styles are heavily influenced by hydrological processes and they also integrate all of the key physical as well as many of the chemical factors that control the potential biotic composition of river corridors (Ward et al., 2002). As planform styles reflect dynamic interactions among river flows, sediments, and vegetation in different topographic settings, they can be highly dynamic in space as well as time, with downstream sequences of river

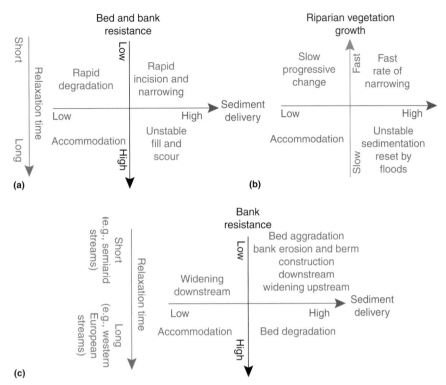

Figure 6 Domains of channel change in response to changes of discharge and sediment load. Responses to a dominant reduction in sediment load (a) are compared with those to a dominant reduction in floods (b); and the special case of channel change below a tributary confluence in a river dominated by flood reduction (c). Adapted from Petts GE and Gurnell AM (2005) Dams and geomorphology: Research progress and future directions. *Geomorphology* 71: 27–47.

styles expanding, contracting, and switching in response to changes in climate, catchment land use and management, and river management. These spatial and temporal controls and responses are embedded in the spatially hierarchical river-styles framework, proposed by Brierley and Fryirs (2005), which provides a geoecological structure within which to explore river-environment functioning and develop management and restoration strategies.

2.10.3.3 The River-Corridor Habitat Mosaic

Naturally functioning river corridors support a variety of river and floodplain styles (Section 2.10.3.1) and therefore are characterized by a diverse array of both vegetated and unvegetated landforms that are subject to continuously varying patterns of water inundation (that connect and disconnect water bodies), moisture content (from precipitation, surface water, and alluvial groundwater dynamics), vegetation development (colonization, growth and establishment, and uprooting and breakage), and morphological change (sediment deposition, sorting, and erosion). Indeed, Tockner and Stanford (2002) identified the maintenance of a high shoreline (water edge) length across the annual range of river flows as a simple but powerful index of habitat quality, noting that dynamic, morphologically complex river corridors maintain shoreline lengths well above the minimum 2 km km^{-1}. For example, in some island-braided reaches, the Tagliamento River, Italy, sustains shoreline lengths of up to 25 km km^{-1} across all but the lowest flows. Furthermore, the length of the edge of patches of riparian vegetation, which is in part a product of inundation dynamics, also displays very high values in naturally functioning multithread rivers (Bertoldi et al., 2009). Typical vegetation-edge lengths within the active tract of island-braided reaches of the Tagliamento River have been found to be 5–6 km per river km (Zanoni et al., 2008). Such naturally functioning open systems are in a state of continual biophysical change where individual landscape elements turnover rapidly as a result of interplay between fluvial disturbances and ecological succession, but their relative abundance tends to remain fairly constant, and therefore predictable (Ward et al., 2002; Zanoni et al., 2008). This phenomenon has been described as the 'shifting habitat mosaic' (Stanford et al., 2005) that drives bio-complexity across spatial and temporal scales (Thorp et al., 2006). Shifting refers specifically to the fact that, although the overall abundance of various landscape elements may remain approximately constant, the individual landscape elements change their location, size, and configuration over time.

As water, sediment, and vegetation dynamics can be viewed across all spatial and temporal scales, research on the river-corridor habitat mosaic is usually based on a spatial, hierarchical, typology of landscape units. This places the units or habitats within segments or sectors of the river channel and floodplain system and then positions each river channel and floodplain sector within the river network or catchment (**Figure 3**). Frissell et al. (1986) were the first to formalize a spatial, hierarchical approach to river-habitat classification, whereby the catchment river network was subdivided into segments (sometimes called sectors), usually terminating at tributary junctions, which were in turn subdivided into river reaches, and then into landforms such as pool-riffle couplets, bars, wood jams, and vegetated patches or islands (habitats or mesohabitats), which in turn supported microhabitats (or patches) of, for example, different sediment caliber or vegetation cover. Since 1986, researchers have reviewed and proposed hierarchical schemes (e.g., Naiman et al., 1992) to structure description, investigation, and management of river systems. These schemes place landscape elements and habitats into their reach and network context (e.g., Montgomery, 1999; Poole, 2002; Snelder and Biggs, 2002; Benda et al., 2004; Thorp et al., 2006). Thorp et al. (2006) proposed a river ecosystem synthesis (RES) of preexisting models, which links the hierarchical structure of river systems and their functioning. The RES is based on 14 tenets relating to distribution of species, community regulation, and ecosystem and river-landscape processes (**Table 2**). The fundamental role of hydrology and fluvial geomorphology is apparent in virtually all of these tenets, with widespread reference to the importance of hydrological/hydraulic and landform/physical habitat and the functional process zones in which they are located.

Three broad properties of these physical habitats have ecological significance: their spatial distribution, their hydrological connectivity, and their dynamics or turnover. These properties are considered in the next two sections.

2.10.3.4 Distribution and Connectivity of Physical Habitats

In naturally functioning river corridors, the pattern of physical habitats reflects interactions between fluvial disturbance and ecological succession. Hydrological connectivity and inundation are the key discriminators of the aggregate and time-varying properties of different habitat types and also of the species and life-cycle stages that the habitats support.

Within the riparian zone, habitats can be defined according to their sediment caliber and vegetation cover and composition but, in addition, they may or may not support surface water during low river flows, and therefore, riparian-habitat characteristics under low-flow conditions reflect the form and water-retention characteristics of the landforms on which they are superimposed. For example, Ward et al. (2002) identified the following spatial elements of river landscapes that underpin the habitat mosaic: (1) geomorphological features or landforms such as channels, floodplains, terraces, levees, bars and islands, and ridges and swales; (2) a spectrum of standing to running surface water bodies; (3) subsurface water bodies showing a spectrum of surface-water influences; and (4) vegetation communities of wetlands, meadows, and alluvial forests. As river dynamics construct floodplains, the nature and proportional mix of riparian habitats reflect river planform style and valley confinement (Nanson and Croke, 1992). This association was supported by Gurnell et al. (2000), who used topographic map data to show how the extent of some easily defined riparian habitats (exposed sediments, vegetated islands, and surface water bodies), varied along the Tagliamento River, in association with changes in lateral confinement, total stream power (5-year flood magnitude and downstream valley-floor slope), and river planform style (single thread, single thread with backwaters, bar braided, bar braided with occasional islands, and island braided) (**Figure 7**).

Increases in river stage progressively extend the river corridor's aquatic zone into the riparian zone, leading to an increase in the extent of inundated habitats and a decrease in exposed sediments and vegetated areas. In complex multi-thread reaches, inundation causes major changes in the degree of connection between surface water bodies, with ponds and backwaters being integrated into bodies of flowing water (e.g., Bertoldi et al., 2009, **Figure 8(a)**) and also in the level of the water table in the alluvial aquifer, and thus water availability to habitats such as islands that may not have been inundated.

Contrasts in hydrological connectivity between subsurface and surface water and between adjacent surface water bodies, as well as the degree of shading by vegetation, strongly affect surface-water quality and temperature. For example, the average daily summer-water temperature in isolated surface water bodies (parafluvial ponds) along the reach of the Tagliamento River depicted in **Figure 8(b)** can range from 13 to 22 °C and the daily (24 h) range in water temperature within a single pond can vary from less than 1 to more than 12 °C (Karaus et al., 2005). Even connected water bodies display strongly varying temperature regimes and, regardless of water-body type, a more stable temperature regime indicates strong connectivity with subsurface waters.

In gravel-bed systems, such vertical connectivity can be high, leading to distinct habitats associated with the surface and subsurface in areas of downwelling and upwelling as well as with the degree of connection between surface water bodies (Arscott et al., 2001). Indeed, Arscott et al. (2001) found that in summer and autumn, thermal variation between lowland floodplain aquatic habitats within the same reach exceeded thermal variation observed in the main channel of the Tagliamento River along the entire studied river corridor (c. 120 km length and c. 1100 m relative relief). The thermal regime is almost as important as the flow regime for aquatic biota (Section 2.10.4). The impacts of surface water–groundwater interactions extend from the aquatic into the riparian zone, with riparian-vegetation growth rates varying between downwelling and upwelling reaches (Harner and Stanford, 2003). The aquatic and terrestrial sides of the dynamic surface-water shoreline also provide important habitats for many organisms. For example, Paetzold et al. (2005) found that on exposed sediments within 2 m of the water's edge of the Tagliamento River, there was a particularly high abundance of ground beetles and spiders, and gut-content analysis revealed that the diet of these terrestrial animals was predominately aquatic insects.

Hydrological connectivity also involves the transfer of many other materials apart from water between river landscape units, controlling the spatial distribution and character of habitats as well as their connectivity. Mineral and organic (coarse particulate organic matter (CPOM) and fine particulate organic matter (FPOM)) sediment particles as well as dissolved chemicals, particularly nutrients, are transferred within and between land and water bodies, driving the construction and turnover of physical habitats and complex biogeochemical processes that are reviewed in **Chapter 2.11 Hydrology and Biogeochemistry Linkages** (see also Craig et al. (2008), who relate nitrogen dynamics to the

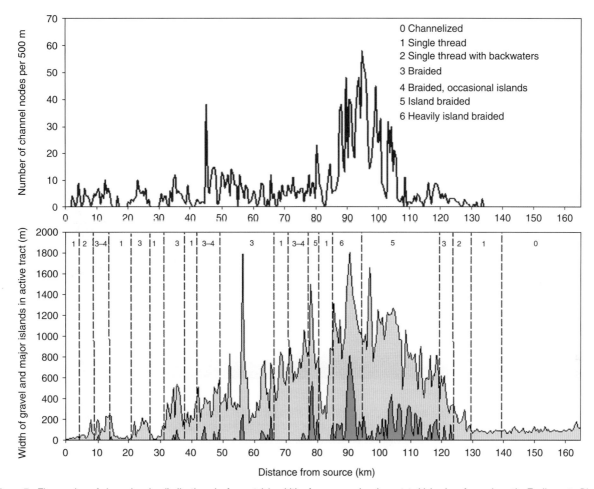

Figure 7 The number of channel nodes (indicating planform style), width of open gravel and vegetated island surfaces along the Tagliamento River, Italy, based on the analysis of information from topographic maps. A classification of six channel styles is superimposed on the lower graph (see key on upper graph for explanation). Adapted from Gurnell AM, Petts GE, Hannah DM, et al. (2000) Wood storage within the active zone of a large European gravel-bed river. *Geomorphology* 34(1–2): 55–72.

morphological complexity and connectivity of the river environment). Viable propagules (e.g., seeds, spores, rhizomes, and eggs), as well as mineral and dead organic matter, are also transported and redistributed by water, supporting colonization of habitats constructed by the erosion, transport, and deposition of mineral sediment. The process of seed dispersal by water, which is known as hydrochory, has been shown to play a major role not only in transporting and depositing freshly produced seeds along river corridors (e.g., Boedeltje et al., 2003; Vogt et al., 2004; Truscott et al., 2006), but also in remobilizing seeds (Pettit and Froend, 2001; Goodson et al., 2003; Gurnell et al., 2008), and structuring riparian plant communities (Nilsson et al., 1991; Johansson et al., 1996; Andersson et al., 2000; Goodson et al., 2002). Large floods may transport mineral and organic particles long distances, but transport of seeds of many species is further facilitated by properties such as their low density or the presence of air-filled seed casing or appendages (Murray, 1986) that can maintain buoyancy and thus transport for prolonged periods, even when river levels and flow velocities are low (Danvind and Nilsson, 1997; Nilsson et al., 2002; Boedeltje et al., 2004).

2.10.3.5 Habitat Dynamics and the Role of Plants as Ecosystem Engineers

In Section 2.10.3.2, it was noted that river styles and their dynamics reflect the interaction of three key ingredients: hydrological processes; sediment supply, transport, and caliber; and vegetation performance. Hydrological and sediment transfer processes provide the broad environmental controls on river styles and their physical habitats, but dead and living vegetation provide key local controls on the type, density, and rate of development of physical habitats. Sediments are redistributed by hydrological processes, causing vertical accretion of river-corridor surfaces (e.g., Asselman and Middelkoop, 1995; Gomez et al., 1998; Benedetti, 2003) and also avulsion, lateral erosion, and lateral/oblique accretion of flow pathways within the river's active tract (e.g., Hickin and Nanson, 1984; Brewer and Lewin, 1998; Mack and Leeder, 1998; Hooke, 2003; Page et al., 2003). However, dead and living vegetation and propagules, which are also transferred and redistributed by hydrological processes, drive much of the detail and complexity of habitats along river corridors (e.g., Gurnell et al., 2001, 2005).

Figure 8 (a) Changes in wetted shoreline length (left graph), proportion of the active tract under disconnected ponds, connected water bodies, bare sediment, and open water (right graph) with increasing river levels in the stretch of the Tagliamento River shown in (b). Adapted from Bertoldi W, Gurnell A, Surian N, et al. (2009) Understanding reference processes: Linkages between river flows, sediment dynamics and vegetated landfroms along the Tagliamento River, Italy. *River Research and Applications* 25: 501–516.

Large wood is transported and deposited to form a habitat in its own right as well as a food source (Gurnell et al., 2002). Wood habitats vary from individual wood pieces to large accumulations of pieces that provide storage reservoirs for leaves and fine sediment particles filtered from flowing water. Wood can snag on landform surfaces such as bars during flood recessions, but retention of large quantities of wood usually depends on the presence of preexisting retention structures such as rocks and trees around which it becomes snagged. Thus, hydrological connection with retention structures is a crucial factor in the distribution of wood across river channels and floodplains, particularly because wood from the majority of tree species floats (Gurnell, 2003). Once wood becomes retained within a river's active tract, its complex hydraulic resistance causes the development of other landforms. In small rivers, large wood can snag easily in many different locations (e.g., between the channel banks, behind trees, or against other large wood pieces) causing the development of downstream plunge pools, upstream dammed pools and bars, and many other sediment and flow-related habitats. In larger rivers, where the wood pieces are significantly smaller than the channel width, snagging depends on floodwaters interacting with hydraulically rough surfaces and objects (Gurnell et al., 2002). Gurnell et al. (2000) estimated that relatively small quantities of wood were stored on smooth open-gravel surfaces along the large Tagliamento River despite their frequent inundation (estimates ranged from 1 to $21 \, t \, ha^{-1}$); intermediate quantities were associated with established islands ($24–186 \, t \, ha^{-1}$), which although morphologically rough (vegetated surfaces and topographic complexity) were located at high elevations where inundation was relatively infrequent; but the largest quantities of wood were associated with small vegetated patches or pioneer islands ($293–1664 \, t \, ha^{-1}$) that were both morphologically rough and at intermediate elevations subject to relatively frequently inundation (Francis, 2007). These observations support the view that although wood can engineer ecosystems in narrower channels, where it can form a significant hydraulic obstruction, its potential to create physical habitats in larger systems is dependent upon the presence of other roughness elements, particularly rooted, living vegetation.

In larger river systems, interactions among trees, wood, river flows, and sediments produce a wide variety of structures, which are fully reviewed by Gurnell (2010). In brief, in forested landscapes characterized by large trees with low decomposition rates that are delivered to the river in large quantities, accumulations of dead wood can block sizable river channels and persist as extensive hard points within alluvial sediments for prolonged periods (e.g., Brooks and Brierley, 2002; Montgomery and Abbe, 2006; Arsenault et al., 2007). In such landscapes, a complex mosaic of habitats develops, each associated with particular soil, vegetation, and large wood assemblages and turned over at characteristic rates to form a shifting mosaic through biophysical feedbacks between geomorphological processes, and large wood and living vegetation (e.g., the Queets River, USA, Latterell et al., 2006). In such systems, low wood-decomposition rates, large piece sizes, and the high rates of recruitment allow wood to act as a river-ecosystem engineer driving a fluvial-biogeomorphic succession (Corenblit et al., 2007). Wood interacts with river flows and fluvial sediments to construct landforms; trap seeds, and shelter them while they germinate and grow; and in the longer term, it protects these sheltered forest patches from erosion.

In large river systems where trees are relatively smaller, often have high rates of decay, or are not supplied to the river system in large quantities, fluvial ecosystem engineering becomes heavily dependent upon living wood – wood pieces or entire shrubs and trees capable of developing root systems and sprouting a canopy following erosion, transport, and deposition. Uprooted trees become snagged on river bars during the falling limb of flood events, typically with their root wad oriented upstream and with smaller wood pieces braced or snagged against the root wad. As in dead wood systems, the hydraulic impact of such trees creates a suite of habitats on the bar surface that is characteristic of bar apex jams (Abbe and Montgomery, 2003). Landforms include deep scour hollows where flows diverge around the root wad of the tree, often exposing coarse lag deposits, and bars of finer sediment in the shelter of the root wad and along the stem and canopy (**Figure 9(a)**). In living wood systems under suitable environmental conditions (Gurnell and Petts, 2006), these features are fundamental to the development of vegetated patches and subsequent islands or floodplain extensions. Many species of riparian willows and poplars, characteristic of river margins across much of the Northern Hemisphere, reproduce vegetatively (Karrenberg et al., 2002). Uprooted trees and fragments produce roots that can extend at rates $> 2.5\,\mathrm{cm\,d^{-1}}$ to track the falling water table following seasonal high flows (Barsoum and Hughes, 1998; Francis et al., 2005). Shoots also grow rapidly (typically $5-10\,\mathrm{mm\,d^{-1}}$ for deposited specimens of *Populus nigra* and *Salix eleagnos* during the summer on the Tagliamento River, Francis et al., 2006), producing a canopy that increases flow resistance above that of an equivalent dead wood deposit, and thereby enhance trapping of sediment from both inundations and wind storms (Gurnell et al., 2008). As a consequence, if the deposited tree survives early uprooting floods and continues to grow, the area of hydraulically induced scour, sedimentation, and growing vegetation enlarges to form a pioneer island (Edwards et al., 1999; **Figure 9(b)**). Pioneer islands usually support a diverse vegetation cover produced by sprouting of trapped wood pieces

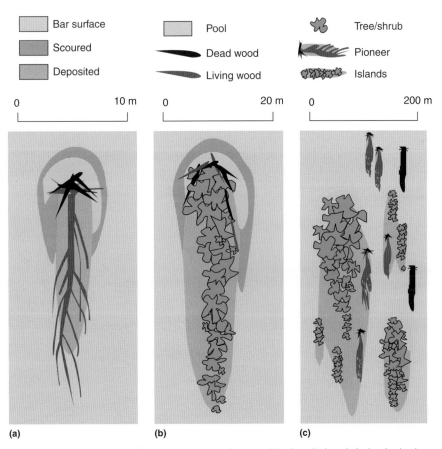

Figure 9 Island development from uprooted, deposited, and resprouting riparian trees: (a) a deposited tree inducing the development of a scour hole and pool and an accumulation of fine sediment in the lee of the root bole; (b) a tree sprouting to produce a line of young shrubs and further scour, deposition of fine sediment, and trapping of wood pieces to form a pioneer island; and (c) an island complex with deposited trees, pioneer islands, and established islands produced by coalescence of smaller islands, distributed across an extensive gravel bar surface. Adapted from Gurnell AM, Tockner K, Edwards PJ, and Petts GE (2005) Effects of deposited wood on biocomplexity of river corridors. *Frontiers in Ecology and Environment* 3(7): 377–382.

and germination of seeds rafted in on the root bole or deposited with mineral and dead organic particles from water or wind transport (Kollmann et al., 1999; Francis et al., 2008). If they survive flood events, pioneer islands continue to trap sediment and plant propagules, and grow and coalesce, forming wooded islands or floodplain extensions (Figure 9(c) Gurnell et al., 2001) that support the same suite of habitats (scour hollows and ponds, coarse lag deposits, and aggraded islands of finer sediment), which form the template for biocomplex river corridors (Gurnell et al., 2005). Vegetated patches also facilitate germination and establishment of tree seedlings on open bar surfaces in their lee (Moggridge and Gurnell, 2009), increasing the potential for woodland to extend rapidly across areas of exposed riverine sediments and to link isolated vegetated patches. On the Tagliamento River, these processes occur rapidly. Figure 10 shows the same area of bar surface between 1999 and 2008, with loss of islands and widespread deposition of wood apparent following a major flood in 2001, and subsequent development and coalescence of pioneer islands to support the establishment of an extensive area of new woodland by 2008.

Ongoing research reveals that there are many plant species or groups of species that are capable of acting as river ecosystem engineers. To date, research has largely focused on riparian trees (e.g., Gurnell et al., 2005; Corenblit et al., 2007), but there is increasing evidence that riparian herbs and grasses (e.g., Corenblit et al., submitted) and aquatic macrophytes (e.g., Gurnell et al., submitted) can induce river-channel adjustments and thereby create or facilitate new habitats that strongly affect the character of the river-habitat mosaic. These engineering plant species not only have a direct effect on the rate and style of physical habitat construction, but also, through their flow resistance, they are very effective in trapping hydrochorously dispersed plant propagules (Gurnell et al., 2008), which in turn can germinate to support diverse plant assemblages on the evolving landforms as well as additional vegetation roughness.

2.10.4 Aquatic Ecosystems

2.10.4.1 Instream Flows and Flow Regimes

In Section 2.10.2, we described some key hydrological characteristics of river networks and briefly illustrated some of the contexts in which these have been shown to have ecological importance. In this section, we revisit these properties, emphasizing how their importance has been revealed in research on aquatic ecosystems.

2.10.4.1.1 The significance of multidimensional variations in flow characteristics

Flowing water along rivers has a number of advantages over still water because it is constantly mixed by turbulence providing nutrients, exchange of respiratory gases, and removal of wastes, and is vital for both downstream movement and upstream migrations of species throughout a drainage network. The communities of animals and plants at any point along a river reflect the species pool of that bioclimatic region modified by location (altitude and distance downstream) within each stream network and the historical sequence of flows, especially the incidence of unpredictable floods and droughts, which disturb biological populations. The spatial distribution of species with shorter life cycles, such as the benthic macro-invertebrates, typically reflect flow conditions at each site along a river (Figure 11(a)) and many species are associated with specific conditions of flow velocity and water depth

Figure 10 The same section of the active tract of the Tagliamento River, Italy, at different dates over an 8-year period, showing strong contrasts in the extent of islands as a result of interactions between tree growth and destructive floods.

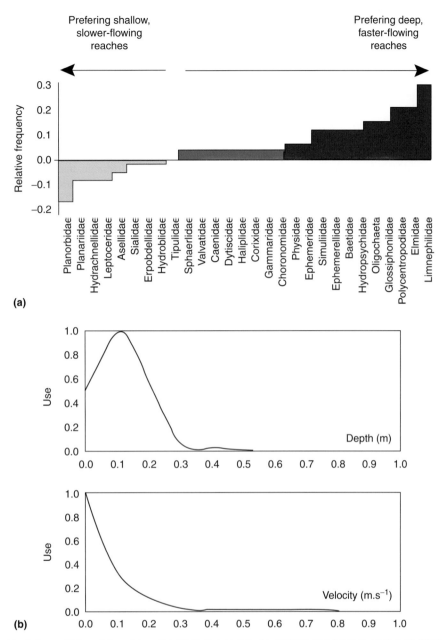

Figure 11 Relationships between biota and flow: (a) spatial distribution of macroinvertebrate taxa between reaches of differing flow depth and velocity along the River Wissey, UK (provided by M.A. Bickerton, School of Geography, Earth and Environmental Sciences, University of Birmingham, UK); and (b) habitat-suitability criteria for bullhead (*Cottus gobio*) showing habitat use in terms of water depth and velocity. Adapted from Gosselin (2008).

(Figure 11(b)). Fundamentally, the ecological integrity of riverine ecosystems depends on their natural dynamic character (Petts and Amoros, 1996; Poff *et al.*, 1997); flow variability maintains habitat complexity and promotes species diversity by providing recruitment opportunities and refuge from competition.

Studies concerned with temporal variability in river flows and, in particular, the impact of fluctuations in the river-flow regime on a river's health have often focused on benthic (riverbed dwelling) macroinvertebrates, because their short life cycles make them sensitive to habitat changes. The impacts of flow-regime properties and change are illustrated by many studies on benthic macroinvertebrate communities, undertaken in many different environments. For example, in a study of summer benthic macroinvertebrate communities within the temperate maritime climate region of England and Wales, Monk *et al.* (2006) illuminated the significance of monthly flows and the magnitude and duration of annual extreme flows through the analysis of a 20-year data set from 83 rivers. Wood *et al.* (2001) demonstrated the particular significance of late winter/spring high flows in a groundwater-dominated chalk stream, especially the absence of these high flows in drought years, which were associated with low invertebrate community abundance. Similarly, for 856 monitoring sites in

New Jersey, USA, Kennen et al. (2008) isolated the importance of disturbance regime, indexed by the average number of annual storms and flashiness of the hydrological regime, in driving the aquatic-invertebrate assemblage. Changes in hydrological regime reflecting land-use changes were associated with the retention of highly tolerant aquatic species and the loss of more sensitive species.

Rivers show a characteristic zonation of biota and the RCC (Vannote et al., 1980; see Section 2.10.2.2) describes progressive changes of stream conditions. The fauna of relatively cool, leaf-litter-dominated headwater streams shaded by trees along the river banks contrasts with that of the relatively wide and shallow midsectors where light and nutrients favor algal production on the channel bed, and the lower river where food chains are based on high levels of fine particulate organic matter from upstream and from floodplain inputs. In forested headwater streams, large wood plays an important role in sustaining the diversity of habitats, in regulating flood flows, and in controlling sediment movement (Maser and Sedell, 1994), whereas in downstream floodplain reaches, the main channel food web can be significantly influenced by dissolved organic carbon and detritus (wood, leaves, seeds, etc.) delivered by the recession limb of floods.

In the middle and downstream reaches of larger rivers, temporal variations in river flows, particularly the seasonal high flows of the annual flow regime, enable lateral movements of plants and animals from main channels into floodplain lakes and backwaters (Petts and Amoros, 1996; Ward and Stanford, 1995; Welcomme, 1979). Furthermore, vertical exchanges of water between the river and the alluvial aquifer and especially the oxygenated subsurface flows where rivers have significant gravel-fills (known as the hyporheos) can also provide important habitats for many species (Stanford and Ward, 1988; Gibert et al., 1990). Indeed, especially along once-glaciated valleys, sequences of alluvial basins separated by rock steps create the sequential downwelling and upwelling of river water into and out of the deep valley-fill sediments. These lateral and vertical flow paths are juxtaposed with longitudinal changes from headwaters to mouth, forming three-dimensional river corridors (Stanford and Ward, 1993; Figure 4a).

Some streams may be adequately described as linear, longitudinal systems but many, especially larger river systems, are more appropriately seen as three-dimensional systems modified by biological hotspots and embedded within a branched network. Thus, the actual spatial pattern of habitats and biota deviates from the idealistic RCC to reflect a number of factors such as history over the Holocene, including glacial-valley deepening and channel incision by sea-level lowering and then valley aggradation, leading to significant vertical fluxes, hydraulic transition zones, and the character of the drainage network. The resilience in regional production of riverine species can be enhanced by the existence of core populations that can buffer metapopulations against environmental change. Metapopulation theory proposes that regional populations have core–satellite structures and the core populations – large populations occupying high-quality habitat – are critical for the persistence of the metapopulation, providing stable sources of dispersers to recolonize peripheral habitats following a major disturbance. Metapopulation linkages allow for local extinction of populations, which reestablish via colonization from adjacent populations. Alluvial zones with active lateral and vertical connections may be particularly important biological hotspots (Stanford et al., 1996). Other hotspots are hydraulic transition zones and major tributary confluences.

The manner in which water velocity, water depth, and channel substratum interact influences the distribution of aquatic biota both along the length of a river and within any given reach (Statzner and Higler, 1986; Statzner et al., 1988). Along the river profile, changes in slope, channel shape, and substratum create hydraulic transition zones between reaches of low and high hydraulic stress, and these zones tend to be associated with high diversity of biota because of the wide variety of hydraulic conditions. The disruption of a simple downstream continuum and its importance for aquatic organisms have been associated with hydraulic discontinuities (Statzner and Higler, 1986), often induced by hydrological and sediment discontinuities at stream network junctions (e.g., Gomi et al., 2002; Benda et al., 2004), which in turn are related to contrasts in the geomorphological characteristics of reaches. Thus, Statzner and Higler (1986) found high benthic invertebrate diversity at transitional zones between those of low and high hydraulic stress for 14 streams worldwide.

The branching and hierarchical drainage network imposes a spatial and temporal organization on river systems (Benda et al., 2004). The influence of the drainage network is seen particularly clearly along recently deglaciated valleys where a diverse community of invertebrates, for example, inhabitat snowmelt or groundwater-fed first- and second-order tributaries flowing through wooded slopes and with stable channels. These tributaries provide sources for rapid colonization of the main channel following further ice retreat or physical disturbance by either downstream migration or drift, or aerial oviposition (Petts and Bickerton, 1994). Tributary junctions represent locations in a network where channel and valley morphology can change and where habitat heterogeneity (in space and time) can be enhanced, potentially leading to increased species richness. Thus, Scarnecchia and Roper (2000) identified tributary mouths as thermal refugia for fish and Rice et al. (2001) found changes in abundance and composition of macroinvertebrate species at confluences. Enhanced channel dynamics at tributary confluences leads to increased width of the active tract and the creation of low-energy backwaters for specialized aquatic species or life stages including fish-rearing habitat. Furthermore, aggradation at confluences can lead to enhanced hyporheic flow with warmer and more nutrient-rich water emerging and supporting increased primary production and microhabitat benefits for some fish (Baxter and Hauer, 2000).

2.10.4.1.2 Adaptations of biota to running water

Adaptations to the flow regime include behaviors to avoid floods, and life-history strategies synchronized with long-term, predictable flow patterns. However, rivers also present hostile environments to biota; natural disturbances, including floods and droughts, and recovery mechanisms contribute to regulating population sizes and species diversity (Milner, 1994; Lake, 2007). Recovery is the natural process by which an ecosystem returns to a condition that closely resembles unstressed rivers in the same region following disturbance. The changes of the biological community that occur at a site following disturbance is known as 'succession' (Fisher, 1990). Except in the severest of floods and droughts, organisms can find refuge locally in the hyporheic zone, pools, and backwaters of the three-dimensional corridor, and these act as colonizing sources to drive succession and recovery along with other sources, especially hotspots, across the drainage network via aquatic and aerial pathways.

The viability of biotic populations links to recruitment and survival rates during early life stages that are determined by external forcing mechanisms, including changes in the flow regime and feedbacks among system components that may also depend on flow (Anderson et al., 2006). Discharge variations drive habitat conditions but large fluctuations in population abundance may be decoupled from long-term availability of usable habitat. Most river fauna are ectotherms, where growth and reproduction are vitally influenced by river temperature, and thus temperature is a critical habitat attribute (Ward, 1985). However, day length also appears to influence the life cycles of many riverine species. Thus, temperature and day length appear to synchronize hatching, maturation of larvae, emergence, and mating of adult aquatic insects (Hynes, 1970; Ward and Stanford, 1982). For aquatic plants of the river food web, availability of light and nutrients is critical. Rivers that flood frequently maintain different food webs than rivers that have more stable flow regimes. Furthermore, biotic interactions (e.g., competition, predation, and parasitism) that occur continually in all habitats are particularly important in spring brooks and lake-outlet streams with naturally stable flow regimes (Ward and Stanford, 1983b). In other rivers, biological interactions tend to become more important with time since the last major flood or drought disturbance.

Over evolutionary time, floods and droughts exert primary selective pressure for adaptation and many organisms have evolved traits that enable them to survive, exploit, and in some cases depend upon, disturbances. Thus, Lytle and Poff (2004) identified three modes of adaptation that plants and animals use to survive floods and droughts. These relate to different hydrological phenomena: the timing of events (calendar day possibly linked to temperature or day length) is important for many life-history adaptations; predictability (the strength and regularity of the seasonal flow cycle) influences behavioral adaptations which may be triggered by linked environmental signals such as rainfall events, seasonal temperature extremes, or sudden changes in flow; and the magnitude and frequency of events of relatively short duration are associated with morphological adaptations. Lytle and Poff (2004) reviewed life-history adaptations that include the timing of reproduction in fish, and diapause and emergence into an aerial adult stage in aquatic invertebrates. The timing of reproduction to coincide with optimal conditions enhances the fitness of offspring and this adaptation appears particularly common in rivers with predictable flow regimes such as those associated with spring/early summer snowmelt or the tropical monsoon.

Although unpredictable floods can cause significant disturbance to river communities, a regular, annual flood can be an advantage to aquatic systems. This is particularly so along large tropical floodplain rivers (Junk et al., 1989) where many species are adapted to the annual flood pulse which connects the main channel to floodplain backwaters and food resources. Agostinho et al. (2004) demonstrate that the flood regime is the primary factor influencing biological processes in neotropical rivers. In the Upper Parana River, the annual variation in the hydrograph affects species with distinct life-history strategies differently, and influences the composition and structure of fish assemblages. Large migratory fish species that spawn in the upper stretches of the basin and use flooded areas as nurseries were favored by annual floods at the beginning of summer that lasted more than 75 days, with longer floods yielding larger populations. Aquatic organisms colonize the floodplain at rising and high water levels because of the breeding and feeding opportunities that arise. Wetland species also benefit from floods. Kingsford and Auld (2005) used 25 years of breeding data for 10 species of colonial waterbirds in the Macquarie Marshes, Australia, to show that the number of waterbird nests was positively related to flow prior to breeding and area inundated, and that breeding was triggered by a threshold flow.

In rivers with unpredictable flows, flood benefits are less obvious (Bayley, 1991). Nevertheless, in these rivers, Lytle and Poff (2004) suggest that bet-hedging strategies might evolve where a parent produces diverse offspring types which correspond to different possible environmental states. Behavioral adaptations often involve reaction on an individual-event basis to a hydrological (change in water level) or hydraulic (change in velocity) cue, often also influenced by a seasonal factor such as temperature. This has been observed in many fish species in seeking refuge from flood flows and in spawning, for example, stimulated by high-flow pulses in warm months or with the onset of declining temperatures. Thus, Schramm and Eggleton (2006), with reference to catfish growth in the lower Mississippi, USA, demonstrated that the FPC applies more strongly to temperate floodplain–river ecosystems when thermal aspects are considered.

2.10.4.2 Ecohydraulics and Mesohabitats

2.10.4.2.1 Hydraulic stream ecology and the mesohabitat template

Statzner et al. (1988) introduced the concept of hydraulic stream ecology to focus on the energy budget of an organism and particularly the relative difference in speed between an organism and the medium in which it lives. Current velocity is significant for lotic organisms influencing respiration and other measures of metabolism, feeding biology, and behavioral characteristics including rheotaxis, locomotory activity, schooling, and territoriality. A common assumption is that biological communities have evolved to exploit the full range of mesohabitats available along rivers (e.g., the structures defined at the habitat scale of Figure 3), the variability of flows determining when and for how long mesohabitats are available to different species at different locations throughout the stream network.

Considerable research effort has been devoted to the categorization of mesohabitats within river channels. These reflect the habitat scale of the hierarchical structure illustrated in Figure 3 and therefore are essentially channel geomorphic units (CGUs, Hawkins et al., 1993) with particular structural and hydraulic properties. These classifications have been developed by some researchers from multivariate statistical associations between the biota and properties of the physical environment; for example, the functional habitats of Harper et al. (1992, 1995), and the mesohabitats of Pardo and Armitage (1997) and Tickner et al. (2000). These classifications have tended to discriminate habitats on the basis of substrate caliber and structural elements of dead and living plants rather than explicitly incorporate topographic features/landforms found on the riverbed and margins. For example, Harper et al. (1995) recognized the following functional habitats: exposed rock, cobbles, gravel, sand, silt, emergent macrophytes, floating-leaved macrophytes, submerged broadleaved macrophytes, submerged fine-leaved macrophytes, mosses, macroalgae, marginal plants, leaf litter, wood debris, and tree roots. An alternative approach develops hydraulically based habitat classifications that reflect flow structures that are influenced by bed roughness (mainly sediment caliber) and morphology (landforms) such as the physical biotopes of Padmore (1997) and the hydraulic biotopes of Wadeson (1994). Newson and Newson (2000) describe physical biotopes based on a simple visual assessment of water-surface flow types. They combined eight flow types (free fall, chute, broken standing waves, unbroken standing waves, rippled, upwelling, smooth boundary turbulent, and scarcely perceptible flow) with other, largely sedimentary, evidence to identify 10 physical biotopes (waterfall, spill, cascade, rapid, riffle, run, boil, glide, pool, and marginal dead water). Harvey et al. (2008) explored the interdependence of these groupings using the UK national database of River Habitat Surveys (Raven et al., 1997), illustrating how vegetation and sediment caliber functional habitats map onto biotopes (**Figure 12**).

This linkage was also developed by Kemp et al. (2000), who found strong associations between flow depth, flow velocity, and the occurrence of functional habitats. The same linkage was also demonstrated at a national scale by Gurnell et al. (2010), who showed discrimination between bed-sediment caliber classes (**Figure 13(a)**) and the presence and abundance of macrophyte morphotypes (**Figure 13(b)**) when they were superimposed on Q (median annual flood) – S (channel gradient) plots for a sample of 467 British river reaches.

The flooding regime leads to a particular configuration of aquatic and riparian habitats but the process of habitat creation and destruction results from the balance between rejuvenating flooding events and habitat stabilization and decay. Habitat turnover may be high along natural river corridors but at the sector scale (a geomorphologically distinctive river segment often of c. 10 km in length, **Figure 3**), the composition and configuration of habitats remain relatively stable (Arscott et al., 2002), providing a continuity of habitat associations that are available to sustain biotic populations. However, there has been much debate about the identification and parameterization of physical habitat at the mesoscale. A major problem is the dynamic relationship between hydraulic parameters and discharge with changes in number and arrangement of mesohabitats (Emery et al., 2003). Thus, attempts to argue the biological significance of mesoscale hydraulic habitat surveys appear premature (Petts, 2009), although the practicality of the mesohabitat approach, and prohibitive cost of microscale surveys, makes it attractive for managers (Newson et al., 1998).

Hydrological change leads to changes of channel morphology and the array of physical habitats available for biota.

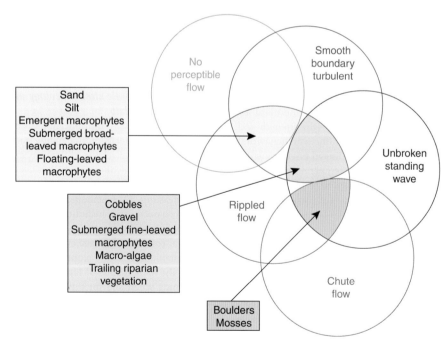

Figure 12 Venn diagram to illustrate the intersection (represented by the yellow-shaded overlap areas) of flow biotopes (no perceptible flow, smooth boundary turbulent flow, rippled flow, and unbroken standing waves) associated with assemblages of functional habitats (listed in the yellow text boxes), based on an analysis of data from the UK Environment Agency's River Habitat Survey. Adapted from Harvey GL, Clifford NJ, and Gurnell AM (2008) Towards an ecologically meaningful classification of the flow biotope for river inventory, rehabilitation, design and appraisal purposes. *Journal of Environmental Management* 88(4): 638–650.

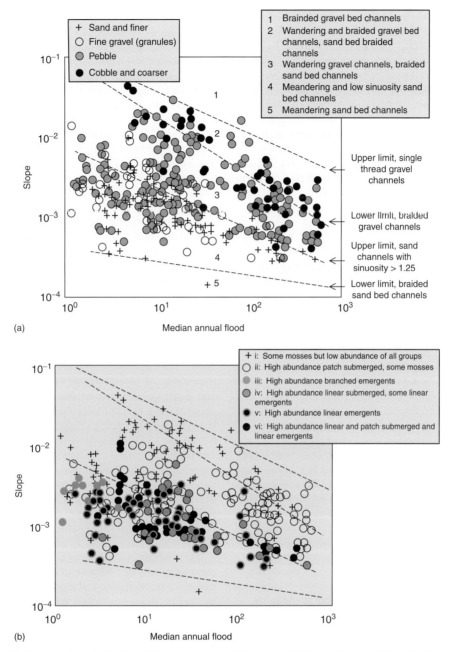

Figure 13 Mean bed-sediment size (a) and the assemblage of macrophyte morphotypes (b) found on a sample of 467 British river reaches in relation to valley gradient (slope in m m^{-1}), the median annual flood (in m^3 s^{-1}), and thresholds of channel style identified by Church M (2002) Geomorphic thresholds in riverine landscapes. *Freshwater Biology* 47: 541–557. Adapted from Gurnell AM, O'Hare JM, O'Hare MT, Dunbar MJ, and Scarlett PD (2010) Associations between assemblages of aquatic plant morphotypes and channel geomorphological properties within British rivers. *Geomorphology* 116: 135–144.

Thus, channel changes to flow regulation below dams involves the complex interaction of sediment-transport processes and riparian-vegetation growth (e.g., Petts and Gurnell, 2005, Section 2.10.3.2). One major concern is the impact of flow regulation on channel sedimentation at salmon-spawning grounds, which could impact upon the intra-gravel environment for egg development over winter and for fry emergence in late winter (Milhous, 1982, 1998; Reiser *et al.*, 1989). In particular, deoxygenated conditions in spawning gravels can cause poor egg survival (Malcolm *et al.*, 2005). Consequently, there have been considerable efforts to quantify the volume, magnitude, duration, and timing of sediment-maintenance flows that flush fines without eroding the important underlying gravels (Wu and Chou, 2004).

2.10.4.2.2 Habitat-suitability criteria

Medium and low flows, together experienced for about 90% of the time along most rivers, sustain a diversity of hydraulic habitats. The complex channel morphology of natural rivers creates a heterogeneous flow environment, described most simply by patterns of velocity and depth. These hydraulic habitats map onto the mesohabitats defined above, but represent the varying velocity-depth conditions within different parts of the river channel to which mobile species can respond as discharge varies. Different species of animals have been observed to have preferred habitats or tolerate different habitats in terms of velocities and depths, to which the caliber of the substratum has been added as a third key habitat characteristic in defining habitat suitability (Bovee, 1978; Gore, 1978). Thus, these associations between the three habitat properties and the preferences of biota have been developed into habitat-suitability criteria (**Figure 11(b)**), which define the probability of use of habitats by a specific life stage or a particular species.

The underlying premise behind the concept of habitat-suitability criteria is that populations, and thus biodiversity in rivers, are limited by habitat events (Stalnaker et al., 1996). Habitat-suitability criteria describe how individuals of a species select the most favorable conditions in a stream but will also use less favorable conditions, with the preference for use decreasing where conditions are less favorable. Simple indices are based on the frequency of occurrence of actual habitat conditions used by a target organism in a particular reach. The ratio of the proportion of habitat utilized to available habitat area within the reach defines the habitat preference. More complex, composite indices may be defined (e.g., Vadas and Orth, 2001; Ahmadi-Nedushan et al., 2006) but these involve several assumptions (Bovee, 1986) not least that all physical variables are equally important and independent.

This habitat-suitability-criteria concept has been challenged over the past 30 years because of the lack of concordance between changes in suitable habitat and fish populations, its simplified approach to hydraulic habitat characterization (e.g., Gore and Nestler, 1988), and lack of biological realism (e.g., Orth, 1987), but it remains central to biological-response models that seek to explain and predict spatial and temporal distributions of instream biotic populations.

2.10.4.2.3 Models of biological responses to changing flows

A very widely used model called Physical HABitat SIMulation (PHABSIM, Tharme, 2003) integrates the changing hydraulic conditions associated with variations in discharge with the habitat preferences of one or more selected species (**Figure 11(b)**). The method relies on three principles (Stalnaker, 1994): the chosen species exhibits preferences within a range of habitat conditions that it can tolerate; these ranges can be defined for each species; and the area of stream providing these conditions can be quantified as a function of discharge and channel structure. The primary approach uses a simple 1-D hydraulic model, but this fails to predict spatial patterns of velocity in natural rivers, although it is useful for determining average velocity variations with changing discharge. This weakness has been overcome by the increasing use of 2-D hydraulic models that can describe the spatial and temporal heterogeneity of hydraulic conditions and provide a link to mesohabitat patterns (Bovee, 1996; Hardy, 1998; Stewart et al., 2005; Crowder and Diplas, 2006).

Considerable efforts have been spent on attempts to assess the ecological credibility of PHABSIM by demonstrating the biological significance of carrying capacity as a limiting factor of population size (Lamouroux et al., 1999; Kondolf et al., 2000). However, validation of the approach in biological terms has proved difficult not least in establishing discrete relationships between biological populations and the weighted usable area (WUA) from empirically derived habitat-suitability curves. From a practical perspective, there is no doubt that the accumulated experience of using PHABSIM means that its strengths and weaknesses are well understood.

Parasiewicz (2003) advanced a PHABSIM derivative, MesoHABSIM. By mapping mesohabitats at different flows along extensive sections of a river and establishing the suitability of each mesohabitat for the dominant members of the fish community, it is possible to derive rating curves to describe changes in relative areas of suitable habitat in response to flow. MesoHABSIM focuses on mesoscale approaches to build on strengths of PHABSIM protocols while providing options for addressing large spatial scales appropriate for water-resource planning (Jacobson, 2008). A rational framework for modeling fish-community response to changing habitat conditions developed by Bain and Meixler (2008) is appropriate for integrating with physical-habitat modeling (Parasiewicz, 2008). The fish-collection survey is the most effort-intensive component of MesoHABSIM, but literature-based evidence and expert opinion can be used, and a regional approach allows transfer of habitat-use models among rivers of similar type (Parasiewicz, 2007). However, the challenge to relate habitat use to changing flows remains elusive (Petts, 2009). The temporal dynamics of habitat quantity may be a major factor determining fish-population responses in riverine environments (Stalnaker et al., 1996), but there is limited evidence that this is manifest by different patterns of habitat use and a large number of empirical case studies have been unable to develop general relationships (Poff and Zimmerman, 2009).

The biomass of a species or a particular life stage within a community can vary because of biological processes such as reproduction, energetics, and mortality that may be influenced by one or more unspecified environmental factors, which undoubtedly blur any simple relationships between species abundance and habitat criteria. Considering trout, for example, recruitment has been shown to be strongly influenced by winter flows (Cattanéo et al., 2002; Lobon-Cervia, 2003; Mitro et al., 2003), but Sabaton et al. (1997) and Gouraud et al. (2001) demonstrated the impact of summer low flows that limit adult-trout biomass and spring flows that limit young-of-the-year numbers between emergence and their first summer, supporting the findings of Capra et al. (2003) that post-emergence high flows have a major impact on the density of 0+ fishes. For unionid mussels, Morales et al. (2006) predicted community development as a function of individual growth and reproduction, biotic interaction involving host fish and intra- and inter-species food competition, and habitat criteria (substrate stability), and demonstrated that for low-density species, even a small level of habitat modification could have a substantial impact on population survival. For common floodplain fishes, Halls and Welcomme (2004) advanced an age-structured population-dynamics model incorporating density-dependent growth, mortality, and recruitment to show the importance of high flood duration, large area of inundation, and a slow rate of flood recession. Thus, biological interactions and flow variability, especially the length of time since the last major flood or drought, may confound attempts to demonstrate simple relationships between habitat availability and fish stocks (Sabaton et al., 2008) and provide a challenge to the future development of predictive tools.

2.10.5 Managing River Flows to Protect Riverine Ecosystems

In this chapter, we have developed three interdependent themes. In Section 2.10.2, we identified those hydrological characteristics of river networks that have been found to be of high ecological importance, primarily focusing on flow regimes, flow extremes, and hydrological connectivity. In Section 2.10.3, we demonstrated that these hydrological characteristics form the main control on the geomorphological style of river corridors and thus their shifting habitat mosaic. Complex interactions occur among river flows, fluvial sediments, and vegetation within naturally adjusting corridors, providing resilient, bio-complex river environments. In Section 2.10.4, we showed how the flow regime and the style and dynamics of the river corridor and river system define the ecohydraulic and mesohabitat complexity of the aquatic ecosystem and its three-dimensional connectivity. Throughout, we have touched on the effects of human interventions. In this section, we conclude our discussion of hydrology and ecology by considering how river flows can be managed to protect riverine ecosystems as well as support human needs, an area that is attracting enormous attention from researchers, managers, and policymakers as the world's rivers come under increasing human pressure (e.g., Annear et al., 2004; Naiman et al., 2002). Many rivers today have flow regimes that differ markedly from the climate-driven regime because of impoundments; the magnitude, frequency, and timing of floods have altered with land-use change; and moderate- to low-flow percentiles have been changed in various ways along the length of a river as a consequence of both abstractions and discharges from wastewater-treatment works (**Figure 14**).

River-flow regulation to control flooding and provide water for human use has had deep-seated impacts on river-corridor ecosystems (e.g., Ward and Stanford, 1979; Petts 1984; Tockner and Stanford, 2002). Direct surface-water abstractions and structural flood-alleviation measures, the construction of all types of surface reservoirs, and development of groundwater resources, including the conjunctive management of surface and groundwater, change the river flow regime and thus induce changes in river-channel characteristics (size, form, and style) and the river corridor and channel-habitat mosaic. The importance of managing flows to sustain riverine ecosystems and especially populations of native species has been demonstrated by the impacts of flow regulation below dams upon river-channel characteristics (Petts and Gurnell, 2005, Section 2.10.3.2) and biota (Petts, 1984, 2007). The deep impact of flow regulation is supported by the observation that regulated rivers regain normative attributes with sufficient distance below a dam and that depressed populations of native species can recover if the natural magnitudes and variations in flows are reestablished. Thus, Armitage's (2006) long-term study of the impacts of Cow Green Dam on the River Tees in Northern England revealed that a narrower range of environmental conditions and increased flow stability led to a dynamically fragile community (indicated by observed changes in community diversity and abundance) which is very susceptible to perturbations because it has developed in their absence. Periphyton and reservoir plankton play an important role in structuring the faunal composition by creating an environment where biotic interactions are more likely. This does not require restoration to some pristine state, but the recovery of some large portion of the lost capacity to sustain native biodiversity and bioproduction is possible by management of processes that maintain normative habitat conditions (Stanford et al., 1996).

Naiman et al. (2002) summarized the fundamental ecological principles for understanding hydrology–ecology relationships along rivers, focusing on the climatically driven variability of flows at least from season to season and from year to year. The two linked general principles are

1. that the natural-flow regime shapes the evolution of aquatic biota and ecological processes; and
2. that every river has a characteristic flow regime and an associated biotic community.

Four further principles were elaborated by Bunn and Arthington (2002):

1. Flow is a major determinant of physical habitat in rivers, which in turn is a major determinant of biotic composition.
2. Maintenance of the natural patterns of connectivity between habitats (a) along a river and (b) between a river and its riparian zone and floodplain is essential to the viability of populations of many riverine species.
3. Aquatic species have evolved life-history strategies primarily in response to the natural-flow regime and the habitats that are available at different times of the year and in both wet and dry years.
4. The invasion and success of exotic and introduced species along river corridors is facilitated by regulation of the flow regime, especially with the loss of natural wet–dry cycles.

These principles underpin three elements of regulated river management: the determination of (1) benchmark flows, (2) ecologically acceptable hydrographs, and (3) ecologically acceptable flow-duration curves (**Figure 15**). These three elements inform short term and local operational rules; seasonal and short series of annual flow management; and long-term water-resource planning, respectively.

The science and application of environmental flows has attracted considerable attention and Tharme (2003) identified over 200 approaches that have been described for advising on environmental flows in 44 countries. These range from reconnaissance-level assessments relying on ecologically informed hydrological methodologies to approaches using complex hydrodynamic habitat modeling. In some areas, such as Australia and southern Africa, a lack of ecological data and

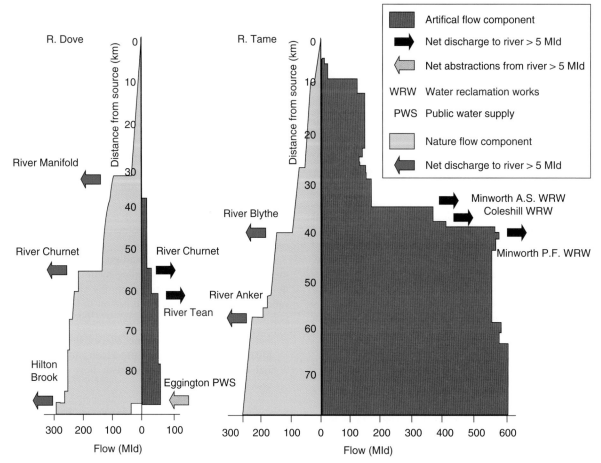

Figure 14 Residual flow diagrams representing the river condition under dry weather flow (DWF) (the average of the annual series of the minimum weekly/7 consecutive days) for a relatively natural (River Dove) and heavily influenced river (River Tame) within the River Trent catchment, UK. The diagrams show downstream flow increments from tributaries (natural flow component) and water-reclamation works and abstractions (artificial flow component). The total discharge at any point along the river is the combination of the natural and artificial flow components. Adapted from Pirt J (1983) The Estimation of Riverflows in Ungauged Catchments. Unpublished Doctoral Thesis, 2 volumes, Loughborough University of Technology.

process models, and political pressure to deliver environmental flow recommendations in short timeframes, often less than 1 year, has led to the use of scientific panels to set environmental flows (Cottingham et al., 2002).

For more than 40 years, tools have been advanced defining benchmark flows to allocate flow to meet in-river needs (Petts and Maddock, 1994) of which PHABSIM has been most widely used. In the 1960s and 1970s, early attempts to set instream flows for rivers focused on the annual minimum flow expressed as a hydrological statistic, commonly as either a flow-duration statistic (such as the 95th percentile flow) or as a fixed percentage of the average daily flow (ADF), with several studies proposing 20% ADF to protect aquatic habitat in streams (e.g., Tennant, 1976). However, recognition of the threat to fisheries of confining flow management to annual minimum flows led to more complex and hydrologically rational approaches (Stalnaker 1979, 1994; Stalnaker et al., 1996). By the early 1990s, the science and management of regulated rivers had expanded from the determination of instream flows to environmental flows and many schemes applied more complex flow-habitat models to address wider issues than the instream needs (the hydraulic habitats) of a single species.

Three general approaches to the allocation of flows to support river-ecosystem needs are being advanced and have achieved some success (Arthington et al., 2003): hydrological methods, hydraulic models, and holistic approaches. These approaches address the sustainability of communities and ecosystems, the access of aquatic biota to seasonal floodplain and riparian habitats as well as the need for high flows to sustain the geomorphological dynamics of the river corridor and floodplain habitats (RRA, 2003). They enabled advancement of an ecologically acceptable flow regime concept (**Figure 15**; Petts, 1996; Petts et al., 1999). This recognized that different life stages and different species benefit from different flows at different times of the year, and in different years. Rivers must be protected in wet years as well as drought years because high flows provide optimum conditions for some species and are also responsible for sustaining the quality and diversity of in-channel and riparian habitats.

Societal demands for river-ecosystem protection have accelerated the development of innovative, locally applicable

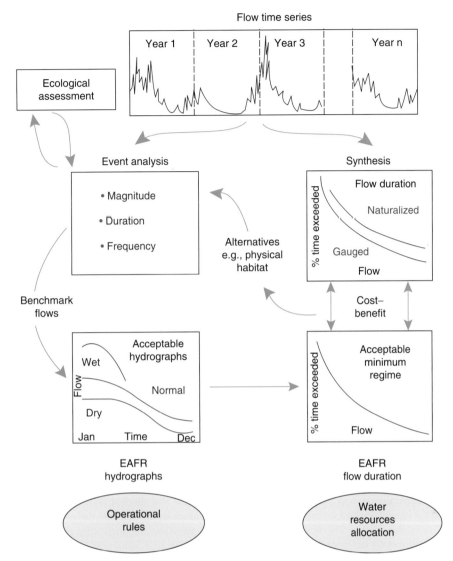

Figure 15 A general procedure for deriving an ecologically acceptable flow regime represented as one or more (e.g., wet and dry years) hydrographs for defining operational rules and as a flow-duration curve for assessing abstractable volumes. The procedure allows the evaluation of alternatives including physical-habitat improvements as part of the decision-making process. Adapted from Petts GE and Amoros C (eds.) (1996) *Fluvial Hydrosystems*, 308pp. London: Chapman and Hall.

methods and tools especially within regions having limited databases. However, there are also many examples where sophisticated, science-based models are being applied to specific problems. For example, Grand *et al.* (2006) used a cell-based model of backwater geometry, a pond-based temperature model, and a model of invertebrate production to investigate the effects of within-day flow fluctuations caused by hydropower operations on nursery habitats for larval and juvenile Colorado pike minnow (*Ptychocheilus lucius*) along the Green River below Flaming Gorge dam, Utah, USA. As noted by Parasiewicz (2001), if community structure reflects habitat structure, then securing habitat for the most common species might preserve the most profound characteristics of the ecosystem and provide survival conditions for the majority of the aquatic community. Progress in developing models that link physical-habitat dynamics and population biology of large organisms such as fish may have been constrained by the difficulty in merging the space- and timescales appropriate to both physical and biological sciences (Petts *et al.*, 2006).

In the hydrological approaches, flow is considered as a simple proxy for a number of related parameters that may have a key influence on the range of aquatic, wetland, and riparian habitats along the river corridor. Thus, Extence *et al.* (1999) developed a scoring system as an indicator of hydrological stress based upon surveys of macroinvertebrate fauna that has been shown to be sensitive to particular hydrological indices (Wood *et al.*, 2001a; Monk *et al.*, 2008). A range of hydrologic parameters for each year of flow record can be used to characterize inter-annual variation before (reference period) and after flow regulation/abstraction (Richter *et al.*, 1996, 1997). Of the hydrological approaches, White *et al.* (2005) used wavelet analysis to assess dam operations in reconstructing desired flow characteristics. This method provides an easy-to-interpret approach for investigating hydrological change when the management history is uncertain and timescales of important cycles are unknown. White *et al.* (2005) used wavelet analysis to detect the hydrological implications of management practices over a range of timescales and suggested that the method could provide a powerful data-mining technique for assessing hydrological changes. The statistical characterization of ecologically relevant hydrograph parameters could be used to define the variability of the dimensions of the flow regime within which artificial influences should be contained (Richter *et al.*, 1997). To date, such approaches have been used in water resources and environmental management in the USA (e.g., Richter *et al.*, 1998, 2006; Mathews and Richter, 2007) and elsewhere (e.g., Shiau and Wu, 2004, 2006). Thus, Galat and Lipkin (2000) recommended changes in reservoir management to return regulated flows to within the pattern of natural variability, thereby simulating a natural riverine ecosystem. They argued that naturalization of the flow regime would benefit not only the ecological system but also the economic value of the Missouri River, once the products of agriculture, electric-power generation, and transportation are integrated with the socio-ecological benefits of a naturalized flow regime. In a follow-up study, Jacobson and Galat (2008) focused on developing a flow regime to support the endangered pallid sturgeon (*Scaphirhynchus albus*). Specific hydrograph requirements for pallid sturgeon reproduction were unknown; so much of the design process was based on hydrological parameters extracted from the reference natural-flow regime.

Three issues often hinder the apparently simple and reasonable application of such hydrological approaches. First, standards need to be set to apply an appropriate record length. At least 12 years data are required for statistical integrity and longer records are needed to incorporate variable weather patterns over decadal timescales and to provide for actual scales of variability in the magnitude and timing of flows and the natural frequencies of these flows. The flow regime is a complex concept. Flow regimes typical of each hydro-climatic region across the globe represent average conditions created by combining a small number of flow-regime types, particular to each hydro-climatic region. Variations of the flow regime from year to year within the British temperate maritime hydro-climatic zone based upon analysis of 80 station-years of flow data (1977–97) from four major rivers across the United Kingdom showed that the typical flow regime with a high-flow season from December to February and a low-flow season from June to August occurred in only 51% of the years (Harris *et al.*, 2000). Of the other flow years, three variants of the typical flow regime were differentiated:

- *Subtype A*. Twenty-eight percent had a dominant peak in November often with a secondary peak in April.
- *Subtype B*. Sixteen percent had a dominant peak in March with a secondary peak in December.
- *Subtype C*. Six percent were characterized by winter drought with no dominant peak and typically a very dry January.

Second is the issue of naturalizing the gauged flow regime. In many areas, the pristine catchment has no relevance to the modern day. The hydrology of catchments characterized by long-term human interference – such as urban conurbations and intensive agriculture – bears little resemblance to the hydrologic character of unmodified catchments in a given hydro-climatic region. The concept for such catchments may be to produce functionally diverse, self-regulating ecological systems (Petts *et al.*, 2000). In reality, this requires determination of the flow regime that would be sustained under current or future catchment conditions in the absence of existing dams, reservoirs, diversions, and abstractions.

Third, the linkages between flow regime and ecological health are complex in both time and space. The natural dynamic character relates not only to flow variability but also to water quality, especially temperature variations; sediment dynamics and channel dynamics (that are also influenced by patterns of woody vegetation growth); changes in food/energy supply; and interactions between biological populations. Across the UK, most regulated rivers are supported in summer by compensation flows that maintain minimum flows or they may even experience enhanced flows during dry summers where the river supports abstractions from the lower river. Under one scenario, the main ecological impact of flow regulation below reservoirs would not be during a summer drought but during the late summer, autumn, and early winter following a subtype C flow year when the need to fill reservoir storage could eliminate high flows along the main river. Under such circumstances, there would be inadequate river flows to stimulate up-river fish migrations and spawning grounds could be impacted by siltation caused by fine-sediment loaded tributary spates.

Water temperature is a particularly important parameter and a river's thermal regime is a key component of environmental flows. Harris *et al.* (2000) and Olden and Naiman (2009) have encouraged ecologists and water managers to broaden their perspective on environmental flows to include both flow and thermal regimes in assessing e-flow needs. Assessments should include the comprehensive characterization of seasonality and variability in stream temperatures in the face of artificial influences on flow and potential impacts of climate change. From a scientific perspective, we need to more clearly elucidate the relative roles of altered flow and temperature in shaping ecological patterns and processes in riverine ecosystems.

In the absence of universal relationships between flows and biotic responses, King *et al.* (2003) advanced a value-based system, Downstream Response to Imposed Flow Transformation (DRIFT). This provided a data-management tool for many types and sources of information, predictive models, theoretical principles, and expert knowledge of a panel of scientists. The approach was developed to link the productivity of large floodplain rivers to their flow characteristics in countries or river basins where data scarcity constrains prediction of ecological responses to flow regulation. It was also produced in a developing region with severe water shortage and uncertainties about river-linked ecological processes and where riparian subsistence populations are important in the decision-making process. DRIFT supports the scientific-panel approach to recommend environmental flows within an adaptive management framework. It is based around four modules: (1) the biophysical module describes the present nature and functioning of the ecosystem; (2) the sociological module identifies subsistence users at risk from flow abstraction or regulation; (3) a module that combines the outputs from the first two to develop biophysical and subsistence scenarios; and (4) a module to address mitigation and compensation costs. Arthington *et al.* (2003) used DRIFT to establish the environmental flow requirements of rivers in Lesotho and contended that the methodology can provide a best-practice framework for conducting scientific-panel studies.

Linking environmental variables to dam-release rules has been shown to achieve significant water savings (Harman and Stewardson, 2005). However, there is no single best method or approach (IUCN, 2003). Given the variety of water resource contexts, the range of environmental settings and variety of species concerned, there has been an increasing tendency to use evidence-based expert judgment. In reality, all environmental flow assessments provide the evidence available to larger or smaller, expert panels, – the decision-making process that evaluates tradeoffs between water users. From a flow-management perspective, these tradeoffs include those between magnitude, frequency, duration, timing, and rate of change (Poff and Ward, 1989), and the evidence is often hierarchically structured to include three orders (Petts, 1980) or levels (Young *et al.*, 2004), broadly linking primary processes, habitat structure, and biota. Moreover, Jacobson and Galat's (2008) experience with flow-regime design using a hydrological approach demonstrated lack of confidence by stakeholders in the value of the natural-flow regime as a measure of ecosystem benefit. The lack of confidence resulted from the lack of fundamental scientific documentation, as might have been provided by a more complex hydrological–hydraulic–biological model. Stakeholders desired proof of ecological benefits commensurate with certainty of economic losses. This conflict between demands for more biologically accountable models and political actions to set environmental flows has also been highlighted by Arthington *et al.* (2006).

Despite considerable progress in understanding how flow variability sustains river ecosystems, there is a growing temptation to ignore natural-system complexity in favor of simplistic, static, environmental flow rules to resolve pressing river-management issues. Arthington *et al.* (2006) argue that such approaches are misguided and will ultimately contribute to further degradation of river ecosystems. In the absence of detailed empirical information of environmental flow requirements for rivers, they proposed a generic approach that incorporates essential aspects of natural-flow variability shared across particular classes of rivers that can be validated with empirical biological data and other information in a calibration process. They further argue that this approach can bridge the gap between simple hydrological rules of thumb and more comprehensive environmental flow assessments and experimental flow-restoration projects. Thus, in the USA, Poff *et al.* (2009) achieved a consensus view from a panel of international scientists on a framework for assessing environmental flow needs that combines a regional hydrological approach and ecological response relationships for each river type based initially on the literature, existing data, and expert knowledge. Stakeholders and decision makers then explicitly evaluate acceptable risk as a balance between perceived value of the ecological goals, the economic costs involved, and the scientific uncertainties. The main risk is a perceived lack of incentive for what could be considered to be costly monitoring and longer-term research to develop evidence of biota–flow relationships for supporting adaptive management.

In conclusion, it is impossible to return regulated rivers to an unimpacted state or even to define what such an unimpacted state might be, given the long history of human intervention across the Earth's surface superimposed on a background of global environmental change. However, it is possible to combine scientific understanding and expert judgment to establish river flows that can support river ecosystems, where the river channel and riparian zone are also managed. At least in a set of reaches distributed across the river network, river flows, sediment, and vegetation need to interact relatively freely to provide refugia for species and resilient sites from which other areas of the river network can be recolonized. Sustaining river ecosystems needs to be based on the maintenance of appropriate river flows coupled with restoration of the potential for reaches within the river network to adjust to those flows.

References

Abbe TB and Montgomery DR (2003) Patterns and processes of wood debris accumulation in the Queets river basin, Washington. *Geomorphology* 51: 81–107.

Agostinho AA, Gomes LC, Verissimo S, and Okada EK (2004) Flood regime, dam regulation and fish in the Upper Parana River: Effects on assemblage attributes, reproduction and recruitment. *Reviews in Fish Biology and Fisheries* 14(1): 11–19.

Ahmadi-Nedushan B, St-Hilaire A, Bérubé M, Robichaud E, Thiémonge N, and Bobée B (2006) A review of statistical methods for the evaluation of aquatic habitat suitability for instream flow assessment. *River Research and Applications* 22(5): 503–523.

Allan JD (2004) Landscapes and riverscapes: The influence of land use on stream ecosystems. *Annual Review of Ecology, Evolution and Systematics* 35: 257–284.

Allen MR and Ingram WJ (2002) Constraints on future changes in climate and the hydrologic cycle. *Nature* 419(6903): 224–232.

Amoros C and Roux AL (1988) Interaction between water bodies within the floodplains of large rivers: Function and development of connectivity. In: Schreiber KL (ed.) *Connectivity in Landscape Ecology*, Proceedings of the 2nd International Seminar of the International Association for Landscape Ecology, Münster, 1987, Munstersche Geographische Arbeiten, pp. 125–130. Paderborn: Schoningh.

Anderson KE, Paul AJ, McCauley E, Jackson LJ, Post JR, and Nisbet RM (2006) Instream flow needs in streams and rivers: The importance of understanding ecological dynamics. *Frontiers in Ecology and the Environment* 4(6): 309–318.

Andersson E, Nilsson C, and Johansson ME (2000) Plant dispersal in boreal rivers and its relation to the diversity of riparian flora. *Journal of Biogeography* 27: 1095–1106.

Amitage PD (2006) Long-term faunal changes in a regulated and an unregulated stream – cow green thrity years on. *River Research and Applications* 22(9): 947–966.

Annear T, Chisholm I, Beecher H, *et al.* (2004) *Instream Flow for Riverine Resource Stewardship*, revised edition. Cheyenne, WY: Instream Flow Council.

Arnell NW (2003) Effects of IPCC SRES emissions scenarios on river runoff: A global perspective. *Hydrology and Earth System Sciences* 7(5): 619–641.

Arora VK and Boer GJ (2001) Effects of simulated climate change on the hydrology of major river basins. *Journal of Geophysical Research – Atmospheres* 106(D4): 3335–3348.

Arscott DB, Tockner K, van der Nat D, and Ward JV (2002) Aquatic habitat dynamics along a braided alpine river ecosystem (Tagliamento River, Northeast Italy). *Ecosystems* 5(8): 802–814.

Arscott DB, Tockner K, and Ward JV (2001) Thermal heterogeneity along a braided floodplain river (Tagliamento River, northeastern Italy). *Canadian Journal of Fisheries and Aquatic Sciences* 58(12): 2359–2373.

Arscott DB, Tockner K, and Ward JV (2003) Spatio-temporal patterns of benthic invertebrates along the continuum of a braided Alpine river. *Archiv für Hydrobiologie* 158(4): 431–460.

Arsenault D, Boucher E, and Bouchon E (2007) Asynchronous forest-stream coupling in a fire-prone boreal landscape: Insights from woody debris. *Journal of Ecology* 95: 789–801.

Arthington AH, Bunn SE, Poff NL, and Naiman RJ (2006) The challenge of providing environmental flow rules to sustain river ecosystems. *Ecological Applications* 16(4): 1311–1318.

Arthington AH, Rall JL, Kennard MJ, and Pusey BJ (2003) Environmental flow requirements of fish in Lesotho rivers using the DRIFT methodology. *River Research and Applications* 19(5–6): 641–666.

Arthurton R, Barker S, Rast W, and Huber M (2007) Water. In: Ashton P (ed.) *Global Environmental Outlook 4 (GEO-4)*, p. 42pp. Malta: UNEP.

Asselman NEM and Middelkoop H (1995) Floodplain sedimentation: Quantities, patterns and processes. *Earth Surface Processes and Landforms* 20: 481–499.

Bain MB and Meixler MS (2008) A target fish community to guide river restoration. *River Research and Applications* 24(4): 453–458.

Barsoum N and Hughes FMR (1998) Regeneration response of black poplar to changing river levels. In: Wheater H and Kirby C (eds.) *Hydrology in a Changing Environment*, pp. 397–412. Chichester: Wiley

Baxter CV and Hauer FR (2000) Geomorphology, hyporheic exchange, and selection of spawning habitat by bull trout (*Salvelinus confluentus*). *Canadian Journal of Fish and Aquatic Sciences* 57: 1470–1481.

Bayley PB (1991) The flood-pulse advantage and the restoration of river-floodplain systems. *Regulated Rivers: Research and Management* 6: 75–86.

Begin ZB (1981) The relationship between flow-shear stress and stream pattern. *Journal of Hydrology* 52(3–4): 307–319.

Benda L, Poff NL, Miller D, *et al.* (2004) The network dynamics hypothesis: How channel networks structure riverine habitats. *BioScience* 54(5): 413–427.

Benedetti MM (2003) Controls on overbank deposition in the Upper Mississippi River. *Geomorphology* 56: 271–290.

Bertoldi W, Gurnell A, Surian N, *et al.* (2009) Understanding reference processes: Linkages between river flows, sediment dynamics and vegetated landforms along the Tagliamento River, Italy. *River Research and Applications* 25: 501–516.

Birken AS and Cooper DJ (2006) Processes of *Tamarix* invasion and floodplain development along the lower Green River, Utah. *Ecological Applications* 16(3): 1103–1120.

Bledsoe BP and Watson CC (2001) Logistic analysis of channel pattern thresholds: Meandering, braiding, and incising. *Geomorphology* 38(3–4): 281–300.

Boedeltje G, Bakker JP, Bekker RM, Van Groenendael JM, and Soesbergen M (2003) Plant dispersal in a lowland stream in relation to occurrence and three specific life-history traits of the species in the species pool. *Journal of Ecology* 91(5): 855–866.

Boedeltje G, Bakker JP, Ten Brinke A, Van Groenendael JM, and Soesbergen M (2004) Dispersal phenology of hydrochorous plants in relation to discharge, seed release time and buoyancy of seeds: The flood pulse concept supported. *Journal of Ecology* 92(5): 786–796.

Bovee KD (1978) The incremental method of assessing habitat potential for coolwater species, with management implications. *American Fisheries Society Special Publication* 11: 340–346.

Bovee KD (1986) Development and evaluation of habitat suitability criteria for use in the instream flow incremental methodology. *Instream Flow Information Paper*, 21, US Fish and Wildlife Service Biological Report No. 86.

Bovee KD (1996) Perspectives on two-dimensional river habitat models: The PHABSIM experience *Proceedings of the Second International Symposium on Habitat Hydraulics*, pp. 149–162. Québec, QC: INRS-Eau.

Bray DI (1982) Regime equations for gravel-bed rivers. In: Hey RD, Bathurst JC, and Thorne CR (eds.) *Gravel-bed Rivers: Fluvial Processes, Engineering and Management*, pp. 517–552. Chichester: Wiley.

Brewer PA and Lewin J (1998) Planform cyclicity in an unstable reach: Complex fluvial response to environmental change. *Earth Surface Processes and Landforms* 23: 989–1008.

Brierley GJ and Fryirs KA (2005) *Geomorphology and River Management: Application of the River Styles Framework*. 398pp. Malden, MA: Blackwell.

Brooks AP and Brierley GJ (2002) Mediated equilibrium: The influence of riparian vegetation and wood on the long-term evolution and behaviour of a near-pristine river. *Earth Surface Processes and Landforms* 27(4): 343–367.

Brooks AP, Brierley GJ, and Millar RG (2003) The long-term control of vegetation and woody debris on channel and flood-plain evolution: Insights from a paired catchment study in southeastern Australia. *Geomorphology* 51(1–3): 7–29.

Brunke M and Gonser T (1997) The ecological significance of exchange processes between rivers and groundwater. *Freshwater Biology* 37: 1–33.

Bruns DA, Minshall GW, Gushing CE, Cummins KW, Brock JT, and Vannote RL (1984) Tributaries as modifiers of the river-continuum concept: Analysis by polar ordination and regression models. *Archiv für Hydrobiologie* 99: 208–220.

Bunn SE and Arthington AH (2002) Basic principles and ecological consequences of altered flow regimes for aquatic biodiversity. *Environmental Management* 30(4): 492–507.

Caissie D (2006) The thermal regime of rivers: A review. *Freshwater Biology* 51(8): 1389–1406.

Capra H, Sabaton C, Gouraud V, Souchon Y, and Lim P (2003) A population dynamics model and habitat simulation as a tool to predict brown trout demography in natural and bypassed stream reaches. *River Research and Applications* 19(5–6): 551–568.

Carson MA (1984a) Observations on the meandering–braided river transition, the Canterbury Plains, New Zealand: Part one. *New Zealand Geographer* 40(1): 12–19.

Carson MA (1984b) Observations on the meandering–braided river transition, the Canterbury Plains, New Zealand: Part two. *New Zealand Geographer* 40(2): 89–99.

Cattanéo F, Lamouroux N, Breil P, and Capra H (2002) The influence of hydrological and biotic processes on brown trout (*Salmo trutta*) population dynamics. *Canadian Journal of Fisheries and Aquatic Sciences* 59(1): 12–22.

Church M (1992) Channel morphology and typology. In: Callow C and Petts G (eds.) *The Rivers Handbook: Hydrological and Ecological Principles*, pp. 126–143. Oxford: Blackwell.

Church M (2002) Geomorphic thresholds in riverine landscapes. *Freshwater Biology* 47: 541–557.

Corenblit D, Steiger J, Gurnell AM, Tabacchi E, and Roques L (2010) Control of sediment dynamics by vegetation as a key ecological function within fluvial corridors. *Earth Surface Processes and Landforms* 34(13): 1790–1810.

Corenblit D, Tabacchi E, Steiger J, and Gurnell AM (2007) Reciprocal interactions and adjustments between fluvial landforms and vegetation dynamics in river corridors: A review of complementary approaches. *Earth-Science Reviews* 84(1–2): 56–86.

Cottingham P, Thoms MC, and Quinn GP (2002) Scientific panels and their use in environmental flow assessment in Australia. *Australian Journal of Water Resources* 5: 103–111.

Coulthard TJ (2005) Effects of vegetation on braided stream pattern and dynamics. *Water Resources Research* 41: W04003.

Craig LS, Palmer MA, Richardson DC, *et al.* (2008) Stream restoration strategies for reducing river nitrogen loads. *Frontiers in Ecology and the Environment* 6(10): 529–538.

Crowder DW and Diplas P (2006) Applying spatial hydraulic principles to quantify stream habitat. *River Research and Applications* 22(1): 79–89.

Dakova S, Uzunov Y, and Mandadjiev D (2000) Low flow – the river's ecosystem limiting factor. *Ecological Engineering* 16: 167–174.

Danielopol DL (1989) Groundwater fauna associated with riverine aquifers. *Journal of the North American Benthological Society* 8(1): 18–35.

Danvind M and Nilsson C (1997) Seed floating ability and distribution of alpine plants along a northern Swedish river. *Journal of Vegetation Science* 8(2): 271–276.

Décamps H and Tabacchi E (1994) Species richness in vegetation along river margins. In: Giller PS, Hildrew AG, and Raffaelli DG (eds.) *Aquatic Ecology: Scale, Pattern and Process*, pp. 1–20. Oxford: Blackwell.

Desloges JR and Church MA (1989) Wandering gravel-bed rivers. *Canadian Geographer* 33(4): 360–364.

Dettinger MD and Diaz HF (2000) Global characteristics of stream flow seasonality and variability. *Journal of Hydrometeorology* 1: 289–310.

Doyle MW, Stanley EH, Strayer DL, Jacobson RB, and Schmidt JC (2005) Effective discharge analysis of ecological processes in streams. *Water Resources Research* 41: W11411.

Edwards PJ, Kollmann J, Gurnell AM, Petts GE, Tockner K, and Ward JV (1999) A conceptual model of vegetation dynamics on gravel bars of a large Alpine river. *Wetlands Ecology and Management* 7: 141–153.

Elwood JW, Newbold JD, O'Neill RV, and Van Winkle W (1983) Resource spiraling: An operational paradigm for analyzing lotic ecosystems. In: Fontaine TD and Bartell SM (eds.) *Dynamics of Lotic Ecosystems*, pp. 3–27. Ann Arbor, MI: Ann Arbor Science.

Emery JC, Gurnell AM, Clifford NJ, Petts GE, Morrissey IP, and Soar PJ (2003) Classifying the hydraulic performance of riffle-pool bedforms for habitat assessment and river rehabilitation design. *River Research and Applications* 19(5–6): 533–549.

Extence CA, Balbi DM, and Chadd RP (1999) River flow indexing using British benthic macroinvertebrates: A framework for setting hydroecological objectives. *Regulated Rivers* 15: 543–574.

Ferguson RI (1987) Hydraulic and sedimentary controls of channel pattern. In: Richards K (ed.) *River Channels*, pp. 129–158. Oxford: Blackwell.

Fisher SG (1990) Recovery processes in lotic ecosystems: Limits of successional theory. *Environmental Management* 14: 725–736.

Francis RA (2007) Size and position matter: Riparian plant establishment from fluvially deposited trees. *Earth Surface Processes and Landforms* 32(8): 1239–1243.

Francis RA, Gurnell AM, Petts GE, and Edwards PJ (2005) Survival and growth responses of *Populus nigra*, *Salix elaeagnos* and *Alnus incana* cuttings to varying levels of hydric stress. *Forest Ecology and Management* 210: 291–301.

Francis RA, Gurnell AM, Petts GE, and Edwards PJ (2006) Riparian tree establishment on gravel bars: Interactions between plant growth strategy and the physical environment. In: Sambrook Smith GH, Best JL, Bristow CS, and Petts GE (eds.) *Braided Rivers: Process, Deposits, Ecology and Management*, International Association of Sedimentologists Special Publication 36, pp. 361–380. VIC: Blackwell.

Francis RA, Tibaldeschi P, and McDougall L (2008) Fluvially-deposited large wood and riparian plant diversity. *Wetlands Ecology and Management* 16(5): 371–382.

Frissell CA, Liss WJ, Warren CE, and Hurley MD (1986) A hierarchical framework for stream habitat classification: Viewing streams in a watershed context. *Environmental Management* 10: 199–214.

Galat DL and Lipkin R (2000) Restoring ecological integrity of great rivers: Historical hydrographs aid in defining reference conditions for the Missouri river. *Hydrobiologia* 422: 29–48.

Gibert J, Dole-Olivier M-J, Marmonier P, and Vervier P (1990) Surface water–groundwater ecotones. In: Naiman RJ and Décamps H (eds.) *The Ecology and Management of Aquatic–Terrestrial Ecotones*, pp. 199–226. Paris: UNESCO.

Glenn EP and Nagler PL (2005) Comparative ecophysiology of *Tamarix ramosissima* and native trees in western US riparian zones. *Journal of Arid Environments* 61(3): 419–446.

Gomez B, Eden DN, Peacock DH, and Pinkney EJ (1998) Floodplain construction by recent, rapid vertical accretion: Waipaoa river, New Zealand. *Earth Surface Processes and Landforms* 23: 405–413.

Gomi T, Sidle RC, and Richardson JS (2002) Understanding processes and downstream linkages of headwater systems. *BioScience* 52(10): 905–916.

Goodson JM, Gurnell AM, Angold PG, and Morrissey IP (2002) Riparian seed banks along the lower River Dove, UK: Their structure and ecological implications. *Geomorphology* 47(1): 45–60.

Goodson JM, Gurnell AM, Angold PG, and Morrissey IP (2003) Evidence for hydrochory and the deposition of viable seeds within winter flow-deposited sediments: The River Dove, Derbyshire, UK. *River Research and Applications* 19(4): 317–334.

Gore JA (1978) A technique for predicting in-stream flow requirements of benthic macroinvertebrates. *Freshwater Biology* 8(2): 141–151.

Gore JA and Nestler JM (1988) Instream flow studies in perspective. *Regulated Rivers: Research and Management* 2: 93–101.

Gouraud V, Baglinière JL, Baran P, Sabaton C, Lim P, and Ombredane D (2001) Factors regulating brown trout populations in two French rivers: Application of a dynamic population model. *Regulated Rivers: Research and Management* 17(4–5): 557–569.

Gradzinski R, Baryla J, Doktor M, *et al.* (2003) Vegetation-controlled modern anastomosing system of the upper Narew River (NE Poland) and its sediments. *Sedimentary Geology* 157(3–4): 253–276.

Graf WL (1978) Fluvial adjustments to the spread of tamarisk in the Colorado Plateau region. *Geological Society of America Bulletin* 89: 1491–1501.

Gran K and Paola C (2001) Riparian vegetation controls on braided stream dynamics. *Water Resources Research* 37: 3275–3283.

Grand TC, Railsback SF, Hayse JW, and LaGory KE (2006) A physical habitat model for predicting the effects of flow fluctuations in nursery habitats of the endangered Colorado pikeminnow (*Ptychocheilus lucius*). *River Research and Applications* 22(10): 1125–1142.

Gregory KJ and Walling DE (1973) *Drainage Basin Form and Process*. 456pp. London: Edward Arnold.

Gregory SV, Swanson FJ, McKee WA, and Cummins KW (1991) An ecosystem perspective of riparian zones. *BioScience* 41(8): 540–551.

Gurnell AM (1995) Vegetation along river corridors: Hydrogeomorphological interactions. In: Gurnell AM and Petts GE (eds.) *Changing River Channels*, pp. 237–260. Chichester: Wiley.

Gurnell AM (2003) Wood storage and mobility. In: Gregory SV, Staley K, and Gurnell AM (eds.) *The Ecology and Management of Wood in World Rivers*, American Fisheries Society American Fisheries Society Symposium 37, pp. 75–91. Bethesda, MD.

Gurnell AM (2003) Wood storage and mobility. In: Gregory SV, Staley K, Gurnell AM (eds.) *The Ecology and Management of Wood in World Rivers*, American Fisheries Society American Fisheries Society Symposium 37, Bethesda, Maryland, 75–91.

Gurnell AM (2010) Wood in fluvial systems. In: Wohl, E (ed.) *Treatise in Geomorphology*, Volume 9, Fluvial Geomorphology, chap. 9.11. Elsevier (in press).

Gurnell AM, Blackall TD, and Petts GE (2008) Characteristics of freshly deposited sand and finer sediments along an island-braided, gravel-bed river: The roles of water, wind and trees. *Geomorphology* 99: 254–269.

Gurnell AM, O'Hare JM, O'Hare MT, Dunbar MJ, and Scarlett PD (2010) Associations between assemblages of aquatic plant morphotypes and channel geomorphological properties within British rivers. *Geomorphology* 116: 135–144.

Gurnell AM, Lee M, and Souch C (2007) Urban rivers: Hydrology, geomorphology, ecology and opportunities for change. *Geography Compass* 1(5): 1118–1137.

Gurnell AM and Petts G (2006) Trees as riparian engineers: The Tagliamento River, Italy. *Earth Surface Processes and Landforms* 31(12): 1558–1574.

Gurnell AM, Petts GE, Hannah DM, *et al.* (2000) Wood storage within the active zone of a large European gravel-bed river. *Geomorphology* 34(1–2): 55–72.

Gurnell AM, Petts GE, Hannah DM, *et al.* (2001) Riparian vegetation and island formation along the gravel-bed Fiume Tagliamento, Italy. *Earth Surface Processes and Landforms* 26(1): 31–62.

Gurnell AM, Piegay H, Swanson FJ, and Gregory SV (2002) Large wood and fluvial processes. *Freshwater Biology* 47(4): 601–619.

Gurnell AM, Thompson K, Goodson J, and Moggridge H (2008) Propagule deposition along river margins: Linking hydrology and ecology. *Journal of Ecology* 96: 553–565.

Gurnell AM, Tockner K, Edwards PJ, and Petts GE (2005) Effects of deposited wood on biocomplexity of river corridors. *Frontiers in Ecology and Environment* 3(7): 377–382.

Gurnell AM, Surian N, and Zanoni L (2009) Multi-thread river channels: A perspective on changing European alpine river systems. *Aquatic Sciences* 71(3): 253–265.

Hack JT (1960) Interpretation of erosional topography in humid temperate regions. *American Journal of Science* 258A: 80–97.

Haines AT, Finlayson BL, and McMahon TA (1988) The global classification of flow regimes. *Applied Geography* 8: 255–272.

Halls AS and Welcomme RL (2004) Dynamics of river fish populations in response to hydrological conditions: A simulation study. *River Research and Applications* 20(8): 985–1000.

Hardy TB (1998) The future of habitat modeling and instream flow assessment techniques. *Regulated Rivers: Research and Management* 14(5): 405–420.

Harman C and Stewardson M (2005) Optimizing dam release rules to meet environmental flow targets. *River Research and Applications* 21(2–3): 113–129.

Harner MJ and Stanford JA (2003) Differences in cottonwood growth between a losing and a gaining reach of an alluvial floodplain. *Ecology* 84(6): 1453–1458.

Harper DM, Smith CD, and Barham PJ (1992) Habitats as the building blocks for river conservation assessment. In: Boon PJ, Calow P, and Petts GE (eds.) *River Conservation and Management*, pp. 311–319. Chichester: Wiley.

Harper DM, Smith CD, and Barham PJ (1995) The ecological basis for management of the natural river environment. In: Harper DM and Erguson AGD (eds.) *The Ecological Basis for River Management*, pp. 219–238. Chichester: Wiley.

Harris NM, Gurnell AM, Hannah DM, and Petts GE (2000) Classification of river regimes: A context for hydroecology. *Hydrological Processes* 14(16–17): 2831–2848.

Harvey GL, Clifford NJ, and Gurnell AM (2008) Towards an ecologically meaningful classification of the flow biotope for river inventory, rehabilitation, design and appraisal purposes. *Journal of Environmental Management* 88(4): 638–650.

Hawkins CP, *et al.* (1993) A hierarchical approach to classifying stream habitat features. *Fisheries* 18: 3–12.

Hickin EJ and Nanson G (1984) Lateral migration rates of river bends. *Journal of Hydraulic Engineering* 110: 1557–1567.

Hooke JM (2003) River meander behaviour and instability: A framework for analysis. *Transactions of the Institute of British Geographers NS* 28: 238–253.

Horton RE (1945) Erosional development of streams and their drainage basins: Hydro-physical approach to quantitative morphology. *Geological Society of America Bulletin* 56(3): 275–370.

Huang HQ and Nanson GC (2006) Why some alluvial rivers develop an anabranching pattern. *Water Resources Research* 43: W07441.

Huet M (1959) Profiles and biology of Western European streams as related to fish management. *Transactions of the American Fisheries Society* 88(3): 19–26.

Huntington TG (2006) Evidence for intensification of the global water cycle: Review and synthesis. *Journal of Hydrology* 319(1–4): 83–95.

Hynes HBN (1970) *The Ecology of Running Waters*. Toronto: University of Toronto.

Hynes HBN (1975) The stream and its valley. *Vereinigung fur Theoretische und Angewandte Limnologie Verhandlungen* 19: 1–15.

Hynes HBN (1983) Groundwater and stream ecology. *Hydrobiologia* 100: 93–99.

Illies J and Botosaneanu L (1963) Problèmes et méthodes de la classification et de la zonation écologique des eaux courantes, considérées surtout du point de vue faunistique. *Mitteilungen der IVL* 12: 57pp.

IUCN (2003) *Flow: The Essentials of Environmental Flows*. Gland: IUCN.

Jacobson RA (2008) Applications of MesoHABSIM using fish community targets. *River Research and Applications* 24(4): 434–438.

Jacobson RB and Galat DL (2008) Design of a naturalized flow regime – an example from the lower Missouri River, USA. *Ecohydrology* 1(2): 81–104.

Johansson ME, Nilsson C, and Nilsson E (1996) Do rivers function as corridors for plant dispersal? *Journal of Vegetation Science* 7(4): 593–598.

Johnson WC (1997) Equilibrium response of riparian vegetation to flow regulation in the Platte river, Nebraska. *Regulated Rivers: Research and Management* 13(5): 403–415.

Johnson WC (2000) Tree recruitment and survival in rivers: Influence of hydrological processes. *Hydrological Processes* 14(16–17): 3051–3074.

Junk WJ, Bayley PB, and Sparks RE (1989) The flood-pulse concept in river-floodplain systems. In: Dodge DP (ed.) *Proceedings of the International Large River Symposium*, Canadian Special Publication of Fisheries and Aquatic Sciences, vol. 106, pp. 110–127. Ottawa, ON: Canadian Government Publishing Centre.

Junk WJ and Wantzen KM (2004) The flood pulse concept: New aspects, approaches, and applications – an update. In: Welcomme RL and Petr T (eds.) *Proceedings of the Second International Symposium on the Management of Large Rivers for Fisheries*, Food and Agriculture Organization & Mekong River Commission, vol. 2, pp. 117–149. FAO Regional Office for Asia and the Pacific, Bangkok. RAP Publication.

Karaus U, Alder L, and Tockner K (2005) Concave islands: Habitat heterogeneity of parafluvial ponds in a gravel-bed river. *Wetlands* 25(1): 26–37.

Karrenberg S, Edwards PJ, and Kollmann J (2002) The life history of Salicaceae living in the active zone of floodplains. *Freshwater Biology* 47: 733–748.

Kemp JL, Harper DM, and Crosa GA (2000) The habitat-scale ecohydraulics of rivers. *Ecological Engineering* 16: 17–29.

Kennen JG, Kauffman LJ, Ayers MA, Wolcock DM, and Colarullo SJ (2008) Use of an integrated flow model to estimate ecologically relevant hydrologic characteristics at stream biomonitoring sites. *Ecological Modelling* 211: 57–76.

King J, Brown C, and Sabet H (2003) A scenario-based holistic approach to environmental flow assessments for rivers. *River Research and Applications* 19(5–6): 619–639.

Kingsford RT and Auld KM (2005) Waterbird breeding and environmental flow management in the Macquarie Marshes, arid Australia. *River Research and Applications* 21(2–3): 187–200.

Knighton D (1998) *Fluvial Forms and Processes: A New Perspective*. London: Arnold.

Kollmann J, Vieli M, Edwards PJ, Tockner K, and Ward JV (1999) Interactions between vegetation development and island formation in the Alpine river Tagliamento. *Applied Vegetation Science* 2: 25–36.

Kondolf GM, Larsen EW, and Williams JG (2000) Measuring and modelling the hydraulic environment for assessing instream flows. *North American Journal of Fisheries Management* 20: 1016–1028.

Krasovskaia I (1996) Sensitivity of the stability of river flow regimes to small fluctuations in temperature. *Hydrological Sciences Journal des Sciences Hydrologiques* 41(2): 251–264.

Krasovskaia I and Saelthun NR (1997) Sensitivity of the stability of Scandinavian river flow regimes to a predicted temperature rise. *Hydrological Sciences Journal des Sciences Hydrologiques* 42(5): 693–711.

Kundzewicz ZW, Mata LJ, Arnell NW, et al. (2007) Freshwater resources and their management. In: Parry ML, Canziani OF, Palutikof JP, van der Linden PJ, and Hanson CE (eds.) *Climate Change 2007: Impacts, Adaptation and Vulnerability*, Contribution of Working Group II to the Fourth Assessment Report of the Intergovernmental Panel on Climate Change, pp. 173–210. Cambridge: Cambridge University Press.

Lake PS (2003) Ecological effects of perturbation by drought in flowing waters. *Freshwater Biology* 48(7): 1161–1172.

Lake PS (2007) Flow-generated disturbances and ecological responses: Floods and droughts. In: Wood PJ, Hannah DM, and Sadler JP (eds.) *Hydroecology and Ecohydrology: Past, Present and Future, pp*, pp. 75–92. Chichester: Wiley.

Lamouroux N, Capra H, Pouilly M, and Souchon Y (1999) Fish habitat preferences in large streams of southern France. *Freshwater Biology* 42(4). 673–673

Lane EW (1957) A study of the shape of channels formed by natural streams flowing in erodible material. US Army Corps of Engineers, Missouri River Division, Sediment Series 9: 106.

Latterell JJ, Bechtold JS, O'Keefe TC, Van Pelt R, and Naiman RJ (2006) Dynamic patch mosaics and channel movement in an unconfined river valley of the Olympic Mountains. *Freshwater Biology* 51: 523–544.

Leopold LB and Maddock T (1953) The hydraulic geometry of stream channels and some physiographic implications. *US Geological Survey Professional Paper* 252: 57pp.

Leopold LB and Wolman MG (1957) River channel patterns – braided, meandering and straight. *US Geological Survey Professional Paper* 282B: 39–85.

Leopold LB, Wolman MG, and Miller JP (1963) *Fluvial Processes in Geomorphology*. 522pp. San Francisco: W.H. Freeman.

Lite SJ and Stromberg JC (2005) Surface water and ground-water thresholds for maintaining Populus-Salix forests, San Pedro River, Arizona. *Biological Conservation* 125: 153–167.

Lobon-Cervia J (2003) Spatio-temporal dynamics of brown trout production in a Cantabrian stream: Effects of density and habitat quality. *Transactions of the American Fisheries Society* 132: 621–637.

Lytle DA and Poff NL (2004) Adaptation to natural flow regimes. *Trends in Ecology and Evolution* 19(2): 94–100.

Mack GH and Leeder MR (1998) Channel shifting of the Rio Grande, southern Rio Grande Rift; implications for alluvial stratigraphic models. *Sedimentary Geology* 117: 207–219.

Malard F, Ferreira D, Dole'dec S, and Ward JV (2003) Influence of groundwater upwelling on the distribution of the hyporheos in a headwater river flood plain. *Archiv für Hydrobiologie* 157: 89–116.

Malard F, Uehlinge U, Zah R, and Tockner K (2006) Flood-pulse and riverscape dynamics in a braided glacial river. *Ecology* 87(3): 704–716.

Malcolm IA, Soulsby C, Youngson AF, and Hannah DM (2005) Catchment-scale controls on groundwater–surface water interactions in the hydorheic zone: Implications for salmon embryo survival. *River Research and Applications* 21: 977–990.

Maser C and Sedell JR (1994) *From the Forest to the Sea: The Ecology of Wood in Streams, Rivers, Estuaries and Oceans*. Florida: St. Lucie, Delray Beach.

Mathews R and Richter BD (2007) Application of the Indicators of Hydrologic Alteration software in environmental flow setting. *Journal of the American Water Resources Association* 43: 1400–1413.

McMahon TA and Finlayson BL (2003) Droughts and anti-droughts: The low flow hydrology of Australian rivers. *Freshwater Biology* 48(7): 1147–1160.

Merritt DM and Cooper DJ (2000) Riparian vegetation and channel change in response to river regulation: A comparative study of regulated and unregulated streams in Green River basin, USA. *Regulated Rivers: Research and Management* 16: 543–564.

Milhous RT (1982) Effect of sediment transport and flow regulation on the ecology of gravel-bed rivers. In: Hey RD, Bathurst JC, and Thorne CR (eds.) *Gravel-Bed Rivers: Fluvial Processes, Engineering and Management*, pp. 819–842. Chichester: Wiley.

Milhous RT (1998) Modelling of instream flow needs: The link between sediment and aquatic habitat. *Regulated Rivers: Research and Management* 14(1): 79–94.

Milly PCD, Wetherald RT, Dunne KA, and Delworth TL (2002) Increasing risk of great floods in a changing climate. *Nature* 415(6871): 514–517.

Milner AM (1994) System recovery. In: Calow P and Petts GE (eds.) *Rivers Handbook*, pp. 76–98. Oxford: Blackwell.

Mitro MG, Zale AV, and Rich BA (2003) The relation between age-0 rainbow trout (*Oncorhynchus mykiss*) abundance and winter discharge in a regulated river. *Canadian Journal of Fisheries and Aquatic Sciences* 60: 135–139.

Moggridge HL and Gurnell AM (2009) Controls on the sexual and asexual regeneration of Salicaceae along a highly dynamic, braided river system. *Aquatic Sciences* 71(3): 305–317.

Mollard JD (1973) Air photo interpretation of fluvial features. In: *Proceedings of the 7th Canadian Hydrology Symposium*, pp. 341–380.

Monk WA, Wood PJ, Hannah DM, and Wilson DA (2007) Selection of river flow indices for the assessment of hydroecological change. *River Research and Applications* 23(1): 113–122.

Monk WA, Wood PJ, Hannah DM, and Wilson DA (2008) Macroinvertebrate community response to inter-annual and regional flow regime dynamics. *River Research and Applications* 24: 988–1001.

Monk WA, Wood PJ, Hannah DM, Wilson DA, Extence CA, and Chadd RP (2006) Flow variability and macroinvertebrate community response within riverine systems. *River Research and Applications* 22(5): 595–615.

Montgomery DR (1999) Process domains and the river continuum. *Journal of the American Water Resources Association* 35: 397–410.

Montgomery DR and Abbe TB (2006) Influence of logjam-formed hard points on the formation of valley-bottom landforms in an old-growth forest valley, Queets River, Washington, USA. *Quaternary Research* 65(1): 147–155.

Morales Y, Weber LJ, Mynett AE, and Newton TJ (2006) Mussel dynamics model: A hydroinformatics tool for analyzing the effect of different stressors on the dynamics of freshwater mussel communities. *Ecological Modelling* 197: 448–460.

Murray DR (1986) Seed dispersal by water. In: Murray DR (ed.) *Seed Dispersal*, pp. 49–85. San Diego, CA: Academic Press.

Naiman RJ, Bunn SE, Nilsson CE, Petts GE, Pinay G, and Thompson LC (2002) Legitimizing fluvial ecosystems as users of water: An overview. *Environmental Management* 30(4): 455–467.

Naiman RJ and Décamps H (eds.) (1990) *The Ecology and Management of Aquatic–Terrestrial Ecotones*. Paris: UNESCO.

Naiman RJ, Lonzarich DG, Beechie TJ, and Ralph SC (1992) General principles of classification and the assessment of conservation potential in rivers. In: Boon PJ, Calow P, and Petts GE (eds.) *River Conservation and Management*, pp. 93–123. Chichester: Wiley.

Nanson GC and Knighton AD (1996) Anabranching rivers: Their causes, character and classification. *Earth Surface Processes and Landforms* 21(3): 217–239.

Nanson GC and Croke JC (1992) A genetic classification of floodplains. *Geomorphology* 4(6): 459–486.

Newbold JD, Elwood JW, O'Neill RV, and Van Winkle W (1981) Nutrient spiralling in streams: The concept and its field measurement. *Canadian Journal of Fisheries and Aquatic Sciences* 38: 860–863.

Newbold JD, Mulholland PJ, Elwood JW, and O'Neill RV (1982) Organic spiralling in stream ecosystems. *Oikos* 38(3): 266–272.

Newson MD, Harper DM, Padmore CL, Kemp JL, and Vogel B (1998) A cost-effective approach for linking habitats, flow types and species requirements. *Aquatic Conservation: Marine and Freshwater Ecosystems* 8(4): 431–446.

Newson MD and Newson CL (2000) Geomorphology, ecology and river channel habitat: Mesoscale to basin-scale challenges. *Progress in Physical Geography* 24. 195–217.

Nilsson C, Andersson E, Merritt DM, and Johansson ME (2002) Differences in riparian flora between riverbanks and river lakeshores explained by dispersal traits. *Ecology* 83(10): 2878–2887.

Nilsson C, Gardfjell M, and Grelsson G (1991) Importance of hydrochory in structuring plant communities along rivers. *Canadian Journal of Botany Revue Canadienne de Botanique* 69(12): 2631–2633.

Nilsson C and Svedmark M (2002) Basic principles and ecological consequences of changing water regimes: Riparian plant communities. *Environmental Management* 30(4): 468–480.

Olden JD and Naiman RJ (2009) Incorporating thermal regimes into environmental flows assessments: Modifying dam operations to restore freshwater ecosystem integrity. *Freshwater Biology* 55(1): 86–107.

Olden JD and Poff NL (2003) Redundancy and the choice of hydrologic indices for characterizing streamflow regimes. *River Research and Applications* 19(2): 101–121.

Orghidan T (1959) Ein neuer Lebensraum des unterirdischen Wassers: Der hyporheische Biotop. *Archiv für Hydrobiologie* 55: 392–414.

Orsborn JF and Allman CH (eds.) (1976) *Proceedings of the Symposium and Specialty Conference on Instream Flow Needs, I and II*. Bethesda, MD: American Fisheries Society.

Orth DJ (1987) Ecological considerations in the development and application of instream flow-habitat models. *Regulated Rivers: Research and Management* 1(2): 171–181.

Osmundson DB, Ryel RJ, Lamarra VL, and Pitlick J (2002) Flow-sediment-biota relations: Implications for river regulation effects on native fish abundance. *Ecological Applications* 12(6): 1719–1739.

Osterkamp WR (1978) Gradient, discharge and particle-size relations of alluvial channels in Kansas, with observations of braiding. *American Journal of Science* 278: 1253–1268.

Padmore CL (1997) Biotopes and their hydraulics: A method for determining the physical component of freshwater habitat quality. In: Boon PJ and Howell DL (eds.) *Freshwater Quality: Defining the Indefinable*, pp. 251–257. Edinburgh: HMSO.

Paetzold A, Schubert CJ, and Tockner K (2005) Aquatic terrestrial linkages along a braided-river: Riparian arthropods feeding on aquatic insects. *Ecosystems* 8: 748–759.

Page KJ, Nanson GC, and Frazier PS (2003) Floodplain formation by oblique accretion on the Murrumbidgee River, Australia. *Journal of Sedimentary Research* 73: 5–14.

Parasiewicz P (2001) MesoHABSIM: A concept for application of instream flow models in river restoration planning. *Fisheries* 26(9): 6–13.

Parasiewicz P (2003) Upscaling: Integrating habitat model into river management. *Canadian Water Resources Journal* 28: 283–299.

Parasiewicz P (2007) The MesoHABSIM model revisited. *River Research and Applications* 23(8): 893–903.

Parasiewicz P (2008) Application of MesoHABSIM and Target Fish Community approaches to restoration of the Quinebaug River, Connecticut and Massachusetts, USA. *River Research and Applications* 24(4): 459–471.

Pardé M (1955) *Fleuves et rivières*. 224pp. Paris: Armand Colin.

Pardo I and Armitage PD (1997) Species assemblages as descriptors of mesohabitats. *Hydrobiologia* 344: 111–128.

Patel JA (2007) Evaluation of low flow estimation techniques for ungauged catchments. *Water and Environment Journal* 21(1): 41–46.

Pettit NE and Froend RH (2001) Availability of seed for recruitment of riparian vegetation: A comparison of a tropical and a temperate river ecosystem in Australia. *Australian Journal of Botany* 49(4): 515–528.

Petts GE (1980) Long-term consequences of upstream impoundment. *Environmental Conservation* 7: 325–332.

Petts GE (1984) *Impounded Rivers*. Chichester: Wiley.

Petts GE (1996) Water allocation to protect river ecosystems. *Regulated Rivers: Research and Management* 12: 353–365.

Petts GE (2009) Instream-flow science for sustainable river management. *Journal of the American Water Resources Association* 45: 1071–1086.

Petts GE and Amoros C (eds.) (1996) *Fluvial Hydrosystems*, 308pp. London: Chapman and Hall.

Petts GE and Bickerton MA (1994) Influence of water abstraction on macroinvertebrate community gradients within a glacial stream system: La Borge d'Arolla, Valais, Switzerland. *Freshwater Biology* 32: 375–386.

Petts GE, Bickerton MA, Crawford C, Lerner DN, and Evans D (1999) Flow management to sustain groundwater-dominated stream ecosystems. *Hydrological Processes* 13(3): 497–513.

Petts GE and Gurnell AM (2005) Dams and geomorphology: Research progress and future directions. *Geomorphology* 71: 27–47.

Petts GE and Maddock I (1994) Flow allocation for in river needs *Rivers Handbook*, pp. 289–307. Oxford: Blackwell.

Petts GE, Nestler J, and Kennedy R (2006) Advancing science for water resources management. *Hydrobiologia* 565(1): 277–288.

Petts GE, Sparks R, and Campbell I (2000) River restoration in developed economies. In: Boon PJ, Davies BR, and Petts GE (eds.) *Global Perspectives on River Conservation*, pp. 493–508. Chichester: Wiley.

Poff NL, Allan D, Bain MB, et al. (1997) The natural flow regime: A paradigm for river conservation and restoration. *BioScience* 47(11): 769–784.

Poff NL, Arthington AH, Bunn SE, et al. (2009) The ecological limits of hydrologic alteration (ELOHA): A new framework for developing regional environmental flow standards. *Freshwater Biology* 55(1): 147–170.

Poff NL, Bledsoe BP, and Cuhaciyan CO (2006) Hydrologic variation with land use across the contiguous United States: Geomorphic and ecological consequences for stream ecosystems. *Geomorphology* 79: 264–285.

Poff NL and Ward JV (1989) Implication of streamflow variability and predictability for lotic community structure: A regional analysis of streamflow patterns. *Canadian Journal of Fisheries and Aquatic Science* 46: 1805–1818.

Poff NL and Zimmerman JKH (2009) Ecological responses to altered flow regimes: A literature review to inform environmental flows science and management. *Freshwater Biology* 55: 194–205.

Poole GC (2002) Fluvial landscape ecology: Addressing uniqueness within the river discontinuum. *Freshwater Biology* 47: 641–660.

Poole GC, O'Daniel SJ, Jones KL, et al. (2008) Hydrologic spiralling: The role of multiple interactive flow paths in stream ecology. *River Research and Applications* 24(7): 1018–1031.

Poole GC, Stanford JA, Running SW, and Frissell CA (2006) Multiscale geomorphic drivers of groundwater flow paths: Subsurface hydrologic dynamics and hyporheic habitat diversity. *Journal of the North American Benthological Society* 25: 288–303.

Power ME, Sun A, Parker G, Dietrich WE, and Wootton JT (1995) Hydraulic food-chain models. *BioScience* 45(3): 159–167.

Pringle CM (2001) Hydrologic connectivity and the management of biological reserves: A global perspective. *Ecological Applications* 11(4): 981–998.

Pringle CM (2003) What is hydrologic connectivity and why is it ecologically important? *Hydrological Processes* 17(13): 2685–2689.

Pringle CM, Naiman RJ, Bretschko G, et al. (1988) Patch dynamics in lotic systems: The stream as a mosaic. *Journal of the North American Benthological Society* 7(4): 503–524.

Raven PJ, Fox P, Everard M, Holmes NTH, and Dawson FH (1997) River habitat survey: A new system for classifying rivers according to their habitat quality. In: Boon PJ and Howell DL (eds.) *Freshwater Quality: Defining the Indefinable?*, pp. 215–234. Edinburgh: HMSO.

Reiser DW, Ramey MP, and Wesche TA (1989) Flushing flows. In: Gore JA and Petts GE (eds.) *Alternatives in Regulated River Management*, pp. 91–138. Boca Raton, FL: CRC.

Resh VH, Brown AV, Covich AP, et al. (1988) The role of disturbance in stream ecology. *Journal of the North American Benthological Society* 7: 443–455.

Rice SP, Greenwood MT, and Joyce CB (2001) Tributaries, sediment sources and the longitudinal organization of macroinvertebrate fauna along river systems. *Canadian Journal of Fisheries and Aquatic Sciences* 58: 824–840.

Richter BD, Baumgartner JV, Braun DP, and Powell J (1998) A spatial assessment of hydrologic alteration within a river network. *Regulated Rivers: Research and Management* 14(4): 329–340.

Richter BD, Baumgartner JV, Powell J, and Braun DP (1996) A method for assessing hydrologic alteration within ecosystems. *Conservation Biology* 10(4): 1163–1174.

Richter BD, Baumgartner JV, Wigington R, and Braun DP (1997) How much water does a river need? *Freshwater Biology* 37(1): 231–249.

Richter BD, Warner AT, Meyer JL, and Lutz K (2006) A collaborative and adaptive process for developing environmental flow recommendations. *River Research and Applications* 22(3): 297–318.

RRA (2003) Environmental flows for river systems. *River Research and Applications, Special Issue* 19(5–6): 375–681.

Sabaton C, Siegler L, Gouraud V, Bagliniere JL, and Manne S (1997) Presentation and first applications of a dynamic population model for brown trout (*Salmo trutta* L.): Aid to river management. *Fisheries Management and Ecology* 4: 425–438.

Sabaton C, Souchon Y, Capra H, Gouraud V, Lascaux J-M, and Tissot L (2008) Long-term brown trout populations responses to flow manipulation. *River Research and Applications* 24(5): 476–505.

Scarnecchia DL and Roper BB (2000) Large scale differential summer habitat use of three anadromous salmonids in a large river basin in Oregon, USA. *Fisheries Management and Ecology* 7: 197–209.

Schramm HL and Eggleton NA (2006) Applicability of the flood-pulse concept to temperate floodplain river ecosystems: Thermal and temporal components. *River Research and Applications* 22: 543–553.

Schumm SA (1977) *The Fluvial System.* 338pp. New York: Wiley.

Schumm SA (1985) Patterns of alluvial rivers. *Annual Reviews of Earth and Planetary Science* 13: 5–27.

Sedell JR, Richey JE, and Swanson FJ (1989) The river continuum concept: A basis for the expected ecosystem behavior of very large rivers? In: Dodge DP (ed.) *Proceedings of the International Large River Symposium*, Canadian Special Publication of Fisheries and Aquatic Sciences, vol. 106, pp. 49–55. Ottawa, ON: Canadian Government Publishing Centre.

Sherrard JJ and Erskine WD (1991) Complex response of a sand-bed stream to upstream impoundment. *Regulated Rivers: Research and Management* 6: 53–70.

Shiau J-T and Wu FC (2004) Assessment of hydrologic alterations caused by Chi-Chi diversion weir in Chou-Shui Creek, Taiwan: Opportunities for restoring natural flow conditions. *River Research and Applications* 20(4): 401–412.

Shiau J-T and Wu FC (2006) Compromise programming methodology for determining instream flow under multiobjective water allocation criteria. *Journal of the American Water Resources Association* 42: 1179–1191.

Smith JD (2004) The role of riparian shrubs in preventing floodplain unravelling along the Clark Fork of the Columbia River in Deer Lodge Valley, Montana. In: Bennett SJ and Simon A (eds.) *Riparian Vegetation and Fluvial Geomorphology, Water Science and Application*, pp. 71–85. Washington, DC: American Geophysical Union.

Snelder TH and Biggs BJF (2002) Multiscale river environment classification for water resources management. *Journal of the American Water Resources Association* 38(5): 1225–1239.

Snelder TH, Biggs BJF, and Woods RA (2005) Improved eco-hydrological classification of rivers. *River Research and Applications* 21(6): 609–628.

Stalnaker CB (1979) The use of habitat structure preferenda for establishing flow regimes necessary for maintenance of fish habitat. In: Ward JV and Stanford JA (eds.) *The Ecology of Regulated Streams*, pp. 321–338. New York: Plenum.

Stalnaker CB (1994) Evolution of instream flow habitat modelling. In: Calow P and Petts GE (eds.) *Rivers Handbook*, pp. 276–288. Oxford: Blackwell.

Stalnaker CB, Bovee KD, and Waddle TJ (1996) Importance of the temporal aspects of habitat dynamics to fish population studies. *Regulated Rivers: Research and Management* 12(2–3): 145–153.

Stanford JA, Lorang MS, and Hauer FR (2005) The shifting habitat mosaic of river ecosystems. *Proceedings – International Association of Theoretical and Applied Limnology* 29: 123–136.

Stanford JA and Ward JV (1988) The hyporheic habitat of river ecosystems. *Nature* 335: 64–66.

Stanford JA and Ward JV (1993) An ecosystem perspective of alluvial rivers: Connectivity and the hyporheic corridor. *Journal of the North American Benthological Society* 12: 48–60.

Stanford JA, Ward JV, Liss WJ, et al. (1996) A general protocl for restoration of regulated rivers. *Regulated Rivers: Research and Management* 12: 391–413.

Statzner B, Gore JA, and Resh VH (1988) Hydraulic stream ecology – observed patterns and potential applications. *Journal of the North American Benthological Society* 7(4): 307–360.

Statzner B and Higler B (1986) Stream hydraulics as a major determinant of benthic invertebrate zonation patterns. *Freshwater Biology* 16: 127–139.

Stewart G, Anderson R, and Wohl E (2005) Two-dimensional modelling of habitat suitability as a function of discharge on two Colorado rivers. *River Research and Applications* 21(10): 1061–1074.

Stromberg JC, Beauchamp VB, Dixon MD, Lite SJ, and Paradzick C (2007) Importance of low-flow and high-flow characteristics to restoration of riparian vegetation along rivers in arid south-western United States. *Freshwater Biology* 52(4): 651–679.

Tal M, Gran K, Murray AB, Paola C, and Hicks DM (2004) Riparian vegetation as a primary control on channel characteristics in multi-thread rivers In: Bennett SJ and Simon A (eds.) *Riparian Vegetation and Fluvial Geomorphology, Water Science and Application*, pp. 43–58., Washington, DC: American Geophysical Union.

Tennant DL (1976) Instream flow requirements for fish, wildlife, recreation and related environmental resources. *Fisheries Management and Ecology* 1: 6–10.

Tharme RE (2003) A global perspective on environmental flow assessment: Emerging trends in the development and application of environmental flow methodologies for rivers. *River Research and Applications* 19(5–6): 397–441.

Thorp JH, Thoms MC, and Delong MD (2006) The riverine ecosystem synthesis: Biocomplexity in river networks across space and time. *River Research and Applications* 22: 123–147.

Tickner D, Armitage PD, Bickerton MA, and Hall KA (2000) Assessing stream quality using information on mesohabitat distribution and character. *Aquatic Conservation – Marine and Freshwater Ecosystems* 10(3): 179–196.

Tockner K, Malard F, and Ward JV (2000) An extension of the flood pulse concept. *Hydrological Processes* 14(16–17): 2861–2883.

Tockner K and Stanford JA (2002) Riverine flood plains: Present state and future trends. *Environmental Conservation* 29(3): 308–330.

Tooth S and Nanson GC (1999) Anabranching rivers on the Northern Plains of arid central Australia. *Geomorphology* 29: 211–233.

Tooth S and Nanson GC (2000) The role of vegetation in the formation of anabranching channels in an ephemeral river, Northern Plains, arid central Australia. *Hydrological Processes* 14: 3099–3117.

Townsend CR (1989) The patch dynamics concept of stream community ecology. *Journal of the North American Benthological Society* 8: 36–50.

Truscott A-M, Soulsby C, Palmer SCF, Newell L, and Hulme PE (2006) The dispersal characteristics of the invasive plant *Mimulus guttatus* and the ecological significance of increased occurrence of high-flow events. *Journal of Ecology* 94: 1080–1091.

Vadas RL and Orth DJ (2001) Formulation of habitat suitability models for stream fish guilds: Do the standard methods work. *Transactions of the American Fisheries Society* 130: 217–235.

van den Berg JH (1995) Prediction of alluvial channel pattern of perennial rivers. *Geomorphology* 12(4): 259–279.

Vannote RL, Minshall GW, Cummins KW, Sedell JR, and Cushing CE (1980) The river continuum concept. *Canadian Journal of Fisheries and Aquatic Sciences* 37: 130–137.

Vogt K, Rasran L, and Jensen K (2004) Water-borne seed transport and seed deposition during flooding in a small river-valley in Northern Germany. *Flora* 199(5): 377–388.

Vörösmarty CJ and Sahagian D (2000) Anthropogenic disturbance of the terrestrial water cycle. *BioScience* 50(9): 753–765.

Vörösmarty CJ, Sharma KP, Fekete BM, et al. (1997) The storage and aging of continental runoff in large reservoir systems of the world. *Ambio* 26(4): 210–219.

Wadeson LA (1994) A geomorphological approach to the identification and classification of instream flow environments. *South African Journal of Aquatic Sciences* 20: 1–24.

Ward JV (1985) Thermal characteristics of running waters. *Hydrobiologia* 125: 31–46.

Ward JV (1989) The four-dimensional nature of lotic ecosystems. *Journal of the North American Benthological Society* 8: 2–8.

Ward JV, Malard F, Tockner K, and Uehlinger U (1999) Influence of ground water on surface water conditions in a glacial flood plain of the Swiss Alps. *Hydrological Processes* 13: 277–293.

Ward JV and Stanford JA (Eds.) (1979) *The Ecology of Regulated Streams.* New York: Plenum Press.

Ward JV and Stanford JA (1982) Thermal responses in the evolutionary ecology of aquatic insects. *Annual Review of Entomology* 27: 97–117.

Ward JV and Stanford JA (1983b) The intermediate disturbance hypothesis: An explanation for biotic diversity patterns in lotic ecosystems. In: Fontaine TD and Bartell SM (eds.) *Dynamics of Lotic Ecosystems*, pp. 347–356. Ann Arbor, MI: Ann Arbor Science.

Ward JV and Stanford JA (1983a) The serial discontinuity concept of lotic systems. In: Fontaine TD and Bartell SM (eds.) *Dynamics of Lotic Ecosystems*, pp. 29–42. Ann Arbor, MI: Ann Arbor Science.

Ward JV and Stanford JA (1995) Ecological connectivity in alluvial river ecosystems and its disruption by flow regulation. *Regulated Rivers: Research and Management* 11: 105–120.

Ward JV, Tockner K, Arscott DB, and Claret C (2002) Riverine landscape diversity. *Freshwater Biology* 47(4): 517–539.

Webster JR, Wade JB, and Patten BC (1975) Nutrient recycling and the stability of ecosystems. In: Howell FG, Gentry JB, and Smith MH (eds.) *Mineral Cycling in Southeastern Ecosystems*, pp. 1–27. Springfield, VA: National Technical Information Service.

Welcomme RL (1979) *Fisheries Ecology of Floodplain Rivers*. London: Longman.

Welcomme RL, Winemiller KO, and Cowx IG (2006) Fish environmental guilds as a tool for assessment of ecological condition of rivers. *River Research and Applications* 22(3): 377–396.

White MA, Schmidt JC, and Topping DJ (2005) Application of wavelet analysis for monitoring the hydrologic effects of dam operation: Glen Canyon Dam and the Colorado River at Lees Ferry, Arizona. *River Research and Applications* 21(5): 551–565.

Williams GP and Wolman MG (1984) Downstream effects of dams on alluvial rivers. *US Geological Survey Professional Paper* 1286: 83.

Wood PJ, Hannah DM, Agnew MD, and Petts GE (2001) Scales of hydroecological variability within a groundwater-dominated stream. *Regulated Rivers: Research and Management* 17(4–5): 347–367.

Wu F-C and Chou Y-J (2004) Tradeoffs associated with sediment-maintenance flushing flows: A simulation approach to exploring non-inferior options. *River Research and Applications* 20(5): 591–604.

Young WJ, Chessman BC, Erskine WD, *et al.* (2004) Improving expert panel assessments through the use of a Composite River Condition Index – the case of the rivers affected by the Snowy Mountains hydro-electric scheme, Australia. *River Research and Applications* 20(6): 733–750.

Zanoni L, Gurnell AM, Drake N, and Surian N (2008) Island dynamics in a braided river from an analysis of historical maps and air photographs. *River Research and Applications* 24(8): 1141–1159.

2.11 Hydrology and Biogeochemistry Linkages

NE Peters, US Geological Survey, Atlanta, GA, USA
JK Böhlke, US Geological Survey, Reston, VA, USA
PD Brooks, University of Arizona, Tucson, AZ, USA
TP Burt, Durham University, Durham, UK
MN Gooseff, Pennsylvania State University, University Park, PA, USA
DP Hamilton, University of Waikato, Hamilton, New Zealand
PJ Mulholland, Oak Ridge National Laboratory, Oak Ridge, TN, USA
NT Roulet, McGill University, Montreal, QC, Canada
JV Turner, CSIRO Land and Water, Wembley, WA, Australia

© 2011 Elsevier B.V. All rights reserved.

2.11.1	Introduction	271
2.11.2	Hydrological Pathways on Drainage Basin Slopes	272
2.11.3	Mountain Environments	274
2.11.3.1	Precipitation	275
2.11.3.1.1	Snow	275
2.11.3.2	Change in Storage	276
2.11.3.3	Evaporation and Transpiration	276
2.11.3.4	Stream Flow	278
2.11.3.4.1	Nitrate isotopes in stream water	278
2.11.3.4.2	Transit time and residence time	279
2.11.3.5	Groundwater Recharge	282
2.11.4	Within-River Processes	283
2.11.5	Wetland Processes	285
2.11.6	Lakes	288
2.11.7	Groundwater	291
2.11.8	Acidic Atmospheric Deposition – Acid Rain	293
2.11.9	Summary and Future Considerations	294
2.11.10	Additional Reading	295
References		296

2.11.1 Introduction

Biota depend on water, energy, and nutrient transfers within and between ecosystems, which result in complex interactions between hydrology and biogeochemistry. The hydrological cycle and the variation in rates and magnitudes of water transfers along pathways in turn affect biogeochemical interactions. Nutrient uptake by biota varies markedly depending on availability of water, the pathways by which it moves through ecosystems, and the ecosystem type (aquatic, terrestrial), climate, and many geomorphological factors, such as slope and soil type. Variations in flow rates along pathways, reaction rates, composition (mineralogy, chemistry, and biology) and characteristics of interacting materials, chemical composition of the water, and temperature are major factors affecting biogeochemical processes. Most chemical cycles either affect or are affected by biological activity. Research since the early 1990s has revealed that even mobile and conservative elements, such as chlorine, are actively cycled by biota (Öberg et al., 1996; White and Broadley, 2001; Öberg, 2003; Lovett et al., 2005; Öberg and Sandén, 2005). However, the cycles of nutrients, carbon, and other biogeochemical components are intricately linked to hydrology and biogeochemistry, and are the foci for most of the discussion in this chapter.

Scientists have been challenged by the task of determining and quantifying the background processes that affect hydrological and biogeochemical linkages because human activities that accompany population growth and the associated requirements of obtaining and consuming natural resources have accelerated landscape changes (Peters and Meybeck, 2000; Meybeck, 2001; Peters et al., 2005). Human impacts on ecosystems are evident everywhere on Earth. For example, deforestation, channelization, dams and river regulation, land drainage, agriculture, energy generation, and urbanization and management of these activities have had major impacts on hydrology and biogeochemistry (Poff et al., 1997; Friedman et al., 1998; Peters and Meybeck, 2000; Meybeck 2001; Paul and Meyer, 2001; Meyer et al., 2005; Peters et al., 2005; Poff et al., 2006; Palmer et al., 2008; Peters, 2008). Human activities and resource management have evolved and it has been widely recognized that while point-source pollution has become more manageable, diffuse or nonpoint sources of pollutants are dominating contamination of ecosystems, and are much more difficult to identify, quantify, and control (Novotny, 2003; Campbell and Novotny, 2004; Loague and

Corwin, 2005). Freshwaters are experiencing declines in biodiversity far greater than those in the most affected terrestrial ecosystems and human threats to global freshwater biodiversity and ecosystem services include overexploitation, water pollution, flow modification, destruction or degradation of habitat, and invasion by exotic species; the combined and interacting influences have resulted in population declines and range reduction of freshwater biodiversity worldwide (Dudgeon et al., 2005). Ecosystem processes, including water, nitrogen, carbon, and phosphorus cycling, changed more rapidly in the second half of the twentieth century than at any time in recorded human history. Human modifications of ecosystems have changed not only the structure of the systems (such as what habitats or species are present in a particular location), but their processes and functioning as well (Millenium Ecosystem Assessment, 2005). The capacity of ecosystems to provide services derives directly from the operation of natural biogeochemical cycles, which in some cases have been modified substantially. Ecosystem services are the human benefits provided by ecosystems.

Furthermore, invasive and exotic species continue to affect ecosystems (Mooney and Hobbs, 2000; Mooney et al., 2005), and the increasing presence of recalcitrant endocrine disruptors and pharmaceuticals in stream water has the potential to alter ecosystems and change biogeochemical cycles (McMaster, 2001; Boyd et al., 2003; Stackelberg et al., 2004). Although our understanding of how invasive species affect ecosystem processes is not well understood (Gordon, 1998; Levine et al., 2003), researchers have reported a wide range of effects on hydrology and biogeochemistry, such as by invasive earthworms in northern temperate forests (Bohlen et al., 2004), invasive vegetation in riparian zones (Décamps et al., 2004) caused by damming and river regulation (Nilsson and Berggren, 2000), and invasive species in grasslands (Scott et al., 2001; Hook et al., 2004). However, it is beyond the scope of this chapter to provide detailed information about hydrological and biogeochemical linkages for these and the myriad of other human activities that affect the landscape.

The objective of this chapter is to provide an overview of the linkages between hydrology and biogeochemistry in terrestrial and aquatic systems by tracking water flow from headwaters to rivers in larger drainage basins, including groundwater, wetlands, and lakes. The selection of foci was arbitrary and determined largely by the expertise of the co-authors, but provides continuity from a hydrological-cycle perspective and with a bias toward a northern temperate hydroclimate, a geographic region with some of the most detailed process research. To focus the discussion, the chapter is sectioned topically and these topics include hydrological pathways on drainage basin slopes, mountain environments, within-river (or in-stream) processes, wetlands, lakes, and groundwater (and groundwater–surface water interactions). In particular, this chapter provides a view of the linkages among the hydrosphere, biosphere, lithosphere, and chemsphere of processes that affect nutrient cycles, particularly nitrogen and carbon. In addition to the general discussion of nutrient cycling, an example is given of the effects of human activities on these linkages through the widespread impacts of acidic atmospheric deposition. Topics discussed in Chapters (see also **Chapter 1.10 Predicting Future Demands for Water** and **Chapter 1.09 Implementation of Ambiguous Water-Quality Policies**) overlap with some of the material in this chapter, hence will not be discussed here in detail.

2.11.2 Hydrological Pathways on Drainage Basin Slopes

> Although the river and the hill-side waste sheet do not resemble each other at first sight, they are only the extreme members of a continuous series, and when this generalization is appreciated, one may fairly extend the 'river' all over its basin and up to its very divides. Ordinarily treated, the river is like the veins of a leaf; broadly viewed, it is like the entire leaf (Davis, 1899).

The drainage basin (also known as watershed in the USA and catchment elsewhere) has long been recognized as the fundamental unit of analysis for the sciences of hydrology and geomorphology. It is usually a clearly defined and unambiguous topographic unit, which acts as an open system for inputs of precipitation and outputs of river discharge and evaporation (Chorley, 1971). The topographic, hydraulic, and hydrological unity of the drainage basin provided the basis of the morphometric stream ordering system of Horton (1945), as elaborated by Strahler (1964). Schumm (1977) had generalized sediment transport within the drainage basin into three zones: source area, transfer zone, and sediment sink or area of deposition. This is also a convenient subdivision for analyzing solute transport through the drainage basin, including in-stream cycling, and thus closely accords with the concepts of river continuum (Vannote et al., 1980) and nutrient spiraling (Webster and Patten, 1979). The source zone includes low-order headwater basins, which comprise most of the basin area, and where stream biogeochemical dynamics are primarily controlled by flushing of solutes and organic matter into the stream. Further downstream, in higher-order reaches, the channel becomes increasingly isolated from the surrounding land. Although concentrated flow may occasionally occur on hillslopes in rills and gullies and in the subsurface through cracks and pipes, there is generally a clear division between diffuse flow on slopes and concentrated flow in the river channel. The nature of these diffuse flows has important implications for the residence time of water in the catchment and the transit time of water moving to the stream channel.

In headwater basins, the occurrence of runoff leads to sediment and solute removal from hillslopes. Therefore, detailing these processes, source areas, and pathways is relevant. Stream flow may be divided into base flow, that is, stream discharge that is not attributable to direct runoff from precipitation or melting snow and generally is maintained by groundwater discharge during rain- and snowmelt-free periods, and runoff (rainstorm and snowmelt) events. The physical characteristics of the soil and bedrock determine the pathways by which hillslope runoff will reach the channel. The paths taken by water (**Figure 1**) determine many of the characteristics of the landscape, human uses of the land, and strategies required for wise land-use management (Dunne and Leopold, 1978). There are essentially two models of storm

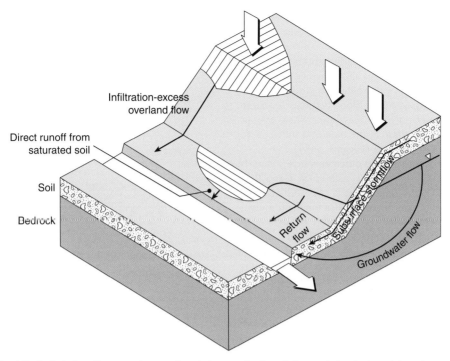

Figure 1 Schematic of the hydrologic pathways and connections between uplands and streams in headwater catchments.

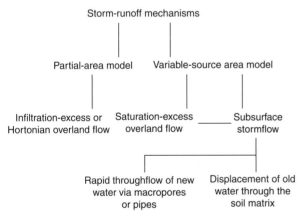

Figure 2 Relations among hydrological pathways/mechanisms delivering stormflow from hillslopes to streams.

runoff generation (**Figure 2**): the partial area and variable-source area models. Horton (1933, 1945) argued that storm runoff is mainly produced by infiltration excess-overland flow. When rainfall intensity exceeds the infiltration capacity of the soil, the excess begins to fill up surface depressions; once these are full, water overflows downslope and surface runoff begins. Horton claimed that infiltration-excess overland flow would occur uniformly across the catchment area, but Betson (1964) showed that this is not necessarily widespread even within a small basin. Betson proposed the partial area model of overland flow generation in which surface runoff is produced only from small areas of the basin during any given storm event. Despite its localized extent, infiltration-excess overland flow can generate large flood peaks and is often associated with soil erosion (Herwitz, 1986). Overland flow moves rapidly and because its residence time is short, it is usually characterized as new water.

Hewlett (1961) proposed the variable-source area model to describe runoff production in areas of permeable soil where subsurface runoff can account for much if not all the storm runoff leaving a basin. The production of significant quantities of subsurface storm flow (otherwise called 'throughflow' or 'interflow') requires the development of saturated conditions within the soil profile, in effect an ephemeral perched water table. Subsurface storm flow may be generated by movement of water through the soil matrix, by flow in macropores (large diameter conduits in the soil, created by agents such as plant roots, soil cracks, or soil fauna), or by a mixture of both (Beven and Germann, 1982). Given its longer residence time within the soil matrix, throughflow usually has a high solute content and therefore appears relatively old compared to precipitation, whereas macropore flow can be sufficiently rapid to retain the characteristics of new water as evidenced from laboratory (Wildenschild et al., 1994; McIntosh et al., 1999) and field studies (Richard and Steenhuis, 1988; Everts et al., 1989; Jardine et al., 1990; Luxmoore et al., 1990; Peters and Ratcliffe, 1998). If the soil profile becomes completely saturated, saturation-excess overland flow (Dunne and Black, 1970a) will be produced, consisting of a mixture of return flow (exfiltrating old soil water) and direct runoff (new rainfall unable to infiltrate the saturated surface). As the extent of the zone of saturation varies seasonally and during storms, Hewlett (1961) coined the phrase variable-source area to describe his model of storm flow generation. Subsurface storm flow dominates the storm hydrograph where deep-permeable soils overlie less-permeable soil or bedrock, and where steep hillslopes abut the stream. Soils may also contain an impeding layer with can cause a perched water table. Soil saturation is more likely to occur in soils of moderate hydraulic conductivity, in areas of reduced soil-moisture storage, and on lower slope angles.

The role of topography is particularly important in determining where soil saturation occurs and is favored in hillslope hollows, in places where the soil profile is shallow, and at the foot of slopes especially where the slope becomes less steep. Considerable erosion can occur in these areas where groundwater discharges to the surface at the base of a hillslope, referred to as groundwater sapping (Higgins, 1984).

Flow in stream channels reflects different source waters. In large seasonally snow-covered basins, such as the Fraser River in western Canada or rivers draining the Rocky Mountains, the Himalayas, or the Alps, snowmelt provides the main annual-discharge response. Depending on the nature of the aquifer, peak baseflow discharge may significantly lag behind precipitation inputs. Groundwater provides the prolonged recession flow associated with the falling limb of the annual hydrograph. Buttle (1994) reviewed processes responsible for conveying pre-event (old) water rapidly to the channel during storm events: groundwater ridging, translatory flow (piston-flow displacement of pre-storm groundwater), macropore flow, saturation-excess overland flow, kinematic waves, and release of water from surface storage. Not all processes occur in all catchments, of course, but whichever predominate, the focus seems inevitably on near-stream zones (Cirmo and McDonnell, 1997; Burt and Pinay, 2005), including upwelling and discharge of deeper groundwater (Winter, 1999; Clément et al., 2003; Lischeid et al., 2007). Bishop et al. (2004) argued that the chemistry of water moving downslope is modified at any given point by soil chemistry, and riparian soils have a particularly important influence on stream-water chemistry because they are the last soils to contact the water before it discharges (Hooper et al., 1998; Buttle, 2005). Riparian zones are not chemically inert and, therefore, whatever the nature and sources of water flowing into the near-stream zone, it should rapidly acquire a chemical signature determined by the nature of the riparian zone (Cirmo and McDonnell, 1997) or mix with upwelling deeper groundwater (Clément et al., 2003).

Dilute (new) event water may reach the stream channel quickly as infiltration-excess overland flow, as direct runoff from saturated areas, or where macropores discharge at the channel bank. The extent to which the chemical composition of flow lines is reset within the near-stream zone depends on the residence time of water there, the size of the riparian aquifer storage, and the amount of new water moving through and mixing with the riparian-zone water. In addition, riparian-zone vegetation can alter the hydrologic cycle (**Figure 3**), including water partitioning through plant physiology (uptake). In addition, the vegetation growth function and structure are affected by water quality (Cirmo and McDonnell, 1997; Tabacchi et al., 2000; Décamps et al., 2004).

Because of variable contributions of water from various pathways, differences exist in the delivery of carbon and nutrients to streams during base flow compared to storm flow (Buffam et al., 2001). Much contemporary research is concerned with the ability of riparian zones to buffer rivers from upslope pollution inputs, nitrate in particular. Biological processes tend to affect nitrate more than most other solutes. Nitrate may be produced in the near-stream zone by

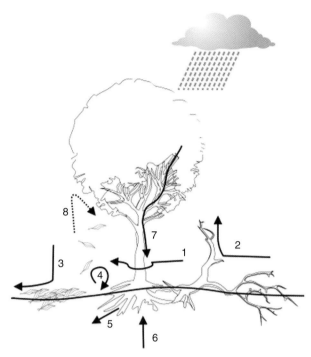

Figure 3 The main physical impacts of riparian vegetation on water cycling: 1, interaction with over-bank flow with stems, branches, and leaves (turbulence); 2, flow diversion by log jams; 3, change in the infiltration rate of flood waters and rainfall by litter; 4, increase of turbulence as a consequence of root exposure; 5, increase of substrate macroporosity by roots; 6, increase of the capillary fringe by fine roots; 7, stem flow (the concentration of rainfall by leaves, branches, and stems); and 8, condensation of atmospheric water and interception of dew by leaves. From Tabacchi E, Lambs L, Guilloy H, PlantyTabacchi AM, Muller E, and Decamps H (2000) Impacts of riparian vegetation on hydrological processes. *Hydrological Processes* 14(16–17): 2959–2976.

mineralization, and rising water tables can then flush this nitrate into the stream (Triska et al., 1989). However, nitrate may also be removed, temporarily by uptake and immobilization, or permanently by denitrification. For forests, stand age affects N-uptake rates and thus N transformation and leaching rates in soils (Stevens et al., 1994; Emmett et al., 1993). Note, however, that channel flow remains a mixture of different source waters, and if significant amounts of water bypass saturated riparian soils, either by flowing across the surface or through permeable strata below the floodplain alluvium, the riparian zone will be ineffective in removing nitrogen (Burt and Pinay, 2005). Contrasts in stream water nitrate response to similar inputs in seemingly comparable watersheds over various timescales also provide insight into the importance of coupling biogeochemical reactions and hydrological pathways (Reynolds et al., 1992; Christopher et al., 2008). Given the several potential mechanisms for stream flow generation, all of these mechanisms point to the fact that streams and groundwater are intricately linked (Winter et al., 1999).

2.11.3 Mountain Environments

The wide ranges of elevation and aspect that characterize mountain environments result in tremendous variability in how hydrology and biogeochemical cycles are linked in space and time. Topographical complexity affects the amount of precipitation, solar radiation, temperature, and the lateral redistribution of water, resulting in highly heterogeneous biogeochemical processes. Energy and water balance at the land surface are intimately related to ecosystem productivity directly through transpiration, growth, acclimation, and assembly (McDowell et al., 2008), and indirectly through changes in surface physical characteristics, such as albedo and roughness (Bonan and Levis, 2006). Together, these factors affect both in situ biogeochemical reactions and the transport of biogeochemical solutes through the landscape and into surface water.

The spatial variability in elevation, aspect, vegetation, soils, and precipitation, including snow cover, associated with mountain systems results in a high degree of both spatial and temporal variability in the coupling of hydrology and biogeochemistry. The coupling of hydrology and biogeochemistry in mountain environments can be observed by addressing a fundamental question, "What happens to precipitation?" (Penman, 1961), which can be evaluated by assessing the components of the water balance equation:

$$P = \Delta S + E + T + Q + R \quad (1)$$

where P is the input of hydrometeors mainly precipitation, and also includes fog, rim ice, and cloud water, ΔS is the change in near-surface water storage, E is the evaporation/sublimation, T is the transpiration, Q is the runoff, and R is the groundwater recharge. The units used for each water-balance component are typically given in depth per unit time, such as millimeter per day or year, for a drainage/catchment/watershed area. These terms are implicitly linked to biogeochemical cycling and suggest mechanisms for explicitly linking hydrology and biogeochemistry in mountain catchments. The remainder of the discussion on mountain environments is organized around these components of the water balance.

2.11.3.1 Precipitation

Precipitation (P) typically increases with elevation, while temperature decreases (Barros and Lettenmaier, 1994; Garcia-Martino et al., 1996). In mountain catchments, soils are wetter for longer at higher elevations for similar landscape positions (Band et al., 2001), for example, riparian zones, stream channels, and hillslopes, and soil types as at lower elevations; but soils generally are thinner at higher elevations. Both increased soil-water availability and decreased temperature at high elevations reduce water stress on vegetation while simultaneously slowing soil heterotrophic activity, resulting in a smaller percentage of fixed carbon being respired from soils (Schlesinger, 1997). Thicker soils and higher temperatures at low elevations favor higher carbon storage and respiration.

Fog and cloud water deposition may be a large percentage of annual water input to some forests in coastal ecosystems (Dawson, 1998; Klemm et al., 2005; Scholl et al., 2007) and to high-elevation forests (Lovett, 1984; Lovett and Kinsman, 1990; Clark et al., 1998; Heath and Huebert, 1999; Herckes et al., 2002; Scholl et al., 2007). Fog and cloud water typically have higher solute concentrations than precipitation (Lovett, 1984; Asbury et al., 1994; Reynolds et al., 1996; Weathers and Likens, 1997; Clark et al., 1998; Oyarzún et al., 2004) and therefore have higher solute-deposition per unit volume of water. Furthermore, fog and cloud water deposition at forest edges is notably higher than within the forest (Weathers et al., 1995; Ewing et al., 2009). The presence of fog or cloud water affects the plant physiology and biogeochemistry (Joslin and Wolfe, 1992; Bruijnzeel and Veneklaas, 1998; Dawson, 1998; Burgess and Dawson, 2004).

Atmospheric washout (scavenging of aerosols and gases by hydrometeors including rain, snow, sleet, freezing rain, and hail (Pruppacher et al., 1983)) affects solute concentrations temporally during rainstorms with concentrations typically decreasing in time. Rainfall also washes off solutes concentrated by evaporation and dry atmospheric deposition that accumulate on vegetation producing much higher solute concentrations in throughfall and stemflow at the onset of rainstorms and decreasing thereafter (Peters and Ratcliffe, 1998).

2.11.3.1.1 Snow

Snow is a special case of precipitation that has important implications for biogeochemistry in mountain environments. Seasonal snow cover has been shown to be an important hydrological control on biogeochemistry in many mountain catchments. Both dry and wet deposition are stored in winter snowpacks and released to soil and surface water in the spring (Jeffries, 1989; Peters and Driscoll, 1989; Bales et al., 1993; Williams et al., 1995). Snow cover also insulates soils from low air temperatures during winter (Peters, 1984; Brooks et al., 1996, 2005), providing an environment where soil microorganisms actively cycle carbon and nutrients during winter. Microbial biomass has been shown to reach annual maximum values under snow cover (Brooks et al., 1997, 1998), and these maximum values are also associated with changes in species composition (Lipson et al., 1999). Overwinter soil respiration may return 20–50% of the carbon fixed during the previous growing season to the atmosphere as CO_2 (Brooks et al., 2005). Variability in the amount of winter CO_2 loss is associated with the timing and amount of snow cover, soil frost, and labile-carbon availability (Brooks and Williams, 1999; Brooks et al., 1999a; Groffman et al., 1999; Grogan and Chapin, 1999; Groffman et al., 2001b). Similarly, overwinter nitrogen mineralization and immobilization in microbial biomass has been shown to be an important source of plant N at the initiation of the growing season. As with CO_2 efflux, the magnitude of overwinter N cycling is related to the timing of snow cover and soil frost (Brooks et al., 1995; Groffman et al., 2001a). Consequently, natural variations in seasonal snow cover and climate change can have major impacts on soil processes (Edwards et al., 2007).

In many forested mountain catchments, snow–vegetation–energy interactions define snow amount and timing of water availability to terrestrial ecosystems during the growing season (Liston, 2004; Molotch and Bales, 2005; Liston and Elder, 2006; Molotch and Bales, 2006; Veatch et al., 2009). Net ecosystem carbon uptake is dominated by fixation during the snowmelt season when water is not limiting (Monson et al., 2002), and soil respiration and N cycling are also strongly controlled by the timing of snow cover (Brooks et al., 1997; Grogan and Chapin, 1999; Groffman et al., 2001b; Grogan and Jonasson, 2003). Finally, snowmelt transports large, but variable amounts of nutrients, organic matter, and inorganic carbon out of the terrestrial ecosystem (Rascher et al., 1987; Hornberger et al., 1994; Campbell et al., 1995; Boyer et al., 1997; Brooks and Williams, 1999; Brooks et al., 1999b; Heuer et al., 1999; Campbell and Law, 2000).

Solute concentrations vary markedly in meltwater and preferentially elute during snowmelt (Johannessen et al., 1975; Johannessen and Henriksen, 1978; Tranter et al., 1986; Berg, 1992). As observed for rainfall and throughfall, early meltwater is high in dissolved solutes that have accumulated in the snowpack and are concentrated in brines around snow crystals during snow metamorphism (Tranter and Jones, 2001). The combination of active abiotic and biotic processes under the snowpack and elution of solutes in meltwater may result in high stream concentrations during snowmelt (Peters and Leavesley, 1995; Mitchell, 2001; Driscoll et al., 2005).

The snowpack is biologically active (Hoham and Duval, 2001), particularly when liquid water is present and temperatures and temperature ranges are optimum (Jones, 1999; Hoham and Duval, 2001). Light is also an important constraint to the distribution and reproduction of some biological components such as snow algae (Hoham et al., 1998, 2000), and these algae in turn affect the timing and magnitude of snowmelt (Hoham and Duval, 2001). In addition to snowpack biological activity, physical characteristics, such as atmospheric conditions (wind, vapor pressure, and temperature), can affect gaseous transfers of nutrients (CO_2 and NO_x) to and from snowpacks (Pomeroy et al., 1999; Tranter and Jones, 2001).

2.11.3.2 Change in Storage

The first-order controls on near-surface storage of water (ΔS) are soil and slope. Soil characteristics and landform are major controls on the partitioning of precipitation into infiltration (Philip, 1967), storage (Beven, 1982), recharge (Gee and Hillel, 1988; Phillips et al., 2004), runoff (see Section 2.11.1), and stream flow (Dunne and Black, 1970b). By affecting the amount and timing of water availability to vegetation and soil microbes, these characteristics interact with climate to affect both potential productivity (carbon fixation) and soil microbial processes. Soils are typically shallower, less weathered, and coarser on ridge lines and higher elevations, and deeper and finer in depressions. Fine-textured soils retain water, reduce diffusion, and result in an environment where anaerobic or facultatively anaerobic processes dominate biogeochemical cycling (Pusch et al., 1998; Hill and Cardaci, 2004). Similarly, topographic depressions and areas of hydrological convergence have higher soil moisture, which may result in either episodic or continuous anaerobic conditions. As aerobic respiration decreases so does carbon mineralization, which can result in increases in denitrification and methanogenesis (Jones et al., 1995; Baker et al., 1999). Several factors affect denitrification across time and space as shown schematically in **Figure 4** and discussed with respect to each of the environments presented in this chapter.

2.11.3.3 Evaporation and Transpiration

Although the absolute magnitude of evaporation and transpiration (E and T) is controlled by climate, the partitioning between E and T is controlled by vegetation water use, and thus is directly linked to carbon fixation and input into the ecosystem. Vegetation serves both as the carbon pump, bringing organic matter into ecosystems, and as the water pump, removing water from ecosystems, and thus controls the amount of chemical energy in organic matter and the amount of water in the environment. Organic matter and water in turn are the primary controls on soil biogeochemical processes, affecting carbon balance, nutrient cycling, and mineral weathering (Amundson et al., 2007).

Both the type and amount of vegetation within an ecosystem vary predictably with elevation and latitude (Holland and Steyn, 1975), as the effect of increasing elevation can generally be equated to that of increasing latitude. Similarly, vegetation varies with aspect in relation to both temperature and water availability (Grace, 1989). For example, the tree line often is limited by temperature and typically extends to a higher elevation on south-facing slopes than north-facing slopes in the Northern Hemisphere (Treml and Banas, 2008). In seasonally water-stressed environments, north-facing and topographically shaded slopes may have more dense vegetation cover and different species assemblages than south-facing or nonshaded slopes (Zhang et al., 2009). These differences arise from the interaction between energy and water. By changing the structure and productivity of the land surface, which control the input of carbon to soil and plant N demand from soil, the potential magnitude of soil biogeochemical processes is affected. Moreover, the root zone and rhizosphere in the soil are particularly active biologically and biogeochemically and the presence of roots affects hydrology. The rhizosphere, which is the dynamic interface among plant roots, soil microbes and fauna, and the soil, is attributed to the evolution of soil, that is, the alteration of primary and secondary soil minerals (Cardon and Whitbeck, 2007).

The redistribution of soil moisture by plants is not typically considered in biogeochemical-cycling research or in typical hydrological-process assessment (Burgess et al., 1998; Caldwell et al., 1998; Jackson et al., 2000; Meinzer et al., 2001). Plant physiologists with the aid of stable isotopes have made major advances in understanding how plants use water (Ehleringer and Dawson, 1992; Dawson, 1993; Emerman and Dawson, 1996; Dawson and Ehleringer, 1998; Moreira et al., 2000). A general pattern is that roots transport water from deep moist horizons to shallow drier surface-soil horizons, particularly at night, by a process called hydraulic lift (Caldwell et al., 1998). It is not surprising that plant redistribution of water would predominate in dry land, as observed in arid areas, for example, for phreatophytes (Hultine et al., 2003) and sage brush (Richards and Caldwell, 1987; Caldwell and Richards, 1989). However, plant redistribution of water has also been documented for sugar maples in northern temperate forests (Dawson, 1996; Emerman and Dawson, 1996), for three species representing each of three canopy niches in Amazonia (Oliveira et al., 2005), and for blue oaks in the foothills of the Sierra Nevada (Millikin Ishikawa and Bledsoe, 2000). Roots not only move water from depth to surface soils, but can

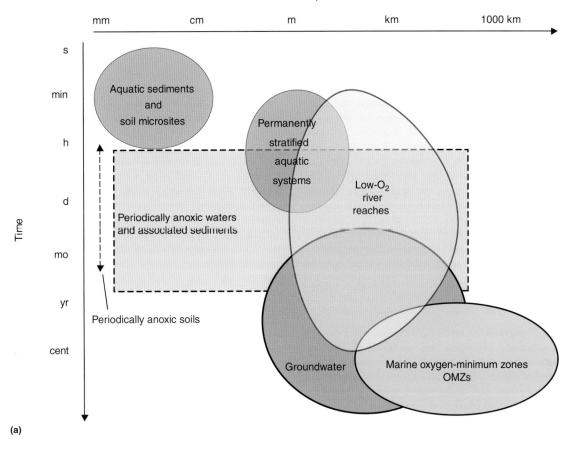

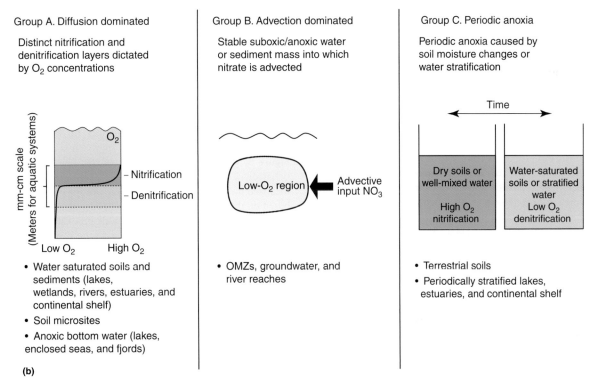

Figure 4 (a) Classification of systems according to the magnitude of temporal and spatial separation between nitrification and denitrification. Diffusion-dominated systems are indicated in gray, advection-dominated systems are indicated with heavy outlines, and systems with periodic anoxia are indicated by dashed lines. (b) Schematic groupings of systems according to mechanism of nitrate delivery to denitrification zone. Vertical profiles of oxygen concentrations are indicated. Adapted from Seitzinger S, Harrison JA, Böhlke JK, et al. (2006) Denitrification across landscapes and waterscapes: A synthesis. *Ecological Applications* 16(5): 2064–2090.

also move water from surface to depth as documented for perennial grasses in northwestern South Africa (Schulze *et al.*, 1998); inverse hydraulic lift may be an important mechanism to facilitate root growth in dry desert soil layers below the upper soil zone to which precipitation penetrates. In addition, meteorological conditions through the effect of shade and clouds on soil water potential can affect hydraulic lift (Williams *et al.*, 1993). Furthermore, water hydraulically lifted by large trees may be used by smaller trees and shrubs as shown by isotope tracers (Dawson, 1996); the authors concluded that small trees and shrubs use soil water and large trees use groundwater. Clearly, these plant–water relations should have major implications for biogeochemistry. However, there is a lack of hard evidence linking biogeochemical cycles to water redistribution by plants, although some of these studies cited above discuss the implications of water redistribution on nutrient cycling.

2.11.3.4 Stream Flow

Stream water integrates the myriad of catchment hydrological and biogeochemical processes. The catchment is a natural spatial domain to study the coupling of hydrological and biogeochemical cycles and simultaneous measurements of catchment discharge (Q) and hydrochemistry capture the integrated catchment hydrological and biogeochemical behavior. Hydrological and biogeochemical processes are tightly coupled at the Earth's surface and hydrological fluxes within catchments provide an opportunity to balance material fluxes at defined spatial and temporal scales. Catchments link the atmosphere, plants, soil surface, subsurface, groundwater, and streams through the convergence and interaction of material and energy flows. Quantifying the variability in the contributions of different water sources to stream flow allows inferences to be drawn about hydrological pathways and biogeochemical processes in these source areas (Peters and Driscoll, 1987b; Hooper *et al.*, 1990; McDonnell *et al.*, 1991). For N export, the magnitude of individual responses to a runoff event appears to be related to soil microbial processes, while seasonal to decadal trends are related to vegetation demand (Bormann and Likens, 1994; Likens and Bormann, 1995). A common observation across mountain catchments is that large fractions of both C and N are exported in stream flow in response to rainstorms and snowmelt runoff (Hood *et al.*, 2006), presumably because of changes in routing through the catchment (McGlynn *et al.*, 2003; Bishop *et al.*, 2004; McGuire and McDonnell, 2006). Shallower soils, higher hydraulic conductivity, and proximity to the stream channel increase the likelihood of increased export of biogeochemically active solutes during runoff events.

2.11.3.4.1 Nitrate isotopes in stream water

The ability to differentiate sources of nitrate in streams has advanced with the application of multiple isotope techniques for $\delta^{15}N$, $\delta^{18}O$, and $\delta^{17}O$ in nitrate (Durka *et al.*, 1994; Kendall, 1998; Burns and Kendall, 2002; Mayer *et al.*, 2002; Fukada *et al.*, 2003; Michalski *et al.*, 2004; Kendall *et al.*, 2007). Nitrate isotopic variations can be related to sources of nitrate, and they also are modified by subsequent reactions such as denitrification and assimilation. Isotope studies emphasizing either source variations or cycling processes may be complicated by these overlapping effects, but they can be useful in some situations. For example, temporal isotopic studies have provided important constraints on the transmission of atmospheric nitrate through watersheds. Oxygen-isotope data indicate that atmospheric nitrate, commonly, is only a minor component of stream nitrate during a range of flow conditions, including snowmelt events, highlighting the importance of nitrification sources in runoff (Burns and Kendall, 2002; Campbell *et al.*, 2002; Buda and DeWalle, 2009). Exceptions include peak flows during storm events, especially in watersheds containing impervious ground cover such as urban areas, where atmospheric nitrate can be a substantial component of stream nitrate (Kendall *et al.*, 2007). Spatially distributed isotopic data in stream networks can provide supporting evidence for varying nitrate sources in different subwatersheds (Mayer *et al.*, 2002; Lindsey *et al.*, 2003). For example, during moderate base-flow conditions in the predominantly agricultural Mahantango WE-38 watershed in Pennsylvania, USA (Lindsey *et al.*, 2003), small streams with low nitrate concentrations and low $\delta^{15}N$ values from forested upland watersheds joined other streams with higher nitrate concentrations and higher $\delta^{15}N$ values draining cropland, whereas a few streams with exceptionally high nitrate concentrations and high $\delta^{15}N$ values drained areas with animal feedlots or pastures. Nitrate apparently was transmitted through the watershed without major isotopic modification after infiltrating through soils.

In contrast to the above, it has been suggested that isotopic indicators could be incorporated into conceptual, analytic, or numerical models of flow, transport, and denitrification to evaluate nitrogen losses within watersheds (Sebilo *et al.*, 2003). This approach requires information or assumptions about the upscaling properties of isotope-fractionation effects from micro- to diffuse scales (scale at which mixing of water and solutes is relatively complete). In principle, the regional-scale status of diffuse denitrification could be evaluated by monitoring $\delta^{15}N_{NO_3}$ and $\delta^{18}O_{NO_3}$ in stream flow to determine the catchment-scale status and dynamics of denitrification. A hypothetical example of such an approach is illustrated in **Figure 5**, which shows predictions of the $\delta^{15}N_{NO_3}$ isotopic behavior of nitrate in stream flow as distributed within an idealized model framework. The stable isotope predictions are based on the riparian nitrate model (RNM), which operates as a filter (plug-in) module within a node-link catchment-scale model (E2); E2 is capable of simulating the hydrological behavior of catchments (Rassam *et al.*, 2006, 2008). The RNM–E2 modeling framework has been augmented with Rayleigh fractionation algorithms at each node within the model domain to track the isotope shifts because of denitrification. The modeling example considers a simple homogeneous riparian-zone soil with decreasing available carbon and related microbial activity with depth and denitrification only occurring in the riparian-zone soils. As denitrification proceeds in the groundwater prior to discharge, the residual nitrate becomes relatively enriched in ^{15}N as nitrate concentration decreases (Marriotti *et al.*, 1988; Kendall, 1998). **Figure 5** illustrates the RNM–E2 simulated dynamics of nitrate concentrations and

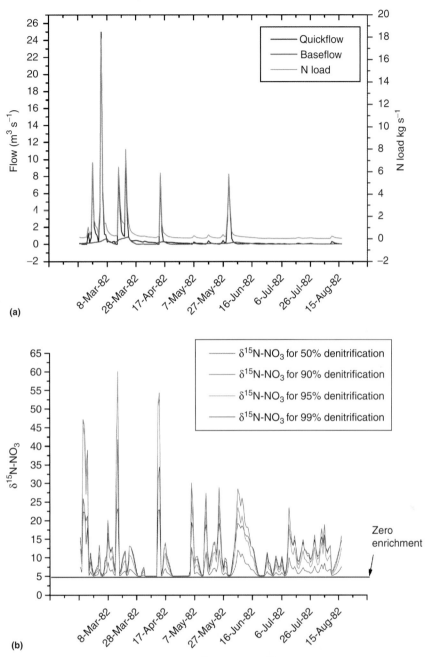

Figure 5 Linkages between stream nitrogen (N) loading, variation of $\delta^{15}N$ of nitrate ($\delta^{15}N$–NO_3) corresponding to varying amounts of dentrification, and hydrology of the Maroochy catchment southeast Queensland, Australia during the wet season 1982; (a) temporal variations in hydrologic response through quick and baseflow components, and N loading; and (b) concurrent predictions of the streamwater $\delta^{15}N$–NO_3. From Rassam DW, Knight JH, Turner J, and Pagendam D (2006) Groundwater surface water interactions: Modelling denitrification and $\delta^{15}N_{NO3}$–$\delta^{18}O_{NO3}$ fractionation during bank storage. In: Institution of Engineers Australia (ed.) *Proceedings of the 30th Hydrology and Water Resources Symposium*, pp. 157–161. Launceston, TAS, Australia, 4–7 December 2006. Sandy Bay, TAS: Conference Design.

$\delta^{15}N$ as the stream flow source switches between base flow and rapid delivery of new water during storm flow. The $\delta^{15}N$ of the stream nitrate increases as the extent of denitrification in the catchment increases. Similar results could be generated for $\delta^{18}O_{NO_3}$ depending on the denitrification fractionation factors used (Böttcher *et al.*, 1990; Fukada *et al.*, 2003; Granger *et al.*, 2008). In these models, isotopic variations in nitrate sources are assumed to be negligible.

2.11.3.4.2 Transit time and residence time

The distributed hydrological response to precipitation encompasses the spatial and temporal variations of water fluxes in landscapes, and, therefore, is directly related to the variability in biogeochemical cycling described earlier. The hydrological response in a mountainous catchment is controlled largely by the near-surface landscape properties (landform and soil characteristics) that function as hydrological filters (Meybeck and

Vörösmarty, 2005); yet, variations are often nonlinear and difficult to represent (Beven and Germann, 1982; Troch et al., 2003). Mean hydrological transit time, or the mean age of water discharging to a stream channel, is an important hydrological descriptor, but its relation to storage, flow pathways, mixing, and sources of water is complex and model dependent (Maloszewski et al., 1983; McGuire et al., 2005; McGuire and McDonnell, 2006; Soulsby et al., 2006). Feedbacks between vegetation structure, soils, biogeochemistry, and landform development result in a range of transit-time distributions throughout a catchment, that is, rates of water movement from the catchment to the channel at various locations along the channel (McGuire and McDonnell, 2006; McDonnell et al., 2007). For example, based on a simple age-distribution model (Tetzlaff et al., 2007), slope was found to be the dominant control of mean transit time in steep Scottish catchments, but soil permeability was a more important control in flat lowland catchments (Tetzlaff et al., 2009). In another study, aspect appeared to be a dominant control on mean transit time, with north-facing slopes having longer transit times than south-facing slopes (Broxton et al., 2009).

From a biogeochemical perspective, the residence time, that is, length of time that water remains in a catchment before it becomes stream flow, has a pronounced effect on potential solute concentrations and export. As water moves from upland areas in the catchment, downslope interactions with other water sources may occur in areas of convergent flow (Anderson and Burt, 1978), promoting physical mixing of waters, exchange of solutes, and increased rates of oxidation–reduction (redox) reactions or hot spots (Fisher et al., 1998; McClain et al., 2003). These hot spots are where high rates of nutrient modifications occur, for example, in riparian zones (Burt and Pinay, 2005), the hyporheic zone, and wetlands (**Figure 6**) with associated chemical transformations (**Figure 7**). Episodic transport during hydrological events, such as snowmelt or heavy rain, can reduce the importance of hot spots by reducing residence time, resulting in increased nutrient flux (Boyer et al., 1997; Stanley et al., 1997).

Even after water has entered a stream channel, its residence time may be greater than that predicted by stream velocity because of hyporheic exchange, that is, the movement of stream water into the subsurface and back to the stream at a location downstream. The typical rough texture of headwater mountain-basin geomorphology (e.g., pool-riffle sequences) often drives exchange of water through the bed and riparian sediments (Harvey and Bencala, 1993; Kasahara and

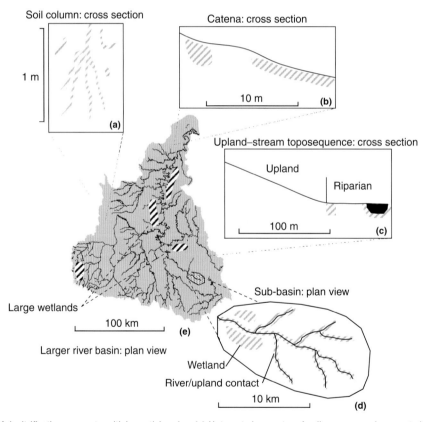

Figure 6 Hot spots of denitrification occur at multiple spatial scales. (a) Hot spots in a meter of soil may occur along root channels where moisture and organic-matter content are high. (b) Topographic depressions that accumulate organic matter and retain moisture may be hot spots within a catena. (c) Along a toposequence from upland to river, the soil–stream interface may represent a hot spot where high-nitrate groundwater intercepts organic-rich soils. (d) At the scale of sub-basins, the occurrence of hot spots may be dictated by the spatial configuration of upland–wetland or upland–river contact zones. (e) The percentage of land occupied by wetlands determines denitrification hot spots at the scale of large river basins. Adapted from McClain ME, Boyer EW, Dent CL, et al. (2003) Biogeochemical hot spots and hot moments at the interface of terrestrial and aquatic ecosystems. *Ecosystems* 6(4): 301–312.

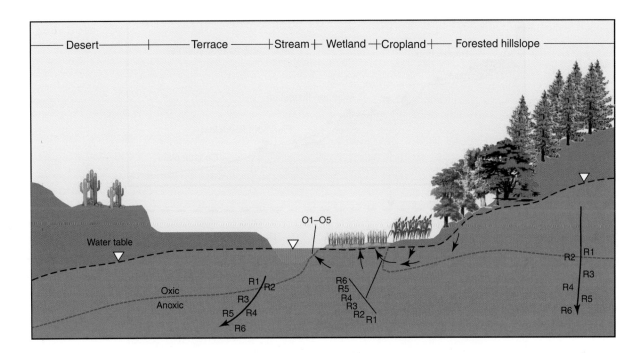
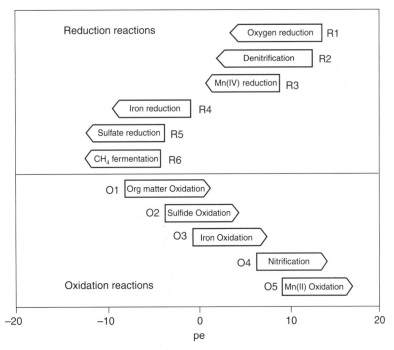

Figure 7 A schematic diagram of some common oxidation and reduction reactions in riparian zones and their relative redox potentials at pH 7. Reaction sequences in waters recharging from, or discharging to, a stream depend on the origin (initial redox status) of the water, the composition of the substrate, and the interaction with biota (bacteria, roots, etc.). Adapted from Dahm CN, Grimm NB, Marmonier P, Valett HM, and Vervier P (1998) Nutrient dynamics at the interface between surface waters and groundwaters. *Freshwater Biology* 40(3): 427–451; and Appelo CAJ and Postma D (2007) *Geochemistry, Groundwater, and Pollution*, 2nd edn., 649pp. Rotterdam: AA Balkema.

Wondzell, 2003). The changing morphology along a river network exerts a strong control on gradients that drive exchange, and therefore, the amount of water that flows through the hyporheic zone (Kasahara and Wondzell, 2003). This exchange not only increases stream water residence time in the basin, but also moves nutrients and other solutes into the subsurface: (1) fueling biogeochemical processes in the shallow subsurface around streams, (2) providing a subsurface habitat that is a mix of surface and groundwater conditions, and (3) generating patches of varying conditions on the streambed in downwelling (where stream water enters the bed) and upwelling (where hyporheic water returns to the

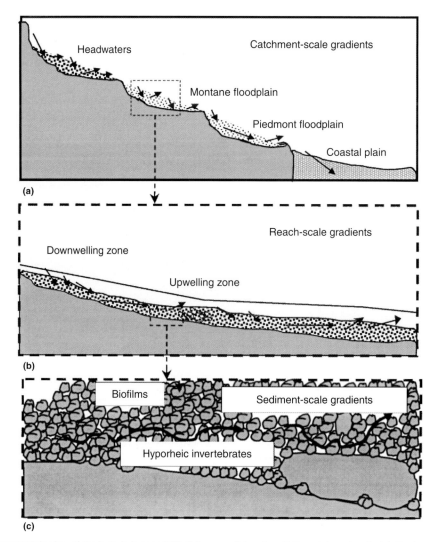

Figure 8 Lateral diagrammatic view of the hyporheic zone (HZ) at three spatial scales. At the catchment scale (a), the hyporheic corridor concept predicts gradients in relative size of the HZ, hydrologic retention, and sediment size (126). At the reach scale (b), upwelling and downwelling zones alternate, generating gradients in nutrients, dissolved gases, and subsurface fauna. At the sediment scale (c), microbial and chemical processes occur on particle surfaces, creating microscale gradients. Arrows indicate water flow paths. From Boulton AJ, Findlay S, Marmonier P, Stanley EH, and Valett HM (1998) The functional significance of the hyporheic zone in streams and rivers. *Annual Review Ecology and Systematics* 29: 59–81.

channel) locations (**Figure 8**). Thus, hyporheic exchange is an important hydrological process that directly contributes to many biological functions such as macroinvertebrate population dynamics (Boulton *et al.*, 1998; Malard *et al.*, 2002) and salmonid spawning (Baxter and Hauer, 2000), and biogeochemically important reactions (**Figure 5**) such as denitrification and retention of dissolved organic carbon (DOC; Baker *et al.*, 1999), water-temperature buffering (Arrigoni *et al.*, 2008), and buffering of heavy metal transport (Fuller and Harvey, 2000).

2.11.3.5 Groundwater Recharge

Groundwater recharge (R) is directly linked to mineral weathering rates through the delivery of water, DOC, and CO_2 to the subsurface; the export of weathering products to surface water; and CO_2 release to the atmosphere (Richey *et al.*, 2002). Consequently, groundwaters with longer transit times typically contain lower DOC concentrations because of microbial degradation, and higher concentrations of conservative alkalinity-associated DIC (Szramek and Walter, 2004), base cations, and silica (Rademacher *et al.*, 2001). Climate affects water–rock interactions. For example, increasing temperature and high rainfall tend to increase rates of chemical weathering (White and Blum, 1995; White and Brantley, 2003). In addition, weathering is controlled by lithology (Meybeck, 1987; Bluth and Kump, 1994; Meybeck and Vörösmarty, 2005). In the long term, linkages or feedbacks between biota and earth materials have modified the near-surface environment of Earth or 'critical zone' (Brantley *et al.*, 2007), and, in turn, the chemistry of streams and rivers shows the evidence of biological processes (Amundson *et al.*, 2007). Groundwater, including recharge and discharge, is an important topic and is presented in a separate section, and is also discussed in each of the following sections where it interfaces with the main topic of the section.

2.11.4 Within-River Processes

Biogeochemical processes within streams and rivers can result in high rates of nutrient uptake, chemical transformation, and release of dissolved materials to water. These biogeochemical processes are largely the result of organisms, such as bacteria and fungi, algae, and higher plants, attached to hard surfaces or to organic and inorganic sediments on the streambed. In larger rivers, biogeochemical processes associated with suspended algae and microbes attached to suspended particles can be important. These processes can significantly alter the flux and chemical form of several biologically active solutes, particularly carbon, nitrogen, and phosphorus.

Most of the organisms responsible for biogeochemical processes in streams and rivers are stationary, associated with the streambed, yet solutes taken up or released to water are under strong advective forces of downstream water flow. Thus, nutrient cycling has a distinctive spatial or longitudinal component along the axis of stream flow. Biological communities generally change from dominance of shredders, associated with leaf litter and woody debris accumulations in turbulent, higher-velocity streams in headwaters, to collector–gatherers in larger more quiescent streams with lower gradients downstream. Nutrient spirals tend to lengthen from upstream where streambed-surface:water-volume ratios tend to be high to downstream where surface:volume ratios are lower (**Figure 9**). The concept of nutrient spiraling was proposed as a framework to study nutrient cycling in streams and explicitly considers the simultaneous processes of biological uptake, transformation, or remineralization and hydrological transport downstream (Webster and Patten, 1979; Newbold et al., 1983). As an example of the nature of these reactions and interactions, a schematic of inorganic nitrogen transformations and interactions with streambed biota and hyporheic zone is shown in **Figure 10** (Peterson et al., 2001). Several metrics quantifying nutrient spiraling have been defined, including uptake length – the average distance traveled by a nutrient atom in water from where it enters the stream to where it is taken up by biota. Methods for field measurement of uptake length have been developed, including the experimental addition of isotopic nutrient tracers, such as ^{15}N for studies of nitrate and ammonium uptake and denitrification (Newbold et al., 1981; Stream Solute Workshop, 1990; Mulholland et al., 2000; Peterson et al., 2001; Böhlke et al., 2004; Mulholland et al., 2004; Böhlke et al., 2009).

Distinctive temporal patterns characterize within-river biogeochemical processes. These patterns are related to seasonality in biological processes and in hydrology. Biological seasonality is largely controlled by the regulation of inputs of light and organic matter by terrestrial ecosystems bordering streams and rivers. In streams draining deciduous forests, nutrient-uptake rates often have two peaks each year: (1) early spring prior to leaf-out when uptake and growth rates of attached algae are high because of high light levels reaching the stream and (2) autumn after leaf-fall when uptake rates by bacteria and fungi associated with leaf decomposition are high (Roberts and Mulholland, 2007). In small, heavily shaded streams, summer is often a period of relatively low rates of nutrient uptake, despite high water temperatures, because algal growth is severely limited by low light levels and activity of heterotrophic microbes is limited by the lack of easily decomposable organic matter in the streambed. However, streams that do not have dense forest canopies (e.g., those in more arid climates and agricultural landscapes) often have late spring or summer peaks in nutrient uptake and cycling because light levels are high and stream flows are more stable (Arango et al., 2008). Late spring and summer peaks in nutrient uptake may also be common in larger rivers because of high light levels and longer water-residence times under lower flows that permit development of larger communities of algae in the water column; however, little is known about rates and temporal dynamics of nutrient uptake in large rivers, that is, with drainage areas greater than 250 000 km².

There is considerable spatial variation in within-river biogeochemical processes, largely controlled by the hydrological regime, channel morphology, catchment land use, and characteristics of stream-bank (riparian) vegetation (Tabacchi et al., 2000). For example, desert streams in monsoonal climates can develop very high rates of nutrient uptake and cycling as algal communities develop during long periods of low, stable stream flow after the monsoon season ends (Grimm et al., 2005). In addition, within-river seasonal variations in macrophyte growth and related geomorphology can affect sediment respiration, nitrification, and denitrification rates (Duff et al., 2002). Furthermore, Duff et al. (2002) showed that as the rooted aquatic macrophyte communities matured, pore water became chemically reduced and nutrient levels increased by one to two orders of magnitude above background in the root zone. These levels were significantly higher than those found in either groundwater or surface water, indicating that streambeds can serve as a nutrient reservoir. Streams with very flashy hydrographs tend to have lower rates of biogeochemical processes because attached organisms are frequently scoured from the streambed during high flows. Cross-site studies have been particularly valuable for identifying broad-scale controls on nutrient uptake and cycling. Stream discharge is often the strongest predictor of nutrient-uptake length in streams, with longer uptake lengths (lower rates of uptake relative to downstream transport) under higher discharge (Peterson et al., 2001; Marti et al., 2004). In relatively undisturbed catchments, land use and riparian vegetation are also important determinants of nutrient uptake, but indirectly via their effects on light regime and stream primary productivity and nutrient inputs (Tabacchi et al., 2000; Hall et al., 2009). Stream algae and microbes are able to increase rates of uptake with increasing nutrient concentrations, although they become somewhat less efficient at removing nutrients from water as concentrations increase (Dodds et al., 2002). As discussed previously, subsurface (hyporheic) zones within streambed sediment accumulations are important sites for biogeochemical processes (Dahm et al., 1998). Water exchange between surface and subsurface zones and between the main channel and backwaters is an important mechanism for increasing rates of nutrient cycling (Triska et al., 1989; Jones and Holmes, 1996), and particularly for rates of denitrification (Duff and Triska, 1990; McMahon and Böhlke, 1996; Mulholland et al., 2009).

In-stream processes can be important for the retention of nutrients within river networks and landscapes, thus reducing the potential for eutrophication and harmful algal blooms in

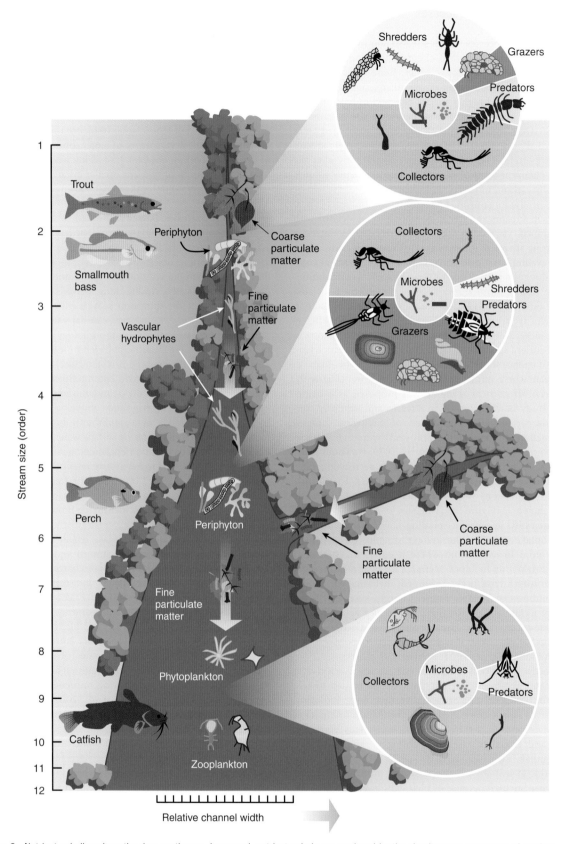

Figure 9 Nutrient spiraling along the river continuum. In general, nutrient spirals are produced by the simultaneous processes of nutrient cycling (uptake from water by biota and subsequent release back to water) and downstream transport. Adapted from Vannote RL, Minshall GW, Cummins KW, Sedell JR, and Cushing CE (1980) The river continuum concept. *Canadian Journal of Fisheries and Aquatic Sciences* 37(1): 130–137.

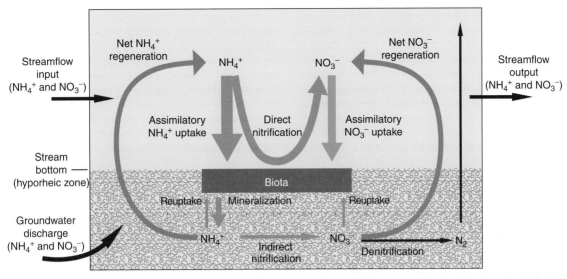

Figure 10 Simplified schematic diagram of selected processes affecting inorganic N cycling in streams. Biota includes bacteria distributed within the sediments. Although not shown, N species can transfer between stream and hyporheic zone without interacting with biota. Modified from Peterson BJ, Wollheim WM, Mulholland PJ, et al. (2001) Control of nitrogen export from watersheds by headwater streams. Science 292(5514): 86–90.

estuaries and coastal waters. Streams receiving relatively low nutrient inputs can retain more than half of their inputs over a 1-km stream length (Peterson et al., 2001). In addition, increasing nitrogen retention by in-stream processes in landscapes recovering from past disturbances has been reported (Bernhardt et al., 2005). Nutrient cycling is a serial process in streams and rivers, and large cumulative stream length results in long residence times and potentially high rates of nutrient retention in river networks (Wollheim et al., 2006; Alexander et al., 2007; Mulholland et al., 2008). Although small, shallow streams in the network can have particularly high rates of nutrient uptake because of high surface:volume ratios (Alexander et al., 2000), nutrient uptake can also be high in larger streams and rivers (Ensign and Doyle, 2006; Alexander et al., 2007; Tank et al., 2008). However, high nutrient inputs can overwhelm the capacity of biological-uptake processes within streams, resulting in low retention efficiency and large losses to downstream ecosystems (Mulholland et al., 2008; Böhlke et al., 2009).

Humans have had large impacts on within-river biogeochemistry. Agriculture and urbanization have been among the most widespread effects on streams. Agriculture often results in stream channelization and other modifications that reduce geomorphologic and hydrodynamic complexity and organic-matter storage and thus the rates of biogeochemical processes; in effect, farmland streams become more of a conduit and less of a barrier to runoff (Burt and Pinay, 2005) and this loss of landscape structure may account, for example, for persistently high concentrations of nitrate in river water (Burt et al., 2008). However, during extended periods of low stable flow in summer, nutrient uptake rates can be high because of high level of light and nutrient concentrations (Bernot et al., 2006). Urbanization can also result in substantial changes to in-stream biogeochemical processes because of many of the same morphological, hydrodynamic, and organic-matter impacts as agriculture, although greater extent of impervious surfaces can result in much flashier urban-stream hydrographs (Boyd et al., 1993; Finkenbine et al., 2000; Rose and Peters, 2001). Rapid runoff or flashiness generally reduces nutrient-uptake rates in urban streams (Paul and Meyer, 2001; Groffman et al., 2004; Meyer et al., 2005; Walsh et al., 2005) and riparian zones (Groffman et al., 2002). In contrast, modifications to stream networks in heavily urbanized areas, such as detention basins and artificial lakes, can enhance nutrient uptake and retention (Grimm et al., 2005).

2.11.5 Wetland Processes

Wetlands are defined differently by country and agency of use, but all definitions recognize the persistence of near-saturated conditions at or above the mineral sediments, hydric soils, and the presence of plants adapted for generally saturated conditions (Mitsch and Gosselink, 2007). Wetlands occur in all climatic and geographic regions of the world but they are more prominent in regions where precipitation exceeds potential evaporation and topographically flat areas where drainage rates are slow. However, given a supply of water, for example, rivers, streams, and groundwater, they can occur even in the most arid regions. The dominant physical factor of wetlands is the presence and persistence of near-saturated conditions. Whether the wetland is in the tropics or the high Arctic, waterlogged conditions are a function of the hydrology of the catchments in which the wetland is located, but the presence of a wetland also alters the hydrology of a catchment. Some wetlands provide temporary storage attenuating flood peaks and sustaining flows during drier periods (Mitsch and Gosselink, 2007); however, wetlands that also accumulate partially decomposed plants, for example, peat, have a poor ability to attenuate runoff because of their hydraulic properties (Bay, 1969; Holden and Burt, 2003; Holden et al., 2006).

The role of wetlands in the hydrology and biogeochemistry of catchments is determined, in part, by the position of the wetland in the landscape. Given topographic depressions and an excess of precipitation over evapotranspiration, wetlands can exist in the headwaters of catchments, in valley floors, or on plateaus adjacent to the drainage divides. While these wetlands might seem isolated from the catchment, their outflow provides water downstream, particularly during periods of storm runoff (Holden and Burt, 2003). The hydrological interaction between streams and wetlands can be very complex, particularly for wetlands adjacent to streams in higher-order drainage basins with low stream gradients. In particular, groundwater can become important in creating saturated conditions that are needed to sustain wetlands (Winter and Woo, 1990). The isolated wetland's hydrology is mainly a function of precipitation and antecedent water storage, making runoff strongly event related (Verry *et al.*, 1988; Evans *et al.*, 1999). In contrast, wetlands that are well connected to groundwater can have a less variable storage dynamic on the short term, for example, wetlands located at break points on hillslopes where they receive groundwater discharge (Winter, 1999).

The magnitude and rate of water input to a wetland relative to the wetland's storage capacity control residence time, and residence time plus the mixture of biological and chemical reactants determine the biogeochemical dynamics. For example, valley-bottom riparian wetlands can receive groundwater discharge that far exceeds precipitation input and this leads to relatively constant levels of water storage and rapid turnover of stored water (Roulet, 1990). Residence time also can be greatly increased by near-shore processes that effectively cycle wetland water from surface water to groundwater and back again. This is because of evapotranspiration (ET) drawing the water table below wetland stage, reversing hydraulic gradients, and allowing wetland water to flow into groundwater troughs that ring the wetlands, a common characteristic of prairie wetlands (Rosenberry and Winter, 1997). In contrast, water exchange in some wetlands, such as raised peat bogs, is confined to a thin hydrologically active layer near the surface (Ingram, 1978) and leads to a two-compartment system – one compartment with a short residence time (hours to days) and the other compartment a deeper groundwater with long residence times (hundreds of years) (Fraser *et al.*, 2001). Water-exchange rates are an important factor affecting wetland biogeochemistry because they affect the magnitude and supply rate of chemical inputs (e.g., DO, SO_4^{2-}, and NO_3^-) relative to the supply of reactants (e.g., Fe^{2+}, Mn^{2+}, NH^{4+}, organic matter, and microbes) in the wetland. For example, headwater and isolated wetlands receive their chemical inputs from precipitation alone; hence, they tend to be nutrient-poor systems, and if they accumulate peat, the surface vegetation can become isolated from the source of minerals in the underlying substrate and result in acidic conditions (Damman, 1986; Wilcox *et al.*, 1986). In contrast, wetlands that receive water that has contacted other land covers and soils, such as marshes, valley-bottom swamps, or prairie pot holes, can receive a large influx of nutrients and cations, which can result in mineral and nutrient-rich productive ecosystems (Bedford *et al.*, 1999).

In wetlands, the temporal and spatial variations in chemical composition and biogeochemical transformations can be evaluated using chemical thermodynamics with a particular emphasis on redox conditions (Hedin *et al.*, 1998). Because of saturated conditions, oxygen is limited in most wetlands. The diffusion of oxygen in water is 10 000 times slower than that in air. Consequently, if there are processes that consume oxygen, such as decomposition of organic matter, the wetland substrate becomes progressively more reduced. In wetlands with short water-residence times, that is, the water is renewed regularly, and wetlands where the surface is flooded, oxic conditions prevail, at least near the surface; biogeochemistry of these wetlands is dominated by oxygen, nitrate, and iron reduction. However, when residence times are longer and oxygen consumption exceeds supply, wetlands become progressively more reduced, and sulfate reduction and eventually methanogenesis become common (Reddy and DeLaune, 2008). In addition to the importance of hydrologic fluxes of oxygenated waters for oxygen input to wetlands, wetland plants oxygenate their roots, and differences in the ability of wetland plants to aerate their submerged tissues under different flooding regimes play a major role in controlling plant distribution (Sorrell *et al.*, 2000; Pezeshki, 2001).

Temporal and spatial variations in saturation are important in controlling temporal and spatial dynamics of wetland biogeochemistry. Spatial and temporal dynamics of the inputs and the distribution of plants are also important controls on wetland biogeochemistry. The wetland setting is based on hydrology, the wetland salinity is based on climate, and biogeochemical response is based on a combination of the two.

Recently, biogeochemists have begun to refer to locations on the landscape that show steep redox gradients, which have high biogeochemical transformation rates, as hot spots and the times when redox conditions change quickly at a location as hot moments (McClain *et al.*, 2003). This conceptualization works well across many scales in wetland settings. For example, in unsaturated wetland sediments, the interface between saturated pores and adjacent air-filled pores could be a hot spot for chemical transformations. At a larger scale, transformations from oxidizing conditions upgradient of a riparian wetland to reducing conditions within a wetland are also likely hot spots. A hot moment may occur when the wetland water table rises rapidly during a hydrological event resulting in a large decrease in oxygen availability.

Wetland ecosystems can also be viewed as a wetland continuum (Euliss *et al.*, 2004), similar to the stream-continuum concept (Vannote *et al.*, 1980). The concept places wetlands in two dimensions (**Figure 11**); one in relation to groundwater interaction, that is, recharge and discharge, and the other with respect to climate condition, that is, atmospheric water, from dry or drought conditions to wet or flood/deluge conditions. The wetland continuum provides a framework for organizing and interpreting biological data by incorporating the dynamic changes these systems undergo as a result of normal climatic variation.

There are many examples of specific linkages between biogeochemistry and hydrology in wetlands. In almost all settings, carbon availability is the main driver of wetland biogeochemistry. This is reflected in the large accumulation of organic matter commonly observed in wetland soils, and in the importance of wetlands as a major source of DOC downstream (Hinton *et al.*, 1998; Freeman *et al.*, 2001; Worrall *et al.*, 2004; Roulet *et al.*, 2007; Nilsson *et al.*, 2008) and

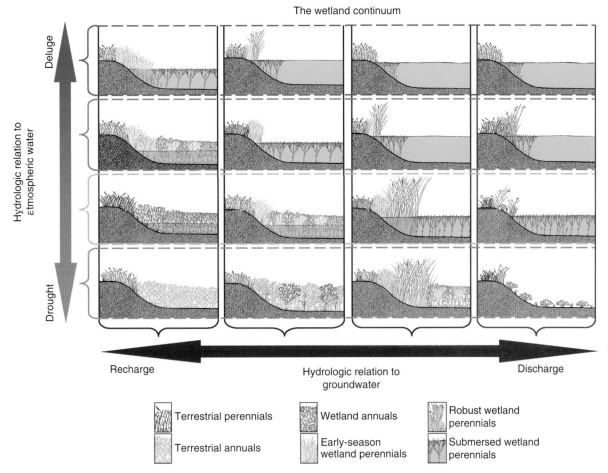

Figure 11 The wetland continuum, a wetland classification based on hydrology with respect to groundwater interactions from recharge to discharge and climate conditions from drought to deluge. Potential plant communities in wetlands at four discrete points along this axis are depicted. Adapted from Euliss NH, Jr., LaBaugh JW, Fredrickson LH, *et al.* (2004) The wetland continuum: A conceptual framework for interpreting biological studies. *Wetlands* 24: 448–458.

emissions of CO_2, methane (CH_4), and N gases (N_2 and N_2O). Wetland CH_4 and N_2O fluxes are extremely high compared to those from other landscapes (Bowden, 1986; Robertson, 2001; Svensson *et al.*, 2001; Rosenberry *et al.*, 2006). Wetland CH_4 fluxes have been linked to water-table depths and temperature (MacDonald *et al.*, 1998), which is the basis for a one-dimensional methane flux model (**Figure 12**) for natural wetlands (Walter and Heimann, 2000; Walter *et al.*, 2001).

Wetlands can accumulate carbon in the form of peat deposits and export water with high DOC concentrations (Moore, 2003); therefore, area of wetlands is often strongly correlated with rates of DOC export (Dillon and Molot, 1997; Xenopoulos *et al.*, 2003). However, this correlation is not universal even in landscapes with significant wetland cover (Frost *et al.*, 2006). Even in catchments where wetlands are only a small fraction of the catchment area, they can still have a profound effect on catchment biogeochemistry producing relatively high DOC concentrations in drainage waters, because they are often the last point of contact before the water enters a stream or river (Bishop *et al.*, 2004). In this case, the role of the wetland can be quite dynamic depending on temporal variations in the hydrological connection of the wetland with the adjacent hillslope (McGlynn *et al.*, 2003; Burt and Pinay, 2005).

Wetlands can be more effective in removing nutrients from circulating waters than lakes or rivers (Saunders and Kalff, 2001), and retaining constituents from atmospheric deposition, but removal is not a universal conclusion. Riparian wetlands, in particular, can be quite effective in removing and retaining nutrients and nitrogen in particular (Jansson *et al.*, 1994), but the dynamics and hydrological setting play a critical role in the effectiveness of nutrient removal (Cirmo and McDonnell, 1997). Nutrients retained under one condition, such as periods when the water table is receding, can be rapidly mobilized upon rewetting, and the retention can be much less throughout the year based on an event or on a seasonal basis (Devito *et al.*, 1989).

Wetlands, through dynamic coupling with uplands and the atmosphere, can sometimes act as biogeochemical hot spot sources instead of sinks. For example, wetlands are sources of methyl-mercury (St. Louis *et al.*, 1996; Babiarz *et al.*, 1998), and the hydrological coupling of wetlands to upland sources of water (Branfireun *et al.*, 1996; St. Louis *et al.*, 1996; Galloway and Branfireun, 2004) and sources of sulfur, either through hydrological input or by atmospheric deposition

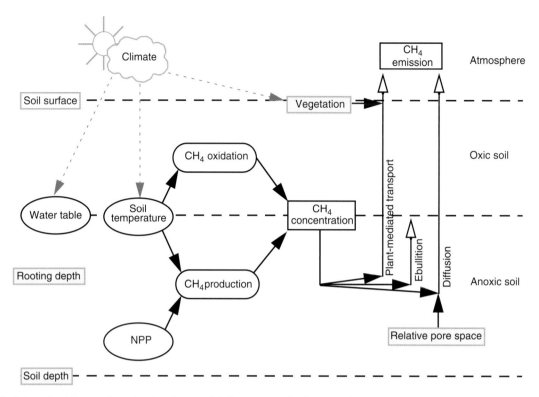

Figure 12 Schematic of the one-dimensional methane model. The processes leading to methane emission to the atmosphere occur in the soil between soil depth and soil surface. Methane production takes place in the anoxic soil below the water table; the methane production rate depends on soil temperature and net primary productivity (NPP). Methane oxidation occurs in the oxic soil above the water table and depends on temperature. The model calculates methane concentrations in each (1 cm thick) soil layer. Transport occurs by diffusion through water/air-filled soil pores, ebullition to the water table, and plant-mediated transport from layers above the rooting depth. From Walter BP, Heimann M, and Matthews E (2001) Modeling modern methane emissions from natural wetlands, 1. Model description and results. *Geophysical Research Letters* 106(D24): 34189–34206.

(Branfireun *et al.*, 1999), is a critical factor. In addition, the reducing conditions in wetlands are active areas for redox reactions involving iron, aluminum, and manganese.

In summary, the key link between hydrology and biogeochemistry in wetlands originates from the dynamics of water storage and related residence time on the supply of oxygen and chemical reactants.

2.11.6 Lakes

Figure 13 shows the major sources, transformations, and sinks of nutrients in lakes. There are three major initial points of entry of nutrients into lakes: surface-water inflows, groundwater inflows, and atmospheric inputs, including wet and dry deposition. Surface-water inflows may include both point sources and nonpoint sources, while atmospheric inputs and most groundwater contributions are nonpoint sources. Occasionally, other nonpoint sources of nutrients, such as guano from abundant bird populations (O'Sullivan, 1995), direct wastewater or pollutant inputs, or salmon-spawning migrations, may make significant contributions to the total lake-input nutrient load (Naiman *et al.*, 2002). In the latter case, the loss of sockeye salmon cadavers from some oligotrophic lakes in British Columbia, Canada, as a result of human effects on spawning runs, was determined to have significantly affected the total load of phosphorus (Stockner and MacIsaac,

1996). An active lake-fertilization program was used in several of these lakes to stimulate primary productivity and initiate a trophic cascade that would ultimately provide improved growth conditions for juvenile salmon (Stockner and MacIsaac, 1996). While reduced levels of primary production in lakes – sometimes termed (re) oligotrophication – may occasionally occur in highland regions or in response to major nutrient load-reduction strategies, the recent history of human influences on lakes has been characterized by nutrient concentrations that are substantially higher than natural or background concentrations. These high nutrient concentrations have led to eutrophication and some of its undesirable consequences such as harmful algal blooms, oxygen depletion of bottom waters, and, occasionally, fish kills.

The species and transformations of nutrients around the point of entry of an inflow to a lake may be highly important in determining the biogeochemical effects of the inflow. Nutrients in groundwater entering a lake may pass through alternating zones of oxidizing and reducing conditions – the latter are often prevalent in the organic-rich sediments deposited in sheltered areas of the lake bed (Schuster *et al.*, 2003). This heterogeneous environment can stimulate a diverse microbial flora associated with the rich array of microenvironments and redox conditions. Reducing conditions can (1) stimulate denitrification of nitrate when this nutrient is present, (2) lead to a buildup of ammonium as nitrification is inhibited, (3) result in dissolution of phosphate

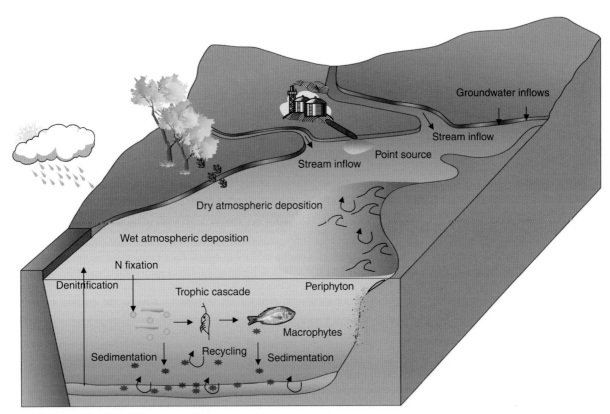

Figure 13 Sources, sinks, and transformations of nutrients in a lake (not including outflows).

in association with reduction of oxidized forms of iron and manganese, and, (4) under strongly reducing conditions, may lead to sulfate reduction and methanogenesis (Reeburgh, 1983). Thus, the rich microbial consortia associated with the land–water transition can strongly influence the availability and species of nutrients, including carbon, nitrogen, and phosphorus.

Processes in the littoral zone of lakes also can profoundly alter the composition of surface inflows. This zone may support a rich benthos and can include emergent and submerged macrophytes and periphyton that may have a large demand for nutrients. Dense beds of submerged macrophytes can also create quiescent conditions suitable for sedimentation of particulate forms of nutrients and associated inorganic sediments (Madsen et al., 2001). The effects of large surface inflows may be strongly dependent on their water-column insertion depths, that is, the depth at which a surface inflow enters the water column of the lake because of density differences, which in turn are regulated by the relative temperature of the inflow and water column, or occasionally also by the salinity or sediment concentration of the inflow (see Chapter 1.08 Managing Agricultural Water). In the case of a thermally stratified lake, an inflow that is warmer than lake water into which it intrudes will promote a surface overflow in which inorganic nutrients in the inflow will be available in the pelagic zone to the resident microscopic suspended plants (i.e., phytoplankton). An inflow that is cooler and therefore denser than water throughout the entire water column of the lake is likely to create an underflow, in which nutrients may not be immediately available to the phytoplankton resident in the photic zone. When these inflows are large relative to the lake volume, they can have important secondary effects such as oxygenation of bottom waters (Hamilton et al., 1995). Many lake inflows have a temperature intermediate between those of the surface and bottom of the water column, and therefore create an interflow that can propagate horizontally at discrete depths in stratified lakes. Thus, nutrients, suspended sediments, contaminants, and microbes (notably pathogens) that are often present at much higher concentrations in storm flow, may be rapidly dispersed across a stratified lake in an interflow (Chung et al., 2009).

In very simple terms, a box-type model can be used to describe steady-state concentrations of nutrients in a lake in which losses are because of sedimentation and outflows. This type of model has been widely used to describe lake-water concentrations of phosphorus (TP_{lake}), but is not commonly used for nitrogen because atmospheric transformations (i.e., N fixation and denitrification) and related gaseous transfers between the lake and the atmosphere are more difficult to quantify. Vollenweider (1969) was first to apply the model in the form

$$TP_{lake} = \frac{L}{\bar{z}(\rho + \sigma)} \qquad (2)$$

where L is the areal loading rate of TP, $\bar{z}$ is the mean lake depth, ρ is the hydraulic flushing rate given by the inflow discharge divided by the lake volume (i.e., $\rho = 1/\tau_w$) where τ_w is the residence time, and σ is a first-order decay rate for TP to account for sedimentation losses. Calculations are typically based on annualized values. Various extensions and simplifications of this model have been made, for example, σ can be approximated by $10\,(m)/\bar{z}$, and Vollenweider and Kerekes (1982) produced the following modified model:

$$\text{TP}_{\text{lake}} = 1.55 \frac{\text{TP}_{\text{in}}}{\left(1 - \sqrt{\tau_w}\right)^{0.82}} \quad (3)$$

where TP_{in} is the inflow total phosphorus concentration. Equation (3) is based on data for 87 lakes, which show hydraulic flushing rate (or residence time) to be the primary factor driving differences in total phosphorus concentrations between the inflows and the lake. Many, mostly empirical, relationships in turn have been used to derive annual mean and peak concentrations of phytoplankton chlorophyll a and primary production as well as Secchi disk transparency from TP_{lake} (Vollenweider and Kerekes, 1982).

The focus on predictive models for total phosphorus reflects a long-held paradigm that phosphorus generally limits productivity in freshwater ecosystems (Likens, 1972; Schindler et al., 2008) and that increases in this nutrient can contribute to lake eutrophication (Lean, 1973). An argument has been made that a shortfall in nitrogen supply compared with the requirements for balanced growth – the Redfield ratio (Sterner and Elser, 2002) – can be compensated for by heterocystous cyanobacteria (blue-green algae) that will fix atmospheric nitrogen (i.e., N_2 dissolved in water) when this nutrient becomes limiting. Earlier work on this subject (Lean, 1973; Smith, 1983) has remained topical but has recently been reignited by Schindler et al. (2008) who declared that controlling nitrogen inputs alone could exacerbate eutrophication by increasing the dominance of N-fixing cyanobacteria and the probability of harmful algal blooms. In contrast, recent research suggests that human activities have changed biogeochemical dynamics (Bergström and Jansson, 2006; Elser et al., 2009a, 2009b) and that phytoplankton biomass yield in most of the lakes in the Northern Hemisphere was limited by N in their natural state. Furthermore, there are advocates for control on both nitrogen and phosphorus loads (Lewis and Wurtsbaugh, 2008) on the basis that N fixation often fails to compensate sufficiently for N limitation in lake phytoplankton, that experimental systems manipulated with additional nutrients have often been found to be similarly controlled by N and P, and that high background loads of P that saturate demand necessarily dictate that another nutrient will be limiting (Lewis et al., 2008). Debate about the relative merits of N versus P control will surely continue into the foreseeable future, but considerations should also be given to the connectivity of inland waters to estuarine and coastal waters for which N limitation is the norm. In some cases, silica has been reported as a limiting nutrient for diatom production (Tilman et al., 1982). In contrast, only rarely is inorganic carbon supply limiting to primary production. These cases may arise for lakes that are poorly buffered and where high rates of photosynthesis remove carbon dioxide and raise pH to a level where bicarbonate (pH $\approx$ 7–9) or even carbonate (pH $\approx \geq 10$) predominate; the latter form is not available to plants, while only some plants can take up bicarbonate (Wetzel, 2001).

The models given by Equations (2) and (3) do not reveal the mechanisms by which nutrients are regenerated and transformed within lakes. As a result of surface wave action, internal waves, or unidirectional currents, the lake bed may be subject to water motions that can disturb pore water and resuspend particulate material, thus increasing concentrations of sediments and nutrients in the overlying water column (Hamilton and Mitchell, 1997). Dissolved nutrients may also be recycled from bottom sediments to the water column as a result of concentration gradients between these two media. These gradients may be enhanced by reducing conditions in the bottom sediments and sometimes in bottom waters, which result in dissolution of iron- or manganese-bound forms of phosphorus and a buildup of ammonium as nitrification shuts down. The pioneering work of Mortimer (1941, 1942) pointed to the key role of dissolved oxygen (DO) and redox potential in waters overlying the sediments. Thus, conditions that stimulate sediment nutrient releases through the benthos also control the benthic macroinvertebrate communities, which burrow deeply into layered sediments and accelerate nutrient cycling through bioturbation and fecal production (Covich et al., 1999). DO concentrations in bottom waters of deep lakes are closely linked to the availability of labile organic matter, the duration of density stratification, and the pool of DO in the bottom waters. Lakes may thermally stratify for periods of minutes to years, creating vertical density gradients that persist only temporarily in shallow lakes, seasonally (monomictic or dimictic lakes) in deeper systems, and not at all in permanently ice-covered lakes or at high altitude (amictic lakes) (Lampert and Sommer, 2007). These mixing patterns, of which there are several variants, dictate the renewal periods of oxygen to bottom waters and therefore play a key role in determining nutrient-release rates from bottom sediments based on their oxidation status. A high rate of supply of organic matter to the bottom waters (hypolimnion) of a stratified lake can completely remove DO from this layer and occurs in deep, eutrophic lakes. In contrast, in oligotrophic lakes that mix seasonally, DO generally remains present in bottom waters between periods of mixing when oxygen is renewed to levels approximating saturation, and nutrient-regeneration rates from bottom sediments are markedly lower than in deep eutrophic lakes. Eutrophic lakes in which the hypolimnion becomes anoxic may support high rates of denitrification and can also emit considerable quantities of both methane and nitrous oxide (Seitzinger et al., 2006). Another situation relevant to eutrophication is when iron and manganese sequester phosphorus under oxic conditions, removing more P to the sediments than in lakes without high iron and manganese concentrations (Dean et al., 2003). Relatively high groundwater discharge to the lake from lithologies containing high amounts of iron and manganese can produce high concentrations of iron and manganese in the lake, that is, a combined lithology and hydrological control on P removal. Redox-sensitive species transformations of iron and manganese mean, however, that under anoxic conditions, P may be released with dissolution of iron and manganese from inorganic sediments.

A key driver of the fluxes of organic matter to the deeper waters and sediments of lakes is the phytoplankton productivity of surface waters, as organic material synthesized in the photic zone eventually falls into bottom waters and sediments if it is not oxidized during settling or removed via outflows. In stratified lakes, most production of phytoplankton biomass occurs in the surface mixed layer (epilimnion) although it can occur at greater depths in oligotrophic lakes with high water clarity. Most nutrients in the epilimnion occur either as organic forms within the biomass, including not only phytoplankton, but also other microorganisms (e.g., bacteria and fungi) and higher trophic levels (e.g., zooplankton or fish), as well as in dead organic matter (detritus) and dissolved nutrient species. Occasionally, in very turbid waters with high concentrations of inorganic suspended solids, concentrations of phosphorus in particulate inorganic form may be the dominant component of the total phosphorus concentration (Grobblelaar and House, 1995). Heterotrophic microorganisms (mostly bacteria and fungi) recycle organic nutrients into forms that can be taken up again by autotrophs. Generally, only a very small fraction of the total nutrients is in a bioavailable inorganic form. The concentration and nature of nonliving particulate organic matter critically influence rates of primary production by controlling rates of regeneration of inorganic nutrients. In large and/or eutrophic lakes, much of this organic matter is generated within the lake itself (autochthonous production), while in smaller lakes, organic matter within the lake may be heavily subsidized by external inputs from the catchment (allochthonous production). Thus, the species and concentrations of nutrients in a lake vary from the interplay of many complex processes, including loading rates, mixing and stratification, redox-associated transformations, and uptake and partitioning of nutrients through the biota.

2.11.7 Groundwater

Groundwater can have various meanings depending on the context and timescale of interest. This section is mainly concerned with the part of the natural hydrological cycle beginning at the top of the saturated zone where groundwater recharge occurs and ending in a discharge area where groundwater becomes surface water (**Figure 14**). Timescales of groundwater movement from recharge to discharge range from minutes (e.g., near-stream response to a rainstorm) to millions of years (e.g., fossil groundwater beneath an arid landscape). The biogeochemistry of groundwater is driven by abiotic and microbially mediated reactions that in part result from the physical transport of aqueous reactants into contact with subsurface materials with which they are not in equilibrium. In this way, groundwater movement and biogeochemistry affect the development and distribution of microbial communities in the subsurface. In turn, groundwater is an important route for delivery of water and solutes, including nutrients and toxins, from the land surface to streams with a range of residence times and chemical compositions that are different from those of surface runoff. The relative proportions of runoff, and shallow and deep groundwater discharge, can change at various time scales, causing marked changes in stream chemistry as a function of flow. Groundwater controls on stream chemistry may be at least as important as in-stream biogeochemical controls in many situations.

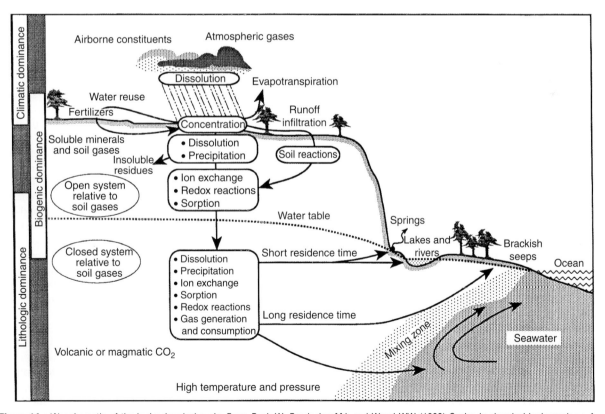

Figure 14 (A) schematic of the hydrochemical cycle. From Back W, Baedecker MJ, and Wood WW (1993) Scales in chemical hydrogeology: A historical perspective. In: Alley WM (ed.) *Regional Ground-Water Quality*, pp. 111–129. New York: van Nostrand Reinhold.

Groundwater recharge beneath an unsaturated zone typically contains oxidants (electron acceptors, e.g., nitrate, sulfate, and dissolved atmospheric oxygen (O_2)), whereas recharge beneath a surface-water body is more likely to contain higher concentrations of reduced species (electron donors, e.g., ammonium, methane, and dissolved organic matter (DOM)). In some cases, electron acceptors and electron donors are carried into the saturated zone together during recharge (e.g., O_2 and DOM), in which case they may react with each other until one or the other is consumed. In many cases, however, solid phases in the aquifer are the predominant sources of electron donors (e.g., organic matter, ferrous iron minerals, and sulfide minerals) or electron acceptors (e.g., ferric iron minerals and sulfate minerals), and the rates and progress of biogeochemical reactions affecting solutes in recharge are largely controlled by the geology of the subsurface (abundance and reactivity of solid phases) and water–rock contact (McMahon and Chapelle, 2008). Many of these reactions are catalyzed by bacteria taking advantage of chemical potential energy caused by the juxtaposition of chemical species that are not in equilibrium (Appelo and Postma, 2007; Stumm and Morgan, 1996). Reaeration of groundwater is limited by downward advection and loss of communication with overlying air in recharge areas. Where groundwater flow is largely unidirectional (advection ± longitudinal dispersion), redox zones generally tend to follow relatively simple progressions with age, for example, if starting with oxic recharge, reduction of O_2, NO_3^- Mn^{4+}, Fe^{2+}, SO_4^{2-}, and CO_2 (**Figure 7**), but the progression through this sequence may be more or less complete depending on groundwater flow time, element availability, and aquifer reactivity (Edmunds et al., 1984). Reduction of O_2 and NO_3^- typically progress more rapidly in organic-rich mudstones and unweathered glacial deposits containing reactive rock fragments, whereas these reactions commonly are slow in highly evolved siliciclastic sediments and some carbonate rocks. Thus, discharging groundwater can be oxic or highly reduced depending on the hydrogeologic setting. Starting with anoxic recharge, other types of reactions may be important and redox processes affecting aqueous species (e.g., oxidation of CH_4, H_2S, NH_4^+, and Fe^{2+}) may be reversed. Reversed redox sequences along groundwater flowpaths are well documented in anthropogenic contaminant plumes, for example, near landfills and organic spill sites (Baedecker et al., 1993; Christensen et al., 2001), and they also occur in aquifers underlying recharge areas in wetlands and lakes containing organic-rich bed sediments (e.g., Katz et al., 1995). Other reactions involving dissolution and precipitation of inorganic solid phases cause concentration gradients in non-redox-sensitive constituents such as SiO_2, Na, Mg, Ca, Sr, and Mg.

For unconsolidated water-table aquifers with typical recharge rates in humid to semiarid environments, groundwater is commonly stratified, being youngest near the water table and progressively older downward. Groundwater chemistry therefore also may be stratified in recharge areas as a result of changing conditions in recharge composition (e.g., changing composition of atmospheric deposition or addition of anthropogenic contaminants) and biogeochemical reactions in the aquifer (Back et al., 1993; Böhlke, 2002). In discharge areas, and where groundwater flowpaths are confined between impermeable units, these patterns can change. In discharge areas, preexisting gradients of groundwater age and chemistry may become horizontal or overturned as flow vectors turn upward (Böhlke et al., 2002). These spatial patterns may be confused with locally generated biogeochemical gradients in the absence of detailed information. Aquifer heterogeneity can result in complex reaction zones including bidirectional transport of reactants and products across aquifer–aquitard contacts (McMahon, 2001). In karst and fractured rock aquifers, groundwater flowpaths and biogeochemical mass transfers may be especially complex because of coexisting high-permeability conduits and massive low-permeability units (Bakalowicz, 2005). Complex patterns of solute transport and redox progression are typical near shallow water tables (Scholl et al., 2006) and near sediment-surface water interfaces, such as hyporheic zones, lake beds, and wetlands, where flow reversals and(or) diffusion are important, for example, beneath forested wetlands (Alewell et al., 2006).

Biogeochemical processes affecting groundwater chemistry operate over a large range of timescales. For example, measured rates of oxygen reduction and denitrification range over at least 8–10 orders of magnitude. Because of this, and because these reactions commonly are limited by the distributions of reactive solid phases, groundwater chemical gradients may be either sharp boundaries (flux-controlled) or gradual transitions (kinetically controlled). Rates of biogeochemical reactions in aquifers have been measured by various techniques, in part reflecting the range of timescales involved. Laboratory experiments with groundwater and aquifer material can be used to study reaction potential on short timescales. In situ tracer injection experiments including isotopically labeled reactants can be used for intermediate timescales. Examples include single-well or push–pull tests and natural-gradient tracer breakthrough experiments (Istok et al., 1997; Smith et al., 2004; Kellogg et al., 2005; Böhlke et al., 2006). Laboratory experiments and single-well injections commonly indicate higher reaction rates than in situ natural gradient measurements, presumably in part because of physical disruption or other forms of biogeochemical stimulation (Smith et al., 1996, 2006). Groundwater dating of reaction-zone chemical gradients may be the only practical empirical method of measurement at longer timescales. Examples include the use of modern atmospheric environmental tracers, for example, tritium and chlorofluorocarbons for reaction zones on 0–60-year timescales (Böhlke et al., 2002; Green et al., 2008) and ^{14}C for reaction zones on 10^3–10^4-year timescales (Vogel et al., 1981; Plummer et al., 1990). Because of aquifer heterogeneity, groundwater ages and reaction rates are evaluated most reliably from field data using solute transport models that account for dispersion and sample mixing (Scholl, 2000; Weissmann et al., 2002; Green et al., 2010).

In discharge areas, groundwater can interact with sediments and plants in riparian wetlands, streambeds, and estuaries, where organic matter and other reactants may be more abundant than elsewhere in the saturated zone. Reactions in these areas are strongly dependent on groundwater flowpaths. Near-stream geomorphology and vertical components of groundwater flow largely determine whether groundwater interacts with shallow riparian soils and plants or bypasses those potential reaction sites before discharging to streams. Slow diffuse flow through reactive material may cause important changes in groundwater chemistry just prior to discharge, whereas rapid flow through permeable layers and macropores may avoid such reactions (Burt et al., 1999; Angier et al., 2005). The relative importance of these flowpaths and reactions for overall mass balance in discharge areas is difficult to assess in watershed-process studies. Direct discharge of groundwater to estuaries is a potential source of land-derived water and nutrients to coastal waters, but is difficult to quantify and may exhibit complex patterns of physical and chemical interaction with salty pore water (Manheim et al., 2004; Andersen et al., 2005). Discharge also may be affected or enhanced by bioirrigation, the augmentation of flow across the sediment–water interface by filter feeders living in estuaries (Martin et al., 2006; Meysman et al., 2006).

Groundwater discharge is an important component of stream flow and solute loads, especially in low-order streams (Alexander et al., 2007). Total stream flow typically is dominated by groundwater discharge except briefly during intense runoff events. Even during flood peaks, the fraction of stream flow delivered from the land surface to the stream without moving into the subsurface typically is small (Buttle, 1994; Buttle and Peters, 1997; Bishop et al., 2004; Burt and Pinay, 2005). As groundwater is stratified in age and chemistry, the age and chemistry of groundwater contributing to stream flow can be quite variable and complex. Old groundwater (decades to millennia) may discharge upward from below the streambed, while younger groundwater (months to years) discharges from shallower flowpaths. Because of variations in subsurface hydraulic properties, age distributions in discharge may be difficult to define and mean age alone may be a poor indicator of the assemblage of watershed transit times. Changing conditions throughout a drainage basin can cause changes in the proportions of different groundwater types and the proportions of groundwater and runoff, contributing to stream flow over timescales ranging from hours to days (storms and snowmelt events) to months (seasonal effects of precipitation and evapotranspiration) to years (interannual and longer climate variation or land-use changes). For example, seasonal variation of nitrate and sulfate concentrations with stream flow may depend on production, consumption, and storage at different timescales in the unsaturated and saturated zones (Lynch and Corbett, 1989; Shanley and Peters, 1993; Huntington et al., 1994; Peters, 1994; Böhlke et al., 2007). At interannual and decadal timescales, responses of streams to changes in loadings of nonpoint-source contaminants can be complex, subdued, or delayed because of changing inputs, groundwater residence times, and water–rock interactions (Böhlke and Denver, 1995; Burt et al., 2008). Surficial aquifers in unconsolidated sediments, such as coastal plains, alluvial valleys, and glacial outwash deposits, commonly have mean groundwater residence times of the order of decades. As changes in land use, agricultural-fertilizer use, and atmospheric deposition commonly occur on decadal timescales, many aquifers contain transient records of anthropogenic nonpoint-source contaminants. This means that the mass flux of a constituent in annual recharge may be different from the mass flux in annual discharge even where the constituent behaves conservatively and the water balance is in steady state. This is a common feature of agricultural drainage basins and an important limitation on watershed nitrogen balances and export predictions (Böhlke, 2002). Commonly, it is difficult to resolve the effects of temporal changes in recharge chemistry from progressive biogeochemical reactions in aquifers without detailed study.

2.11.8 Acidic Atmospheric Deposition – Acid Rain

Many linkages between hydrology and biogeochemistry were revealed from research conducted to understand the effects of acid rain on terrestrial and aquatic ecosystems beginning in the 1970s. Some of the most import linkages were demonstrated through studies of deleterious effects on biota, particularly forests and fish. Decreases in pH and increases in dissolved inorganic aluminum concentrations have diminished species diversity and abundance of plankton, invertebrates, and fish in acid-impacted surface waters (Schindler, 1988).

Acid rain effects on ecosystems include forest decline (Pitelka, 1994; DeHayes et al., 1999), bird population declines and changes (Graveland, 1998), and aquatic-biota declines including algae, macroinvertebrates, and fish (Schindler, 1988). Extremely high deposition of N species (wet and dry deposition) has had a range of effects on forests from fertilization to changes in N mineralization and increased N leaching through soils to surface water (Vitousek et al., 1982; Aber, 1992; Aber et al., 1995, 1998; Emmett et al., 1998a, 1998b; Emmett, 1999; Mitchell, 2001). Vegetation filters atmospheric contaminants and dry deposition can be a major input to ecosystems (Reynolds et al., 1994; Peters et al., 1998). Acidification causes base cations and metals, particularly inorganic aluminum, to be mobilized, which in turn, has deleterious effects on aquatic biota, such as fish (Driscoll et al., 1980). For example, aluminum precipitates on fish gills ultimately affecting blood pH and decreasing the capacity of hemoglobin to transfer oxygen (Fromm, 1980). The loss of nutrient base cations, such as calcium, from soils affects forest growth and health (DeHayes et al., 1999), and subsequent decreases in receiving waters affect aquatic biota (Holt and Yan, 2003; Keller et al., 2003; Jeziorski and Yan, 2006; Jeziorski et al., 2008; Cairns and Yan, 2009). A knock-on effect of the S emissions and deposition associated with acid rain is that increased inputs of sulfate decrease methane production in wetlands (Schimel, 2004).

Hydrology is a major driver that delivers acid rain through terrestrial vegetation, soils, and groundwater to streams and lakes. Acid-neutralizing capacity (or alkalinity) is generated by mineral weathering, but base-poor lithologies for which weathering rates are relatively low, including quartzites, sandstones, and granitoid metamorphic and igneous rocks, are particular susceptible to the addition of acids (Schnoor and Stumm, 1986; Reynolds et al., 2001). Glaciated terrain on these lithologies is susceptible to acidic deposition, particularly where glacial deposits are thin and relatively impermeable. For example, US lakes and streams with comparable-sized drainage basins in the west-central Adirondack Mountains, NY, receiving similar acidic deposition, responded differently in neutralizing the acidity simply because of differences in the thickness and distribution of surficial material (Peters and Murdoch, 1985; Peters and Driscoll, 1987a). The residence time of water combined with the weathering rate of surficial materials determines the amount of alkalinity in the water. Longer residence time associated with long hydrological pathways results in higher alkalinities compared to those of short hydrological pathways. For moderate to high acidic-deposition rates, streams become chronically acidic when residence times are short and similarly for lakes where the drainage basin:lake area is high. However, even drainage basins with large deposits of thick till or stratified drift and generally long residence time may experience episodic acidification during snowmelt, large rainstorms, or sea-salt episodes (Wright *et al.*, 1988; Heath *et al.*, 1992; Hindar *et al.*, 1995; Larssen and Holme, 2005).

Through cleaner fuel technologies and emission-control systems, acid rain generally has decreased throughout North America and Europe since the, 1970s, but acidity remains higher than the inferred pre-industrial conditions (Stoddard *et al.*, 1999; Jeffries *et al.*, 2003; Wright *et al.*, 2005). In addition, liming of lakes and watersheds can restore pH values, but other changes resulting from acid deposition, such as those in soil chemistry and related biota, are not reversible on short timescales (Schindler, 1999). For example, weathering is a primary source for soil base cations and the process of restoring the original soil cation-exchange complex may take several hundred years before full recovery occurs (Cosby *et al.*, 1985; Driscoll *et al.*, 2001). The persistence of surface-water acidification even with reductions in acid deposition has been attributed to losses of exchangeable base cations in the soil (Lawrence *et al.*, 1999; Lawrence, 2002), which is reflected in many soil organisms and other biota, such as birds (Graveland, 1998; Hamburg *et al.*, 2003). A multitracer assessment of red spruce, a species showing recent growth reductions and decreases in plant-available calcium in the northeastern US, suggests a progressive shallowing of effective depth of base-cation uptake by fine roots (Bullen and Bailey, 2005). Forest harvesting also decreases exchangeable base-cation pools (Federer *et al.*, 1989; Watmough *et al.*, 2003). The reductions in atmospheric emissions have targeted S and although substantial – for example, greater than 40% in the northeastern United States and eastern Canada and greater than 60% in Norway – have not been sufficient for surface waters to recover chemically and biologically (Stoddard *et al.*, 1999; Aherne *et al.*, 2003; Jeffries *et al.*, 2003; Skjelkvåle *et al.*, 2003; Watmough and Dillon, 2003; Larssen, 2005; Wright *et al.*, 2005) and the biological recovery may be hampered by other environmental factors such as drought and increasing water temperatures (Arnott and Yan, 2002; Ashforth and Yan, 2008). Remediation has resulted in restoration of some aquatic biota, such as fish, but restoration of surface-water chemistry to pre-industrial conditions may not be possible and the trajectory of biogeochemical and species evolution has likewise changed (Schindler, 1999). The restoration may also be exacerbated by other environmental factors such as climate-change affects on the frequency and severity of sea-salt episodes and drought, turnover of organic carbon, and mineralization of nitrogen (Skjelkvåle *et al.*, 2003; Wright *et al.*, 2006).

2.11.9 Summary and Future Considerations

The hydrosphere, biosphere, lithosphere, and chemosphere are intricately linked through a wide range of spatial and temporal scales. For a comprehensive understanding of biogeochemical cycling, an understanding of the hydrological-processes is required. In addition to the wealth of information linking hydrology and biogeochemistry across different aspects of the hydrological cycle, there is a wealth of information on in-stream hydrological variability and biogeochemical processing in streams and rivers.

Section 2.11.2 provided an overview of hydrological processes in headwaters with respect to stream flow generation. Mechanisms delivering water from hillslopes to stream channels were presented and discussed with respect to the relative contributions of old water, that is, water stored in the basin soils and groundwater, and new water, that is, associated with precipitation and snowmelt. The relative importance of biogeochemical processes along hydrological pathways was highlighted with a particular focus on the importance of near-stream (riparian) saturated zones in resetting the chemical signature of water flowing into the riparian zone. The riparian zone in many basins effectively buffers upslope nutrient inputs, but may also alter nutrient concentrations and fluxes through N cycling processes, such as mineralization, denitrification, and uptake by riparian vegetation.

Section 2.11.3 discussed processes affecting the components of the water budget, snow formation, and ablation processes, and those in the soil below snow-cover overwinter and during snowmelt. Microbes remain active in soils under the snowpack where water is not limited. The coupling of these nutrient transformations and snow-meltwater fluxes can result in delivery of large quantities of nutrients, organic matter, and carbon export from terrestrial ecosystems. Furthermore, solutes in snowpacks preferentially elute during melting, which in turn, can stimulate biological activity within the snowpack, for example, snow algae. The presence and rate of water movement combined with the organic-matter composition and temperature of soils largely determine the nature of the biogeochemical reactions, for example, aerobic versus anaerobic. Vegetation and solar radiation control soil water content through evaporation and transpiration and the vegetation is in part controlled by soil type and thickness, aspect, and elevation. Tree roots can redistribute water in soils affecting nutrient uptake. Plant–soil relations are intricately linked to biogeochemical cycling through the rhizosphere. Downstream mixing affects water and solute transit times, which are intricately linked to hydrological pathways through soils and groundwater and in streams with riparian and hyporheic zones. These pathway contributions, in turn, are controlled by the magnitude and intensity of rainfall and snowmelt.

Section 2.11.4 presented the concept of nutrient spiraling including the concept of nutrient-uptake length and the importance of temperature and stream flow variability on biogeochemistry. The effects of stream–groundwater interactions through hyporheic and riparian zones were also discussed. Hyporheic zone processes tend to have a larger effect per unit area on the water column in shallow upper reaches, but continuing losses through large river networks can have large cumulative effects.. Field studies involving isotopes (including isotopically labeled compounds) have elucidated the within-river transformations of nitrogen species including how these are affected by seasonality, stream flow, light penetration, and terrestrial organic matter and nutrient inputs from near-stream ecosystems. Spatial variations in within-river processes are also controlled by hydrology, channel morphology, catchment land use, and riparian vegetation.

Section 2.11.5 contrasted important processes in hydrologically isolated wetlands with those temporally connected to streams and rivers. The exchange of water, sediments, and nutrients in wetlands with adjacent catchment areas, groundwater, and streams has a major effect on biogeochemical processes. Residence time is a key driver of biogeochemical dynamics ranging from rapid turnover rates in valley-bottom riparian wetlands with high groundwater discharge to extremely slow turnover rates in a thin active layer at the surface of raised peat bogs. The near-saturated conditions of wetlands with typically high organic contents control the redox potential, which drives the biogeochemical processes. Oxygen typically limits degradation rates in wetlands and carbon is the main driver of wetland biogeochemistry. Furthermore, the temporal and spatial variability of residence time and related turnover rates therefore dictate the biogeochemical processes.

Section 2.11.6 discussed atmospheric, stream, and groundwater nutrient inputs, stratification and within-lake processes, interactions with sediments, and limiting nutrients. The nutrients associated with groundwater discharge to lakes are affected by the composition of sediments, which may alternate from oxidized to reduced conditions. Differences in sediment composition control redox conditions and, subsequently, aerobic or anaerobic reactions that affect nutrient transformations and species. Plants in littoral zones, such as emergent and submerged macrophytes and periphyton, can also alter lake nutrient composition by trapping particulates and through nutrient uptake (growth) and release (decay). Phosphorus generally limits productivity in freshwater ecosystems, but with excess phosphorus, nitrogen may be limiting; however, nitrogen can be supplemented by blooms of N-fixing blue-green algae. Although recent research suggests that surface waters were N limited prior to industrialization, the science is still contentious about the relative importance (or limitations) of N and P in controlling biological productivity of freshwaters. Lake stratification controls mixing of top and bottom waters, thus affecting biogeochemical processes. The nutrient status and productivity of surface waters determines light penetration and subsequent supply of organic matter and nutrients to bottom waters.

Section 2.11.7 presented information about typical reactions controlled by hydrological pathways, lithology (mineralogy) and biota, the importance of residence time in biogeochemical evolution, and linkages between groundwater and surface water. Biogeochemistry of groundwater is largely related to microbially mediated redox reactions that result from physical transport of aqueous reactants into contact with subsurface materials with which they are not in equilibrium, where microbial communities develop to catalyze reactions in exchange for energy. Redox conditions in groundwater vary depending on landscape position, with oxidizing conditions prevailing in headwaters and beneath the unsaturated zone and more reducing conditions occurring in lowlands and under streams and lakes. Redox conditions may also be affected by lithology. Consequently, discharging groundwater can be oxic or highly reduced depending on the hydrogeologic setting. Biogeochemical processes affecting groundwater chemistry operate over a large range of timescales (e.g., 8–10 orders of magnitude for oxygen reduction and denitrification). Stream flow is typically dominated by groundwater discharge, even during floods. But because groundwater can vary markedly in age and chemistry, the discharging mixture of groundwater contributed from a wide range of hydrological pathways can cause stream water composition and delivery of nutrients to aquatic ecosystems to likewise vary markedly in time and space.

Examples are given of the effects of human activities on hydrology and biogeochemistry linkages in each of the sections and in a separate section on acidic atmospheric deposition.

Although much research has been conducted in assessing the linkages between hydrology and biogeochemistry, many challenges remain, particularly in linking observations across a wide range of temporal and spatial scales. Vegetation, soils, hydrology, and biogeochemistry develop and respond together; yet, our efforts to study these linkages are often narrowly focused, resulting in high levels of site-specific knowledge, but slower progress in extrapolating to larger spatial scales and in developing meaningful generalizations. The need for more-comprehensive interdisciplinary studies is warranted to link terrestrial vegetation and soils in headwaters through riparian zones/floodplains to streams. These interdisciplinary studies would incorporate in-stream processes including interactions with the hyporheic zone, across scales and hydroclimatic zones. Understanding hydrological and biogeochemical processes also requires knowledge of the biological components and their functioning within these studies.

Advances in technology continue to provide smaller and more robust sensors, smaller data-acquisition packages with innovative data-transmission capabilities, and better analytical instrumentation for accurate and precise measurement of low elemental and solute concentrations on small samples. In addition, new tools are evolving in the areas of nanotechnology, remote sensing, and biosensor technology, which are providing new and innovative ways to evaluate processes linking hydrology and biogeochemistry. In addition, computer-technology advances and new visualization software with much higher computation and processing speeds provide a platform for innovative designs in data analysis and modeling. Interdisciplinary research incorporating some of these new technologies for data collection and processing coupled with the computer processing and visualization may provide new ways of data mining and testing of hydrological, biological, and biogeochemical process interactions.

2.11.10 Additional Reading

The literature is comprehensive with information about hydrology and biogeochemistry linkages. For some additional details about general water-quality characteristics, see Meybeck *et al.* (2005) and Peters *et al.* (2005); for nitrogen cycling, see Burt *et al.* (1993), Heathwaite *et al.* (1996), Cirmo and McDonnell (1997), Edwards and Wetzel (2005), Goulding *et al.* (1998), Lohse *et al.* (2009), Mitchell (2001), Vollenweider and Kerekes (1982), and Wetzel (2001); and for stream-groundwater interactions, see Burt and Pinay (2005), Dahm *et al.* (1998), Jones and Holmes (1996), Jones and Mulholland (2000), Rosenberry and Labaugh (2008), Winter (1999), Winter and Woo (1990), and Winter *et al.* (1999). Finally, Lohse *et al.* (2009) provide an overview of linkages between hydrology and biogeochemistry, and Belnap *et al.* (2005) discuss hydrology and microbial linkages in arid and semiarid watersheds.

References

Aber JD (1992) Nitrogen cycling and nitrogen saturation in temperate forest ecosystems. *Trends in Ecology and Evolution* 7(7): 220–224.

Aber JD, Magill A, Mcnulty SG, *et al.* (1995) Forest biogeochemistry and primary production altered by nitrogen saturation. *Water, Air and Soil Pollution* 85(3): 1665–1670.

Aber JD, McDowell W, Nadelhoffer K, *et al.* (1998) Nitrogen saturation in temperate forest ecosystems. *BioScience* 48(11): 921–934.

Aherne J, Dillon PJ, and Cosby BJ (2003) Acidification and recovery of aquatic ecosystems in south central Ontario, Canada: Regional application of the MAGIC model. *Hydrology and Earth System Science* 7(4): 561–573.

Alewell C, Paul S, Lischeid G, Küsel K, and Gehre M (2006) Characterizing the redox status in three different forested wetlands, with geochemical data. *Environmental Science and Technology* 40(24): 7609–7615.

Alexander RB, Boyer EW, Smith RA, Schwarz GE, and Moore RB (2007) The role of headwater streams in downstream water quality. *Journal of the American Water Resources Association* 43: 41–59.

Alexander RB, Smith RA, and Schwartz GE (2000) Effect of stream channel size on the delivery of nitrogen to the Gulf of Mexico. *Nature* 403: 758–761 (doi:10.1038/35001562).

Amundson R, Richter DD, Humphreys GS, Jobbágy EG, and Jérôme G (2007) Coupling between biota and earth materials in the critical zone. *Elements* 3(5): 327–332 (doi:10.2113/gselements.3.5.327).

Andersen MS, Nyvang V, Jakobsen R, and Postma D (2005) Geochemical processes and solute transport at the seawater/freshwater interface of a sandy aquifer. *Geochimica et Cosmochimica Acta* 69(16): 3979–3994 (doi:10.1016/j.gca.2005.03.017).

Anderson MG and Burt TP (1978) The role of topography in controlling throughflow generation. *Earth Surface Processes* 3(4): 331–344.

Angier JT, McCarty GW, and Prestegaard KL (2005) Hydrology of a first-order riparian zone and stream, mid-Atlantic coastal plain, Maryland. *Journal of Hydrology* 309: 149–166 (doi:10.1016/j.jhydrol.2004.11.017).

Appelo CAJ and Postma D (2007) *Geochemistry, Groundwater, and Pollution*, 2nd edn. Rotterdam: AA Balkema. 649pp.

Arango CP, Tank JL, Johnson LT, and Hamilton SK (2008) Assimilatory uptake rather than nitrification and denitrification determines nitrogen removal patterns in streams of varying land use. *Limnology and Oceanography* 53(6): 2558–2572.

Arnott SE and Yan ND (2002) The influence of drought and re-acidification on zooplankton emergence from resting stages. *Ecological Applications* 12(1): 138–153.

Arrigoni AS, Poole GC, Mertes LAK, O'Daniel SJ, Woessner WW, and Thomas SA (2008) Buffered, lagged, or cooled? Disentangling hyporheic influences on temperature cycles in stream channels. *Water Resources Research* 44: W09418 (doi:09410.01029/02007WR006480).

Asbury CE, Mcdowell WH, Trinidadpizarro R, and Berrios S (1994) Solute deposition from cloud water to the canopy of a Puerto Rican montane forest. *Atmospheric Environment* 28(10): 1773–1780.

Ashforth D and Yan ND (2008) The interactive effects of calcium concentration and temperature on the survival and reproduction of *Daphnia pulex* at high and low food concentrations. *Limnology and Oceanography* 53(2): 420–432.

Babiarz CL, Hurley JP, Benoit JM, Shafer MM, Andren AW, and Webb DA (1998) Seasonal influences on partitioning and transport of total and methylmercury in rivers from contrasting watersheds. *Biogeochemistry* 41: 237–257.

Back W, Baedecker MJ, and Wood WW (1993) Scales in chemical hydrogeology: A historical perspective. In: Alley WM (ed.) *Regional Ground-Water Quality*, pp. 111–129. New York: van Nostrand Reinhold.

Baedecker MJ, Cozzarelli IM, Eganhouse RP, Siegel DI, and Bennett PC (1993) Crude oil in a shallow sand and gravel aquifer – III. Biogeochemical reactions and mass balance modeling in anoxic groundwater. *Applied Geochemistry* 8(6): 569–586.

Bakalowicz M (2005) Karst groundwater: A challenge for new resources. *Hydrogeology Journal* 13(1): 148–160.

Baker MA, Dahm CN, and Valett HM (1999) Acetate retention and metabolism in the hyporheic zone of a mountain stream. *Limnology and Oceanography* 44(6): 1530–1539.

Bales RC, Davis RE, and Williams MW (1993) Tracer release in melting snow: Diurnal and seasonal patterns. *Hydrological Processes* 7(4): 389–401.

Band LE, Tague CL, Groffman P, and Belt K (2001) Forest ecosystem processes at the watershed scale: Hydrological and ecological controls of nitrogen export. *Hydrological Processes* 15(10): 2013–2028.

Barros AP and Lettenmaier DP (1994) Dynamic modelling of orographically induced precipitation. *Reviews of Geophysics* 32(3): 265–284.

Baxter CV and Hauer FR (2000) Geomorphology, hyporheic exchange, and selection of spawning habitat by bull trout (*Salvelinus confluentus*). *Canadian Journal of Fisheries and Aquatic Sciences* 57(7): 1470–1481.

Bay RR (1969) Runoff from small peatland watersheds. *Journal of Hydrology* 9: 90–102.

Bedford BL, Walbridge MR, and Aldous A (1999) Patterns in nutrient availability and plant diversity of temperate North American wetlands. *Ecology* 80(7): 2151–2169.

Belnap J, Welter JR, Grimm NB, Barger N, and Ludwig JA (2005) Linkages between microbial and hydrologic process in arid and semiarid watersheds. *Ecology* 86(2): 298–307.

Berg NH (1992) Ion elution and release sequence from deep snowpacks in the central Sierra Nevada, California. *Water, Air and Soil Pollution* 61: 139–168.

Bergström A-K and Jansson M (2006) Atmospheric nitrogen deposition has caused nitrogen enrichment and eutrophication of lakes in the northern hemisphere. *Global Change Biology* 12(4): 635–643 (doi:10.1111/j.365-486.2006.01129.x).

Bernhardt ES, Likens GE, Hall RO, *et al.* (2005) Can't see the forest for the stream? In-stream processing and terrestrial nitrogen exports. *BioScience* 55(3): 219–230.

Bernot MJ, Tank JL, Royer TV, and David MB (2006) Nutrient uptake in streams draining agricultural catchments of the midwestern United States. *Freshwater Biology* 51(3): 499–509 (doi:10.1111/j.365-2427.2006.01508.x).

Betson RP (1964) What is watershed runoff? *Journal of Geophysical Research* 69(8): 1542–1552.

Beven K (1982) On subsurface stormflow: Predictions with simple theory for saturated and unsaturated flows. *Water Resources Research* 18(6): 1627–1633.

Beven K and Germann P (1982) Macropores and water flow in soils. *Water Resources Research* 18(5): 1311–1325.

Bishop K, Seiber J, Kohler S, and Laudon H (2004) Resolving the double paradox of rapidly mobilised old water with highly variable responses in runoff chemistry. *Hydrological Processes* 18: 185–189.

Bluth GJS and Kump LR (1994) Lithologic and climatologic controls of river chemistry. *Geochimica et Cosmochimica Acta* 58(10): 2341–2359.

Bohlen PJ, Scheu S, Hale CM, *et al.* (2004) Non-native invasive earthworms as agents of change in northern temperate forests. *Frontiers in Ecology and the Environment* 2(8): 427–435.

Bonan GB and Levis S (2006) Evaluating aspects of the community land and atmosphere models (CLM3 and CAM3) using a dynamic global vegetation model. *Journal of Climate* 19(11): 2290–2301.

Bormann FH and Likens GE (1994) *Pattern and Process in a Forested Ecosystem*. New York: Springer. 253pp.

Boulton AJ, Findlay S, Marmonier P, Stanley EH, and Valett HM (1998) The functional significance of the hyporheic zone in streams and rivers. *Annual Review Ecology and Systematics* 29: 59–81.

Bowden WB (1986) Gaseous nitrogen emissions from undisturbed terrestrial ecosystems: An assessment of their impacts on local and global nitrogen budgets. *Biogeochemistry* 2: 249–279.

Boyd GR, Reemtsma H, Grimm DA, and Mitra S (2003) Pharmaceuticals and personal care products (PPCPs) in surface and treated waters of Louisiana, USA and Ontario, Canada. *Science of the Total Environment* 311: 135–149.

Boyd MJ, Bufill MC, and Knee RM (1993) Pervious and impervious runoff in urban catchments. *Hydrological Sciences Journal/Journal des Sciences Hydrologiques* 38(6): 463–479.

Boyer EW, Hornberger GM, Bencala KE, and McKnight DM (1997) Response characteristics of DOC flushing in an alpine catchment. *Hydrological Processes* 11: 1635–1647.

Böhlke JK (2002) Groundwater recharge and agricultural contamination. *Hydrogeology Journal* 10: 153–179 (doi:10.1007/s10040-001-0183-3).

Böhlke JK, Antweiler RC, Harvey JW, et al. (2009) Multi-scale measurements and modeling of denitrification in streams with varying flow and nitrate concentration in the upper Mississippi River basin, USA. *Biogeochemistry* 93: 117–141 (doi:10.1007/s10533-008-9282-8).

Böhlke JK and Denver JM (1995) Combined use of groundwater dating, chemical, and isotopic analyses to resolve the history and fate of nitrate contamination in two agricultural watersheds, Atlantic coastal plain, Maryland. *Water Resources Research* 31(9): 2319–2339.

Böhlke JK, Harvey JW, and Voytek MA (2004) Reach-scale isotope tracer experiment to quantify denitrification and related processes in a nitrate-rich stream, midcontinent United States. *Limnology and Oceanography* 49(3): 821–838.

Böhlke JK, O'Connell ME, and Prestegaard KL (2007) Ground-water stratification and delivery of nitrate to an incised stream in varying flow conditions. *Journal of Environmental Quality* 36(3): 664–680.

Böhlke JK, Smith RL, and Miller DN (2006) Ammonium transport and reaction in contaminated ground water: Application of isotopic tracers and isotope fractionation studies. *Water Resources Research* 42: W05411 (doi:10,1029/2005WR004349).

Böhlke JK, Wanty R, Tuttle M, Delin G, and Landon M (2002) Denitrification in the recharge area and discharge area of a transient agricultural nitrate plume in a glacial outwash sand aquifer, Minnesota. *Water Resources Research* 38(7): 10.1–10.26 (doi:10.1029/2001WR000663).

Böttcher J, Strebel O, Voerkelius S, and Schmidt H-L (1990) Using isotope fractionation of nitrate-nitrogen and nitrate-oxygen for evaluation of microbial denitrification in a sandy aquifer. *Journal of Hydrology* 114: 413–424.

Branfireun B, Heyes A, and Roulet NT (1996) The hydrology and methyl mercury dynamics of a headwater wetland. *Water Resources Research* 32: 1785–1794.

Branfireun BA, Routlet NT, Kelly CA, and Rudd JWM (1999) *In situ* sulphate stimulation of mercury methylation in a boreal peatland: Toward a link between acid rain and methylmercury contamination in remote environments. *Global Biogeochemical Cycles* 13(3): 743–750.

Brantley SL, Goldhaber MB, and Ragnarsdottir KV (2007) Crossing disciplines and scales to understand the critical zone. *Elements* 3(5): 207–314.

Brooks P, McKnight D, and Bencala K (1999a) The relationship between soil heterotrophic activity, soil dissolved organic carbon (DOC) leachate, and catchment-scale DOC export in headwater catchments. *Water Resources Research* 35(6): 1895–1902.

Brooks PD, Campbell DH, Tonnessen KA, and Heuer K (1999b) Natural variability in N export from headwater catchments: Snow cover controls on ecosystem N retention. *Hydrological Processes* 13(14): 2191–2201.

Brooks PD, McKnight D, and Elder K (2005) Carbon limitation of soil respiration under winter snowpacks: Potential feedbacks between growing season and winter carbon fluxes. *Global Change Biology* 11(2): 231–238.

Brooks PD, Schmidt SK, and Williams MW (1997) Winter production of CO_2 and N_2O from alpine Tundra: Environmental controls and relationship to inter-system C and N fluxes. *Oecologia* 110(3): 403–413.

Brooks PD and Williams MW (1999) Snowpack controls on nitrogen cycling and export in seasonally snow-covered catchments. *Hydrological Processes* 13(14–15): 2177–2190.

Brooks PD, Williams MW, and Schmidt SK (1995) Snowpack controls on soil nitrogen dynamics in the Colorado alpine. In: Tonnessen KA, Williams MW, and Tranter M (eds.) *Biogeochemistry of Seasonally Snow-Covered Catchments*, vol. 228, pp. 283–292. Wallingford: IAHS.

Brooks PD, Williams MW, and Schmidt SK (1996) Microbial activity under alpine snowpacks, Niwot Ridge, Colorado. *Biogeochemistry* 32(2): 92–113.

Brooks PD, Williams MW, and Schmidt SK (1998) Inorganic nitrogen and microbial biomass dynamics before and during spring snowmelt. *Biogeochemistry* 43(1): 1–15.

Broxton PD, Troch PA, and Lyon SW (2009) On the role of aspect to quantify water transit times in small mountainous catchments. *Water Resources Research* 45: W08427 (doi:10.1029/2008WR007438).

Bruijnzeel LA and Veneklaas EJ (1998) Climatic conditions and tropical montane forest productivity: The fog has not lifted yet. *Ecology* 79(1): 3–9.

Buda AR and DeWalle DR (2009) Dynamics of stream nitrate sources and flow pathways during stormflows on urban, forest and agricultural watersheds in central Pennsylvania, USA. *Hydrological Processes* 23(22): 3292–3305.

Buffam I, Galloway JN, Blum LK, and McGlathery KJ (2001) A stormflow/baseflow comparison of dissolved organic matter concentrations and bioavailability in an Appalachian stream. *Biogeochemistry* 53(3): 269–306.

Bullen TD and Bailey SW (2005) Identifying calcium sources at an acid deposition-impacted spruce forest: A strontium isotope, alkaline earth element multi-tracer approach. *Biogeochemistry* 74(1): 63–99.

Burgess SSO, Adams MA, Turner NC, and Ong CK (1998) The redistribution of soil water by tree root systems. *Oecologia* 115: 306–311.

Burgess SSO and Dawson TE (2004) The contribution of fog to the water relations of *Sequoia sempervirens* (D. Don): Foliar uptake and prevention of dehydration. *Plant, Cell and Environment* 27(8): 1023–1034.

Burns DA and Kendall C (2002) Analysis of $\delta^{15}N$ and $\delta^{18}O$ to differentiate NO_3^- sources in runoff at two watersheds in the Catskill Mountains of New York. *Water Resources Research* 38(5): 1051 (doi:10.1029/2001WR000292).

Burt TP, Heathwaite AL, and Trudgill ST (1993) *Nitrate: Processes, Patterns and Management*. Chichester: Wiley.

Burt TP, Howden NJK, Worrall F, and Whelan MJ (2008) Importance of long-term monitoring for detecting environmental change: Lessons from a lowland river in south east England. *Biogeosciences* 5(6): 1529–1535.

Burt TP, Matchett LS, Goulding KWT, Webster CP, and Haycock NE (1999) Denitrification in riparian buffer zones: The role of floodplain sediments. *Hydrological Processes* 13(10): 1431–1463.

Burt TP and Pinay G (2005) Linking hydrology and biogeochemistry in complex landscapes. *Progress In Physical Geography* 29(3): 297–316.

Buttle JM (1994) Isotope hydrograph separations and rapid delivery of pre-event water from drainage basins. *Progress in Physical Geography* 18(1): 16–41.

Buttle JM (2005). Isotope hydrograph separation of runoff sources. In: Anderson MG, McDonnell JJ (eds.) *Encyclopedia of Hydrological Sciences*. vol. 3, Part 8, ch. 116, pp. 1763–1774. Chichester: Wiley.

Buttle JM and Peters DL (1997) Inferring hydrological processes in a temperate basin using isotopic and geochemical hydrograph separation: A re-evaluation. *Hydrological Processes* 11(6): 557–573.

Cairns A and Yan ND (2009) A review of the influence of low ambient calcium concentrations on freshwater daphniids, gammarids, and crayfish. *Environmental Reviews* 17: 67–79.

Caldwell MM and Richards JH (1989) Hydraulic lift: Water efflux from upper roots improves effectiveness of water uptake by deep roots. *Oecologia* 79: 1–5.

Caldwell MM, Dawson TE, and Richards JH (1998) Hydraulic lift: Consequences of water eflux from the roots of plants. *Oecologia* 113: 151–161.

Campbell DH, Clow DW, Ingersoll GP, Mast MA, Spahr NE, and Turk JT (1995) Nitrogen deposition and release in alpine watersheds, Loch Vale, Colorado, USA. In: Tonnessen KA, Williams MW, and Tranter M (eds.) *Biogeochemistry of Seasonally Snow-Covered Catchments*, vol. 228, pp. 243–253. Wallingford: IAHS.

Campbell DH, Kendall C, Change CCY, Silva SR, and Tonnessen KA (2002) Pathways for nitrate release from an alpine watershed: Determination using $\partial15N$ and $\partial18O$. *Water Resources Research* 38(5). (doi:10.1029/2001WR000294).

Campbell JL and Law BE (2000) Dissolved organic nitrogen budgets for upland, forested ecosystems in New England. *Biogeochemistry* 49(2): 123–142.

Campbell NS and Novotny V (2004) *Water Quality – Prevention, Identification, and Management of Diffuse Pollution*. New York: IWA Publishing. 322pp.

Cardon ZG and Whitbeck JL (2007) *The Rhizosphere: An Ecological Perspective*. New York, NY: Elsevier. 212pp.

Chorley RJ (1971) The drainage basin as the fundamental geomorphic unit. In: Chorley RJ (ed.) *Introduction to Physical Hydrology*, pp. 37–59. London: Methuen.

Christensen TH, Kjeldsen P, Bjerg PL, et al. (2001) Biogeochemistry of landfill leachate plumes. *Applied Geochemistry* 16(7–8): 659–718.

Christopher SF, Mitchell MJ, McHale MR, Boyer EW, Burns DA, and Kendall C (2008) Factors controlling nitrogen release from two forested catchments with contrasting hydrochemical responses. *Hydrological Processes* 22(1): 46–62.

Chung SW, Hipsey MR, and Imberger J (2009) Modelling the propagation of turbidity density inflows into a stratified lake: Daecheong Reservoir, Korea. *Environmental Modelling and Software* 24(12): 1467–1482 (doi:10.1016/j.envsoft.2009.05.016).

Cirmo CP and McDonnell JJ (1997) Linking the hydrologic and biogeochemical controls of nitrogen transport in near-stream zones of temperate-forested catchments: A review. *Journal of Hydrology* 199(12): 88–120.

Clark KL, Nadkarni NM, Schaefer D, and Gholz HL (1998) Cloud water and precipitation chemistry in a tropical montane forest, Monteverde, Costa Rica. *Atmospheric Environment* 32(9): 1595–1603.

Clément J-C, Aquilina L, Bour O, Plaine K, Burt TP, and Pinay G (2003) Hydrological flowpaths and nitrate removal rates within a riparian floodplain along a fourth-order stream in Brittany (France). *Hydrological Processes* 17(6): 1177–1195.

Cosby BJ, Wright RF, Hornberger GM, and Galloway JN (1985) Modeling the effects of acid deposition: Estimation of long-term water quality responses in a small forested catchment. *Water Resources Research* 21(11): 1591–1601.

Covich AP, Palmer MA, and Crowl TA (1999) The role of benthic invertebrate species in freshwater ecosystems: Zoobenthic species influence energy flows and nutrient cycling. *BioScience* 49(2): 119–127.

Dahm CN, Grimm NB, Marmonier P, Valett HM, and Vervier P (1998) Nutrient dynamics at the interface between surface waters and groundwaters. *Freshwater Biology* 40(3): 427–451.

Damman A (1986) Hydrology, development, and biogeochemistry of ombrogenous peat bogs with special reference to nutrient relocation in a western Newfoundland bog. *Canadian Journal of Botany* 64(2). 384–294

Davis WM (1899) The geographical cycle. *Geographical Journal* 14: 481–504.

Dawson TE (1993) Hydraulic lift and water use by plants: Implications for water balance, performance and plant–plant interactions. *Oecologia* 95(4): 565–574.

Dawson TE (1996) Determining water use by trees and forests from isotopic, energy balance and transpiration analyses: The roles of tree size and hydraulic lift. *Tree Physiology* 16: 263–272.

Dawson TE (1998) Fog in the California redwood forest: Ecosystem inputs and use by plants. *Oecologia* 117(4): 476–485.

Dawson TE and Ehleringer JR (1998) Plants, isotopes and water use: A cachment-scale perspective. In: Kendall C and McDonnell JJ (eds.) *Isotope Tracers in Catchment Hydrology*, pp. 165–202. New York: Elservier.

Dean WE, Neff BP, Rosenberry DO, Winter TC, and Parkhurst RS (2003) The significance of ground water to the accumulation of iron and manganese in the sediments of two hydrologically distinct lakes in North-Central Minnesota: A geological perspective. *Ground Water* 41: 951–963.

DeHayes DH, Schaberg PG, Hawley GJ, and Strimbeck GR (1999) Acid rain impacts on calcium nutrition and forest health. *BioScience* 49(10): 789–800.

Décamps H, Pinay G, Naiman RJ, et al. (2004) Riparian zones; where biogeochemistry meets biodiversity in management practice. *Polish Journal of Ecology* 52(1): 3–18.

Devito KJ, Dillon PJ, and Lazerte BJ (1989) Phosphorus and nitrogen retention in five Precambrian shield wetlands. *Biogeochemistry* 8(3): 185–204.

Dillon PJ and Molot LA (1997) Dissolved organic and inorganic carbon mass balances in central Ontario lakes. *Biogeochemistry* 36(1): 29–42.

Dodds WK, Lobez AJ, Bowden WB, et al. (2002) Nitrogen uptake as a function of concentration in streams. *Journal of the North American Benthological Society* 21: 206–220.

Driscoll CT, Baker JP, Bisogni JJJ, and Schofield CL (1980) Effect of aluminium speciation on fish in dilute acidified waters. *Nature* 284: 161–164.

Driscoll CT, Fallon-Lambert K, Chen L (2005) Acidic deposition: Sources and effects. In: Anderson MG, McDonnell JJ (eds), *Encyclopedia of Hydrological Sciences*, vol. 3, Part 8, ch. 95, pp. 1441–1457. Chichester: Wiley.

Driscoll CT, Lawrence GB, Bulger AJ, et al. (2001) Acidic deposition in the northeastern United States: Sources and inputs, ecosystem effects, and management strategies. *BioScience* 51(3): 180–198.

Dudgeon D, Artheington AH, Gessner MO, et al. (2005) Freshwater biodiversity: Importance, threats, status and conservation challenges. *Biological Reviews* 81: 163–182.

Duff JH, Hendricks SP, Jackman AP, and Triska FJ (2002) The effect of *Elodea canadensis* beds on porewater chemistry, microbial respiration, and nutrient retention in the Shingobee River, Minnesota, North America. *Verhandlungen Internationale Vereinigung fur Theoretische und Angewandte Limnologie* 28: 1–9.

Duff JH and Triska FJ (1990) Denitrification in sediments from the hyporheic zone adjacent to a small forested stream. *Canadian Journal of Fisheries and Aquatic Sciences* 47(6): 1140–1147.

Dunne T and Black RD (1970a) An experimental investigation of runoff production in permeable soils. *Water Resources Research* 6(2): 478–490.

Dunne T and Black RD (1970b) Partial area contributions to storm runoff in a small New-England watershed. *Water Resources Research* 6: 1296–1305.

Dunne T and Leopold LB (1978) *Water in Environmental Planning*. New York, NY: WH Freeman. 818pp.

Durka W, Schulze E-D, Gebauer G, and Voerkeliust S (1994) Effects of forest decline on uptake and leaching of deposited nitrate determined from ^{15}N and ^{18}O measurements. *Nature* 372(6508): 765–767 (doi:10.1038/372765a0).

Edmunds WM, Miles DL, and Cook JM (1984) A comparative study of sequential redox processes in three British aquifers from the United Kingdom. In: Eriksson E (ed.) *Hydrochemical Balances of Freshwater Systems*, vol. 150, pp. 55–70. Wallingford: IAHS.

Edwards AC, Scalenghe R, and Freppaz M (2007) Changes in the seasonal snow cover of alpine regions and its effect on soil processes: A review. *Quarternary International* 162–163: 172–181.

Edwards AC and Wetzel RL (2005) Nutrient cycling. In: Anderson MG and McDonnell JJ (eds.), *Encyclopedia of Hydrological Sciences*, vol. 3, Part 8, ch. 96, pp. 1459–1477. Chichester: Wiley.

Ehleringer JR and Dawson TE (1992) Water uptake by plants – perspectives from stable isotope composition. *Plant, Cell and Environment* 15(9): 1073–1082.

Elser JJ, Andersen T, Baron JS, et al. (2009a) Shifts in lake N:P stoichiometry and nutrient limitation driven by atmospheric nitrogen deposition. *Science* 326(5954): 835–837.

Elser JJ, Kyle M, Steger L, Nydick KR, and Baron JS (2009b) Nutrient availability and phytoplankton nutrient limitation across a gradient of atmospheric nitrogen deposition. *Ecology* 90(11): 3062–3073.

Emerman SH and Dawson TE (1996) Hydraulic lift and its influence on the water content of the rhizosphere: An example from sugar maple. *Acer saccharum*. *Oecologia,* 108: 273–278.

Emmett BA (1999) The impact of nitrogen on forest soils and feedbacks on tree growth. *Water, Air and Soil Pollution* 116(1–2): 65–74.

Emmett BA, Boxman D, Bredemeier M, et al. (1998a) Predicting the effects of atmospheric nitrogen deposition in conifer stands: Evidence from the NITREX ecosystem-scale experiments. *Ecosystems* 1(4): 352–360.

Emmett BA, Reynolds B, Silgram M, Sparks TH, and Woods C (1998b) The consequences of chronic nitrogen additions on N cycling and soilwater chemistry in a Sitka spruce stand, North Wales. *Forest Ecology and Management* 101(1–3): 165–175.

Emmett BA, Reynolds B, Stevens PA, et al. (1993) Nitrate leaching from afforested Welsh catchments: Interaction between stand age and nitrogen deposition. *AMBIO* 22(6): 386–394.

Ensign SH and Doyle MW (2006) Nutrient spiraling in streams and river networks. *Journal of Geophysical Research* 111: G04009 (doi:10.1029/2005JG000114).

Euliss NH Jr., LaBaugh JW, Fredrickson LH, et al. (2004) The wetland continuum: A conceptual framework for interpreting biological studies. *Wetlands* 24: 448–458.

Evans MG, Burt TP, Holden J, and Adamson JK (1999) Runoff generation and water table fluctuations in blanket peat: Evidence from UK data spanning the dry summer of 1995. *Journal of Hydrology* 221: 141–160.

Everts CJ, Kanwar RS, Alexander ECJ, and Alexander SC (1989) Comparison of tracer mobilities under laboratory and field conditions. *Journal of Environmental Quality* 18(4): 491–498.

Ewing HA, Weathers KC, Templer PH, et al. (2009) Fog water and ecosystem function: Heterogeneity in a California Redwood Forest. *Ecosystems* 12(3): 417–433.

Federer CA, Hornbeck JW, Tritton LM, Martin CW, Pierce RS, and Smith CT (1989) Long-term depletion of calcium and other nutrients in eastern US forests. *Environmental Management* 13(5): 593–601.

Finkenbine JK, Atwater JW, and Mavinic DS (2000) Stream health after urbanization. *Journal of the American Water Resources Association* 36(5): 1149–1160.

Fisher SG, Grimm NB, Marti E, Holmes RM, and Jones JB (1998) Material spiraling in stream corridors: A telescoping ecosystem model. *Ecosystems* 1(1): 19–34.

Fraser CJD, Roulet NT, and Lafleur M (2001) Groundwater flow patterns in a large peatland. *Journal of Hydrology* 246: 142–154.

Freeman C, Evans CD, Monteith DT, Reynolds B, and Fenner N (2001) Export of organic carbon from peat soils. *Nature* 412(6849): 785 (doi:10.1038/35090628).

Friedman JM, Osterkamp WR, Scott ML, and Auble GT (1998) Downstream effects of dams on channel geometry and bottomland vegetation: Regional patterns in the Great Plains. *Wetlands* 18(4): 619–633.

Fromm PO (1980) A review of some physiological and toxicological responses of freshwater fish to acid stress. *Environmental Biology of Fishes* 5(1): 79–93.

Frost PC, Larson JH, Johnston CA, et al. (2006) Landscape predictors of stream dissolved organic matter concentration and physicochemistry in a Lake Superior river watershed. *Aquatic Sciences* 68(1): 40–51 (doi:10.1007/s00027-005-0802-5).

Fukada T, Hiscock KM, Dennis PF, and Grischek T (2003) A dual isotope approach to identify denitrification in groundwater at a river-bank infiltration site. *Water Research* 37: 3070–3078.

Fuller CC and Harvey JW (2000) Reactive uptake of trace metals in the hyporheic zone of a mining-contaminated stream, Pinal Creek, Arizona. *Environmental Science and Technology* 34(7): 1150–1155.

Galloway ME and Branfireun BA (2004) Mercury dynamics of a temperate forest wetland. *Science of the Total Environment* 325: 239–254.

Garcia-Martino AR, Warner GS, Scatena FN, and Civco DL (1996) Rainfall, runoff and elevation relationships in the Luquillo Mountains of Puerto Rico. *Caribbean Journal of Science* 32(4): 413–424.

Gee GW and Hillel D (1988) Groundwater recharge in arid regions – review and critique of estimation methods. *Hydrological Processes* 2(3): 255–266.

Gordon DR (1998) Effects of invasive, non-indigenous plant species on ecosystem processes: Lessons from Florida. *Ecological Applications* 8(4): 975–989.

Goulding KWT, Bailey NJ, Bradbury NJ, et al. (1998) Nitrogen deposition and its contribution to nitrogen cycling and associated processes. *New Phytologist* 139: 49–58.

Grace J (1989) Tree lines. *Philosophical Transactions of the Royal Society, London B* 324: 233–245.

Granger J, Sigman DM, Lehmann MF, and Torell PD (2008) Nitrogen and oxygen isotope fractionation during dissimilatory nitrate reduction by denitrifying bacteria. *Limnology and Oceanography* 53(6): 2533–2545.

Graveland J (1998) Effects of acid rain on bird populations. *Environmental Reviews* 6(1): 41–54.

Green CT, Böhlke JK, Bekins B and Phillips, S. (2010) Mixing effects on apparent reaction rates and isotope fractionation during denitrification in a heterogeneous aquifer. *Water Resources Research* 46 (doi:0.1029/2009WR008903).

Green CT, Puckett LJ, Böhlke JK, et al. (2008) Limited occurrence of denitrification in four shallow aquifers in agricultural areas of the United States. *Journal of Environmental Quality* 37: 994–1007.

Grimm NB, Sheibley RW, Crenshaw CL, Dahm CN, Roach WJ, and Zeglin LH (2005) N retention and transformation in urban streams. *Journal of the North American Benthological Society* 24: 626–642.

Grobblelaar JU and House WA (1995) Phosphorus as a limiting resource in inland waters; interactions with nitrogen. In: Tiessen H (ed.) *Phosphorus in the Global Environment: Transfers, Cycles and Management*, pp. 255–276. Chichester: Wiley.

Groffman PM, Boulware NJ, Zipperer WC, Puyat RV, Band LE, and Colosimo MF (2002) Soil nitrogen cycle processes in urban riparian zones. *Environmental Science and Technology* 36(21): 4547–4552.

Groffman PM, Driscoll CT, Fahey TJ, Hardy JP, Fitzhugh RD, and Tierney GL (2001a) Colder soils in a warmer world: A snow manipulation study in a northern hardwood forest ecosystem. *Biogeochemistry* 56(2): 135–150.

Groffman PM, Driscoll CT, Fahey TJ, Hardy JP, Fitzhugh RD, and Tierney GL (2001b) Effects of mild winter freezing on soil nitrogen and carbon dynamics in a northern hardwood forest. *Biogeochemistry* 56(2): 191–213.

Groffman PM, Hardy JP, Nolan SS, Fitzhugh RD, Driscoll CT, and Fahay TJ (1999) Snow depth, soil frost and nutrient loss in a northern hardwood forest. *Hydrological Processes* 13(14): 2275–2286.

Groffman PM, Law NL, Belt KT, Band LE, and Fisher GT (2004) Nitrogen fluxes and retention in urban watershed ecosystems. *Ecosystems* 7(4): 393–403.

Grogan P and Chapin FS (1999) Arctic soil respiration: Effects of climate and vegetation depend on season. *Ecosystems* 2(5): 451–459.

Grogan P and Jonasson S (2003) Controls on annual nitrogen cycling in the understory of a subarctic birch forest. *Ecology* 84(1): 202–218.

Hall ROJ, Tank JL, Sobota DJ, et al. (2009) Nitrate removal in stream ecosystems measured by 15N addition experiments: Total uptake. *Limnology and Oceanography* 54: 653–665.

Hamburg SP, Yanai RD, Arthur MA, Blum JD, and Siccama TG (2003) Biotic control of calcium cycling in northern hardwood forests: Acid rain and aging forests. *Ecosystems* 6(4): 399–406 (doi:10.1007/s10021-002-0174-9).

Hamilton DP and Mitchell SF (1997) Wave-induced shear stresses, plant nutrients and chlorophyll in seven shallow lakes. *Freshwater Biology* 38: 159–168.

Hamilton DP, Schladow SG, and Fisher IH (1995) Controlling the indirect effects of flow diversions on water quality in an Australian reservoir. *Environment International* 21(5): 583–590.

Harvey JW and Bencala KE (1993) The effect of streambed topography on surface–subsurface water exchange in mountain catchments. *Water Resources Research* 29(1): 89–98.

Heath JA and Huebert BJ (1999) Cloudwater deposition as a source of fixed nitrogen in a Hawaiian montane forest. *Biogeochemistry* 44(2): 119–134.

Heath RH, Kahl SA, Noerton SA, and Fernandez IJ (1992) Episodic stream acidification caused by atmophseric deposition of sea salts at Acadia National Park, Maine, United States. *Water Resources Research* 28(4): 1081–1088.

Heathwaite AL, Johnes PJ, and Peters NE (1996) Trends in nutrients. *Hydrological Processes* 10(2): 263–293.

Hedin LO, von Fischer JC, Ostrom NE, Kennedy BP, Brown MG, and Robertson GP (1998) Thermodynamic constraints on nitrogen transformations and other biogeochemical processes at soil–stream interfaces. *Ecology* 79(2): 684–703.

Herckes P, Wendling R, Sauret N, Mirabel P, and Wortham H (2002) Cloudwater studies at a high elevation site in the Vosges Mountains (France). *Environmental Pollution* 117(1): 169–177.

Herwitz SR (1986) Infiltration-excess caused by stemflow in a cyclone-prone tropical rainforest. *Earth Surface Processes and Landforms* 11(4): 401–412.

Heuer K, Brooks PD, and Tonnessen KA (1999) Nitrogen dynamics in two high elevation catchments during spring snowmelt 1996, Rocky Mountains, Colorado. *Hydrological Processes* 13(14): 2203–2214.

Hewlett JD (1961) Watershed management. *Report for 1961 Southeastern Forest Experiment Station*. Asheville, NC: US Department of Agriculture, Forest Service.

Higgins CG (1984) Piping and sapping: Development of landforms by groundwater outflow. In: La Fleur RG (ed.) *Groundwater as a Geomorphic Agent*, pp. 18–58. London: Allen and Unwin.

Hill AR and Cardaci M (2004) Denitrification and organic carbon availability in riparian wetland soils and subsurface sediments. *Soil Science Society of America Journal* 68(1): 320–325.

Hindar A, Henriksen A, Kaste Ø, and Tørseth K (1995) Extreme acidification in small catchments in southwestern Norway associated with a sea salt episode. *Water, Air and Soil Pollution* 85(2): 547–552.

Hinton MJ, Schiff SL, and English MC (1998) Sources and flowpaths of dissolved organic carbon during storms in two forested watersheds of the Precambrian Shield. *Biogeochemistry* 41(2): 175–197.

Hoham RW and Duval B (2001) Microbial ecology of snow and freshwater ice with emphasis on snow algae. In: Jones HG, Pomeroy JW, Walker DA, and Hoham RW (eds.) *Snow Ecology: An Interdisciplinary Examination of Snow-Covered Ecosystems*, pp. 166–226. Cambridge: Cambridge University Press.

Hoham RW, Marcarelli AM, Rogers HS, et al. (2000) The importance of light and photoperiod in sexual reproduction and geographical distribution in the green snow alga, *Chloromonas* sp.-D (Chlorophyceae, Volvocales). *Hydrological Processes* 14(18): 3309–3321.

Hoham RW, Schlag EM, Kang JY, et al. (1998) The effects of irradiance levels and spectral composition on mating strategies in the snow alga, *Chloromonas* sp.-D, from the Tughill Plateau, New York State. *Hydrological Processes* 12(10–11): 1627–1639.

Holden J and Burt TP (2003) Runoff production in blanket peat covered catchments. *Water Resources Research* 39(7). SWC 6.1–6.9.

Holden J, Burt TP, Evans MG, and Horton M (2006) Impact of land drainage on peatland hydrology. *Journal of Environmental Quality* 35: 1764–1778.

Holland PG and Steyn DG (1975) Vegetational responses to latitudinal variations in slope angle and aspect. *Journal of Biogeography* 2(3): 179–183.

Holt C and Yan ND (2003) Recovery of crustacean zooplankton communities from acidification in Killarney Park, Ontario, 1971–2000: pH 6 as a recovery goal. *AMBIO* 32(3): 203–207.

Hood EW, Gooseff MN, and Johnson SL (2006) Changes in the character of stream water dissolved organic carbon during flushing in three small watersheds, Oregon. *Journal of Geophysical Research – Biosciences* 111: G01007 (doi:10.1029/2005JG000082).

Hook PB, Olson BE, and Wraith JM (2004) Effects of the invasive forb *Centaurea maculosa* on grassland carbon and nitrogen pools in Montana, USA. *Ecosystems* 7(6): 686–694.

Hooper RP, Aulenbach BT, Burns DA, et al. (1998) Riparian control of stream-water chemistry: Implications for hydrochemical basin models. In: Kovar K, Tappeiner U, Peters NE, and Craig RG (eds.) *Hydrology, Water Resources and Ecology in Headwaters*, vol. 248, pp. 451–458. Wallingford: IAHS.

Hooper RP, Christophersen N, and Peters NE (1990) Modelling streamwater chemistry as a mixture of soilwater end-members – an application to the Panola Mountain catchment, Georgia, U. S. A. *Journal of Hydrology* 116(1): 321–343.

Hornberger GM, Bencala KE, and Mcknight DM (1994) Hydrological controls on dissolved organic carbon during snowmelt in the Snake River near Montezuma, Colorado. *Biogeochemistry* 25(3): 147–165.

Horton RE (1933) The role of infiltration in the hydrologic cycle. *Transactions of the American Geophysical Union* 14: 446–460.

Horton RE (1945) Erosional development of streams and their drainage basins: Hydrophysical approach to quantitative morphology. *Bulletin of the Geological Society of America* 56: 275–370.

Hultine KR, Williams DG, Burgess SSO, and Keefer TO (2003) Contrasting patterns of hydraulic redistribution in three desert phreatophytes. *Oecologia* 135: 167–175.

Huntington TG, Hooper RP, and Aulenbach BT (1994) Hydrologic processes controlling sulfate mobility in a small forested watershed. *Water Resources Research* 30(2): 283–295.

Ingram HAP (1978) Soil layers in mires: Function and terminology. *Journal of Soil Science* 29: 224–227.

Istok JD, Humphrey MD, Schroth MH, Hyman MR, and O'Reilly KT (1997) Single-well, "Push-Pull" test for *in situ* determination of microbial activities. *Ground Water* 35(4): 619–631.

Jackson RB, Sperry JS, and Dawson TE (2000) Root water uptake and transport: Using physiological processes in global predictions. *Trends in Plant Science* 5(11): 482–488.

Jansson M, Andersson R, Berggren H, and Leonardson L (1994) Wetlands and lakes as nitrogen traps. *AMBIO* 23(6): 320–326.

Jardine PM, Wilson GV, and Luxmoore RJ (1990) Unsaturated solute transport through a forest soil during rain storm events. *Geoderma* 46: 103–118.

Jeffries DS (1989) Snowpack storage of pollutants, release during melting, and impact on receiving waters. In: Norton SA, Lindberg SE, and Page AL (eds.) *Acidic Precipitation, Volume 4: Soils, Aquatic Processes and Lake Acidification*, pp. 107–154. New York: Springer.

Jeffries DS, Clair TA, Couture S, et al. (2003) Assessing the recovery of lakes in southeastern Canada from the effects of acidic deposition. AMBIO 32(3): 176–182.

Jeziorski A and Yan ND (2006) Species identity and aqueous calcium concentrations as determinants of calcium concentrations of freshwater crustacean zooplankton. Canadian Journal of Fisheries and Aquatic Sciences 63(5): 1007–1013.

Jeziorski A, Yan ND, Paterson AM, et al. (2008) The widespread threat of calcium decline in fresh waters. Science 322(5902): 1374–1377.

Johannessen M, Dale T, Gjessing ET, Henriksen A, and Wright RF (1975) Acid precipitation in Norway: The regional distribution of contaminants in snow and the chemical concentration processes during snowmelt. In: Nye JF, Miiller F, and Oeschger H (eds.) Isotopes and Impurities in Snow and Ice, vol. 118, pp. 116–120. Wallingford: IAHS.

Johannessen M and Henriksen A (1978) Chemistry of snow meltwater: Changes in concentration during melting. Water Resources Research 14(4): 615–619.

Jones HG (1999) The ecology of snow-covered systems: A brief overview of nutrient cycling and life in the cold. Hydrological Processes 13: 2135–2147.

Jones JB, Holmes RM, Fisher SG, Grimm NB, and Greene DM (1995) Methanogenesis in Arizona, USA dryland streams. Biogeochemistry 31(3): 155–173.

Jones JB and Mulholland PJ (2000) Streams and Ground Waters. San Diego, CA: Academic Press. 425pp.

Jones JBJ and Holmes RM (1996) Surface–subsurface interactions in stream ecosystems. Trends in Ecology and Evolution 11(6): 239–242.

Joslin JD and Wolfe MH (1992) Red spruce soil solution chemistry and root distribution across a cloud water deposition gradient. Canadian Journal of Forest Research 22(6): 893–904.

Kasahara T and Wondzell SM (2003) Geomorphic controls on hyporheic exchange flow in mountain streams. Water Resources Research 39: (doi, 10.1029/2002WR001386).

Katz BG, Plummer LN, Busenberg E, Revesz KM, Jones BF, and Lee TM (1995) Chemical evolution of groundwater near a sinkhole lake, northern Florida, 2. Chemical patterns, mass transfer modeling, and rates of mass transfer reactions. Water Resources Research 31(6): 1565–1584.

Keller W, Yan ND, Somers KM, and Heneberry JH (2003) Crustacean zooplankton communities in lakes recovering from acidification. Canadian Journal of Fisheries and Aquatic Sciences 60(11): 1307–1313 (doi: 10.1139/f03-111).

Kellogg DQ, Gold AJ, Groffman PM, Addy K, Stolt MH, and Blazejewski G (2005) In situ ground water denitrification in stratified, permeable soil underlying riparian wetlands. Journal of Environmental Quality 34(2): 524–533.

Kendall C (1998) Tracing nitrogen sources and cycling in catchments. In: Kendall C and McDonnell JJ (eds.) Isotope Tracers in Catchment Hydrology, pp. 519–576. New York: Elsevier.

Kendall C, Elliott EM, and Wankel SD (2007) Tracing anthropogenic inputs of nitrogen to ecosystems. In: Michener RH and Lajtha K (eds.) Stable Isotopes in Ecology and Environmental Science, 2nd edn., pp. 375–449. Malden, MA: Blackwell.

Klemm O, Wrzesinsky T, and Scheer C (2005) Fog water flux at a canopy top: Direct measurement versus one-dimensional model. Atmospheric Environment 39(29): 5375–5386.

Lampert W and Sommer U (2007) Limnoecology. The Ecology of Lakes and Streams. Oxford: Oxford University Press. 336pp.

Larssen T (2005) Model prognoses for future acidification recovery of surface waters in Norway using long-term monitoring data. Environmental Science and Technology 39: 7970–7979.

Larssen T and Holme J (2005) Afforestation, seasalt episodes and acidification – a paired catchment study in western Norway. Environmental Pollution 139(3): 440–450.

Lawrence GB (2002) Persistent episodic acidification of streams linked to acid rain effects on soil. Atmospheric Environment 36(10): 1589–1598.

Lawrence GB, David MB, Lovett GM, et al. (1999) Soil calcium status and the response of stream chemistry to changing acidic deposition rates. Ecological Applications 9(3): 1059–1072.

Lean DRS (1973) Phosphorus dynamics in lake water. Science 79: 678–680.

Levine JM, Vilà M, D'Antonio CM, Dukes JS, Grigulis K, and Lavorel S (2003) Mechanisms underlying the impacts of exotic plant invasions. Proceedings of the Royal Society of London B 270(1517): 775–781.

Lewis WM, Saunders JS, and McCutchan JH (2008) Application of a nutrient saturation concept to the control of algal growth in lakes. Lake and Reservoir Management 24: 41–46.

Lewis WM and Wurtsbaugh WA (2008) Control of lacustrine phytoplankton by nutrients: Erosion of the phosphorus paradigm. International Review of Hydrobiology 93(4–5): 446–465.

Likens GE (1972) Nutrients and eutrophication – the limiting-nutrient controversy. Limnology and Oceanography (Special Symposium) 1: 1–328.

Likens GE and Bormann FH (1995) Biogeochemistry of a Forested Ecosystem, 2nd edn. New York: Springer. 170pp.

Lindsey BD, Phillips SW, Donnelly CA, et al. (2003) Residence times and nitrate transport in ground water discharging to streams in the Chesapeake Bay Watershed. US Geological Survey Water Resources Invesigations Report 03-4035, 201pp. Denver, CO: US Geological Survey Branch of Information Services.

Lipson GE, Schmidt SK, and Monson RK (1999) Links between microbial population dynamics and nitrogen availability in an alpine ecosystem. Ecology 80(5): 1623–1631.

Lischeid G, Kolb A, Alewell C, and Paul S (2007) Impact of redox and transport processes in a riparian wetland on stream water quality in the Fichtelgebirge region, southern Germany. Hydrological Processes 21(1): 123–132.

Liston GE (2004) Representing subgrid snow cover heterogeneities in regional and global models. Journal of Climate 17(6): 1381–1397.

Liston GE and Elder K (2006) A distributed snow-evolution modeling system (SnowModel). Journal of Hydrometeorology 7(6): 1259–1276.

Loague K and Corwin DL (2005) Point and nonpoint source pollution. In: Anderson, MG and McDonnell JJ (eds.) Encyclopedia of Hydrological Sciences, vol. 3, Part 8, ch. 95 (doi:10.1002/0470848944.hsa097). Chichester: Wiley.

Lohse KA, Brooks PD, McIntosh JC, Meixner T, and Huxman TE (2009) Interactions between biogeochemistry and hydrologic systems. Annual Review of Environment and Resources 34: 65–96.

Lovett GM (1984) Rates and mechanisms of cloud water deposition to a subalpine balsam fir forest. Atmospheric Environment 18(2): 361–371.

Lovett GM and Kinsman JD (1990) Atmospheric pollutant deposition to high elevation ecosystems. Atmospheric Environment 24A: 2767–2786.

Lovett GM, Likens GE, Buso DC, Driscoll CT, and Bailey SW (2005) The biogeochemistry of chlorine at Hubbard Brook, New Hampshire, USA. Biogeochemistry 72: 191–232.

Luxmoore RJ, Jardine PM, Wilson GV, Jones JR, and Zelazny LW (1990) Physical and chemical controls of preferred path flow through a forested hillslope. Geoderma 46: 139–154.

Lynch JA and Corbett ES (1989) Hydrologic control of sulfate mobility in a forested watershed. Water Resources Research 25(7): 1695–1703.

MacDonald JA, Fowler D, Hargreaves KJ, Skiba U, Leith ID, and Murray MB (1998) Methane emission rates from a northern wetland; response to temperature, water table and transport. Atmospheric Environment 32(1): 3219–3227.

Madsen JD, Chambers PA, James WF, Koch EW, and Westlake DF (2001) The interaction between water movement, sediment dynamics and submerged macrophytes. Hydrobiologia 444: 71–84.

Malard F, Tockner K, Dole-Olivier M-J, and Ward JV (2002) A landscape perspective of surface–subsurface hydrological exchanges in river corridors. Freshwater Biology 47(4): 621–640.

Maloszewski P, Rauert W, Stichler W, and Herrmann A (1983) Application of flow models in an alpine catchment area using tritium and deuterium data. Journal of Hydrology 66: 319–330.

Manheim FT, Krantz DE, and Bratton JF (2004) Studying ground water under Delmarva coastal bays using electrical resistivity. Ground Water 42(7): 1052–1068.

Mariotti A, Landreau A, and Simon B (1988) ^{15}N isotope biogeochemistry and natural denitrification process in groundwater: Application to the chalk aquifer of northern France. Geochimica et Cosmochimica Acta 52(7): 1869–1878.

Marti E, Aumatell J, Gode L, Poch M, and Sabater F (2004) Nutrient retention efficiency in streams receiving inputs from wastewater treatment plants. Journal of Environmental Quality 33(1): 285–293.

Martin JB, Cable JE, Jaeger J, Hartl K, and Smith CG (2006) Thermal and chemical evidence for rapid water exchange across the sediment–water interface by bioirrigation in the Indian River Lagoon, Florida. Limnology and Oceanography 51(3): 1332–1341.

Mayer B, Boyer EW, Goodale C, et al. (2002) Sources of nitrate in rivers draining sixteen watersheds in the northeastern U.S.: Isotopic constraints. Biogeochemistry 57/58: 171–192.

McClain ME, Boyer EW, Dent CL, et al. (2003) Biogeochemical hot spots and hot moments at the interface of terrestrial and aquatic ecosystems. Ecosystems 6(4): 301–312.

McDonnell JJ, Sivapalan M, Vaché K, et al. (2007) Moving beyond heterogeneity and process complexity: A new vision for watershed hydrology. Water Resources Research 43: W07301 (doi: 10.1029/2006WR005467).

McDonnell JJ, Stewart MK, and Owen IF (1991) Effect of catchment-scale subsurface mixing on stream isotopic response. Water Resources Research 27(12): 3065–3073.

McDowell NG, White S, and Pockman WT (2008) Transpiration and stomatal conductance across a steep climate gradient in the southern Rocky Mountains. Ecohydrology 1(3): 193–204.

McGlynn BL, McDonnell J, Stewart M, and Seibert J (2003) On the relationships between catchment scale and streamwater mean residence time. Hydrological Processes 17(1): 175–181.

McGuire KJ and McDonnell JJ (2006) A review and evaluation of catchment transit time modeling. *Journal of Hydrology* 330(3–4): 543–563.

McGuire KJ, McDonnell JJ, Weiler M, *et al.* (2005) The role of topography on catchment-scale water residence time. *Water Resource Research* 41: W05002 (doi: 10.1029/2004WR003657).

McIntosh J, McDonnell JJ, and Peters NE (1999) Tracer and hydrometric study of preferential flow in large undisturbed soil cores from the Georgia Piedmont, USA. *Hydrological Processes* 13(2): 139–155.

McMahon PB and Böhlke JK (1996) Denitrification and mixing in a stream–aquifer system: Effects on nitrate loading to surface water. *Journal of Hydrology* 186: 105–128.

McMahon PB and Chapelle FH (2008) Redox processes and water quality of selected principal aquifer systems. *Ground Water* 46: 259–271.

McMahon PG (2001) Aquifer/aquitard interfaces: Mixing zones that enhance biogeochemical reactions. *Hydrogeology Journal* 9(1): 34–43.

McMaster ME (2001) A review of the evidence for endocrine disruption in Canadian aquatic ecosystems. *Water Quality Research Journal of Canada* 36(2): 215–231.

Meinzer FC, Clearwater MJ, and Goldstein G (2001) Water transport in trees: Current perspectives, new insights and some controversies. *Environmental and Experimental Botany* 45: 239–262.

Meybeck M (1987) Global chemical weathering of surficial rocks estimated from river dissolved loads. *American Journal of Science* 287: 401–428.

Meybeck M (2001) River basins under anthropocene conditions. In: von Bodungen B and Turner RK (eds.) *Science and Integrated Coastal Management*, pp. 275–294. Berlin: Dahlem University Press.

Meybeck M, Peters NE, and Chapman DV (2005) Water quality. In: Anderson MG and McDonnell JJ (eds.) *Encyclopedia of Hydrological Sciences,* vol. 3, part 8, ch. 91, pp. 1373–1385. Chichester: Wiley.

Meybeck M and Vörösmarty CJ (2005) Fluvial filtering of land-to-ocean fluxes: From natural Holocene variations to Anthropocene. *CR Geoscience* 337: 107–123 (doi: 10.1016/j.crte.2004.09.016).

Meyer JL, Paul MJ, and Taulbee WK (2005) Stream ecosystem function in urbanizing landscapes. *Journal of the North American Benthological Society* 24(3): 602–612.

Meysman FJR, Galaktionov OS, Gribsholt B, and Middelburg JJ (2006) Bioirrigation in permeable sediments: Advective pore-water transport induced by burrow ventilation. *Limnology and Oceanography* 51(1): 142–156.

Michalski G, Meixner T, Fenn M, *et al.* (2004) Tracing atmospheric nitrate depsition in a complex semiarid ecosystem using $\Delta 17O$. *Environmental Science and Technology* 38(7): 2175–2181.

Millenium Ecosystem Assessment E (2005) *Ecosystems and Human Well-Being: Synthesis.* Washington, DC: Island Press. 137pp.

Millikin Ishikawa C and Bledsoe CS (2000) Seasonal and diurnal patterns of soil water potential in the rhizosphere of blue oaks: Evidence for hydraulic lift. *Oecologia* 125: 459–465.

Mitchell MJ (2001) Linkages of nitrate losses in watersheds to hydrological processes. *Hydrological Processes* 15(17): 3305–3307.

Mitsch WJ and Gosselink J (2007) *Wetlands.* Hoboken, NJ: Wiley.

Molotch NP and Bales RC (2005) Scaling snow observations from the point to the grid element: Implications for observation network design. *Water Resources Research* 41: W11421 (doi: 10.1029/2005WR004229).

Molotch NP and Bales RC (2006) SNOTEL representativeness in the Rio Grande headwaters on the basis of physiographics and remotely sensed snow cover persistence. *Hydrological Processes* 20(4): 723–739.

Monson RK, Turnipseed AA, Sparks JP, *et al.* (2002) Carbon sequestration in a high-elevation, subalpine forest. *Global Change Biology* 8(5): 459–478.

Mooney HA and Hobbs RJ (2000) *Invasive Species in a Changing World.* Washington, DC: Island Press. 457pp.

Mooney HA, Mack RN, McNeely JK, *et al.* (eds.) (2005) *Invasive Alien Species: A New Synthesis.* Washington, DC: Island Press.

Moore TR (2003) Dissolved organic carbon in a boreal landscape. *Global Biogeochemical Cycles* 17(4): 1109 (doi: 10.1029.2003GB002050).

Moreira MZ, Sternberg LdSL, and Nepstad DC (2000) Vertical patterns of soil water uptake by plants in a primary forest and an abandoned pasture in the eastern Amazon: An isotopic approach. *Plant and Soil* 222: 95–107.

Mortimer CH (1941) The exchange of dissolved substances between mud and water in lakes. I and II. *Journal of Ecology* 29(2): 280–329.

Mortimer CH (1942) The exchange of dissolved substances between mud and water in lakes. III and IV. *Journal of Ecology* 30: 147–201.

Mulholland PJ, Hall ROJ, Sobota DJ, *et al.* (2009) Nitrate removal in stream ecosystems measured by 15N addition experiments: Denitrification. *Limnology and Oceanography* 54: 666–680.

Mulholland PJ, Helton AM, Poole GC, *et al.* (2008) Excess nitrate from agricultural and urban areas reduces denitrification efficiency in streams. *Nature* 452: 202–205.

Mulholland PJ, Tank JL, Sanzone DM, *et al.* (2000) Nitrogen cycling in a forest stream determined by a N-15 tracer addition. *Ecological Monographs* 70(3): 471–493.

Mulholland PJ, Valett HM, Webster JR, *et al.* (2004) Stream denitrification and total nitrate uptake rates measured suing a field 15N isotope tracer approach. *Limnology and Oceanography* 49: 809–820.

Naiman RJ, Bilby RE, Schindler DE, and Helfield JM (2002) Pacific salmon, nutrients, and the dynamics of freshwater and riparian ecosystems. *Ecosystems* 5(4): 399–417.

Newbold JD, Elwood JW, O'Neill RV, and Sheldon AL (1983) Phosphorus dynamics in a woodland stream ecosystem: A study of nutrient spiralling. *Ecology* 64: 1249–1265.

Newbold JD, Elwood JW, O'Neill RV, and VanWinkle W (1981) Measuring nutrient spiralling in streams. *Canadian Journal of Fisheries and Aquatic Sciences* 38: 860–863.

Nilsson C and Berggren K (2000) Alterations of riparian ecosystems caused by river regulation. *BioScience* 50(9): 783–792.

Nilsson M, Sagerfors J, Buffman I, *et al.* (2008) Contemporary carbon accumulation in a boreal oligotrophic minerogenic mire – a significant sink after accounting for all C-fluxes. *Global Change Biology* 14(10): 2317–2332.

Novotny V (2003) *Water Quality: Diffuse Pollution and Watershed Management*, 2nd edn. New York: Wiley. 864pp.

O'Sullivan PE (1995) Eutrophication. *International Journal of Environmental Studies* 47: 173–195.

Oliveira RS, Dawson TE, Burgess SSO, and Nepstad DC (2005) Hydraulic redistribution in three Amazonian trees. *Oecologia* 145: 354–363.

Oyarzún CE, Godoy R, Schrijver AD, Staelens J, and Lust N (2004) Water chemistry and nutrient budgets in an undisturbed evergreen rainforest of southern chile. *Biogeochemistry* 71(1): 107–123.

Öberg G (2003) The biogeochemistry of chlorine in soil. In: Gribble GW (ed.) *The Handbook of Environmental Chemistry, Volume 3: Natural Production of Organohalogen Compounds*, pp. 43–62. Berlin: Springer.

Öberg G, Nordlund E, and Berg B (1996) *In situ* formation of organically bound halogens during decomposition of Norway spruce needles: Effects of fertilization. *Canadian Journal of Forest Research* 26(6): 1040–1048.

Öberg G and Sandén P (2005) Retention of chloride in soil and cycling of organic matter-bound chlorine. *Hydrological Processes* 19(11): 2123–2136.

Palmer MA, Reidy Liermann CA, Nilsson C, *et al.* (2008) Climate change and the world's river basins: Anticipating management options. *Frontiers in Ecology and the Environment* 6: (doi:10.1890/060148).

Paul MJ and Meyer JL (2001) Streams in the urban landscape. *Annual Review Ecology and Systematics* 32: 333–365.

Penman HL (1961) Weather, plant and soil factors in hydrology. *Weather* 16: 207–219.

Peters NE (1984) Comparison of air and soil temperatures at forested sites in the west-central Adirondack Mountains. *Northeastern Environmental Science* 3(2): 67–72.

Peters NE (1994) Water-quality variations in a forested Piedmont catchment, Georgia, USA. *Journal of Hydrology* 156: 73–90.

Peters NE (2008) Water-quality monitoring and process understanding in support of environmental policy and management. In: Tchiguirinskaia I, Demuth S, and Hubert P (eds.) *9th Kovacs Colloquium: River Basins – From Hydrological Science to Water Management/Les bassins versant – de la science hydrologique à la gestion des eaux*, vol. 323, pp. 93–109. Paris: International Association of Hydrological Sciences.

Peters NE, Cerny J, and Havel M (1998) Factors controlling streamwater nitrate concentrations in a forested headwater catchment, Krusné hory Mountains, Czech Republic. In: Haigh MJ, Krecek J, Rajwar GS, and Kilmartin MP (eds.) *Headwaters: Water Resources and Soil Conservation*, pp. 147–157. Rotterdam: AA Balkema.

Peters NE and Driscoll CT (1987a) Hydrogeologic controls of surface-water chemistry in the Adirondack region of New York State. *Biogeochemistry* 3(1/3): 163–180.

Peters NE and Driscoll CT (1987b) Sources of acidity during snowmelt at a forested site in the west-central Adirondack Mountains, New York. In: Swanson RH, Bernier PY, and Woodard PD (eds.) *Forest Hydrology and Watershed Management*, vol. 167, pp. 99–108. Wallingford: IAHS.

Peters NE and Driscoll CT (1989) Temporal variations in solute concentrations of meltwater and forest floor leachate at a forested site in the Adirondacks, New York. In: Lewis JE (ed.) *46th Annual Eastern Snow Conference, Eastern Snow Conference*, pp. 45–56. Quebec City, Canada.

Peters NE and Leavesley GH (1995) Biotic and abiotic processes controlling water chemistry during snowmelt at Rabbit Ears Pass, Rocky Mountains, Colorado, USA. *Water, Air, and Soil Pollution* 79(1/4): 171–190.

Peters NE and Meybeck M (2000) Water quality degradation effects on freshwater availability: Impacts of human activities. *Water International* 25(2): 185–193.

Peters NE, Meybeck M and Chapman DV (2005) Effects of human activities on water quality. In: Anderson MG and McDonnell JJ (eds.) *Encyclopedia of Hydrological Sciences,* vol. 3, Part 8, ch. 93, pp. 1409–1425. Chichester: Wiley.

Peters NE and Murdoch PS (1985) Hydrogeologic comparison of an acidic-lake basin with a neutral-lake basin in the west-central Adirondack Mountains. New York. *Water, Air and Soil Pollution*, 26: 387–402.

Peters NE and Ratcliffe EB (1998) Tracing hydrologic pathways using chloride at the Panola Mountain Research Watershed, Georgia, USA. *Water, Air, and Soil Pollution* 105(1–2): 263–275.

Peterson BJ, Wollheim WM, Mulholland PJ, et al. (2001) Control of nitrogen export from watersheds by headwater streams. *Science* 292(5514): 86–90.

Pezeshki SR (2001) Wetland plant responses to soil flooding. *Environmental and Experimental Botany* 46(3): 299–312.

Philip JR (1967) Sorption and infiltration in heterogeneous media. *Australian Journal of Soil Research* 5(1): 1–10.

Phillips FM, Hogan JF, and Scanlon BR (2004) Section I: Introduction and overview. In: Hogan JF, Phillips FM, and Scanlon BR (eds.) *Groundwater Recharge in a Desert Environment: The Southwestern United States*, pp. 1–14. Washington, DC: American Geophysical Union.

Pitelka LF (1994) Air pollution and terrestrial ecosystems. *Ecological Applications* 4(4): 627–628.

Plummer LN, Busby JF, Lee RW, and Hanshaw BB (1990) Geochemical modeling of the Madison Aquifer in parts of Montana, Wyoming, and South Dakota. *Water Resources Research* 26(9): 1981–2014.

Poff NL, Allan JD, Bain MB, et al. (1997) The natural flow regime – a paradigm for river conservation and restoration. *BioScience* 47(11): 769–784.

Poff NL, Bledsoe BP, and Cuhaciyan CO (2006) Hydrologic variation with land use across the contiguous United States: Geomorphic and ecological consequences for stream ecosystems. *Geomorphology* 79: 264–285.

Pomeroy JW, Davies TD, Jones HG, Marsh P, Peters NE, and Tranter M (1999) Transformations of snow chemistry in the boreal forest: Accumulation and volatilization. *Hydrological Processes* 13(14–15): 2257–2273.

Pruppacher HR, Semonin RG, and Slinn WGN (1983) *Precipitation Scavenging, Dry Deposition, and Resuspension, Volume1: Precipitation Scavenging* London: Elsevier. 729pp.

Pusch M, Fiebig D, Brettar I, et al. (1998) The role of micro-organisms in the ecological connectivity of running waters. *Freshwater Biology* 40(3): 453–495.

Rademacher LK, Clark JF, Hudson GB, Erman DC, and Erman NA (2001) Chemical evolution of shallow groundwater as recorded by springs, Sagehen basin; Nevada County, California. *Chemical Geology* 179: 37–51.

Rascher CM, Driscoll CT, and Peters NE (1987) Concentration and flux of solutes from snow and forest floor during snowmelt in the west-central Adirondack Region of New York. *Biogeochemistry* 3: 209–224.

Rassam DW, Knight JH, Turner J, and Pagendam D (2006) Groundwater surface water interactions: Modelling denitrification and $\delta^{15}NNO_3$–$\delta^{18}ONO_3$ fractionation during bank storage. In: Institution of Engineers Australia (ed.) *Proceedings of the 30th Hydrology and Water Resources Symposium*, pp. 157–161. Launceston, TAS, Australia, 4–7 December 2006. Sandy Bay, TAS: Conference Design.

Rassam DW, Pagendam DE, and Hunter HM (2008) Conceptualisation and application of models for groundwater–surface water interactions and nitrate attenuation potential in riparian zones. *Environmental Modelling and Software* 23(7): 859–875 (doi:10.1016/j.envsoft.2007.11.003).

Reddy KR and DeLaune RD (2008) *Biogeochemistry of Wetlands: Science and Applications*. Boca Raton, FL: Taylor and Francis.

Reeburgh WS (1983) Rates of biogeochemical processes in anoxic sediments. *Annual Review of Earth and Planetary Sciences* 11: 269–298.

Reynolds B, Emmett BA, and Woods C (1992) Variations in streamwater nitrate concentrations and nitrogen budgets over 10 years in a headwater catchment in mid-Wales. *Journal of Hydrology* 136(1–4): 155–175.

Reynolds B, Fowler D, and Thomas S (1996) Chemistry of cloud water at an upland site in mid-Wales. *Science of the Total Environment* 188: 115–125.

Reynolds B, Neal C, and Norris DA (2001) Evaluation of regional acid sensitivity predictions using field data: Issues of scale and heterogeneity. *Hydrology and Earth System Science* 5(1): 75–81.

Reynolds B, Ormerod SJ, and Gee AS (1994) Spatial patterns in stream nitrate concentrations in upland Wales in relation to catchment forest cover and forest age. *Environmental Pollution* 84(1): 27–33.

Richard TL and Steenhuis TS (1988) Tile drain sampling of preferential flow on a field scale. *Journal of Contaminant Hydrology* 3: 307–325.

Richards JH and Caldwell MM (1987) Hydraulic lift: Substantial nocturnal water transport between soil layers by *Artemisia tridentata* roots. *Oecologia* 73: 486–489.

Richey JE, Melack JM, Aufdenkampe AK, Ballester VM, and Hess LL (2002) Outgassing from Amazonian rivers and wetlands as a large tropical source of atmospheric CO_2. *Nature* 416(6881): 617–620.

Roberts BJ and Mulholland PJ (2007) In-stream biotic control on nutrient biogeochemistry in a forested headwater stream, West Fork of Walker Branch. *Journal of Geophysical Research* 112: G04002 (doi:10.1029/2007JG000422).

Robertson K (2001) Emissions of N_2O in Sweden – natural and anthropogenic sources. *AMBIO* 20(3–4): 151–155.

Rose S and Peters NE (2001) Effects of urbanization on streamflow in the Atlanta area (Georgia, USA): A comparative hydrological approach. *Hydrological Processes* 15(8): 1441–1457.

Rosenberry DO, Glaser PH, and Siegel DI (2006) The hydrology of northern peatlands as affected by biogenic gas: Current developments and research needs. *Hydrological Processes* 20(17): 3601–3610.

Rosenberry DO and Labaugh JW (2008) Field techniques for estimating water fluxes between surface water and ground water. In: US Geological Survey (ed.) *US Geological Survey Technical Water Resources Investigations*, Book 4, ch. D2, 128pp. Washington, DC: USGS.

Rosenberry DO and Winter TC (1997) Dynamics of water-table fluctuations in an upland between two prairie-pothole wetlands in North Dakota. *Journal of Hydrology* 191: 266–269.

Roulet N, Lafleur P, Richard P, Moore T, Humphreys E, and Bubier J (2007) Contemporary carbon balance and late Holocene carbon accumulation in a northern peatland. *Global Change Biology* 13(2): 379–411.

Roulet NT (1990) Hydrology of a headwater basin wetland: Groundwater discharge and wetland maintenance. *Hydrological Processes* 4(4): 387–400.

Saunders DL and Kalff J (2001) Nitrogen retention in wetlands, lakes and rivers. *Hydrobiologia* 443: 205–212.

Schimel J (2004) Playing scales in the methane cycle: From microbial ecology to the globe. *Proceedings of the National Academy of Sciences of the United States of America* 101(34): 12400–12401.

Schindler DW (1988) Effects if acid rain on freshwater ecosystems. *Science* 239(4836): 149–157.

Schindler DW (1999) From acid rain to toxic snow. *AMBIO* 28(4): 350–355.

Schindler DW, Hecky RE, Findlay DL, et al. (2008) Eutrophication of lakes cannot be controlled by reducing nitrogen input: Results of a 37-year whole-ecosystem experiment. *Proceedings of the National Academy of Sciences of the United States of America* 105(32): 11254–11258.

Schlesinger WH (1997) *Biogeochemistry: An Analysis of Global Change*, 2nd edn. San Diego, CA: Academic Press.

Schnoor JL and Stumm W (1986) The role of chemical weathering in the neutralization of acidic deposition. *Aquatic Sciences* 48(2): 171–195.

Scholl MA (2000) Effects of heterogeneity in aquifer permeability and biomass on biodegradation rate calculations – results from numerical simulations. *Ground Water* 38(5): 702–712.

Scholl MA, Cozzarelli IM, and Christenson SC (2006) Recharge processes drive sulfate reduction in an alluvial aquifer contaminated with landfill leachate. *Journal of Contaminant Hydrology* 86(3–4): 239–261.

Scholl MA, Giambelluca TW, Gingerich SB, Nullet MA, and Loope LL (2007) Cloud water in windward and leeward mountain forests: The stable isotope signature of orographic cloud water. *Water Resources Research* 43: W12411.

Schulze E-D, Caldwell MM, Canadell J, et al. (1998) Downward flux of water through roots (i.e. inverse hydraulic lift) in dry Kalahari sands. *Oecologia* 115: 460–462.

Schumm SA (1977) *The Fluvial System*. New York: Wiley. 338pp.

Schuster PF, Reddy MM, LaBaugh JW, et al. (2003) Characterization of lake water and ground water movement in the littoral zone of Williams Lake, a closed-basin lake in north central Minnesota. *Hydrological Processes* 17(4): 823–838.

Scott NA, Sagger S, and McIntosh PD (2001) Biogeochemical impact of *Hieracium* invasion in New Zealand's grazed tussock grasslands: Sustainability implications. *Ecological Applications* 11(5): 1311–1322.

Sebilo M, Billen G, Grably M, and Mariotti A (2003) Isotopic composition of nitrate–nitrogen as a marker of riparian and benthic denitrification at the scale of the whole Seine River system. *Biogeochemistry* 65: 35–51.

Seitzinger S, Harrison JA, Böhlke JK, et al. (2006) Denitrification across landscapes and waterscapes: A synthesis. *Ecological Applications* 16(5): 2064–2090.

Shanley JB and Peters NE (1993) Variations in aqueous sulfate concentrations at Panola Mountain, Georgia. *Journal of Hydrology* 146: 361–382.

Skjelkvåle BL, Evans C, Larssen T, Hindar A, and Raddam GG (2003) Recovery from acidification in European surface waters: A view to the future. *AMBIO* 32(3): 170–175.

Smith RL, Baumgartner LK, Miller DN, Repert DA, and Böhlke JK (2006) Assessment of nitrification potential in ground water using short term, single-well injection experiments. *Microbial Ecology* 51: 22–35.

Smith RL, Böhlke JK, Garbedian SP, Revesz KM, and Yoshinari T (2004) Assessing denitrification in groundwater using natural gradient tracer tests with ^{15}N: *In situ* measurement of a sequential multistep reaction. *Water Resources Research* 40: W07101 (doi:10.1029/2003WR002919).

Smith RL, Garabedian SP, and Brooks MH (1996) Comparison of denitrification activity measurements in groundwater suing cores and natural-gradient tracer tests. *Environmental Science and Technology* 30(12): 3448–3456.

Smith VH (1983) Low nitrogen to phosphorus ratios favor dominance by blue-green algae in lake phytoplankton. *Science* 221: 669–671.

Sorrell BK, Mendelssohn IA, McKee KL, and Woods RA (2000) Ecophysiology of wetland plant roots: A modelling comparison of aeration in relation to species distribution. *Annals of Botany* 86(3): 675–685.

Soulsby C, Tetzlaff D, Rodgers P, Dunn S, and Waldron S (2006) Runoff processes, stream water residence times and controlling landscape characteristics in a mesoscale catchment: An initial evaluation. *Journal of Hydrology* 325: 197–221.

St. Louis VL, Rudd JWM, Kelly CA, Beaty KG, Flett RJ, and Roulet NT (1996) Production and loss of methylmercury and loss of total mercury from boreal forest catchments containing different types of wetlands. *Environmental Science and Technology* 30(9): 2719–2729.

Stackelberg PE, Furlong ET, Meyer MT, Zaugg SD, Henderson AK, and Reissman DB (2004) Persistence of pharmaceutical compounds and other organic wastewater contaminants in a conventional drinking-water treatment plant. *Science of the Total Environment* 339: 99–113.

Stanley EH, Fisher SG, and Grimm NB (1997) Ecosystem expansion and contraction in streams. *BioScience* 47(7): 427–435.

Sterner RW and Elser JJ (2002) *Ecological Stoichiometry: The Biology of Elements from Molecules to the Biosphere*. Princeton, NJ: Princeton University Press. 466pp.

Stevens PA, Norris DA, Sparks TH, and Hodgson AL (1994) The impacts of atmospheric N inputs on throughfall, soil and stream water interactions for different aged forest and moorland catchments in Wales. *Water, Air and Soil Pollution* 73(1): 297–317.

Stockner JG and MacIsaac EA (1996) British Columbia lake enrichment programme: Two decades of habitat enhancement for sockeye salmon. *Regulated Rivers: Research and Management* 12: 547–561.

Stoddard JL, Jeffries DS, Lukewille A, et al. (1999) Regional trends in aquatic recovery from acidification in North America and Europe. *Nature (London)* 401: 575–578.

Strahler AN (1964) Quantitative geomorphology of drainage basins and channel networks. In: Chow VT (ed.) *Handbook of Applied Hydrology*, pp. 4–11. New York: McGraw-Hill.

Stream Solute Workshop (1990) Concepts and methods for assessing solute dynamics in stream ecosystems. *Journal of the North American Benthological Society* 9: 95–119.

Stumm W and Morgan JJ (1996) *Aquatic Chemistry*, 3rd edn. New York: Wiley. 1022pp.

Svensson BH, Lantsheer JC, and Rodhe H (2001) Sources and sinks of methane in Sweden. *AMBIO* 20(3–4): 155–160.

Szramek K and Walter LM (2004) Impact of carbonate precipitation on riverine inorganic carbon mass transport from a mid-continent, forested watershed. *Aquatic Geochemistry* 10(1–2): 99–137.

Tabacchi E, Lambs L, Guilloy H, PlantyTabacchi AM, Muller E, and Decamps H (2000) Impacts of riparian vegetation on hydrological processes. *Hydrological Processes* 14(16–17): 2959–2976.

Tank JL, Rosi-Marshall EJ, Baker MA, and Hall RO (2008) Are rivers just big streams? A pulse method to quantify nitrogen demand in a large river. *Ecology* 89(10): 2935–2945 (doi:10.1890/07-1315.1).

Tetzlaff D, Seibert J, and Soulsby C (2009) Inter-catchment comparison to assess the influence of topography and soils on catchment transit times in a geomorphic province; the Cairngorm mountains. *Scotland. Hydrological Processes*, 23(13): 1874–1886.

Tetzlaff D, Soulsby C, Waldron S, et al. (2007) Conceptualization of runoff processes using a geographical information system and tracers in a nested mesoscale catchment. *Hydrological Processes* 2(110): 1289–1307.

Tilman D, Kilham SS, and Kilham P (1982) Phytoplankton community ecology: The role of limiting nutrients. *Annual Review Ecology and Systematics* 13: 349–372.

Tranter M, Brimblecombe P, Davies TD, Vincent CE, and Abrahams PW (1986) The composition of snowfall, snowpack and meltwater in the Scottish Highlands – evidence for preferential elution. *Atmospheric Environment* 20(3): 517–525.

Tranter M and Jones HG (2001) The chemistry of snow: Processes and nutrient cycling. In: Jones HG, Pomeroy JW, Walker DA, and Hoham RW (eds.) *Snow Ecology: An Interdisciplinary Examination of Snow-Covered Ecosystems*, pp. 127–165. Cambridge: Cambridge University Press.

Treml V and Banas M (2008) The effect of exposure on alpine treeline position: A case study from the High Sudetes, Czech Republic. *Arctic, Antarctic, and Alpine Research* 40(4): 751–760.

Triska FJ, Kennedy VC, Avanzino RJ, Zellweger GW, and Bencala KE (1989) Retention and transport of nutrients in a third-order stream in northwestern California: Hyporheic processes. *Ecology* 70(6): 1893–1905.

Troch PA, Paniconi C, and van Loon EE (2003) Hillslope-storage Boussinesq model for subsurface flow and variable source areas along complex hillslopes: 1. Formulation and characteristic response. *Water Resources Research* 39(11): 1316 (doi:10.1029/2002WR001728).

Vannote RL, Minshall GW, Cummins KW, Sedell JR, and Cushing CE (1980) The river continuum concept. *Canadian Journal of Fisheries and Aquatic Sciences* 37(1): 130–137.

Veatch W, Brooks PD, Gustafson JR, and Molotch NP (2009) Quantifying the effects of forest canopy cover on net snow accumulation at a continental, mid-latitude site. *Ecohydrology* 2(2): 115–128.

Verry ES, Brooks KN, and Barten PK (1988) Streamflow response from an ombrotorhic mire. In: *International Symposium on the Hydrology of Wetlands in Temperate and Cold Regions, Joensuu, Finland*, pp. 52–59. Helsinki, Finland: Academy of Finland.

Vitousek PM, Gosz JR, Grier CC, Melillo JM, and Reiners WA (1982) A comparative analysis of potential nitrification and nitrate mobility in forest ecosystems. *Ecological Monographs* 52(2): 155–177.

Vogel JC, Talma AS, and Heaton THE (1981) Gaseous nitrogen as evidence for denitrification in groundwater. *Journal of Hydrology* 50: 191–200.

Vollenweider RA (1969) Possibilities and limits of elementary models concerning the budget of substances in lakes. *Archiv für Hydrobiologie* 66: 1–36.

Vollenweider RA and Kerekes J (1982) *Eutrophication of Waters*. Paris: OECD. 154pp.

Walsh CJ, Roy AH, and Feminella JW (2005) The urban stream syndrome: Current knowledge and the search for a cure. *Journal of the North American Benthological Society* 24(3): 706–723.

Walter BP and Heimann M (2000) A process-based, climate-sensitive model to derive methane emissions from natural wetlands: Application to five wetland sites, sensitivity to model parameters, and climate. *Global Biogeochemical Cycles* 14(3): 745–765.

Walter BP, Heimann M, and Matthews E (2001) Modeling modern methane emissions from natural wetlands, 1. Model description and results. *Geophysical Research Letters* 106(D24): 34189–34206.

Watmough SA, Aherne J, and Dillon PJ (2003) Potential impact of forest harvesting on lake chemistry in south-central Ontario at current levels of acid deposition. *Canadian Journal of Fisheries and Aquatic Sciences* 60(9): 1095–1103.

Watmough SA and Dillon PJ (2003) Base cation and nitrogen budgets for seven forested catchments in central Ontario, 1983–1999. *Forest Ecology and Management* 177: 155–177.

Weathers KC and Likens GE (1997) Clouds in southern Chile: An important source of nitrogen to nitrogen-limited ecosystems? *Environmental Science and Technology* 31: 210–213.

Weathers KC, Lovett GM, and Likens GE (1995) Cloud deposition to a spruce forest edge. *Atmospheric Environment* 29(6): 65–672.

Webster JR and Patten BC (1979) Effects of watershed perturbation on stream potassium and calcium dynamics. *Ecological Monographs* 49(1): 51–72.

Weissmann GS, Zhang Y, LaBille EM, and Fogg GE (2002) Dispersion of groundwater age in an alluvial aquifer system. *Water Resources Research* 38(1). (doi:10.1029/2001WR000907).

Wetzel RL (2001) *Limnology: Lake and River Ecosystems*, 3rd edn. San Diego, CA: Academic Press. 1006pp.

White AF and Blum AE (1995) Effects of climate on chemical weathering in watersheds. *Geochimica et Cosmochimica Acta* 59(9): 1729–1747.

White AF and Brantley SL (2003) The effect of time on the weathering of silicate minerals: Why do weathering rates differ in the laboratory and field? *Chemical Geology* 202: 479–506.

White PJ and Broadley MR (2001) Chloride in soils and its uptake and movement within the plant: A review. *Annals of Botany* 88(6): 967–988.

Wilcox DA, Shedlock RJ, and Hendrickson WH (1986) Hydrology, water chemistry and ecological relations in the raised mound of Cowles Bog. *Journal of Ecology* 74(4): 1103–1117.

Wildenschild D, Jensen KH, Villholth K, and Illangasekare TH (1994) A laboratory analysis of the effect of macropores on solute transport. *Ground Water* 32(3): 381–389.

Williams K, Caldwell MM, and Richards JH (1993) The influence of shade and clouds on soil water potential: The buffered behavior of hydraulic lift. *Plant and Soil* 157: 83–95.

Williams MW, Bales RC, Brown AD, and Melack JM (1995) Fluxes and transformations of nitrogen in a high-elevation catchment, Sierra Nevada. *Biogeochemistry* 28(1): 1–31.

Winter TC (1999) Relation of streams, lakes and wetlands to groundwater flow systems. *Hydrogeology Journal* 7: 28–45.

Winter TC, Harvey JW, Franke OL, and Alley WM (1999) Ground water and surface water – a single resource. *US Geological Survey Circular* 1139: 1–79.

Winter TC and Woo M-K (1990) Hydrology of lakes and wetlands. In: Wolman MG and Riggs HC (eds.) *Surface Water Hydrology. The Geology of North America*, vol. 0–1, pp. 159–187. Boulder, CO: Geological Society of America.

Wollheim WM, Vörösmarty CJ, Peterson BJ, Seitzinger SP, and Hopkinson CS (2006) Relationship between river size and nutrient removal. *Geophysical Research Letters* 33: L06410 (doi:10.1029/2006GL025845).

Worrall F, Harriman R, Evans CD, *et al.* (2004) Trends in dissolved organic carbon in UK rivers and lakes. *Biogeochemistry* 70(3): 369–402.

Wright RF, Aherne J, Bishop K, *et al.* (2006) Modelling the effect of climate change on recovery of acidified freshwaters: Relative sensitivity of individual processes in the MAGIC model. *Science of the Total Environment* 365: 154–166.

Wright RF, Larssen T, Camarero L, *et al.* (2005) Recovery of European acidfied surface waters. *Environmental Science and Technology* 39(3): 64A–72A (doi: 10.1021/es0531778).

Wright RF, Norton SA, Brakke DF, and Frogner T (1988) Experimental verification of episodic acidification of freshwaters by sea salts. *Nature* 334: 422–424.

Xenopoulos MA, Lodge DM, Fretress J, *et al.* (2003) Regional comparisons of watershed determinants of dissolved organic carbon in temperate lakes from the Upper Great Lakes region and selected regions globally. *Limnology and Oceanography* 48(6): 2321–2334.

Zhang Q, Zhang Y, Peng S, Yirdaw E, and Wu N (2009) Spatial structure of alpine trees in Mountain Baima Xueshan on the southeast Tibetan Plateau. *Silva Fennica* 42(2): 197–208.

2.12 Catchment Erosion, Sediment Delivery, and Sediment Quality

DE Walling, University of Exeter, Exeter, UK
SN Wilkinson, CSIRO Land and Water, Townsville, QLD, Australia
AJ Horowitz, US Geological Survey, Atlanta, GA, USA

© 2011 Elsevier B.V. All rights reserved.

2.12.1	A Changing Context	305
2.12.2	Sediment Budgets	307
2.12.2.1	The Sediment Budget as an Integrating Concept	307
2.12.2.2	The Functioning of the Sediment Budget	308
2.12.2.3	The Global Sediment Budget	311
2.12.3	Documenting Catchment Sediment Budgets	312
2.12.3.1	The Background	312
2.12.3.2	The Use of Fallout Radionuclides	313
2.12.3.3	Sediment Source Fingerprinting	314
2.12.3.4	The Future	317
2.12.4	Modeling the Catchment Sediment Budget	317
2.12.4.1	The Requirement	317
2.12.4.2	Model Development	317
2.12.4.2.1	Modeling approaches and model complexity	317
2.12.4.2.2	Empirical modeling of catchment sediment yield	318
2.12.4.2.3	Conceptual process modeling of catchment sediment budgets	319
2.12.4.2.4	Mechanistic, physically based modeling of hillslope processes	320
2.12.4.3	SedNet – A Sediment Budget Model for River Networks	320
2.12.4.3.1	Model outline	320
2.12.4.3.2	Management applications	320
2.12.4.4	Current Status and Future Directions	322
2.12.4.4.1	Modeling across scales for planning and management	322
2.12.4.4.2	Directions in modeling erosion and deposition processes	322
2.12.4.4.3	Model uncertainty considerations	323
2.12.5	The Quality Dimension	323
2.12.5.1	Introduction	323
2.12.5.2	Basic Sediment Geochemistry	324
2.12.5.3	Major Issues Associated with Sediment Quality	324
2.12.5.3.1	Background/baseline sediment-associated constituent concentrations	324
2.12.5.3.2	The collection of representative sediment samples and the issues of spatial and temporal variability	325
2.12.5.3.3	The chemical analysis of suspended and bed sediments	329
2.12.5.3.4	Bioavailability and toxicity	330
2.12.5.4	Future Directions	331
References		331

2.12.1 A Changing Context

Although it has attracted the interest of fluvial geomorphologists, geologists, sedimentologists, and hydrologists, the study of erosion and sediment transport by rivers has traditionally been largely the preserve of the agricultural engineer and the hydraulic or civil engineer (e.g., ASCE, 1975; Schwab et al., 1981; Lal, 1994; Julien, 1995, 2002; Morgan, 1995; Yang, 1996; US Department of Agriculture, 1997; Chien and Wan, 1999; Fangmeier et al., 2006; US Bureau of Reclamation, 2006; Garcia, 2008). In the case of erosion, attention focused largely on soil erosion on agricultural land and emphasized the assessment of rates of soil loss and their implications for crop productivity and the sustainability of land use practices, as well as the design of soil conservation measures.

The well-known Universal Soil Loss Equation (USLE) developed in the USA (see Wischmeier and Smith, 1978) and modified for application elsewhere (e.g., Schwertmann et al., 1990; Larionov, 1993) was a product of this interest in erosion processes, providing a basis for predicting the spatial variation of rates of soil loss in response to their control by rainfall, topography, and land use practices, and for assessing the potential impact of improved management and cropping practices. Hydraulic engineers directed attention to the study of sediment transport by rivers and related problems associated with the management of river channels for navigation and flood control and reservoir sedimentation, as well as to the design of hydraulic structures able to cope with high sediment loads. Such work commonly emphasized the coarser fractions of the sediment load, as this was most important in terms of river morphology. It was also more readily predicted from a knowledge of sediment properties and flow conditions, than the finer washload, which was generally a noncapacity load and therefore a supply-controlled load. The washload was frequently viewed as being of limited importance, since it was readily transported through a river system and had limited morphological impact. The emphasis on the transport of coarser sediment either as bedload or as suspended bed material is well demonstrated by the large number of sediment transport formulas that were developed in the middle years of the twentieth century for predicting these components of the sediment load (see ASCE, 1975). As a noncapacity or supply-controlled load, the washload of a river was seen as something that was not easy to predict using sediment transport formulas and, if it was of interest, it therefore needed to be measured.

Against this background, attention traditionally focused on soil loss from upstream areas, with emphasis on on-site problems of soil degradation and loss of productivity and on downstream sediment transport and sediment yield. There was often only limited contact between those working on these two aspects. Stated very simply, soil loss from fields was lost from the farm and its ultimate fate was of limited interest to the agricultural engineer. Equally, eroded soil generally contributed primarily to the finer washload of a river, which was seen as being of limited importance in terms of river morphology and hydraulics, although it was an important cause of reservoir sedimentation and some knowledge of downstream suspended sediment yields was therefore needed. In general, the degree of attention given to the study of erosion and sediment transport was broadly proportional to the magnitude of erosion rates and sediment loads. In countries such as the USA, there was considerable activity in these areas, whereas in countries such as the UK, where erosion rates were low and soil erosion was not perceived to be a problem, and sediment yields were also low and rivers relatively small, activity and interest were limited.

A major change in the above situation occurred in the latter part of the twentieth century. Changing perspectives on erosion and sediment transport promoted increased interest in this general field and emphasized the need for a more multidisciplinary perspective and a more integrated approach that directed attention to the functioning of the entire catchment system. Greater emphasis was therefore placed on the hydrological context. Several key drivers of these changes can be identified. One was the recognition of the importance of fine sediment, both as a key water-quality parameter in terms of its physical presence and also as an important control on river water quality more generally. Many pollutants and contaminants, including heavy metals, pesticides and other organic contaminants, as well as nutrients such as phosphorus, are transported primarily in association with sediment and interactions between the solid (sediment) and liquid (water) phases exert an important influence on water quality (e.g., Golterman, 1977; Shear and Watson, 1977; UNESCO, 1978; Allan, 1979; Förstner and Wittmann, 1981; Hart, 1982; Salomons and Förstner, 1984; Horowitz, 1991; Ongley et al., 1992; Santiago et al., 1994; US Environmental Protection Agency, 1997; House and Warwick, 1999; Warren et al., 2003). In addition to considering the amount of sediment transported by a river, there was also an increasing need to consider the quality of that sediment. Another important driver was the increasing evidence of the detrimental effects of fine sediment in degrading aquatic habitats and ecosystems, through, for example, the siltation of fish spawning gravels, the smothering of aquatic vegetation and increased nutrient inputs to floodplains, riparian areas, and other water bodies, through sediment deposition (see Ritchie, 1972; Clark, 1985; Clark et al., 1985; Waters, 1995; Newcombe and Jensen, 1996; Wood and Armitage, 1997; Soulsby et al., 2001; Suttle et al., 2004; Cavalcanti and Lockaby, 2005).

Both in terms of its physical presence and its quality, fine sediment is frequently an important cause of environmental degradation and it has been widely referred to as the world's number one pollutant. Diffuse source pollution was increasingly recognized as an important cause of water pollution, and sediment, which can be mobilized from throughout a river basin, is a major component of such pollution. Around the Great Lakes of North America, for example, concern for the eutrophication and pollution of these water bodies and particularly Lake Erie, directed attention to the need to control diffuse source pollution and sediment assumed a central role in the Pollution from Land Use Activities Reference Group (PLUARG) program developed by the International Joint Commission on the Great Lakes (Coote et al., 1982; Ongley, 1982). In many ways, this program was ahead of its time in recognizing the importance of sediment and the role of land use in influencing sediment mobilization and transfer to water bodies. Sediment has also assumed considerable importance in the recent EU Water Framework and Habitats Directives (Förstner, 2002, 2003) aimed at improving land management practices, protecting aquatic habitats, and maintaining conditions of good ecological status in rivers. In addition, increasing interest in the functioning of the Earth's system has highlighted the important role of land–ocean sediment transfer in global geochemical cycling, and particularly the carbon cycle (Ludwig et al., 1996; Lyons et al., 2002; Beusen et al., 2005; Seitzinger et al., 2005; Gislason et al., 2006; Van Oost et al., 2007; Saenger et al., 2008). River sediment loads have been shown to be very sensitive to the various drivers of global change (e.g., Walling and Fang, 2003; Walling, 2008) and to exert an important influence on the health of receiving waters in the coastal zone. Within the International Geosphere Biosphere Programme (IGBP), launched in 1987 by the International Council for Scientific Unions (ICSU) particular attention was directed to land–ocean sediment and material transfers through its Land–Ocean Interactions in the Coastal Zone (LOICZ) core project.

Significant outcomes of the evolution of these new perspectives on erosion and sediment transport include the following. First there has been an increasing emphasis on fine sediment (see Owens et al., 2005). This is the most significant fraction of the sediment load in terms of pollution and sediment-associated transport of nutrients and contaminants, since contaminants are in most cases preferentially associated with the finer ($<63\,\mu m$) particles (Horowitz, 1991). Equally, it is generally fine sediment which is of greatest importance in terms of the degradation of aquatic ecosystems and habitats. Since the fine sediment load of a river is commonly a noncapacity load and supply-limited, interest in fine sediment transport has necessarily shifted the emphasis of sediment transport studies. It has moved away from a hydraulic approach, that emphasized hydraulic controls and the transport conditions in the channel toward developing an improved understanding of sediment mobilization and transfer within the entire catchment and thus the supply of sediment to the river system. Clearly, this demands a hydrological approach. Second, with greater emphasis being directed to sediment quality, and increasing recognition that sediment quality is in most instances closely related to sediment source, there has been a need to develop an improved understanding of potential sediment sources and transfer pathways. This was also a key requirement for any attempt to develop sediment management or control programs and to implement mitigation measures. Resources need to be targeted to those sources or parts of a catchment that provide the main source of the sediment transported by a river and which need to be controlled. Again, therefore, a distributed hydrological approach for understanding and modeling the sediment dynamics of a catchment or river system has been increasingly required. Third, the shift of emphasis away from concern for sediment problems linked primarily to the amount of sediment and thus the magnitude of erosion rates and sediment yields to the more wide ranging environmental significance of fine sediment has broadened the relevance of the study of erosion and sediment transport to include most areas of the world. Paradoxically, it is often areas with low erosion rates and low sediment yields where the environmental impacts of sediment are potentially greatest and the need to develop an improved understanding of the processes of sediment mobilization and transfer is therefore strongest. Finally, as indicated above, these new perspectives and requirements have created the need to integrate studies of erosion and sediment transport. The off-site problems of soil erosion, which relate to the onward transfer of the mobilized sediment through a drainage basin and the impacts of this sediment, are frequently seen as being equally, if not more, important than the on-site problems of soil loss. Equally, the increased emphasis on fine, rather than coarse, sediment transport has focused attention on sediment supply to river channels and the need to look beyond the river channel and to consider the processes of sediment mobilization and transfer within the entire upstream catchment area.

A hydrological perspective is a key requirement for the new perspectives on erosion and sediment transport outlined above. This has in turn strengthened the position of the study of erosion and sediment transport as an important branch of hydrology. This position has long been recognized by the International Association of Hydrological Sciences (IAHS), through the activities of its Commission on Continental Erosion which was established in the middle years of the twentieth century. Equally, the Hydrology section of the American Geophysical Union and the Hydrological Sciences division of the European Geosciences Union both include groups devoted to the study of erosion and sedimentation. The need for a broader multidisciplinary perspective is also demonstrated by the emergence of specialist groups focusing on sediment studies, such as the International Association for Sediment Water Science (IASWS), which was established in 1984, and the World Association for Sedimentation and Erosion Research (WASER), which was founded in 2004.

This contribution reviews some of the key developments associated with the changing focus of studies of erosion and sediment yield outlined above. Attention is directed first to the sediment budget concept, which provides a valuable framework for studying, modeling, and managing erosion and sediment yield in catchments. Second, approaches to modeling catchment sediment budgets are considered. Finally, several of the key contemporary issues associated with sediment quality are discussed.

2.12.2 Sediment Budgets

2.12.2.1 The Sediment Budget as an Integrating Concept

Although there is undoubtedly still a place for a reductionist approach, which focuses attention on the dynamics of a particular process associated with erosion and sediment transport, recognition of the wide ranging environmental significance of fine sediment and the need to link information on sediment output from drainage basins with information on sediment sources and sediment mobilization, transfer, and storage has resulted in a general acceptance of the sediment budget as a central integrating concept for the study of erosion and sediment yield. In addition to integrating the various components of sediment mobilization, transfer, storage, and output and providing a valuable scientific framework for research investigations, the sediment budget concept also provides an essential management tool (Walling and Collins, 2008). It identifies the key sediment sources and transfer pathways within a catchment, which are likely to represent the focus of any management strategy. Furthermore, it emphasizes the sensitivity of the sediment response of a catchment to environmental change and the potentially complex links between changing erosion rates and changes in sediment yield, which must be recognized when planning and implementing sediment management and control strategies.

Figure 1, which is based on the classic work of Trimble (1983) in the 360 km^2 catchment of Coon Creek, Wisconsin, USA, provides a useful demonstration of the catchment sediment budget concept and the way in which it integrates consideration of sources, sinks, and output and thus sediment mobilization, transport, deposition and storage, as well as the dynamic interaction of these components. In the Coon Creek study, two separate budgets were developed. The first was for the period of poorly managed agriculture and severe erosion that followed land clearance and the expansion of agriculture in the latter half of the nineteenth century and the early part of the twentieth century. The second was for the subsequent period, when soil conservation measures were introduced to control erosion and soil degradation. An important feature of the budgets for both periods is that only a relatively small proportion of the total mass of sediment mobilized within the basin by erosion reaches the basin outlet (i.e., ~5–7%). This emphasizes that information on the sediment yield at a basin outlet may provide a poor indication of the overall amount of sediment mobilized and moved through a basin and emphasizes that the key to understanding the system frequently lies in identifying and quantifying the sediment sinks or stores. Whereas attention has traditionally focused on erosion processes and sediment transport, the sinks and stores can dominate the functioning of a catchment sediment budget, with much of the sediment mobilized in catchments being deposited on hillslopes, in riverbeds, on floodplains, and in reservoirs, rather than contributing to catchment export (e.g., Dunne et al., 1998; Trimble and Crosson, 2000). Figure 1 demonstrates that during both periods, large amounts of sediment were being stored in the colluvial deposits associated with the hillslopes within the upland areas and in alluvial sinks within both the tributary valleys and the main valley of Coon Creek. Comparison of the sediment budgets for the two periods shows that although the implementation of soil conservation measures after 1938 greatly reduced upland erosion rates, producing a substantial (i.e., $\sim 25\%$) reduction in sediment mobilization from the slopes, the sediment yield at the basin outlet changed very little, due to the increased efficiency of sediment transfer through the channel system (i.e., reduced deposition) and the remobilization of sediment that had accumulated within the middle valley during the preceding period of accelerated erosion.

From a management perspective, a sediment budget, such as that presented in Figure 1, provides valuable information for use in developing effective catchment-based sediment management and control strategies. It identifies the most important sediment sources that would need to be targeted in any attempt to reduce downstream sediment fluxes and thus facilitates the optimum use of the resources available for implementing sediment control measures. Equally, it also emphasizes that reduction of upstream erosion may not necessarily result in a significant reduction of the downstream sediment yield. Much of the sediment generated upstream may have previously been deposited and stored before reaching the catchment outlet, and reduction in upstream sediment mobilization could be offset by remobilization of sediment from intervening stores. An understanding of the sediment budget of a drainage basin is clearly also important for predicting the likely impact of future climate change on downstream sediment response. This could change significantly, if hydrological changes resulted in the remobilization of sediment from existing sediment sinks, for example, through changing channel morphology and increased channel migration and erosion.

2.12.2.2 The Functioning of the Sediment Budget

As indicated above, the sediment budget concept provides a valuable integrating framework for studying the various processes of sediment mobilization and delivery operating within a catchment. Table 1 lists most of the key processes involved and some of the recent research aimed at developing an improved understanding of these processes. Although, as shown in Figure 1, the emphasis is commonly placed on the magnitude of the fluxes and stores, and thus on the quantities of sediment involved, it is also important to recognize that the properties of the sediment associated with different

Table 1 Examples of recent research on the component processes of the sediment budget

Process	References
Sediment mobilization	
Interrill or sheet erosion	Abrahams et al. (2001), Giménez and Govers (2002), Prosser and Rustomji (2000), Valmis et al. (2005), Wei et al. (2009), Zhang et al. (2009a, 2009b).
Rill erosion	Cerdan et al. (2006), Govers et al. (2007), Lei et al. (2006), Merten et al. (2001), Schiettecatte et al. (2008).
Gully erosion	Gomez et al. (2003), Gordon et al. (2008), Poesen et al. (2003), Rustomji (2006), Valentin et al. (2005).
Mass movements	Brayshaw and Hassan (2009), Chappell et al. (2004), Hassan et al. (2005), Heimsath et al. (2002), Lavigne and Suwa (2003),Wemple et al. (2001), Schuerch et al. (2006).
Channel bank erosion	Atkinson et al. (2003), Florsheim et al. (2008), Fox et al. (2007), Rinaldi et al., (2008), Hupp et al. (2009), Jeffries et al. (2003), Laubel et al. (2003), Wynn and Mostaghimi (2006).
Sediment transfer or delivery	
Slope to channel	Rustomji and Prosser (2001), Croke et al. (2005), Deasy et al. (2009), Haygarth et al. (2006), Preston and Schmidt (2003), Smith and Dragovich (2008).
Channel	Droppo et al. (2001, 2004), Forbes and Lamoreux (2005), Malmon et al. (2005), Petticrew et al. (2007), Rubin and Topping (2001), Simon et al. (2004), Stone et al. (2008).
Sediment deposition and storage	
Colluvial	Brardinoni et al. (2009), Cochrane and Flanagan (2006), Croke et al. (1999), de Moor and Verstraeten (2008), Macaire et al. (2002), Rommens et al. (2006).
Channel	Collins and Walling (2007), Hart (2002), Macnab et al. (2006), Petticrew et al. (2007), Smith et al. (2003), Steiger et al. (2003).
Alluvial fans	Field (2001), Harvey (2002), Harvey et al. (2005), Leeder and Mack (2001), Ritter et al. (2000), Staley et al. (2006).
Floodplain	Aalto et al. (2003, 2008), Hughes et al. (2009), Kronvang et al. (2009), Lauer and Parker (2008), Sweet et al. (2003), Swanson et al. (2008), Jeffries et al. (2003), Thonon (2006), Thonon et al. (2007).
Sediment yield	
	Ali and de Boer (2008), Evans and Slaymaker (2004), Haregeweyn et al. (2008), Steegen et al. (2001), Syvitski and Milliman (2007), Molina et al. (2007), Tamene et al. (2006), Verstraeten and Poesen (2002), Verstraeten et al. (2003).

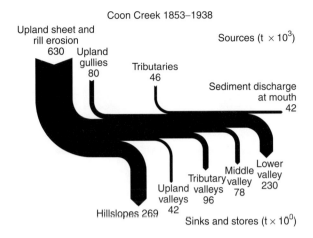

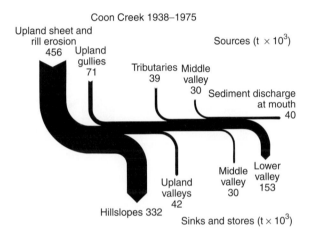

Figure 1 The sediment budgets for Coon Creek, Wisconsin, USA for the periods 1853–1938 and 1938–1975 produced by Trimble (1983). The fluxes shown are mean annual values. From Trimble (1983) A sediment budget for Coon Creek basin in the Driftless Area, Wisconsin, 1853–1977. *American Journal of Science* 283: 454–474.

components of the budget may change, as sediment is transferred from source to sink. Mobilization, transfer, and deposition processes will frequently involve selectivity related to both particle size and particle density, resulting in contrasts between the composition of sources and sediment associated with different components of the budget (e.g., Fontaine *et al.*, 2000; Stone and Droppo, 1996). In the case of variations in particle size, the contrast between the effective or *in situ* grain size and the ultimate or absolute grain size of the sediment, which reflects the existence of composite particles (i.e., aggregates and flocs), can introduce further complexity and exert an important influence on the properties of sediment associated with individual components of the sediment budget (see Stone and Saunderson, 1992; Stone and Walling, 1997; Walling *et al.*, 2000; Blake *et al.*, 2005; Woodward and Walling, 2007). In the case of sinks, for example, coarser particles may be preferentially deposited, but if these coarser particles comprise aggregates or flocs, they may contain a significant proportion of fine particles, the deposition of which might otherwise be unexpected (Nicholas and Walling, 1996). Droppo (2001) has called for a rethinking of conventional approaches to investigating suspended sediment dynamics to reflect the existence and importance of such composite particles. Equally, sediment mobilized from different sources may be characterized by different properties, and the sediment associated with different components of the budget may change according to its source or the relative contribution from different sources. In this context, contrasts in the properties of sediment derived from different sources may reflect both the source type (e.g., sheet and rill erosion vs. gully and channel bank erosion) and the spatial variability of source material properties caused by variations in geology, soil type, or land use across a catchment.

Key characteristics of the functioning of a sediment budget include its connectivity and thus the extent to which the slopes or upstream parts of a catchment are linked to the channel system or the catchment outlet. The connectivity of a system will depend on the incidence and efficiency of the transfer pathways and the magnitude of the stores. The connectivity of the system is clearly of fundamental importance when investigating and attempting to control sediment-induced diffuse source pollution. Detailed assessment of connectivity necessitates consideration of the transfer pathways involved and their efficiency in transferring sediment through the sediment delivery system. By providing a clearer representation of the links between erosion or sediment mobilization and sediment yield, the catchment sediment budget represents an important advance over the sediment delivery ratio concept. The latter simple blackbox concept (e.g., Roehl, 1962; Walling, 1983) recognized that the sediment output was likely to be less than the gross sediment mobilization and represented the ratio of the former to the latter. The magnitude of this ratio was in turn linked to the size of the catchment, with its magnitude commonly decreasing as the scale of the catchment increased. Many limitations of the sediment delivery ratio concept have been widely debated (e.g., Walling, 1983; Parsons *et al.*, 2006; de Vente *et al.*, 2007) and Beven *et al.* (2005) provide a useful overview of the problems to be faced in conceptualizing sediment delivery to stream channels.

The precise form taken by the sediment budget of a catchment will reflect a wide range of controls, including the local topography and the hydrological regime, as well as the size of the catchment. **Figure 2** provides an indication of the potential nature and extent of such variability by indicating the key characteristics of the sediment budgets of four small drainage basins on the Russian Plain documented by Golosov *et al.* (1992). These are all relatively small basins, heavily impacted by agricultural land use and associated soil erosion. The investigation aimed to establish the proportion of the sediment mobilized within the catchments by different erosion processes that reached the basin outlets. In this environment, three key sediment mobilization processes were identified. These comprised sheet erosion (i.e., widespread erosion of the surface by surface wash), rill erosion (i.e., erosion by concentrated flow in micro-channels developed on the slopes), and gully erosion (i.e., erosion within deeper ephemeral channels that dissect the landscape and where sediment is mobilized by mass movements on the gully sides as well as by the flow through the gully). In this environment, sheet and rill erosion are generally more important than gully erosion as a sediment source and there is little evidence of sediment storage on the lower parts of the slopes. Slopes are frequently convex, terminating at the margins of balkas (flat floored, gully-like features), that dissect the landscape. Even within this relatively homogeneous area, the proportion of the sediment mobilized by erosion within the individual catchments that reaches the basin outlet ranges from 0% to 89%. In most of the catchments, both the balka bottoms and the river floodplains constitute major sinks for sediment moving through the system and as with Coon Creek (see **Figure 1**), the sinks represent a very important component of their sediment budgets.

As the scale of the drainage basin increases, deposition of sediment on the river floodplains in the lower parts of the basin will commonly assume increasing importance. Work within the catchments of the Rivers Ouse (3315 km^2) and Wharfe (818 km^2) in Yorkshire, UK, reported by Walling et al. (1998) has, for example, shown that as much as 30–40% of the sediment delivered to the main channel system is deposited on the adjacent floodplains during overbank flood events and does not reach the basin outlet. At a larger scale, Bobrovitskaya et al. (1996) provided information on the sediment budget of the lower River Ob which drains a vast catchment of 2 950 000 km^2 in Siberia to the Arctic Ocean. The available information on the suspended load of this river provided by two gauging stations on its lower reaches separated by an 870-km reach indicates that in its lower reaches approximately 50% of its suspended sediment load is deposited on the well-developed floodplain bordering the river and fails to reach the lowest measuring station. At this larger scale, tectonic subsidence within the interior of a river basin can also promote the development of major sediment sinks which reduce the downstream sediment flux. This is well illustrated by the Rio Madeira, a major tributary of the Amazon in Bolivia and information reported by Baby et al. (2009). The upper basin of this river, which extends to ~170 000 km^2, drains the Andean Cordillera where erosion rates are high and consequently transports a very high annual suspended sediment load of ~500–600 Mt. However, on leaving the Andes, the downstream course of the Rio Madeira passes through a subsiding foreland basin where of the order of 270 Mt of sediment is deposited each year. As a result, only about 45% of the upstream load of the Rio Madeira is transported downstream into the main Amazon river system.

The operation of sediment budgets, such as those depicted in **Figures 1–3**, can be considered over several different timescales. Several recent sediment budget investigations that have taken a longer-term perspective have emphasized the importance of sediment storage, with the majority of the sediment mobilized being stored within the catchment over long periods. For example, Rommens et al. (2005) reported a Holocene sediment budget for a small 103 ha agricultural catchment in the Belgian loess belt that shows that 58–80% of the sediment mobilized within the catchment had been stored near its source and not delivered to downstream rivers. Similarly, Prosser et al. (2001a) estimated that as much as 80% of the sediment eroded from large coastal catchments in Eastern Australia in historical times remains stored in their channels and floodplains. Many of the sediment sinks associated with a sediment budget are likely to be long-term sinks. For example, the sediment deposited on the lower parts of a slope will

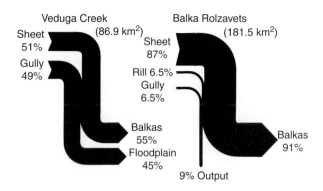

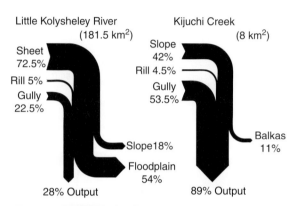

Figure 2 The sediment budgets for four small drainage basins on the Russian Plain, established by Golosov et al. (1992). From Golosov et al. (1992) Sediment budgets of river catchments and river channel aggradation on the Russian plain. *Geomorphology (Moscow)* 4: 62–71 (in Russian).

commonly remain in near-permanent storage, unless there is a significant change in the pattern of erosion. River floodplains will frequently also represent longer-term sinks, with bank erosion perhaps causing some loss, which is balanced by point bar formation and deposition elsewhere. Floodplain sinks could, however, be rapidly remobilized by changes in channel pattern and increased channel migration associated with changes in the flow regime caused by human activity in the upstream catchment or climate change. Some sinks will, however, operate as shorter-term stores. This was the case with the middle valley sink within the Coon Creek catchment depicted in **Figure 1**. Furthermore, at the annual timescale, sediment deposited within the channel system may accumulate during one period of the year, only to be remobilized and flushed out during a subsequent period (e.g., Collins and Walling, 2007). In this situation, storage is clearly temporary and in their study of three groundwater-dominated lowland catchments in the UK, Collins and Walling (2007) demonstrated that fine sediment accumulated within the channel during the winter period, when most sediment was transported through the system, and was subsequently remobilized during the summer period. Estimates of the average mass of sediment stored in the channel systems of the three catchments during the 2-year study period demonstrated that this was equivalent to between 21% and 38% of the mean annual

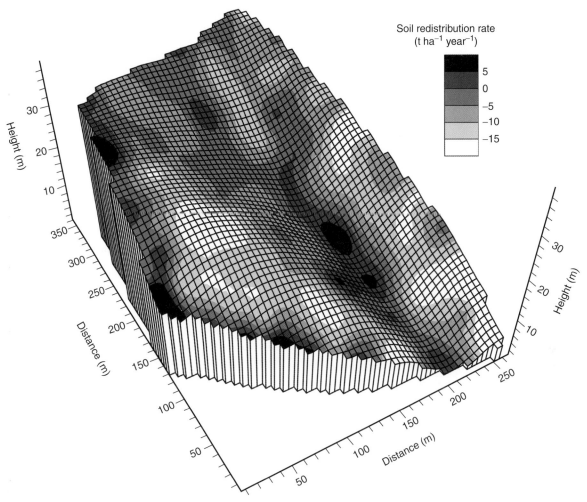

Figure 3 The pattern of soil redistribution within a 7.5 ha field at Butsford Barton near Colebrooke, Devon, UK established by Walling and his co-workers using ^{137}Cs measurements. More than 200 bulk cores were collected from the field at the intersections of a 20 m grid. The soil redistribution rates depicted represent mean annual values for a ~40 year period prior to the mid-1990s.

suspended sediment export from the three catchments. By documenting changes in storage through time, it was also possible to estimate the total amount of sediment entering and leaving channel storage within the three catchments over the study period. The amounts of sediment entering and leaving channel storage within the three catchments were equivalent to between ~20% and 75% and between 25% and 70%, respectively, of the mean annual sediment yield, demonstrating that a significant proportion of the sediment flux passed through this short-term sink.

2.12.2.3 The Global Sediment Budget

Although the application is somewhat different, in terms of both scale and the nature of the budget, a sediment budget approach has also been used to assess the impact of recent changes in the sediment loads of the world's rivers on the global land–ocean sediment transfer. Such changes have important implications for global geochemical cycling. In this case the emphasis has been on the total land–ocean sediment flux and the magnitude of the changes in this flux that have occurred as a result of human activity. Although reliable information on sediment loads is unavailable for many world rivers, there is a general consensus that the contemporary land–ocean sediment flux is of the order of 15 Gt yr^{-1}, and a recent study reported by Syvitski et al. (2005) suggests that the value may be somewhat lower around 12.6 Gt yr^{-1}. However, it is known that this flux is changing as a result of human impact (Walling, 2006a; Syvitski and Milliman, 2007). In some rivers it is increasing, due to land clearance and the expansion of agricultural land use and associated increases in erosion, whereas in others it is declining due to the trapping of sediment by dams. In some river basins both drivers may be operating and the net effect will depend on their relative importance. A key issue is the extent to which the global sediment budget has been perturbed by human influence. This involves establishing the likely natural land–ocean sediment flux and then assessing the extent to which it has been increased and reduced by land disturbance and dam construction respectively.

Syvitski et al. (2005) have used a regression model incorporating the main controls on natural river loads to estimate the pre-human sediment loads of the world's major rivers as being $\sim 14\,\mathrm{Gt\,yr}^{-1}$. This is $1.4\,\mathrm{Gt\,yr}^{-1}$ greater than their estimate of the contemporary land–ocean sediment flux ($12.6\,\mathrm{Gt\,yr}^{-1}$), which will have been influenced by both increases and decreases relative to the pre-human flux, as a result of land disturbance and sediment trapping by dams, respectively. Lack of sediment load data for many rivers in the developing world, where sediment loads are likely to have increased as a result of population growth, makes it difficult to estimate the magnitude of any increase, but more information is available on the impact of dams in reducing sediment fluxes. Although the impact of dams in reducing the sediment loads of the world's rivers is widely recognized (see Milliman et al., 1984; Vörösmarty et al., 1997, 2003; Walling and Fang, 2003; Walling, 2006a), there is currently considerable uncertainty associated with existing estimates of the likely amount of sediment sequestered behind dams on the world's rivers and the resulting reduction in the global land–ocean sediment flux. Vörösmarty et al. (2003) estimated that more than 40% of the global river discharge is currently intercepted by large ($\geq 0.5\,\mathrm{km}^3$ maximum storage capacity) reservoirs, and by coupling this information with estimates of reservoir trap efficiency they estimated that reservoirs are currently sequestering $\sim 4\text{-}5\,\mathrm{Gt\,yr}^{-1}$ of sediment, with the potential for this value to be considerably higher if the large number of smaller reservoirs are also taken into account. Using a similar approach, Syvitski et al. (2005) suggested that the contemporary land–ocean sediment flux is being reduced by $\sim 3.6\,\mathrm{Gt\,yr}^{-1}$ as a result of sediment trapping by dams. These values are, however, an order of magnitude lower than the estimate of the current sedimentation behind the world's major dams provided by a recent study involving the $\sim 33\,000$ dams included in the ICOLD World Register of Dams (ICOLD, 2006), undertaken by the ICOLD Reservoir Sedimentation Committee and reported by Basson (2008). The data provided by this study suggest that sedimentation behind the world's major dams is currently equivalent to an annual sequestration of $\sim 60\,\mathrm{Gt\,yr}^{-1}$ (see Walling, 2008).

It is, however, important to recognize that the estimate of the current rate of sediment sequestration in the world's reservoirs of $\sim 60\,\mathrm{Gt\,yr}^{-1}$ presented above represents the mass of sediment sequestered behind the dams and this does not equate to the associated reduction in the land–ocean sediment flux. Much of this sediment would previously not have reached the oceans, due to deposition and storage within the river system, and particularly on river floodplains. The conveyance loss associated with sediment movement through a river system can clearly be expected to vary according to the magnitude of the sediment flux, the sediment transport and flood regime of the river, and the morphology of the channel system, and is likely to decrease in heavily managed channels, where the flow is constricted and flood inundation restricted. It is therefore difficult to propose a typical value for the conveyance loss likely to be associated with the $\sim 60\,\mathrm{Gt\,yr}^{-1}$ of sediment currently being sequestered behind the dams constructed on the world's rivers. However, Walling, (2008) has suggested a value of 60% as a first-order estimate. Use of this value would mean that 40% of the total $\sim 60\,\mathrm{Gt\,yr}^{-1}$ might be expected to have previously reached the oceans and that dam construction is currently reducing the global land–ocean

Table 2 A comparison of the estimates of the major components of the global sediment budget and their modification by human activity provided by Syvitski et al. (2005) with those generated by Walling (2008), using a different estimate of the reduction in the contemporary sediment flux caused by sediment trapping

Component	Syvitski et al. (2005)	Walling (2008)
Pre-human land–ocean flux (Gt yr^{-1})	14.0	14.0
Contemporary land–ocean sediment Flux (Gt yr^{-1})	12.6	12.6
Reduction in flux associated with reservoir trapping (Gt yr^{-1})	3.6	24
Contemporary flux in the absence of reservoir trapping (Gt yr^{-1})	16.2	36.6
Increase over pre-human flux due to human activity (%)	22	160
Reduction in contemporary gross flux due to reservoir trapping (%)	16	66

sediment flux by about $24\,\mathrm{Gt\,yr}^{-1}$, a value that is considerably in excess of the likely contemporary global land–ocean sediment flux. This value of $24\,\mathrm{Gt\,yr}^{-1}$ is approaching an order of magnitude greater than the values of $3.6\,\mathrm{Gt\,yr}^{-1}$ suggested by Syvitski et al. (2005) as representing the reduction in the contemporary global annual land–ocean flux resulting from sediment trapping by reservoirs.

Taking the above information on the likely magnitude of the contemporary and pre-human land–ocean sediment flux and the potential impact of sediment trapping by dams, it is possible to speculate further on the possible nature of the global sediment budget and the extent to which it has been perturbed by human activity (see Table 2). If the contemporary land–ocean sediment flux is taken to be $12.6\,\mathrm{Gt\,yr}^{-1}$, but it is assumed that this has been reduced by $3.6\,\mathrm{Gt\,yr}^{-1}$ as a result of reservoir trapping, the contemporary flux in the absence of reservoir trapping would be $16.2\,\mathrm{Gt\,yr}^{-1}$. This represents a $\sim 16\%$ increase over the pre-human flux, with this contemporary flux being reduced by $\sim 22\%$. As such, the perturbation associated with human activity is fairly limited. If, however, the same value is used for the contemporary land–ocean flux ($12.6\,\mathrm{Gt\,yr}^{-1}$), but it is assumed that this has been reduced by $24\,\mathrm{Gt\,yr}^{-1}$ as a result of reservoir trapping, the contemporary flux in the absence of reservoir trapping would be $36.6\,\mathrm{Gt\,yr}^{-1}$. This represents a $\sim 169\%$ increase over the pre-human flux and this has, in turn, been reduced by $\sim 66\%$ as a result of reservoir trapping. Under this scenario, human activity must be seen to have had a major influence on the global sediment budget. Further research is clearly required to confirm the magnitude of human impact on the global sediment budget.

2.12.3 Documenting Catchment Sediment Budgets

2.12.3.1 The Background

Traditionally, fine sediment monitoring programs in river basins have focused on measuring the sediment load at the outlet of the catchment or river basin under investigation. This enables the sediment yield ($t\,yr^{-1}$) and specific sediment yield ($t\,km^{-2}\,yr^{-1}$) to be quantified. Such traditional measuring programs are commonly based on manual suspended sediment sampling, using samplers designed to collect representative suspended sediment samples from the measuring cross section (e.g., Guy and Norman, 1970; Gray et al., 2008). These samples are subsequently filtered to determine the suspended sediment concentration ($mg\,l^{-1}$). The suspended sediment flux at the time of sampling is computed as the product of the water discharge and the sediment concentration, taking account of the variation of both sediment concentration and flow velocity in the cross section. If frequent samples are collected, it is possible to interpolate the record of sediment flux between the sampling times, to compute the total load for the study period. Where, as in many situations, fewer samples are collected, rating curves representing relationships between suspended sediment concentration or discharge and water discharge are established using the available samples and the rating curve is used in conjunction with the record of water discharge to estimate the load for the period of interest. The use of rating curves introduces the potential for significant errors in the estimate of sediment flux (see Walling, 1977; Ferguson, 1986; Walling and Webb, 1988). Recent technological advances have greatly expanded the potential for obtaining more reliable estimates of suspended sediment load. Programmable automatic samplers can be used to increase the sampling frequency and to ensure that samples are collected at key times during a flood event (e.g., Lewis and Eads, 2001; Alexandrov et al., 2003). *In situ* sensors can also be deployed to record surrogate information that can be used to provide a continuous or near continuous record of suspended sediment concentration. Turbidity measurements obtained using both optical backscatter and transmission sensors have proved to be particularly valuable for this purpose (e.g., Gippel, 1995; Glysson and Gray, 2002; Schoellhamer and Wright, 2003), although their use is generally limited to relatively low levels of suspended sediment concentration. Other principles, involving lasers and ultrasonic sensors, have also been successfully used to collect information on the variation of both suspended sediment concentration and its grain size composition through time (e.g., Melis and Topping, 2003; Thonon et al., 2005; Topping et al., 2005).

The requirement for information on the overall sediment budget of a catchment, rather than simply an estimate of the sediment load at the catchment outlet, has, however, necessarily introduced a need to develop new approaches capable of documenting rates of sediment mobilization, quantifying the storage elements of sediment budgets and obtaining information on sediment sources and transfer pathways. Traditional techniques provide some scope for assembling such information, but in a recent paper Walling (2006b) suggested that there was a need for a new paradigm which focused on tracing rather than monitoring. Monitoring was seen as continuing to be important, and indeed to be an essential component of any comprehensive measurement program, but the use of tracing techniques was viewed as representing the only effective means of assembling much of the information required to establish a sediment budget.

Two key advances in the application of tracer techniques for investigating catchment sediment budgets can usefully be highlighted. The first is the use of fallout radionuclides to obtain information on soil and sediment redistribution rates within a catchment and the second is the use of sediment source fingerprinting techniques to provide information on the relative contribution of a range of potential sources to the sediment output from a catchment. Both applications are briefly considered below.

2.12.3.2 The Use of Fallout Radionuclides

The use of fallout radionuclides to obtain information on soil and sediment redistribution rates within a catchment is founded on the existence of a number of natural and man-made radionuclides that reach the land surface as fallout, primarily as wet fallout in association with rainfall, and are rapidly and strongly fixed by the surface soil or sediment. The subsequent redistribution of these radionuclides within a catchment or river system is a direct reflection of the movement of the soil or sediment particles to which the radionuclides are attached. By studying the post-fallout redistribution and fate of the selected fallout radionuclide, it is possible to obtain information on soil and sediment redistribution and, therefore, on erosion and deposition rates. The fallout radionuclide most widely used for this purpose is cesium-137 (^{137}Cs) (see Ritchie and Ritchie, 2008). Cesium-137 is a man-made radionuclide that was produced by the testing of thermonuclear weapons in the 1950s and early 1960s. The ^{137}Cs released by these bomb tests was carried up into the stratosphere and transported around the globe. Significant fallout occurred in most areas of the world during the period extending from the mid-1950s through to the 1970s, although the depositional fluxes were much greater in the northern than the southern hemisphere. In the absence of further bomb tests after the Nuclear Test Ban Treaty in 1963, fallout effectively ceased in the mid-1970s. However, in some areas of the world a further fallout input occurred in 1986 as a result of the Chernobyl accident. Fallout from that accident was short-lived, but in some neighboring regions the total fallout associated with the Chernobyl accident exceeded the earlier bomb fallout. Cesium-137 has a half-life of 30.2 years and much of the original fallout still remains within the upper horizons of the soils and sediments of a catchment. By investigating the current distribution of the radionuclide in the landscape, it is possible to obtain information on the net effect of soil and sediment redistribution processes operating over the past ~ 50 years (see Zapata, 2002). When sampling the soils and sediments in a catchment, attention is usually directed to both the inventory or the total amount of ^{137}Cs contained in the soil or sediment ($Bq\,m^{-2}$) and its depth distribution. However, emphasis is frequently placed on the collection of bulk soil cores and their use to determine the inventory at the sampling point, since the sectioning of a core to determine the depth distribution necessitates the analysis of a much larger number of samples, which can prove to be costly and time-consuming. Samples are analyzed by gamma spectrometry and count times of 12–24 h may be required when activities are low. Mean soil redistribution rates over the past ~ 50 years, since the main period of fallout, are established by comparing the inventories measured at individual sampling points with the reference inventory for the study site.

The latter is commonly based on cores collected from an adjacent undisturbed area with minimum slope that can be expected to have experienced neither erosion nor deposition over the past ~50 years. Points with inventories less than the reference inventory are indicative of eroding areas, whereas those with inventories in excess of the reference value indicate deposition. A range of conversion models have been developed for use in estimating erosion and deposition rates, based on the degree of departure of the measured inventory from the reference inventory (e.g., Walling and He, 1999a; Walling et al., 2002; Li et al., 2009). Using a similar approach, ^{137}Cs measurements have also been successfully used to estimate deposition rates on river floodplains over the past ~50 years (Walling and He, 1993, 1997; Terry et al., 2002; Ritchie et al., 2004).

Cesium-137 has now been successfully used in many areas of the world to obtain hitherto essentially unavailable information on medium-term rates of soil and sediment redistribution (see **Figures** 3 and 4, **Table** 3; Ritchie and Ritchie, 2008) and its value as a tracer has been strongly promoted by the International Atomic Energy Agency (IAEA) (see Zapata, 2002). Most applications have involved relatively small areas, since this permits the collection of sufficient samples to obtain representative information on the spatial patterns of soil and sediment redistribution involved. There is, nevertheless, a need for further work to establish procedures for using the approach in a reconnaissance mode, in order to obtain information from larger areas without a major increase in the number of samples that need to be collected and analyzed. Key advantages of the approach include the ability to obtain retrospective information on medium-term soil redistribution rates, the need for only a single sampling campaign, the provision of spatially distributed information relating to the individual sampling points, and the ability to collect information from the natural landscape, without the need to install plots or to otherwise constrain the location of the measuring points.

Although most studies employing fallout radionuclides have been based on ^{137}Cs, both excess lead-210 (^{210}Pb$_{ex}$) and beryllium-7 (^{7}Be) have also been used in a similar manner (see Mabit et al., 2008). Lead-210 is a natural geogenic radionuclide produced as a product of the uranium decay series. Radium-226 (^{226}Ra) is found in most soils and rocks and this decays to produce gaseous radon-222 (^{222}Rn), which in turn decays to ^{210}Pb. Some of the ^{222}Rn diffuses upward through the soil and escapes into the atmosphere where it decays to ^{210}Pb and is deposited as fallout. As with ^{137}Cs, the ^{210}Pb fallout reaching the land surface is rapidly fixed by the soil and its subsequent redistribution is governed by the movement of soil and sediment particles. The fallout ^{210}Pb is termed excess or unsupported ^{210}Pb, to distinguish it from the ^{210}Pb produced by in situ decay, which will be in equilibrium with, or supported by, the parent ^{226}Ra. The use of ^{210}Pb$_{ex}$ to document soil and sediment redistribution within the landscape employs similar assumptions and procedures to those used with ^{137}Cs. Walling and He (1999b) discuss its use in soil erosion studies and He and Walling (1996) provide examples of its application for estimating rates of overbank sedimentation on river floodplains. The half-life of ^{210}Pb is 22.3 years and therefore similar to that of ^{137}Cs. However, because ^{210}Pb is a natural geogenic radionuclide, the fallout has been essentially constant through time and the activity in the soil will reflect fallout receipt and subsequent decay over the past ~100 years. In the case of soil redistribution on slopes, the influence of past erosion on the present inventory will increase toward the present. Measurements of ^{210}Pb$_{ex}$ activity can therefore provide information on longer-term soil and sediment redistribution rates over the past ~100 years and use of both ^{137}Cs and ^{210}Pb$_{ex}$ in combination can provide additional information on the erosional or depositional behavior of a study area (He and Walling, 1996; Walling et al., 2003a) (see also **Figure** 4). Beryllium-7 is a natural cosmogenic radionuclide formed in the upper atmosphere by its bombardment with cosmic rays. In contrast to ^{137}Cs and ^{210}Pb, ^{7}Be has a very short half-life of only 53 days and, because of this, it can be used to provide information on soil and sediment redistribution rates associated with individual events or short periods of heavy rainfall extending over a few weeks (e.g., Walling et al., 1999; Blake et al., 2002; Wilson et al., 2003; Schuller et al., 2006; Sepulveda et al., 2007). The principles involved in applying ^{7}Be measurements are similar to those for ^{137}Cs and ^{210}Pb$_{ex}$, but for most approaches it is important to ensure that the period of interest conforms to a number of requirements, to avoid carry-over effects from previous periods of heavy rain, which could influence the magnitude and spatial distribution of ^{7}Be inventories across the study area. Walling et al. (2009) have recently described a refined procedure for employing ^{7}Be measurements, which largely overcomes this constraint and makes the approach more generally applicable.

2.12.3.3 Sediment Source Fingerprinting

Sediment source fingerprinting techniques can provide important information on the source of the suspended sediment transported by a stream. In simple terms, the techniques attempt to match the properties of the sediment to those of potential sources within the catchment and to establish the relative contribution of those sources to a given sediment sample. Source can be interpreted in terms of both spatial sources, representing different parts of the catchment, perhaps different tributaries or areas underlain by different rock types, and source types, representing sediment mobilized by different processes or from areas with different land use. A set of potential source types could, for example, include surface erosion from cultivated areas and areas of permanent pasture or range, gully erosion, and channel erosion. A wide of sediment properties including color (e.g., Krein et al., 2003), geochemistry (e.g., Collins and Walling, 2002), mineral magnetic properties (e.g., Caitcheon, 1993; Hatfield and Maher, 2009), radionuclide content (e.g., Wallbrink et al., 1998; Matisoff et al., 2002), and stable isotopes (e.g., Fox and Papanicolaou, 2007) have been used to fingerprint potential sources and in most cases a composite fingerprint incorporating several properties is required to discriminate between potential sediment sources. A mixing (or unmixing) model is used to estimate the relative contribution of the potential sediment sources to the sediment sample under consideration. Walling (2005) provides an overview of the development of source tracing techniques and their potential, emphasizing many complexities that need to be taken into account in order

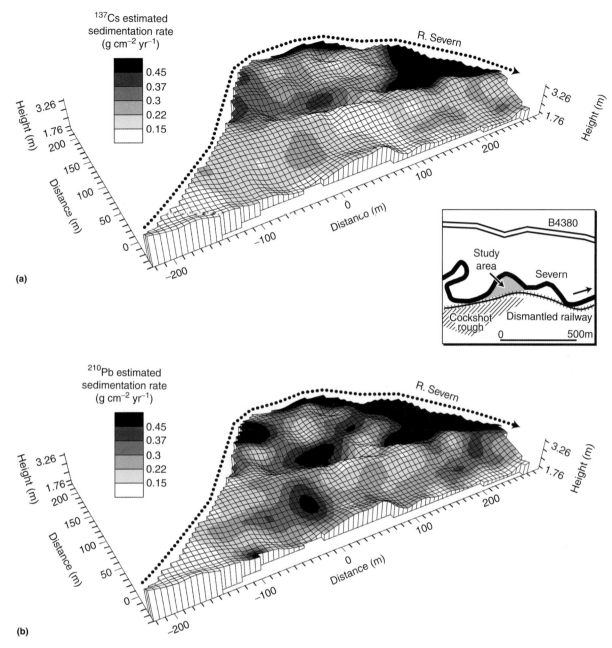

Figure 4 The patterns of overbank deposition of fine sediment on a portion of the floodplain of the River Severn near Buildwas, Shropshire, UK established by Walling and his co-workers using ^{137}Cs and ^{210}Pb$_{ex}$ measurements. 124 bulk cores were collected from the floodplain at the intersections of a 25-m grid. The estimates of mean annual sedimentation rate estimated using the ^{137}Cs measurements relate to the past ~40 years, whereas those based on the ^{210}Pb$_{ex}$ measurements relate to the past ~100 years.

to obtain meaningful and reliable results. Of particular importance are the need to verify statistically the discriminatory power of the fingerprints employed (e.g., Collins et al., 1997a), to recognize many sources of uncertainty incorporated into the approach (e.g., Rowan et al., 2000) and to express and interpret the results accordingly, and to take account of possible differences between source material and sediment samples in terms of grain size composition and organic matter content (e.g., Collins et al., 1997b). The success of source fingerprinting techniques depends heavily on identifying a range of elements and/or isotopes that are capable of discriminating potential sources with a high degree of reliability. Fallout radionuclides have frequently been successfully incorporated into fingerprints used to discriminate between surface sources under different land use (e.g., cultivation, pasture, and forest) and channel/subsurface sources (e.g., Walling et al., 2008) but a new generation of fingerprint properties based on compound specific stable isotopes (CSSIs) associated with the fatty acids produced by plants appears to offer the potential to discriminate between source areas supporting different vegetation covers. Gibbs (2008) reports a study undertaken in North Island New Zealand where CSSIs were used to establish the relative importance of areas under sheep pasture, indigenous forest, and exotic forest plantations as sources of the sediment deposited in a downstream estuary.

Most existing fingerprinting studies focus on establishing the source of the suspended sediment transported by a river. In early work this commonly involved collecting individual samples of sediment, but more recent studies have frequently made use of time-integrating sediment traps (e.g., Phillips et al., 2000), in order to provide a single sample representative of the sediment transported during the period of sample collection. In addition to suspended sediment, the fingerprinting approach has also been applied to overbank sediment deposits on floodplains (e.g., Bottrill et al., 2000) and fine sediment recovered from salmon spawning gravels (e.g., Walling et al., 2003c) and it is possible to generate a temporal perspective and to investigate changes in sediment source through time by applying the same approach to a sediment core collected from a lake or floodplain and interpreting downcore changes in sediment properties in terms of source fingerprints (e.g., Collins et al., 1997b; Walling et al., 2003b; Pittam et al., 2009).

Taken together and combined with more traditional monitoring techniques for obtaining information on the sediment flux at the catchment outlet, these two sets of sediment tracing techniques afford a valuable means of obtaining much of the information required to establish a catchment sediment budget (e.g., Walling et al., 2001, 2006). Thus, for example, it is possible to link information on the source of the sediment load at the catchment outlet provided by source fingerprinting techniques with information on rates of sediment redistribution within those source areas and rates of accretion in sediment sinks such as river floodplains provided by fallout radionuclides, to quantify the key sources and sinks within the sediment budget of a catchment. **Figure 5** depicts

Table 3 A spatially integrated assessment of soil redistribution within the 7.5 ha field at Butsford Barton near Colebrooke, Devon, UK, based on the estimates of soil redistribution rates provided by ^{137}Cs measurements presented in **Figure 3**

Parameter	Value
Percentage area with erosion (%)	79
Percentage area with deposition (%)	20
Mean erosion rate for eroding area (t ha^{-1} yr^{-1})	10
Mean deposition rate for deposition zones (t ha^{-1} yr^{-1})	7.5
Net erosion rate for the field (t ha^{-1} yr^{-1})	6.5
Sediment delivery ratio for the field (%)	81

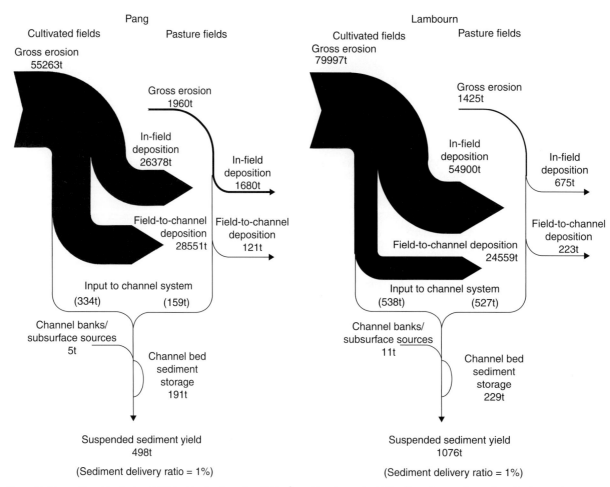

Figure 5 The sediment budgets established for the Pang (~166 km^2) and Lambourn (~234 km^2) catchments in southern England by Walling et al. (2006).

the sediment budgets of the Pang ($\sim 166\,km^2$) and Lambourn ($\sim 234\,km^2$) catchments (located on the chalk of southern England), which were established using this approach. In this case, the highly permeable strata underlying the catchments and the resulting dominance of groundwater in the runoff from the catchment mean that storm runoff is limited and that little sediment leaves the catchment. However, there is evidence of relatively high rates of sediment mobilization and redistribution within the catchments, and in this environment the functioning of their sediment budgets is dominated by the internal sediment sinks.

2.12.3.4 The Future

Recent years have seen important advances in the development of improved methods for characterizing and establishing catchment sediment budgets. The methods now available are able to provide information and understanding to support the development of sediment management programs. However, further advances are required to meet future information requirements, which are likely to place increasing emphasis on the targeting of sediment control strategies, in order to maximize the benefits achieved by investment in such strategies. The use of new sediment source fingerprints, such as CSSIs, can be expected to provide significant improvements in tracing sediment from specific sources and assessing the importance of those sources. In the case of fallout radionuclide applications, most existing work has focused on small areas and there is an important need to upscale their use to larger areas. Because of the limitations on sample numbers commonly associated with sample counting facilities, this upscaling cannot be achieved by simply increasing the number of samples collected. Attention needs to be directed to the development of reconnaissance sampling strategies capable of maximizing the information supplied by a small number of samples. In turn, there is a need to integrate the use of fallout radionuclide techniques with numerical modeling and geographical information systems (GISs), in order to optimize the spatial extrapolation of the resulting information. Advances in sensor technology will undoubtedly bring new and improved methods for monitoring sediment fluxes at catchment outlets that increase the temporal resolution of the records obtained and extend the scope of the data obtained. Scope undoubtedly exists to obtain valuable information on the grain size composition of the sediment load, as well as its magnitude. Such developments are likely to prove to be important in compensating for the progressive reduction in field staff dedicated to sediment sampling and other related monitoring activities that have occurred in many countries in recent years.

2.12.4 Modeling the Catchment Sediment Budget

2.12.4.1 The Requirement

Supporting national and international programs that aim to reduce the threat soil erosion poses to agricultural production and downstream aquatic environments require systematic assessments of sediment sources and their connectivity to downstream impacts (Phillips, 1986). Assessments can provide a technical basis to assist targeting of limited resources to maximize benefits (NLWRA, 2001; Bohn and Kershner, 2002). At river basin scale, managers are often faced with a paucity of erosion data or uneven distribution of measurements across the assessment area. Measurements will have been made at different times and various scales for a range of purposes. Modeling can enable systematic assessment of erosion severity over much larger areas than can be practically covered by measurement alone (Reid and Dunne, 2003). Modeling can also enable assessment over longer time periods with a wider range of climatic conditions, including potential future climates, which is critical for long-term planning, given the high temporal variability of erosion and sediment delivery. Modeling periods of decades may be required to represent adequately the aggregate effects of climatic variability.

Several requirements for modeling erosion and sediment delivery can be identified from a management perspective:

1. Models should explicitly represent the primary erosion processes occurring, so that priorities can be developed, effective interventions identified, and the effect of alternative management scenarios simulated.
2. Spatial patterns in erosion rates should be identified by representing the primary environmental drivers of erosion and sediment delivery.
3. The connectivity of upstream erosion sources to downstream sediment loads should be represented, which requires consideration of sediment sinks as well as sources, and the potential for source connectivity to vary spatially depending on the location of sediment sinks.
4. Assessing erosion and sediment delivery at national or continental scales requires models with modest data requirements.
5. Where sediment-associated pollutants, such as phosphorus and agro-chemicals, are a focus of management, then sediment particle size fractions should be explicitly represented. Pollutants preferentially attach to fine sediment fractions (Section 2.12.5.2), and erosion and transport process behavior will differ between fine and coarse fractions.

These requirements, particularly the first three, suggest that a sediment budget is a suitable framework for modeling, because it accounts for the sources, transport and sinks of material, with a river basin being the confining domain (Trimble, 1993; Reid and Dunne, 2003). The following subsections describe three aspects of sediment budget modeling: (1) the evolution and development of models, (2) an example of integrated sediment budget modeling, and (3) the current status and future directions of model development and application.

2.12.4.2 Model Development

2.12.4.2.1 Modeling approaches and model complexity

Erosion and sediment delivery modeling can be based upon interpolation and synthesis of measurements, or upon physical reasoning and identification of the environmental factors that control the key processes. Typically, a combination is used (DeRoo, 1996), with process models providing spatial resolution between measurement points and helping to identify the upstream erosion and delivery processes, and measurement points being used to constrain model predictions. Three approaches to erosion and sediment delivery modeling can be identified, representing different weightings between measurement and mathematical process description (Beck, 1987; Merritt et al., 2003):

1. Empirical modeling, usually based on a small number of causal variables, often spatially and temporally lumped, and calibrated to measurements. These include catchment-specific relationships between catchment area and sediment yield for example.
2. Conceptual process modeling, which represents generation, routing and storage processes within landscape units or catchments using simple representation of their controlling parameters, and without process interactions. They may be semi-lumped into units or subcatchments with time-steps of days to decades (see Merritt et al., 2003).
3. Mechanistic physical-process modeling provides detailed representation of runoff-generation processes, and usually for application at point, field, or small watershed scale, at finer temporal and spatial resolution than employed in conceptual modeling.

Model complexity generally increases from empirical through conceptual to mechanistic approaches, as the level of process representation and the level of spatial and temporal resolution required increase. There is interdependence between process complexity and the temporal and spatial resolution of a model; finer spatial and temporal resolution requires more complexity in process representation, and vice versa. For example, conceptual rainfall-runoff models require more storage terms and model parameters for predicting daily or monthly runoff than for long-term average runoff (Jothityangkoon et al., 2001). The interdependence between process complexity and model resolution scale means that most empirical models focus on lumped process representations of catchment sediment yield or individual erosion processes, and can be implemented over large areas and long time periods.

Most mechanistic models are spatially distributed and focus on hillslopes or small watersheds for individual events. Conceptual models are often semi-lumped, and focus on simple representation of catchment erosion and deposition processes.

The most appropriate model design considers each component of the erosion and sediment system with a level of complexity that is appropriate for the problem at hand and for the data available (Reid and Dunne, 2003). This is important because, for a given data availability and process knowledge, there is a maximum model complexity which fully exploits the information provided by input data and above which predictive capacity is reduced (Grayson and Blöschl, 2000; Figure 6). Above this level of complexity, it also becomes difficult to identify appropriate parameter values (Beck, 1987). The optimization of model complexity to maximize predictive capacity has contributed to the number of models which have been developed in recent decades.

The evolution of each modeling approach is briefly described below, where some of the more widely known models are used as examples and model-specific reviews are cited.

2.12.4.2.2 Empirical modeling of catchment sediment yield

Empirical modeling of catchment sediment yield emphasizes available data rather than process representations, and this feature can represent both a strength and weakness, depending on the problem being addressed. On the one hand, sensitivity to data constrains model output and enhances the opportunity for new system understanding, making empirical modeling particularly useful where system understanding is weak. Model empiricism is also vital for investigating or calibrating models of sediment yield processes for which the fundamental physical constraints are not sufficiently well known or described. Empirical models are generally relatively simple, and consequently have modest data requirements.

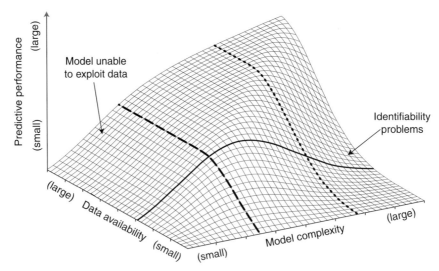

Figure 6 Schematic diagram of the relationship between hydrologic model complexity, data availability and predictive performance. Reproduced from Grayson R and Blöschl G (2000) Spatial modelling of catchment dynamics. In: Grayson R and Blöschl G (eds.) *Spatial Patterns in Catchment Hydrology*, ch. 3, pp. 51–81. Cambridge: Cambridge University Press.

Examples of empirical sediment yield modeling include:

- Modeling the sediment load at a river station, using discharge and sediment concentration data. Commonly, frequent discharge measurements ($L^3 T^{-1}$) are available, with occasional measurements of suspended sediment concentration ($M L^{-3}$). There is considerable short-term temporal variability in sediment concentration (Nistor and Church, 2005). The key modeling challenge is to explain and predict the variation in sediment concentrations between observations, with sediment load in a given time period ($M T^{-1}$) then being the product of discharge and concentration. A common approach is to model concentration using functions of discharge, or sediment rating curves (Asselman, 2000). Rating curves assume steady-state behavior, although records can be divided into multiple time windows to represent system changes. More recently, neural network and other modeling techniques have been employed, which consider the influence of antecedent as well as present discharge and concentration on sediment load (Kisi, 2005).
- Lumped models of sediment yield, based on upstream catchment area (Wasson, 1994), include other basin metrics such as relief, runoff, climate zone, lithology, and anthropogenic factors (Syvitski and Milliman, 2007).

Weaknesses of empirical modeling include: (1) the spatial and temporal resolution and extent are limited by the data available; (2) their lack of explicit process representations can limit predictive capacity outside the study area or measured range of environmental characteristics; (3) the heterogeneity of catchment characteristics such as rainfall, topography, lithology, and land use is not usually represented in spatially lumped models; this reduces predictive capacity, given the significant spatial correlations, and nonlinear dependencies, between slope gradient, runoff, and other driving variables of erosion (Van Rompaey et al., 2001); (4) the absence of source and sink process representations in empirical sediment yield models can limit the number of different types of data which can be meaningfully assembled. Data sets that may enhance the model may instead be used to interpret model results (e.g., Singer and Dunne, 2001).

2.12.4.2.3 Conceptual process modeling of catchment sediment budgets

Conceptual process modeling generally represents sediment routing through a catchment, using a semi-lumped structure of subcatchments or management units. Source (erosion) and sink (deposition) processes are commonly represented for each defined spatial unit, although not all models calculate complete sediment budgets. Feedbacks between processes will not usually feature. Models that identify sediment sources and delivery through the river network are more consistent with field-based evidence of sediment fluxes (Trimble and Crosson, 2000), and are more suitable than lumped models of sediment yield for the modeling requirements identified in Section 2.12.4.1 (Phillips, 1986).

Individual erosion and deposition components of conceptual process models are based on prior field measurements and studies, which provide the basis to identify and mathematically formulate the environmental controls and physical constraints on process behavior. Thus, conceptual models evolve as the process understanding improves. Process component models are usually somewhat empirical in nature, requiring some calibration to match observed erosion and deposition rates, but providing predictive capacity where measurements are not available.

There are several types of conceptual process models of erosion and sediment budgets, which employ different types of input data:

1. A catchment sediment budget developed from a combination of air-photo interpretation, mapping of geomorphic units or zones, field erosion measurements, and dating of sediment deposits (Trimble, 1983; Wasson et al., 1998; Curtis et al., 2005). As the amount of data increase, these budget models become increasingly complex in terms of the sources and sinks represented, and the spatial and temporal resolutions and extents. These models can help guide rehabilitation efforts (Trimble, 1993). Prediction outside the study area is limited where measurements dominate model inputs.
2. A reach sediment budget generated from load estimates based on sediment monitoring data, which can be used to identify reaches of net erosion or deposition (Singer and Dunne, 2001).
3. A source fingerprinting analysis based on sediment tracer properties, which assesses the relative contribution of erosion processes and/or source areas to river sediment without requiring direct measurement of erosion or deposition rates (Walling, 2005; Walling and Collins, 2008; Davis and Fox, 2009).
4. Semi-lumped spatial models of sediment budgets which commonly use GIS functions to divide river basins into subcatchments or watersheds, each draining to a river link, and route sediment through river networks (Benda and Dunne, 1997; Prosser et al., 2001b; Wilkinson et al., 2006, 2009).
5. Distributed spatial models which compute hillslope runoff, erosion, and sediment delivery at the resolution of input data sets, with sediment routed according to topography. Examples include AGNPS (Agricultural Nonpoint Source model; Young et al., 1989) and SEDEM (SEdiment DElivery Model; Van Rompaey et al., 2001). More precisely, these models should be seen as partially distributed, with some lumped elements remaining.

Spatial lumping of model domains into morphological units or subcatchments, and application of lumped parameter values across units and timescales is common in conceptual models. However, the scale of lumping can strongly influence the model predictions. For example, spatial patterns of land use within units may be far from random, which can affect sediment delivery.

For practical reasons related to the time and effort involved, more complex semi-lumped and distributed models were originally limited to small hillslope applications. By the late 1990s, the wide availability of GIS software and techniques enabled more complex models with spatially varying input data to be easily applied to larger areas. Today, data requirements and the ability to realistically describe processes are the dominant limitations on modeling.

2.12.4.2.4 Mechanistic, physically based modeling of hillslope processes

Available mechanistic models span a range of complexity in the process representations used to model runoff and erosion. Some models are more physically based, representing soil infiltration and runoff routing analytically, for example, using kinematic waves, while others use empirical approaches such as runoff curve numbers. Surface erosion is modeled using USLE-based detachment equations, or more complex shear stress and stream power functions (DeRoo et al., 1996; Srivastava et al., 2007). Some models predict net soil loss considering both detachment and deposition (Nearing et al., 1989).

Indicating their focus toward hillslope or small watershed scales, mechanistic models do not commonly include gully and riverbank erosion and channel deposition processes. Several stand-alone models predict ephemeral gully erosion using flow shear stress and sediment transport capacity, although their predictive capacity has received little testing (Poesen et al., 2003). Several mechanistic models of permanent gullies have also been developed, describing the evolution of morphology during the early stages of gully development and the final morphological characteristics (Poesen et al., 2003). Again, however, their applicability has not been widely tested.

Available physically based models cover a wide variety of spatial and temporal scales, with the former ranging from individual hillslopes to fields. Some models are pixel-based, and accommodate GIS data to facilitate modeling small watersheds (DeRoo et al., 1996; Srivastava et al., 2007). The majority of physically based models are designed to simulate individual events rather than long time periods (Aksoy and Kavvas, 2005). Many mechanistic and distributed models have input data requirements that are technically or financially unattainable over large river basins or continents or multiyear time periods (Van Rompaey et al., 2001).

The distributed or hillslope unit structure of mechanistic models addresses problems associated with spatial lumping, such as spatial correlation between inputs. However, considerable spatial variability in surface roughness, slope gradient, and other variables often exist even within model spatial units. Distributed models are sensitive to errors in surface slope and topography. Consequently, it is common for mechanistic physics-based models to require calibration to reproduce observed behavior, which potentially introduces error in process modules.

2.12.4.3 SedNet – A Sediment Budget Model for River Networks

2.12.4.3.1 Model outline

To illustrate considerations in design and application of sediment budget models, this section describes the SedNet (Sediment budget river Network) model. SedNet constructs budgets of the primary sources and sinks of sediment for each link in a river network (Prosser et al., 2001b; Wilkinson et al., 2004, 2009). This model structure may be considered generic to all semi-lumped sediment budget models. The structure enables spatial representation of the connectivity of upstream sources to downstream yields, including the role of floodplains and impoundments within catchments (Prosser et al., 2001c). Predicted suspended sediment yield is supply limited in the long term, which is consistent with observations. The river network is defined from a digital elevation model (DEM). Separate budgets are constructed for sand and gravel bed material (Wilkinson et al., 2006), for suspended sediment (Wilkinson et al., 2009), and for particulate and dissolved phosphorus and nitrogen (Wilkinson et al., 2004). The sediment yield from the downstream end of each link accounts for material sourced from hillslope and gully erosion in the subcatchment which drains directly to the link, bank erosion along the link, and from upstream tributaries. Deposition is accounted for on floodplains, in impoundments or reservoirs, and accumulation of bed material in the river channel. The processes of land-sliding, debris flow, and hillslope soil creep are not significant sediment sources in the Australian environment (for which SedNet was developed), although they could be added for model application elsewhere. The net change in channel storage of suspended sediment over decades is assumed to be negligible relative to other terms, and so in-channel deposition and re-entrainment of suspended sediment is ignored.

The budget is reported as mean annual values for a set of conditions. The effects of temporal variability in climate and hydrology on each source and sink are modeled by regionalizing statistics of daily discharge. The process representations are generally conceptual in nature, designed to show the primary physical controls to provide predictive capacity in low-data environments. For example, hillslope erosion is represented by the Revised Universal Soil Loss Equation (RUSLE; Renard et al., 1997), with a sediment delivery ratio accounting for deposition within hillslopes. Gully sediment yield is constrained by the estimated volume of gully networks and their period of development. Spatial variation in riverbank erosion is estimated as a function of stream power, and the extent of erodible soil and riparian vegetation (Wilkinson et al., 2009). Parameter values for a given environment are specified based on the knowledge of erosion and deposition processes developed through field measurement (Bartley et al., 2007), reconstruction of erosion histories (Wasson et al., 1998), sediment tracing (Wallbrink et al., 1998), and independent sediment yield estimates (Rustomji et al., 2008).

2.12.4.3.2 Management applications

SedNet was developed for the Australian National Land and Water Resources Audit, which investigated the spatial patterns in erosion processes, and the offsite impacts of agriculture across the Australian continent (NLWRA, 2001). The modeling indicated marked differences in sediment supply between regions, with gully and river bank erosion dominating sediment supply in temperate regions, and hillslope erosion dominating in tropical regions, due to the higher rainfall intensity. Only 25% of fine sediment delivered to streams was predicted to be delivered to estuaries overall (Prosser et al., 2001a).

SedNet has since been applied at regional scale using higher-resolution datasets, to better support catchment planning. For example, increased riverine sediment exports from the catchments draining to the Great Barrier Reef (GBR) threaten to degrade near-shore coral reef and benthic ecosystems (De'ath and Fabricius, 2010). Modeling predicted that 70% of sediment export comes from just 20% of the total catchment area (**Figure 7**). The spatial pattern of contribution to export was highest in near-coastal areas with high rainfall intensity, steep slopes, and more intensive land management (McKergow et al., 2005). The model has also been used to compare scenarios of future management. **Figure 8** demonstrates that targeting erosion control to areas and erosion sources in descending order of their contribution to sediment export can achieve reductions in export several times larger than would be achieved by spatial random changes in land management.

Evaluation of SedNet against yield estimates from suspended sediment rating curves, and against sediment tracer data, indicates that the model can reliably differentiate between the areas contributing most and least to basin yield, provided that input data are of good quality (Wilkinson, 2008; Wilkinson et al., 2009). SedNet has also been used to assess the location and extent of accumulations of sand and gravel bed material within river networks, indicating that up to 25% of the river network is affected in some river basins, particularly downstream of gully and riverbank erosion hotspots. This approach provides improved predictive capacity over spatially lumped models (Wilkinson et al., 2006). River links predicted to have bed material accumulation have impaired biological health, with lower abundance

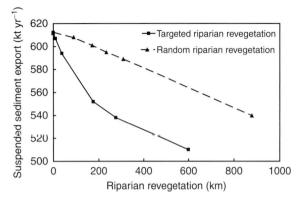

Figure 8 Simulated reductions in suspended sediment export from the Murrumbidgee River catchment, showing that spatially-targeted control of gully and river bank erosion can achieve larger reductions in export than spatially-random control measures. Reproduced from Wilkinson SN, Prosser IP, Olley JM, and Read A (2005) Using sediment budgets to prioritise erosion control in large river systems. In: Batalla RJ and Garcia C (eds.) *Geomorphological Processes and Human Impacts in River Basins*, IAHS Publication 299, pp. 56–64. Solsona, Catalonia: IAHS Press, with permission from IAHS press.

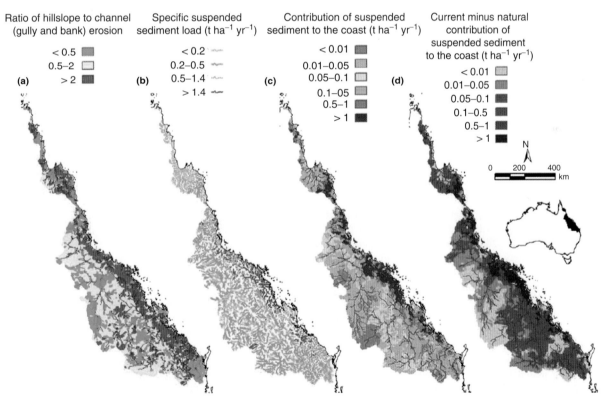

Figure 7 SedNet results for the catchments draining to the Great Barrier Reef (a) estimated ratio of hillslope erosion to channel erosion (gully plus riverbank) in each subcatchment, (b) predicted specific suspended sediment load, (c) predicted contribution of suspended sediment to the coast under current conditions, and (d) the difference between estimated current and natural contribution to suspended sediment export. Reproduced from McKergow et al. (2005) Sources of sediment to the Great Barrier Reef World Heritage Area. *Marine Pollution Bulletin* 51: 200–211, with permission from Elsevier.

of habitat-sensitive macro-invertebrate taxa (Harrison et al., 2008).

2.12.4.4 Current Status and Future Directions

2.12.4.4.1 Modeling across scales for planning and management

There is an increasing demand for land management and environmental stewardship to be underpinned by technical assessments. Land management and planning is increasingly driven by, and administered through, national and international programs, rather than locally. The challenge for erosion and sediment modeling is to provide robust assessments of the effects of local and dynamic changes in land management on erosion and sediment yield outcomes over large areas ($100-100\,000\,km^2$) and long time periods (10–100 years). Modelling and measurement are becoming more integrated in the study of catchment erosion and sediment delivery. Remote sensing has provided a rapid increase in the quality of spatial data of topography, soil, and vegetation cover (Vrieling, 2006). A range of data types are now being used to constrain the modeled sediment budget, including independent load estimates and sediment tracing (Rustomji et al., 2008; Wilkinson et al., 2009), and dating of sediment deposits (de Moor and Verstraeten, 2008).

There are no models that simultaneously operate at all scales from point land management practices to basin sediment yields, and such a model would require large increases in input data resolution (Srivastava et al., 2007). The present response to the need for information at multiple scales is to apply erosion and sediment delivery models of different scales in a more closely coupled and integrated fashion. For example, conceptual river basin models can be used to identify priority source areas, where mechanistic models are applied at the field scale to optimize management practices, the outputs of which are then represented in the basin-scale model. Given the considerable resources and expertise required to operate models and improve data inputs, having a clear strategy for model integration will help achieve the most effective outcomes. The current status of empirical, conceptual, and mechanistic modeling approaches is summarized below.

Empirical models of sediment yield can help to identify areas or catchments with more intense erosion or deposition (Singer and Dunne, 2001). Because they do not usually identify sediment source processes, their ability to simulate climate or land management scenarios is limited. Empirical sediment yield models are often used to validate conceptual and mechanistic sediment budget process models (Takken et al., 1999; Wilkinson et al., 2009). In this context, methods to quantify the uncertainty associated with empirical sediment load estimates are important (Rustomji and Wilkinson, 2008).

Conceptual and mechanistic models provide frameworks for routing material from sources to sinks and downstream environments, to identify erosion hotspots and to estimate the effects of practice changes. The benefit of conceptual and mechanistic modeling approaches are realized only when all of the important erosion and deposition processes are represented. Many models provide in-depth treatment of surface erosion but omit gully and riverbank erosion and channel deposition processes, despite their importance at basin scale (de Vente and Poesen, 2005). Inappropriate or omitted process representations can lead to a model predicting the right river basin sediment yield for the wrong reasons, jeopardizing investment priorities (Boomer et al., 2008).

The application of mechanistic models across the large areas and long time-periods of interest for land management is often constrained by limited data with which to specify parameter values, and simplified versions of mechanistic models have emerged (Van Rompaey et al., 2001; Borah and Bera, 2002; Brasington and Richards, 2007). Stochastic or probabilistic descriptions of hydrology and soil properties have also been proposed (Aksoy and Kavvas, 2005). Spatial interfaces have been developed to facilitate application of hillslope profile models to broader areas with complex topography (e.g., Ascough et al., 1997; Renschler, 2003).

Simulating the effects of global warming-induced climate change on erosion and sediment yield is likely to become more common, but requires careful consideration of all model inputs, including changes in variability and mean condition.

2.12.4.4.2 Directions in modeling erosion and deposition processes

A fundamental principle guiding further developments in modeling erosion and sediment delivery processes is that the complexity and the spatial and temporal scales should be appropriate to the depth of process knowledge for the study area, the input data available, and the modeling objective. Four less-well developed areas can be identified as foci for further development of process modeling:

Overland sediment transport capacity. It is now well recognized that models of catchment sediment delivery should separate surface erosion and overland sediment transport from erosion and deposition processes occurring at larger spatial scales, such as gully erosion and floodplain deposition (Trimble and Crosson, 2000; de Vente et al., 2007). Predicting the spatial variations in overland transport capacity is challenging due to local variability in terrain, soil properties, and vegetation cover, and consequently predictions are usually calibrated to match observed hillslope sediment yields (Verstraeten et al., 2007). Recent approaches include: (1) pixel-based sediment transport capacity estimation as a function of the erosion rate (Van Rompaey et al., 2001), (2) functions of stream power (Young et al., 1989; Prosser and Rustomji, 2000), and (3) subcatchment delivery ratio estimation using functions of storm duration and runoff travel time, considering the distance to stream, surface slope, and roughness (Ferro and Minacapilli, 1995; Lu et al., 2005).

Hillslope mass failures are an important sediment source in many mountainous and hilly areas. Modeling the spatial controls on hillslope mass failure has received less attention to date than sediment mobilization processes that are more common in lowland areas. Rainfall intensity and duration, slope gradient, and soil properties are important determining factors (Chang and Chiang, 2009). Predicting the occurrence of mass failure at given locations is difficult over broad areas, and the temporal patterns of sediment delivery to streams can be described stochastically (Benda and Dunne, 1997).

Channel network erosion. Available models of the extent and erosion rate of gully networks are empirical, which is an appropriate given the strong random component to the upstream extent of incised channel networks (Shreve, 1966). Modeling gully extent across large river basins can be based on manual mapping of sample areas. Soil type, slope, and climate variability provide useful, but not powerful, explanatory variables (Hughes *et al.*, 2001; Kuhnert *et al.*, 2010). Where gully extent estimates are available, gully volume and age provide constraints on long-term gully sediment yield (Prosser *et al.*, 2001b; Wilkinson *et al.*, 2006). Gully sediment yield also declines with gully age (Prosser and Winchester, 1996). There are no widely validated models available for assessing gully sediment yield dynamics over shorter time periods, although runoff, land use, drainage area, slope gradient, and soil properties are factors commonly applied to explain local variability in measured rates (Poesen *et al.*, 2003; Valentin *et al.*, 2005).

The most common drivers used to predict spatial variation in bank erosion rates are bankfull discharge (Rutherfurd, 2000), stream power (Finlayson and Montgomery, 2003), riparian vegetation, and bank erodibility (Wilkinson *et al.*, 2009). Recent work is improving ability to represent the mechanisms driving bank erosion, including subaerial weathering, scour, mass failure, and channel meandering (e.g., Lawler, 1995; Sun *et al.*, 1996; Abernethy and Rutherfurd, 1998; Langendoen and Simon, 2008).

Achieving robust predictions at finer temporal resolution is a particular challenge, but is not always required for management purposes. There have been limited data on the extent and rates of the above erosion processes, which have constrained model development. However, high-resolution DEMs from laser altimetry are now providing data over much larger spatial extents on gully dimensions (Ritchie, 1996) and on bank erosion rates (Notebaert *et al.*, 2009). Pixel resolution is a key factor influencing the utility of laser altimetry DEMs for mapping erosion features and quantifying erosion rates (Notebaert *et al.*, 2009).

River channel and floodplain deposition. Representing deposition remains essential to modeling sediment delivery at river basin scale in most environments. However, hydraulic controls on sediment deposition cannot generally be resolved in basin-scale modeling (Nicholas *et al.*, 2006).

Conceptually, the primary controls on floodplain deposition are the sediment delivery to floodplains, controlled by the overbank discharge and concentration, and the residence time of overbank flow determined by floodplain size (Prosser *et al.*, 2001c). More complex models consider shallow-water hydraulics more explicitly, and the uncertainty associated with model parameters (Nicholas *et al.*, 2006). Dating of floodplain sediments is useful to verify model predictions (Nicholas *et al.*, 2006; de Moor and Verstraeten, 2008). Long-term fine sediment deposition within river channels, such as lateral point-bar accretion, is controlled by similar fluid mechanics principles to vertical floodplain deposition. Accounting for temporary sediment storage within river channels, which is remobilized in subsequent flow events, can be important for modeling fine sediment yield in shorter time periods, and accounting for the progression of bed material through river networks (Viney and Sivapalan, 1999; Wilkinson *et al.*, 2006).

However, few river basin models currently represent temporary channel storage, and further development is warranted. Analytic representations of channel storage and sediment wave routing are under active development, but data availability has prevented application or verification at basin scale (Cui *et al.*, 2005; Lauer and Parker, 2008).

2.12.4.4.3 Model uncertainty considerations

Uncertainty analysis methods are covered in the Chapter Chapter 2.17 Uncertainty of Hydrological Predictions. Many of these methods are applied in erosion and sediment modeling, especially with more complex models where parameter values may be less well defined. The importance of quantifying model uncertainty, relative to generating best estimates of catchment function, depends on whether the modeling purpose is to formulate hypotheses for further investigation, or for practical application such as developing and justifying investment priorities (Sivapalan, 2009). Understanding the relative contributions of sources of uncertainty is useful for guiding efforts to improve in model performance. This is determined by model sensitivity to changes in parameter values, but also to the levels of uncertainty in each parameter (Reid and Dunne, 2003).

Calibration data on erosion and deposition rates are sparser than data on rainfall and runoff, and there is potential for calibration of erosion and sediment models to distort predicted spatial patterns and sediment source contributions. The method of model calibration should ideally align with the modeling purpose. For example, sediment yield data are useful for calibrating sediment yield predictions, but if the modeling purpose is to predict relative contributions of erosion processes to yield then sediment tracing data may be more useful. The requirement for evaluating model predictions is especially important when modeling environments in which the erosion and deposition processes are less well understood.

2.12.5 The Quality Dimension

2.12.5.1 Introduction

Although, as indicated in Section 2.12.1, fluvial/lacustrine suspended and bed sediments have traditionally been treated as a physical issue (e.g., reservoir sedimentation, channel and harbor silting, bridge scour, and soil erosion and loss), they also can pose a significant chemical/toxicological (water-quality) problem (Vanoni, 1977; Walling, 1977; Baker, 1980; Förstner and Wittmann, 1981; Salomons and Förstner, 1984; Ferguson, 1986; Horowitz, 1991, 1995; de Vries and Klavers, 1994; Stumm and Morgan, 1996; US Environmental Protection Agency, 1997; Horowitz *et al.*, 2001; Walling *et al.*, 2003d; Blum *et al.*, 2004; Cinque *et al.*, 2004; Reed *et al.*, 2004; de Vente and Poesen, 2005; Radaone and Radaone, 2005; Walling, 2005; de Araújo *et al.*, 2006; Black *et al.*, 2007; Domenici *et al.*, 2007; Horowitz and Stephens, 2008). Chemical constituents that primarily are sediment-associated fall into a general class called hydrophobes or hydrophobic compounds (e.g., Förstner and Wittmann, 1981; Luthy *et al.*, 1997; Warren *et al.*, 2003). This group includes heavy metals/trace elements (e.g., Cu, Pb, Zn, As, and Hg), nutrients (e.g., N, P, Si, and C), and persistent organic compounds such as polycyclic aromatic hydrocarbons (PAHs), polychlorinated biphenyls (PCBs), dioxin, kepone, and chlorinated pesticides (e.g., Aldrin, Chlordane, Mirex, DDT and its breakdown products DDD and DDE; e.g., US Environmental Protection Agency, 1997; Simpson et al., 2005).

Even in relatively pristine environments, barely detectable dissolved-constituent concentrations occurring in the water column can simultaneously be detected at levels 3–5 orders of magnitude higher in association with naturally occurring suspended and bed sediments (Förstner and Wittmann, 1981; Förstner, 1989; Horowitz, 1991; Chapman, 1992; Foster and Charlesworth, 1996; Horowitz and Stephens, 2008). Further, bed sediments can make substantial chemical contributions to interstitial water, usually as the result of changing redox conditions that trigger post-depositional mineralogical changes and subsequent chemical remobilization, often into a more bioavailable form.

Numerous studies have demonstrated that sediment-associated chemical constituents can affect aquatic organisms. The organisms can range from small zooplankton (near the base of the food chain), through benthic organisms that live in intimate contact with bed sediment and its surrounding interstitial water, to humans, who may be affected ultimately as constituent levels increase and bioaccumulate up the food chain (Förstner and Wittmann, 1981; Salomons and Förstner, 1984; Chapman, 1992; Förstner and Heise, 2006).

Ever since the publication of the Hawkes and Webb (1962) treatise on geocemical exploration, as well as the subsequent publication of numerous geochemical atlases (e.g., Webb et al., 1978; Fauth et al., 1985; Ottesen et al., 2000), there is a widely accepted perception that suspended and bed sediments reflect local environmental inputs. This is the result of both physical (e.g., grain size, surface area) and chemical (e.g., unbalanced surface charges, presence of oxyhydroxide or organic coatings) factors that make aquatic sediments akin to chemical sponges (Förstner and Wittmann, 1981; Horowitz, 1991; US Environmental Protection Agency, 1997). Hence, sediment-associated chemical levels can increase or decrease in response to natural environmental processes (e.g., changes in Eh, pH, and grain-size distribution) and interactions (e.g., changes in local geology, volcanic activity). For similar reasons, sediment chemistry also tends to reflect anthropogenically derived contributions, for example, due to land use changes, from both point and nonpoint sources (e.g., Reimann and Garrett, 2005; Horowitz and Stephens, 2008).

2.12.5.2 Basic Sediment Geochemistry

The majority of sediment-associated chemical constituents are found on or near the surface of sediment particles, and usually are held by sorption or complexation as a result of unbalanced surface charges (e.g., Förstner and Wittmann, 1981; Salomons and Förstner, 1984; Horowitz, 1991). As such, grain size and particle surface area play a significant role in controlling sediment chemical-associated concentrations (e.g., Horowitz, 1991). Generally, as particle size decreases, total surface area increases, as do the chemical levels; hence, elevated concentrations are more likely to be found associated with silt- and/ or clay-sized particles ($\leq 63\,\mu m$) than coarser sand-sized material ($\geq 63\,\mu m$). Although some contaminants can attach directly to the surfaces of sediment particles (e.g., clay minerals), it is more typical to find them associated with particle coatings that are composed of either organic matter or Fe and/ or Mn oxides and oxyhydroxides (e.g., Förstner and Wittmann, 1981; Förstner, 1989; Horowitz, 1991; Foster and Charlesworth, 1996).

Authigenic minerals that form *in situ*, and which exist as separate particles, may entrain chemical constituents within their crystalline or cryptocrystalline structure as a result of either chemical bonding or physical trapping (e.g., Föstner and Wittmann, 1981; Horowitz, 1991). As with surface coatings, sorption/desorption processes may increase or reduce associated chemical levels depending on changing physicochemical conditions. Less commonly, substantial concentrations of inorganic constituents may be held within mineral lattices. This is most likely to occur in association with mining or mining related and some industrial activities, and occurs as a result of the discharge of ore minerals and/or mining/industrial waste (typically sulfides such as pyrite (Fe), arsenopyrite (As), galena (Pb), sphalerite (Zn), etc.), through the physical erosion of exposed mine tailings (e.g., Horowitz et al., 1988, 1993; Pope, 2005), or industrial discharges.

2.12.5.3 Major Issues Associated with Sediment Quality

Despite the potential environmental impacts of sediment-associated chemical constituents, only a very limited number of countries currently (e.g., Canada, The Netherlands, Australia, New Zealand, and Germany) have established sediment-chemical regulatory limits; however, many have established guidelines (e.g., Persaud et al., 1993; US Environmental Protection Agency, 2005; Simpson et al., 2005). This situation reflects a number of long-standing arguments associated with sediment chemical quality that have yet to be fully resolved. Four of the most significant ones are: (1) what are the background/baseline concentrations for a variety of sediment-associated constituents; (2) how best to collect representative suspended and bed sediment samples for subsequent chemical analysis; (3) how best to determine the concentrations of sediment-associated constituents; and (4) how to estimate/determine bioavailability?

2.12.5.3.1 Background/baseline sediment-associated constituent concentrations

Unlike the vast majority of sediment-associated synthetic organic compounds that have no natural source(s), unless they are manufactured copies of natural substances, sediment-associated inorganic constituents typically do occur in the environment. As a result, a background/baseline level must be established to determine the presence of contamination and/ or the impact of variations in land use (e.g., Goldschmidt, 1958; Hawkes and Webb, 1962; Plant et al., 1997; Reimann and Garrett, 2005). Background and baseline are concepts that tend to be used interchangeably, and often are qualified by terms such as geochemical, natural, or ambient. However, background concentrations usually refer to chemical levels that imply the exclusion of anthropogenic influence whereas baseline concentrations are typically determined at a particular point in space and/or time; albeit, it may imply limited anthropogenic effects (e.g., Gough, 1993; Reimann and

Garrett, 2005). The occurrence of natural geological phenomena (e.g., volcanic eruptions, unworked ore deposits) and changing local geology, combined with the advent of the industrial revolution, and the concomitant eolian and fluvial distribution/redistribution of a variety of materials, and their associated chemical constituents means that it is unlikely that background concentrations can be determined from any current surficial material. It also means that background/baseline concentrations can change spatially and temporally. This leads to a major issue: how to define the natural inorganic chemical composition of sediments, and has led many geochemists/environmental chemists to accept two precepts: (1) background chemical composition is neither spatially nor temporally static, and should be viewed as a range rather than as a single value; and (2) chemical changes induced by natural processes should not be viewed as contamination, even when the source may be an unmined mineralized zone. Hence, only sediment-chemical enhancements derived from anthropogenic activities/sources should be viewed as contamination. On the other hand, regulatory agencies tend to take a broader view, and define contaminated sediment as that which "contains chemical substances in excess of appropriate geochemical, toxicological, or sediment quality criteria or measures, or is otherwise considered to pose a threat to human health or the environment" (US Environmental Protection Agency, 1997).

For many years, geochemists attempted to provide a single set of chemical values in an effort to define the natural background or baseline for the inorganic composition of sediments, and used average crustal chemical abundances, or constructed values for a so-called average shale for that purpose (e.g., Clarke and Washington, 1924; Poldervaart, 1955; Turekian and Wedepohl, 1961; Taylor, 1964; Krauskopf, 1967; Bowen, 1979; Wedepohl, 1995). In turn, these background/baseline concentrations were then used to determine the presence and extent of sediment-chemical contamination. A recent US Geological Survey (USGS) National Water Quality Assessment (NAWQA) Program study has provided a fairly comprehensive and up-to-date (1990–2000) continental-scale assessment of baseline values for a wide variety of sediment-associated inorganic constituents (Horowitz and Stephens, 2008). The results are based on the chemical analyses of the ≤63-μm fraction of nearly 450 bed-sediment samples obtained from undeveloped or agricultural sites within the conterminous US (Table 4). The baseline values associated with these samples do not appear markedly different from those generated from other studies performed in the US and globally (Bowen, 1979; Shacklette and Boerngen, 1984; Horowitz, et al., 1991; Gustavsson, et al., 2001; Manheim and Hayes, 2002). Hence, they probably represent a useful benchmark for identifying anthropogenically enhanced (contaminated) sediment-associated constituents and levels. The same study also indicates that with the exception of mining, urbanization, and population density (Horowitz and Stephens, 2008), land use does not exercise substantive controls on sediment-associated inorganic chemical concentrations (Table 4).

2.12.5.3.2 The collection of representative sediment samples and the issues of spatial and temporal variability

It should be accepted as given in any sediment-associated chemical study that no amount of high-quality analytical work can overcome poor and/or nonrepresentative sampling (e.g., Horowitz, 1997). Consequently, the number, location, and density of sampling locations need to be carefully evaluated within the context of study objectives, acceptable levels of associated error, and some knowledge of the expected levels of chemical variation. In other words, what is the minimum level of acceptable error that still permits sound management decisions and or data interpretations (e.g., Keith, 1988; Mudroch and MacKnight, 1991; Horowitz, 2008). Both bed and suspended sediments have unique characteristics and distribution patterns that must be understood prior to designing an adequate sampling and analysis program (e.g., Keith, 1988; Horowitz, 1991, 2008; Mudroch and MacKnight, 1991).

Studies have provided a clear picture regarding the distribution of suspended sediment in fluvial cross sections, as well as the impacts of those distributions on sediment-associated chemical concentrations (Horowitz et al., 1989, 1992; Horowitz, 1991, 1995, 2008). Suspended sediment and suspended sediment-associated chemical concentrations can exhibit marked short-term spatial and temporal variability (e.g., Horowitz, 1991, 1995, 2008). These distributions tend to support the use of equal-width increment (EWI) or equal-discharge increment (EDI), depth- and width-integrated isokinetic sampling. In other words, collecting a dip sample at a single location in a normal fluvial cross section is unlikely to generate a representative sample of suspended sediment or sediment-associated chemical constituents (Horowitz et al., 1989, 1990, 1992; Horowitz, 1991, 1995, 2008). There are exceptions, such as during low flow (baseflow) conditions when suspended sediment concentrations are very low (≤ 10 mg l^{-1}), or at elevated energy levels in relatively narrow channels when suspended sediment concentrations can be very high, but such cases require careful sampling and subsequent analysis to justify the use of localized (e.g., a dip sample, an autosampler) rather than depth- and width-integrated sampling (e.g., Horowitz, 2008).

When both sand- (>63 μm) and silt/clay-sized (<63 μm) particles are present in a stream, the concentration of suspended sediment tends to increase with increasing distance from the riverbanks (Figure 9). This results from increasing stream velocity (discharge) due to decreasing frictional resistance away from the riverbanks and the riverbed (in shallow water) (e.g., Vanoni, 1977). Note that the typical cause of an increase in suspended sediment concentration is an increase in the amount of sand-sized (>63 μm) material (Figure 9). These concentration changes can occur over relatively short distances (e.g., <3 m). There is a concomitant decrease in the concentrations of most sediment-associated chemical constituents with increasing distance from the riverbanks (Figure 9). This decrease occurs as a direct result of the increase in sand-sized (>63 μm) particles because the coarser material typically contains markedly lower chemical concentrations (e.g., trace elements, nutrients) than the finer silt/clay-sized material (Förstner and Wittmann, 1981; Salomons and Förstner, 1984; Horowitz, 1991). Note that Al does not follow

Table 4 A continental scale assessment of baseline sediment chemistry for the conterminous US. The table presents information on minimum, maximum, mean, median, and median absolute deviations (MAD) chemical concentrations for background, as well as various land-use categories

	Al (%)	Sb (mg kg⁻¹)	As (mg kg⁻¹)	Ba (mg kg⁻¹)	Be (mg kg⁻¹)	Cd (mg kg⁻¹)	Ca (%)	Ce (mg kg⁻¹)	Cr (mg kg⁻¹)	Co (mg kg⁻¹)	Cu (mg kg⁻¹)	Fe (%)	La (mg kg⁻¹)	Pb (mg kg⁻¹)	Li (mg kg⁻¹)	Mg (%)	Mn (mg kg⁻¹)	Hg (mg kg⁻¹)	Mo (mg kg⁻¹)	Ni (mg kg⁻¹)	P (%)	K (%)	Se (mg kg⁻¹)	Ag (mg kg⁻¹)	Na (%)	Sr (mg kg⁻¹)	S (%)	Sn (mg kg⁻¹)	V (mg kg⁻¹)	Zn (mg kg⁻¹)	Ti (%)	OC (%)	TC (%)
Background chemical concentrations – all data																																	
Count	448	446	447	447	448	445	447	447	447	448	448	448	447	448	448	447	448	448	448	447	447	447	447	445	447	448	439	433	448	448	444	425	426
Min	0.2	0.1	0.1	7.0	0.1	0.1	0.1	12.0	6.3	0.5	1.0	0.2	6.3	2.0	3.0	0.04	15	0.01	0.3	1.0	0.02	0.03	0.1	0.1	0.02	17	0.03	1.2	5.1	5.2	0.04	0.01	0.7
Max	13.0	3.7	60	1300	7.0	2.8	28	360	270	78	150	10	190	200	97	4.3	9000	3.1	13	160	0.47	3.1	5.6	4.3	2.2	970	1.5	54	380	430	1.9	25	25
Mean	6.0	0.8	8.1	470	1.6	0.5	3.0	79	66	14	24	3.3	42	24	33	1.0	1100	0.08	1.1	28	0.11	1.4	0.8	0.3	0.7	160	0.12	2.8	92	100	0.38	3.7	4.5
Median	5.9	0.7	6.6	490	1.8	0.4	1.8	69	58	12	20	2.9	39	20	30	0.9	840	0.04	1.0	23	0.10	1.5	0.7	0.2	0.6	150	0.08	2.5	83	91	0.33	2.4	3.3
MAD	1.0	0.2	2.2	110	0.8	0.2	1.3	15	13	4.0	6.0	0.7	8.0	6.0	10	0.4	360	0.02	0.0	7.0	0.02	0.3	0.2	0.1	0.3	60	0.04	<0.1	21	20	0.08	1.1	1.6
Agricultural sites (≥50%)																																	
Count	237	237	237	237	237	237	237	237	237	237	237	237	237	237	237	237	237	237	237	237	237	237	237	237	237	237	236	222	237	237	234	220	220
Min	3.2	0.1	2.4	9.0	0.5	0.1	0.1	35	34	5.0	6.0	0.3	19	6.0	20	0.4	190	0.01	0.3	12	0.04	0.1	0.1	0.1	0.03	24	0.03	1.4	47	30	0.14	0.02	0.8
Max	10.5	3.0	60	860	6.0	2.8	12.0	360	200	78	86	10.0	150	310	110	4.4	8400	1.0	6.5	160	0.31	3.1	4.8	1.0	1.8	470	0.75	5.0	260	190	1.1	18	18
Mean	6.0	0.8	8.8	490	1.4	0.4	3.3	74	65	14	24	3.3	40	24	33	1.2	1200	0.06	1.2	30	0.11	1.4	0.8	0.2	0.6	160	0.10	2.7	96	99	0.37	2.7	3.6
Median	5.8	0.7	7.2	500	1.0	0.4	2.6	66	58	12	22	2.9	36	20	30	1.0	870	0.04	1.0	25	0.10	1.5	0.7	0.2	0.6	140	0.07	2.5	88	93	0.31	2.2	3.2
MAD	0.9	0.2	1.9	70	0.3	0.1	1.8	12	11	3.0	6.0	0.6	6.0	5.0	8.0	0.4	260	0.02	0.1	5.0	0.02	0.3	0.2	0.1	0.2	40	0.04	0.0	21	19	0.06	0.8	1.4
Forest sites (≥50%)																																	
Count	286	284	285	285	286	283	285	285	285	286	286	286	285	286	286	285	286	286	286	285	285	285	285	283	285	286	278	275	286	285	282	269	270
Min	1.4	0.1	0.1	7.0	0.5	0.1	0.1	13	13.0	2.0	1.0	0.7	9.0	2.0	6.0	0.1	20	0.01	0.3	6.0	0.02	0.1	0.1	0.1	0.0	17	0.03	1.2	14	21	0.05	0.01	0.7
Max	13.0	3.7	41	1300	7.0	4.2	26	350	270	64	250	8.5	190	200	97	4.2	20000	3.1	13	140	0.39	2.7	8.6	4.3	2.2	660	1.5	54	380	440	1.9	25	25
Mean	6.4	0.8	7.9	420	1.9	0.6	1.9	89	71	18	29	3.7	47	34	38	0.8	1400	0.12	1.2	30	0.13	1.3	0.9	0.3	0.6	130	0.13	3.1	91	130	0.43	4.8	5.2
Median	6.5	0.7	6.8	420	2.0	0.4	1.0	81	65	17	24	3.6	43	28	39	0.6	1000	0.07	1.0	29	0.11	1.4	0.7	0.2	0.4	100	0.09	2.5	86	110	0.41	3.3	3.6
MAD	0.9	0.2	2.0	80	0.6	0.1	1.0	14	12	4.0	8.0	0.8	7.0	7.0	14	0.2	400	0.02	0.0	6.5	0.04	0.3	0.3	0.1	0.2	38	0.03	0.0	21	31	0.06	0.9	1.0
Rangeland sites (≥50%)																																	
Count	59	59	59	58	59	59	59	59	58	59	59	59	59	59	59	59	59	59	59	58	59	59	59	59	59	59	59	59	59	59	59	59	59
Min	2.1	0.1	1.9	10	0.5	0.1	1.0	28	18	4.0	4.0	1.6	17	6.0	10	0.4	170	0.01	0.3	7.0	0.04	0.5	0.2	0.1	0.1	110	0.03	1.4	28	38	0.14	0.5	0.7
Max	11.0	3.7	24	1100	3.0	1.6	25	130	220	32	79	6.2	100	330	80	3.8	3200	4.7	1.4	160	0.19	2.1	2.6	1.0	1.7	970	0.48	5.0	200	150	0.72	3.7	9.9
Mean	5.9	1.1	6.9	590	1.5	0.4	5.0	69	59	11	24	2.7	40	24	33	1.2	660	0.12	0.9	24	0.11	1.7	0.8	0.3	0.8	290	0.12	2.5	82	85	0.32	1.6	2.9
Median	5.7	0.6	5.2	590	1.6	0.3	3.7	69	48	9.0	20	2.4	40	18	30	1.1	490	0.03	1.0	19	0.10	1.7	0.6	0.2	0.8	250	0.08	2.5	72	84	0.30	1.4	2.6
MAD	0.6	0.1	1.4	85	0.6	0.1	2.0	10	11	2.0	5.0	0.4	5.0	4.0	8.0	0.3	120	0.02	0.0	5.0	0.02	0.2	0.2	0.1	0.3	50	0.04	0.0	15	17	0.04	0.6	1.2
Urban sites (≥50%)																																	
Count	94	94	94	93	94	94	94	94	94	94	94	94	94	94	94	94	94	94	94	94	94	94	94	94	94	94	94	88	94	94	93	88	88
Min	3.9	0.2	1.3	33	0.5	0.2	0.2	27	45	5.0	9.0	1.6	12	8.0	10	0.2	130	0.03	0.9	11	0.04	0.2	0.2	0.1	0.04	39	0.03	2.0	52	45	0.18	0.01	0.9
Max	13.0	10	140	920	4.3	7.3	19.0	270	700	64	420	11.0	120	590	100	4.7	12000	2.2	11	130	1.3	2.5	4.1	17	2.0	1000	1.0	69	180	1700	0.85	16	19
Mean	6.6	1.6	13	430	1.8	1.4	2.9	85	97	18	76	4.2	45	110	33	1.2	1600	0.25	2.2	39	0.20	1.4	1.0	1.1	0.7	180	0.20	7.4	99	330	0.42	4.4	5.5
Median	5.8	1.1	9.1	450	1.7	1.0	1.7	78	81	16	53	3.9	41	76	30	0.9	1100	0.13	1.0	36	0.14	1.4	0.7	0.5	0.5	120	0.15	2.5	94	270	0.40	3.3	4.9
MAD	0.8	0.4	3.0	70	0.6	0.5	1.0	22	18	4.0	24	0.8	12	35	10	0.4	570	0.07	0.0	7.0	0.04	0.3	0.3	0.1	0.3	40	0.08	0.1	18	120	0.12	1.4	1.7
Population density sites (≥50 Percentile; ≥27 p km⁻²)																																	
Count	505	504	504	501	505	503	504	505	504	505	505	505	505	505	505	505	505	504	505	504	505	505	504	503	505	505	504	471	505	504	492	468	467
Min	1.7	0.1	1.2	6	0.5	0.1	0.1	15	11	3	6	0.9	10	8	6	0.1	74	0.01	0.3	4.2	0.03	0.1	0.1	0.1	0	31	0.03	1.1	24	2	0.1	0.01	0.7
Max	14	24	140	920	12	18	20	270	700	170	620	21	130	590	110	4.7	12000	14.5	34	170	1.8	2.8	13	17	2.6	1600	1.7	92	240	1700	1.1	29	29
Mean	6.4	1.2	10	460	1.8	0.9	3	83	79	17	51	3.8	45	64	37	1.1	1400	0.22	1.6	34	0.15	1.5	0.9	0.7	0.6	160	0.16	5	91	200	0.4	3.9	4.8
Median	6	0.7	7.9	460	2	0.5	1.7	74	69	15	36	3.6	40	39	34	0.9	1000	0.09	1	30	0.13	1.5	0.7	0.4	0.5	120	0.11	2.5	86	150	0.36	2.9	3.9
MAD	1	0.2	2.3	80	0.5	0.2	1.2	20	15	5	13	0.8	11	16	6	0.4	400	0.04	0	7	0.04	0.3	0.2	0.2	0.3	40	0.06	0	16	51	0.1	1.1	1.5

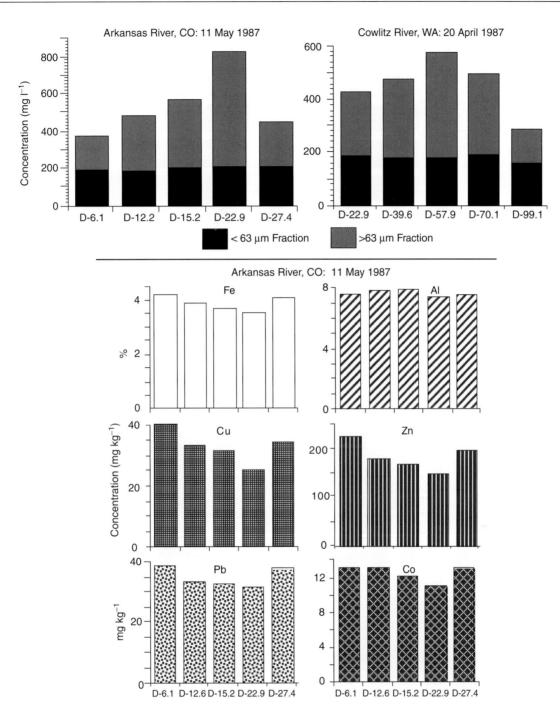

Figure 9 Horizontal cross sectional changes in suspended sediment concentration for the Arkansas and Cowlitz rivers based on isokinetic depth-integrated vertical samples. The numbers following the D are distances, in meters, from the left bank (upper). Horizontal cross sectional variations in selected suspended sediment-associated trace element in depth-integrated isokinetic vertical samples from the Arkansas River on May 11, 1987. The numbers following the D are distances, in meters, from the left bank of the river (lower).

this pattern, because the majority of this element is lattice-held rather than sorbed to mineral surfaces, and both fractions contain substantial quantities of aluminosilicates (Horowitz, 2008).

Vertical concentrations of suspended sediment in fluvial systems tend to increase with increasing depth; this also is due to an increase in sand-sized material (**Figure 10**). This occurs because the velocity (discharge) in most rivers, under normal flow conditions, is insufficient to homogeneously distribute the coarser material. The majority of sand-sized particles tend to be transported on or near the riverbed. The increase in sand-sized particle concentration, from top to bottom, also leads to a concomitant decrease in sediment-associated chemical concentrations (**Figure 10**). As with the horizontal variations noted previously, these changes can occur over relatively short distances (<1.5 m).

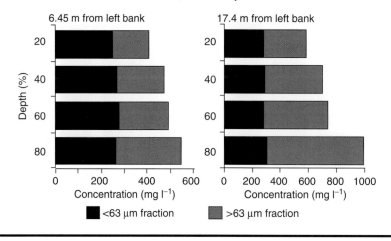

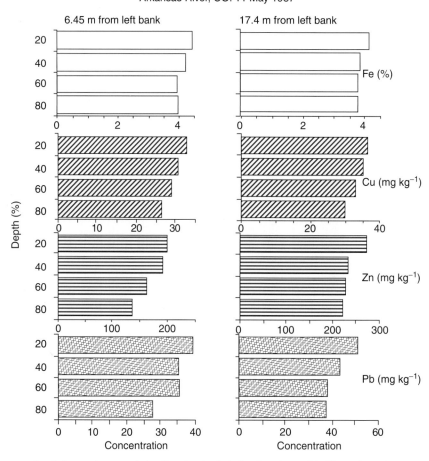

Figure 10 Vertical cross sectional changes in suspended sediment and selected sediment-associated trace element concentrations for the Arkansas River on May 11, 1987, based on isokinetic point samples collected at 20%, 40%, 60%, and 80% of depth. One vertical was 6.45 m and the other was 17.4 m from the left bank.

Sediment chemistry, especially suspended sediment chemistry is markedly affected by hydrology and can display substantial changes in concentration over relatively short as well as relatively longer timescales (e.g., Horowitz, 1995, 2008; Horowitz et al., 2008). This accrues for two reasons, and means that in order to delimit the range of sediment-associated constituent concentrations at any particular location, samples must be collected over a range of flow conditions and temporal scales. The hydrologic linkage can be both direct and indirect. The direct linkage results from the changes in suspended sediment grain-size distribution mentioned earlier. As velocity (discharge) increases, the median grain size of suspended sediment tends to increase that normally produces a concomitant decrease in sediment-associated constituent concentrations (e.g., Horowitz, 1991, 1995, 2008; Horowitz, et al., 2008). Although sediment-associated chemical concentrations decline as discharge increases, the fluxes (loads) of these same constituents usually increase. This occurs because the increase in discharge, in conjunction with increasing amounts of suspended sediment, although coarser, typically more than compensates for the decline in actual chemical concentrations (e.g., Horowitz, 1995, 2008; Old et al., 2003, 2006; Lawler et al., 2006; Horowitz et al., 2008).

The indirect linkage between hydrology and sediment-associated chemical concentrations occurs as a result of changing sediment sources. Although it is something of an over-simplification, there is a generally accepted perception that under baseflow, sediment chemistry tends to be dominated by point sources, whereas during high flow (stormflow), sediment chemistry tends to be dominated by nonpoint (diffuse) sources (e.g., Horowitz, 1995, 2008; Old, et al., 2003, 2006; Horowitz, et al., 2008). Sediment from nonpoint sources in urban (e.g., trace elements, nutrients, PAHs, and pesticides), mining (e.g., trace/major elements), and agricultural areas (e.g., nutrients, agricultural chemicals) is particularly enriched in a wide variety of both organic and inorganic constituents (e.g., Horowitz, 1995; Horowitz, et al., 2008). At least in the US, baseflow point-source sediment-associated chemical effects tend to be limited in terms of both amount and chemical concentration by controlling end-of-pipe discharges through permitting processes such as National Pollutant Discharge Elimination System (NPDES) under the Clean Water Act of 1972. On the other hand, nonpoint-source discharges are much harder to limit/control, and entail the application of a variety of best-management practices (BMPs) such as increasing the width of riparian zones, procurement of additional green space, reforestation, low or no tillage in agricultural areas, installation of highway runoff settling ponds, etc.

As a result, of all the foregoing, accurate assessments of suspended sediment-associated chemical concentrations may require sampling over several different temporal scales (e.g., sampling over the course of a storm to determine concentration changes associated with the rising limb, the peak, and the falling limb of the hydrograph, sampling between storms to determine the impact of different lengths of antecedent dry conditions, and seasonal sampling to deal with the application of various agricultural chemicals (e.g., fertilizers, pesticides), or to evaluate the impact of, for example, deicing salts). On the other hand, such factors as changing demographics (e.g., population density) and land use factors (e.g., urbanization) can affect sediment-associated chemical concentrations over longer temporal scales, of the order of years or decades, depending on the size of the hydrologic system.

Bed sediments can exhibit marked spatial variability, but rarely short-term temporal variability (e.g., Horowitz, 1991, 1995). In addition, unlike suspended sediment samples, bed sediments rarely pose a problem relative to collecting sufficient masses to meet any requisite analytical needs. As such, it usually is far easier to collect representative bed sediment rather than suspended sediment samples. As long as the goal of a study is not intended to determine relative levels of localized spatial variability, the sampling issues associated with bed sediments usually can be addressed through the production of composite samples generated by combining a sufficient number of spatially separated equal-volume aliquots. The number of requisite aliquots typically is predicated on the level of local spatial variability in conjunction with acceptable levels of chemical variance. Normally, the more subsamples that are combined, the more representative the composite, and the smaller the level of associated chemical variance (e.g., Hakanson, 1984; Garner et al., 1988; Horowitz, 1991; Mudroch and MacKnight, 1991).

2.12.5.3.3 The chemical analysis of suspended and bed sediments

As noted previously, sediment-associated constituent concentrations typically occur at levels three to five orders of magnitude higher than found in solution (e.g., Horowitz, 1991, 1995, 2008). As such, analytical sensitivity normally is not an issue except when samples masses are very small; this rarely is the case with bed sediment, but can be an issue with suspended sediment. However, even in the case of suspended sediment, there are a number of techniques available, such as flow-through centrifugation or filtration that permit the collection/concentration of sufficient amounts of suspended matter for subsequent chemical analyses (e.g., Horowitz, 1995, 2008).

Although there are a number of direct techniques for the chemical analysis of solid-phase materials, the majority, with limited exceptions, are performed on either liquid extracts (organic constituents) or liquid digests (inorganic constituents). In the past, the method of choice for the solubilization of sediment-associated organic constituents was the soxhlet extraction (e.g., Amalric and Mouvet, 1997). However, modern production laboratories have begun to switch to either microwave extraction (e.g., Morozova et al., 2008; Forster et al., 2009) or accelerated solvent extraction (especially useful for the determination of hydrophobic persistent organic pollutants; e.g., Jacobsen et al., 2004; Silvia Diaz-Cruz et al., 2006; Berrada et al., 2008). The latter two procedures are markedly faster than the soxhlet extraction, and can be modified to be compound and/or compound-class specific through the selection of an appropriate solvent, and by programming the temperature and/or pressure, and the length of time for the extraction (e.g., Raynie, 2004, 2006). The analytical instruments of choice currently in use are: (1) gas chromatography using various detectors; (2) gas chromatography-mass spectroscopy; and (3) liquid chromatography-mass spectroscopy. Extraction efficiencies are normally determined by the concomitant analysis of reference materials, and analytical recoveries are typically determined by spiking extracts with known amounts of the analyte(s) of interest.

The analytical instrumentation of choice for a wide variety of inorganic constituents including trace elements (e.g., Cu, Zn, Cd, and Pb), major elements (e.g., Fe, Al, Na, and K), and some nutrients (e.g., P, S) is some type of inductively coupled plasma (ICP)-based system. The two most common devices are ICP-atomic emission spectroscopy (ICP-AES), sometimes called ICP-optical emission spectroscopy (ICP-OES), and ICP-mass spectroscopy (ICP-MS). The former is a purely optical system that measures emitted light at a specific (constituent-specific) wavelength whereas the latter is based on determining the concentrations of specific individual stable isotopes, and converting that value to a total concentration in the digestate based on fixed isotopic percentages. Both systems require sample solubilization before quantitation. Several common nutrient determinations (e.g., total carbon, total organic carbon, total nitrogen, and total sulfur) are determined directly on dried sediment by combusting the sample at high temperatures, in the presence of oxygen, and quantitating the evolved gases using a variety of different detectors.

There are numerous choices for the solubilization of various sediment-associated inorganic constituents for subsequent chemical analysis (e.g., Johnson and Maxwell, 1981; van Loon, 1985; Batley, 1989; ASTM, 2008). Essentially, these methods fall into one of the three categories: total analyses, total recoverable analyses, or selective extractions (e.g., Horowitz, 1995, 2008). Geochemists normally define a total analysis as one in which $\geq 95\%$ of the analyte of concern is quantified. This approach usually entails the complete breakdown of mineral lattices, and requires a mixture of concentrated mineral acids and relatively high temperatures, and/or some type of fusion with various fluxes (e.g., sodium carbonate, lithium tetraborate/metaborate) at elevated temperatures, with the solubilization of the resulting bead (e.g., Johnson and Maxwell, 1981). Analytical precision and bias are normally determined by the concomitant analysis of appropriate reference materials. Analyses of this type are unambiguous because they are independent of sediment-associated mineralogical/petrological variation; hence, they are comparable across spatial and/or temporal scales, and represent a known end-member because of the level of recovery ($\geq 95\%$).

Total recoverable digestions, typically favored by regulatory agencies at least in the US, are nonspecific partial extractions usually employing mineral acids and some level of heating. The levels of constituent solubilization and subsequent quantitation are highly dependent on variations in mineralogy/petrology, as such, concentrations determined using this procedure may not be comparable across spatial and/or temporal scales, and can be very difficult to interpret. It sometimes has been claimed that this type of digestion produces a measure of bio- and/or environmental-availability but that interrelation has never been demonstrated (e.g., Horowitz, 1991, 2008). Another difficulty associated with this approach is a lack of certified reference materials, so while it is possible to determine analytical precision (reproducibility) by replicate analyses of the same material, it is difficult to evaluate analytical bias as well as percent recoveries. Spiking digestates is not a typical sediment-associated inorganic analytical procedure because the major source of analytical variance does not result from the quantitation step, but from the solubilization step.

Selective extractions represent a special category of partial and/or recoverable digestions that are intended to provide specific information about sediment-chemical partitioning (e.g., Kersten and Förstner, 1987; Horowitz, 1991). Sediment-chemical partitioning entails all the various procedures that can be used to determine how (mechanistic approach; e.g., adsorption, complexation; within mineral lattices) and/or where (phase approach; e.g., iron oxides, manganese oxides, organic matter; carbonates) various chemical constituents are associated with sediment particles (e.g., Chao, 1984; Bately, 1989; Horowitz, 1991; Hall and Pelchat, 1999). Unfortunately, all these procedures can be categorized as operational definitions and there are numerous procedures that purport to provide similar information (e.g., bound to iron oxides), but which do not provide equivalent analytical results (e.g., Horowitz, 1991). These procedures usually also have additional limitations such as they can only be used on sediment collected in oxidized environments (e.g., Tessier et al., 1979).

2.12.5.3.4 Bioavailability and toxicity

By far, the single biggest barrier to the widespread development and acceptance of sediment quality guidelines/regulatory limits is the continuing controversy regarding the bioavailability, and potential toxicity of sediment-associated chemical constituents. Unlike dissolved constituents, where the total concentration is presumed to be bioavailable, it has always been assumed that only limited portions of sediment-associated constituents (e.g., non-lattice held) fall into the same category (e.g., Allen et al., 1993; Hansen et al., 2005; Simpson et al., 2005). Additional concerns stem from disagreements over the interpretation of toxicity tests in terms of exposure rates, selection of appropriate test organisms, and relative measures of lethality, as well as an understanding of sediment-chemical partitioning and potential bioaccumulation (e.g., Lahr et al., 2003; Hansen et al., 2005).

Current sediment quality guidelines (SQGs) are typically provided at two concentrations. The first level is the lower of the two, and normally reflects concentrations at which little or no biological/ecological effects are expected. This level has been given a variety of names such as lowest effect level (Persaud et al., 1993), threshold-effects concentration (TEC; MacDonald et al., 2000), and threshold-effects level (TEL; US Environmental Protection Agency, 1997). The second level is the higher of the two, and normally reflects concentrations at which biological/ecological effects are expected. This level has also been given a variety of names such as severe effect level (Persaud et al., 1993), probable-effects concentration (PEC; MacDonald, et al., 2000), and probable-effects level (PEL; US Environmental Protection Agency, 1997). Although these various concentrations tend to overlap, there currently appears to be no consensus on a single set of values for either category or for various sediment-associated constituents (e.g., Persaud et al., 1993; US Environmental Protection Agency, 1997; MacDonald et al., 2000; Hansen et al., 2005; Simpson et al., 2005). To further complicate this issue, other terms have been used to indicate the potential effects of sediment-associated chemical constituents. For example, in 1997, the US EPA evaluated sediment chemical data from over 21 000 locations in the US and found that 26% had a higher probability and 49% had an intermediate probability of adverse effects on aquatic life and human health. The chemical constituents most often associated with these increased probabilities were PCBs, Hg, DDT, Cu, Ni, and Pb (US Environmental Protection Agency, 1997).

Other than the results from various toxicity tests, there appear to be two basic approaches associated with establishing SQGs; one is based on equilibrium partitioning models whereas the other is based on some type of sediment-chemical partitioning (e.g., Hansen et al., 2005; Simpson et al., 2005). In either case, evaluations have to be made on a site-specific basis (e.g., Simpson et al., 2005). The equilibrium partitioning approach requires measuring chemical concentrations in interstitial water; the underlying assumption being that whatever concentrations in the porewater represent the bioavailable component of the equivalent sediment-associated constituent (e.g., Simpson et al., 2005). The actual guideline concentrations are then based on those established for dissolved constituents. There are three major issues with this approach. The first is that it obviously does not apply to suspended sediment. The second issue is associated with those organisms that actually ingest sediment. The physicochemical conditions in the gut of an organism are obviously not the same as those that exist in the interstitial water (e.g., they are likely to be more acidic and less oxygen will be present); hence, bioavailable concentrations derived from interstitial water may be inappropriate. The third issue is associated with specific size fractions from a bulk sediment sample. Examination of the gut contents of various aquatic organisms indicates that they tend to limit their intake to specific grain-size ranges. As such, determining equilibrium partitioning concentrations based on interstitial water derived from a bulk sediment sample may be inappropriate.

The sediment partitioning approach is predicated on the view that some form of chemical extraction/digestion can be found that functions as a measure of, or surrogate for bioavailability (e.g., Di Toro et al., 1992; Allen et al., 1993). One such procedure that has received a good deal of attention is acid volatile sulfides-simultaneously extracted metals (AVS-SEM). This extraction employs cold dilute HCl; those metals that exceed the concentration of available sulfide, on a molar basis, are considered bioavailable (Di Toro et al., 1992; Allen et al., 1993; Hansen et al., 2005; Simpson et al., 2005). However, a number of studies have indicated that the AVS-SEM method can over- or under-predict bioavailability when compared to other approaches, and may not be appropriate for all environments and/or organisms (e.g., Chen and Mayer, 1999; Lahr et al., 2003; Meador et al., 2005; Simpson and Batley, 2007; Prica et al., 2008).

2.12.5.4 Future Directions

While the foregoing summary, covering the current status of sediment quality, clearly indicates that a great deal is known about the subject, there still remains much to do with respect to achieving scientific consensus in a variety of areas. While the scientific community has begun to reach some level of consensus relative to background/baseline values for sediment-associated constituents, more refinement is needed. It would also be appropriate to begin to delineate those constituents that are particularly sensitive to, or indicative of various types of land use or source material. The general lack of broad scale agreement on appropriate sampling and analytical procedures is likely to continue to make transboundary/multinational studies difficult as a result of potential data incompatibilities.

Lastly, it appears that a great deal of more work needs to be done in the areas of sediment-associated constituent bioavailability and toxicity before there can even be a modicum of consensus that could lead to fairly universal sediment-quality guidelines.

References

Aalto R, Lauer JW, and Dietrich WE (2008) Spatial and temporal dynamics of sediment accumulation and exchange along Strickland River floodplains (Papue New Guinea) over decadal-to-centennial timescales. *Journal of Geophysical Research Earth Surface* 113: F01F04.

Aalto R, Maurice-Bourgoin L, Dunne T, Montgomery DR, Nittrouer CA, and Guyot J-L (2003) Episodic sediment accumulation on Amazonian floodplains influenced by El Nino/Southern Oscillation. *Nature* 425: 493–497.

Abernethy B and Rutherfurd I (1998) Where along a river's length will vegetation most effectively stabilise stream banks? *Geomorphology* 23: 55–75.

Abrahams AD, Li G, Krishnan C, and Atkinson JF (2001) A sediment transport equation for interrill overland flow. *Earth Surface Processes and Landforms* 26: 1443–1459.

Aksoy H and Kavvas ML (2005) A review of hillslope and watershed scale erosion and sediment transport models. *Catena* 64: 247–271.

Alexandrov Y, Laronne JB, and Reid I (2003) Suspended sediment concentration and its variation with water discharge in a dryland ephemeral channel, Northern Negev, Israel. *Journal of Arid Environments* 53: 73–84.

Ali KF and De Boer DH (2008) Factors controlling sediment yield in the upper Indus River basin, northern Pakistan. *Hydrological Processes* 22: 3102–3114.

Allan RJ (1979) *Sediment-Related Fluvial Transmission of Contaminants: Some Advances by 1979*, Inland Waters Directorate, Environment Canada, Scientific Series No. 107. Ottawa: Environment Canada.

Allen HE, Fu G, and Deng B (1993) Analysis of acid-volatile sulfide (AVS) and simultaneously extracted metals (SEM) for the estimation of potential toxicity in aquatic sediments. *Environmental Toxicology and Chemistry* 12: 1441–1453.

Amalric L and Mouvet C (1997) Comparison of SFE, soxhlet and sonication for the extraction of organic contaminants from sandstone. *International Journal of Environmental Analytical Chemistry* 68: 171–186.

ASCE (1975) *Sedimentation Engineering*. American Society of Civil Engineers Manuals and Reports on Engineering Practice No. 54. New York: ASCE.

Ascough JC, Baffaut C, Nearing MA, and Liu BY (1997) The WEPP watershed model. 1. Hydrology and erosion. *Transactions of the ASAE* 40: 921–933.

Asselman NEM (2000) Fitting and interpretation of sediment rating curves. *Journal of Hydrology* 234: 228–248.

ASTM (2008) *Annual Book of ASTM Standards*, vol. 11.02, pp. 384–387. West Conshohocken, PA: ASTM International.

Atkinson PM, German SE, Sear DA, and Clark MJ (2003) Exploring the relations between riverbank erosion and geomorphological controls using geographically weighted logistic regression. *Geographical Analysis* 35: 58–82.

Baby P, Guyot J-L, and Hérail G (2009) Tectonic control of erosion and sedimentation in the Amazon Basin of Bolivia. *Hydrological Processes* 23: 3225–3229 (doi:10.1002/hyp.7391).

Baker RA (ed.) (1980) *Contaminants and Sediments, Vol. 1: Fate and Transport, Case Studies, Modeling, Toxicity* Ann Arbor, MI: Ann Arbor Science Publishers.

Bartley R, Hawdon A, Post DA, and Roth CH (2007) A sediment budget for a grazed semi-arid catchment in the Burdekin Basin, Australia. *Geomorphology* 87: 302–321.

Basson G (2008) Reservoir sedimentation – an overview of global sedimentation rates and predicted sediment deposition. In: *Erosion, Transport and Deposition*. Workshop Berne, Switzerland, April 2008, Abstracts, pp. 74–79.

Batley GE (1989) *Trace Element Speciation: Analytical Methods and Problems*. Boca Raton, FL: CRC Press.

Beck MB (1987) Water-quality modeling – a review of the analysis of uncertainty. *Water Resources Research* 23: 1393–1442.

Benda L and Dunne T (1997) Stochastic forcing of sediment routing and storage in channel networks. *Water Resources Research* 33: 2865–2880.

Berrada H, Borrull F, Guillermina F, and Marce RM (2008) Determination of macrolide antibiotics in meat and fish using pressurized liquid extraction and liquid chromatography-mass spectrometry. *Journal of Chromatography, A* 1208: 83–89.

Beusen AHW, Dekkers ALM, Bouwman AF, Ludwig W, and Harrison J (2005) Global river export of suspended solids and particulate carbon and nutrients. *Global Biochemical Cycles* 19: GB4S05.

Beven K, Heathwaite L, Haygarth P, Walling D, Brazier R, and Withers P (2005) On the concept of delivery of sediment and nutrients to stream channels. *Hydrological Processes* 19: 551–556.

Black KS, Athey S, Wilson P, and Evans D (2007) The use of particle tracking in sediment transport studies: A review. *Geological Society, Special Publication* 274: 73–91.

Blake WH, Droppo IG, Wallbrink PJ, Doerr SH, Shakesby RA, and Humphreys GS (2005) Impacts of wildfire on effective sediment particle size: Implications for post-fire sediment budgets. In: Walling DE and Horowitz AJ (eds.) *Sediment Budgets 1*, IAHS Publication No. 236, pp. 115–123. Wallingford: IAHS Press.

Blake WH, Walling DE, and He Q (2002) Using cosmogenic Beryllium-7 as a tracer is sediment budget investigations. *Geografiska Annala* 84A: 89–102.

Blum WEH, Büsing J, and Montanarella L (2004) Research needs in support of the European thematic strategy for soil protection. *TrAC – Trends in Analytical Chemistry* 23: 680–685.

Bobrovitskaya NN, Zubkova C, and Meade RH (1996) Discharges and yields of suspended sediment in the Ob and Yenesey Rivers of Siberia. In: Walling DE and Webb BW (eds.) *Erosion and Sediment Yield: Global and Regional Perspectives*, IAHS Publication No. 325, pp. 323–338. Wallingford: IAHS Press.

Bohn BA and Kershner JL (2002) Establishing aquatic restoration priorities using a watershed approach. *Journal of Environmental Management* 64: 355–363.

Bottrill LJ, Walling DE, and Leeks GJL (2000) Using recent overbank deposits to investigate contemporary sediment sources in large river systems. In: Foster IDL (ed.) *Tracers in Geomorphology*, pp. 369–387. Chichester: Wiley.

Bowen HJM (1979) *Environmental Chemistry of the Elements*. London: Academic Press.

Brardinoni F, Hassan MA, Rollerson T, and Maynard D (2009) Colluvial sediment dynamics in mountain drainage basins. *Earth and Planetary Science Letters* 284: 310–319.

Boomer KB, Weller DE, and Jordan TE (2008) Empirical models based on the universal soil loss equation fail to predict sediment discharges from Chesapeake Bay catchments. *Journal of Environmental Quality* 37: 79–89.

Borah DK and Bera M (2002) Watershed-scale hydrologic and nonpoint-source pollution models: Review of mathematical bases. In: *ASAE International Meeting*, pp. 1553–1566. Chicago, IL, USA.

Brasington J and Richards K (2007) Reduced-complexity, physically-based geomorphological modelling for catchment and river management. *Geomorphology* 90: 171–177.

Brayshaw D and Hassan MA (2009) Debris flow initiation and sediment recharge in gullies. *Geomorphology* 109: 122–131.

Caitcheon GG (1993) Sediment source tracing using environmental magnetism – a new approach with examples from Australia. *Hydrological Processes* 7(4): 349–358.

Cavalcanti GG and Lockaby BG (2005) Effects of sediment deposition on fine root dynamics in riparian forests. *Soil Science Society of America Journal* 69: 729–737.

Cerdan O, Poesen J, Govers G, et al. (2006) Sheet and rill erosion rates in Europe. In: Boardman J and Poesen J (eds.) *Soil Erosion in Europe*, pp. 501–514. Chichester: Wiley.

Chang KT and Chiang SH SH (2009) An integrated model for predicting rainfall-induced landslides. *Geomorphology* 105: 366–373.

Chao TT (1984) Use of partial dissolution techniques in geochemical exploration. *Journal of Geochemical Exploration* 20: 101–135.

Chapman D (ed.) (1992) *Water Quality Assessments, a Guide to the Use of Biota, Sediments, and Water in Environmental Monitoring*. London: Chapman and Hall.

Chappell NA, Douglas I, Hanapi JM, and Tych W (2004) Sources of suspended sediment within a tropical catchment recovering from selective logging. *Hydrological Processes* 18: 685–702.

Chen Z and Mayer LM (1999) Assessment of sedimentary Cu availability: A comparison of biometric and AVS approaches. *Environmental Science and Technology* 33: 650–652.

Chien N and Wan Z (1999) *Mechanics of Sediment Transport*. New York: ASCE.

Cinque K, Stevens MA, Roser DJ, Ashbolt NJ, and Leeming R (2004) Assessing the health implications of turbidity and suspended particles in protected catchments. *Water Science and Technology* 50: 205–210.

Clark EH (1985) The off-site costs of soil erosion. *Journal of Soil and Water Conservation* 40: 19–22.

Clark EH, Haverkamp JA, and Chapman W (1985) *Eroding Soils: the Off-Farm Impacts*. Washington, DC: The Conservation Foundation.

Clarke FW and Washington HS (1924) The composition of the earth's crust. *U.S. Geological Survey Professional Paper 127*. Washington, DC: US Geological Survey.

Cochrane T and Flanagan D (2006) Sediment deposition in a simulated rill under shallow flow conditions. *Transactions of the ASAE* 49: 893–903.

Collins AL and Walling DE (2002) Selecting fingerprint properties for discriminating potential suspended sediment sources in river basins. *Journal of Hydrology* 261: 218–244.

Collins AL and Walling DE (2007) Fine-grained bed sediment storage within the main channel system of the Frome and Piddle catchments, Dorset, UK. *Hydrological Processes* 21: 1448–1459.

Collins AL, Walling DE, and Leeks GJL (1997a) Source type ascription for fluvial suspended sediment based on a quantitative composite fingerprinting technique. *Catena* 29: 1–27.

Collins AL, Walling DE, and Leeks GJL (1997b) Sediment sources in the Upper Severn catchment: A fingerprinting approach. *Hydrology and Earth Systems Science* 1: 509–521.

Coote DR, MacDonald EM, Dickinson WT, Ostry RC, and Frank R (1982) Agriculture and water quality in the Canadian Great Lakes Basin: I. Representative agricultural watersheds. *Journal of Environmental Quality* 11: 473–481.

Croke J, Hairsine P, and Fogarty P (1999) Sediment transport, redistribution and storage on logged forest hillslopes in south-eastern Australia. *Hydrological Processes* 13: 2705–2720.

Croke J, Mockler S, Fogarty P, and Takken I (2005) Sediment concentration changes in runoff pathways from a forest road network and the resultant spatial pattern of catchment connectivity. *Geomorphology* 68: 257–268.

Cui Y, Parker G, Lisle TE, Pizzuto JE, and Dodd AM (2005) More on the evolution of bed material waves in alluvial rivers. *Earth Surface Processes and Landforms* 30: 107–114.

Curtis JA, Flint LE, Alpers CN, and Yarnell SM (2005) Conceptual model of sediment processes in the upper Yuba River watershed, Sierra Nevada, CA. *Geomorphology* 68: 149–166.

Davis CM and Fox JF (2009) Sediment Fingerprinting: Review of the method and future improvements for allocating nonpoint source pollution. *Journal of Environmental Engineering* 135: 490–504.

Deasy C, Brazier RE, Heathwaite AL, and Hodgkinson R (2009) Pathways of runoff and sediment transfer in small agricultural catchments. *Hydrological Processes* 23: 1349–1358.

de Araújo JC, Güntner A, and Bronstert A (2006) Loss of reservoir volume by sediment deposition and its impact on water availability in semiarid Brazil. *Hydrological Sciences Journal* 51: 157–170.

De'ath G and Fabricius K (2010) Water quality as a regional driver of coral biodiversity and macroalgae on the Great Barrier Reef. *Ecological Applications* 20(3): 840–850.

de Moor JJW and Verstraeten G (2008) Alluvial and colluvial sediment storage in the Geul River catchment (The Netherlands) – combining field and modelling data to construct a Late Holocene sediment budget. *Geomorphology* 95: 487–503.

de Vente J and Poesen J (2005) Predicting soil erosion and sediment yield at the basin scale: Scale issues and semi-quantitative models. *Earth Science Reviews* 71: 95–125.

de Vente J, Poesen J, Arabkhedri J, and Verstraeten G (2007) The sediment delivery problem revisited. *Progress in Physical Geography* 31: 155–178.

de Vries A and Klavers HC (1994) Riverine fluxes of pollutants: Monitoring strategy first, calculation methods second. *European Journal of Water Pollution Control* 4: 12–17.

DeRoo APJ (1996) Soil erosion assessment using G.I.S. In: Singh VP and Fiorentino M (eds.) *Geographical Information Systems in Hydrology*, pp. 339–356. Dordrecht: Kluwer.

DeRoo APJ, Wesseling CG, and Ritsema CJ (1996) LISEM: A single-event physically based hydrological and soil erosion model for drainage basins. 1: Theory, input and output. *Hydrological Processes* 10: 1107–1117.

Di Toro DM, Mahony JD, Hansen DJ, Scott KJ, Carlson AR, and Ankley GT (1992) Acid volatile sulfide predicts the acute toxicity of cadmium and nickel in sediments. *Environmental Science and Technology* 26: 96–101.

Domenici P, Claireaux G, and McKenzie DJ (2007) Environmental constraints upon locomotion and predator–prey interactions in aquatic organisms: An introduction. *Philosophical Transactions of the Royal Society B: Biological Sciences* 362: 1929–1936.

Droppo IG (2001) Rethinking what constitutes suspended sediment. *Hydrological Processes* 15: 1551–1564.

Droppo IG, Lau YL, and Mitchell C (2001) The effect of depositional history on contaminated bed sediment stability. *Science of the Total Environment* 266: 7–13.

Droppo IG, Nackaerts K, Walling DE, and Williams N (2004) Can flocs and water stable soil aggregates be differentiated within fluvial systems. *Catena* 60: 1–18.

Dunne T, Mertes LAK, Meade RH, Richey JE, and Forsberg BR (1998) Exchanges of sediment between the flood plain and channel of the Amazon River in Brazil. *Geological Society of America Bulletin* 110: 450–466.

Edwards TK and Glysson GD (1988) Field methods for measurement of fluvial sediment. U.S. Geological Survey Open-file Report 86–531, 118pp.

Evans M and Slaymaker O (2004) Spatial and temporal variation of sediment delivery from alpine lake basins: Cathedral Provincial Park, Southern British Columbia. *Geomorphology* 61: 209–224.

Fangmeier DD, Elliot WJ, Workman SR, Huffman RL, and Schwab GO (2006) *Soil and Water Conservation Engineering*. New York: Delmar Thompson.

Fauth H, Hindel R, Siewers U, and Zinner J (1985) *Geochemischer Atlas Bundesrepublik Deutschland*. Stuttgart: BGR Hannover; Schweizerbart'sche Verlagsbuchhandlung.

Ferguson RI (1986) River loads underestimated by rating curves. *Water Resources Research* 22: 74–76.

Ferro V and Minacapilli M (1995) Sediment delivery processes at basin scale. *Hydrological Sciences Journal (Des Sciences Hydrologiques)* 40: 703–717.

Field J (2001) Channel avulsion on alluvial fans in southern Arizona. *Geomorphology* 37: 93–104.

Finlayson DP and Montgomery DR (2003) Modeling large-scale fluvial erosion in geographic information systems. *Geomorphology* 53: 147–164.

Florsheim JL, Mount JF, and Chin A (2008) Bank erosion as a desirable attribute of rivers. *BioScience* 58: 519–529.

Fontaine TA, Moore TD, and Burgoa B (2000) Distributions of contaminant concentrations and particle size in fluvial sediment. *Water Research* 34: 3473–3477.

Forbes AC and Lamoureux SF (2005) Climatic controls on streamflow and suspended sediment transport in three large middle Arctic catchments, Boothia Peninsula, Nunavut, Canada. *Arctic, Antarctic and Alpine Research* 37: 304–315.

Forster M, Laabs V, Lamshoft M, et al. (2009) Sequestration of manure-applied sulfadiazine residues in soils. *Environmental Science and Technology* 43: 1824–1830.

Förstner U (1989) *Contaminated Sediments. Lecture on Environmental Aspects of Particle-Associated Chemicals in Aquatic Systems*. New York: Springer.

Förstner U (2002) Sediments and the European water framework directive. *Journal of Soils and Sediments* 2: 54.

Förstner U (2003) Sediments and the European water framework directive. *Journal of Soils and Sediments* 3: 138.

Förstner U and Heise S (2006) Assessing and managing contaminated sediments: Requirements on data quality – from molecular to river basin scale. *Croatica Chemica Acta* 79: 5–14.

Förstner U and Wittmann GTW (1981) *Metal Pollution in the Aquatic Environment*, 2nd edn. New York: Springer.

Foster IDL and Charlesworth SM (1996) Heavy metals in the hydrocycle: Trends and explanation. *Hydrological Processes* 10: 227–261.

Fox GA, Wilson GV, Simon A, Langendoen EJ, Akay O, and Fuchs JW (2007) Measuring streambank erosion due to groundwater seepage: Correlation to bank pore water pressure, precipitation and stream stage. *Earth Surface Processes and Landforms* 32: 1558–1573.

Fox JF and Papanicolaou AN (2007) The use of carbon and nitrogen isotopes to study small watershed erosion processes. *Journal of the American Water Resources Association* 43: 1047–1064.

Garcia MH (ed.) (2008) *Sedimentation Engineering: Processes, Measurements, Modeling and Practice*, American Society of Civil Engineers Manuals and Reports on Engineering Practice No. 110. New York: ASCE.

Garner FC, Stapanian MA, and Williams LR (1988) Composite sampling for environmental monitoring. In: Keith LH (ed.) *Principles of Environmental Sampling*, pp. 363–374. Washington, DC: American Chemical Society.

Gibbs MM (2008) Identifying source soils in contemporary estuarine sediments: A new compound-specific isotope method. *Estuaries and Coasts* 31: 344–359.

Giménez R and Govers G (2002) Flow detachment by concentrated flow on smooth and irregular beds. *Soil Science Society of America Journal* 66: 1475–1483.

Gippel CJ (1995) Potential of turbidity monitoring for measuring the transport of suspended solids in streams. *Hydrological Processes* 9: 83–97.

Gislason SR, Oelkers EH, and Snorrason A (2006) Role of river-suspended material in the global carbon cycle. *Geology* 34: 49–52.

Glysson GD and Gray JR (eds.) (2002) *Proceedings of the Federal Interagency Workshop on Turbidity and Other Sediment Surrogates, U.S. Geological Survey Circular 1250*. Washington, DC: US Geological Survey.

Goldschmidt VM (1958) *Geochemistry*. Oxford: Clarendon.

Golosov VN, Ivanova NN, Litvin LF, and Sidorchuk AYu (1992) Sediment budgets of river catchments and river channel aggradation on the Russian plain. *Geomorphology (Moscow)* 4: 62–71 (in Russian).

Golterman HL (ed.) (1977) *Interactions between Sediment and Fresh Water*. The Hague: Dr. W Junk Publishers.

Gomez B, Banbury K, Marden M, Trustrum NA, Peacock DH, and Hoskin PJ (2003) Gully erosion and sediment production: Te Weraroa stream, New Zealand. *Water Resources Research* 39: 1187.

Gordon LM, Bennett SJ, Alonso CV, and Bingner RL (2008) Modeling long-term soil losses on agricultural fields due to ephemeral gully erosion. *Journal of Soil and Water Conservation* 63: 173–189.

Gough LP (1993) *Understanding our Fragile Environment: Lessons from Geochemical Studies, U.S. Geological Survey Circular 1105*. Washington, DC: US Geological Survey.

Govers G, Giménez R, and van Oost K (2007) Rill erosion: Exploring the relationship between experiments, modelling and field observations. *Earth-Science Reviews* 84: 87–102.

Gray J, Glysson D, and Edwards T (2008) Suspended-sediment samplers and sampling methods. In: *Sedimentation Engineering: Processes, Measurements, Modeling and Practice*, American Society of Civil Engineers Manuals and Reports on Engineering Practice No. 110. New York: ASCE.

Grayson R and Blöschl G (2000) Spatial modelling of catchment dynamics. In: Grayson R and Blöschl G (eds.) *Spatial Patterns in Catchment Hydrology*, ch. 3, pp. 51–81. Cambridge: Cambridge University Press. (http://www.hydro.tuwien.ac.at/downloads.html).

Gustavsson N, Bolviken B, Smith DB, and Severson RC (2001) Geochemical landscapes of the conterminous United States – new map presentations for 22 elements. U.S. Geological Survey Professional Paper 1648. Washington, DC: US Geological Survey.

Guy HP and Norman VW (1970) Field Methods for Measurement of Fluvial Sediment, U.S. Geological Survey Techniques of Water Resources Investigations, Book 3, ch. C2. Washington, DC: US Geological Survey.

Hakanson L (1984) Sediment sampling in different aquatic environments – statistical aspects. *Water Resources Research* 20: 41–46.

Hall GEM and Pelchat P (1999) Comparability of results obtained by the use of different selective extraction schemes for the determination of element forms in soil. *Water, Air, and Soil Pollution* 112: 41–53.

Hansen DJ, Di Toro DM, Berry WJ, et al. (2005) Procedures for the Derivation Equilibrium Partitioning Sediment Benchmarks (ESBs) for the Protection of Benthic Organisms: Metal Mixtures (Cadmium, Copper, Lead, Nickel, Silver and Zinc). U.S. EPA/600/R-02/011. Washington, DC: US Environmental Protection Agency, Office of Research and Development.

Haregeweyn N, Poesen J, Nyssen J, et al. (2008) Sediment yield variability in Northern Ethiopia: A quantitative analysis of its controlling factors. *Catena* 75: 65–78.

Harrison ET, Norris RH, and Wilkinson SN (2008) Can a catchment-scale sediment model be related to an indicator of river health? *Hydrobiologia* 600: 49–64.

Hart BT (1982) Uptake of trace metals by sediments and suspended particulates: A review. *Hydrobiologia* 91–92: 299–313.

Hart EA (2002) Effects of woody debris on channel morphology and sediment storage in headwater streams in the Great Smoky Mountains, Tennessee-North Carolina. *Physical Geography* 23: 492–510.

Harvey AM (2002) The role of base-level change in the dissection of alluvial fans: Case studies from southeast Spain and Nevada. *Geomorphology* 45: 67–87.

Harvey AM, Mather AE, and Stokes M (2005) Alluvial fans: Geomorphology, sedimentology, dynamics – introduction. A review of alluvial fan research. *Geological Society London Special Publication* 252: 1–7.

Hassan MA, Church M, Lisle TE, Brardinoni F, Benda L, and Grant GE (2005) Sediment transport and channel morphology of small forested streams. *Journal of the American Water Resources Association* 41: 853–876.

Hatfield RG and Maher BA (2009) Fingerprinting upland sediment sources: Particle size-specific magnetic linkages between soils, lake sediments and suspended sediments. *Earth Surface Processes and Landforms* 34(10): 1359–1373 (doi:10.1002/esp.1824).

Hawkes HE and Webb JS (1962) *Geochemistry in Mineral Exploration*. New York, NY: Harper and Row.

Haygarth PM, Bilotta GS, Bol R, et al. (2006) Processes affecting transfer of sediment and colloids with associated phosphorus, from intensively farmed grasslands: An overview of key issues. *Hydrological Processes* 20: 4407–4413.

He Q and Walling DE (1996) Use of fallout Pb-210 measurements to investigate longer-term rates and patterns of overbank sediment deposition on the floodplains of lowland rivers. *Earth Surface Processes and Landforms* 21: 141–154.

Heimsath AM, Chappell J, Spooner NA, and Questiaux DG (2002) Creeping soil. *Geology* 30: 111–114.

House WA and Warwick MS (1999) Interactions of phosphorus with sediments in the River Swale, Yorkshire, UK. *Hydrological Processes* 13: 1103–1115.

Horowitz AJ (1991) *A Primer on Sediment-Trace Element Chemistry*, 2nd edn. Chelsea, MI: Lewis.

Horowitz AJ (1995) *The Use of Suspended Sediment and Associated Trace Elements in Water Quality Studies*. IAHS Special Publication No. 4. Wallingford: IAHS Press.

Horowitz AJ (1997) Some thoughts on problems associated with various sampling media used for environmental monitoring. *Analyst* 122: 1193–1200.

Horowitz AJ (2008) Contaminated sediments: Inorganic constituents. In: Anderson MG (ed.) *Encyclopedia of Hydrological Sciences*, Wiley Interscience, published online at http://mrw.interscience.wiley.com/emrw/9780470848944/ehs/article/hsa317/current/html, DOI: 10.1002/0470848944.hsa317, 27pp.

Horowitz AJ (2008) Determining annual suspended sediment and sediment-associated trace element and nutrient fluxes. *Science of the Total Environment* 400: 315–343.

Horowitz AJ, Elrick KA, and Callender E (1988) The effect of mining on the sediment-trace element geochemistry of cores from the Cheyenne River arm of Lake Oahe, South Dakota, U.S.A. *Chemical Geology* 67: 17–33.

Horowitz AJ, Elrick KA, and Cook RB (1993) The effect of mining and related activities on the sediment-trace element geochemistry of Lake Coeur d'Alene, Idaho, U.S.A. – Part I: Surface sediments. *Hydrological Processes* 7: 403–423.

Horowitz AJ, Elrick KA, Demas CR, and Demcheck DK (1991) The use of sediment-trace element geochemical models for the identification of local fluvial baseline concentrations. In: Peters NE and Walling DE (eds.) *Sediment and Stream Water Quality in a Changing Environment: Trends and Explanation*, IAHS Publication 203, pp. 339–348. Wallingford: IAHS Press.

Horowitz AJ, Elrick KA, and Hooper RC (1989) A comparison of instrumental dewatering methods for the separation and concentration of suspended sediment for subsequent trace element analysis. *Hydrological Processes* 3: 163–184.

Horowitz AJ, Elrick KA, and Smith JJ (2001) Annual suspended sediment and trace element fluxes in the Mississippi, Columbia, Colorado, and Rio Grande drainage basins. *Hydrological Processes* 15: 1169–1207.

Horowitz AJ, Elrick KA, and Smith JJ (2008) Monitoring urban impacts on suspended sediment, trace element, and nutrient fluxes within the City of Atlanta, Georgia, U.S.A.: Program design, methodological considerations, and initial results. *Hydrological Processes* 22: 1473–1496.

Horowitz AJ, Elrick KA, von Guerard PB, Young NO, Buell GR, and Miller TL (1992) The use of automatically collected point samples to estimate suspended sediment and associated trace element concentrations for determining annual mass transport. In: Bogen J, Walling D, and Day T (eds.) *Proceedings of the IAHS Symposium Erosion and Sediment Transport Monitoring Programmes in River Basins*, IAHS Publication No. 210, pp. 209–218. Oslo: IAHS Press.

Horowitz AJ, Rinella F, Lamothe P, *et al.* (1990) Variations in suspended sediment and associated trace element concentrations in selected riverine cross sections. *Environmental Science and Technology* 24: 1313–1320.

Horowitz AJ and Stephens VC (2008) The effects of land use on fluvial sediment chemistry for the conterminous U.S. – results from the first cycle of the NAWQA Program: Trace and major elements, phosphorus, carbon, and sulfur. *Science of the Total Environment* 400: 290–314.

Hughes AO, Olley JM, Croke JC, and Webster IT (2009) Determining floodplain sedimentation rates using ^{137}Cs in a low fallout environment dominated by channel- and cultivation-derived sediment inputs, central Queensland, Australia. *Journal of Environmental Radioactivity* 100(10): 858–865 (doi:10.1016/j.jenrad.2009.06.011).

Hughes AO, Prosser IP, Stevenson J, *et al.* (2001) Gully erosion mapping for the National Land and Water Resources Audit. *Technical Report 26/01*. Canberra: CSIRO Land and Water.

Hupp CR, Schenk ER, Richter JM, Peet RK, and Townsend PA (2009) Bank erosion along the dam-regulated lower Roanoke River, North Carolina. In: *Management and Restoration of Fluvial Systems with Broad Historical Changes and Human Impacts*, The Geological Society of America Special Paper 451, pp. 97–108. Boulder, CO: Geological Society of America.

ICOLD (2006) *World Register of Dams*. ICOLD http://www.icold-cigb.net (accessed April 2010).

Jacobsen AM, Halling-Sorenson B, Ingerslev F, and Hansen SH (2004) Simultaneous extraction of tetracycline, macrolide and sulfonamide antibiotics from agricultural soils using pressurized liquid extraction, followed by solid-phase extraction and liquid chromatography-tandem mass spectrometry. *Journal of Chromatography A* 1038: 157–170.

Jeffries R, Darby SE, and Sear DA (2003) The influence of vegetation and organic debris on flood-plain sediment dynamics: Case study of a low-order stream in the New Forest England. *Geomorphology* 51: 61–80.

Johnson WM and Maxwell JA (1981) *Rock and Mineral Analysis*. New York, NY: Wiley.

Jothityangkoon C, Sivapalan M, and Farmer DL (2001) Process controls on water balance variability in a large semi-arid catchment: Downward approach to hydrological model development. *Journal of Hydrology* 254: 174–198.

Julien P (1995) *Erosion and Sedimentation*. Cambridge: Cambridge University Press.

Julien P (2002) *River Mechanics*. Cambridge: Cambridge University Press.

Keith LH (ed.) (1988) *Principles of Environmental Sampling, American Chemical Society*. York, PA: Maple Press.

Kersten M and Förstner U (1987) Effect of sample pretreatment on the reliability of solid speciation data of heavy metals: Implications for the study of early diagenetic processes. *Marine Chemistry* 22: 299–312.

Kisi O (2005) Suspended sediment estimation using neuro-fuzzy and neural network approaches. *Hydrological Sciences Journal (Des Sciences Hydrologiques)* 50: 683–696.

Krauskopf KB (1967) *Introduction to Geochemistry*. New York, NY: McGraw-Hill.

Krein A, Petticrew E, and Udelhoven T (2003) The use of fine sediment fractal dimensions and color to determine sediment sources in a small watershed. *Catena* 53: 165–179.

Kronvang B, Hoffmann CC, and Dröge R (2009) Sediment deposition and net phosphorus retention in a hydraulically restored lowland river floodplain in Denmark: Combining field and laboratory experiments. *Marine and Freshwater Research* 60: 638–646.

Kuhnert PM, Henderson A, Bartley R, and Herr A (2010) Incorporating uncertainty in gully erosion calculations using the Random Forests modelling approach. *Environmetrics* 21(5): 493–509. doi:10.1002/env.999.

Lahr J, Maas-Diepeveen JL, Stuijfzand SC, *et al.* (2003) Responses in sediment bioassays used in the Netherlands: Can observed toxicity be explained by routinely monitoring priority pollutants? *Water Research* 37: 1691–1710.

Lal R (ed.) (1994) *Soil Erosion Research Methods*. Ankeny, IA: Soil and Water Conservation Society.

Langendoen EJ and Simon A (2008) Modeling the evolution of incised streams. II: Streambank erosion. *Journal of Hydraulic Engineering-ASCE* 134: 905–915.

Larionov GA (1993) *Erosion and Wind Blown Soil*. Moscow: Moscow State University Press.

Laubel A, Kronvang B, Hald AB, and Jensen C (2003) Hydromorphological and biological factors influencing sediment and phosphorus loss via bank erosion in small lowland rural streams in Denmark. *Hydrological Processes* 17: 3443–3463.

Lauer JW and Parker G (2008) Modeling framework for sediment deposition, storage, and evacuation in the floodplain of a meandering river: Theory. *Water Resources Research* 44: W04425.

Lavigne F and Suwa H (2003) Contrasts between debris flows, hyperconcentrated flows and stream flows at a channel of Mount Semeru, East Java, Indonesia. *Geomorphology* 61: 41–58.

Lawler DM (1995) The impact of scale on the processes of channel-side sediment supply: A conceptual model. In: *Effects of Scale on Interpretation and Management of Sediment and Water Quality*, IAHS Publication 226, pp. 175–184. Boulder, NC: IAHS Press.

Lawler DM, Petts GE, Foster IDL, and Harper S (2006) Turbidity dynamics during spring storm events in an urban headwater river system: The Upper Tame, West Midlands, UK. *Science of the Total Environment* 360: 109–126.

Leeder MR and Mack GH (2001) Lateral erosion ('toe-cutting') of alluvial fans by axial rivers: Implications for basin analysis and architecture. *Journal of the Geological Society* 158: 885–893.

Lei TW, Zhang QW, Zhao J, and Nearing MS (2006) Tracing sediment dynamics and sources with rare earth elements. *European Journal of Soil Science* 57: 287–294.

Lewis J and Eads RE (2001) Turbidity threshold sampling for suspended sediment load estimation. In: *Proceedings of the 7th Federal Interagency Sedimentation Conference*, pp. III-110–III-117. Reno, Nevada. Federal Interagency Project, Technical Committee of the Subcommittee on Sedimentation.

Li S, Lobb DA, Tiessen KHD, and McConkey BG (2009) Selecting and applying cesium-137 conversion models to estimate soil erosion rates in cultivated fields. *Journal of Environmental Quality* 39: 204–219.

Lu H, Moran CJ, and Sivapalan M (2005) A theoretical exploration of catchment-scale sediment delivery. *Water Resources Research* 41: W09415 (doi:10.1029/2005WR004018).

Ludwig W, Probst J-L, and Kempe S (1996) Predicting the oceanic input of organic carbon by continental erosion. *Global Biogeochemical Cycles* 10: 23–41.

Luthy RG, Aiken GR, Brusseau ML, *et al.* (1997) Sequestration of hydrophobic organic contaminants by geosorbents. *Environmental Science and Technology* 31: 3341–3347.

Lyons WB, Nezat CA, Carey AE, and Hicks DM (2002) Organic carbon fluxes to the ocean from high-standing islands. *Geology* 30: 443–446.

Mabit L, Benmansour M, and Walling DE (2008) Comparative advantages and limitations of fallout radionuclides (^{137}Cs, ^{210}Pb and ^{7}Be) to assess soil erosion and sedimentation. *Journal of Environmental Radioactivity* 99: 1799–1807.

Macaire J-J, Bellemlih S, Di-Giovanni C, De Luca P, Visset L, and Bernard J (2002) Sediment yield and storage variations in the Negron River catchment (south western Parisian Basin, France) during the Holocene period. *Earth Surface Processes and Landforms* 27: 991–1009.

MacDonald DD, Ingersoll CG, and Berger TA (2000) Development and evaluation of consensus-based sediment quality guidelines for freshwater ecosystems. *Archives of Environmental Contamination and Toxicology* 39: 20–31.

Macnab K, Jacobson C, and Brierley G (2006) Spatial variability of controls on downstream patterns of sediment storage: A case study in the Lane Cove Catchment, New South Wales, Australia. *Geographical Research: Journal of the Institute of Australian Geographers* 44: 255–271.

Malmon DV, Reneau SL, Dunne T, Katzman D, and Drakos PG (2005) Influence of sediment storage on downstream delivery of contaminated sediment. *Water Resources Research* 41: W05008.

Manheim FT and Hayes LH (2002) Lake Ponchartrain basin, bottom sediments and related resources. *U.S. Geological Survey Professional Paper 1634*, CD-ROM.

Matisoff G, Bonniwell EC, and Whiting PJ (2002) Soil erosion and sediment sources in an Ohio watershed using beryllium-7, cesium-137, and lead-210. *Journal of Environmental Quality* 31: 54–61.

McKergow L, Prosser I, Hughes A, and Brodie J (2005) Sources of sediment to the Great Barrier Reef World Heritage Area. *Marine Pollution Bulletin* 51: 200–211.

Meador JP, Ernest DW, and Kagley AN (2005) A comparison of the non-essential elements cadmium, mercury, and lead found in fish and sediment from Alaska and California. *Science of the Total Environment* 339: 189–205.

Melis TS and Topping DG (2003) Testing laser based sensors for continuous in situ monitoring of suspended sediment in the Colorado River, Arizona. In: Bogen J, Fergus T, and Walling DE (eds.) *Erosion and Sediment Transport Measurement in Rivers: Technological and Methodological Advances*, IAHS Publication No. 283, pp. 21–27. Wallingford: IAHS Press.

Merritt WS, Letcher RA, and Jakeman AJ (2003) A review of erosion and sediment transport models. *Environmental Modelling and Software* 18: 761–799.

Merten GHG, Nearing MA, and Borges ALO (2001) Effect of sediment load on soil detachment and deposition in rills. *Soil Science Society of America Journal* 65: 861–868.

Milliman JD, Quirashee GS, and Beg MAA (1984) Sediment discharge from the Indus River to the ocean: Past, present and future. In: Haq BU and Milliman JD (eds.) *Marine Geology and Oceanography of Arabian Sea and Coastal Pakistan*, pp. 65–70. New York: Van Nostran Rheinhold.

Molina A, Govers G, Poesen J, Van Hemelryck H, De Bièvre B, and Vanecker V (2007) Environmental factors controlling spatial variation in sediment yield in a Central Andean mountain area. *Geomorphology* 98: 178–186.

Morgan RPC (1995) *Soil Erosion and Conservation*. Harlow: Longman Scientific and Technical.

Morozova VS, Eremin SA, Nesterenko PN, Klyuev NA, Shelepchikov AA, and Kubrakova IA (2008) Microwave and ultrasonic extraction of chlorophenoxy acids from soil and their determination by fluorescence polarization immunoassay. *Journal of Analytical Chemistry* 63: 127–134.

Mudroch A and MacKnight SD (1991) *Handbook of Techniques for Aquatic Sediment Sampling*. Boca Raton, FL: CRC Press.

Nearing MA, Foster GR, Lane LJ, and Finkner SC (1989) A process-based soil-erosion model for USDA-Water Erosion Prediction Project technology. *Transactions of the ASAE* 32(5): 1587–1593.

Newcombe CP and Jensen JOT (1996) Channel suspended sediment and fisheries: A synthesis for quantitative assessment of risk and impact. *North American Journal of Fish Management* 16: 693–727.

Nicholas AP and Walling DE (1996) The significance of particle aggregation in the overbank deposition of suspended sediment on river floodplains. *Journal of Hydrology* 186: 275–293.

Nicholas AP, Walling DE, Sweet RJ, and Fang X (2006) Development and evaluation of a new catchment-scale model of floodplain sedimentation. *Water Resources Research* 42: W10426 (doi:10.1029/2005WR004579).

Nistor CJ and Church M (2005) Suspended sediment transport regime in a debris-flow gully on Vancouver Island, British Columbia. *Hydrological Processes* 19: 861–885.

NLWRA (2001) *Australian Agriculture Assessment 2001*. National Land and Water Resources Audit, Canberra. http://www.anra.gov.au/topics/publications (accessed April 2010).

Notebaert B, Verstraeten G, Govers G, and Poesen J (2009) Qualitative and quantitative applications of LiDAR imagery in fluvial geomorphology. *Earth Surface Processes and Landforms* 34: 217–231.

Old GH, Leeks GJL, Packman JC, Smith BPG, Lewis S, and Hewitt EJ (2006) River flow and associated transport of sediments and solutes through a highly urbanised catchment, Bradford, West Yorkshire. *Science of the Total Environment* 360: 98–108.

Old GH, Leeks GJL, Packman JC, *et al.* (2003) The impact of a convectional summer rainfall event on river flow and fine sediment transport in a highly urbanized catchment: Bradford, West Yorkshire. *Science of the Total Environment* 314/316: 495–512.

Ongley ED (1982) The PLUARG experience: Scientific implications for diffuse source management. In: Hart BT (ed.) *Water Quality Management: Monitoring Programs and Diffuse Runoff*, pp. 87–101. Melbourne: Water Studies Centre, Chisholm Institute of Technology and Australian Society for Limnology.

Ongley ED, Krishnappan BG, Droppo I, Rao SS, and Maguire RJ (1992) Cohesive sediment transport: Emerging issues for toxic chemical management. *Hydrobiologia* 235–236: 177–187.

Ottesen RT, Bogen J, Bolviken B, Volden T, and Haugland T (2000) *Geokjemisk Atlas for Norge*, del. 1. Trondheim, Norway: Norges geologiske undersokelse, Norges vassdrags–og energidirektorat.

Owens PN, Batalla RJ, Collins AJ, *et al.* (2005) Fine-grained sediment in river systems: Environmental significance and management issues. *River Research and Applications* 21: 693–717.

Parsons AJ, Wainwright J, Brazier RE, and Powell DM (2006) Is sediment delivery a fallacy? *Earth Surface Processes and Landforms* 31: 1325–1328.

Persaud D, Jaagumagi R, and Hayton A (1993) *Guidelines for the Protection and Management of Aquatic Sediment Quality in Ontario*. Ontario: Water Resources Branch, Ontario Ministry of the Environment and Energy.

Petticrew EL, Krein A, and Walling DE (2007) Evaluating fine sediment mobilisation and storage in a gravel-bed river using controlled reservoir releases. *Hydrological Processes* 21: 198–210.

Phillips JD (1986) The utility of the sediment budget concept in sediment pollution-control. *Professional Geographer* 38: 246–252.

Phillips JM, Russell MA, and Walling DE (2000) Time-integrated sampling of fluvial suspended sediment: A simple methodology for small catchments. *Hydrological Processes* 14: 2589–2602.

Pittam NJ, Foster IDL, and Mighall TM (2009) An integrated lake-catchment approach for determining sediment source changes at Aqualate Mere, Central England. *Journal of Paleolimnology* 42(2): 215–232 (doi:10.1007/s10933-008-9272-9).

Plant JA, Klaver G, Locutura J, Salminen R, Vrana K, and Fordyce FM (1997) The forum of European geological surveys geochemistry task group inventory 1994–1996. *Journal of Geochemical Exploration* 59: 123–146.

Poesen J, Nachtergaele J, Verstraeten G, and Valentin C (2003) Gully erosion and environmental change: Importance and research needs. *Catena* 50: 91–133.

Poldervaart A (ed.) (1955) Crust of the earth. *Geological Society of America Special Paper 62*. Denver, CO: Geological Society of America.

Pope LM (2005) Assessment of contaminated streambed sediment in the Kansas part of the historic Tri-State Lead and Zinc Mining District, Cherokee County, 2004. *U.S. Geological Survey Scientific Investigations Report 2005–5251*. Lawrence, KS: US Geological Survey.

Preston N and Schmidt J (2003) Modelling sediment fluxes at large spatial and temporal scales. *Lecture Notes in Earth Sciences* 101: 53–72.

Prica M, Dalmacija B, Roncevic S, Krcmar D, and Becelic M (2008) A comparison of sediment quality results with acid volatile sulfides (AVS) and simultaneously extracted metals (SEM) ratio in Vojvodina (Serbia) sediments. *Science of the Total Environment* 389: 235–244.

Prosser IP, Hughes A, Rustomji P, Young WJ, and Moran CJ (2001a) Predictions of the sediment regime of Australian rivers. In: Rutherfurd ID, Sheldon F, Brierley G, and Kenyon C (eds.) *Third Australian Stream Management Conference*, pp. 529–533. Brisbane: Cooperative Research Centre for Catchment Hydrology.

Prosser IP and Rustomji P (2000) Sediment capacity relationships for overland flow. *Progress in Physical Geography* 24: 179–193.

Prosser IP, Rustomji P, Young B, Moran C, and Hughes A (2001b) Constructing river basin sediment budgets for the National Land and Water Resources Audit. *Technical Report 15/01*. Canberra: CSIRO Land and Water.

Prosser IP, Rutherfurd ID, Olley JM, Young WJ, Wallbrink PJ, and Moran CJ (2001c) Large-scale patterns of erosion and sediment transport in river networks, with examples from Australia. *Marine and Freshwater Research* 52: 81–99.

Prosser IP and Winchester SJ (1996) History and processes of gully initiation and development in eastern Australia. *Zeitschrift fur Geomorphologie, Supplement Bund* 105: 91–109.

Radoane M and Radoane N (2005) Dams, sediment sources and reservoir silting in Romania. *Geomorphology* 71: 112–125.

Raynie DE (2004) Modern extraction techniques. *Analytical Chemistry* 76: 4659–4664.
Raynie DE (2006) Modern extraction techniques. *Analytical Chemistry* 78: 3997–4004.

Reed T, Wielgus SJ, Barnes AK, Schiefelbein JJ, and Fettes AL (2004) Refugia and local controls: Benthic invertebrate dynamics in Lower Green Bay, Lake Michigan following zebra mussel invasion. *Journal of Great lakes Research* 30: 390–396.

Reid LM and Dunne T (2003) Sediment budgets as an organizing framework in fluvial geomorphology. In: Kondolf MG and Piégay H (eds.) *Tools in Fluvial Geomorphology*, pp. 463–500. New York: Wiley.

Reimann C and Garrett RG (2005) Geochemical background – concept and reality. *Science of the Total Environment* 350: 12–27.

Renard KG, Foster GA, Weesies DK, McCool DK, and Yoder DC (1997) *Predicting Soil Erosion by Water: A Guide to Conservation Planning with the Revised Universal Soil Loss Equation. Agriculture Handbook 703*. Washington, DC: United States Department of Agriculture.

Renschler CS (2003) Designing geo-spatial interfaces to scale process models: the GeoWEPP approach. *Hydrological Processes* 17: 1005–1017.

Rinaldi M, Mengoni B, Luppi L, Darby SE, and Mosselman E (2008) Numerical simulation of hydrodynamics and bank erosion in a river bend. *Water Resources Research* 44: W09428.

Ritchie JC (1996) Remote sensing applications to hydrology: Airborne laser altimeters. *Hydrological Sciences Journal (Des Sciences Hydrologiques)* 41: 625–636.

Ritchie JC (1972) Sediment, fish, and fish habitat. *Journal of Soil and Water Conservation* 27: 124–125.

Ritchie JC, Finney VL, Oster KJ, and Ritchie CA (2004) Sediment deposition in the flood plain of Stemple Creek watershed, northern California. *Geomorphology* 61: 347–360.

Ritchie JC and Ritchie CA (2008) Bibliography of publications of ^{137}cesium studies related to erosion and sediment deposition. *USDA-ARS Hydrology and Remote Sensing Laboratory Occasional Paper HRSL-2008-02*. http://www.ars.usda.gov/Main/docs.htm?docid = 17939 (accessed April 2010).

Ritter JB, Miller JR, and Husek-Wulforst J (2000) Environmental controls on the evolution of alluvial fans in Buena Vista Valley, North Central Nevada, during late quaternary time. *Geomorphology* 36: 63–87.

Roehl JE (1962) Sediment source areas, delivery ratios and influencing morphological factors. In: *Commission of Land Erosion (Proceedings Symposium of Bari, October, 1962)*, IAHS Publication No. 59, pp. 202–213. Wallingford: IAHS Press.

Rommens T, Verstraeten G, Bogman P, et al. (2006) Holocene alluvial sediment storage in a small river catchment in the loess area of Central Belgium. *Geomorphology* 77: 187–201.

Rommens T, Verstraeten G, Lang A, et al. (2005) Soil erosion and sediment deposition in the Belgian loess belt during the Holocene: Establishing a sediment budget for a small agricultural catchment. *Holocene* 15: 1032–1043.

Rowan JS, Goodwill P, and Franks SW (2000) Uncertainty estimation in fingerprinting suspended sediment sources. In: Foster IDL (ed.) *Tracers in Geomorphology*, pp. 279–290. Chichester: Wiley.

Rubin DM and Topping DJ (2001) Quantifying the relative importance of flow regulation and grain size regulation of suspended sediment transport and tracking changes in grain size of bed sediment. *Water Resources Research* 37: 133–146.

Rustomji P (2006) Analysis of gully dimensions and sediment texture for southeast Australia for catchment sediment budgeting. *Catena* 67: 119–127.

Rustomji P, Caitcheon G, and Hairsine P (2008) Combining a spatial model with geochemical tracers and river station data to construct a catchment sediment budget. *Water Resources Research* 44: W01422.

Rustomji P and Prosser IP (2001) Spatial patterns of sediment delivery to valley floors: Sensitivity to sediment transport capacity and hillslope hydrology relations. *Hydrological Processes* 15: 1003–1018.

Rustomji P and Wilkinson SN (2008) Applying bootstrap resampling to quantify uncertainty in fluvial suspended sediment loads estimated using rating curves. *Water Resources Research* 44: W09434.

Rutherfurd ID (2000) Some human impacts on Australian stream channel morphology. In: Brizga S and Finlayson B (eds.) *River Management: The Australasian Experience*, pp. 11–49. London: Wiley.

Saenger C, Cronin TM, Willard D, Halka J, and Kerhin R (2008) Increased terrestrial to ocean sediment and carbon fluxes in the northern Chesapeake Bay associated with twentieth century land alteration. *Estuaries and Coasts* 31: 492–500.

Salomons W and Förstner U (1984) *Metals in the Hydrocycle*. New York, NY: Springer.

Santiago S, Thomas RL, Larbaigt G, et al. (1994) Nutrient, heavy metal and organic pollutant composition of suspended and bed sediments in the Rhone River. *Aquatic Sciences* 56: 220–242.

Schiettecatte W, Gabriels D, Cornelis WM, and Hofman G (2008) Enrichment of organic carbon in sediment transport by interrill and rill erosion processes. *Soil Science Society of America Journal* 72: 50–55.

Schoellhamer DH and Wright SA (2003) Continuous measurement of suspended-sediment discharge in rivers by use of optical backscatterance sensors. In: Bogen J, Fergus T, and Walling DE (eds.) Erosion and Sediment Transport Measurement in Rivers: Technological and Methodological Advances, IAHS Publication No. 283, pp. 28–36. Wallingford: IAHS Press.

Schuerch P, Densmore AL, McArdell BW, and Molnar P (2006) The influence of landsliding on sediment supply and channel change in a steep mountain catchment. *Geomorphology* 78: 222–235.

Schuller P, Iroume A, Walling DE, Mancilla HB, Castillo A, and Trumper RE (2006) Use of beryllium-7 to document soil redistribution following forest harvest operations. *Journal of Environmental Quality* 35: 1756–1763.

Schwab GO, Frevert RK, Edminster TW, and Barnes KK (1981) *Soil and Water Conservation Engineering*, 3rd edn. New York: Wiley.

Schwertmann U, Vogel W, and Kainz M (1990) *Bodenerosion Durch Wasser*. Stuttgart: Eugen Ulmer GmbH.

Seitzinger SP, Harrison JA, Dumont E, Beusen AHW, and Bouwman AF (2005) Sources and delivery of carbon, nitrogen and phosphorus to the coastal zone: An overview of global nutrient export from watersheds (NEWS) models and their application. *Global Biogeochemical Cycles* 19: GB4S01.

Sepulveda A, Schuller P, Walling DE, and Castillo A (2007) Use of ^{7}Be to document soil erosion associated with a short period of extreme rainfall. *Journal of Environmental Radioactivity* 99: 35–49.

Shacklette HT and Boerngen JG (1984) Elemental concentrations in soils and other surficial materials of the conterminous United States. *U.S. Geological Survey Professional Paper 1270*. Washington, DC: US Geological Survey.

Shear H and Watson AEP (eds.) (1977) *The Fluvial Transport of Sediment-Associated Nutrients and Contaminants*. Windsor, ON: International Joint Commission.

Shreve RL (1966) Statistical law of stream numbers. *Journal of Geology* 74: 17–37.

Silvia Diaz-Cruz M, Lopex de M, and Barcelo D (2006) Determination of antimicrobials in sludge from infiltration basins at two artificial recharge plants by pressurized liquid extraction–liquid chromatography–tandem mass spectrometry. *Journal of Chromatography A* 1130: 72–82.

Simon A, Dickerson W, and Heine A (2004) Suspended-sediment transport rates at the 1.5-year recurrence interval for ecoregions of the United States: Transport conditions at the bankfull and effective discharge? *Geomorphology* 58: 243–262.

Simpson SL and Batley GE (2007) Predicting metal toxicity in sediments: A critique of current approaches. *Integrated Environmental Assessment and Management* 3: 18–31.

Simpson SL, Bately GE, Charlton AA, et al. (2005) *Handbook for Sediment Quality Assessment*. Lucas Heights, NSW: CSIRO, Centre for Environmental Contaminants Research.

Singer MB and Dunne T (2001) Identifying eroding and depositional reaches of valley by analysis of suspended sediment transport in the Sacramento River, California. *Water Resources Research* 37: 3371–3381.

Sivapalan M (2009) The secret to doing better hydrological science?: Change the question!. *Hydrological Processes* 23: 1391–1396.

Smith BPG, Naden PS, Leeks GJL, and Wass PD (2003) The influence of storm events on fine sediment transport, erosion and deposition in a reach of the River Swale, Yorkshire, UK. *Science of the Total Environment* 314–316: 451–474.

Smith HG and Dragovich D (2008) Sediment budget analysis of slope-channel coupling and in-channel sediment storage in an upland catchment, southeastern Australia. *Geomorphology* 101: 643–654.

Soulsby C, Youngson AF, Moira HJ, and Malcolm IA (2001) Fine sediment influence on salmonid spawning habitat in a lowland agricultural stream: A preliminary assessment. *Science of the Total Environment* 265: 295–307.

Srivastava P, Migliaccio KW, and Simunek J (2007) Landscape models for simulating water quality at point, field, and watershed scales. *Transactions of the ASABE* 50: 1683–1693.

Staley DM, Wasklewicz TA, and Blaszczynski JS (2006) Surficial patterns of debris flow deposition on alluvial fans in Death Valley, CA using airborne laser swath mapping data. *Geomorphology* 74: 152–163.

Steegen A, Govers G, Takken I, Nachtergaele J, Posen J, and Merckx R (2001) Factors controlling sediment and phosphorus export from two Belgian agricultural catchments. *Journal of Environmental Quality* 30: 1249–1258.

Steiger J, Gurnell AM, and Goodson JM (2003) Quantifying and characterizing contemporary riparian sedimentation. *River Research and Applications* 19: 335–352.

Stone M and Droppo IG (1996) Distribution of lead, copper and zinc in size-fractionated river bed sediment in two agricultural catchments of southern Ontario, Canada. *Environmental Pollution* 93: 353–362.

Stone M, Krishnappan BG, and Emelko MB (2008) The effect of bed age and shear stress on the particle morphology of eroded cohesive river sediment in an annular flume. *Water Research* 42: 4179–4187.

Stone M and Saunderson H (1992) Particle size characteristics of suspended sediment in Southern Ontario rivers tributary to the Great Lakes. *Hydrological Processes* 6: 189–198.

Stone PM and Walling DE (1997) Particle size selectivity considerations in suspended sediment budget investigations. *Water, Air and Soil Pollution* 99: 63–70.

Stumm W and Morgan JJ (1996) *Aquatic Chemistry: Chemical Equilibria and Rates in Natural Waters*. New York, NY: Wiley.

Sun T, Meakin P, Jossang T, and Schwarz K (1996) A simulation model for meandering rivers. *Water Resources Research* 32: 2937–2954.

Suttle KB, Powers ME, Levine JM, and McNeely C (2004) How fine sediment in riverbeds impairs growth and survival of juvenile salmonids. *Ecological Applications* 14: 969–974.

Swanson KM, Watson E, Aalto R, et al. (2008) Sediment load and floodplain deposition rates: Comparison of the Fly and Strickland Rivers, Papue New Guinea. *Journal of Geophysical Research, Earth Surface* 113: F01S03.

Sweet RJ, Nicholas AP, Walling DE, and Fang X (2003) Morphological controls on medium-term sedimentation rates on British lowland river floodplains. *Hydrobiologia* 494: 177–183.

Syvitski JPM and Milliman JD (2007) Geology, geography and humans battle for dominance over delivery of fluvial sediment to the coastal ocean. *Journal of Geology* 115: 1–19.

Syvitski JPM, Vörösmarty CJ, Kettner AJ, and Green P (2005) Impact of humans on the flux of terrestrial sediment to the global coastal ocean. *Science* 308: 376–380.

Takken I, Beuselinck L, Nachtergaele J, Govers G, Poesen J, and Degraer G (1999) Spatial evaluation of a physically-based distributed erosion model (LISEM). *Catena* 37: 431–447.

Tamene L, Park SJ, Dikau R, and Vlek PLJ (2006) Analysis of factors determining sediment yield variability in the highlands of northern Ethiopia. *Geomorphology* 76: 76–91.

Taylor SR (1964) Abundance of chemical elements in the continental crust: A new table. *Geochimica et Cosmochimica Acta* 28: 1273–1285.

Terry JP, Garimella S, and Kostaschuk RA (2002) Rates of floodplain accretion in a tropical island river system impacted by cyclones and large floods. *Geomorphology* 42: 171–182.

Tessier A, Campbell PGC, and Bisson M (1979) Sequential extraction procedure for the speciation of particulate trace metals. *Analytical Chemistry* 51: 844–851.

Thonon I (2006) Deposition of Sediment and Associated Heavy Metals on Floodplains. PhD Thesis, University of Utrecht.

Thonon I, de Jong K, van der Perk M, and Middelkoop H (2007) Modelling floodplain sedimentation using particle tracking. *Hydrological Processes* 21: 1402–1412.

Thonon I, Roberti JR, Middelkoop H, van der Perk M, and Burrough P (2005) In situ measurement of sediment settling characteristics in floodplains using a LISST-ST. *Earth Surface Processes and Landforms* 30: 1327–1343.

Topping D, Wright S, Melis T, and Rubin D (2005) High-resolution monitoring of suspended-sediment concentration and grain size in the Colorado River using laser-diffraction instruments and a three-frequency acoustic system. In: *Proceedings of the Federal Interagency Sediment Monitoring Instrument and Analysis Research Workshop*. Flagstaff, AZ, USA, 9–11 September 2003. US Geological Survey Circular 1276.

Trimble SW (1983) A sediment budget for Coon Creek basin in the Driftless Area, Wisconsin, 1853–1977. *American Journal of Science* 283: 454–474.

Trimble SW (1993) The distributed sediment budget model and watershed management in the Paleozoic plateau of the upper midwestern United States. *Physical Geography* 14: 285–303.

Trimble SW and Crosson P (2000) U.S. soil erosion rates – myth and reality. *Science* 289: 248–250.

Turekian KK and Wedepohl KH (1961) Distribution of the elements in some major units of the earth's crust. *Bulletin of the Geological Society of America* 72: 175–192.

UNESCO (1978) *Monitoring of Particulate Matter Quality in Rivers and Lakes*. Paris: UNESCO.

US Bureau of Reclamation (2006) *Erosion and Sedimentation Manual*. Denver: US Bureau of Reclamation.

US Department of Agriculture (1997) *Predicting Soil Erosion by Water, a Guide to Conservation Planning with the Revised Universal Soil Loss Equation (RUSLE)*, USDA Agricultural Research Service, Agricultural Handbook No. 703. Washington, DC: USDA.

US Environmental Protection Agency (1997) *The Incidence and Severity of Sediment Contamination in Surface Waters of the United States*, vol. 1, EPA-823-R-97-006. Washington, DC: National Sediment Quality Survey, Office of Science and Technology.

US Environmental Protection Agency (2005) Procedures for the Derivation of Equilibrium Partitioning Sediment Benchmarks (ESBs) for the Protection of Benthic Organisms: Metal Mixtures (Cadmium, Copper, Lead, Nickel, Silver, and Zinc), EPA-600-R-02-011. Washington, DC: US Environmental Protection Agency, Office of Research and Development.

VA (ed.) (1977) *Sedimentation Engineering*, pp. 17–316. New York, NY: American Society of Civil Engineers, Task Committee for the Preparation of the Manual on Sedimentation of the Hydraulics Division.

Valentin C, Poesen J, and Li Y (2005) Gully erosion: Impacts, factors and control. *Catena* 63: 132–153.

Valmis S, Dimoylannis D, and Danalatos NG (2005) Assessing interrill erosion rate from soil aggregate instability index, rainfall intensity and slope angle on cultivated soils in central Greece. *Soil and Tillage Research* 80: 139–147.

van Loon JC (1985) *Selected Methods of Trace Metal Analysis: Biological and Environmental Samples*. New York, NY: Wiley.

Van Oost K, Quine TA, Govers G, et al. (2007) The impact of agricultural soil erosion on the global carbon cycle. *Science* 318: 626–629.

Van Rompaey AJJ, Verstraeten G, Van Oost K, Govers G, and Poesen J (2001) Modelling mean annual sediment yield using a distributed approach. *Earth Surface Processes and Landforms* 26: 1221–1236.

Verstraeten G and Poesen J (2002) Regional scale variability in sediment and nutrient delivery from small agricultural watersheds. *Journal of Environmental Quality* 31: 870–879.

Verstraeten G, Poesen J, de Vente J, and Koninckx X (2003) Sediment yield variability in Spain: A quantitative and semiqualitative analysis using reservoir sedimentation rates. *Geomorphology* 50: 327–348.

Verstraeten G, Prosser IP, and Fogarty P (2007) Predicting the spatial patterns of hillslope sediment delivery to river channels in the Murrumbidgee catchment, Australia. *Journal of Hydrology* 334: 440–454.

Viney NR and Sivapalan M (1999) A conceptual model of sediment transport: Application to the Avon River Basin in Western Australia. *Hydrological Processes* 13: 727–743.

Vörösmarty CJ, Meybeck M, Fekete B, and Sharma K (1997) The potential impact of neo-Castorization on sediment transport by the global network of rivers. In: *Human Impact on Erosion and Sedimentation*, IAHS Publication No. 245, pp. 261–273. Wallingford: IAHS Press.

Vörösmarty CJ, Meybeck M, Fekete B, Sharma K, Green P, and Syvitski JPM (2003) Anthropogenic sediment retention: Major global impact from registered river impoundments. *Global and Planetary Change* 39: 169–190.

Vrieling A (2006) Satellite remote sensing for water erosion assessment: A review. *Catena* 65: 2–18.

Wallbrink PJ, Murray AS, Olley JM, and Olive LJ (1998) Determining the sources and transit times of suspended sediment in the Murrumbidgee River, New South Wales, Australia, using fallout 137 Cs and 210Pb. *Water Resources Research* 34: 879–887.

Walling DE (1977) Assessing the accuracy of suspended sediment rating curves for a small basin. *Water Resources Research* 12: 1869–1894.

Walling DE (1983) The sediment delivery problem. *Journal of Hydrology* 65: 209–237.

Walling DE (2005) Tracing suspended sediment sources in catchments and river systems. *Science of the Total Environment* 344: 159–184.

Walling DE (2006a) Human impact on land–ocean sediment transfer by the world's rivers. *Geomorphology* 79: 192–216.

Walling DE (2006b) Tracing versus monitoring: New challenges and opportunities in erosion and sediment delivery research. In: Owens PN and Collins AJ (eds.) *Soil Erosion and Sediment Redistribution in River Catchments*, pp. 13–27. Wallingford: CABI.

Walling DE (2008) The changing sediment loads of the world's rivers. In: Schmidt J, Cochrane T, Phillips C, Elliott S, Davies T, and Basher L (eds.) *Sediment Dynamics in Changing Environments*, IAHS Publication No. 325, pp. 323–338. Wallingford: IAHS Press.

Walling DE and Collins AL (2008) The catchment sediment budget as a management tool. *Environmental Science and Policy* 11: 136–143.

Walling DE, Collins AL, and McMellin GK (2003c) A reconnaissance survey of the source of interstitial fine sediment recovered from salmonid spawning gravels in England and Wales. *Hydrobiologia* 497: 91–108.

Walling DE, Collins AL, and Sichingabula HM (2003a) Using unsupported lead-210 measurements to investigate soil erosion and sediment delivery in a small Zambian catchment. *Geomorphology* 52: 193–213.

Walling DE, Collins AL, Jones PA, Leeks GJL, and Old G (2006) Establishing fine-grained sediment budgets for the Pang and Lambourn LOCAR catchments, UK. *Journal of Hydrology* 330: 126–141.

Walling DE, Collins AL, Sichingabula HW, and Leeks GJL (2001) Integrated assessment of catchment sediment budgets. *Land Degradation and Development* 12: 387–415.

Walling DE, Collins AL, and Stroud RW (2008) Tracing suspended sediment and particulate phosphorus sources in catchments. *Journal of Hydrology* 350: 274–289.

Walling DE and Fang D (2003) Recent trends in the suspended sediment loads of the world's rivers. *Global and Planetary Change* 39: 111–126.

Walling DE and He Q (1993) Use of cesium-137 as a tracer in the study of rates and patterns of floodplain sedimentation. In: Peters NE, Hoehn E, Leibundgut Ch, Tase N, and Walling DE (eds.) *Tracers in Hydrology*, IAHS Publication No. 215, pp. 319–328. Wallingford: IAHS Press.

Walling DE and He Q (1997) Use of fallout ^{137}Cs in investigations of overbank deposition on floodplains. *Catena* 29: 263–282.

Walling DE and He Q (1999a) Improved models for estimating spoil erosion rates from cesium-137 measurements. *Journal of Environmental Quality* 28: 611–622.

Walling DE and He Q (1999b) Using fallout lead-210 measurements to estimate soil erosion on cultivated land. *Soil Science Society of America Journal* 63: 1404–1412.

Walling DE, He Q, and Appleby PG (2002) Conversion models for use in soil-erosion, soil redistribution and sedimentation investigations. In: Zapata F (ed.) *Handbook for the Assessment of Soil Erosion and Sedimentation Using Environmental Radionuclides*, pp. 111–164. Dordrecht: Kluwer.

Walling DE, He Q, and Blake WH (1999) Use of ^{7}Be and ^{137}Cs measurements to document short- and medium-term rates of water-induced soil erosion on agricultural land. *Water Resources Research* 35: 3865–3874.

Walling DE, Owens PN, Carter J, et al. (2003d) Storage of sediment-associated nutrients and contaminants in river channel and floodplain systems. *Applied Geochemistry* 18: 195–220.

Walling DE, Owens PN, Foster IDL, and Lees JA (2003b) Changes in the fine sediment dynamics of the Ouse and Tweed basins over the last 100–150 years. *Hydrological Processes* 17: 3245–3269.

Walling DE, Owens PN, and Leeks GJL (1998) The role of channel and floodplain storage in the suspended sediment budget of the River Ouse, Yorkshire, UK. *Geomorphology* 22: 225–242.

Walling DE, Owens PN, Waterfall BD, Leeks GJL, and Wass PD (2000) The particle size characteristics of fluvial suspended sediment in the Humber and Tweed catchments, UK. *Science of the Total Environment* 251–252: 205–222.

Walling DE, Schuller P, Zhang Y, and Iroumé A (2009) Extending the timescale for using beryllium 7 measurements to document soil redistribution by erosion. *Water Resources Research* 45: W02418 (doi:10.1029/2008WR007143).

Walling DE and Webb BW (1988) The reliability of rating curve estimates of suspended sediment yield. Some further comments. In: *Sediment Budgets, Proceedings of the Porto Alegre Symposium*. IAHS Publication No. 174, pp. 337–350. December 1988. Wallingford: IAHS Press.

Warren N, Allan IJ, Carter JE, House WA, and Parker A (2003) Pesticides and other micro-organic contaminants in freshwater sedimentary environments – a review. *Applied Geochemistry* 18: 159–194.

Wasson RJ (1994) Annual and decadal variation of sediment yield in Australia, and some global comparisons. In: Olive LJ, Loughran RJ, and Kesby JA (eds.) *Variability in Stream Erosion and Sediment Transport*, IAHS Publication 224, pp. 269–279. Canberra: IAHS Press.

Wasson RJ, Mazari RK, Starr B, and Clifton G (1998) The recent history of erosion and sedimentation on the Southern Tablelands of southeastern Australia: sediment flux dominated by channel incision. *Geomorphology* 24: 291–308.

Waters TF (1995) *Sediment in Streams: Sources, Biological Effects and Control*. Bethesda: American Fisheries Society.

Webb JS, Thornton I, Thompson M, Howarth RJ, and Lowenstein PL (1978) *The Wolfson Geochemical Atlas of England and Wales*. Oxford: Oxford University Press.

Wedepohl KH (1995) The composition of the continental crust. *Geochimica et Cosmochimica Acta* 59: 1217.

Wei H, Nearing MA, Stone JJ, et al. (2009) A new splash and sheet erosion equation for rangelands. *Soil Science Society of America Journal* 73: 1386–1392.

Wemple BC, Swanson FJ, and Jones JA (2001) Forest roads and geomorphic process interactions, Cascade Range Oregon. *Earth Surface Processes and Landforms* 26: 191–204.

Wilkinson S, Henderson A, Chen Y, and Sherman B (2004) SedNet User Guide, Version 2. *Client Report*. Canberra: CSIRO Land and Water. http://www.toolkit.net.au/sednet (accessed April 2010).

Wilkinson SN, Prosser IP, Olley JM, and Read A (2005) Using sediment budgets to prioritise erosion control in large river systems. In: Batalla RJ and Garcia C (eds.) *Geomorphological Processes and Human Impacts in River Basins*, IAHS Publication 299, pp. 56–64. Solsona, Catalonia: IAHS Press.

Wilkinson SN, Prosser IP, and Hughes AO (2006) Predicting the distribution of bed material accumulation using river network sediment budgets. *Water Resources Research* 42: W10419.

Wilkinson SN (2008) Testing the capability of a sediment budget model to target remediation measures to reduce suspended-sediment yield. In: Schmidt J, Cochrane T, Phillips C, et al. (eds.) *Sediment Dynamics in Changing Environments*, IAHS Publication 325, pp. 559–566. Christchurch: IAHS Press.

Wilkinson SN, Prosser IP, Rustomji P, and Read AM (2009) Modelling and testing spatially distributed sediment budgets to relate erosion processes to suspended sediment yields. *Environmental Modelling and Software* 24: 489–501.

Wilson CG, Matisoff G, and Whiting PJ (2003) Short-term erosion rates from a 7Be inventory balance. *Earth Surface Processes and Landforms* 28: 967–977.

Wischmeier WH and Smith DD (1978) *Predicting Rainfall Erosion Losses. A Guide to Conservation Planning*. Agriculture Handbook No. 537. Washington, DC: US Department of Agriculture.

Wood PJ and Armitage PD (1997) Biological effects of fine sediment in the lotic environment. *Environmental Management* 21: 203–217.

Woodward JC and Walling DE (2007) Composite suspended sediment particles in river systems: Their incidence, dynamics and physical characteristics. *Hydrological Processes* 21: 3601–3614.

Wynn T and Mostaghimi S (2006) The effects of vegetation and soil type on streambank erosion, southwestern Virginia, USA. *Journal of the American Water Resources Association* 42: 69–82.

Yang CT (1996) *Sediment Transport: Theory and Practice*. New York: McGraw-Hill.

Young RA, Onstad CA, Bosch DD, and Anderson WP (1989) AGNPS – a nonpoint-source pollution model for evaluating agricultural watersheds. *Journal of Soil and Water Conservation* 44: 168–173.

Zapata F (2002) *Handbook for the Assessment of Soil Erosion and Sedimentation Using Environmental Radionuclides*. Dordrecht: Kluwer.

Zhang G, Liu Y, Han Y, and Zhang XC (2009a) Sediment transport and soil detachment on steep slopes: I. Transport capacity estimation. *Soil Science Society of America Journal* 73: 1291–1297.

Zhang G, Liu M, Han Y, and Zhang XC (2009b) Sediment transport and soil detachment on steep slopes: II. Sediment feedback relationship. *Soil Science Society of America Journal* 73: 1298–1304.

Relevant Websites

http://www.loicz.org
    Land Ocean Interactions in the Coastal Zone.

2.13 Field-Based Observation of Hydrological Processes

M Weiler, Albert-Ludwigs University of Freiburg, Freiburg, Germany

© 2011 Elsevier B.V. All rights reserved.

2.13.1	Runoff Generation Processes	339
2.13.1.1	Early Research	339
2.13.1.2	Defining the Pathways of Storm Runoff	340
2.13.1.3	Current Directions	342
2.13.2	Quantifying the Processes	343
2.13.2.1	Field-Based Observations	343
2.13.2.2	Quantifying the Processes: Hydrometric Observations	343
2.13.2.3	Quantifying the Processes: Tracers	345
2.13.3	Conclusion	347
References		348

2.13.1 Runoff Generation Processes

2.13.1.1 Early Research

Infiltration was the first process recognized as being significant to runoff generation during a precipitation event. In the early part of the twentieth century, Robert Elymer Horton first described quantitatively the process of infiltration into the soil surface and introduced terminology still used by hydrologists today (Horton, 1933). Following Horton, others recognized that surface runoff was often not the dominant process responsible for increased stream discharge observed during precipitation events. Beginning with Hursh and Brater's (1941) work at Coweeta (North Carolina, USA), subsurface flow became recognized as a potentially important component of storm flow. Later, studies identified the concepts of variable runoff source areas and the importance of subsurface flow as a contributor to event stream flow response (Betson, 1964; Hewlett and Hibbert, 1963, 1967). Shortly after these developments, old water (pre-event water stored in the watershed as soil water or/and groundwater) was identified as being a significant contributor to runoff (e.g., Pinder and Jones, 1969; Sklash and Farvolden, 1979). Indeed, it is now widely accepted that old water constitutes the majority of stormflow in humid watersheds (e.g., Pearce et al., 1986). However, new water may still be an important contribution to storm runoff in urbanized watersheds or many arid and mountainous watersheds.

Horton (1933) defined infiltration as a result of the need to describe the physical process by which water moves into the soil, distinct from other terms sometimes used such as percolation or absorption. Horton defined infiltration capacity as "the maximum rate at which rain can be absorbed by a given soil at a given condition" (Horton, 1933: 453). Horton attributed surface runoff to rainfall intensities that exceeded the infiltration capacity of the soil. This is widely known as Hortonian overland flow or infiltration excess overland flow. However, Horton was not working in forested environments and therefore probably concluded incorrectly that runoff for an individual storm event was mainly or wholly surface runoff. The storm hydrograph response in a forested watershed was shown to consist of subsurface flow and channel precipitation by Hursh and Brater (1941). Engler (1919) already recognized the importance of subsurface stormflow after making detailed measurements of infiltration and physical properties of soil, including porosity, water content, soil texture, and hydraulic conductivity. Subsequently, soil depth, topography, and hydrologic characteristics associated with different elevations were shown to influence peak discharge (Hoover and Hursh, 1943). Hewlett and Hibbert (1963) first recognized the importance of unsaturated flow and concluded that unsaturated flow could not be ignored in hydrograph analysis. Utilizing a concrete trough to observe unsaturated flow at the Coweeta experimental watershed, they coined the term 'translatory flow' to describe unsaturated flow and attributed it to the thickening of water films surrounding soil particles, which results in a pulse of water.

Substantial amounts of runoff can be generated on areas which have been saturated with water (Dunne and Black, 1970). Furthermore, not only water quantity but also water chemistry and quality are affected by runoff from saturated soils (Molénat et al., 2002). Cappus (1960) recognized that saturated areas often occur at specific locations in a watershed which led to the development of the partialcontribution area concept (Betson, 1964). Runoff-generating areas are frequently located in valley floors and on particularly shaped slopes (Amerman, 1965). Even though extent and location of runoff generation areas can vary notably, it has been demonstrated that generally only a small part of a watershed contributes to storm runoff from saturated areas (Ragan, 1968).

Cappus (1960) characterized a catchment in terms of its runoff-generating areas. His research showed that it is parts of the watershed, not the whole area, that contribute to runoff (partial contribution area concept). Concerning the involved parts of the catchment, he differentiated between infiltration excess (roads and compacted soil) and saturation areas (valley bottoms). The variable source area concept was developed in the early 1960s and is largely attributed to Betson (1964). He found out that contributions made by different parts of the catchment depend on the precipitation intensity, but the variability is so small that the contributing areas remain almost constant for the duration of one event in his study area. Betson (1964) demonstrated that contributing areas were almost constant during heavy rainfalls. For such events, infiltration excess overland flow was observed.

Research by Ragan (1968) supports the partial contribution area concept, when he states that only a small part of the basin, less than 3% of the total watershed, contributes an appreciable amount to the storm hydrograph. Amerman (1965) found out that these areas are often located on ridge tops, in valley floors, and on valley slopes. Dunne and Black (1970) collected subsurface flow and saturation overland flow (SOF) with a large trench and could demonstrate that the partial area concept can be extended from infiltration excess overland flow (Betson, 1964) to saturation excess overland flow. Furthermore, Weyman (1970) described that the concept of partial contributing areas, approved by his experiments, can be extended to subsurface flow. He observed that subsurface throughflow and saturation excess surface runoff mainly occurred in specific parts of his watershed.

Subsurface stormflow was finally recognized as being an important contributor to event-based stream discharge. In addition, it was previously observed that preferential subsurface stormflow could occur in forest soils (i.e., water moving faster than the soil matrix should allow, typically through some form of soil pipe) (Whipkey, 1965). Whipkey was the father of trench studies, in which trenches are commonly excavated along the base of a hillslope down to the impermeable layer and flow from the soil horizons is collected and measured. Figure 1 shows an example of how permanent trenches can be built to collect runoff at the soil surface and in the subsurface.

2.13.1.2 Defining the Pathways of Storm Runoff

Development of runoff theory proceeded rapidly in the 1960s and 1970s and the studies conducted by Dunne and Black (1970) set a precedent that was rarely exceeded during the next two decades. Dunne and Black used intensive instrumentation across various hillslope types to observe subsurface processes in a wet, mountainous area of Vermont, USA. Three hillslopes consisting of well-drained sandy loams over glacial till, with convex, concave, and straight contours, were instrumented with wells and piezometers to measure water-table elevation and pressure potential, and a nuclear depth probe was used to measure soil moisture along a transect up the middle of each slope. A trench was excavated along the base to the hillslopes to measure runoff at various levels and weirs were installed above and below the reach of river channel running at the

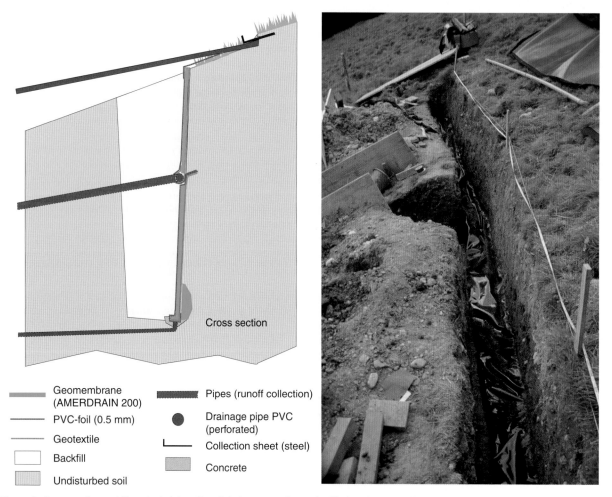

Figure 1 Cross section and the actual picture (trench is in construction and refilled to the upper observation depth) of a permanent trench for measuring surface runoff and subsurface runoff in two different depths.

base of the study site. Subsurface stormflow was found to occur only during large events and SOF occurred in significant quantities only on the concave (hollow) hillslope. Overland flow occurring on the concave hillslope during large precipitation events was the only flow measured in large enough quantities to account for the measured stream flow.

Other important contributions during this decade include Weyman's (1973) study, which advocated the theory of a saturated wedge developing from riparian margins and moving upslope with increasing precipitation. Groundwater hydrologists, such as Alan Freeze, were developing theories on regional groundwater flow in the early 1970s (e.g., Freeze and Witherspoon, 1967). Freeze et al. (1972) suggested that the majority of event hydrograph response could be attributed to subsurface stormflow. Near the end of the 1970s, a series of studies focused on searching for the mechanism that could explain this process (e.g., Sklash and Farvolden, 1979).

Until this time, subsurface flow was considered to be a function of measurable physical properties of soil, namely hydraulic conductivity. However, measurement of soil hydraulic conductivity could not account for the rates of flow necessary to deliver water, via the subsurface, to the stream channel in order to affect the observed stream response. This quandary is resolved by separately considering the flow in the soil matrix described by Darcy's law (where flow is dependent on soil hydraulic conductivity) and preferential flow pathways via soil pipes and macropores (e.g., Harr, 1977). Studies such as Mosley (1979) showed that rates of subsurface flow could be large enough to account for the observed hydrograph response in a steep headwater catchment with very moist conditions (M8 catchment, Maimai, New Zealand). Large peak flow rates observed at concentrated locations of soil pit faces were found to coincide with stream hydrograph peaks and dye-tracing experiments were used to quantify the rate of water movement through the profile. Mosley's dye experiments led him to conclude that the majority of the stream flow response was from the contribution of event or new water (water contributed by the current precipitation event). Significant debate over the source of water that generates the storm hydrograph response followed Mosley's (1979) published work. Mosley believed that the new water was entering macropores and the soil surface, and flowing by lateral macropores without interacting with the soil matrix.

Around the time Mosley was working in Maimai, a forested headwater catchment in New Zealand, a new method of examining the source of stormflow was conceived. Pinder and Jones (1969) were the first to use the two-component mixing model to separate event water on the basis of chemical signatures by measuring various ions in rainwater, storm discharge, and stream baseflow. However, it would be almost 20 years before hydrochemical observations were combined with hydrometric observations. Pinder and Jones (1969) concluded that up to 42% of event stream flow might be old water in the Nova Scotia catchment studies. Later, Sklash and Farvolden (1979) measured tritium, oxygen-18 ($\delta^{18}O$), and deuterium isotope ratios across various watersheds and concluded that groundwater was the main contributor to the event hydrograph. The process responsible for the transfer of old water to the stream in sufficient quantities to explain their observations was attributed to groundwater ridging near the riparian margins via the rapid conversion of the tension saturated capillary fringe to phreatic water (i.e., saturation occurred soon after the commencement of an event). Gillham (1984) studied these groundwater ridging processes and further realized the importance for stream flow generation in watersheds with extending riparian zones.

Following Mosley (1979), Pearce et al. (1986) and Sklash et al. (1986) published the results of studies in which they examined the relative concentrations of chloride, deuterium, and $\delta^{18}O$ in addition to the electrical conductivity of samples of rainfall, streamflow, and soil water flowing from pit faces in the Maimai catchment, New Zealand. Generally, old (preevent) water and new (event) water were thought to be mixing in the soil profile and then discharging to the stream in a fairly uniform mixture in terms of isotopic and chemical composition (Pearce et al., 1986, Sklash et al., 1986). Groundwater ridging and saturated wedge development from the rapid conversion of tension-saturated zones to positive potentials were cited as the mechanisms responsible for the delivery of stormflow, although hydrometric data were not available to augment these findings. If conversion of tension-saturated zones to positive potentials was occurring, rapid transmission of new water was not needed to explain stormflow. The majority of stormflow would be contributed by old water already stored in the soil and only a small amount of new water would be needed as input (Pearce et al., 1986; Sklash et al., 1986).

To solve the old-water new-water dichotomy, a unification of hydrochemical and hydrometric measurements was necessary and McDonnell (1990) did just that in the same catchment (Maimai-M8) as studied by Mosley (1979), Pearce et al. (1986), and Sklash et al. (1986). Using the same soil pits excavated in the previous studies, a combination of isotope and chemical tracing, and an extensive tensiometer network, McDonnell (1990) observed that water tables arising at the soil bedrock interface were not maintained but correlated well with throughflow rates. Soil piping (connection of macropores) was suggested to explain the rapid dissipation of the water table and pore water pressures (McDonnell, 1990). To explain his observations, McDonnell (1990) suggested that rapidly infiltrating new water perched at the impermeable layer and mixed with larger volumes of old water and subsequently drained as the saturated areas in the hillslope expanded creating continuous saturated areas thus affecting rapid stormflow. The formation of these saturated areas, then, largely depends on topography. McGlynn et al. (2002) provided a thorough review of the experiments to date at the Maimai research area. Another explanation of the old-water dominance was provided by the transmissivity feedback mechanism (Rodhe, 1989; Seibert et al., 2003). The process, which was mainly observed in glacial till soils in Scandinavia and Canada, describes the rapid rise of the water table into the more transmissive (permeable) topsoil and the resulting higher subsurface flow. As water is stored below field capacity in the soils, the new water mixes with the old water and the resulting runoff is characterized by dominance in old water (Laudon et al., 2004). Beven (2006) compiled many relevant original papers about runoff generation that are also introduced in this chapter and provided the historical context in detail.

2.13.1.3 Current Directions

A range of mechanisms facilitates subsurface stormflow and it is useful to separate these into various subareas in order to examine the controls on subsurface stormflow. From the above discussion, we know that increases in subsurface flow are a result of increasing hydraulic gradient, cross-sectional area, rise in the water table into more transmissive soil layers, and the linking of isolated saturated areas across the hillslope (variable source area). To what degree each of these phenomena influences subsurface stormflow appears to depend on various conditions such as antecedent moisture conditions and the morphology of the defined basin or hillslope. Subsurface stormflow initiation and the mechanisms for preferential flow are still debatable issues. Topographic control is being examined in greater detail as it becomes apparent how complex the influence of morphology is on subsurface flow. The aim to better understand the subsurface structures focused on using different geophysical methods. It is also recognized that there are thresholds to the occurrence of subsurface storm flow. Again, these thresholds seem to vary with individual site conditions and it is recognized that the response is nonlinear (Weiler et al., 2005). Research in the last decade focuses on any one or on a number of these issues, either directly or indirectly.

Preferential flow has been shown to be important for both flow initiation and rapid lateral transport of water downslope (Mosley, 1979; McDonnell, 1990). In the first case, preferential infiltration has been identified as being significant enough for rapid development of saturated areas and water tables in more permeable soils (e.g., De Vries and Chow, 1978; McDonnell, 1990; Weiler and Naef 2003). In order for rapid stream flow response to be facilitated by subsurface storm flow, water must infiltrate and move down slope at rates greater than the estimates based on soil matrix properties would predict. Preferential flow can occur via macropores, cracks and soil pipes, and in areas of higher permeability in the soil, including highly permeable layers (Bonell, 1998; McGlynn et al., 2002). It has also been recognized that the permeability of macropore and crack walls may be lower than that of the soil matrix, which would allow for rapid unimpeded flow once water fills these conduits (Calver and Cammeraat, 1993; McDonnell, 1990). Rates of pipe flow are largely determined by their diameter, and it has been recognized that there are certain precipitation thresholds that must be exceeded before pipe flow will dominate the subsurface flow (Weiler and McDonnell, 2007). In Bonell's (1998) review of runoff generation, he states "reconciling their [soil pipes] hydrochemistry coupled with the need for more sophisticated hydrometric studies to address the pipeflow issue, stands out as one of the principle research challenges connected with storm hydrograph separations." We are still lacking the knowledge to address these pipe flow issues; however, there have been several attempts to perform more sophisticated tracer-based and hydrometric studies (Anderson et al. 2009a, 2009b; Anderson et al., 2010) or to use other approaches to understand the flow pathways along the soil bedrock interface (Graham, 2009).

Geophysical methods offer the opportunity to rapidly collect subsurface information in a noninvasive or minimally invasive manner, which may be a key information to see flow pathways and hence understand runoff generation processes. These techniques are sensitive to different physical properties (e.g., magnetic, elastic, and electrical properties) of subsurface materials. In near-surface environments, techniques such as ground-penetrating radar (GPR), electrical resistivity tomography (ERT), electromagnetic induction surveying, or different seismic methods have been proven to provide valuable data for a variety of applications (Butler, 2005). Much of the related work and progress made in the field of hydrology within the past decade is documented in Rubin and Hubbard (2005) and Vereecken et al. (2006). Especially, GPR and seismic methods may provide structural information to characterize the subsurface. In sandy sediments, GPR allows imaging of subsurface geometries up to depths of ∼10 m with a resolution at the dm-to-m scale (Beres et al., 1999). However, translation of geophysical observations into the relevant subsurface state and properties, such as moisture content, or hydraulic conductivity, remains a challenging task. The relations between geophysical and the hydrological target variables are usually complex, nonunique, and site specific (Schön, 1998). Uncertainty in analyzing and interpreting geophysical data may be reduced by multimethod approaches. For example, time-lapse imaging of subsurface flow processes is possible by combining geophysical techniques with artificial tracers, as for instance, salt tracers in the unsaturated zone imaged by ERT or cross-hole radar attenuation tomography (Johnson et al., 2007). Kienzler (2007) developed a nice example of the potential combining artificial salt tracer injection to visualize preferential lateral flow pathways (**Figure 2**).

Another interesting idea relies on extracting structural information such as correlation lengths from geophysical images. The basic assumption is that the statistical properties of geophysical structures and parameter variations, respectively, can be used as a proxy for the statistical characteristics of the target hydrological parameter field. Until now, this idea has primarily been used with GPR reflection data (e.g., Knight et al., 2007); more experience, using realistic synthetic scenarios and other geophysical data, is clearly needed. Nevertheless, such concepts can be extremely useful to investigate and characterize geophysical data and hydrological systems to understand runoff generation mechanisms.

Until today, experimental studies to understand runoff generation processes and flow pathways have been conducted basically worldwide in various climatic and geological settings, leading to an advanced and refined perception of rainfall runoff processes (Graham, 2009; Kienzler, 2007; Scherrer et al., 2006). Despite the large variety of observed flow processes, one commonality is the strong nonlinearity of hillslope response to rainfall (Tromp-van Meerveld and Weiler, 2008). One possibility to explain this behavior is the sudden connection of different areas in the watershed that are locally generating runoff (either at the soil surface or as subsurface runoff) by different flow pathways (macropores, pipes, and channels). Tromp-van Meerveld and McDonnell (2006) proposed the so-called fill-and-spill mechanisms to explain the threshold behavior at the Panola watershed experiment, USA. Bachmair and Weiler (2010) extended this concept to the connect-and-react hypothesis to generally describe the sudden connection of runoff generation processes in the watershed

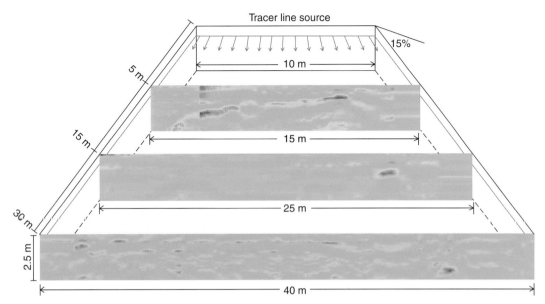

Figure 2 Subsurface flow paths detected with ground-penetrating radar (GPR) at the Koblenz experimental slope, Switzerland. The profiles have been taken before as well as 4 h after the start of a salt tracer injection. Displayed are the differences between these two measurements, large differences (yellow–red colors) indicate flow path locations. From Kienzler P (2007) Experimental Study of Subsurface Stormflow Formation. Combining Tracer, Hydrometric and Geophysical Techniques. Diss. ETH Nr. 17330, Eidgenössische Technische Hochschule (ETH), Zürich, Switzerland.

and the resulting nonlinearity or threshold behavior in the runoff response. However, explaining and predicating the threshold behavior continue to be a challenging task (Zehe and Sivapalan, 2009).

2.13.2 Quantifying the Processes

2.13.2.1 Field-Based Observations

Many different techniques and methods have been developed in the last 100 years to directly or indirectly observe runoff generation processes in the field. Many of these methods require only simple instruments or observations, but their power can be increased if enough spatial and temporal explicit observations (e.g., Trubilowicz et al., 2009) are being taken to understand the complex spatial–temporal processes during storm runoff generation. In Table 1, measurement methods are listed together with the spatial scale and the processes which can be observed. Many of these methods are explained in more detail in the following chapter by showing the potential and issues within sample application and past studies. The references listed in Table 1 are only possible sources of information.

2.13.2.2 Quantifying the Processes: Hydrometric Observations

The contribution of SOF to storm runoff has been repeatedly studied since the first work of Dunne and Black (1970). A central aspect of the SOF estimation is the delineation of the contributing saturated areas. Soil saturation can be detected with remote sensing (e.g., Mohanty and Skaggs, 2001). However, saturation patterns cannot be captured efficiently with remote sensing in dense forests (e.g., Kite and Pietroniro, 1996). Soil saturation under forests was therefore mapped based on soil and vegetation characteristics in order to evaluate the models for saturation predictions (Güntner et al., 2004). However, only few studies have evaluated the mapping criteria with direct saturation measurements (Rosin et al., 2009). Moreover, saturated areas have been mapped and modeled in a single climate (e.g., Blazkova et al., 2002), but only few investigations have been made to compare mapping and modeling in different climates, mostly using different data sources (Mérot et al., 2003). It still seems to be important to improve our methods to better monitor the spatial dynamics and connectivity of saturated areas in particular in watersheds that are dominated by SOF. The observations in the Miniflet catchment in Sweden (Myrabo, 1997) are a nice example of the space–time variations of water-table response and saturation areas due to subsurface flow and flow accumulation (**Figure 3**).

Observing subsurface runoff is still a challenge. Woods and Rowe (1996) excavated a trench in the Maimai M8 catchment and measured flow rate and quantity in a series of troughs along a trench face coupled with tensiometer and piezometer measurements. They found that bedrock topography was responsible for flow routing as saturated areas developed in hollows and converged as ribbons of concentrated flow. Flow volumes were highly variable and were not well predicted by the surface topography and flow accumulation. In a follow-up study, Woods et al. (1997) concluded that variability in runoff depends on both topography and soil moisture conditions. Freer et al. (1997) excavated another similar trench in the Panola watershed close to Atlanta, USA. They determined that bedrock topography improved predictions as subsurface flow routing is dependent on the morphology of the impeding layer or bedrock. Freer et al. (1997) also noted that antecedent soil moisture conditions were a significant control on the occurrence of saturation.

Table 1 Measurement methods to observe runoff generation processes at different scales

Measurement	Spatial scales	Processess	Issues	References
Infiltrometer	Plot, hillslope	Infiltration	Boundary conditions	Mertens et al. (2002)
Sprinkling experiments	Plot, hillslope	Infiltration, overland, and subsurface flow	Drop size distribution and intensity. Trench is necessary for subsurface runoff	Weiler and Naef (2003)
Soil moisture measurement	Point	Percolation, saturation, evapotranspiration	Point measurement, high spatial, and temporal variability	Jost et al. (2005)
Lysimeter	Plot	Percolation, evapotranspiration, groundwater recharge	High costs, disturbed soil monolith	Scanlon et al. (2002), Aboukhaled et al. (1982)
Soil water potential	Point	Flow direction, saturation	Point measurement, high spatial, and temporal variability	Anderson and Burt (1978)
Dye staining experiments	Plot, hillslope	Flow pathways, preferential flow	Destructive but very visually informative, artificial experiment	Weiler and Naef (2003), Anderson et al. (2005)
Artificial tracer experiments (1-D)	Plot, hillslope	Flow velocity, preferential flow	Sampling method influences results (destructive, flow, suction cups, etc.)	Weiler et al. (1998), McGuire et al. (2005)
Artificial tracer experiments (2-D)	Plot	Flow velocity, flow direction, preferential flow	Stationary conditions	Roth et al. (1991)
Trenching	Hillslope	Subsurface flow	Interruption of low pathways, artificial drainage	Woods and Rowe (1996)
Chemical hydrograph separation	Hillslope, catchment	Water sources (spatially or temporally)	Concentration of end members	Hooper and Shoemaker (1986)
Mapping of saturated areas	Hillslope, catchment	Potential saturation	Depends on climate	Merot et al. (1995)
Saturation collectors	Plot, hillslopes	Saturation	Point observations	Rosin et al. (2009)

Utilizing smaller troughs to collect discharge from the soil profile of a hillslope and a network of piezometers, Hutchinson and Moore (2000) examined how throughflow is related to surface topography and basal till/confining layer. The troughs were oriented such that the hillslope was divided up into units so they could be compared. They found that at the lowest flow, the subsurface flow distribution was correlated well with upslope contributing area as calculated from the basal till layer topography. However, at the highest flows, subsurface flow was more closely related to the contributing area of the surface topography. In other words, the saturated area, or water table, shifted from being parallel to the confining layer to being parallel to the surface. Moreover, they observed macropores which can deliver enough discharge to negate the topography as a control on subsurface flow. It was suggested that macropores can route water laterally which questions the validity of models which use topographic controls to predict subsurface flow (Hutchinson and Moore, 2000). Finally, it is of great relevance to this study that topographic models usually assume a quasi-steady-state for throughflow, which Hutchinson and Moore (2000) did not find appropriate at their site.

Scherrer et al. (2006) surpassed, in number of trenches, all other experimental trench studies to understand runoff generation processes. They performed sprinkling experiments at $60\,m^2$ hillslopes and also measured, in addition to surface runoff, subsurface runoff in a 6-m-wide trench (**Figure 4**). In order to quantify the internal processes, they also instrumented the slope with many time-domain reflectometer (TDR) probes, tensiometers, and piezometers. This combined hydrometric observation setup allowed a detailed description of flow processes within the hillslope, resulting in runoff generation processes. Tromp-van Meerveld and Weiler (2008) also instrumented a hillslope in detail, similar as Scherrer et al. (2006) did, but with the focus of a longer-term observation to explore the wet and dry season and its processes. In particular, the combination of soil moisture measurements, maximum water-table observations, and sap-flow measurements enabled them to observe the spatial–temporal patterns of flow processes in the soil and transpiration processes.

Jost et al. (2005) presented another approach, focusing on soil moisture measurements with TDRs to observe the spatial–temporal patterns of transpiration, recharge, and soil moisture storage. They measured soil moisture dynamic at 198 locations in a 0.5-ha forest site and were able to use these observations together with geostatistical methods to predict the patterns of water fluxes at the soil vegetation atmosphere. In future, the potential to use a large number of sensors to better observe the spatial–temporal patterns of fluxes and storage changes in the saturated and unsaturated zone will increase when using wireless sensor techniques. The wireless sensors will only provide a more efficient way to collect the data, but the development of cost-effective sensors to observe the relevant fluxes states the need to go hand in hand. Trubilowicz

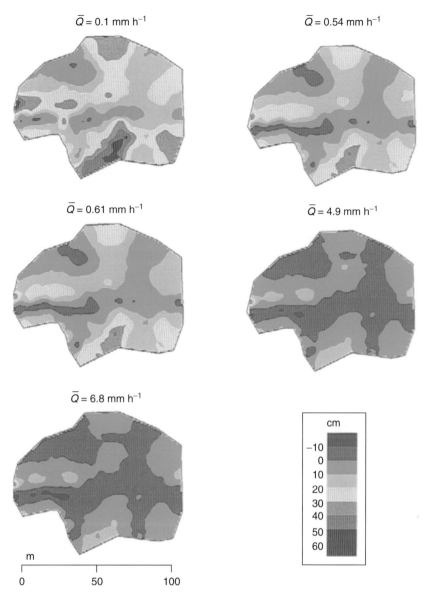

Figure 3 Water table variations and saturation (blue areas) at the Minifelt watershed in Sweden for different stream runoff conditions. From Myrabo S (1997) Temporal and spatial scale of response area and groundwater variation in till. *Hydrological Processes* 11: 1861–1880.

et al. (2009) tested the potential of low-cost, low-power wireless sensor networks (mote networks) for monitoring throughfall, temperature, humidity, soil water content, and water-table dynamics using 41 motes in a small forested watershed (**Figure 5**). They found that while motes gave the ability to monitor a catchment at resolution levels that were previously impossible, they still need to evolve into an easier-to-use, more reliable platform before they can replace traditional data collection methods.

2.13.2.3 Quantifying the Processes: Tracers

Hydrologists have used tracers to study water movement for several decades and there are a number of different tracers available some being better suited to specific applications (see also **Chapter 2.09 Tracer Hydrology**). There are two basic types of tracers, the one considered natural and the other artificial. Natural tracers are ones that can be found in the natural environment such as oxygen isotope 18, tritium, or weathered materials like silicates. These can be measured from water samples in the soil, precipitation, groundwater, and the stream. Artificial tracers are applied to the system; this includes various types of dyes and anions (e.g., chloride). Of course, chloride is naturally occurring but it is often applied in much larger quantities so that it overrides the natural background concentrations. Natural and artificial tracers both have advantages and disadvantages and neither are necessarily better, even of these two types, one tracer may be completely inappropriate where another is very useful. For example, Rhodamine WT is not a very useful soil–water tracer in lab application, whereas Lissamine FF is (Trudgill, 1987).

The first tracers to be used in runoff generation studies were naturally occurring. By measuring the relative concentrations in the different sources (soil water, precipitation, and

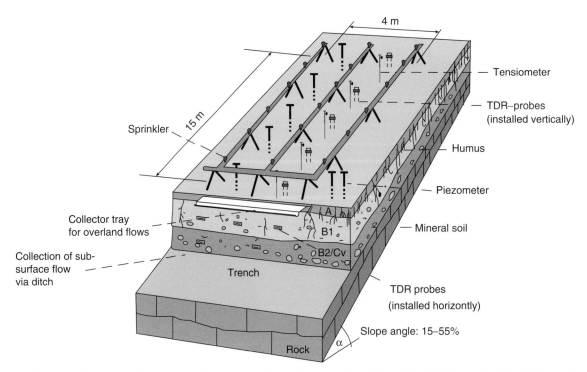

Figure 4 Experimental setup to study runoff generation processes at the hillslope scale. From Scherrer S (1997) Abflussbildung bei Starkniederschlägen, Identifikationvon Abflussprozessen mittels künstlicher Niederschläge. Versuchsanstalt fürWasserbau, Hydrologie und Glaziologie der ETH Zürich, Zürich, 147pp.

stream), researchers could separate the storm hydrograph into different component sources (e.g., Pinder and Jones, 1969). For example, McGuire et al. (2005) used $\delta^{18}O$ to measure the residence time of water falling on eight different catchments in the HJ Andrews experimental forest, Coastal Range Oregon. They measured $\delta^{18}O$ in the various input sources and in the stream water to determine the source of the stormflow and how long it has been in the catchment. By accounting for variations in isotope ratios with changes in elevation and also values for snowpack melt water, they were able to determine residence times for the catchments and compare how it varied across scale. Interestingly, residence time was not dependent on scale but was more closely related to simple topographic measures such as median flow path length and gradient.

Other types of natural tracing include measurements of silica content and alkalinity. For example, Soulsby et al. (2004) used alkalinity and silica content measurements to determine the main sources of runoff in the Scottish highlands. Natural tracers lend themselves well to catchment scale and larger studies while artificial tracers are convenient for hillslope and plot scale applications.

The main drawback of natural tracers is the uncertainty associated with characterizing the sources (e.g., Didszun and Uhlenbrook, 2008). We know that chemical signatures are variable in both space and time. The chemical signature of soil moisture varies spatially and temporally depending on the length of time that moisture has resided in the soil. In turn, residence time of soil water is dependent on the length of time since the last precipitation event and the size of that event. This dependence makes it difficult to account for the soil water signature. In addition, the influence of interception and the spatial variation in chemical composition of precipitation at the catchment scale has received little attention to date. Most hydrochemical studies have focused on the hillslope scale and often use only a single rain gauge to determine the chemical signature of rainwater. As McGuire et al. (2005) pointed out, the isotope signature of rainfall varies with elevation. In addition, deposition of minerals and soil physical properties can vary over small spatial scales, which will alter the chemical signature of water flowing through various areas. It would be nearly impossible, or at least very labor intensive, to account for such variations. Nonetheless, naturally occurring tracers continue to be used and do provide certain advantages over artificial tracers, mainly that they can be used on a larger scale and are ubiquitous.

Artificial tracers overcome the uncertainty associated with characterizing the naturally occurring tracers, in that we can control when, where, and how much to apply. Artificial tracers have been used around for some time as early as the late 1960s (e.g., radioactive tracers used by Pilgrim (1966)). Exploration of the utility of tracers in hydrological study continued through the 1970s (e.g., Pilgrim and Huff, 1978; Smart and Laidlaw, 1977). Pilgrim and Huff (1978) demonstrated the usefulness of artificial tracers for monitoring water movement in the subsurface and observed irregular patterns of movement despite a uniform surface.

More recently, dye experiments have been used to examine infiltration in greater detail (Weiler and Flühler, 2004; Weiler and Naef, 2003). Weiler and Flühler (2004) used simulated rainfall with Brilliant Blue dye followed by soil pit excavation and image analysis to examine infiltration (example images in Figure 6). Extended vertically stained sections of macropore

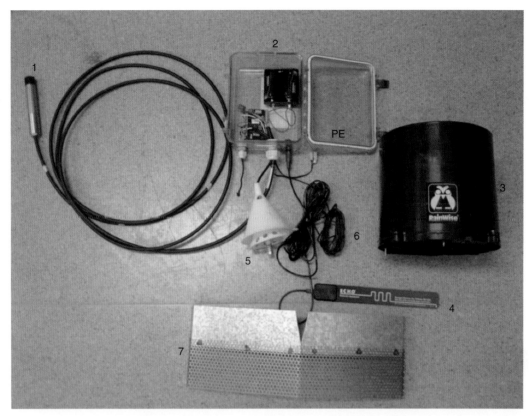

Figure 5 Wireless sensor network with measurement station ready for deployment: (1) pressure transducer; (2) mote with power supply; (3) tipping bucket; (4) soil moisture probe; (5) air temperature/humidity probe with radiation shield; (6) ground temperature thermistor; (7) overland flow weir. From Trubilowicz J, Cai K, and Weiler M (2009) Viability of motes for hydrological measurement. *Water Resources Research* 45: W00D22 (doi:10.1029/2008WR007046).

flow were observed which initiated close to the soil surface. This is in accordance with the conclusions of McDonnell (1990). Weiler and Naef (2003) concluded that although macropores make up a much smaller fraction of the total porosity (<1%), they account for the majority of saturated flow and preclude the use of Darcy's law or the Richards' equation to predict flow rates. However, dye-staining experiments at the hillslope scale seem to be possible. Anderson *et al.* (2009a) were able to reconstruct lateral preferential flow networks by staining a 30-m-long hillslope and excavating the pathways. The experiment revealed that larger contributing areas coincided with highly developed and hydraulically connected preferential flow paths that had flow with little interaction with the surrounding soil matrix. They found evidence of subsurface erosion and deposition of soil and organic material laterally and vertically within the soil (see detailed information about the experimental setup, results, and interpretation in the electronic supplements). These dye-staining results are important because they add to the understanding of the runoff generation, infiltration, solute transport, and slope stability of preferential flow-dominated soils.

Artificial tracers have also been used at the hillslope scale often injected through piezometers at specific depth (e.g., Talamba *et al.* (2000) or Weiler *et al.* (1998)) or done as a line application at the top of a hillslope or hillslope plot (Weiler *et al.*, 1998). It seems that the time has come for artificial tracers to be tested at the small catchment scale. Application is

the biggest limitation with artificial tracers. It is either labor intensive or expensive in the case of sprinkler systems. Rodhe *et al.* (1996) and Lange *et al.* (1996) conducted studies in catchment which have been covered below the canopy so that chemical signature of the input water could be controlled, an impressive undertaking. Rodhe *et al.* (1996) used $\delta^{18}O$ ratios while Lange *et al.* (1996) used LiBr as their tracers. Of greatest interest are the results of Lange *et al.* (1996) who had low recovery and concluded that residence time was long enough to permit equilibrium exchange between the soil water and soil matrix. They believed that hydrochemical processes related to catchment runoff are underestimated because they are often based on soil column studies that do not account for lateral movement. While artificial tracers have been used in hydrology for some time, their usefulness as a tool for studying runoff generation has not been extensively explored. It could be possible in future to use artificial tracer more extensively if instrumentation techniques to detect the tracers are becoming better and smaller amounts need to be applied to observe the movement of tracers in the watershed.

2.13.3 Conclusion

In the 1960s and 1970s, the focus was on observing hydrological processes in the field. There have been many groundbreaking studies and experiments observing the

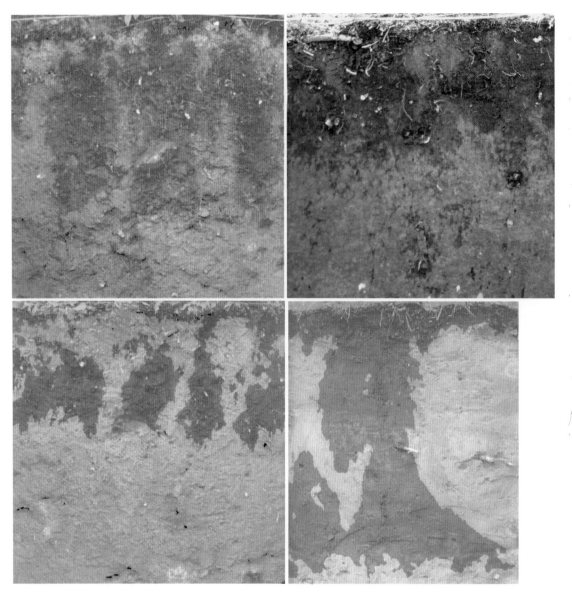

Figure 6 Examples of four dye patterns in forest soils after sprinkling 60 mm dyed water in 3 h. The soil types range from sandy soils to loamy soils; however, macropores, root channels, and hydrophobicity at the soil surface are more relevant for generating different infiltration patterns than soil type.

spatial–temporal dynamics of water and solute fluxes on the plot, hillslope, and catchment to understand the interplay of different hydrological processes. Unfortunately, the focus on hydrology shifted toward modeling and computer simulations in the last 20 years. Field-based observations are demanding and time consuming and the rewards are often not as pronounced compared to developing or applying a new hydrological model at the watershed or even at the continental scale. However, we have forgotten many important lessons that we have learned about the functioning of watersheds. Our current hydrological models are all very similar and most of them do not incorporate the hydrological processes and flow pathways that have been observed in the field. As pleaded, for example, by Weiler and McDonnell (2004), a new area of more connections and discussions between field hydrologists and hydrological modeler is needed to overcome the current deficit in hydrological model development. It is also believed that new techniques and new sensors need to be developed, tested, and implemented into field-based observation to enhance the possibility and understanding of processes, in particular flow processes in the subsurface and surface–groundwater interaction.

References

Aboukhaled A, Alfaro A, and Smith M (1982) *Lysimeters, Irrigation and Drainage*, Paper 39, 68pp. Rome: FAO.

Amerman CR (1965) The use of unit-source watershed data for runoff prediction. *Water Resources Research* 1(4): 499–507.

Anderson AE, Weiler M, Alila Y, and Hudson RO (2009a) Dye staining and excavation of a lateral preferential flow network. *Hydrology and Earth System Sciences* 13: 935–944.

Anderson AE, Weiler M, Alila Y, and Hudson RO (2009b) Subsurface flow velocities in a hillslope with lateral preferential flow. *Water Resources Research* 45: W11407 (doi:10.1029/2008WR007121).

Anderson AE, Weiler M, Alila Y, and Hudson RO (2010) Water table response in zones of a watershed with lateral preferential flow as a first order control on subsurface flow. *Hydrological Processes* (in press).

Anderson MG and Burt TP (1978) The role of topography in controlling throughflow generation. *Earth Surfaces Processes and Landforms* 3: 331–334.

Bachmair S and Weiler M (2010) New dimensions of hillslope hydrology. In: Levia D, Carlyle-Moses D, and Tanaka T (eds.) *Forest Hydrology and Biogeochemistry: Synthesis of Research and Future Directions.* New York, NY: Springer.

Beres M, Huggenberger P, Green AG, and Horstmeyer H (1999) A study of glaciofluvial architectures using two- and three-dimensional georadar methods. *Sedimentary Geology* 129: 1–24.

Betson RP (1964) What is watershed runoff? *Journal of Geophysical Research* 69(8): 1541–1551.

Beven K (2006) Streamflow generation processes. In: McDonnell JJ (ed.) *IAHS Benchmark Papers in Hydrology Series*, 432pp. Wallingford: IAHS.

Blazkova S, Beven KJ, and Kulasova A (2002) On constraining TOPMODEL hydrograph simulations using partial saturated area information. *Hydrological Processes* 16(2): 441–458.

Bonell M (1998) Selected challenges in runoff generation research in forests from the hillslope to headwater drainage basin scale. *Journal of the American Water Resources Association* 34(4): 765–786.

Butler DK (ed.) (2005) *Near-Surface Geophysics.* Tulsa, OK: Society of Exploration Geophysicists.

Calver A and Cammeraat LH (1993) Testing a physically based runoff model against field observations on a Luxembourg hillslope. *Catena* 20: 273–288.

Cappus P (1960) Etude des lois de l'eÀ coulement, application au calcul et a la prevision des de bits. *La Houille Blanche A* 493–520.

De Vries J and Chow TL (1978) Hydrologic behavior in a forested mountain soil in coastal British Columbia. *Water Resources Research* 14(5): 935–942.

Didszun J and Uhlenbrook S (2008) Scaling of dominant runoff generation processes: Nested catchments approach using multiple tracers. *Water Resources Research* 44: W02410 (doi:101029/2006WR005242).

Dunne T and Black RD (1970) An experimental investigation of runoff production in permeable soils. *Water Resources Research* 6(2): 478–490.

Engler A (1919) Untersuchungen über den Einfluss des Waldes auf den Stand der Gewässer. *Mitteilung der Schweizerischen Anstalt für fortsliches Versuchswesen* 12: 1–626.

Freer J, McDonnell JJ, Beven KJ, et al. (1997) Topographic controls on subsurface storm flow at the hillslope scale for two hydrologically distinct small catchments. *Hydrological Processes* 11(9): 1347–1352.

Freeze AR, McDonnell JJ, Beven KJ, et al. (1972) Role of subsurface flow in generating surface runoff: 2 Upstream source areas. *Water Resources Research* 8(5): 1272–1283.

Freeze AR and Witherspoon PA (1967) Theoretical analysis of regional groundwater flow: 2. Effect of water-table configuration and subsurface permeability variation. *Water Resources Research* 3: 623–634.

Gillham RW (1984) The capillary fringe and its effect on water-table response. *Journal of Hydrology* 67: 307–324.

Graham C (2009) A Macroscale Measurement and Modeling Approach to Improve Understanding of the Hydrology of Steep, Forested Hillslopes. PhD Thesis, Oregon State University, USA.

Güntner A, Seibert J, and Uhlenbrook S (2004) Modeling spatial patterns of saturated areas: An evaluation of different terrain indices. *Water Resources Research* 40: W05114 (doi:10.1029/2003wr002864).

Harr RD (1977) Water flux in soil and subsoil on a steep forested slope. *Journal of Hydrology* 33: 37–58.

Hewlett JD and Hibbert AR (1963) Moisture and energy conditions within a sloping soil mass during drainage. *Journal of Geophysical Research* 68: 1081–1087.

Hewlett JD and Hibbert AR (1967) Factors affecting the response of small watersheds to precipitation in humid areas. In: Sopper WE and Lull HW (eds.) *Forest Hydrology*, pp. 275–291. New York, NY: Pergamon.

Hooper RP and Shoemaker CA (1986) A comparison of chemical and isotopic hydrograph separation. *Water Resources Research* 22(10): 1444–1454.

Hoover MD and Hursh CR (1943) Influence of topography and soil depth on runoff from forest land. *Transactions of the American Geophysical Union* 2: 693–698.

Horton RE (1933) The role of infiltration in the hydrological cycle. In: *Transactions of the American Geophysical Union, Fourteenth Annual Meeting*, pp. 445–460. Washington, DC.

Hursh CR and Brater EF (1941) Separating storm-hydrographs from small drainage-areas into surface- and subsurface-flow. *Transactions of the American Geophysical Union* 3: 863–871.

Hutchinson DG and Moore RD (2000) Throughflow variability on a forested slope underlain by compacted glacial till. *Hydrological Processes* 14(10): 1751–1766.

Johnson TC, Routh PS, Barrash W, and Knoll MD (2007) A field comparison of Fresnel zone and ray-based GPR attenuation-difference tomography for time-lapse imaging of electrically anomalous tracer or contaminant plumes. *Geophysics* 72: G21–G29.

Jost G, Heuvelink GBM, and Papritz A (2005) Analysing the space-time distribution of soil water storage of a forest ecosystem using spatio-temporal kriging. *Geoderma* 128(3–4): 258–273.

Kienzler P (2007) Experimental Study of Subsurface Stormflow Formation. Combining Tracer, Hydrometric and Geophysical Techniques. Diss. ETH Nr. 17330, Eidgenössische Technische Hochschule (ETH), Zürich, Switzerland.

Kite GW and Pietroniro A (1996) Remote sensing applications in hydrological modelling. *Hydrological Sciences* 563–591.

Knight R, Irving J, Tercier P, Freeman G, Murray C, and Rockhold M (2007) A comparison of the use of radar images and neutron probe data to determine the horizontal correlation length of water content. In: Hyndman DW, Day-Lewis FD, and Singha K (eds.) *Subsurface Hydrology: Data Integration for Properties and Processes*, Geophysical Monograph Series, vol. 171, pp. 31–44. Washington, DC: AGU.

Lange H, Lischeid G, Hoch R, and Hauhs M (1996) Water flow paths and residence times in a small headwater catchment in Gårdsjön, Sweden, during steady state storm flow conditions. *Water Resources Research* 32: 1689–1698.

Laudon H, Seibert J, Kohler S, and Bishop K (2004) Hydrological flow paths during snowmelt: Congruence between hydrometric measurements and oxygen 18 in meltwater, soil water, and runoff. *Water Resources Research* 40: W03102 (doi:10.1029/2003WR002455).

McDonnell JJ (1990) A rationale for old water discharge through macropores in a steep, humid catchment. *Water Resources Research* 26(11): 2821–2832.

McGlynn BL, McDonnell JJ, and Brammer DD (2002) A review of the evolving perceptual model of hillslope flowpaths at the Maimai catchments, New Zealand. *Journal of Hydrology* 257: 1–26.

McGuire KJ, McDonnell M, Weiler M, et al. (2005) The role of topography on catchment-scale water residence time. *Water Resources Research* 41: W05002.

Merot Ph, Ezzehar B, Walter C, and Aurousseau P (1995) Mapping waterlogging of soils using digital terrain models. *Hydological Processes* 9: 27–34.

Mérot P, Squividant H, Aurousseau P, et al. (2003) Testing a climato-topographic index for predicting wetlands distribution along an European climate gradient. *Ecological Modelling* 163(1–2): 51–71.

Mertens J, Jacques D, Vanderborght J, and Feyen J (2002) Characterisation of the field-saturated hydraulic conductivity on a hillslope: In situ single ring pressure infiltrometer measurements. *Journal of Hydrology* 263(1–4): 217–229.

Mohanty BP and Skaggs TH (2001) Spatio-temporal evolution and time-stable characteristics of soil moisture within remote sensing footprints with varying soil, slope, and vegetation. *Advances in Water Resources* 24(9–10): 1051–1067.

Molénat J, Durand P, Gascuel-Odoux C, Davy P, and Gruau G (2002) Mechanisms of nitrate transfer from soil to stream in an agricultural watershed of French Brittany. *Water, Air, and Soil Pollution* 133(1–4): 161–183.

Mosley MP (1979) Streamflow generation in a forested watershed. *Water Resources Research* 15: 795–806.

Myrabo S (1997) Temporal and spatial scale of response area and groundwater variation in till. *Hydrological Processes* 11: 1861–1880.

Pearce AJ, Stewart MK, and Sklash MG (1986) Storm runoff generation in humid headwater catchments: 1. Where does the water come from? *Water Resources Research* 22: 1263–1272.

Pilgrim DH (1966) Radioactive tracing of storm runoff on a small catchment. *Journal of Hydrology* 4: 289–326.

Pilgrim DH and Huff DD (1978) A field evaluation of subsurface and surface runoff: I. Tracer studies. *Journal of Hydrology* 38: 299–318.

Pinder GF and Jones JF (1969) Determination of the groundwater component of peak discharge for the chemistry of total runoff. *Water Resources Research* 5(2): 438–445.

Ragan RM (1968) An experimental investigation of partial area contributions. In: *Proceedings of the Berne Symposium*, IAHS Publ. 76, pp. 241–249.

Rodhe A (1989) On the generation of stream runoff in till soils. *Nordic Hydrology* 20: 1–8.

Rodhe A, Nyberg L, and Bishop K (1996) Transit times for water in a small till catchment from a step shift in the oxygen 18 content of the water input. *Water Resources Research* 32: 3497–3511.

Rosin K, Weiler M, and Smith R (2009) Evaluating soil saturation models in forests in different climates. *Journal of Hydrology* (in review).

Roth K, Jury WA, Flühler H, and Attinger W (1991) Transport of chloride through an unsaturated field soil. *Water Resources Research* 27(10): 2533–2541.

Rubin Y and Hubbard SS (2005) *Hydrogeophysics*. Dordrecht: Springer.

Scanlon BR, Healy RW, and Cook PG (2002) Choosing appropriate techniques for quantifying groundwater recharge. *Hydrogeology Journal* 10(1): 18–39.

Scherrer S (1997) Abflussbildung bei Starkniederschlägen, Identifikationvon Abflussprozessen mittels künstlicher Niederschläge. Versuchsanstalt fürWasserbau, Hydrologie und Glaziologie der ETH Zürich, Zürich, 147pp.

Scherrer S, Naef F, Faeh AO, and Cordery I (2006) Formation of runoff at the hill-slope scale during intense precipitation. *Hydrology and Earth System Sciences* 11(2): 907–922.

Schön JH (1998) *Physical Properties of Rocks: Fundamentals and Principles of Petrophysics*. Oxford: Pergamon.

Seibert J, Bishop K, Rodhe A, and McDonnell JJ (2003) Groundwater dynamics along a hillslope: A test of the steady state hypothesis. *Water Resources Research* 39(1). 2-1–2-9 (doi:1029/2002WR001404 2003).

Sklash MG, Beven KJ, Gilman K, and Darling WG (1996) Isotope studies of pipeflow at Plynlimon, Wales, UK. *Hydrological Processes* 10(7): 921–944.

Sklash MG and Farvolden RN (1979) The role of groundwater in storm runoff. *Journal of Hydrology* 43: 45–65.

Sklash MG, Stewart MK, and Pearce AJ (1986) Storm runoff generation in humid headwater catchments: 2. A case study of hillslope and low-order stream response. *Water Resources Research* 22(8): 1273–1282.

Smart PL and Laidlaw IMS (1977) An evaluation of some fluorescent dyes for water tracing. *Water Resources Research* 13: 15–33.

Soulsby C, Rodgers PJ, Petry J, Hannah DM, Malcolm IA, and Dunn SM (2004) Using tracers to upscale flow path understanding in mesoscale mountainous catchments: Two examples from Scotland. *Journal of Hydrology* 291: 174–196.

Talamba D, Joerin C, and Musy A (2000) Study of subsurface flow using environmental and artificial tracers: The Haute-Mentue case, Switzerland. In: *Tracers and Modelling in Hydrogeology*, IAHS Publication No. 262, pp. 559–264.

Tromp-van Meerveld HJ and McDonnell JJ (2006) Threshold relations in subsurface stormflow: 2. The fill and spill hypothesis. *Water Resources Research* 42: W02411 (doi:10.1029/2004WR003800).

Tromp-van Meerveld I and Weiler M (2008) Hillslope dynamics modeled with increasing complexity. *Journal of Hydrology* 361(1–2): 24–40.

Trubilowicz J, Cai K, and Weiler M (2009) Viability of motes for hydrological measurement. *Water Resources Research* 45: W00D22 (doi:10.1029/2008WR007046).

Trudgill ST (1987) Soil water dye tracing, with special reference to the use of rhodamine WT, Lissamine FF and amino G acid. *Hydrological Processes* 1: 149–170.

Vereecken H, Binley A, Cassiani G, Revil A, and Titov K (2006) *Applied Hydrogeophysics*. Dordrecht: Springer.

Weiler M and Flühler H (2004) Inferring flow types from dye patterns in macroporous soils. *Geoderma* 120(1–2): 137–153.

Weiler M and McDonnell J (2007) Conceptualizing lateral preferential flow and flow networks and simulating the effects on gauged and ungauged hillslopes. *Water Resources Research* 43: W03403 (doi:10.1029/2006WR004867).

Weiler M, McDonnell J, Tromp-van Meerveld HJ, and Uchida T (2005) Subsurface stormflow. In: Anderson MG and Jeffrey JJ (eds.) *Encyclopedia of Hydrological Sciences*, vol. 3, ch. 112, pp. 1719–1732. Chichester: Wiley.

Weiler M and Naef F (2003) An experimental tracer study of the role of macropores in infiltration in grassland soils. *Hydrological Processes* 17(2): 477–493.

Weiler M, Naef F, Leibundgut C (1998) Study of runoff generation on hillslopes using tracer experiments and physically based numerical model. IAHS Publication No. 248, pp. 353–360

Weyman DR (1970) Throughfall on hillslopes and its relation to the stream-hydrograph. *Bulletin of the International Association of the Scientific Hydrology* 15: 23–25.

Weyman DR (1973) Measurements of the downslope flow of water in a soil. *Journal of Hydrology* 20: 267–288.

Whipkey RZ (1965) Subsurface storm flow from forested slopes. *Bulletin of the International Association of the Scientific Hydrology* 2: 74–85.

Woods R and Rowe L (1996) The changing spatial variability of subsurface flow across a hillside. *Journal of Hydrology (NZ)* 35(1): 51–86.

Woods RA, Sivapalan M, and Robinson JS (1997) Modelling the spatial variability of subsurface runoff using a topographic index. *Water Resources Research* 31: 2097–2110.

Zehe E and Sivapalan M (2009) Threshold behavior in hydrological systems as (human) geo-ecosystems: Manifestations, controls, implications. *Hydrology and Earth System Sciences* 13: 1273–1297.

2.14 Observation of Hydrological Processes Using Remote Sensing

Z Su, University of Twente, Enschede, The Netherlands
RA Roebeling, Royal Netherlands Meteorological Institute, De Bilt, The Netherlands
J Schulz, Deutscher Wetterdienst, Offenbach, Germany
I Holleman, Royal Netherlands Meteorological Institute, De Bilt, The Netherlands
V Levizzani, ISAC-CNR, Bologna, Italy
WJ Timmermans, University of Twente, Enschede, The Netherlands
H Rott, University of Innsbruck, Innsbruck, Austria
N Mognard-Campbell, OMP/LEGOS, Toulouse, France
R de Jeu, VU University Amsterdam, Amsterdam, The Netherlands
W Wagner, Vienna University of Technology, Vienna, Austria
M Rodell, NASA/GSFC, Greenbelt, MD, USA
MS Salama, GN Parodi, and L Wang, University of Twente, Enschede, The Netherlands

© 2011 Elsevier B.V. All rights reserved.

2.14.1	General introduction	352
2.14.1.1	Water Cycle and Water Resources Management	352
2.14.1.2	Water and Energy Balance of the Earth	353
2.14.1.3	From Radiometric Observations to Object Properties	356
2.14.2	Water in the Atmosphere: Clouds and Water Vapor	357
2.14.2.1	Introduction	357
2.14.2.2	Satellite RS	357
2.14.2.2.1	Observing water vapor	358
2.14.2.2.2	Observing clouds	360
2.14.2.3	Retrieval Algorithms	360
2.14.2.3.1	Water vapor	360
2.14.2.3.2	Total column water vapor	360
2.14.2.3.3	Water vapor profiles	361
2.14.2.3.4	Upper tropospheric humidity	361
2.14.2.4	Cloud Detection	361
2.14.2.4.1	Cloud property retrievals	361
2.14.2.5	Validation	362
2.14.2.5.1	Water vapor	362
2.14.2.5.2	Cloud properties	362
2.14.2.6	Data Sets	363
2.14.2.6.1	Water vapor products	363
2.14.2.6.2	Cloud products	363
2.14.3	Water from the Atmosphere: Precipitation	363
2.14.3.1	Introduction	363
2.14.3.2	Precipitation Measurements from Weather Radars	363
2.14.3.3	Precipitation Measurements from Satellite	364
2.14.3.3.1	Retrievals from VIS–IR sensors	364
2.14.3.3.2	Retrievals from passive MW sensors	364
2.14.3.4	Validation	365
2.14.3.4.1	Weather radar retrievals	365
2.14.3.4.2	Satellite retrievals	366
2.14.3.5	Applications	366
2.14.4	Water to the Atmosphere – Evaporation	367
2.14.4.1	Introduction and Historic Development	367
2.14.4.2	Current State of Science	367
2.14.4.2.1	Statistical approaches	367
2.14.4.2.2	Variability approaches	368
2.14.4.2.3	Physical approaches	368
2.14.4.3	Future Research Needs	369
2.14.4.3.1	Scaling	369
2.14.4.3.2	Feedbacks	369
2.14.4.3.3	Validation	370
2.14.5	Water on the Land – Snow and Ice	370

2.14.5.1	Introduction	370
2.14.5.2	Techniques for Retrieval of Extent and Physical Properties of Snow and Ice	370
2.14.5.3	Examples of Products and Applications	372
2.14.5.4	Future Research Needs	373
2.14.6	**Water on the Land – Surface Water, River Flows, and Wetlands (Altimetry)**	**373**
2.14.6.1	Introduction	373
2.14.6.2	*In Situ* Measurements	374
2.14.6.3	RS Techniques	374
2.14.6.4	Validation and Synergy of RS Techniques	375
2.14.6.5	Availability of the Satellite Data Sets	376
2.14.6.6	SWOT: The Future Satellite Mission Dedicated to Surface Hydrology	376
2.14.7	**Water in the Ground – Soil Moisture**	**377**
2.14.7.1	Introduction	377
2.14.7.2	State of the Art	377
2.14.7.3	Data Sets BBB	378
2.14.7.3.1	Active MW data sets	378
2.14.7.3.2	Passive MW data sets	379
2.14.7.4	Validation	380
2.14.8	**Water in the Ground – Groundwater (Gravity Observations)**	**380**
2.14.8.1	Introduction	380
2.14.8.2	GRACE Data Processing	380
2.14.8.3	Retrievals of Groundwater Storage with GRACE Data	381
2.14.8.4	GRACE Data Access	381
2.14.8.5	Concluding Remarks and Future Perspective	381
2.14.9	**Optical RS of Water Quality in Inland and Coastal Waters**	**381**
2.14.9.1	Introduction	381
2.14.9.2	Atmospheric Correction	382
2.14.9.3	Retrieval Algorithms	383
2.14.9.4	Uncertainty Estimates	383
2.14.9.5	Concluding Remarks and Future Perspective	384
2.14.10	**Water Use in Agro- and Ecosystems**	**384**
2.14.10.1	Introduction	384
2.14.10.2	Continuous Evaluation of Crop Water Use with Support from RS	385
2.14.10.3	Drought Indices and Soil Moisture Monitoring	386
2.14.10.4	Algorithm Retrievals and Operability	387
2.14.10.5	SEBS Algorithm	387
2.14.10.6	Evaluation Example	388
Acknowledgment		**389**
References		**389**

2.14.1 General introduction

2.14.1.1 Water Cycle and Water Resources Management

The United Nations (UN) Millennium Declaration called on all members "to stop the unsustainable exploitation of water resources by developing water management strategies at the regional, national and local levels which promote both equitable access and adequate supplies." Improving water management can make a significant contribution to achieving most of the Millennium Development Goals established by the UN General Assembly in 2000, especially those related to poverty, hunger, and major diseases. The World Summit on Sustainable Development (WSSD) in 2002 recognized this need. Water and sanitation in particular received great attention from the Summit. The Johannesburg Plan of Implementation recommended to improve water resources management and scientific understanding of the water cycle through joint cooperation and research. For this purpose, it is recommended to promote knowledge sharing, provide capacity building, and facilitate the transfer of technology including remote-sensing (RS) and satellite technologies, especially to developing countries and countries with economies in transition, and to support these countries in their efforts to monitor and assess the quantity and quality of water resources, for example, by establishing and/or further developing national monitoring networks and water resources databases and by developing relevant national indicators. The Johannesburg Plan also adopted integrated water resources management as the overarching concept in addressing and solving water-related issues. As a result of the commitments made in the Johannesburg Plan of Implementation, several global and regional initiatives have emerged.

Current international initiatives such as the Global Monitoring for Environment and Security (GMES) program of the European Commission and the European Space Agency (ESA), and the Global Earth Observation System of Systems (GEOSS)

10-Year Implementation Plan (GEO, 2005), have all identified Earth observation (EO) of the water cycle as the key in helping to solve the world's water problems. More specifically, the 10-year implementation plan states that "Enhanced prediction of the global water cycle variation based on improved understanding of hydrological processes and its close linkage with the energy cycle and its sustained monitoring capability is a key contribution to mitigation of water-related damages and sustainable human development. Improved monitoring and forecast information, whether of national or global origin, if used intelligently, can provide large benefits in terms of reduced human suffering, improved economic productivity, and the protection of life and property. In many cases, the combination of space-based data and high-resolution in-situ data provides a powerful combination for effectively addressing water management issues. Information on water quantity and quality and their variation is urgently needed for national policies and management strategies, as well as for UN conventions on climate and sustainable development, and the achievement of the Millennium Goals" (GEO, 2005).

The availability of spatial information on water quantity and quality will also enable closure of the water budget at river basin and continental scales to the point where effective water management is essential (e.g., as requested by the European Union's Water Framework Directive (WFD), as well as national policies). Geo-information science and EO are vital in achieving a better understanding of the water cycle and better monitoring, analysis, prediction, and management of the world's water resources.

Subject to climate change, the security of freshwater resources has emerged as one of the key societal problems. According to a report prepared under the auspices of the Intergovernmental Panel on Climate Change (IPCC, 2008), "Observational records and climate projections provide abundant evidence that freshwater resources are vulnerable and have the potential to be strongly impacted by climate change, with wide-ranging consequences on human societies and ecosystems." Floods, droughts, water scarcity, water usage, water quality, water and ecosystem interactions, and water and climate interactions are all issues of direct importance to our human society. The only key to safeguard the security of water resources is better water resources management. This in turn requires better understanding of the water cycle, water climate interactions, and water ecosystem interactions in the Earth's climate system. To achieve such an understanding, it is essential to be able to measure hydroclimatic variables at different spatial and temporal scales, such as radiation, precipitation, evaporation and transpiration (or evapotranspiration (ET)), soil moisture, clouds, water vapor, surface water and runoff, vegetation state, albedo and surface temperature, etc.

The major components of the water cycle of the Earth system and their possible observations are presented in Figure 1. Such observations are essential to understand the global water cycle and its variability, both spatially and temporally, and can only be achieved consistently by means of EOs. Additionally, such observations are essential to advance our understanding of coupling between the terrestrial, atmospheric, and oceanic branches of the water cycle, and how this coupling may influence climate variability and predictability. Figure 1 also shows the proportion of the water-cycle flux components in the ocean (including evaporation of the ocean water into the atmosphere and condensation of the water vapor falling as precipitation into the ocean again), the proportion of the terrestrial water-cycle components (including precipitation as a consequence of condensation of water vapor generated by evaporation and transpiration from land and water vapor transported from the ocean, the river discharge, and groundwater discharge returning water into the ocean), water and ocean ice sheets in the ocean, permafrost and snow, soil moisture and groundwater on land, and atmospheric water vapor. Water resources management directly interferes with the natural water cycle in the forms of building dams, reservoirs, water transfer systems, and irrigation systems that divert and redistribute part of the water storages and fluxes on land. The water cycle is mainly driven and coupled to the energy cycle in terms of phase changes of water (changes among liquid, water vapor, and solid phases) and transport of water by winds in addition to gravity and diffusion processes. The water-cycle components can be observed with *in situ* sensors as well as airborne and satellite sensors in terms of radiative quantities. Processing and conversion of these radiative signals are necessary to retrieve the water-cycle components.

To enhance prediction of the global water-cycle variation, based on improved understanding of hydrological processes and its close linkage with the energy cycle and its sustained monitoring capability, is a key contribution to mitigation of water-related damages and sustainable human development. In many cases, the combination of space-based data and high-resolution *in situ* data in a modeling system using data assimilation provides a powerful tool for effectively addressing water management issues.

2.14.1.2 Water and Energy Balance of the Earth

The Sun is the primary source of energy of Earth's climate system and its five major components: the atmosphere, the biosphere, the cryosphere, the hydrosphere, and the land surface (ESA, 2006). In Earth's energy balance, the shortwave (solar) radiation is redistributed by different radiative climate forcing components. In the long term, the amount of incoming solar radiation absorbed by land, ocean, and atmosphere is balanced by releasing the same amount of outgoing longwave (terrestrial) radiation from Earth to space. About half of the incoming solar radiation is absorbed by the Earth's surface. This energy is transferred to the atmosphere by warming the air in contact with the surface (thermals), by ET and by longwave radiation that is absorbed by clouds and greenhouse gases. The atmosphere in turn radiates longwave radiation back to the Earth's surface as well as out to space. Changes in greenhouse gas concentrations cause altering the longwave radiation from the Earth out to space. The climate system will respond directly to such changes, as well as indirectly, through a variety of feedback mechanisms. For example, an increased concentration of water vapor enhances the amount of thermal radiation absorbed by the atmosphere and consequently leads to an increase of the surface temperature, but will also likely lead to an increase of cloud amount and precipitation. Simplified schematic representations of the annual mean energy flux budgets for the Earth,

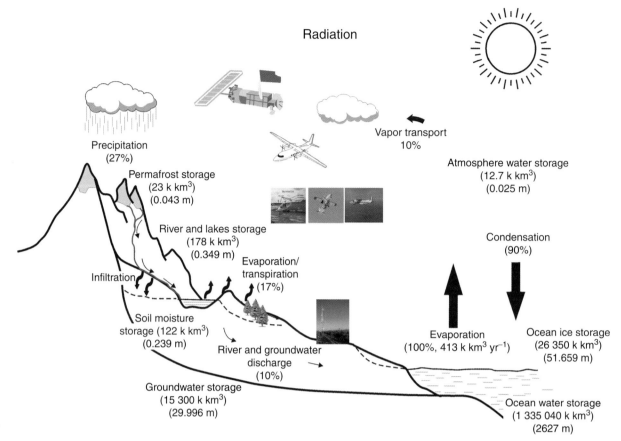

Figure 1 Global water cycle of the Earth system and their possible observations with *in situ*, airborne instruments (low altitude and high altitude), and satellites. The flux components (condensation, water vapor transport, precipitation, evaporation and transpiration, and river and groundwater discharge) are normalized with the total ocean evaporation of 413 000 km³ yr⁻¹ (100%). The storage components are also converted to water depth using the total surface area of earth 510 072 000 km². Data from Trenberth *et al.* (2007).

land, and ocean are presented in **Figure 2**, using data reported by Trenberth *et al.* (2009). For the Earth energy budget, the incoming solar radiation at the top of atmosphere (TOA) is 341.3 W m⁻², equivalent to one-quarter of the solar constant 1365.2 W m⁻², of which 101.9 W m⁻² is reflected (79 W m⁻² by clouds and 23 W m⁻² by the Earth's surface) to space resulting in a planetary albedo (or TOA albedo) of 29.8%. The surface albedo 14.3% (at the bottom of atmosphere, BOA) is the ratio of the reflected solar radiation (23 W m⁻²) to the absorbed solar radiation (161 W m⁻²). In addition, 78 W m⁻² of the incoming solar radiation is absorbed by the atmosphere. The atmosphere emits 333 W m⁻² downward to the surface, while the surface emits 396 W m⁻² upward to the atmosphere, resulting in a net upward surface longwave of 93 W m⁻². Part of the emitted surface longwave radiation passing through the atmospheric window (40 W m⁻²), together with the upward longwave radiation from the atmosphere (169 W m⁻²) and that from clouds (30 W m⁻²), makes up the outgoing longwave radiation to space (238.5 W m⁻²). The sum of the net radiation at the surface 98 W m⁻² (net downward solar radiation 161 W m⁻² less net surface upward longwave radiation 63 W m⁻²) is balanced by the thermals (i.e., sensible heat flux 17 W m⁻²) and latent heat flux for evaporation/transpiration (80 W m⁻²), with 0.9 W m⁻² absorbed by the surface. At the TOA, the radiation balance is also 0.9 W m⁻² (incoming solar radiation 341.3 W m⁻² less reflected solar radiation 101.9 W m⁻² and outgoing longwave radiation 238.5 W m⁻²), indicating a net gain of 0.9 W m⁻² in energy, which may be conceived as a possible warming of the Earth system. However, this quantity is derived only for the Clouds and the Earth's Radiant Energy System (CERES) (Wielicki *et al.*, 1996) period from March 2000 to May 2004 and cannot be taken as long-term evidence.

Similar explanations can be made for the energy budgets for land and ocean separately. The differences in land and ocean energy budget components are caused mainly by different albedo over land and water as well as the different thermodynamic properties of land and water.

EO of water cycle primarily uses information in the optical, thermal, and microwave (MW) regions of the electromagnetic spectrum to retrieve water-cycle components, though other measurement using, for example, gravity measurement has also shown great promise for monitoring mass changes. One example of EO of water cycle is the Water Cycle Multi-mission Observation Strategy (WACMOS) project initiated by the European Space Agency (ESA) and the Global Energy and Water Cycle Experiment (GEWEX) of the World Climate Research Programme (WCRP). The WACMOS project aims to develop and validate novel and improved multimission-based global water-cycle data sets using multimission satellite data.

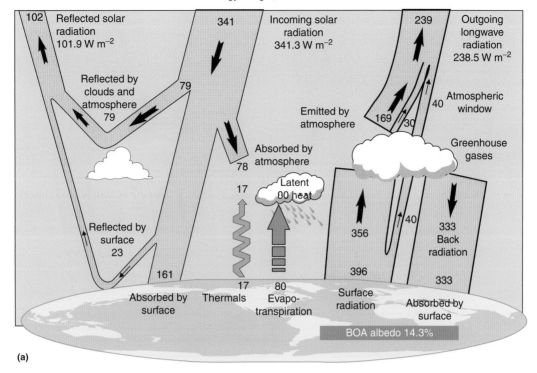

(a)

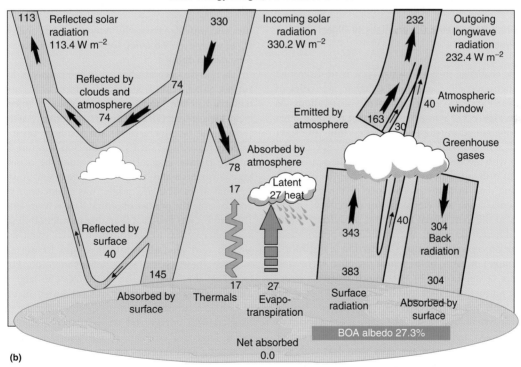

(b)

Figure 2 Schematic representation of the mean annual energy budgets for the earth, land, and ocean. The surface emissivity is assumed to be 1.0. Scheme and primary data from Trenberth KE, Fasullo JT, and Kiehl J (2009) Earth's global energy balance. *Bulletin of the American Meteorological Society* 311–323: doi:10.1175/2008BAMS2634.1.

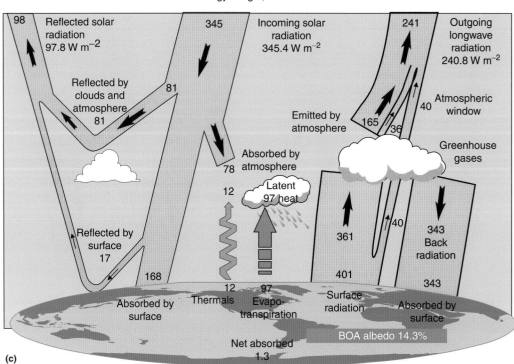

Figure 2 Continued.

2.14.1.3 From Radiometric Observations to Object Properties

The scientific challenge in EO of water cycle is to determine turbulent, thermodynamic, and fluid dynamic properties of the whole water cycle by using radiometric observations. As illustrated in **Figure 3**, a sensor with a certain response function measuring radiation reflected or emitted from an object can have different geometric arrangements with respect to the object, each with always atmosphere between the sensor and the object. In order to retrieve the properties of the object in concern using data in terms of its range to the sensor, its combined temperature and emissivity (or reflectivity) at different times, at different spatial resolution, at different wavelengths, at different direction, and at different polarization, detailed data processing is needed (see Section 2.14.10 for a detailed example). In terms of the sensor response, we need to ask two types of questions: (A) How much radiation is detected at the sensor? (B) When and how does it arrive? If the answers are only relevant to question (A), then we have a passive sensor system, otherwise if the answers are relevant to both questions (A) and (B), then we have an active sensor system. Many excellent textbooks exist that deal with the theoretical aspects of the sensor–object relationships and the practical issues related to retrievals of object properties (e.g., Rees, 2001; and Liang, 2004). Applications of RS in hydrology and climate studies can be found in related chapters in Anderson and McDonnell (2005) and in Bolle (2003); the current chapter is a continuation of these earlier efforts on RS in hydrology and water resources management. Many Internet sites also provide very useful resources, data, and examples of EO of water-cycle variables, some of the most relevant ones are provided in **Table 1**.

There have been excellent field campaigns in hydrology – HAPEX Sahel, HAPEX Mobily (Goutorbe et al., 1997), International Satellite Land Surface Climatology Project (ISLSCP) Field Experiment (FIFE; Sellers et al., 1992), MONSOON'90 (Kustas and Goodrich, 1994), the Southern Great Plains Hydrology Experiment (SGP), and the Soil Moisture Experiment (SMEX; Jackson et al., 1999) – to study the complex hydrological processes and land–atmosphere interactions at local to regional scales. The FIFE project was a large-scale climatology project set in the prairies of central Kansas from 1987 to 1989. This project was designed to improve understanding of carbon and water cycles; to coordinate data collected by satellites, aircraft, and ground instruments; and to use satellites to measure these cycles. More information on FIFE can be found on the Internet. The MONSOON'90 large-scale interdisciplinary field experiment was conducted in the summer of 1990 in southeastern Arizona to investigate the utility of RS coupled with energy and water-balance modeling for providing large-area estimates of fluxes in semiarid rangelands. Large-scale field experiments related to soil moisture are the series of the SGP and the SMEX series, focusing on validation of retrieval algorithms and demonstration of technological feasibilities of RS of soil moisture.

Some examples of both sensor systems and the retrievals of the relevant geo-biophysical parameters can be found in some recent large-scale field experiments, the SPAR 2004 and SEN2FLEX 2005 campaigns (Sobrino et al., 2008, 2009;

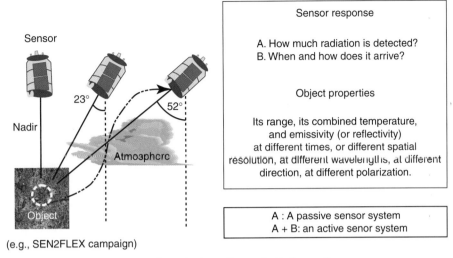

Figure 3 Schematic representation of links between radiometric observations and object properties.

Su et al., 2008) as well as the EAGLE 2006 campaign (Su et al., 2009). The data collected from these field experiments are open to the scientific community for collaborative investigations and are accessible at the European Space Agency's Principle Investigators portal or by contacting the authors directly. As an example, the spectra of bright sand and of a young pine tree collected during the EAGLE 2006 campaign are shown in **Figure 4**, indicating the sensor responses to different properties of the object.

Some most relevant web links to data portals, software tools, and training courses related to water and energy balance of the Earth are given **Table 1**.

In the following sections, we discuss details of the different components of the water cycles from the perspectives of EO.

2.14.2 Water in the Atmosphere: Clouds and Water Vapor

2.14.2.1 Introduction

Accurate information on the distribution of water vapor and clouds in the atmosphere is essential for water and energy balance studies. The atmosphere acts as a medium for the transport of water around the globe. Water vapor is brought into the atmosphere through evaporation from liquid water bodies (~90%) and transpiration from plants (~10%). Clouds are formed in lifting air parcels, in which water vapor condensates into cloud particles due to the cooler temperatures. Once in the atmosphere, clouds are moved around the globe by strong winds and either evaporate back into water vapor or disappear as precipitation to replenish the earth-bound parts of the water cycle.

The presence of water vapor and clouds in the atmosphere warms the Earth's troposphere and surface, and acts as a partial blanket for the longwave radiation coming from the surface. Water vapor and clouds absorb and emit infrared (IR) radiation and thus contribute to warming the Earth's surface. For clouds, this effect is counterbalanced by the reflection of visible (VIS) radiation, which reduces the amount of short-wave (solar) incoming radiation at the Earth's surface and has a cooling effect on the climate system. The net average effect of the Earth's cloud cover in the present climate is a slight cooling because the reflection of radiation more than compensates for the blanketing effect of clouds. Information on the distribution of water vapor and clouds in the atmosphere is also relevant for studying the hydrological cycle. The shortwave and longwave radiation that reach the Earth's surface directly affect the evaporation (latent) and sensible heat fluxes. The part of the radiation that is used to evaporate soil moisture (evaporation) or crop moisture (transpiration) is released to the atmosphere as water vapor. The evaporated water vapor, in turn, is carried upward where it condenses into cloud droplets, ice crystals, or precipitation.

Ground-based measurements are inadequate to observe the large spatial and temporal variations in water vapor and cloud properties (Rossow and Schiffer, 1999). The advent of satellite RS has changed this situation. Satellites can provide the required information at adequate temporal and spatial scales. Satellite observations can be used to retrieve integrated water vapor amounts and cloud physical properties from passive MW radiometers or spectral radiances, respectively. The accuracies and precisions of these satellite retrievals have been well determined within various validation studies.

2.14.2.2 Satellite RS

Since the 1960s, many satellites have been providing continuous observations of the state of the atmosphere over very large regions or even for the entire globe. The satellite instruments of most interest for observing water vapor and clouds are MW radiometers that measure emitted MW radiation of the Earth's surface, atmosphere, and clouds, and the

Table 1 Some resources for Earth observation program, satellites and data, software, and training courses related to earth observation of water cycle

Organization/program	Web links	Comments
Group on Earth Observations (GEO)	http://www.earthobservations.org/	GEO coordinates international efforts to build a Global Earth Observation System of Systems (GEOSS). In its water societal benefit area it aims at improving water resource management through better understanding of the water cycle.
GEO portal	http://www.geoportal.org/	The GEO portal provides an entry point to access remote sensing, geospatial static, and *in situ* data, information and services. Water is one of the nine societal benefit areas.
GEO applications	http://www.earthobservations.org/documents/the_full_picture.pdf	'The Full Picture' provides an overview of the progresses in applications of GEOSS in the nine societal benefit areas till 2007.
European space agency (esa)	http://www.esa.int/esaeo/ http://www.esa.int/esalp/	ESA's Earth observation programs include the Global Monitoring for Environment and Security (GMES) and the Living Planet Programme. Observation of the hydrosphere is one of the foci of this program.
European Space Agency (ESA) Dragon programme	http://earth.esa.int/dragon/	The ESA-MOST (Ministry of Science and Technology, China) Dragon program includes a dedicated training program to provide training in data processing, algorithm and product development from ESA Earth Observation (EO) data in land, ocean, and atmospheric applications.
National Aeronautics and Space Administration (NASA)	http://neptune.gsfc.nasa.gov/hsb/	NASA's Hydrological Sciences focuses on the interpretation of remotely sensed data and land surface hydrological, meteorological, and climate modeling.
Global land data assimilation system (gldas)	http://ldas.gsfc.nasa.gov/	GLDAS generates optimal fields of land surface states and fluxes (Rodell *et al.*, 2004) by assimilating satellite- and ground-based observational data products into advanced land surface models. The high-quality, global land surface fields provided by GLDAS support several current and proposed weather and climate prediction, water resources applications, and water-cycle investigations. GLDAS has resulted in a massive archive of modeled and observed, global, surface meteorological data, parameter maps, and output which include 1° and 0.25° resolution 1979–present simulations of the Noah, CLM, and Mosaic land surface models.
Land surface analysis satellite applications facility (lsa saf)	http://landsaf.meteo.pt/	The main purpose of the Land SAF is to increase the benefits from Meteosat Second Generation (MSG) and European Polar Satellite (EPS) data related to land, land–atmosphere interactions and biophysical applications by developing techniques, products and algorithms that will allow a more effective use of data from the two satellites of the European Organisation for the Exploitation of Meteorological Satellites (EUMETSAT).
Committee on Earth Observation Satellites (CEOS)	http://www.eohandbook.com	The Earth Observation Handbook presents the main capabilities of satellite Earth observations, their applications, and a systematic overview of present and planned CEOS agency Earth observation satellite missions and their instruments. It also explores society's increasing need for information on our planet.

passive imagers that measure VIS, near-IR, and IR radiances. The importance of satellite observations is determined by the spatial and temporal sampling resolution of their instruments. Frequent sampling is especially required for parameters that are highly variable in space and time, such as clouds.

2.14.2.2.1 Observing water vapor

The strong variations of water vapor in space and time lead to the necessity of monitoring this quantity globally from satellites. Absorption lines of water vapor are present in almost every part of the electromagnetic spectrum. A great variety of space-borne sensors are used to retrieve atmospheric profiles of humidity or the column amount, even if they were not designed for it. These sensors observe the interaction of radiation with water vapor in the different parts of the spectrum (MW, IR, optical, and ultraviolet (UV)). The number of available instruments is further increased due to the need for measurements at different observation geometries (nadir view, limb scanning, occultation, and day or night) and at different orbit orientations. Here, only some of the observation systems for tropospheric water vapor are described.

MW radiometers observe the radiation close to the 22-GHz water vapor absorption line that is closely related to the total column content of water vapor. These observations can be used over oceans in clear sky and cloudy conditions. The conically scanning special sensor microwave/imager (SSM/I) on the DMSP satellites is available since 1987 and is continued with the special sensor microwave imager sounder (SSMIS) instrument into the future. Among others, this type of radiometers is flown on the US TRMM satellite (TRMM Microwave Imager (TMI)) and on the US Aqua mission

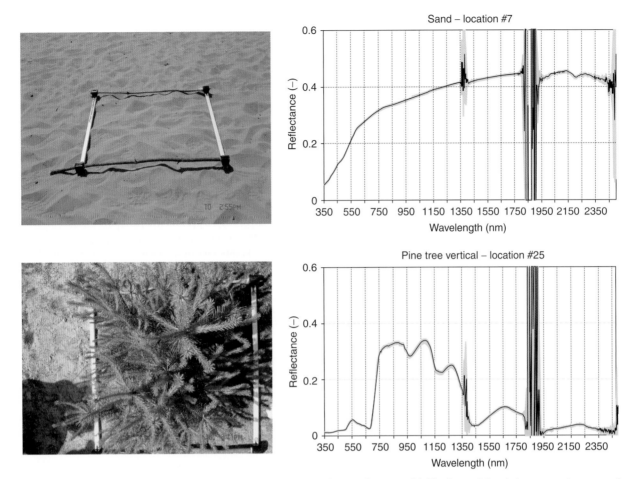

Figure 4 Example spectra of bright sand (upper panels) and of a young pine tree (lower panels). The frame of the photos represents an area of $1 \times 1\,m^2$. On the graphs, the gray lines show the measured spectra, the thick black line is the spectrum accepted as the characteristic spectrum of the site under consideration (Su et al., 2009).

(Advanced Microwave Scanning Radiometer for the Earth Observing System (AMSR-E)). In addition, the advanced microwave sounding unit (AMSU) makes its observations at this frequency but is a cross-track scanner.

Imaging spectrometers, such as ESA's medium-resolution imaging spectrometer (MERIS), are used to retrieve the total column content of water vapor at a very high spatial resolution (~300 m) from near-IR observations during daytime (Rast et al., 1999). MERIS is especially useful over land surfaces. Such observations are also available from the moderate-resolution imaging spectroradiometer (MODIS) flown on the NASA Aqua and Terra satellites (NASA, National Aeronautics and Space Administration).

Since 1977, the Meteosat Visible and Infrared Radiation Imager (MVIRI) and Spinning-Enhanced Visible and Infrared Imager (SEVIRI) instruments in geostationary orbit observe radiation at 6.3 and 7.2 μm (only SEVIRI), and allow the retrieval of upper tropospheric humidity (UTH) with a very high temporal resolution (up to 15 min) that allows for studies of atmospheric dynamics (Schmetz et al., 2002). Also in geostationary orbit, humidity sounders similar to the HIRS instrument are flown on the US Geostationary Operational Environmental Satellites (GOES).

UV/VIS spectrometers, such as the Global Ozone Monitoring Experiment (GOME) and Scanning Imaging Absorption Spectrometer for Atmospheric Cartography (Sciamachy), are used for the retrieval of total column water vapor over land and ocean surfaces with approximately the same accuracy as the SSM/I but only under daylight and clear sky conditions at much coarser spatial resolution (Burrows et al., 1999).

Since 1978, the observations of the Advanced Television and Infrared Observation Satellite (TIROS) Operational Vertical Sounder (ATOVS) on NOAA and MetOp satellites, with its IR spectrometer (High resolution Infrared Radiation Sounder (HIRS)), and MW radiometers (AMSU-A/B and Microwave Humidity Sounder (MHS)), have been combined to derive atmospheric temperature and humidity profiles. Since 2007, EUMETSATs MetOp satellite has been carrying the Infrared Atmospheric Sounding Interferometer (IASI) instrument. This new generation of IR sounding instruments is capable of observing about 15 independent pieces of information on the vertical profile by performing observations over a large part of the IR spectrum (4–50 μm) (Simeoni et al., 1997). A similar instrument, called Atmospheric Infrared Sounder (AIRS), is flown since 2002 on NASAs Aqua mission (Aumann et al., 2003).

Finally, temperature and humidity profiles can also be retrieved from Radio Occultation measurements, performed by, for example, the GRAS (Loiselet et al., 2000) instrument onboard the *MetOp* or the COSMIC fleet (Anthes et al., 2008).

2.14.2.2.2 Observing clouds

During the last 25 years, observations from passive imaging satellites have been successfully used for the retrieval of cloud cover and cloud physical properties. Rossow and Garder (1993) used observations from the Advanced Very High Resolution Radiometer (AVHRR) instrument onboard the NOAA series of polar-orbiting satellites to derive global cloud climatology since 1982. Recently, several more sophisticated instruments for EOs have been launched. These include the instruments that are flown onboard the NASA Earth Observing System (EOS) geosynchronous orbiting satellites, which were launched in 1999 (*Terra*) and in 2002 (*Aqua*). The MODIS instruments on both satellites operate the required spectral channels for the retrieval of cloud properties at high spatial resolutions (better than $1 \times 1 \, \text{km}^2$) globally, but at low temporal resolutions (revisit time 1 day or more). The unprecedented sampling frequency of geostationary satellites (better than 30 min) allows for monitoring the diurnal variations in cloud properties. The SEVIRI instrument on board *METEOSAT-8*, which was launched in 2002, is the first instrument that can be used for the retrieval of these properties from a geostationary orbit (**Figure 5**). SEVIRI constitutes a valuable source of data for water and energy balance studies.

Another types of instruments for cloud observations are passive MW radiometers. These instruments measure emitted MW radiances from the Earth surface and the overlaying atmosphere. Greenwald et al. (1993) showed that the radiances observed by these instruments can be used for a simultaneous retrieval of atmospheric water vapor and cloud liquid water.

Figure 5 METEOSAT-8/SEVIRI image of the visible channel (0.6 μm) for the SEVIRI field of view for 17 January 2006 at 11:45 UTC.

Recently, even more advanced satellite systems are available for observing clouds. The most advanced cloud measurements are provided by the radar on the *Cloudsat* satellites and lidar on the *Calipso* satellites, which were launched in 2006 and fly in the A-train constellation. These instruments measure vertical profiles of cloud reflectivity of large particles (radar) and small particles (lidar).

2.14.2.3 Retrieval Algorithms

2.14.2.3.1 Water vapor

Retrieval methods have to correspond to the instrument spectral range and observation geometry. Processes in the atmosphere complicate the retrieval task, for example, the coexistence of the three thermodynamic phases of water on the Earth, interaction with aerosols, and varying surface emissivity. The number of retrieval algorithms is much larger than the number of sensors.

Retrieval methods generally depend on *a priori* information, which could be the coefficients of a regression, the constraints for retrieval based on inversion, or the training set of a neural network. The quality of a retrieval scheme depends on the applicability of the *a priori* information to the prevailing environmental conditions, that is, surface properties, clouds, etc.

2.14.2.3.2 Total column water vapor

Major instruments utilized for the retrieval of total column water vapor are MW radiometers (SSM/I), UV/VIS spectrometers (GOME), and VIS and near-IR imaging spectrometers (MERIS).

Retrieval schemes for MW radiometer can be distinguished in semiphysical and physical schemes. In both cases, observations of the instrument are simulated using a radiative transfer model. Input to the model is the atmospheric state vector and instrument parameters. The semiphysical schemes then retrieve the water vapor content by applying a statistical scheme (linear regression or neural networks) based on the training data (Schlüssel and Emery, 1990).

The physical schemes mostly use a first guess, often coming from a numerical weather forecast model (NWP), as the basis for the forward computation and then vary the first guess until the used set of observed radiances is best matched (e.g., Wentz, 1997). The latter requires much more computer power but has generally replaced statistical methods in the past 10 years.

The basic principle in retrieval applied to GOME is the Differential Optical Absorption Spectroscopy (DOAS) method to calculate the difference between the Sun normalized measured Earthshine radiance and absorption cross sections at wavelengths where water vapor absorbs radiation and relate this absorption depth to the water vapor column concentration (e.g., Noel et al., 1999). The DOAS method provides a global (land and ocean) completely independent data set, because it does not rely on any additional external information.

Near-IR MERIS algorithms are based on radiative transfer simulations, where the radiance ratio between the MERIS channels 15 (900 nm) and 14 (885 nm) are used in an inversion procedure based on regression (Bennartz and Fischer, 2001). Near-IR and IR algorithms were also developed for the MODIS instrument by Huang et al. (2004).

2.14.2.3.3 Water vapor profiles

In general, the so-called 1 D-VAR technique is employed for water profile retrievals. 1 D-VAR schemes use the variational principle to solve the retrieval problem (Eyre, 1989), and invert the radiances to simultaneously retrieve the temperature and humidity profile, the surface temperature and MW emissivity, as well as cloud amount and cloud-top pressure. It employs an iterative method, which finds the maximum probability solution to a nonlinear retrieval/analysis problem. Li et al. (2000) applied a 1 D-VAR scheme to TOVS/ATOVS observations. Moreover, 1 D-VAR schemes are also applied to atmospheric sounders, such as AIRS and IASI and to Radio Occultation instruments.

Semiphysical schemes, such as neural networks, can also be used to simultaneously retrieve temperature and water vapor profiles (e.g., Kuligowski and Barros, 2001).

2.14.2.3.4 Upper tropospheric humidity

The relative humidity (RH) of the upper troposphere has a strong influence on the amount of outgoing longwave radiation. It is often derived employing IR and MW instruments as HIRS and MVIRI/SEVIRI as well as AMSU-B/MHS. The brightness temperature of one channel, 6.3 μm for IR and 183 GHz for MW, is related to the RH of the upper troposphere. A typical physical retrieval method for Meteosat is described in Schmetz and Turpeinen (1988). The retrieval is confined to areas with neither medium- nor high-level clouds. Similar schemes, using Jacobian vertical weighting, have been developed by Buehler and John (2005) for AMSU-B, Brogniez et al. (2007) for Meteosat, and Jackson and Bates (2001) for HIRS.

2.14.2.4 Cloud Detection

In general, cloud detection methods are based on the fact that clouds have a higher reflectance and a lower temperature than the underlying Earth surface. In addition, cloudy scenes have a higher spatial and temporal variability than clear sky scenes. However, difficulties in cloud detection appear when the contrast between the cloud and underlying surface is small. At VIS wavelengths, it is difficult to detect clouds over high reflecting surfaces such as snow or desert. At IR wavelengths, it is difficult to discriminate low clouds from clear sky land surfaces during the night, when surface temperatures may drop below cloud-top temperatures. In these cases, testing the spatial coherence in IR radiances in cloudy and clear skies is an effective manner to identify cloudy areas (Coakley and Bretherton, 1982). Moreover, cloud edges are difficult to detect, as the satellite pixels at these edges are only partly cloudy. Part of the difficulties touched on above may be alleviated by the combined use of the multi-spectral observations from satellite (Saunders and Kriebel, 1988; Ackerman et al., 1998).

2.14.2.4.1 Cloud property retrievals

The spectral variations in cloud reflectances provide a powerful diagnostic tool to identify differences in cloud micro- and macro-physical properties, such as cloud optical thickness, cloud thermodynamic phase, and cloud particle size. The retrieval of cloud physical properties from reflectances at VIS (0.6 or 0.8 μm) and near-IR (1.6, 2.2, or 3.9 μm) wavelengths is based on the principle that the reflectance of clouds at a nonabsorbing wavelength in the VIS region is strongly related to the optical thickness and has very little dependence on particle size, whereas the reflectance of clouds at an absorbing wavelength in the near-IR region is primarily related to particle size (Nakajima and King, 1990; Han et al., 1995). The example simulated TOA solar reflectance spectra presented in **Figure 6** shows the substantial differences in the absorption properties of water and ice in the near-IR solar region ($0.7\,\mu m < \lambda < 4\,\mu m$). Especially at 1.6, 2.2, and 3.9 μm, ice exhibits stronger absorption than water. Due to differences in cloud optical thickness, the ice cloud is somewhat brighter than the water cloud in the VIS region ($\lambda < 0.7\,\mu m$).

An inversion procedure is used to relate observed radiances to cloud thermodynamic phase, optical thickness, and particle size. A radiative transfer model is used to prepare lookup tables of simulated reflectances for clouds with different optical thicknesses, thermodynamic phases, and particle sizes for a wide variety of solar and satellite viewing geometries. Liquid and ice water path are computed from retrieved cloud optical thickness and particle size. Note that the retrieval of particle size from near-IR reflectances is weighted toward the upper part of the cloud (Platnick, 2001).

At thermal IR wavelengths, the retrieval of cloud micro- and macro-physical properties is based on the interpretation spectral variations in emitted radiances at the cloud top. Absorption and emission dominate cloud radiative transfer at these wavelengths, because cloud particles have a low single scattering albedo and a large asymmetry parameter. For clouds with an optical thickness smaller than 4, the amount of observed upwelling radiance at the top of the atmosphere will be affected by cloud properties, such as optical thickness, particle size, and thermodynamic phase (Baum et al., 1994). By selecting two (or more) appropriate wavelengths, it is feasible to infer the emissivity and temperature at the cloud top, and deduce information on cloud optical thickness, particle size, and thermodynamic phase. A major drawback of IR retrievals techniques is that cloud emissivities saturate at relatively low optical thicknesses.

Passive MW radiometers measure radiances, expressed as brightness temperatures, at various frequencies between 10 and 100 GHz. These radiances have distinct atmospheric

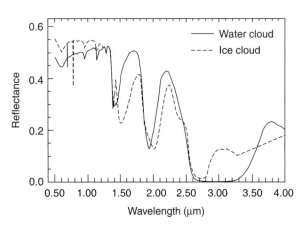

Figure 6 Radiative transfer model simulated top of atmosphere reflectance spectra for water and ice clouds.

absorption characteristics, and can be used for a simultaneous retrieval of atmospheric water vapor and cloud liquid water (Greenwald et al., 1993). The liquid water path (LWP) retrievals from MW radiometers provide a measurement of the integrated LWP, and only represent the liquid droplets volume in the cloud.

Numerous algorithms have been developed to estimate the incoming shortwave radiation from satellite radiances (Pinker et al., 1995). Some methods calculate the incoming shortwave radiance by directly interpreting the TOA albedo in terms of atmospheric transmission (Mueller et al., 2009), while others calculate the transmission for a clear and cloudy atmosphere separately, using atmospheric water vapor and cloud microphysical properties (Deneke et al., 2008).

2.14.2.5 Validation

Validation is prerequisite to generate accurate data sets of water vapor and cloud properties for water and energy balance studies.

2.14.2.5.1 Water vapor

The validation of atmospheric water vapor retrieval schemes is very difficult because classical observations are only sparsely available; for example, radiosondes are mostly available over land surfaces and their observation time does not match overpass times of polar orbiting satellites. Ground-based global positioning system (GPS) observations are available more often over land surfaces, but they are only suitable to validate total column water vapor estimates. As aircraft observations are only available along major flight paths, and the accuracy of their instruments often insufficient for validation, the upper troposphere and lower stratosphere are hard to validate. Instead, satellite systems are compared among themselves or to atmospheric reanalysis. Such comparison can also help to uncover specific instrumental and retrieval problems. For instance, the comparison of the passive MW AMSR-E and IR AIRS estimates of total water vapor content revealed some systematic differences due to the treatment of clouds in the AIRS retrievals (Fetzer et al., 2006).

The most comprehensive comparison of SSM/I-based retrievals among themselves and to radiosondes, performed by Sohn and Smith (2003), revealed that differences in statistical retrievals are mostly caused by differences in the training data that were used. It was also found that statistical algorithms outperform physical ones because of simplifying assumptions on tangential factors, such as near-surface wind speed, sea-surface temperature, and residual cloud liquid water. On a seasonal scale (3 months means), the differences between satellite and radiosondes are $\sim 1\,\mathrm{kg\,m^{-2}}$ bias and $\sim 2.5\,\mathrm{kg\,m^{-2}}$ rms.

Sensitivity of instruments influences the satellite comparisons. Fetzer et al. (2008) compared AIRS UTH with the Microwave Limb Sounder (MLS) data. The mean values agree well within 10% and standard deviations of their differences are 30% or less. Differences in wet and dry regimes were found to be caused by different sensitivities of the two instruments.

Milz et al. (2009) compared monthly mean distributions of UTH products from AMSU-B, Humidity Sounder Brazil (HSB), and AIRS for January 2003. The UTH, based on simulated AMSU-B brightness temperatures from AIRS profiles, has a slight moist bias of up to 4% in RH. This bias is small compared to the differences in UTH observations from radiosondes and nadir-looking IR sounders, which were between 10% and 15%, depending on the type of radiosondes (Soden and Lanzante, 1996). It is also small compared to the large differences in UTH between different climate models (John and Soden, 2007). Thus, most of the existing UTH data sets are suitable as benchmark for improving climate model representations of humidity.

Li et al. (2000) reported for temperature profile retrievals from ATOVS, an accuracy of 2 K for temperatures at 1-km resolution and 3–6 K for dew-point temperatures. IASI profile retrievals have recently been evaluated by Pougatchev et al. (2009). Besides the very much improved temperature retrieval, they found that the instantaneous RH retrievals have a bias of about ±10%, and a standard error lower than 10% in the 800–300-hPa range.

2.14.2.5.2 Cloud properties

The validation data of cloud properties are obtained from flight measurements or special observatory sites. During flight measurement campaigns, heavily instrumented aircrafts collect very detailed measurements of cloud micro- and macrophysical properties over a limited period of time, providing valuable information to obtain a better understanding of cloud microphysics (e.g., EUCAARI over Europe, AMMA over Africa, and RICO over the Caribbean). Special observatory sites aim to measure the physical state of the (cloudy) atmosphere over longer periods of time (years). These sites are equipped with a suite of RS instruments to measure radiation, water vapor, and cloud properties. The number of these sites is limited, and comprises the three America Atmospheric Radiation Measurement (ARM) sites and the four Cloudnet sites in Northern Europe. The measurements of the above-described observatory sites play a key role in the continuous validation of cloud properties.

Recently, measurements from *Cloudsat* (radar) and *Calipso* (lidar) can be used for validation as well. The combined use of radar and lidar observations allows the retrieval of vertical profiles of cloud optical thickness, cloud phase, particle size, and cloud water content (Delanoë and Hogan, 2008). These retrievals are of great value for the validation of cloud property retrievals or for deriving global cloud climatology.

Validation studies confirmed that LWP can be retrieved with high accuracy from both AVHRR (Han et al., 1995; Jolivet and Feijt, 2005) and SEVIRI (Roebeling et al., 2006). Although some retrieval algorithms use the 0.6-, 3.8-, and 10.5-μm radiances (Han et al., 1995), while others use the 0.6- and 1.6-μm radiances (Jolivet and Feijt 2005; Roebeling et al. 2008), similar accuracies ($\sim 15\,\mathrm{g\,m^{-2}}$) and precisions ($\sim 30\,\mathrm{g\,m^{-2}}$ for thin clouds and up to $100\,\mathrm{g\,m^{-2}}$ for thick clouds) were found. The above-mentioned accuracies suggest that LWP retrievals from satellite could be an appropriate source of information for the evaluation of climate-model-predicted LWP values. For nonprecipitating water clouds, Van Meijgaard and Crewell (2005) found differences up to $50\,\mathrm{g\,m^{-2}}$ between climate-model-predicted and MWR-inferred LWP values. During the FIRE Arctic cloud experiment, Curry et al. (2000) compared large-scale model LWP values to MWR-inferred LWP values.

They found that all models underestimate the mean LWP by 20–30 g m^{-2}, which corresponded to a relative accuracy worse than 60%.

2.14.2.6 Data Sets

2.14.2.6.1 Water vapor products

In the framework of the GEWEX Water Vapor Project, the NVAP total column water vapor product (Randel *et al.*, 1996) was derived from a combination of SSM/I, TOVS, and radiosonde data for the years 1988–2001. This product was partly renewed by the additional use of AMSU and TRMM data, but this covers only the years 2000–01.

Over ocean, the total column water vapor derived from SSM/I (**Figure 7**) is available from the EUMETSAT Satellite Application Facility on Climate Monitoring (CM-SAF) (Schulz *et al.*, 2009) and from RS systems (Wentz, 1997). These data sets have been successfully used for climate analysis, the evaluation of climate models, model-based reanalysis, trend studies (Trenberth *et al.*, 2005), and investigations of the human impact on the water vapor distribution (Santer *et al.*, 2007). Moreover, GOME/SCIAMACHY data sets have also been used to compute trends of total column water vapor (Mieruch *et al.*, 2008).

High-quality data sets of atmospheric profiles for climate studies, based on TOVS, have been derived by Scott *et al.* (1999). These profiles are highly correlated to corresponding ATOVS profiles. Operationally processed data from ATOVS, AIRS, and IASI exist at various places, such as NOAA, NASA, EUMETSAT, and the CM-SAF.

UTH data sets are derived from AMSU-B and described in Buehler *et al.* (2008). UTH data sets derived from geostationary satellites have been used for the evaluation of climate models (Brogniez *et al.*, 2005).

2.14.2.6.2 Cloud products

The International Satellite Cloud Climatology Project (ISCCP) provided the first global climatology of cloud cover at an acceptable spatial resolution of 30 × 30 km (Rossow and Garder, 1993). For a limited area, Karlsson (2003) presented a cloud climatology from AVHRR observations for Scandinavia. With more advanced retrieval algorithms, MODIS continues the survey of cloud cover (Ackerman *et al.*, 1998).

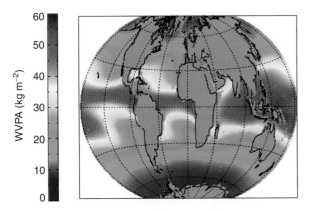

Figure 7 EUMETSAT CM-SAF SSM/I-derived total column water vapor.

The ISCCP data have been successfully used to derive information on cloud physical properties, such as cloud phase, cloud optical depth, or cloud particle size (Rossow and Schiffer, 1999). In turn, these properties have been applied to derive parameters, such as the shortwave radiation budget (Gupta *et al.*, 1999). Other global cloud climatologies are derived from AVHRR, such as the PATMOS climatology (Jacobowitz *et al.*, 2003), or MODIS (Minnis *et al.*, 2003; Platnick *et al.*, 2003).

2.14.3 Water from the Atmosphere: Precipitation

2.14.3.1 Introduction

Precipitation can be considered the most crucial link between the atmosphere and the surface in weather and climate processes. Quantitative precipitation estimates (QPEs) on high spatial and temporal resolutions are of increasing importance for water resources management, for improving the precipitation prediction scores in numerical weather prediction (NWP) models, and for monitoring seasonal to interannual climate variability. Although operational networks of weather radars are expanding over Europe and North America, large areas remain where information on the occurrence and intensity of rainfall is missing. Rain rate estimates from passive MW or VIS and IR imaging sensors on polar and/or geostationary orbiting satellites may bridge this gap, and provide quasi-global information on the spatial extent and intensity of rain.

2.14.3.2 Precipitation Measurements from Weather Radars

Weather radars employ scattering of radio-frequency waves (5.6 GHz for C-band) to measure precipitation and other particles in the atmosphere (Rinehart, 2004). The intensity of the atmospheric echoes is converted to the so-called radar reflectivity factor Z using the Rayleigh scattering approximation (Probert-Jones, 1962):

$$Z = \sum_i D_i^6 \qquad (1)$$

where D_i is the diameter of raindrop i and the summation is over all drops in a unit volume. Marshall and Palmer (1948) proposed a simple exponential form of the drop size distribution $N(D)$ which is widely accepted:

$$N(D) = N_0 \exp(-\Lambda D) \qquad (2)$$

where the drop density $N_0 = 8 \times 10^3$ mm^{-1} m^{-3} and $\Lambda = 4.1 R^{-0.21}$ mm^{-1} depends on the rain rate R in mm h^{-1}. The radar reflectivity factor can be estimated from the sixth moment of the drop size distribution:

$$Z = \int N(D) D^6 \mathrm{d}D = 720 \, N_0/\Lambda^7 = 296 R^{1.47} \qquad (3)$$

with Z in mm^6 m^{-3}. Many different Z–R power laws are used as the appropriate power law depends on climatic and actual meteorological circumstances (e.g., stratiform vs. convective precipitation). Apart from variations in the drop size distribution, several other factors impact the quality of radar-based QPE (Rossa et al., 2005).

The vertical profile of reflectivity (VPR) is, especially at higher latitudes, the major source of error in QPE deduced from weather radar observations (Joss and Waldvogel, 1990; Koistinen, 1991). At longer ranges, the height of observation will increase and in the presence of a significant gradient in the VPR this will typically generate an underestimation of the accumulated precipitation. Many different techniques have been developed to estimate the VPR and to subsequently correct the radar QPEs for this profile. The VPR can be estimated from weather radar data using climatological profiles, mean reflectivity profiles, or local profiles obtained at short ranges (Vignal and Krajewski, 2001). On the other hand, gauge adjustment techniques have been developed which correct the radar precipitation estimates using a second-order polynomial in range (Michelson et al., 2000).

The radio frequency radiation transmitted and received by weather radar is scattered by precipitation. During very intense precipitation, scattering can become so strong that the radar beam is attenuated causing underestimation of precipitation intensity or even disappearance of the rain cells behind very strong cells. The observed radar reflectivity may be corrected for the attenuation when the one-way attenuation due to rainfall is approximated by a power law. However, the correction algorithm for attenuation is potentially highly unstable (Hitschfeld and Bordan, 1954). For the (near) future, dual-polarization weather radars offer promising new possibilities to correct for attenuation during intense rainfall (Bringi and Chandrasekar, 2001).

2.14.3.3 Precipitation Measurements from Satellite

The reader can find an up-to-date review of satellite rainfall retrieval methods in Kidd et al. (2009) and Levizzani et al. (2007).

2.14.3.3.1 Retrievals from VIS–IR sensors

Over the past decades, several rain rate retrieval methods based on observations from VIS and IR sensors were developed. The methods based on geostationary (GEO) satellites often use thermal IR observations and relate daily minimum cloud-top temperatures (Adler and Negri, 1988; Anagnostou et al., 1999) or cold cloud durations (CCD) to rain rates (Todd et al., 1995). These methods tend to perform reasonably well over areas where rainfall is governed by deep convection, but are less effective at higher latitudes, where precipitation originates from both convective and stratiform systems. A major limitation of the CCD methods is that rain rates are proportional to cloud duration, which is an assumption that fails in case high rain intensities occur over a short time period (Alemseged and Rientjes, 2007).

Several methods have been developed that relate cloud physical properties, retrieved from passive imaging sensors, to precipitation. The GOES Multi-Spectral Rainfall Algorithm (GMSRA; Ba and Gruber, 2001) utilizes data from five channels, covering the VIS, near IR, water vapor, and two thermal channels, to extract information on the cloud and rain extent. Nauss and Kokhanovsky (2007) showed that cloud LWP retrievals from MODIS daytime observations are directly proportional to the probability of rainfall. On the other hand, Rosenfeld and Gutman (1994) and Rosenfeld and Lensky (1998) found that clouds require droplets with effective radii > 14 µm for the onset of precipitation. This is consistent with the findings of Twomey (1977), who reported that the precipitation efficiency of a given cloud depends on the size of the cloud droplets and the amount of aerosols in the air. Roebeling and Holleman (2009) present a novel approach, which uses information on cloud condensed water path, particle effective radius, and cloud thermodynamic phase to detect precipitating clouds, while information on condensed water path and cloud top height is used to estimate rain rates. The fact that their approach can be applied to GEO observations from the SEVIRI potentially allows for the provision of precipitation observations over large parts of the globe every 15 min. **Figure 8** shows the effect of increasing threshold condensed water path and droplet effective radius values on the spatial extent of precipitation over the Netherlands as retrieved from SEVIRI.

2.14.3.3.2 Retrievals from passive MW sensors

A more direct measurement of precipitation from satellite is made possible by the use of the MW frequencies as in this part of the spectrum precipitation-sized particles are the main source of atmospheric attenuation. Over ocean, the signal is mainly due to the increased emission of radiation from rain droplets so that rain areas appear warmer over the radiometrically 'cold' water background. Over land, rainfall is associated with scattering of the upwelling surface radiation by precipitation-related ice particles. The main problem of the passive-MW-based techniques is that the instruments are currently only available on low-Earth orbiting (LEO) satellites, and thus observations are available only twice per day per satellite (at best). Moreover, the resolutions of the measurements are for ocean rainfall products of the order of $50 \times 50\ km^2$, while over land they are typically no better than $10 \times 10\ km^2$.

MW-based estimation techniques belong to two main groups: empirical techniques that calibrate the observations with surface data sets and physical techniques that minimize the difference between a modeled atmospheric rainfall event and the observation. An example of the physical techniques is the Goddard profiling (GPROF) technique (Kummerow et al., 2001) that uses a database of model-generated atmospheric profiles to which the observed satellite measurements are compared, and the best profile match is selected. The advantage of such a technique, first conceived for the TRMM TMI, is that it provides more information about the precipitation system than techniques that just provide information on surface rainfall.

With the launch of sensors such as the Advanced Microwave Sounding Unit-B (AMSU-B, cross-track scanner) or the Special Sensor Microwave Imager/Sounder (SSMIS, conical scanner), higher frequency channels in the strong water vapor absorption lines at 183 GHz became available. The response functions of these channels peak at altitudes higher than 2 km and thus are much less influenced by ground emissivity features that represent a large portion of the errors in precipitation estimation over land. Several

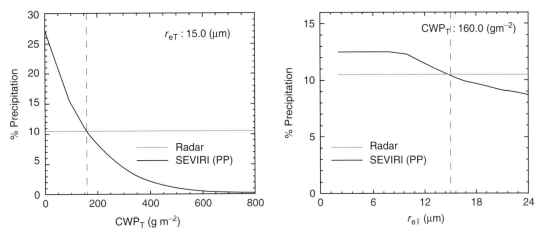

Figure 8 Relationship between spatial extents of precipitation and threshold condensed water path values (CWP$_T$) for clouds with particle sizes larger than 15 μm (left), and threshold particle sizes (r_{eT}) for clouds with condensed water path values larger than 160 g m^{-2} (right). The horizontal gray line indicates the spatial extent of precipitation derived from weather radar observations that were collocated and synchronized with the SEVIRI retrievals. Note that the optimum thresholds for the detection of precipitation are 160 gm^{-2} for CWP and 15 μm for r_e.

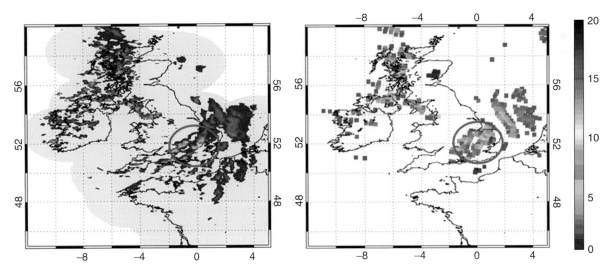

Figure 9 22 November 2008. Precipitation retrieval (in mm h^{-1}) over NW France, the Channel and the UK using high-frequency AMSU-B MW channels for a mixed-type precipitating system (right) and radar retrieval from the NIMROD network (left). The circle delimits the area where both radar and satellite sense snowfall. Image courtesy of S. Laviola, ISAC-CNR.

algorithms are now available for operational applications, including detection of cloud droplets, snowfall, and snow on the ground (e.g., Ferraro et al., 2005; Laviola and Levizzani, 2008; Surussavadee and Staelin, 2008; Weng et al., 2003). An example of mixed-phase precipitation retrieval is shown in Figure 9.

2.14.3.4 Validation

2.14.3.4.1 Weather radar retrievals

Surface networks of rain gauges can be used for both reduction of the gross errors and validation of quantitative precipitation estimates. Wilson (1970) pioneered with the integration of radar and rain gauge data and showed that this can improve the area rainfall measurements. A real-time calibration of radar-based surface rainfall estimates by telemetering rain gauges was performed by Collier (1983) and an improved accuracy was seen on most locations. Nowadays, mean-field bias adjustment of radar-based quantitative precipitation estimates is widely used.

At the Royal Netherlands Meteorological Institute (KNMI), mean-field bias adjustment with gauges is used operationally for an hourly updated QPE product (Holleman, 2007). An extensive spatial and temporal verification of the bias-adjusted radar composites over a 6-year period (2000–05) using dependent and independent gauge data is performed. It is found that the real-time adjustment scheme effectively removes the mean-field bias from the raw accumulations over a large area and that it substantially reduces the daily standard deviation. The adjustment method cannot correct for a range-dependent bias and it is recommended to use a simple VPR adjustment procedure for that.

2.14.3.4.2 Satellite retrievals

Weather radar observation can be used to validate retrievals of the occurrence and intensity of precipitation from passive imaging satellites. Roebeling and Holleman (2009) compared 15-min SEVIRI retrievals of spatial extent of precipitation and rain rates against weather radar observations (**Figure 10**). The instantaneous rain rates from SEVIRI are retrieved with a high accuracy of about 0.1 mm h^{-1}, and a satisfactory precision of about 0.8 mm h^{-1}. SEVIRI is very accurate in detecting percentages of precipitation over larger domains (the Netherlands), which is shown by the high correlation of about 0.90 between spatial extents of precipitation from SEVIRI and weather radar. Similarly, the rain rates retrievals from SEVIRI correlate reasonably well with the weather radar observations (corr. = 0.63).

An international effort is being conducted by the International Precipitation Working Group (IPWG) to obtain reasonably homogeneous validation figures for the various satellite rainfall estimation algorithms over the various continents. Ebert *et al.* (2007) argued that the results of such a validation exercise so far confirm that the performance of satellite precipitation estimates is highly dependent on the rainfall regime and generally opposed to those of the NWP model Quantitative Precipitation Forecasts (QPF). Satellite estimates of rainfall occurrence and amount are most accurate during summer and at lower latitudes, whereas the NWP models show greatest skill during winter and at higher latitudes. In general, the more the precipitation regime tends toward deep convection, the more (less) accurate the satellite (model) estimates are.

2.14.3.5 Applications

The third phase of the Network of European Meteorological Services (EUMETNET) Operational Program on the Exchange of Weather Radar Information (OPERA) is a joint effort of 30 European countries, which runs from 2007 till 2011, and is managed by KNMI. OPERA-3 is designed to firmly establish the Program as the host of the European Weather Radar Network. Currently, OPERA's operational network consists of more than 175 weather radars, of which roughly 100 systems have Doppler processing and about 15 systems have dual-polarization capability. In the coming years, the number of dual-polarization systems is expected to increase dramatically, thus offering new opportunities for quantitative precipitation estimation (Bringi and Chandrasekar, 2001).

During this program phase, an OPERA Data Center (ODC) for the weather radar network should be specified, developed, and operated. This data center is crucial for reaching the main objective of OPERA-3, that is, establishing the weather radar networking as a solid element of the European infrastructure. The ODC will enhance and monitor availability of radar data, facilitate quality control of single-site radar data, stimulate exchange of volume radar data and quality information, and produce a homogeneous European radar composite. Furthermore, the ODC will deliver radar data to users inside and outside the National Meteorological Services. In May 2009, the EUMETNET Council has approved further development of the ODC. Start of operation of the ODC is planned for early 2011. More information on OPERA can be found in Holleman *et al.* (2008) and on the website.

Satellite rainfall products, in spite of their intrinsic problems and not completely defined quality figures, are characterized by a global perspective that no other measurement method has. Because of this, their applications are numerous and cover very different fields such as meteorology, hydrology, civil protection, and climate. We will only mention a few examples without pretending of being complete.

As is the case of radar precipitation measurement, the first field of application is in nowcasting, when a larger coverage than the one ensured by the radar is necessary. The satellite, in fact, ensures a mesoscale perspective, which becomes synoptic when LEO and GEO orbits are used. Another very important meteorological application is in data assimilation for NWP. Several physical (nudging) and variational methods have been developed in time at all scales. It is generally accepted that precipitation assimilation (e.g., Davolio and Buzzi, 2003) is more suited at the mesoscale, where current models start to incorporate the appropriate cloud parametrizations that general circulation models often lack.

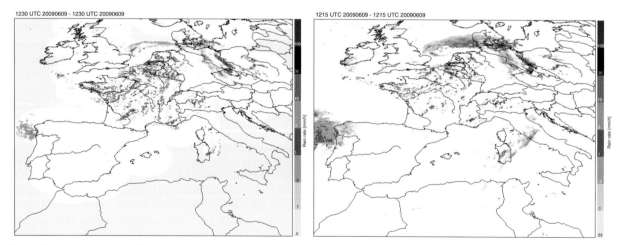

Figure 10 9 June 2009, 12:30 UTC. An example of Opera rain rate composite (left) and MSG-SEVIRI rain rate retrievals (right) for Europe, presented in the projection of MSG.

Hydrological applications of satellite rainfall products span from the assimilation into hydrological models for basin management to global hydrological predictions. In all cases, uncertainty definition is the key to successfully use satellite data in this field (e.g., Voisin et al., 2008). Another important problem of current global products is the effect of orography on the retrieval (Adam et al., 2006). An upcoming very interesting application that merges a hydrological and a civil protection perspective is the one that uses satellite global rainfall data for landslide prediction (Hong and Adler, 2008); their methodology identifies landslide-prone areas on the basis of morphological information and rainfall for providing a hazard map.

The number of climatological applications is expected to increase substantially over the next few years, given the global character of satellite data. The Global Precipitation Climatology Project (GPCP; Adler et al., 2003) has gathered global satellite rainfall estimations since 1979. GPCP products are now used to evaluate models and verify scenarios on the impacts of the various phenomena (ENSO, volcanic eruptions, etc.) on climate. The most important scenario is to verify whether global warming produces an acceleration of the global water cycle with more extremes (droughts on one side and extreme floods on the other) or not (e.g., Curtis et al., 2007). Finally, regional studies are conducted to examine the structure of propagating convective episodes in the warm season for their better forecasting and their modification in a climate perspective (e.g., Carbone and Tuttle, 2008; Laing et al., 2008).

2.14.4 Water to the Atmosphere – Evaporation

2.14.4.1 Introduction and Historic Development

In the middle of the last century, ET from well-watered land surfaces was thought to be controlled by meteorological conditions, and only in the 1970s it was recognized that spatially and temporally dynamic feedback mechanisms between ET and land surface (e.g., albedo, rooting depth, and temperature) play an important role. This invoked the first applications of RS-based approaches, which mainly made use of airborne scanners (Bartholic et al., 1972; Idso et al., 1975; Jackson et al., 1977; Stone and Horton, 1974). It was only in the following decade that the first use of thermal data obtained from satellites to estimate ET was seen (Price, 1982; Seguin and Itier, 1983). They comprised of statistical approaches using linear relationships between daily totals of ET and net radiation and the difference between near-midday observations of radiant temperature and near-surface air temperature. Naturally, these linear relations needed local calibration, and the influences such as wind velocity, thermal stratification, and surface roughness were incorporated in later work (Riou et al., 1988), as such trying to bridge the gap already recognized by Seguin and Itier (1983) between sophisticated models useful for understanding basic processes and for performing informative simulations on the one hand and real estimation of ET on the other.

These developments led to the general acceptance of the idea that spatial variability in ET is important, which in turn stimulated the development of effective methods for determining landscape-scale ET. However, still a need was noticed for models that can realistically simulate the distributed nature of land surface processes and for techniques capable of upscaling estimates that are based on point-scale observations (Shuttleworth, 1988; Kalma and Calder, 1994), mainly used for validation. As observed by Kalma et al. (2008), this viewpoint developed more or less simultaneously with the use of airborne eddy correlation measurements (Schuepp et al., 1992; Mann and Lenschow, 1994), the development of scintillometry (de Bruin et al., 1995; Green et al., 2001), and an increased use of RS techniques. The main attraction of the last technique probably is the possibility of integration over a heterogeneous area at different resolutions and of routinely generating operational ET estimates. From 1990 onward until now, a vast amount of models have been developed and tested in a large number of multidisciplinary large-scale field experiments (Kustas and Goodrich, 1994; Shuttleworth et al., 1989; Kabat et al., 1997; Hollinger and Daughtry, 1999; Su et al., 2008, 2009). The increased understanding of the observed processes resulted in a number of excellent overview papers on both these processes and their typical impediments (Moran and Jackson, 1991; Becker and Li, 1995) as well as on the methodologies to estimate ET themselves by Kustas and Norman (1996) and Quattrochi and Luvall (1999) and more recently by Kalma et al. (2008).

The models that have evolved mainly differ in type or purpose of the application which basically determines the type of RS data used and to which extent ancillary data are needed. What they all have in common is that the main input originating from RS is thermal information. It is obvious that no method or algorithm will outperform all other methods under all conditions and that a selection has to be based on the scale and purpose of the application as well as on the availability of the required data.

2.14.4.2 Current State of Science

There are currently several methods being used, which can roughly be divided into three categories: they are either based on statistics and empirics, on spatial variability using either within image hydrological contrasts or some kind of index, or they are physically based, more specific on the energy balance at the Earth's surface. As this chapter deals with the observation of hydrological processes, we will focus on the last category. For the sake of completeness, first, we briefly discuss the statistical and spatial variability methodologies followed by a description of current frequently used physical, or analytical, approaches.

2.14.4.2.1 Statistical approaches

The methods using mainly statistical and empirical relations, also the first that were developed, make use of quasi-linear relationships between difference in daily amounts of ET and net radiation on one side and observed instantaneous differences between radiometric temperature and near-surface air temperature on the other. A prerequisite here is the use of near-midday temperature differences on clear days as these are representative of the entire day due to the regular course of climatic parameters during cloud-free days. These methods all originate from the work of Jackson et al. (1977), who derived a single statistical regression constant for the relation between the instantaneous temperature differences and daily ET and net radiation. Although later work incorporated aerodynamic surface properties on atmospheric stability effects (Seguin and Itier, 1983; Riou et al., 1988) and, as such, moved into the direction of a physically based energy balance approach (Nieuwenhuis et al., 1985; Soer, 1980; Lagouarde and McAneney, 1992), these types of approaches still require local calibration. Therefore, they are currently more often used in combination with scaled indices derived from scatterplots of midday temperature versus normalized difference vegetation index (NDVI; Carlson et al., 1995) as such reducing the need for local calibration and ancillary data input, making them more suitable for operational monitoring of ET, reaching accuracies of around 1 mm on a daily basis (Kustas and Norman, 1996).

2.14.4.2.2 Variability approaches

This brings us to the methods using the spatial variability within the image. They use either two-dimensional scatterplots of surface radiant temperature versus NDVI, the so-called triangle methods (Nemani and Running, 1989; Price, 1990) or the within-image relation between surface temperature and surface reflection (Bastiaanssen et al., 1998a; Roerink et al., 2000; Su et al., 1999), both based on the original work of Menenti and Choudhury (1993) to determine the hydrological wet and dry extremes. In the triangle approaches basically the observed position of a pixel within the scatterplot determines the ratio between actual and potential ET. The methods using temperature and reflective properties use a scaling between the observed wet and dry edges along the surface temperature, using either solely temperature such as in the S-SEBI approach (Roerink et al., 2000) or in combination with the use of a local surface roughness estimate as in the SEBAL approach (Bastiaanssen et al., 1998a). Whereas these methods do not need very accurate atmospheric correction techniques nor detailed meteorological inputs, they are limited by the fact that hydrological contrast needs to be present within the observed scene. In the case of the triangle methods, this is circumvented by comparison with a theoretically derived scatter triangle (Jiang and Islam, 2001; Carlson et al., 1994; Gillies et al., 1997; Venturini et al., 2004), whereas a temperature scaling was coupled to surface resistance by inverting an energy balance model by Boegh et al. (2002) and McVicar and Jupp (2002) to overcome this shortcoming.

2.14.4.2.3 Physical approaches

This brings us to the physically based RS algorithms to derive ET estimates. They are all based on the idea that ET is a change of the liquid state of water to the gaseous state, hereby using available energy in the environment for vaporization. The available energy is the net radiation, which is the budget of all shortwave and longwave incoming and outgoing radiation at the Earth–atmosphere interface, less the heat used for heating up that interface, that is, the Earth's surface, commonly referred to as the soil heat. The available energy is then thought to be used either for heating up the atmosphere, the so-called sensible heat, or for changing the state of water, the latent heat. Soil heat is generally considered a fraction of net radiation (Su, 2002; Norman et al., 1995; Anderson et al., 1997) depending on vegetation characteristics, and several studies have indicated that net radiation can be accurately determined from RS data (Timmermans et al., 2007; Boegh et al., 1999; Kustas and Norman, 1999; Su et al., 2001); the main remaining task is the division of the available energy between sensible and latent heat. The most widespread approach, also commonly used in land-surface modeling (Overgaard et al., 2006), is to consider the Earth–atmosphere interface, the Earth's surface, as an electrical analog. Basically, this means that the rate of exchange (i.e., flux) of a quantity (e.g., temperature or vapor pressure) between two media (e.g., the Earth and the atmosphere) is driven by a difference in potential of that quantity, and controlled by a number of resistances that depend on the local climate as well as on the internal properties of the two media. The remote determination of vapor pressure is not feasible with the current state of technology. Therefore, the approach is to determine the rate of exchange of temperature between the Earth and the atmosphere, that is, the sensible heat flux, and determine the latent heat flux as a rest term. Dividing the latent heat flux by the latent heat of vaporization then yields ET. This means that current research efforts aim at the proper determination of the sensible heat flux. The different approaches to this problem are sketched in **Figure 11**.

Basically, they differ in whether or not they discriminate between soil and canopy components. In **Figure 11(a)**, the

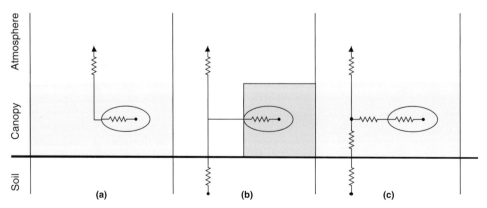

Figure 11 Sketch of different resistance schemes.

sensible heat flux is driven by the difference between the aerodynamic surface temperature at the canopy source/sink height and the near-surface air temperature and controlled by a single aerodynamic resistance to sensible heat transfer between the canopy source/sink height and the air/atmosphere at a reference height above the canopy. The aerodynamic resistance is generally calculated from local wind speed, surface roughness length, and atmospheric stability (Brutsaert, 1982, 1992). Although these single-source models are known to give good results under a variety of conditions and environments (Kustas, 1990; Kustas et al., 1996; Bastiaanssen et al., 1998; Su, 2002; Jia et al., 2003), their main problem is that the necessary aerodynamic surface temperature at the mean canopy air stream is different from the radiometric surface temperature obtained from RS observations. This is usually corrected for by introducing an extra resistance that mainly depends on the inverse Stanton number, a dimensionless heat transfer coefficient that originates from the difference in source/sink heights for momentum and for heat transport. Although robust models exist to estimate this parameter (Massman, 1999; Su et al., 2001), it is known to vary widely, especially over sparse vegetation. This has led to the so-called dual-source models that treat the soil and canopy separately.

Two different approaches are noticed. When the surface, or pixel, is divided into different fractions of bare soil and vegetation, the soil and vegetation components do not interact. These models, first introduced by Avissar and Pielke (1989), are known as patch models, or parallel resistance network as they are more frequently named in the RS community (see **Figure 11(b)**). However, when sparse vegetation is present, the soil and vegetation components are known to interact and a so-called series resistance network is more appropriate. In this case, the canopy consists of a semi-transparent layer located above the soil surface such that heat and moisture have to enter or leave the surface layer through the canopy layer, whereby the component fluxes are allowed to interact (see **Figure 11(c)**). The structure proposed by Shuttleworth and Wallace (1985) is most widely used and incorporates a bulk stomata resistance for the vegetation as well as a resistance controlling the soil fluxes. It is assumed that aerodynamic mixing within the canopy invokes a mean canopy-airflow where fluxes from the components are allowed to interact after which they are exchanged with the atmosphere, controlled by a third aerodynamic resistance. Both structures require component temperatures, whereas a remote sensor only observes the effective radiometric surface temperature, which is a combination of the component temperatures, depending on viewing angle and fractional vegetation cover. To derive the component temperatures from the effective temperature, additional information is required. Several methods have been developed, ranging from empirical relationships (Lhomme et al., 1994), via coupling to a crop growth model (Chehbouni et al., 1996; Chehbouni et al., 1997) and the NDVI–surface temperature relationship (Boegh et al., 1999), to dual viewing angle approaches (Francois et al., 1997; Kustas and Norman, 1997; Merlin and Chehbouni, 2004). A different approach was developed by Norman et al. (1995) where transpiration initially is estimated through the Priestley–Taylor equation, whereby they were able to relate the canopy temperature to air temperature. This allowed initial guesses of component temperatures, which were then used in an iterative procedure to derive soil evaporation and canopy transpiration that satisfied the energy balance. In their original paper, Norman et al. (1995) described an operational procedure both for the series as well as for the parallel approach that, in different improved versions (Kustas et al., 2001; Anderson et al., 1997, 2005; Kustas and Norman 1999, 2000), is nowadays widely used (Sanchez et al., 2008; French et al., 2002; Li et al., 2005).

By now it may be clear that using RS observations to derive latent heat fluxes, or ET, requires a certain amount of assumptions depending on the model and data used as well as on the purpose of the application. As such, it is no surprise that different techniques lead to deviating estimates (Zhan et al., 1996; French et al., 2005; Timmermans et al., 2007) and works still need to be undertaken to minimize those deviations.

2.14.4.3 Future Research Needs

From the previous some residual challenges and thus future directions of research follow. They relate to spatial and temporal scaling issues, coupling and feedback issues, and, last but not least, validation issues. Although they may be categorized, they are discussed here in a coherent manner, as most of them are interrelated.

2.14.4.3.1 Scaling

Depending on the application purpose, models describing the land–atmosphere interaction assume that both processes and variables are scale invariant (Menenti et al., 2004), which means that it is assumed that the relation of observations with model variables is the same at all spatial scales. Intermodel variability of predicted fluxes is therefore often large and causes are difficult to pinpoint (Menenti et al., 2004), which is probably the reason why only a few pixel-by-pixel flux comparisons (Timmermans et al., 2007; French et al., 2005; Boegh et al., 2004; Timmermans et al., 2009; De Lathauwer et al., 2009) are made (Overgaard et al., 2006). An in-depth analysis is needed of the nature of feasible observations in the soil–vegetation–atmosphere system at different (Tol et al., 2009; Timmermans et al., 2009) and multiple scales (McCabe et al., 2006) to detect and understand inconsistencies in model variables and parametrizations.

There is also a need for improved temporal scaling procedures to extrapolate instantaneous estimates of ET derived from RS platforms to hourly, daily or longer periods (Kalma et al., 2008). Concepts most widely used so far (Shuttleworth et al., 1989; Batra et al., 2006) to extrapolate to daily values yield unsatisfying results, especially over drying surfaces (Chehbouni et al., 2008; Gentine et al., 2007). Given the fact that cloudy conditions hamper the frequent remote observation of ET, alternative approaches have to be explored, especially on timescales longer than 1 day.

2.14.4.3.2 Feedbacks

Describing transfer of energy into the atmosphere using the energy balance methods generally invokes assuming homogeneous atmospheric properties. This requires neglecting fast changes in air temperature and humidity, and thus in fluxes thereof, due to turbulence. These changes occur at small spatial scales implying that spatial variability of the atmospheric properties at a given time is significant. A more realistic description of the structure and dynamics of the atmosphere is obtained by large eddy simulation (Albertson, 1996; Albertson et al., 2001). Opportunities to improve current RS-based ET estimates therefore include testing the spatial validity of the meteorological data used (Gowda et al., 2007). In addition, there is a critical need to understand the feedbacks between the land surface and the atmosphere at various scales (Wood, 1998). Feedback between land and atmosphere arises from the fact that the fluxes of heat and water from the land surface to the atmosphere will change the properties of the atmosphere, which in turn will change the fluxes. Therefore, more work is required in the line of Bertoldi et al. (2007) and Timmermans et al. (2008), who examined feedback effects at multiple scales using an RS-based energy balance model dynamically coupled to a large eddy simulation model.

2.14.4.3.3 Validation

Apart from the ongoing discussion on the mismatch between available energy observations and turbulent flux measurements from eddy correlation (Foken, 2008) resulting in uncertainties of up to 30% in validation data, there is also considerable doubt on the applicability of scintillometry over very heterogeneous terrain (Timmermans et al., 2009; Ezzahar et al., 2007). Moreover, RS-based energy balance models tend to be validated versus a handful of tower-based measurements, which does not ensure a reliable performance over the broader landscape. To address this uncertainty, intercomparisons of flux model output need to be performed as reported by Timmermans et al. (2007) and French et al. (2005). In addition, a dynamic coupling of distributed hydrological and atmospheric models through an RS-based surface energy balance model, such as Timmermans et al. (2008), Velde et al. (2009), and Bertoldi et al. (2007), is vital for future applications and probably improves possibilities for making a more spatially detailed evaluation (Overgaard et al., 2006).

To summarize, advances in improving parametrization and validation of physically based ET models will rely heavily on the understanding of physical processes at different scales as well as on the ability to obtain distributed physical information. In order to achieve this, satellite EO will prove to be of paramount importance in the future.

2.14.5 Water on the Land – Snow and Ice

2.14.5.1 Introduction

The seasonal and perennial snow and ice masses (the cryosphere) cover a major part of the land surfaces. They are essential or dominating elements of the hydrological cycle in mid- and high latitudes, as well as in many mountain areas. The terrestrial cryosphere comprises the seasonal snow cover, lake and river ice, permafrost, seasonally frozen ground, glaciers, ice caps, and the large ice sheets of Greenland and Antarctica. Of these, seasonal snow cover and frozen ground on land dominate in spatial extent and temporal variability, covering at maximum about 50% of the land area in the Northern Hemisphere. Due to feedbacks with the atmosphere and other elements of the hydrosphere, the cryosphere responds very sensitively to climate warming, as reports on past and ongoing changes of the snow and ice masses confirm (Lemke et al., 2007).

Due to the large spatial extent and temporal variability of snow and ice coverage, RS techniques provide the only feasible means for timely and comprehensive observation of these elements of the Earth system. The potential of RS for monitoring snow and ice has been recognized already in the 1960s, applying optical imaging sensors of the NOAA satellites to mapping the global snow cover on a weekly basis (Robinson et al., 1993). Thanks to advancements in sensor technology, the 1970s brought in a big step forward in satellite-borne RS, including observations of the cryosphere. Optical sensors of improved spatial and spectral resolution and new active and passive MW sensors opened up the opportunity to monitor all the individual elements of the global cryosphere. Already at that time, RS became an indispensable tool for snow and ice monitoring and research that further evolved over the years, thanks to advancements in sensor technology and data processing (Key et al. 2007). Airborne sensors play an important role in the development of techniques for data processing and analysis, as well as in local to regional surveys of snow and ice. However, due to the near-global coverage and the regular repeat capabilities, satellite-borne sensors are the main tool for snow and ice monitoring. Therefore, in this chapter, we focus on applications of satellite sensors.

2.14.5.2 Techniques for Retrieval of Extent and Physical Properties of Snow and Ice

Sensors in the VIS, infrared, and MW part of the electromagnetic spectrum are employed to monitor the extent and physical properties of snow and ice. In order to explain the information content of the various sensor types, the main features affecting the radiance reflected or emitted by snow and ice are summarized below.

Electromagnetic waves, incident on a snow or ice medium, are subject to scattering at volume inhomogeneities (snow grains and air bubbles in ice) and absorption along the propagation path. In the case of melting snow, water adds as a third component of the mixture. The absorption and scattering characteristics are determined by the dielectric and structural properties of the medium and the sensor wavelength.

At VIS wavelengths, the dielectric losses of ice and water are small, but increase considerably in the near- and mid-IR. In the thermal IR, snow is almost a black body (emissivity 0.99). Consequently, clean fresh snow has a high reflectance in the visible part of the spectrum ($0.9 < R_{VIS} < 0.99$), dropping to $R < 0.1$ in the shortwave IR at wavelengths $\geqslant 1.5\,\mu m$. The spectral reflectance in the visible decreases significantly with aging of snow due to pollution. In the near IR, between 0.9 and $1.3\,\mu m$, the reflectance decreases with increasing size of the snow grains, which is used to estimate this parameter from satellite measurements (Dozier and Painter, 2004). The direct effect of liquid water in a snow pack on near-IR reflectance is small, although the reflectivity decreases because melt metamorphosis causes snow grains to grow.

The decrease of reflectance in the IR is employed by the normalized difference snow index (NDSI) for discriminating snow cover and snow-free surfaces:

$$\text{NDSI} = \frac{R_{\text{VIS}} - R_{\text{SWIR}}}{R_{\text{VIS}} + R_{\text{SWIR}}} \quad (4)$$

The automated MODIS snow-mapping algorithm uses at satellite reflectances in MODIS bands 4 (0.545–0.565 μm) and 6 (1.628–1.652 μm) to calculate the NDSI (Hall et al., 2002). Different thresholds of the NSDI are used to detect snow in forested areas and open land (Salminen et al., 2009). For excluding cloud-covered pixels, the quality flag from the MODIS cloud-masking algorithm is applied which uses visible, SWIR, and thermal IR channels to detect clouds (Ackerman et al., 1998). In the case of patchy snow cover, the binary classification shows a trend of overestimating the total snow area. To account for these effects, spectral unmixing techniques, using VIS and near-IR channels to map snow cover fraction at subpixel scale, are applied (Vikhamar and Solberg, 2003; Dozier and Painter, 2004; Sirguey et al., 2009).

Active MW sensors (synthetic aperture radar, SAR) and passive MW sensors (radiometers) are widely applied for mapping the extent and physical properties of the snow cover. For interpreting and analyzing MW measurements of snow, it is essential to consider the layers contributing to the observed signal. The penetration depth, d_p, can be computed from the complex permittivity ($\varepsilon = \varepsilon' - i\varepsilon''$) by

$$d_p = \frac{\lambda}{2\pi} \frac{\sqrt{\varepsilon'}}{\varepsilon''} \quad (5)$$

The imaginary part of the permittivity of snow, ε'', and, therefore, also d_p, shows a strong dependence on the liquid water content (Mätzler, 1987). The dielectric losses in dry snow are small, and the penetration depth is of the order of several hundred wavelengths (e.g., about 20 m for C-band SAR with $\lambda = 5.6$ cm). On the other hand, the penetration depth in wet snow is only about one wavelength or less due to the high dielectric losses of water. In the C-band, for example, the penetration d_p in snow with 5% by volume of liquid water is only 3 cm. This has an important impact on the signal observed by MW sensors. For wet snow, the MW signal reflected or emitted from a melting snow pack originates from a thin top snow layer and the snow surface, whereas for dry snow both the snow volume and the medium below the snow pack contribute to the observed signal (Rott, 1997).

These properties cause distinct differences in the information provided by the various MW sensors. Imaging radars for snow mapping typically operate in the C- and X-band ($\lambda = 5.6$ cm, $\lambda = 3$ cm). At these wavelengths, the scattering contribution of a dry winter snow pack is small, and the backscatter contribution of the ground surface dominates. On the other hand, due to the high absorption losses, the radar reflectivity of melting snow is rather low. These characteristics enable to map melting snow areas by means of C- and X-band SAR, applying a change detection algorithm using SAR image time series (Nagler and Rott, 2000). Combining optical and SAR sensors for snow area mapping helps to overcome the cloud handicap of optical sensors which is a particular problem for updating snow extent in the ephemeral melting snow zones (Solberg et al., 2008).

For observing snow water equivalent (SWE), a critical parameter for snow hydrology, shorter MW wavelengths need to be employed to obtain a distinct signal of the snow volume. The radiance emitted by the ground is attenuated in the dry snow pack by scattering at the snow grains. The attenuation due to volume scattering increases inversely to the third power of the wavelength ($\sim \lambda^{-3}$) (Hallikainen et al., 1987). The scattering losses depend on snow depth, density, and grain size. Currently, no satellite-borne imaging radar systems are available at short wavelengths, but MW radiometers are applied to map the depth and water equivalent of the snow pack. Retrieval of SWE from passive MW data is conventionally based on empirically determined relationships between SWE and emitted brightness temperature (T_B). Standard procedures apply the difference in T_B at 37 GHz ($\lambda = 0.8$ cm) and 19 GHz ($\lambda = 1.6$ cm) to estimate SWE (Foster et al., 2005). In order to compensate for effects of grain size, the parameters of the retrieval algorithms need to be tuned to regional snow conditions (Derksen et al., 2003). Another option for compensating grain size effects is the assimilation of in situ snow measurements in the SWE processing line (Pulliainen, 2006). Due to the coarse resolution of the sensors and the saturation of the signal in deep snow, radiometric SWE retrievals are subject to major errors in mountain areas and forests.

Satellite-borne RS is widely applied for mapping the extent, surface topography, and motion of glaciers. For glacier mapping, spectral ratios in optical imagery are applied, similar to the techniques for snow mapping (Kargel et al., 2005; Paul et al., 2002). Manual post-processing is required to correct for debris-covered glacier surfaces. Stereo-optical satellite imagery (ASTER, SPOT-5) is applied to map surface topography, but the limited radiometric contrast reduces the accuracy in the snow areas (Berthier and Toutin, 2008). This problem can be overcome by radar interferometry (InSAR). Single-pass interferometry with two antennas on a platform, as on the Shuttle Radar Topography Mission (SRTM), avoids the problem of temporal decorrelation of the radar signal. The SRTM data set, acquired in February 2000, is the basis of a freely available DEM (90 m grid) covering the land surfaces between 60°N and 56°S (Rodriguez et al., 2005).

Repeat-pass SAR images enable the mapping of ice motion at high accuracy by means of differential processing techniques. Differential InSAR processing techniques are applied to separate the phase contributions of surface motion and topography (Hanssen, 2001). However, decorrelation of the radar phase due to snowfall, wind drift, or melt in the time interval between the image acquisitions severely limits the application of repeat-pass interferometry over snow and ice. A unique InSAR data set for glacier studies, less affected by decorrelation, was acquired during the concurrent ('tandem') operation of the satellites ERS-1 and ERS-2 in the years 1995–99 (Weydahl, 2001). During the tandem phase, the two satellites imaged the same swath on the Earth's surface at a time difference of 24 h.

If stable features are apparent on a glacier surface, image correlation techniques can be applied to map glacier motion from repeat-pass images of high-resolution optical sensors and SAR. This technique is less sensitive to motion than InSAR, but does not require phase coherence (Strozzi et al., 2002). The new very high resolution SAR systems of TerraSAR-X and COSMO-SkyMed are very useful for this application (Floricioiu et al., 2008).

2.14.5.3 Examples of Products and Applications

Satellite-derived products on snow and land ice have been widely used in research and are increasingly applied also for operational applications in hydrology and water management. **Table 2** provides an overview on some key snow and land ice products, including a few links to sample data sets.

Medium-resolution optical sensors (e.g., AVHRR on NOAA and MODIS on the *Terra* and *Aqua* platforms) are the main data sources for snow mapping at national to global scales. These data are widely applied for studies in climate research and hydrology of snow-covered regions (e.g., Brown et al., 2008; Pu et al., 2007; Rodell and Houser, 2004; Shamir and Georgakakos, 2006). An example for a snow map derived from MODIS data is shown in **Figure 12** for the Ötztal basin in the Austrian Alps. The inset shows the area-altitude distribution of the snow cover, which is used as input to a semi-distributed model for simulating and forecasting snowmelt runoff (Nagler et al., 2008). Daily snow maps are often rather fragmentary due to cloud cover, so that for some applications (e.g., climate studies) weekly composites are preferred. MODIS daily snow maps, 8-day composites, and monthly fractional snow cover can be found on the Internet. Sensors at higher spatial and/or spectral resolution are used for regional studies of snow physical properties and snow distribution (Dozier and Painter, 2004; Molotoch, 2009), but usually lack the temporal sequence required for real-time runoff forecasting applications.

SAR data are used for regional snow mapping, with emphasis on snow depletion during the melt period, exploiting the sensitivity of the sensors for detecting melting snow. Preferably, SAR data of the wide swath mode (ScanSAR) are used, providing a swath width of 400 km (Envisat ASAR) and 500 km (Radarsat) (Luojus et al., 2007; Nagler and Rott, 2005). SAR-derived snow maps are applied for snowmelt runoff modeling and forecasting (Nagler et al., 2008), for snow cover modeling linked to regional meteorological models (Longépé et al., 2009), and for climate studies.

Global maps of snow depth and water equivalent, derived from satellite-borne multichannel MW radiometer data reaching back to 1979, are available for climate studies (Foster and Chang, 1993). However, in many regions, the data show systematic differences to *in situ* measurements, requiring further improvement of retrieval algorithms (Foster et al., 2005). In western Canada, weekly SWE maps retrieved from satellite MW radiometer data are produced on an operational basis since the 1980s (Derksen et al., 2003). An example of such a product for the Canadian Prairies is shown in **Figure 13**. Because the retrieval parameters are tuned for regional snow morphology, these SWE maps provide better accuracy than the global products.

EO satellite data are widely applied for compiling and updating glacier inventories and provide key input data for models of glacier mass balance, hydrology, and ice dynamics. The main satellite products for glacier research and monitoring applications are maps of glacier area, topography, surface velocity, diagenetic facies, and albedo. The Global Land Ice Measurements from Space (GLIMS) initiative is aimed at compiling a global data base of glacier outlines in digital format from optical satellite data (Raup et al., 2007). The database, available to the public, includes satellite image glacier maps, vector outlines and related metadata, and, optionally, also snow lines, center flow lines, hypsometry data, and surface velocity fields. Observations of the temporal evolution of the snowline during the ablation period are used as input for modeling glacier mass balance and runoff (Rott et al., 2008). Satellite observations of changes in ice surface elevation and ice fluxes are also very relevant to mass balance studies (Bamber and Rivera, 2007). SAR interferometry is an

Table 2 Overview on selected snow and land ice products derived from satellite observations

Product type	Sensor type	Spatial resolution (typical range)	Sensors (examples)	Selected data sets
Snow area (total)	Multispectral optical imager	30 m–1 km	Modis, avhrr, landsat	Global snow area: http://modis-snow-ice.gsfc.nasa.gov/
Snow area (melting)	SAR, scatterometer	30–100 m	Asar, radarsat	
Snow albedo	Multispectral optical imager	250 m–1 km	Modis, meris	http://www-modis.bu.edu/brdf/userguide/albedo.html
Snow water equivalent	Imaging microwave radiometer	25 km	Ssm/i, amsr	Amsr-e/aqua daily l3 global snow water equivalent: http://www.nsidc.org/data/ae_dysno.html
Lake and river ice extent and concentration	Multispectral optical imager, SAR	30–250 m	Modis, asar, radarsat	http://www.polarview.org/services/lim.htm http://www.polarview.org/services/rim.htm
Glacier outlines	Multispectral optical imager	5–30 m	Spot hrv, aster, landsat	Glims glacier data base http://nsidc.org/glims/
Glacier surface topography	Interferometric SAR, optical stereo imager	10–100 m	Spot hrv, aster, srtm, asar	SRTM data products: http://www2.jpl.nasa.gov/srtm/cbanddataproducts.html
Glacier motion	SAR, optical imager	3–30 m	ASAR, radarsat, terrasar-X, optical	

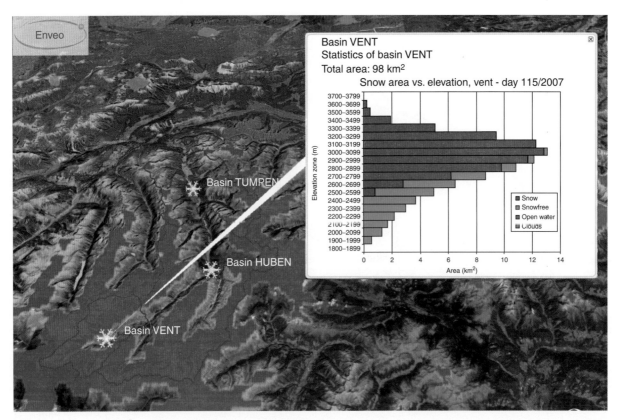

Figure 12 Snow map of the Ötzal Alps, Austria, derived from MODIS image data, 25 April 2007. Superimposed to Google Earth. (Inset) Area-altitude distribution of snow-covered and snow-free surfaces in the sub-basin Vent.

important tool for studying glacier hydraulics (Magnússon et al., 2007) and provides detailed maps of ice flow, which can be used for estimating the ice export of calving glaciers (Stuefer et al., 2007).

2.14.5.4 Future Research Needs

Currently available satellite missions and sensors are providing important information on the extent and physical properties of snow and ice from local to regional and global scales. This potential has been utilized so far mainly for dedicated research studies in the fields of water balance and hydrology, surface energy fluxes, land surface processes, and Earth surface/atmosphere interactions. However, the potential of remotely sensed cryosphere data for process modeling has so far been rarely exploited. Fostering the use of spatially distributed snow data requires further advancements of data assimilation techniques, a topic that has gained in importance over the last years (e.g., Clark et al., 2006; Kolberg et al., 2006; Nagler et al., 2008; Rodell and Houser, 2004; Slater and Clark, 2006).

Regarding sensors and satellites, a large variety of imaging sensors in the optical and MW spectral range is available, many of which can be employed for snow and ice observations. However, many sensors lack continuity, which represents an obstacle for operational use in hydrology and water management. New initiatives will provide better continuity of observations, such as the *Sentinel* satellites within the GMES initiative of the ESA and the European Union. A major observational deficit is the lack of a sensor for spatially detailed observations of the snow mass (SWE), a key parameter of the water balance. The feasibility of a satellite mission for SWE mapping with dual frequency (X-band and Ku-band) SAR is presently studied by ESA (Kern et al., 2008).

2.14.6 Water on the Land – Surface Water, River Flows, and Wetlands (Altimetry)

2.14.6.1 Introduction

Terrestrial surface water is absolutely essential to life, economies, environment, climate, and weather. Both national and local economies rely on flowing rivers to transport storm waters, sewage, and other effluents away from cities besides offering major shipping lanes to inland areas. The ecologies of wetlands and floodplains depend on surface water flows to deliver nutrients and to exchange carbon and sediments. Surface waters play a role in global climate through energy and water mass exchange with the lower atmosphere. Moreover, local weather is strongly affected by the surface area of nearby water bodies. Runoff is a strong indicator of accumulated precipitation throughout a watershed, and large, periodically flooded wetlands provide vast surfaces for evaporation as well as water storage. Earth's 6 billion people critically rely upon surface water availability for domestic use, agriculture, and industry, while human health is impacted by waterborne diseases (e.g., disease-vector-related malaria). National defense issues are related to surface water, particularly via politically charged water impoundment projects. The global

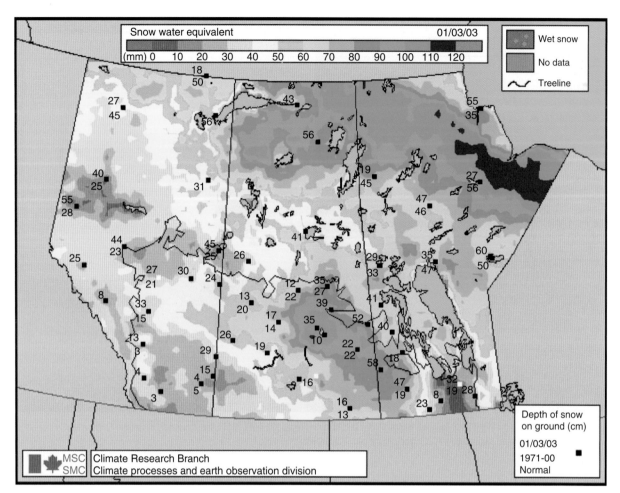

Figure 13 Map of snow water equivalent (color coded) over the Canadian Prairies, derived from satellite microwave radiometer measurements. The numbers refer to SWE measured at snow stations. Courtesy: Meteorological Service of Canada.

water issues will have large effects on many of the world's major decisions in the next decades and will require operational monitoring tools to support water policies. The following sections provide a review of the surface-water measurements available on the ground, onboard satellites, and a description of the future satellite mission Surface Water and Ocean Topography (SWOT), first satellite mission dedicated to the hydrology of continental surface water.

2.14.6.2 *In Situ* Measurements

In situ gauging networks have been installed for several decades in many river basins, distributed nonuniformly throughout the world. *In situ* measurements provide time series of water levels and discharge rates, which are used for studies of regional climate variability as well as for socioeconomic applications (e.g., water resources allocation, navigation, land use, infrastructures, hydroelectric energy, and flood hazards) and environmental studies (rivers, lakes, wetlands, and floodplain ecohydrology). *In situ* methods are essentially a one-dimensional, point-based sampling of the water surface that relies on well-defined channel boundaries to confine the flow. Yet, water flow and storage changes across wetlands and floodplains are spatially complex with both vast diffusive flows and narrow confined hydraulics. This complexity is fundamentally a three-dimensional process varying in space and time, which cannot be adequately sampled with one-dimensional approaches. In addition, gauging stations are scarce or even absent in parts of large river basins due to geographical, political, or economic limitations. For example, over 20% of the freshwater discharge to the Arctic Ocean is ungauged and surface water across much of Africa and portions of the Arctic either is not measured or has experienced the loss of over two-thirds of the gauges (Stokstad, 1999). Therefore, our ability to measure, monitor, and forecast global supplies of freshwater using *in situ* methods is essentially impossible because of (1) the decline in the numbers of gauges worldwide (Vörösmarty *et al.*, 2001), (2) the poor economic and infrastructure problems that exist for nonindustrialized nations, and (3) the physics of water flow across vast lowlands.

2.14.6.3 RS Techniques

During the past decade, RS techniques (satellite altimetry, radar and optical imagery, active and passive MW techniques, InSAR, and space gravimetry) have been used to monitor some components of the water cycle in large river basins (Cazenave *et al.*, 2004). Radar altimetry, in particular, has been used extensively in the recent years to monitor water levels of lakes, rivers, inland seas, floodplains, and wetlands (e.g., Birkett, 1995, 1998; Birkett *et al.*, 2002; Mercier *et al.*, 2002, Maheu *et al.*, 2003; Kouraev *et al.*, 2004). A few examples of altimetry-derived water level time series over rivers are presented in Figure 14.

Nadir-viewing altimetry has a number of limitations over land because radar waveforms (e.g., raw radar altimetry echoes after reflection on the land surface) are complex and multi-peaked due to interfering reflections from water, vegetation canopy, and rough topography. These effects result in less valid data than over oceans. Systematic reprocessing of raw radar waveforms with optimized algorithms provides decade-long time series of terrestrial water levels, at least over large (>1 km width) rivers. Repeat-pass SAR interferometry has been shown to offer important information about floodplains in measuring small water-level changes (Alsdorf *et al.*, 2000). Poor temporal resolutions are associated with repeat-pass interferometric SAR. Off-nadir single-pass interferometric SAR does not work over open water; instead, it requires special hydro-geomorphologies of flooded vegetation (Alsdorf *et al.*, 2000; Lu *et al.*, 2005; Kim *et al.*, 2005).

Optical sensors are used to provide estimates of surface water extent under favorable conditions when there are few or no clouds. The GRACE gravimetry mission provides estimates of water volume but its resolution, on the order of 400 km, is poor (Tapley *et al.*, 2004). Although the Shuttle Radar Topography Mission (SRTM) produced a high spatial resolution image of heights, the errors over water surfaces are quite large and the mission was active for a sampling period of only 11 days in February 2000, preventing temporal change studies (e.g., ±5.5 m; LeFavour and Alsdorf, 2005).

2.14.6.4 Validation and Synergy of RS Techniques

Surface water levels estimated from conventional nadir altimetry have been compared to those obtained from *in situ* gauges located along the satellite tracks and in the proximity of the altimeter swath over many of the largest river basins. The rms differences between *in situ* and altimetry-derived water levels have been computed and are usually in the order of a few to several tens of centimeters (Kouraev *et al.*, 2004).

Combining nadir altimetry-derived water levels with satellite imagery provides a new method for remotely measuring surface water volumes over large floodplains. Figure 15 shows an example of the interannual surface water volume signal variability obtained with a combination of altimetry and NDVI data from the SPOT-4/Vegetation instrument over the lower Mekong River Basin compared with the GRACE signal (black) that integrates surface and underground water

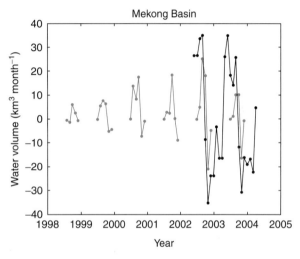

Figure 15 Variation of surface-water volume from ERS-2/ENVISAT altimetry and SPOT-VGT imagery (green) and total land water volume from GRACE (black) (Frappart *et al.*, 2006).

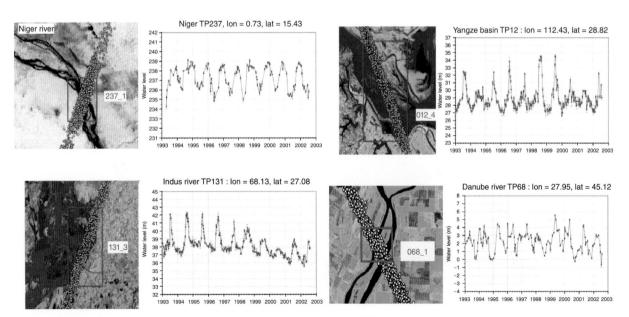

Figure 14 Water-level time series over the Niger (upper panel; left), Yangtze (upper panel; right), Indus (lower panel; left), and Danube (lower panel; right) based on *Topex–Poseidon* altimetry. From http://www.legos.obs-mip.fr/en/soa/hydrologie/hydroweb/.

(Frappart *et al.*, 2006). In the near future, when GRACE observations improve in terms of geographical resolution, it will be possible to estimate change in water volumes stored in soil and underground reservoirs by using in synergy GRACE, altimetry, and imagery data.

2.14.6.5 Availability of the Satellite Data Sets

A recently developed water-level database for major rivers, lakes, and wetlands using altimetry measurements from *Topex/Poseidon*, *Jason-1*, *ERS-2*, *ENVISAT*, and *GFO* satellites can be accessed through the Internet. The database includes water levels for over 130 lakes and man-made reservoirs, 250 virtual stations on rivers, and about 100 sites on flooded areas. The time series are regularly updated and the number of sites increases regularly. Users have access to associated errors.

For optical sensors, several databases are available through the web. For instance, the SPOT VEGETATION products can be found on the website where the NDVI products are available; for the MERIS instrument, the data can be found on the Internet.

2.14.6.6 SWOT: The Future Satellite Mission Dedicated to Surface Hydrology

The currently operating radar altimeters built to sample the surface of the open ocean miss numerous water bodies between orbital tracks. Optical sensors cannot penetrate the canopy of inundated vegetation and typically fail to image water surfaces when clouds or smoke is present (e.g., Smith, 1997). The prevalent vegetation and atmospheric conditions in the tropics lead to much reduced performances for technologies operating in and near the optical spectrum. Hydraulic measurements with repetitive global coverage of the continental surface water are needed to accurately model the water cycle and to guide water management (Alsdorf *et al.*, 2003; Alsdorf and Lettenmaier, 2003).

The future satellite mission SWOT dedicated to continental surface hydrology in cooperation between NASA and CNES will contribute to a fundamental understanding of the global water cycle by providing for the first time global measurements of terrestrial surface water storage changes and discharge, which are critical for present and future climate modeling (Mognard and Alsdorf, 2006; Alsdorf *et al.*, 2007; Mognard *et al.*, 2007; Fu *et al.*, 2009). The Ka-band Radar Interferometer (KaRIN) (**Figure 16**) is the technology capable of supplying the required imaging capability of water level (h) with global coverage at least twice every 21 days. KaRIN has two Ka-band SAR antennae at opposite ends of a 10-m boom with both antennae transmitting and receiving the emitted radar pulses along both sides of the orbital track. Look angles are limited to less than 4.5° providing a 120-km-wide swath. The 200-MHz bandwidth achieves cross-track ground

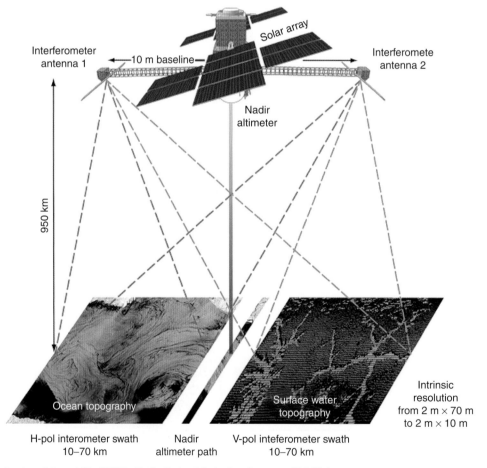

Figure 16 Artistic view of the satellite SWOT with the Ka-band Radar Interferometer (KaRIN) instrument.

resolutions varying from about 10 m in the far swath to about 60 m in the near swath. A resolution of about 2 m in the along track direction is derived by means of synthetic aperture processing.

SWOT will contribute to a fundamental understanding of the global water cycle by providing global measurements of terrestrial surface-water storage changes and discharge, which are critical for present and future climate modeling. SWOT will facilitate societal needs by (1) improving our understanding of flood hazards by measuring flood waves and water elevations, which are critical for hydrodynamic models; (2) freely providing water volume information to countries that critically rely on rivers that cross political borders; and (3) mapping the variations in water bodies that contribute to disease vectors (e.g., malaria).

2.14.7 Water in the Ground – Soil Moisture

2.14.7.1 Introduction

Soil moisture is defined as the amount of water in the rooting zone, or any other depth in the unsaturated zone and is usually expressed in volumetric percentage (Hillel, 1998). It is a variable that has always been required in many disciplinary and cross-cutting scientific and operational applications such as numerical weather prediction, ecology, biogeochemical cycles, flood forecasting, etc. (Jackson et al., 1999). With increasing evidence of climate change, it becomes even more urgent to be able to elucidate the critical role of soil moisture. Unfortunately, soil moisture is notoriously difficult to observe at large (landscape to global) scale due to its high spatial and temporal variability. Most of our limited understanding of the role of soil moisture in meteorology, hydrology, ecology, and biogeochemistry has been developed from point to field-scale studies, where the emphasis was typically on the variation of soil moisture with depth. Our failure to translate this small-scale understanding to natural landscapes can be attributed largely to our lack of understanding of soil moisture variability at larger spatial scales. As a parallel consequence, most models have been designed around the available point data and do not reflect spatial variability (Leese et al., 2000).

The potential to use MW RS for measuring soil moisture has been recognized early (Eagleman and Ulaby, 1975). The theoretical basis for measuring soil moisture at MW frequencies lies in the large contrast between the dielectric properties of liquid water and dry soil material. The large dielectric constant of water is the result of the water molecule's alignment of its permanent electric dipole in response to an applied electromagnetic field. Therefore, when water is added to the soil matrix, the effective dielectric constant of the soil increases strongly (Hipp, 1974). As the emission and scattering properties of the soil are strongly influenced by the soil dielectric constant, both active and passive MW measurements are highly sensitive to soil moisture (Ulaby, 1976; Schmugge et al., 1974).

Methodological problems, lack of validation, and limitations in computing have frequently delayed the research process to retrieve soil moisture from space observations (Wagner et al., 2007). But research in these fields evolved, resulting in several global-and continental-scale soil moisture data sets (e.g., (Wagner et al., 2003; Owe et al., 2008; Njoku et al., 2003). This section gives a brief overview of the state of science on satellite soil moisture.

2.14.7.2 State of the Art

Since the early days, satellite RS was seen as a potential tool to provide spatial and temporal continuous soil moisture measurements (Engman and Chaunhan, 1995). In particular, MW sensors are attractive because they can acquire imagery day and night unimpeded by cloud cover. However, even more important is the fact that many MW sensors are operated at frequencies below the relaxation frequency of water (9–17 GHz, depending on temperature) where the dielectric constant of soil changes strongly with the soil moisture content. For example, at 1.4 GHz, the dielectric constant of dry soil is around 3, while it is around 20–25 for a wet soil depending on soil texture (Wang and Schmugge, 1980). Given the strong effect of the soil dielectric properties on the emission and scattering of electromagnetic waves, both passive and active MW sensors provide a relatively direct means for assessing soil moisture when the soil is not frozen or snow covered. Further, sensors operating in the VIS and IR parts of the electromagnetic spectrum have been used for mapping soil moisture (Verstraeten et al., 2008). These methods use remotely sensed surface variables such as surface temperature or vegetation to constrain the surface energy and water balances to indirectly infer soil moisture. These methods essentially belong to the group of methods used for estimating evaporation and are hence discussed elsewhere.

Active MW sensors used for soil moisture retrieval include synthetic aperture radars (SARs) for local- to regional-scale mapping and scatterometers for global monitoring (GM). These instruments transmit an electromagnetic pulse and measure the energy scattered back from the Earth's surface. On the other hand, passive MW sensors (radiometers) merely record the radiation emitted by the Earth surface itself, which is related to the physical temperature of the emitting layer and the emissivity of the surface (Ulaby et al., 1981). Even though one might expect that active and passive sensors observe very different surface properties due to their different measurements principles, several land surface parameters, such as soil moisture, surface roughness, or vegetation biomass, have a comparable impact on both active and passive measurements. The fundamental reason for this is Kirchhoff's law which, applied to the problem of RS of the Earth's surface, states that the emissivity is one minus the hemisphere integrated reflectivity (Schanda, 1986). Therefore, soil moisture observed by active or passive sensors can be directly compared, particularly when the sensors are operated at the same frequency.

The basic challenge for both active and passive soil moisture retrieval methods is that other surface variables, such as vegetation water content, vegetation structure, and surface roughness, also have a strong impact on the MW signal. Therefore, successful retrieval methods must be able to account for all these confounding land surface parameters. This might suggest that one should use models that describe the interaction of the MWs with the Earth's surface in as much details as possible. Yet, such models become very complex and it is in general not possible to invert them. Even more problematic is that one generally does not have enough experimental observations to falsify complex models, simply because different model structures and parameter sets may explain the observations equally well (Wagner *et al.*, 2009). This so-called equifinality problem (Beven, 2001) is possibly the major reason why it is often not possible to transfer algorithms calibrated over one region to another. These considerations show that it is essential to develop models that capture the main physical phenomena, yet be simple enough to allow falsification and inversion. This implies that models may differ depending on the spatial resolution of the satellite system, because the dominant processes often change with scale.

Equally important as the retrieval algorithm is the selection of MW instruments. A suitable sensor exhibits a high sensitivity to soil moisture while minimizing instrument noise and the perturbing impacts of other surface variables on the measured signal (Wagner *et al.*, 2007a). Many RS studies conducted in the 1970s, 1980s, and 1990s indicated that low-frequency MW radiometers should offer the best performance because of the minimal influence of surface roughness and vegetation on these measurements (Jackson *et al.*, 1999). Therefore, the first satellite mission dedicated to measuring soil moisture on a global scale uses an MW radiometer operated at a frequency of 1.4 GHz (L-band), that is, the Soil Moisture and Ocean Salinity (SMOS) launched on November 2009 (**Figure 17**). To improve its spatial resolution, SMOS uses a passive interferometric design inspired from the very large baseline antenna concept in radio astronomy (Kerr *et al.*, 2001). Yet, its spatial resolution will only be in the order of about 40 km, which limits its use to large-scale applications such as numerical weather prediction or climate change. To enlarge the number of potential applications, the Soil Moisture Active/Passive (SMAP) mission foreseen for launch in 2014 uses both active radar and passive radiometer instruments at L-band. It will use a 6-m large rotating mesh antenna shared by the radar and radiometer to cover a 1000-km-wide swath (**Figure 17**). Thus, SMAP will offer a 40-km soil moisture product derived from its passive observations and a 10-km product derived from the combined active and passive observations (Entekhabi *et al.*, 2010).

SMOS and SMAP employ novel measurement concepts with the goal to measure soil moisture with unprecedented accuracy, and also existing MW sensors operated at frequencies below about 10 GHz can provide valuable soil moisture information. Particularly, in recent years, several soil moisture data sets derived from both active and passive MW sensors have become freely available, which demonstrate the advances made in algorithmic research. Wagner *et al.* (2007) suggested that this initially less visible revolution became possible, thanks to the increasing availability of computer power, disk space, and powerful programming languages at affordable costs. This has allowed more students and researchers to develop and test algorithms on regional to global scales, which lead to a greater diversity of methods and, consequently, to more successful algorithms.

2.14.7.3 Data Sets BBB

The soil moisture data sets described in this section are all available for user download via file transfer protocol (FTP) or web portals. **Table 3** gives an overview of the different products. Most of these data sets have a rather coarse spatial resolution in the order of 20–50 km because they are derived from MW radiometers or scatterometers. In addition, a first continental-scale 1-km soil moisture data set derived from ENVISAT Advanced Synthetic Aperture Radar (ASAR) operated in GM mode has recently been published.

2.14.7.3.1 Active MW data sets

Investigations into the potential of active MW sensors for soil moisture retrieval began already in the 1960s and gained momentum in the 1990s due to the launch of several satellites that carried a synthetic aperture radar (SAR) on board. Unfortunately, there is still no widely accepted method that delivers SAR-derived soil moisture data at fine spatial scales (10–100 m). This is to some extent surprising given that a large number of backscatter models and retrieval approaches were proposed and successfully applied within pilot studies (Dubois *et al.*, 1995; Zribi *et al.*, 2005). Unfortunately, independent testing and transferring of the methods to other

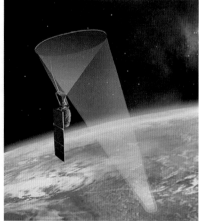

Figure 17 Artist impressions of ESAs Soil Moisture and Ocean Salinity Mission (left) and NASAs Soil Moisture Active Passive Mission (right). Both satellites will be used for global soil moisture mapping. (Left) Image courtesy: ESA. (Right) image courtesy NASA.

Table 3 A short list of accessible satellite derived soil moisture products using active and passive microwave instruments

Product name	Satellite	Spatial resolution	Temporal resolution	Period	Url	Reference
NSIDC L3 soil moisture	Amsr-e	0.25°	Sub daily	2002–now	http://nsidc.org/	Njoku et al. (2003)
LPRM soil moisture	Amsr-e, trmm-tmi, ssm/i, smmr	0.25°	Sub daily	1978–now	http://geoservices.falw.vu.nl	Owe et al. (2008)
Windsat soil moisture	Windsat	0.25°	Daily	2003–now	http://www.nrl.navy.mil/windsat/	Li et al. (2009)
Scat	Ers-1, ers-2	50 km	~ 6 days	1991–now	http://www.ipf.tuwien.ac.at/radar/	Wagner et al. (2003)
Ascat	Metop	25/50 km	Daily	2006–now	http://www.eumetsat.int/	Bartalis et al. (2007)
Asar	Envisat	1–5 km	Weekly	2005–now	http://www.ipf.tuwien.ac.at/radar/	Pathe et al. (2009)

regions or data sets often did not yield the hoped-for results (Walker et al., 2004). The major problem appears to be the failure to accurately model surface roughness and vegetation effects at fine spatial scales (Verhoest et al., 2008), besides the technical characteristics of most SARs (revisit time, frequency, etc.) are not well suited for the task of soil moisture monitoring.

Parallel to the work on SAR, some research groups started to investigate the potential of the ERS scatterometer for land applications in the 1990s (Pulliainen et al., 1998; Woodhouse and Hoekman, 2000). Despite scatterometers were designed for monitoring winds over the oceans, these studies quickly demonstrated the potential of the ERS scatterometer for soil moisture monitoring at a 50-km scale (Wagner et al., 1999; Wen and Su, 2003). From an algorithmic point of view, the advantage of working at a scale to 50 km is that surface roughness and land cover can reasonably be assumed to be constant. The major technical benefits of the ERS scatterometer are its short revisit time and its high radiometric accuracy. In addition, its three antennas acquire three quasi-instantaneous backscatter measurements from different azimuth and incidence angles, which is important for separating vegetation and soil moisture effects on the signal.

The first global soil moisture data set was derived from ERS scatterometer data using a change detection algorithm (Wagner et al., 2003). It was released in 2003 and has since then been used in several validation and application studies (e.g. Scipal et al., 2008). Using the same algorithm, the European Organisation for the Exploitation of Meteorological Satellites (EUMETSAT) has developed the first operational, near-real-time soil moisture monitoring system based upon the Advanced Scatterometer (ASCAT) flown on board of the Meteorological Operational (METOP) satellite series. ASCAT is the successor instrument of the ERS scatterometer and offers a twofold improved temporal and spatial resolution (Bartalis et al., 2007).

The change-detection algorithm developed for the scatterometer has been adapted to 1-km GM mode data as acquired by the Advanced Synthetic Aperture (ASAR) on board of ENVISAT (Pathe et al., 2009). This particular SAR mode has a rather poor radiometric resolution, but requires less energy as high-resolution SAR modes. Thus, it offers a good temporal coverage suitable for studying soil moisture dynamics.

2.14.7.3.2 Passive MW data sets

In the passive domain, soil moisture research already started in the 1970s and one of the first soil moisture retrieval algorithms was developed by Njoku and Kong (1977). This algorithm used a simple regression technique on multifrequency MW observations to obtain soil moisture from a controlled bare soil site. In time, this modeling approach started to become more complex with the addition of a surface roughness module (Choudhury et al., 1979; Wang and Choudhury, 1981; Wigneron et al., 2001), a vegetation module (Kirdiashev et al., 1979; Meesters et al., 2005), and a dielectric mixing module to convert the soil dielectric properties to soil moisture (Wang and Schmugge, 1980; Dobson et al., 1985; Mironov et al., 2004).

On a later stage, an atmosphere module (Pellarin et al., 2003; Liebe, 2004) and snow module (Pulliainen et al., 1999) were introduced to obtain a better description of the MW emission as measured by the satellite.

Most of the global soil moisture data sets from passive MW observations are based on a selection of the given modules and the differences between the different products vary on the choice of modules. In this section, we describe the two most commonly used global soil moisture data sets.

The first global soil moisture product was developed by Njoku et al. (2003) and uses X-band AMSR-E MW observations to retrieve soil moisture. This model uses a multichannel iterative forward-model optimization method to solve simultaneously for surface temperature, soil moisture, and vegetation water content (Njoku et al., 2003). In the forward mode, the retrieval algorithm iteratively adjusts values of the retrieval parameters using Fresnel relations adjusted for surface roughness and attenuation by vegetation cover using time-invariant parameters based on land cover type (Njoku and Chan, 2006). The modeled brightness temperature is then compared to the observed at-sensor brightness temperature until an iterative least-squared minimized solution is obtained. Polarization ratios are used instead of absolute brightness temperature because these minimize the effects of surface temperature (Sahoo et al., 2008). The model uses the X-band frequency to minimize the effects of radio-frequency interference (RFI) on the at-sensor brightness temperature (Njoku et al., 2005). The final soil moisture data set is screened to remove data over large water bodies, dense vegetation, snow, and permanent ice. This product is distributed by the

National Snow and Ice Data Center (NSIDC) in EASE–GRID format with a nominal grid spacing of 25 km.

The second global soil moisture product was obtained from different satellites sensors, including Nimbus SMMR, TRMM TMI, SSM/I, and AMSR-E (Owe et al., 2008). It used the Land Parameter Retrieval Model (LPRM) to retrieve soil moisture from passive MW observations. The soil moisture retrievals from LPRM were based on the solution of an MW radiative transfer model and solved simultaneously for the surface soil moisture and vegetation optical depth without *a priori* information of land surface characteristics (Meesters et al., 2005). The flexible approach created the possibility to retrieve soil moisture from a variety of frequencies. LPRM produced volumetric (approximately in $m^3 m^{-3}$) soil moisture of approximately the first 1^{-2} cm for C-band MW observations. For X band, the penetration depth is a bit smaller, resulting in soil moisture values of the first centimeter. The data are distributed in a rectangular grid with a pixel spacing of $0.25°$.

2.14.7.4 Validation

Soil moisture products from passive and active MW satellite observations were extensively validated. In the absence of a homogeneous global soil moisture station network, the data sets were validated over regional networks (Ceballos et al., 2005; Wagner et al., 2007; Draper et al., 2009), intercompared (De Jeu et al., 2008; Rudiger et al., 2009; Mladenova et al., 2009), and evaluated against model data (Wagner et al., 2003). These studies found high correlations with *in situ* observations in semi-arid regions and somewhat lower correlations in agricultural areas.

On average, the current active and passive MW soil moisture products have an accuracy of about $0.06 m^3 m^{-3}$ for sparse-to-moderate vegetated regions (De Jeu et al., 2008). For denser vegetation classes such as forests, the soil moisture retrievals start to become less accurate and at an LAI of about 4, no reliable soil moisture can be retrieved from the current passive MW sensors (De Jeu, 2003). Nevertheless, recent assimilation studies have demonstrated the potential use of these existing data sets for the regions where they can obtain reliable soil moisture.

The assimilation of soil moisture observations from operational satellite systems was found to improve the model performance in agro-meteorology (de Wit and van Diepen, 2007), hydrology (Parajka et al., 2006), meteorology (Drusch et al., 2009; Zhao et al., 2006; Scipal et al., 2008), and climate (Liu et al., 2007; Loew et al., 2009).

With the anticipated launch of the new satellites with more innovative sensors and the continuous scientific movement in algorithm development, an improvement on the quality of satellite soil moisture is expected. Furthermore, the use of satellite soil moisture in environmental research is not yet fully exploited, and further research is necessary to fully demonstrate the potential of these new data sets.

2.14.8 Water in the Ground – Groundwater (Gravity Observations)

2.14.8.1 Introduction

Groundwater is vital for meeting agricultural, domestic, and industrial water needs, particularly in parts of the world where the climate or topography does not allow for a reliable supply of surface water. It is also by far the most abundant form of fresh, unfrozen water on the Earth (Shiklomanov, 1993). Groundwater storage does not vary as rapidly as soil moisture or surface water, but it does exhibit significant seasonal and interannual variability (Rodell and Famiglietti, 2001) and it is susceptible to overexploitation (Alley et al., 2002). The slow process of groundwater recharge acts like a low-pass filter on transient weather conditions, so that multiannual water-table fluctuations in a natural setting may be a useful indicator of climate variations. Hence, groundwater storage observations are valuable for both practical and scientific applications.

As with other water-cycle variables, monitoring groundwater storage at regional scales using *in situ* measurements is expensive and problematic, and at the global scale it is simply not feasible. RS has propelled global hydrology forward in the past 30 years, but because groundwater is hidden deep beneath the surface, it was the last component of the terrestrial water cycle to benefit from the technology. Near-surface stocks and fluxes of the water cycle can be inferred based on electromagnetic radiation (various wavelengths of light) emitted or reflected from the land surface and atmosphere. Satellites can only sense groundwater by the effect it has on Earth's time-varying gravity field. Redistributions of water and other forms of mass cause changes in gravitational potential, which is imperceptible to human beings yet strong enough to perturb satellite orbits. This is the concept behind one of the most innovative Earth-observing satellite systems yet launched, the Gravity Recovery and Climate Experiment (GRACE).

2.14.8.2 GRACE Data Processing

The primary goal of GRACE is to map the static and time-varying components of the Earth's gravity field with better spatial resolution and accuracy than ever before (Tapley et al., 2004). GRACE comprises two satellites in a tandem, near-polar orbit, approximately 200 km apart and 500 km above the Earth. As they orbit, a K-band MW ranging system continuously measures the distance between the two satellites, which is affected by heterogeneities in the Earth's gravity field. These measurements, along with precise location information, can be used to construct a mathematical model of the shape of the gravity field, nominally on a monthly basis. Each gravity field solution is delivered as a set of spherical harmonic coefficients, rather than a gridded map. Wahr et al. (1998) and Rowlands et al. (2005) described two of the techniques available for converting the GRACE gravity data to mass anomalies (deviations from the long-term mean field). Further, in order to isolate changes in terrestrial water storage mass (groundwater, soil moisture, snow and ice, surface water, and biomass) one must remove the effects of atmosphere and ocean circulations using atmospheric analysis and ocean model outputs. Glacial isostatic adjustment must also be considered in certain regions, and a major earthquake can produce a gravitational anomaly, but the timescales of most solid earth processes are too long to be an issue (Dickey et al., 1997). Because of the nature of the measurements, GRACE has no 'footprint' or pixel resolution. Rather, there is a tradeoff between resolution and accuracy, so that the effective limit of resolution for estimating changes in terrestrial water storage is approximately 160 000 km² (Rodell and Famiglietti, 1999; Rowlands et al., 2005; Swenson et al., 2006).

2.14.8.3 Retrievals of Groundwater Storage with GRACE Data

Despite its origins in the field of geodesy, GRACE's greatest contributions have been in the cryospheric and hydrologic sciences. GRACE has monitored the melting of the Greenland and Antarctic ice sheets as never before possible (Luthcke et al., 2006; Velicogna and Wahr, 2006) and quantified glacier melt in the Gulf of Alaska (Chen et al., 2006). GRACE terrestrial water-storage data have been used to constrain regional ET rates (Rodell et al., 2004), river discharge (Syed et al., 2005), soil moisture variations (Swenson et al., 2008), and surface-water-storage variations (Han et al., 2009), and to describe intercontinental teleconnections (Crowley et al., 2006). GRACE is also the first satellite system to observe regional scale variations in aquifer storage.

Isolating groundwater from GRACE-derived terrestrial water-storage data requires knowledge of the other water-storage variables, because gravimeters provide no indication of the sources or stratification of the mass changes affecting the time-variable gravity field. In polar and alpine regions, terrestrial water-storage variability is often dominated by changes in snow and ice (Niu et al., 2007). In humid tropical regions, such as the Amazon, surface water can be the major variable (Han et al., 2009). In the rest of the world, soil water typically exhibits the largest fluctuations on daily-to-seasonal timescales, whereas groundwater storage amplitudes can be as large or larger on seasonal and longer timescales (Rodell and Famiglietti, 2001). Biomass variations are near or below GRACE's limit of detectability (Rodell et al., 2005).

Following the approach suggested by Rodell and Famiglietti (2001), Yeh et al. (2006) and Rodell et al. (2007) demonstrated that groundwater storage variations can be isolated from GRACE terrestrial water-storage data using in situ root zone soil moisture observations or numerically modeled soil moisture fields. They verified their results using data from groundwater monitoring networks in Illinois and the Mississippi River Basin. Strassberg et al. (2007) achieved good results using the model-supported technique to estimate groundwater storage changes in the High Plains aquifer, likewise verified by monitoring well observations. Rodell et al. (2009) applied the technique to determine that groundwater beneath the Indian states of Rajasthan, Punjab, and Haryana (including Delhi) is being depleted at a rate of 17.7 km³ yr⁻¹ due to withdrawals for irrigation.

Zaitchik et al. (2008) presented a more sophisticated approach for disaggregating GRACE-derived terrestrial water storage into its components, whereby an ensemble Kalman smoother is used to assimilate the GRACE data into a numerical land surface model. This approach has several advantages. First, physical equations of hydrologic and energetic processes, integrated within the model, provide a basis for synthesizing GRACE and other relevant observations such as precipitation in a physically consistent manner. Second, the model fills spatial and temporal data gaps, while observations anchor the results in reality. Third, in addition to separating groundwater, soil moisture, and other component contributions, the assimilated output has much higher spatial and temporal resolutions than the original GRACE data. Zaitchik et al. validated the technique in the Mississippi River Basin using groundwater data from a network of wells, and showed significant improvement in both the timing and amplitude of modeled groundwater variations.

2.14.8.4 GRACE Data Access

GRACE gravity data are produced and distributed by three centers that support the mission: the University of Texas Center for Space Research, NASA's Jet Propulsion Laboratory (JPL), and the German Research Centre for Geosciences (GFZ). GRACE terrestrial water-storage products have been developed by many groups. Visualization and data portals include those provided by NASA/JPL, NASA/Goddard Space Flight Center (GSFC) and Stinger Ghaffarian Technologies, and the University of Colorado.

2.14.8.5 Concluding Remarks and Future Perspective

Although other RS data can provide clues as to the location and characteristics of aquifers (Becker, 2006), satellite gravimetry is the only technology currently available for measuring regional-scale groundwater storage changes from space. In addition to GRACE, two other advanced gravity-monitoring satellites have been launched: GFZ's Challenging Minisatellite Payload (CHAMP) in 2000 and the European Space Agency's Gravity Field and Steady-State Ocean Circulation Explorer (GOCE) in 2009. CHAMP was a major advance in gravimetry at the time of launch, but it was not accurate enough to infer water-storage changes, and it was quickly made obsolete by GRACE. GOCE will map the static gravity field with significantly higher spatial resolution than GRACE, but it is not well suited for monitoring the time-variable gravity field and inferring changes in groundwater storage (Han and Ditmar, 2008).

GRACE is in its extended mission phase, beyond its initial 5-year goal. It could potentially continue through 2012. NASA, ESA, and many independent reports (e.g., NRC, 2007) have recognized the importance of the data provided by GRACE and the need for a follow-on mission to enable continued monitoring of terrestrial water and ice as only satellite gravimetry can. Technology upgrades, such as a laser ranging system, a lower Earth orbit with drag-free propulsion, or more satellites and different orbital configurations, could increase the accuracy and spatial resolution of the products. However, at the time of writing, a next-generation time-variable gravity mission had not yet been approved.

2.14.9 Optical RS of Water Quality in Inland and Coastal Waters

2.14.9.1 Introduction

Inland and coastal waters are important natural resources yet they are seriously threatened by eutrophication, salinization, and heavy metal contamination. Excessive concentrations of suspended particulate matter (SPM) influence the productivity and thermodynamic stability of inland and coastal waters (Muller-Krager, 2005: 348). Traditional measurements of water quality are costly, time consuming, and are limited in their spatial and temporal coverage. EO data, on the other hand, provide rapid and repeated information over large and often inaccessible areas. EO, in conjunction with modeling and strategic *in situ* sampling, can play a crucial role in determining the current status of water-quality conditions and helps anticipate, mitigate, and even avoid future water catastrophes (DiGiacomo *et al.*, 2007).

The primary measurement of EO data over water is the visible light leaving the water column, hereafter called the water-leaving reflectance. In inland and coastal waters, this water-leaving reflectance is strongly affected by different materials, for example, terrigenous particulate and dissolved materials, resuspended sediment, or highly concentrated phytoplankton bloom. The majority of inland and coastal waters can therefore be classified as case 2 waters (Gordon and Morel, 1983). In case 2 waters, the constituents are independent of each other and do not covary with chlorophyll *a* as in case 1 waters.

RS of inland and coastal waters is quite challenging due to the complicated signals from turbid water, substrate reflectance, and adjacent land surfaces (**Figure 18**).

Consistent EO estimates of water-quality parameters in inland and coastal waters require three components: (1) a reliable atmospheric correction method; (2) an accurate retrieval algorithm; and (3) an objective method to estimate the uncertainty budget based on their sources.

Because of limitation in length, the scope of this chapter has been narrowed to confine some of the recent developments in each of the above-mentioned areas. Knowledge of the basic concepts of aquatic optics is assumed available.

2.14.9.2 Atmospheric Correction

Most of the atmospheric correction procedures fail over inland and coastal waters, that is, case 2 waters. The failure of atmospheric correction might be attributed to the complexity of the recorded reflectance. The standard approach by Gordon and Wang (1994), for example, assumes a zero water-leaving reflectance in the near-infrared (NIR). In case 2 waters, this water-leaving reflectance has distinctive values at the NIR part of the spectrum (Siegel *et al.*, 2000). The non-negligible value of water-leaving reflectance at the NIR was accounted by many authors (Carder *et al.*, 1999; Gould *et al.*, 1999; Ruddick *et al.*, 2000; Hu *et al.*, 2000; Salama *et al.*, 2004).

Coupled approaches are increasingly used to retrieve the optical properties of both water and atmosphere simultaneously. For each atmosphere–water setup, a TOA reflectance is simulated at variable viewing-illumination conditions. The parameters that define each media are tuned until the best convergence to the recorded reflectance is found (Chomko *et al.*, 2003; Stamnes *et al.*, 2003; Gordon *et al.*, 1997; Zhao and Nakajima, 1997).

However, most of these algorithms were developed for case 1 waters, that is, assuming known and spatially homogeneous water-leaving reflectance at the NIR. Newly developed algorithms are emerging for case 2 waters (Kuchinke *et al.*, 2009a, 2009b). The spectral optimization method (Kuchinke *et al.*,

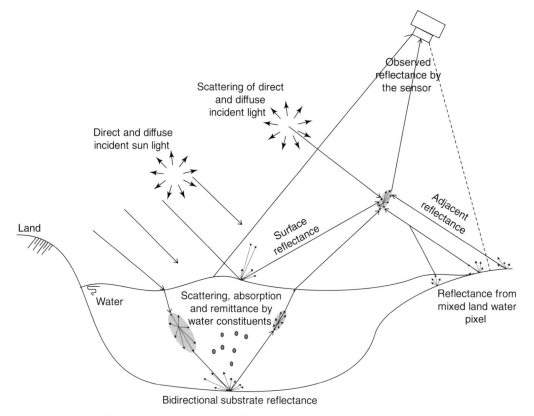

Figure 18 Schematic diagram of the different processes that contribute to the observed remote-sensing reflectance at a pixel size in inland and coastal waters.

2009b) was constrained to $0.1\,\mathrm{m\,m^{-1}}$ as a maximum value of backscattering coefficient of SPM at $0.443\,\mu m$. This value of backscattering is equivalent to $12\,\mathrm{g\,m^{-3}}$ concentration of suspended particles using the specific backscattering coefficient of Albert and Gege (2006). On the other hand, artificial neural network techniques (Doerffer and Schiller, 2007) are usually limited to the range of their training sets. Most of estuarine and coastal waters have high loads of SPM, exceeding $12\,\mathrm{g\,m^{-3}}$. For instance, the Yangtze estuarine water is extremely turbid with SPM concentration ranging between 80 and $500\,\mathrm{g\,m^{-3}}$ (Shen et al., 2010).

The spatial variabilities of the aerosol and water signals at the NIR part of the spectrum are characteristic features of turbid inland and coastal waters. These variabilities can be attributed to the different aerosol types that may coexist in this transaction zone as well as to the distinctive shape of the water-leaving reflectance. Salama and Shen (2009) proposed an analytical approach to consider and quantify this variability (**Figure 19**). Their method was validated with *in situ* measurements and successfully applied on data obtained from orbital ocean color and geostationary sensors. Deriving water-quality parameters from geostationary satellite in open coastal areas (Salama and Shen, 2009; Neukermans et al., 2008) is of unprecedented benefit. It will facilitate resolving the temporal dynamic of marine bio-geophysical parameters and overcome cloud covers.

2.14.9.3 Retrieval Algorithms

Most developed algorithms for water-quality retrievals in inland and coastal waters are empirical in nature. This empiricism limits their application to a specific range of concentrations, area, and season.

Kallio et al. (2001) studied different algorithms to estimate chlorophyll *a* in lakes. These algorithms were empirical and estimated one variable using band ratio of approximately 0.675 and $0.705\,\mu m$ (Dekker et al., 1992; Gitelson et al., 1993). A generalized retrieval algorithm is, however, hindered by the large natural variability of inland waters. Significant efforts on improving the accuracy of air- and space-borne-derived water-quality parameters are therefore required for inland and near-coastal waters. Many studies have used semi-analytical models to derive water-quality parameters in lakes (Hoogenboom et al., 1998; Gons et al., 2002). Derived water-quality parameters from multivariable inversion methods are ambiguous and not unique (Sydor et al., 2004). Other promising methods are used when the inversion employs lookup tables (Van Der Woerd and Pasterkamp, 2008). Salama et al. (2009) showed that the inversion is very sensitive to the spectral shape parameters of SPM backscattering and absorption of dissolved organic matter. Including these two parameters has enhanced the retrieval in inland waters (**Figure 20**).

2.14.9.4 Uncertainty Estimates

Reliable methods for uncertainty quantification of water-quality EO products are important for sensor and algorithm validation, assessment, and operational monitoring. High accuracy in both observations and algorithms may reduce considerable ranges of errors. EO-derived water-quality parameters have, however, an inherent stochastic component. This is due to the dynamic nature of water, intrinsic fluctuations, model approximations, correction schemes, and inversion methods. Quantitative measures of uncertainty support water quality and ocean-color product validation, especially with the introduction of the new AERONET-OC network (Zibordi, 2006). Due to stochasticity of the measurements, as well as model approximations and inversion ambiguity, the retrieved inherent optical properties (IOPs) are not the only possible set that caused the observed spectrum (Duarte et al., 2003; Sydor et al., 2004). Instead, many other sets of IOPs may be derived. Each of these sets has an unknown probability of being the derived ocean-color product. The probability distribution of the estimated IOPs provides, therefore, all the necessary information about the variability and uncertainties of derived water-quality parameters.

Several efforts have been carried out to resolve the uncertainty of the derived IOPs. Duarte et al. (2003) analyzed the sensitivity of the observed RS reflectance due to variable concentrations of water constituents. Salama (2003) proposed a stochastic technique to quantify and separate the source of errors of IOPs derived from hyperspectral airborne measurements. Maritorena and Siegel (2005) employed a nonlinear regression technique for consistent merging of different ocean-color-derived products. Wang et al. (2005) performed a detailed study on the uncertainties of ocean-color model inversion related to fluctuations in each of the IOPs and their spectral shape. In general, these studies used the method of Bates and Watts (1988) to construct the confidence interval around the derived IOPs following different approaches, however. It is adequate as long as model inversion has a well-conditioned Jacobian matrix of the minimum cost function. Recently, Salama and Stein (2009) developed a generic method to quantify the uncertainties in the derived water-quality products based on their sources, namely, model approximations, measurement noise, and atmosphere correction. The method was evaluated and validated using ocean-color data sets (**Figure 21**). The method has promising applications for inland and coastal water. Moreover, it provides vital input to SPM assimilation models (Eleveld et al., 2008) and EO product merging (Pottier et al., 2006).

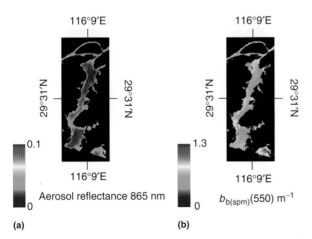

Figure 19 (a) Derived aerosol reflectance above the Poyang Lake. (b) Derived SPM concentration in the Poyang Lake. Notice that they are totally uncorrelated.

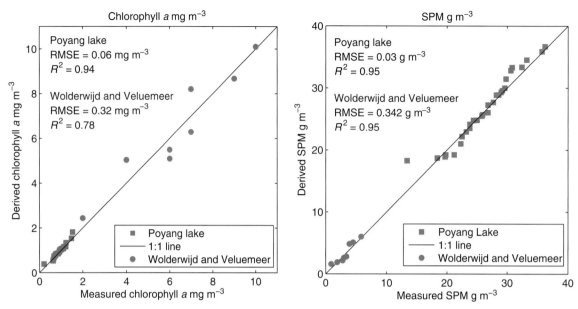

Figure 20 Results from remote-sensing inversion in inland waters. The spectral shape parameters were also derived from the inversion. (a) Derived chlorophyll a and (b) derived suspended particulate matter.

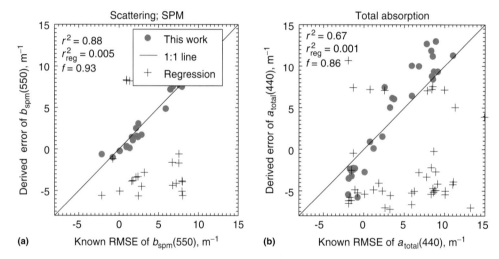

Figure 21 Estimated total error (dot symbols) on IOPs derived from SeaWiFS spectra of the Nomad data set. Nonlinear regression results are superimposed as plus symbols; r^2 and r^2_{reg} are the correlation coefficients for dot and plus symbols respectively; f is the fraction of successful retrievals. (a) Errors of scattering coefficient; (b) error of total absorption coefficient. From Salama MS and Stein A (2009) Error decomposition and estimation of inherent optical properties. *Applied Optics* 48: 4947–4962.

2.14.9.5 Concluding Remarks and Future Perspective

We have summarized the three requirements for reliable retrievals of water-quality parameters from RS data in inland and coastal waters. Although EO operational products are still under development, there are few issues that need extra attention in the future:

1. The red, NIR, and even shortwave IR bands, with sufficient signal-to-noise ratio, are necessary for RS of inland waters. They improve the accuracy of derived IOPs.
2. Improved parametrization of IOPs is needed for inland waters. The improvement should (a) account for different phytoplankton species and (b) deconvolve the overlapped absorptions at the blue.
3. Reliable methods to account for absorbing aerosol and adjacency effect in inland waters.
4. Studying the effects of climate change on water quality and a better understanding of the role of water quality of large inland lakes on the radiative energy budget on a subcatchment scale.

2.14.10 Water Use in Agro- and Ecosystems

2.14.10.1 Introduction

A comprehensive review of reflective (and partly thermal) RS techniques applied to agro-hydrology and ecological systems was given by Dorigo *et al.* (2007). Understanding the opportunistic nature of RS acquisitions, the traditional approach selected in most cases for the evaluation of RS-derived biophysical variables consists of the statistical comparison between field ancillary data and a corresponding RS data subset. The analysis ends by evaluating the strength of the correlation between these data sets.

Actual water use, transpired by the crops and evaporated from the soil, is the main output of a vast number of soil–vegetation–atmosphere transfer (SVAT) algorithms and surface energy balance (SEB) models. In the SVAT and SEB sequence, a great number of submodels are required to retrieve land properties from reflective optical measurable from RS instruments. As such, crop-water requirements from RS (AET-RS) processes are heavily demanding in terms of data input and modeling. Along with the image process and analysis, a dedicated number of intermediate products are elaborated which are simultaneously essential for many other eco-hydrological applications. We consider that a good overview of the use of water in agro- and ecosystem can be tackled by reviewing SVAT and SEB models.

Efforts focused on the use of AET-RS and its pre- and post-elaborated products are not only to improve water management in irrigated lands, but also associated with this, in irrigation planning, and irrigation monitoring, leading to performance indicators (Bos et al., 2005), water competition and water strategy at basin level (Bos et al., 2009), soil moisture retrievals (Wang and Qu, 2009), and several other categories of hydrological modeling benefiting from this approach.

AET-RS estimation opens the essential spatial dimension to a diversity of agro-ecological models on the one hand, and welcomes continuous model output to cover the typical temporal gaps in RS imagery. Increasingly, data assimilation techniques are used to integrate RS information into continuous modeling with success (Loew, 2007).

In this section, the topic of AET-RS products and the suggested link to FAO crop factors and irrigation (FAO, Food and Agriculture Organization; Allen et al., 1998) is selected as an example of a variety of opportunities that congregates much of the knowledge of RS in agro- and ecosystem.

2.14.10.2 Continuous Evaluation of Crop Water Use with Support from RS

Despite the intensive research in the field of RS to evaluate energy fluxes on an instantaneous basis, setting an operational scheme for practical agronomical purposes proves more troublesome. The techniques to elaborate energy flux maps from RS are temporally intermittent, as ephemeral as the image opportunity. They depend on the availability of the image that is affected by spatial resolution (nadir and off-nadir views), revisiting time, and, mainly, cloudiness. High-resolution thermal sensors are less available today (2009) than it was in the past, probably after the failure in establishing operational sequences for use in the market. As such, low-resolution thermal sensors are the only available RS source with adequate temporal revisiting time. The spatial resolution is, at the same time, the main limitation of these sensors for applications at field scale.

The elaboration of a single flux map requires the highest level of expertise and continuous upgrading. Updated preprocessing including sensor calibration, atmospheric correction, narrow-to-broadband conversions, emissivity retrievals, angular-geometrical effects, and instantaneous-to-daily integration are some of the necessary steps to be performed. The processing time is much slower than the dynamics of the ET process, such that this might be the cause of the few operational efforts beyond the framework of international projects.

The selected criteria for classification of the approaches to evaluate AET-RS regarding the FAO Kc (crop factor) link is based on the relative weight that the RS and modeling components have on the final product. In view of the extensive bibliography available, only a few main references are indicated here.

Continuous monitoring of the crop-water requirements using RS images exclusively can be attempted in areas where both clear days and image coverage are frequent. Daily AET is estimated from SEB models (see Section 2.14.4) on the available cloud-free images. Auxiliary ground data are collected from meteorological stations on daily (or shorter) basis to complement the database required for the SEB.

The net radiation 'Rn' is calculated on a daily basis from a combination of RS and meteorological stations (Hurtado and Sobrino, 2001) or from ground stations when images are not available (Allen et al., 1998; Irmak et al., 2003).

For the days when images are available, surface albedo (Liang et al., 2003), surface temperature (sensor dependent) (Coll et al., 1994, 2005; Gillespie et al., 1998), emissivity (Valor and Caselles, 1996), and fractional vegetation coverage (Su, 2002) can be evaluated. Considering the dynamic behavior of these parameters in the diurnal cycle, the intention is to describe a continuous pixel-based temporal evolution of them from day to day, as good as possible. If the daily values of each parameter are known on a daily basis, a 'potential image' of these parameters is obtained. After that, the evaporative fraction 'EFi' can be evaluated (Su, 2002). This method also considers the evaporative fraction being conservative during the daily course (Brutsaert and Sugita, 1992), providing a method to scale instantaneous evaporative fraction to daily evaporative fraction. However, several authors have shown the variability of the evaporative fraction during the day (Chehbouni et al., 2008; Crago, 1996) to the point that the original hypothesis stated by the one-time AET-RS must be reviewed for the particular situation considered.

The soil heat flux averaged over the day is usually assumed zero (Brutsaert, 2008), so the AET for any pixel at any day 'i' is calculated as: $Rn_i * EF_i$.

As AET and potential evapotranspiration (PET) are obtained on a daily basis, direct application to irrigation becomes possible. The approach is suited for irrigated lands in areas of data scarcity where clear skies are common as shown in a case study in Morocco (Jacobs et al., 2008).

A second approach estimates AET replacing the Penman–Monteith method with the Priesley and Taylor (PT) method that has proved to be reasonable for wet irrigated areas (de Bruin, 1987; McNaughton and Spriggs, 1989). Under these circumstances, the evaluation of $Kc * Ks = AET/ET0$ can be resolved from the balance of radiative fluxes at the surface only (Mekonnen and Bastiaanssen, 2000). If the area is under irrigation, then the water availability factor $Ks = 1$, and a locally adjusted value of Kc can be evaluated directly from RS measurements.

This method of evaluating a Kc becomes also suitable for heterogeneous cropping patterns in the same land, typical of many agricultures in Asia and Africa. The fractional cover of each crop can be evaluated and the average Rn, G, and H used in the evaluation of a 'composite Kc' can be obtained from the homogeneous covers by using their fractions as weights.

As AET is estimated from PT, the values could be verified by using an SEB method from RS, which allows monitoring. The reference ET0 in the above method is obtained with the PT method for grasslands. This approach is used in irrigated areas with data scarcity in order to improve the estimations of Kc and reduce the need of images.

As Kc varies in terms of weeks, this approach is suitable to build adjusted Kc factors for irrigated crops after some seasons of analysis. After the Kc is evaluated, it can be applied for crop specifics without the requirement of images.

The main deficiency of the purely RS-based models is the uncertainty of clear image acquisition in the right time window. As such, major attention is turned to dual approaches where a model does the calculations on crop-water requirements and RS images are used to feed the model. Difference in these methods is centered on how the images are used as model input.

The first group of approaches using SVAT models such as SWAT (Arnold and Fohrer, 2005; Arnold et al., 1998), CRIWAR (Bos et al., 2009), SWATRE (Belmans, 1983), or SIMGRO (Querner and van Bakel, 1989) is applied independently or combined with ground data to make the actual evaluation of the crop-water requirements. As the models are based on water balance, soil moisture and AET are estimated at every time step and for each land-use location.

Due to the balance agreement between soil moisture and evaporation, a good calibration of the soil moisture ensures a good estimate of the ET from the model, mainly in the case of irrigation (surplus).

After the model is calibrated, AET is a product that can be used on daily basis. In well-managed irrigation schemes, the assessment of water efficiency and uniformity of water distribution is preferred over the strict use of the Kc factor for crop-water requirements estimates. These two items are considered essential in the evaluation of irrigation performances (Bos et al., 2005).

In this type of appraisal model, the concept of Kc is replaced for the 'relative ET' defined as $RE = AET/PET$, where PET can be obtained from standard meteorological measurements (Doorenbos and Pruitt, 1977). Irrigation efficiency in the command area is then evaluated as the combination of three measurable properties:

- Optimal water requirement ensures that no stress occurs. In general, when $AET/PET \geq 0.75$, this condition prevails. The value of 0.75 is generally adopted, although it might be adjusted locally.
- Water efficiency is defined as the ratio of AET to irrigation water depth given to each unit of parcel or crop.
- Uniformity refers to the homogeneity of the distribution of water in space and time inside the command area. As AET and RE are obtained on a daily and pixel basis, the coefficient of variation of these variables is used to evaluate it. A threshold is set to evaluate uniformity.

As presented here, the approach does not use AET derived from imagery as part of the procedure. Then, images obtained on clear days are used to derive AET establishing a strong basis of comparison with the information produced by the model. AET estimated by the image is used to evaluate the three indicators mentioned earlier, allowing real-time monitoring and the detection of flaws in the model that then can be corrected.

This approach was used very successfully by Roerink et al. (1997) in the irrigation fields in Mendoza, Argentina, with irrigation performance methods developed by Bos et al. (1994).

For operational purposes, the most accepted approach is to allow a continuous SVAT model to perform the calculations of AET and crop-water requirement. The model is calibrated mainly against soil moisture and groundwater table, both key variables in the process that can also be easily monitored at point scale. No SEB-RS model is used in this case. In this sense, this approach is similar to the previous method. The difference occurs on the use of the imagery as input to the model.

RS imagery is used only to evaluate crop patterns and land parameters or properties that are required for the model. From a sequence of high-resolution images, a selected vegetation index evolution in time is estimated at pixel level (usually the Normalized Difference Vegetation index (NDVI) is used in many SEB). From the individual points in time, smoothing, filtering, or mainly harmonic techniques are normally adopted to achieve a time-continuous evolution (Verhoef, 1996).

The NDVI time series and up-to-date land-use maps suffice to evaluate basis Kc (Kcb) described in Allen et al. (1998) and the fractional cover (fc), both needed to input in the model (Bausch and Neale, 1987; Bausch, 1995; Gonzalez Piqueras, 2006; Heilman et al., 1982; Ray and Dadhwal, 2001; Valor and Caselles, 1996). The model 'HidroMORE' was the result of the application of Irrigation Advisory Services (IAS) in Mediterranean areas, and validation of the results are given by Rubio et al. (2003). This approach was successfully used in a demonstration European project called 'Demeter' (Jochum and Calera, 2006) and in worldwide international projects such as 'Pleiades'.

In general, AET derived from SEB algorithms using RS imagery needs strict field (on-site) validation and local fine-tuning. On wet conditions on typically well-covered irrigated fields, the actual evaporation approaches to the potential one and in this situation most of the SEB methods work well (Bastiaanssen et al., 1998; Norman et al., 1995b; Su, 2002). Kite and Droogers (2000) reviewed some models and RS retrieval under the same conditions and warned about the great variability in the results between FAO-24 and satellite methods. There is a general agreement in the circle of experts that some high degree of expertise is required for the application of AET-RS SEB models, a warning that must not be overlooked.

2.14.10.3 Drought Indices and Soil Moisture Monitoring

The partition of available energy at the Earth surface is largely controlled by the available soil moisture. The relation between soil moisture and ET is very tight, to the point that irrigation supply (moisture) can be estimated directly from accumulated ET as well as from monitoring soil moisture changes. However, from the water management point of view these two processes are very different in the sense that ET can be monitored by RS but cannot be controlled easily and the soil moisture can be controlled but it is not directly observable in the rooting depth. As we can only adjust the water allocated to crops, the soil moisture becomes the most relevant of all parameters in the water cycle for agro-systems.

In watershed management, there are situations requiring continuous monitoring of soil moisture as during droughts, which is the most devastating form of agricultural deficiency. The evaluation of droughts is done through methods that account for the endurance of low soil moisture values through time on a pixel basis. RS in the visible and thermal wavelengths is unable to directly measure soil moisture on the ground. As such, indirect methods were designed to approach it.

There are three approaches to monitor droughts purely from standard passive RS. The initial methods were based on the fact that during drought conditions a sudden change in soil moisture would be followed by a distinctive jump in the spectral reflectance of the observed pixel. In a soil, more humidity implies less reflectance. Bowers and Hanks (1965) and Bowers and Smith (1972) linearly related the soil moisture and the spectral bands where soil moisture is an energy absorber.

The property that bodies oppose to temperature changes is called 'thermal inertia'. It can be evaluated from multitemporal imagery (day and night) from the visible and thermal bands. Pratt and Ellyet (1979) presented a modification of the model to map soil moisture, and later Price (1985) used the energy-balance concept to add certainty to the model. All these approaches were successful under controlled conditions in areas of sparse vegetation, as soil contrast needs to be only affected by moisture.

More appropriate was the attempt to monitor soil moisture under vegetated conditions. Moisture depletion affects plant physiology and, in particular, the reflective properties of the vegetation. In the absence of vegetation water as incoming radiation needs dissipation, vegetation reacts by increasing both the reflectivity of the leaves (albedo) and the sensible heat (surface temperature). The accounting of the difference between the surface and air temperature in time leads to the design of the crop-water stress index (Idso et al., 1981, 1975; Jackson et al., 1981), which was suitable for evaluating stress for full cover situations. Moran et al. (1994) developed a water deficit index (WDI) using vegetation indices to account for partially vegetation covered areas, using a composite of the surface temperature for vegetation and land, an approach that was later on extended to SEB models. As air temperature was needed in this type of method and the availability of ground meteorological measurements was restrictive, the use of the variability of the canopy surface temperature allowed a pure RS approach, especially applicable to nonfully covered areas (Gonzalez Dugo et al., 2005).

Vegetation indices and surface temperature were also used in the development of indices dedicated to drought monitoring (Carlson et al., 1991, 1994; Kogan, 1990; Sandholt et al., 2002). Indices are rarely absolute, as they might not directly compare to a similar degree of drought severity, which leads to the designs of regional indices that can be compared.

RS-derived SEB subproducts were earlier involved in the calculation of drought indices. Using air and surface temperature and its variation, albedo, and net radiation in a consistent framework allowed quantifying droughts spatially. It was only after AET-SEB product became operational, for example, SEBI (Menenti and Choudhury, 1993), S-SEBI (Roerink et al., 2000), SEBAL (Bastiaanssen, 1995), TSEB (Kustas and Norman, 1999; Kustas et al., 2004; Norman et al., 1995a), ALEXI (Anderson et al., 1997; Mecikalski et al., 1999), and Dis-ALEXI (Norman et al., 1995a), the derived energy fluxes, directly conditioned by soil moisture, could be used for the evaluation of drought indices.

Su et al. (2003) used outputs of SEBS (Su, 2002) to propose a fully RS-derived Drought Severity Index (DSI) for the North China Plain for low-resolution imagery. The results showed that the relative evaporation (actual latent heat flux over potential latent heat flux) can be used to predict soil moisture within one standard deviation, but the effects of cloud-contaminated pixels highly condition the applicability of the approach.

Soil moisture derived from MW RS techniques can also be used for drought monitoring; however, the applicability is limited to continental scale as the most available products have rather coarse spatial resolution (tens of kilometers) and hence not applicable for most agricultural applications. More details on MW-derived soil moisture can be found in Section 2.14.7. More recently, it has been shown that time series of soil moisture derived from high-resolution MW sensors (tens of meters), such as ASAR (Van der Velde and Su, 2009). However, at the time of writing, such data are restricted to selected experimental areas where access of data is guaranteed.

2.14.10.4 Algorithm Retrievals and Operability

AET estimates from RS are now available from a number of empirical, physically based, and mixed-approach methods. The scientific community agrees that there is no one single approach that best suits all cases, but some methods are preferred over others according to specific land-cover patterns and wetness characteristics of the site.

In order to make RS information operational, continuous modeling, testing, and validation are required. The Demeter and Pleiades projects are examples of European efforts in using the latest observation and communication technologies to narrow the gap between research and operational needs. The reader is referred to the cited literature for detailed information on the different retrieval algorithms; we will only briefly describe the practical issues used in the SEBS algorithm for practical applications.

2.14.10.5 SEBS Algorithm

The SEBS algorithm (Su, 2002) is a single-source physical model for the evaluation of the energy-balance fluxes from RS imagery. A full description of the model can be found in Su (2002, 2005) and the corresponding open-source software SEBS4ILWIS from the International Institute for Geo-information Science and Earth Observation.

SEBS is a column model, which means that information on adjacent pixels is not affecting the pixel where the calculations are done. This means that, in the ideal case, the information required as input needs to be measured independently at the site. However, in practice, this is never the case and interpolation techniques are always required. The examples presented at the end of the section were evaluated at measuring points. In the following, the values between brackets are examples taken from one point during this analysis. In the case that SEBS applied to imagery, the input indicated with (*) should desirably be a map.

SEBS for ILWIS Open Source software also provides a set of routine for bio-geophysical parameter extraction. It uses satellite EO data, in combination with meteorological information as inputs, to produce the evaporative fraction, net radiation, and soil heat flux parameters. The main steps using the MODIS data as an example are as follows (SEBS core is theoretically independent of the sensor used):

1. Reprojecting and converting MODIS level-1 B data with the ModisSwathTool software.
2. Importing images into ILWIS.
3. Preprocessing for SEBS:

   - raw data to radiance/reflectance conversion;
   - brightness temperature computation;
   - SMAC for atmospheric correction;
   - land surface albedo computation;
   - land surface emissivity, NDVI, vegetation proportion, and emissivity difference computation; and
   - land surface temperature computation.

4. SEBS core model for bio-geophysical parameter extraction.

Inputs in SEBS. Meteorological information from meteorological stations or atmospheric model out fields is used at the time of the satellite pass:

- Reference height (z_{ref}): height from the ground where measurements of temperature, wind, pressure, and specific humidity are made [m].
- Specific humidity: [kg kg^{-1}].
- Wind speed (u_{ref}): [m s^{-1}].
- Air temperature at reference height (T_a): [°C].
- Air pressure at reference height: [Pa].
- Air pressure at land surface: [Pa].
- PBL height (hi): height of the planetary boundary layer (PBL) in [m] that can be estimated by radiosounding or using atmospheric model outputs. (default hi = 1000 m).
- Incoming global solar radiation: [W m^{-2}] (636 W m^{-2}).

Input normally derived after image preprocessing:

- Atmospherically corrected broadband albedo (α) [–] (0.18).
- Surface emissivity (ε_0) (0.98).
- Atmospherically corrected surface temperature [K] (296 K).

Input derived from land-cover properties. Land properties affect roughness, the proven most sensitive information in all SEB-SVAT methods. A good estimation of aerodynamic roughness is the key for success:

- The percentage of fractional vegetation cover (fc) – it controls the partition of energy fluxes between vegetation and bare soil.
- The land use that contains roughness classes (z_{om}) and displacement zero heights (d_0), all in (m), associated with vegetation height values.
- The leaf area index is included in the deduction of the aerodynamic roughness for heat transport (z_{oh}).

Outputs of SEBS. After the successful completion of the SEBS operation in ILWIS, following raster maps are generated:

- sebs_evap: evaporative fraction [–]
- sebs_daily_evap: daily evaporation [mm d^{-1}]
- sebs_evap_relative: relative evaporation [–]
- sebs_G0: soil heat flux [W m^{-2}]
- sebs_H_dry: sensible heat flux at the dry limit [W m^{-2}]
- sebs_H_i: sensible heat flux [W m^{-2}]
- sebs_H_wet: sensible heat flux at the wet limit [W m^{-2}]
- sebs_Rn: net radiation [W m^{-2}]
- sebs_LE: latent heat flux [W m^{-2}]

More details on the operation of the software SEBS4ILWIS can also be obtained from the online help.

2.14.10.6 Evaluation Example

A simple application example is presented for part of the Guareñas catchment in the Duero Basin, Spain, which is being monitored with the Network of Soil Moisture Measurement Stations of the University of Salamanca (REMEDHUS). Measurements started from June 1999 to the present. The network consists of a series of 23 soil moisture stations, three meteorological stations, and discharge gages.

First, RS-derived soil moisture estimates were compared to ground-truth data. The RS soil moisture retrievals were obtained indirectly from the SEBS RS model. AET retrievals using SEBS (Su, 2002) were calculated using the software SEBS4IL-WIS (ITC, 2008; Wang et al., 2008).

The top soil moisture values of the 23 stations were recorded simultaneously for 13 clear-day MODIS images taken during 2007. All VIS and NIR bands were radiometrically and atmospherically corrected using the SMAC (Rahman and Dedieu, 1994) version implemented in the Integrated Land and Water Information System (ILWIS, a GIS system). Surface temperature was obtained using a split window technique (Sobrino and Raissouni, 2003) and albedo estimates following (Liang, 2001).

Due to the very low resolution ($\sim$1000 m in the thermal band) of the imagery, the comparison between the ground measurement and the RS soil moisture derived at the pixel where each ground station was located was futile, because the soil moisture at the spot did not represent the one of the pixel.

However, a more representative and realistic approach was to make use of the simultaneous information of all 23 stations and the corresponding RS-derived soil moisture at the pixels where the stations were located. The RS-derived soil moisture average was obtained for the network.

To estimate RS soil moisture, it was considered that the relative soil moisture (the ratio of the actual soil moisture to the soil moisture at the limiting wet case) was equal to the relative ET that could be evaluated after SEBS (Su et al., 2003). The soil moisture at limiting wet case was evaluated after laboratory analysis. The results illustrate that there is a good

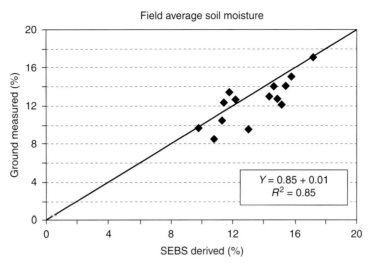

Figure 22 SEBS-derived vs. ground-measured soil moisture for the REMEDHUS project. Data courtesy: The University of Salamanca.

Table 4 Comparison between FAO 56 and remote-sensing-derived Kc factors after SEBS analysis

Crop Stage	Image date	Daily ET_{act} (mm d^{-1})	Average daily ET_{act} (mm d^{-1})	Ten-day average ET_0 (mm d^{-1})	Kc average calculated ET_{act}/ET_0	Kc FAO guide lines
Initial	(14 Nov. 2007)	1.09	1.09	1.57	0.70	0.70
Crop development	16 Dec. 2007)	0.99	0.99	0.83	1.19	0.7–1.15
	(8 Mar. 2007)	2.17	2.17	2.18	1.00	0.7–1.15
Mid-season	(27 Apr. 2008)	4.47	4.06	4.23	1.19	1.15
	(1 May. 2008)	3.65				
Late season	(18 Jun. 2008)	4.98	4.98	5.75	0.87	1.15–0.25

agreement between SEBS-estimated soil moisture values and ground-measured values on the field scale level with a strong correlation. Figure 22 shows the results of the comparison. The general dry condition for the soils in the region was captured by the model. This information suffices for modeling purposes of the surface water that was done with the HVB model (Bergström, 1995).

As a second example, SEBS was used to calculate AET from the available imagery. Then, RS-derived Kc factor for four stages of wheat development was derived and compared with the tabulated values of the FAO guidelines. The single crop coefficient (Allen et al., 1998) is used to calculate crop ET.

With the available imageries, the Kc was computed for four stages of wheat development, namely, the initial, the crop development, the mid-season, and the late season and then compared with the tabulated values of the FAO guidelines. In the study area, the sowing dates for winter wheat vary from October to November and the harvesting dates vary from June to July. The subset was selected in a way of having a clear wheat zone close to meteorological stations. The nearest two meteorological stations, Canizal and VA_02, were selected to calculate a 10-day average crop reference ET. The 10-day average ET0 was calculated based on the estimations of 5 consecutive days before and after the imagery date.

Table 4 shows the results of the comparison. The RS values of Kc are in agreement with the values given in the FAO guidelines; however, the procedure allows local fitting of the Kc. The procedure is universally applicable mainly in irrigated areas, as it is recommended for spatial irrigation performance studies (Bos et al., 2005).

Acknowledgment

This work was partially funded by EUMETSAT Satellite Application Facility on Climate Monitoring.

References

Ackerman SA, Strabala KI, Menzel WP, Frey RA, Moeller CC, and Gumley LE (1998) Discriminating clear sky from clouds with MODIS. *Journal of Geophysical Research* 103(D24): 32141–32157.

Adam JC, Clark EA, Lettenmaier DP, and Wood EF (2006) Correction of global precipitation products for orographic effects. *Journal of Climate* 19: 15–38.

Adler RF, Huffman GJ, Chang A, et al. (2003) The version-2 Global Precipitation Climatology Project (GPCP) monthly precipitation analysis (1979–present). *Journal of Hydrometeorology* 4: 1147–1167.

Adler RF and Negri AJ (1988) A satellite IR technique to estimate tropical convective and stratiform rainfall. *Journal of Applied Meteorology* 27: 30–51.

Albert A and Gege P (2006) Inversion of irradiance and remote sensing reflectance in shallow water between 400 and 800 nm for calculations of water and bottom properties. *Applied Optics* 45: 2331–2343.

Albertson JD (1996) Large Eddy Simulation of Land–Atmosphere Interaction. PhD Thesis, University of California, Davis, 185pp.

Albertson JD, Kustas WP, and Scanlon TM (2001) Large-eddy simulation over heterogeneous terrain with remotely sensed land surface conditions. *Water Resources Research* 37: 1939–1953.

Alemseged TH and Rientjes THM (2007) Spatio-temporal rainfall mapping from space: Setbacks and strengths. In: *Proceedings of the 5th International Symposium on Spatial Data Quality SDQ 2007, Modelling Qualities in Space and Time*, 9pp. Enschede, The Netherlands, 13–15 June. Enschede: ITC.

Allen RG, Pereira LS, Raes D, and Smith M (1998) Crop Evapotranspiration. Guidelines for Computing Crop Water Requirements. In: *Irrigation and Drainage Paper*, FAO 56, 300pp. FAO, Rome.

Alley WM, Healy RW, LaBaugh JW, and Reilly TE (2002) Flow and storage in groundwater systems. *Science* 296: 1985–1990.

Alsdorf DE, Fu L-L, Mognard NM, et al. (2007) Measuring the global oceans and terrestrial fresh water from space. *EOS, Transactions, American Geophysical Union* 18(24): 253–257.

Alsdorf DE and Lettenmaier DP (2003) Tracking fresh water from space. *Science* 301: 1485–1488.

Alsdorf DE, Lettenmaier DP, Vörösmarty C, and the NASA Surface Water Working Group (2003) The need for global, satellite-based observations of terrestrial surface waters. *EOS, Transactions, American Geophysical Union* 84(269): 275–276.

Alsdorf DE, Melack JM, Dunne T, Mertes LAK, Hess LL, and Smith LC (2000) Interferometric radar measurements of water level changes on the Amazon floodplain. *Nature* 404: 174–177.

Anagnostou EN, Negri AJ, and Adler RF (1999) A satellite infrared technique for diurnal rainfall variability studies. *Journal of Geophysical Research* 104: 31477–31488.

Anderson MC, Norman JM, Diak GR, Kustas WP, and Mecikalski JR (1997) A two-source time-integrated model for estimating surface fluxes using thermal infrared remote sensing. *Remote Sensing of Environment* 60: 195–216.

Anderson MC, Norman JM, Kustas WP, Li F, Prueger JH, and Mecikalski JR (2005) Effects of vegetation clumping on two-source model estimates of surface energy fluxes from an agricultural landscape during SMACEX. *Journal of Hydrometeorology* 6: 892–909.

Anderson MG and McDonnell JJ (eds.) (2005) *Encyclopedia of Hydrological Sciences, Vol. 2, Part 5, Remote Sensing*, pp. 657–996. Chichester: Wiley.

Anthes RA, Bernhardt PA, Chen Y, et al. (2008) The COSMIC/FORMOSAT-3 mission: Early results. *Bulletin of the American Meteorological Society* 89: 313–333.

Arnold JG and Fohrer N (2005) SWAT2000: Current capabilities and research opportunities in applied watershed modeling. *Hydrological Processes* 19(3): 563–572.

Arnold JG, Srinivasan RS, Muttiah RS, and Williams JR (1998) Large-area hydrologic modeling and assessment: Part I. Model development. *Journal of the American Water Resources Association* 34(1): 73–89.

Aumann HH, Chahine MT, Gautier C, et al. (2003) AIRS/AMSU/HSB on the aqua mission: Design, science objectives, data products, and processing systems. *IEEE Transactions on Geoscience and Remote Sensing* 41: 253–264.

Avissar R and Pielke RA (1989) A parameterization of heterogeneous land surfaces for atmospheric numerical models and its impact on regional meteorology. *Monthly Weather Review* 117: 2113–2136.

Ba MB and Gruber A (2001) GOES Multispectral Rainfall Algorithm (GMSRA). *Journal of Applied Meteorology* 40: 1500–1514.

Bamber JL and Rivera A (2007) A review of remote sensing methods for glacier mass balance determination. *Global and Planetary Change* 59: 138–148.

Bartalis Z, Wagner W, Naeimi V, et al. (2007) Initial soil moisture retrievals from the METOP-A advanced scatterometer (ASCAT). *Geophysical Research Letters* 34: L20401 (1–5, doi:10.1029/2007GL031088).

Bartholic JF, Namken LN, and Wiegand CL (1972) Aerial thermal scanner to determine temperatures of soil and of crop canopy differing in water stress. *Agronomy Journal* 64: 603–608.

Bastiaanssen WGM (1995) Regionalization of Surface Flux Densities and Moisture Indicators in Composite Terrain – a Remote Sensing Approach under Clear Skies in Mediterranean Climates, 273pp. PhD Thesis, Wageningen Agricultural University, Wageningen.

Bastiaanssen WGM, Menenti M, Feddes RA, and Holtslag AAM (1998a) A remote sensing surface energy balance algorithm for land (SEBAL): 1. Formulation. *Journal of Hydrology* 212–213: 198–212.

Bastiaanssen WGM, Pelgrum H, Wang J, et al. (1998b) A remote sensing surface energy balance algorithm for land (SEBAL): 2. *Validation. Journal of Hydrology* 212–213: 213–229.

Bates D and Watts D (1988) *Nonlinear Regression Analysis and Its Applications*. New York, NY: Wiley.

Batra N, Islam S, Venturini V, Bisht G, and Jiang L (2006) Estimation and comparison of evapotranspiration from MODIS and AVHRR sensors for clear sky days over the Southern Great Plains. *Remote Sensing of Environment* 103: 1–15.

Baum BA, Arduini RF, Wielicki BA, Minnis P, and Tsay S-C (1994) Multilevel cloud retrieval using multispectral HIRS and AVHRR data: Nighttime oceanic analysis. *Journal of Geophysical Research* 99: 5499–5514.

Bausch WC (1995) Remote sensing of crop coefficients for improving the irrigation scheduling of corn. *Agricultural Water Management* 27: 55–68.

Bausch W and Neale CMU (1987) Crop coefficients derived from reflected canopy radiation: A concept. *Transactions of the ASAE* 30(3): 703–709.

Becker F and Li Z-L (1995) Surface temperature and emissivity at various scales: Definition, measurement and related problems. *Remote Sensing Reviews* 12: 225–253.

Becker MW (2006) Potential for satellite remote sensing of groundwater. *Groundwater* 44: 306–318.

Belmans C (1983) Simulation model of the water balance of a cropped soil: SWATRE. *Journal of Hydrology* 63(3–4): 271–286.

Bennartz R and Fischer J (2001) Retrieval of columnar water vapor over land from backscattered solar radiation using the Medium Resolution Imaging Spectrometer. *Remote Sensing of Environment* 78: 274–283.

Bergström S (1995) The HBV model. In: Sing VP (ed.) *Computer Models of Watershed Hydrology*, pp. 443–476. Highlands Ranch, CO: Water Resources Publications.

Berthier E and Toutin T (2008) SPOT5-HRS digital elevation models and the monitoring of glacier elevation changes in North-West Canada and South-East Alaska. *Remote Sensing of Environment* 112(5): 14–28.

Bertoldi G, Albertson JD, Kustas WP, Li F, and Anderson MC (2007) On the opposing roles of air temperature and wind speed variability in flux estimation from remotely sensed land surface states. *Water Resources Research* 43: W10433 (doi:10.1029/2007WR005911).

Beven K (2001) How far can we go in distributed hydrological modeling. *Hydrology and Earth System Sciences* 5: 1–12.

Birkett CM (1995) The contribution of TOPEX/POSEIDON to the global monitoring of climatically sensitive lakes. *Journal of Geophysical Research* 100: 25179–25204.

Birkett CM (1998) Contribution of the TOPEX NASA radar altimeter to the global monitoring of large rivers and wetlands. *Water Resources Research* 34: 1223–1239.

Birkett CM, Mertes LAK, Dunne T, Costa M, and Jasinski J (2004) Altimetric remote sensing of the Amazon: Application of satellite radar altimetry. *Journal of Geophysical Research* 107(D20): 8059 (doi:10.1029/2001JD000609).

Boegh E, Soegaard H, Hanan N, Kabat P, and Lesch L (1999) A remote sensing study of the NDVI-Ts relationship and the transpiration from sparse vegetation in the Sahel based on high-resolution satellite data. *Remote Sensing of Environment* 69: 224–240.

Boegh E, Soegaard H, and Thomsen A (2002) Evaluating evapotranspiration rates and surface conditions using Landsat TM to estimate atmospheric resistance and surface resistance. *Remote Sensing of Environment* 79: 329–343.

Boegh E, Thorsen M, Butts MB, et al. (2004) Incorporating remote sensing data in physically based ditributed agro-hydrological modeling. *Journal of Hydrology* 287: 279–299.

Bolle H-J (ed.) (2003). *Mediterranean Climate – Variability and Trends*, 372pp. Berlin, Springer.

Bos MG, Burton MA, and Molden DJ (2005) *Irrigation and Drainage Performance Assessment. Practical Guidelines*, 158pp. Trowbridge: Cromwell Press.

Bos MG, Kselik RAL, Allen RG, and Molden DJ (2009) *Water Requirements for Irrigation and the Environment*. Dordrecht: Springer.

Bos MG, Murray-Rust DH, Merrey D, Johnson HG, and Snellen WB (1994) Methodologies for assessing performance of irrigation and drainage management. *Irrigation and Drainage System* 7: 231–261.

Bowers SA and Hanks RJ (1965) Reflection of radiant energy from soil. *Soil Science* 100(2): 130–138.

Bowers SA and Smith SJ (1972) Spectrophotometric determination of soil water content. *Soil Science Society of America Proceedings* 36: 978–980.

Bringi VN and Chandrasekar V (2001) *Polarimetric Doppler Weather Radar*, 636pp. New York, NY: Cambridge University Press.

Brogniez H, Lémond J, Roca R, and Picon L (2007) Variability of free tropospheric humidity from Meteosat over the tropics: 1983–2005. In: *EUM P.50 – Joint 2007 EUMETSAT and 15th AMS Conference*, 6pp. http://www.eumetsat.int/Home/Main/AboutEUMETSAT/Publications/ConferenceandWorkshopProceedings/2007/groups/cps/documents/document/pdf_conf_p50_s7_02_brogniez_v.pdf (accessed September 2010).

Brogniez H, Roca R, and Picon L (2005) Evaluation of the distribution of subtropical free tropospheric humidity in AMIP-2 simulations using METEOSAT water vapor channel data. *Geophysical Research Letters* 32: L19708 (doi:10.1029/2005GL024341).

Brown L, Thorne R, and Woo M-K (2008) Using satellite imagery to validate snow distribution simulated by a hydrological model in large northern basins. *Hydrological Processes* 22(15): 2777–2787.

Brutsaert W (1982) *Evaporation into the Atmosphere*. 299pp. Dordrecht: Reidel.

Brutsaert W (1992) Stability correction functions for the mean wind speed and temperature in the unstable surface layer. *Geophysical Research Letters* 19: 469–472.

Brutsaert W (2008) *Hydrology. An Introduction*. New York, NY: Cambridge University Press.

Brutsaert W and Sugita M (1992) Application of self-preservation in the diurnal evolution of the surface energy budget to determine daily evaporation. *Journal of Geophysical Research* 97(18): 377–382.

Buehler S and John V (2005) A simple method to relate microwave radiances to upper tropospheric humidity. *Journal of Geophysical Research* 110: D02110 (doi:10.1029/2004JD005111).

Buehler SA, *et al*. (2008) An upper tropospheric humidity data set from operational satellite microwave data. *Journal of Geophysical Research* 113: D14110 (doi:10.1029/2007JD009314).

Burrows JP, *et al*. (1999) The global ozone monitoring experiment (GOME): Mission concept and first scientific results. *Journal of the Atmospheric Sciences* 56: 151–175.

Carbone RE and Tuttle J (2008) Rainfall occurrence in the U.S. warm season: The diurnal cycle. *Journal of Climate* 21: 4132–4146.

Carder K, Cattrall C, and Chen F (1999) MODIS Clear Water Epsilons Algorithm Theoretical Basis Document ATBD, 72pp. http://modis.gsfc.nasa.gov/data/atbd/atbd_mod21.pdf (accessed September 2010).

Carlson TN, Capehart WJ, and Gillies RR (1995) A new look at the simplified method for remote sensing of daily evapotranspiration. *Remote Sensing of Environment* 54: 161–167.

Carlson TN, Gillies RR, and Perry EM (1994) A method to make use of thermal infrared temperature and NDVI measurements to infer surface soil water content and fractional vegetation cover. *Remote Sensing Reviews* 9: 161–173.

Carlson TN, Perry EM, and Schmugge TJ (1991) Remote sensing estimation of soil moisture availability and fractional vegetation cover for agricultural fields. *Agricultural and Forest Meteorology* 52: 45–69.

Cazenave A, Milly PCD, Douville H, Benveniste J, Kosuth P, and Lettenmaier DP (2004) Space techniques used to measure change in terrestrial waters. *EOS, Transactions, American Geophysical Union* 85(6): 58–59.

Ceballos A, Scipal K, Wagner W, and Martinez-Fernandez J (2005) Validation of ERS scatterometer-derived soil moisture data in the central part of the Duero Basin, Spain. *Hydrological Processes* 19(8): 1549–1566.

CEOS (2009) *The Earth Observation Handbook*. http://www.eohandbook.com (accessed September 2010).

Chehbouni A, Hoedjes JCB, Rodriguez J-C, *et al*. (2008) Using remotely sensed data to estimate area-averaged daily surface fluxes over a semi-arid mixed agricultural land. *Agricultural and Forest Meteorology* 148: 330–342.

Chehbouni A, Seen DL, Njoku EG, Lhomme JP, Monteny B, and Kerr YH (1997) Estimation of sensible heat flux over sparsely vegetated surfaces. *Journal of Hydrology* 188–189: 855–868.

Chehbouni A, Seen DL, Njoku EG, and Monteney BM (1996) Examination of the difference between radiative and aerodynamic surface temperatures over sparsely vegetated surfaces. *Remote Sensing of Environment* 58: 177–186.

Chen JL, Tapley BD, and Wilson CR (2006) Alaskan mountain glacial melting observed by satellite gravimetry. *Earth and Planetary Science Letters* 248: 368–378.

Chomko R, Gordon H, Maritorena S, and Siegel D (2003) Simultaneous retrieval of oceanic and atmospheric parameters for ocean color imagery by optimization: a validation. *Remote Sensing of Environment* 84: 208–220.

Choudhury BJ, Schmugge TJ, Chang A, and Newton RW (1979) Effect of surface roughness on the microwave emission from soils. *Journal of Geophysical Research* 84: 5699–5706.

Clark MP, Slater AG, Barrett AP, *et al*. (2006) Assimilation of snow covered area information into hydrologic and land-surface models. *Advances in Water Resources* 29: 1209–1221.

Coakley JA Jr. and Bretherton FP (1982) Cloud cover from high-resolution scanner data: Detecting and allowing for partially filled fields of view. *Journal of Geophysical Research* 87: 4917–4932.

Coll C, Caselles V, Sobrino JA, and Valor E (1994) On the atmospheric dependence of the split-window equation for land surface temperature. *International Journal of Remote Sensing* 15(1): 105–122.

Coll C, *et al*. (2005) Ground measurements for the validation of land surface temperatures derived from AATSR and MODIS data. *Remote Sensing of Environment* 97(3): 288–300.

Collier CG (1983) A weather radar correction procedure for real-time estimation of surface rainfall. *Quarterly Journal of the Royal Meteorological Society* 109: 589–608.

Crago R (1996) Conservation and variability of the evaporative fraction during the daytime. *Journal of Hydrology* 180(1–5): 173–194.

Crowley JW, Mitrovica JX, Bailey RC, Tamisiea ME, and Davis JL (2006) Land water storage within the Congo Basin inferred from GRACE satellite gravity data. *Geophysical Research Letters* 33: L19402.

Curry JA, Hobbs PV, King MD, *et al*. (2000) FIRE Arctic clouds experiment. *Bulletin of the American Meteorological Society* 81: 5–29.

Curtis S, Salahuddin A, Adler RF, Huffman GJ, Gu G, and Hong Y (2007) Precipitation extremes estimated by GPCP and TRMM: ENSO relationships. *Journal of Hydrometeorology* 8: 678–689.

Davolio S and Buzzi A (2003) A nudging scheme for the assimilation of precipitation data into a mesoscale model. *Weather Forecasting* 19: 855–871.

de Bruin HAR (1987) A model for the Priestley–Taylor parameter. *Journal of Climate and Applied Meteorology* 12: 572–578.

de Bruin HAR, van den Hurk BJJM, and Kohsiek W (1995) The scintillation method tested over a dry vineyard area. *Boundary-Layer Meteorology* 76: 25–40.

De Jeu R (2003) *Retrieval of Land Surface Parameters Using Passive Microwave Remote Sensing*, 122pp. PhD Dissertation, VU University Amsterdam, Amsterdam.

De Jeu R, *et al*. (2008) Global soil moisture patterns observed by space borne microwave radiometers and scatterometers. *Surveys in Geophysics* 28: 399–420.

De Lathauwer E, Timmermans WJ, Satalino G, Mattia F, Loew A, and Pauwels VRN (2009) Assessment of the potential use of remotely sensed soil moisture values for spatially distributed latent heat flux estimation. *JSTARS-IEEE* (submitted).

De Wit A and van Diepen C (2007) Crop model data assimilation with the Ensemble Kalman Filter for improving regional crop yield forecasts. *Agricultural and Forest Meteorology* 146(1–2): 38–56.

Dekker A, Malthus T, Wijnen M, and Seyhan E (1992) Remote sensing as a tool for assessing water quality in Loosdrecht lakes. *Hydrobiologia* 233: 137–159.

Delanoë J and Hogan RJ (2008) A variational scheme for retrieving ice cloud properties from combined radar, lidar, and infrared radiometer. *Journal of Geophysical Research* 113: D07204 (doi:10.1029/2007JD009000).

Deneke HM, Roebeling RA, and Feijt A (2008) Estimation surface solar irradiance from METEOSAT SEVIRI-derived cloud properties. *Remote Sensing of Environment* 112: 3131–3141.

Derksen C, Walker A, and Goodison B (2003) A comparison of 18 winter seasons of in situ and passive microwave derived snow water equivalent estimates in Western Canada. *Remote Sensing of Environment* 88(3): 271–282.

Dickey JO, Bentley CR, Bilham R, *et al*. (1997) *Satellite gravity and the geosphere*. National Research Council Report, 112pp. Washington, DC: National Academies Press.

DiGiacomo P, Hook S, Neumann A, *et al*. (2007) GEO inland and nearshore coastal water quality remote sensing. In: Bauer M, Dekker A, DiGiacomo P, *et al*. (eds.), *GEO Group on Earth Observation*, 31pp. http://www.earthobservations.org/meetings/20070327_29_water_quality_workshop_report.pdf (accessed September 2010).

Dobson MC, Ulaby FT, Hallikainen MT, and Elrayes MA (1985) Microwave dielectric behavior of wet soil. 2. Dielectric mixing models. *IEEE Transactions on Geoscience and Remote Sensing* 23(1): 35–46.

Doerffer R and Schiller H (2007) The MERIS case 2 water algorithm. *International Journal of Remote Sensing* 28: 517–535.

Doorenbos J and Pruitt WO (1977). *Guidelines for Predicting Crop Water Requirements*, Irrigation and Drainage Paper, FAO 33, 193pp. FAO, Rome.

Dorigo WA, *et al*. (2007) A review on reflective remote sensing and data assimilation techniques for enhanced agroecosystem modeling. *International Journal of Applied Earth Observation and Geoinformation* 9: 165–193.

Dozier J and Painter TH (2004) Multispectral and hyperspectral remote sensing of alpine snow properties. *Annual Review of Earth and Planetary Science* 32: 465–494.

Draper C, Walker J, Steinle P, De Jeu R, and Holmes T (2009) Evaluation of AMSR-E derived soil moisture over Australia. *Remote Sensing of Environment* 113: 703–710.

Drusch M, Holmes TRH, Rosnay PD, and Balsamo G (2009) Comparing ERA-40 based L-band brightness temperatures with Skylab observations: A preliminary calibration/validation study in preparation for the SMOS mission. *Journal of Hydrometeorology* 10(1): 213–226.

Duarte J, Vélez-Reyes M, Tarantola S, Gilbes F, and Armstrong R (2003) A probabilistic sensitivity analysis of water-leaving radiance to water constituents in coastal shallow waters. In: Frouin RJ, Gilbert GD, and Pan D (eds.) *Ocean Remote Sensing and Imaging*, Proceedings of SPIE, vol. 5155, pp. 162–173.

Dubois PC, van Zyl J, and Engman ET (1995) Measuring soil moisture with imaging radars. *IEEE Transactions on Geoscience and Remote Sensing* 33: 915–926.

Eagleman JR and Ulaby FT (1975) Remote sensing of soil moisture by SKYLAB radiometer and scatterometer sensors. *Journal of the Astronautical Science* 23: 147–159.

Ebert EE, Janowiak J, and Kidd C (2007) Comparison of near-real-time precipitation estimates from satellite observations and numerical models. *Bulletin of the American Meteorological Society* 88: 47–64.

Eleveld M, van der Woerd H, El Serafy G, Blaas M, van Kessel T, and de Boer G (2008) Assimilation of remotely sensed observation in a sidement transport model. In: *Ocean Optics*. Barga, Italy. Halifax: Lewis Conference Services International. http://oceanopticsconference.org/abstracts (accessed September 2010).

Engman E and Chaunhan N (1995) Satus of microwave soil moisture measurements with remote sensing. *Remote Sensing of Environment* 51: 189–198.

Entekhabi D, Njoku E, O'Neill P, et al. (2010) The soil moisture active and passive (SMAP) mission. *Proceedings of the IEEE* 98: 704–716 (doi:10.1109/JPROC.2010.2043918).

ESA (2006) The Changing Earth, ESA SP-1304, 85pp. http://esamultimedia.esa.int/docs/SP-1304.pdf (accessed September 2010).

Eyre R (1989) Inversion of cloudy satellite sounding radiances by nonlinear optimal estimation. I: Theory and simulation for TOVS. *Quarterly Journal of the Royal Meteorological Society* 115: 1001–1026.

Ezzahar J, Chehbouni A, Hoedjes JCB, and Chehbouni A (2007) On the application of scintillometry over heterogeneous grids. *Journal of Hydrology* 334: 493–501.

Ferraro RR, Weng F, Grody NC, et al. (2005) NOAA operational hydrological products derived from the Advanced Microwave Sounding Unit. *IEEE Transactions on Geoscience and Remote Sensing* 43: 1036–1049.

Fetzer EJ, Lambrigtsen BH, Eldering A, Aumann HH, and Chahine MT (2006) Biases in total precipitable water vapor climatologies from Atmospheric Infrared Sounder and Advanced Microwave Scanning Radiometer. *Journal of Geophysical Research* 111: D09S16 (doi:10.1029/2005JD006598).

Fetzer EJ, et al. (2008) Comparison of upper tropospheric water vapor observations from the Microwave Limb Sounder and Atmospheric Infrared Sounder. *Journal of Geophysical Research* 113: D22110 (doi:10.1029/2008JD010000).

Floricioiu D, Eineder M, Rott H, and Nagler T (2008) Velocities of major outlet glaciers of the Patagonia Icefield observed by TerraSAR-X. In: *Proceedings of the IEEE International Geoscience and Remote Sensing Symposium 2008*, paper no. 2241, pp. 31–34 (ISBN 978-1-4244-2808).

Foken T (2008) The energy balance closure problem: an overview. *Ecological Application* 18: 1351–1367.

Foster JL and Chang ATC (1993) Snow cover. In: Gurney RJ, et al. (eds.) *Atlas of Satellite Observations Related to Global Change*, pp. 361–370. Cambridge: Cambridge University Press.

Foster JL, Sun C, Walker JP, et al. (2005) Quantifying the uncertainty in passive microwave snow water equivalent observations. *Remote Sensing of Environment* 94: 187–203.

Francois C, Ottle C, and Prevot L (1997) Analytical parameterization of canopy directional emissivity and directional radiance in the thermal infrared. Application on the retrieval of soil and foliage temepratures using two directional measurements. *International Journal of Remote Sensing* 18: 2587–2621.

Frappart F, DoMinh K, L'Hermitte J, et al. (2006) Water volume change in the lower Mekong from satellite altimetry and imagery data. *Geophysical Journal International* 167: 570–584.

French A, Schmugge TJ, and Kustas WP (2002) Estimating evapotranspiration over El Reno, Oklahoma with ASTER imagery. *Agronomie* 22: 105–106.

French AN, Jacob F, Anderson MC, et al. (2005) Surface energy fluxes with the Advanced Spaceborne Thermal Emission and Reflection radiometer (ASTER) at the Iowa 2000 SMACEX site (USA). *Remote Sensing of Environment* 99: 55–65.

Fu L-L, Alsdorf DE, Rodriguez E, et al. (2009) The SWOT (surface water and ocean topography) mission: Spaceborne radar interferometry for oceanographic and hydrological applications. soumis à OCEANOBS'09.

Gentine P, Entekhabi D, Chehbouni A, Boulet G, and Duchemin B (2007) Analysis of evaporative fraction diurnal behaviour. *Agricultural and Forest Meteorology* 143: 13–29.

GEO (2005) GEOSS. 10-Year Implementation Plan. Reference Document, GEO 1000R, GEO 1000R/ESA SP-1284, February 2005, Noordwijk, The Netherlands, 73pp.

GEO (2007) The Full Picture. http://www.earthobservations.org/documents/the_full_picture.pdf.

Gillies RR, Carlson TN, Cui J, Kustas WP, and Humes KS (1997) A verification of the triangle method for obtaining surface soil water content and energy fluxes from remote measurements of Normalized Difference Vegetation Index (NDVI) and surface radiant temperature. *International Journal of Remote Sensing* 18: 3145–3166.

Gillespie AR, et al. (1998) A temperature and emissivity separation algorithm for advanced spaceborne thermal emission and reflection radiometer ASTER images. *IEEE Transactions on Geoscience and Remote Sensing* 21: 2127–2132.

Gitelson A, Garbuzov G, Szilagyi F, Mittenzwey K, Karnieli A, and Kaiser A (1993) Quantitative remote sensing methods for real-time monitoring of inland waters quality. *International Journal of the Remote Sensing* 14: 1269–1295.

Gons H, Rijkeboer M, and Ruddick K (2002) A chlorophyll-retrieval algorithm for satellite imagery (Medium Resolution Imaging Spectrometer) of inland and coastal waters. *Journal of Plankton Research* 24: 947–951.

Gonzalez Dugo MP, Moran MS, Mateos L, and Bryant R (2005) Canopy temperature variability as an indicator of crop water stress severity. *Irrigation Science* 24: 233–240.

Gonzalez Piqueras J (2006) Evapotranspiracion de la cubierta vegetal mediante la determinacion del coeficiente de cultivo por teledeccion. PhD Thesis, Universidad de Valencia, Valencia.

Gordon H, Du T, and Zhang T (1997) Remote sensing of ocean color and aerosol properties: Resolving the issue of aerosol absorption. *Applied Optics* 36: 8670–8684.

Gordon H and Morel A (1983) *Remote Assessment of Ocean Color for Interpretation of Satellite Visible Imagery: A Review*. New York, NY: Springer.

Gordon H and Wang M (1994) Retrieval of water-leaving radiance and aerosol optical thickness over the ocean with SeaWiFS: A preliminary algorithm. *Applied Optics* 33: 443–452.

Gould R, Arnone R, and Martinolic P (1999) Spectral dependence of scattering coefficients in case I and case II waters. *Applied Optics* 38: 2377–2383.

Goutorbe JP, Dolman AJ, Gash JHC, et al. (eds.) (1997) HAPEX-Sahel. *Journal of Hydrology* 188–189: 1–1079.

Gowda PH, Chavez JL, Colaizzi PD, Evett SR, Howell TA, and Tolk JA (2007) Remote sensing based energy balance algorithms for mapping ET: Current status and future challenges. *Transactions of the ASABE* 50: 1639–1168.

Green AE, Astill MS, McAneney KJ, and Nieveen JP (2001) Path-averaged surface fluxes determined from infrared and microwave scintillometers. *Agricultural and Forest Meteorology* 109: 233–247.

Greenwald TJ, Stephens GL, Vonder Haar TH, and Jackson DL (1993) A physical retrieval of cloud liquid water over the global oceans using SSM/I observations. *Journal of Geophysical Research* 98: 18471–18488.

Gupta SK, Ritchey AC, Wilber AC, Whitlock CH, Gibson GG, and Stackhouse PW (1999) A climatology of surface radiation budget derived from satellite data. *Journal of Climate* 12: 2691–2710.

Hall DK, Riggs GA, Salomonson VV, DiGirolamo NE, and Bayr KJ (2002) MODIS snow-cover products. *Remote Sensing of Environment* 83: 181–194.

Hallikainen M, Ulaby FT, and Deventer T (1987) Extinction behavior of dry snow in the 18- to 90-GHz range. *IEEE Transactions on Geoscience and Remote Sensing* 25: 737–745.

Han Q, Rossow W, Welch R, White A, and Chou J (1995) Validation of satellite retrievals of cloud microphysics and liquid water path using observations from FIRE. *Journal of the Atmospheric Sciences* 52: 4183–4195.

Han SC and Ditmar P (2008) Localized spectral analysis of global satellite gravity fields for recovering time-variable mass redistributions. *Journal of Geodesy* 82: 423–430.

Han SC, Kim H, Yeo I-Y, et al. (2009) Dynamics of surface water storage in the Amazon inferred from measurements of inter-satellite distance change. *Geophysical Research Letters* 36: L09403.

Hanssen RF (2001) *Radar Interferometry*. 308 pp. Dordrecht: Kluwer Academic.

Heilman JL, Heilman WE, and Moore DG (1982) Evaluating the crop coefficient using spectral reflectance. *Agronomy Journal* 74: 967–971.

Hillel D (1998) *Environmental Soil Physics*. Boston: Academic Press.

Hipp J (1974) Soil electromagnetic parameters as functions of frequency, soil density, and soil moisture. *Proceedings of the IEEE* 62: 98–103.

Hitschfeld W and Bordan J (1954) Errors inherent in the radar measurement of rainfall at attenuating wavelengths. *Journal of Meteorology* 11: 58–67.

Holleman I (2007) Bias adjustment and long-term verification of radar-based precipitation estimates. *Meteorological Applications* 14: 195–203.

Holleman I, Delobbe L, and Zgonc A (2008). Update on the European weather radar network (OPERA). In: *Proceedings of ERAD, 2008*, pp. 1–5. http://www.wmo.int/pages/prog/www/IMOP/publications/IOM-96_TECO-2008/1(11)_Holleman_The_Netherlands.pdf (accessed September 2010).

Hollinger SE and Daughtry CST (1999) Southern Great Plains 1997 hydrological experiment: Vegetation sampling and data documentation. *Technical Report, Atmosphere Environment*, Section III. State Water Survey, Champaign, 1999.

Hong Y and Adler RF (2008) Predicting global landslide spatiotemporal distribution: Integrating landslide susceptibility zoning techniques and real-time satellite rainfall estimates. *International Journal of Sediment Research* 23: 249–257.

Hoogenboom H, Dekker A, and De Haan J (1998) Retrieval of chlorophyll a and suspended matter in inland waters from CASI data by matrix inversion. *Canadian Journal of Remote Sensing* 24: 144–152.

Hu C, Carder K, and Muller-Karger F (2000) Atmospheric correction of SeaWiFS imagery over turbid coastal waters: A practical method. *Remote Sensing of Environment* 74: 195–206.

Huang H-L, et al. (2004) International MODIS and AIRS processing package (IMAPP): A direct broadcast software package for the NASA Earth Observing System. *Bulletin of the American Meteorological Society* 85: 159–161.

Hurtado R and Sobrino JA (2001) Daily net radiation estimated from air temperature and NOAA–AVHRR data: A case study for the Iberian Peninsula. *International Journal of Remote Sensing* 22(8): 1521–1533.

Idso SB, Kackson RD, and Pinter PJ (1981) Normalizing the stress degree of water for environmental variability. *Agricultural and Forest Meteorology* 24: 45–55.

Idso SB, Schmugge TJ, and Jackson RD (1975) The utility of surface temperature measurements for remote sensing of soil water studies. *Journal of Geophysical Research* 80: 3044–3049.

Idso SB, Schmugge TJ, Jackson RD, and Reginato RJ (1975) The utility of surface temperature measurements for the remote sensing of the soil water status. *Journal of Geophysical Research* 80: 3044–3049.

IPCC (2008) Technical Paper on Climate Change and Water. IPCC, April 2008, 240pp. Geneva, Switzerland: World Meteorological Organization.

Irmak S, et al. (2003) Predicting daily net radiation using minimum climatological data. *Journal of Irrigation and Drainage Engineering* 129(4): 256–269.

ITC (2008) *Integrated Land and Water Information System* 3.6. Open Source. 52 North. ITC.

Jackson D and Bates J (2001) Upper tropospheric humidity algorithm assessment. *Journal of Geophysical Research* 106: 32259–32270.

Jackson RD, Idso SB, and Reginato RJ (1981) Canopy temperature as a crop water stress indicator. *Water Resources Research* 17: 1133–1138.

Jackson RD, Reginato RJ, and Idso SB (1977) Wheat canopy temperature: A practical tool for evaluating water requirements. *Water Resources Research* 13: 651–656.

Jackson TJ, et al. (1999a) Soil moisture mapping at regional scales using microwave radiometry: The Southern Great Plains hydrology experiment. *IEEE Transactions on Geoscience and Remote Sensing* 37(8): 2136–2149.

Jackson TJ, Le Vine DM, Hsu AY, et al. (1999b) Soil moisture mapping at Regional Scales using microwave radiometry: The Southern Great Plains hydrology experiment. *IEEE Transactions on Geoscience and Remote Sensing* 37(5): 2136–2151.

Jacobowitz H, et al. (2003) The advanced very high resolution radiometer pathfinder atmosphere (PATMOS) climate dataset: A resource for climate research. *Bulletin of the American Meteorological Society* 84(6): 785–793.

Jacobs C, Roerink G, and Hammani A (2008) Crop water stress detection from remote sensing using the SSEBI-2 algorithm: A case study in Morocco. Unpublished lecture, Alterra.

Jia L, Su Z, van den Hurk B, et al. (2003) Estimation of sensible heat flux using the Surface Energy Balance System (SEBS) and ATSR measurements. *Physics and Chemistry of the Earth* 28: 75–88.

Jiang L and Islam S (2001) Estimation of surface evaporation map over southern Great Plains using remote sensing data. *Water Resources Research* 37: 329–340.

Jochum MAO and Calera A (2006) Operational space-assisted irrigation advisory services: Overview of and lessons learned from the project DEMETER, AIP conference. *Earth Observation for Vegetation Monitoring and Water Management* 852: 3–13.

John VO and Soden BJ (2007) Temperature and humidity biases in global climate models and their impact on climate feedbacks. *Geophysical Research Letters* 34: L18704 (doi:10.1029/2007GL030429).

Jolivet D and Feijt A (2005) Quantification of the accuracy of LWP fields derived from NOAA-16 advanced very high resolution radiometer over three ground stations using microwave radiometers. *Journal of Geophysical Research* 110: D11204 (doi:10.1029/2004JD005205).

Joss J and Waldvogel A (1990) Precipitation measurement and hydrology. In: Atlas D (ed.) *Radar in Meteorology*, pp. 577–606. Boston, MA: American Meteorological Society.

Kabat P, Prince SD, and Prihodko L (eds.) (1997) *HAPEXnSahel West Central Supersite: Methods, Measurements and Selected Results, Report 130*. Wageningen: DLO Winand Staring Center.

Kallio K, Kutser T, Hannonen T, et al. (2001) Retrieval of water quality from airborne imaging spectrometry of various lake types in different seasons. *Science of the Total Environment* 208: 59–77.

Kalma JD and Calder IR (1994) Land surface processes in large scale hydrology. *Operational Hydrology Report No. 40*, 60pp. Geneva, Switzerland: World Meteorological Organization.

Kalma JD, McVicar TR, and McCabe MF (2008) Estimating land surface evaporation: A review of methods using remotely sensed surface temperature data. *Surveys in Geophysics* 29: 421–469.

Kargel JS, et al. (2005) Multispectral imaging contributions to global land ice measurements from space. *Remote Sensing of Environment* 99: 187–219.

Karlsson K (2003) A 10 year cloud climatology over Scandinavia derived from NOAA advanced very high resolution radiometer imagery. *International Journal of Climatology* 23: 1023–1044.

Kern M, et al. (2008) CoReH$_2$O: Candidate earth explorer core missions. *Reports for Assessment*, ESA SP-1313(3), 104pp. (ISSN 0379-6566). Noordwijk: Mission Science Division, ESA-ESTEC.

Kerr YH, et al. (2001) Soil moisture retrieval from space: The Soil Moisture and Ocean Salinity (SMOS) mission. *IEEE Transactions on Geoscience and Remote Sensing* 39: 1729–1735.

Key J, et al. (2007) *IGOS Cryosphere Theme Report*. WMO/TD No. 1405, 100pp., August 2007. http://cryos.ssec.wisc.edu/docs/cryos_theme_report.pdf (accessed September 2010).

Kidd C, Levizzani V, Turk FJ, and Ferraro RR (2009) Satellite precipitation measurements for water resource monitoring, *Journal of the American Water Resources Association* 45: 567–579.

Kim SW, Hong SH, and Won JS (2005) An application of L-band synthetic aperture radar to tide height measurement. *IEEE Transactions on Geoscience and Remote Sensing* 4: 2708–2710.

Kirdiashev KP, Chukhlantsev AA, and Shutko AM (1979) Microwave radiation of the earth's surface in the presence of vegetation cover. *Radiotekhnika i Elektronika* 24(8): 256–264.

Kite GW and Droogers P (2000) Comparing evapotranspiration estimates from satellites, hydrological models and field data. *Journal of Hydrology* 229: 3–18.

Kogan FN (1990) Remote sensing of weather impacts on vegetation in non-homogeneous areas. *International Journal of Remote Sensing* 11: 1405–1419.

Koistinen J (1991) Operational correction of radar rainfall errors due to the vertical profile of reflectivity. In: *25th Radar Meteorological Conference*, pp. 91–96. Paris, France. Reading: American Meteorological Society.

Kolberg S, Rue H, and Gottschalk L (2006) A Bayesian spatial assimilation scheme for snow coverage observations in a gridded snow model. *Hydrology and Earth System Sciences* 10: 369–381.

Kouraev AV, Zakharova EA, Samain O, Mognard NM, and Cazenave A (2004) Ob' river discharge from TOPEX/Poseidon satellite altimetry data. *Remote Sensing of Environment* 93: 238–245.

Kuchinke C, Gordon H, and Franz B (2009a) Spectral optimization for constituent retrieval in case 2 waters I: Implementation and performance. *Remote Sensing of Environment* 113: 571–587.

Kuchinke C, Gordon H, Harding L, Jr, and Voss K (2009b) Spectral optimization for constituent retrieval in case 2 waters II: Validation study in the Chesapeake Bay. *Remote Sensing of Environment* 113: 610–621.

Kuligowski R and Barros A (2001) Combined IR-microwave satellite retrieval of temperature and dewpoint profiles using artificial neural networks. *Journal of Applied Meteorology* 40: 2051–2067.

Kummerow CD, Hong Y, Olson WS, et al. (2001) The evolution of the Goddard Profiling Algorithm (GPROF) for rainfall estimation from passive microwave sensors. *Journal of Applied Meteorology* 40: 1801–1820.

Kustas WP (1990) Estimates of evapotranspiration within a one- and two-layer model of heat transfer over partial vegetation cover. *Journal of Applied Meteorology* 29: 704–715.

Kustas WP, Diak GR, Norman JM. (2001). Time difference methods for monitoring regional scale heat fluxes with remote sensing. In: Lakshmi V, Albertson J, and Schaake J. (eds.), *Observations and Modeling of the Land Surface Hydrological Processes*, American Geophysical Union Water Science and Application Series, vol. 3, pp. 15–29. Washington, DC: American Geophysical Union.

Kustas WP and Goodrich DC (1994) Monsoon'90 multidisciplinary experiment, preface. *Water Resources Research* 30: 1211–1225.

Kustas WP, Humes KS, Norman JM, and Moran MS (1996) Single- and dual-source modeling of surface energy fluxes with radiometric surface temperature. *Journal of Applied Meteorology* 35: 110–121.

Kustas WP and Norman JM (1996) Use of remote sensing for evapotranspiration monitoring over land surfaces. *Hydrological Sciences* 41: 495–516.

Kustas WP and Norman JM (1997) A two-source approach for estimating turbulent fluxes using multiple angle thermal infrared observations. *Water Resources Research* 33: 1495–1508.

Kustas WP and Norman JM (1999) Evaluation of soil and vegetation heat flux predictions using a simple two-source model with radiometric temperatures for partial canopy cover. *Agricultural and Forest Meteorology* 94: 13–29.

Kustas WP and Norman JM (2000) A two-source energy balance approach using directional radiometric temperature observations for sparse canopy covered surfaces. *Agronomy Journal* 92: 847–854.

Kustas WP, Norman JM, Schmugge TJ, and Anderson MC (2004) Mapping surface energy fluxes with radiometric temperature. In: Quattrochi DA and Luvall JC (eds.) *Thermal Remote Sensing in Land Surface Processes*, pp. 205–253. Boca Raton, FL: CRC Press.

Lagouarde J-P and McAneney KJ (1992) Daily sensible heat flux estimation from a single measurement of surface temperature and maximum air temperature. *Boundary-Layer Meteorology* 59: 341–362.

Laing AG, Carbone RE, Levizzani V, and Tuttle J (2008) The propagation and diurnal cycles of deep convection in northern tropical Africa. *Quarterly Journal of the Royal Meteorological Society* 134: 93–109.

Laviola S and Levizzani V (2008) Rain retrieval using 183 GHz absorption lines. In: *IEEE Proceedings of the MicroRad 2008, 10th Specialist Meeting of Microwave Radiometry and Remote Sensing of the Environment*, pp. 1–4 (doi: 10.1109/MICRAD.2008.4579505). Firenze, 11–14 March.

Leese J, Jackson TJ, Pitman A, and Dirmeyer P (2000) Meeting summary, GEWEX/BAHC International Workshop on soil moisture monitoring, analysis, and prediction for hydrometeorological and hydroclimatological applications. *Bulletin of the American Meteorological Society* 39: 1248–1268.

LeFavour G and Alsdorf D (2005) Water slope and discharge in the Amazon River estimated using the shuttle radar topography mission digital elevation model. *Geophysical Research Letters* 32: L17404 (doi:10.1029/2005GL023836).

Lemke P, Ren J, et al. (2007) Observations: Changes in snow, ice and frozen ground. In: *Climate Change 2007: The Physical Science Basis, Contribution of Working Group I to the Fourth Assessment Report of the Intergovernmental Panel on Climate Change*, pp. 337–383. Cambridge and New York, NY: Cambridge University Press.

Levizzani V, Bauer P, and Turk FJ (eds.) (2007) *Measuring Precipitation from Space*. Dordrecht: Springer., 722pp.

Lhomme JP, Monteney B, and Amadou M (1994) Estimating sensible heat flux from radiometric temperature over sparse millet. *Agricultural and Forest Meteorology* 68: 77–91.

Li F, Kustas WP, Prueger JH, Neale CMU, and Jackson TJ (2005) Utility of remote sensing based two-source energy balance model under low and high vegetation cover conditions. *Journal of Hydrometeorology* 6: 878–891.

Li J, Wolf WW, Menzel WP, Zhang WJ, Huang HL, and Achtor TH (2000) Global soundings of the atmosphere from ATOVS measurements: The algorithm and validation. *Journal of Applied Meteorology* 39: 1248–1268.

Liang S (2001) Narrowband to broadband conversions of land surface albedo I: Algorithms. *Remote Sensing of Environment* 76(2): 213–238.

Liang S (2004) *Quantitative Remote Sensing of Land Surface*. Hoboken, NJ: Wiley.

Liang S, et al. (2003) Narrowband to broadband conversions of land surface albedo: II. Validation. *Remote Sensing of Environment* 84(1): 25–41.

Liebe HJ (2004) MPM – an atmospheric millimeter-wave propagation model. *International Journal of Infrared and Millimeter Waves* 10(8): 631–650.

Liu Y, De Jeu RAM, van Dijk AIJM, and Owe M (2007) TRMM-TMI satellite observed soil moisture and vegetation density (1998–2005) show strong connection with El Nino in eastern Australia. *Geophysical Research Letters* 34 doi:10.1029/2007GL030311.

Loew A (2007) Remote sensing data assimilation. *Special Issue. Remote Sensing of Environment* 112(4): 1257.

Loew A, Holmes T, and De Jeu R (2009) The European heat wave 2003: Early indicators from multisensoral microwave sensing. *Journal of Geophysical Research* 114: D05103 (doi:10.1029/2008JD010533).

Loiselet M, Stricker N, Menard Y, and Luntama JP (2000) GRAS – Metop's GPS-based atmospheric sounder. *ESA Bulletin* 102: 38–44.

Longépé N, Allain S, Ferro-Famil L, Pottier E, and Durand Y (2009) Snowpack characterization in mountainous regions using C-band SAR data and a meteorological model. *IEEE Transactions on Geoscience and Remote Sensing* 47(2): 406–418.

Lu Z, Crane M, Kwoun O, Wells C, Swarzenski C, and Rykhus R (2005) C-band radar observes water-level change in coastal Louisiana swamp forests. *EOS, Transactions, American Geophysical Union* 86: 141–144.

Luojus KP, Pulliainen JT, Metsämäki SJ, and Hallikainen MT (2007) Snow-covered area estimation using satellite radar wide-swath images. *IEEE Transactions on Geoscience and Remote Sensing* 45(4): 978–989.

Luthcke SB, Zwally HJ, Abdalati W, et al. (2006) Recent Greenland ice mass loss by drainage system from satellite gravity observations. *Science* 314: 1286–1289.

Magnússon E, Rott H, Björnsson H, and Pálsson F (2007) The impact of jökulhlaups on basal sliding observed by SAR interferometry on Vatnajökull. *Journal of Glaciology* 35(181): 232–240.

Maheu C, Cazenave A, and Mechoso C (2003) Water level fluctuations in the Plata Basin (South America) from Topex/Poseidon Satellite Altimetry. *Geophysical Research Letters* 30: 1143–1146.

Mann J and Lenschow DH (1994) Errors in airborne flux measurements. *Journal of Geophysical Research* 99: 14519–14526.

Maritorena S and Siegel D (2005) Consistent merging of satellite ocean color data sets using a bio-optical model. *Remote Sensing of Environment* 94: 429–440.

Marshall J and Palmer W (1948) The distribution of raindrops with size. *Journal of Meteorology* 5: 165–166.

Massman WJ (1999) A model study of kB-1 H for vegetated surfaces using 'localized near-field' Lagrangian theory. *Journal of Hydrology* 223: 27–43.

Mätzler C (1987) Applications of the interaction of of microwaves with the natural snow cover. *Remote Sensing Reviews* 2: 259–387.

McCabe MF and Wood EF (2006) Scale influences on the remote estimation of evapotranspiration using multiple satellite sensors. *Remote Sensing of Environment* 105: 271–285.

McNaughton KG and Spriggs TW (1989) *An Evaluation of the Priestley and Taylor Equation and the Complementary Relationship Using Results from a Mixed-Layer Model of the Convective Boundary Layer*, pp 89–104. Wallingford: IAHS Red Book.

McVicar TR and Jupp DLB (2002) Using covariates to spatially interpolate moisture availability in the Murray–Darling Basin: A novel use of remotely sensed data. *Remote Sensing of Environment* 79: 199–212.

Mecikalski JR, Diak GR, Anderson MC, and Norman JM (1999) Estimating fluxes on continental scales using remotely-sensed data in an atmospheric–land exchange model. *Journal of Applied Meteorology* 38: 1352–1369.

Meesters AGCA, De Jeu RAM, and Owe M (2005) Analytical derivation of the vegetation optical depth from the microwave polarization difference index. *IEEE Geoscience and Remote Sensing Letters* 2(2): 121–123.

Mekonnen G and Bastiaanssen WGM (2000) A new simple method to determine crop coefficients for water allocation planning from satellites: Results from Kenya. *Irrigation and Drainage Systems* 14: 237–256.

Menenti M and Choudhury BJ (1993) Parametrization of land surface evapotranspiration using a location-dependent potential evapotranspiration and surface temperature range. In: Bolle HJ, Feddes RA, and Kalma JD (eds.) *Exchange Processes at the Land Surface for a Range of Space and Time Scales*, Proceedings of the Yokohama Symposium, July 1993, pp. 561–568. Yokohama: IAHS.

Menenti M, Jia L, and Bastiaanssen WGM (2004) Energy and water flow through the soil–vegetation–atmosphere system: The fiction of measurements and the reality of models. In: Feddes RA, de Rooij GH, and van Dam JC (eds.) *Frontis Workshop on Un-Saturated Zone Modeling: Progress, Challenges and Applications*, pp. 211–229. Wageningen, The Netherlands, 3–5 October 2004.

Mercier F, Cazenave A, and Maheu C (2002) Interannual lake level fluctuations in Africa from Topex–Poseidon: Connections with ocean–atmosphere interactions over the Indian Ocean. *Global and Planetary Change* 32: 141–163.

Merlin O and Chehbouni A (2004) Different approaches in estimating heat flux using dual angle observations of radiative surface temperature. *International Journal of Remote Sensing* 25: 275–289.

Michelson DB, Andersson T, Koistinen J, et al. (2000) BALTEX Radar Data Centre Products and their methodologies. Report RMK No. 90, Swedish Meteorological and Hydrological Institute.

Mieruch S, Noel S, Bovensmann H, and Burrows JP (2008) Analysis of global water vapor trends from satellite measurements in the visible spectral range. *Atmospheric Chemistry and Physics* 8: 491–504.

Milz M, Buehler SA, and John VO (2009) Comparison of AIRS and AMSU-B monthly mean estimates of upper tropospheric humidity. *Geophysical Research Letters* 36: L10804 (doi:10.1029/2008GL037068).

Minnis P, et al. (2003) A global cloud database from VIRS and MODIS for CERES. *Proceedings of the SPIE* 4891: 115–126.

Mironov VL, Dobson C, Kaupp VH, Komarov VA, and Kleshchenko VN (2004) Generalized refractive mixing dielectric model for moist soils. *IEEE Transactions on Geoscience and Remote Sensing* 42: 773–785.

Mladenova I, Lakshmi V, Walker J, Long D, and De Jeu R (2009) Observations for disaggregation of radiometer derived soil moisture estimates over the NAFE'06 study area. *IEEE Geoscience and Remote Sensing Letters* 6(4): 640–643 (doi:10.1109/LGRS.2009.2021492).

Mognard NM and Alsdorf DE (2006) Why a hydrology mission needs two-dimensional acquisitions of water surface elevations? In: *Proceedings of the 15 Years of Progress in Radar Altimetry Symposium Venice*, ESA Special Publication SP-614. Italy, 13–18 March 2006.

Mognard NM, Alsdorf DE, Cazenave A, and Rodriguez E (2007) The water and terrestrial elevation recovery hydrosphere mapper (WATER-HM): A dedicated surface water mission. In: *Proceedings of the 2nd Space for Hydrology Workshop*, ESA Workshop Proceedings Publication, WPP-280. 11–14 November 2007. Geneva: WMO.

Molotoch NP (2009) Reconstructing snow water equivalent in the Rio Grande headwaters using remotely sensed snow cover data and a spatially distributed snowmelt model. *Hydrological Processes* 23: 1076–1089.

Moran MS, Clarke T, Inoue Y, and Vidal A (1994) Estimating crop water deficit using the relation between surface-air temperature and spectral vegetation index. *Remote Sensing of Environment* 49: 246–263.

Moran MS and Jackson RD (1991) Assessing the spatial distribution of evapotrnaspiration using remotely sensed inputs. *Journal of Environmental Quality* 20: 725–737.

Mueller RW, Matsoukas C, Gratzki A, Behr HD, and Hollmann R (2009) The CM-SAF operational scheme for the satellite based retrieval of solar surface irradiance — a LUT based eigenvector hybrid approach. *Remote Sensing of Environment* 113: 1012–1024.

Muller-Krager F, Hu C, Andréfouët S, Varela R, and Thunell R (2005) The color of the coastal ocean and applications in the solution of research and management problems. In: Miller R, Del Castillo C, and McKee B (eds.) *Remote Sensing of Coastal Aquatic Environments: Technologies, Techniques and Applications*, pp. 101–127. Dordrecht: Springer.

Nagler T and Rott H (2000) Retrieval of wet snow by means of multitemporal SAR data. *IEEE Transactions on Geoscience and Remote Sensing* 38(2): 754–765.

Nagler T and Rott H (2005) Snow classification algorithm for Envisat ASAR. In: *Proceedings of the Envisat & ERS Symposium*, ESA SP-572. Salzburg, Austria, 6–10 September 2004. http://earth.esa.int/workshops/salzburg04/papers_posters/4C4_nagler_169.pdf (accessed September 2010).

Nagler T, Rott H, Malcher P, and Müller F (2008) Assimilation of meteorological and remote sensing data for snowmelt runoff forecasting. *Remote Sensing of Environment* 112: 1408–1420.

Nakajima T and King MD (1990) Determination of the optical thickness and effective particle radius of clouds from reflected solar radiation measurements. Part I. Theory. *Journal of the Atmospheric Science* 47: 1878–1893.

Nauss T and Kokhanovsky AA (2007) Assignment of rainfall confidence values using multispectral satellite data at mid-latitudes: First results. *Advanced Geosciences* 10: 99–102.

Nemani RR and Running SW (1989) Estimation of regional surface resistance to evapotranspiration from NDVI and thermal-IR AVHRR data. *Journal of Applied Meteorology* 28: 276–284.

Neukermans G, Nechad B, and Ruddick K (2008) Optical remote sensing of coastal waters from geostationary platforms: A feasibility study – Mapping Total Suspended Matter with SEVIRI. In: *Ocean Optics*. Barga, Italy, 6–10 October 2008. Halifax: Lewis Conference Services International.

Nieuwenhuis GJA, Smidt EH, and Thunissen HAM (1985) Estimation of regional evapotranspiration of arable crops from thermal infrared images. *International Journal of Remote Sensing* 6: 1319–1334.

Niu GY, Seo KW, Yang ZL, et al. (2007) Retrieving snow mass from GRACE terrestrial water storage change with a land surface model. *Geophysical Research Letters* 34: L15704.

Njoku E, Ashcroft P, Chan T, and Li L (2005) Global survey and statistics of radio-frequency interference in AMSR-E land observations. *IEEE Transactions on Geoscience and Remote Sensing* 43(5): 938–947.

Njoku EG and Chan SK (2006) Vegetation and surface roughness effects on AMSR-E land observations. *Remote Sensing of Environment* 100(2): 190–199.

Njoku EG, Jackson TJ, Lakshmi V, Chan TK, and Nghiem SV (2003) Soil moisture retrieval from AMSR-E. *IEEE Transactions on Geoscience and Remote Sensing* 41(2): 215–229.

Njoku EG and Kong J (1977) Theory for passive microwave remote sensing of near-surface soil moisture. *Journal of Geophysical Research* 82(20): 3108–3118.

Noel S, Buchwitz M, Bovensmann H, Hoogen R, and Burrows J P (1999) Atmospheric water vapor amounts retrieved from GOME satellite data. *Geophysical Research Letters* 26: 1841–1844.

Norman J, Kustas WP, and Humes KS (1995) A two-source approach for estimating soil and vegetation energy fluxes in observations of directional surface temperature. *Agricultural and Forest Meteorology* 77: 263–293.

NRC (*National Research Council*) (2007) Earth Science and Applications from Space, Earth Science and Applications from Space: National Imperatives for the Next Decade and Beyond, 456pp. Washington, DC: National Academies Press.

Overgaard J, Rosjberg D, and Butts MB (2006) Land-surface modelling in hydrological perspective – a review. *Biogeosciences* 3: 229–241.

Owe M, De Jeu R, and Holmes T (2008) Multisensor historical climatology of satellite-derived global land surface moisture. *Journal of Geophysical Research* 113: F01002 (doi:10.1029/2007JF000769).

Parajka J, et al. (2006) Assimilating scatterometer soil moisture data into conceptual hydrologic models at the regional scale. *Hydrology and Earth System Sciences* 10(3): 353–368.

Pathe C, Wagner W, Sabel D, Doubkova M, and Basara J (2009) Using ENVISAT ASAR Global Mode data for surface soil moisture retrieval over Oklahoma, USA. *IEEE Transactions on Geoscience and Remote Sensing* 47(2): 468–480.

Paul F, Kääb A, Maisch M, Kellenberger T, and Haeberli W (2002) The new remote sensing-derived Swiss glacier inventory: I. Methods. *Annals of Glaciology* 34: 355–361.

Pellarin T, Wigneron JP, Calvet JC, and Waldteufel P (2003) Global soil moisture retrieval from a synthetic L-band brightness temperature data set. *Journal of Geophysical Research* 108: 9–1.

Pinker R, Frouin R, and Li Z (1995) A review of satellite methods to derive surface shortwave irradiance. *Remote Sensing of Environment* 51: 108–124.

Platnick S (2001) A superposition technique for deriving mean photon scattering statistics in plane-parallel cloudy atmospheres. *Journal of Quantitative Spectroscopy and Radiative Transfer* 68: 57–73.

Platnick S, et al. (2003) The MODIS cloud products: algorithms and examples from Terra. *IEEE Transactions on Geoscience and Remote Sensing* 41: 459–473.

Pottier C, Garçon V, Larnicol G, Sudre J, Schaeffer P, and Le Traon P-Y (2006) Merging SeaWiFS and MODIS/Aqua ocean color data in North and Equatorial Atlantic using weighted averaging and objective analysis. *IEEE Transactions on Geoscience and Remote Sensing* 44: 3436–3451.

Pougatchev P, et al. (2009) IASI temperature and water vapor retrievals – error assessment and validation. *Atmospheric Chemistry and Physics Dicussions* 9: 7971–7989.

Pratt DA and Ellyet CD (1979) The thermal inertia approach to mapping of soil moisture and geology. *Remote Sensing of Environment* 8: 147–158.

Price JC (1982) Estimation of regional scale evapotranspiration through analysis of satellite thermal-infrared data. *IEEE Transactions on Geoscience and Remote Sensing* GE-20: 286–292.

Price JC (1985) On the analysis of thermal infrared imagery, the limited utility of apparent thermal inertia. *Remote Sensing of Environment* 18: 59–73.

Price JC (1990) Using spatial context in satellite data to infer regional scale evapotranspiration. *IEEE Transactions on Geoscience and Remote Sensing* 28: 940–948.

Probert-Jones JR (1962) The radar equation in meteorology. *Quarterly Journal of the Royal Meteorological Society* 88: 486–495.

Pu Z, Xu L, and Salomonson VV (2007) MODIS/Terra observed seasonal variations of snow cover over the Tibetan Plateau. *Geophysical Research Letters* 34(6): L06706.

Pulliainen J (2006) Mapping of snow water equivalent and snow depth in boreal and sub-arctic zones by assimilating space-borne microwave radiometer data and ground-based observations. *Remote Sensing of Environment* 101: 257–269.

Pulliainen JT, Hallikainen MT, and Grandell J (1999) HUT snow emission model and its applicability to snow water equivalent retrieval. *IEEE Transactions on Geoscience and Remote Sensing* 37(8): 1378–1390.

Pulliainen JT, Manninen T, and Hallikainen M (1998) Application of ERS-1 Wind Scatterometer data to soil frost and soil moisture monitoring in boreal forest zone. *IEEE Transactions on Geoscience and Remote Sensing* 36(3): 849–863.

Quattrochi DA and Luvall JC (1999) Thermal infrared remote sensing for analysis of landscape ecological processes: Methods and applications. *Landscape Ecology* 14: 577–598.

Querner EP and van Bakel PJT (1989) *Description of the Regional Groundwater Flow Model SIMGRO*. Wageningen: DLO-Staring Centrum.

Rahman H and Dedieu G (1994) SMAC: A simplified method for the atmospheric correction of satellite measurements in the solar spectrum. *International Journal of Remote Sensing* 15(1): 123–143.

Randel DL, Vonder TH, Haar MA, et al. (1996) A new global water vapor dataset. *Bulletin of the American Meteorological Society* 77: 1233–1246.

Rast M, Bezy JL, and Bruzzi S (1999) The ESA Medium Resolution Imaging Spectrometer MERIS – a review of the instrument and its mission. *International Journal of Remote Sensing* 20: 1681–1702.

Raup BH, et al. (2007) Remote sensing and GIS technology in the Global Land Ice. Measurements from Space (GLIMS) Project. *Computers and Geosciences* 33: 104–125.

Ray SS and Dadhwal VK (2001) Estimation of crop evapotranspiration of irrigation command area using remote sensing and GIS. *Agricultural Water Management* 49(3): 239–249.

Rees WG (2001) *Physical Principles of Remote Sensing*. Cambridge: Cambridge University Press.

Rinehart RE (2004) *Radar for Meteorologists*, 4th edn., 482pp. Columbia, MO: Rinehart Publications

Riou C, Itier B, and Seguin B (1988) The influence of surface roughness on the simplified relationship between daily evaporation and surface temperature. *International Journal of Remote Sensing* 9: 1529–1533.

Robinson DA, Dewey KF, and Heim RR (1993) Global snow cover monitoring: an update. *Bulletin of the American Meteorological Society* 74: 1689–1696.

Rodell M, Chao BF, Au AY, Kimball J, and McDonald K (2003) Global biomass variation and its geodynamic effects, 1982–1998. *Earth Interactions* 9(2): 1–19.

Rodell M, Chen J, Kato H, Famiglietti JS, Nigro J, and Wilson CR (2007) Estimating ground water storage changes in the Mississippi River Basin (USA) using GRACE. *Hydrogeology Journal* 15: doi:10.1007/s10040-006-0103-7.

Rodell M and Famiglietti JS (1999) Detectability of variations in continental water storage from satellite observations of the time dependent gravity field. *Water Resources Research* 35: 2705–2723.

Rodell M and Famiglietti JS (2001) An analysis of terrestrial water storage variations in Illinois with implications for the Gravity Recovery and Climate Experiment (GRACE). *Water Resources Research* 37: 1327–1340.

Rodell M, Famiglietti JS, Chen J, et al. (2004a) Basin scale estimates of evapotranspiration using GRACE and other observations. *Geophysical Research Letters* 31: L20504 (doi:10.1029/2004GL020873).

Rodell M and Houser PR (2004) Updating a land surface model with MODIS-derived snow cover. *Journal of Hydrometeorology* 5: 1064–1075.

Rodell M, Houser PR, Jambor U, et al. (2004b) The Global Land Data Assimilation System. *Bulletin of the American Meteorological Society* 85(3): 381–394.

Rodell M, Velicogna I, and Famiglietti JS (2009) Satellite-based estimates of groundwater depletion in India. *Nature*, doi:10.1038/nature08238.

Rodriguez E, Morris C, Belz J, et al. (2005) An assessment of the SRTM topographic products. *Technical Report JPL D-31639, Jet Propulsion Laboratory*, Pasadena, CA, 143pp.

Roebeling RA, Deneke HM, and Feijt AJ (2008) Validation of cloud liquid water path retrievals from SEVIRI using one year of CloudNET observations. *Journal of Applied Meteorology and Climatology* 47: 206–222.

Roebeling RA, Feijt AJ, and Stammes P (2006) Cloud property retrievals for climate monitoring: Implications of differences between SEVIRI on METEOSAT-8 and AVHRR on NOAA-17. *Journal of Geophysical Research* 111: D20210 (doi:10.1029/2005JD006990).

Roebeling RA and Holleman I (2009) Development and validation of rain rate retrievals from SEVIRI using weather radar observations. *Journal of Geophysical Research* (submitted).

Roerink GJ, Bastiaanssen WGM, Chambouleyron J, and Menenti M (1997) Relating crop water consumption to irrigation water supply by remote sensing. *Water Resources Management* 11: 445–465.

Roerink GJ, Su Z, and Menenti M (2000) S-SEBI: A simple remote sensing algorithm to estimate the surface energy balance. *Physics and Chemistry of the Earth (B)* 25: 147–157.

Rosenfeld D and Gutman G (1994) Retrieving microphysical properties near the tops of potential rain clouds by multispectral analysis of AVHRR data. *Atmospheric Research* 34: 259–283.

Rosenfeld D and Lensky IM (1998) Satellite-based insights into precipitation formation processes in continental and maritime clouds. *Bulletin of the American Meteorological Society* 79: 2457–2476.

Rossa A, Bruen M, Macpherson B, Holleman I, Michelson D, and Michaelides S (eds.) (2005) COST Action 717: Use of Radar *Observations in Hydrology and NWP Models*, EUR 21954, 292pp. Brussels: European Union.

Rossow WB and Garder L (1993) Cloud detection using satellite measurements of infrared and visible radiances for ISCCP. *Journal of Climate* 6: 2341–2369.

Rossow WB and Schiffer RA (1999) Advances in understanding clouds from ISCCP. *Bulletin of the American Meteorological Society* 80: 2261–2287.

Rott H (1997) Capabilities of microwave sensors for monitoring areal extent and physical properties of the snowpack. In: Sooroshian S, Gupta HV, and Rodda JC (eds.) *Land Surface Processes in Hydrology, NATO ASI Series I: Global Environmental Change*, vol. 46, pp. 135–167. Berlin: Springer.

Rott H, Nagler T, Malcher P, and Müller F (2008) A satellite-based information system for glacier monitoring and modelling. In: Gomarasca MA (ed.) *GeoInformation in Europe, Proceedings of the 27th EARSeL Symposium*, pp. 395–402 (ISBN 9789059660618). Rotterdam: Millpress.

Rowlands DD, Luthcke SB, Klosko SM, et al. (2005) Resolving mass flux at high spatial and temporal resolution using GRACE intersatellite measurements. *Geophysical Research Letters* 32: L04310 (doi:10.1029/2004GL021908).

Rubio E, Mejuto M, Calera A, Vela A, Castaño S, and Moratalla A (2003) Validation of an operational modelling of evapotranspiration and direct recharge. In: *Proceedings of the 10th International Symposium*. Barcelona: SPIE.

Ruddick K, Ovidio F, and Rijkeboer M (2000) Atmospheric correction of SeaWiFS imagery for turbid coastal and inland waters. *Applied Optics* 39: 897–912.

Rudiger C, et al. (2009) An Intercomparison of ERS-Scat, AMSR-E soil moisture observations with model simulations over France. *Journal of Hydrometeorology* 10: 431–447.

Sahoo AK, et al. (2008) Evaluation of AMSR-E soil moisture results using the *in-situ* data over the Little River Experimental Watershed, Georgia. *Remote Sensing of Environment* 112(6): 3142–3152.

Salama MS and Shen F (2009) Simultaneous atmospheric correction and quantification of suspended particulate matters from orbital and geostationary earth observation sensors. *Estuarine, Coastal and Shelf Science* (in press).

Salama MS and Stein A (2009) Error decomposition and estimation of inherent optical properties. *Applied Optics* 48: 4947–4962.

Salama MS, Su Z, Mannaerts CM, and Verhoef W (2009) Deriving inherent optical properties and associated uncertainties for the Dutch inland waters during the eagle campaign. *Hydrology and Earth System Sciences* 13: 1113–1121.

Salama S (2003) Optical remote sensing for the estimation of marine bio-geophysical quantities. In: *Hydraulics*, p. 175. Leuven: Katholieke Universiteit Leuven.

Salama S, Monbaliu J, and Coppin P (2004) Atmospheric correction of Advanced Very High Resolution Radiometer imagery. *International Journal of Remote Sensings* 25: 1349–1355.

Salminen M, Pulliainen J, Metsämäki S, Kontu A, and Suokanerva H (2009) The behaviour of snow and snow-free surface reflectance in boreal forests: Implications to the performance of snow covered area monitoring. *Remote Sensing of Environment* 113: 907–918.

Sanchez JM, Kustas WP, Caselles V, and Anderson MC (2008) Modelling surface energy fluxes over maize using a two-source patch model and radiometric soil and canopy temperature observations. *Remote Sensing of Environment* 112: 1130–1143.

Sandholt I, Rasmussen K, and Andersen J (2002) A simple interpretation of the surface temperature/vegetation index space for assessment of surface moisture status. *Remote Sensing of Environment* 79: 213–224.

Santer BD, et al. (2007) Identification of human-induced changes in atmospheric moisture content. *Proceedings of the National Academy of Sciences* 104(39): 15248–15253.

Saunders RW and Kriebel KT (1988) An improved method for detecting clear sky and cloudy radiances from AVHRR radiances. *International Journal of Remote Sensings* 9: 123–150.

Schanda E (1986) *Physical Fundamentals of Remote Sensing*, 187pp. Berlin and Heidelberg: Springer.

Schmetz J, Pili P, Tjemkes S, et al. (2002) An introduction to Meteosat Second Generation (MSG). *Bulletin of the American Meteorological Society* 83: 977–992.

Schmetz J and Turpeinen O (1988) Estimation of the upper tropospheric relative humidity field from METEOSAT water vapor image data. *Journal of Applied Meteorology* 27: 889–899.

Schmugge T, Gloersen P, Wilheit T, and Geiger F (1974) Remote-sensing of soil-moisture with microwave radiometers. *Journal of Geophysical Research* 79(2): 317–323.

Schuepp PH, MacPherson JI, and Desjardins RL (1992) Adjustment of footprint correction for airborne flux mapping over the FIFE site. *Journal of Geophysical Research* 97: 18455–18466.

Schulz J, et al. (2009) The EUMETSAT satellite application facility on climate monitoring (CM-SAF): Mission concept and first results. *Atmospheric Chemistry and Physics* 9: 1687–1709.

Scipal K, Drusch M, and Wagner W (2008) Assimilation of a ERS scatterometer derived soil moisture index in the ECMWF numerical weather prediction system. *Advances in Water Resources* 31(8): 1101–1112.

Scott N, et al. (1999) Characteristics of the TOVS Pathfinder Path-B dataset. *Bulletin of the American Meteorological Society* 80: 2679–2701.

Seguin B and Itier B (1983) Using midday surface temperature to estimate daily evporation from satellite thermal IR data. *International Journal of Remote Sensings* 4: 371–383.

Sellers PJ, Hall FG, Asrar G, Strebel DE, and Murphy RE (1992) An overview of the First International Satellite Land Surface Climatology Project (ISLSCP) Field Experiment (FIFE). *Journal of Geophysical Research* 97(D17): 18345–18371.

Shamir E and Georgakakos KP (2006) Distributed snow accumulation and ablation modeling in the American River basin. *Advances in Water Resources* 29: 558–570.

Shen F, Salama MS, Zhou Y, Li J, Su Z, and Kuang D (2010) Remote-sensing reflectance characteristics of highly turbid estuarine waters – a comparative experiment of the Yangtze River and the Yellow River. *International Journal of Remote Sensing* 31(10): 2639–2654.

Shiklomanov IA (1993) World fresh water resources. In: Gleick PH (ed.) *Water in Crisis*. New York, NY: Oxford University Press.

Shuttleworth JW and Wallace JS (1985) Evaporation from sparse crops – an energy combination theory. *Quarterly Journal of the Royal Meteorological Society* 111: 839–855.

Shuttleworth WJ (1988) Macrohydrology – the new challenge for process hydrology. *Journal of Hydrology* 100: 31–56.

Shuttleworth WJ, Gurney RJ, Hsu AY, and Ormsby JP (1989) *FIFE: The Variation in Energy Partition at Surface Flux Sites*, vol. 186, pp. 67–74. Wallingford: IAHS Publication.

Siegel D, Wang M, Maritorena S, and Robinson W (2000) Atmospheric correction of satellite ocean color imagery: The black pixel assumption. *Applied Optics* 39: 3582–3591.

Simeoni D, Singer C, and Chalon G (1997) Infrared atmospheric sounding interferometer. *Acta Astronautica* 40: 113–118.

Sirguey P, Mathieu R, and Arnaud Y (2009) Subpixel monitoring of the seasonal snow cover with MODIS at 250 m spatial resolution in the Southern Alps of New Zealand: Methodology and accuracy assessment. *Remote Sensing of Environment* 113: 160–181.

Slater AG and Clark MP (2006) Snow data assimilation via an ensemble Kalman filter. *Journal of Hydrometeorology* 7: 478–493.

Smith LC (1997) Satellite remote sensing of river inundation area, stage, and discharge: A review. *Hydrological Processes* 11: 1427–1439.

Sobrino JA, Jiménez-Muñoz JC, Sòria G, et al. (2008) Thermal remote sensing in the framework of the SEN2FLEX project: Field measurements, airborne data and applications. *International Journal of Remote Sensing* 29(17): 4961–4991.

Sobrino JA, Jimenez-Munoz JC, Zarco-Tejada PJ, et al. (2009) Thermal remote sensing from Airborne Hyperspectral Scanner data in the framework of the SPARC and SEN2FLEX projects: An overview. *Hydrology and Earth System Sciences* 13: 2031–2037.

Sobrino JA and Raissouni N (2003) Surface temperature and water vapour retrieval from MODIS data. *International Journal of Remote Sensings* 24(24): 5162–5182.

Soden BJ and Lanzante JR (1996) An assessment of satellite and radiosonde climatologies of upper-tropospheric water vapor. *Journal of Climate* 9: 1235–1250.

Soer GJR (1980) Estimation of regional evapotranspiration and soil moisture conditions using remotely sensed crop surface temperatures. *Remote Sensing of Environment* 9: 27–45.

Sohn BJ and Smith EA (2003) Explaining sources of discrepancy in SSM/I water vapor algorithms. *Journal of Climate* 16: 3229–3255.

Solberg R, Huseby RB, Koren H, and Malnes E (2008) Time-series fusion of optical and SAR data for snow cover area mapping. In: *Proceedings of the 5th EARSeL LIS-SIG Workshop: Remote Sensing of Land Ice and Snow*. Bern, 11–13 February 2008. http://www.conferences.earsel.org/abstract/show/395 (accessed September 2010).

Stamnes K, Li W, Yan B, et al. (2003) Accurate and self-consistent ocean color algorithm: Simultaneous retrieval of aerosol optical properties and chlorophyll concentrations. *Applied Optics* 42: 939–951.

Stokstad E (1999) Scarcity of rain, stream gages threatens forecasts. *Science* 285: 1199.

Stone LR and Horton ML (1974) Estimating evapotrnapsiration using canopy temperatures: Field evaluation. *Agronomy Journal* 66: 450–454.

Strassberg G, Scanlon BR, and Rodell M (2007) Comparison of seasonal terrestrial water storage variations from GRACE with groundwater-level measurements from the High Plains Aquifer (USA). *Geophysical Research Letters* 34: L14402 (doi:10.1029/2007GL030139).

Strozzi T, Luckman A, Murray T, Wegmüller U, and Werner CL (2002) Glacier motion estimation using SAR offset-tracking procedures. *IEEE Transactions on Geoscience and Remote Sensing* 40(119): 2384–2391.

Stuefer M, Rott H, and Skvarca P (2007) Glaciar Perito Moreno, Patagonia: Climate sensitivities and glacier characteristics preceding the 2003/2004 and 2005/2006 damming events. *Journal of Glaciology* 53(180): 1–13.

Su Z (2002) The Surface Energy Balance System (SEBS) for estimation of turbulent heat fluxes. *Hydrology and Earth System Sciences* 6: 85–99.

Su Z (2005) Estimation of the surface energy balance. In: Anderson MG and McDonnell JJ (eds.) *Encyclopedia of Hydrological Sciences*, pp. 731–752. Chichester: Wiley.

Su Z and Pelgrum H (1999) and Menenti MAggregation effects of surface heterogeneity in land surface processes. *Hydrology and Earth System Sciences* 3: 549–563.

Su Z, Schmugge T, Kustas WP, and Massman WJ (2001) An evaluation of two models for estimation of the roughness height for heat transfer between the land surface and the atmosphere. *Journal of Applied Meteorology* 40: 1933–1951.

Su Z, Timmermans WJ, Gieske A, et al. (2008) Quantification of land–atmosphere exchanges of water, energy and carbon dioxide in space and time over the heterogeneous Barrax site. *International Journal of Remote Sensing* 29(17): 5215–5235.

Su Z, Timmermans WJ, Tol Cvd, et al. (2009) EAGLE 2006 – multi-purpose, multi-angle and multi-sensor in-situ, airborne and space borne campaigns over grassland and forest. *Hydrology and Earth System Sciences* 13: 833–845.

Su Z, et al. (2003) Assessing relative soil moisture with remote sensing data: Theory and experimental validation. *Physics and Chemistry of the Earth* 28(1–3): 89–101.

Surussavadee C and Staelin DH (2008) Global millimeter-wave precipitation retrievals trained with a cloud-resolving numerical weather prediction model, part I: Retrieval design. *IEEE Transactions on Geoscience and Remote Sensing* 46: 99–108.

Swenson S, Famiglietti J, Basara J, and Wahr J (2008) Estimating profile soil moisture and groundwater variations using GRACE and Oklahoma Mesonet soil moisture data. *Water Resources Research* 44: W01413.

Swenson S, Yeh PJF, Wahr J, and Famiglietti J (2006) A comparison of terrestrial water storage variations from GRACE with in situ measurements from Illinois. *Geophysical Research Letters* 33: L08607 (doi:10.1029/2005GL025489).

Sydor M, Gould R, Arnone R, Haltrin V, and Goode W (2004) Uniqueness in remote sensing of the inherent optical properties of ocean water. *Applied Optics* 43: 2156–2162.

Syed TH, Famiglietti JS, Chen J, et al. (2005) Total basin discharge for the Amazon and Mississippi River basins from GRACE and a land–atmosphere water balance. *Geophysical Research Letters* 32: L24404 (doi:10.1029/2005GL024851).

Tapley BD, Bettadpur S, Ries JC, Thompson PF, and Watkins MM (2004) GRACE measurements of mass variability in the Earth system. *Science* 305: 503–505.

Timmermans J, Verhoef W, Tol Cvd, and Su Z (2009a) Retrieval of canopy component temperatures through Bayesian inversion of directional thermal measurements. *Hydrology and Earth System Sciences Discussions* 6: 3007–3040.

Timmermans WJ, Bertoldi G, Albertson JD, Olioso A, Su Z, and Gieske ASM (2008) Accounting for atmospheric boundary layer variability on flux estimation from RS observations. *International Journal of Remote Sensing* 29: 5275–5290.

Timmermans WJ, Jiménez-Muñoz JC, Hidalgo V, et al. (2009b) Surface energy fluxes from high resolution data during AgriSAR-2006: A model inter-comparison. *Journal of Selected Topics in Applied Earth Observations and Remote Sensing – IEEE* (in review).

Timmermans WJ, Kustas WP, Anderson MC, and French AN (2007) An intercomparison of the surface energy balance algorithm for land (SEBAL) and the two-source energy balance (TSEB) modeling schemes. *Remote Sensing of Environment* 108: 369–384.

Timmermans WJ, Su Z, and Olioso A (2009c) Footprint issues in scintillometry over heterogeneous landscapes. *Hydrology and Earth System Sciences Discussions* 6: 2099–2127.

Todd MC, Barrett EC, Beaumont MJ, and Green JL (1995) Satellite identification of rain days over the upper Nile river basin using an optimum infrared rain/no-rain threshold temperature model. *Journal of Applied Meteorology* 34: 2600–2611.

Tol Cvd, Tol Svd, Verhoef A, et al. (2009) A Bayesian approach to estimate sensible and latent heat over vegetation. *Hydrology and Earth System Sciences* 13: 749–758.

Trenberth KE, Fasullo JT, and Kiehl J (2009) Earth's global energy balance. *Bulletin of the American Meteorological Society* 311–323: doi:10.1175/2008BAMS2634.1.

Trenberth KE, Fasullo J, and Smith L (2005) Trends and variability in column-integrated atmospheric water vapor. *Climate Dynamics* 24: 741–758 (doi:10.1007/s00382-005-0017-4).

Trenberth KE, Smith L, Qian T, Dai A, and Fasullo J (2007) Estimates of the global water budget and its annual cycle using observational and model data. *Journal of Hydrometeorology* 8: 758–769.

Twomey S (1977) *Atmospheric Aerosols*, 302pp. Elsevier.

Ulaby FT (1976) Passive microwave remote-sensing of Earth's surface. *IEEE Transactions on Antennas and Propagation* 24(1): 112–115.

Ulaby FT, Moore RK, and Fung AK (eds.) (1981) *Microwave Remote Sensing: Active and Passive*, Vol. I: *Microwave Remote Sensing Fundamentals and Radiometry*, 456pp. Reading, MA: Addison-Wesley. Advanced Book Program.

Valor E and Caselles V (1996) Mapping land surface emissivity from NDVI: Application to European, African and South American areas. *Remote Sensing of Environment* 57(3): 167–184.

Van der Velde R and Su Z (2009) Dynamics in land surface conditions on the Tibetan Plateau observed by advanced synthetic aperture radar, ASAR. *Hydrological Sciences Journal* 54(6): 1079–1093.

Van Der Woerd H and Pasterkamp R (2008) HYDROPT: A fast and flexible method to retrieve chlorophyll-*a* from multispectral satellite observations of optically complex coastal waters. *Remote Sensing of Environment* 112: 1795–1807.

van Meijgaard E and Crewell S (2005) Comparison of model predicted liquid water path with ground-based measurements during CLIWA-NET. *Atmospheric Research, Special Issue: CLIWA-NET: Observation and Modelling of Liquid Water Clouds* 75(3): 201–226 (doi:10.1016/j.atmosres.2004.12.006).
Velde R, Su Z, Ek M, Rodell M, and Ma Y (2009) Influence of thermodynamic soil and vegetation parameterizations on the simulation of soil temperature states and surface fluxes by the Noah LSm over a Tibetan Plateau site. *Hydrology and Earth System Sciences* 13: 759–777.
Velicogna I and Wahr J (2006) Measurements of time-variable gravity show mass loss in Antarctica. *Science* 311: 1754–1756.
Venturini V, Bisht G, Islam S, and Jiang L (2004) Comparison of evaporative fractions estimated from AVHRR and MODIS sensors over South Florida. *Remote Sensing of Environment* 93: 77–86.
Verhoef W (1996) *Application of Harmonic Analysis of NDVI Time Series*, Report No. 108. Wageningen, NL: DLO Winnand Staring Centre.
Verhoest NEC, Lievens H, Wagner W, Álvarez-Mozos J, Moran MS, and Mattia F (2008) On the soil roughness parameterization problem in soil moisture retrieval of bare surfaces from synthetic aperture radar. *Sensors* 8: 4213–4248.
Verstraeten W, Veroustraete F, and Feyen J (2008) Assessment of evapotranspiration and soil moisture content across different scales of observation. *Sensors* 8: 70–117.
Vignal B and Krajewski W (2001) Large-sample evaluation of two methods to correct range-dependent error for WSR-88 D rainfall estimates. *Journal of Hydrometeorology* 2: 490–504.
Vikhamar D and Solberg R (2003) Snow-cover mapping in forests by constrained linear spectral unmixing of MODIS data. *Remote Sensing of Environment* 88(3): 309 – 323.
Voisin N, Wood AW, and Lettenmaier DP (2008) Evaluation of precipitation products for global hydrological prediction. *Journal of Hydrometeorology* 9: 388–407.
Vörösmarty CJ, Askew A, Barry R, et al. (2001) Global water data: A newly endangered species. *EOS, Transactions, American Geophysical Union* 82(5): 54–58.
Wagner W, et al. (2003) Evaluation of the agreement between the first global remotely sensed soil moisture data with model and precipitation data. *Journal of Geophysical Research – Atmospheres* 108(D19): 4611.
Wagner W, et al. (2007a) Operational readiness of microwave remote sensing of soil moisture for hydrologic applications. *Nordic Hydrology* 38(1): 1–20.
Wagner W, Lemoine G, and Rott H (1999) A method for estimating soil moisture from ERS scatterometer and soil data. *Remote Sensing of Environment* 70: 191–207.
Wagner W, Naeimi V, Scipal K, De Jeu R, and Martinez-Fernandez J (2007b) Soil moisture from operational meteorological satellites. *Hydrogeology Journal* 15(1): 121–131.
Wagner W, Verhoest NEC, Ludwig R, and Tedesco M (2009) Editorial, Remote sensing in hydrological sciences. *Hydrology and Earth System Sciences* 13: 813–817.
Wahr J, Molenaar M, and Bryan F (1998) Time-variability of the Earth's gravity field: Hydrological and oceanic effects and their possible detection using GRACE. *Journal of Geophysical Research* 103(30): 205–230.
Walker JP, Houser PR, and Willgoose GR (2004) Active microwave remote sensing for soil moisture measurement: A field evaluation using ERS-2. *Hydrological Processes* 18: 1975–1997.
Wang JR and Choudhury BJ (1981) Remote sensing of soil moisture content over bare field at 1. GHz frequency. *Journal of Geophysical Research* 4(86): 5277–5287.
Wang JR and Schmugge TJ (1980) An empirical model for the complex dielectric permittivity of soils as a function of water content. *IEEE Transactions on Geoscience and Remote Sensing* 18: 288–295.
Wang L, Parodi GN, and Su Z (2008) SEBS module BEAMS: A practical tool for surface energy balance estimates from remote sensing data. In: Agency ES (ed.) *Proceedings of the 2nd MERIS/(A)ATSR Workshop*, ESA SP-666, November 2008, Tool and Services, 8pp. Frascati, Italy, 22–26 September 2008. Noordwijk: ESA Communication Production Office ESTEC.
Wang L and Qu JJ (2009) Satellite remote sensing applications for surface soil moisture monitoring: A review. *Frontiers of Earth Science in China* 3(2): 237–247.
Wang P, Boss E, and Roesler C (2005) Uncertainties of inherent optical properties obtained from semianalytical inversions of ocean color. *Applied Optics* 44: 4074–4084.
Wen J and Su Z (2003) A time series based method for estimating relative soil moisture with ERS wind scatterometer data. *Geophysical Research Letters* 30(7): 1397 (doi:10.1029/ 2002GL016557).
Weng F, Zhao L, Ferraro RR, Poe G, Li X, and Grody NC (2003) Advanced microwave sounding unit cloud and precipitation algorithms. *Radio Science* 38: 8068 (doi:10.1029/2002RS002679).
Wentz FJ (1997) A well-calibrated ocean algorithm for special sensor microwave/imager. *Journal of Geophysical Research* 102: 8703–8718.
Weydahl DJ (2001) Analysis of ERS tandem coherence from glaciers, valleys and fjord ice in Svalbard. *IEEE Transactions on Geoscience and Remote Sensing* 39: 2029–2039.
Wielicki BA, Barkstrom BR, Harrison EF, Lee RB, Smith GL, and Cooper JE (1996) Clouds and the Earth's radiant energy system (CERES): An earth observing system experiment. *Bulletin of the American Meteorological Society* 77: 853–868.
Wigneron JP, Laguerre L, and Kerr Y (2001) A simple parmeterization of the L-band microwave emission from rough agricultural soils. *IEEE Transactions on Geoscience and Remote Sensing* 39: 1697–1707.
Wilson JW (1970) Integration of radar and raingage data for improved rainfall measurement. *Journal of Applied Meteorology* 9: 489–497.
Wood EF (1998) Hydrologic measurements and observations: An assessment of needs. In: *Hydrologic Sciences: Taking Stock and Looking Ahead*, pp. 67–86. Washington, DC: National Academies Press.
Woodhouse IH and Hoekman DH (2000) A model-based determination of soil moisture trends in Spain with the ERS-scatterometer. *IEEE Transactions on Geoscience and Remote Sensing* 38(4): 1783–1793.
Yeh PJ-F, Swenson SC, Famiglietti JS, and Rodell M (2006) Remote sensing of groundwater storage changes in Illinois using the Gravity Recovery and Climate Experiment (GRACE). *Water Resources Research* 42: doi:10.1029/2006WR005374.
Zaitchik BF, Rodell M, and Reichle RH (2008) Assimilation of GRACE terrestrial water storage data into a land surface model: Results for the Mississippi River Basin. *Journal of Hydrometeorology* 9(3): 535–548 (doi:10.1175/2007JHM951.1).
Zhan X, Kustas WP, and Humes KS (1996) An intercomparison study on models of sensible heat flux over partial canopy surfaces with remotely sensed surface temperature. *Remote Sensing of Environment* 58: 242–256.
Zhao D, Su ZB, and Zhao M (2006) Soil moisture retrieval from satellite images and its application to heavy rainfall simulation in eastern China. *Advances in Atmospheric Sciences* 23(2): 299–316.
Zhao F and Nakajima T (1997) Simultaneous determination of water-leaving reflectance and aerosol optical thickness from Coastal Zone Color Scanner measurements. *Applied Optics* 36: 6949–6956.
Zibordi G (2006) A network for standardized ocean color validation measurements. *EOS, Transactions, American Geophysical Union* 87: 293–297.
Zribi M, Baghdadi N, Holah N, and Fafin O (2005) New methodology for soil moisture estimation and its application to ENVISAT-ASAR multi-incidence data inversion. *Remote Sensing of Environment* 96: 485–496.

Relevant Websites

http://grace.sgt-inc.com
 Access Program; Mass Anomalies.
http://www.cloud-net.org
 Cloudnet.
http://www.demeter-ec.net
 DEMETER.
http://earth.esa.int
 ESA Earthnet.
http://www.eumetnet.eu.org
 EUMETNET, the Network of European Meteorological Services.
http://www.knmi.nl
 EUMETNET, the Network of European Meteorological Services; Opera.
http://ec.europa.eu
 European Commission; GMES.
http://geoid.colorado.edu
 Fedora Core Test Page; Grace.
http://www-app2.gfz-potsdam.de
 GFZ Potsdam, Department 1: The Grace Mission.
http://www.glims.org
 GLIMS: Global Land Ice Measurements from Space.
http://www.gewex.org
 Global Energy and Water Cycle Experiment: GEWEX.
http://www.isac.cnr.it
 ISAC; CGMS, IPWG; IPWG-5, Hamburg, Germany.
http://www.itc.nl
 ITC, Faculty of Geo-Information Science and Earth Observation.
http://www.legos.obs-mip.fr
 LEGOS; Hydrology from Space.
http://grace.jpl.nasa.gov
 NASA; Grace Tellus.

http://daac.ornl.gov
　ORNL, DAAC; The First ISLSCP Field Experiment (FIFE).
http://www.pleiades.es
　PLEIADES.
http://postel.mediasfrance.org
　Postel, December 2008: Globcover
　Validation Report and New Regional Land Cover Products Available.
http://modis-snow-ice.gsfc.nasa.gov
　The Modis Snow/Ice Global Mapping Project.

http://www.ars.usda.gov
　United States Department of Agriculture; Agricultural Research Service; Monsoon'90; Soil Moisture Experiments.
http://www.csr.utexas.edu
　University of Texas at Austin, Center for Space Research.
http://free.vgt.vito.be
　Vegetation, Free Vegetation Products.
http://www.wacmos.org
　WACMOS.

2.15 Hydrogeophysics

SS Hubbard, Lawrence Berkeley National Laboratory, Berkeley, CA, USA
N Linde, University of Lausanne, Lausanne, Switzerland

© 2011 Elsevier B.V. All rights reserved.

2.15.1	**Introduction to Hydrogeophysics**	402
2.15.2	**Geophysical Methods**	404
2.15.2.1	Electrical Resistivity Methods	404
2.15.2.2	IP Methods	405
2.15.2.3	SP Methods	405
2.15.2.4	Controlled-Source Inductive EM Methods	406
2.15.2.5	GPR Methods	406
2.15.2.6	Seismic Methods	407
2.15.2.7	Surface Nuclear Magnetic Resonance	407
2.15.2.8	Gravity	407
2.15.2.9	Magnetics	408
2.15.2.10	Well Logging	408
2.15.3	**Petrophysical Models**	408
2.15.3.1	Electrical Conductivity	408
2.15.3.1.1	Archie's law	409
2.15.3.1.2	Waxman–Smits law	409
2.15.3.1.3	The Johnson, Koplik, and Schwartz model	409
2.15.3.1.4	Self-similar models	410
2.15.3.2	Dielectric Permittivity	410
2.15.3.2.1	Volume averaging	410
2.15.3.2.2	Topp's equations	410
2.15.3.3	Complex Conductivity	410
2.15.3.3.1	Cole–Cole model	410
2.15.4	**Parameter Estimation/Integration Methods**	411
2.15.4.1	Key Components, Constraints, Metrics, and Steps in Parameter Estimation	411
2.15.4.1.1	Model space and initial model	411
2.15.4.1.2	Objective function (systems of equations)	411
2.15.4.1.3	Inversion step or model proposal	414
2.15.4.1.4	Geophysical model or model population	414
2.15.4.2	Example Parameter Estimation Approaches	414
2.15.4.2.1	Direct mapping approaches	414
2.15.4.2.2	Integration approaches (geostatistical, Bayesian)	414
2.15.4.2.3	Joint inversion or fully coupled hydrogeophysical inversion	415
2.15.5	**Case Studies**	415
2.15.5.1	Subsurface Architecture Delineation	415
2.15.5.1.1	3D resistivity mapping of a Galapagos volcano aquifer	415
2.15.5.1.2	High-resolution GPR imaging of alluvial deposits	417
2.15.5.1.3	Subsurface flow architecture delineation using seismic methods	417
2.15.5.1.4	Fracture zonation characterization using azimuthal electrical methods	419
2.15.5.2	Delineation of Anomalous Fluid Bodies	420
2.15.5.2.1	Electrical resistivity to delineate high-ionic-strength plume boundaries	420
2.15.5.2.2	SP imaging of redox potentials associated with contaminated plumes	420
2.15.5.3	Hydrological Process Monitoring	421
2.15.5.3.1	Soil moisture monitoring	422
2.15.5.3.2	Saline tracer monitoring in fractured rock using time-lapse GPR methods	423
2.15.5.3.3	Seasonal changes in regional saltwater dynamics using time-lapse EM methods	424
2.15.5.4	Hydrogeological Parameter or Zonation Estimation for Improving Flow Predictions	425
2.15.5.4.1	Hydraulic conductivity and zonation estimation using GPR and seismic methods	425
2.15.5.4.2	Joint modeling to estimate temporal changes in moisture content using GPR	427
2.15.6	**Summary and Outlook**	429
Acknowledgments		430
References		431

2.15.1 Introduction to Hydrogeophysics

The shallow subsurface of the Earth is an extremely important geological zone, one that yields our water resources, supports our agriculture and ecosystems, influences our climate, and serves as the repository for our contaminants. The need to develop sustainable water resources for increasing population, agriculture, and energy needs and the threat of climate and land-use change on ecosystems contribute to an urgency associated with improving our understanding of flow and transport processes in the shallow subsurface.

Developing a predictive understanding of subsurface flow and transport is complicated by the disparity of scales across which controlling hydrological properties and processes span (e.g., Gelhar, 1993). For example, the distributions of microfractures and geological formations both influence the hydraulic conductivity and thus subsurface flow, albeit over dramatically different spatial scales. Similarly, different hydrological processes may exert varying degrees of control on subsurface flow and transport as a function of the scale: the overall system response of the particular problem may be dominated by seasonal precipitation patterns or by surface–groundwater interactions at the catchment scale; by the influence of groundwater pumping wells, gradients, and heterogeneity-induced mixing at the local scale; and by microbe–mineral interactions and diffusion at the grain scale (**Figure 1**). The level of subsurface characterization required for a particular problem depends therefore on many factors,

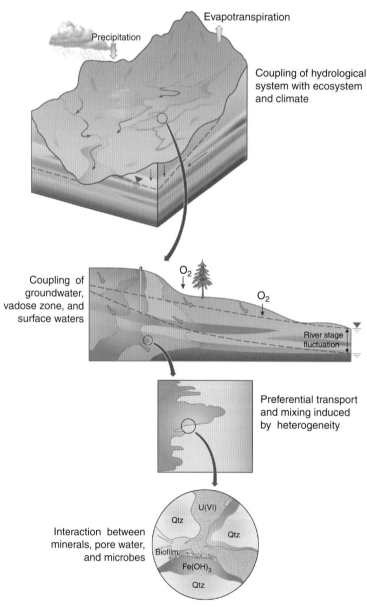

Figure 1 Subsurface flow and transport is impacted by coupled processes and properties that preferentially exert influence over a wide range of spatial scales, rendering characterization based on borehole data only challenging. Modified from US DOE (2010) Complex Systems Science for Subsurface Fate and Transport, Report from the August 2009 Workshop, DOE/SC-0123, U.S. Department of Energy Office of Science (www.science.doe.gov/ober/BER_workshops.html).

including: the level of heterogeneity relative to the characterization objective, the spatial and temporal scales of interest, and regulatory or risk drivers. In some cases, reconnaissance efforts that delineate major characteristics of the study site may be sufficient, while other investigations may require a much more intensive effort.

Conventional techniques for characterizing or monitoring the hydrogeological properties that control flow and transport typically rely on borehole access to the subsurface. For example, established hydrological characterization methods (such as pumping, slug, and flowmeter tests) are commonly used to measure hydraulic conductivity in the vicinity of the wellbore (e.g., Freeze and Cherry, 1979; Butler, 2005; Molz et al., 1994), and wellbore fluid samples are often used for water-quality assessment (e.g., Chapelle, 2001). Unfortunately, data obtained using borehole methods may not capture sufficient information away from the wellbore to describe the key controls on subsurface flow. The inability to characterize controlling properties at a high-enough spatial resolution and over a large-enough volume for understanding and predicting flow and transport processes using borehole methods often hinders our ability to predict and optimally manage associated resources.

The field of hydrogeophysics has developed in recent years to explore the potential that geophysical methods have for characterization of subsurface properties and processes relevant for hydrological investigations. Because geophysical data can be collected from many different platforms (such as from satellites and aircrafts, at the ground surface of the Earth, and within and between wellbores), integration of geophysical data with direct hydrogeological or geochemical measurements can provide characterization information over a variety of spatial scales and resolutions. The main advantage of using geophysical data over conventional measurements is that geophysical methods can provide spatially extensive information about the subsurface in a minimally invasive manner at a comparatively high resolution. The greatest disadvantage is that the geophysical methods only provide indirect proxy information about subsurface hydrological properties or processes relevant to subsurface flow and transport.

Hydrogeophysical investigations strive to provide information that can be used to (1) develop insights about complex hydrological processes, (2) serve as input data to construct flow and transport models, and (3) guide the management of subsurface water resources and contaminants. The field of hydrogeophysics builds on many decades of experience associated with the mining and petroleum industries, which have relied heavily on geophysical methods to guide the exploration of ore and hydrocarbons, respectively. Because geophysics has been used as a tool in these industries for so long, there is a relatively good understanding about methods and optimal data acquisition approaches for given problems, as well as about petrophysical relationships associated with the consolidated, high-pressure, and high-temperature subsurface environments common to those industries. However, such subsurface conditions are quite different from the shallow, low-temperature, low-pressure, and weakly consolidated environments that typify most hydrogeological investigation sites. The parameter values and the functional form of the petrophysical relationships that link geophysical properties to subsurface parameters, as well as the geophysical response itself, can vary dramatically between different types of environments.

In the last decade, many advances have been made that facilitate the use of geophysical data for shallow subsurface hydrogeological characterization. These advances include those associated with instrument development, interpretation procedures, petrophysical relationships relevant to near subsurface environments, integration or joint inversion approaches for combining multiple data sets, and coupled hydrological and geophysical modeling. Simultaneously, the number of publications related to hydrogeophysics has dramatically increased and are now common contributions to hydrological and geophysical journals such as *Water Resources Research*, *Journal of Hydrology*, *Vadose Zone Journal*, and *Geophysics*. Most hydrological and Earth science professional meetings (such as American Geophysical Union, Geological Society of America, and European Geosciences Union) now commonly host one or more hydrogeophysical special sessions at their annual meetings. These meetings have created an active environment where geophysicists and hydrologists can interact to learn about each other's methods and challenges. Many groundbreaking hydrogeophysical studies have now been published by researchers with a formal hydrological training and it is becoming more common for geophysicists to strive to gain hydrological insights in addition to focusing primarily on advancing geophysical instrumentation and methodology. Some of the fairly recent hydrogeophysical advances are summarized in two edited books *Hydrogeophysics* (Rubin and Hubbard, 2005) and *Applied Hydrogeophysics* (Vereecken et al., 2006), as well as by numerous individual publications.

Generally, hydrogeophysical characterization and monitoring objectives can often be categorized into the following three categories:

1. hydrological mapping of subsurface architecture or features (such as interfaces between key geological units, water table, or contaminant plume boundaries);
2. estimating subsurface properties or state variables that influence flow and transport (such as hydraulic conductivity or soil moisture); and
3. monitoring subsurface processes associated with natural or engineered *in situ* perturbations (such as infiltration through the vadose zone and tracer migration).

There are several components that are common to most hydrogeophysical studies. First and foremost, it is critical to collect high-quality geophysical data sets using the geophysical method or methods that are most likely to provide data that can help to resolve the hydrogeological characterization or monitoring objective and that work well in the given environment. Although the corresponding geophysical properties (such as electrical conductivity/resistivity from electrical and electromagnetic (EM) methods or dielectric constant from ground penetrating radar (GPR) methods) can be used to infer hydrogeological properties or structures, petrophysical relationships must be developed and invoked at some stage to link the geophysical properties or data with the property or variable of interest (such as hydraulic conductivity or water content). Integration or joint inversion methodologies are used to systematically integrate or fuse disparate data sets (geophysical and hydrogeological) to obtain a meaningful interpretation that honors all data and physical laws. The ultimate step is the use of the integrated hydrogeophysical property or state model to interpret complex subsurface system processes or to guide the optimal management of subsurface water resources and contaminants.

The key objectives of this chapter are to familiarize hydrogeologists and water-resource professionals with the state of the art as well as the existing challenges associated with hydrogeophysics. We provide a review of the key components of many hydrogeophysical studies as well as example case studies that are relevant to understanding of hydrological behavior at the field scale. The remainder of this chapter is organized as follows. A brief description of some of the key geophysical methods that are used in hydrogeophysics is provided in Section 2.15.2. Descriptions of theoretical and empirical petrophysical relationships that can be used to link the geophysical attributes to the hydrogeological property of interest are discussed in Section 2.15.3. Section 2.15.4 reviews parameter estimation and integration methods that are used to combine disparate data sets for a consistent interpretation of critical flow and transport properties. Finally, in Section 2.15.5 we present various case studies that illustrate the use of geophysical data sets, petrophysics, and estimation methods to investigate near subsurface systems, with a particular emphasis on case studies that are conducted over field scales relevant to water resources and contaminant remediation.

2.15.2 Geophysical Methods

The purpose of this section is to introduce some of the geophysical techniques that are most commonly used for hydrogeological studies, including electrical resistivity tomography (ERT), induced polarization (IP), electromagnetic induction (EMI), self-potential (SP), GPR, seismic, surface nuclear magnetic resonance (SNMR), gravity, magnetics, and wellbore logging techniques. For each method, we provide a brief description of the underlying physical principles and instrumentation, common acquisition strategies, and general data reduction and interpretation methods. We restrict our discussions to practical use and limitations of common geophysical methods; geophysical theory (e.g., Telford et al., 1990) is beyond the scope of this discussion. For detailed information, references are given for each geophysical method. This discussion of classical geophysical methods is envisioned to compliment existing literature on what are typically considered to be hydrological sensors or measurement approaches, even though they rely on geophysical mechanisms; examples include soil moisture probes (time domain reflectometer and capacitance probes), EM wellbore flowmeters, and various remote-sensing sensors deployed from airborne or space-borne platforms. Reviews of these methods are provided by Vereecken et al. (2008) and Butler (2005). The discussion of geophysical methods provided here is augmented by Section 2.15.3, where several petrophysical relationships are provided that may permit the transfer of geophysical measurements and models into estimates of hydrological parameters.

2.15.2.1 Electrical Resistivity Methods

For groundwater studies, electrical resistivity methods have perhaps been more frequently used than any other geophysical method. Resistivity is a measure of the ability to resist electrical current flow through materials; it is the inverse of electrical conductivity and is an intrinsic property of the material. In electrical resistivity methods, a typically low-frequency (<1 Hz) current is injected into the ground between two current electrodes, while one or more pairs of potential electrodes are used to measure electrical potential differences. At the low frequencies measured, energy loss via ionic and electronic conduction dominates. Ionic conduction results from the electrolyte filling the interconnected pore space (Archie, 1942) as well as from surface conduction via the formation of an electrical double layer at the grain-fluid interface (e.g., Revil and Glover, 1997, 1998). Electronic conduction resulting from the formation of continuous conductive pathways by metallic minerals is typically not important for most environmental applications. The current distribution can be visualized by equipotential surfaces, with current flow lines running perpendicular to these surfaces. The fraction of total current flow that penetrates to a particular depth is a function of the current electrode spacing and location, the electrical resistivity distribution of the subsurface materials, and the topography.

Most resistivity surveys utilize a four-electrode measurement approach. To obtain a value for subsurface resistivity, two potential electrodes are placed at some distance from the current electrodes, and the difference in electrical potential or voltage is measured. This measurement, together with the injected current and the geometric factor which is a function of the particular electrode configuration and spacing, can be used to calculate resistivity for uniform subsurface conditions following Ohm's law. Common electrode configurations include the Wenner, the Schlumberger, and the dipole–dipole arrays. In real heterogeneous (nonuniform) subsurface environments, the more general term 'apparent resistivity' is used, which refers to the resistivity of an equivalent uniform media.

There are several modes of acquiring electrical data. Profiling is undertaken by moving the entire array laterally along the ground surface by a fixed distance after each reading to obtain apparent resistivity measurements over a relatively constant depth as a function of distance. As profiles give lateral variations in electrical conductivity but not information about vertical distribution, the interpretation of profile data is generally qualitative, and the primary value of the data is to delineate sharp lateral contrasts associated with vertical/near vertical contacts. Vertical electrical sounding (VES) curves give information about the vertical variations in electrical conductivity at a single ground surface location assuming an idealized one-dimensional (1D) resistivity structure. For example, soundings with the Wenner array are obtained by expanding the array along a straight line so that the spacing between the individual electrodes remains equal for each measurement, but increases after each measurement.

The depth of investigation for a given measurement is a function of the electrode spacing as well as the subsurface resistivity contrasts; as the electrode spacing is increased, the data are increasingly sensitive to deeper structures.

Modern multichannel geoelectrical equipment now includes multiplexing capabilities and automatic and autonomous computer acquisition, which greatly facilitate data acquisition within acceptable timeframes. Such surface imaging, now commonly called electrical resistivity tomography or ERT, allows the electrodes (tens to hundreds) to be used alternatively as both current and potential electrodes to obtain 2D or 3D electrical resistivity models (e.g., Günther et al., 2006). In fact, when performing ERT, it is limiting to restrict the measurement sequence to a given configuration type, since optimal data sets often consist of a combination of traditional and nontraditional configuration types (Stummer et al., 2004; Wilkinson et al., 2006). With the development of advanced and automated acquisition systems, robust inversion routines, and the capability of recording tens of thousands of measurements per hour, ERT has proved to be useful for dynamic process monitoring using electrodes placed at the ground surface or in wellbores. A review of surface and crosshole ERT methods for hydrogeological applications is given by Binley and Kemna (2005), and discussion of petrophysical relationships that link the electrical properties with hydrological properties of interest is described in Section 2.15.3.1.

2.15.2.2 IP Methods

IP methods measure both the resistive and capacitive properties of subsurface materials. IP measurements can be acquired using the same four-electrode geometry that is conventionally used for electrical resistivity surveys, although IP surveys typically employ nonpolarizing electrodes. Surveys can be conducted in the time domain as well as in the frequency domain. In the time domain, the current is injected and the decay of the voltage over time is measured. Frequency-domain methods measure the impedance magnitude and phase shift of the voltage relative to an injected alternating current. Spectral induced polarization (SIP) methods measure the polarization relaxation over many frequencies (typically over the range of 0.1–1000 Hz). The voltage decay (in the time domain) and spectral response (in the frequency domain) are caused by polarization of ions in the electrical double layer at the mineral–fluid interface, by accumulation of electrical charges at pore space constrictions (e.g., pore throats), and by conduction in the pore fluid and along the fluid-grain boundaries. More information about IP methods is provided by Binley and Kemna (2005) and Leroy and Revil (2009).

The linkage between IP attributes, granulometric properties, and interfacial phenomena suggests that it also holds significant potential for exploring hydrogeological properties (e.g., Slater and Lesmes, 2002) as well as complex biogeochemical processes associated with contaminant remediation (e.g., Williams et al., 2005, 2009; Slater et al., 2007). Section 2.15.3.3 provides discussion of petrophysical models associated with SIP data sets.

2.15.2.3 SP Methods

SP is a passive method where naturally occurring electric fields (voltage gradients) are measured at the ground surface or in wellbores using nonpolarizable electrodes and a high-impedance voltmeter. Electrical potentials measured with the SP method obey a Poisson's equation with a source term given by the divergence of an electrical source current density (e.g., Minsley et al., 2007). The source current density has several possible contributors, including those associated with ground water flow, redox phenomena, and ionic diffusion. The electrokinetic contribution associated with the flow of ground water in a porous medium (or more precisely, with the drag of charges contained in the diffuse layer that surrounds mineral surfaces) has been recognized for many decades and has been used to qualitatively interpret SP signals in terms of seepage beneath dams or to map groundwater flow (e.g., Poldini, 1938). However, only more recently have such data sets been used to quantify hydrological properties by coupling equations that represent volumetric fluid flux and volume current density, which are linked by a coupling coefficient (e.g., Sill, 1983; Revil et al., 2003). The underlying physics of the redox and ionic diffusion contributions are now better understood and current research is advancing our ability to use SP for quantitative hydrogeochemical characterization (such as for characterizing field-scale redox gradients; refer to case study provided in Section 2.15.5.2).

The SP method is the only geophysical method that is directly sensitive to hydrological fluxes (e.g., Sill, 1983). Even if several alternative formulations exist to describe electrokinetic phenomena, we consider here the case where the SP sources are expressed in terms of Q_v where Q_v is expressed as excess charge in the diffuse double layer per saturated pore volume. The relative movement of an electrolyte with respect to mineral grains with a charged surface area results in so-called streaming currents (e.g., Sill, 1983). These currents are intimately linked to the Darcy velocity $\mathbf{U}$ and an excess charge Q_v along the hydrological flow paths. A practical formulation of the streaming currents that corresponds to this parametrization is (e.g., Revil and Linde, 2006)

$$J_s = Q_v \mathbf{U} \qquad (1)$$

This equation is only strictly valid when the size of the double layer is comparable to the size of the pores (see Revil and Linde (2006) for a description of chemico-electromechanical coupling under such conditions) and when internal permeability variations within the averaged volume are small. Equation (1) can be used in heterogeneous media or in coarse sediments when we replace Q_v with an effective Q_v^{eff} that is scaled with the relative contributions to permeability of all flow paths in the media (Linde, 2009). It is straightforward to deduce Q_v^{eff} of aquifer materials in the laboratory using the relationship (Revil and Leroy, 2004)

$$C_{sat} = \frac{Q_v^{eff} k}{\mu_w \sigma} \qquad (2)$$

where k is the permeability and μ_w is the dynamic water viscosity. The voltage coupling C_{sat} can at moderate to high permeabilities be obtained using a simple experimental setup, for example, using the type of column experiment presented by Suski et al. (2006).

The dependence of $Q_{v,sat}^{eff}$ with water content depends on the geological media considered, but it is to a first order inversely related to water content (Linde et al., 2007; Linde, 2009). The source current that is responsible for observed electrical potential signals associated with these processes is given by the divergence of Equation (1) (e.g., Linde et al., 2007).

2.15.2.4 Controlled-Source Inductive EM Methods

Controlled-source inductive EM methods use a transmitter to pass a time- or frequency-varying current through a coil or dipole placed on the Earth's surface, in boreholes, mounted on an aircraft or towed behind a ship. Governed by Maxwell's equations and typically operating in the 1–15 kHz range, this alternating current produces a time-varying primary magnetic field, which in turn interacts with the conductive subsurface to induce time-varying eddy currents. These eddy currents give rise to a secondary EM field. Attributes of this secondary magnetic field, such as amplitude, orientation, and/or phase shift, can be measured by a receiver coil. By comparing these attributes with those of the primary field, information about the presence of subsurface electrical conductors or the subsurface electrical conductivity distribution can be inferred. Because a conductive subsurface environment or target is required to set up the secondary field measured with inductive EM methods, EM methods are best suited for use when attempting to detect the presence of high-conductivity subsurface targets, such as saltwater saturated sediments or clay layers. However, because coils do not require contact with the ground, EM methods are often more successful on electrically resistive or paved ground than the classical DC resistivity method, which requires electrode contact.

As with ERT and SIP data, EM induction data can often be collected in profile or sounding mode. The mode of acquisition and the resolution and depth penetration of the data are dictated by the electrical conductivity distribution of the subsurface and the coil spacing and source configuration. For frequency domain systems, high transmitter frequencies permit high-resolution investigation of subsurface conductors at shallow depths, while lower transmitter frequencies permit deeper observations but at a loss in resolution. Time domain systems measure the secondary magnetic field as a function of time, and early-time measurements yield information about the near surface, while later-time measurements are increasingly influenced by the electrical properties at larger depths. The depth of penetration and resolution are also governed by coil configuration; the measurements from larger coil separations are influenced by electrical properties at greater depths, while smaller coil spacings sample from the near surface. A review and discussion of the use of controlled-source EM methods for hydrogeological investigations is given by Everett and Meju (2005). It should be noted that it is also possible to use civilian and military radio transmitters, operating in the 10–250 kHz frequency range, as the source signal. These are the signals used in the popular very low frequency (VLF) (e.g., Pedersen et al., 1994) and radio magnetotelluric (RMT) (e.g., Linde and Pedersen, 2004) techniques.

2.15.2.5 GPR Methods

GPR methods use EM energy at frequencies of ∼ 10 MHz to 1 GHz to probe the subsurface. At these frequencies, the separation (polarization) of opposite electric charges within a material that has been subjected to an external electric field dominates the electrical response. GPR systems consist of an impulse generator which repeatedly sends a particular voltage and frequency source to a transmitting antenna. When the source antenna is placed on or above the ground surface, waves are radiated downward into the soil. In general, GPR performs better in unsaturated coarse or moderately coarse textured soils; GPR signal strength is strongly attenuated in electrically conductive environments (such as systems dominated by the presence of clays or high ionic strength pore fluids). Together, the electrical properties of the host material and the frequency of the GPR signal primarily control the resolution and the depth of penetration of the signal. Increasing the frequency increases the resolution but decreases the depth of penetration.

GPR data sets can be collected in the time or in the frequency domain. Time-domain systems are most commonly used in near-surface investigations. Generally, one chooses a radar center frequency that yields both sufficient penetration and resolution; for field applications this is often between 50 and 250 MHz. However, significant advances have been made in the development of frequency domain systems. Lambot et al. (2004a) describe a stepped-frequency continuous-wave radar deployed using an off-ground horn antenna over the frequency range of 0.8–3.4 GHz. The wide bandwidth and off-ground configuration permits more accurate modeling of the radar signal, thus potentially leading to improved estimates of subsurface parameters (Lambot et al., 2004b, 2006).

The most common ground surface GPR acquisition mode is surface common-offset reflection, in which one (stacked) trace is collected from a transmitter–receiver antenna pair pulled along the ground surface. With this acquisition mode, GPR antennas can be pulled along or above the ground surface at walking speed. When the EM waves in the ground reach a contrast in dielectric constants, part of the energy is reflected and part is transmitted deeper into the ground. The reflected energy is displayed as 2D profiles that indicate the travel time and amplitude of the reflected arrivals; such profiles can be displayed in real time during data collection and can be stored digitally for subsequent data processing. An example of the use of GPR profiles for interpreting subsurface stratigraphy is provided in Section 2.15.5.1.

The velocity of the GPR signal can be obtained by measuring the travel time of the signal over a known distance between the transmitter and the receiver. The propagation phase velocity (V) and signal attenuation are controlled by the dielectric constant (κ) and the electrical conductivity of the subsurface material through which the wave travels. At the high-frequency range used in GPR, the velocity in a low electrical conductivity material can be related to the dielectric constant, also known as the dielectric permittivity, as (Davis and Annan, 1989)

$$\kappa \approx \left(\frac{c}{V}\right)^2 \quad (3)$$

where c is the propagation velocity of EM waves in free space $(3 \times 10^8 \text{ m s}^{-1})$.

Approaches that facilitate EM velocity analysis include surface common-midpoint (CMP), crosshole tomography acquisition, as well as analysis of the groundwave arrival recorded using common-offset geometries. Full-waveform inversion approaches have recently been developed (e.g., Ernst et al., 2007; Sassen and Everett, 2009) that offer potential for improved subsurface property characterization over methods based on travel times alone. Discussion of petrophysical relationships that link dielectric permittivity with hydrological properties of interest is described in Section 2.15.3.2. A review of GPR methods applied to hydrogeological applications is given by Annan (2005).

2.15.2.6 Seismic Methods

Seismic methods common to hydrological investigations use high-frequency (~ 100–5000 Hz) pulses of acoustic energy to probe the subsurface. These pulses are generally artificially produced (using weight drop, hammers, explosives, piezoelectric transducers, etc.) and propagate outward as a series of wavefronts. The passage of the wavefront creates a motion that can be detected by a sensitive geophone or hydrophone. According to the theory of elasticity upon which seismic wave propagation is based, several different waves are produced by a disturbance; these waves travel with different propagation velocities that are governed by the elastic constants and density of the material. The P-wave energy is transmitted by a back-and-forth particle movement in the direction of the propagating wave. Transverse waves, also called S (secondary or shear)-waves, have lower velocities than the P-wave and thus arrive later in the recording. P-wave arrivals are the easiest to detect and most commonly used arrival; we focus here exclusively on information available from P-waves. The principles of seismic reflection, refraction, and tomographic methods are briefly described below.

The surface reflection technique is based on the return of reflected P-waves from boundaries where velocity and density (or seismic impedance) contrasts exist. Processing of seismic reflection data generally produces a wiggle-trace profile that resembles a geologic cross section. However, due to the lack of well-defined velocity contrasts and strong signal interference in shallow unconsolidated and unsaturated materials, seismic reflection approaches to image near subsurface architecture can be challenging. With refraction methods, the incident ray is refracted along the target boundary before returning to the surface. The refracted energy arrival times are displayed as a function of distance from the source, and interpretation of this energy can be accomplished by using simple software or forward modeling techniques. As with GPR methods, the arrival times and distances can be used to obtain velocity information directly. More advanced applications include multi-dimensional inversion for the subsurface velocity distribution using many first arrival travel times corresponding to refracted energy for many combinations of transmitter and receiver locations. Refraction techniques are most appropriate when there are only a few shallow (< 50 m) targets of interest, or where one is interested in identifying gross lateral velocity variations or changes in interface dip. Seismic refraction methods yield much lower resolution than seismic reflection and crosshole methods. However, because refraction methods are inexpensive and acquisition may be more successful in unsaturated and unconsolidated environments, they are often chosen over reflection methods for applications such as determining the depth to the water table and to the top of bedrock, the gross velocity structure, or for locating significant faults. With crosshole seismic tomographic data, the multiple sampling of the inter-wellbore area via raypaths that emanate from instruments lowered down boreholes permits very detailed estimation of the velocity structure that can be used to estimate hydrogeological properties. A review of shallow seismic acquisition and processing techniques is given by Steeples (2005).

2.15.2.7 Surface Nuclear Magnetic Resonance

SNMR is a geophysical method that takes advantage of the NMR response of hydrogen protons, which are components of water molecules, to estimate water content. This method involves the use of a transmitting and a receiving loop to induce and record responses to an EM excitation induced at the resonance frequency of protons (the Larmor frequency). Under equilibrium conditions, the protons of water molecule hydrogen atoms have a magnetic moment that is aligned with the Earth's local magnetic field. Upon excitation, the axis of the precession is modified. When the external field is removed, relaxation occurs as a function of the spatial distribution, amount, and mobility of water; this relaxation manifests itself as an EM signal that decays over time. Through use and analysis of different excitation intensities, initial amplitudes, and decay time, approaches have been proposed to estimate the density distribution of hydrogen atoms as well as associated pore and grain size and water content.

Although SNMR holds significant potential for directly investigating subsurface hydrological properties, it is still in an early stage of development and its resolving power is rather limited. As described by Yaramanci et al. (2005), advances are needed to overcome induction effects and inversion errors associated with multi-dimensional heterogeneities and regularization. A further problem with this method is that it is very sensitive to cultural EM noise and that the measured signals are often weak. Hertrich (2008) provides a review of SNMR for groundwater applications, and describes recent algorithm and method development.

2.15.2.8 Gravity

Measurements of changes in gravitational acceleration can be used to obtain information about subsurface density variations that can in turn be related to variations in lithology or moisture content. The common measuring device for this potential field method is a gravimeter, an instrument which is portable and easy to use. An extremely sensitive spring balance inside the gravimeter measures differences in the weight of a small internal object from location to location; the weight differences are attributed to changes in the acceleration of gravity due to lateral variations in subsurface density. Measurements can be collected at a regional or local scale depending on the station spacing, which is usually less than half of the depth of interest.

The theoretical response to the gravitational field due to factors such as the datum, latitude, terrain, drift, and regional gradient is typically compensated for prior to interpretation of the remaining gravity anomaly. Qualitative interpretation usually consists of constraining a profile or contoured anomaly map with other known geologic information to delineate, for example, the boundary of a sedimentary basin that overlies denser bedrock. A general review of the gravity technique and applications to environmental studies is given by Hinze (1990). More recently, microgravity studies have recently been performed in an attempt to quantify changes in water storage associated with hydrological processes (e.g., Krause et al., 2009) and to characterize cavities in karstic terrains (Styles et al., 2005).

2.15.2.9 Magnetics

Magnetic methods obtain information related to the direction, gradient, or intensity of the Earth's magnetic field. The intensity of the magnetic field at the Earth's surface is a function of the location of the observation point in the primary earth magnetic field as well as from contributions from local or regional variations of magnetic material such as magnetite, the most common magnetic mineral. After correcting for the effects of the Earth's natural magnetic field, magnetic data can be presented as total intensity, relative intensity, and vertical or horizontal gradient anomaly profiles or contour maps. Interpretation of magnetic surveys generally involves forward modeling or mapping of the anomalies correlating them with other known geologic information. As magnetic signatures depend to a large extent on magnetic mineral content, which is low in most sediments that comprise aquifers, magnetics is not commonly employed for hydrological investigations, but it can be a very powerful technique to locate lateral boundaries of landfills. Exceptions include mapping subsurface structures (basement topography, faults, and paleochannels), provided that a sufficient magnetic signature or contrast exists. A review of magnetic methods as applied to environmental problems is given by Hinze (1990).

2.15.2.10 Well Logging

Well logging refers to the process of recording and analyzing measurements collected discretely or continually within wellbores. Borehole measurements are made by lowering a probe into the borehole on the end of an electric cable. The probe, generally 2.5–10.0 cm in diameter and 0.5–10.0 m in length, typically encloses sources, sensors, and the electronics necessary for transmitting and recording signals. A variety of different types of wellbore probes are available; perhaps the most common for hydrological studies include: SP, electrical, EM, gamma–gamma, natural gamma, acoustic, temperature, flowmeter, neutron–neutron, televiewer, and caliper logs. The volume of investigation of the borehole measurement is related to the log type, source–detector spacing, the borehole design, and the subsurface material. The well log measurements can be compared with each other and with direct measurements (such as from core samples) to develop site-specific petrophysical relationships. Log data are also useful to tie hydrological and geological data collected at the wellbore location with geophysical signatures of property variations collected using surface or crosshole geophysical data. References for borehole geophysics applied to hydrogeologic investigations are given by Keys (1989) and Kobr et al. (2005).

2.15.3 Petrophysical Models

To be useful in hydrology, geophysical data and hence the corresponding geophysical properties need to be sensitive to hydrological primary (e.g., total and effective porosity, and permeability) or state variables (e.g., salinity, water content, and pressure gradients). In this section, we introduce different petrophysical models that link hydrological and geophysical properties. We focus on models related to electrical properties, since they dominate hydrogeophysical applications through methods such as ERT, SIP, EM, and GPR (see Section 2.15.2). Pertinent models related to gravity, seismics, and borehole geophysical data are not considered here for brevity, but can be found in references such as Mavko et al. (1998), Schön (1996), Guéguen and Palciauskas (1994), and Carcione et al. (2007). A wealth of models for electrical properties in porous media has been proposed and only main results are summarized below; the reader is referred to Lesmes and Friedman (2005), Keller (1987), and Slater (2007) for more information. The petrophysical models discussed below were chosen because they are fairly general, and also because most of them share a similar parametrization.

Purely mathematical models are useful to define bounds on properties, such as the classical Hashin–Shtrikman bounds (Hashin and Shtrikman, 1962). More common in hydrogeophysical studies is the use of semi-empirical models that partly incorporate geometrical and physical properties of the components that comprise the porous media. Examples of such models are Archie's law (Archie, 1942) or the complex refractive index model (Birchak et al., 1974). In many cases, purely empirical relationships are obtained by fitting polynomial functions (e.g., Topp et al., 1980). Below, we briefly review petrophysical models associated with electrical conductivity, dielectric permittivity, complex conductivity, and electrokinetics.

2.15.3.1 Electrical Conductivity

The conductive and capacitive properties of an isotropic and homogeneous media can be represented by a complex conductivity (σ^*), a complex resistivity (ρ^*), or a complex permittivity (ε^*):

$$\sigma^*(\omega) = \frac{1}{\rho^*(\omega)} = i\omega\varepsilon^*(\omega) \tag{4}$$

where ω is the angular frequency and $i = \sqrt{-1}$. It is common practice to refer to the real-valued component of $\sigma^*(\omega) = \sigma'(\omega) + i\sigma''(\omega)$ at low frequencies (say 0–250 kHz) as σ and the real-valued relative permittivity at high frequencies (10–1000 MHz) as $\kappa = \varepsilon'/\varepsilon_0$, where ε' is the effective permittivity of the media and ε_0 is the permittivity of vacuum. It is important to note that in these frequency ranges both properties, σ and κ, have a weak frequency dependency (e.g., see Figure 4.1 in Lesmes and Friedman (2005)) that needs to be taken into account for quantitative comparisons. Low-frequency polarization $\sigma''(\omega)$ is discussed in Section 2.15.3.3.

2.15.3.1.1 Archie's law

The aggregated empirical Archie's first and second law (Archie, 1942), expressed here in terms of electrical conductivity, is probably the most commonly used model to interpret electrical conductivity in hydrological studies:

$$\sigma = \sigma_w S_w^n \phi^m = \sigma_w S_w^n F^{-1} \qquad (5)$$

where σ is the bulk electrical conductivity of the media, σ_w is the electrical conductivity of the pore fluid, S_w is the water saturation, n is the water saturation exponent, ϕ is the porosity, and m is the cementation exponent. The electrical formation factor F is defined in the absence of surface conductivity σ_s as (e.g., Revil et al., 1998)

$$\frac{1}{F} \equiv \lim_{\sigma_s \to 0}\left(\frac{\sigma}{\sigma_w}\right) = \phi^m \qquad (6)$$

The attraction of Archie's law in hydrological applications is obvious since it includes key properties, namely the electrical conductivity of the pore fluid related to salinity and the inverse of the electrical formation factor, which can be thought of as an effective interconnected porosity (Revil and Cathles, 1999). Archie's law not only explains a lot of experimental data, but also is physically justified when surface conduction is negligible (Sen et al., 1981). In the vadose zone, the water saturation exponent may display significant hysteresis (Knight, 1991). Archie's law is only valid for a continuous water phase, which might break down in dry areas where evaporation is significant (e.g., Shokri et al., 2009). Another more serious problem with this model is that surface conduction, which plays a role when significant clay and silt fractions are present in the media, is ignored.

2.15.3.1.2 Waxman–Smits law

A number of models have been proposed to incorporate surface conduction. One of the most commonly used models that includes surface conduction in saturated media is the model of Waxman and Smits (1968):

$$\sigma = \frac{1}{F}(\sigma_w + BQ_v) \qquad (7)$$

where B is the equivalent conductance per ion and Q_v is the density of counter ions per unit pore volume. Electrical conduction is here modeled as being composed of an electrical path in the pore volume and another parallel path at the mineral–water interface. This equation has been extensively used in the oil industry and it provides normally a good fit to experimental data when the electrolytic conductivity term dominates over the surface conductivity term (e.g., Waxman and Smits, 1968; Johnson et al., 1986; Sen et al., 1988). One problem with Waxman and Smits' model is that it uses an average Q_v determined by titration while only the excess charge located along the conducting paths in the pore space will contribute to electrical flow. Another problem arises when surface conductivity becomes more important, since the electrical conduction paths change and can no longer be expressed by F (see Equation (6)) only (Johnson et al., 1986; Revil et al., 1998).

2.15.3.1.3 The Johnson, Koplik, and Schwartz model

A fundamental length-scale parameter Λ was introduced by Johnson et al. (1986) as

$$\frac{2}{\Lambda} = \frac{\int |\nabla \psi_0(\mathbf{r})|^2 dS}{\int |\nabla \psi_0(\mathbf{r})|^2 dV_p} \qquad (8)$$

where $\nabla \psi_0(\mathbf{r})$ is the electrical potential gradient at position $\mathbf{r}$ in the absence of surface conductivity from a current source imposed from the sides and where the integration is performed over the mineral–water interface (S) and the pore volume (V_p), respectively. It follows that $2/\Lambda$ is an effective surface-to-pore-volume ratio weighted by the local strength of the electric field. This weighting eliminates contributions from dead-end pores (Johnson et al., 1986). Johnson et al. (1986) use a perturbation technique to derive the following equation:

$$\sigma = \frac{1}{F}\left(\sigma_w + \frac{2\Sigma_s}{\Lambda}\right) + O(\Sigma_s^2) \qquad (9)$$

where the specific surface conductivity is given by (e.g., Schwartz et al., 1989)

$$\Sigma_s = \int_0^\infty [\sigma(\varepsilon) - \sigma_w]d\varepsilon \qquad (10)$$

where ε measures the distance along a normal directed into the pore space from the grain boundary. The contributions to Σ_s become insignificant for values much larger than the Debye screening length that is at most some 100 Å. The Σ_s is fairly well-known and is much less variable than Λ (Leroy and Revil, 2004).

Neglecting second-order terms in Equation (9), $O(\Sigma_s^2)$, is only valid in the vicinity of the high-salinity limit. Schwartz et al. (1989) extended the theory of Johnson et al. (1986) to the low-salinity limit in which the electrical flow paths are determined by regions with significant surface conduction. They showed that Padé approximants (a ratio of two polynomials) are effective to interpolate between the high- and low-salinity limits.

Johnson et al. (1986) also show that Λ can be used to predict permeability k with a high predictive power using the relation (see also Bernabé and Revil, 1995)

$$k \approx \frac{\Lambda^2}{4F} \qquad (11)$$

2.15.3.1.4 Self-similar models

Another approach to model electrical conductivity is based on self-similar models with electrolytic conduction only (Sen et al., 1981) or with surface conductivity included (Bussian, 1983). Revil et al. (1998) extended the model of Bussian (1983) to explicitly model the different conduction paths taken by anions and cations. Tortuosity affecting the migration of the anions is given by $F\phi$, but the dominant conduction paths for the cations shift toward the conduction paths defined by the distribution of Q_v at the mineral–water interfaces as the salinity decreases.

The ubiquitous presence of surface conductivity in geological porous media makes models of electrical conductivity alone uncertain tools in hydrological studies, since a moderately high electrical conductivity can be explained by either a fairly high σ_w with a well-connected pore space (i.e., low F) without any clay particles, or a low σ_w and a poorly connected pore space (i.e., high F) with a moderate clay fraction. The hydrological behaviors of these two types of media are fundamentally different and electrical conductivity data alone may not offer even qualitative information about the dominant hydrological properties (e.g., Purvance and Andricevic, 2000). To make quantitative predictions, it is therefore often necessary to have access to other types of geophysical (such as IP) or geological data or to perform time-lapse experiments, where temporal variations in the geophysical data are recorded (e.g., Binley et al., 2002).

2.15.3.2 Dielectric Permittivity

2.15.3.2.1 Volume averaging

Variations of electrical polarization at the frequencies used in ground-penetrating radar (GPR; 10–1000 MHz) are mainly determined by water content and less by mineralogy, even if polarizations of mineral grains need to be considered. Due to the need for complimentary data in many hydrogeophysical applications, it is common to use estimates of both electrical conductivity and the relative permittivity (e.g., Binley et al., 2002; Linde et al., 2006a). When explaining relative permittivity data, it can therefore be useful to use a relative permittivity model that shares a similar parametrization of the pore geometry as the one used to explain electrical conductivity. Such an approach was presented by Pride (1994) who used a volume-averaging approach to derive the following equation for relative permittivity

$$\kappa = \left[\frac{1}{F}(\kappa_w - \kappa_s) + \kappa_s\right] \quad (12)$$

where κ_w is the relative permittivity of water ($\kappa_w \approx 80$) and κ_s is the relative permittivity of the solid ($\kappa_s = 3-8$). This equation was extended by Linde et al. (2006a) to incorporate partial saturations as

$$\kappa = \frac{1}{F}\left[S_w^n \kappa_w + (1 - S_w^n)\kappa_a + (F-1)\kappa_s\right] \quad (13)$$

where κ_a is the relative permittivity of air ($\kappa_a = 1$) (see Linde et al. (2006a) for a corresponding model for electrical conductivity with surface conductivity included).

One of the most common petrophysical models used to estimate water content from relative permittivity data is the so-called Lichteneker–Rother model (e.g., Guéguen and Palciauskas, 1994):

$$\kappa^\alpha = \sum_{i=1}^{n} \phi_i \kappa_i^\alpha \quad (14)$$

where the subscript i indicates the contribution of each phase (e.g., rock matrix, water, and air). Equation (14) with $\alpha = 0.5$ is referred to as the complex refractive index model (Birchak et al., 1974)

$$\sqrt{\kappa} = \theta\sqrt{\kappa_w} + (\phi - \theta)\sqrt{\kappa_a} + (1-\phi)\sqrt{\kappa_s} \quad (15)$$

where θ is water content. Brovelli and Cassiani (2008) showed that this commonly used model is only valid when the cementation exponent $m \sim 2$ and when the dielectric contrast between phases are large. This means that Equation (15) is based on an implicit assumption about the connectedness of the pore space that in reality varies (e.g., m is typically ~ 1.5 in unconsolidated aquifer materials; Lesmes and Friedman, 2005). Recently, Brovelli and Cassiani (2010) showed convincingly that an appropriately weighted combination of the lower and upper Hashin–Shtrikman bounds using the cementation factor could predict permittivity measurements very well.

2.15.3.2.2 Topp's equations

A set of models that are purely empirical but have high predictive power in soils are the so-called Topp's equations that were derived at high frequencies for different soil types. The general Topp equation (Topp et al., 1980) when the soil type is unknown is

$$\kappa = 3.03 + 9.3\theta + 146\theta^2 - 76.7\theta^3 \quad (16)$$

2.15.3.3 Complex Conductivity

2.15.3.3.1 Cole–Cole model

We now focus on the frequency behavior of the imaginary component of the complex electrical conductivity (Equation (4)) $\sigma''(\omega)$ at low frequencies. Recent experiments suggest that the electrochemical polarization of a grain is dominated by the mineral/water interface of the Stern layer and Maxwell–Wagner effects associated with accumulation of electrical charges at pore throats (Leroy et al., 2008). SIP data (also referred to as complex conductivity; Kemna, et al., 2000) have been identified as the most promising method to develop robust inferences of polarization processes (Ghorbani et al., 2007) and potentially permeability in hydrological studies (Slater and Lesmes, 2002; Binley et al., 2005). The most common petrophysical model used in SIP is the phenomenological Cole–Cole model (Cole and Cole, 1941) or combinations of several Cole–Cole models. The Cole–Cole model can be expressed as

$$\sigma^*(\omega) = \sigma_0\left[1 + m\left(\frac{(i\omega\tau)^c}{1+(i\omega\tau)^c(1-m)}\right)\right] \quad (17)$$

where σ_0 is the conductivity at the DC limit, τ is the mean relaxation time, c is an exponent that typically takes values in the range of 0.1–0.6, and m is the chargeability ($m = 1 - \sigma_0/\sigma_\infty$, where σ_∞ is the electrical conductivity at high frequency). Parameters of this model might be sensitive to specific surface area (Börner and Schön, 1991; Slater et al., 2006), dominant pore-throat sizes (Scott and Barker, 2003), or effective grain sizes (Slater and Lesmes, 2002). Laboratory measurements on sandstone suggest a strong correlation between the relaxation time and the permeability ($r^2 = 0.78$) (Binley et al., 2005) and promising results have been reported from field applications (Hördt et al., 2007). It is likely that new physical models based on a more physical parametrization of the pore space that is consistent with the ones developed for other electrical properties (e.g., Leroy et al., 2008; Leroy and Revil, 2009) will help to gain a better understanding of the low-frequency polarization response and its sensitivity to hydrological parameters. In particular, it is important to develop a theory that holds at any frequency and that takes the characteristics of the electrical double layer and the surface chemistry into account.

2.15.4 Parameter Estimation/Integration Methods

This section addresses how geophysical data and models can be used together with hydrological data and models to improve the imaging of hydrological properties or monitoring of hydrological processes. The approaches that have been presented in the literature differ mainly in how they represent the model parameter space; what importance and representation is given to *a priori* information; at what stage different data types are coupled; how uncertainties in the observations, the forward models, and the petrophysical models are treated. **Figure 2** provides a schematic view of how geophysical and hydrological data and models can be integrated at different stages in the inversion process. The figure can also be seen as a general representation of how joint inversion can be carried out for the case of two data types. For an in-depth treatment of inversion theory, the reader is referred to Menke (1984), Parker (1994), McLaughlin and Townley (1996), and Tarantola (2005).

Study objectives and the available budget will determine many of the choices made throughout the inversion process. These aspects are not incorporated in **Figure 2**, since it mainly serves to illustrate where interactions between geophysical and hydrological components of the inversion process might take place. In a given hydrogeophysical inversion method only a fraction of the links between the geophysical and hydrological compartments in **Figure 2** is likely to be used.

2.15.4.1 Key Components, Constraints, Metrics, and Steps in Parameter Estimation

2.15.4.1.1 Model space and initial model
A key choice in any inversion is to decide on the model parametrization used to represent the subsurface, the permissible ranges of model parameters, and the initial model (see box 'Model space and initial model'). These choices will mainly be based upon prior knowledge (see box 'Prior knowledge'). Prior knowledge is information about characteristics of the system that we have obtained from other sources of information than the actual geophysical or hydrogeological data that we try to invert. Prior knowledge might, in this case, be related to information about the geological setting and previous exploratory or detailed studies. The link from box 'Geophysical (or hydrological) system property data' to box 'Prior knowledge' indicates also estimates of system properties that have been made outside the parameter estimation procedure (e.g., sonic log data transformed to P-wave velocities, neutron–neutron data transformed into porosity, and EM flowmeter data translated into relative variations in permeability) and we assume that these properties are known with an associated uncertainty.

2.15.4.1.2 Objective function (systems of equations)
The number of independent parameters that can be inferred from hydrological and geophysical data depend on the data type, the experimental design, the number of data available, the data quality, and the forward model used. There exists however an upper limit of how many parameters one can independently estimate from a given data set. For this reason, we must find ways to constrain model space in order to obtain meaningful results, in addition to simply decrease computing time and memory use. In practice, it is necessary to explicitly constrain the parameter space by solving either an overdetermined problem with few model parameters or an underdetermined problem where a unique solution defined as the model that fits the data with the least model structure as defined by the regularization constraints used to stabilize the inverse solution. The zonation approach to model parametrization is to assume that the subsurface can be represented by a number of zones with similar physical properties, where the boundaries are either assumed to be known or updated during the inversion process. Possible applications where a zonation approach could be justified are the delineation of sand from interbedded clay layers or sediments from the underlying bedrock. The advantage of the zonation approach is that the number of model parameters can be kept relatively small and smoothness constraints across boundaries in the inversion may thus be avoided. The geostatistical approach is based on the assumption that the parameter field can be explained by a known or estimated spatial random variable with a certain correlation structure and deterministic trend. This parametrization is probably preferable when the parameters of interest vary in more or less random fashion and there is no clearly defined structure (see further discussion in Mclaughlin and Townley (1996)). Geophysical inversion is typically performed using a very fine model discretization where the aim of the inversion is to fit the data to a certain error level while minimizing deviations from an assumed prior model or spatial variability between neighboring cells. Regardless of the parametrization used, it is clear that prior knowledge should affect the objective function as indicated in **Figure 2** (see arrow to box 'Objective function (systems of equations)').

After defining the model parametrization, it is necessary to define a metric that defines what constitutes a good model and an algorithm that can be used to find such models. There are two main groups of inversion strategies: (1) deterministic

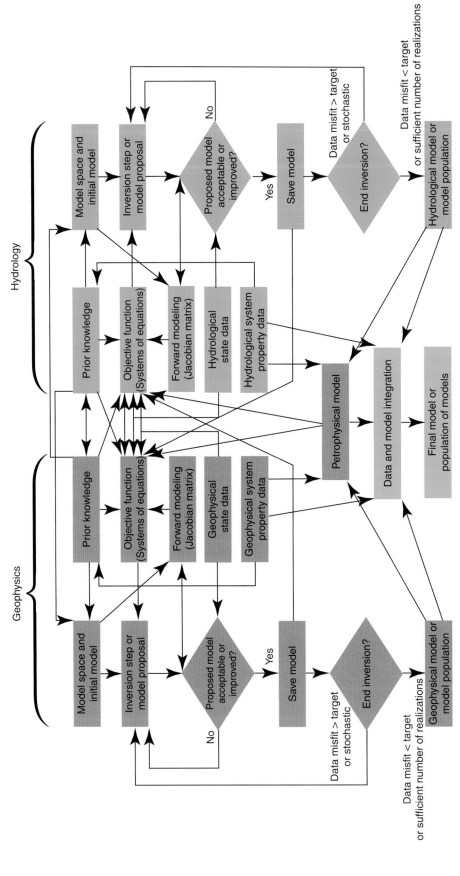

Figure 2 This flowchart illustrates possible ways that geophysics and hydrology can be integrated in hydrogeophysical studies. Recent hydrogeophysical research indicates that it is important to tightly couple hydrology and geophysics throughout the inversion and modeling process; see possible connections from blue hydrological boxes to green geophysics boxes, and vice versa. Please refer to text for details.

inversion where one unique model is sought that describes the subsurface in some average sense (Menke, 1984; Parker, 1994) and (2) stochastic inversion where a probabilistic description of the model space is used and where a large population of possible models are sampled without specifying which is the best model, only how likely they are to explain the available data and any prior knowledge (Tarantola, 2005). Regardless of the inversion approach, the definition of the objective function will, to a large degree, determine the type of models that will be obtained for a given data set.

Objective functions, at least in deterministic inversions, often include two different terms: (1) one data misfit term that characterizes how well a model explains the observed data and (2) one model misfit term that defines how a model corresponds with prior knowledge or any assumptions about how the model is likely to vary spatially. The most common approach is to quantify these two terms by using a least-squares formulation, where a weighted sum of the two squared misfit terms is penalized simultaneously, which typically works well when system properties are expected to vary smoothly and when data errors have an approximately Gaussian distribution. In this case, the data misfit is expressed as

$$\chi_d^2 = (\mathbf{d} - \mathbf{F}[\mathbf{m}])^T \mathbf{C}_d^{-1} (\mathbf{d} - \mathbf{F}[\mathbf{m}]) \quad (18)$$

where $\mathbf{d}$ is an $N \times 1$ data vector (e.g., electrical resistances or observed drawdown at a pumping well); $\mathbf{F}[\mathbf{m}]$ is a forward model operator response for a given model vector $\mathbf{m}$ of size $M \times 1$; superscript T indicates transposition; $\mathbf{C}_d^{-1}$ is the inverse of the data covariance matrix. It is commonly assumed that data errors are uncorrelated, rendering $\mathbf{C}_d^{-1}$ a diagonal matrix that contains the inverses of the estimated variances of the data errors; thus, more reliable data carry larger weight when evaluating the data fit. The corresponding model norm is

$$\chi_m^2 = (\mathbf{m} - \mathbf{m}_0)^T \mathbf{C}_m^{-1} (\mathbf{m} - \mathbf{m}_0) \quad (19)$$

where $\mathbf{m}_0$ is a reference model of size $M \times 1$; $\mathbf{C}_m^{-1}$ is the inverse of the model covariance matrix, which characterizes the expected variability and correlation of model parameters (Maurer et al., 1998; Linde et al., 2006a). It should be noted that it is often common to neglect the term $\mathbf{m}_0$ and replace $\mathbf{C}_m^{-1}$ with a regularization term that approximates the square of a first or second derivative of the model. The objective function for a classical geophysical deterministic inversion is in the general least-squares case:

$$W_\lambda(\mathbf{m}) = (\mathbf{m} - \mathbf{m}_0)^T \mathbf{C}_m^{-1} (\mathbf{m} - \mathbf{m}_0) \\ + \lambda^{-1} \{ (\mathbf{d} - F[\mathbf{m}])^T \mathbf{C}_d^{-1} (\mathbf{d} - F[\mathbf{m}]) \} \quad (20)$$

where λ^{-1} acts as a trade-off parameter between the smooth well-conditioned problem defined by a heavy penalty on deviations from the predefined model behavior (i.e., λ is large) and the ill-conditioned problem defined by the data misfit term (i.e., λ is small).

In order to obtain models that display sharper contrasts between geological units or when data noise has a non-Gaussian distribution, it is possible to use a method called iteratively reweighted least-squares based on the Ekblom l_p-norm (Farquharson, 2008), thereby approaching a formulation of the inverse problem where only absolute differences in misfit are penalized, while maintaining the numerical advantages of least-squares formulations. A number of alternative data and model norms have been proposed to obtain models that provide closer representations of the expected model behavior; these methods are all based on iterative reweighting of the model misfit terms (Zhdanov, 2009; Ajo-Franklin et al., 2007; Minsley et al., 2007).

In order to evaluate the performance of a proposed model for a given objective function, it is necessary to have access to a forward model (see box 'Forward modeling (Jacobian matrix)'), which is the model that numerically solves the governing partial differential equation for a given model, boundary conditions, and excitation (e.g., current injection, detonation of explosives, or water injection in a wellbore). The accuracy of the forward model is of key importance in any inversion scheme. In deterministic inversions, it is also important to have access to the Jacobian or sensitivity matrix that defines how sensitive the modeled data are to a given small perturbation of each model parameter.

The objective function offers many opportunities to couple different data types to perform joint inversion by simply augmenting $\mathbf{d}$ and $\mathbf{m}$ with new data and model types, respectively. In order to perform joint inversion, it is necessary to define some sort of constraint such that the different data types and models interact in a meaningful way. These constraints can be of many types, such as structural constraints that penalize dissimilarity between two types of models (see arrows from box 'Save model'). One possible approach to structural joint inversion is to assume that the gradients in two models should be parallel or anti-parallel, thereby providing models that are structurally similar (e.g., Gallardo and Meju, 2003, 2004; Linde et al., 2006a, 2008). When performing joint inversion, only one objective function is used. To decrease the number of model unknowns when solving the corresponding system of equations, it is also possible to use an iterative sequential approach where two different objective functions are used as indicated in **Figure 2**. Another approach is to use the final model from one method to define spatial statistics that can be used to constrain the other model (Saunders et al., 2005), or the models can be constrained by using system property data from another method (Dafflon et al., 2009) or key interfaces can be incorporated, such as the depth to bedrock determined by seismic refraction in hydrogeological modeling of hillslope processes. Such information enters the objective function through the box 'Prior knowledge'.

Another approach when an accurate petrophysical relation is known to exist is to couple different model or data types by directly assuming that a given petrophysical model (known or with a given functional form with unknown parameter values) exists (see box 'Petrophysical models'). In this way, a geophysical model can be defined by a number of hydrological properties and state variables, and the geophysical data can thereby be directly incorporated into the hydrological inversion without the need to construct a geophysical model. In this case, only a geophysical forward model and a petrophysical model is needed to interpret the geophysical data within a hydrological inverse framework. These types of inversion methods are often referred to as fully coupled hydrogeophysical inversion (Kowalsky et al., 2005; Pollock and Cirpka, 2010). One problem with such an approach is that not only the parameter values used in petrophysical models might change within the study area, but also their functional form if they are too simplified.

2.15.4.1.3 Inversion step or model proposal

The next step corresponds to the box 'Inversion step or model proposal'. For the deterministic case, the system of equations are solved for a given trade-off λ of the different data and model misfit terms of the objective function. A large amount of numerical methods are available to solve this problem and a review of the most common methods is outside the scope of this chapter, but a good starting point is Golub and van Loan (1996). In a typical deterministic inversion, the inversion process continues until $\chi_d^2 \approx \chi_*^2$, where χ_*^2 is a predefined target data misfit. A new inversion step using the model obtained in the previous model is carried out if $\chi_d^2 > \chi_*^2$, where typically also the value of λ is decreased. If $\chi_d^2 < \chi_*^2$, it is customary to repeat the previous inversion step with a larger λ until $\chi_d^2 \approx \chi_*^2$ (see boxes 'End inversion?'). In cases where no convergence is obtained, one needs to change the inversion settings.

In stochastic inversions, a proposed model is evaluated based on prior knowledge and the so-called likelihood function, which is closely related to the data misfit term. Bayesian theory offers a consistent and general framework to sequentially condition models to different data types. The posterior distribution of the model parameters **m** given data **d** is given by Bayes' theorem

$$p(\mathbf{m}|\mathbf{d}) = Cp(\mathbf{d}|\mathbf{m})p(\mathbf{m}) \quad (21)$$

where C is a normalizing coefficient, $p(\mathbf{d}|\mathbf{m})$ is the likelihood function, and $p(\mathbf{m})$ is the prior distribution of the model parameters (permissible range and the distribution within the range for each parameter). The likelihood functions provide information about how likely it is that a given model realization is responsible for the observed data.

A main attraction of Bayesian sampling methods is that virtually any formulation of the likelihood function and the prior model can be used and it can differ between data and model types if performing joint inversion. The functional form of the petrophysical relationship can also be chosen in a flexible manner. The aim of Bayesian methods is generally to explore $p(\mathbf{d}|\mathbf{m})$ and this is often done by using Monte Carlo Markov Chain (MCMC) methods (e.g., Hastings, 1970; Mosegaard and Tarantola, 1995; Chen et al., 2006; Vrugt et al., 2009).

2.15.4.1.4 Geophysical model or model population

Assessment of the uncertainty in the final inversion images obtained from deterministic inversion is often limited to classical linear uncertainty estimates based on the posterior model covariance matrix and resolution measures based on the resolution matrix. These estimates bear a strong imprint of the regularization used to create a stable solution (Alumbaugh and Newman, 2000). The estimated uncertainty of individual model parameters is therefore often vastly underestimated.

Different approaches have been proposed in the literature to address the variance and resolution properties of deterministic inversion models. One popular approach is simply to perform several inversions where the regularization operators or the initial model vary. This approach provides a qualitative assessment of parameters that are well resolved by the geophysical data (Oldenburg and Li, 1999). Another approach is to perform a most-squares inversion (Jackson, 1976), where the bounds within which a model parameter can vary are sought for a given small increase in data misfit. Kalscheuer and Pedersen (2007) present a nonlinear variance and resolution analysis that investigate for a given variance of a model parameter, the resulting resolution properties of this estimate. The advantage compared with classical resolution analysis (e.g., Alumbaugh and Newman, 2000; Friedel, 2003) is that resolution properties are calculated for the same model variance and that regularization operators do not influence resolution estimates. Nonlinearity is partly handled by introducing nonlinear semi-axis that takes nonlinearity in the model eigenvectors into account. Even if these methods provide a qualitative assessment of model resolution and parameter uncertainty, they provide limited insight with respect to the probability distribution of the underlying model parameters and their multi-dimensional cross-correlations.

2.15.4.2 Example Parameter Estimation Approaches

2.15.4.2.1 Direct mapping approaches

The simplest application of geophysical data in quantitative hydrology is direct mapping (Linde et al., 2006b). In its simplest case, all boxes and arrows related to hydrology in **Figure 2** are removed and the inversion is performed using a standard geophysical inversion method. It is assumed that a known petrophysical model exists and that it can be used to map the final geophysical model into a hydrological model. Such transformations can be useful, but it is important to understand that geophysical models are only smoothed descriptions of the real property distribution and that the estimates might be biased. Day-Lewis and Lane (2004) and Day-Lewis et al. (2005) have developed a framework to describe how resolution in geophysical images degrade as a function of experimental design and data errors for linear and linearized nonlinear problems. They also show how it is possible to establish apparent petrophysical models from a known intrinsic petrophysical model that take this smoothing into account and thereby transform the geophysical model into a more realistic hydrological model. Direct mapping approaches can be made more effective when defining the model space and initial model, as well as the objective function, using prior knowledge related to the hydrology.

2.15.4.2.2 Integration approaches (geostatistical, Bayesian)

A more advanced approach is to combine site-specific hydrological system property data with geophysical models. We refer to this group of models as integration approaches and they are often based on concepts from geostatistics (Linde et al., 2006b). In this case, the geophysical inversion is performed in the same way as for direct mapping, but the petrophysical model and the model integration differ. One example of this approach would be to update a model of hydraulic conductivity observed at observation wells with geophysical models that are partly sensitive to hydraulic conductivity (e.g., Chen et al., 2001). Such models can incorporate some of the uncertainty in the geophysical and petrophysical relationships in the resulting hydrological models, but they are bound to use either petrophysical models with parameters determined from laboratory measurements (which are often unsuitable in this context; Moysey et al., 2005) or empirical field-specific relationships (which may be invalid away from calibration points; Linde et al., 2006c).

Direct mapping and integration approaches are useful routine tools, but they share several main limitations: (1) laboratory-based or theoretical petrophysical models often cannot be used directly, (2) the estimation of site-specific parameter values of the petrophysical models are not included within the inversion process, (3) there is no information-sharing between different data types during the inversion, (4) resulting uncertainty estimates are qualitative at best, and (5) they often provide physically impossible models (e.g., mass is not conserved when performing tracer tests, e.g., Singha and Gorelick, 2005).

2.15.4.2.3 Joint inversion or fully coupled hydrogeophysical inversion

The hydrogeophysical research community has in the recent years developed approaches that do not suffer from some of the limitations of direct mapping and integration methods by using both hydrological and geophysical state data during the inversion process and by coupling the hydrological and geophysical models during the inversion. We refer to such approaches as joint inversion (Linde et al., 2006b) or, alternatively, as fully coupled hydrogeophysical inversion. These approaches often include one or more of the following: (1) hydrological flow and transport modeling form together with geophysical forward modeling an integral part of the parameter estimation process; (2) petrophysical relationships are inferred during the inversion process; (3) nonuniqueness is explicitly recognized and a number of equally possible models are evaluated. This type of approach has at least four main advantages: (1) mass conservation can be assured in time-lapse studies and physically impossible flow fields are avoided when incorporating flow and transport simulations within the inversion framework; (2) data sharing during the inversion makes it often possible to obtain more realistic models with a higher resolution; (3) unknown parameters of petrophysical models can be estimated during the inversion process; (4) and physically implausible model conceptualization might make it impossible to fit the data to a realistic error level. This last point is important, since it makes joint inversion well suited to distinguish not only between possible realistic models and inconsistent parameter distributions, but also between competing conceptual models. Joint inversion comes at a price since it is necessary to develop new inversion codes that are suitable to the available data and model objectives; recent hydrogeophysical joint or fully coupled inversion methodologies include Kowalsky et al. (2005, 2006); Chen et al. (2006, 2010), Linde et al. (2006a, 2008),Lambot et al. (2009), Pollock and Cirpka (2009), and Huisman et al. (2010). Such developments can in practice be greatly facilitated by incorporating freely available forward codes or by using commercial multiphysics modeling packages. Despite this, the amount of work involved is typically more significant compared with direct mapping and data integration approaches.

2.15.5 Case Studies

Several case studies are presented to illustrate the use of geophysical methods for delineating subsurface architecture (Section 2.15.5.1), delineating anomalous subsurface fluid bodies (Section 2.15.5.2), monitoring hydrological processes (Section 2.15.5.3), and estimating hydrological properties (Section 2.15.5.4). The examples are based primarily on published hydrogeophysical studies that were conducted to gain insights about field-scale system behavior, improve flow and transport predictions, or to provide input to water resources or contaminant remediation management decisions. Examples were chosen to illustrate the utility for a variety of different characterization objectives, geophysical methods, and hydrogeophysical estimation approaches. Each example provides a brief background of the study as well as references for readers interested in more information.

2.15.5.1 Subsurface Architecture Delineation

Because geophysical properties are often sensitive to contrasts in physical and geochemical properties, geophysical methods can be useful for mapping subsurface architecture, defined here as a distribution of hydrogeologically distinct units. Using geophysical methods for subsurface mapping is perhaps the most well-developed application in hydrogeophysics, and it is often commonly performed using surface-based geophysical techniques. Examples of common mapping objectives in hydrogeological applications include the mapping of stratigraphy or the depth to bedrock or the water table. The ability to distinguish hydrogeologically meaningful boundaries using geophysical data depends on their sensitivity to subsurface physical properties, contrasts in these properties, and the resolution of the geophysical method at the characterization target depth.

In this section, we describe the use of geophysical data for mapping subsurface architecture or features by presenting several case studies that differ in their choice of geophysical method, the scales involved, characterization objective, and interpretation or integration approach. These examples include the use of airborne EM data to map lithofacies relevant for water resources at an island; high-resolution GPR imaging of braided river deposits; seismic methods to estimate subsurface architecture in contaminated environments; and fracture characterization using azimuthal SP measurements.

2.15.5.1.1 3D resistivity mapping of a Galapagos volcano aquifer

A fundamental limitation of most geophysical methods used in hydrogeophysics is that they have a limited ability to cover areas larger than ~ 1 km^2 within a reasonable time and budget at a high resolution. If funding permits, airborne geophysics can be very useful for gaining an overall view of the geological structure at the watershed scale. Most airborne data that are collected by geological surveys around the world provide only a limited depth resolution (e.g., Pedersen et al., 1994), whereas systems developed by the mining industry are designed for deeper targets and for more pronounced anomalies such that data quality demands are lower than in hydrogeological applications.

An exception is the helicopter-borne SkyTEM system that was purposefully developed for mapping of geological structures in the near surface for groundwater and environmental applications (Sørensen and Auken, 2004). This is a transient electromagnetic (TEM) system that uses a strong current flowing in the transmitter coil to induce weak secondary subsurface currents whose resulting magnetic fields are subsequently measured with a receiver coil. SkyTEM measurements provide similar data quality and resolution as ground-based TEM systems, but with the advantage that measurements are carried out at speeds exceeding 15 km h^{-1} corresponding to a typical station spacing of 35–45 m. This system operates normally at altitudes of 15–20 m with the helicopter located at an altitude of 50 m. The system is a stand-alone system that can be attached to the cargo hook of any helicopter. It uses a four-turn 12.5×12.5 m^2 transmitter loop with a low moment using one turn only and a high moment using all four turns. The receiver coil (0.5×0.5 m^2) is located 1.5 m above a corner of the quadratic and rigidly fixed transmitter loop.

D'Ozouville et al. (2008) illustrate the tremendous amount of information that airborne EM can provide in hydrological studies in remote areas where only limited prior geological and geophysical work have previously been carried out. This study was motivated by the need to better understand the groundwater resources on the volcanic island of Santa Cruz in the Galapagos island, which is experiencing challenges in meeting the water demands of the island's population and its many visitors. In order to provide an overall view of potential groundwater resources at the island, a SkyTEM survey of 900 km covering 190 km^2 was carried out to obtain a detailed view of the island's internal 3D electrical resistivity structure.

Figure 3 shows the resulting electrical resistivity models from two profiles that cross the island. 3D inversion of TEM data is computationally infeasible for large data sets, and the inversions were performed using 1 D forward modeling. The strong lateral continuity of these models is the result of lateral model constraints that are imposed during the inversion (Viezzoli et al., 2008). Four hydrogeological units were interpreted and they are indicated as I–IV in **Figure 3**. Unit I represents unsaturated fractured basalts with resistivities above 800 Ω m; unit II is the other resistivity end-member with resistivities smaller than 10 Ω m representing fractured basalt invaded by seawater. Unit III is a near-surface unit and unit IV a buried unit with resistivity values that range between 50 and 200 Ω m. These units might correspond to weathered zones or fractured basalts saturated with freshwater.

The top of unit II images the saltwater wedge to distances approximately 9 km inland and its slope is in perfect agreement with predictions based on the hydraulic gradient observed in one borehole and the density contrast between

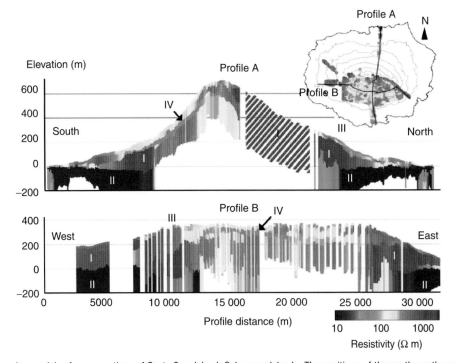

Figure 3 Two inversion models of cross sections of Santa Cruz Island, Galapagos Islands. The positions of the south–north and west–east profiles across the island are shown in the inset over a background of near-surface average resistivity showing extent of mapped area. The four units of hydrological interest are: (I) high-resistivity unsaturated basalts; (II) seawater intrusion wedge underlying the brackish basal aquifer; (III) near-surface, low-resistivity units consisting of colluvial deposits; and (IV) internal, low-resistivity unit of saturated basalts overlying an impermeable substratum. From figure 3 in D'Ozouville N, Auken E, Sorensen K, et al. (2008) Extensive perched aquifer and structural implications revealed by 3D resistivity mapping in a Galapagos volcano. *Earth and Planetary Science Letters* 269: 518–522.

salt- and freshwater. Of significant hydrological interest is unit IV that displays electrical resistivities similar to those of freshwater-saturated basalts on other islands. It forms an internal low-resistivity zone that is present only in the upper section of the southern side of the volcano. This wedge-shaped unit covers $50\,km^2$ and it has a thickness that varies between 10 and 80 m. It is quasi-parallel to the topography and coincides with the area of maximum precipitation. D'Ozouville et al. (2008) interpret unit IV as being composed of a similar basalt as unit I, but underlain by an impermeable layer that prohibits further downward percolation.

2.15.5.1.2 High-resolution GPR imaging of alluvial deposits

When surface conductivity is insignificant and pore water salinity is reasonably low, one of the primary tools in near-surface (up to tens of meters or so) hydrogeophysical studies is ground-penetrating radar (Davis and Annan, 1989). This method can be used to image interfaces of the 3D water content distribution and can therefore be very useful to gain information about variations in water saturation in the vadose zone (Irving et al., 2009) and porosity in the saturated zone (Beres and Haeni, 1991). GPR can also be used to image fractures (Grasmueck, 1996) or investigate the depositional setting (van Overmeeren, 1998; Beres et al., 1999). The widespread use of GPR is mainly due to its superior vertical resolution (in the order of 0.1–1.0 m depending on the antenna frequency and the velocity of the subsurface) and the very fast data acquisition, which makes it possible to routinely obtain high-quality data at close to walking speed.

Gravelly, braided river deposits form many aquifers and hydrocarbon reservoirs. These deposits typically display a hierarchical architecture where permeability varies over a multitude of scales (e.g., Ritzi et al., 2004), since permeability is linked to the sediment texture, geometry, and spatial distribution of sedimentary stata. Detailed characterization of such systems is difficult, but at least their statistical properties need to be known prior to attempting reservoir or aquifer management. In order to improve the understanding of such systems, Lunt et al. (2004) developed a 3D depositional model of the gravelly braided Sagavanirktok River in northern Alaska. The data used to construct this model were obtained from cores, wireline logs, trenches, and some 90 km of GPR profiles using different antenna frequencies (110, 225, 450, and 900 MHz) with corresponding depths of penetration varying from 7 to 1.5 m. The 17 boreholes only provided limited sampling, and the 1.3 km of destructive trenches provided information down to the water table only.

The GPR data provide continuous coverage over the whole thickness of the deposits and provide information about the depositional setting both across stream and along stream over the whole channel-belt width of 2.4 km. The GPR data made it possible to locate channel fill, unit bars, side bar deposits, confluence scour, compound braid bar deposits, and other depositional features that are of importance to understand the depositional setting. Figure 4 displays a comparison between a sedimentary log and a collocated GPR profile. Not only reflections corresponding to large-scale compound bar boundaries, but also certain unit bar boundaries are clearly imaged.

2.15.5.1.3 Subsurface flow architecture delineation using seismic methods

The use of both GPR and seismic data sets typically entails the processing of the geophysical measurements into estimates of geophysical properties, such as reflectivity or velocity, followed by a comparison of the attributes with direct measurements often available from wellbores (e.g., lithological boundaries). Figure 4 illustrated a comparison between GPR reflectivity and wellbore lithological information. Although this two-step method often provides useful information and takes advantage of expert knowledge, the qualitative approach can limit our ability to quantify errors associated with the interpretation and it can lead to dramatically different interpretations of subsurface heterogeneity depending on the interpreter and the processing steps employed.

To circumvent these limitations, Chen et al. (2010) developed a joint inversion method that simultaneously considers surface seismic refraction travel times and wellbore data sets for delineating watershed architecture that may exert an influence on contaminant plume mobility at the Oak Ridge National Laboratory site in Tennessee. The groundwater at this site includes uranium, nitrate, and other contaminants that emanated from a seepage basin (S-3 ponds, Figure 5). Underlying the seepage basin is weathered and fractured saprolite that overlies bedrock. Flow is expected to preferentially occur through the more intensely fractured and weathered zones. It is impossible to image individual fractures on a 100-m scale. However, because the fractures occur in discrete zones at this site and because the P-wave velocity in weathered and fractured zones should be lower than the surrounding more competent rock (e.g., Mair and Green, 1981, Chen et al., 2006, Juhlin and Stephens, 2006), seismic methods hold potential for aiding in the delineation of preferential flow zones.

A Bayesian joint inversion approach was developed and tested at two locations within the watershed to delineate architecture that may be important for controlling plume scale transport. Within the developed framework, the seismic first-arrival times and wellbore information about key interfaces were considered as input. A staggered-grid finite-difference method was used to forward model the full seismic waveform in 2D with subsequent automated travel-time picking. Seismic slowness and indicator variables of key interfaces are considered as unknown variables in the framework. By conditioning to the seismic travel times and wellbore information, Chen et al. (2010) estimated the probability of encountering key interfaces (i.e., between fill, saprolite, weathered low velocity zone, and consolidated bedrock) as a function of location and depth within the watershed. An example of the results obtained from two surface seismic data sets collected along the watershed reveals the presence of a distinct low velocity zone that is coincident with the trend of the plume axis (Figure 5). This example illustrates how the joint inversion approach can explicitly incorporate wellbore data into the inversion of the seismic travel time data in the estimation of aquifer architecture. Although not shown,

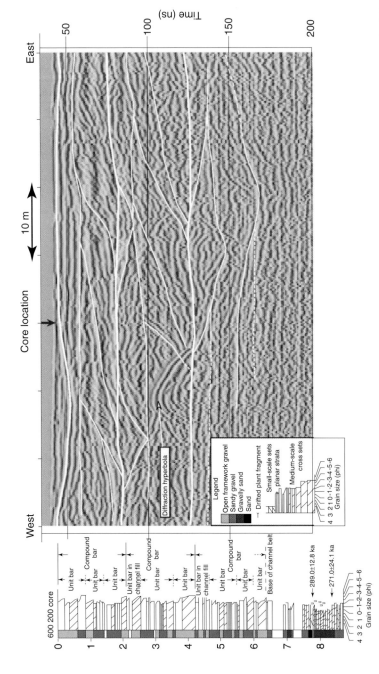

Figure 4 Comparison between a sedimentary log and a GPR profile, see arrow that indicates the core location on the GPR profile. GPR profile has a vertical exaggeration of 5:1. Sedimentary core log shows three compound bar deposits, each comprising two or three unit-bar deposits. The additional information offered by the continuous GPR profiles is evident. Modified from Figure 12 in Lunt IA, Bridge JS, and Tye RS (2004) A quantitative, three-dimensional depositional model of gravelly braided rivers. *Sedimentology* 51: 377–414.

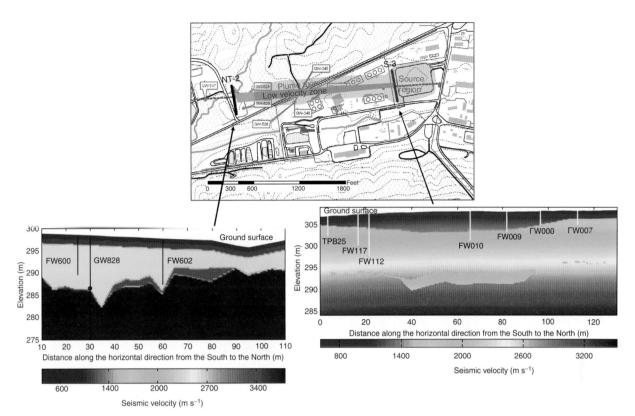

Figure 5 Bottom: Examples of seismic velocity models and subsurface architecture obtained through joint stochastic inversion of wellbore and surface-based seismic refraction data sets, which reveal distinct low velocity zones that are laterally persistent along the plume axis. Top: Superposition of low velocity zone region (shown in purple) on top of plume distribution (shown in pink), suggesting the control of the low velocity zone on plume migration. Modified from Chen J, Hubbard S, Gaines D, Korneev V, Baker G, and Watson D (2010) Stochastic estimation of aquifer geometry using seismic refraction data with borehole depth constraints. Water Resources Research (in press).

the approach also provides estimates of uncertainty about the location of the interfaces.

2.15.5.1.4 Fracture zonation characterization using azimuthal electrical methods

A significant body of literature has developed on using azimuthal electrical resistivity soundings to determine anisotropic electrical properties in fractured media (e.g., Taylor and Fleming, 1988; Lane et al., 1995). Electrical anisotropy in such systems is due to preferential fracture orientations, variable aperture distributions with azimuth, or clay-filled fractures. Field data suggest that directions of electrical anisotropy can, under certain conditions, be linked to anisotropy in hydraulic transmissivity (e.g., Taylor and Fleming, 1988), and it could therefore serve as an important data source in hydrogeological applications in fractured rock systems. Watson and Barker (1999) show that many of the electrode configurations that have been employed in past azimuthal resistivity surveys cannot discriminate between electrical anisotropy and heterogeneity. This problem can be avoided by using certain specialized electrode configurations. Unfortunately, data collection is very slow and no inversion for anisotropic parameters using such surveys has been performed to date. Linde and Pedersen (2004) demonstrate how these problems can be avoided by employing a frequency-domain EM method,

namely RMT. Despite these methodological developments to estimate azimuthal electrical anisotropy, it is not guaranteed that electrical anisotropy coincides with preferred hydrological flow directions and any such relationship is likely to be site specific.

Wishart et al. (2006, 2008) introduced the azimuthal self-potential gradient (ASPG) method, which provides data that may be sensitive to dominant hydrological flow directions in fractured media. In ASPG, one electrode is successively moved in steps on the order of 10° on the perimeter of an inner circle while the reference electrode moves with steps of the same size on the perimeter of an outer circle with the same midpoint as the inner circle. If the underground is predominantly anisotropic, the data will display a 180° symmetry except for data errors, while a 360° symmetry appears for measurements where lateral heterogeneity dominates. Wishart et al. (2008) applied this technique to four different fractured rock field sites in the New Jersey Highlands and found that three sites showed ASPG responses that compared well with observed fracture patterns at the sites. Figure 6 shows an example from one of the sites where the ASPG data display a significant 180° symmetry indicating that large-scale fracture anisotropy is responsible for the observed ASPG data. It is also seen that the direction of the anomalies corresponds well with the mapped fracture directions at outcrops within 100 m of the measurement locations. The data from an azimuthal resistivity survey

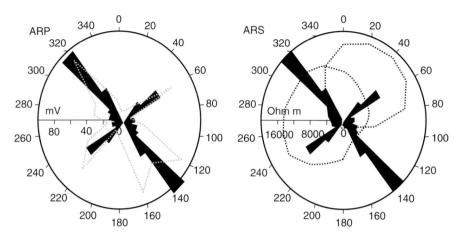

Figure 6 Azimuthal self-potential (SP) and resistivity data superimposed on rose diagrams of fracture strike sets mapped at the Wawayanda State Park, New Jersey. From Figure 4 in Wishart DN, Slater LD, and Gates A (2008) Fracture anisotropy characterization in crystalline bedrock using field-scale azimuthal self potential gradient. *Journal of Hydrology* 358: 35–45.

using an asymmetric arrow type array (Bolshakov et al., 1997) do not seem to correspond to either of the dominant fracture directions; the data are strongly nonsymmetric, indicating that lateral variations in the electrical conductivity structure dominate. It is reasonable to assume that the positive lobes of the ASPG data indicate groundwater flow directions, but no detailed field evidence is available to confirm this even if regional flow considerations point in this direction. The magnitudes of the SP signals presented by Wishart et al. (2008) are rather large (e.g., up to 300 mV over distances of 36 m) and the variations of ASPG signals with offset are sharp. The large magnitudes can partly be explained by the shallow water table (~ 0.5 m) and only a few meters of till overlying the highly resistive fractured rock mass. Applications in locations with thicker and more conductive overburden and with a deeper location of the source current will likely result in much smaller magnitude and a less clear-cut interpretation even for identical flow and fracture conditions. The usefulness of this technique in other field-settings needs to be assessed, but it appears that the ASPG technique can be a very rapid method to nonintrusively map preferential flow paths in fractured rock at shallow depths where overburden thickness is thin.

2.15.5.2 Delineation of Anomalous Fluid Bodies

Geophysical methods, particularly those collected from the ground surface or from aircrafts (e.g., Paine, 2003), have been successfully used to identify anomalous subsurface fluid bodies, such as contaminant plume boundaries and regions impacted by saltwater intrusion. Here, we illustrate the use of surface electrical approaches for delineating high ionic strength plumes and for characterizing redox gradients associated with contaminant plumes.

2.15.5.2.1 Electrical resistivity to delineate high-ionic-strength plume boundaries

Because of the strong positive correlation between total dissolved solids (TDSs) and the electrical conductivity of the pore fluid, electrical methods are commonly used to delineate subsurface plumes having high ionic strength (e.g., Watson et al., 2005; Adepelumi et al., 2005; Titov et al., 2005). As shown in Equation (5), electrical resistivity responds to porosity and surface conduction (often linked to lithology) as well as to saturation and pore fluid ionic strength. As described by Atekwana et al. (2004), activity of the natural microbial population can also impact the electrical resistivity through facilitating processes such as mineral etching, which appear to be more prevalent at the fringes of organic plumes. If the contrast between the concentration of the groundwater and the plume is great enough so that other contributions are considered to be negligible, electrical methods can be used, at least in the absence of significant clay units, to indicate contrasts in pore water electrical conductivity, or to delineate approximate boundaries of high-ionic-strength plumes.

An example of the use of inverted surface electrical resistivity data to delineate a deep (~ 50 m) nitrate plume at the contaminated Department of Energy (DOE) Hanford Reservation in Washington is given by Rucker and Fink (2007). They collected six ERT transects (each at least 200 m long), inverted the data to estimate the electrical resistivity distribution in the contaminated region, and compared their results with wellbore borehole measurements of pore water electrical conductivity and contaminant concentrations. They found a strong, negative correlation between electrical resistivity and nitrate concentration above a threshold value, which was used with the electrical models to delineate the plume. **Figure 7** shows several of the inverted transects as well as the correlation between electrical resistivity and nitrate concentration obtained from co-located electrical and wellbore measurements.

2.15.5.2.2 SP imaging of redox potentials associated with contaminated plumes

The traditional application of the SP method has been in mineral exploration where large negative SP anomalies are

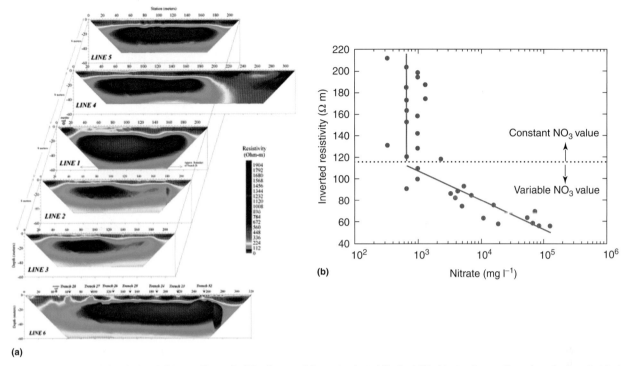

Figure 7 (a) Inverted electrical resistivity profiles at the BC crib area of the contaminated Hanford, Washington Reservation, where the low electrical resistivity (high electrical conductivity) regions were interpreted as the plume. (b) Observed relationship between electrical conductivity and nitrate concentration at a single wellbore location. Modified from Rucker DF and Fink JB (2007) Inorganic plume delineation using surface high resolution electrical resistivity at the BC Cribs and Trenches Site, Hanford. *Vadose Zone Journal* 6: 946–958.

typically associated with mineral veins (Fox, 1830). As an extreme example, Goldie (2002) presents a peak anomaly of −10.2 V associated with highly resistive high-sulfidation oxide gold deposits. The main contribution of such anomalies is thought to be related to electrochemical half-reactions (Sato and Mooney, 1960; Bigalke and Grabner, 1997), even if it has been suggested that some field data contradict this model (Corry, 1985).

Naudet *et al.* (2003, 2004) observed large negative SP anomalies over the Entressen domestic landfill outside Marseille, France. Redox potential, or Eh, indicates the tendency for oxidation–reduction reactions to occur. Strong redox gradients often become established adjacent to contaminant plumes. They found that the residual SP data (**Figure 8(a)**), where the effects of streaming currents had been filtered out, were strongly correlated with the difference in redox potential between groundwater samples in the contaminated landfill and uncontaminated areas. They invoked an explanation in analogy with the models of Sato and Mooney (1960) and Bigalke and Grabner (1997).

To remotely map variations in redox potential, Linde and Revil (2007) developed a linear inversion model where the difference in redox potential is retrieved from the residual SP data assuming a known 1 D electrical resistivity model and a known depth at which electrochemical reactions take place. They created a simplified representation of the electrical conductivity structure of the Entressen landfill based on ERT models and they assumed that source currents are located at the water table. **Figure 8(b)** shows a comparison between the simulated and observed residual SP data of Naudet *et al.* (2003, 2004). **Figure 8(c)** displays the retrieved redox potentials assuming a known background value outside of the contaminated area. By comparing these estimates and measured redox potentials in the wells (**Figure 8(d)**), they found that the inversion results can retrieve the measured redox potentials quite well given the simplifying assumptions involved. It should be noted that equally good data fits between the simulated and observed SP data could have been achieved by shifting the depth at which the sources are imposed or by assuming that the vertical dipole sources are distributed over a volume and not over an area (Blakely, 1996). The interpretation of SP data must therefore be treated with caution, and significant prior constraints must be imposed. Nevertheless, it appears that SP mapping and monitoring may provide a cheap and reliable method for monitoring field scale distribution of redox potential at contaminated sites. It is necessary that this approach is tested at other research sites before its applicability can be properly assessed.

2.15.5.3 Hydrological Process Monitoring

A particularly powerful component in hydrogeophysics is the use of a suite of geophysical data sets, collected at the same locations as a function of time, to monitor hydrological processes. Such repeated studies are often referred to as time-lapse geophysics, and their advantages compared to static images are

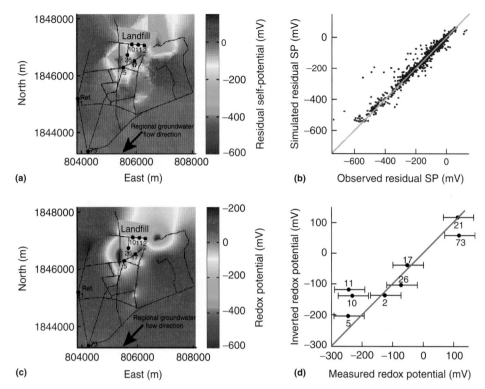

Figure 8 (a) Residual SP map at the Entressen landfill, in which the black lines indicate the SP profiles (2417 SP measurements). (b) Comparison of simulated SP with the residual SP estimated from the measured SP data. The response of the inverted model fits the residual SP to the estimated standard deviation. (c) Inverted redox potential in the aquifer at Entressen. (d) Comparison of inverted redox potentials in the aquifer with *in situ* measurements from Entressen (the correlation coefficient is 0.94). Modified from Linde N and Revil A (2007) Inverting self-potential data for redox potentials of contaminant plumes. *Geophysical Research Letters* 34: L14302 (doi: 10.1029/2007GL030084).

significant for process monitoring. First of all, changes in well-designed time-lapse inversions are most often primarily related to changes in state variables only (e.g., temperature, pressure, partial saturations of different phases, and the electrical conductivity of the pore fluid) and not to characteristics of the rock matrix itself. Time-lapse imaging has also the advantage that errors in the forward model tend to cancel (e.g., LaBrecque and Yang, 2001; Lien and Mannseth, 2008). It should be noted that subsurface engineered manipulations, such as those associated with environmental remediation, aquifer storage and recovery, and carbon sequestration can indeed alter the physical properties of the material. For example, remediation treatments can induce biogeochemical transformations that in turn alter the pore geometry and ultimately the fluid flow characteristics (Englert et al., 2009; Li et al., 2009). The geophysical responses to such processes are currently under intense study in the research area of biogeophysics (Atekwana et al., 2006; Williams et al., 2005; Chen et al., 2009; Slater et al., 2009), but are not covered in this chapter.

In what follows, we present several examples that illustrate the use of time-lapse geophysics, including: GPR to monitor the spatiotemporal distribution of soil water content in agricultural and hillslope settings; the use of GPR to monitor the distribution of saline tracers in fractured rock; and the use of EM methods to monitor seasonal changes in freshwater–seawater dynamics.

2.15.5.3.1 Soil moisture monitoring

The vadose zone mediates many of the processes in the hydrological cycle, such as the partitioning of precipitation into infiltration and runoff, groundwater recharge, contaminant transport, plant growth, evaporation, and sensible and latent energy exchanges between the Earth's surface and its atmosphere. As an example, in catchment hydrology, the readiness of an area to generate surface runoff during storm rainfall is related to its surface storage capacity. Given the predominant effects of soil moisture on the production of crops, soil salinization, carbon cycling, and climate feedback, development of methods for monitoring moisture content over field-relevant scales is desirable (e.g., Vereecken et al., 2008). Equations (5) and (13) indicate that both the dielectric constant and electrical conductivity are sensitive to water content. Because of this sensitivity, geophysical methods that are sensitive to these properties (e.g., GPR and ERT) have been used fairly extensively to monitor the spatiotemporal distribution of soil moisture.

As described by Huisman et al. (2003) and Lambot et al. (2008), GPR is commonly used in hydrogeophysical studies to estimate water content. Various GPR waveform components and configurations have been used to estimate water content, including: crosshole radar velocity (Hubbard et al., 1997; Binley et al., 2002), surface ground wave velocity (Grote et al., 2003), subsurface reflection (Greaves et al., 1996; Lunt et al., 2005), and air-launched ground-surface reflection approaches

(Lambot et al., 2006). An example of the use of time-lapse surface reflection GPR coupled with a Bayesian method to estimate seasonal changes in water content in the root zone of an agricultural site is given by Hubbard et al. (2006). Within a 90 m × 220 m section of this agricultural site, a thin (~0.1 m), low-permeability clay layer was identified from borehole samples and logs at a depth of 0.8–1.3 m below ground surface. GPR data were collected several times during the growing season using 100 MHz surface antennas; these data revealed that the thin clay layer was associated with a subsurface channel. Following equations (3) and (15), as the bulk water content in the unit above the GPR reflector increased, the dielectric constant increased, which lowered the velocity and lengthened the two-way travel time to the reflector. As a result, the GPR reflections revealed seasonal changes in the travel time to the clay layer as a function of average root zone moisture content. At the wellbore locations, a site-specific relationship between the dielectric constant and volumetric water content was used with the radar travel times to the clay reflector to estimate the depth-averaged volumetric water content of the soils above the reflector. Compared to average water content measurements from calibrated neutron probe logs collected over the same depth interval, the estimates obtained from GPR reflections at the borehole locations had an average error of 1.8% (Lunt et al., 2005). To assess seasonal variations in the root zone water content between the wellbores, the travel time picks associated with all GPR data sets, the wellbore information about the depths to the clay layer, and the site-specific petrophysical relationship were used within a Bayesian procedure (Hubbard et al., 2006). **Figure 9** illustrates the estimated volumetric water content for the zone located above the reflecting clay layer at different times during the year. The figure indicates seasonal variations in mean water content and also that the channel-shaped feature influences water content distribution: within this area the soils are consistently wetter than the surrounding soils. The soil moisture variations appeared to play a significant role in the crop performance: crops located within the channel region had consistently higher crop weight relative to the surrounding regions. These results suggest that the two-way GPR reflection travel times can be used to obtain estimates of average soil layer water content when GPR reflectors are present and when sufficient borehole control is available.

Several studies have also explored the use of surface ERT data sets for characterizing moisture infiltration and redistribution at the hillslope and watershed scales. For example, Berthold et al. (2004) compared electrical conductivity estimates from surface ERT images with groundwater electrical conductivity measurements to evaluate the roles of wetlands and ponds on depression-focused groundwater recharge within a Canadian wildlife region. The surface electrical data revealed a complex pattern of salt distribution that would have been difficult to understand given point measurements alone. Koch et al. (2009) collected surface electrical profiles over time along 18 transects within a German hillslope environment, and used the images together with conventional measurements to interpret flow pathways and source areas of runoff.

2.15.5.3.2 Saline tracer monitoring in fractured rock using time-lapse GPR methods

Hydrogeophysical applications in fractured media are challenging because of the large and discrete variations between the physical properties of the intact rock mass and the fractures (NRC, 1996). Time-lapse imaging of geophysically detectable tracers has been used in recent years to improve the understanding of fracture distribution and connectivity. The best adapted geophysical technique to image individual

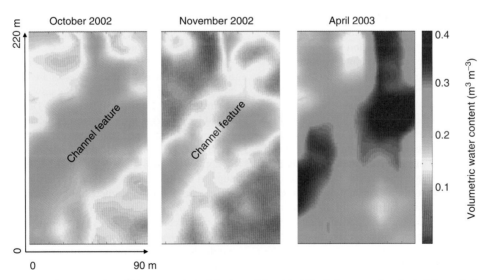

Figure 9 Plan-view map of average volumetric water content of the top soil layer (<1.5 m below ground surface) at the agricultural study site, estimated using 100 MHz GPR reflection travel-time data and borehole neutron probe data within a Bayesian estimation approach. Color key at right indicates relative volumetric water content, from red (drier) to blue (wetter). Modified from Hubbard S, Lunt I, Grote K, and Rubin Y (2006) Vineyard soil water content: mapping small scale variability using ground penetrating radar. In: Macqueen RW and Meinert LD (eds.) *Fine Wine and Terroir – The Geoscience Perspective*. Geoscience Canada Reprint Series Number 9 (ISBN 1-897095-21-X; ISSN 0821-381X). St. John's, NL: Geological Association of Canada.

fractures away from boreholes, up to few tens of meters away from the boreholes, is probably single-hole radar reflection measurements. This method has been shown to be a useful tool in site characterization efforts to determine possible orientations and lengths of fractures in nuclear waste repository laboratories (Olsson et al., 1992) and in characterizing unstable rock masses (Spillmann et al., 2007). By stimulating individual fractures by adding a saline tracer, it is possible to image tracer movement from the surface (Tsoflias et al., 2001; Talley et al., 2005; Tsoflias and Becker, 2008) and in-between boreholes (Day-Lewis et al., 2003, 2006) by investigating how the amplitude of GPR signals varies over time for a given transmitter–receiver geometry while the saline tracer migrates in the rock fractures. One problem with such studies is that the data acquisition time is often comparable to the timescale of the hydrological flow processes in the fractures where fluid flow velocities might be very high, creating large inversion artifacts if the data acquisition time is ignored in the inversion process.

Day-Lewis et al. (2002, 2003) present an innovative inversion method for difference-attenuation crosshole GPR data where the data acquisition time is included within the inversion. Synthetic (Day-Lewis et al., 2002) and field-based (Day-Lewis et al., 2003, 2006) inversion results show significant improvements compared with classical time-lapse inversion algorithms. The research of Day-Lewis et al. (2003, 2006) was carried out at the Forest Service East (FSE) well field at the US Geological Survey (USGS) Fractured Rock Hydrology Research Site located near Mirror Lake, New Hampshire. This well field consists of 14 boreholes distributed over an area of $120 \times 80 \, m^2$. Saline injection tests were carried out at 45 m depth where four boreholes, with side lengths of approximately 10 m located in a square-like shape seen from above, are hydraulically connected (Hsieh and Shapiro, 1996). These tracer tests were performed using weak-doublet tracer tests, where fluid was pumped out of one borehole at a rate of $3.8 \, l \, min^{-1}$ and water was injected in another borehole at $1.9 \, l \, min^{-1}$. After achieving steady-state flow, the injection fluid was changed from freshwater to a sodium chloride (NaCl) concentration of $50 \, g \, l^{-1}$ NaCl. Injection of freshwater was resumed after 10 min. The electrical conductivity ratio of these two fluids was estimated to be close to 170. A conventional packer system was used in the pumping well, whereas a special PVC packer system that allowed measurements while preventing vertical flow and the saline solution from entering the boreholes was used in the injection well and in two neighboring wells where GPR measurements were also conducted.

The energy of a GPR signal that arrives at the receiving antenna depends to a large degree on the electrical conductance of the media in between the transmitting and receiving antenna. It is expected that the magnitude of the signal at the receiving antenna decreases significantly when a saline tracer passes the ray path. Difference-attenuation inversion is a linear problem since electrical conductivity has no significant effect on the actual ray path. **Figure 10** displays variations in the ray energy that arrives in the receiver antennas normalized by the ray length for different transmitter and receiver separations during the tracer experiment of Day-Lewis et al. (2003) for a borehole plane roughly perpendicular to the injection and pumping borehole. **Figure 12** also shows the corresponding chloride concentration in the pumping well. The geophysical difference-attenuation data and the chloride data seem to agree qualitatively and a quicker breakthrough in the GPR data is observed because they were acquired over an area halfway between the injection and pumping boreholes.

This type of data was later inverted by Day-Lewis et al. (2003) and they showed that it was possible to remotely monitor the tracer movement relatively well given that only three 2D slices through the 3D volume could be imaged. It appears that difference-attenuation data might provide the resolution needed to study fluid flow in fractured rock with only limited hydrological point sampling.

2.15.5.3.3 Seasonal changes in regional saltwater dynamics using time-lapse EM methods

Falgàs et al. (2009) present one of few published time-lapse hydrogeophysical studies at the km scale (see Ogilvy et al. (2009) and Nguyen et al. (2009) for seawater intrusion studies using ERT). They used a frequency-domain EM method, namely Controlled-Source Audiomagnetotellurics (CSAMT) (Zonge and Hughes, 1991), to monitor freshwater–seawater interface dynamics in the deltaic zone of the Tordera River in northeastern Spain. Monitoring of saltwater intrusion in coastal aquifers is important due to population growth and since most of the World's population is concentrated along coastal areas. The CSAMT data were collected over an ancient paleochannel that controls seawater intrusion in a part of the delta. During 2 years, a profile of seven soundings was acquired along a 1700-m N–S trending line. Due to agricultural activity, they could not recover previous site locations with accuracy higher than 100 m when performing the repeated measurements.

The resulting individually inverted resistivity models are shown in **Figure 11** together with a weighted root-mean-square (RMS) data misfit calculated with an assumed error level of 5%. To better distinguish temporal changes, the inversions used the inversion results of the first survey as initial model for the subsequent inversions. The changes of the electrical resistivity models over time clearly indicate saltwater encroachment in the low-resistivity layer at approximately 50 m depth. These dynamic processes are best imaged in the northern part of the profile where the seawater wedge retreated toward the sea from April 2004 until December 2004, followed by progression until August 2005, and finally followed by a new retreat until May 2006. Multilevel sampling of a piezometer (W06) in April 2004 displayed a saltwater content of approximately 8% at a 50-m depth. Additional evidence to support the interpretation of the geoelectrical models in terms of seawater intrusion is offered by the piezometric levels that were the lowest in August 2005 when the seawater intrusion was interpreted to be at its maximum. Another zone displaying seawater intrusion dynamics is shown in a shallow aquifer located in the upper tenths of meters close to the sea located to the South.

Even if the study of Falgàs et al. (2009) had certain limitations, namely rather few stations, long periods between measurements, and not identical measurement locations between surveys, it still shows the potential of EM methods to monitor seawater intrusion processes on a scale that is relevant for water-resource planning.

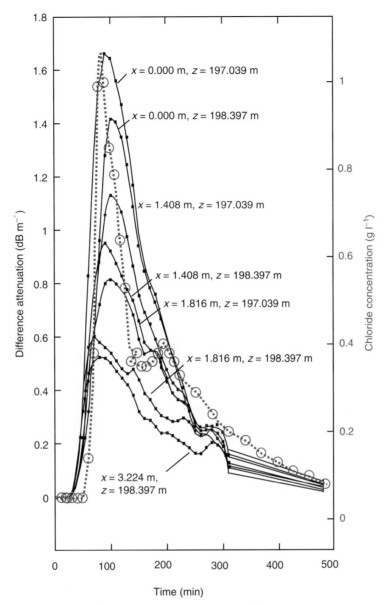

Figure 10 Nodal difference-attenuation histories between two boreholes (in black) and measured chloride concentration (in red). From Day-Lewis FD, Lane JW, and Gorelick SM (2006) Combined interpretation of radar, hydraulic, and tracer data from a fractured-rock aquifer near Mirror Lake, New Hampshire, USA. *Hydrogeology Journal* 14: 1–14.

EM methods have a higher sensitivity to conductors (e.g., the seawater plume) than the more commonly used ERT method, even if they have a poor resolving power in defining the lower boundary of conductors. This limitation can partly be resolved by combining this type of geophysical data with other types of geophysical data, such as seismic refraction data during the inversion (Gallardo and Meju, 2007).

2.15.5.4 Hydrogeological Parameter or Zonation Estimation for Improving Flow Predictions

Developing a predictive understanding of subsurface flow is complicated by the inaccessibility of the subsurface, the disparity of scales across which controlling processes dominate (e.g., Gelhar, 1993), and the sampling bias associated with different types of measurements (e.g., Scheibe and Chien, 2003). In this section, we describe the use of geophysical methods to improve flow predictions, through improved parametrization of flow and transport models as well as through fully coupled hydrogeophysical inversion. Although the examples provided here have been conducted at the local scale, joint or fully coupled hydrogeophysical inversion at the watershed scale is a research area that we expect to become more advanced in the coming years.

2.15.5.4.1 Hydraulic conductivity and zonation estimation using GPR and seismic methods

Several studies have described the use of geophysical data for estimating hydraulic conductivity (e.g., Cassiani *et al.*, 1998; Hyndman *et al.*, 2000; Hubbard *et al.*, 2001; Chen *et al.*, 2001;

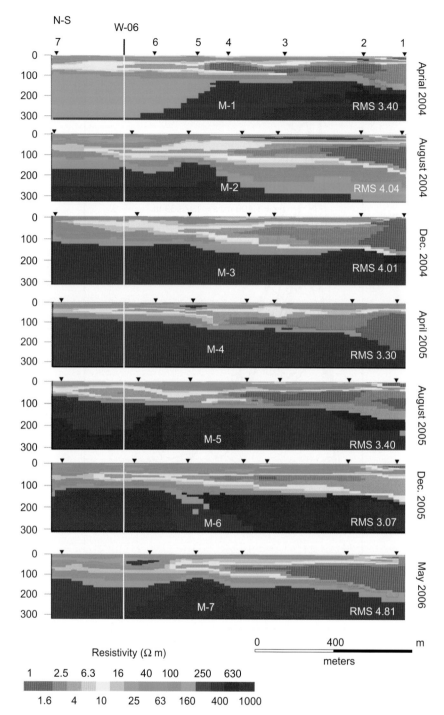

Figure 11 A series of two-dimensional (2D) electrical conductivity models as a function of date, with the triangles indicating survey location and the y-axis depth in m. From Falgàs E, Ledo J, Marcuello A, and Queralt P (2009) Monitoring freshwater–seawater interface dynamics with audiomagnetotellurics data. *Near Surface Geophysics* 7(5–6): 391–399.

Gloaguen et al., 2001; Slater, 2007; Linde et al., 2008). A few studies have also illustrated the value of geophysically obtained information for improving flow and transport predictions (Scheibe and Chien, 2003; Bowling et al., 2006; Scheibe et al., 2006). One such example is provided by the linked hydrogeophysical-groundwater modeling study performed at the DOE Oyster Site in Virginia. At this site, tomographic data were used together with borehole flowmeter logs to develop a site-specific petrophysical relationship that linked radar and seismic velocity with hydraulic conductivity. Using a Bayesian approach, a prior probability of hydraulic conductivity was first obtained through geostatistical interpolation (i.e., kriging) of the hydraulic conductivity values obtained at the wellbore location using the flowmeter logs. Within the Bayesian framework, these estimates were then updated using the developed petrophysical relationship and estimates of radar and seismic velocity were obtained along the tomographic transects (**Figure 12**). The method yielded posterior estimates of hydraulic conductivity (and their uncertainties) along the geophysical transects that honored the wellbore measurements (Chen *et al.*, 2001; Hubbard *et al.*, 2001). Examples of mean values of the geophysically obtained hydraulic conductivity estimates are shown in **Figure 12**, where the transects are oriented parallel and perpendicular to geological strike. The estimates were obtained at the spatial resolution of the geophysical model, which had pixel dimensions of 0.25 m × 0.25 m.

The geophysically obtained estimates were then used to develop a synthetic aquifer model (Scheibe and Chien, 2003). Other types of data sets were also used to develop other aquifer models, including interpolated core hydraulic conductivity measurements and interpolated flowmeter data. The breakthrough of a bromide tracer through these different aquifer models was simulated and subsequently compared with the breakthrough of the bromide tracer measured at the Oyster site itself (Scheibe and Chien, 2003). Even though this site was fairly homogeneous (the hydraulic conductivity varied over one order of magnitude) and had extensive borehole control (i.e., wellbores every few meters), it was difficult to capture the variability of hydraulic conductivity using borehole data alone with sufficient accuracy to ensure reliable transport predictions. Scheibe and Chien (2003) found that "conditioning to geophysical interpretations with larger spatial support significantly improved the accuracy and precision of model predictions" relative to wellbore-based data sets. This study suggested that the geophysically based methods provided information at a reasonable scale and resolution for understanding field-scale processes. This is an important point, because it is often difficult to take information gained at the laboratory scale or even from discrete wellbore samples and apply it at the field scale.

The level of detail shown in the hydraulic conductivity estimates of **Figure 12** may not always be necessary to adequately describe the controls on transport; in some cases, defining only contrasts between hydraulic units (Hill, 2006) or the hydraulic zonation may be sufficient to improve flow predictions. Several studies have illustrated the utility of tomographic methods for mapping zonation of lithofacies or hydrologically important parameters. Hyndman and Gorelick (1996) jointly used tracer and seismic tomographic data to map hydrological zonation within an alluvial aquifer. Linde *et al.* (2006c) used tomographic zonation constraints in the inversion of tracer test data and found that the constraints improved hydrogeological site characterization. Hubbard *et al.* (2008) used a discriminant analysis approach to estimate hydraulic conductivity zonation at the contaminated Hanford 100 H site, and found that the identified heterogeneity controlled the distribution of remedial amendments injected into the subsurface for bioremediation purposes as well as the subsequent biogeochemical transformations.

2.15.5.4.2 Joint modeling to estimate temporal changes in moisture content using GPR

In this example, we illustrate the value of the joint inversion approach for taking advantage of the complementary nature of geophysical and hydrological data and for circumventing some of the obstacles commonly encountered in other types of integration approaches (see Section 2.15.4.2). As was previously discussed, the use of GPR methods for mapping water content distributions in the subsurface is now well established. However, in general, GPR measurements cannot be directly related to the soil hydraulic parameters needed to make hydrological predictions in the vadose zone (such as the permeability and the parameters describing the relative permeability and capillary pressure functions). On the other hand, time-lapse GPR data often contain information that can be indirectly related to the soil hydraulic properties, since these soil hydraulic properties influence the time- and space-varying changes in water distribution, which in turn affect GPR data.

Kowalsky *et al.* (2004, 2005) illustrated an approach for incorporating time-lapse GPR and hydrological measurements into a hydrological–geophysical joint inversion framework for estimating soil hydraulic parameter distributions. Coupling between the hydrological and GPR simulators was accomplished within the framework of an inverse model (iTOUGH2, Finsterle, 1999). The inversion was performed using a maximum *a posteriori* (MAP) approach that utilized concepts from the pilot point method. One of the benefits of this approach was that it directly used the GPR travel times rather than radar velocity tomograms, which circumvented some of the problems that were discussed in Section 2.15.4.2. The approach also accounted for uncertainty in the petrophysical function that related water content and dielectric permittivity.

The approach was applied to data collected at the 200 East Area of the US Department of Energy (DOE) Hanford site in Washington. The Hanford subsurface is contaminated with significant quantities of metals, radionuclide, and organics; contaminants are located in the saturated as well as in a thick vadose zone. Gaining an understanding of vadose zone hydraulic parameters, such as permeability, is critical for estimating plume infiltration at the site and the ultimate interception with groundwater and the nearby Columbia River. To gain information about the vadose zone hydraulic parameters, an infiltration test was performed by ponding water on the ground surface and subsequently measuring the subsurface moisture distribution over time using neutron probe data collected within wellbores and radar tomographic data collected between boreholes (**Figures 13(a)** and **13(b)**).

Because water infiltration behavior is a function of the permeability distribution, the joint inversion procedure could be used with the time-lapse moisture data to estimate log permeability. The inversion procedure was also used to estimate other parameters of the petrophysical relationship, porosity, and the injection rate, none of which were measured precisely at the site. Figure 13(c) illustrates the permeability values estimated from the joint inversion procedure, which have been conditioned to GPR travel times and to the measured hydrological properties. The obtained permeability values were then used to predict fluid flow at future times. The accuracy of predictions for future times was evaluated through comparison with data collected at later times but not used in the inversion. In the first case, inversion was performed using only neutron probe data collected in two wells at three different times. In the second case, inversion was performed

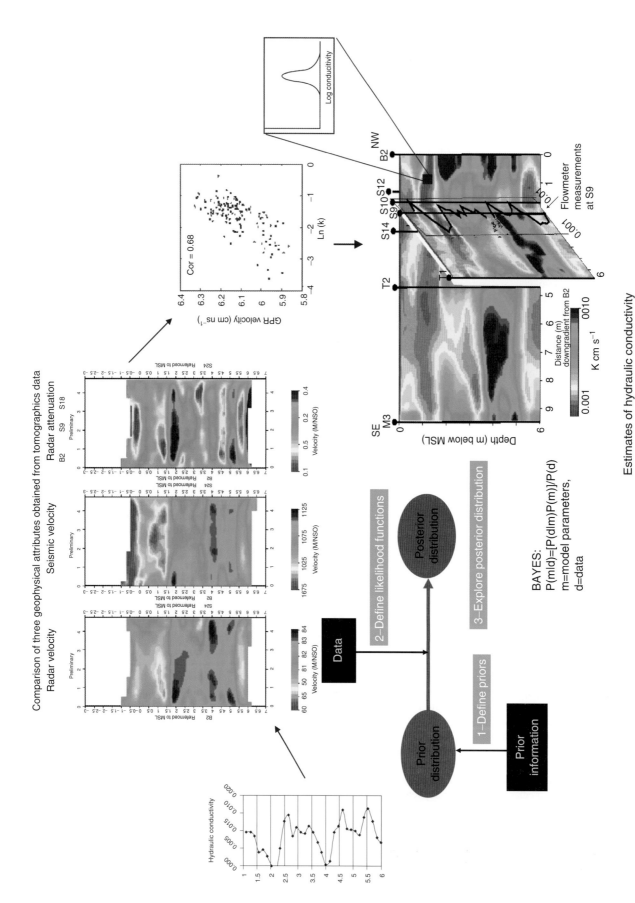

Figure 12 Example of Bayesian approach for integrating disparate data sets for the estimation of hydraulic conductivity distributions is shown on the bottom right. Modified from Hubbard S, Chen J, Peterson J, et al. (2001) Hydrogeological characterization of the DOE bacterial transport site in Oyster, Virginia, using geophysical data. *Water Resources Research* 37(10): 2431–2456.

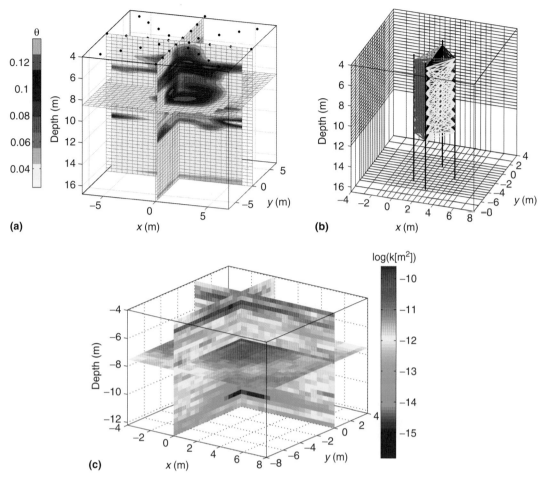

Figure 13 Time-lapse data sets collected during water injection at Hanford site, including (a) interpolated water content inferred from dense neutron-probe measurements and (b) ground-penetrating radar acquisition geometry. Estimates of log permeability (c) obtained using the coupled inversion approach. Modified from Kowalsky et al. (2006).

using GPR data collected at two times in addition to the neutron probe data used in the first case. Compared to predictions made through inversion of only neutron probe data, inclusion of GPR data in the joint inversion resulted in more accurate estimates of water content at later times.

2.15.6 Summary and Outlook

This chapter has reviewed several case studies that illustrated how hydrogeophysical methods can be used to: map subsurface architecture, estimate subsurface hydrological properties or state variables, and monitor subsurface processes associated with natural or engineered *in situ* perturbations to the subsurface system. These and many other studies have now demonstrated that hydrogeophysical approaches can successfully be used to gain insight about subsurface hydrological processes, provide input that improves flow and transport predictions, and provide information over spatial scales that are relevant to the management of water resources and contaminant remediation. Critical to the success of hydrogeophysical studies are several factors: (1) the acquisition of high-quality geophysical data; (2) the availability of petrophysical relationships that can link geophysical properties to the parameters or processes relevant for the hydrological study; and (3) the use of inversion approaches that allow for reliable and robust estimation of hydrological parameters of interest. Here, we briefly comment on each of these important factors and associated research needs.

Section 2.15.2 reviewed many of the geophysical methods that are common or are being increasingly employed in hydrogeophysical studies, including: electrical resistivity, IP, controlled-source inductive EM, SP, GPR, seismic, SNMR, gravity, magnetics, and well logging methods. We stressed that acquisition of high-quality data is critical to a successful hydrogeophysical study. The choice of which geophysical data to invoke for a particular investigation must be made based on the expected sensitivity of the geophysical attribute to the properties associated with the characterization objective (or the contrast of the target properties with the surrounding sediments or rocks). Different geophysical methods perform optimally in different environments and have different resolving capabilities. It is thus necessary, when deciding on which geophysical method to use to consider the general geological setting and the size/depth/contrast magnitude of the characterization target. Although these characteristics should be considered prior to choosing a method (and ideally considered through synthetic modeling), commonly the performance of a geophysical method cannot be truly assessed until it is tested at a particular field site. This is because factors that influence its performance (such as clay content, depth of particular target, contrast in characterization target properties with surrounding material, and presence of cultural features such as underground pipes) are often not known with sufficient certainty prior to field testing. For this reason, geophysical campaigns are ideally performed in an iterative manner, starting first with reconnaissance campaigns that involve testing the geophysical responses of a few different methods prior to choosing the method for further high-resolution investigation.

Section 2.15.3 described common petrophysical models that link electrical conductivity, dielectric permittivity, complex conductivity, and SP measurements to hydrological variables, which are commonly based on theoretical considerations or on laboratory- or field-based experiments. Unfortunately, all of these petrophysical model types pose challenges for hydrogeophysical studies. Theoretical models often invoke assumptions or simplifications that deviate from heterogeneous, *in situ* conditions. Problems with laboratory-based measurements (e.g., Ferré *et al.*, 2005) are that it is very difficult to (1) acquire undisturbed samples that adequately represent conditions in the near subsurface and (2) upscale developed relationships from the laboratory to the field scale (Moysey *et al.*, 2005). Application of field-scale relationships (e.g., using co-located hydrogeophysical wellbore data sets; Hubbard *et al.*, 2001) can also be problematic if the petrophysical relationship differs at locations away from the wellbore (Linde *et al.*, 2006c). Finally, because most geophysical attributes are sensitive to more than one property that typically varies substantially in the subsurface, methods must be developed to handle nonuniqueness in geophysical responses to property variations (Hubbard and Rubin, 1997). The development and testing of petrophysical relationships that describe the linkages between field-scale geophysical responses to variably saturated, semi- to unconsolidated, low-pressure materials that typify many of our shallow subsurface environments continues to be a need in hydrogeophysics. Embedded in that need is the development of methods that can adequately handle scale effects, nonuniqueness, and uncertainties associated with petrophysical relationships.

The importance of parameter estimation/integration methods that honor available hydrogeological and geophysical data in the interpretation procedure was described in Section 2.15.4. We defined three different parameter estimation processes, namely: (1) direct mapping; (2) integration approaches (geostatistical and Bayesian); and (3) joint inversion or fully coupled hydrogeophysical inversion. Each of these has advantages and limitations, and the decision about which approach to use is a function of the data available, the characterization objective and project budget, and the experience of the interpreter with the different methods. Clearly, the motivation exists to take advantage of the complimentary nature of hydrological and geophysical data and modeling to improve experimental design and interpretation while recognizing that each of these approaches has associated uncertainty. We thus believe that one of the most important developments in hydrogeophysical research in the coming years will arise from data integration schemes that provide a flexible way to couple different hydrological and geophysical data and model types in a framework that explicitly assesses uncertainty in the final model or model predictions. An important challenge will be the development of methods that disregard models that are inconsistent with our available data and *a priori* conceptions while retaining a representative subset of models that are consistent with available data. It is expected that joint inversion approaches can provide more significant improvements compared to other approaches, especially when working with time-lapse data and when the hydrological dynamics of the geophysical and hydrological forward responses display strong nonlinearities. Although inversion approaches have been developed to meet some of these criteria, for the most part they have been tested in conjunction with specific research projects and are not generally accessible for use by nonspecialists or flexible enough to be applied to other problems and data sets. An existing need is thus the development of software that will facilitate the transfer of the state-of-the-art inversion algorithms, which allow joint consideration of geophysical and hydrological measurements and phenomena and that provide meaningful assessments of uncertainty, into practice.

Related to all three key factors in hydrogeophysical studies (high-quality geophysical data sets, petrophysics, and integration methods) is the need to better advance our capabilities to improve the characterization of subsurface hydrological parameters and processes at the larger watershed scale. The majority of the hydrogeophysical studies that have focused on quantitative hydrological parameter estimation or model coupling have been performed at the local scale (typically with length scales < 10 m), where the disparity in measurement support scale between wellbore (direct) measurements and geophysical measurements is smaller and where stationarity of petrophysical relationships can often be reasonably assumed. Although these studies have illustrated the power of hydrogeophysical methods for improving the resolution and understanding of subsurface properties or processes at the local scale, they are often still limited in their ability to inform about behavior that may be most relevant at the larger scales where water resources or environmental contaminants are managed. As described in Section 2.15.4, although a handful of case studies have now illustrated the potential that geophysical methods hold for providing quantitative information over large spatial scales, additional effort is needed to continue to advance this area of watershed hydrogeophysics (Robinson *et al.*, 2008). In particular, there is a great need to develop petrophysical models and integration schemes that permit the coupling of different hydrological and geophysical data and model types within a framework that explicitly assesses uncertainty in the final model or model predictions over watershed- or plume-relevant scales.

Acknowledgments

Support for Susan Hubbard was provided by the US Department of Energy, Biological and Environmental Research Program as part of the Oak Ridge Integrated Field Research

Center (ORIFRC) project and through DOE Contract DE-AC0205CH11231 to the LBNL Sustainable Systems Scientific Focus Area. We thank Lee Slater and Sébastien Lambot whose constructive reviews helped to substantially improve the text.

References

Adepelumi AA, Ako BD, Afolabi O, and Arubayi JB (2005) Delineation of contamination plume around oxidation sewage ponds in southwestern Nigeria. *Environmental Geology* 48: 1137–1146.

Ajo-Franklin JB, Minsley BJ, and Daley TM (2007) Applying compactness constraints to differential traveltime tomography. *Geophysics* 72: R67–R75.

Alumbaugh DL and Newman GA (2000) Image appraisal for 2-D and 3-D electromagnetic inversion. *Geophysics* 65: 1455–1467.

Annan AP (2005) GPR methods for hydrogeological studies. In: Rubin Y and Hubbard S (eds.) *Hydrogeophysics*, ch. 7, pp. 185–214. Dordrecht: Springer.

Archie GE (1942) The electrical resistivity log as an aid in determining some reservoir parameters. *Transactions of the American Institute of Mining, Metallurgy, and Petroleum Engineering* 146: 54–62.

Atekwana EA, Werkema DD, Allen JP, et al. (2004) Evidence of microbial enhanced electrical conductivity in hydrocarbon-contaminated sediments. *Geophysics Research Letters* 31: L2350.

Atekwana EA, Werkema DD, and Atekwana EA (2006) Biogeophysics: The effects of microbial processes on geophysical properties of the shallow subsurface. In: Vereecken H, Binley A, Cassiani G, Revil A, and Titov K (eds.) *Applied Hydrogeophysics*, ch. 6, pp. 161–193. Dordrecht: Springer.

Beres M and Haeni FP (1991) Application of ground-penetrating radar methods in hydrogeological studies. *Ground Water* 29: 375–386.

Beres M, Huggenberger P, Green AG, and Horstmeyer H (1999) Using two- and three-dimensional georadar methods to characterize glaciofluvial architecture. *Sedimentary Geology* 129: 1–24.

Bernabé Y and Revil A (1995) Pore-scale heterogeneity, energy dissipation and the transport properties of rocks. *Geophysical Research Letters* 22: 1529–1532.

Berthold S, Bentley L, and Hayashi M (2004) Integrated hydrogeological and geophysical study of depression-focused groundwater recharge in the Canadian prairies. *Water Resources Research* 40: W06505.

Bigalke J and Grabner EW (1997) The geobattery model: A contribution to large scale electrochemistry. *Electrochemica Acta* 42: 3443.

Binley A, Cassiani G, Middleton R, and Winship P (2002) Vadose zone flow model parameterisation using cross-borehole radar and resistivity imaging. *Journal of Hydrology* 267: 147–159.

Binley A and Kemna A (2005) DC resistivity and induced polarization methods. In: Rubin Y and Hubbard S (eds.) *Hydrogeophysics*, ch. 5, pp. 129–156. Dordrecht: Springer.

Binley A, Slater LD, Fukes M, and Cassiani G (2005) Relationship between spectral induced polarization and hydraulic properties of saturated and unsaturated sandstone. *Water Resources Research* 41: W12417.

Birchak JR, Gardner CG, Hipp JE, and Victor JM (1974) High dielectric constant microwave probes for sensing soil moisture. *Proceedings of the IEEE* 62: 93–98.

Blakely RJ (1996) *Potential Theory in Gravity and Magnetic Applications.* Cambridge: Cambridge University Press.

Bolshakov DK, Modin IN, Pervage EV, and Shevnin VA (1997) Separation of anisotropy and inhomogeneity influence by the spectral analysis of azimuthal resistivity diagrams. Paper presented at 3rd EEGS-ES Meeting, Aarhus, Denmark.

Bowling JC, Zheng C, Rodriguez AB, and Harry D (2006) Geophysical constraints on contaminant modeling in a heterogeneous fluvial aquifer. *Journal of Contaminant Hydrology* 85: 72–88.

Brovelli A and Cassiani G (2008) Effective permittivity of porous media: A critical analysis of the complex refractive index model. *Geophysical Prospecting* 56: 715–727.

Brovelli A and Cassiani G (2010) A combination of the Hashin–Shtrikman bounds aimed at modelling electrical conductivity and permittivity of variably saturated porous media. *Geophysical Journal International* 180: 225–237.

Bussian AE (1983) Electrical conductance in a porous medium. *Geophysics* 48: 1258–1268.

Butler J (2005) Hydrogeological methods for estimation of spatial variations in hydraulic conductivity. In: Rubin Y and Hubbard S (eds.) *Hydrogeophysics*, ch. 2, pp. 23–58. Dordrecht: Springer.

Börner FD and Schön JH (1991) A relation between the quadrature component of electrical conductivity and the specific surface area of sedimentary rocks. *Log Analyst* 32: 612–613.

Carcione JM, Ursin B, and Nordskag JI (2007) Cross-property relations between electrical conductivity and the seismic velocity of rocks. *Geophysics* 72: E193–E204.

Cassiani G, Bohm G, Vesnaver A, and Nicholich R (1998) A geostatistical framework for incorporating seismic tomography auxillary data into hydraulic conductivity estimation. *Journal of Hydrology* 206(1–2): 58–74.

Chapelle FH (2001) *Groundwater Microbiology and Geochemistry.* New York: Wiley.

Chen J, Hubbard S, Gaines D, Korneev V, Baker G, and Watson D (2010) Stochastic estimation of aquifer geometry using seismic refraction data with borehole depth constraints. *Water Resources Research* (in press).

Chen J, Hubbard S, Peterson J, et al. (2006) Development of a joint hydrogeophysical inversion approach and application to a contaminated fractured aquifer. *Water Resources Research* 42: W06425 (doi: 10.1029/ 2005WR004694).

Chen J, Hubbard S, and Rubin Y (2001) Estimating the hydraulic conductivity at the South Oyster Site from geophysical tomographic data using Bayesian techniques based on the normal linear regression model. *Water Resources Research* 37: 1603–1616.

Chen J, Hubbard SS, Williams KH, Pride S, Li L, and Slater L (2009) A state-space Bayesian framework for estimating biogeochemical transformations using time-lapse geophysical data. *Water Resources Research* 45: W08420 (doi: 10.1029/ 2008WR007698).

Cole KS and Cole RH (1941) Dispersion and absorption in dielectrics, 1. Alternating current characteristics. *Journal of Chemical Physics* 9: 341–351.

Corry CE (1985) Spontaneous polarization associated with porphyry sulfide mineralization. *Geophysics* 50: 1020.

Dafflon B, Irving J, and Holliger K (2009) Simulated-annealing-based conditional simulation for the local-scale characterization of heterogeneous aquifers. *Journal of Applied Geophysics* 68: 60–70.

Davis JL and Annan AP (1989) Ground-penetrating radar for high-resolution mapping of soil and rock stratigraphy. *Geophysical Prospecting* 37: 531–551.

Day-Lewis FD, Harris JM, and Gorelick SM (2002) Time-lapse inversion of crosswell radar data. *Geophysics* 67: 1740–1752.

Day-Lewis FD and Lane JW, Jr. (2004) Assessing the resolution-dependent utility of tomograms for geostatistics. *Geophysical Research Letters* 31: L07503.

Day-Lewis FD, Lane JW, and Gorelick SM (2006) Combined interpretation of radar, hydraulic, and tracer data from a fractured-rock aquifer near Mirror Lake, New Hampshire, USA. *Hydrogeology Journal* 14: 1–14.

Day-Lewis FD, Lane JW, Jr., Harris JM, and Gorelick SM (2003) Time-lapse imaging of saline-tracer transport in fractured rock using difference-attenuation radar tomography. *Water Resources Research* 39: 1290.

Day-Lewis FD, Singha K, and Binley A (2005) Applying petrophysical models to radar travel time and electrical resistivity tomograms: Resolution-dependent limitations. *Journal of Geophysical Research* 110: B08206.

D'Ozouville N, Auken E, Sorensen K, et al. (2008) Extensive perched aquifer and structural implications revealed by 3D resistivity mapping in a Galapagos volcano. *Earth and Planetary Science Letters* 269: 518–522.

Englert A, Hubbard SS, Williams KH, Li L, and Steefel CI (2009) Feedbacks between hydrological heterogeneity and bioremediation induced biogeochemical transformations. *Environmental Science and Technology* 43(13): 5197–5204.

Ernst JR, Maurer H, Green AG, and Holliger K (2007) Full-waveform inversion of crosshole radar data based on 2-D finite-difference time-domain solutions of Maxwell's equations. *IEEE Transactions on Geoscience and Remote Sensing* 45(9): 2807–2828.

Everett ME and Meju MA (2005) Near-surface controlled source electromagnetic induction: Background and recent advances. In: Rubin Y and Hubbard S (eds.) *Hydrogeophysics*, ch. 6, pp. 157–183. Dordrecht: Springer.

Falgàs E, Ledo J, Marcuello A, and Queralt P (2009) Monitoring freshwater–seawater interface dynamics with audiomagnetotellurics data. *Near Surface Geophysics* 7 (5–6): 391–399.

Farquharson CG (2008) Constructing piecewise-constant models in multidimensional minimum-structure inversions. *Geophysics* 73: K1–K9.

Ferré TPA, Binley A, Geller J, Hill E, and Illangasekare T (2005) Hydrogeophysical methods at the laboratory scale. In: Rubin Y and Hubbard S (eds.) *Hydrogeophysics*, pp. 441–463. Dordrecht: Springer.

Finsterle S (1999) iTOUGH2 User Guide, Lawrence Berkeley National Laboratory LBNL-40040.

Fox RM (1830) On the electromagnetic properties of metalliferous veins in the mines of Cornwall. *Philosophical Transactions of the Royal Society* 130: 399.

Freeze RA and Cherry JA (1979) *Groundwater.* Upper Saddle River, NJ: Prentice Hall.

Friedel S (2003) Resolution, stability and efficiency of resistivity tomography estimated from a generalized inverse approach. *Geophysical Journal International* 153: 305–316.

Gallardo LA and Meju MA (2003) Characterization of heterogeneous near-surface materials by joint 2D inversion of dc resistivity and seismic data. *Geophysical Research Letters* 30: 1658.

Gallardo LA and Meju MA (2004) Joint two-dimensional DC resistivity and seismic travel time inversion with cross-gradient constraints. *Journal of Geophysical Research* 109: B03311.

Gallardo LA and Meju MA (2007) Joint two-dimensional cross-gradient imaging of magnetotelluric and seismic traveltime data for structural and lithological classification. *Geophysical Journal International* 169: 1261–1271.

Gelhar LW (1993) *Stochastic Subsurface Hydrology*. New York: Prentice Hall.

Ghorbani A, Camerlynck C, Florsch N, Cosenza P, and Revil A (2007) Bayesian inference of the Cole–Cole parameters from time- and frequency-domain induced polarization. *Geophysical Prospecting* 55: 589–605.

Gloaguen E, Chouteau M, Marcotte D, and Chapuis R (2001) Estimation of hydraulic conductivity of an unconfined aquifer using cokriging of GPR and hydrostratigraphic data. *Journal of Applied Geophysics* 47(2): 135–152.

Goldie M (2002) Self-potentials associated with the Yanacocha high-sulfidation gold deposit in Peru. *Geophysics* 67: 684–689.

Golub GH and van Loan CF (1996) *Matrix Computations*, 3rd edn. Baltimore, MD: Johns Hopkins Press.

Grasmueck M (1996) 3-D ground-penetrating radar applied to fracture imaging in gneiss. *Geophysics* 61: 1050–1064.

Greaves RJ, Lesmes DP, Lee JM, and Toksov MN (1996) Velocity variations and water content estimated from multi-offset, ground-penetrating radar. *Geophysics* 61: 683–695.

Grote K, Hubbard SS, and Rubin Y (2003) Field-scale estimation of volumetric water content using GPR ground wave techniques. *Water Resources Research* 39: 1321–1333.

Guéguen Y and Palciauskas V (1994) *Introduction to the Physics of Rocks*. Princeton, NJ: Princeton University Press.

Günther T, Rücker C, and Spitzer K (2006) Three-dimensional modelling and inversion of dc resistivity data incorporating topography – II. Inversion. *Geophysical Journal International* 166: 506–517.

Hashin Z and Shtrikman S (1962) A variational approach to the theory of the effective magnetic permeability of multiphase materials. *Journal of Applied Physics* 33: 3125–3131.

Hastings WK (1970) Monte Carlo sampling methods using Markov chains and their applications. *Biometrika* 57: 97–109.

Hertrich M (2008) Imaging of groundwater with nuclear magnetic resonance. *Progress in Nuclear Magnetic Resonance Spectroscopy* 53: 227–248.

Hill M (2006) The practical use of simplicity in developing groundwater models. *Ground Water* 44(6): 775–781.

Hinze WJ (1990) The role of gravity and magnetic methods in engineering and environmental studies. In: Ward S (ed.) *Geotechnical and Environmental Geophysics: Review and Tutorial*, SEG Investigations in Geophysics No. 5, vol. 1, pp. 75–126. Tulsa, OK: Society of Exploration Geophysicists.

Hsieh PA and Shapiro AM (1996) Hydraulic characteristics of fractured bedrock underlying the FSE well field at the Mirror Lake Site, Grafton Country, New Hampshire. In: Morganwalp DW and Aronson DA (eds.) *U.S. Geological Survey Toxic Substances Hydrology Program: Proceedings of the Technical Meeting*, US Geological Survey Water Resources Investigative Report 94-4015, vol. 1, pp. 127–130, Colorado Springs, CO, USA. 20–24 September 1993.

Hördt A, Blaschlek R, Kemna A, and Zisser N (2007) Hydraulic conductivity estimation from induced polarisation data at the field scale – the Krauthausen case history. *Journal of Applied Geophysics* 62: 33–46.

Hubbard S, Chen J, Peterson J, et al. (2001) Hydrogeological characterization of the DOE bacterial transport site in Oyster, Virginia, using geophysical data. *Water Resources Research* 37(10): 2431–2456.

Hubbard S, Lunt I, Grote K, and Rubin Y (2006) Vineyard soil water content: mapping small scale variability using ground penetrating radar. In: Macqueen RW and Meinert LD (eds.) *Fine Wine and Terroir – The Geoscience Perspective*. Geoscience Canada Reprint Series Number 9 (ISBN 1-897095-21-X; ISSN 0821-381X). St. John's, NL: Geological Association of Canada.

Hubbard S, Peterson JE, Majer EL, et al. (1997) Estimation of permeable pathways and water content using tomographic radar data. *The Leading Edge of Exploration* 16(11): 1623–1628.

Hubbard S and Rubin Y (1997) Ground penetrating radar assisted saturation and permeability estimation in bimodal systems. *Water Resources Research* 33(5): 971–990.

Hubbard SS, Williams K, Conrad M, et al. (2008) Geophysical monitoring of hydrological and biogeochemical transformations associated with Cr(VI) biostimulation. *Environmental Science and Technology* (doi: 10.1021/es071702s)

Huisman S, Hubbard SS, Redman D, and Annan P (2003) Monitoring soil water content with ground-penetrating radar: A review. *Vadose Zone Journal* 2: 476–491.

Huisman JA, Rings J, Vrugt JA, Sorg J, and Vereecken H (2010) Hydraulic properties of a model dike from coupled Bayesian and multi-criteria hydrogeophysical inversion. *Journal of Hydrology* 380: 62–73.

Hyndman DW and Gorelick SM (1996) Estimating lithologic and transport properties in three dimensions using seismic and tracer data: The Kesterson aquifer. *Water Resources Research* 32(9): 2659–2670.

Hyndman DW, Harris JM, and Gorelick SM (2000) Inferring the relationship between seismic slowness and hydraulic conductivity in heterogeneous aquifers. *Water Resources Research* 36(8): 2121–2132.

Irving J, Knight R, and Holliger K (2009) Estimation of the lateral correlation structure of subsurface water content from surface-based ground-penetrating radar reflection images. *Water Resources Research* 45: W12404.

Jackson D (1976) Most squares inversion. *Journal of Geophysical Research* 81: 1027–1030.

Johnson DL, Koeplik J, and Schwartz LM (1986) New pore-size parameter characterizing transport in porous media. *Physical Review Letters* 57: 2564–2567.

Juhlin C and Stephens MB (2006) Gently dipping fracture zones in Paleoproterozoic metagranite, Sweden: Evidence from reflection seismic and cored borehole data and implications for the disposal of nuclear waste. *Journal of Geophysical Research* 111: B09302.

Kalscheuer T and Pedersen LB (2007) A non-linear truncated SVD variance and resolution analysis of two-dimensional magnetotelluric models. *Geophysical Journal International* 169: 435–447.

Keller GV (1987) Rock and mineral properties. In: Nabighian MN (ed.) *Electromagnetic Methods in Applied Geophysics, Vol. 1 – Theory*, Investigations in Geophysics 3, pp. 13–51. Tulsa, OK: Society of Exploration Geophysicists (SEG).

Kemna A, Binley A, Ramirez A, and Daily W (2000) Complex resistivity tomography for environmental applications. *Chemical Engineering Journal* 77: 11–18.

Keys SW (1989) *Borehole Geophysics Applied to Ground-Water Investigations*. Dublin, OH: National Water Well Association.

Knight R (1991) Hysteresis in the electrical resistivity of partially saturated sandstone. *Geophysics* 56: 2139–2147.

Kobr M, Mares S, and Paillet F (2005) Geophysical well logging: Borehole geophysics for hydrogeological studies: Principles and applications. In: Rubin Y and Hubbard S (eds.) *Hydrogeophysics*, ch. 10, pp. 291–332. Dordrecht: Springer.

Koch K, Wenninger J, Uhlenbrook S, and Bonell M (2009) Joint interpretation of hydrological and geophysical data: Electrical resistivity tomography results from a process hydrological research site in the Black Forest Mountains, Germany. *Hydrological Processes* 23: 1501–1513.

Kowalsky MB, Finsterle S, Peterson J, et al. (2005) Estimation of field-scale soil hydraulic and dielectric parameters through joint inversion of GPR and hydrological data. *Water Resources Research* 41: W11425.

Kowalsky MB, Finsterle SA, and Rubin Y (2004) Estimating flow parameter distributions using ground-penetrating radar and hydrological measurements during transient flow in the vadose zone. *Advances in Water Resources* 27(6): 583–599.

Krause P, Naujoks M, Fink M, and Kroner C (2009) The impact of soil moisture changes on gravity residuals obtained with a superconducting gravimeter. *Journal of Hydrology* 373: 151–163.

LaBrecque DJ and Yang X (2001) Difference inversion of ERT data: A fast inversion method for 3-D *in situ* monitoring. *Journal of Environmental and Engineering Geophysics* 6: 83–89.

Lambot S, Binley A, Slob E, and Hubbard S (2008) Ground penetrating radar in hydrogeophysics. *Vadose Zone Journal* 7: 137–139 (doi: 10.2136/vzj2007.0180)

Lambot S, Rhebergen J, van den Bosch I, Slob EC, and Vanclooster M (2004b) Measuring the soil water content profile of a sandy soil with an off-ground monostatic ground penetrating radar. *Vadose Zone Journal* 3: 1063–1071.

Lambot S, Slob E, Rheberger J, Lopera O, Jadoon KZ, and Vereecken H (2009) Remote estimation of the hydraulic properties of a sand using full-waveform integrated hydrogeophysical inversion of time-lapse, off-ground GPR data. *Vadose Zone Journal* 8: 743–754.

Lambot S, Slob EC, van den Bosch I, Stockbroeckx B, and Vanclooster M (2004a) Modeling of ground penetrating radar for accurate characterization of subsurface electric properties. *IEEE Transactions on Geoscience and Remote Sensing* 42: 2555–2568.

Lambot S, Weihermüller L, Huisman JA, Vereecken H, Vanclooster M, and Slob EC (2006) Analysis of air-launched ground-penetrating radar techniques to measure the soil surface water content. *Water Resources Research* 42: W11403 (doi: 10.1029/2006WR005097).

Leroy P Jr. and Revil P (2004) A triple-layer model of the surface electrochemical properties of clay minerals. *Journal of Colloid and Interface Science* 270: 371–380.

Leroy P and Revil A (2009) A mechanistic model for the spectral induced polarization of clay materials. *Journal of Geophysical Research* 114: B10202 (doi: 10.1029/2008JB006114).

Leroy P, Revil A, Kemna A, Cosenza P, and Ghorbani A (2008) Complex conductivity of water-saturated packs of glass beads. *Journal of Colloid and Interface Sciences* 321: 103–117.

Lesmes DP and Friedman SP (2005) Relationships between the electrical and hydrogeological properties of rocks and soils. In: Rubin Y and Hubbard SS (eds.) *Hydrogeophysics*, pp. 87–128. Dordrecht: Springer.

Li L, Steefel CI, Williams KH, Wilkins MJ, and Hubbard SS (2009) Mineral transformation and biomass accumulation during uranium bioremediation at Rifle, Colorado. *Environmental Science and Technology* 43(14): 5429–5435.

Lien M and Mannseth T (2008) Sensitivity study of marine CSEM data for reservoir production monitoring. *Geophysics* 73: F151–F163.

Linde N (2009) A comment on "Characterization of multiphase coupling using a bundle of capillary tubes model" by M. D. Jackson (*Journal of Geophysical Research*, 113: B04201). *Journal of Geophysical Research* 114: B06209.

Linde N, Binley A, Tryggvason A, Pedersen LB, and Revil A (2006a) Improved hydrogeophysical characterization using joint inversion of cross-hole electrical resistance and ground-penetrating radar traveltime data. *Water Resources Research* 42: W12404.

Linde N, Chen J, Kowalsky MB, and Hubbard S (2006b) Hydrogeophysical parameter estimation approaches for field scale characterization. In: Vereecken H, Binley A, Cassiani G, Revil A, and Titov K (eds.) *Applied Hydrogeophysics*, ch. 2, pp. 9–44. Dordrecht: Springer.

Linde N, Finsterle S, and Hubbard S (2006c) Inversion of tracer test data using tomographic constraints. *Water Resources Research* 42: W04410.

Linde N, Jougnot D, Revil A, *et al.* (2007) Streaming current generation in two-phase flow conditions. *Geophysical Research Letters* 34: L03306.

Linde N and Pedersen LB (2004) Evidence of electrical anisotropy in limestone formations using the RMT technique. *Geophysics* 69: 909–916.

Linde N and Revil A (2007) Inverting self-potential data for redox potentials of contaminant plumes. *Geophysical Research Letters* 34: L14302 (10.1029/2007GL030084).

Linde N, Tryggvason A, Peterson J, and Hubbard S (2008) Joint inversion of crosshole radar and seismic traveltimes. *Geophysics* 73: G29–G37.

Lunt IA, Bridge JS, and Tye RS (2004) A quantitative, three-dimensional depositional model of gravelly braided rivers. *Sedimentology* 51: 377–414.

Lunt IA, Hubbard SS, and Rubin Y (2005) Soil moisture content estimation using ground-penetrating radar reflection data. *Journal of Hydrology* 307: 254–269.

Mair JA and Green AG (1981) High-resolution seismic-reflection profiles reveal fracture-zones within a homogeneous granite batholith. *Nature* 294: 439–442.

Maurer H, Holliger K, and Boerner DE (1998) Stochastic regularization: Smoothness or similarity? *Geophysical Research Letters* 25: 2889–2892.

Mavko G, Mukerji T, and Dvorkin J (1998) *The Rock Physics Handbook*. Cambridge: Cambridge University Press.

McLaughlin D and Townley LR (1996) A reassessment of the groundwater inverse problem. *Water Resources Research* 32: 1131–1161.

Menke W (1984) *Geophysical Data Analysis–Discrete Inverse Theory*. New York: Academic Press.

Minsley BJ, Sogade J, and Morgan FD (2007) Three-dimensional source inversion of self-potential data. *Journal of Geophysical Research* 112: B02202.

Molz F, Boman GK, Young SC, and Waldrop WR (1994) Borehole flowmeters: Field applications and data analysis. *Journal of Hydrology* 163: 347–371.

Mosegaard K and Tarantola A (1995) Monte-Carlo sampling of solutions to inverse problems. *Journal of Geophysical Research* 100: 12431–12447.

Moysey S, Singha K, and Knight R (2005) A framework for inferring field-scale rock physics relationships through numerical simulation. *Geophysical Research Letters* 32: L08304.

Naudet V, Revil A, Bottero JY, and Begassat P (2003) Relationship between self-potential (SP) signals and redox conditions in contaminated groundwater. *Geophysical Research Letters* 30: 2091.

Naudet V, Revil A, Rizzo E, Bottero JY, and Begassat P (2004) Groundwater redox conditions and conductivity in a contaminant plume from geoelectrical investigations. *Hydrology and Earth Systems Sciences* 8: 8–22.

Nguyen F, Kemna A, Antonsson A, *et al.* (2009) Characterization of seawater intrusion using 2D electrical imaging. *Near Surface Geophysics* 7: 377–390.

NRC (1996) *Rock Fractures and Fluid Flow*. Washington, DC: National Academy Press.

Ogilvy RD, Meldrum PI, Kuras I, *et al.* (2009) Automated monitoring of coastal aquifers with electrical resistivity tomography. *Near Surface Geophysics* 7: 367–375.

Oldenburg DW and Li Y (1999) Estimating the depth of investigation in dc resistivity and IP surveys. *Geophysics* 64: 403–416.

Olsson O, Falk L, Forslund O, Lundmark L, and Sandberg E (1992) Borehole radar applied to the characterization of hydraulically conductive fracture zones in crystalline rock. *Geophysical Prospecting* 40: 109–142.

Paine JG (2003) Determining salinization extent, identifying salinity sources, and estimating chloride mass using surface, borehole, and airborne electromagnetic induction methods. *Water Resources Research* 39(3): 3–1–3–10.

Parker RL (1994) *Geophysical Inverse Theory*. Princeton, NJ: Princeton University Press.

Pedersen LB, Qian W, Dynesius L, and Zhang P (1994) An airborne tensor VLF system – from concept to realization. *Geophysical Prospecting* 42: 863–883.

Poldini E (1938) Geophysical exploration by spontaneous polarization methods. *Mining Magazine, London* 59: 278–282.

Pollock D and Cirpka OA (2010) Fully coupled hydrogeophysics inversion of synthetic salt tracer experiments. *Water Resources Research* 46: W07501.

Pride S (1994) Governing equations for the coupled electromagnetics and acoustics of porous media. *Physical Review B* 50: 15678–15696.

Purvance D and Andricevic R (2000) On the electrical-hydraulic conductivity correlation in aquifers. *Water Resources Research* 36: 2905–2913.

Revil A and Cathles LM (1999) Permeability of shaly sands. *Water Resources Research* 35: 651–662.

Revil A, Cathles LM, Losh S, and Nunn JA (1998) Electrical conductivity in shaly sands with geophysical applications. *Journal of Geophysical Research* 103: 23925–23936.

Revil A and Glover PWJ (1997) Theory of ionic-surface electrical conduction in porous media. *Physical Review B* 55: 1757–1773.

Revil A and Glover PWJ (1998) Nature of surface electrical conductivity in natural sands, sandstones, and clays. *Geophysical Research Letters* 25: 691–694.

Revil A and Leroy P (2004) Constitutive equations for ionic transport in porous shale's. *Journal of Geophysical Research* 109: B03208.

Revil A and Linde N (2006) Chemico-electromechanical coupling in microporous media. *Journal of Colloid and Interface Sciences* 302: 682–694.

Revil A, Naudet V, Nouzaret J, and Pessel M (2003) Principles of electrography applied to self-potential electrokinetic sources and hydrogeological applications. *Water Resources Research* 39(5): 1114.

Ritzi RW, Dai ZX, Dominic DF, and Rubin YN (2004) Spatial correlation of permeability in cross-stratified sediment with hierarchical architecture. *Water Resources Research* 40: W03513.

Robinson DA, Binley A, Crook N, *et al.* (2008) Advancing process-based watershed hydrological research using near-surface geophysics: A vision for, and review of, electrical and magnetic geophysical methods. *Hydrological Processes* 22: 3604–3635.

Rubin Y and Hubbard S (2005) *Hydrogeophysics*, Water and Science Technology Library, vol. 50. Dordrecht: Springer.

Rucker DF and Fink JB (2007) Inorganic plume delineation using surface high resolution electrical resistivity at the BC Cribs and Trenches Site, Hanford. *Vadose Zone Journal* 6: 946–958.

Sassen DS and Everett ME (2009) 3D polarimetric GPR coherency attributes and full- waveform inversion of transmission data for characterizing fractured rock. *Geophysics* 74: J23–J34 (doi: 10.1190/1.3103253).

Sato M and Mooney HM (1960) The electrochemical mechanism of sulfide self-potentials. *Geophysics* 25: 226–249.

Saunders JH, Herwanger JV, Pain CC, Worthington MH, and de Oliveira CRE (2005) Constrained resistivity inversion using seismic data. *Geophysical Journal International* 160: 785–796.

Scheibe T, Fang Y, Murray CJ, *et al.* (2006) Transport and biogeochemical reactions of metals in a physically and chemically heterogeneous aquifer. *Geosphere* 2(4): 220–235 (doi: 10.1130/GES00029.1).

Scheibe TD and Chien YJ (2003) An evaluation of conditioning data for solute transport prediction. *Ground Water* 41(2): 128–141.

Schwartz LM, Sen PN, and Johnson DL (1989) Influence of rough surfaces on electrolytic conduction in porous media. *Physical Review B* 40: 2450–2458.

Schön JH (1996) *Physical Properties of Rocks – Fundamentals and Principles of Petrophysics*. Amsterdam: Elsevier.

Scott JBT and Barker RD (2003) Determining pore-throat size in Permo-Triassic sandstones from low-frequency electrical spectroscopy. *Geophysical Research Letters* 30: 1450.

Sen PS, Goode PA, and Sibbit A (1988) Electrical conduction in clay bearing sandstones at low and high salinities. *Journal of Applied Geophysics* 63: 4832–4840.

Sen PN, Scala C, and Cohen MH (1981) A self-similar model for sedimentary rocks with application to the dielectric constant of fused glass beads. *Geophysics* 46: 781–795.

Shokri N, Lehmann P, and Or D (2009) Characteristics of evaporation from partially wettable porous media. *Water Resources Research* 45: W02415.

Sill WR (1983) Self-potential modeling from primary flows. *Geophysics* 48: 76–86.
Singha K and Gorelick SM (2005) Saline tracer visualized with three-dimensional electrical resistivity tomography: Field-scale spatial moment analysis. *Water Resources Research* 41: W05023.
Slater L (2007) Near surface electrical characterization of hydraulic conductivity: From petrophysical properties to aquifer geometries – a review. *Surveys in Geophysics* 28: 167–197.
Slater L, Atekwana E, Brantley S, et al. (2009) Exploring the geophysical signatures of microbial processes. *Eos Transactions AGU* 90: 83.
Slater L and Lesmes DP (2002) Electrical-hydraulic relationships observed for unconsolidated sediments. *Water Resources Research* 38: 1213.
Slater L, Ntarlagniannis D, Personna Y, and Hubbard S (2007) Pore-scale spectral induced polarization (SIP) signatures associated with FeS biomineral transformations. *Geophysical Research Letters* 34: L21404 (doi: 10.1029/2007GL031840).
Slater L, Ntarlagniannis D, and Wishard D (2006) On the relationship between induced polarization and surface area in metal–sand and clay–sand mixtures. *Geophysics* 71: A1–A5.
Sørensen KI and Auken E (2004) SkyTEM – a new high-resolution helicopter transient electromagnetic system. *Exploration Geophysics* 35: 191–199.
Spillmann T, Maurer H, Willenberg H, et al. (2007) Characterization of an unstable rock mass based on borehole logs and diverse borehole radar data. *Journal of Applied Geophysics* 61: 16–38.
Steeples D (2005) Shallow seismic methods. In: Rubin Y and Hubbard S (eds.) *Hydrogeophysics*, ch. 7, pp. 215–252. Dordrecht: Springer.
Stummer P, Maurer H, and Green AG (2004) Experimental design: Electrical resistivity data sets that provide optimum subsurface information. *Geophysics* 69: 120–139.
Styles P, McGrath R, Thomas E, and Cassidy NJ (2005) The use of microgravity for cavity characterization in Karstic terrains. *Quarterly Journal of Engineering Geology and Hydrogeology* 38(2): 155–169.
Suski B, Revil A, Titov K, et al. (2006) Monitoring of an infiltration experiment using the self-potential method. *Water Resources Research* 42: W08418.
Talley J, Baker GS, Becker MW, and Beyrle N (2005) Four dimensional mapping of tracer channelization in subhorizontal bedrock fractures using surface ground penetrating radar. *Geophysical Research Letters* 32: L04401.
Tarantola A (2005) *Inverse Problem Theory and Methods for Model Parameter Estimation*. Philadelphia, PA: Society for Industrial and Applied Mathematics (SIAM).
Taylor RW and Fleming AH (1988) Characterizing jointed systems by azimuthal resistivity techniques. *Ground Water* 26: 464–474.
Telford WM, Geldart LP, and Sheriff RE (1990) *Applied Geophysics*. Cambridge: Cambridge University Press.
Titov KV, Levitski A, Knoosavski PK, Tarasov AV, Ilyin YT, and Bues MA (2005) Combined application of surface geoelectrical methods for groundwater-flow modeling: A case history. *Geophysics* 70(5): H21–H31.
Topp GC, Davis JL, and Annan AP (1980) Electromagnetic determination of soil water content: Measurements in coaxial transmission lines. *Water Resources Research* 16: 574–582.
Tsoflias GP and Becker MW (2008) Ground-penetrating-radar response to fracture-fluid salinity: Why lower frequencies are favorable for resolving salinity changes. *Geophysics* 73: J25–J30.
Tsoflias GP, Halihan T, and Sharp JM Jr. (2001) Monitoring pumping test response in a fractured aquifer using ground-penetrating radar. *Water Resources Research* 37: 1221–1229.
US DOE (2010) Complex Systems Science for Subsurface Fate and Transport, Report from the August 2009 Workshop, DOE/SC-0123, U.S. Department of Energy Office of Science (www.science.doe.gov/ober/BER_workshops.html).
van Overmeeren RA (1998) Radar facies of unconsolidated sediments in The Netherlands: A radar stratigraphy interpretation method for hydrogeology. *Journal of Applied Geophysics* 40: 1–18.
Vereecken H, Binley A, Cassiani G, Revil A, and Titov K (2006) *Applied Hydrogeophysics*, NATO Science Series, Earth and Environmental Science, vol. 71. Dordrecht: Springer.
Vereecken H, Huisman JA, Bogena H, Vanderborght J, Vrugt JA, and Hopmans JW (2008) On the value of soil moisture measurements in vadose zone hydrology: A review. *Water Resources Research* 44: W00D06.
Viezzoli A, Christiansen AV, Auken E, and Sørensen K (2008) Quasi-3D modeling of airborne TEM data by spatially constrained inversion. *Geophysics* 73: F105–F113.
Vrugt JA, ter Braak CJF, Diks CGH, et al. (2009) Accelerating Markov chain Monte Carlo simulation by differential evolution with self-adaptive randomized subspace sampling. *International Journal of Nonlinear Sciences and Numerical Simulation* 10: 273–290.
Watson DB, Doll WE, Gamey TJ, Sheehan JR, and Jardine PM (2005) Plume and lithologic profiling with surface resistivity and seismic tomography. *Ground Water* 43(2): 169–177.
Watson KA and Barker RD (1999) Differentiating anisotropy and lateral effects using azimuthal resistivity offset Wenner soundings. *Geophysics* 64: 739–745.
Waxman MH and Smits LJM (1968) Electrical conductivities in oil-bearing shaly sands. *Transactions AIME* 243: 107–122.
Wilkinson PB, Meldrum PI, Chambers JE, Kuras O, and Ogilvy R (2006) Improved strategies for the automatic selection of optimized sets of electrical resistivity tomography measurement configurations. *Geophysical Journal International* 167: 1119–1126.
Williams KH, Kemna A, Wilkins M, et al. (2009) Geophysical monitoring of microbial activity during stimulated subsurface bioremediation. *Environmental Science and Technology* (doi 10.1021/es900855j).
Williams KH, Ntarlagiannis D, Slater LD, Dohnalkova A, Hubbard SS, and Banfield JF (2005) Geophysical imaging of stimulated microbial biomineralization. *Environmental Science and Technology* 39(19): 7592–7600.
Wishart DN, Slater LD, and Gates A (2008) Fracture anisotropy characterization in crystalline bedrock using field-scale azimuthal self potential gradient. *Journal of Hydrology* 358: 35–45.
Wishart DN, Slater LD, and Gates AE (2006) Self potential improves characterization of hydraulically-active fractures from azimuthal geoelectrical measurements. *Geophysical Research Letters* 33: L17314.
Yaramanci U, Kemna A, and Vereecken H (2005) Emerging technologies in hydrogeophysics. In: Rubin Y and Hubbard S (eds.) *Hydrogeophysics*, ch. 16, pp. 467–486. Dordrecht: Springer.
Zhdanov MS (2009) New advances in regularized inversion of gravity and electromagnetic data. *Geophysical Prospecting* 57: 463–478.
Zonge KL and Hughes LJ (1991) Controlled source audio-frequency magnetotellurics. In: Nabighian MN (ed.) *Electromagnetic Methods in Applied Geophysics, Vol. 2 – Applications*, pp. 713–810. Tulsa, OK: Society of Exploration Geophysicists.

2.16 Hydrological Modeling

DP Solomatine, UNESCO-IHE Institute for Water Education and Delft University of Technology, Delft, The Netherlands
T Wagener, The Pennsylvania State University, University Park, PA, USA

© 2011 Elsevier B.V. All rights reserved.

2.16.1	**Introduction**	435
2.16.1.1	What Is a Model	435
2.16.1.2	History of Hydrological Modeling	436
2.16.1.3	The Modeling Process	436
2.16.2	**Classification of Hydrological Models**	438
2.16.2.1	Main types of Hydrological Models	438
2.16.3	**Conceptual Models**	439
2.16.4	**Physically Based Models**	440
2.16.5	**Parameter Estimation**	441
2.16.6	**Data-Driven Models**	444
2.16.6.1	Introduction	444
2.16.6.2	Technology of DDM	444
2.16.6.2.1	Definitions	444
2.16.6.2.2	Specifics of data partitioning in DDM	445
2.16.6.2.3	Choice of the model variables	446
2.16.6.3	Methods and Typical Applications	446
2.16.6.4	DDM: Current Trends and Conclusions	448
2.16.7	**Analysis of Uncertainty in Hydrological Modeling**	449
2.16.7.1	Notion of Uncertainty	449
2.16.7.2	Sources of Uncertainty	449
2.16.7.3	Uncertainty Representation	450
2.16.7.4	View at Uncertainty in Data-Driven and Statistical Modeling	450
2.16.7.5	Uncertainty Analysis Methods	451
2.16.8	**Integration of Models**	452
2.16.8.1	Integration of Meteorological and Hydrological Models	452
2.16.8.2	Integration of Physically Based and Data-Driven Models	452
2.16.8.2.1	Error prediction models	452
2.16.8.2.2	Integration of hydrological knowledge into DDM	453
2.16.9	**Future Issues in Hydrological Modeling**	453
References		454

2.16.1 Introduction

Hydrological models are simplified representations of the terrestrial hydrological cycle, and play an important role in many areas of hydrology, such as flood warning and management, agriculture, design of dams, climate change impact studies, etc. Hydrological models generally have one of two purposes: (1) to enable reasoning, that is, to formalize our scientific understanding of a hydrological system and/or (2) to provide (testable) predictions (usually outside our range of observations, short term vs. long term, or to simulate additional variables). For example, catchments are complex systems whose unique combinations of physical characteristics create specific hydrological response characteristics for each location (Beven, 2000). The ability to predict the hydrological response of such systems, especially stream flow, is fundamental for many research and operational studies.

In this chapter, the main principles of and approaches to hydrological modeling are covered, both for simulation (process) models that are based on physical principles (conceptual and physically based), and for data-driven models. Our intention is to provide a broad overview and to show current trends in hydrological modeling. The methods used in data-driven modeling (DDM) are covered in greater depth since they are probably less widely known to hydrological audiences.

2.16.1.1 What Is a Model

A model can be defined as a simplified representation of a phenomenon or a process. It is typically characterized by a set of variables and by equations that describe the relationship between these variables. In the case of hydrology, a model represents the part of the terrestrial environmental system that controls the movement and storage of water. In general terms, a system can be defined as a collection of components or elements that are connected to facilitate the flow of information, matter, or energy. An example of a typical system considered in hydrological modeling is the watershed or catchment. The extent of the system is usually defined by the control volume or modeling domain, and the overall

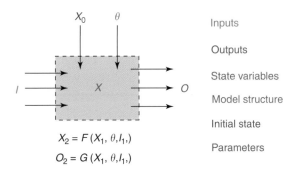

Figure 1 Schematic of the main components of a dynamic mathematical model. I, inputs; O, outputs; X, state variables; X_0, initial states; and θ, parameters.

modeling objective in hydrology is generally to simulate the fluxes of energy, moisture, or other matter across the system boundaries (i.e., the system inputs and outputs). Variables or state variables are time varying and (space/time) averaged quantities of mass/energy/information stored in the system. An example would be soil moisture content or the discharge in a stream [L3/T]. Parameters describe (usually time invariant) properties of the specific system under study inside the model equations. Examples of parameters are hydraulic conductivity [L/T] or soil storage capacity [L].

A dynamic mathematical model has certain typical elements that are discussed here briefly for consistency in language (**Figure 1**). Main components include one or more inputs I (e.g., precipitation and temperature), one or more state variables X (e.g., soil moisture or groundwater content), and one or more model outputs O (e.g., stream flow or actual evapotranspiration). In addition, a model typically requires the definitions of initial states X_0 (e.g., is the catchment wet or dry at the beginning of the simulation) and/or the model parameters θ (e.g., soil hydraulic conductivity, surface roughness, and soil moisture storage capacity).

Hydrological models (and most environmental models in general) are typically based on certain assumptions that make them different from other types of models. Typical assumptions that we make in the context of hydrological modeling include the assumption of universality (i.e., a model can represent different but similar systems) and the assumption of physical realism (i.e., state variables and parameters of the model have a real meaning in the physical world; Wagener and Gupta, 2005). The fact that we are dealing with real-world environmental systems also carries certain problems with it when we are building models. Following Beven (2009), these problems include the fact that it is often difficult to (1) make measurements at the scale at which we want to model; (2) define the boundary conditions for time-dependent processes; (3) define the initial conditions; and (4) define the physical, chemical, and biological characteristics of the modeling domain.

2.16.1.2 History of Hydrological Modeling

Hydrological models applied at the catchment scale originated as simple mathematical representations of the input-response behavior of catchment-scale environmental systems through parsimonious models such as the unit hydrograph (for flow routing) (e.g., Dooge, 1959) and the rational formula (for excess rainfall calculation) (e.g., Dooge, 1957) as part of engineering hydrology. Such single-purpose event-scale models are still widely used to estimate design variables or to predict floods. These early approaches formed a basis for the generation of more complete, but spatially lumped, representations of the terrestrial hydrological cycle, such as the Stanford Watershed model in the 1960s (which formed the basis for the currently widely used Sacramento model (Burnash, 1995)). This advancement enabled the continuous time representation of the rainfall–runoff relationship, and models of this type are still at the heart of many operational forecasting systems throughout the world. While the general equations of models (e.g., the Sacramento model) are based on conceptualizing plot (or smaller) scale hydrological processes, their spatially lumped application at the catchment scale means that parameters have to be calibrated using observations of rainfall–runoff behavior of the system under study. Interest in predicting land-use change leads to the development of more spatially explicit representations of the physics (to the best of our understanding) underlying the hydrological system in form of the Systeme Hydrologique Europeen (SHE) model in the 1980s (Abbott et al., 1986). The latter is an example of a group of highly complex process-based models whose development was driven by the hope that their parameters could be directly estimated from observable physical watershed characteristics without the need for model calibration on observed stream flow data, thus enabling the assessment of land cover change impacts (Ewen and Parkin, 1996; Dunn and Ferrier, 1999).

At that time, these models were severely constrained by our lack of computational power – a constraint that decreases in its severity with increases in computational resources with each passing year. Increasingly available high-performance computing enables us to explore the behavior of highly complex models in new ways (Tang et al., 2007; van Werkhoven et al., 2008). This advancement in computer power went hand in hand with new strategies for process-based models, for example, the use of triangular irregular networks (TINs) to vary the spatial resolution throughout the model domain, that have been put forward in recent years; however, more testing is required to assess whether previous limitations of physically based models have yet been overcome (e.g., the lack of full coupling of processes or their calibration needs) (e.g., Reggiani et al., 1998, 1999, 2000, 2001; Panday and Huyakorn, 2004; Qu and Duffy, 2007; Kollet and Maxwell, 2006, 2008).

2.16.1.3 The Modeling Process

The modeling process, that is, how we build and use models is discussed in this section. For ease of discussion, the process is divided into two components. The first component is the model-building process (i.e., how does a model come about), whereas the second component focuses on the modeling protocol (i.e., a procedure to use the model for both operational and research studies).

The model-building process requires (at least implicitly) that the modeler considers four different stages of the model (see also Beven, 2000). The first stage is the perceptual model. This model is based on the understanding of the system in the modeler's head due to both the interaction with the system and the modeler's experience. It will, generally, not be formalized on paper or in any other way. This perceptual model forms the basis of the conceptual model. This conceptual model is a formalization of the perceptual model through the definition of system boundaries, inputs–states–outputs, connections of system components, etc. It is not to be mistaken with the conceptual type of models discussed later. Once a suitable conceptual model has been derived, it has to be translated into mathematical form. The mathematical model formulates the conceptual model in the form of input (–state)–output equations. Finally, the mathematical model has to be implemented as computer code so that the equation can be solved in a computational model.

Once a suitable model has been built or selected from existing computer codes, a modeling protocol is used to apply this model (Wagener and McIntyre, 2007). Modeling protocols can vary widely, but generally contain some or most of the elements discussed below (**Figure 2**). A modeling protocol – at its simplest level – can be divided into model identification and model evaluation parts. The model identification part mainly focuses on identifying appropriate parameters (one set or many parameter sets if uncertainty in the identification process is considered), while the latter focuses on understanding the behavior and performance of the model.

The starting point of the model identification part should be a detailed analysis of the data available. Beven (2000) provided suggestions on how to assess the quality of data in the context of hydrological modeling. This is followed by the model selection or building process. The model-building process has already been outlined previously. In many cases, it is likely that an existing model will be selected though, either because the modeler has extensive experience with a particular model or because he/she has applied a model to a similar hydrological system with success in the past. The universality of models, as discussed above, implies that a typical hydrological model can be applied to a range of systems as long as the basic physical processes of the system are represented within the model. Model choice might also vary with the intended modeling purpose, which often defines the required spatio-temporal resolution and thus the degree of detail with which the system has to be modeled.

Once a model structure has been selected, parameter estimation has to be performed. Parameters, as defined above, reflect the local physical characteristics of the system. Parameters are generally derived either through a process of calibration or by using *a priori* information, for example, of soil or vegetation characteristics. For calibration, it is necessary to assess how closely simulated and observed (if available) output time series match. This is usually done by the use of an objective function (sometimes also called loss function or cost function), that is, a measure based on the aggregated differences between observed and simulated variables (called residuals). The choice of objective function is generally closely coupled with the intended purpose of the modeling study. Sometimes this problem is posed as a multiobjective optimization problem. Methods for calibration (parameter estimation) are covered later in Section 2.16.5. Further, the model

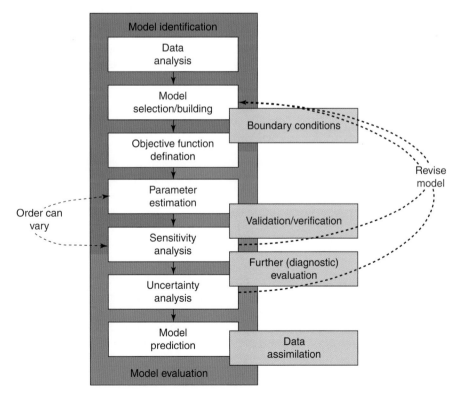

Figure 2 Schematic representation of a typical modeling protocol.

should be evaluated with respect to whether it provides the right result for the right reason. Parameter estimation (calibration) is followed by the model evaluation, including validation (checking model performance on an unseen data set, thus imitating model operation), sensitivity, and uncertainty analysis. A comprehensive framework for model evaluation (termed diagnostic evaluation) is proposed by Gupta et al. (2008).

One tool often used in such an evaluation is sensitivity analysis, which is the study of how variability or uncertainty in different factors (including parameters, inputs, and initial states) impacts the model output. Such an analysis is generally used either to assess the relative importance of model parameters in controlling the model output or to understand the relative distributions of uncertainty from the different factors. It can therefore be part of the model identification as well as the model evaluation component of the modeling protocol.

The subsequent step of uncertainty analysis – the quantification of the uncertainty present in the model – is increasingly popular. It usually includes the propagation of the uncertainty into the model output so that it can be considered in subsequent decision making (see Section 2.16.7).

When a model is put into operation, the data progressively collected can be used to update (improve) the model parameters, state variables, and/or model predictions (outputs), and this process is referred to as data assimilation.

One aspect needs mentioning here. Due to the lack of information about the modeled process, a modeler may decide not to try to build unique (the most accurate) model, but rather consider many equally acceptable model parametrizations. Such reasoning has led to a Monte-Carlo-like method of uncertainty analysis called Generalised Likelihood Uncertainty Estimator (GLUE) (Beven and Binley, 1992), and to research into the development of the (weighted) ensemble of models, or multimodels (see e.g., Georgakakos et al., 2004).

2.16.2 Classification of Hydrological Models

2.16.2.1 Main types of Hydrological Models

A vast number of hydrological model structures has been developed and implemented in computer code over the last few decades (see, e.g., Todini (1988) for a historical review of rainfall–runoff modeling). It is therefore helpful to classify these structures for an easier understanding of the discussion.

Many authors present classification schemes for hydrological models (see, e.g., Clarke, 1973; Todini, 1988; Chow et al., 1988; Wheater et al., 1993; Singh, 1995b; and Refsgaard, 1996). The classification schemes are generally based on the following criteria: (1) the extent of physical principles that are applied in the model structure and (2) the treatment of the model inputs and parameters as a function of space and time. According to the first criterion (i.e., physical process description), a rainfall–runoff model can be attributed to two categories: deterministic and stochastic (see **Figure 3**). A deterministic model does not consider randomness; a given input always produces the same output. A stochastic model has outputs that are at least partially random.

Deterministic models can be classified based on whether the model represents a lumped or distributed description of the considered catchment area (i.e., second criterion) and whether the description of the hydrological processes is empirical, conceptual, or more physically based (Refsgaard, 1996). With respect to deterministic models, we will distinguish three classes: (1) data-driven (also called data-based, metric, empirical, or black box models), (2) conceptual (also called parametric, explicit soil moisture accounting or gray box models), and (3) physically based (also called physics-based, mechanistic, or white box models) models. The two latter classes are sometimes referred to as simulation (or process) models. **Figure 4** provides some guidelines on estimation of structure and parameters for various types of deterministic models.

Note that the distinction between deterministic and stochastic models is not clear-cut. In many modeling studies, it is

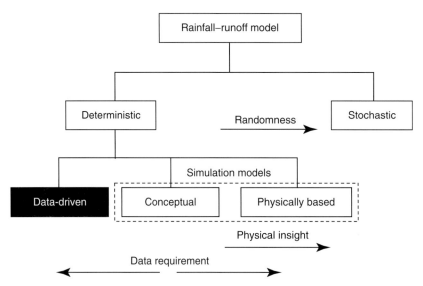

Figure 3 Classification of hydrological models based on physical processes. Adapted from Refsgaard JC (1996) Terminology, modelling protocol and classification of hydrological model codes. In: Abbott MB and Refsgaard JC (eds.) *Distributed Hydrological Modelling*, pp. 17–39. Dordrecht: Kluwer.

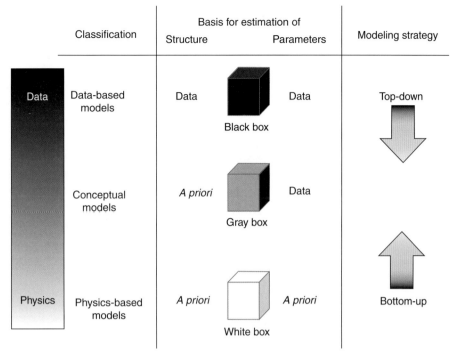

Figure 4 Estimation of structure and parameters for various types of deterministic models.

assumed that the modeled variables are not deterministic, but still more developed apparatus of deterministic modeling is used. To account for stochasticity, additional uncertainty analysis is conducted assuming probability distributions for at least some of the variables and parameters involved.

2.16.3 Conceptual Models

Conceptual modeling uses simplified descriptions of hydrological processes. Such models use storage elements as the main building component. These stores are filled through fluxes such as rainfall, infiltration, or percolation, and emptied through processes such as evapotranspiration, runoff, and drainage. Conceptual models generally have a structure that is specified *a priori* by the modeler, that is, it is not derived from the observed rainfall–runoff data. In contrast to empirical models, the structure is defined by the modeler's understanding of the hydrological system. However, conceptual models still rely on observed time series of system output, typically stream flow, to derive the values of their parameters during the calibration process. The parameters describe aspects such as the size of storage elements or the distribution of flow between them. A number of real-world processes are usually aggregated (in space and time) into a single parameter, which means that this parameter can therefore often not be derived directly from field measurements. Conceptual models make up the vast majority of models used in practical applications. Most conceptual models consider the catchment as a single homogeneous unit. However, one common approach to consider spatial variability is the segmentation of the catchment into smaller subcatchments, the so-called semi-distributed approach.

One typical example of a conceptual model – Hydrologiska Byråns Vattenbalansavdelning (HBV) – (Bergström, 1976) as rainfall–runoff model is given below. The HBV model was developed at the Swedish Meteorological and Hydrological Institute (Hydrological Bureau Water balance section). The model was originally developed for Scandinavian catchments, but has been applied in more than 30 countries all over the world (Lindström *et al.*, 1997).

A schematic diagram of the HBV model (Lindström *et al.*, 1997) is shown in **Figure 5**. The model of one catchment comprises subroutines for snow accumulation and melt, soil moisture accounting procedure, routines for runoff generation, and a simple routing procedure. The soil moisture accounting routine computes the proportion of snowmelt or rainfall P (mm h^{-1} or mm d^{-1}) that reaches the soil surface, which is ultimately converted to runoff. If the soil is dry (i.e., small value of SM/CF), the recharge R, which subsequently becomes runoff, is small as a major portion of the effective precipitation P is used to increase the soil moisture. Whereas if the soil is wet, the major portion of P is available to increase the storage in the upper zone.

The runoff generation routine transforms excess water R from the soil moisture zone to runoff. The routine consists of two conceptual reservoirs. The upper reservoir is a nonlinear reservoir whose outflow simulates the direct runoff component from the upper soil zone, while the lower one is a linear reservoir whose outflow simulates the base flow component of the runoff. The total runoff Q is computed as the sum of the outflows from the upper and the lower reservoirs. The total runoff is then smoothed using a triangular transformation function.

Input data are observations of precipitation and air temperature, and estimates of potential evapotranspiration. The

440 Hydrological Modeling

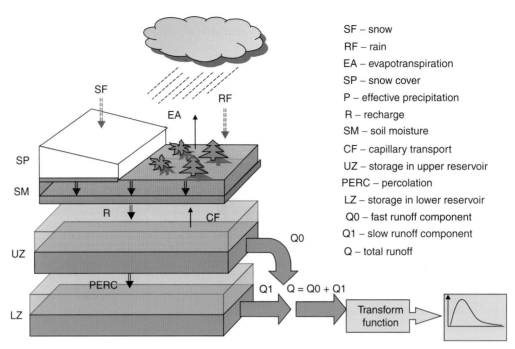

Figure 5 Schematic representation of the HBV-96 model with routines for snow, soil, and runoff response. Modified from Lindström G, Johansson B, Persson M, Gardelin M, and Bergström S (1997) Development and test of the distributed HBV-96 hydrological model. *Journal of Hydrology* 201: 272–228.

time step is usually 1 day, but it is possible to use shorter time steps. The evaporation values used are normally monthly averages, although it is possible to use the daily values. Air temperature data are used for calculations of snow accumulation and melt. It can also be used to adjust potential evaporation when the temperature deviates from normal values, or to calculate potential evaporation.

Note that the software IHMS-HBV allows for linking several lumped models and thus making it possible to build separate models for sub-basins, which are integrated, so that the overall model is the semi-distributed model.

The HBV model is an example of a typical lumped conceptual model. Other examples of such models differ in the details of describing the catchment hydrology. The following examples can be mentioned: Sugawara's tank model (Sugawara, 1995), Sacramento model (Burnash, 1995), Xinanjiang model (Zhao and Liu, 1995), and Tracer Aided Catchment (TAC) model (Uhlenbrook and Leibundgut, 2002).

2.16.4 Physically Based Models

Physically based models (e.g., Freeze and Harlan, 1969; Beven, 1996, 1989, 2002; Abbott *et al.*, 1986; Calver, 1988) use much more detailed and rigorous representations of physical processes and are based on the laws of conservation of mass, momentum, and energy. They became practically applicable in 1980s, as a result of improvements in computer power. The hope was that the degree of physical realism on which these models are based would be sufficient to relate their parameters, such as soil moisture characteristic and unsaturated zone hydraulic conductivity functions for subsurface flow or friction coefficients for surface flow, to physical characteristics of the catchment (Todini, 1988), thus eliminating the need for model calibration. However, mechanistic models suffer from high data demand, scale-related problems (e.g., the measurement scales differ from the simulation model (parameter) scales), and from over-parametrization (Beven, 1989).

One consequence of the problems of scale is that (at least not all of) the model parameters cannot be derived through measurements; physically based models structures, therefore, still require calibration, usually of a few key parameters (Calver, 1988; Refsgaard, 1997; Madsen and Jacobsen, 2001). The expectation that these models could be applied to ungauged catchments has, therefore, not yet been fulfilled (Parkin *et al.*, 1996; Refsgaard and Knudsen, 1996). They are typically rather applied in a way that is similar to conceptual models (Beven, 1989), thus demanding continued research into new approaches to merge these models with data. Physically based models often use spatial discretizations based on grids, triangular irregular networks, or some type of hydrologic response unit (e.g., Uhlenbrook *et al.*, 2004). A typical model of this kind is, for example, a physically based model based on triangular irregular networks – the Penn State Integrated Hydrologic Model (PIHM) (Qu and Duffy, 2007); its simplified structure is presented in **Figure 6**.

Such models are therefore particularly appropriate when a high level of spatial detail is important, for example, to estimate local levels of soil erosion or the extent of inundated areas (Refsgaard and Abbott, 1996). However, if the main interest simply lies in the estimation of stream flow at the catchment scale, then simpler conceptual or data-driven models often perform well and the high complexity of physically based models is not required (e.g., Loague and Freeze, 1985; Refsgaard and Knudsen, 1996). Regarding the results of

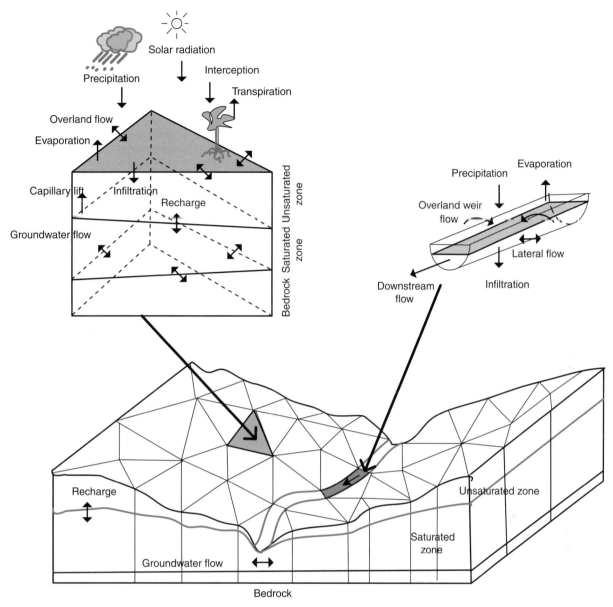

Figure 6 Schematic representation of the PIHM model, an example of a TIN-based physically based hydrological model (Qu and Duffy, 2004).

a comprehensive experiment to compare lumped and distributed models, the reader is referred to Reed *et al.* (2004). This experiment has shown that due to difficulties in calibrating distributed models, in many cases, conceptual models are in fact more accurate in reproducing the resulting catchment stream flow than the distributed ones.

2.16.5 Parameter Estimation

Many, if not most, rainfall–runoff model structures currently used to simulate the continuous hydrological response can be classified as conceptual, if this classification is based on two criteria (Wheater *et al.*, 1993): (1) the model structure is specified prior to any modeling being undertaken and (2) (at least some of) the model parameters do not have a direct physical interpretation, in the sense of being independently measurable, and have to be estimated through calibration against observed data. Calibration is a process of parameter adjustment (automatic or manual), until catchment and model behavior show a sufficiently (to be specified by the hydrologist) high degree of similarity. The similarity is usually judged by one or more objective functions accompanied by visual inspection of observed and calculated hydrographs (Gupta *et al.*, 2005).

The choice of such objective functions has itself been the subject of extensive research over many years. Traditionally, measures based on the mean squared error (MSE) criterion were used, for example, root mean squared error (RMSE) or Nash–Sutcliff efficiency (NSE). Appearance of the squared errors in many formulations is the result of an assumption of the normality (Gaussian distribution) of model errors, and using the principle of maximum likelihood to derive the error function. Calibration includes the process of finding a set of parameters providing the minimum of RMSE or the maximum value of NSE. For more information on the formulations of these and other error function, the reader is referred to, for example, Gupta et al. (1998). In a recent paper, Gupta et al. (2009) showed certain deficiencies of MSE-based objective functions and suggested possible remedies.

Often a single measure may not be enough to capture all the aspects of the system response that the model is supposed to reproduce, and several criteria (objective functions) have to be considered simultaneously so that multiobjective optimization algorithms have to be used. Examples of such multiple objectives include the RMSE calculated separately on low and high flows, timing errors, and error in reproducing the water balance. The models constituting the Pareto in criteria space should be seen as the best models, since there are no models better than these on all criteria.

If a single model is to be selected from this set, it is done either by a decision maker (who would use some additional criteria that are difficult to formalize), or by measuring the distance of the models to the ideal point in criteria space, or by using the (weighted) sum of objective functions values. This section covers single-objective optimization; for the use of multiobjective methods, the reader is directed to the papers by Gupta et al. (1998), Khu and Madsen (2005), and Tang et al. (2006) (with the subsequent discussion).

Hydrological model structures of the continuous watershed response (mainly stream flow) became feasible in the 1960s. They were usually relatively simple lumped, conceptual mathematical representations of the (perceived to be important) hydrological processes, with little (if any) consideration of issues such as identifiability of the parameters or information content of the watershed response observations. It became quickly apparent that the parameters of such models could not be directly estimated through measurements in the field, and that some sort of adjustment (fine-tuning) of the parameters was required to match simulated system responses with observations (e.g., Dawdy and O'Donnell 1965). Adjustment approaches were initially based on manual perturbation of the parameter values and visual inspection of the similarity between simulated and observed time series. Over the years, a variety of manual calibration procedures have been developed, some having reached very high levels of sophistication allowing hydrologists to achieve very good performing and hydrologically realistic model parameters and predictions, that is, a well-calibrated model (Harlin, 1991; Burnash, 1995). This hydrological realism is still a problem for most automated procedures as discussed in van Werkhoven et al. (2008).

Necessary conditions for a hydrological model to be well calibrated are that it exhibits (at least) the following three characteristics (Wagener et al. 2003; Gupta et al. 2005):

1. the input–state–output behavior of the model is consistent with the measurements of watershed behavior;
2. the model predictions are accurate (i.e., they have negligible bias) and precise (i.e., the prediction uncertainty is relatively small); and
3. the model structure and behavior are consistent with a current hydrological understanding of reality.

This last characteristic is often ignored in operational settings, where the focus is generally on useful rather than realistic models. This will be an adequate approach in many cases, but will eventually lead to limitations of potential model uses. This problem is exemplified in the current attempts to modeling watershed residence times and flow paths (McDonnell, 2003). This aspect of the hydrologic system, though often not crucial for reliable quantitative flow predictions, is however relevant for many of today's environmental problems, but cannot be simulated by many of the currently available models.

The high number of nonlinearly interacting parameters present in most hydrological models makes manual calibration a very labor-intensive and a difficult process, requiring considerable experience. This experience is time consuming to acquire and cannot be easily transferred from one hydrologist to the next. In addition, manual calibration does not formally incorporate an analysis of uncertainty, as is required in a modern decision-making context. The obvious advantages of computer-based automatic calibration procedures began to spark interest in such approaches as soon as computers became more easily available for research.

In automatic calibration, the ability of a parameter set to reproduce the observed system response is measured (summarized) by means of an objective function (also sometimes called loss or cost function). As discussed above, this objective function is an aggregated measure of the residuals, that is, the differences between observed and simulated responses at each time step. An important early example of automatic calibration is the dissertation work by Ibbitt (1970) in which a variety of automated approaches were applied to several watershed models of varying complexity (see also Ibbitt and O'Donnell, 1971). The approaches were mainly based on local-search optimization techniques, that is, the methods that start from a selected initial point in the parameter space and then walk through it, following some predefined rule system, to iteratively search for parameter sets that yield progressively better objective function values. Ibbitt (1970) found that it is difficult to conclude when the best parameter set has been found, because the result depends both on the chosen method and on the initial starting parameter set. The application of local-search calibration approaches to all but the most simple watershed models has been largely unsuccessful. In reflection of this, Johnston and Pilgrim (1976) reported the failure of their 2-year quest to find an optimal parameter set for a typical conceptual rainfall–runoff (RR) model. Their honesty in reporting this failure ultimately led to a paradigm shift as researchers started to look closely at the possible reasons for this lack of success.

The difficulty of the task at hand, in fact, only became clear in the early 1990s when Duan et al. (1992) conducted a detailed study of the characteristics of the response surface that any search algorithm has to explore. Their studies showed that the specific characteristics of the response surface, that is, the $(n+1)$-dimensional space of n model parameters and an objective function, of hydrological models give rise to conditions that make it extremely difficult for local optimization strategies to be successful. They listed the following characteristics commonly associated with the response surface of a typical hydrological model:

- it contains more than one main region of attraction;
- each region of attraction contains many local optima;
- it is rough with discontinuous derivatives;
- it is flat in many regions, particularly in the vicinity of the optimum, with significantly different parameter sensitivities; and
- its shape includes long and curved ridges.

Concluding that optimization strategies need to be powerful enough to overcome the search difficulties presented by these response surface characteristics, Duan et al. (1992) developed the shuffled complex evolution (SCE-UA) global optimization method (UA, University of Arizona). The SCE-UA algorithm has since been proved to be highly reliable in locating the optimum (where one exists) on the response surfaces of typical hydrological models. However, in a follow-up paper, Sorooshian et al. (1993) used SCE-UA to show that several different parameter combinations of the relatively complex Sacramento model (13 free parameters) could be found which produced essentially identical objective function values, thereby indicating that not all of the parameter uncertainty can be resolved through an efficient global optimizer (see discussion in Wagener and Gupta (2005)). Similar observations of multiple parameter combinations producing similar performances have also been made by others (e.g., Binley and Beven, 1991; Beven and Binley, 1992; Spear, 1995; Young et al., 1998; Wagener et al., 2003). Part of this problem had been attributed to overly complex models for the information content of the system response data available, usually stream flow (e.g., Young, 1992, 1998).

It is worth mentioning that practically any direct search optimization algorithm can be used for model calibration. The reason of using direct search (i.e., the search based purely on calculation of the objective function values for different points in the search space) is that for most calibration problems computation of the objective function gradients is not possible, so the efficient gradient-based search cannot be used. (Another name for this class of algorithms is global optimization algorithms since they are focused on finding the global minimum rather than a local one.) For example, in many studies a popular genetic algorithm (GA) is used.

If a model is simple and fast running, then it is really not important how efficient the optimization algorithm is. Here, efficiency is measured by the number of the model runs needed by an optimization algorithm to find a more-or-less accurate estimate of the parameter vector leading to the minimum value of the model error. However, if a model is computationally complex, as is the case for physically based and distributed models, efficiency of the optimization algorithm used becomes an issue. With this in mind, the so-called adaptive cluster-covering algorithm (ACCO) was developed (Solomatine, 1995; Solomatine et al., 1999, 2001), and it was shown that on a number of calibration problems it is more efficient than GA and several other algorithms.

A large number of other algorithms has been applied to hydrological models, including a multialgorithm genetically adaptive method (AMALGAM) (Vrugt and Robinson, 2007) and epsilon-NSGA-II (NSGA, nondominated sorting genetic algorithm; Tang et al., 2006), and different algorithms have come out as most effective or most efficient depending on the study. A range of algorithms for model calibration can be obtained from the Hydroarchive website.

As mentioned above (**Figure 2**), once parameters are estimated, the model has to be validated, that is, the degree to which a model is an accurate representation of the modeled process has to be determined.

Case of ungauged basins. A different problem has to be solved in the case of the so-called ungauged basins, that is, watersheds for which none or insufficiently long observations of the hydrological response variable of interest (usually stream flow) are available. The above-discussed strategy of model calibration cannot be used under those conditions. Early attempts to model ungauged catchments simply used the parameter values derived for neighboring catchments where stream flow data were available, that is, a geographical proximity approach (e.g., Mosley, 1981; Vandewiele and Elias, 1995). However, this seems to be insufficient since nearby catchments can even be very different with respect to their hydrological behavior (Post et al., 1998; Beven, 2000). Others propose the use of parameter estimates directly derived from, among others, soil properties such as porosity, field capacity, and wilting point (to derive model storage capacity parameters); percentage forest cover (evapotranspiration parameters); or hydraulic conductivities and channel densities (time constants) (e.g., Koren et al., 2000; Duan et al., 2001; Atkinson et al., 2002).

The main problem here is that the scale at which the measurements are made (often from small soil samples) is different from the scale at which the model equations are derived (often laboratory scale) and at which the model is usually applied (catchment scale). The conceptual model parameters represent the effective characteristics of the integrated (heterogeneous) catchment system (e.g., including preferential flow), which are unlikely to be easily captured using small-scale measurements since there is generally no theory that allows the estimation of the effective values within different parts of a heterogeneous flow domain from a limited number of small-scale or laboratory measurements (Beven, 2000). It seems unlikely that conceptual model parameters, which describe an integrated catchment response, usually aggregating significant heterogeneity (including the effect of preferential flow paths, different soil and vegetation types, etc.), can be derived from catchment properties that do not consider all influences on water flow through the catchment. Further fine-tuning of these estimates using locally observed flow data is needed because the physical information available to estimate *a priori* parameters is not adequate to define local physical properties of individual basins for accurate hydrological forecasts (Duan et al., 2001). However, useful initial values might be derived in this way (Koren et al., 2000). The advantages of this approach are that the assumed physical basis of the parameters is preserved and (physical) parameter dependence can be accounted for, as shown by Koren et al. (2000).

Probably, the most common apprnoach to ungauged modeling is to relate model parameters and catchment characteristics in a statistical manner (e.g., Jakeman et al., 1992; Sefton et al., 1995; Post et al., 1998; Sefton and Howarth, 1998; Abdullah and Lettenmaier, 1997; Wagener et al., 2004;

Merz and Bloschl, 2004; Lamb and Kay, 2004; Seibert, 1999; Lamb *et al.*, 2000; Post and Jakeman, 1996; Fernandez *et al.*, 2000), assuming that the uniqueness of each catchment can be captured in a distinctive combination of catchment characteristics. The basic methodology is to calibrate a specific model structure, here called the local model structure, to as large a number of (gauged) catchments as possible and derive statistical (regression) relationships between (local) model parameters and catchment characteristics. These statistical relationships, here called regional models, and the measurable properties of the ungauged catchment can then be used to derive estimates of the (local) model parameters. This procedure is usually referred to as regionalization or spatial generalization (e.g., Lamb and Calver, 2002). While this approach has been widely applied, it still does not constrain existing uncertainty sufficiently in many cases (Wagener and Wheater, 2006).

Recent approaches also used regionalized information about stream flow characteristics to further reduce this uncertainty (Yadav *et al.*, 2007; Zhang *et al.*, 2008). It seems as if the most promising strategies for the future lie in combining as much information as possible to reduce predictive uncertainty, rather than relying on a single approach.

2.16.6 Data-Driven Models

2.16.6.1 Introduction

Along with the physically based and conceptual models, the empirical models based on observations (experience) are also popular. Such models involve mathematical equations that have been assessed not from the physical process in the catchment but from analysis of data – concurrent input and output time series. Typical examples here are the unit hydrograph method and various statistical models – for example, linear regression, multilinear, ARIMA, etc. During the last decade, the area of empirical modeling received an important boost due to developments in the area of machine learning (ML). It can be said that it now entered a new phase and deserves a special name – DDM.

DDM is based on the analysis of all the data characterizing the system under study. A model can then be defined based on connections between the system state variables (input, internal and output variables) with only a limited number of assumptions about the physical behavior of the system. The methods used nowadays can go much further than the ones used in conventional empirical modeling: they allow for solving prediction problems, reconstructing highly nonlinear functions, performing classification, grouping of data, and building rule-based systems.

It is worth mentioning that among some hydrologists there is still a certain skepticism about the use of DDM. In their opinion, such models do not relate to physical principles and mathematical reasoning, and view building models from data sets as a purely computational exercise. This is true, and indeed DDM cannot be a replacement of process-based modeling, but should be used in situations where data-driven models are capable of generating improved forecasts of hydrological variables.

There are cases where the traditional statistical models (typically linear regression or ARIMA-class models) are accurate enough, and there is no need of using sophisticated methods of ML. Some of the concerns of this nature are discussed, for example, by Gaume and Gosset (2003), See *et al.* (2007), Han *et al.* (2007), and Abrahart *et al.*, 2008. Abrahart and See (2007) also addressed some of these problems, however, demonstrated that the existing nonlinear hydrological relationships, which are so important when building flow forecasting models, are effectively captured by a neural network, the most widely used DDM method. In this respect, positioning of data-driven models is important: they should be seen as complementary to process-based simulation models; they cannot explain reality but could be effective predictive tools.

2.16.6.2 Technology of DDM

2.16.6.2.1 Definitions

One may identify several fields that contribute to DDM: statistical methods, ML, soft computing (SC), computational intelligence (CI), data mining (DM), and knowledge discovery in databases (KDDs). ML is the area concentrating on the theoretical foundations of learning from data and it can be said that it is the major supplier of methods for DDM. SC is emerging from fuzzy logic, but many authors attribute to it many other techniques as well. CI incorporates two areas of ML (neural networks and fuzzy systems), and, additionally, evolutionary computing that, however, can be better attributed to the field of optimization than to ML. DM and KDDs used, in fact, the methods of ML and are focused typically at large databases being associated with banking, financial services, and customer resources management.

DDM can thus be considered as an approach to modeling that focuses on using the ML methods in building models that would complement or replace the physically based models. The term modeling stresses the fact that this activity is close in its objectives to traditional approaches to modeling, and follows the steps traditionally accepted in (hydrological) modeling. Examples of the most common methods used in data-driven hydrological modeling are linear regression, ARIMA, artificial neural networks (ANNs), and fuzzy rule-based systems (FRBSs).

Such positioning of DDM links to learning which incorporates determining the so far unknown mappings (or dependencies) between a system's inputs and its outputs from the available data (Mitchell, 1997). By data, we understand the known samples (data vectors) that are combinations of inputs and corresponding outputs. As such, a dependency (mapping or model) is discovered (induced), which can be used to predict (or effectively deduce) the future system's outputs from the known input values.

By data, we usually understand a set K of examples (or instances) represented by duple $\langle x_k, y_k \rangle$, where $k = 1, \ldots, K$, vector $x_k = \{x_1, \ldots, x_n\}_k$, vector $y_k = \{y_1, \ldots, y_m\}_k$, n = number of inputs, and m = number of outputs. The process of building a function (or mapping, or model) $y = f(x)$ is called training. If only one output is considered, then $m = 1$. (In relation to hydrological and hydraulic models, training can be seen as calibration.)

In the context of hydrological modeling, the inputs and outputs are typically real numbers ($x_k, y_k \in \geq \mathscr{R}^n$), so the main learning problem solved in hydrological modeling is numerical prediction (regression). Note that the problems of clustering and classification are rare but there are examples of it as well (see, e.g., Hall and Minns, 1999; Hannah et al., 2000; Harris et al., 2000).

As already mentioned, the process of building a data-driven model follows general principles adopted in modeling: study the problem, collect data, select model structure, build the model, test the model, and (possibly) iterate. There is, however, a difference with physically based modeling: in DDM not only the model parameters but also the model structure are often subject to optimization. Typically, simple (or parsimonious) models are valued (as simple as possible, but no simpler). An example of such parsimonious model could be a linear regression model versus a nonlinear one, or a neural network with the small number of hidden nodes. Such models would automatically emerge if the so-called regularization is used: the objective function representing the overall model performance includes not only the model error term, but also a term that increases in value with the increase of model complexity represented, for example, by the number of terms in the equation, or the number of hidden nodes in a neural network.

If there is a need to build a simple replica of a sophisticated physically based hydrological model, DDM can be used as well: such models are called surrogate, emulation, or meta-models (see, e.g., Solomatine and Torres, 1996; Khu et al., 2004). They can be used as fast-working approximations of complex models when speed is important, for example, in solving the optimization or calibration problems.

2.16.6.2.2 Specifics of data partitioning in DDM

Obviously, data analysis and preparation play an important role in DDM. These steps are considered standard by the experts in ML but are not always given proper attention by hydrologists building or using such models.

Three data sets for training, cross-validation, and testing. Once the model is trained (but before it is put into operation), it has to be tested (or verified) by calculating the model error (e.g., RMSE) using the test (or verification) data set. However, during training often there is a need to conduct tests of the model that is being built, so yet another data set is needed – the cross-validation set. This set serves as the representative of the test set. As a model gradually improves as a result of the training process, the error on the training data will be gradually decreasing. The cross-validation error will also be first decreasing, but as the model starts to reproduce the training data set better and better, this error will start to increase (effect of over fitting). This typically means that the training should be stopped when the error on cross-validation data set starts to increase. If these principles are respected, then there is a hope that the model will generalize well, that is, its prediction error on unseen data will be small. (Note that the test data should be used only to test the final model, but not to improve (optimize) the model.)

One may see that this procedure is more complex than the standard procedure of the hydrological model calibration – when no data are allocated for cross-validation, and, worse, often the whole data set is used to calibrate the model.

Note that in an important class of ML models – support vector machines (SVMs) – a different approach is taken: it is to build the model that would have the best generalization ability possible without relying explicitly on the cross-validation set (Vapnik, 1998).

In connection to the issues covered above, there are two common pitfalls, especially characteristic of DDM applications where time series are involved, that are worth mentioning here.

The desired properties of the three data sets. It is desired that the three sets are statistically similar. Ideally, this could be automatically ensured by the fact that data sets are sufficiently large and sampled from the same distribution (typical assumption in machine and statistical learning). However, in reality of hydrological modeling, such situations are rare, so normally a modeler should try to ensure at least some similarity in the distributions, or, at least, similar ranges, mean and variance. Statistical similarity can be achieved by careful selection of examples for each data set, by random sampling data from the whole data set, or employing an optimization procedure resulting in the sets with predefined properties (Bowden et al., 2002).

One of the approaches is to use the 10-fold validation method when a model is built 10 times, trained each time on 9/10th of the whole set of available data and validated on 1/10th (number of runs is not necessarily 10). A version of this method is the leave-one-out method when K models are built using $K - 1$ examples and not using one (every time different). The modeler is left with 10 or K trained models, so the resulting model to be used is either one of these models, or an ensemble of all the built models, possibly with the weighted outputs.

Strictly speaking, for generation of the statistically similar training data sets for building a series of similar but different models, one should typically rely on the well-developed statistical (re)sampling methods such as bootstrap originated by Tibshirani in the 1970s (see Efron and Tibshirani, 1993) where (in its basic form) K data are randomly selected from K original data.

For many hydrologists, there could be a visualization (or even a psychological) problem. If one of these procedures is followed, the data will not be always contiguous: it would not be possible to visualize a hydrograph when the model is fed with the test set. There is nothing wrong with such a model if the time structure of all the data sets is preserved. Such models, however, may be rejected by practitioners, since they are so different from the traditional physically based models that always generate contiguous time series. A possible solution here is to consider the hydrological events (i.e., contiguous blocks of data), to group the data accordingly, and to try to ensure the presence of statistically similar events in all the three data sets.

This is all possible of course, if there is enough data. In the situations when the data set is not large enough to allow for building all three sets of substantial size, modelers could be forced not to build cross-validation set at all with the hope that the model trained on training set would perform well on the test set as well. An alternative could be performing 10-fold cross-validation but it is somehow rarely used.

2.16.6.2.3 Choice of the model variables

Apart from dividing the data into several subsets, data preparation also includes the selection of proper variables to represent the modeled process, and, possibly, their transformation (Pyle, 1999). A study on the influence of different data transformation methods (linear, logarithmic, and seasonal transformations, histogram equalization, and a transformation to normality) was undertaken by Bowden et al. (2003). On a (limited) case study (forecasting salinity in a river in Australia 14 days ahead), they found that the model using the linear transformation resulted in the lowest RMSE and more complex transformations did not improve the model. Our own experience shows that it is sometimes also useful to apply the smoothing filters to reduce the noise in the hydrological time series.

Choice of variables is an important issue, and it has to be based on taking the physics of the underling processes into account. State variables of data-driven models have nothing to do with the physics, but their inputs and outputs do have. In DDM, the physics of the process is introduced mainly via the justified and physically based choice of the relevant input variables.

One may use visualization to identify the variables relevant for predicting the output value. There are also formal methods that help in making this choice more justified, and the reader can be directed to the paper by Bowden et al. (2005) for an overview of these.

Mutual information which is based on Shannon's entropy (Shannon, 1948) is used to investigate linear and nonlinear dependencies and lag effects (in time series data) between the variables. It is the measure of information available from one set of data having knowledge of another set of data. The average mutual information (AMI) between two variables X and Y is given by

$$\text{AMI} = \sum_{i,j} P_{XY}(x_i, y_j) \log_2 \left[\frac{P_{XY}(x_i, y_j)}{P_X(x_i) P_Y(y_j)} \right] \quad (1)$$

where $P_X(x)$ and $P_Y(y)$ are the marginal probability density functions (PDFs) of X and Y, respectively, and $P_{XY}(x,y)$ the joint PDFs of X and Y. If there is no dependence between X and Y, then by definition the joint probability density $P_{XY}(x,y)$ would be equal to the product of the marginal densities ($P_X(x) P_Y(y)$). In this case, AMI would be zero (the ratio of the joint and marginal densities in Equation (1) being 1, giving the logarithm a value of 0). A high value of AMI would indicate a strong dependence between two variables. Accurate estimate of the AMI depends on the accuracy of estimation of the marginal and joint probabilities density in Equation (1) from a finite set of examples. The most widely used approach is estimation of the probability densities by histogram with the fixed bin width. More stable, efficient, and robust probability density estimator is based on the use of kernel density estimation techniques (Sharma, 2000).

It is our hope that the adequate data preparation and the rational and formalized choice of variables will become a standard part of any hydrological modeling study.

2.16.6.3 Methods and Typical Applications

Most hydrological modeling problems are formulated as simulation of forecasting of real-valued variables. In terminology of machine (statistical) learning, this is a regression problem. A number of linear and (sometimes) nonlinear regression methods have been used in the past. Most of the methods of ML can also be seen as sophisticated nonlinear regression methods. Many of them, instead of using very complex functions, use combinations of many simple functions. During training, the number of these functions and the values of their parameters are optimized, given the functions' class. Note that ML methods typically do not assume any special kind of distribution of data, and do not require the knowledge of such distribution.

Multilayer perceptron (MLP) is a device (mathematical model) that was originally referred to as an ANN (Haykin, 1999). Later ANN became a term encompassing other connectionist models as well. MLP consists of several layers of mutually interconnected nodes (neurons), each of which receives several inputs, calculates the weighted sum of them, and then passes the result to a nonlinear squashing function. In this way, the inputs to an MLP model are subjected to a multiparameter nonlinear transformation so that the resulting model is able to approximate complex input–output relationships. Training of MLP is in fact solving the problem of minimizing the model error (typically, MSE) by determining the optimal set of weights.

MLP ANNs are known to have several dozens of successful applications in hydrology. The most popular application was building rainfall–runoff models: Hsu et al. (1995), Minns and Hall (1996), Dawson and Wilby (1998), Dibike et al. (1999), Abrahart and See (2000), Govindaraju and Rao (2001), Coulibaly et al. (2000), Hu et al. (2007), and Abrahart et al. (2007b). They were also used to model river stage–discharge relationships (Sudheer and Jain, 2003; Bhattacharya and Solomatine, 2005). ANNs were also used to build surrogate (emulation, meta-) models for replicating the behavior of hydrological and hydrodynamic models: in model-based optimal control of a reservoir (Solomatine and Torres, 1996), calibration of a rainfall–runoff model (Khu et al., 2004), and in multiobjective decision support model for watershed management (Muleta and Nicklow, 2004).

Most theoretical problems related to MLP have been solved, and it should be seen as a quite reliable and well-understood method.

Radial basis functions (RBFs) could be seen as a sensible alternative to the use of complex polynomials. The idea is to approximate some function $y = f(x)$ by a superposition of J functions $F(x, \sigma)$, where σ is a parameter characterizing the span or width of the function in the input space. Functions F are typically bell shaped (e.g., a Gaussian function) so that they are defined in the proximity to some representative locations (centers) w_j in n-dimensional input space and their values are close to zero far from these centers. The aim of learning here is to find the positions of centers w_j and the parameters of the functions $f(x)$. This can be accomplished by building an RBF neural network; its training allows the identification of these unknown parameters. The centers w_j of the RBFs can be chosen using a clustering algorithm, the parameters of the Gaussian can be found based on the spread (variance) of data in each cluster, and it can be shown that the weights can be found by solving a system of linear equations. This is done for a certain number of RBFs, with the exhaustive optimization run across the number of RBFs in a certain range.

The areas of RBF networks applications are the same as those of MLPs. Sudheer and Jain (2003) used RBF ANNs for modeling river stage–discharge relationships and found out that on the considered case study RBF ANNs were superior to MLPs; Moradkhani *et al.* (2004) used RBF ANNs for predicting hourly stream flow hydrograph for the daily flow for a river in USA as a case study, and demonstrated their accuracy if compared to other numerical prediction models. In this study, RBF was combined with the self-organizing feature maps used to identify the clusters of data. Nor *et al.* (2007) used RBF ANN for the same purpose, however, for the hourly flow and considering only storm events in the two catchments in Malaysia as case studies.

Regression trees and M5 model trees. These models can be attributed simultaneously to (piece-wise) linear regression models, and to modular (multi)models. They use the following idea: progressively split the parameter space into areas and build in each of them a separate regression model of zero or first order (**Figure 7**). In M5 trees models in leaves are first order (linear). The Boolean tests a_i at nodes have the form $x_i < C$ and are used to progressively split the data set. The index of the input variable i and value C are chosen to minimize the standard deviation in the subsets resulting from the split. Mn are models built for subsets filtered down to a given tree leaf. The resulting model can be seen as a set of linear models being specialized on the certain subsets of the training set – belonging to different regions of the input space. M5 algorithm to build such model trees was proposed by Quinlan (1992).

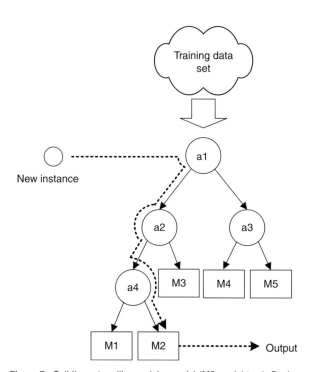

Figure 7 Building a tree-like modular model (M5 model tree). Boolean tests a_i have the form $x_i < C$ and split data set during training. Mn are linear regression models built on data subsets, and applied to a new instance input vector in operation.

Combination of linear models was already used in dynamic hydrology in the 1970s (e.g., multilinear models by Becker and Kundzewicz (1987)). Application of the M5 algorithm to build such models adds rigor to this approach and makes it possible to build the models automatically and to generate a range of models of different complexity and accuracy.

MTs are often almost as accurate as ANNs, but have some important advantages: training of MTs is much faster than ANNs, it always converges, and the results can be easily understood by decision makers.

An early (if not the first) application of M5 model trees in river flow forecasting was reported by Kompare *et al.* (1997). Solomatine and Dulal (2003) used M5 model tree in rainfall–runoff modeling of a catchment in Italy. Stravs and Brilly (2007) used M5 trees in modeling the precipitation interception in the context of the Dragonja river basin case study.

Genetic programming (GP) and evolutionary regression. GP is a symbolic regression method in which the specific model structure is not chosen *a priori*, but is a result of the search (optimization) process. Various elementary mathematical functions, constants, and arithmetic operations are combined in one function and the algorithm tries to construct a model recombining these building blocks in one formula. The function structure is represented as a tree and since the resulting function is highly nonlinear, often nondifferentiable, it is optimized by a randomized search method – usually a GA. Babovic and Keijzer (2005) gave an overview of GP applications in hydrology. Laucelli *et al.* (2007) presented an application of GP to the problem of forecasting the groundwater heads in the aquifer in Italy; in this study, the authors also employed averaging of several models built on the data subsets generated by bootstrap.

One may limit the class of possible formulas (regression equations), allowing for a limited class of formulas that would *a priori* be reasonable. In evolutionary regression (Giustolisi and Savic, 2006), a method similar to GP, the elementary functions are chosen from a limited set and the structure of the overall function is fixed. Typically, a polynomial regression equation is used, and the coefficients are found by GA. This method overcomes some shortcomings of GP, such as the computational requirements – the number of parameters to tune and the complexity of the resulting symbolic models. It was used, for example, for modeling groundwater level (Giustolisi *et al.*, 2007a) and river temperature (Giustolisi *et al.*, 2007b), and the high accuracy and transparency of the resulting models were reported.

FRBSs. Probability is not the only way to describe uncertainty. In his seminal paper, Lotfi Zadeh (1965) introduced yet another way of dealing with uncertainty – fuzzy logic, and since then it found multiple successful applications, especially in application to control problems.

Fuzzy logic can be used in combining various models, as done previously, for example, by See and Openshaw (2000) and Xiong *et al.* (2001), building the so-called fuzzy committees of models (Solomatine, 2006), and also the instrumentarium of fuzzy logic can be used for building the so-called FRBSs which are effectively regression models. FRBS can be built by interviewing human experts, or by processing historical data and thus forming a data-driven model. These rules are patches of local models overlapped throughout the parameter space, using a sort of interpolation at a lower level to represent patterns in complex nonlinear relationships. The basics of the data-driven approach and its use in a number of water-related applications can be found in Bárdossy and Duckstein (1995).

Typically, the following rules are considered:

$$IF\ x_1\ is\ A_{1,r}\ AND\ldots AND\ x_n\ is\ A_{n,y}\ THEN\ y\ is\ B$$

where $\{x_1,\ldots,x_n\} = x$ is the input vector; A_{im} the fuzzy set; r the index of the rule, $r = 1,\ldots,R$. Fuzzy sets A_{ir} (defined as membership functions with the values ranging from 0 to 1) are used to partition the input space into overlapping regions (for each input these are intervals). The structure of B in the consequent could be either a fuzzy set (then such model is called a Mamdani model) or a function $y = f(x)$, often linear, and then the model is referred to as Takagi–Sugeno–Kang (TSK) model. The model output is calculated as a weighted combination of the R rules' responses. Output of the Mamdani model is fuzzy (a membership function of irregular shape), so the crisp output has to be calculated by the so-called defuzzification operator. Note that in TSK model, each of the r rules can be interpreted as local models valid for certain regions in the input space defined by the antecedent and overlapping fuzzy sets A_{ir}. Resemblance to the RBF ANN is obvious.

FRBSs were effectively used for drought assessment (Pesti et al., 1996); reconstruction of the missing precipitation data by a Mamdani-type system (Abebe et al., 2000b); control of water levels in polder areas (Lobbrecht and Solomatine et al., 1999); and modeling rainfall–discharge dynamics (Vernieuwe et al., 2005; Nayak et al., 2005). Casper et al. (2007) presented an interesting study where TSK type of FRBS has been developed using soil moisture and rainfall as input variables to predict the discharge at the outlet of a small catchment, with the special attention to the peak discharge. One of the limitations of FRBS is that the demand for data grows exponentially with an increase in the number of input variables.

SVMs. This ML method is based on the extension of the idea of identifying a hyperplane that separates two classes in classification. It is closely linked to the statistical learning theory initiated by V. Vapnik in the 1970s at the Institute of Control Sciences of the Russian Academy of Science (Vapnik, 1998). Originally developed for classification, it was extended to solving prediction problems, and, in this capacity, was used in hydrology-related tasks. Dibike et al. (2001) and Liong and Sivapragasam (2002) reported using SVMs for forecasting the river water flows and stages. Bray and Han (2004) addressed the issue of tuning SVMs for rainfall–runoff modeling. In all reported cases, SVM-based predictors have shown good results, in many cases superseding other methods in accuracy.

Instance-based learning (IBL). This method allows for classification or numeric prediction directly by combining some instances from the training data set. A typical representative of IBL is the k-nearest neighbor (k-NN) method. For a new input vector x_q (query point), the output value is calculated as the mean value of the k-nearest neighboring examples, possibly weighted according to their distance to x_q. Further extensions are known as locally weighted regression (LWR) when the regression model is built on k nearest instances: the training instances are assigned weights according to their distance to x_q and the regression equations are generated on the weighted data.

In fact, IBL methods construct a local approximation to the modeled function that applies well in the immediate neighborhood of the new query instance encountered. Thus, it describes a very complex target function as a collection of less complex local approximations, and often demonstrates competitive performance when compared, for example, to ANNs.

Karlsson and Yakowitz (1987) introduced this method in the context of water, focusing however only on (single-variate) time-series forecasts. Galeati (1990) demonstrated the applicability of the k-NN method (with the vectors composed of the lagged rainfall and flow values) for daily discharge forecasting and favorably compared it to the statistical autoregressive model with exogenous input (ARX) model, and used the k-NN method for adjusting the parameters of the linear perturbation model for river flow forecasting. Toth et al. (2000) compared the k-NN approach to other time-series prediction methods in a problem of short-term rainfall forecasting. Solomatine et al. (2007) explored a number of IBL methods, tested their applicability in rainfall–runoff modeling, and compared their performance to other ML methods.

To conclude the coverage of the popular data-driven methods, it can be mentioned that all of them are developed in the ML and CI community. The main challenges for the researchers in hydrology and hydroinformatics are in testing various combinations of these methods for particular water-related problems, combining them with the optimization techniques, developing the robust modeling procedures able to work with the noisy data, and in developing the methods providing the model uncertainty estimates.

2.16.6.4 DDM: Current Trends and Conclusions

There are a number of challenges in DDM: development of the optimal model architectures, making models more robust, understandable, and ready for inclusion into existing decision support systems. Models should adequately reflect reality, which is uncertain, and in this respect developing the methods of dealing with the data and model uncertainty is currently an important issue.

One of the interesting questions that arise in case of using a data-driven model is the following one: to what extent such models could or should incorporate the expert knowledge into the modeling process. One may say that a typical ML algorithm minimizes the training (cross validation) error seeing it as the ultimate indicator of the algorithms performance, so is purely data-driven – and this is what is expected from such models. Hydrologists, however, may have other consideration when assessing the usefulness of a model, and typically wish to have a certain input to building a model rightfully hoping that the direct participation of an expert may increase the model accuracy and trust in the modeling results. Some of the examples of merging the hydrological knowledge and the concepts of process-based modeling with those of DDM are mentioned in Section 2.16.8.2.

Data-driven models are seen by many hydrologists as tools complementary to process-based models. More and more practitioners are agreeing to that, but many are still to be convinced. Research is now oriented toward development of the optimal model architectures and avenues for making data-driven models more robust, understandable, and very useful for practical applications. The main challenge is in the inclusion of DDM into the existing decision-making frameworks, while taking into consideration both the system's physics, expert judgment, and the data availability. For example, in operational hydrological forecasting, many practitioners are trained in using process-based models (mainly conceptual ones) that serve them reasonably well, and adoption of another modeling paradigm with inevitable changes in their everyday practice could be a painful process. Making models capable of dealing with the data and model uncertainty is currently an important issue as well.

It is sensible to use DDM if (1) there is a considerable amount of observations available; (2) there were no considerable changes to the system during the period covered by the model; and (3) it is difficult to build adequate process-based simulation models due to the lack of understanding and/or to the ability to satisfactorily construct a mathematical model of the underlying processes. Data-driven models can also be useful when there is a necessity to validate the simulation results of physically based models.

It can be said that it is practically impossible to recommend one particular type of a data-driven model for a given problem. Hydrological data are noisy and often of poor quality; therefore, it is advisable to apply various types of techniques and compare and/or combine the results.

2.16.7 Analysis of Uncertainty in Hydrological Modeling

2.16.7.1 Notion of Uncertainty

Webster's Dictionary (1998) defines uncertain as follows: not surely or certainly known, questionable, not sure or certain in knowledge, doubtful, not definite or determined, vague, liable to vary or change, not steady or constant, varying. The noun uncertainty results from the above concepts and can be summarized as the state of being uncertain. However, in the context of hydrological modeling, uncertainty has a specific meaning, and it seems that there is no consensus about the very term of uncertainty, which is conceived with differing degrees of generality (Kundzewicz, 1995).

Often uncertainty is defined with respect to certainty. For example, Zimmermann (1997) defined certainty as "certainty implies that a person has quantitatively and qualitatively the appropriate information to describe, prescribe or predict deterministically and numerically a system, its behaviour or other phenomena." Situations that are not described by the above definition shall be called uncertainty. A similar definition has been given by Gouldby and Samuels (2005): "a general concept that reflects our lack of sureness about someone or something, ranging from just short of complete sureness to an almost complete lack of conviction about an outcome."

In the context of modeling, uncertainty is defined as a state that reflects our lack of sureness about the outcome of a physical processes or system of interest, and gives rise to potential difference between assessment of the outcome and its true value. More precisely, uncertainty of a model output is the state or condition that the output cannot be assessed uniquely. Uncertainty stems from incompleteness or imperfect knowledge or information concerning the process or system in addition to the random nature of the occurrence of the events. Uncertainty resulting from insufficient information may be reduced if more information is available.

2.16.7.2 Sources of Uncertainty

Uncertainties that can affect the model predictions stem from a variety of sources (e.g., Melching, 1995; Gupta et al., 2005), and are related to our understanding and measurement capabilities regarding the real-world system under study:

1. Perceptual model uncertainty, that is, the conceptual representation of the watershed that is subsequently translated into mathematical (numerical) form in the model. The perceptual model (Beven, 2001) is based on our understanding of the real-world watershed system, that is, flow-paths, number and location of state variables, runoff production mechanisms, etc. This understanding might be poor, particularly for aspects relating to subsurface system characteristics, and therefore our perceptual model might be highly uncertain (Neuman, 2003).

2. Data uncertainty, that is, uncertainty caused by errors in the measurement of input (including forcing) and output data, or by data processing. Additional uncertainty is introduced if long-term predictions are made, for instance, in the case of climate change scenarios for which as per definition no observations are available. A hydrological model might also be applied in integrated systems, for example, connected to a socioeconomic model, to assess, for example, impacts of water resources changes on economic behavior. Data to constrain these integrated models are rarely available (e.g., Letcher et al., 2004). An element of data processing, that is, uncertainty, is introduced when a model is required to interpret the actual measurement. A typical example is the use of radar rainfall measurements. These are measurements of reflectivity that have to be transformed to rainfall estimates using a (empirical) model with a chosen functional relationship and calibrated parameters, both of which can be highly uncertain.

3. Parameter estimation uncertainty, that is, the inability to uniquely locate a best parameter set (model, i.e., a model structure parameter set combination) based on the available information. The lack of correlation often found between conceptual model parameters and physical watershed characteristics will commonly result in significant prediction uncertainty if the model is extrapolated to predict the system behavior under changed conditions (e.g., land-use change or urbanization) or to simulate the behavior of a similar but geographically different watersheds for which no observations of the variable of interest are available (i.e., the ungauged case). Changes in the represented system have to be considered through adjustments of the model parameters (or even the model structure), and the degree of adjustment has so far been difficult to determine without measurements of the changed system response.

4. Model structural uncertainty introduced through simplifications, inadequacies, and/or ambiguity in the description

of real-world processes. There will be some initial uncertainty in the model state(s) at the beginning of the modeled time period. This type of uncertainty can usually be taken care of through the use of a warm-up (spin-up) period or by optimizing the initial state(s) to fit the beginning of the observed time series. Errors in the model (structure and parameters) and in the observations will also commonly cause the states to deviate from the actual state of the system in subsequent time periods. This problem is often reduced using data assimilation techniques as discussed later.

Figure 8 presents how different sources of the uncertainty might vary with model complexity. As the model complexity (and the detailed representation of the physical process) increases, structural uncertainty decreases. However, with the increasing complexity of model, the number of inputs and parameters also increases and consequently there is a good chance that input and parameter uncertainty will increase. Due to the inherent trade-off between model structure uncertainty and input/parameter uncertainty, for every model there is the optimal level of model complexity where the total uncertainty is minimum.

2.16.7.3 Uncertainty Representation

For many years, probability theory has been the primary tool for representing uncertainty in mathematical models. Different methods can be used to describe the degree of uncertainty. The most widely adopted methods use PDFs of the quantity, subject to the uncertainty. However, in many practical problems the exact form of this probability function cannot be derived or found precisely.

When it is difficult to derive or find PDF, it may still be possible to quantify the level of uncertainty by the calculated statistical moments such as the variance, standard deviation, and coefficient of variation. Another measure of the uncertainty of a quantity relates to the possibility to express it in terms of the two quantiles or prediction intervals. The prediction intervals consist of the upper and lower limits between which a future uncertain value of the quantity is expected to lie with a prescribed probability. The endpoints of a prediction interval are known as the prediction limits. The width of the prediction interval gives us some idea about how uncertain we are about the uncertain entity.

Although useful and successful in many applications, probability theory is, in fact, appropriate for dealing with only a very special type of uncertainty, namely random (Klir and Folger, 1988). However, not all uncertainty is random. Some forms of uncertainty are due to vagueness or imprecision, and cannot be treated with probabilistic approaches. Fuzzy set theory and fuzzy measures (Zadeh, 1965, 1978) provide a nonprobabilistic approach for modeling the kind of uncertainty associated with vagueness and imprecision.

Information theory is also used for representing uncertainty. Shannon's (1948) entropy is a measure of uncertainty and information formulated in terms of probability theory. Another broad theory of uncertainty representation is the evidence theory introduced by Shafer (1976). Evidence theory, also known as Dempster–Shafer theory of evidence, is based on both the probability and possibility theory. In hydrological modeling, the primary tool for handling uncertainty is still probability theory, and, to some extent, fuzzy logic.

2.16.7.4 View at Uncertainty in Data-Driven and Statistical Modeling

In DDM, the sources of uncertainty are similar to those for other hydrological models, but there is an additional focus on data partitioning used for model training and verification. Often data are split in a nonoptimal way. A standard procedure for evaluating the performance of a model would be to split the data into training set, cross-validation set, and test set. This approach is, however, very sensitive to the specific sample splitting (LeBaron and Weigend, 1994). In principle, all these splitting data sets should have identical distributions, but we do not know the true distribution. This causes uncertainty in prediction as well.

The prediction error of any regression model can be decomposed into the following three sources (Geman et al., 1992): (1) model bias, (2) model variance, and (3) noise. Model bias and variance may be further decomposed into the contributions from data and training process. Furthermore, noise can also be decomposed into target noise and input noise. Estimating these components of prediction error (which is however not always possible) helps to compute the predictive uncertainty.

The terms bias and variance come from a well-known decomposition of prediction error. Given N data points and M models, the decomposition is based on the following equality:

$$\frac{1}{NM}\sum_{i=1}^{N}\sum_{j=1}^{M}(t_i - y_{ij})^2 = \frac{1}{N}\sum_{i=1}^{N}(t_i - \bar{y}_i)^2 + \frac{1}{NM}\sum_{i=1}^{N}\sum_{j=1}^{M}(\bar{y}_i - y_{ij})^2 \quad (2)$$

where t_i is the ith target, y_{ij} the ith output of the jth model, and $\bar{y}_i = 1/M \sum_{j=1}^{M} y_{ij}$ the average model output calculated for input i.

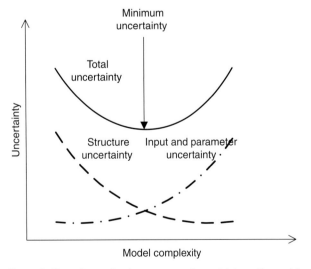

Figure 8 Dependency of various sources of uncertainty on the model complexity.

The left-hand-side term of Equation (2) is the well-known MSE. The first term on the right-hand side is the square of bias and the last term is the variance. The bias of the prediction errors measures the tendency of over- or under-prediction by a model, and is the difference between the target value and the model output. From Equation (2), it is clear that the variance does not depend on the target, and measures the variability in the predictions by different models.

2.16.7.5 Uncertainty Analysis Methods

Once the uncertainty in a model is acknowledged, it should be analyzed and quantified with the ultimate aim to reduce the impact of uncertainty. There is a large number of uncertainty analysis methods published in the academic literature. Pappenberger et al. (2006) provided a decision tree to help in choosing an appropriate method for a given situation. Uncertainty analysis process in hydrological models varies mainly in the following: (1) type of hydrological models used; (2) source of uncertainty to be treated; (3) the representation of uncertainty; (4) purpose of the uncertainty analysis; and (4) availability of resources. Uncertainty analysis has comparatively a long history in physically based and conceptual modeling (see, e.g., Beven and Binley, 1992; Gupta et al., 2005).

Uncertainty analysis methods in all the above cases should involve: (1) identification and quantification of the sources of uncertainty, (2) reduction of uncertainty, (3) propagation of uncertainty through the model, (4) quantification of uncertainty in the model outputs, and (6) application of the uncertain information in decision-making process.

A number of methods have been proposed in the literature to estimate model uncertainty in rainfall–runoff modeling. Reviews of various methods of uncertainty analysis on hydrological models can be found in, for example, Melching (1995), Gupta et al. (2005), Montanari (2007), Moradkhani and Sorooshian (2008), and Shrestha and Solomatine (2008). These methods are broadly classified into several categories (most of them result in probabilistic estimates):

1. analytical methods (see, e.g., Tung, 1996);
2. approximation methods, for example, first-order second moment method (Melching, 1992);
3. simulation and sampling-based (Monte Carlo) methods leading to probabilistic estimates that may also use Bayesian reasoning (Kuczera and Parent, 1998; Beven and Binley, 1992);
4. methods based on the analysis of the past model errors and either using distribution transforms (Montanari and Brath, 2004) or building a predictive ML of uncertainty (Shrestha and Solomatine, 2008; Solomatine and Shrestha, 2009); and
5. methods based on fuzzy set theory (e.g., Abebe et al., 2000a; Maskey et al., 2004).

Analytical and approximation methods can hardly be applicable in case of using complex computer-based models. Here, we will present only the most widely used probabilistic methods based on random sampling – Monte Carlo simulation method. The GLUE method (Beven and Binley, 1992), widely used in hydrology, can be seen as a particular case of MC approach. MC simulation is a flexible and robust method capable of solving a great variety of problems. In fact, it may be the only method that can estimate the complete probability distribution of the model output for cases with highly nonlinear and/or complex system relationship (Melching, 1995). It has been used extensively and also as a standard means of comparison against other methods for uncertainty assessment. In MC simulation, random values of each of uncertain variables are generated according to their respective probability distributions and the model is run for each of the realizations of uncertain variables. Since we have multiple realizations of outputs from the model, standard statistical technique can be used to estimate the statistical properties (mean, standard deviation, etc.) and empirical probability distribution of the model output. MC simulation method involves the following steps:

1. randomly sample uncertain variables X_i from their joint probability distributions;
2. run the model $y = g(x_i)$ with the set of random variables x_i;
3. repeat the steps 1 and 2 s times, storing the realizations of the outputs $y_1, y_2, \ldots, y_s$; and
4. from the realizations $y_1, y_2, \ldots, y_s$, derive the cdf and other statistical properties (e.g., mean and standard deviation) of Y.

When MC sampling is used, the error in estimating PDF is inversely proportional to the square root of the number of runs s and, therefore, decreases gradually with s. As such, the method is computationally expensive, but can reach an arbitrarily level of accuracy. The MC method is generic, invokes fewer assumptions, and requires less user input than other uncertainty analysis methods. However, the MC method suffers from two major practical limitations: (1) it is difficult to sample the uncertain variables from their joint distribution unless the distribution is well approximated by a multinormal distribution (Kuczera and Parent, 1998) and (2) it is computationally expensive for complex models. Markov chain Monte Carlo (MCMC) methods such as Metropolis and Hastings (MH) algorithm (Metropolis et al., 1953; Hastings, 1970) have been used to sample parameter from its posterior distribution. In order to reduce the number of samples (model simulations) necessary in MC sampling, more efficient Latin Hypercube sampling has been introduced (McKay et al., 1979). Further, the following methods in this row can be mentioned: Kalman filter and its extensions (Kitanidis and Bras, 1980), the DYNIA approach (Wagener et al., 2003), the BaRE approach (Thiemann et al., 2001), the SCEM-UA algorithm (Vrugt et al., 2003), and the DREAM algorithm (Vrugt et al., 2008b), a version of the MCMC scheme.

Most of the probabilistic techniques for uncertainty analysis treat only one source of uncertainty (i.e., parameter uncertainty). Recently, attention has been given to other sources of uncertainty, such as input uncertainty or structure uncertainty, as well as integrated approach to combine different sources of uncertainty. The research shows that input or structure uncertainty is more dominant than the parameter uncertainty. For example, Kavetski et al. (2006) and Vrugt et al. (2008a), among others, treat input uncertainty in hydrological modeling using Bayesian approach. Butts *et al.* (2004) analyzed impact of the model structure on hydrological modeling uncertainty for stream flow simulation. Recently, new schemes have emerged to estimate the combined uncertainties in rainfall–runoff predictions associated with input, parameter, and structure uncertainty. For instance, Ajami *et al.* (2007) used an integrated Bayesian multimodel approach to combine input, parameter, and model structure uncertainty. Liu and Gupta (2007) suggested an integrated data assimilation approach to treat all sources of uncertainty.

Regarding the sources of uncertainty, Monte-Carlo-type methods are widely used for parameter uncertainty, Bayesian methods and/or data assimilation can be used for input uncertainty and Bayesian model averaging method is suitable for structure uncertainty. The appropriate uncertainty analysis method also depends on whether the uncertainty is represented as randomness or fuzziness. Similarly, uncertainty analysis methods for real-time forecasting purposes would be different from those used for design purposes (e.g., when estimating design discharge hydraulic structure design).

It should be noted that the practice of uncertainty analysis and the use of the results of such analysis in decision making are not yet widely spread. Some possible misconceptions are stated by Pappenberger and Beven (2006):

> a) uncertainty analysis is not necessary given physically realistic models, b) uncertainty analysis is not useful in understanding hydrological and hydraulic processes, c) uncertainty (probability) distributions cannot be understood by policy makers and the public, d) uncertainty analysis cannot be incorporated into the decision-making process, e) uncertainty analysis is too subjective, f) uncertainty analysis is too difficult to perform and g) uncertainty does not really matter in making the final decision.

Some of these misconceptions however have explainable reasons, so the fact remains that more has to be done in bringing the reasonably well-developed apparatus of uncertainty analysis and prediction to decision-making practice.

2.16.8 Integration of Models

2.16.8.1 Integration of Meteorological and Hydrological Models

Water managers demand much longer lead times in the hydrological forecasts. Forecasting horizon of hydrological models can be extended if along with the (almost) real-time measurements of precipitation (radar and satellite images, gauges), their forecasts are used. The forecasts can come only from the numerical weather prediction (NWP; meteorological) models.

Linking of meteorological and hydrological models is currently an adopted practice in many countries. One of the examples of such an integrated approach is the European flood forecasting system (EFFS), in which development started in the framework of EU-funded project in the beginning of the 2000s. Currently, this initiative is known as the European flood alert system (EFAS), which is being developed by the EC Joint Research Centre (JRC) in close collaboration with several European institutions. EFAS aims at developing a 4–10-day in-advance EFFS employing the currently available medium-range weather forecasts. The framework of the system allows for incorporation of both detailed models for specific basins as well as a broad scale for entire Europe. This platform is not supposed to replace the national systems but to complement them.

The resolution of the existing NWP models dictates to a certain extent the resolution of the hydrological models. LISFLOOD model (Bates and De Roo, 2000) and its extension module for inundation modeling LISFLOOD-FP have been adopted as the major hydrological response model in EFAS. This is a rasterized version of a process-based model used for flood forecasting in large river basins. LISFLOOD is also suitable for hydrological simulations at the continental scale, as it uses topographic and land-use maps with a spatial resolutions up to 5 km.

It should be mentioned that useful distributed hydrological models that are able to forecast floods at meso-scales have grid sizes from dozens of meters to several kilometers. At the same time, the currently used meteorological models, providing the quantitative precipitation forecasts, have mesh sizes from several kilometers and higher. This creates an obvious inconsistency and does not allow to realize the potential of the NWP outputs for flood forecasting – see, for example, Bartholmes and Todini (2005). The problem can partly be resolved by using downscaling (Salathe, 2005; Cannon, 2008), which however may bring additional errors. As NWP models use more and more detailed grids, this problem will be becoming less and less acute.

One of the recent successful software implementations of allowing for flexible combination of various types of models from different suppliers (using XML-based open interfaces) and linking to the real-time feeds of the NWP model outputs is the Delft-FEWS (FEWS, flood early warning system) platform of Deltares (Werner, 2008). Currently, this platform is being accepted as the integrating tool for the purpose of operational hydrological forecasting and warning in a number of European countries and in USA. The other two widely used modeling systems (albeit less open in the software sense) that are also able to integrate meteorological inputs are (1) the MIKE FLOOD by DHI Water and Environment, based on the hydraulic/hydrologic modeling system MIKE 11 and (2) FloodWorks by Wallingford Software.

2.16.8.2 Integration of Physically Based and Data-Driven Models

2.16.8.2.1 Error prediction models

Consider a model simulating or predicting certain water-related variable (referred to as a primary model). This model's outputs are compared to the recorded data and the errors are calculated. Another model, a data-driven model, is trained on the recorded errors of the primary model, and can be used to correct errors of the primary model. In the context of river modeling, this primary model would be typically a physically based model, but can be a data-driven model as well.

Such an approach was employed in a number of studies. Shamseldin and O'Connor (2001) used ANNs to update runoff forecasts: the simulated flows from a model and the current and previously observed flows were used as input, and the corresponding observed flow as the target output. Updates of daily flow forecasts for a lead time of up to 4 days were made, and the ANN models gave more accurate improvements than autoregressive models. Lekkas et al. (2001) showed that error prediction improves real-time flow forecasting, especially when the forecasting model is poor. Abebe and Price (2004) used ANN to correct the errors of a routing model of the River Wye in UK. Solomatine et al. (2006) built an ANN-based rainfall–runoff model whose outputs were corrected by an IBL model.

2.16.8.2.2 Integration of hydrological knowledge into DDM

An expert can contribute to building a DDM by bringing in the knowledge about the expected relationships between the system variables, in performing advanced correlation and mutual information analysis to select the most relevant variables, determining the model structure based on hydrological knowledge (allowed, e.g., by the M5flex algorithm by Solomatine and Siek (2004)), and in deciding what data should be used and how it should be structured (as it is done by most modelers).

It is possible to mention a number of studies where an attempt is made to include a human expert in the process of building a data-driven model. For solving a flow forecasting problem, See and Openshaw (2000) built not a single overall ANN model but different models for different classes of hydrological events. Solomatine and Xue (2004) introduced a human expert to determine a set of rules to identify various hydrological conditions for each of which a separate specialized data-driven model (ANN or M5 tree) was built. Jain and Srinivasulu (2006) and Corzo and Solomatine (2007) also applied decomposition of the flow hydrograph by a threshold value and then built the separate ANNs for low and high flow regimes. In addition, Corzo and Solomatine (2007) were building two separate models related to base and excess flow which were identified by the Ekhardt's (2005) method, and used overall optimization of the resulting model structure. All these studies demonstrated the higher accuracy of the resulting models where the hydrological knowledge and, wherever possible, models were directly used in building data-driven models.

2.16.9 Future Issues in Hydrological Modeling

Natural and anthropogenic changes constantly impact the environment surrounding us. Available moisture and energy change due to variability and shifts in climate, and the separation of precipitation into different pathways on the land surface are altered due to wildfires, beetle infestations, urbanization, deforestation, invasive plant species, etc. Many of these changes can have a significant impact on the hydrological regime of the watershed in which they occur (e.g., Milly et al., 2005; Poff et al., 2006; Oki and Kanae, 2006; Weiskel et al., 2007). Such changes to water pathways, storage, and subsequent release (the blue and green water idea of Falkenmark and Rockstroem (2004)) are predicted to have significant negative impacts on water security for large population groups as well as for ecosystems in many regions of the world (e.g., Sachs, 2004). The growing imbalances among freshwater supply, its consumption, and human population will only increase the problem (Vörösmarty et al., 2000). A major task for hydrologic science lies in providing predictive models based on sound scientific theory to support water resource management decisions for different possible future environmental, population, and institutional scenarios.

But can we provide credible predictions of yet unobserved hydrological responses of natural systems? Credible modeling of environmental change impact requires that we demonstrate a significant correlation between model parameters and watershed characteristics, since calibration data are, by definition, unavailable. Currently, such *a priori* or regionalized parameters estimates are not very accurate and will likely lead to very uncertain prior distributions for model parameters in changed watersheds, leading to very uncertain predictions. Much work is to be done to solve this and to provide the hydrological simulations with the credibility necessary to support sustainable management of water resources in a changing world.

The issue of model validation has to be given much more attention. Even if calibration and validation data are available, the historical practice of validating the model based on calculation of the Nash–Sutcliffe coefficient or some other squared error measure outside the calibration period is inadequate. Often low or high values of these criteria cannot clearly indicate whether or not the model under question has descriptive or predictive power. The discussion on validation has to move on to use more informative signatures of model behavior, which allow for the detection of how consistent the model is with system at hand (Gupta et al., 2008). This is particularly crucial when it comes to the assessment of climate and land-use change impacts, that is, when future predictions will lie outside the range of observed variability of the system response.

Another development is expected with respect to modeling technologies, mainly in the more effective merging of data into models. One of the aspects here is the optimal use of data for model calibration and evaluation. In this respect, more rigorous approach adopted in DDM (e.g., use of cross-validation and optimal data splitting) could be useful. Modern technology allows for accurate measurements of hydraulic and hydrologic parameters, and for more and more accurate precipitation forecasts coming from NWP models. Many of these come in real time, and this permits for a wider use of data-driven models with their combination with the physically based models, and for wider use of updating and data assimilation schemes. With more data being collected and constantly increasing processing power, one may also expect a wider use of distributed models. It is expected that the way the modeling results are delivered to the decision makers and public will also undergo changes. Half of the global population already owns mobile phones with powerful operating systems, many of which are connected to wide-area networks, so the Information and communication technology (ICT) for the quick dissemination of modeling results, for example, in the form of the flood alerts, is already in place. Hydrological models will be becoming more and more integrated into hydroinformatics systems that support full information cycle, from data gathering to the interpretation and use of modeling results by decision makers and the public.

References

Abbott MB, Bathurst JC, Cunge JA, O'Connell PE, and Rasmussen J (1986) An introduction to the European Hydrological System – Systeme Hydrologique Europeen, SHE. 1. History and philosophy of a physically-based, distributed modelling system. *Journal of Hydrology* 87: 45–59.

Abdulla FA and Lettenmaier DP (1997) Development of regional parameter estimation equations for a macroscale hydrologic model. *Journal of Hydrology* 197: 230–257.

Abebe AJ, Guinot V, and Solomatine DP (2000a) Fuzzy alpha-cut vs. Monte Carlo techniques in assessing uncertainty in model parameters. In: *Proceedings of the 4th International Conference on Hydroinformatics.* Cedar Rapids, IA, USA.

Abebe AJ and Price RK (2004) Information theory and neural networks for managing uncertainty in flood routing. *ASCE: Journal of Computing in Civil Engineering* 18(4): 373–380.

Abebe AJ, Solomatine DP, and Venneker R (2000b) Application of adaptive fuzzy rule-based models for reconstruction of missing precipitation events. *Hydrological Sciences Journal* 45(3): 425–436.

Abrahart B, See LM, and Solomatine DP (eds.) (2008) *Hydroinformatics in Practice: Computational Intelligence and Technological Developments in Water Applications.* Heidelberg: Springer.

Abrahart RJ, Heppenstall AJ, and See LM (2007a) Timing error correction procedure applied to neural network rainfall–runoff modelling. *Hydrological Sciences Journal* 52(3): 414–431.

Abrahart RJ and See L (2000) Comparing neural network and autoregressive moving average techniques for the provision of continuous river flow forecast in two contrasting catchments. *Hydrological Processes* 14: 2157–2172.

Abrahart RJ and See LM (2007) Neural network modelling of non-linear hydrological relationships. *Hydrology and Earth System Sciences* 11: 1563–1579.

Abrahart RJ, See LM, Solomatine DP, and Toth E (eds.) (2007b) Data-driven approaches, optimization and model integration: Hydrological applications. *Special Issue: Hydrology and Earth System Sciences* 11: 1563–1579.

Ajami NK, Duan Q, and Sorooshian S (2007) An integrated hydrologic Bayesian multimodel combination framework: Confronting input, parameter, and model structural uncertainty in hydrologic prediction. *Water Resources Research* 43: W01403 (doi:10.1029/2005WR004745).

Atkinson SE, Woods RA, and Sivapalan M (2002) Climate and landscape controls on water balance model complexity over changing timescales. *Water Resources Research* 38(12): 101029/2002WR001487.

Babovic V and Keijzer M (2005) Rainfall runoff modelling based on genetic programming. In: Andersen MG (ed.) *Encyclopedia of Hydrological Sciences*, vol. 1, ch. 21. New York: Wiley.

Bárdossy A and Duckstein L (1995) *Fuzzy Rule-Based Modeling with Applications to Geophysical, Biological and Engineering Systems.* Boca Raton, FL: CRC Press.

Bartholmes J and Todini E (2005) Coupling meteorological and hydrological models for flood forecasting. *Hydrology and Earth System Sciences* 9: 55–68.

Bates PD and De Roo APJ (2000) A simple raster-based model for floodplain inundation. *Journal of Hydrology* 236: 54–77.

Becker A and Kundzewicz ZW (1987) Nonlinear flood routing with multilinear models. *Water Resources Research* 23: 1043–1048.

Bergström S (1976) Development and application of a conceptual runoff model for Scandinavian catchments. *SMHI Reports RHO, No. 7.* Norrköping, Sweden: SMHI.

Beven K (1989) Changing ideas in hydrology: The case of physically-based models. *Journal of Hydrology* 105: 157–172.

Beven KJ (1996) A discussion on distributed modelling. In: Reefsgard CJ and Abbott MB (eds.) *Distributed Hydrological Modelling*, pp. 255–278. Kluwer: Dodrecht.

Beven K (2000) *Rainfall–Runoff Modeling – the Primer.* Chichester: Wiley.

Beven KJ (2001) Uniqueness of place and process representations in hydrological modeling. *Hydrology and Earth System Sciences* 4: 203–213.

Beven KJ (2002) Towards an alternative blueprint for a physically based digitally simulated hydrologic response modelling system. *Hydrological Processes* 16: 189–206.

Beven KJ (2009) *Environmental Modeling: An Uncertain Future?* London: Routledge.

Beven K and Binley AM (1992) The future role of distributed models: Model calibration and predictive uncertainty. *Hydrological Processes* 6: 279–298.

Bhattacharya B and Solomatine DP (2005) Neural networks and M5 model trees in modelling water level – discharge relationship. *Neurocomputing* 63: 381–396.

Bowden GJ, Dandy GC, and Maier HR (2003) Data transformation for neural network models in water resources applications. *Journal of Hydroinformatics* 5: 245–258.

Bowden GJ, Dandy GC, and Maier HR (2005) Input determination for neural network models in water resources applications. Part 1 – background and methodology. *Journal of Hydrology* 301: 75–92.

Bowden GJ, Maier HR, and Dandy GC (2002) Optimal division of data for neural network models in water resources applications. *Water Resources Research* 38(2): 1–11.

Bray M and Han D (2004) Identification of support vector machines for runoff modelling. *Journal of Hydroinformatics* 6: 265–280.

Burnash RJC (1995) The NWS river forecast system – catchment model. In: Singh VP (ed.) *Computer Models of Watershed Hydrology*, pp. 311–366. Highlands Ranch, CO: Water Resources Publications.

Butts MB, Payne JT, Kristensen M, and Madsen H (2004) An evaluation of the impact of model structure on hydrological modelling uncertainty for streamflow simulation. *Journal of Hydrology* 298: 222–241.

Calder IR (1990) *Evaporation in the Uplands.* Chichester: Wiley.

Calver A (1988) Calibration, validation and sensitivity of a physically-based rainfall–runoff model. *Journal of Hydrology* 103: 103–115.

Cannon AJ (2008) Probabilistic multisite precipitation downscaling by an expanded Bernoulli-gamma density network. *Journal of Hydrometeorology* 9(6): 1284–1300.

Casper MPO, Gemmar M, Gronz J, and Stüber M (2007) Fuzzy logic-based rainfall-runoff modelling using soil moisture measurements to represent system state. *Hydrological Sciences Journal (Special Issue: Hydroinformatics)* 52(3): 478–490.

Chow V, Maidment D, and Mays L (1988) *Applied Hydrology.* New York, NY: McGraw-Hill.

Clarke R (1973) A review of some mathematical models used in hydrology, with observations on their calibration and use. *Journal of Hydrology* 19(1): 1–20.

Corzo G and Solomatine DP (2007) Baseflow separation techniques for modular artificial neural network modelling in flow forecasting. *Hydrological Sciences Journal* 52(3): 491–507.

Coulibaly P, Anctil F, and Bobbe B (2000) Daily reservoir inflow forecasting using artificial neural networks with stopped training approach. *Journal of Hydrology* 230: 244–257.

Dawdy DR and O'Donnell T (1965) Mathematical models of catchment behavior. *Journal of the Hydraulics Division, American Society of Civil Engineers* 91: 113–137.

Dawson CW and Wilby R (1998) An artificial neural network approach to rainfall–runoff modelling. *Hydrological Sciences Journal* 43(1): 47–66.

Dibike Y, Solomatine DP, and Abbott MB (1999) On the encapsulation of numerical-hydraulic models in artificial neural network. *Journal of Hydraulic Research* 37(2): 147–161.

Dibike YB, Velickov S, Solomatine DP, and Abbott MB (2001) Model induction with support vector machines: Introduction and applications. *ASCE Journal of Computing in Civil Engineering* 15(3): 208–216.

Dooge JCI (1957) The rational method for estimating flood peaks. *Engineering* 184: 311–313.

Dooge JCI (1959) A general theory of the unit hydrograph. *Journal of Geophysical Research* 64: 241–256.

Dunn SM and Ferrier RC (1999) Natural flow in managed catchments: A case study of a modelling approach. *Water Research* 33: 621–630.

Duan Q, Gupta VK, and Sorooshian S (1992) Effective and efficient global optimization for conceptual rainfall–runoff models. *Water Resources Research* 28: 1015–1031.

Duan Q, Schaake J, and Koren V (2001) *A priori* estimation of land surface model parameters. In: Lakshmi V, Albertson J, and Schaake J (eds.) *Land Surface Hydrology, Meteorology, and Climate: Observations and Modelling.* American Geophysical Union, Water Science and Application, vol. 3, pp. 77–94.

Efron B and Tibshirani RJ (1993) *An Introduction to the Bootstrap.* New York: Chapman and Hall.

Ekhardt K (2005) How to construct recursive digital filters for baseflow separation. *Hydrological Processes* 19: 507–515.

Ewen J and Parkin G (1996) Validation of catchment models for predicting land-use and climate change impacts. 1: Method. *Journal of Hydrology* 175: 583–594.

Falkenmark M and Rockstroem J (2004) *Balancing Water for Humans and Nature. The New Approach in Ecohydrology.* (ISBN 1853839272).. Sterling, VA: Stylus Publishing.

Fernandez W, Vogel RM, and Sankarasubramanian S (2000) Regional calibration of a watershed model. *Hydrological Sciences Journal* 45(5): 689–707.

Freeze RA and Harlan RL (1969) Blueprint for a physically-based, digitally-simulated hydrologic response model. *Journal of Hydrology* 9: 237–258.

Galeati G (1990) A comparison of parametric and non-parametric methods for runoff forecasting. *Hydrological Sciences Journal* 35(1): 79–94.

Gaume E and Gosset R (2003) Over-parameterisation, a major obstacle to the use of artificial neural networks in hydrology? *Hydrology and Earth System Sciences* 7(5): 693–706.

Geman S, Bienenstock E, and Doursat R (1992) Neural networks and the bias/variance dilemma. *Neural Computation* 4(1): 1–58.

Georgakakos KP, Seo D-J, Gupta H, Schaake J, and Butts MB (2004) Towards the characterization of streamflow simulation uncertainty through multimodel ensembles. *Journal of Hydrology* 298: 222–241.

Giustolisi O, Doglioni A, Savic DA, and di Pierro F (2007a) An evolutionary multiobjective strategy for the effective management of groundwater resources. *Water Resources Research* 44: W01403.

Giustolisi O, Doglioni A, Savic DA, and Webb BW (2007b) A multi-model approach to analysis of environmental phenomena. *Environmental Modelling Systems Journal* 22(5): 674–682.

Giustolisi O and Savic DA (2006) A symbolic data-driven technique based on evolutionary polynomial regression. *Journal of Hydroinformatics* 8(3): 207–222.

Gouldby B and Samuels P (2005) Language of risk. Project definitions. FLOODsite Consortium Report T32-04-01. http://www.floodsite.net (accessed April 2010)

Govindaraju RS and Rao RA (eds.) (2001) *Artificial Neural Networks in Hydrology*. Dordrecht: Kluwer.

Gupta HV, Beven KJ, and Wagener T (2005) Model calibration and uncertainty estimation. In: Andersen M (ed.) *Encyclopedia of Hydrological Sciences*, pp. 2015–2031. New York, NY: Wiley.

Gupta HV, Kling H, Yilmaz KK, and Martinez GF (2009) Decomposition of the mean squared error and NSE performance criteria: Implications for improving hydrological modelling. *Journal of Hydrology* 377: 80–91.

Gupta HV, Sorooshian S, and Yapo PO (1998) Toward improved calibration of hydrological models: Multiple and noncommensurable measures of information. *Water Resources Research* 34(4): 751–763.

Gupta HV, Wagener T, and Liu Y (2008) Reconciling theory with observations: Elements of a diagnostic approach to model evaluation. *Hydrological Processes* 22(18): 3802–3813 (doi:10.1002/hyp.6989).

Hall MJ and Minns AW (1999) The classification of hydrologically homogeneous regions. *Hydrological Sciences Journal* 44: 693–704.

Han D, Kwong T, and Li S (2007) Uncertainties in real-time flood forecasting with neural networks. *Hydrological Processes* 21(2): 223–228.

Hanna SR (1988) Air quality model evaluation and uncertainty. *Journal of the Air Pollution Control Association* 38: 406–442.

Hannah DM, Smith BPG, Gurnell AM, and McGregor GR (2000) An approach to hydrograph classification. *Hydrological Processes* 14: 317–338.

Harlin J (1991) Development of a process oriented calibration scheme for the HBV hydrological model. *Nordic Hydrology* 22: 15–36.

Harris NM, Gurnell AM, Hannah DM, and Petts GE (2000) Classification of river regimes: A context for hydrogeology. *Hydrological Processes* 14: 2831–2848.

Hastings WK (1970) Monte Carlo sampling methods using Markov chains and their applications. *Biometrika* 57(1): 97–109.

Haykin S (1999) *Neural Networks: A Comprehensive Foundation*. New York: McMillan.

Hsu KL, Gupta HV, and Sorooshian S (1995) Artificial neural network modelling of the rainfall–runoff process. *Water Resources Research* 31(10): 2517–2530.

Hu T, Wu F, and Zhang X (2007) Rainfall–runoff modeling using principal component analysis and neural network. *Nordic Hydrology* 38(3): 235–248.

Ibbitt RP (1970) Systematic parameter fitting for conceptual models of catchment hydrology. PhD dissertation, Imperial College of Science and Technology, University of London.

Ibbitt RP and O'Donnell T (1971) Fitting methods for conceptual catchment models. *Journal of the Hydraulics Division, American Society of Civil Engineers* 97(9): 1331–1342.

Jain A and Srinivasulu S (2006) Integrated approach to model decomposed flow hydrograph using artificial neural network and conceptual techniques. *Journal of Hydrology* 317: 291–306.

Jakeman AJ, Hornberger GM, Littlewood IG, Whitehead PG, Harvey JW, and Bencala KE (1992) A systematic approach to modelling the dynamic linkage of climate, physical catchment descriptors and hydrological response components. *Mathematics and Computers in Simulation* 33: 359–366.

Johnston PR and Pilgrim DH (1976) Parameter optimization for watershed models. *Water Resources Research* 12(3): 477–486.

Karlsson M and Yakowitz S (1987) Nearest neighbour methods for non-parametric rainfall runoff forecasting. *Water Resources Research* 23(7): 1300–1308.

Kavetski D, Kuczera G, and Franks SW (2006) Bayesian analysis of input uncertainty in hydrological modeling: 1. Theory. *Water Resources Research* 42: W03407 (doi:10.1029/2005WR004368).

Khu ST and Madsen H (2005) Multiobjective calibration with Pareto preference ordering: An application to rainfall–runoff model calibration. *Water Resources Research* 41: W03004 (doi:10.1029/2004WR003041).

Khu S-T, Savic D, Liu Y, and Madsen H (2004) A fast evolutionary-based meta-modelling approach for the calibration of a rainfall–runoff model. In: Pahl-Wostl C, Schmidt S, Rizzoli AE, and Jakeman AJ (eds.) *Transactions of the 2nd Biennial Meeting of the International Environmental Modelling and Software Society*, vol. 1, pp. 141–146. Manno, Switzerland: iEMSs.

Kitanidis PK and Bras RL (1980) Real-time forecasting with a conceptual hydrologic model: Analysis of uncertainty. *Water Resources Research* 16(6): 1025–1033.

Klir G and Folger T (1988) *Fuzzy Sets, Uncertainty, and Information*. Englewood Cliffs, NJ: Prentice Hall.

Kollet S and Maxwell R (2006) Integrated surface–groundwater flow modeling: A free-surface overland flow boundary condition in a parallel groundwater flow model. *Advances in Water Resources* 29: 945–958.

Kollet S and Maxwell R (2008) Demonstrating fractal scaling of baseflow residence time distributions using a fully-coupled groundwater and land surface model. *Geophysical Research Letters* 35(7): L07402 (doi:10.1029/2008GL033215).

Kompare B, Steinman F, Cerar U, and Dzeroski S (1997) Prediction of rainfall runoff from catchment by intelligent data analysis with machine learning tools within the artificial intelligence tools. *Acta Hydrotechnica* 16/17: 79–94 (in Slovene).

Koren VI, Smith M, Wang D, and Zhang Z (2000) Use of soil property data in the derivation of conceptual rainfall–runoff model parameters. *15th Conference on Hydrology, Long Beach*, American Meteorological Society, USA, Paper 2.16.

Kuczera G and Parent E (1998) Monte Carlo assessment of parameter uncertainty in conceptual catchment models: The Metropolis algorithm. *Journal of Hydrology* 211: 69–85.

Kundzewicz Z (1995) Hydrological uncertainty in perspective. In: Kundzewicz Z (ed.) *New Uncertainty Concepts in Hydrology and Water Resources*, pp. 3–10. Cambridge: Cambridge University Press.

Lamb R and Calver A (2002) Continuous simulation as a basis for national flood frequency estimation. In: Littlewood I (ed.) *Continuous River Flow Simulation: Methods, Applications and Uncertainties*, pp. 67–75. BHS–Occasional Papers, No. 13, Wallingford.

Lamb R, Crewett J, and Calver A (2000) Relating hydrological model parameters and catchment properties to estimate flood frequencies from simulated river flows. BHS 7th National Hydrology Symposium, Newcastle-upon-Tyne, vol. 7, pp. 357–364.

Lamb R and Kay AL (2004) Confidence intervals for a spatially generalized, continuous simulation flood frequency model for Great Britain. *Water Resources Research* 40: W07501 (doi:10.1029/2003WR002428).

Laucelli D, Giustolisi O, Babovic V, and Keijzer M (2007) Ensemble modeling approach for rainfall/groundwater balancing. *Journal of Hydroinformatics* 9(2): 95–106.

LeBaron B and Weigend AS (1994) Evaluating neural network predictors by bootstrapping. In: *Proceedings of the International Conference on Neural Information Processing (ICONIP'94)*. Seoul, Korea.

Lekkas DF, Imrie CE, and Lees MJ (2001) Improved non-linear transfer function and neural network methods of flow routing for real-time forecasting. *Journal of Hydroinformatics* 3(3): 153–164.

Letcher RA, Jakeman AJ, and Croke BFW (2004) Model development for integrated assessment of water allocation options. *Water Resources Research* 40: W05502.

Lindström G, Johansson B, Persson M, Gardelin M, and Bergström S (1997) Development and test of the distributed HBV-96 hydrological model. *Journal of Hydrology* 201: 272–228.

Liong SY and Sivapragasam C (2002) Flood stage forecasting with SVM. *Journal of American Water Resources Association* 38(1): 173–186.

Liu Y and Gupta HV (2007) Uncertainty in hydrologic modeling: Toward an integrated data assimilation framework. *Water Resources Research* 43: W07401 (doi:10.1029/2006WR005756).

Loague KM and Freeze RA (1985) A comparison of rainfall–runoff techniques on small upland catchments. *Water Resources Research* 21: 229–240.

Lobbrecht AH and Solomatine DP (1999) Control of water levels in polder areas using neural networks and fuzzy adaptive systems. In: Savic D and Walters G (eds.) *Water Industry Systems: Modelling and Optimization Applications*, pp. 509–518. Baldock: Research Studies Press.

Madsen H and Jacobsen T (2001) Automatic calibration of the MIKE SHE integrated hydrological modelling system. *4th DHI Software Conference*, Helsingør, Denmark, 6–8 June 2001.

McDonnell JJ (2003) HYPERLINK "http://www.cof.orst.edu/cof/fe/watershd/pdf/2003/McDonnellHP2003.pdf" Where does water go when it rains? Moving beyond the variable source area concept of rainfall–runoff response. *Hydrological Processes* 17: 1869–1875.

Maskey S, Guinot V, and Price RK (2004) Treatment of precipitation uncertainty in rainfall–runoff modelling: A fuzzy set approach. *Advances in Water Resources* 27(9): 889–898.

McKay MD, Beckman RJ, and Conover WJ (1979) A comparison of three methods for selecting values of input variables in the analysis of output from a computer code. *Technometrics* 21(2): 239–245.

Melching CS (1992) An improved first-order reliability approach for assessing uncertainties in hydrological modelling. *Journal of Hydrology* 132: 157–177.

Melching CS (1995) Reliability estimation. In: Singh VP (ed.) *Computer Models of Watershed Hydrology*, pp. 69–118. Highlands Ranch, CO: Water Resources Publications.

Merz B and Blöschl G (2004) Regionalisation of catchment model parameters. *Journal of Hydrology* 287(1–4): 95–123.

Metropolis N, Rosenbluth AW, Rosenbluth MN, and Teller AH (1953) Equations of state calculations by fast computing machines. *Journal of Chemical Physics* 26(6): 1087–1092.

Milly PCD, Dunne KA, and Vecchia AV (2005) Global patterns of trends in streamflow and water availability in a changing climate. *Nature* 438: 347–350.

Minns AW and Hall MJ (1996) Artificial neural network as rainfall–runoff model. *Hydrological Sciences Journal* 41(3): 399–417.

Mitchell TM (1997) *Machine Learning*. New York: McGraw-Hill.

Montanari A (2007) What do we mean by uncertainty? The need for a consistent wording about uncertainty assessment in hydrology. *Hydrological Processes* 21(6): 841–845.

Montanari A and Brath A (2004) A stochastic approach for assessing the uncertainty of rainfall–runoff simulations. *Water Resources Research* 40: W01106 (doi:10.1029/2003WR002540).

Moradkhani H, Hsu KL, Gupta HV, and Sorooshian S (2004) Improved streamflow forecasting using self-organizing radial basis function artificial neural networks. *Journal of Hydrology* 295(1): 246–262.

Moradkhani H and Sorooshian S (2008) General review of rainfall–runoff modeling: Model calibration, data assimilation, and uncertainty analysis. In: Sorooshian S, Hsu K, Coppola E, Tomassetti B, Verdecchia M, and Visconti G (eds.) *Hydrological Modelling and the Water Cycle*, pp. 1–24. Heidelberg: Springer.

Mosley MP (1981) Delimitation of New Zealand hydrologic regions. *Journal of Hydrology* 49: 173–192.

Muleta MK and Nicklow JW (2004) Joint application of artificial neural networks and evolutionary algorithms to watershed management. *Journal of Water Resources Management* 18(5): 459–482.

Nayak PC, Sudheer KP, and Ramasastri KS (2005) Fuzzy computing based rainfall–runoff model for real time flood forecasting. *Hydrological Processes* 19: 955–968.

Neuman SP (2003) Maximum likelihood Bayesian averaging of uncertain model predictions. *Stochastic Environmental Research and Risk Assessment* 17(5): 291–305.

Nor NA, Harun S, and Kassim AH (2007) Radial basis function modeling of hourly streamflow hydrograph. *Journal of Hydrologic Engineering* 12(1): 113–123.

Oki T and Kanae S (2006) Global hydrological cycles and world water resources. *Science* 313(5790): 1068–1072.

Panday S and Huyakorn PS (2004) A fully coupled physically-based spatially-distributed model for evaluating surface/subsurface flow. *Advances in Water Resources* 27: 361–382.

Pappenberger F and Beven KJ (2006) Ignorance is bliss: Or seven reasons not to use uncertainty analysis. *Water Resources Research* 42: W05302 (doi:10.1029/2005WR004820).

Pappenberger F, Harvey H, Beven K, Hall J, and Meadowcroft I (2006) Decision tree for choosing an uncertainty analysis methodology: A wiki experiment. *Hydrological Processes* 20: 3793–3798.

Parkin G, O'Donnell G, Ewen J, Bathurst JC, O'Connell PE, and Lavabre J (1996) Validation of catchment models for predicting land-use and climate change impacts. 1. Case study for a Mediterranean catchment. *Journal of Hydrology* 175: 595–613.

Pesti G, Shrestha BP, Duckstein L, and Bogárdi I (1996) A fuzzy rule-based approach to drought assessment. *Water Resources Research* 32(6): 1741–1747.

Poff NL, Bledsoe BP, and Cuhaciyan CO (2006) Hydrologic variation with land use across the contiguous United States: Geomorphic and ecological consequences for stream ecosystems. *Geomorphology* 79: 264–285.

Post DA and Jakeman AJ (1996) Relationships between physical attributes and hydrologic response characteristics in small Australian mountain ash catchments. *Hydrological Processes* 10: 877–892.

Post DA, Jones JA, and Grant GE (1998) An improved methodology for predicting the daily hydrologic response of ungauged catchments. *Environmental Modeling and Sofware* 13: 395–403.

Pyle D (1999) *Data Preparation for Data Mining*. San Francisco, CA: Morgan Kaufmann.

Qu Y and Duffy CJ (2007) A semi-discrete finite-volume formulation for multi-process watershed simulation. *Water Resources Research* 43: W08419 (doi:10.1029/2006WR005752).

Quinlan JR (1992) Learning with continuous classes. In: Adams N and Sterling L (eds.) *AI'92 Proceedings of the 5th Australian Joint Conference on Artificial Intelligence*, pp. 343–348. World Scientific: Singapore.

Reed S, Koren V, Smith M, *et al*. (2004) Overall distributed model intercomparison project results. *Journal of Hydrology* 298(1–4): 27–60.

Refsgaard JC (1996) Terminology, modelling protocol and classification of hydrological model codes. In: Abbott MB and Refsgaard JC (eds.) *Distributed Hydrological Modelling*, pp. 17–39. Dordrecht: Kluwer.

Refsgaard JC (1997) Validation and intercomparison of different updating procedures for real-time forecasting. *Nordic Hydrology* 28(2): 65–84.

Refsgaard JC and Abbott MB (1996) The role of distributed hydrological modeling in water resources management. In: *Distributed Hydrologic Modeling*, chap. 1. Dordrecht, The Netherlands: Kluwer Academic Publishers.

Refsgaard JC and Knudsen J (1996) Operational validation and intercomparison of different types of hydrological models. *Water Resources Research* 32(7): 2189–2202.

Reggiani P, Hassanizadeh EM, Sivapalan M, and Gray WG (1999) A unifying framework for watershed thermodynamics: Constitutive relationships. *Advances in Water Resources* 23(1): 15–39.

Reggiani P, Sivapalan M, Hassanizadeh M, and Gray WG (2001) Coupled equations for mass and momentum balance in a stream network: Theoretical derivation and computational experiments. *Proceedings of the Royal Society of London. Series A. Mathematical, Physical and Engineering Sciences* 457: 157–189.

Reggiani P, Sivapalan M, and Hassanizadeh SM (1998) A unifying framework for watershed thermodynamics: Balance equations for mass, momentum, energy and entropy, and the second law of thermodynamics. *Advances in Water Resources* 22(4): 367–398.

Reggiani P, Sivapalan M, and Hassanizadeh SM (2000) Conservation equations governing hillslope responses: Exploring the physical basis of water balance. *Water Resources Research* 36(7): 1845–1863.

Sachs JS (2004) Sustainable development. *Science* 304(5671): 649 (doi: 10.1126/science.304.5671.649).

Salathe EP (2005) Downscaling simulations of future global climate with application to hydrologic modelling. *International Journal of Climatology* 25: 419–436.

See LA, Solomatine DP, Abrahart R, and Toth E (2007) Hydroinformatics: Computational intelligence and technological developments in water science applications – editorial. *Hydrological Sciences Journal* 52(3): 391–396.

See LM and Openshaw S (2000) A hybrid multi-model approach to river level forecasting. *Hydrological Sciences Journal* 45(4): 523–536.

Sefton CEM and Howarth SM (1998) Relationships between dynamic response characteristics and physical descriptors of catchments in England and Wales. *Journal of Hydrology* 211: 1–16.

Sefton CEM, Whitehead PG, Eatherall A, Littlewood IG, and Jakeman AJ (1995) Dynamic response characteristics of the Plynlimon catchments and preliminary analysis of relationships to physical catchment descriptors. *Environmetrics* 6: 465–472.

Seibert J (1999) Regionalisation of parameters for a conceptual rainfall–runoff model. *Agricultural and Forest Meteorology* 98–99: 279–293.

Shafer G (1976) *Mathematical Theory of Evidence*. Princeton, NJ: Princeton University Press.

Shamseldin AY and O'Connor KM (2001) A non-linear neural network technique for updating of river flow forecasts. *Hydrology and Earth System Sciences* 5(4): 557–597.

Shannon CE (1948) A mathematical theory of communication. *Bell System Technical Journal* 27: 379–423. and 623–656.

Sharma A (2000) Seasonal to interannual rainfall probabilistic forecasts for improved water supply management: Part 3 – a nonparametric probabilistic forecast model. *Journal of Hydrology* 239(1–4): 249–258.

Shrestha DL and Solomatine D (2008) Data-driven approaches for estimating uncertainty in rainfall–runoff modelling. *International Journal of River Basin Management* 6(2): 109–122.

Singh VP (1995b) Watershed modeling *Computer Models of Watershed Hydrology*, pp. 1–22. Highlands Ranch, CO: Water Resources Publication.

Solomatine D, Dibike YB, and Kukuric N (1999) Automatic calibration of groundwater models using global optimization techniques. *Hydrological Sciences Journal* 44(6): 879–894.

Solomatine D and Shrestha DL (2009) A novel method to estimate total model uncertainty using machine learning techniques. *Water Resources Research* 45: W00B11 (doi:10.1029/2008WR006839).

Solomatine DP (1995) The use of global random search methods for models calibration *Proceedings of the 26th Congress of the International Association for Hydraulic Research*, vol. 1, pp. 224–229. London: Thomas Telford.

Solomatine DP (2006) Optimal modularization of learning models in forecasting environmental variables. In: Voinov A, Jakeman A, and Rizzoli A (eds.) *Proceedings of the iEMSs 3rd Biennial Meeting: "Summit on Environmental Modelling and*

Software". Burlington, VT, USA, July 2006. Manno, Switzerland: International Environmental Modelling and Software Society.

Solomatine DP and Dulal KN (2003) Model tree as an alternative to neural network in rainfall–runoff modelling. *Hydrological Sciences Journal* 48(3): 399–411.

Solomatine DP, Maskey M, and Shrestha DL (2006) Eager and lazy learning methods in the context of hydrologic forecasting *Proceedings of the International Joint Conference on Neural Networks.* Vancouver, BC: IEEE.

Solomatine DP, Shrestha DL, and Maskey M (2007) Instance-based learning compared to other data-driven methods in hydrological forecasting. *Hydrological Processes* 22(2): 275–287 (doi:10.1002/hyp.6592).

Solomatine DP and Siek MB (2004) Flexible and optimal M5 model trees with applications to flow predictions In: Liong, Phoon, and Babovic (eds.) *Proceedings of the 6th International Conference on Hydroinformatics.* Singapore: World Scientific.

Solomatine DP and Torres LA (1996) Neural network approximation of a hydrodynamic model in optimizing reservoir operation. In: Muller A (ed.) *Proceedings of the 2nd International Conference on Hydroinformatics*, pp. 201–206. Rotterdam: Balkema.

Solomatine DP, Velickov S, and Wust JC (2001) Predicting water levels and currents in the North Sea using chaos theory and neural networks. In: Li, Wang, Pettijean, and Fisher (eds.) *Proceedings of the 29th IAHR Congress.* Beijing, China.

Solomatine DP and Xue Y (2004) M5 model trees and neural networks: Application to flood forecasting in the upper reach of the Huai River in China. *ASCE Journal of Hydrologic Engineering* 9(6): 491–501.

Sorooshioan S, Duan Q, and Gupta VK (1993) Calibration of rainfall–runoff models: Application of global optimization to the Sacramento Soil Moisture Accounting Model. *Water Resources Research* 29: 1185–1194.

Spear RC (1998) Large simulation models: Calibration, uniqueness, and goodness of fit. *Environmental Modeling and Sofware* 12: 219–228.

Stravs L and Brilly M (2007) Development of a low-flow forecasting model using the M5 machine learning method. *Hydrological Sciences–Journal–des Sciences Hydrologiques (Special Issue: Hydroinformatics)* 52(3): 466–477.

Sudheer KP and Jain SK (2003) Radial basis function neural network for modeling rating curves. *ASCE Journal of Hydrologic Engineering* 8(3): 161–164.

Sugawara M (1995) Tank model. In: Singh VP (ed.) *Computer Models of Watershed Hydrology*, pp. 165–214. Highlands Ranch, CO: Water Resources Publication.

Tang Y, Reed P, Van Werkhoven K, and Wagener T (2007) Advancing the identification and evaluation of distributed rainfall–runoff models using global sensitivity analysis. *Water Resources Research* 43: W06415.1–W06415.14 (doi:10.1029/2006WR005813).

Tang Y, Reed P, and Wagener T (2006) How effective and efficient are multiobjective evolutionary algorithms at hydrologic model calibration? *Hydrology and Earth System Sciences* 10: 289–307.

Thiemann M, Trosser M, Gupta H, and Sorooshian S (2001) Bayesian recursive parameter estimation for hydrologic models. *Water Resources Research* 37(10): 2521–2536.

Todini E (1988) Rainfall–runoff modeling – past, present and future. *Journal of Hydrology* 100(1–3): 341–352.

Toth E, Brath A, and Montanari A (2000) Comparison of short-term rainfall prediction models for real-time flood forecasting. *Journal of Hydrology* 239: 132–147.

Tung Y-K (1996) Uncertainty and reliability analysis. In: Mays LW (ed.) *Water Resources Handbook*, pp. 7.1–7.65. New York, NY: McGraw-Hill.

Uhlenbrook S and Leibundgut Ch (2002) Process-oriented catchment modeling and multiple-response validation. *Hydrological Processes* 16: 423–440.

Uhlenbrook S, Roser S, and Tilch N (2004) Hydrological process representation at the meso-scale: The potential of a distributed, conceptual catchment model. *Journal of Hydrology* 291(3–4): 278–296.

Vandewiele GL and Elias A (1995) Monthly water balance of ungauged catchments obtained by geographical regionalization. *Journal of Hydrology* 170(1–4,8): 277–291.

Van Werkhoven K, Wagener T, Tang Y, and Reed P (2008) Understanding watershed model behavior across hydro-climatic gradients using global sensitivity analysis. *Water Resources Research* 44: W01429 (doi:10.1029/2007WR006271).

Vapnik VN (1998) *Statistical Learning Theory.* New York: Wiley.

Vernieuwe H, Georgieva O, De Baets B, Pauwels VRN, Verhoest NEC, and De Troch FP (2005) Comparison of data-driven Takagi–Sugeno models of rainfall-discharge dynamics. *Journal of Hydrology* 302(1–4): 173–186.

Vörösmarty CJ, Green P, Salisbury J, and Lammers RB (2000) Global water resources: Vulnerability from climate change and population growth. *Science* 289: 284–288 (doi:10.1126/science.289.5477.284).

Vrugt J, ter Braak C, Clark M, Hyman J, and Robinson B (2008a) Treatment of input uncertainty in hydrologic modeling: Doing hydrology backward with Markov chain Monte Carlo simulation. *Water Resources Research* 44: W00B09 (doi:10.1029/2007WR006720).

Vrugt JA, Gupta HV, Bouten W, and Sorooshian S (2003) A shuffled complex evolution metropolis algorithm for optimization and uncertainty assessment of hydrologic model parameters. *Water Resources Research* 39(8): 1201 (doi:10.1029/2002WR001642).

Vrugt JA and Robinson BA (2007) Improved evolutionary optimization from genetically adaptive multimethod search. *Proceedings of the National Academy of Sciences* 104(3): 708–711.

Vrugt JA, ter Braak CJF, Gupta HV, and Robinson BA (2008b) Equifinality of formal (DREAM) and informal (GLUE) Bayesian approaches in hydrologic modeling? *Stochastic Environmental Research and Risk Assessment* 23: 1011–1026 (doi:10.1007/s00477-008-0274-y).

Wagener T and Gupta HV (2005) Model identification for hydrological forecasting under uncertainty. *Stochastic Environmental Research and Risk Assessment* 19: 378–387 (doi:10.1007/s00477-005-0006-5).

Wagener T and McIntyre N (2007) Tools for teaching hydrological and environmental modeling. *Computers in Education Journal* XVII(3): 16–26.

Wagener T, McIntyre N, Lees MJ, Wheater HS, and Gupta HV (2003) Towards reduced uncertainty in conceptual rainfall–runoff modelling: Dynamic identifiability analysis. *Hydrological Processes* 17(2): 455–476.

Wagener T and Wheater HS (2006) Parameter estimation and regionalization for continuous rainfall–runoff models including uncertainty. *Journal of Hydrology* 320(1–2): 132–154.

Wagener T, Wheater HS, and Gupta HV (2004) *Rainfall–Runoff Modelling in Gauged and Ungauged Catchments.* 332pp. London: Imperial College Press.

Weiskel PK, Vogel RM, Steeves PA, DeSimone LA, Zarriello PJ, and Ries KG III (2007) Water-use regimes: Characterizing direct human interaction with hydrologic systems. *Water Resources Research* 43: W04402 (doi:10.1029/2006WR005062).

Wheater HS, Jakeman AJ, and Beven KJ (1993) Progress and directions in rainfall–runoff modelling. In: Jakeman AJ, Beck MB, and McAleer MJ (eds.) *Modelling Change in Environmental Systems*, pp. 101–132. Chichester: Wiley.

Werner M (2008) Open model integration in flood forecasting. In: Abrahart B, See LM, and Solomatine DP (eds.) *Practical Hydroinformatics: Computational Intelligence and Technological Developments in Water Applications*, p. 495. Heidelberg: Springer.

Widen-Nilsson E, Halldin S, and Xu C-Y (2007) Global water-balance modelling with WASMOD-M: Parameter estimation and regionalization. *Journal of Hydrology* 340: 105–118.

Xiong LH, Shamseldin AY, and O'Connor KM (2001) A non-linear combination of the forecasts of rainfall–runoff models by the first-order Takagi–Sugeno fuzzy system. *Journal of Hydrology* 245(1–4): 196–217.

Yadav M, Wagener T, and Gupta HV (2007) Regionalization of constraints on expected watershed response behavior for improved predictions in ungauged basins. *Advances in Water Resources* 30: 1756–1774 (doi:10.1016/j.advwatres.2007.01.005).

Young PC (1992) HYPERLINK "http://eprints.lancs.ac.uk/21104/" Parallel processes in hydrology and water quality: A unified time-series approach. *Water and Environment Journal* 6(6): 598–612.

Young PC (1998) Data-based mechanistic modelling of environmental, ecological, economic and engineering systems. *Environmental Modeling and Sofware* 13: 105–122.

Zadeh LA (1965) Fuzzy sets. *Information and Control* 8(3): 338–353.

Zadeh LA (1978) Fuzzy set as a basis for a theory of possibility. *Fuzzy Sets and Systems* 1(1): 3–28.

Zhang Z, Wagener T, Reed P, and Bushan R (2008) Ensemble streamflow predictions in ungauged basins combining hydrologic indices regionalization and multiobjective optimization. *Water Resources Research* 44: W00B04 (doi:10.1029/2008WR006833).

Zhao RJ and Liu XR (1995) The Xinanjiang model. In: Singh VP, et al. (eds.) *Computer Models of Watershed Hydrology*, pp. 215–232. Littelton, CO: Water Resources Publication.

Zimmermann H-J (1997) A fresh perspective on uncertainty modeling: Uncertainty vs. uncertainty modeling. In: Ayyub BM and Gupta MM (eds.) *Uncertainty Analysis in Engineering and Sciences: Fuzzy Logic, Statistics, and Neural Network Approach*, pp. 353–364. Heidelberg: Springer.

Relevant Websites

http://www.deltares.nl
  Deltares.
http://www.dhigroup.com
  DHI; DHI software.
http://efas.jrc.ec.europa.eu
  European Commission Joint Research Centre.
http://www.sahra.arizona.edu
  SAHRA; Hydroarchive.

2.17 Uncertainty of Hydrological Predictions

A Montanari, University of Bologna, Bologna, Italy

© 2011 Elsevier B.V. All rights reserved.

2.17.1	**Introduction**	459
2.17.2	**Definitions and Terminology**	461
2.17.2.1	Probability	461
2.17.2.2	Randomness	461
2.17.2.3	Random Variable	461
2.17.2.4	Stochastic Process	461
2.17.2.5	Stationarity	461
2.17.2.6	Ergodicity	462
2.17.2.7	Uncertainty	462
2.17.2.8	Global Uncertainty and Individual Uncertainties	462
2.17.2.9	Uncertainty Assessment	462
2.17.2.10	Probabilistic Estimate/Estimation/Assessment of Uncertainty (Probabilistic Uncertainty)	462
2.17.2.11	Nonprobabilistic Estimate/Estimation/Assessment of Uncertainty (Nonprobabilistic Uncertainty)	462
2.17.2.12	Confidence Band	463
2.17.2.13	Equifinality	463
2.17.2.14	Behavioral Model	463
2.17.3	**Classification of Uncertainty and Reasons for the Presence of Uncertainty in Hydrology**	463
2.17.3.1	Inherent Randomness	464
2.17.3.2	Model Structural Uncertainty	464
2.17.3.3	Model Parameter Uncertainty	464
2.17.3.4	Data Uncertainty	465
2.17.3.5	Operation Uncertainty	465
2.17.4	**Uncertainty Assessment**	465
2.17.5	**Classification of Approaches to Uncertainty Assessment**	465
2.17.5.1	Research Questions about Uncertainty in Hydrology	465
2.17.5.2	An Attempt of Classification	466
2.17.6	**Assessment of the Global Uncertainty of the Model Output**	467
2.17.6.1	Analytical Methods	467
2.17.6.2	The Generalized Likelihood Uncertainty Estimation	467
2.17.6.3	The Bayesian Forecasting System	468
2.17.6.4	Techniques Based on the Statistical Analysis of the Model Error	469
2.17.6.5	Bayesian Model Averaging	470
2.17.6.6	Machine Learning Techniques	471
2.17.7	**Assessment of Data Uncertainty**	472
2.17.7.1	Precipitation Uncertainty	472
2.17.7.2	River Discharge Uncertainty	473
2.17.8	**Assessment of Parameter Uncertainty**	474
2.17.8.1	The MOSCEM-UA Method	474
2.17.8.2	The AMALGAM Method	474
2.17.9	**Assessment of Model Structural Uncertainty**	474
2.17.10	**Uncertainty Assessment as a Learning Process**	475
2.17.11	**Conclusions**	475
Acknowledgments		476
References		476

2.17.1 Introduction

Hydrological modeling is receiving increasing attention from researchers and practitioners. The increasing availability of mathematical tools and computing power together with an improved understanding of the dynamics of hydrological processes has favored the continuous development of new modeling approaches in the past few decades (see **Chapter 2.16 Hydrological Modeling**). Hydrological modeling is an attractive option today for solving many practical problems of environmental engineering, flood protection, water resource management, and applied hydrology in general.

Setting up a hydrological model in order to solve a practical problem requires the application of proper procedures of model identification, parameter calibration, hypothesis testing, model testing (also called model validation), and uncertainty assessment. The above procedures are often strictly related and are the subject of an increasing research activity by hydrologists. In particular, uncertainty estimation is very much related with parameter calibration and model validation. It consists of a verification of the hydrological model appropriateness and performances finalized to providing a quantitative assessment of its reliability.

As a matter of fact, uncertainty estimation in hydrological surface and subsurface modeling is today one of the most important subfields of hydrology, according to the numerous contributions in recent scientific literature. Uncertainty reduction is also one of the main goals of the Prediction in Ungauged Basins (PUB) initiative promoted by the International Association of Hydrological Sciences.

While quantitative uncertainty assessment in hydrology is often considered a relatively new topic, it is worth noting that hydrologists were aware of uncertainty and used to deal with it because the first hydrological studies and applications were carried out. In particular, empirical techniques were used to compensate for the lack of information about model reliability. For instance, hydrologists are well used to adopt safety factors or allowance for freeboard, which are usually set basing on consensus, expert opinion, and empirical evidence. These safety factors were the first and very useful tools to take into account inherent uncertainty and imperfect knowledge of hydrological processes in hydrological design. However, expert knowledge is by itself subjective and referred to specific contexts and situations. The call for a generalized and systematic approach to uncertainty estimation in hydrology is the motivation for the renewed interest in the past few years.

One of the reasons why uncertainty assessment in hydrology was not much investigated on theoretical basis until the recent past is that hydrological modeling itself is a relatively young discipline. In fact, the first hydrological models were the rational formula proposed by Kuichling (1889) (although the principles of the method were introduced by Mulvaney (1851)), and the unit hydrograph model proposed by Sherman (1932). Most of the hydrological models that are used today were proposed after the 1960s. The interest in new techniques for uncertainty assessment was stimulated by Spear and Hornberger (1980), who introduced the generalized sensitivity analysis methodology, also known as regional sensitivity analysis. Their work inspired the development of the generalized likelihood uncertainty estimation (GLUE; Beven and Binley, 1992; see Section 2.17.6.2), which works under the hypothesis that different sets of model parameters/structures may be equally likely as simulators of the real system. In the 1990s the emerging need for reliable techniques for uncertainty estimation, for the multitude of modeling situations and approaches that are experienced in hydrology, stimulated the development of many methods (for a long, though still incomplete, list one can refer to Liu and Gupta (2007) and Matott et al. (2009)).

Another reason limiting the use of uncertainty assessment methods is that the transfer of the know-how about uncertainty in hydrology from scientists to end-users was and still is, difficult, notwithstanding the relevant research activity mentioned above. Pappenberger and Beven (2006) provided an extensive analysis of this issue. A relevant problem today is that uncertainty assessment in hydrology suffers from the lack of a coherent terminology and a systematic approach. The result of this situation is that it is extremely difficult (if not impossible) to obtain a coherent picture of the available methods. This lack of clarity is an example of linguistic uncertainty (Regan et al., 2003). Therefore, much is still to be done to reach a coherent treatment of the topic.

Quantitative uncertainty assessment in conditions of data scarcity is a very difficult task, if not impossible, in some cases. Usually, uncertainty estimation in applied scientific modeling is dealt with by comparing the model output with observed data, by borrowing concepts from statistics. According to this procedure, model reliability is quantified in a probabilistic framework. However, statistical testing becomes not as reliable in situations of data scarcity and therefore the use of statistical concepts for uncertainty assessment in hydrology sometimes may not be appropriate. This is one of the reasons why hydrologists are looking for different procedures that can be complementary or alternative to statistics. Moreover, uncertainty in hydrology might arise from limited knowledge (epistemic uncertainty, see Section 2.17.3) or from natural variability. In the former case, we deal with uncertainties that might not be aleatory in nature. They can be treated with statistical methods (e.g., the BATEA method, see Section 2.17.7.1), but many authors question the validity of statistics in this case and prefer nonstatistical approaches. These procedures are generally conceived in order to allow incorporation of expert knowledge in a theoretically based framework. They are characterized by a certain degree of subjectivity, which needs to be reduced as much as possible in order to allow their application in situations where knowledge is lacking.

Therefore, different philosophies and approaches for quantifying the reliability of hydrological models were recently proposed. As a result, an active debate recently began about the relative advantages of each of them. Such debate in many cases assumed a philosophical behavior, because the philosophy underlying each method is one of the main subjects of the discussion. On the one hand, such a debate stimulated additional developments and insights in itself; on the other, it is still not clear which approach is most appropriate given the needs of the user. For this reason, the hydrologic scientific community still calls for more pragmatism in uncertainty estimation.

On the one hand, hydrology is a science where uncertainty is very significant. Progress in monitoring techniques, process understanding, and modeling will certainly reduce uncertainty in the future but will never eliminate it. On the other hand, hydrologists are in charge of providing design variables that play a fundamental role in water engineering, civil protection, and water resource management. Therefore, it is clear that the efficient real world use of an uncertain design variable should necessarily be based on uncertainty assessment.

This chapter aims at presenting a comprehensive introduction to the subject of uncertainty assessment in hydrology. After presenting a brief glossary and a discussion about the reasons for the presence of uncertainty in hydrology, a review of the most-used approaches to uncertainty assessment is presented.

2.17.2 Definitions and Terminology

There is currently a linguistic uncertainty affecting the topic of uncertainty assessment in hydrology (Regan *et al.*, 2003; Beven, 2009), meaning that an agreed terminology is still lacking. Some basic definitions are provided in the following.

2.17.2.1 Probability

Probability can be defined in different ways. In fact, probability is currently interpreted according to two broad and distinguished views.

The classical frequentist view of probability defines the probability of an event occurring in a particular trial as the frequency with which it occurs in a long sequence of similar trials. In a Bayesian or subjectivist view, the probability of an event is dependent upon the state of information available and this information can include expert opinion. Probability theory forms the basis of classical statistics, which has estimators based on a likelihood function that represents how likely an observed data sample is for a given model and parameter set.

2.17.2.2 Randomness

Randomness is a term that is used within science with different meanings. In statistics, and hydrology as well, a random process is such that its outcome cannot be predicted deterministically. Randomness does not imply lack of knowledge about the process dynamics or impossibility to set up a deterministic model for it. However, if a deterministic model can be set up for a process, randomness implies that such a model cannot perfectly predict the process outcome.

For instance, in the case of a roulette wheel, if the geometric and dynamic behaviors of the system are perfectly known, then the number on which the ball will stop would be a certainty. However, one is fully aware that even a small imperfection in the description of the geometry of the system and/or its initial conditions makes the outcome of the experiment unpredictable. A probabilistic description can thus be more useful than a deterministic one for analyzing the pattern of outcomes of repeated rolls of a roulette wheel. Physicists face the same situation in kinetic theory of gases, where the system, while deterministic in principle, is very complex so that only a statistical description of its properties is feasible.

Another example is the experiment of dropping balls into a spiked sieve. Here, the geometry of the system is perfectly known as well as the initial and boundary conditions. However, once a ball is dropped in the sieve, it is impossible to predict deterministically its trajectory, because no one can predict which way the ball will follow after hitting a spike. However, the distribution of the balls at the bottom of the sieve is well known to be Gaussian. In this case, the full comprehension of the geometry and dynamics of the system does not allow one to set up a deterministic description, while a stochastic description can provide a satisfactory model. Actually, one cannot exactly predict the number of balls in each bar, but the probabilistic prediction will have a small uncertainty.

An important discovery of the twentieth-century physics was the random character of all physical processes that occur at subatomic scales and are governed by the laws of quantum mechanics. This means that probability theory is required to describe nature. This type of interpretation was questioned by many scientists, as the famous quote by Albert Einstein, from a letter to Max Born, clearly testifies: "I am convinced that He does not play dice."

A similar controversy currently occurs in hydrology (for an interesting discussion, see Koutsoyiannis *et al.* (2009)). The trend toward the so-called physically based models induced in the last few decades the inspiration to pursue a completely deterministic description of hydrological systems, through a better understanding of the internal dynamics of hydrological processes. However, such deterministic description is so complicated that only a probabilistic treatment is possible. This does not mean that knowledge is unuseful. On the contrary, it allows one to set up a plausible probabilistic description of the random outcome.

2.17.2.3 Random Variable

A random variable maps all possible outcomes from a random event into the real numbers. As such, it is affected by uncertainty and cannot be deterministically predicted. Random variables can assume discrete and continuous values.

2.17.2.4 Stochastic Process

A stochastic process can be defined as a collection of random variables. For instance, if we assume that the river flow at time t is a random variable, then the time series of river flow observations during an assigned observation period is a realization of a stochastic process. While a deterministic process gives only one possible value of its output under assigned initial and boundary conditions (as it is the case, e.g., for the solution of an ordinary differential equation), the output of a stochastic process is affected by some uncertainty that is described by the corresponding probability distributions. This means that there are many possible paths for the evolution of the process, with some of them being more likely and others less. A stochastic process can assume discrete or continuous values. Although the random variables of a stochastic process may be independent, in most commonly considered situations in hydrology, they exhibit statistical correlations. A stochastic process can include a deterministic representation but always includes a random component which makes its output uncertain.

2.17.2.5 Stationarity

A stochastic process is strictly stationary when the joint probability distribution of an arbitrary number of its random variables does not change when shifted in time or space. As a result, parameters such as the statistics of the process also do not change over time or position. Stationarity is a property of the mathematical representation of the system, or an ensemble of outcomes from a repeatable experiment, and therefore does not constitute an actual property of the natural process itself. This latter follows just one trajectory and therefore its outcome is unique, because nature and life do not enable repeatability. Stationarity is a property that is used in statistics in order to make inference about the physical process and therefore does not imply any assumption on the natural process itself.

It is interesting to mention that the opposite of stationarity is nonstationarity, which implies that the above statistics change accordingly to deterministic functions of time, where deterministic means that the above-mentioned functions should be known independently of the data and should apply to any time, past, present, and future (Papolulis, 1991). Conversely, if the above functions are random (i.e., realizations of stationary stochastic processes), then the process is stationary.

The concept of stationarity is a way to find invariant properties in complex natural systems. In view of what was anticipated above, it is important to note that stationarity does not imply that the statistics of a realization of a process are constant in time. Actually, such statistics are affected by sampling variability and therefore they certainly change after a time shift. The crucial issue is to detect if such a change exists in the process and can be expressed through a deterministic function of time.

Recently, the scientific literature presented contributions stating that stationarity is dead because of hydrological change and climate change. Actually, stationarity is an assumtpion and therefore can hardly be dead.

2.17.2.6 Ergodicity

A stochastic process is said to be ergodic if its statistical properties can be deduced from a single, sufficiently long sample (realization) of the process.

2.17.2.7 Uncertainty

Uncertainty can be defined as an attribute of information (Zadeh, 2005; Montanari, 2007). In the context of hydrology, uncertainty is generally meant to be a quantitative indication of reliability for a given hydrological quantity, either observed or inferred by using models. The indication of reliability can be provided by estimating the error affecting the quantity or the expected range of variability (due to uncertainty) for the quantity itself. Uncertainty can be broadly grouped into two major categories, namely, aleatory and epistemic uncertainty (see Section 2.17.3), and can be inferred by using probabilistic or nonprobabilistic methods.

2.17.2.8 Global Uncertainty and Individual Uncertainties

Global uncertainty can be defined as the discrepancy between the model output and the true value of the corresponding variable. Different uncertainties can compensate each other in the formation of the global uncertainty; for instance, parameter errors can compensate, at least in part, for data errors and model structural errors. These different uncertainties are termed individual uncertainties and are specifically referred to with a terminology which recall their causal origin, such as parameter uncertainty and model structural uncertainty (see Section 2.17.3.2 for an extended description). The terms above are not formally defined and therefore some linguistic uncertainty is present. For instance, the terms parameter uncertainty, input uncertainty, and model structural uncertainty should be used to indicate the uncertainty affecting the model parameters, input, and structure, respectively. Hereafter this is the meaning that will be used in this chapter. However, these terms are sometimes used to indicate the part of uncertainty in the model output that is caused by imperfect parameters, input, and model structure, respectively.

While global uncertainty is relatively easy to estimate *a posteriori*, for instance, by computing the difference between the model output and the corresponding observed variable (under the assumption that this latter is correct), the identification of the contribution of individual uncertainties above is impossible, unless assumptions are introduced or independent observations are available (see Section 2.17.4). This means that it is usually difficult, if not impossible, to assess whether the model performance is affected by, say, a parameter error rather than a model structural error.

2.17.2.9 Uncertainty Assessment

In what follows, we refer to uncertainty assessment to mean a quantitative evaluation of uncertainty affecting a hydrological variable, parameter, or model. Uncertainty estimation and uncertainty quantification will be considered synonymous with uncertainty assessment, which is different from uncertainty analysis and uncertainty modeling. The former is a preliminary step of uncertainty assessment aimed at identifying the reasons for the presence of uncertainty and the nature of uncertainty itself, while the latter term refers to the tools that are used for uncertainty assessment.

2.17.2.10 Probabilistic Estimate/Estimation/Assessment of Uncertainty (Probabilistic Uncertainty)

We will use the term probabilistic estimate of uncertainty to mean that uncertainty estimation for a given hydrological quantity has been carried out consistently with formal probability theory. In the probabilistic approach, uncertainties are characterized by the probabilities associated with events. Therefore, if one refers to the output of a hydrological model, the related probability distribution should actually provide an estimate of the frequency with which the true values fall within a given range.

2.17.2.11 Nonprobabilistic Estimate/Estimation/Assessment of Uncertainty (Nonprobabilistic Uncertainty)

Nonprobabilistic methods to uncertainty estimation in hydrology are frequently applied. Nonprobabilistic methods are various generalizations of probability theory that have emerged since the 1950s, including random set theory, evidence theory, fuzzy set theory, and possibility theory (Jacquin and Shamseldin, 2007). In particular, fuzzy set theory and possibility theory have received considerable attention from hydrologists, because much human reasoning about hydrological systems is possibilistic rather than strictly probabilistic. We reason about whether a given scenario could happen, without necessarily endeavoring to attach probabilities to the likelihood of it happening, particularly in situations of very scarce information.

In a more general context, we will refer to nonprobabilistic uncertainty when the estimation is carried out by using other approaches than formal probabilistic ones. This category includes probabilistic methods where some of the underlying assumptions are relaxed.

2.17.2.12 Confidence Band

A range around an estimated quantity that encompasses the true value with a probability $1 - \alpha$, where α is the significance level and $1 - \alpha$ is the confidence level. It is worth pointing out that the terminology is sometimes ambiguous. Some authors use the term confidence band or confidence interval when referring to the distribution of estimates that cannot be observed (e.g., a model parameter), while the term prediction interval is used when referring to the distribution of future values. Moreover, some authors indicate with the term tolerance interval a range in the observations that encompasses a $1 - \alpha$ proportion of the population of the related random variable. For more details, the reader is reffered to Hahn and Meeker (1991).

Figure 1 shows an example of confidence bands computed with the meta-Gaussian approach (Montanari and Brath, 2004; see Section 2.17.6.2) for river flow simulations referred to the Samoggia River at Calcara (Italy). It is interesting to note that the shape of the confidence bands themselves provides indications about the goodness of the fit provided by the model. Moreover, the skew in the prediction distribution results indicates that a systematic error is likely to be present.

2.17.2.13 Equifinality

Equifinality implies that in a system interacting with its environment a given end state can be reached by more than one potential mean. The term is due to von Bertalanffy (1968), the founder of general systems theory. The idea of equifinality suggests that similar results may be achieved with different initial conditions, different model parameters, and different model structures. In hydrology the concept of equifinality was introduced by Beven (1993) as an unavoidable effect of the presence of uncertainty. For an extended discussion, see Beven (2006a). Equifinality leads to the idea of multimodeling solutions in hydrology (see Section 2.17.6.5).

2.17.2.14 Behavioral Model

Within the context of equifinality, a behavioral model is one that provides an acceptable simulation of observed natural processes. In a multimodel approach, the collection of behavioral models provides a means for assessing the uncertainty of their output (see Section 2.17.6.5).

2.17.3 Classification of Uncertainty and Reasons for the Presence of Uncertainty in Hydrology

There have been many attempts presented by the literature to classify uncertainty in hydrology. The proposed solutions were not always in agreement because, given the uncertain nature of hydrological processes, it is sometimes impossible to unambiguously decipher the reason for the presence of errors. It is generally agreed that uncertainties can be grouped into two major categories: (1) natural variability (also called structural uncertainty, aleatory, external, objective, inherent, random, irreducible, or stochastic uncertainty) and (2) knowledge uncertainty (also called epistemic, functional, internal, reducible, or subjective uncertainty (**Table 1** in NRC, 2000; Koutsoyiannis et al., 2009; Hall and Solomatine, 2008). These two categories have different ramifications. In fact, the global uncertainty of a given model or variable may be characterized in three ways: purely structural, partly epistemic and partly structural, and purely epistemic (Cullen and Frey, 1999). When evaluating model performances and when possible, the different types of uncertainty should be separated (Cullen and Frey, 1999; Hoffman and Hammonds, 1994; Nauta, 2000; Sonich-Mullin, 2001). However, this is not always possible and therefore epistemic uncertainty and natural variability are often dealt with in an integrated fashion.

Other classifications were proposed. According to the causes for the presence of uncertainty in hydrology (which nevertheless are not always identifiable), one may identify the following categories: (1) inherent randomness (the geometry

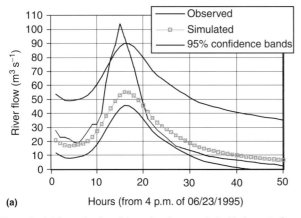
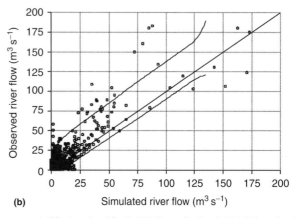

Figure 1 (a) Example of confidence bands computed with the meta-Gaussian approach (Montanari and Brath (2004); see Section 2.17.6.2) for a flood event occurred in the Samoggia River at Calcara (Italy) in 1995. (b) Example of confidence bands computed with the meta-Gaussian approach (Montanari and Brath (2004); see Section 2.17.6.2) drawn on a scatterplot of observed versus simulated hourly river flows for the Samoggia River at Calcara (Italy) during the years 1995–97.

Table 1 Uncertainty assessment methods in hydrology, along with their classification (see Section 2.17.5) and purpose (see Sections 45.6–45.10)

Assessment method	Classification	Type of uncertainty estimated
AMALGAM	Nonprobabilistic, parameter estimation	Parameter
BATEA	Probabilistic, parameter estimation, uncertainty assessment, sensitivity analysis	Precipitation induced
BFS	Probabilistic, Bayesian	Global
BMA	Probabilistic, multimodel	Global
DYNIA	Nonprobabilistic, identifiability analysis	Parameter
GLUE	Nonprobabilistic (when an informal likelihood is used), parameter estimation, uncertainty assessment, sensitivity analysis	Global, parameter, data, structural
IBUNE	Probabilistic, parameter estimation, uncertainty assessment, sensitivity analysis	Global, precipitation induced, model structure induced
Machine learning	Nonprobabilistic	Usually global, in principle all
Meta-Gaussian	Probabilistic, data analysis	Global
MOSCEM-UA	Nonprobabilistic, parameter estimation, sensitivity analysis	Parameter
SCE-UA	Probabilistic, parameter estimation	Parameter

Classification is ambiguous in some cases; it distinguishes between probabilistic and nonprobabilistic methods, as well as among the seven categories introduced by Matott *et al.* (2009) (see Section 2.17.5.2).
AMALGAM, a multialgorithm genetically adaptive method for multiobjective optimization; BATEA, Bayesian total error analysis; BFS, Bayesian forecasting system; BMA, Bayesian multimodel analysis; DYNIA, dynamic identifiability analysis; GLUE, generalized likelihood uncertainty estimation; IBUNE, integrated Baysian uncertainty estimator; MOSCEM-UA, multiobjective shuffled complex evolution University of Arizona; SCE-UA, shuffled complex evolution university of Arizona. References for the methods are in the text.

of the control volumes, the weather, etc.); (2) model structural uncertainty that reflects the inability of a model to represent precisely the true behavior of the system; (3) model parameter uncertainty; and (4) data uncertainty. When using models to make engineering or management decisions about hydrologic systems we also have to deal with (5) operation uncertainties (associated with construction and maintenance; Loucks and Van Beek, 2005). The above sources of uncertainty are briefly discussed in the following.

It is generally agreed that uncertainty in hydrology cannot be eliminated, no matter if it is epistemic in nature or induced by inherent randomness. For instance, rainfall inputs to a catchment might be highly structured, with different structures in different events that lead to nonrandom errors in estimates of areal rainfall. This type of error can be reduced with new measurement techniques but cannot be fully removed.

2.17.3.1 Inherent Randomness

Inherent randomness is one of the main reasons for the presence of uncertainty and is a intrinsic behavior of hydrological processes. For instance, a deterministic description of subsurface flow paths is impossible. Different soils and rocks, irregular macropores, faults and cracks with their heterogeneous patterns in both space and time, combined with two phase flows, varying wetting fronts, form a extremely complex system, for which a deductive description is impossible (Koutsoyiannis *et al.*, 2009). Inherent randomness emerges also from meteorology, variability of the surface flow paths, and so on. It is not only related to a coarse description of the system (that also induces uncertainty, which is nevertheless epistemic at least in part; in fact, it could be reduced by an increased capability to monitor the processes at finer spatial and temporal scales) but rather related to the effective impossibility to describe deterministically the inherent variability of the process. Inherent randomness has been long discussed by the hydrologic community in the recent past (Koutsoyiannis *et al.*, 2009; Koutsoyiannis, 2009). It has been argued that dynamical systems theory has well shown that uncertainty can emerge even from an insignificant perturbation of the initial conditions of a pure, simple, and fully known deterministic (chaotic) dynamics.

2.17.3.2 Model Structural Uncertainty

In the ideal situation in which perfect input data are available and model parameters are perfect, model structural uncertainty is defined as the uncertainty in the model output induced by the inhability of the hydrological model to perfectly reproduce the dynamics of hydrological systems. This means that the model output would still be uncertain even in the ideal situation in which no other uncertainties are present. Model structural uncertainty can be induced by imperfect model structure or lack of computational power. If the reason for the presence of uncertainty is an incorrect selection of the model or the computational tools, then model structural uncertainty is epistemic; on the other hand, the effective impossibility to describe the system with a mathematical model induces the presence of irreducible uncertainty. Given that model structural uncertainty is epistemic at least in part, the search for improved modeling tools has been the main focus of the hydrologic scientific community in the last few decades.

2.17.3.3 Model Parameter Uncertainty

Model parameter uncertainty is the result of the lack of a sufficiently extended database of good quality, or the inefficiency of the optimization algorithm and/or the related objective function, which induce parameter estimates to be significantly uncertain even if a perfect model and a perfect knowledge of the system were available. This is a relevant problem in all the hydrological applications and motivates the intense efforts that were dedicated to parameter estimation and related uncertainty assessment (e.g., Ibbitt and O'Donnell, 1974; Alley, 1984; Kleisenn et al., 1990; Gan and Burges, 1990; Duan et al., 1992; Brath et al., 2004; Vrugt et al., 2003a; 2003b; Vrugt and Robinson, 2007). Parameter estimation can be coupled with sensitivity analysis and model diagnostic to identify the most sensitive parameters in periods of model failures, thus gaining insights into the reasons for model inadequacy (Sieber and Uhlenbrook, 2005). To this end, Wagener et al. (2003) proposed the dynamic identifiability analysis (DYNIA).

Usually, an objective function is used to calibrate the model parameters to observed data. Independently of the objective function and the tools employed to optimize the model parameters, most hydrological models suffer from the existence of multiple optima of the objective function itself and the presence of high interaction or correlation among the parameters. These problems make parameter calibration uncertain even when a relatively large database is at disposal (Kuczera and Mroczkowski, 1998). Parameter uncertainty also arises when the parameters are not calibrated but rather estimated on the basis of field surveys or expert knowledge, for instance, while defining land-cover parameters. Parameter uncertainty can be epistemic at least in part.

2.17.3.4 Data Uncertainty

Data uncertainty is an emerging problem that is gaining renewed attention by hydrologists in the recent past (see, e.g., Di Baldassarre and Montanari, 2009; Dottori et al., 2009; Koussis, 2009; Petersen-Øverleir and Reitan, 2009). In fact, even modern technologies cannot avoid the presence of a significant approximation in observations of, say, rainfall, river flows (for both low and high flows), and so forth. Data uncertainty emerges from limitation of the monitoring techniques (instrumentation error, rating curve approximations, etc.) or variability of the spatial and temporal distribution of the observed hydrological variables (spatial variability of rainfall, time variability of streamflow, etc.). It follows that hydrological models are optimized against imperfect data and therefore an error is induced in hydrological simulations. Data uncertainty has both epistemic and aleatory components and therefore it is particularly important how observation errors are treated. Some authors claim that treating data error with purely statistical approaches may induce overconditioning in hydrological modeling (Beven, 2006a).

2.17.3.5 Operation Uncertainty

Operation uncertainty arises when hydrological models are used in the real world. In fact, it is well known that in real-time applications uncertainties of different nature are present that do not affect off-line exercises. Often the data and the initial and boundary conditions cannot be preliminary checked, the computational time might become a relevant constraint, the end-users operate under stress and therefore the human error becomes more likely, there is a weak ability to identify decision criteria, communication becomes difficult, and so forth. Operation uncertainty is difficult to assess, is rarely considered by researchers, and represents an emerging awareness among hydrological modelers and end-users. As a matter of fact, the identification of the most suitable model should be carried out in view of operation uncertainty as well. Data assimilation can be used to constrain uncertainty during model application.

2.17.4 Uncertainty Assessment

It is well known that uncertainty assessment in hydrology is a topical issue. Already in 1905, W.E. Cooke, who was issuing daily weather forecasts in Australia, stated: "It seems to me that the condition of confidence or otherwise form a very important part of the prediction, and ought to find expression." Uncertainty assessment in hydrology involves the analysis of multiple sources of error, the main ones being outlined in Section 2.17.3. The contribution of these latter to the formation of the global uncertainty cannot be quantified independently, unless (1) one is willing to introduce subjective assumptions about the nature of the individual error components or (2) independent observations are available for estimating each source of error. As an example for the latter solution, the reader is referred to Winsemius et al. (2006, 2008) where gravity and evaporation measurements are used to constrain the water balance and the land surface parameters, respectively, for a rainfall–runoff model.

However, in some hydrological applications it is not necessary to separate different sources of error. For this reason in many cases, uncertainty is assessed in an aggregated solution, therefore quantifying global uncertainty.

2.17.5 Classification of Approaches to Uncertainty Assessment

This section aims to propose a classification of uncertainty assessment methods in hydrology. Classifying the methods is useful to clarify their behavior and operational purpose. However, it should be premised that such a classification might be subjective, because some methods lend themselves to different interpretations of their nature and scope.

2.17.5.1 Research Questions about Uncertainty in Hydrology

The uncertain nature of hydrology has pushed hydrologists to raise many questions related to uncertainty assessment. The most urgent ones are those related to quantifying the reliability of the output variables of hydrological models (forecasts, simulations, etc.). Hydrological simulation is often used in real-time prediction systems for natural hazards or for assessing long-term effects of climate change or for assessing the reliability of proposed water resource management strategies. In these cases, quantifying the uncertainty of the hydrological model response is extremely important from a societal point of view.

Uncertainty assessment in hydrology includes additional research issues. Among them, there is the call for assessing the uncertainty of observed data, model parameters, and model structure. These issues are also significant for gaining further insight into the dynamics of hydrological processes. Indeed, to identify the most appropriate model is a means to provide support to hydrological theory. Therefore, uncertainty assessment became strongly related to parameter estimation, multi-objective optimization, model identification, model building, model diagnostics, model averaging, data collection, and information theory in general. All topics in this list have gained the attention of researchers in recent years and are often allocated under the one umbrella of uncertainty assessment in hydrology. Indeed, it would be helpful for end-users to formally identify such subtopics and the related research questions.

2.17.5.2 An Attempt of Classification

The traditional way of dealing with uncertainty in science is through statistics and probability (see, e.g., Montanari et al., 2009) but, as mentioned above, nonprobabilistic approaches to uncertainty analysis are also popular in hydrology.

In some cases, it is not easy to classify an approach as either probabilistic or not. In fact, there are some methods that are based on probability theory, but in real-word applications simplifying assumptions are often introduced which finally lead to a nonprobabilistic estimation of the likelihood of a given scenario. Such assumptions are introduced in order to overcome operational problems, for instance, due to lack of enough data to support a statistical application.

The decision to use probabilistic or nonprobabilistic methods is currently the most controversial issue in hydrologic uncertainty analysis. This debate has raised the very relevant question about the capability of probabilistic and nonprobabilistic methods to correctly infer the frequency properties of hydrological simulations and predictions (see, e.g., Beven, 2006a; Montanari, 2005, 2007; Mantovan and Todini, 2006; Beven et al., 2007, 2008). Criticism about probabilistic methods is focused on the concern that for many data sets it is not clear if the assumptions of classical statistics (e.g., stationarity) can be justified. The main reason for criticism of nonprobabilistic methods is that they are subjective and not necessarily coherent from a statistical point of view (see, e.g., the criticism of Mantovan and Todini (2006) with respect to GLUE). Moreover, on known problems for which the data do support the necessary probabilistic assumptions, probabilistic and nonprobabilistic methods provide different answers (e.g., Stedinger et al., 2008).

The suitability of probabilistic versus nonprobabilistic methods and the difference in their response are dictated by the knowledge that the user has about the structure of the error model. Using a correctly based inference should lead to similar results in uncertainty assessment. Conversely, some authors claim that with unknown error structure it is dangerous to rely on statistical methods based on simple assumptions about the nature of the errors themselves.

There is an increasing consensus about the opportunity to use probabilistic approaches, as a way to efficiently summarize the information content of the data, when sufficient information is available to support statistical hypotheses with appropriate statistical tests (Montanari et al., 2009). Conversely, data scarcity calls for expert knowledge to support uncertainty assessment. Above all, data scarcity calls for the integration of different types of information, within a framework that is unavoidably subjective, given that the information itself is often soft.

Besides the above, additional classifications were recently proposed for uncertainty assessment methods. For instance, Matott et al. (2009) identified seven categories of models: (1) Data analysis methods, including analytical and statistical procedures for evaluating the accuracy of data. These include also parametrization of probability distributions. (2) Identifiability analysis, aiming at detecting data inadequacy and suggesting model improvements. (3) Parameter estimation methods, quantifying uncertain model parameters. (4) Uncertainty analysis techniques, meaning methods to propagate sources of uncertainty through the model to generate probability distributions for the model output. These methods include approximation and sampling methods. (5) Sensitivity analyses, investigating to what extent different sources of variation in the input of a mathematical model affect the variation of the output. Sensitivity analysis aims at identifying what source of uncertainty weights more on the model output (see, e.g., Van Griensven et al. (2006); Götzinger and Bárdossy, 2008). Sensitivity analysis and uncertainty estimation are well distinguished. Their results can be comparable, because a probability distribution of model outputs corresponding to different inputs can be similar to the analogous distribution derived through the analysis of probabilistic uncertainty. This similarity of results has originated a confusion of terms in some applications. (6) Multimodel analysis, consisting of generating multiple possible outputs accordingly to different models, parameters, and boundary conditions. (7) Bayesian methods, which were previously defined (this category could be joined with category 4 above). The seven categories above are not strictly separated, meaning that a method can belong to more than one of them.

Another classification for uncertainty assessment methods for the model output was recently proposed by Shrestha and Solomatine (2008) who consider the following categories: (1) analytical methods, using derived distribution methods to compute the probability distribution function of the model output; (2) approximation methods, providing only the moments of the distribution of the uncertain output variable; (3) simulation and sampling–based methods, estimating the full distribution of the model output via simulation; (4) Bayesian methods, which combine Bayes' theorem and various simulation approaches to either estimate or update the probability distribution function of the parameters of the model and consequently estimate the uncertainty of the model output; (5) methods based on the analysis of the model errors, such as the meta-Gaussian approach described in Section 2.17.6.4; and (6) fuzzy-theory-based methods, providing a nonprobabilistic approach for modeling the kind of uncertainty associated with vagueness and imprecision.

Whatever approach is chosen to uncertainty assessment, the end-user should be made fully aware of the assumptions and drawbacks of the method that is being used. The presence of subjectivity should be clearly stated and the limitations of the underlying hypotheses, both in the probabilistic and nonprobabilistic approaches, clearly described and discussed. An appropriate terminology should also be used to make the meaning of the provided confidence bands clear. Whenever a subjective method is adopted, the user should be made aware that the uncertainty bands reflect user belief instead of providing a frequentist assessment of the probability of the true value to fall between them. Appropriate use of the methods being proposed by the scientific community, depending on the user needs and data availability, would allow us to successfully reach a better communication between scientists and end-users. It is as important to communicate uncertainty as communicate the assumptions on which an assessment has been based.

2.17.6 Assessment of the Global Uncertainty of the Model Output

Assessment of the global uncertainty for the model output is by far the application that is most frequently presented by the hydrological literature, as a means for quantifying model reliability and providing end-users with operational indications. Several methods are available to this end, ranging from statistically based to subjective approaches.

2.17.6.1 Analytical Methods

The most direct method to assess the uncertainty of a system output is to derive its statistics from a knowledge of the statistical properties of the system itself and the input data (Langley, 2000). However, this approach may be limited by two main problems: first, the derivation of the statistics of the output can imply significant mathematical and numerical difficulties; and, second, the statistical properties of the system and the input may not be known in detail.

The first difficulty has stimulated the development of a first type of uncertainty assessment technique, namely, the approximate analytical methods. An example is the asymptotic reliability analysis, like the first-order reliability method (FORM) and second-order reliability method (SORM). Examples of applications in hydrology are given by Melching (1992) and Vrugt and Bouten (2002). Point estimate methods are an interesting option too, in view of their computational efficiency (Tsai and Franceschini, 2005).

The second problem mentioned above may be even more difficult to deal with. For instance, the definition of the statistics of the system is a delicate step of the uncertainty assessment method recently proposed by Huard and Mailhot (2006) in a hydrological context.

2.17.6.2 The Generalized Likelihood Uncertainty Estimation

GLUE was introduced by Beven and Binley (1992), who were inspired by the generalized sensitivity analysis methodology proposed by Spear and Hornberger (1980).

GLUE rejects the concept of an optimum model and parameter set and assumes that, prior to input of data into a model, all models and parameter sets have an equal likelihood of being acceptable. The acceptance of the existence of multiple likely models has been called equifinality (Beven, 1993) to suggest that this should be accepted as a generic problem in hydrological modeling rather than simply reflecting the problem of identifying the true model in the face of uncertainty.

GLUE is performed by first selecting different modeling options (different hydrological models and different parameters). In order to reduce the computational requirements of the procedure, it might be necessary to limit the dimension of the sample space of the parameters and models. Then, a high number N of simulation is generated by sampling the model and parameter spaces accordingly to a prior probability distribution. In the absence of prior knowledge, uniform sampling can be used. By increasing N one increases the probability of trying all of the most relevant solutions. The different models are then run for each of the parameter sets and the model output is then compared to a record of observed data (e.g., for observed hydrographs or annual maximum peak flows, see Cameron et al. (1999); another interesting example is given by Blazkova and Beven (2009)). The performance of each trial is assessed via likelihood measures, either formal or informal. This includes rejecting some parameter sets as nonbehavioral. For instance, the Nash and Sutcliffe (1970) efficiency can be used as informal likelihood measure of the simulation of a continuous hydrograph. All parameter sets that lead to obtaining an efficiency above a subjective threshold are retained. Finally, likelihood weighted uncertainty bounds are calculated depending on the likelihood (Freer et al., 1996). For instance, the calculated likelihoods can be rescaled to produce a cumulative sum of 1.0, thereby obtaining informal weights. A cumulative distribution function of simulated discharges is then constructed using the rescaled weights. Linear interpolation is used to extract the discharge estimates corresponding to cumulative probabilities of $\alpha/2$, 0.5, and $1.0 - \alpha/2$. This allows $100(1 - \alpha)\%$ uncertainty bounds to be derived, in addition to a median simulation.

If either (or both) the likelihood measure or the procedure for computing the rescaled weights is informal, the probabilities computed with GLUE do not possess the classical frequentist meaning. Therefore, strictly speaking, it is inappropriate to refer to them with the term probability and many authors classify GLUE as a nonprobabilistic approach. Conversely, if formal statistical procedures are used, GLUE assumes the behavior of a probabilistic methodology. For extended discussions, the reader is referred to Beven et al. (2008) and Stedinger et al. (2008).

GLUE could be applied in principle even in the absence of observed historical data, in those real-world applications in which the likelihood measure is estimated on the basis of expert knowledge. GLUE is highly computationally demanding, especially if the number of significant model parameters is high. This problem may prevent the application of GLUE when dealing with complex models.

Beven (2006a) formally introduced a different procedure for the identification of behavioral models, by following previous practical experiments by Pappenberger and Beven (2004) and Page et al. (2007). A recent interesting application is presented by Liu et al. (2009). In this approach, limits of acceptability are preliminarily identified for the model output or selected performance measures. All the models that meet the limits of acceptability are retained so that an envelope of behavioral model simulations can be identified. Finally, a likelihood weighted cumulative density function for the model output can be computed as previously in GLUE so that simulation quantiles can be estimated (see also Blazkova and Beven, 2009).

There are many other variants of GLUE; for example, Tolson and Shoemaker (2008) and Mugunthan and Shoemaker (2006) combined optimization methods with a non-probabilistic GLUE-like approach to increase computational efficiency of nonprobabilistic uncertainty analysis.

The hydrological literature presented many applications to GLUE to numerous hydrological problems, including rainfall–runoff modeling (Cameron *et al.*, 1999), groundwater modeling (Christensen, 2003), inundation modeling (Aronica *et al.*, 1998, 2002), and urban water-quality modeling (Freni *et al.*, 2008, 2009).

2.17.6.3 The Bayesian Forecasting System

The Bayesian Forecasting System (BFS) was proposed by Krzysztofowicz (1999, 2001a, 2002), Krzysztofowicz and Kelly (2000), and Krzysztofowicz and Herr (2001). The purpose is to produce a probabilistic river stage forecast (PRSF) based on a probabilistic quantitative precipitation forecasting (PQPF) as an input to a hydrological model that is in charge of simulating the response of a river basin to precipitation. It can be adapted to produce a probabilistic river discharge forecast.

The BFS assumes that the dominant source of uncertainty derives from the imperfect knowledge of the future precipitation, so that it can be assumed that all other sources of uncertainty play a minor role. The system can work with any hydrological model and aims at estimating the global uncertainty of the forecast, which is considered to be caused by: (1) precipitation uncertainty, which is dominant and quantified by the probability distribution of the future rainfall specified by the PQPF and (2) hydrologic uncertainty, which is the aggregate of all uncertainties arising from sources other than precipitation uncertainty. In particular, it aggregates the model uncertainty and parameter uncertainty.

The BFS has three structural components: the precipitation uncertainty processor (PUP; Kelly and Krzysztofowicz, 2000), the hydrologic uncertainty processor (HUP; Krzysztofowicz and Kelly, 2000), and the integrator (INT; Krzysztofowicz, 2001b). **Figure 2** reports a sketch of the BFS structure adapted from Krzysztofowicz (2002). The PUP has the purpose of mapping precipitation uncertainty to output uncertainty under the hypothesis that there is no hydrologic uncertainty. This involves running the hydrological model for a set of specified quantiles of the probability distribution of the future rainfall. The HUP quantifies hydrologic uncertainty under the hypothesis that there is no precipitation uncertainty. Finally, the INT integrates the two uncertainties in order to produce a PRSF. For extended details on the PUP and INT, the interested reader is invited to refer to Kelly and Krzysztofowicz (2000) and Krzysztofowicz (2001b, 2002). Next, we provide a brief description of the HUP for the purpose of illustrating the meta-Gaussian approach adopted by BFS.

Let h_n denote the true river stage on day n, counting from day $n=0$ when the forecast is issued. At the forecast time the actual river stage on day n is unknown and thus uncertain.

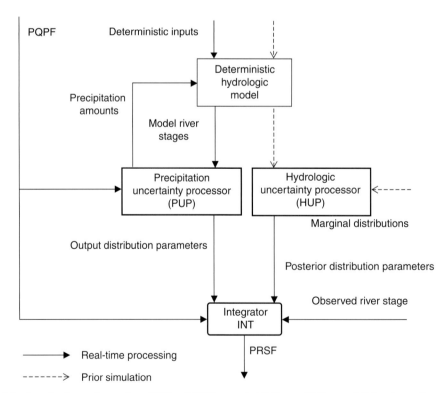

Figure 2 Sketch of the Bayesian forecasting system. PQPF and PRSF are probabilistic quantitative precipitation forecasting and probabilistic river stage forecasting, respectively. Adapted from Krzysztofowicz R (2002) Bayesian system for probabilistic river stage forecasting. *Journal of Hydrology* 268: 16–40.

Therefore, it is treated as a random variable which we refer to with the symbol H_n. Let s_n be an estimate of H_n from the hydrologic model based on all the input variables and the true precipitation amount. Estimate s_n is treated as a realization of the random variable S_n. One would observe $h_n = s_n$ if there were no hydrologic uncertainty (let us remark that HUP is developed under the assumptions that there is no precipitation uncertainty). The presence of hydrologic uncertainty gives rise to a probability distribution of the actual river stage H_n, conditional on a realization of the model river stage $S_n = s_n$. Therefore, we can treat H_n as a random variable whose probability distribution is conditioned on the corresponding realization of the model river stage $S_n = s_n$. The purpose of the HUP is to provide an estimate for such probability distribution. This is achieved by applying a Bayesian technique.

First of all, H_n and S_n are transformed to the Gaussian variables W_n and X_n respectively, by applying the standard normal quantile transform (NQT; see Kelly and Krzysztofowicz, 1997) and assuming that all conditional and joint densities are Gaussian. Then, it is assumed that the actual river stage process is well represented by a Markov stochastic process of order 1 and strictly stationary. This allows one to derive the prior probability distribution $g_n(w_n|w_0)$ of w_n conditional on $W_0 = w_0$. The prior density is derived under the assumption that the following normal linear equation applies in the transformed domain:

$$W_n = rW_{n-1} + \Theta_n \tag{1}$$

where r is a parameter and Θ_n is a random variable stochastically independent of W_{n-1} and normally distributed with mean zero and variance $\sigma^2(\Theta_n)$. Therefore, the probability distribution of w_n is Gaussian with mean equal to rw_{n-1} and variance $\sigma^2(\Theta_n)$. The recursive application of the above derivation allows one to estimate $g_n(w_n|w_0)$.

Subsequently, a probability distribution of the normalized model river stage x_n conditioned on w_n and w_0 is built, and is denoted as $f_n(x_n|w_n, w_0)$. This is derived under the hypothesis that the stochastic dependence between the transformed variables is governed by the following normal linear equation:

$$X_n = a_n W_n + d_n W_0 + b_n + \Phi_n \tag{2}$$

in which a_n, b_n, and d_n are parameters and Φ_n is a variate stochastically independent of (W_n, W_0) and normally distributed with mean zero and variance $\sigma^2(\Phi_n)$. It follows that the probability distribution $f_n(x_n|w_n, w_0)$ is Gaussian with mean and variance (see Krzysztofowicz and Kelly, 2000):

$$E(X_n|W_n = w_n, W_0 = w_0) = a_n w_n + d_n w_0 + b_n \tag{3}$$

$$\mathrm{Var}(X_n|W_n = w_n, W_0 = w_0) = \sigma_n^2(\Phi) \tag{4}$$

The coefficients r, a_n, b_n, and d_n are derived by running the hydrological model for an extended record of the model input with a perfect forecast of precipitation amount, thus obtaining joint realizations of the model-actual river stage process that are transformed to the Gaussian probability distribution through the NQT. These joint realizations are used to estimate the parameters of the above regressions (1) and (2). This design of the analysis assures that there is no precipitation uncertainty, but only hydrologic uncertainty.

Once $g_n(w_n|w_0)$ and $f_n(x_n|w_n, w_0)$ are known, the total probability law allows one to derive the distribution $k_n(x_n|w_0)$ of the transformed model river stage conditioned on w_0, while the Bayes theorem allows one to derive the posterior density of w_n conditioned on x_n and w_0, namely:

$$r(w_n|x_n, w_0) = \frac{f_n(x_n|w_n, w_0) g_n(w_n, w_0)}{k_n(x_n|w_0)} \tag{5}$$

where

$$k(x_n|w_0) = \int_{-\infty}^{\infty} f_n(x_n|w_n, w_0) g_n(w_n|w_0) \mathrm{d}w_n \tag{6}$$

Finally, the inverse of the NQT allows one to derive the posterior density of h_n conditioned on s_n and h_0. Such distribution allows one to quantify the hydrologic uncertainty. More details are provided by Krzysztofowicz and Kelly (2000).

Despite a theoretical development that may appear complicated, the BFS has the advantage of being easy to apply and allowing rapid implementation in real time. However, it was conceived for estimating the uncertainty of forecasted variables only.

2.17.6.4 Techniques Based on the Statistical Analysis of the Model Error

Several methods for uncertainty assessment were proposed based on the statistical analysis of the model error. Accordingly, the model error is treated as a stochastic process for which realizations are obtained by performing off-line simulations which are matched with the corresponding observations. Of course, observed data are themselves uncertain and therefore the model reliability analysis could not be correct in absolute terms (in the ideal situation of a perfect model, if we compared its response with uncertain output observations, that we assumed to be correct, we would wrongly conclude that the model is uncertain). However, in any case, from a practical point of view, the difference between the model response and what we measure in the field gives an important information for the sake of inferring reality based on the model output (Refsgaard et al., 2006).

A technique for global uncertainty assessment based on the analysis of the model error is the meta-Gaussian approach proposed by Montanari and Brath (2004) for the case of hydrological simulations and extended by Montanari and Grossi (2008) for hydrological forecasting. Next, the latter methodology is presented, therefore making reference to real-time flood forecasting systems. The meta-Gaussian approach is probabilistic.

In order to estimate the uncertainty of a hydrological forecast, it is assumed that the forecast error is a stationary and ergodic stochastic process, denoted with the symbol $E(t)$. Its statistical properties are inferred by analyzing a past realization $e_{\mathrm{obs}}(t) = Q_{\mathrm{obs}}(t) - Q_{\mathrm{pred}}(t)$ that it is assumed to be available, where $Q_{\mathrm{obs}}(t)$ and $Q_{\mathrm{pred}}(t)$ are true and forecasted river flows, respectively. The use of a meta-Gaussian model is then proposed to derive the time-varying probability distribution of the forecast error. In particular, the probability distribution of $E(t)$ is inferred on the basis of its dependence on M selected explanatory random variables. The statistical inference is performed in the Gaussian domain, by preliminarily transforming $E(t)$ and the explanatory variables to the Gaussian probability distribution. The above transformation is operated through the NQT.

The probabilistic model for $E(t)$ is built as follows. First of all, it is assumed that positive and negative errors come from two different statistical populations $E^{(+)}(t)$ and $E^{(-)}(t)$. Therefore, the probability model for $E(t)$ is given by a mixture of two probability distributions, one for $E^{(+)}(t)$ and one for $E^{(-)}(t)$. The mixture is composed such that the area of the probability distribution of $E^{(+)}(t)$ is equal to the percentage, $P^{(+)}$, of positive errors over the total sample size of the available past realization $e_{obs}(t)$ of the forecast error.

The two realizations $e^{(+)}{}_{obs}(t)$ and $e^{(-)}{}_{obs}(t)$ are transformed through the NQT, therefore obtaining the normalized realizations $Ne^{(+)}{}_{obs}(t)$ and $Ne^{(-)}{}_{obs}(t)$. Then, M explanatory variables, $X^{(i)}(t)$ with $i = 1, \ldots, M$ (which should be readily available at the forecast time), are selected in order to explain the variability in time of the marginal statistics of $E^{(+)}(t)$ and $E^{(-)}(t)$. The values of such explanatory variables for the realizations $e^{(+)}{}_{obs}(t)$ and $e^{(-)}{}_{obs}(t)$ above are estimated and then transformed by using the NQT, therefore obtaining the normalized explanatory variables $Nx^{(i)}{}_{obs}(t)$ with $i = 1, \ldots, M$.

In the Gaussian domain, it is assumed that the forecast error can be expressed as a linear combination of the selected explanatory variables. Let us focus on the positive error. The linear combination can be expressed through the following relationship:

$$Ne^{(+)}(t_j) = C_1^{(+)} Nx^{(1)}(t_j) + C_2^{(+)} Nx^{(2)}(t_j) + \cdots + C_M^{(+)} Nx^{(M)}(t_j) + \varepsilon^{(+)}(t_j) \quad (7)$$

where $\varepsilon^{(+)}(t_j)$ is an outcome of a homoscedastic and Gaussian random variable and t_j is an assigned time step. An analogous relationship holds for $Ne^{(-)}(t)$. It is assumed that positive and negative errors are conditioned by the same explanatory variables, but the fit of the linear regression (7) leads to a different set of coefficient values. Such coefficients are estimated by inserting in (7) the past realizations of transformed forecast error, $Ne^{(+)}_{obs}(t)$, and explanatory variables, $Nx^{(i)}_{obs}(t)$, and then by identifying the coefficient values that lead to the best fit (for instance by minimizing the sum of the squares of $\varepsilon^{(+)}(t_j)$). The goodness of the fit provided by (7) can be verified by drawing a normal probability plot and a residual plot for $\varepsilon^{(+)}(t_j)$ as in Montanari and Brath (2004).

Once the linear regression (7) has been calibrated, for positive and negative errors, the probability distribution of the transformed positive forecast error can be easily derived for real-time and real-world applications. Such distribution is Gaussian and is expressed by the following relationship:

$$Ne^{(+)}(t_j) \sim G[\mu[Ne^{(+)}(t_j)], \sigma[Ne^{(+)}(t_j)]] \quad (8)$$

where '$\sim$' means equality in probability distribution and G indicates the Gaussian distribution whose parameters are given by

$$\mu[Ne^{(+)}(t_j)] = C_1^{(+)} Nx^{(1)}(t_j) + C_2^{(+)} Nx^{(2)}(t_j) + \cdots + C_M^{(+)} Nx^{(M)}(t_j) \quad (9)$$

$$\sigma[Ne^{(+)}(t_j)] = \sigma[\varepsilon^{(+)}(t_j)] \quad (10)$$

Analogous relationships (from (8) to (10)) hold for the negative error. Therefore, the confidence bands for the transformed forecast and an assigned significance level can be straightforwardly derived. In detail, the upper confidence band of the transformed forecast at the α significance level is given by the $1 - \alpha/(2 \cdot P^{(+)})$ quantile of the Gaussian distribution given by (8), (9), and (10). Given that $P^{(+)}$ can be arbitrarily close to 0, in the technical computation one may obtain values greater than 1 of $\alpha/(2 \cdot P^{(+)})$. This means that the probability of getting a positive forecast error is small enough to make equal to 0 the width of the upper confidence band at the α significance level.

For instance, if $P^{(+)} = 0.5$ and $\alpha = 10\%$, the transformed upper confidence band is given by the well-known relationship:

$$Ne^{(+)}_{90\‰}(t_j) = \mu[Ne^{(+)}(t_j)] + 1.96\sigma[Ne^{(+)}(t_j)] \quad (11)$$

Finally, by applying back the NQT one obtains the confidence bands for the assigned significance level in the untransformed domain.

The reason why positive and negative errors are treated separately is that a good fit is frequently not achieved through the linear regression (7) when the errors are pooled together. In fact, in this case, it appears that the NQT is not effective in making the errors homoscedastic and therefore the assumption of linearity does not hold. The reason for this result is that the NQT is not efficient in assuring homoscedasticity if the mean of the model error is not significantly changing across the range of the error itself, as it often happens when dealing with hydrological models. By treating positive and negative errors separately, the problem disappears and the assumptions of the linear regression are met. Finally, it is important to note that the only assumption made about the sign of the future forecast error is that it has a probability equal to $P^{(+)}$ to be positive. Therefore, no inference is made on the sign of the forecast error on the basis of the explanatory variables.

Figure 3 shows the confidence bands computed with the meta-Gaussian approach for the forecast with 1-h lead time of two flood events occurred on the Toce River at Candoglia, in Italy.

2.17.6.5 Bayesian Model Averaging

Bayesian model averaging (BMA, Hoeting et al., 1999) is a statistical way of postprocessing model output ensembles to derive predictive probability density functions for hydrological variables. It represents the predictive probability distribution as a weighted average of the individual predictive probabilities of each model, where the weights are posterior probabilities of the models themselves and reflect the models' relative contributions to predictive skill over a training period. The combination of multiple models is an important component of

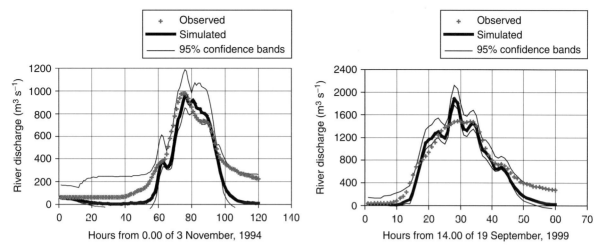

Figure 3 95% confidence bands computed with the meta-Gaussian approach for the forecast with 1-h lead time of two flood events occurred on the Toce River at Candoglia, on 3 November 1994 (left) and 19 September 1999 (right). From Montanari A and Grossi G (2008) Estimating the uncertainty of hydrological forecasts: A statistical approach. *Water Resources Research* 44: W00B08 (doi:10.1029/2008WR006897).

model validation (Burnham and Anderson, 2002). Multi-modeling solutions are often applied in real time forecasting (ensemble forecasting). See, for instance, the activity carried out in the framework of the HEPEX Project (Franz *et al.*, 2005; Zappa *et al.*, 2008).

BMA is applied in a Bayesian framework. Let $\mathbf{M} = \{M_i; i = 1, 2, \ldots, N\}$ be a set of N hydrological models for obtaining the vector $\hat{z}$ of hydrological variables. Given a set of data, $\mathbf{D}$, the posterior probability $\Pr(\hat{z}|\mathbf{D})$ of $\hat{z}$ is obtained through the BMA according to the law of total probability:

$$\Pr(\hat{z}|\mathbf{D}) = E_\mathbf{M}[\Pr(\hat{z}|M_i, \mathbf{D})] = \sum_{i=1}^{N} \Pr(\hat{z}|M_i, \mathbf{D}) \Pr(M_i|\mathbf{D}) \quad (12)$$

where $\Pr(\hat{z}, \mathbf{D})$ is the posterior probability of $\hat{z}$ for the given data set $\mathbf{D}$, $\Pr(\hat{z}|M_i, \mathbf{D})$ is the posterior probability of $\hat{z}$ for given data set $\mathbf{D}$ and model M_i, $\Pr(M_i, \mathbf{D})$ is the posterior model probability for model M_i, and $E_\mathbf{M}$ is the expectation operator over simulation models. Essentially, Equation (12) says that the probability distribution given by the model ensemble for the output variable is a weighted mixture of the individual distributions given by each model, where the weights are the posterior model probabilities. Therefore, Equation (12) presupposes that individual probability distributions for the output from each model, conditioned on the model itself and the available data set, are available. According to Bayes' rule, the posterior model weight is

$$\Pr(M_i, \mathbf{D}) = \frac{\Pr(\mathbf{D}|M_i)\Pr(M_i)}{\sum_{i=1}^{N} \Pr(\mathbf{D}|M_i)\Pr(M_i)} \quad (13)$$

where $\Pr(\mathbf{D}|M_i)$ is the marginal model likelihood function for model M_i, $\Pr(M_i)$ is the prior model probability for model M_i, and $\sum_p \Pr(M_i) = 1$. A uniform distribution can be assumed for the priors if better information is not available. Equation (13) implies the total model weight $\sum_p \Pr(M_i|\mathbf{D}) = 1$. The marginal model likelihood function $\Pr(\mathbf{D}|M_i)$ plays an important role in the determination of the degree of importance for each model, given the same data set. For noninformative model priors, higher posterior model weights reflect better agreement between results and observed data.

According to Equation (12), the law of total expectation allows one to obtain the means of the predicted $\hat{z}$ over the models for given data $\mathbf{D}$:

$$E(\hat{z}|\mathbf{D}) = E_\mathbf{M}[E|\hat{z}|M^{(p)}, \mathbf{D}] = \sum_p E[\hat{z}|M^{(p)}, \mathbf{D}]\Pr(M^{(p)}|\mathbf{D}) \quad (14)$$

where E is the expectation operation over $\hat{z}$. Analogous relationships allow one to obtain the covariance matrix of the predicted $\hat{z}$, therefore allowing a quantification of uncertainty. For more details, and an application that refers to the prediction of groundwater heads, the reader is referred to Li and Tsai (2009).

There are plenty of applications of BMA in hydrology (see, for instance, Ajami *et al.* (2006), Duan *et al.* (2007), Zhang *et al.* (2009), Reggiani *et al.* (2009), and Li and Tsai (2009)). **Figure 4** shows confidence bands computed with BMA for simulations of river flows obtained with the soil and water assessment tool (SWAT) model in the Yellow River Headwater Basin (from Zhang *et al.*, 2009)

BMA tends to be computationally demanding and relies heavily on prior information about models. Neuman (2003) proposed a maximum likelihood version (MLBMA) of BMA to render it computationally feasible and to allow dealing with cases where reliable prior information is lacking (Ye *et al.*, 2004). BMA is also used within the Integrated Bayesian Uncertainty Estimator (IBUNE) proposed by Ajami *et al.* (2007).

2.17.6.6 Machine Learning Techniques

In the recent past, there has been an increased interest about machine learning technique for global uncertainty assessment (see, for instance, Shrestha *et al.*, 2009; Solomatine and Shrestha, 2009; Hall and Solomatine, 2008). These methods are frequently used as a mean to approximate complex models for uncertainty assessment, therefore obtaining a less computationally intensive approach.

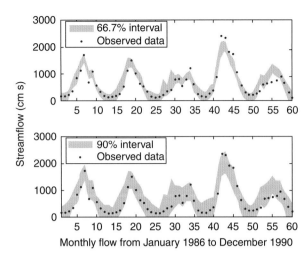

Figure 4 Confidence bands computed with BMA for simulations of river flows obtained with the SWAT model in the Yellow River Headwater Basin. From Zhang X, Srinivasan R, and Bosch D (2009) Calibration and uncertainty analysis of the SWAT model using genetic algorithms and Bayesian model averaging. *Journal of Hydrology* 374: 307–317.

Machine learning is concerned with the design and development of algorithms that allow computers to learn based on data, such as from sensor data or databases. A major focus of machine learning research is to automatically learn to recognize complex patterns and make intelligent decisions based on data. Machine learning techniques include, among others, approaches that have been widely used in hydrology, such as neural networks, nearest-neighbor methods, and statistical methods.

2.17.7 Assessment of Data Uncertainty

Among the sources of uncertainty in hydrology, data uncertainty is often believed to play a marginal role. Accordingly, only a few attempts have been made to quantify the effects of uncertainty in observations on hydrological modeling (see, for instance, Clarke, 1999).

Different types of observations are currently used in hydrological modeling. The most common applications usually refer to precipitation as input and river flows as output data, although very often solar radiation, temperature, wind speed, soil moisture, groundwater levels, geomorphological features, land use, and others are also employed. Some of the above variables are affected by a limited uncertainty with respect to the others. In particular, uncertainty in precipitation and river flow is often considered to be dominant, because of the spatial variability of rainfall and snowfall on the one hand, and the errors in the determination of the rating curve on the other.

The presence of uncertainty in input and output data induces two types of problems to hydrologists: the first is related to its estimation (to what extent the observed data are uncertain?), whereas the second is connected to accounting for such uncertainty in hydrological modeling.

2.17.7.1 Precipitation Uncertainty

Hydrologists are well aware that a multitude of problems and research issues are related to precipitation uncertainty, which are connected to precipitation monitoring and prediction (Chua and Bras, 1982; Gottschalk and Jutman, 1982).

Precipitation monitoring is carried out through direct measurements (raingauges and snowgauges) or remote sensing (satellite, radar, and microsensors). The uncertainty of gauge measurements is typically limited and therefore the estimation error of the precipitation field is mainly induced by spatial variability. When remote-sensed data are used, spatial variability is generally better estimated but the uncertainty in point measurements is relevant. There is a large body of literature about uncertainty assessment for precipitation, starting from the pioneeristic work of Thiessen (1911).

Uncertainty of gauge measurements of precipitation has been the subject of numerous case studies (see, for instance, Morrissey et al., 1995; Brath et al., 2004). These studies proved that the estimation error of mean areal precipitation significantly depends on the climatic conditions, the spatial structure of the precipitation itself, the morphology of the catchment, and the gauging network.

The task of quantifying remotely sensed precipitation uncertainty has proved to be difficult. A fundamental problem is the lack of a term of comparison (Habib and Krajewski, 2002). Numerous studies compared remotely sensed and gauged data and showed significant disagreement. For instance, Austin (1987) found that for individual storms, radar and raingauge measurements can differ of a factor of 2 or more. In a more recent investigation, Brandes et al. (1999) found that radar-to-gauge ratios of storm totals were in the range of about 0.7–1.9. These differences become even more significant when satellite versus raingauge comparisons are carried out.

In general, estimating precipitation uncertainty is a difficult task and no general rule exists. Integrating different monitoring techniques is certainly a potentially valuable solution.

Turning to the purpose of accounting for precipitation uncertainty in hydrological modeling, different methods were proposed by the literature. The BFS described in Section 2.17.6.3 is a relevant example. GLUE can be applied as well, by introducing an input error model and then generating many different realizations of the input data themselves, with which one can derive a likelihood weighted model output.

Kavetski et al. (2002, 2006a, 2006b) introduced the Bayesian total error analysis (BATEA), that is, a method for explicitly accounting for sampling and measurement uncertainty in both input and output data. In view of the inability to build a formal and sufficiently representative input error model in many real-world applications, BATEA is based on the use of vague error models, with the awareness that such an approach can cause a degeneration of the reliability of the inference equations.

The basic working hypothesis of BATEA consists of assuming that the input uncertainty is multiplicative Gaussian and independent of each storm, even though its general framework allows alternative uncertainty models. The multiplier approach assumes that the storm depth is the only quantity in error, whereas the rainfall pattern is correct up to a multiplicative constant m_i, such that $d_i^t = m_i d_i$, where d_i^t and d_i are the observed and true precipitation depths for the ith storm. Accordingly, the parameter vector of the hydrologic model is extended to include the parameters of the uncertainty models. Parameter inference is then carried out within a Bayesian framework, which requires indentifying a prior distribution of the model parameters that is subsequently updated by using Bayes' theorem in view of the available observations.

The above treatment of input uncertainty implies that the dimensionality of the parameter vector is increased with latent variables, whose number depends on the sample size of the observed data (and the type of error model assumed). Moreover, if both the input and the output data are observed with large uncertainty, the utility of any parameter estimation methodology becomes questionable. Finally, one might be concerned that rainfall multipliers can possibly interact with other sources of error and therefore separation of errors in BATEA is conditional on other sources of uncertainty. For instance, classic underprediction by a hydrological model after a long dry period can be compensated by increasing the rainfall multiplier. A similar treatment for precipitation uncertainty is used in IBUNE (Ajami et al., 2007).

2.17.7.2 River Discharge Uncertainty

Pelletier (1987) reviewed 140 publications dealing with uncertainty in the determination of the river discharge, thereby providing an extensive summary. He referred to the case where the river discharge at a given cross section is measured by using the velocity–area method, that is,

$$Q_{obs}(t) = A(t) \cdot v(t) \qquad (15)$$

where t is the sampling time, $Q_{obs}(t)$ the measured river discharge, $A(t)$ the cross-sectional area of the river, and $v(t)$ the velocity of the river flow averaged over the cross section. Errors in $Q_{obs}(t)$ are originated by uncertainties in both $A(t)$ and $v(t)$, which in turn are originated by uncertainty in the current meter, variability of the river flow velocity over the cross section, and uncertainty in the estimation of the cross-section geometry. Pelletier (1987) highlighted that the overall uncertainty in a single determination of river discharge, at the 95% confidence level, can vary in the range (8–20%), mainly depending on the exposure time of the current meter, the number of sampling points where the velocity is measured, and the value of $v(t)$. Another interesting contribution was provided by the European ISO EN Rule 748 (1997) that quantified the expected errors in the determination of the river discharge with the velocity–area method. The conclusions were similar to those of Pelletier (1987).

In some cases, including the usual practice in many countries of Europe, river discharge values are estimated by using the rating curve method, which is very easy to apply. According to the rating curve method, observations of river stage are converted into river discharge by means of a rating curve, which is preliminarily estimated by using observations collected using the velocity–area method. Hence, an additional error is induced by imperfect estimation of the rating curve.

Di Baldassarre and Montanari (2009) proposed a model for estimating the error affecting river flow observations derived by the rating curve method. The model aims at taking into account the main sources of uncertainty within a simplified approach. The most important assumptions underlying the model are as follows. (1) The uncertainty induced by imperfect observation of the river stage is negligible. This is consistent with the fact that these errors are usually very limited (around 1–2 cm; e.g., Schmidt, 2002; Pappenberger et al., 2006) and therefore of the same order of magnitude as standard topographic errors. (2) The geometry of the river is assumed to be invariant, which means that the rating curve changes in time only because of seasonal variation of roughness (see below). This assumption has been made because the uncertainty induced by possible variations of the river geometry is heavily dependent on the considered case study and no general rule can be suggested. However, it is worth noting that, using this assumption, the study neglects one of the most relevant sources of uncertainty that may affect river discharge observations where relevant sediment transport and erosion processes are present. (3) Uncertainty in the rating curve derives from the following causes: errors in the river discharge measurements that are used to calibrate the rating curve itself; interpolation and extrapolation error of the rating curve; unsteady flow conditions; and seasonal changes of roughness.

The uncertainty affecting the river discharge measurements was estimated by Di Baldassarre and Montanari (2009) according to the guidelines reported by the European ISO EN Rule 748 (1997), which lead to an estimate of about 5–6% when the measurements are collected in ideal conditions. This outcome matches the indications reported in Leonard et al. (2000) and Schmidt (2002).

The remaining sources of uncertainty were evaluated by Di Baldassarre and Montanari (2009) by developing a numerical simulation study for a 330-km reach of the Po River (Italy). The study focused on 17 cross sections and found that the estimation of river discharge using the rating curve method is affected by an increasing error for increasing river discharge values. At the 95% confidence level, the error ranges from 6.2% to 42.8% of the observation, with an average value of 25.6%. Furthermore, the uncertainty induced by the extrapolation of the rating curve is dominating the other errors in high flow conditions. In fact, previous contributions in hydrology (e.g., Rantz et al., 1982) do not recommend extrapolating rating curves beyond a certain range. Nevertheless, several hydrological applications are unavoidably based on flood flow observations (e.g., calibration and validation of rainfall–runoff models, flood frequency analysis, and boundary conditions of flood inundation models) and therefore one needs to extrapolate the rating curve beyond the measurement range (Pappenberger et al., 2006).

The above analysis proved that river flow uncertainty can indeed be very significant and therefore should be accounted for in practical applications. An interesting opportunity is offered by the application of GLUE according to the limits of acceptability concept (Blazkova and Beven, 2009). Once the uncertainty in the river flows is estimated, it is possible to fix limits of acceptability for the observed river flows, and the models that do not respect them can be rejected as nonbehavioral. The collection of the behavioral outputs allows the user to obtain an envelope of likely model simulations. The above approach is nonprobabilistic.

2.17.8 Assessment of Parameter Uncertainty

Calibration in hydrology is increasingly done automatically, while manual calibration (through trial and error procedures) is used only when dealing with complex models requiring high computational costs. Parameter calibration techniques lead to either a single solution or multiple solutions (i.e., parameter sets). The approaches leading to a single solution are basically optimization problems, while techniques leading to multiple likely solutions can serve as tools for uncertainty assessment. Many search algorithms have been successfully devised and applied to automatically find the optimal parameter set for hydrological models, which can be subdivided into local, global, and hybrid search techniques (Duan et al., 1992; Mugunthan and Shoemaker, 2006; Tolson and Shoemaker, 2008; Thyer et al., 2009; Tonkin and Doherty, 2009).

Approaches for multiple-solution parameter estimation can be broadly divided into two categories: importance sampling and Markov chain Monte Carlo (MCMC) sampling (Kuczera and Parent, 1998). With this approach full parameter distributions rather than simple point estimates can be obtained. Methods based on importance sampling aim to identify a set of behavioral model parameter configurations according to a selected objective function. Then, parameter distributions are estimated using a weighted combination of the behavioral parameter sets. GLUE is perhaps the most-used method based on importance sampling.

MCMC parameter estimation incorporates importance sampling into a procedure for evaluating conditional probability distributions. Prior parameter distributions are selected (for instance, by assigning a uniform distribution or a distribution derived through expert knowledge) and the sampler evolves them into posterior distributions that are estimated by using the observed data.

Thus, multiple-solution approaches can be used to assess parameter uncertainty. A relevant example within this respect is the SCEM-UA algorithm by Vrugt et al. (2003a). Once the uncertainty in the parameters is known, simulation approaches can be applied to estimate the related uncertainty induced in the model output. An example is given by Thorsen et al. (2001) who assessed the uncertainty in simulations of nitrate leaching induced by using model parameters obtained from databases at the European level.

End-users frequently experience the case where multiple or competing objectives need to be optimized. According to this need, numerous multiobjective optimization algorithms have been devised, with numerous developments in the recent past (see, for instance, Zhang et al., 2008). Relevant examples are the MOSCEM-UA and AMALGAM methods (Vrugt et al., 2003b; Vrugt and Robinson, 2007). These two methods are briefly described in the following.

2.17.8.1 The MOSCEM-UA Method

Multiobjective calibration problems can be dealt with by defining more than one optimization criteria (objective functions) that correspond to different performance measures of the selected model. Then, a multicriteria optimization method can be used to identify the set of nondominated, efficient, or Pareto optimal solutions (Gupta et al., 1998). The Pareto solutions represent tradeoffs among the different performance measures that are often conflicting. As such, moving from one solution to another results in the improvement of one objective and deterioration in one or more others.

A simple way to deal with multiobjective calibration is to weigh the different criteria into a single objective function and to run a large number of independent single-criteria optimization runs using different values for the weights (Madsen, 2000). This method is simple to implement, but has the drawback that a complete single-objective optimization is to be solved to obtain each discrete Pareto solution. Moreover, maintaining the independence of the various criteria will allow the user to analyze the tradeoffs among the different criteria, therefore enabling an improved understanding of the limitations of the model structure.

MOSCEM-UA (Vrugt et al., 2003b) is an effective and efficient MCMC sampler, which is capable of generating a fairly uniform approximation of the Pareto frontier within a single optimization run. The algorithm is closely related to the SCEM-UA algorithm (Vrugt et al., 2003a). In addition, MOSCEM-UA uses a newly developed, improved concept of Pareto dominance, thereby also containing the single-criteria solutions at the extremes of the Pareto solution set. For more details, the interested reader is invited to refer to Vrugt et al. (2003b).

The ensemble of the models lying on the Pareto frontier allows the user to identify an envelope of model outputs corresponding to the nondominated solutions.

2.17.8.2 The AMALGAM Method

AMALGAM (Vrugt and Robinson, 2007) is a follow-up of MOSCEM-UA and is specifically designed to take full advantage of the power of distributed computer networks. AMALGAM runs multiple different search strategies simultaneously for population evolution and adaptively updates the weights of these individual methods based on their reproductive success. This ensures a fast, reliable, and computationally efficient solution to multiobjective optimization problems.

2.17.9 Assessment of Model Structural Uncertainty

Model structural uncertainty is induced by inadequateness of the hydrological model to represent the hydrological system. This situation is also characterized by reduced model identifiability, because the imperfectness of the modeling solutions makes many of them potentially suboptimal, regardless of the different values of the selected performance measure. In the presence of model, structural uncertainty a performance measure becomes less effective and therefore the highest of its value does not necessarily identify the best model. For instance, a performance measure that lays emphasis on floods may be biased toward a (imperfect) model that could not be as reliable in reproducing the low flows. This is the reason why multiobjective calibration is frequently applied in hydrology.

A statistical and rigorous evaluation of model structural uncertainty is not possible in practical hydrological applications, at least because it should necessarily be performed with perfect data. The literature proposed approximate techniques for estimating the uncertainty in the model output induced by model structural uncertainty. The most popular of them are based on multimodel analysis. In fact, the variability of the response provided by different models, if other uncertainty sources are negligible, provides indications on the uncertainty induced by a wrong model structure.

Multimodel analysis is based on the use of many different plausible models that may consider, for instance, alternative processes and alternative simplified approximations. An example of model combination is the BMA presented in Section 2.17.6.5.

Another quantitative approach for performing multimodel application was presented by Burnham and Anderson (2002) and Ye *et al.* (2008). It is implemented by assigning performance scores and importance weights to each candidate model with which the ensemble of model outputs can be constructed basing on the importance of each model.

Multimodel applications can be performed also by applying GLUE, with which different models can be considered and evaluated according to a single likelihood measure or one or more likelihood measures. An example of application of GLUE with different modeling solutions is provided by Rojas *et al.* (2009).

From a practical point of view, the above techniques are often applied for assessing the global uncertainty in the model output instead of model structural uncertainty only, because it is impossible to carry out such techniques in the absence of data and parameter uncertainty. As such, the combination of different models with uncertain parameters and uncertain data bases does not allow one to separate the above sources of uncertainty, unless one makes heavy assumptions (see, e.g., the IBUNE method; Ajami *et al.* (2007); see also Clark *et al.* (2008)).

2.17.10 Uncertainty Assessment as a Learning Process

Uncertainty assessment is an effective mean to quantitatively assess model reliability and therefore perform model diagnostic and evaluation. These latter, in turn, provide indications about the model ability to simulate hydrology at a given place and therefore about the correctness of our understanding of the hydrological processes at that place. Thus, uncertainty assessment plays a fundamental role in the learning process.

In the past, the learning process was mainly linked to parameter estimation for a given model. The optimal parameter values, actually, provide information about the conditions of the system. Treating modeling more explicitly as a learning process allows one to follow a new approach to this problem based on a methodology that will link models, databases, and parameters with the areas of interest, thereby providing information on the dominant hydrological processes (see Beven (2007); applications are presented in Montanari *et al.* (2006), Fenicia *et al.* (2008), and Schoups *et al.* (2008)). This is part of the downward modeling approach that recently gained increased attention within the context of PUB.

One of the most exciting future perspectives is the possibility to implement many different models as a process of learning about specific places (Beven, 2007). The representation will be uncertain so that this learning process should be implemented within a framework of uncertainty estimation. Indeed, uncertainty estimation, providing quantitative information about model reliability, if coupled with a multimodel approach, could provide indications about the dominant hydrological processes and their dynamics. Within this framework it is also necessary to set up a mechanism for model rejection (e.g., the model providing the simulations presented in **Figure 1** could be rejected because it is too biased). There is the potential problem that model rejection is not by default embedded in uncertainty assessment methods. In particular, it is not embedded in statistical approaches, which in many cases do not assess the motivation for the presence of uncertainty. Model rejection is often based on expert knowledge, which is subjective but indeed necessary in the context of a learning process (see, for instance, Merz and Blöschl (2008a, 2008b)). This implies that the use of statistical methods for uncertainty assessment in a learning process should be based on including in the statistical representation the available information about the underlying physical process (for an extended discussion, see Koutsoyiannis (2009)).

2.17.11 Conclusions

Uncertainty assessment in hydrology is a relevant practical problem and still a research challenge. The limited extension of hydrological databases and the complexity of hydrological processes, whose dynamics and domains are to a great extent nonobservable, make the interpretation of the results of hydrological modeling studies not easy. The intense research activity recently done on uncertainty resulted in the development of many new techniques for uncertainty assessment, which differ in behavior and scope. It is essential to formally define a terminology and make clear the prerogatives of each method in order to make clear to end-users the meaning of uncertainty in hydrology and convey them a useful information.

In order to provide a contribution to this end, we provide in **Table 1** a brief summary of the most-used uncertainty assessment methods, including those presented here, by also providing an attempt of classification and by specifying their purpose.

Uncertainty assessment in hydrology will represent a research challenge for a long time to come. Uncertainty is an inherent property of hydrological processes which in principle will not prevent gaining a much better understanding of how water flows downstream. Uncertainty in hydrology should not be viewed as a limitation to be eliminated but rather as a intrinsic feature that needs to be properly and objectively quantified, whenever possible, with scientific method, that is, through the collection of data by means of observation and experimentation, and the formulation and testing of hypotheses.

Communicating uncertainty to end-users should not undermine their confidence in models (Beven, 2006b; Pappenberger and Beven, 2006; Faulkner *et al.*, 2007), but rather increase it through an improved perception of the underlying natural processes and an increased awareness of model reliability. Uncertainty does not mean lack of knowledge or lack of modeling capability but that the predicted value of a hydrological variable is uncertain. A proper estimation of uncertainty is the way forward to a reliable hydrological design and therefore a proper management of the environment and water resources.

Acknowledgments

The author is grateful to Demetris Koutsoyiannis, Jasper Vrugt, Keith Beven, Simone Castiglioni, an anonymous referee and the Editor Stefan Uhlenbrook for providing very useful comments on the text. The support of the Italian Government, through the National Research Project "Uncertainty estimation for precipitation and river discharge data. Effects on water resources planning and flood risk management" is also acknowledged.

References

Ajami NK, Duan Q, Gao X, and Sorooshian S (2006) Multi-model combination techniques for hydrological forecasting: Application to distributed model intercomparison project results. *Journal of Hydrometeorology* 8: 755–768.

Ajami NK, Duan Q, and Sorooshian S (2007) An integrated hydrologic Bayesian multimodel combination framework: Confronting input, parameter, and model structural uncertainty in hydrologic prediction. *Water Resources Research* 43: W01403 (doi:10.1029/2005WR004745).

Alley WM (1984) On the treatment of evapotranspiration, soil moisture accounting and aquifer recharge in monthly water balance models. *Water Resources Research* 20: 1137–1149.

Aronica G, Bates PD, and Horritt MS (2002) Assessing the uncertainty in distributed model predictions using observed binary pattern information within GLUE. *Hydrological Processes* 16: 2001–2016.

Aronica G, Hankin B, and Beven KJ (1998) Uncertainty and equifinality in calibrating distributed roughness coefficients in a flood propagation model with limited data. *Advances in Water Resources* 22: 349–365.

Austin PM (1987) Relation between measured radar reflectivity and surface rainfall. *Monthly Weather Review* 115: 1053–1070.

Beven KJ (1993) Prophesy, reality and uncertainty in distributed hydrological modeling. *Advances in Water Resources* 16: 41–51.

Beven KJ (2006a) A manifesto for the equifinality thesis. *Journal of Hydrology* 320: 18–36.

Beven KJ (2006b) On undermining the science? *Hydrological Processes* 20: 3141–3146.

Beven KJ (2007) Towards integrated environmental models of everywhere: Uncertainty, data and modeling as a learning process. *Hydrology and Earth System Sciences* 11: 460–467.

Beven KJ (2009) *Environmental Modelling: An Uncertain Future?: An Introduction to Techniques for Uncertainty Estimation in Environmental Prediction.* London: Taylor and Francis.

Beven KJ and Binley AM (1992) The future of distributed models: Model calibration and uncertainty prediction. *Hydrological Processes* 6: 279–298.

Beven KJ, Smith PJ, and Freer JE (2007) Comment on "Hydrological forecasting uncertainty assessment: Incoherence of the GLUE methodology" by Pietro Mantovan and Ezio Todini. *Journal of Hydrology* 338: 315–318.

Beven KJ, Smith PJ, and Freer JE (2008) So just why would a modeller choose to be incoherent? *Journal of Hydrology* 354: 15–32.

Blazkova S and Beven KJ (2009) A limits of acceptability approach to model evaluation and uncertainty estimation in flood frequency estimation by continuous simulation: Skalka catchment, Czech Republic. *Water Resources Research* 45: W00B16 (doi:10.1029/2007WR006726).

Brandes EA, Vivekanandan J, and Wilson JW (1999) A comparison of radar reflectivity estimates of rainfall from collocated radars. *Journal of Atmospheric Oceanic Technology* 16: 1264–1272.

Brath A, Montanari A, and Toth E (2004) Analysis of the effects of different scenarios of historical data availability on the calibration of a spatially-distributed hydrological model. *Journal of Hydrology* 291: 272–288.

Burnham KP and Anderson DR (2002) *Model Selection and Multimodel Inference*, 2nd edn. New York: Springer.

Cameron DS, Beven KJ, Tawn J, Blazkova S, and Naden P (1999) Flood frequency estimation by continuous simulation for a gauged upland catchment (with uncertainty). *Journal of Hydrology* 219: 169–187.

Christensen S (2003) A synthetic groundwater modelling study of the accuracy of GLUE uncertainty intervals. *Nordic Hydrology* 35: 45–59.

Chua S-H and Bras RL (1982) Optimal estimations of mean areal precipitation in regions of orographic influence. *Journal of Hydrology* 57: 23–48.

Clark MP, Slater AG, Rupp DE, et al. (2008) Framework for understanding structural errors (FUSE): A modular framework to diagnose differences between hydrological models. *Water Resources Research* 44: W00B02 (doi:10.1029/2007WR006735).

Clarke RT (1999) Uncertainty in the estimation of mean annual flood due to rating curve indefinition. *Journal of Hydrology* 222: 185–190.

Cullen AC and Frey HC (1999) *Probabilistic Techinques in Exposure Assessment: A Handbook for Dealing with Variability and Uncertainty in Model and Inputs.* New York: Plenum.

Di Baldassarre G and Montanari A (2009) Uncertainty in river discharge observations: A quantitative analysis. *Hydrology and Earth System Sciences* 13: 913–921.

Dottori F, Martina MLV, and Todini E (2009) A dynamic rating curve approach to indirect discharge measurement. *Hydrology and Earth System Sciences* 13: 847–863.

Duan Q, Ajami NK, Gao X, and Sorooshian S (2007) Multi-model ensemble hydrologic prediction using Bayesian model averaging. *Advances in Water Resources* 30: 1371–1386.

Duan Q, Sorooshian S, and Gupta VK (1992) Effective and efficient gobal optimization for conceptual rainfall-runoff models. *Water Resources Research* 28: 1015–1031.

European ISO EN Rule 748 (1997) Measurement of liquid flow in open channels – velocity-area methods, Reference number ISO 748:1997 (E), International Standard.

Faulkner H, Parker D, Green C, and Beven KJ (2007) Developing a translational discourse to communicate uncertainty in flood risk between science and the practitioner. *AMBIO* 36: 692–704.

Fenicia F, Savenije HHG, Matgen P, and Pfister L (2008) Understanding catchment behavior through stepwise model concept improvement. *Water Resources Research* 44: W01402 (doi:10.1029/2006WR005563).

Franz K, Ajami N, Schaake J, and Buizza R (2005) Hydrologic ensemble prediction experiment focuses on reliable forecasts. *Eos* 86: 239.

Freer J, Beven KJ, and Ambroise B (1996) Bayesian uncertainty in runoff prediction and the value of data: An application of the GLUE approach. *Water Resources Research* 32: 2163–2173.

Freni G, Mannina G, and Viviani G (2008) Uncertainty in urban stormwater quality modelling: The effect of acceptability threshold in the GLUE methodology. *Water Research* 42: 2061–2072.

Freni G, Mannina G, and Viviani G (2009) Uncertainty assessment of an integrated urban drainage model. *Journal of Hydrology* 373: 392–404.

Gan TY and Burges SJ (1990) An assessment of a conceptual rainfall-runoff model's ability to represent the dynamics of small hypothetical catchments 2. Hydrologic responses for normal and extreme rainfall. *Water Resources Research* 26: 1605–1619.

Gottschalk L and Jutman T (1982) Calculation of areal means of meteorologic variables for watersheds. *7th Nordic Hydrological Conference.* NHP-Report No. 2, pp. 720–736.

Götzinger J and Bárdossy A (2008) Generic error model for calibration and uncertainty estimation of hydrological models. *Water Resources Research* 44: W00B07 (doi:10.1029/2007WR006691).

Gupta HV, Sorooshian S, and Yapo PO (1998) Toward improved calibration of hydrologic models: Multiple and noncommensurable measures of information. *Water Resources Research* 34: 751–763.

Habib E and Krajewski WF (2002) Uncertainty analysis of the TRMM ground-validation radar-rainfall products: Application to the TEFLUN-B field campaign. *Journal of Applied Meteorology* 41: 558–572.

Hahn GJ and Meeker WQ (1991) *Statistical Intervals; A Guide for Practitioners.* New York: Wiley.

Hall J and Solomatine DP (2008) A framework for uncertainty analysis in flood risk management decisions. *International Journal of River Basin Management* 6: 85–98.

Hoeting JA, Madigan D, Raftery AE, and Volinsky CT (1999) Bayesian model averaging: A tutorial. *Statistical Science* 14: 382–401 (doi:10.1214/ss/1009212519).

Hoffman FO and Hammonds JS (1994) Propagation of uncertainty in risk assessments: The need to distinguish between uncertainty due to lack of knowledge and uncertainty due to variability. *Risk Analysis* 14: 707–712.

Huard D and Mailhot A (2006) A Bayesian perspective on input uncertainty in model calibration: Application to hydrological model "abc". *Water Resources Research* 42: W07416 (doi:10.1029/2005WR004661).

Ibbitt RP and O'Donnell T (1974) Designing conceptual catchment models for automatic fitting methods. In: *Mathematical Models in Hydrology Symposium*, IAHS-AISH, Publication No. 2, pp. 461–475.

Jacquin AP and Shamseldin AY (2007) Development of a possibilistic method for the evaluation of predictive uncertainty in rainfall-runoff modeling. *Water Resources Research* 43: W04425. doi:10.1029/2006WR005072.

Kavetski D, Franks SW and Kuczera G (2002) Confronting input uncertainty in environmental modelling. In: *Calibration of Watershed Models*. Duan Q, Sorooshian S, Gupta H, Rosseau H, and Turcotte R (eds.). Water Science and Application, vol. 6, pp. 49–68. Washington, DC: AGU.

Kavetski D, Kuczera G, and Franks SW (2006a) Bayesian analysis of input uncertainty in hydrological modeling: 1. Theory. *Water Resources Research* 42: W03407 (doi:10.1029/2005WR004368).

Kavetski D, Kuczera G, and Franks SW (2006b) Bayesian analysis of input uncertainty in hydrological modeling: 2. Application. *Water Resources Research* 42: W03408 (doi:10.1029/2005WR004376).

Kelly KS and Krzysztofowicz R (1997) A bivariate meta-Gaussian density for use in hydrology. *Stochastic Hydrology and Hydraulics* 11: 17–31.

Kelly KS and Krzysztofowicz R (2000) Precipitation uncertainty processor for probabilistic river stage forecasting. *Water Resources Research* 36: 2643–2653.

Kleisenn FM, Beck MB, and Weather HS (1990) The identifiability of conceptual hydrochemical models. *Water Resources Research* 26: 2979–2992.

Koussis A (2009) Comment on "A dynamic rating curve approach to indirect discharge measurement" by Dottori *et al.* (2009). *Hydrology and Earth System Sciences Discussions* 6: 7429–7437.

Koutsoyiannis D (2009) A random walk on water. *Hydrology and Earth System Sciences Discussions* 6: 6611–6658.

Koutsoyiannis D, Makropoulos C, Langousis A, *et al.* (2009) HESS opinions: "Climate, hydrology, energy, water: Recognizing uncertainty and seeking sustainability". *Hydrology and Earth System Sciences* 13: 247–257.

Krzysztofowicz R (1999) Bayesian theory of probabilistic forecasting via deterministic hydrologic model. *Water Resources Research* 35: 2739–2750.

Krzysztofowicz R (2001a) The case for probabilistic forecasting in hydrology. *Journal of Hydrology* 249: 2–9.

Krzysztofowicz R (2001b) Integrator of uncertainties for probabilistic river stage forecasting: Precipitation-dependent model. *Journal of Hydrology* 249: 69–85.

Krzysztofowicz R (2002) Bayesian system for probabilistic river stage forecasting. *Journal of Hydrology* 268: 16–40.

Krzysztofowicz R and Herr HD (2001) Hydrologic uncertainty processor for probabilistic river stage forecasting: Precipitation-dependent model. *Journal of Hydrology* 249: 46–68.

Krzysztofowicz R and Kelly KS (2000) Hydrologic uncertainty processor for probabilistic river stage forecasting. *Water Resources Research* 36: 3265–3277.

Kuczera G and Mroczkowski M (1998) Assessment of hydrologic parameter uncertainty and the worth of multiresponse data. *Water Resources Research* 34: 1481–1489.

Kuczera G and Parent E (1998) Monte Carlo assessment of parameter uncertainty in conceptual catchment models: The Metropolis algorithm. *Journal of Hydrology* 211: 69–85.

Kuichling E (1889) The relation between rainfall and the discharge of sewers in populous districts. *Transactions ASCE* 20: 1–60.

Langley RS (2000) Unified approach to probabilistic and possibilistic analysis of uncertain systems. *Journal of Engineering Mechanics* 126: 1163–1172.

Leonard J, Mietton M, Najib H, and Gourbesville P (2000) Rating curve modelling with Manning's equation to manage instability and improve extrapolation. *Hydrological Sciences Journal* 45: 739–750.

Li X and Tsai FT-C (2009) Bayesian model averaging for groundwater head prediction and uncertainty analysis using multimodel and multimethod. *Water Resources Research* 45: W09403 (doi:10.1029/2008WR007488).

Liu Y, Freer J, Beven KJ, and Matgen P (2009) Towards a limits of acceptability approach to the calibration of hydrological models: Extending observation error. *Journal of Hydrology* 367: 93–103.

Liu Y and Gupta HV (2007) Uncertainty in hydrologic modeling: Toward an integrated data assimilation framework. *Water Resources Research* 43: W07401 (doi:10.1029/2006WR005756).

Loucks DP and Van Beek E (2005) Water resources systems planning and management an introduction to methods, models and applications. In: *Studies and Reports in Hydrology*, United Nations Educational, Scientific and Cultural Organization (ISBN 92-3-103998-9). Paris, France: UNESCO.

Madsen H (2000) Automatic calibration of a conceptual rainfall-runoff model using multiple objectives. *Journal of Hydrology* 235: 276–288.

Mantovan P and Todini E (2006) Hydrological forecasting uncertainty assessment: Incoherence of the GLUE methodology. *Journal of Hydrology* 330: 368–381.

Matott LS, Babendreier JE, and Purucker ST (2009) Evaluating uncertainty in integrated environmental models: A review of concepts and tools. *Water Resources Research* 45: W06421 (doi:10.1029/2008WR007301).

Melching CS (1992) An improved first-order reliability approach for assessing uncertainties in hydrologic modeling. *Journal of Hydrology* 132: 157–177.

Merz R and Blöschl G (2008a) Flood frequency hydrology: 1. Temporal, spatial, and causal expansion of information. *Water Resources Research* 44: W08432, doi:10.1029/2007WR006744.

Merz R and Blöschl G (2008b) Flood frequency hydrology: 2. Combining data evidence. *Water Resources Research* 44: W08433, doi:10.1029/2007WR006745.

Montanari A (2005) Large sample behaviors of the generalized likelihood uncertainty estimation (GLUE) in assessing the uncertainty of rainfall-runoff simulations. *Water Resources Research* 41: W08406 (doi:10.1029/2004WR003826).

Montanari A (2007) What do we mean by 'uncertainty'? The need for a consistent wording about uncertainty assessment in hydrology. *Hydrological Processes* 21: 841–845.

Montanari A and Brath A (2004) A stocastic approach for assessing the uncertainty of rainfall-runoff simulations. *Water Resources Research* 40: W01106 (doi:10.1029/2003WR002540).

Montanari A and Grossi G (2008) Estimating the uncertainty of hydrological forecasts: A statistical approach. *Water Resources Research* 44: W00B08 (doi:10.1029/2008WR006897).

Montanari A, Shoemaker CA, and Van de Giesen NC (2009) Uncertainty assessment in surface and subsurface hydrology: An overview. *Water Resources Research* 45: W00B00 (doi:10.1029/2009WR008471).

Montanari L, Sivapalan M, and Montanari A (2006) Investigation of dominant hydrological processes in a tropical catchment in a monsoonal climate via the downward approach. *Hydrology and Earth System Sciences* 10: 769–782.

Morrissey ML, Maliekal JA, Greene JS, and Wang J (1995) The uncertainty of simple spatial averages using rain gauge networks. *Water Resources Research* 31: 2011–2017.

Mugunthan P and Shoemaker CA (2006) Assessing the impacts of parameter uncertainty for computationally expensive groundwater models. *Water Resources Research* 42: W10428 (doi:10.1029/2005WR004640).

Mulvaney TJ (1851) On the use of self-registering rain and flood gauges in making observations of the relations of rainfall and of flood discharges in a given catchment. *Proceeding of the Institute of Civil Engineers of Ireland* 4: 18–31.

Nash JE and Sutcliffe JV (1970) River flow forecasting through conceptual models. 1: A discussion of principles. *Journal of Hydrology* 10: 282–290.

Nauta MJ (2000) Separation of uncertainty and variability in quantitative microbial risk assessment models. *International Journal of Food Microbiology* 57: 9–18.

Neuman SP (2003) Maximum likelihood Bayesian averaging of alternative conceptual-mathematical models. *Stochastic Environmental Resources Risk Assessment* 17: 291–305.

NRC (National Research Council) (2000) *Risk Analysis and Uncertainty in Flood Damage Reduction Studies*. Washington, DC: National Academy Press.

Page T, Beven KJ, Freer J, and Neal C (2007) Modelling the chloride signal at Plynlimon, Wales, using a modified dynamic TOPMODEL incorporating conservative chemical mixing (with uncertainty). *Hydrological Processes* 21: 292–307.

Papoulis A (1991) *Probability, Random Variables, and Stochastic Processes*, 3rd edn. New York: McGraw-Hill.

Pappenberger F and Beven KJ (2004) Functional classification and evaluation of hydrographs based on Multicomponent Mapping (M^x). *International Journal of River Basin Management* 2: 89–100.

Pappenberger F and Beven KJ (2006) Ignorance is bliss: Or seven reasons not to use uncertainty analysis. *Water Resources Research* 42: W05302 (doi:10.1029/2005WR004820).

Pappenberger F, Matgen P, Beven KJ, Henry JB, Pfister L, and de Fraipont P (2006) Influence of uncertain boundary conditions and model structure on flood inundation predictions. *Advances in Water Resources* 29: 1430–1449.

Pelletier MP (1987) Uncertainties in the determination of river discharge: A literature review. *Canadian Journal of Civil Engineering* 15: 834–850.

Petersen-Øverleir A and Reitan T (2009) Accounting for rating curve imprecision in flood frequency analysis using likelihood-based methods. *Journal of Hydrology* 366: 89–100.

Rantz SE, *et al.* (1982) *Measurement and Computation of Streamflow*, US Geological Survey, Water Supply Paper 2175. http://water.usgs.gov/pubs/wsp/wsp2175/ (accessed March 2010).

Refsgaard JC, van der Sluijs JP, Brown J, and Van der Keur P (2006) A framework for dealing with uncertainty due to model structure error. *Advances in Water Resources* 29: 1586–1597.

Regan HM, Akcakaya HR, Ferson S, Root KV, Carroll S, and Ginzburg LR (2003) Treatments of uncertainty and variability in ecological risk assessment of single-species populations. *Human Ecology Risk Assessment* 9: 889–906.

Reggiani P, Renner M, Weerts AH, and van Gelder PAHJM (2009) Uncertainty assessment via Bayesian revision of ensemble streamflow predictions in the operational river Rhine forecasting system. *Water Resources Research* 45: W02428 (doi:10.1029/2007WR006758).

Rojas R, Batelaan O, Feyen L, and Dassargues A (2009) Assessment of conceptual model uncertainty for the regional aquifer Pampa del Tamarugal – North Chile. *Hydrology and Earth System Sciences Discussion* 6: 5881–5935.

Schmidt AR (2002) Analysis of stage-discharge relations for open channel flow and their associated uncertainties. PhD Thesis, University of Illinois.

Schoups G, Van de Giesen NC, and Savenije HHG (2008) Model complexity control for hydrologic prediction. *Water Resources Research* 44: W00B03 (doi:10.1029/2008WR006836).

Sherman LK (1932) Streamflow from rainfall by the unit-graph method. *Engineering News Record* 108: 501–505.

Shrestha DL, Kayastha N, and Solomatine DP (2009) A novel approach to parameter uncertainty analysis of hydrological models using neural networks. *Hydrology and Earth System Sciences* 13: 1235–1248.

Shrestha DL and Solomatine DP (2008) Data-driven approaches for estimating uncertainty in rainfall-runoff modelling. *International Journal of River Basin Management* 6: 109–122.

Sieber A and Uhlenbrook S (2005) Sensitivity analyses of a distributed catchment model to verify the model structure. *Journal of Hydrology* 310: 216–235.

Solomatine DP and Shrestha DL (2009) A novel method to estimate model uncertainty using machine learning techniques. *Water Resources Research* 45: W00B11 (doi:10.1029/2008WR006839).

Sonich-Mullin C (2001) Harmonizing the incorporation of uncertainty and variability into risk assessment: An international approach. *Human and Ecological Risk Assessment* 7: 7–13.

Spear RC and Hornberger GM (1980) Eutrophication in Peel Inlet: II. Identification of critical uncertainties via generalized sensitivity analysis. *Water Research* 14: 43–49.

Stedinger JR, Vogel RM, Lee SU, and Batchelder R (2008) Appraisal of the generalized likelihood uncertainty estimation (GLUE) method. *Water Resources Research* 44: W00B06 (doi:10.1029/2008WR006822).

Thiessen AH (1911) Precipitation averages for large areas. *Monthly Weather Report* 39: 1082–1084.

Thorsen M, Refsgaard JC, Hansen S, Pebesma E, Jensen JB, and Kleeschulte S (2001) Assessment of uncertainty in simulation of nitrate leaching to aquifers at catchment scale. *Journal of Hydrology* 242: 210–227.

Thyer M, Renard B, Kavetski D, Kuczera G, Franks SW, and Srikanthan S (2009) Critical evaluation of parameter consistency and predictive uncertainty in hydrological modeling: A case study using Bayesian total error analysis. *Water Resources Research* 45: W00B14 (doi:10.1029/2008WR006825).

Tolson BA and Shoemaker CA (2008) Efficient prediction uncertainty approximation in the calibration of environmental simulation models. *Water Resources Research* 44: W04411 (doi:10.1029/2007WR005869).

Tonkin M and Doherty J (2009) Calibration-constrained Monte Carlo analysis of highly parameterized models using subspace techniques. *Water Resources Research* 45: W00B10 (doi:10.1029/2007WR006678).

Tsai CW and Franceschini S (2005) Evaluation of probabilistic point estimate methods in uncertainty analysis for environmental engineering applications. Journal of Environmental Engineering 131: 387–395.

Van Griensven A, Meixner T, Grunwald S, Bishop T, Di luzio M, and Srinivasan R (2006) A global sensitivity analysis tool for the parameters of multi-variable catchment models. *Journal of Hydrology* 324: 10–23.

Von Bertalanffy L (1968) *General System Theory*. New York: George Braziller.

Vrugt JA and Bouten W (2002) Validity of first-order approximations to describe parameter uncertainty in soil hydrologic models. *Soil Science Society of America Journal* 66: 1740–1752.

Vrugt JA, Gupta HV, Bastidas LA, Bouten W, and Sorooshian S (2003b) Effective and efficient algorithm for multiobjective optimization of hydrologic models. *Water Resources Research* 39: 1214 (doi:10.1029/2002WR001746).

Vrugt JA, Gupta HV, Bouten W, and Sorooshian S (2003a) *A Shuffled Complex Evolution Metropolis algorithm for optimization and uncertainty assessment of hydrologic model parameters*, Water Resources Research 39: 1201 (doi:10.1029/2002WR001642).

Vrugt JA and Robinson BA (2007) Improved evolutionary optimization from genetically adaptive multimethod search. *Proceedings of the National Academy of Science of the United States of America* 104: 708–711.

Wagener T, McIntyre N, Lees MJ, Wheater HS, and Gupta HV (2003) Towards reduced uncertainty in conceptual rainfall-runoff modelling: Dynamic identifiability analysis. *Hydrological Processes* 17: 455–476.

Winsemius HC, Savenije HHG, and Bastiaanssen WGM (2008) Constraining model parameters on remotely sensed evaporation: Justification for distribution in ungauged basins? *Hydrology and Earth System Sciences* 12: 1403–1413.

Winsemius HC, Savenije HHG, Gerrits AMJ, Zapreeva EA, and Klees R (2006) Comparison of two model approaches in the Zambezi river basin with regard to model reliability and identifiability. *Hydrology and Earth System Sciences* 10: 339–352.

Ye M, Meyer PD, and Neumann SP (2008) On model selection criteria in multimodel analysis. *Water Resources Research* 44: W03428 (doi:10.1029/2008WR006803).

Ye M, Neuman SP, and Meyer PD (2004) Maximum likelihood Bayesian averaging of spatial variability models in unsaturated fractured tuff. *Water Resources Research* 40: W05113 (doi:10.1029/2003WR002557).

Zadeh LA (2005) Toward a generalized theory of uncertainty (GTU)—an outline. *Information Sciences* 172: 1–40.

Zappa M, Rotach MV, Arpagaus M, *et al.* (2008) MAP D5-PHASE: Real-time demonstration of hydrological ensemble prediction systems. *Atmospheric Science Letters* 9: 80–87.

Zhang X, Srinivasan R, and Bosch D (2009) Calibration and uncertainty analysis of the SWAT model using genetic algorithms and bayesian model averaging. *Journal of Hydrology* 374: 307–317.

Zhang Z, Wagener T, Reed P, and Bhushan R (2008) Reducing uncertainty in predictions in ungauged basins by combining hydrologic indices regionalization and multiobjective optimization. *Water Resources Research* 44: W00B04 (doi:10.1029/2008WR006833).

Relevant Websites

http://education.mit.edu
　Gaussian Distribution.
http://www.itia.ntua.gr
　Presentation: Hurst-Kolmogorov dynamics and uncertainity.
http://www.agu.org
　Special issue on uncertainty.

2.18 Statistical Hydrology

S Grimaldi, Università degli Studi della Tuscia, Viterbo, Italy
S-C Kao, Oak Ridge National Laboratory, Oak Ridge, TN, USA
A Castellarin, Università degli Studi di Bologna, Bologna, Italy
S-M Papalexiou, National Technical University of Athens, Zographou, Greece
A Viglione, Technische Universität Wien, Vienna, Austria
F Laio, Politecnico di Torino, Torino, Italy
H Aksoy and A Gedikli, Istanbul Technical University, Istanbul, Turkey

© 2011 Elsevier B.V. All rights reserved.

2.18.1	Introduction	480
2.18.2	Analysis and Detection of Nonstationarity in Hydrological Time Series	480
2.18.2.1	The Common Nonstationarity Analysis Methods	481
2.18.2.1.1	Randomness test	481
2.18.2.1.2	Detection of trend	482
2.18.2.1.3	Simple regression on time	482
2.18.2.1.4	Mann–Kendall test	482
2.18.2.1.5	Spearman rank order correlation test	482
2.18.2.1.6	Detection of shifts (segmentation)	483
2.18.2.1.7	t-Test	483
2.18.2.1.8	Mann–Whitney test	483
2.18.2.2	A New Method of Segmentation	484
2.18.3	Extreme Value Analysis: Distribution Functions and Statistical Inference	485
2.18.3.1	Probability Distributions for Extreme Events	486
2.18.3.1.1	Normal distribution	486
2.18.3.1.2	Lognormal distribution	486
2.18.3.1.3	Exponential distribution	487
2.18.3.1.4	Gamma distribution	488
2.18.3.1.5	Pearson type 3 distribution	488
2.18.3.1.6	Log-Pearson type 3 distribution	488
2.18.3.1.7	Extreme value distributions	489
2.18.3.1.8	Generalized Pareto distribution	490
2.18.3.1.9	Generalized logistic distribution	490
2.18.3.2	Parameter Estimation Methods	490
2.18.3.2.1	Method of moments	490
2.18.3.2.2	Method of L-moments	490
2.18.3.2.3	Method of the maximum-likelihood and Bayesian methods	491
2.18.3.3	Model Verification: Goodness-of-Fit Tests	491
2.18.4	IDF Curves	493
2.18.4.1	Definition of IDF Curves and Clarifications	494
2.18.4.2	Empirical Methods	494
2.18.4.2.1	Parameter estimation	495
2.18.4.2.2	Application in a real-world data set	496
2.18.4.3	Theoretically Consistent Methods	499
2.18.4.3.1	Parameter estimation	499
2.18.4.3.2	Application in a real-world data set	500
2.18.5	Copula Function for Hydrological Application	501
2.18.5.1	Concepts of Dependence Structure and Copulas	502
2.18.5.2	Copulas in Hydrologic Applications	505
2.18.5.3	Remarks on Copulas and Future Research	506
2.18.6	Regional Frequency Analysis	506
2.18.6.1	Index-Flood Procedure, Extensions and Evolutions	506
2.18.6.2	Classical Regionalization Approach	507
2.18.6.2.1	Estimation of the index flood	507
2.18.6.2.2	Estimation of the regional dimensionless quantile	508
2.18.6.2.3	Homogeneity testing	508
2.18.6.2.4	Choice of a frequency distribution	509
2.18.6.2.5	Estimation of the regional frequency distribution	510

2.18.6.2.6	Validation of the regional model	510
2.18.6.3	Open Problems and New Advances	511
References		512

2.18.1 Introduction

Hydrological phenomena such as precipitation, floods, and droughts are inherently random by nature. Due to the complexity of the hydrologic system, these physical processes are not fully understood and reliable deterministic mathematical models are still to be developed. Therefore, in order to provide useful analyses for designing hydraulic facilities and infrastructures, statistical approaches have been commonly adopted.

In literature and in the practical hydrological applications, many statistical methods are considered with different aims. Simulation, forecasting, uncertainty analysis, spatial interpolation, and risk analysis are some of the most important ones. The use of statistical analyses is strongly related to the data availability and to the quality of observations. Particular emphasis is given to the case of ungauged area where the statistical approach is particularly important to develop hydrological analyses without direct observations (the relevance of this issue is well documented by the Decade on Prediction in Ungauged Basins (PUB) promoted by the International Association of Hydrological Sciences (IAHS, Sivapalan et al., 2003).

This chapter describes some statistical topics widely used in hydrology. Among the large number of subjects available in literature, the attention is focalized on some of them particularly useful either for innovative hydrological analyses or for an appropriate application of common procedures.

Many statistical methods are strongly affected by specific conditions to be verified on the available data set. Indeed, for instance, complex procedures, used for different important applications, usually need a very common and simple hypothesis: the stationarity. This condition, simple to define but very difficult to verify, probably is the most important in statistical hydrology. For this reason, the first section of this chapter provides a short review of this topic and a detailed description of the segmentation method that is a promising procedure for time series trend detection.

Another primary topic, described here, is the univariate extreme value (EV) analysis. The EV approach is the widest used in hydrology (i.e., for the derivation of return levels for extreme rainfall and flood estimates) and it should be carefully and correctly applied in order to avoid dangerous under- or overestimation of the analyzed design variables (rainfall, runoff). With this aim in the second section, a detailed distribution functions used with hydrological variables are described; moreover, the approaches to develop the parameter estimation and the goodness-fit-test steps are reviewed.

Since rainfall is the most-observed hydrological phenomenon, a peculiar section is included in this chapter providing an update description of EV-IDF (intensity–duration–frequency) procedure. IDF curves are an invaluable tool in hydrology having a crucial role in the safe and efficient design of major or minor infrastructures (e.g., water dams, urban hydraulic works, flood design, etc.) that affect human lives. IDF curves are in use almost for a century, and the many different forms and methods proposed and studied through the years underline their importance. During all those years, IDF curves have evolved from purely empirical forms to theoretically more consistent, while today, their study still remains an active field of research. In this text, some of the most commonly used forms and techniques have been presented and applied in a real world data set. The search of the literature and the application presented here reveals that some commonly used techniques and forms of IDF curves may result in underestimating the rainfall intensity, especially for large return periods, and thus should be used with caution. More advanced forms and estimation procedures are described and compared to the most commonly used ones in practical applications.

Until now the efforts of hydrologists were primarily devoted to analyze single parameters (flood peak, rainfall intensity, etc.), not because it is not important to consider other variables (i.e., flood duration and flood volume, or rainfall duration and volume, etc.) but because of the absence of a flexible approach to jointly analyze these different but useful variables. However, this is now finally possible, thanks to the relatively recent introduction of copula function. This statistical and mathematical method is quickly evolving and numerous applications are described in literature. Since this approach is promising and it could change and improve many hydrological procedures, a specific section on copula function is considered in this chapter, providing an updated review useful for hydrological applications.

As mentioned at the beginning of this section, the ungauged basin is a sensitive problem. Most of the little basins ($<150\,km^2$) are characterized by poor hydrological observations (usually few raingauges are available) that stimulated an intense research on statistical methods for regional frequency analysis. Therefore, in the last section, it is essential to include a review and a specific description of this important topic.

This chapter is written by a group of researcher members of the Statistics in Hydrology – STAHY Working Group recently launched by the International Association of Hydrological Sciences – IAHS with the purpose of sharing knowledge and stimulating research activities on statistical hydrology.

2.18.2 Analysis and Detection of Nonstationarity in Hydrological Time Series

Hydrological time series used in water resources planning studies are very often supposed to meet the stationary hypothesis. Under steady-state natural conditions, time series exhibit regular fluctuations around a mean value; however, when the natural conditions change markedly, they may form trends or exhibit jumps. Hydrological data series frequently show this type of significant nonstationarity due to several reasons (human activities, climate change, etc.).

A random process is an indexed family $(x_t)_{t \in I}$ of random variables, which may be discrete time if I is a set of integers. A discrete random time process $\mathbf{x} = (x_1, x_2, \ldots, x_n)$ is said to be stationary if, for every k and n, the distribution of $x_{k+1}, x_{k+2}, \ldots, x_{k+n}$ is the same as the distribution of $x_1, x_2, \ldots, x_n$ (Baseville and Nikiforov, 1993). In other words, a random process or variable is said to be strictly stationary if its statistical properties do not vary with time, and hence independent of changes of time origin.

Trends, jumps/shifts, seasonality/periodicity, or non-randomness in a hydrological time series can be referred to as components of the time series. Presence of these components makes the time series nonstationary. Indeed, nonstationarity is under the effect of persistency and scaling issues (Koutsoyiannis, 2006).

Hydrological time series frequently exhibit nonstationary behavior, for example; flow and precipitation or rainfall stay below or above the mean long-term average (Rao and Yu, 1986), although they are generally assumed to be stationary at annual scale. When the time interval used is shorter than a year (month, week, or day), the stationarity assumption in the hydrological time series then becomes nonvalid simply because of the annual cycle of the Earth around the Sun.

Trends in a time series can result from gradual natural and human-induced disruptive and evolutionary changes in the environment, whereas a jump may result from sudden catastrophic natural events (Haan, 2002). Any change in the time series is most reliable if it is detected by statistical tests and also has physical and historical evidences (Salas et al., 1980). Therefore, it is considered an important issue to identify (detect), describe (test), and remove these components.

2.18.2.1 The Common Nonstationarity Analysis Methods

A number of parametric and nonparametric tests have been suggested in literature for the detection of trend and jumps, and for checking randomness. These tests are considered to be important for scientific purposes as well as for practicing hydrologists.

In what follows, a combination of the above-mentioned tests has been briefly described.

2.18.2.1.1 Randomness test

An adapted version of a simple nonparametric run test, reported by Adeloye and Montaseri (2002), is given below. The test consists of the following steps (**Figure 1**):

1. The median of the observation is determined.
2. Each data item is examined to find out if it exceeds the median. If a data item exceeds the median, this is defined as a case of success, S, if not, this is defined as a case of failure, F. Cases that are exactly equal to the median are excluded.
3. Successes and failures are counted and denoted by n_1 and n_2, respectively.
4. The total number of runs (R) in the data set is determined. A run is a continuous sequence of successes until it is interrupted by a failure or vice versa.
5. The test statistics is computed by

$$z = \frac{R - \left(\dfrac{2n_1 n_2}{n_1 + n_2} - 1\right)}{\sqrt{\dfrac{2n_1 n_2(2n_1 n_2 - n_1 - n_2)}{(n_1 + n_2)^2 (n_1 + n_2 - 1)}}} \quad (1)$$

where z has a standard normal distribution under the null hypothesis, H_0, that the sequence of successes and failures is random.

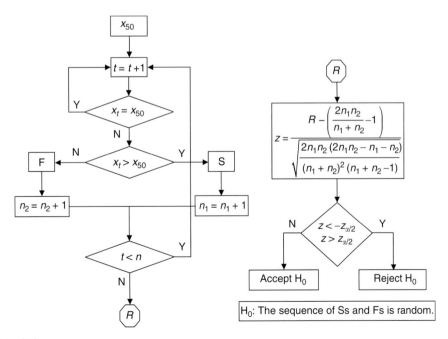

Figure 1 Randomness test.

6. Critical values of the standard normal distribution are obtained for the chosen significance level, α, and denoted by $\pm z_{\alpha/2}$.
7. Computed statistics z is compared to the critical values $\pm z_{\alpha/2}$. H_0 is rejected if $z < -z_{\alpha/2}$ or $z > z_{\alpha/2}$.

2.18.2.1.2 Detection of trend

A number of parametric and nonparametric trend detection tests are available in the literature (Berryman et al., 1988; Cluis et al., 1989; Helsel and Hirsch, 1992; Salas, 1993; Fanta et al., 2001; Yue et al., 2002; Burn and Elnur, 2002; Adeloye and Montaseri, 2002; Xiong and Guo, 2004; Koutsoyiannis, 2006). Among these, one parametric and two nonparametric tests are supplied below.

2.18.2.1.3 Simple regression on time

The simple linear trend line between the variable (x) and time (t) can be written as

$$x_t = a + bt \qquad (2)$$

where a and b are parameters of the regression model. A linear trend exists when the null hypothesis that $b=0$ is rejected. The null hypothesis is rejected if the test statistics, T_c, satisfies

$$T_c = \left| \frac{\sqrt{n-2}}{r\sqrt{1-r^2}} \right| > T_{1-\alpha/2, \nu} \qquad (3)$$

where r is the cross-correlation coefficient between the variable x ($x_1, x_2, \ldots, x_n$) and time $t = 1, 2, \ldots, n$, and $T_{1-\alpha/2,\nu}$ is the $1 - \alpha/2$ quantile of the Student t distribution with $\nu = n - 2$ degrees of freedom.

2.18.2.1.4 Mann–Kendall test

The Mann–Kendall test checks the existence of a trend without specifying if the trend is linear or nonlinear. It is widely reported as in Libiseller and Grimwall (2002).

The univariate statistics for monotone trend in a time series $x_t (t = 1, 2, \ldots, n)$ is defined as

$$S = \sum_{j<i} \text{sgn}(x_i - x_j) \qquad (4)$$

where

$$\text{sgn}(x) = \begin{cases} 1, & \text{if } x > 0 \\ 0, & \text{if } x = 0 \\ -1, & \text{if } x < 0 \end{cases} \qquad (5)$$

If no ties are present and the values of $x_1, x_2, \ldots, x_n$ are randomly ordered, the test statistics has expectation zero and variance

$$V(S) = \frac{n(n-1)(2n+5)}{18} \qquad (6)$$

In the case of presence of tied groups, equations are modified (Salas, 1993).

2.18.2.1.5 Spearman rank order correlation test

The Spearman rank order correlation nonparametric test is used to investigate the existence of a trend that might be found in the time series. The step-by-step explanation of the test for a time series x_t ($t = 1, \ldots, n$) observed in time t (**Figure 2**) is as follows:

1. Ranks, R_{x_t}, are assigned to x_t, such that the rank 1 is assigned to the largest x_t and the rank n to the least x_t. Where there are ties in the x_t, then a rank equal to the average of

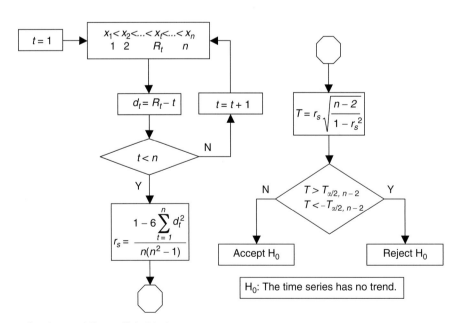

Figure 2 Spearman rank order correlation coefficient test.

the ranks which would have been used had there been no ties is assigned to each of the ties.

2. The difference

$$d_t = R_{x_t} - t \qquad (7)$$

is computed.

3. The coefficient of trend, r_s, is computed by

$$r_s = \frac{1 - 6\sum_{t=1}^{n} d_t^2}{n(n^2 - 1)} \qquad (8)$$

Under the null hypothesis that the time series has no trend, the variable

$$T = r_s \sqrt{\frac{n-2}{1-r_s^2}} \qquad (9)$$

has a Student's t-distribution with $n-2$ degrees of freedom.

4. The critical values of the t-distribution for the chosen significance level, α, and $n-2$ degrees of freedom are obtained. For a two-tailed test, the critical values are denoted by $\pm T_{\alpha/2,\ n-2}$.
5. The values of T are compared to the critical values. H_0 is rejected if $T > T_{\alpha/2,\ n-2}$ or $T < -T_{\alpha/2,\ n-2}$.

2.18.2.1.6 Detection of shifts (segmentation)

Segmentation of a time series is the first step of jump analysis also called change point detection problem (or detection of shifts) for which statistical tests such as the Pettitt (1979) and Alexandersson (1986) tests are available in the literature. The simplest case is the segmentation with regression by constant in which it is aimed to determine the change points or boundaries where the average of the current segment is statistically different than the average of the next segment as well as that of the previous one. This shift or jump may be either positive or negative. By using a proper algorithm, the time series is first divided into segments with different mean values. Then the significance of the difference in the mean is tested. A number of tests are available in the literature to test the significance, that is to detect whether the time series is consistent. The tests are either parametric or nonparametric as in the trend detection tests (Hirsch et al., 1993; Chen and Rao, 2002; Fanta et al., 2001; Xiong and Guo, 2004; Wong et al., 2006). Here, the t-test and Mann–Whitney test are described.

2.18.2.1.7 t-Test

A segmentation algorithm can be used for splitting the sample into segments with significantly different means. The segmentation algorithm divides the time series into as many segments as possible. Then, if two or more segments are identified, the starting year of the last segment is chosen as the first year for splitting the time series. The comparison is made between the segments before and after the chosen year. Once segmentation is completed, the jump analysis is performed by

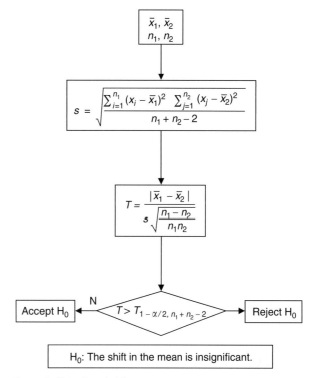

Figure 3 Detection of shift: t-test.

using the parametric t-test for which details are given below (Figure 3).

1. The time series is divided into several segments by a segmentation algorithm.
2. The average of two consecutive segments ($\bar{x}_1$ and $\bar{x}_2$) is calculated and the length of the segments (n_1 and n_2) is determined.
3. The t-statistics is calculated by

$$T = \frac{|\bar{x}_1 - \bar{x}_2|}{s\sqrt{\frac{n_1 + n_2}{n_1 n_2}}} \qquad (10)$$

with $n_1 + n_2 - 2$ degrees of freedom. s in Equation (10) is the pooled variance given by

$$s = \sqrt{\frac{\sum_{i=1}^{n_1}(x_i - \bar{x}_1)^2 + \sum_{j=1}^{n_2}(x_j - \bar{x}_2)^2}{n_1 + n_2 - 2}} \qquad (11)$$

4. The null hypothesis that shift in the average value is insignificant is rejected if the sample T statistics in Equation (10) is greater than the critical value of Student's t-distribution $T_{1-\alpha/2,\ n_1+n_2-2}$ with $n_1 + n_2 - 2$ degrees of freedom.

2.18.2.1.8 Mann–Whitney test

The Mann–Whitney test is used when a time series x_t ($t = 1, 2, \ldots, n$) can be divided into two segments $x_1, x_2, \ldots, x_{n_1}$ and $x_{n_1+1}, x_{n_1+2}, \ldots, x_n$ such that $n_2 = n - n_1$. This is a widely reported test and briefly given below as reported in Salas (1993): a new series z_t ($t = 1, 2, \ldots, n$) is defined by rearranging the original data x_t at increasing order of magnitude. The hypothesis that the mean of the first segment is equal to the mean of the second segment is tested by

$$u_c = \frac{\sum_{t=1}^{n_1} R(x_t) - n_1(n_1 + n_2 + 1)/2}{[n_1 n_2 (n_1 + n_2)/12]^{1/2}} \quad (12)$$

where $R(x_t)$ is the rank of the observation x_t in ordered series z_t. The hypothesis of equal means is rejected at the significance level α when $|u_c| > u_{1-\alpha/2}$ where $u_{1-\alpha/2}$ is the $1-\alpha/2$ quantile of the standard normal distribution.

2.18.2.2 A New Method of Segmentation

In addition to the classical tests, various segmentation algorithms have been developed to determine stationary segments and estimate parameters characterizing each segment. The usual criterion to decide if a change point exists is based on the segmentation cost defined as the sum of squared deviation of the data from the means of their respective segments. The number of segments has the lowest limit of 1 and the highest n, the length of the time series, and determines the order of segmentation, that is, a fifth-order segmentation, for instance, means the time series is divided into five segments.

The segmentation procedure of Hubert et al. (1989) and Hubert (2000) used also by Cluis and Laberge (2001), Fortin et al. (2004), Aksoy (2006), Dahamsheh and Aksoy (2007), and Aksoy et al. (2008b), among others, is available in the literature. Some earlier examples developed to determine stationary segments and estimate the parameters characterizing each segment can be found in Appel and Brandt (1983) and Imberger and Ivey (1991). Kehagias (2004) and Kehagias et al. (2006) developed segmentation algorithms based on dynamic programming (DP) and hidden Markov model (HMM). Gedikli et al. (2008) made the segmentation algorithm – denoted as AUG – available. Gedikli et al. (2010b) modified the DP algorithm (mDP). The HMM, DP, AUG, and mDP algorithms are all motivated from the segmentation algorithm of Hubert (2000). In the following subsection, the AUG algorithm is briefed with the aim of piecewise-stationarity analysis of hydrological time series.

Following definitions are required to explain the formulation behind the segmentation algorithm. For details, the reader is referred to Gedikli et al. (2008) and Aksoy et al. (2008a).

Assume that a time series $\mathbf{x} = (x_1, x_2, \ldots, x_n)$ is given. Segmentation of such a series can be described by a sequence $\mathbf{t} = (t_0, t_1, \ldots, t_K)$ to satisfy $0 = t_0 < t_1 < \ldots < t_{k-1} < t_k = n$. The intervals of integers $[t_0 + 1, t_1][t_1 + 1, t_2]\ldots,[t_{K-1} + 1, \ldots, t_K]$ are called segments, the times $t_0, t_1, \ldots, t_K$ are called segment boundaries and K, the number of segments, is called the order of the segmentation. In other words, the time points where changes take place are called change points; the interval included between two change points is a segment (of the time series); and the procedure by which the segments of a time series are determined is called time series segmentation.

The set of all segmentations of $\{1, 2, \ldots, n\}$ is denoted by $\mathbf{N}$ and the set of all segmentations of order K by $\mathbf{N}_K$. Clearly, $\mathbf{N} = \cup_{K=1}^{n} \mathbf{N}_K$. The number of all possible segmentations of $\{1, 2, \ldots, n\}$ is 2^{n-1}. This can be formulated as an optimization problem. In other words, the optimal segmentation

depends on $\mathbf{x}$. The segmentation cost $J(\mathbf{t})$ is defined by

$$J(\mathbf{t}) = \sum_{k=1}^{K} d_{t_{k-1}+1, t_k} \quad (13)$$

where $d_{s,t}$ (for $0 \leq s < t \leq T$) is the segment error corresponding to segment $[s,t]$. The segment error depends on the data vector $\{x_s, x_{s+1}, \ldots, x_t\}$. A variety of $d_{s,t}$ functions can be used. In this study,

$$d_{s,t} = \sum_{\tau=s}^{t} (x_\tau - \mu_{s,t})^2 \quad (14)$$

is used where the segment mean is given by

$$\mu_{s,t} = \frac{\sum_{\tau=s}^{t} x_\tau}{t - s + 1} \quad (15)$$

The optimal segmentation, denoted by $\hat{\mathbf{t}} = (\hat{t}_0, \hat{t}_1, \ldots, \hat{t}_K)$, is defined as $\hat{\mathbf{t}} = \arg\min_{\mathbf{t} \in \mathbf{N}} J(\mathbf{t})$ and the optimal segmentation of order K, denoted by $\hat{\mathbf{t}}^{(K)} = (\hat{t}_0^{(K)}, \hat{t}_1^{(K)}, \ldots, \hat{t}_K^{(K)})$, is defined as $\hat{\mathbf{t}}^{(K)} = \arg\min_{\mathbf{t} \in \mathbf{N}_K} J(\mathbf{t})$. The optimal segmentation can be found by exhaustive enumeration of all possible segmentations (and computation of the corresponding $d_{s,t}$). In computational sense, this is an infeasible way as the total number of segmentations increases exponentially with T. In order to obtain fast algorithms, a fast method for computing the costs $d_{s,t}$ is first required. For this aim, the recursive formulation of

$$d_{s,t+1} = d_{s,t} + (t - s + 1)(\mu_{s,t} - \mu_{s,t+1})^2 + (x_{t+1} - \mu_{s,t+1})^2 \quad (16)$$

is easily proved where

$$\mu_{s,t+1} = \frac{(t - s + 1)\mu_{s,t} + x_{t+1}}{t - s + 2} \quad (17)$$

The segmentation algorithm is based on the branch-and-bound-type technique. The branches are the possible segments of the kth-order segmentation. As suggested by Hubert (2000), the upper bound, u, of the kth segment in the Kth-order segmentation can trivially be given as

$$t_k \leq u = n - K + k \quad (18)$$

In the segmentation algorithm, the term 'upper bound' is the possible maximum value that t_k can take. The basic idea of the algorithm is to enumerate (branch into) the possible solutions of the segmentation problem but, at the same time, to avoid exhaustive enumeration by eliminating clearly suboptimal solutions (bounds). It is possible to eliminate segmentations by reducing the upper bound of the segments as defined in Equation (18). It is also easy to check that

$$c_{t+1}^k \geq c_t^k \geq \left(c_t^{k+1} \text{ and } c_{t+1}^{k+1}\right) \quad (19)$$

is valid for $t = 2, \ldots, N-1$ and $k = 1, 2, \ldots, t$. Equation (19) is rather obvious; a detailed derivation of it can be found in Gedikli et al. (2008). In order to reduce the upper bound, u, the remaining cost concept is defined as

$$R_{n,t}^{K,k} = c_n^K - c_t^k \quad (20)$$

where $k \leq K$ and $t \leq n$. This is a unique concept developed to make the algorithm fast.

The segmentation algorithm computes a sequence of optimal segmentations $\hat{t}_1, \hat{t}_2, \ldots, \hat{t}_k$, where $\hat{t}_k$ is the kth-order optimal segmentation. For a given segmentation ($\hat{t}_k$ for instance), the hypothesis that the means of consecutive segments are significantly different is tested. Determining the optimal order of segmentation, that is, selecting the number of segments, is a subsequent step in the segmentation procedure to be performed for which the Scheffe (1959) test is employed. The test is run on the optimal segmentations $\hat{t}^{(1)}, \hat{t}^{(2)}, \ldots, \hat{t}^{(K)}$. Hubert (2000) accepts $\hat{t}^{(k)}$ as the optimal segmentation when $\hat{t}^{(k+1)}$ is the first lowest order segmentation which is rejected by the Scheffe test (i.e., the first segmentation for which at least two consecutive segments do not show a statistically significant difference in their means). In the AUG algorithm, however, not the first lowest but the highest order segmentation which is accepted by the Scheffe test is considered instead.

The application of the segmentation algorithm was performed by using a previously used data set: the annual mean streamflow data of Senegal River originating from Hubert (2000) and used by Kehagias (2004), Kehagias et al. (2006), and Gedikli et al. (2008). The data set is available on the Internet. A user-friendly software (the AUG-Segmenter version – 1.1) based on the above algorithm is now available (Gedikli et al., 2010a). The software is able to segment time series efficiently and fast. Using this software the Senegal River annual mean streamflow data set is segmented. The length of the data is 84 years for the period 1903–86. The fifth-order segmentation is found to be optimal after the execution of the algorithm (Table 1 and Figure 4).

2.18.3 Extreme Value Analysis: Distribution Functions and Statistical Inference

The study of the statistics of extreme events is the first step for most of the hydrological studies. In many situations, historical records containing observations from the past are the only reliable source of information. In the flood contest, the analysis of extreme events was introduced at the beginning of the twentieth century (e.g., Fuller, 1914) to replace the earlier design flood procedures, such as envelope curves and empirical formulas, by more objective estimation methods. When longer flood records became available by the middle of the twentieth century and with further theoretical developments such as extreme value theory of Gumbel (1958), the method rapidly became what Klemeš (1993) termed 'the standard approach to frequency analysis'.

As stressed in Stedinger et al. (1993), "frequency analysis is an information problem." If one had a sufficiently long record of flood flows, rainfall, low flows, etc., then a frequency distribution for a site could be precisely determined, so long as change over time due to urbanization or natural processes did not alter the relationships of concern. However in most situations, available data are not enough to precisely define the risk of large floods, rainfall, or low flows. This forces hydrologists to use practical knowledge of the processes involved, and efficient and robust statistical techniques, to develop the best estimates of risk that they can. These techniques are generally restricted, with 10–100 sample observations, to estimate events exceeded with a chance of at least 1 in 100, corresponding to exceedance probabilities of 1% or more. In some cases, they are used to estimate the rainfall exceeded with a chance of 1 in 1000 (the rainfall with return period of 1000 years), and even the flood flows for spillway design exceeded with a chance of 1 in 10 000 (the 10 000 years flood).

In essence, the extreme value analysis consists of fitting distribution functions to ordered sequences of observed data and extrapolating the tails of the distribution to low exceedance probabilities. The immediate problem pertains to the way in which the probabilities are estimated and what level of accuracy is associated with such probabilities. The hydrologist should be aware that in practice the true probability distributions of the phenomena in question are not known. Even if they were, their functional representation would likely have too many parameters to be of much practical use. The practical issues are: how to select a reasonable and simple distribution to describe the phenomenon of interest, finding the correct trade-off between estimation bias and variance, that respectively decreases and increases as the number of model parameters increases; to estimate the distribution's parameters;

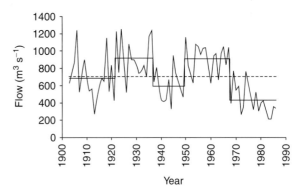

Figure 4 The fifth-order segmentation of the Senegal River annual streamflow data.

Table 1 Change points of annual streamflow data of Senegal River data (1903–86)

Segmentation order	Change points					
2	1902	1967	1986			
3	1902	1949	1967	1986		
4	1902	1938	1949	1967	1986	
5	1902	1921	1936	1949	1967	1986

and thus to obtain risk estimates of satisfactory accuracy for the problem at hand.

2.18.3.1 Probability Distributions for Extreme Events

In this section, several distributions commonly used in hydrology are briefly described. Tables 2 and 3 provide a summary of the probability density functions (PDFs) or cumulative distribution functions (CDFs) of these probability distributions. The moments and L-moments for these distributions are reported in Tables 4 and 5 (see Section 2.18.3.2 for more details).

2.18.3.1.1 Normal distribution

The normal distribution (N) arises from the central limit theorem, which states that if a sequence of random variables X_i are independently and identically distributed, then the distribution of the sum of n such random variables tends toward the normal distribution as n becomes large. The important point is that this is true no matter what the probability distribution function of X is. Hydrologic variables, such as annual precipitation, calculated as the sum of the effects of many independent events tend to follow the normal distribution. The main limitations of the normal distribution for describing hydrological variables are that it varies over a continuous range $(-\infty, +\infty)$, while most hydrologic variables are non-negative, and that it is symmetric about the mean, while hydrologic data tend to be skewed. Because of its definition, the normal distribution is not suitable for extreme value analysis.

2.18.3.1.2 Lognormal distribution

If the random variable $Y = \log(X)$ is normally distributed, then X is said to be lognormally distributed (LN). This distribution is applicable to hydrologic variables formed as the products of other variables, because of the central limit theorem, provided that these are independent and identically distributed (see, e.g., Sangal and Biswas (1970), Martins and Stedinger (2001), and Kroll and Vogel (2002) for applications in hydrology). The lognormal distribution has the advantages over the normal distribution that it is bounded ($X > 0$) and that the log transformation tends to reduce the positive skewness commonly found in hydrologic data (especially in extremes), because taking logarithms reduces large numbers proportionately more than small numbers. Some limitations of the lognormal distribution are that it has only two parameters

Table 2 Commonly used frequency distributions in hydrology

Distribution	PDF, fx(x), CDF, Fx(x), and quantile function, x_F	Range
Normal (N)	$f_X(x) = \dfrac{1}{\theta_2\sqrt{2\pi}}\exp\left[-\dfrac{1}{2}\left(\dfrac{x-\theta_1}{\theta_2}\right)^2\right]$ $x_F = \theta_1 + \theta_2 \Phi^{-1}(F)$	$-\infty < x < \infty$ $\theta_2 > 0$
Lognormal (LN)	$f_X(x) = \dfrac{1}{x\theta_2\sqrt{2\pi}}\exp\left[-\dfrac{1}{2}\left(\dfrac{\log(x)-\theta_1}{\theta_2}\right)^2\right]$ $x_F = \exp[\theta_1 + \theta_2 \Phi^{-1}(F)]$	$x > 0$ $\theta_2 > 0$
3-par Lognormal (LN3)	$f_X(x) = \dfrac{1}{(x-\theta_1)\theta_3\sqrt{2\pi}}\exp\left[-\dfrac{1}{2}\left(\dfrac{\log(x-\theta_1)-\theta_2}{\theta_3}\right)^2\right]$ $x_F = \theta_1 + \exp[\theta_2 + \theta_3 \Phi^{-1}(F)]$	$x > \theta_1$ $\theta_3 > 0$
Exponential (E)	$f_X(x) = \dfrac{1}{\theta_2}\exp\left[-\dfrac{x-\theta_1}{\theta_2}\right]$ $F_X(x) = 1 - \exp\left[-\dfrac{x-\theta_1}{\theta_2}\right]$ $x_F = \theta_1 - \theta_2 \ln(1-F)$	$x > \theta_1$ for $\theta_2 > 0$
Gamma (G)	$f_X(x) = \dfrac{1}{|\theta_1|\Gamma(\theta_2)}\left(\dfrac{x}{\theta_1}\right)^{\theta_2-1}\exp\left[-\dfrac{x}{\theta_1}\right]$	$x \geq 0$
Pearson type 3 (P3)	$f_X(x) = \dfrac{1}{|\theta_2|\Gamma(\theta_3)}\left(\dfrac{x-\theta_1}{\theta_2}\right)^{\theta_3-1}\exp\left[-\dfrac{x-\theta_1}{\theta_2}\right]$	$\theta_1 < x < \infty$ if $\theta_2 > 0$ $-\infty < x < \theta_1$ if $\theta_2 < 0$ $\theta_3 > 0$
Log-Pearson type 3 (LP3)	$f_X(x) = \dfrac{1}{x|\theta_2|\Gamma(\theta_3)}\left(\dfrac{\log(x)-\theta_1}{\theta_2}\right)^{\theta_3-1}\exp\left[-\dfrac{\log(x)-\theta_1}{\theta_2}\right]$	$\exp(\theta_1) < x < \infty$ if $\theta_2 > 0$ $0 < x < \exp(\theta_1)$ if $\theta_2 < 0$ $\theta_3 > 0$

θ_1, θ_2, and θ_3 are distribution parameters, Φ is the standard normal CDF, and Γ is the gamma function.

Table 3 Commonly used frequency distributions in hydrology

Distribution	PDF, fx(x), CDF, Fx(x), and quantile function, x_F	Range
Gumbel (EV1)	$f_X(x) = \dfrac{1}{\theta_2} \exp\left\{-\dfrac{x-\theta_1}{\theta_2} - \exp\left[-\dfrac{x-\theta_1}{\theta_2}\right]\right\}$	$-\infty < x < \infty$
	$F_X(x) = \exp\left\{-\exp\left[-\dfrac{x-\theta_1}{\theta_2}\right]\right\}$	
	$x_F = \theta_1 - \theta_2 \ln[-\ln(F)]$	
Fréchet (EV2)	$f_X(x) = \dfrac{\theta_2}{\theta_1}\left(\dfrac{\theta_1}{x}\right)^{\theta_2+1} \exp\left[-\left(\dfrac{\theta_1}{x}\right)^{\theta_2}\right]$	$x > 0;\ \theta_1,\ \theta_2 > 0$
	$F_X(x) = \exp\left[-\left(\dfrac{\theta_1}{x}\right)^{\theta_2}\right]$	
	$x_F = \theta_1[-\ln(F)]^{-1/\theta_2}$	
Weibull (EV3)	$f_X(x) = \dfrac{\theta_2}{\theta_1}\left(\dfrac{x}{\theta_1}\right)^{\theta_2-1} \exp\left[-\left(\dfrac{x}{\theta_1}\right)^{\theta_2}\right]$	$x > 0;\ \theta_1,\ \theta_2 > 0$
	$F_X(x) = 1 - \exp\left[-\left(\dfrac{x}{\theta_1}\right)^{\theta_2}\right]$	
	$x_F = \theta_1[-\ln(1-F)]^{1/\theta_2}$	
GEV	$F_X(x) = \exp\left\{-\left[1 - \theta_3\dfrac{(x-\theta_1)}{\theta_2}\right]^{1/\theta_3}\right\}$	$x < \left(\theta_1 + \dfrac{\theta_2}{\theta_3}\right)$ if $\theta_3 > 0$
	$x_F = \theta_1 + \dfrac{\theta_2}{\theta_3}[1 - (-\ln(F)^{\theta_3})]$	$x > \left(\theta_1 + \dfrac{\theta_2}{\theta_3}\right)$ if $\theta_3 < 0$
Generalized Pareto (GP)	$f_X(x) = \dfrac{1}{\theta_2}\left[1 - \theta_3\dfrac{(x-\theta_1)}{\theta_2}\right]^{1/\theta_3-1}$	$\theta_1 \leq x < \infty$ if $\theta_3 < 0$
	$F_X(x) = 1 - \left[1 - \theta_3\dfrac{(x-\theta_1)}{\theta_2}\right]^{1/\theta_3}$	$\theta_1 \leq x \leq \theta_1 + \dfrac{\theta_2}{\theta_3}$ if $\theta_3 > 0$
	$x_F = \theta_1 + \dfrac{\theta_2}{\theta_3}[1 - (1-F)^{\theta_3}]$	
Generalized Logistic (GL)	$F_X(x) = \dfrac{1}{1 + \left[1 - \dfrac{\theta_3}{\theta_2}(x-\theta_1)\right]^{1/\theta_3}}$	$x > \left(\theta_1 + \dfrac{\theta_2}{\theta_3}\right)$ if $\theta_3 < 0$
	$x_F = \theta_1 + \dfrac{\theta_2}{\theta_3}\left[1 - \left(\dfrac{1-F}{F}\right)^{\theta_3}\right]$	$x < \left(\theta_1 + \dfrac{\theta_2}{\theta_3}\right)$ if $\theta_3 > 0$

θ_1, θ_2, and θ_3 are distribution parameters, Φ is the standard normal CDF, Γ is the gamma function.

and that it requires the logarithms of the data to be symmetric about their mean. Moreover, the lognormal distribution cannot be used when dealing with variables that can assume null values (e.g., discharge in ephemeral rivers).

The three-parameter lognormal distribution (LN3) differs from the LN2 distribution by the introduction of a lower bound (indicated as θ_1 in **Table 2**) so that if X follows the LN3 distribution, $\log(X - \theta_1)$ is normally distributed.

2.18.3.1.3 Exponential distribution

Some sequences of hydrologic events, such as the occurrence of precipitation, may be considered Poisson processes, in which events occur instantaneously and independently on a time horizon, or along a line. The time between such events, or interarrival time, is described by the exponential distribution (E) whose parameter θ_2 is the mean rate of occurrence of the events. The exponential distribution is used to describe the interarrival times of random shocks to hydrologic systems, such as slugs of polluted runoff entering streams as rainfall washes the pollutants off the land surface. The advantage of the exponential distribution is that it is easy to estimate θ_2 from observed data and the exponential distribution lends itself well to theoretical studies, such as a probability model for the linear reservoir ($\theta_2 = 1/k$, where k is the storage constant in the linear reservoir). Its disadvantage is that it requires the occurrence of each event to be completely independent of its neighbors, which may not be a valid assumption for the process under study (e.g., the arrival of a front may generate many showers of rain) and this has led investigators to study

Table 4 Moments and L-moments of commonly used frequency distributions in hydrology

Distribution	Moments	L-moments
Normal (N)	$\mu = \theta_1$, $\sigma = \theta_2$ $\gamma = 0$, $\kappa = 3$	$\lambda_1 = \theta_1$, $\lambda_2 = \pi^{-1/2}\theta_2$ $\tau_3 = 0$, $\tau_4 = 0.1226$
Lognormal (LN)	$\mu = \exp(\theta_1 + \theta_2^2/2)$ $\sigma^2 = [\exp(\theta_2^2) - 1]\exp(2\theta_1 + \theta_2^2)$ $\gamma = [\exp(\theta_2^2) + 2]\sqrt{\exp(\theta_2^2) - 1}$ $\kappa = e^{4\theta_2^2} + 2e^{3\theta_2^2} + 3e^{2\theta_2^2} - 3$	$\lambda_1 = \exp(\theta_1 + \theta_2^2/2)$ $\lambda_2 = \exp(\theta_1 + \theta_2^2/2)\,[2\Phi(\theta_2/\sqrt{2}) - 1]$ τ_3: NA, see HW, eq. (A72) τ_4: NA, see HW, eq. (A73)
3-par Lognormal (LN3)	$\mu = \theta_1 + \exp(\theta_2 + \theta_3^2/2)$ $\sigma^2 = [\exp(\theta_3^2) - 1]\exp(2\theta_2 + \theta_3^2)$ $\gamma = [\exp(\theta_3^2) + 2]\sqrt{\exp(\theta_3^2) - 1}$ $\kappa = e^{4\theta_3^2} + 2e^{3\theta_3^2} + 3e^{2\theta_3^2} - 3$	$\lambda_1 = \theta_1 + \exp(\theta_2 + \theta_3^2/2)$ $\lambda_2 = \exp(\theta_2 + \theta_3^2/2)\,[2\Phi(\theta_3/\sqrt{2}) - 1]$ τ_3: NA, see HW, eq. (A72) τ_4: NA, see HW, eq. (A73)
Exponential (E)	$\mu = \theta_1 + \theta_2$, $\sigma^2 = \theta_2^2$ $\gamma = 2$, $\kappa = 9$	$\lambda_1 = \theta_1 + \theta_2$, $\lambda_2 = \theta_2/2$ $\tau_3 = 1/3$, $\tau_4 = 1/6$
Gamma (G)	$\mu = \theta_1\theta_2$ $\sigma^2 = \theta_2\theta_1^2$ $\gamma = 2\,\text{sign}(\theta_1)/\sqrt{\theta_2}$ $\kappa = 6/\theta_2 + 3$	$\lambda_1 = \theta_1\theta_2$ $\lambda_2 = \pi^{-1/2}\theta_1\Gamma(\theta_2 + 1/2)/\Gamma(\theta_2)$ τ_3: NA, see HW, eq. (A86) and (A88) τ_4: NA, see HW, eq. (A87) and (A89)
Pearson type 3 (P3)	$\mu = \theta_1 + \theta_3\theta_2$ $\sigma^2 = \theta_3\theta_2^2$ $\gamma = 2\,\text{sign}(\theta_2)/\sqrt{\theta_3}$ $\kappa = 6/\theta_3 + 3$	$\lambda_1 = \theta_1 + \theta_2\theta_3$ $\lambda_2 = \pi^{-1/2}\theta_2\Gamma(\theta_3 + 1/2)/\Gamma(\theta_3)$ τ_3: NA, see HW, eq. (A86) and (A88) τ_4: NA, see HW, eq. (A87) and (A89)
Log-Pearson type 3 (LP3)	$\mu = e^{\theta_1}[(1 - \theta_2)]^{-\theta_3}$ $\sigma^2 = e^{2\theta_1}\left[(1 - 2\theta_2)^{-\theta_3} - (1 - \theta_2)^{-2\theta_3}\right]$ γ: NA, see ST, page 18.21 κ: NA, see ST, page 18.21	$\lambda_1 = e^{\theta_1}[(1 - \theta_2)]^{-\theta_3}$ λ_2: NA τ_3: NA τ_4: NA

θ_1, θ_2, and θ_3 are distribution parameters, Φ is the standard normal CDF, and Γ is the gamma function. NA indicates that the moment or L-moment is very complicated or not available in analytical form, with reference to Hosking and Wallis (1997) (HW in the table) or Stedinger et al. (1993) (ST in the table) when formulas or approximations are available.

various forms of compound Poisson processes, in which θ_2 is considered a random variable instead of a constant. The exponential distribution has been used in extreme value analysis as a simple model of the flood or rainfall exceedances over high thresholds in peak over threshold analyses (see, e.g., Todorovic, 1978).

2.18.3.1.4 Gamma distribution

The time taken for a number of events, n, to occur in a Poisson process is described by the gamma distribution (G), which is the distribution of a sum of n independent and identical exponentially distributed random variables. The gamma distribution has a smoothly varying form and is useful for describing skewed hydrologic variables without the need for log transformation. It has been applied, for example, to describe the distribution of depth of precipitation in storms (see, e.g., Sivapalan et al., 2005; Viglione and Blöschl, 2009). The two-parameter gamma distribution has a lower bound at zero, which is a disadvantage for application to hydrologic variables that have a lower bound larger than zero.

2.18.3.1.5 Pearson type 3 distribution

The Pearson type 3 distribution (P3), also called the three-parameter gamma distribution, introduces a third parameter, the lower bound. This is a very flexible distribution, assuming a number of different shapes as the parameters vary. The normal distribution is a special case of the Pearson type 3 distribution, describing a nonskewed variable. The Pearson type 3 distribution was first applied in hydrology by Foster (1924) to describe the probability distribution of annual maximum flood peaks. When the data are very positively skewed, a log transformation is used to reduce the skewness.

Examples of use of the Pearson type 3 distribution in extreme value analysis are Matalas and Wallis (1973), Bobée and Rasmussen (1995), and Kroll and Vogel (2002) among others.

2.18.3.1.6 Log-Pearson type 3 distribution

If log(X) follows a Pearson type 3 distribution, then X is said to follow a log-Pearson type 3 distribution (LP3). This distribution is the standard distribution for frequency analysis of annual maximum floods in the United States (Benson, 1968; Stedinger and Griffis, 2008). As a special case, when log(X) is symmetric about its mean, the log-Pearson type 3 distribution reduces to the lognormal distribution. The location of the bound θ_1 in the log-Pearson type 3 distribution depends on the skewness of the data. If the data are positively skewed, then $\log(X) > \theta_1$ and θ_1 is a lower bound, whereas if the data are negatively skewed, $\log(X) > \theta_1$ and θ_1 is an upper bound.

Table 5 Moments and L-moments of commonly used frequency distributions in hydrology

Distribution	Moments	L-moments
Gumbel (EV1)	$\mu = \theta_1 + 0.5772\,\theta_2,\ \sigma_2 = \pi^2\theta_2^2/6$ $\gamma = 1.1396,\ \kappa = 5 + 2/5$	$\lambda_1 = \theta_1 + 0.5772\,\theta_2,\ \lambda_2 = \theta_2 \ln(2)$ $\tau_3 = 0.1699,\ \tau_4 = 0.1504$
Fréchet (EV2)	$\mu = \theta_1 \Gamma(1 - 1/\theta_2)$ $\sigma^2 = \theta_1^2 \left[\Gamma(1 - 2/\theta_2) - \Gamma^2(1 - 1/\theta_2)\right]$	$\lambda_1 = \theta_1 \Gamma(1 - 1/\theta_2)$ $\lambda_2 = \theta_1 \Gamma(1 - 1/\theta_2)\ (2^{1/\theta_2} - 1)$
Weibull (EV3)	$\mu = \theta_1 \Gamma(1 + 1/\theta_2)$ $\sigma^2 = \theta_1^2 \left[\Gamma(1 + 2/\theta_2) - \Gamma^2(1 + 1/\theta_2)\right]$	$\lambda_1 = \theta_1 \Gamma(1 + 1/\theta_2)$ $\lambda_2 = \theta_1 \Gamma(1 + 1/\theta_2)\ (2^{-1/\theta_2} - 1)$
GEV	$\mu = \theta_1 + \theta_2[1 - \Gamma(1 + \theta_3)]/\theta_3$ $\sigma^2 = \left(\dfrac{\theta_2}{\theta_3}\right)^2 \left[\Gamma(1 + 2\theta_3) - \Gamma^2(1 + \theta_3)\right]$ γ: NA, see ST, eq. (18.2.19) κ: NA	$\lambda_1 = \theta_1 + \theta_2[1 - \Gamma(1 + \theta_3)]/\theta_3$ $\lambda_2 = \theta_2(1 - 2^{-\theta_3})\Gamma(1 + \theta_3)/\theta_3$ $\tau_3 = 2(1 - 3^{-\theta_3})/(1 - 2^{-\theta_3}) - 3$ $\tau_4 = \dfrac{5(1 - 4^{-\theta_3}) - 10(1 - 3^{-\theta_3}) + 6(1 - 2^{-\theta_3})}{1 - 2^{-\theta_3}}$
Generalized Pareto (GP)	$\mu = \theta_1 + \theta_2/(1 + \theta_3)$ $\sigma^2 = \theta_2^2 / \left[(1 + \theta_3)^2 (1 + 2\theta_3)\right]$ $\gamma = 2\sqrt{1 + 2\theta_3}(1 - \theta_3)/(1 + 3\theta_3)$ $\kappa = \dfrac{3(1 + 2\theta_3)(3 - \theta_3 + 2\theta_3^2)}{(1 + 3\theta_3)(1 + 4\theta_3)}$	$\lambda_1 = \theta_1 + \theta_2/(1 + \theta_3)$ $\lambda_2 = \theta_2/[(1 + \theta_3)(2 + \theta_3)]$ $\tau_3 = (1 - \theta_3)/(3 + \theta_3)$ $\tau_4 = (1 - \theta_3)(2 - \theta_3)/[(3 + \theta_3)(4 + \theta_3)]$
Generalized Logistic (GL)	$\mu = \theta_1 + \theta_2(1/\theta_3 - \pi/\sin(\pi\theta_3))$ $\sigma^2 = \pi\theta_2^2 \left(\dfrac{2}{\theta_3 \sin(2\pi\theta_3)} - \dfrac{\pi}{\sin^2(\pi\theta_3)}\right)$ γ: NA, see JO, eq. (23.71) κ: NA, see JO, eq. (23.71)	$\lambda_1 = \theta_1 + \theta_2(1/\theta_3 - \pi/\sin(\pi\theta_3))$ $\lambda_2 = \theta_2 \theta_3 \pi/\sin(\pi\theta_3)$ $\tau_3 = -\theta_3$ $\tau_4 = (1 + 5\theta_3^2)/6$

$\theta_1,\ \theta_2$, and θ_3 are distribution parameters, Φ is the standard normal CDF, and Γ is the gamma function. NA indicates that the moment or L-moment is very complicated or not available in analytical form, with reference to Hosking and Wallis (1997) (HW in the table) or Stedinger et al. (1993) (ST in the table) or Johnson et al. (1994) (JO in the table) when formulas or approximations are available.

The log transformation reduces the skewness of the transformed data and may produce transformed data which are negatively skewed from original data which are positively skewed. In this case, the application of the log-Pearson type 3 distribution would impose an artificial upper bound on the data.

Depending on the values of the parameters, the log-Pearson type 3 distribution can assume many different shapes. Its use is justified by the fact that it has been found to yield good results in many applications, particularly for flood peak data (e.g., Bobée, 1975).

2.18.3.1.7 Extreme value distributions

Extreme values are selected maximum or minimum values of sets of data. For example, the annual maximum discharge at a given location is the largest recorded discharge value during a year, and the annual maximum discharge values for each year of historical record make up a set of extreme values that can be analyzed statistically. Distributions of the extreme values selected from sets of samples of any probability distribution have been shown by Fisher and Tippett (1928) to converge to one of three forms of extreme value distributions, called types I, II, and III, respectively, when the number of selected extreme values is large. Unfortunately, for many hydrologic variables this convergence is too slow for this argument alone to justify adoption of an extreme value distribution as a model of annual maxima and minima. The properties of the three limiting forms were further developed by Gumbel (1941) for the extreme value type I (EV1) distribution, Fréchet (1927) for the extreme value type II (EV2), and Weibull (1939) for the extreme value type III (EV3). The three limiting forms were shown by Jenkinson (1955) to be special cases of a single distribution called the generalized extreme value (GEV) distribution. The three limiting cases are: (1) for $\theta_3 = 0$, the EV1 distribution for which x is unbounded; (2) for $\theta_3 < 0$, the EV2 distribution for which x is bounded from below by $\theta_1 + \theta_2/\theta_3$; (3) for $\theta_3 > 0$, the EV3 distribution for which x is bounded from above by $\theta_1 + \theta_2/\theta_3$.

The EV1 and EV2 distributions are also known as the Gumbel and Fréchet distributions, respectively. Note that the Gumbel and Fréchet distributions are mutually related through the logarithmic transformation, that is, if X is a Fréchet-distributed variable, then $\log(X)$ is distributed as a Gumbel. If a variable x is described by the EV3 distribution, then $-x$ is said to have a Weibull distribution. The Gumbel model is widely applied, often gives satisfactorily results, and is parsimonious (two parameters), but may underestimate the design rainfall depth or discharge for large return periods (e.g., Koutsoyiannis, 2004a, 2004b). By contrast, the GEV model encounters difficulties when its parameters are estimated using small to medium-size samples, due to the large estimation variance of the shape parameter θ_3.

Examples of using the extreme value distribution in extreme value analysis are of course very numerous; an extensive list of references would be out of the scope of this chapter.

2.18.3.1.8 Generalized Pareto distribution

The generalized Pareto (GP) distribution is a simple distribution useful for describing events which exceed a specified lower bound, such as all floods above a threshold or daily flows above zero. The GP distribution allows a continuous range of possible shapes that includes both the exponential and Pareto distributions as special cases. The GP distribution is commonly used in the peaks over threshold (POT) approach. Examples of use of the generalized Pareto in extreme value analysis are Hosking and Wallis (1987), Rosbjerg *et al.* (1992), Madsen *et al.* (1997a, 1997b), Lang *et al.* (1999), and Claps and Laio (2003).

2.18.3.1.9 Generalized logistic distribution

The generalized logistic distribution (GL) has been used extensively for maximum rainfall modeling, and in the UK and elsewhere is used in hydrological risk analysis as the standard model for flood frequency estimation (Institute of Hydrology, 1999; Atiem and Harmancioglu, 2006).

2.18.3.2 Parameter Estimation Methods

Fitting a distribution to data sets provides a compact and smoothed representation of the frequency distribution revealed by the available data, and leads to a systematic procedure for extrapolation to frequencies beyond the range of the data set. When flood flows, low flows, rainfall, or water-quality variables are well described by some family of distributions, a task for the hydrologist is to estimate the parameters Θ of that distribution so that required quantiles and expectations can be calculated with the fitted model. Appropriate choices for distribution functions can be based on examination of the data using probability plots, moment ratios and goodness-of-fit tests (discussed in Section 2.18.3.3), the physical origins of the data, previous experience, and administrative guidelines.

Several approaches are available for estimating the parameters of a distribution. Some commonly used approaches are described in the following subsections.

2.18.3.2.1 Method of moments

The method of moments was first developed by Karl Pearson in 1902. He considered that good estimates of the parameters of a probability distribution are those for which moments of the PDF about the origin are equal to the corresponding moments of the sample data. Pearson originally considered only moments about the origin, but later it became customary to use the variance as the second central moment and the coefficient of skewness as the standardized third central moment, to determine second and third parameters of the distribution if required.

Given a distribution function $f_X(x)$, the mean is defined as

$$\mu = E[X] = \int_{-\infty}^{\infty} x f_X(x) \mathrm{d}x \quad (21)$$

The second moment about the mean is the variance defined as $\sigma^2 = E[(X-\mu)^2]$. The standard deviation σ is the square root of the variance and describes the width or scale of a distribution. These are examples of product moments because they depend upon powers of X. A dimensionless measure of the variability in X, appropriate for use with positive random variables $X \geq 0$, is the coefficient of variation, defined as σ/μ. The relative asymmetry of a distribution is described by the coefficient of skewness $\gamma = E[(X-\mu)^3]/\sigma^3$ while the coefficient of kurtosis $\kappa = E[(X-\mu)^4]/\sigma^4$ describes the thickness of distribution's tails. These four moments are tabled for different distributions in Tables 4 and 5.

From a set of observations $(X_1, \ldots, X_n)$, the unbiased estimators of the mean, variance, and coefficient of skewness are

$$\hat{\mu} = \bar{X} = \frac{1}{n} \sum_{i=1}^{n} X_i \quad (22)$$

$$\hat{\sigma}^2 = S^2 = \frac{1}{n-1} \sum_{i=1}^{n} (X_i - \bar{X})^2 \quad (23)$$

$$\hat{\gamma} = \frac{n}{(n-1)(n-2)S^3} \sum_{i=1}^{n} (X_i - \bar{X})^3 \quad (24)$$

The method of moments consists of inverting the equations in Tables 4 and 5 so as to express the parameters of the distributions in terms of their moments and then using the sample moments to estimate the distribution moments.

2.18.3.2.2 Method of L-moments

L-moments are another way to summarize the statistical properties of hydrologic data. L-moments were introduced by Sillitto (1969) and formalized by Hosking (1990) and are linear combinations of the probability-weighted moments defined by Greenwood *et al.* (1979). The first L-moment estimator is again the mean $\lambda_1 = E[X]$. Let $X_{i:n}$ be the *i*th smallest observation in a sample of size n ($i=1$ corresponds to the smallest). Then, for any distribution, the second L-moment is a description or scale based on the expected difference between two randomly selected observations:

$$\lambda_2 = \tfrac{1}{2} E[X_{2:2} - X_{1:2}] \quad (25)$$

Similarly, the third and fourth L-moments are

$$\lambda_3 = \tfrac{1}{3} E[X_{3:3} - 2X_{2:3} + X_{1:3}] \quad (26)$$

$$\lambda_4 = \tfrac{1}{4} E[X_{4:4} - 3X_{3:4} + 3X_{2:4} - X_{1:4}] \quad (27)$$

and in general

$$\lambda_r = r^{-1} \sum_{j=0}^{r-1} (-1)^j \binom{r-1}{j} E[X_{r-j:r}] \quad (28)$$

The coefficient of L-variation (L-CV) is defined by the ratio of two L-moments as $\tau = \lambda_2/\lambda_1$. Other L-moment ratios are $\tau_3 = \lambda_3/\lambda_2$ and $\tau_4 = \lambda_4/\lambda_2$ that measure the skewness and the kurtosis of the distributions.

Analogously to the method of moments, the method of L-moments consists of inverting the equations of **Table 2** so as to express the parameters of the distributions in terms of their L-moments and to use the sample L-moments as estimators of distribution L-moments. Sample L-moments are defined as

$$l_r = \sum_{k=0}^{r-1} p^*_{r-1,k} b_k \qquad (29)$$

where the coefficients

$$p^*_{r,k} = (-1)^{r-k} \binom{r}{k}\binom{r+k}{k} \qquad (30)$$

are those of the 'shifted Legendre polynomials' (see Hosking and Wallis, 1997) and b_k are the sample probability-weighted moments. These are computed from the ordered statistics $X_{1:n}$, $X_{2:n}, \ldots, X_{n:n}$, that is, the data values arranged in increasing order, as

$$b_k = n^{-1} \binom{n-1}{k}^{-1} \sum_{j=k+1}^{n} \binom{j-1}{k} X_{j:n} \qquad (31)$$

where n is the sample length and k the order of the probability-weighted moment. Since L-moment estimators are linear functions of the sample values, they should be virtually unbiased and have relatively small sampling variance.

The sample L-CV is defined by the ratio $t = l_2/l_1$, where l_1 is the sample mean and l_2 a measure of the dispersion around the mean value. Other sample L-moment ratios are $t_3 = l_3/l_2$ and $t_4 = l_4/l_2$ that measure the skewness and the kurtosis of data. According to Hosking (1990), also the L-moment ratio estimators are asymptotically normally distributed and have small bias and variance, especially if compared with the classical coefficients of variation, skewness, and kurtosis (Hosking and Wallis, 1997).

In many hydrological applications an occasional event may be several times larger than other values; when product moments are used, such values can mask the information provided by the other observations, while product moments of the logarithms of sample values can overemphasize small values. In a wide range of hydrologic applications, L-moments provide simple and reasonably efficient estimators of the characteristics of hydrologic data and of a distribution's parameters.

2.18.3.2.3 Method of the maximum-likelihood and Bayesian methods

Still another method that has strong statistical motivation is the maximum-likelihood method. Maximum-likelihood estimators (MLEs) have very good statistical properties in large samples, and experience has shown that they generally do well with records available in hydrology. MLEs sometimes perform poorly when the distribution of the observations deviates in significant ways from the distribution being fitted.

MLE methods provide a computationally convenient way to fit frequency distributions by using different sources of information. In flood frequency analysis, for example, systematic records can be combined with historical events through a proper formulation of the likelihood function in which also uncertainties (particularly measurement errors) are taken into account (see, e.g., Stedinger and Cohn, 1986; O'Connell et al., 2002; O'Connell, 2005). The best parametrization of the assumed flood frequency distribution can then be obtained maximizing the likelihood function.

The Bayesian inference, in addition to the maximum likelihood method, combines prior information (or, for example, regional hydrologic information) with the likelihood function in a posterior probability model that quantifies the belief in the hypothesis (i.e., the flood frequency distribution with a given parameter set) after evidence (the flood data) has been observed. Another advantage of the Bayesian method over the method of maximum likelihood is that it allows the explicit modeling of uncertainty in the parameters of the frequency distribution, which can be used to assign confidence bounds to the estimated flood quantiles.

2.18.3.3 Model Verification: Goodness-of-Fit Tests

Goodness-to-fit criteria are useful for gaining an appreciation for whether the lack of fit is likely to be due to sample-to-sample variability, or whether a particular departure of the data from a model is statistically significant. Model testing and verification are basic steps of statistical inference, and several testing techniques, borrowed from applied statistics, have been applied in the hydrologic field. However, none of these tests has reached a broad consensus in the hydrologic community, possibly due to some complications that inevitably arise when the parameters of the hypothetical distribution are unknown. In order to evaluate which is the best distribution for a specific sample, a first simple step can be the graphical representation of the functions.

The tail behavior of the distributions described in the previous paragraphs is analyzed in **Figure 5**, which shows the quantile function versus the return period for three-parameters distributions with equal L-moments of order 1, 2, and 3. It is evident that the distributions follow a similar behavior for return periods up to 50 years, but they tend to diverge for larger return periods. This highlights the necessity to consider methods for the choice of the distribution function from a set of candidate distributions. Graphical procedures form a useful visual method of verifying whether a theoretical distribution fits an empirical distribution (a data sample). Among these procedures, probability plots are commonly used in hydrology (Powell, 1943). In most cases, probability plots are constructed to suit the CDF of a particular distribution. Thus, when the distribution function is plotted against the variate, a linear relationship is obtained if the observations are from the hypothetical distribution. **Figure 5** does not refer to one particular distribution but is simply a convenient probability plot that highlights the shape of different distributions for very low exceedance probabilities (very high return periods). If one plots a data sample (which,

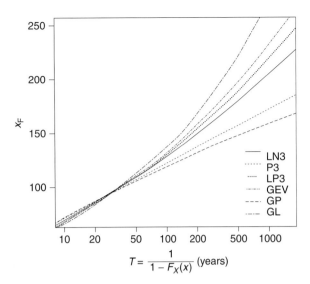
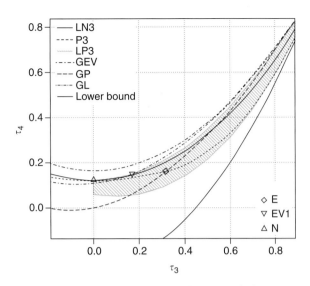

Figure 5 Frequency plot of the quantile function x_F vs. the return period T for some three-parameter distributions with $\lambda_1 = 40$, $\tau = 0.3$, and $\tau_3 = 0.3$.

Figure 6 Probability distributions in the L-kurtosis vs. L-skewness diagram.

for example, has sample L-moments $l_1 = 40$, $t = 0.3$ and $t_3 = 0.3$) using a plotting position on such a graph, one can have an idea of which distribution is better suited to the data. Since the method is subjective, it should always be supplemented by objective goodness-of-fit tests and/or by the application of model selection techniques. Objective procedures for the probabilistic model selection can be found in the specific literature. The subject was first proposed in the Akaike (1973) work, where the principle of maximum entropy was introduced as the theoretical basis for model selection, and by Schwartz (1978) who, by developing a similar idea in a Bayesian context, proposed the Bayesian information criterion for model selection. Applications of objective model selection techniques in statistical hydrology can be found in Strupczewski et al. (2002), Mitosek et al. (2006), Di Baldassarre et al. (2008), and Laio et al. (2009).

Perhaps the most common approach to the choice of the probabilistic model in hydrology is based on the use of L-moments plots, which are used to determine the probability distribution closer to the available sample of data (see, e.g., Hosking, 1990; Chowdhury et al., 1991; Stedinger et al., 1993; Hosking and Wallis, 1997; Peel et al., 2001). **Figure 6** represents the distributions treated in this chapter on the L-moment ratio diagram τ_3–τ_4. Two-parameter distributions are shown as points while three-parameter distributions are represented as curves. Because the LP3 distribution has two shape parameters, the L-moment ratio diagram covers a two-dimensional area (see Griffis and Stedinger, 2007). The diagram is convenient for plotting sample at-site or regional average L-moments ratios for comparison with the population values. Also this approach, however, is not fully objective, because the goodness-of-fit of a distribution to the data is often based only upon graphic judgment. Again, the choice of the distribution should be based on goodness-of-fit test.

In the following some goodness-of-fit tests are described. It will be referred to as 'case p', in analogy to the Stephens (1986) use of 'case 0' for the case when the parameters are fully specified a priori. In case p, the distribution test statistics depend on the so-called null hypothesis H_0, that is, on the probability distribution that is being tested (e.g., Stephens, 1986). This means that the percentage points, that is, the $100(1-\alpha)$ percentiles of the distributions of the test statistics (α is the significance level of the test), have to be recalculated for each H_0. The method of parameter estimation, the presence of a shape parameter, and the sample size also have an influence on percentage points, and this further complicates the analysis. There is thus the necessity to have a different table for each distribution, and tables of percentage values for some tests and distributional families are still lacking.

Some testing techniques commonly used in the hydrologic field are listed in the following, with reference to available tables of percentage points when applicable.

1. *Tests of chi-squared type*. The use of the classical Pearson test requires that the range of x is partitioned in classes; a convenient procedure to avoid arbitrariness and maximize the power of the test entails the choice of k equiprobable classes under the hypothesized distribution, with $k = 2n^{0.4}$ (Moore, 1986). The test statistic distribution in case 0 is the chi-squared distribution with $k - 1$ degrees of freedom. In case p the distribution is not completely known, since there is a partial recovery of degrees of freedom of the chi-squared distribution with respect to the commonly recommended value of $k - p - 1$ (p is the number of model parameters), when efficient estimators are used (e.g., Kendall and Stuart, 1977: 455; Moore, 1986). When using maximum likelihood to estimate model parameters, the critical points fall between those of $\chi^2(k-1)$ and those of $\chi^2(k-p-1)$, and not even this can be said when moments or L-moments estimators are employed.
2. *Tests using the linearity of the probability plot for measuring the goodness of fit*. A probability plot is a graph of the ranked observations $x_{(i)}$ versus an approximation of their expected value, $F^{-1}(1-q_i)$, where q_i is the plotting position, which

can be written as

$$q_i = \frac{i-a}{n+1-2a}$$

where $a<0.5$ is a coefficient (see Stedinger et al., 1993, table 18.3.1 for standard a values). Appropriate critical values for probability plot tests for the EV1 and normal distributions are tabled by Stedinger et al. (1993). No such tables exist for the GEV with the three parameters estimated from the sample. For the P3 distribution, the testing procedure is described by Vogel and McMartin (1991).

3. *Tests based on the comparison of empirical and hypothetical L-moments ratios.* The appropriate testing procedure and percentage points are found in Fill and Stedinger (1995) for the EV1 distribution, in Stedinger et al. (1993) for the normal distribution, and in Wang (1998) for the GEV distribution. The test is not available for the P3 distribution.

4. *Tests based in the empirical distribution function (EDF).* EDF tests are based on the comparison between the hypothetical and empirical distribution function, $F_n(x)$, a cumulative probability distribution function that concentrates probability $1/n$ at each of the n values in a sample. The discrepancy between the two distributions can be measured either with statistics of the form $\max|F_n(x) - F_X(x)|$ (Kolmogorov–Smirov (KS) test), or using quadratic statistics, $Q^2 = n \int_{\text{all } x} [F_n(x) - F_X(x)]^2 \Psi(x)\, dx$, where $\Psi(x)$ is a weight function. When $\Psi(x) = 1$, one has the Cramer–von Mises statistic, usually called W^2, which is a measure of the mean square difference between the empirical and hypothetical CDF; when $\Psi(x) = [F_X(x)(1 - F_X(x))]^{-1}$, the tails of the distribution are weighted more than the central part, and one has the Anderson–Darling statistic, called A^2. W^2 and A^2 are estimated in practice as (e.g., Stephens, 1986)

$$W^2 = \frac{1}{12n} + \sum_{i=1}^{n} \left\{ F_X(x_{(i)}) - \frac{(2i-1)}{2n} \right\}^2 \quad (32)$$

and

$$A^2 = -n - \frac{1}{n}\sum_{i=1}^{n}\{(2i-1)\ln(F_X(x_{(i)})) + (2n+1-2i)\ln(1-(F_X(x_{(i)})))\} \quad (33)$$

respectively, where $x_{(i)}$ represents the ith element in the ordered sample.

Suitable tables of percentage points for the KS test in the p-case are found in Stephens (1986) for the EV1 and normal distributions. For the GEV and GAM distributions with all the parameters estimated, the appropriate percentage points are instead not known. Percentage points in the p-case for EV1, NORM, GAM, and GEV distributions can be calculated following the procedure described by Laio (2004) for the Cramer–von Mises and Anderson–Darling tests.

2.18.4 IDF Curves

IDF curves are probably one of the most commonly used tools in engineering practice. IDF curves are simple functions between the rainfall intensity i, the timescale k at which the rainfall process is studied, and the return period T (see Section 2.18.4.1 for definitions). Nevertheless, the concept of IDF curves is often misinterpreted, mainly because of the imprecise terms used in its definition. It will be apparent in the following analysis that both terms, 'duration' and 'frequency', are misleading.

Specifically, the term 'duration' is often misinterpreted as the actual time duration of a rainfall episode, while the term is meant for the time interval k, or else the timescale k, over which the rainfall process is averaged. For example, the actual duration of a rainfall episode may be only a fraction of the time interval k, or may be equal to many time intervals k.

In addition, the term 'frequency' traditionally is meant for the number of occurrences of a periodic event during a time unit, and is reciprocal to the period which is defined as the exact time between two successive occurrences of the event. Thus, this may falsely lead to the belief that the rainfall intensity value assigned to a return period T will occur once every T years. Of course, this is wrong, as the correct interpretation is that the rainfall intensity value i, assigned to a return period T, will on average be exceeded once every T years. This means that during a particular period of T years the value i may not be exceeded at all, or exceeded several times, and only on average it will be once every T years. Thus, frequency in the IDF curves, it could be said, is referred to an 'average frequency'.

Given the above, a more correct term for IDF curves would be intensity–timescale–return period curves. Recently, the term 'ombrian curves' has been coined (Papalexiou and Koutsoyiannis, 2008) based on the ancient Greek word 'όμβρος' (pronounced ómbros) meaning rain (send by the Olympian god Zeus). Nevertheless, as the prevailing term at the moment is IDF curves, this term will be used in the rest of the text.

It seems that Kuichling (1889) was the first, who studied the rainfall in relation with timescale; however, IDF curves, on a basis that is still in use, were established by Bernard (1932). Since then, the importance of IDF curves in engineering necessitated the study and the construction of IDF curves in several parts of world. For example, in the USA, the US Weather Bureau created a rainfall frequency Atlas (Hershfield, 1961); in addition, the NOOA developed maps for the Western US (Miller, 1973) and the Eastern and Central US (Frederick, 1977). Similar maps were constructed for Sri Lanka (Baghirathan, 1978), Namibia (Pitman, 1980), areas of Brazil (Brasil Vieira and Zink de Souza, 1985), Australia (Canterford, 1986), Pennsylvania (Gert, 1987), India (Kothyari, 1992), and many more. More recently, IDF relationships have been studied for Southeast Asia (Dairaku et al., 2004), Quebec in Canada (Mailhot et al., 2007), the Netherlands (Overeem et al., 2008), and for Denmark (Madsen et al., 2009).

IDF curves have a great variety of applications, as they are a very convenient and useful tool used in the hydraulic design of flood protection infrastructures and in flood risk management in general. They provide the basic input in models that convert the rainfall to flood discharge, for example, they provide the rainfall rate for the so-called rational method. Essentially, their usefulness is in predicting the average rainfall intensity value, for a given timescale that depends on the infrastructure's characteristics, and for a given return period that depends on the infrastructure's importance and the aimed safety.

2.18.4.1 Definition of IDF Curves and Clarifications

IDF curves are mathematical formulas that relate the rainfall intensity i with the timescale k and the return period T, that is, formulas that establish a one-to-one correspondence among rainfall intensity i, timescale k, and return period T.

To clarify, the rainfall intensity is, in general, a continuous time stochastic process, that is, at every time instant t, the rainfall intensity has a value $i(t)$ that can be either zero or positive. Of course, the instantaneous rainfall intensity $i(t)$ cannot be known, however; the rainfall depth can be measured at consecutive time intervals of duration k each, and thus, dividing the rainfall depth by the time duration k results in the average rainfall intensity time series. The time duration k over which the rainfall intensity is averaged is called the timescale k.

Obviously, as in reality only the average rainfall intensity is used, and for the sake of brevity, instead of the term average rainfall intensity the term rainfall intensity will be used. The rainfall intensity at timescale k can be regarded as a random variable (r.v.), denoted as $I(k)$, following a probability distribution $F_{I(k)}(i)$, whereas the timescale k is a specified quantity and not an r.v. Moreover, it is well known that the return period T assigned to a value of an r.v. is defined as the average time needed for this value to be exceeded, and in a discrete time process is explicitly related to the probability distribution F of the r.v. by

$$T = \frac{1}{1-F} \quad (34)$$

It is noted that T is expressed in the same time units as the timescale k of the discrete time process, for example, if the timescale k is 1 h then T is expressed in hours. Therefore, if the probability distribution $F_{I(k)}(i)$ is known, the rainfall intensity at timescale k and for a return period T can be estimated, given Equation (34), by the quantile function $Q_{I(k)}(T)$ of the distribution, $F_{I(k)}(i)$:

$$i(k,T) = F^{-1}_{I(k)}\left(1 - \frac{1}{T}\right) = Q_{I(k)}(T) \quad (35)$$

Nevertheless, the estimation of rainfall intensity for a given return period T and for an arbitrary timescale k within a desired interval – as it is often the demand in engineering practice – would require knowledge of the distribution $F_{I(k)}(i)$ for every timescale k within this interval. Undoubtedly, this is hard to accomplish, if not impossible, as in reality, the distribution $F_{I(k)}(i)$ can only be estimated for a few discrete timescales. In fact, the construction of IDF curves remedies this problem by using the few estimated distributions $F_{I(k)}(i)$ to establish a function that assigns a rainfall intensity value to any given timescale k and any return period T.

In the literature, there are several different techniques for constructing IDF curves that vary significantly. Regarding the starting series, or the historical samples used for the construction of IDF curves, some methods use annual maxima series (AMS) of rainfall intensity, that is, the annual maximum values of every timescale, and others use partial duration series (PDS), that is, the series of values above a threshold (for comparison see e.g., Langbein, 1949; Cunnane, 1973; Takeuchi, 1984; Buishand, 1989; Madsen, 1997a, 1997b; Begueria, 2005; Ben-Zvi, 2009). Nevertheless, the use of AMS is by far more popular as the AMS are usually readily available, or can be easily prepared. In addition, the use of AMS offers computational simplicity (it will be apparent in the next sections), as it can be assumed that the probability distribution of the annual maximum rainfall intensity at each timescale k_j belongs to the same family of distributions, that is, the extreme value distributions.

Methodologies for constructing IDF curves do not only vary in the samples used, but also may be based on different approaches. For example, the classical empirical forms are presented in Chow et al. (1988: 459); a more general approach applied in United States has been proposed by Chen (1983); general forms consistent with the probability theory are given by Koutsoyiannis et al. (1998); forms in relation with L-moments by Hosking and Wallis (2005); approaches in relation with multifractals are given by Bendjoudi et al. (1997) and Veneziano et al. (2007); and in relation with copula functions by Singh and Zhang (2007).

Nevertheless, most of forms of IDF curves can be combined in the following general expression:

$$i(k,T) = \frac{g(T)}{h(k)} \quad (i \text{ in mm h}^{-1}, k \text{ in h}, T \text{ in years}) \quad (36)$$

where $g(T)$ is a function of the return period T and $h(k)$ is a function of the timescale k. Clearly, this expression implies the separable function dependence of the rainfall intensity i on the return period T and on the timescale k, and even though the theoretical consistency of this assumption has been recently disputed, for moderate and large return periods provides a close approximation sufficient for practical purposes (Papalexiou and Koutsoyiannis, 2008). Of course, as it is obvious form Equation (36) that the rainfall intensity is a monotonically increasing function of the return period T, and a monotonically decreasing function of the timescale k.

2.18.4.2 Empirical Methods

Empirical forms of IDF curves, due to their long history, as they date back to 1932 (Bernard, 1932), are those mostly studied and used in practice, while are still the most popular forms covered in existing text books (e.g., Chow et al., 1988: 459; Wanielista, 1990: 61; Shaw and Shaw, 1998: 228; Mays, 2004: 219). In general, compared to other forms of IDF curves, their expressions are characterized by simplicity, while their parameters are easy to estimate, at least for the most simple forms among them. The most commonly used empirical expression for the return period function is $g(T) = aT^\beta$, while others have also been used. In addition, the timescale function can be found in many variations that, however, can all be combined to the general expression $h(k)=(k^\gamma+\delta)^\varepsilon$ (for a comparison see, e.g., García-Bartual and Schneider, 2001; Di Baldassarre et al., 2006a). For convenience, the different variations of the return period function $g(T)$ and the timescale function $h(k)$ used in this text are distinctly named:

$$g(T): \quad g_1(T)=aT^\beta, \; g_2(T)=a+\beta\ln T \qquad (37)$$

$$h(k): \quad h_1(k)=k^\gamma, \; h_2(k)=k^\gamma+\delta,$$
$$h_3(k)=(k+\delta)^\varepsilon, \; h_4(k)=(k^\gamma+\delta)^\varepsilon \qquad (38)$$

where α, β, γ, δ, and ε are the parameters to be estimated. Of course, all the different variations of the $g(T)$ and the $h(k)$ functions may be used, thus resulting in several different empirical forms of IDF curves.

2.18.4.2.1 Parameter estimation

The typical estimation procedure of IDF curves' parameters, based on AMS (e.g., Chow et al., 1988: 459), can be summarized in three steps.

Step 1: In the first step, a suitable probability distribution is selected and fitted to each maximum rainfall intensity data set that comprises values of the same timescale k_j, where $j=1$, ..., m, with m denoting the total number of different timescales that data are available.

Clearly, the many distribution choices and the many available distribution fitting methods (e.g., the method of moments and L-moments, or the maximum likelihood and the least-squares error methods) may significantly affect the estimated parameters of the IDF curves.

Nevertheless, a natural choice for the probability distribution to be fitted, given that the groups of rainfall intensity values are annual maximum values, is one of the two maximum extreme value distributions, that is, the Gumbel distribution, given in Equation (39), or the GEV distribution, given in Equation (40) (with parameter $\theta_3<0$ in order to be unbounded form above), and with the latter comprising the Gumbel distribution as a special case for $\theta_3=0$:

$$F_{I(k_j)}(i)=\exp\left[-\exp\left(-\frac{i-\theta_1}{\theta_2}\right)\right] \quad (\theta_1\in\mathbb{R}, \theta_2>0) \quad (39)$$

$$F_{I(k_j)}(i)=\exp\left[-\left(1-\theta_3\frac{i-\theta_1}{\theta_2}\right)^{1/\theta_3}\right]$$
$$(\theta_1\in\mathbb{R}, \theta_2>0, \theta_3>0) \qquad (40)$$

where the symbol $I(k_j)$ stands for the annual maximum rainfall intensity at timescale k_j. Yet, it should be noted that apart from the Gumbel and GEV distributions, other distributions have also been used to describe annual maxima, for example, the log-Pearson III and lognormal distributions.

Although the Gumbel distribution has been the traditional choice for describing maxima, as it is a parsimonious model and often gives good results, new evidence suggests (Gellens, 2002; Ramesh and Davison, 2002; Koutsoyiannis, 2004a, 2004b; Salvadori and De Michele, 2006) that the Gumbel distribution may seriously underestimate the rainfall intensity for large return periods, and thus its use should be avoided. Alternatively, the GEV distribution, which has gained popularity the last decade, can be used taking special care in the estimation of the parameter θ_3. In particular, as the typical rainfall samples are usually small, the estimation of the parameter θ_3 may be highly uncertain. In order to remedy this, Koutsoyiannis (2004a, 2004b) proposed to adopt a global value for θ_3, that is, $\theta_3=-0.15$, as this value resulted from studying many rainfall samples from stations all over the world.

Step 2: In the second step, a set of p characteristic return period values $\{T_1, \ldots, T_l, \ldots, T_p\}$ is defined (e.g., $\{2, 5, 10, 20, 50, \ldots, T_p\}$) and the m fitted distributions form step 1 are used to evaluate the rainfall intensity for the selected return periods and for each of the m timescales k_j. This procedure will result in a set comprising $m\times p$ points of the form $(i_{j,l}, k_j, T_l)$. The evaluation of the maximum rainfall intensities $i_{j,l}$ can be accomplished using the quantile functions $Q_{I(k_j)}(T_l)$ of the fitted distributions expressed in relation with the return period T. For the Gumbel and GEV distributions, the quantile functions are given, respectively:

$$i_{j,l}=Q_{I(k_j)}(T_l)=\theta_1-\theta_2\ln\left[-\ln\left(1-\frac{1}{T_l}\right)\right] \qquad (41a)$$

$$i_{j,l}=Q_{I(k_j)}(T_l)=\theta_1+\frac{\theta_2}{\theta_3}\left\{1-\left[-\ln\left(1-\frac{1}{T_l}\right)\right]^{-\theta_3}\right\} \qquad (41b)$$

Step 3: In this final step the parameters of the selected form of IDF curves are estimated.

The parameter estimation of the most simple and one of the most commonly used empirical forms of IDF curves, that is, $i(k,T)=g_1(T)/h_1(k)$, can be done analytically using the method of multiple linear regression. Clearly, logarithmizing this simple form results in

$$\ln i(k,T)=\ln \alpha+\beta\ln T-\gamma\ln k \qquad (42)$$

which is for the form $y=\xi_0+\xi_1 x_1+\xi_2 x_2$, and consequently, the parameters α, β, and γ can be estimated by performing a multiple linear regression using the set of $(\ln i_{j,l}, \ln k_j, \ln T_l)$ points, evaluated in step 2. Obviously, the rainfall intensity logarithm $\ln i$ is considered as the dependent variable y, whereas the timescale logarithm $\ln k$ and the return period logarithm $\ln T$ as the independent variables x_1 and x_2, respectively. The parameters α, β, and γ will straightforwardly result from the estimated multiple linear regression coefficients ξ_0, ξ_1, and ξ_2, that is, $a=\exp(\xi_0)$, $\beta=\xi_1$, and $\gamma=-\xi_2$.

This technique, however, is not directly applicable in the case of more general forms of IDF curves. Specifically, logarithmizing the $i(k,T)=\alpha T^\beta/(k^\gamma+\delta)^\varepsilon$ results in

$$\ln i(k,T)=\beta\ln T-\varepsilon\ln(k^\gamma+\delta)+\ln\alpha \qquad (43)$$

which is not of the form $y=\xi_0+\xi_1 x_1+\xi_2 x_2$. Nevertheless, inspection of Equation (43) suggests that if the timescale function $h(k)=h_3(k)$, then the term $\ln(k^\gamma+\delta)$ in the equation becomes $\ln(k+\delta)$, and thus, multiple regression can be performed by assuming given values of δ. The estimated parameters α, β, and γ should be those for that δ minimizes a proper norm between the values of the set of points $(\ln i_{j,l}, \ln k_j, \ln T_l)$ and the estimated ones by the selected form of IDF curves. Obviously, if the return period function $g(T) = g_2(T)$, this methodology is not applicable.

In addition, one global way of fitting a function, linear or nonlinear, to a given data set of values, and thus applicable in the parameter estimation of IDF curves, is to minimize the mean square error (MSE) between the values of the given data set and the corresponding values calculated from the function to be fitted. Alternatively, in cases where there are large differences between the values of the given data set, instead of minimizing the MSE, it may be more suitable to minimize the logarithmic SE (log SE). In the case studied here, the MSE and the log SE are given, respectively, by

$$\text{MSE} = \frac{1}{mp} \sum_{j=1}^{m} \sum_{l=1}^{p} \left[i(k_j, T_l) - Q_{I(k_j)}(T_l) \right]^2 \quad (44)$$

$$\log \text{SE} = \sum_{j=1}^{m} \sum_{l=1}^{p} \log^2 \frac{i(k_j, T_l)}{Q_{I(k_j)}(T_l)} \quad (45)$$

where $i(k_j, T_l)$ is the rainfall intensity values calculated from the selected form of IDF curves, for example, one of the forms resulted from the combinations given in Equations (37) and (38) and $Q_{I(kj)}(T_l)$ is the quantile functions of the distributions fitted in step 1, for the rainfall intensity of timescale k_j and of return period T_l.

2.18.4.2.2 Application in a real-world data set

In this section, the methodologies described in Section 2.18.4.2.1 will be applied in a real-world data set of recorded rainfall intensities form 1987 to 2004, in the station Ardeemore in UK. The data set was originally available, by the British Atmospheric Data Centre (BACD), as tipping bucket measurements that were first converted in the 5-min temporal resolution, second, aggregated over several timescales and third, the annual maximum values of each timescale were extracted (British Atmospheric Data Centre, 2006). The summary statistics of the resulted data set of maximum rainfall intensities in several different timescales are presented in Table 6.

As described in Section 2.18.4.2.1, the typical parameter estimation procedure of IDF curves begins with selecting and fitting a theoretical probability distribution to the same timescale data. In this application, for demonstration and comparison, both the Gumbel and the GEV distributions are fitted to the data. In addition, the L-moments method (Hosking, 1990) was selected as a fitting method, as it is robust and easy to apply – especially for the Gumbel and the GEV distributions, it results in analytical equations. The results are presented in Table 7.

The fitted distributions to the empirical data, and the empirical distribution functions according to Weibull plotting position, are depicted in Figure 7. Clearly, both distributions perform very well up to return period values approximately equal to 20 years. Of course, as it was expected, the estimated rainfall intensity difference between the two distributions increases monotonically with the return period, and for return period values higher than 100 years, the difference gets significant, with the GEV distribution resulting in higher rainfall intensity estimates.

Both the Gumbel and the GEV distributions fit equally well to the empirical points (see Figure 7); however, the Gumbel distribution may underestimate the rainfall intensity for high return periods (see step 1 of Section 2.18.4.2.1), and thus, the GEV distribution is preferred to generate the set of $(i_{j,l}, k_j, T_l)$ points described in step 2 of Section 2.18.4.2.1. In addition, this argument is fortified by noticing in Figure 7 that the empirical return period of the higher historical value in the smaller timescales is disproportionally small compared to the one resulted by the Gumbel distribution. For example, the empirical return period of the largest value in the 5 min timescale is 19 years, and while the GEV distribution assigns a theoretical return period to that value approximately equal to 80 years, the corresponding value by the Gumbel distribution is about 180 years.

The selected characteristic return periods T_l are $\{2, 5, 10, 20, 50, 100, 200, 500, 1000\}$ years, a total of $p = 9$ values. It is noted that due to the small recorded sample (18 years), the rainfall intensity estimates in high return periods will be uncertain. Furthermore, the number of the selected timescales k_j is $m = 9$ (see Table 6), and therefore, a set of $m \times p = 99$ points of the form $(i_{j,l}, k_j, T_l)$ is generated. The rainfall intensity values $i_{j,l}$ are calculated using the quantile function of the GEV distribution, given in Equation (41), for the characteristics return periods T_l, with the estimated parameters that correspond to the timescale k_j (see Table 7). The set of generated points is depicted in Figure 8.

Table 6 Summary statistics of maximum rainfall intensity data (mm h^{-1}) at several timescales observed at Ardeemore

Timescale k_j	Sample size	Mean	Standard deviation	Variation coefficient	Skewness coefficient	Minimum	Maximum
5 min	18	44.03	20.77	0.47	2.04	20.30	112.83
10 min	18	33.07	19.14	0.58	2.86	16.20	103.22
20 min	18	23.00	10.03	0.44	1.52	11.35	52.97
30 min	18	17.49	6.62	0.38	1.14	8.27	35.50
60 min	18	11.23	3.64	0.32	0.61	4.90	20.40
2 h	18	7.89	3.26	0.41	1.86	4.19	18.10
3 h	18	5.99	2.05	0.34	2.04	3.13	12.60
6 h	18	4.22	0.91	0.22	0.58	2.70	6.30
12 h	18	2.85	0.70	0.25	0.03	1.50	4.07
24 h	18	1.82	0.40	0.22	−0.09	1.00	2.61
48 h	18	1.25	0.37	0.30	0.94	0.72	2.17

Table 7 Estimated sample L-moments of the maximum rainfall intensity in 11 different timescales and the corresponding estimated parameters of the fitted Gumbel and GEV distributions

Timescale	Sample L-moments		Gumbel parameters		GEV parameters[a]	
	l_1	l_2	θ_1	θ_2	θ_1	θ_2
5 min	44.03	10.29	35.47	14.84	34.54	12.66
10 min	33.07	8.25	26.21	11.90	25.46	10.15
20 min	23.00	5.30	18.59	7.65	18.11	6.53
30 min	17.49	3.62	14.47	5.23	14.14	4.46
60 min	11.23	2.06	9.51	2.97	9.32	2.54
2 h	7.89	1.63	6.53	2.35	6.38	2.01
3 h	5.99	0.97	5.18	1.40	5.09	1.19
6 h	4.22	0.51	3.79	0.74	3.75	0.63
12 h	2.85	0.41	2.51	0.59	2.47	0.50
24 h	1.82	0.23	1.63	0.33	1.61	0.28
48 h	1.25	0.21	1.08	0.30	1.06	0.25

[a]Parameters estimated setting *a priori* $\theta_3 = -0.15$ in all timescales.

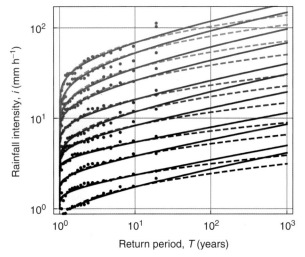

Figure 7 Empirical distribution functions (dots) according to the Weibull plotting position, and fitted by the method of L-moments, Gumbel (dashed lines) and GEV (solid lines) distributions for the data in the timescales given in **Table 8**, ranging form 5 min to 48 h (from above to below).

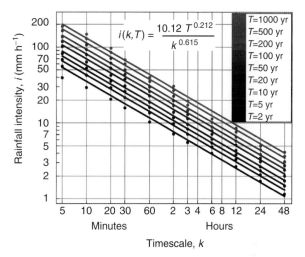

Figure 8 IDF curves for the Ardeemore station in UK constructed using the typical parameter estimation procedure, and the set of rainfall intensity points generated form the fitted GEV distribution to the empirical data.

Finally, the parameters of the most simple form of IDF curves (i.e., $i(k,T) = aT^\beta/k^\gamma$) are estimated (see step 3 in Section 2.18.4.2.1) by performing a multiple linear regression to the set of $(\ln i_{j,l}, \ln k_j, \ln T_l)$ points. The estimated multiple linear regression coefficients are $\xi_0 = 2.315$, $\xi_1 = 0.211$, and $\xi_2 = -0.615$, and consequently, the resulted parameters of the IDF curves are $\alpha = 10.12$, $\beta = 0.211$, and $\gamma = 0.615$. The resulted IDF curves are depicted in **Figure 8**.

In addition to the previous simple form of IDF curves, some more complicated empirical forms – and in order to demonstrate and compare their performance – were constructed for the Ardeemore station data set, by numerically minimizing the logSE given in Equation (45) (see step 3 in Section 2.18.4.2.1). The logSE minimization was performed using one of the many software packages that include numerical minimization routines. It is worth noting that estimating the parameters of the simple $i(k,T) = aT^\beta/k^\gamma$ by minimizing the logSE is actually the same as performing the multilinear regression method. The fitted IDF curves the estimated parameters, and the resulted logSE are presented in **Table 8**.

As expected, the additional parameters result in smaller logSE, and thus, in a better fit. Nevertheless, especially in the particular case studied here, the difference in logSE among the different forms of IDF curves is not substantial, and therefore, according to the principle of parsimony, a more parsimonious form should be preferred than the most general five-parameter case. The last argument is also fortified by **Figure 9**, where the more general IDF curves of **Table 9** are depicted. Clearly, the IDF curves with return period function $g_1(T)$ and the three-parameter timescale function $h_4(k)$, compared with the ones with the two-parameter functions $h_2(k)$ and $h_3(k)$, are not significantly different.

Table 8 Five different empirical forms of IDF curves fitted by minimizing the logarithmic square error

IDF curves $i(k,T)$	Estimated parameters					Log SE
	α	β	γ	δ	ε	
aT^β/k^γ	10.12	0.212	0.615			0.99
$aT^\beta/(k^\gamma + \delta)$	10.47	0.212	0.627	0.020		0.97
$aT^\beta/(k + \delta)^\varepsilon$	10.42	0.212		0.015	0.626	0.95
$aT^\beta/(k^\gamma + \delta)^\varepsilon$	10.39	0.212	1.810	0.005	0.347	0.91
$(a + \beta \ln T)/(k^\gamma + \delta)^\varepsilon$	7.33	4.66	3.258	0.0003	0.192	0.92

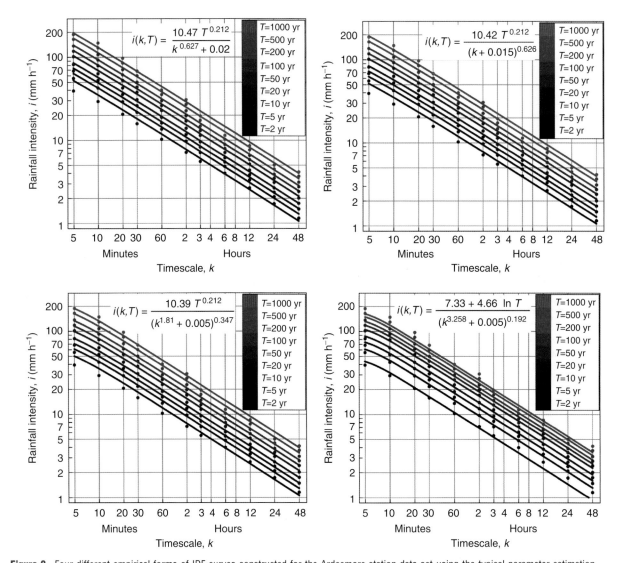

Figure 9 Four different empirical forms of IDF curves constructed for the Ardeemore station data set using the typical parameter estimation procedure.

Moreover, it is important to note that all IDF curves shown in **Figure 9**, compared to the simple form depicted in **Figure 8**, exhibit a slight curvature in small timescales – which is more apparent in the IDF curves with the timescale function $h_4(k)$. This is obviously the effect of the parameter δ, a very important parameter that allows a better fit of the IDF curves in small timescales, or equally in the high rainfall intensities. Although in this particular case the resulted curvature is very slight, and thus not important, this is not the general rule, as for a different data set this curvature may be very strong and

Table 9 Theoretically consistent forms of IDF curves fitted by two different methods for the Ardeemore station data set

Method	Parameters	IDF curves $i(k,T)$				
		$\dfrac{Q_{\text{GEV}}(T)}{(k^\gamma + \delta)^\varepsilon}$	$\dfrac{Q_{\text{GEV}}(T)}{(k + \delta)^\varepsilon}$	$\dfrac{Q_{\text{GEV}}(T)^a}{(k + \delta)^\varepsilon}$	$\dfrac{Q_{\text{Gumb}}(T)}{(k^\gamma + \delta)^\varepsilon}$	$\dfrac{Q_{\text{Gumb}}(T)}{(k + \delta)^\varepsilon}$
Two-step robust estimation	θ_1	9.92	10.02	9.95	10.04	10.14
	θ_2	2.82	2.88	2.67	3.09	3.14
	θ_3	0.09	0.09	0.15		
	γ	2.298			2.298	
	δ	0.002	0.017	0.017	0.002	0.017
	ε	0.255	0.593	0.593	0.255	0.593
	K_{KW}	12.08	13.38	13.38	12.08	13.38
One-step log SE minimization	θ_1	9.65	9.83	9.85	9.85	10.04
	θ_2	3.02	3.07	3.10	3.28	3.36
	θ_3	0.16	0.16	0.15		
	γ	3.089			3.089	
	δ	0.0004	0.022	0.022	0.0004	0.022
	ε	0.185	0.577	0.577	0.185	0.577
	log SE	2.13	2.16	2.16	2.26	2.29

[a]The parameters were estimated by setting *a priori* $\theta_3 = -0.15$.

thus essential in engineering practice. Consequently, it is proposed that the selected form of IDF curves should include this parameter.

2.18.4.3 Theoretically Consistent Methods

Apart from the classical empirical forms of IDF curves, described in Section 2.18.4.2, there are forms of IDF curves, based on the general Equation (36), that are theoretically more consistent. Koutsoyiannis *et al.* (1998) proposed that empirically derived return period functions $g(T)$ are unnecessary, as the $g(T)$ can be determined from the probability distribution function of the maximum rainfall intensity $I(k)$. Specifically, this method is based on the fact that the probability distribution $F_I(i)$ of the r.v. $I = I(k)h(k)$ is just a scaled version of the distribution of the r.v. $I(k)$, as the function $h(k)$ for a certain timescale k is just a real number. Thus, instead of estimating the probability distribution $F_{I(k)}(i)$ of the r.v. $I(k)$ for several timescales, only the distribution of the r.v. I should be estimated. Consequently, the form of IDF curves would be

$$i(k,T) = \frac{F_I^{-1}(1 - 1/T)}{h(k)} = \frac{Q_I(T)}{h(k)} \quad (46)$$

where $Q_I(T)$ is the quantile function of the r.v. I, and not some empirically proposed function. Using the extreme value distributions in this framework, as they are the natural choice for describing maxima – although many other distributions have been used – the forms of IDF curves, according to Equation (46) and for the general three-parameter timescale function $h_4(k)$, for the Gumbel and GEV distributions, respectively, become

$$i(k,T) = \frac{Q_{\text{Gumb}}(T)}{h_4(k)} = \frac{\theta_1 - \theta_2 \ln\left[-\ln\left(1 - \dfrac{1}{T_l}\right)\right]}{(k^\gamma + \delta)^\varepsilon} \quad (47)$$

$$i(k,T) = \frac{Q_{\text{GEV}}(T)}{h_4(k)} = \frac{\theta_1 + \dfrac{\theta_2}{\theta_3}\left\{1 - \left[-\ln\left(1 - \dfrac{1}{T_l}\right)\right]^{\theta_3}\right\}}{(k^\gamma + \delta)^\varepsilon} \quad (48)$$

where the symbols $Q_{\text{Gumb}}(T)$ and $Q_{\text{GEV}}(T)$, obviously, denote the quantile functions of the GEV and Gumbel distributions, respectively.

Nevertheless, in this theoretical framework, the return period functions in the empirical forms of IDF curves actually correspond to theoretical probability distributions. Specifically, the return period function $g_1(T) = \alpha T^\beta$ corresponds to

$$\alpha T^\beta = Q_I(T) \Leftrightarrow \alpha \left(\frac{1}{1 - F_I(i)}\right)^\beta = i \Leftrightarrow F_I(i) = 1 - \left(\frac{i}{\alpha^\beta}\right)^{-1/\beta} \quad (49)$$

which is the celebrated two-parameter Pareto distribution with parameters $\alpha > 0$, $\beta > 0$ and support $i \in [\alpha^\beta, \infty)$. In addition, the return period function $g_2(T) = \alpha + \beta \ln T$ corresponds to

$$\alpha + \beta \ln T = Q_I(T) \Leftrightarrow \alpha + \beta \ln \frac{1}{1 - F_I(i)}$$
$$= i \Leftrightarrow F_I(i) = 1 - \exp\left(-\frac{i - \alpha}{\beta}\right) \quad (50)$$

which is the celebrated two-parameter exponential distribution with parameters $\alpha \in \mathbb{R}$, $\beta > 0$, and support $i \in [\alpha, \infty)$.

2.18.4.3.1 Parameter estimation

Two-step robust estimation method. This method (Koutsoyiannis *et al.*, 1998) estimates the parameters of the IDF curves in two steps. First, it estimates the parameters of the scale function $h(k)$, and second, the parameters of the return period function $g(T)$, which, in this framework, is the quantile function of a probability distribution. The method is based on the fact that, the r.v.'s $I_j = I(k_j)h(k_j)$ should be distributed identically.

Given the above, in the first step, multiplying the values of each timescale k_j group, denoted $\{i_{j,1}, \ldots, i_{j,n_j}\}$ by the value $h(k_j)$, should result in groups from the same population. Apparently, the function of $h(k)$ is not *a priori* known. Consequently, the method assumes a set of values for the $h(k)$ parameters, and consecutively uses an appropriate statistic to check that indeed the resulted different timescale groups (i.e., $\{h(k_j) \cdot i_{j,1}, \ldots, h(k_j) \cdot i_{j,n_j}\}$) belong to same population. This naturally leads to the Kruskal–Wallis test (Kruskal and Wallis, 1952), which is a nonparametric test applied to infer whether or not different groups of values belong to the same population. The test static K_{KW} is given by

$$K_{KW} = \frac{12}{N(N+1)} \sum_{j=1}^{m} n_j \left(\bar{r}_j - \frac{N+1}{2} \right)^2, \quad \bar{r}_j = \frac{1}{n_j} \sum_{i=1}^{n_j} r_{j,l} \quad (51)$$

where m is the total number of timescales, n_j is the sample size of the timescale k_j group, N is the total sample size across all groups, $\bar{r}_j$ the average rank of the timescale k_j group, and $r_{j,l}$ the rank (among all data) of the lth data value of the timescale k_j group.

Clearly, different groups from the same population would result in a small value of the KKW statistic. Therefore, the estimated parameters of the scale function $h(k)$ are those that minimize the K_{KW} statistic. Essentially, minimizing the K_{KW} statistic results in forcing the different groups of data to belong to the same population. Unfortunately, this minimization can only be accomplished numerically, but numerical optimization can now be routinely performed with widely spread software packages.

Once the parameters of the timescale function $h(k)$ are estimated, it is straightforward to estimate the parameters of the return period function $g(T)$. Specifically, the values of all the resulted groups $\{h(k_j) \cdot i_{j,1}, \ldots, h(k_j) \cdot i_{j,n_j}\}$ are unified in one sample – at least theoretically should belong to the same population – and the probability distribution that corresponds to the return period function $g(T)$ of the selected form of IDF curve is just fitted to this unified sample. The estimated parameters of the fitted distribution are, evidently, the parameters of the return period function $g(T)$.

One-step least-squares estimation method. The basic difference of this method compared to the least-squares method presented in Section 2.18.4.2.1 is that it uses historical data (Koutsoyiannis *et al.*, 1998). As a result, first, it avoids the procedure of fitting distributions to each timescale k_j group and generates values using a set of characteristic return periods, and second, it does not depend on the range of the characteristic return period set, as it uses the empirical return periods resulting from the historical data.

In particular, to every rainfall intensity value of every timescale k_j group of historical data, an empirical return period can be assigned. Specifically, sorting in decreasing order the values of every timescale k_j group, for example, $i_{j,(1)} > \ldots > i_{j,(l)} > \ldots > i_{j,(n_j)}$, the empirical return period $T_{j,l}$ of the lth largest rainfall intensity value, denoted by $i_{j,l}$, of the timescale k_j group, is given according, to the Weibull plotting position, by

$$T_{j,l} = \frac{n_j + 1}{l}, \quad l = 1, \ldots, n_j \quad (52)$$

where n_j is the sample size of the timescale k_j group. Consequently, the MSE and logSE given in Equations (44) and (45), respectively, are modified to

$$\text{MSE} = \frac{1}{mn} \sum_{j=1}^{m} \sum_{l=1}^{n_j} \left[i(k_j, T_{j,l}) - i_{j,l} \right]^2 \quad (53)$$

$$\log \text{SE} = \sum_{j=1}^{m} \sum_{l=1}^{n_j} \log^2 \frac{i(k_j, T_{j,l})}{i_{j,l}} \quad (54)$$

where $i(k_j, T_{j,l})$ is the rainfall intensity calculated from the selected form of IDF curves for the timescale k_j and the empirical return period $T_{j,l}$ of the historical rainfall intensity value $i_{j,l}$ as defined above.

Therefore, the one-step least-squares error method consists of selecting a form of IDF curves, for example, one of the theoretically consistent forms given in Equations (47) or (48), and numerically minimizes the resulted MSE or logSE between the selected form of IDF curves and the historical data. Again, as noted in Section 2.18.4.2.1 the logSE may be more suitable due to the large differences in the rainfall intensity values in small and large timescales. Evidently, the estimated parameters are the ones that minimize the MSE or the logSE. Alternatively, this method can also be used with the empirical forms of IDF curves or with any other form.

2.18.4.3.2 Application in a real-world data set

This section demonstrates the applicability of the aforementioned methodologies and presents the consistent forms of IDF curves constructed for the Ardeemore station data set used in Section 2.18.4.2.2. Among the several forms of IDF curves that could emerge by combining different return period functions $g(T)$ and timescale functions $h(k)$, the ones presented in Equations (47) and (48) are used for the two- and three-parameter timescale functions $h_2(k)$ and $h_4(k)$, respectively. Each form of IDF curves, for comparison and demonstration, was fitted using both methods described in Section 2.18.4.3.1, that is, the two-step robust estimation method, and the one-step least-squares estimation method. The results are presented in Table 9.

It seems that the one-step least-squares error method is the most straightforward method to apply. Simply, the desired form of IDF curves is selected for an arbitrary set of parameters, and is used to estimate the rainfall intensity values $i(k_j, T_{j,l})$ that correspond to the empirical return periods $T_{j,l}$ given by Equation (52). The numerical minimization of the MSE or of the logSE, given in Equations (53) and (54), respectively, between the historical data and the ones predicted by the selected form of IDF curves, results in the estimated parameters. The selected forms that were fitted by minimizing the logSE as the estimated parameters are presented in Table 9.

Among the several variations of fitted IDF curves presented in Table 9, Figure 10 depicts the $Q_{GEV}(T)/h_2(k)$ for

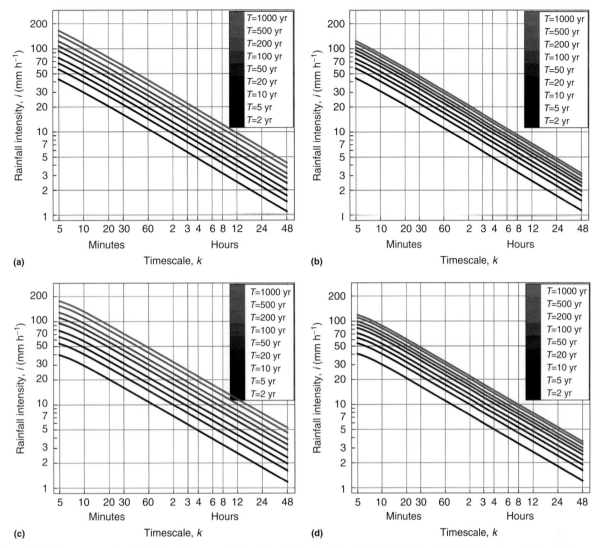

Figure 10 IDF curves constructed for the Ardeemore station data set. Graphs (a) and (b) depict the $Q_{GEV}(T)/h_2(k)$ for $\theta_3 = -0.15$ and the $Q_{Gumb}(T)/h_2(k)$, respectively, fitted with the robust estimation method, while graphs (c) and (d) depict the same IDF curves fitted with the one-step LSE method. The parameters for each case are given in **Table 9**.

$\theta_3 = -0.15$, and the $Q_{Gumb}(T)/h_2(k)$, fitted with robust estimation and by minimizing the log SE. The comparison of the same form of IDF curves fitted by different methods reveals that there are small, albeit noticeable, differences between them. Specifically, the IDF curves fitted with the one-step LSE method, especially for large return periods, are slightly more conservative, that is, the predicted rainfall intensity is higher compared to the one predicted by the other method. In addition, the one-step LSE method results in stronger curvature in the area of small timescales. This behavior can be explained by the presence of a very large value in the small timescales of the historical data set (see **Figure 7**), and it is well known that the LSE methods, in general, are sensitive to outliers. Of course, as **Figure 10** demonstrates, the major difference is between the two forms of IDF curves: the form that uses the quantile of the Gumbel distribution as a return period function, especially for the large return periods, predicts significantly smaller rainfall intensity values compared to one that uses the quantile of the GEV distribution. Obviously, from a mathematical point of view this was expected, but given that the log SE is smaller in the form $Q_{GEV}(T)/h_2(k)$ than in the $Q_{Gmnb}(T)/h_2(k)$, this may suggest that adoption of the $Q_{Gumb}(T)/h_2(k)$ for the design purposes may be a dangerous choice.

2.18.5 Copula Function for Hydrological Application

Since most of the hydrologic phenomena involved multiple variables across various temporal and spatial scales with significant inter-dependencies and non-Gaussian-like behaviors, univariate approaches with the assumption of normality or independence among variables may cause significant over-simplification. In order to address the interwoven dependencies between hydrologic variables, multivariate joint probability distribution needs to be properly modeled. In the past there were attempts focusing on preserving the correct correlation relationship (e.g., Goel et al., 2000; Singh and Singh, 1991; Yue, 2001), but they usually required more assumptions and case-specific restrictions (types of marginal distributions, variables, and selected fixed-form joint distributions) may apply. Thus, though the univariate approach may be less realistic, it is sometimes a necessary trade-off between complexity and applicability.

With the need to characterize multidimensional randomness in nature, a flexible approach with general applicability is of desire. Such a method should be able to model different types of probability distributions for hydrologic variables governed by various physical mechanisms (e.g., rainfall intensity, flood peak, and drought severity), while also being able to faithfully describe their dependence structures. Delightfully, these challenges can now be addressed by using a novel statistical tool – copulas. Copulas got the name as functions that couples arbitrary univariate distributions to form the multivariate joint distribution. In order words, they are the mathematical formulations of the entire dependence space rather than a single correlation or dependence measure (e.g., Pearson's linear correlation coefficient). Since all multivariate probability functions (such as multivariate Gaussian and bivariate exponential distributions) can be re-expressed into the combinations of their marginal distributions and the corresponding copulas, the use of copulas does not conflict with the existing multivariate techniques, but endow them with more possibility. In practice, conceptually similar to the selection of a most appropriate PDF for each individual variable, the most suitable dependence structure between variables of interest can be identified by testing various candidate copula functions. Associated with the identified marginal distributions, together a general joint distribution can be formed.

Though the core theorem supporting copulas was proposed by Sklar early in 1959, the growing number of copula applications was not found until recently, mostly in the field of finance (see Cherubini et al., 2004). In the hydrologic community, it is still a relatively new concept. Nevertheless, copulas were soon found useful in various types of water resources problems due to their great feasibility in modeling multivariate dependence structure. Generally speaking, there are several advantages that make copulas an appealing method for hydrologic topics: (1) it can model non-Gaussian-like variables; (2) the assumption of statistical independence is not a prerequisite; (3) it proceeds in a parallel fashion and all the existing univariate techniques hold; (4) it is less mathematical challenging compared to the conventional multivariate statistical approach; and (5) it helps generate sets of random vectors with prescribed marginal distributions and dependence levels conveniently. Expect that copulas will gradually play a more important role in the future hydrologic study; this section aims to provide the general hydrologic audience with the introduction, state-of-the-art applications, limitations, and future research needs of the copula techniques.

2.18.5.1 Concepts of Dependence Structure and Copulas

As Gaussian distribution has been the most commonly used statistical model for probability distributions and uncertainties, to some engineers and hydrologists, the term correlation (or dependence) refers directly to Pearson's linear correlation coefficient ρ. For random variables, X and Y with means as $\bar{x}$ and $\bar{y}$, ρ is defined as $E[(X-\bar{x})(Y-\bar{y})]/\text{Std}[X]\text{Std}[Y]$, in which $E[\cdot]$ and $\text{Std}[\cdot]$ are the operators of expectation and standard deviation. Though ρ is widely adopted, its limitations are less emphasized: (1) ρ tends to be highly affected by outliers and hence is not suitable for extreme value analysis; (2) the value of ρ may change if X and/or Y are transformed monotonically (such as exponentiation) while their rank correlations remain the same; (3) most important of all, ρ is only adequate for Gaussian (or elliptical) distributions (Nelsen, 2006).

An example is illustrated in **Figure 11**, where two bivariate distributions and the corresponding realizations are presented. **Figure 11(a)** shows the bivariate Gaussian distribution with $\rho = 0.8$. The two-dimensional surface represents the joint PDF $h_{XY}(x,y)$. When integrating either one of the variables over the entire domain $(-\infty, \infty)$, the marginals $f_X(x)$ and $f_Y(y)$ can be obtained, which are plotted on the two sides. In **Figure 11(a)**, both marginals are the typical bell-shape Gaussian densities. In the other case, the joint distribution shown in **Figure 11(b)** is clearly not bivariate Gaussian, as $h_{XY}(x,y)$ has a different shape and the realizations reveal dissimilar patterns. However, the marginals shown in **Figure 11(b)** can still be univariate Gaussian (identical to **Figure 11(a)**), and the correlation coefficient ρ is again the same as 0.8. This example indicates that: (1) the joint distribution cannot be determined only by known marginals and (2) the correlation coefficient ρ is not a sufficient measurement of dependence for non-Gaussian distributions.

As a matter of fact, a single dependence measure (e.g., besides ρ, Kendall's concordance measure τ and Spearman's rank correlation ρ) may not be sufficient to describe the entire dependence space, just as the statistical moments are only the summary of a univariate PDF. Hence, it motivates the use of copulas. The first usage of copula is attributed to Sklar (1959) in a theorem describing how one-dimensional distribution functions can be combined to form multivariate distributions. For d-dimensional continuous random variables $\{X_1, \ldots, X_d\}$ with marginal CDFs $u_j = F_{x_j}(x_j)$, $j = 1, \ldots, d$, Sklar showed that there exists one unique d-copula $C_{U_1, \ldots, U_d}$ such that

$$C_{U_1,\ldots,U_d}(u_1, \ldots, u_d) = H_{X_1,\ldots,X_d}(x_1, \ldots, x_d) \quad (55)$$

where u_j is the jth marginal and $H_{X_1,\ldots,X_d}$ is the joint CDF of $\{X_1, \ldots, X_d\}$. Copulas $C_{U_1,\ldots,U_d}$ can be regarded as a transformation of $H_{X_1,\ldots,X_d}$ from $[-\infty, \infty]^d$ to $[0,1]^d$. The consequence of this transformation is that the marginal distributions are segregated from $H_{X_1,\ldots,X_d}$. Hence, $C_{U_1,\ldots,U_d}$ becomes only relevant to the association between variables, and it gives a complete description of the entire dependence structure. In other words, the characterization of joint distributions can be performed separately for the marginal distributions and for the dependence structure (described by copulas), and therefore the dependence between variables can be clearly revealed.

Among various types of copula function, one-parameter Archimedean copulas have attracted the most attention owing to their several convenient properties. For an Archimedean copula, there exists a generator φ such that the following

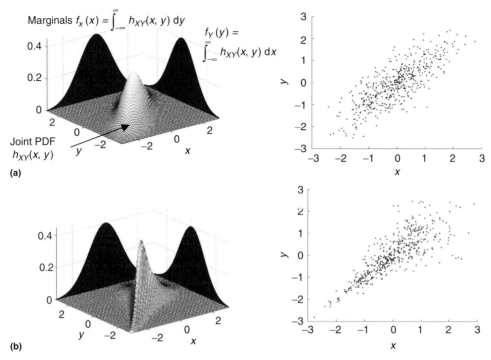

Figure 11 Illustration of bivariate joint distributions: (a) bivariate Gaussian distribution with $\rho = 0.8$ and the corresponding samples and (b) joint distribution with Gaussian marginals and Clayton copulas and the corresponding samples ($\rho = 0.8$).

relationship holds:

$$\varphi(C(u,v)) = \varphi(u) + \varphi(v) \quad (56)$$

where the generator φ is a continuous, strictly decreasing function defined in $[0,1]$, and $\varphi(1) = 0$. When the generator $\varphi(t) = -\ln t$, the copula in (56) is $C(u,v) = uv$, which is the special case when the variables are independent. Some commonly used families of one-parameter Archimedean copulas are listed in **Table 10**, in which θ is the dependence parameter. It should be noted that not every family of Archimedean copulas can accommodate the entire range of dependencies (from perfectly positive dependence to perfectly negative dependence). The choice of copulas depends on the range of dependence levels they can describe. For instance, Gumbel-Hougaard can only be applied for positive dependence, Ali-Mikhail-Haq is only suitable for weaker dependence ($-0.1807 < \tau < 0.3333$), while Clayton, Frank, and Genest-Ghoudi are suitable for both positive and negative dependencies. **Figure 12** shows an example of using Frank family of copulas in computing random samples with various levels of dependence.

Archimedean copulas find wide applications because they are easy to construct and possess several nice features. For example, several statistical properties can be simply expressed in terms of φ, such as the distribution function K_C of copulas (i.e. $K_C(t) = P[C(U,V) \leq t]$):

$$K_C(t) = t - \frac{\varphi(t)}{\varphi'(t)}, \quad t \in [0, 1] \quad (57)$$

The distribution function K_C offers a cumulative probability measure for the set $\{(u,v) \in [0,1]^2 | C(u,v) \leq t\}$, and it can help project multivariate information onto a single axis. The quantity K_C was also used by Salvadori and De Michele (2004b) for defining secondary return period for bivariate copulas. Another statistic that can be related to φ is the concordance measure Kendall's τ, which is defined as $\tau = P[(X_1 - X_2)(Y_1 - Y_2) > 0] - P[(X_1 - X_2)(Y_1 - Y_2) < 0]$, where (X_1, Y_1) and (X_2, Y_2) are independent and identically distributed random vectors with the same joint CDF $H_{XY}(x,y)$. Kendall's τ can be interpreted as the difference between probability of concordance $P[(X_1 - X_2)(Y_1 - Y_2) > 0]$ (for positive dependence) and probability of discordance $P[(X_1 - X_2)(Y_1 - Y_2) < 0]$ (for negative dependence). The value of Kendall's τ falls in $[-1,1]$, where 1 represents total concordance, -1 represents total discordance, and 0 represents concordance. To obtain the sample estimator of Kendall's τ, let (x_1, y_1) and (x_2, y_2) be two observations from a size-n sample space, and then $\hat\tau$ can be estimated by

$$\hat\tau = \frac{(c-d)}{\binom{n}{2}} \quad (58)$$

where c denotes concordant pairs $((x_2 - x_1)(y_2 - y_1) > 0)$, and d denotes disconcordant pairs $((x_2 - x_1)(y_2 - y_1) < 0)$. By using generator φ, the theoretical Kendall's τ can be expressed as

$$\tau = 1 + 4 \int_0^1 \frac{\varphi(t)}{\varphi'(t)} \, dt \quad (59)$$

This useful property leads to the nonparametric procedure of estimating dependence parameter θ by equating $\hat\tau = \tau$. This nonparametric estimator does not rely on prior information of

Table 10 Some commonly used one-parameter Archimedean copulas

Family	$\varphi(t)^a$	Range of θ^a	$K_C(t)^b$	$\tau(t)^{b,c}$
Ali-Mikhail-Haq	$\ln\dfrac{1-\theta(1-t)}{t}$	$[-1, 1)$	$t + \dfrac{t(1-\theta+\theta t)}{1-\theta}\ln\dfrac{1-\theta+\theta t}{t}$	$1 - \dfrac{2}{3\theta} - \dfrac{2}{3}\left(1 - \dfrac{1}{\theta}\right)^2 \ln(1-\theta)^c$
Clayton	$\dfrac{1}{\theta}(t^{-\theta} - 1)$	$[-1, 0) \cup (0, \infty)$	$t\left(1 + \dfrac{1}{\theta}\right) - \dfrac{t^{\theta+1}}{\theta}$	$\dfrac{\theta}{\theta + 2}$
Frank	$-\ln\dfrac{e^{-\theta t} - 1}{e^{-\theta} - 1}$	$(-\infty, 0) \cup (0, \infty)$	$t + \dfrac{e^{\theta t} - 1}{\theta}\ln\dfrac{e^{-\theta} - 1}{e^{-\theta t} - 1}$	$1 + \dfrac{4}{\theta}[D_1(\theta) - 1]^d$
Genest-Ghoudi	$(1 - t^{1/\theta})^\theta$	$[1, \infty)$	$t^{1-1/\theta}$	$\dfrac{2\theta - 3}{2\theta - 1}$
Gumbel-Hougaard	$(-\ln t)^\theta$	$[1, \infty)$	$t\left(1 - \dfrac{\ln t}{\theta}\right)$	$\dfrac{\theta - 1}{\theta}$

^{a}Column $\varphi(t)$, range of θ adapted from Nelsen (2006).
^{b}Column $K_C(t)$ and (t) of the Genest-Ghoudi family adopted from Kao and Govindaraju (2007b).
$^c\tau(t)$ of the Ali-Makhail-Haq, Frank, and Gumbel-Hougaard families adopted from Zhang and Singh (2007a), $\tau(t)$ of the Clayton family adopted from Grimaldi and Serinaldi (2006b).
$^dD_1$ is the Debye function of order 1, $D_1(\theta) = \int_0^\theta (t/\theta(e^t - 1))\, dt$.

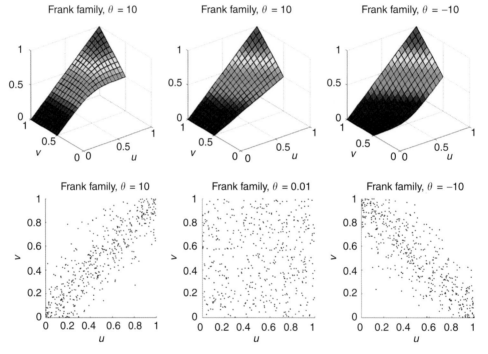

Figure 12 Frank family of Archimedean copulas with various levels of dependencies.

marginal distributions, and provides a more objective measure of dependence structure. Therefore, the existence of outliers, while affecting the estimation of marginal distributions, would not affect the determination of copulas. This aspect of being able to determine the dependence structure independent of marginals can be an advantage. Further information on statistical interference, detailed theoretical background, and descriptions of copulas can be found in Genest and Rivest (1993), Genest et al. (1995), Joe (1997), Nelsen (2006), and Salvadori et al. (2007).

To examine the appropriateness of a selected copula, one can construct the empirical copula (i.e., the observed probabilities) and apply it to perform goodness-of-fit tests. Similar to the concept of plotting position formula used in univariate statistical analysis (e.g. Weibull formula), empirical copulas are rank-based empirically joint cumulative probability measures. For sample size n, the d-dimensional empirical copula C_n is

$$C_n\left(\frac{k_1}{n}, \frac{k_2}{n}, \ldots, \frac{k_d}{n}\right) = \frac{a}{n} \qquad (60)$$

where a is the number of samples $\{x_1,\ldots,x_d\}$ with $x_1 \leq x_{1(k_1)}$, $\ldots, x_d \leq x_{d(k_d)}$, and $x_{1(k_1)}, \ldots, x_{d(k_d)}$ with $1 \leq k_1,\ldots,k_d \leq n$ are the order statistics from the sample. In an analogous fashion, the empirical distribution function K_{C_n} can be expressed as

$$K_{C_n}\left(\frac{l}{n}\right) = \frac{b}{n} \qquad (61)$$

where b is the number of samples $\{x_1,\ldots,x_d\}$ with $C_n(k_1/n,\ldots,k_d/n) \leq l/n$. Empirical copulas C_n and empirical distribution function K_{C_n} are mostly applied for model verification and are treated as the observed (real) dependence structure. Currently, the development of goodness-of-fit tests for copulas remains a major interest. Several applicable tests include: multidimensional KS test (Saunders and Laud, 1980), tests based on the probability integral transformation (Breymann et al., 2003; Genest et al., 2006; Dobric and Schmid, 2007; Kojadinovic, in press), kernel-based smoothing techniques (Fermanian, 2005; Panchenko, 2005; Scaillet, 2007), and cross-product ratio model (Wallace and Clayton, 2003).

2.18.5.2 Copulas in Hydrologic Applications

Though being a relatively new method in statistical hydrology, the flexibility offered by copulas for constructing joint distributions is now evident from many applications. Copulas were firstly applied in hydrologic studies by De Michele and Salvadori (2003), and Salvadori and De Michele (2004a) for rainfall frequency analysis. Hourly precipitation data from two raingauges at La Presa (Italy) for 7 years (from 1990 to 1996) were utilized to construct a bivariate model for regular storms, in which the generalized Pareto distribution was selected to describe the marginals, and the Frank family of Archimedean copulas was adopted to construct the dependence structure between rainfall duration and average intensity. This study was further extended to the trivariate level in Salvadori and De Michele (2006), where the dry period between rainfall events was added as the third variable, and again the generalized Pareto distribution and Frank families of Archimedean copulas were adopted.

For studying extreme rainfall behavior, Grimaldi and Serinaldi (2006b) and Serinaldi et al. (2005) discussed the relationship between design rainfall depth (critical depth, obtained from IDF curves by specifying design duration and return period) and the actual features of extreme rainfall events. Half-hourly rainfall data from 10 raingauges at Umbria (Italy) from 1995 to 2001 were combined with the assumption of regional homogeneity to form a 70-year annual maximum series for analysis. The trivariate model containing critical depth, actual total depth, and peak intensity was constructed via copulas. By providing critical depth, it was expected that important features of extreme rainfall could be obtained. Zhang and Singh (2007a, 2007b) performed multivariate analysis for extreme rainfall events via copulas. Hourly precipitation data from three raingauges at Amite River basin in Louisiana (US) for 42 years were analyzed. Bivariate rainfall models between total depth (volume), duration, and average intensity were constructed. Several types of conditional and joint return periods were illustrated in their study. Kao and Govindaraju (2007b, 2008) adopted 53 hourly precipitation stations in Indiana, USA with a minimum recording of 50 years. Extreme rainfall events were analyzed using both bivariate Archimedean copulas and trivariate Plackett copulas. The most appropriate definition for the selection of extreme rainfall samples was suggested. Joint distributions of extreme rainfall events were constructed and used to compute design rainfall estimates. Comparisons between the conventional and copula-based rainfall estimates showed that the traditional univariate analysis provides reasonable estimates of rainfall depths for durations greater than 10 h but fails to capture the peak features of rainfall. Further applications of copulas in rainfall analysis can be found in Kuhn et al. (2007), Singh and Zhang (2007), Evin and Favre (2008), Laux et al. (2009), and Serinaldi (2009).

Copulas were adopted in flood related problems as well. Favre et al. (2004) applied copulas for multivariate flood frequency analysis for two watersheds in Quebec, Canada. The combined flooding risk of multiple catchments and the joint distribution of peak flows and volume were discussed. Two families of Archimedean copulas (Frank, Clayton), independent and Farlie-Gumbel-Morgenstern copulas, were investigated. They showed that conditional probability of flood volumes is quite different when compared to the univariate result. Therefore, return periods of design floods would be different when the joint behavior is taken into account. De Michele et al. (2005) used the Gumbel family of Archimedean copulas to model the dependence between flood peaks and flood volumes. These two margins were analyzed by the generalized extreme value distribution. A bivariate model was constructed to calculate the flood hydrographs for a given return period, and was combined with the linear reservoir model to assess the adequacy of dam spillway of Ceppo Modrelli dam in Northern Italy. Zhang and Singh (2006, 2007c) investigated the dependence structure between flood peak, volume, and duration by testing four different families of Archimedean copulas: Gumbel-Hougaard, Ali-Mikhail-Haq, Frank, and Cook-Johnson. The margins were analyzed by using the extreme value type I and log-Pearson type III distributions. They found positive correlations between flood peak and volume, and flood volume and duration, and concluded that the Gumbel-Hougaard family was most appropriate to characterize the dependence structure. They also applied the copula model in calculating conditional return period. Other applications of copulas in flood frequency analysis can be found in Grimaldi and Serinaldi (2006a), Shiau et al. (2006), Genest et al. (2007), Renard and Lang (2007), and Serinaldi and Grimaldi (2007).

Besides rainfall and flood frequency analyses, copulas were adopted in other hydrologic topics as well. Salvadori and De Michele (2004b) discussed the use of copulas to assess the return period of hydrological events using bivariate models and defined the secondary return period. Kao and Govindaraju (2007a) quantified the effect of dependence between rainfall duration and average intensity on surface runoff and demonstrated that the use of copulas could result in simpler, more elegant mathematical treatment of zero runoff probabilities. Copulas were also applied in estimating groundwater parameters using a copula-based geostatistics approach (Bárdossy, 2006; Bárdossy and Li, 2008), drought analysis (Beersma and Buishand, 2004; Shiau, 2006; Shiau et al., 2007; Serinaldi *et al.*, 2009), multivariate L-moment homogeneity test (Chebana and Ouarda, 2007), tail dependence in hydrologic data (de Waal *et al.*, 2007; Poulin *et al.*, 2007), uncertainty quantification in remote-sensed data (Gebremichael and Krajewski, 2007; Pan *et al.*, 2008; Villarini *et al.*, 2008), rainfall IDF curves (Singh and Zhang, 2007), sea storm and wave height analysis (Wist *et al.*, 2004; de Waal and van Gelder, 2005; De Michele *et al.*, 2007), and atmospheric and climatologic studies (Vrac *et al.*, 2005; Maity and Kumar, 2008; Norris *et al.*, 2008). A review of copulas in Genest and Favre (2007) indicated that application of copulas in hydrology is still in its nascent stages, and their full potential for analyzing hydrologic problems is yet to be realized. More detailed theoretical background and descriptions for the use of copulas in problems related to water resources can be found in Dupuis (2007), Salvadori *et al.* (2007), and Salvadori and De Michele (2007).

2.18.5.3 Remarks on Copulas and Future Research

While copulas can help advance hydrologic analysis at multivariate levels and provide broad potential applications, it is important to bear in mind also their limitations. One should always be aware that the reliability of copulas is founded upon the sufficiency and quality of observations. Copulas, like any other statistical methods, can only elucidate the information embedded in the samples. In order to characterize multivariate joint distributions, a much larger sample size is needed. As such, data processing and quality control will play equally important roles. The assumption of stationarity should also be examined. It is particularly necessary since the changing climate may cause fundamental changes in the past climate pattern and invalidate the predictions based on historic observations.

Another major limitation, which is essential for more extensive usage in hydrologic applications, is the curse of dimensionality. Though Sklar's theorem was proposed for a general dimension, most of the current copula functions are valid only on the bivariate level. Choices are limited especially on a higher dimension (>4). It is a major disadvantage particularly while modeling natural phenomenon with a complicated dependence structure such as droughts. Moreover, the mathematical compatibility among various marginals and lower-level dependencies complicates the issue and it remains as an open problem (see Kao and Govindaraju, 2008). Further researches are deemed necessary to explore more choices of copulas, for the expanding possibility in modeling complex natural dependence structures.

2.18.6 Regional Frequency Analysis

Regional frequency analysis is widely employed for estimating design variables in ungauged sites or when dealing with data record lengths that are short as compared to the recurrence interval of interest (see, e.g., Stedinger *et al.*, 1993). Regionalization procedures embody the first principles proposed by the NRC-US (1988) for hydro-meteorological modeling: 'substitute time for space' by using hydrologic information collected at different locations to compensate for the limited (or absence of) information at the site of interest.

The literature proposes several approaches to regionalization (traditional approaches are illustrated for instance in Stedinger *et al.* (1993), Hosking and Wallis (1997), Pandey and Nguyen (1999), and FEH (1999)) and presents applications to different hydrological problems and contexts such as the *T*-year flood estimation (see, e.g., Dalrymple, 1960; Burn, 1990; Gabriele and Arnell, 1991; Castellarin *et al.*, 2001; Merz and Blöschl, 2005), frequency analysis of low flows (see, e.g., Smakhtin, 2001; Castellarin *et al.*, 2004; Laaha and Blöschl, 2006; Vogel and Kroll, 1992; Furey and Gupta, 2000), and rainfall extremes (Schaefer, 1990; Alila, 1999; Faulkner, 1999; Brath *et al.*, 2003; Di Baldassarre *et al.*, 2006b).

This section makes a direct and explicit reference to regional flood frequency analysis (RFFA); nevertheless, concepts and algorithms presented herein are suitable for regionalization of other hydrological extremes (e.g., low flows and rainstorms). More in general, the main features of a regionalization procedure, which are summarized in the remainder of this section, can be easily extended to broader hydrologic problems such as the prediction of within-year variability of streamflow regime in ungauged basins. Examples are reported in the literature that describe how to regionalize annual and long-term flow–duration curves (see, e.g., Fennessey and Vogel, 1990; Smakhtin *et al.*, 1997; Castellarin *et al.*, 2004, 2007) or to predict the streamflow regime in ungauged catchments via simulation by using rainfall-runoff models with regional parameters (see, e.g., Parajka *et al.*, 2007).

2.18.6.1 Index-Flood Procedure, Extensions and Evolutions

A traditional approach to regional frequency analysis is the index-flood procedure (Dalrymple, 1960). The approach has a long history in hydrology and is based on the identification of homogeneous groups of sites (homogeneous regions) for which the frequency distribution of floods (or other hydrological extremes) is the same except for a scale parameter, called index flood (or in general index term, e.g., index storm for the regionalization of rainfall extremes), which reflects the local hydrological conditions.

According to the index-flood procedure, the *T*-year flood at a given site, Q_T, can be expressed as the product of two terms: the index-flood, μ_Q, and the dimensionless growth factor q_T, which describes the relationship between the dimensionless flood and the recurrence interval, *T* (the so-called growth curve):

$$Q_T = \mu_Q \cdot q_T \qquad (62)$$

The index flood μ_Q in (62) is generally a measure of central tendency of the at-site frequency distribution. It is common to refer to the mean of the distribution (see, e.g., Brath *et al.*, 2001), but the literature points out that the median is a valid alternative (see, e.g., Robson and Reed, 1999). The procedure allows the estimation of more reliable growth curves due to the exploitation of information from the entire homogenous region.

The index-flood approach can be applied with respect to AMS or PDS (also known as POT; see, e.g., Madsen *et al.*, 1997a, 1997b). The remainder of this section refers directly to AMS, which are generally more common and easier to obtain,

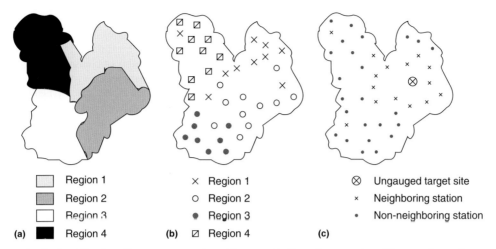

Figure 13 Approaches for the delineation of homogeneous regions: (a) geographically continuous regions; (b) noncontiguous homogeneous regions; and (c) hydrologic neighborhoods. From Ouarda TBMJ, Girard C, Cavadias GS and Bobee B (2001) Regional flood frequency estimation with canonical correlation analysis. *Journal of Hydrology* 254(1–4): 157–173, Fig. 1.

even though the concepts and procedures described can be easily extended for PDS (or POT) series.

The classical implementation of the index-flood procedure is based on the most restrictive fundamental hypothesis of existence of homogeneous regions within which the statistical properties of dimensionless flood flows do not vary with location (e.g., dimensionless statistical moments such as the coefficients of variation and skewness are constant). Nevertheless, after proposing the original procedure the literature reported several extensions and evolutions, which partly relax the above fundamental hypothesis. The hierarchical application of the index-flood procedure, for instance, assumes that statistics of increasing order are constant within a set of nested regions; the larger the order of the statistics, the larger the region (see, e.g., Gabriele and Arnell, 1991). Another relevant example of evolution of the original hypothesis is or region of influence (RoI) approach (see, e.g., Burn, 1990), which replaces the original concept of fixed and contiguous regions with homogeneous pooling groups of sites. The pooling groups are identified in order to maximize the hydrological affinity with the site of interest (focused pooling), and the regionalization procedure enables one to weight the regional hydrological information according to the similarities with the site of interest and to use the at-site information in a very efficient way (see, e.g., Zrinji and Burn, 1996; Castellarin et al., 2001).

Although the techniques for delineating homogeneous pooling group of sites were definitely enhanced and improved since the first studies on regionalization, moving toward more objective and process-related approaches, the definition of homogeneous pooling-group of sites is still a hot topic of regional frequency analysis, highly debated among the scientific community. During the evolution of the regionalization techniques, the definition of boundaries between pooling-group of sites evolved steadily, from administrative boundaries to physiographic and meteo-climatic boundaries (panel a) in **Figure 13**), from geographically identifiable boundaries (panels a) and b) in **Figure 11**) to boundaries associated with the particular site of interest (panel c) in **Figure 13**; RoI, see, e.g., Burn, 1990; Reed et al., 1999; Ouarda et al., 2001.

The latest advances tend to remove completely the concept of boundaries from the definition of homogeneous regions (pooling group of sites). Examples in this direction are the models in which regional parameters vary continuously with geomorfoclimatic indices (see, for instance, Alila, 1999; Di Baldassarre et al., 2006b) or the adaptation of geostatistical interpolation techniques to the problem of regionalization of hydrological information.

2.18.6.2 Classical Regionalization Approach

A few main steps characterize any regionalization procedure. If the index flood is selected as the regionalization scheme and one is interested in estimating the flood quantile at a given site, these steps can be summarized as follows (see, e.g., Hosking and Wallis, 1997): (1) estimation of the index flood, μ_Q; (2) estimation of the regional quantile, q_T, that is, (2a) identification of homogeneous pooling group of sites, (2b) choice of a frequency distribution, and (2c) estimation of the regional frequency distribution; and (3) validation of the regional model.

2.18.6.2.1 Estimation of the index flood

Without lack of generality, let us consider the case in which μ_Q is assumed to be the mean of the distribution. The estimation of the index flood is straightforward when the site of interest is gauged and the record length is sufficiently long. In this case μ_Q can be obtained directly by calculating the arithmetic mean of the available observations. Indirect methods have to be used in ungauged sites, instead. Multiregression models are probably the most common indirect methods (see, e.g., Brath et al., 2001; Castellarin et al., 2007). They link μ_Q to an appropriate set of morphological and climatic descriptors of the basin through statistical relations that, for instance, may read as

$$\hat{\mu}_Q = A_0 \omega_1^{A_2} \omega_2^{A_3} \ldots \omega_i^{A_n} + \vartheta \quad (63)$$

where $\hat{\mu}_Q$ is the index-flood estimate, ω_i, $i = 1, 2, \ldots, n$, are the explanatory variables of the model (i.e., a suitable set of geomorphologic and climatic indexes), A_i, $i=0,1,\ldots,n$, are parameters, and θ is the residual of the model. The structure of (63), that is, selection of the smallest and most efficient set of catchment descriptors, and the values of the parameters can be identified by multivariate stepwise regression analysis (e.g., Wiesberg, 1985; Brath et al., 2001). Instead of stepwise regression analysis, alternative multivariate procedures can also be adopted for this task, such as artificial neural network, principal component or canonical correlation analysis (see, e.g., Shu and Burn, 2004; Ouarda et al., 2001; Chokmani and Ouarda, 2004).

The literature indicates that statistical indirect models such as (63) are generally more accurate than conceptual indirect models for predicting μ_Q in ungauged basins (see, e.g., Brath et al., 2001). The latter models attempt to interpret the dynamics of rainfall–runoff transformation and are characterized by a more rigid structure. As a consequence, conceptual models reduce the influence on model parametrization of the specific information which arrives from any gauged station and therefore are typically more robust than statistical models. The literature also reports that direct estimation may be a preferable alternative to indirect methods when 2–5 years of data are available, especially for basins with physiographic and climatic characteristics that differ significantly from the average characteristics of the set of basins considered for the identification of the indirect estimation models.

2.18.6.2.2 Estimation of the regional dimensionless quantile

The literature reports on different approaches for delineating pooling groups of sites, as well as selecting and estimating the regional frequency distribution (see, e.g., GREHYS, 1996; FEH, 1999). Hosking and Wallis (1997) proposed an integrated approach completely based on the use of L-moments.

The approach summarizes the frequency regime of the pooling group of sites through the regional L-moment ratios. Regional L-moment ratios can be defined as follows:

$$\bar{t}_r = \sum_{i=1}^{R} n_i t_r^{(i)} / \sum_{i=1}^{R} n_i \quad (64)$$

where $\bar{t}_r$ is the regional L-moment ratio of order r (e.g., the L-coefficients of variation, skewness, and kurtosis, L-Cv, L-Cs, and L-Ck, correspond to $r=2$, 3, and 4 in this order), $t_r^{(i)}$ is the sample L-moment ratio of order r for site i that can be computed as described in Section 2.18.3, n_i is the record length for site i, while R is the number of sites in the pooling group. Numerous applications in different contexts and to different regionalization problems (not necessarily confined to the estimation of the design flood) proved the validity and reliability of the approach, whose main steps are briefly illustrated in this section.

2.18.6.2.3 Homogeneity testing

Once a pooling group of sites has been delineated, its homogeneity degree has to be tested. The homogeneity of the group of sites is a fundamental requirement in order to perform an effective estimation of the T-year quantile (e.g., Lettenmaier et al., 1987; Stedinger and Lu, 1995: see Viglione et al. (2007) for a comparison of some homogeneity tests proposed by the literature). Hosking and Wallis (1997) defined a heterogeneity measure that is a standardized measure of the intersite variability of L-moment ratios. This measure is routinely used by hydrologists to test regional homogeneity.

The Hosking and Wallis (1997) heterogeneity measure assesses the homogeneity of a group of basins at three different levels by focusing on three measures of dispersion for different orders of the sample L-moment ratios.

A measure of dispersion for the L-Cv

$$V_1 = \frac{\sum_{i=1}^{R} n_i \left(t_2^{(i)} - \bar{t}_2\right)^2}{\sum_{i=1}^{R} n_i} \quad (65)$$

A measure of dispersion for both the L-Cv and the L-Cs coefficients in the L-Cv–L-Cs space

$$V_2 = \frac{\sum_{i=1}^{R} n_i \left[\left(t_2^{(i)} - \bar{t}_2\right)^2 + \left(t_3^{(i)} - \bar{t}_3\right)^2\right]^{1/2}}{\sum_{i=1}^{R} n_i} \quad (66)$$

A measure of dispersion for both the L-Cs and the L-Ck coefficients in the L-Cs–L-Ck space

$$V_3 = \frac{\sum_{i=1}^{R} n_i \left[\left(t_3^{(i)} - \bar{t}_3\right)^2 + \left(t_4^{(i)} - \bar{t}_4\right)^2\right]^{1/2}}{\sum_{i=1}^{R} n_i} \quad (67)$$

where $\bar{t}_2$, $\bar{t}_3$, and $\bar{t}_4$ are the regional L-Cv, L-Cs, and L-Ck respectively; $t_2^{(i)}, t_3^{(i)}, t_4^{(i)}$, and n_i are the values of L-Cv, L-Cs, L-Ck, and the sample size for site i; and R is the number of sites in the pooling group.

The underlying concept of the test is to measure the sample variability of the L-moment ratios and compare it to the variation that would be expected in a homogeneous group. The expected mean value and standard deviation of these dispersion measures for a homogeneous group, namely μ_{V_k} and σ_{V_k}, are assessed through repeated simulations, by generating homogeneous groups of basins having the same record lengths as those of the observed data. To avoid any unduly commitment to a particular three-parameter distribution, the authors recommend the four-parameter kappa distribution to generate the synthetic groups of flood sequences. The kappa distribution includes, as special cases, several well-known two- and three-parameter distributions (see, e.g., Hosking and Wallis, 1997; Castellarin et al., 2007). The heterogeneity measures are then evaluated using the following expression:

$$H_k = \frac{V_k - \mu_{V_k}}{\sigma_{V_k}}, \quad \text{for } k = 1, 2, 3 \quad (68)$$

Hosking and Wallis suggest that a group of sites may be regarded as 'acceptably homogeneous' if $H_k < 1$, 'possibly heterogeneous' if $1 \leq H_k < 2$, and 'definitely heterogeneous' if $H_k > 2$. According to the authors, these reference values are guidelines. For instance, the amount $H = 1$ can be regarded as the borderline of whether a redefinition of the region may lead to a meaningful increase in the accuracy of the regional quantile estimate.

H_1 is the most selective heterogeneity measure. H_2 and H_3 tend to identify larger homogeneous pooling groups of sites; therefore, the utilization of all three measures well suits the application of a hierarchical regionalization approach (Gabriele and Arnell, 1991; Castellarin et al., 2001).

2.18.6.2.4 Choice of a frequency distribution

Hosking and Wallis suggest to base the selection of the parent distribution (the frequency distribution for all sites in the pooling-group) on the value of regional L-moments. The authors define a goodness-of-fit measure to be used for selecting the candidate distribution among a family of possible three-parameter distributions. The authors consider as possible candidates the generalized logistic (GLO), generalized Pareto (GPA), lognormal (LN3), Pearson type III (PE3), and GEV (see Section 2.18.3). The measure defined for selecting the three-parameter distribution quantifies how well the L-Cs and the L-Ck of the of the fitted distribution match the regional average L-Cs and L-Ck. The fit can be geometrically interpreted on a diagram that reports the values of L-Cs and L-Ck on the x- and y-axis (L-moment ratio diagram) as the vertical distance between the point corresponding to the regional average and the curve representing the theoretical relationship between L-Cs and L-Ck for the considered distribution (see **Figure 14**). Hosking and Wallis (1997), Vogel and Fennessey (1993), and Peel et al. (2001), among others, recommended using the L-moment ratio diagrams to guide the selection of the most suitable parent distribution.

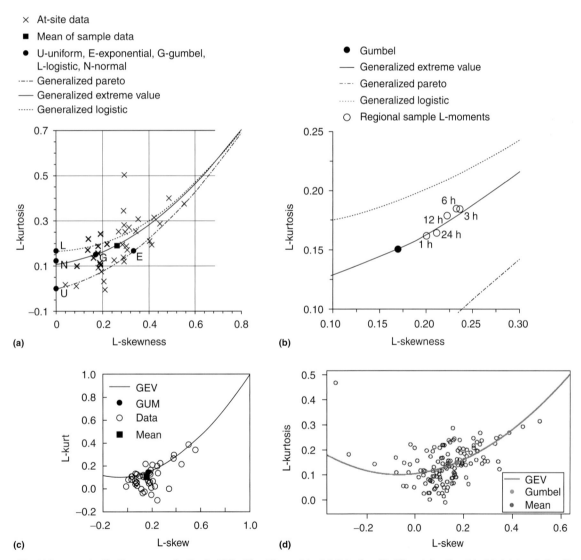

Figure 14 (a) L-moment ratio diagrams: application to AMS of flood flows; (b) rainfall depths with different duration; (c) global data set of earthquake magnitudes; and (d) extreme wind speeds at 129 stations in the contiguous United States. (a) From Castellarin A, Burn DH and Brath A (2001) Assessing the effectiveness of hydrological similarity measures for flood frequency analysis. *Journal of Hydrology* 241: 270–285, Fig. 2. (b) From Brath A, Castellarin A and Montanari A (2003) Assessing the reliability of regional depth-durationfrequency equations for gaged and ungaged sites. *Water Resources Research* 39(12): 1367–1379, Fig. 3. (c) From Thompson EM, Baise LG and Vogel RM (2007) A global index earthquake approach to probabilistic assessment of extremes. *Journal of Geophysical Research* 112: B06314, Fig. 1. (d) Personal communication, Eugene Morgan, Tufts University, Deparment of Civil and Environmental Engineering, 2009.

For the details concerning the goodness-of-fit measure the interested reader is referred to Hosking and Wallis (1997).

The literature documents that the GEV distribution is conceptually appropriate and technically suitable for accurately reproducing the sample frequency distribution of geophysical extremes observed in different geographical contexts around the world (see, e.g., Stedinger et al., 1993; Vogel and Wilson, 1996; Robson and Reed, 1999; Castellarin et al., 2001; Thompson et al., 2007). For instance, **Figure 14** shows by means of L-moment ratios the appropriateness of the GEV distribution for flood flows, rainfall extremes, earthquake magnitudes, and wind speed extremes.

2.18.6.2.5 Estimation of the regional frequency distribution

The estimation of the regional distribution can be performed through the method of L-moments. This method is analogous to the method of moments, which is probably the oldest and widely understood technique for fitting frequency distributions to observed data (Vogel and Fennessey, 1993). The method equates regional L-moment estimates with the theoretical L-moments of the distribution, resulting in a system of nonlinear equations whose variables are the parameters to be estimated.

The use of L-moments instead of conventional moments offers several advantages, for instance, the possibility to characterize a wider range of distributions, smaller bias and higher robustness of the estimators when applied to short samples (see, e.g., Hosking and Wallis, 1997).

For the case of the GEV distribution, Hosking and Wallis (1997) proposed the following system of equations for the application of the method of L-moments:

$$\hat{k}^R \approx 7.8590c + 2.9554c^2, \quad c = \frac{2}{3+\bar{t}_3} - \frac{\ln 2}{\ln 3} \quad (69a)$$

$$\hat{\alpha}^R = \frac{\bar{t}_2 \hat{k}^R}{(1-2^{-\hat{k}^R})\Gamma(1+\hat{k}^R)}, \quad \hat{\xi}^R = 1 - \hat{\alpha}^R[1-\Gamma(1+\hat{k}^R)]\hat{k}^R \quad (69b)$$

where $\hat{k}^R$, $\hat{\alpha}^R$, and $\hat{\xi}^R$ indicates the regional estimates of the GEV parameters, $\bar{t}_2$ and $\bar{t}_3$ are the regional L-Cv and L-Cs, $\Gamma(x) = \int_0^\infty t^{x-1}e^{-t}dt$ is the gamma function, and the regional GEV parent is assumed to have unit mean (i.e., index flood coincides with the mean of the distribution). The empirical polynomial equation in c has accuracy better than 9×10^4 for typical L-Cs values. Once the regional parameters are estimated, the regional dimensionless quantiles can be computed as follows:

$$\hat{q}_T = \hat{\xi}^R + \frac{\hat{\alpha}^R}{\hat{k}^R}\left\{1 - \left[-\ln\left(1-\frac{1}{T}\right)\right]^{\hat{k}^R}\right\} \quad \text{for } k \neq 0 \quad (70a)$$

$$\hat{q}_T = \hat{\xi}^R - \hat{\alpha}^R\left\{\ln\left[-\ln\left(1-\frac{1}{T}\right)\right]\right\} \quad \text{for } k \neq 0 \quad (70b)$$

To improve the accuracy of the regional quantile estimates, the target pooling-group size can be determined according to the 5T guideline (Jakob et al., 1999), which suggests that a pooling group should contain at least 5T station-years of data so as to obtain reasonably accurate estimates of the T-year quantile avoiding undue extrapolations.

2.18.6.2.6 Validation of the regional model

Regional flood frequency models are generally applied for predicting the flood frequency regime in ungauged basin. Therefore, it is fundamental to quantify the accuracy of the models and the uncertainty of regional estimates when no observation is available at the site of interest. A powerful and easy-to-implement cross-validation technique that can be used for this purpose is the jack-knife resampling procedure (see, e.g., Shao and Tu, 1995; Castellarin, 2007; Castellarin et al., 2007).

Regardless of the regionalization approach or structure of the regional model being considered, the jack-knife procedure is a leave-one-out cross-validation technique that can be described as follows:

1. one gauging station (site i) is removed from the set of R stations;
2. the regional model is constructed for site i pooling group by neglecting site i data;
3. the quantity of interest (i.e., T-year flood Q_T) is estimated at site i through the regional model identified at step 2 (jack-knife estimate); and
4. steps 1–3 are repeated $R-1$ times for each one of the remaining gauges.

The R jack-knife estimates are then compared with the corresponding reference values (i.e., regional estimates that do consider the data observed at site i, or at-site estimate if viable), for instance in terms of relative BIAS, MSE, and Nash and Sutcliffe efficiency measure (NSE).

$$\text{BIAS} = \frac{1}{R}\sum_{i=1}^{R}\frac{\hat{x}_i^{jk} - x_i}{x_i}$$

$$\text{MSE} = \frac{1}{R}\sum_{i=1}^{R}\left(\frac{\hat{x}_i^{jk} - x_i}{x_i}\right)^2$$

$$\text{NSE} = 1 - \sum_{i=1}^{R}\left(\frac{\hat{x}_i^{jk} - x_i}{x_i - \bar{x}}\right)^2 \quad (71)$$

where x_i and $\hat{x}_i^{jk}$ are respectively the reference and jack-knife estimates for site i; R is the number of sites in the pooling group; and $\bar{x}$ is the average of the R reference values. NSE varies between $-\infty$ and 1, where NSE = 1 indicates the perfect fit, and NSE = 0 stands for a model that performs as efficiently as a mean regional value.

The general structure of the jack-knife procedure can be applied for the cross-validation of any regional model. The actual implementation of step 2 depends on the particular regional model being considered. For example, if a regional multiregression model for the estimation of the index flood is considered, step 2 will involve the calibration of the statistical coefficients of the model (i.e., coefficients A_j, with $j = 1, 2, \ldots, n$ in Equation (63)). If the estimation of Q_T is considered instead, step 2 will include the whole regionalization process. **Figure 15** reports an example of cross-validation for a regional model adopting the index-flood scheme (see also Castellarin, 2007).

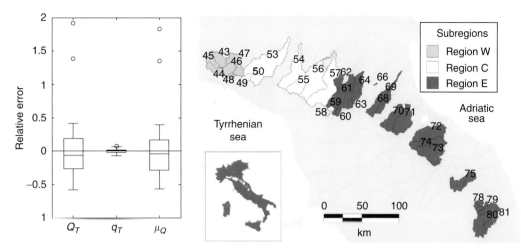

Figure 15 Box plots summarizing the relative error distributions in terms of 25th, 50th, and 75th percentiles, maximum and minimum values and outliers (circles) for cross-validated flood quantiles of given recurrence interval T, Q_T, dimensionless flood quantiles, q_T, and index flood, μ_Q for the three homogeneous regions depicted in the right panel. From Castellarin A, Camorani G and Brath A (2007) Predicting annual and long-term flow-duration curves in ungauged basins. *Advances in Water Resources* 30(4): 937–953, Fig. 6.

2.18.6.3 Open Problems and New Advances

RFFA has been a research sector for more than five decades now, yet the scientific community is still very active on this topic. This results from the existence of open problems, as it is discussed later, but it also can be ascribed to the potentiality of statistical regionalisation for solving a very common problem in hydrology, that is prediction in ungauged basins (see, e.g., Sivapalan *et al.*, 2003). For instance, probabilistic interpretation and regionalization of classical deterministic hydrological tools (e.g., flow duration curves, FDC, regional envelope curves of flood flows, REC, etc.) renewed the scientific appeal of these simple methods, further promoting RFFA among hydrological research topics (see, e.g., Castellarin *et al.* (2004, 2007) for FDC and Castellarin *et al.* (2005) and Castellarin (2007) for REC).

Some issues and aspects associated with RFFA may perhaps be considered to be well studied and the margin of improvement in the accuracy of regional estimates associated with them is probably rather limited. Examples are the choice and estimation of the regional parent distribution or the statistical homogeneity testing (Castellarin and Laio, 2006). Some other issues are still critical, instead, and further analyses may significantly improve the accuracy of regional predictions in ungauged sites.

One of these issues is certainly the estimation of the index flood in ungauged basins. **Figure 15** eloquently shows for a given case study, but this is a widespread condition (see, e.g., Kjeldsen and Jones, 2007) that the largest amount of uncertainty is associated with this step of the regionalization procedure. Investigators are still dedicating a great deal of effort to the improvement of existing methodologies (see, e.g., Shu and Burn, 2004; Kjeldsen and Jones, 2007) and to the definition of guidelines for the identification of the most reliable and suitable ones depending on the problem at hand (see, e.g., Bocchiola *et al.*, 2003).

Also, classical studies document that intersite correlation among flood flows observed at different sites is typically not negligible (see, e.g., Matalas and Langbein, 1962; Stedinger, 1983) and leads to increases in the variance of regional flood statistics (see, for instance, Hosking and Wallis, 1988). Nevertheless, the analysis of the impacts of cross-correlation on regional estimates is still poorly understood. Recent studies have pointed out that cross-correlation may significantly reduce the regional information content in practical applications, quantified in terms of equivalent number of independent observation (see, e.g., Troutman and Karlinger, 2003, Castellarin *et al.*, 2005; Castellarin, 2007). This reduction has an impact on the reliability of regional quantiles, as it increases the variance of regional estimators, and can also severely affect the power of statistical tests for assessing the regional homogeneity degree (Castellarin *et al.*, 2008).

The delineation of homogeneous pooling group of sites, or catchment classification, is still an open and highly debated problem, on which the scientific community is very active (see, e.g., McDonnell and Woods, 2004). Concerning this issue, the main research activities focus on: (1) the identification of the most descriptive and informative physiographic and climatic catchment descriptors to be used as proxies for the flood frequency regime (see, e.g., Castellarin *et al.*, 2001; Merz and Blöschl, 2005) and (2) the development of pooling procedures as objective and nonsupervised as possible. Several objective approaches have been proposed by the scientific literature, such as cluster analysis (Burn, 1989) or unsupervised artificial neural networks (ANNs) (see, e.g., Hall and Minns, 1999; Toth, 2009). Furthermore, the scientific community is dedicating an increasing attention to the possibilities offered by the application of geostatistical techniques to the problem of statistical regionalisation. These techniques have been showed to have a significant potential for regionalization and, for this reason, will be briefly discussed and presented in this section.

Geostatistical procedures were originally developed for the spatial interpolation of point data (see, e.g., kriging: Kitanidis, 1997). The literature proposes two different ways to apply geostatistics to the problem of regionalisation of hydrological

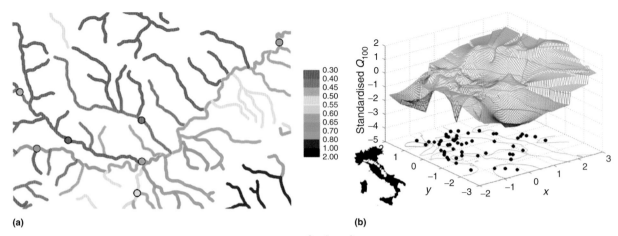

Figure 16 (a) Topkriging: 100-year flood per unit area (color codes in $m^3 s^{-1} km^{-2}$) for a portion of the Mur region (Austria): Topkriging estimates along the stream network and empirical values as circles. From Skøien JO, Merz R and Bloschl G (2006) Top-kriging - geostatistics on stream networks. *Hydrology and Earth System Sciences* 10(2): 277–287, Fig. 7. (b) PSBI: 3D representation of standardised value of 100-year flood over the physiographic space identified for a set of basins in northern central Italy (gauged basins are represented as dots).

information. The first technique is called physiographic space-based interpolation (PSBI) and performs the spatial interpolation of the desired hydrometric variable (e.g., T-year flood, but also annual streamflow, peak flow with a certain return period, low flows, etc.) in the bidimensional space of geomorphoclimatic descriptors (Chokmani and Ouarda, 2004; Castiglioni et al., 2009). The x and y orthogonal coordinates of the bidimensional space are derived from an adequate set of $n > 1$ geomorphologic and climatic descriptors of the river basin (such as drainage area, main channel length, mean annual precipitation, and indicators of seasonality; see Castellarin et al., 2001) through the application of multivariate techniques, such as the principal components or canonical correlation analysis (Shu and Ouarda, 2007). The second technique, named Topological kriging or Topkriging, is a spatial estimation method for streamflow-related variables. It interpolates the streamflow value of interest (i.e., T-year flood, low-flow indices, etc.) along the stream network by taking the area and the nested nature of catchments into account (Skøien et al., 2006; Skøien and Blöschl, 2007).

The philosophy behind these innovative approaches to regionalization is rather interesting because they enable one to regionalize hydrometric variables dispensing with the definition or identification of homogeneous regions or pooling groups of sites (see **Figure 13**). The approaches are particularly appealing for predictions in ungauged basins as they provide a continuous representation of the quantity of interest (e.g., T-year flood) along the stream network (Topkriging) or in the physiographic space (PSBI), providing the user with an estimate of the uncertainty associated with the interpolated value. In particular, the final output for Topkriging is the estimation of the measure of interest (with uncertainty) along the stream network (see **Figure 16**). A little less intuitive is the output of PSBI. With this technique any given basin (gauged or ungauged) can be represented as a point in the x–y space described above; in the same way the set of gauged basins of the study area can be represented by a cloud of points in this space. The empirical values of the quantity of interest (e.g., T-year flood) can be represented along the third dimension z for each gauged catchment, and can then be spatially interpolated (with uncertainty) by applying a standard interpolation algorithm (e.g., ordinary or universal kriging). The spatial interpolation enables one to represent the quantity of interest over the entire portion of the x–y space containing empirical data, and therefore to estimate it at ungauged sites lying within the same portion of the space (see **Figure 16**).

References

Adeloye AJ and Montaseri M (2002) Preliminary streamflow data analyses prior to water resources planning study. *Hydrological Sciences Journal* 47(5): 679–692.

Akaike H (1973) Information theory and an extension of the maximum likelihood principle. In: Petrov BN and Csaki F (eds.) *Second International Symposium on Information Theory.* 281pp. Budapest: Academiai Kiado.

Aksoy H (2006) Hydrological variability of the European part of Turkey. *Iranian Journal of Science and Technology* 31(B2): 225–236.

Aksoy H, Unal NE, Gedikli A, and Kehagias A (2008a) Fast segmentation algorithms for long hydrometeorological time series. *Hydrological Processes* 22: 4600–4608.

Aksoy H, Unal NE, Alexandrov V, Dakova S, and Yoon J (2008b) Hydrometeorological analysis of northwestern Turkey with links to climate change. *International Journal of Climatology* 28: 1047–1060.

Alexandersson H (1986) A homogeneity test applied to precipitation data. *Journal of Climatology* 6: 661–675.

Alila Y (1999) A hierarchical approach for the regionalization of precipitation annual maxima in Canada. *Journal of Geophysical Research* 104: 31645–31655.

Appel U and Brandt AV (1983) Adaptive sequential segmentation of piecewise stationary time series. *Information Sciences* 29: 27–56.

Atiem IA and Harmancioglu NB (2006) Assessment of regional floods using l-moments approach: The case of the river Nile. *Water Resources Management* 20(5): 723–747.

Baghirathan VR (1978) Rainfall depth-duration-frequency studies for Sri Lanka. *Journal of Hydrology* 37(3–4): 223–239.

Bárdossy A (2006) Copula-based geostatistical models for groundwater quality parameters. *Water Resources Research* 42: W11416 (doi:10.1029/2005WR004754).

Bárdossy A and Li J (2008) Geostatistical interpolation using copulas. *Water Resources Research* 44: W07412 (doi:10.1029/2007WR006115).

Baseville M and Nikiforov IV (1993) *Detection of abrupt changes: Theory and application.* Englewood Cliffs, NJ: Prentice Hall.

Beersma JJ and Buishand TA (2004) Joint probability of precipitation and discharge deficits in the Netherlands. *Water Resources Research* 40: W12508 (doi: 10.1029/2004WR003265).

Begueria S (2005) Uncertainties in partial duration series modelling of extremes related to the choice of the threshold value. *Journal of Hydrology* 303(1–4): 215–230.

Bendjoudi H, Hubert P, Schertzer D, and Lovejoy S (1997) Multifractal point of view on rainfall intensity-duration-frequency curves. *Comptes Rendus de l'Academie de Sciences – Serie IIa: Sciences de la Terre et des Planetes* 325(5): 323–326.

Benson M (1968) Uniform flood-frequency estimating methods for federal agencies. *Water Resources Research* 4(5): 891–908.

Ben-Zvi A (2009) Rainfall intensity-duration-frequency relationships derived from large partial duration series. *Journal of Hydrology* 367(1–2): 104–114 (doi:10.1016/j.jhydrol.2009.01.007).

Bernard MM (1932) Formulas for rainfall intensities of long durations. *Transaction of the American Society of Civil Engineers* 96: 592–624.

Berryman D, Bobee B, Cluis D, and Haemmerli J (1988) Nonparametric tests for trend detection in water quality time series. *AWRA Water Resources Bulletin* 24(3): 545–556.

Bobée B (1975) Log Pearson type-3 distribution and its application in hydrology. *Water Resources Research* 11(5): 681–689.

Bobée B and Rasmussen PF (1995) Recent advances in flood frequency-analysis. *Reviews of Geophysics* 33(S1): 1111–1116.

Bocchiola D, De Michele C, and Rosso R (2003) Review of recent advances in index flood estimation. *Hydrology and Earth System Sciences* 7(3): 283–296.

Brasil Vieira D and Zink de Souza C (1985) Analysis of the relation intensity–duration–frequency of heavy rains for Ribeirao Preto. *ICID Bulletin (International Commission on Irrigation and Drainage)* 34(1): 49–55, 64.

Brath A, Castellarin A, and Montanari A (2003) Assessing the reliability of regional depth–duration–frequency equations for gaged and ungaged sites. *Water Resources Research* 39(12): 1367–1379.

Breymann W, Dias A, and Embrechts P (2003) Dependence structures for multivariate high-frequency data in finance. *Quantitative Finance* 3: 1–14 (doi:10.1080/713666155).

British Atmospheric Data Centre (2006) UK Meteorological Office. MIDAS Land Surface Stations data (1853-current). http://badc.nerc.ac.uk/data/ukmo-midas (accessed March 2010).

Buishand TA (1989) The partial duration series method with a fixed number of peaks. *Journal of Hydrology* 109(1–2): 1–9.

Burn DH (1989) Cluster analysis as applied to regional flood frequency. *Journal of Water Resources Planning and Management* 115(5): 567–582.

Burn DH (1990) Evaluation of regional flood frequency analysis with a region of influence approach. *Water Resources Research* 26(10): 2257–2265.

Burn DH and Elnur MAH (2002) Detection of hydrologic trends and variability. *Journal of Hydrology* 255: 107–122.

Canterford R P (1986) Frequency analysis of Australian rainfall data as used for flood analysis and design. In: *Regional Flood Frequency Analysis: Proceedings of the International Symposium on Flood Frequency and Risk Analyses*, pp. 293–302.

Castellarin A (2007) Probabilistic envelope curves for design-flood estimation at ungaged sites. *Water Resources Research* 43: W04406.

Castellarin A, Burn DH, and Brath A (2001) Assessing the effectiveness of hydrological similarity measures for flood frequency analysis. *Journal of Hydrology* 241: 270–285.

Castellarin A, Burn DH, and Brath A (2008) Homogeneity testing: how homogeneous do heterogeneous cross-correlated regions seem? *Journal of Hydrology* 360: 67–76.

Castellarin A, Camorani G, and Brath A (2007) Predicting annual and long-term flow-duration curves in ungauged basins. *Advances in Water Resources* 30(4): 937–953.

Castellarin A, Galeati G, Brandimarte L, Montanari A, and Brath A (2004) Regional flow-duration curves: Reliability for ungauged basins. *Advances in Water Resources* 27: 953–965.

Castellarin A and Laio F (2006) *Regional frequency analysis, obsolete or ongoing? Seminar in Italian within the workshop New Frontiers in Hydrology*. Potenza, Italy: Università degli Studi della Basilicata.

Castellarin A, Vogel RM, and Matalas NC (2005) Probabilistic behavior of a regional envelope curve. *Water Resources Research* 41: W06018.

Castiglioni S, Castellarin A, and Montanari A (2009) Prediction of low-flow indices in ungauged basins through physiographical space-based interpolation. *Journal of Hydrology* 378: 272–280.

Chebana F and Ouarda TBMJ (2007) Multivariate L-moment homogeneity test. *Water Resources Research* 43: W08406 (doi:10.1029/2006WR005639).

Chen C (1983) Rainfall intensity-duration-frequency formulas. *Journal of Hydraulic Engineering-ASCE* 109(12): 1603–1621.

Chen HL and Rao AR (2002) Testing hydrologic time series for stationarity. *ASCE Journal of Hydrologic Engineering* 7(2): 129–136.

Cherubini U, Luciano E, and Vecchiato W (2004) *Copula Methods in Finance*. West Sussex: Wiley.

Chokmani K and Ouarda TBMJ (2004) Physiographical space-based kriging for regional flood frequency estimation at ungauged sites. *Water Resources Research* 40: W12514.

Chow VT, Maidment DR, and Mays LW (1988) *Applied Hydrology*, International edn. New York: McGraw-Hill Higher Education.

Chowdhury JU, Stedinger JR, and Lu L (1991) Goodness-of-fit tests for regional generalized extreme value flood distributions. *Water Resources Research* 27(7): 1765–1776.

Claps P and Laio F (2003) Can continuous streamflow data support flood frequency analysis? An alternative to the partial duration series approach. *Water Resources Research* 39(8): 1216.

Cluis D and Laberge C (2001) Climate change and trend detection in selected rivers within the Asia-Pacific region. *Water International* 26(3): 411–424.

Cluis D, Langlois C, van Coillie R, and Laberge C (1989) Development of a software package for trend detection in temporal series: Application to water and industrial effluent quality data for the St. Lawrence river. *Environmental Monitoring and Assessment* 12: 429–441.

Cunnane C (1973) A particular comparison of annual maxima and partial duration series methods of flood frequency prediction. *Journal of Hydrology* 18(3–4): 257–271.

Dahamsheh A and Aksoy H (2007) Structural characteristics of annual precipitation data in Jordan. *Theoretical and Applied Climatology* 88: 201–212.

Dairaku K, Emori S, and Oki T (2004) Rainfall amount, intensity, duration, and frequency relationships in the Mae Chaem watershed in Southeast Asia. *Journal of Hydrometeorology* 5(3): 458–470.

Dalrymple T (1960) *Flood frequency analyses*. Water Supply Paper 1543–A. Reston, VA, USA: USGS.

De Michele C and Salvadori G (2003) A Generalized Pareto intensity-duration model of storm rainfall exploiting 2-Copulas. *Journal of Geophysical Research* 108(D2): 4067 (doi:10.1029/2002JD002534).

De Michele C, Salvadori G, Canossi M, Petaccia A, and Rosso R (2005) Bivariate statistical approach to check adequacy of dam spillway. *Journal of Hydrologics Engineering* 10(1): 50–57 (doi:10.1061/(ASCE)1084-0699(2005)10:1(50)).

De Michele C, Salvadori G, Passoni G, and Vezzoli R (2007) A multivariate model of sea storms using copulas. *Coastal Engineering* 54(10): 734–751 (doi:10.1016/j.coastaleng.2007.05.007).

de Waal DJ and van Gelder PHAJM (2005) Modelling of extreme wave heights and periods through copulas. *Extremes* 8: 345–356 (doi:10.1007/s10687-006-0006-y).

Di Baldassarre G, Brath A, and Montanari A (2006a) Reliability of different depth–duration–frequency equations for estimating short-duration design storms. *Water Resources Research* 42(12): W12501.

Di Baldassarre G, Castellarin A, and Brath A (2006b) Relationships between statistics of rainfall extremes and mean annual precipitation: An application for design-storm estimation in northern central Italy. *Hydrology and Earth System Sciences* 10: 589–601.

Di Baldassarre G, Laio F, and Montanari A (2008) Design flood estimation using model selection criteria. *Physics and Chemistry of the Earth* 34: 606–611 (doi: 10.1016/j.pce.2008.10.066).

Dobric J and Schmid F (2007) A goodness of fit test for copulas based on the Rosenblatt's transformation. *Computational Statistics and Data Analysis* 51(9): 4633–4642 (doi: 10.1016/j.csda.2006.08.012).

Dupuis DJ (2007) Using copulas in hydrology: Benefits, cautions, and issues. *Journal of Hydrologics Engineering* 12(4): 381–393 (doi: 10.1061/(ASCE) 1084-0699(2007)).

Evin G and Favre A-C (2008) A new rainfall model based on the Neyman-Scott process using cubic copulas. *Water Resources Research* 44: W03433 (doi:10.1029/2007WR006054).

Fanta B, Zaake BT, and Kachroo RK (2001) A study of variability of annual river flow of the southern African region. *Hydrological Sciences Journal* 46(4): 513–524.

Faulkner D (1999) Rainfall frequency estimation. In: *Flood Estimation Handbook (FEH)*, Vol. 2, Wallingford: Institute of Hydrology.

Favre A-C, El Adlouni S, Perreault L, Thiémonge N, and Bobée B (2004) Multivariate hydrological frequency analysis using copulas. *Water Resources Research* 40: W01101 (doi: 10.1029/2003 WR002456).

FEH (1999) *Flood Estimation Handbook*. Wallingford: Institute of Hydrology.

Fennessey NM and Vogel RM (1990) Regional flow-duration curves for ungauged sites in Massachusetts. *Journal of Water Resources Planning and Management, ASCE* 116(4): 531–549.

Fermanian J-D (2005) Goodness-of-fit tests for copulas. *Journal of Multivariate Analysis* 95: 119–152 (doi:10.1016/j.jmva.2004.07.004).

Fill HD and Stedinger JR (1995) L-moment and probability plot correlation coefficient goodness-of-fit tests for the Gumbel distribution and impact of autocorrelation. *Water Resources Research* 31(1): 225–229.

Fisher R and Tippett L (1928) Limiting forms of the frequency distribution of the largest or smallest member of a sample. *Proceedings of the Cambridge Philosophical Society* 24(2): 180–191.

Fortin V, Perreault L, and Salas JD (2004) Restrospective analysis and forecasting of streamflows using a shifting level models. *Journal of Hydrology* 296: 135–163.

Foster H (1924) Theoretical frequency curves and their application to engineering problems. *Transactions of the American Society of Civil Engineers* 87: 142–173.

Fréchet M (1927) Sur ia loi de probabilité de l'écart maximum (On the probability law of maximum values). In: *Annales de la societé' Polonaise de Mathematique*, vol. 6, pp. 93–116. Krakow, Poland.

Frederick RH (1977) Five- to 60-minute precipitation frequency for the eastern and central United States. *NOAA Technical Memorandum NWS HYDRO-35*.

Fuller W (1914) Flood flows. *Transactions of the American Society of Civil Engineers* 77: 564–617.

Furey PR and Gupta VA (2000) Space-time variability of low streamflows in river networks. *Water Resources Research* 36(9): 2679–2690.

Gabriele S and Arnell N (1991) A hierarchical approach to regional flood frequency analysis. *Water Resources Research* 27(6): 1281–1289.

García-Bartual R and Schneider M (2001) Estimating maximum expected short-duration rainfall intensities from extreme convective storms. *Physics and Chemistry of the Earth, Part B: Hydrology, Oceans and Atmosphere* 26(9): 675–681.

Gebremichael M and Krajewski WF (2007) Application of copulas to modeling temporal sampling errors in satellite-derived rainfall estimates. *Journal of Hydrologics Engineering* 12(4): 404–408 (doi:10.1061/(ASCE)1084-0699(2007)12:4(404)).

Gedikli A, Aksoy H, and Unal NE (2008) Segmentation algorithm for long time series analysis. *Stochastic Environmental Research and Risk Assessment* 22: 291–302.

Gedikli A, Aksoy H, and Unal NE (2010a). AUG-segmenter: A user-friendly tool for segmentation of long time series. *Journal of Hydroinformatics* 12(3): 318–328.

Gedikli A, Aksoy H, Unal NE, and Kehagias A (2010b) Modified dynamic programming approach for offline segmentation of long hydrometeorological time series. *Stochastic Environmental Research and Risk Assessment* 24: 547–557.

Gellens D (2002) Combining regional approach and data extension procedure for assessing GEV distribution of extreme precipitation in Belgium. *Journal of Hydrology* 268(1–4): 113–126.

Genest C and Favre A-C (2007) Everything you always wanted to know about copula modeling but were afraid to ask. *Journal of Hydrologics Engineering* 12(4): 347–368 (doi:10.1061/(ASCE) 1084-0699(2007)12:4(347)).

Genest C, Favre A-C, Béliveau J, and Jacques C (2007) Metaelliptical copulas and their use in frequency analysis of multivariate hydrological data. *Water Resources Research* 43: W09401 (doi:10.1029/2006WR005275).

Genest C, Ghoudi K, and Rivest L-P (1995) A semiparametric estimation procedure of dependence parameters in multivariate families of distributions. *Biometrika* 82(3): 543–552.

Genest C, Quessy J-F, and Rémillard B (2006) Goodness-of-fit procedures for copula models based on the probability integral transformation. *Scandinavian Journal of Statistics* 33: 337–366 (doi:10.1111/j.1467-9469.2006.00470.x).

Genest C and Rivest L-P (1993) Statistical inference procedures for bivariate archimedean copulas. *Journal of the American Statistical Association* 88(423): 1034–1043.

Gert A (1987) Regional rainfall intensity-duration-frequency curves for Pennsylvania. *Water Resources Bulletin* 23(3): 479–486.

Goel NK, Kurothe RS, Mathur BS, and Bogel RM (2000) A derived flood frequency distribution for correlated rainfall intensity and duration. *Journal of Hydrology* 228: 56–67 (doi:10.1016/S0022-1694(00)00145-1).

Greenwood J, Landwehr J, Matalas N, and Wallis J (1979) Probability weighted moments: Definition and relation to parameters of several distributions expressible in inverse form. *Water Resources Research* 15: 1049–1054.

GREHYS (1996) Inter-comparison of regional flood frequency procedures for Canadian rivers. *Journal of Hydrology* 186: 85–103.

Griffis V and Stedinger JR (2007) Log-Pearson type 3 distribution and its application in flood frequency analysis. I: Distribution characteristics. *Journal of Hydrologic Engineering* 12(5): 482–491.

Grimaldi S and Serinaldi F (2006a) Asymmetric copula in multivariate flood frequency analysis. *Advances in Water Resources* 29(8): 1155–1167 (doi:10.1016/j.advwatres.2005.09.005).

Grimaldi S and Serinaldi F (2006b) Design hyetograph analysis with 3-copula function. *Hydrological Sciences Journal* 51(2): 223–238 (doi:10.1623/hysj.51.2.223).

Gumbel EJ (1941) The return period of flood flows. *Annals of Mathematical Statistics* 12: 163–190.

Gumbel EJ (1958) *Statistics of Extremes*. New York: Columbia University Press.

Hall MJ and Minns AW (1999) The classification of hydrologically homogeneous regions. *Hydrological Sciences–Journal–des Sciences Hydrologiques* 44(5): 693–704.

Haan CT (2002) *Statistical Methods in Hydrology*, 2nd edn. Ames, Iowa: Iowa State Press.

Helsel DR and Hirsch RM (1992) *Statistical Methods in Water Research*. Amsterdam: Elsevier.

Hershfield DM (1961) Rainfall frequency Atlas of the United States for durations from 30 minutes to 24 hours and return periods from 1 to 100 years. *US Weather Bureau Technical Paper 40*. Washington, DC: US Government Printing Office.

Hirsch RM, Helsel DR, Cohn TA, and Gilroy EJ (1993) Statistical analysis of hydrologic data. In: Maidment DR (ed.) *Handbook of Hydrology*. New York: Mc-Graw Hill.

Hosking J and Wallis J (1987) Parameter and quantile estimation for the generalized Pareto distribution. *Technometrics* 29(3): 339–349.

Hosking JRM (1990) L-moments: Analysis and estimation of distributions using linear combinations of order statistics. *Journal of the Royal Statistical Society, Series B (Methodological)* 52: 105–124.

Hosking JRM and Wallis JR (1988) The effect of intersite dependence on regional flood frequency-analysis. *Water Resources Research* 24(4): 588–600.

Hosking JRM and Wallis JR (1997) *Regional Frequency Analysis: An Approach Based on L-Moments*. Cambridge: Cambridge University Press.

Hosking JRM and Wallis JR (2005) *Regional Frequency Analysis: An Approach Based on L-Moments*. Cambridge: Cambridge University Press.

Hubert P (2000) The segmentation procedure as a tool for discrete modeling of hydrometeorological regimes. *Stochastic Environmental Research and Risk Assessment* 14: 297–304.

Hubert P, Carbonnel JP, and Chaouche A (1989) Segmentation des series hydrométéorologiques – application à des séries de précipitations et de débits de l'afrique de l'ouest. *Journal of Hydrology* 110(3–4): 349–367.

Imberger J and Ivey GN (1991) On the nature of turbulence in a stratified fluid. Part II: Applications to lakes. *Journal of Physical Oceanography* 21(5): 659–679.

Institute of Hydrology (1999) *Flood Estimation Handbook*. Crowmarsh Gifford: Institute of Hydrology.

Jakob D, Reed DW, and Robson AJ (1999) Choosing a pooling-group. In: *Flood Estimation Handbook*, vol. 3, Wallingford: Institute of Hydrology.

Jenkinson AF (1955) The frequency distribution of the annual maximum (or minimum) of meteorological elements. *Quarterly Journal of the Royal Meteorological Society* 81: 158–171.

Joe H (1997) *Multivariate models and dependence concepts*. London: Chapman and Hall.

Johnson NL, Kotz S, and Balakrishnan N (1994) *Continuous Univariate Distributions*. 2nd edn., vol. 1. New York: Wiley.

Kao S-C and Govindaraju RS (2007a) Probabilistic structure of storm surface runoff considering the dependence between average intensity and storm duration. *Water Resources Research* 43: W06410 (doi:10.1029/2006WR005564).

Kao S-C and Govindaraju RS (2007b) A bivariate rainfall frequency analysis of extreme rainfall with implications for design. *Journal of Geophysical Research* 112: D13119 (doi:10.1029/2007JD008522).

Kao S-C and Govindaraju RS (2008) Trivariate statistical analysis of extreme rainfall events via Plackett family of copulas. *Water Resources Research* 44: W02415 (doi:10.1029/2007WR006261).

Kehagias A (2004) A hidden Markov model segmentation procedure for hydrological and environmental time series. *Stochastic Environmental Research and Risk Assessment* 18: 117–130.

Kehagias A, Nidelkou E, and Petridis V (2006) A dynamic programming segmentation procedure for hydrological and environmental time series. *Stochastic Environmental Research and Risk Assessment* 20: 77–94.

Kendall MG and Stuart A (1977) *The Advanced Theory of Statistics; Vol. 2: Inference and Relationship*, 4th edn. London: Griffin and Co.

Kitanidis PK (1997) *Introduction to Geostatistics: Applications to Hydrogeology*. Cambridge: Cambridge University Press.

Kjeldsen TR and Jones D (2007) Estimation of an index flood using data transfer in the UK. *Hydrological Sciences–Journal–des Sciences Hydrologiques* 52(1): 86–98.

Klemeš V (1993) Probability of extreme hydrometeorological events – a different approach. In: Kundzewicz Z (ed.) *Extreme hydrological events: Precipitation, Floods and Droughts*, vol. 213, pp. 167–176. Wallingford: IAHS.

Kojadinovic I (2010) Hierarchical clustering of continuous variables based on the empirical copula process and permutation linkages. *Computational Statistics and Data Analysis* 54(1): 90–108.

Kothyari UC (1992) Rainfall intensity–duration–frequency formula for India. *Journal of Hydraulic Engineering – ASCE* 118(2): 323–336.

Koutsoyiannis D (2004a) Statistics of extremes and estimation of extreme rainfall: I. Theoretical investigation. *Hydrological Sciences Journal* 49(4): 575–590.

Koutsoyiannis D (2004b) Statistics of extremes and estimation of extreme rainfall: II. Empirical investigation of long rainfall records. *Hydrological Sciences Journal* 49(4): 591–610.

Koutsoyiannis D (2006) Nonstationarity versus scaling in hydrology. *Journal of Hydrology* 324: 239–254.

Koutsoyiannis D, Kozonis D, and Manetas A (1998) A mathematical framework for studying rainfall intensity–duration–frequency relationships. *Journal of Hydrology* 206(1–2): 118–135.

Kroll C and Vogel RM (2002) Probability distribution of low streamflow series in the United States. *Journal of Hydrologic Engineering* 7(2): 137–146.

Kruskal WH and Wallis WA (1952) Use of ranks in one-criterion variance analysis. *Journal of the American Statistical Association* 47: 583–621.

Kuhn G, Khan S, Ganguly AR, and Branstetter ML (2007) Geospatial–temporal dependence among weekly precipitation extremes with applications to observations and climate model simulations in South America. *Advances in Water Resources* 30: 2401–2423 (doi:10.1016/j.advwatres.2007.05.006).

Kuichling E (1889) The relation between the rainfall and the discharge of sewers in populous districts. *Transactions of the American Society of Civil Engineers* 20(140): 782–794.

Laaha G and Blöschl G (2006) Seasonality indices for regionalizing low flows. *Hydrological Processes* 20: 3851–3878.

Laio F (2004) Cramer–von Mises and Anderson-Darling goodness of fit tests for extreme value distributions with unknown parameters. *Water Resources Research* 40: W09308 (doi:10.1029/2004WR003204).

Laio F, Di Baldassarre G, and Montanari A (2009) Model selection techniques for the frequency analysis of hydrological extremes. *Water Resources Research* 45: W07416 (doi:10.1029/2007WR006666).

Lang M, Ouarda T, and Bobée B (1999) Towards operational guidelines for over-threshold modeling. *Journal of Hydrology* 225(3–4): 103–117.

Langbein WB (1949) Annual floods and the partial duration flood series. *Transactions of the American Geophysical Union* 30(6): 879–881.

Laux P, Wagner S, Wagner A, Jacobeit J, Bárdossy A, and Kunstmann H (2009) Modelling daily precipitation features in the Volta Basin of West Africa. *International Journal of Climatology* (doi:10.1002/joc.1852).

Lettenmaier DP, Wallis JR, and Wood EF (1987) Effect of regional heterogeneity on flood frequency estimation. *Water Resources Research* 23(2): 313–323.

Libiseller C and Grimwall A (2002) Performance of partial Mann–Kendall tests for trend detection in the presence of covariates. *Environmetrics* 13: 71–84.

Madsen H (1997a) Comparison of annual maximum series and partial duration series methods for modeling extreme hydrologic events 1. At-site modeling. *Water Resources Research* 33(4): 747–757.

Madsen H (1997b) Comparison of annual maximum series and partial duration series methods for modeling extreme hydrologic events 2. Regional modeling. *Water Resources Research* 33(4): 759–769.

Madsen H, Arnbjerg-Nielsen K, and Mikkelsen P (2009) Update of regional intensity–duration–frequency curves in Denmark: Tendency towards increased storm intensities. *Atmospheric Research* 92(3): 343–349.

Madsen H, Pearson C, and Rosbjerg D (1997a) Comparison of annual maximum series and partial duration series methods for modeling extreme hydrologic events 2. Regional modeling. *Water Resources Research* 33(4): 759–769.

Madsen H, Rasmussen PF, and Rosbjerg D (1997b) Comparison of annual maximum series and partial duration series methods for modeling extreme hydrologic events 1. At-site modeling. *Water Resources Research* 33(4): 747–757.

Mailhot A, Duchesne S, Caya D, and Talbot G (2007) Assessment of future change in intensity–duration–frequency (IDF) curves for Southern Quebec using the Canadian Regional Climate Model (CRCM). *Journal of Hydrology* 347(1–2): 197–210.

Maity R and Kumar DN (2008) Probabilistic prediction of hydroclimatic variables with nonparametric quantification of uncertainty. *Journal of Geophysical Research* 113: D14105 (doi:10.1029/2008JD009856).

Martins E and Stedinger JR (2001) Historical information in a generalized maximum likelihood framework with partial duration and annual maximum series. *Water Resources Research* 37(10): 2559–2567.

Matalas N and Wallis J (1973) Eureka–it fits a Pearson type 3 distribution. *Water Resources Research* 9(2): 281–289.

Matalas NC and Langbein WB (1962) Information content of the mean. *Journal of Geophysical Research* 67(9): 3441–3448.

Mays LW (2004) *Water Resources Engineering*, 2005th edn. New York: Wiley.

McDonnell JJ and Woods RA (2004) On the need for catchment classification. *Journal of Hydrology* 299: 2–3.

Merz R and Blöschl G (2005) Flood frequency regionalisation – spatial proximity vs catchment attributes. *Journal of Hydrology* 302(1–4): 283–306.

Miller JF (1973) *Precipitation Frequency Analysis of the Western United States*, NOAA Atlas 2.

Mitosek HT, Strupczewski WG, and Singh VP (2006) Three procedures for selection of annual flood peak distribution. *Journal of Hydrology* 323: 57–73.

Moore DS (1986) Tests of chi-squared type. In: D'Agostino RB and Stephens AM (eds.) *Goodness-of-Fit Techniques*, pp. 63–96. New York: Dekker.

Nelsen RB (2006) *An Introduction to Copulas*. New York: Springer.

Norris PM, Oreopoulos L, Hou AY, Tao W-K, and Zeng X (2008) Representation of 3D heterogeneous cloud fields using copulas: Theory for water clouds. *Quarterly Journal of the Royal Meteorological Society* 134: 1843–1864 (doi:10.1002/qj.321).

NRC-US (1988) *US National Research Council, Committee on Techniques for Estimating Probabilities of Extreme Floods, Estimating Probabilities of Extreme Floods, Methods and Recommended Research*. Washington DC: National Academies Press.

O'Connell DR (2005) Nonparametric Bayesian flood frequency estimation. *Journal of Hydrology* 313(1–2): 79–96.

O'Connell DR, Ostenaa DA, Levish DR, and Klinger RE (2002) Bayesian flood frequency analysis with paleohydrologic bound data. *Water Resources Research* 38(5): 1058 (doi:10.1029/2000WR000028).

Ouarda TBMJ, Girard C, Cavadias GS, and Bobee B (2001) Regional flood frequency estimation with canonical correlation analysis. *Journal of Hydrology* 254(1–4): 157–173.

Overeem A, Buishand A, and Holleman I (2008) Rainfall depth-duration-frequency curves and their uncertainties. *Journal of Hydrology* 348(1–2): 124–134 (doi:10.1016/j.jhydrol.2007.09.044).

Pan M, Wood EF, Wojcik R, and McCabe MF (2008) Estimation of regional terrestrial water cycle using multi-sensor remote sensing observations and data assimilation. *Remote Sensing Environment* 112(4): 1282–1294 (doi:10.1016/j.rse.2007.02.039).

Panchenko V (2005) Goodness-of-fit test for copulas. *Physica A* 355(1): 176–182 (doi:10.1016/j.physa.2005.02.081).

Pandey GR and Nguyen VTV (1999) A comparative study of regression based methods in regional flood frequency analysis. *Journal of Hydrology* 225(1–2): 92–101.

Papalexiou S and Koutsoyiannis D (2008) Ombrian curves in a maximum entropy framework. In: *European Geosciences Union General Assembly*, p. 00702. http://www.itia.ntua.gr/en/docinfo/851/ (accessed March 2010).

Parajka J, Blöschl G, and Merz R (2007) Regional calibration of catchment models: Potential for ungauged catchments. *Water Resources Research* 43: W06406.

Peel MC, Wang QJ, Vogel RM, and McMahon TA (2001) The utility of L-moment ratio diagrams for selecting a regional probability distribution. *Hydrological Sciences Journal des Sciences Hydrologiques* 46(1): 147–156.

Pettitt AN (1979) A non-parametric approach to the change-point detection. *Applied Statistics* 28: 126–135.

Pitman WV (1980) A depth–duration–frequency diagram for point rainfall in SWA-Namibia. *Water SA* 6(4): 157–162.

Poulin A, Huard D, Favre A-C, and Pugin S (2007) Importance of tail dependence in bivariate frequency analysis. *Journal of Hydrologics Engineering* 12(4): 394–403 (doi:10.1061/(ASCE) 1084-0699(2007)12:4(394)).

Powell R (1943) A simple method of estimating flood frequencies. *Civil Engineering* 13: 105–106.

Ramesh N and Davison A (2002) Local models for exploratory analysis of hydrological extremes. *Journal of Hydrology* 256(1–2): 106–119.

Rao AR and Yu GH (1986) Detection of nonstationarity in hydrologic time series. *Management Science* 32(9): 1206–1217.

Reed DW, Jakob D, Robinson AJ, Faulkner DS, and Stewart EJ (1999) Regional frequency analysis: A new vocabulary. In: *Hydrological Extremes: Understanding, Predicting, Mitigating*, Proc. IUGG 99 Symposium, Publ. no. 255, pp. 237–243. Birmingham: IAHS.

Renard B and Lang M (2007) Use of a Gaussian copula for multivariate extreme value analysis: Some case studies in hydrology. *Advances in Water Resources* 30: 897–912 (doi:10.1016/j.advwatres.2006.08.001).

Robson AJ and Reed DW (1999) *Flood Estimation Handbook, Vol. 3: Statistical Procedures for Flood Frequency Estimation*. Wallingford: Institute of Hydrology.

Rosbjerg D, Madsen H, and Rasmussen PF (1992) Prediction in partial duration series with generalized Pareto distributed exceedances. *Water Resources Research* 28(11): 3001–3010.

Salas JD (1993) Analysis and modeling of hydrologic time series. In: Maidment DR (ed.) *Handbook of Hydrology*. New York: Mc-Graw Hill.

Salas JD, Delleur JW, Yevjevich V, and Lane WL (1980) *Applied Modeling of Hydrologic Time Series*. Littleton, CO: Water Resources Publications.

Salvadori G and De Michele C (2004a) Analytical calculation of storm volume statistics involving Pareto-like intensity-duration marginals. *Geophysical Research Letters* 31: L04502 (doi:10.1029/2003GL018767).

Salvadori G and De Michele C (2004b) Frequency analysis via copulas: Theoretical aspects and applications to hydrological events. *Water Resources Research* 40: W12511 (doi:10.1029/2004WR003133).

Salvadori G and De Michele C (2006) Statistical characterization of temporal structure of storms. *Advances in Water Resources* 29(6): 827–842 (doi:10.1016/j.advwatres.2005.07.013).

Salvadori G and De Michele C (2007) On the use of copulas in hydrology: Theory and practice. *Journal of Hydrologics Engineering* 12(4): 369–380 (doi:10.1061/(ASCE)1084-0699(2007)12:4 (369)).

Salvadori G, De Michele C, Kottegoda NT, and Rosso R (2007). *Extremes in nature–An Approach Using Copulas: Water Sciences and Technology* Library Series. vol. 56. New York: Springer.

Sangal B and Biswas A (1970) 3-parameter lognormal distribution and its applications in hydrology. *Water Resources Research* 6(2): 505–515.

Saunders R and Laud P (1980) The multidimensional Kolmogorov goodness-of-fit test. *Biometrika* 67(1): 237.

Scaillet O (2007) Kernel based goodness-of-fit tests for copulas with fixed smoothing parameters. *Journal of Multivariate Analysis* 98(3): 533–543 (doi:10.1016/j.jmva.2006.05.006).

Schaefer MG (1990) Regional analysis of precipitation annual maxima in Washington State. *Water Resources Research* 26(1): 119–131.

Scheffe M (1959) *The analysis of variance*. 477 pp. New York: Wiley.

Schwartz G (1978) Estimating the dimension of a model. *Annals of Statistics* 6: 461–464.

Serinaldi F (2009) Copula-based mixed models for bivariate rainfall data: An empirical study in regression perspective. *Stochostic Environmental Research and Risk Assessment* 23(5): 677–693. (doi:10.1007/s00477-008-0249-z).

Serinaldi F, Bonaccorso B, Cancelliere A, and Grimaldi G (2009) Probabilistic characterization of drought properties through copulas. *Physics Chemistry of the Earth* (in press) (doi:10.1016/j.pce.2008.09.004).

Serinaldi F and Grimaldi S (2007) Fully nested 3-copula: Procedure and application on hydrological data. *Journal of Hydrologics Engineering* 12(4): 420–430 (doi:10.1061/(ASCE) 1084-0699(2007) 12:4(420)).

Serinaldi F, Grimaldi S, Napolitano F, and Ubertini L (2005) A 3-copula function application for design hyetograph analysis. In: Savic DA, Mariño MA, Savenije HHG, and Bertoni JC (eds.) *Sustainable Water Management Solutions for Large Cities*. Wallingford: IAHS. 293 ISBN 1-901502-97-X. 203-212.

Shao J and Tu D (1995) *The Jackknife and Bootstrap*. New York: Springer.

Shaw E and Shaw EM (1998) *Hydrology in Practice*, 3rd edn. Abingdon, Oxon: Routledge.

Shiau J-T (2006) Fitting drought duration and severity with two-dimensional copulas. *Water Resources Management* 20: 795–815 (doi:10.1007/s11269-005-9008-9).

Shiau J-T, Feng S, and Nadarajah S (2007) Assessment of hydrological droughts for the Yellow River, China, using copulas. *Hydrological Processes* 21(16): 2157–2163 (doi:10.1002/hyp.6400).

Shiau J-T, Wang H-Y, and Tsai C-T (2006) Bivariate frequency analysis of flood using copulas. *Journal of American Water Resources Association* 42(6): 1549–1564 (doi:10.1111/j.1752-1688. 2006.tb06020.x).

Shu C and Burn DH (2004) Artificial neural network ensembles and their application in pooled flood frequency analysis. *Water Resources Research* 40: W09301.

Shu C and Ouarda TBMJ (2007) Flood frequency analysis at ungauged sites using artificial neural networks in canonical correlation analysis physiographic space. *Water Resources Research* 43: W07438.

Singh K and Singh VP (1991) Derivation of bivariate probability density functions with exponential marginals. *Stochastics Hydrology and Hydraulics* 5(1): 55–68.

Singh VP and Zhang L (2007) IDF curves using the Frank Archimedean copula. *Journal of Hydrologics Engineering* 12(6): 651–662 (doi:10.1061/(ASCE)1084-0699(2007)12:6(651)).

Sillitto GP (1969) Derivation of approximants to the inverse distribution function of a continuous univariate population from the order statistics of a sample. *Biometrika* 56(3): 641–650.

Sivapalan M, Blöschl G, Merz R, and Gutknecht D (2005) Linking flood frequency to long-term water balance: Incorporating effects of seasonality. *Water Resources Research* 41(6): W06012 (doi:10.1029/2004WR003439).

Sivapalan, et al. (2003) IAHS decade on predictions in ungauged basins (PUB), 2003–2012: Shaping an exciting future for the hydrological sciences. *Hydrological Sciences Journal des Sciences Hydrologiques* 48(6): 857–880.

Sklar A (1959) Functions de répartition à n dimensions et leurs marges. Publ. Inst. Statist. Univ. Paris 8: 229–231.

Skøien JO and Blöschl G (2007) Spatiotemporal topological kriging of runoff time series. *Water Resources Research* 43: W09419.

Skøien JO, Merz R, and Bloschl G (2006) Top-kriging–geostatistics on stream networks. *Hydrology and Earth System Sciences* 10(2): 277–287.

Smakhtin VU (2001) Low flow in hydrology: A review. *Journal of Hydrology* 240: 147–186.

Smakhtin VY, Hughes DA, and Creuse-Naudine E (1997) Regionalization of daily flow characteristics in part of the Eastern Cape, South Africa. *Hydrological Sciences Journal des Sciences Hydrologiques* 42(6): 919–936.

Stedinger JR (1983) Estimating a regional flood frequency distribution. *Water Resources Research* 19: 503–510.

Stedinger JR and Cohn TA (1986) Flood frequency analysis with historical and paleoflood information. *Water Resources Research* 22(5): 785–793.

Stedinger JR and Griffis VW (2008) Flood frequency analysis in the United States: Time to update. *Journal of Hydrological Engineering* 13(4): 199–204.

Stedinger JR and Lu L (1995) Appraisal of regional and index flood quantile estimators. *Stochastic Hydrology and Hydraulics* 9(1): 49–75.

Stedinger JR, Vogel RM, and Foufoula-Georgiou E (1993) Frequency analysis of extreme events. In: Maidment DR (ed.) *Handbook of Hydrology*, Ch. 18. New York: McGraw-Hill Inc.

Stephens MA (1986) Tests based on EDF statistics. In: D'Agostino RB and Stephens AM (eds.) *Goodness-of-fit techniques*, pp. 97–194. New York: Dekker.

Strupczewski WG, Singh VP, and Weglarczyk S (2002) Asymptotic bias of estimation methods caused by the assumption of false probability distributions. *Journal of Hydrology* 258: 122–148.

Takeuchi K (1984) Annual maximum series and partial-duration series-Evaluation of Langbein's formula and Chow's discussion. *Journal of Hydrology* 68(1–4): 275–284.

Thompson EM, Baise LG, and Vogel RM (2007) A global index earthquake approach to probabilistic assessment of extremes. *Journal of Geophysical Research* 112: B06314.

Todorovic P (1978) Stochastic models of floods. *Water Resources Research* 14(2): 345–356.

Toth E (2009) Classification of hydro-meteorological conditions and multiple artificial neural networks for streamflow forecasting. *Hydrology and Earth System Sciences Discussion* 6: 897–919.

Troutman BM and Karlinger MR (2003) Regional flood probabilities. *Water Resources Research* 39(4): 1095.

Veneziano D, Lepore C, Langousis A, and Furcolo P (2007) Marginal methods of intensity-duration-frequency estimation in scaling and nonscaling rainfall. *Water Resources Research* 43(10): W10418, 14.

Viglione A and Blöschl G (2009) On the role of storm duration in the mapping of rainfall to flood return periods. *Hydrology and Earth System Sciences* 13(2): 205–216.

Viglione A, Laio F, and Claps P (2007) A comparison of homogeneity tests for regional frequency analysis. *Water Resources Research* 43: W03428.

Villarini G, Serinaldi F, and Krajewski WF (2008) Modeling radar-rainfall estimation uncertainties using parametric and non-parametric approaches. *Advances in Water Resources* 31(12): 1674–1686 (doi:10.1016/j.advwatres.2008.08.002).

Vogel RM and Fennessey NM (1993) L-moment diagrams should replace product-moment diagrams. *Water Resources Research* 29(6): 1745–1752.

Vogel RM and Kroll CN (1992) Regional geohydrologic–geomorphic relationships for the estimation of low-flow statistics. *Water Resources Research* 28(9): 2451–2458.

Vogel RM and McMartin DE (1991) Probability plot goodness-of-fit and skewness estimation procedures for the Pearson type 3 distribution. *Water Resources Research* 27(12): 3149–3158.

Vogel RM and Wilson I (1996) Probability distribution of annual maximum, mean, and minimum streamflows in the United States. *Journal of Hydrologic Engineering* 1(2): 69–76.

Vrac M, Chedin A, and Diday E (2005) Clustering a global field of atmospheric profiles by mixture decomposition of copulas. *Journal of Atmospheric and Oceanic Technology* 22(10): 1445–1459 (doi:10.1175/JTECH1795.1).

Wallace C and Clayton D (2003) Estimating relative recurrence risk ratio. *Genetic Epidemiology* 25(4): 293–302 (doi: 10.1002/gepi.10270).

Wang QJ (1998) Approximate goodness-of-fit tests of fitted generalized extreme value distributions using LH moments. *Water Resources Research* 34(12): 3497–3502.

Wanielista M (1990) *Hydrology and Water Quality Control*. New York: Wiley.

Weibull W (1939) A statistical theory of the strength of materials, pp. 5–45. Number 51. Ingeniors Vetenskaps Akademien (The Royal Swedish Institute for Engineering Research).

Wiesberg S (1985) *Applied Linear Regression*, 2nd edn. New York: Wiley.

Wist HT, Myrhaug D, and Rue H (2004) Statistical properties of successive wave heights and successive wave periods. *Applied Ocean Research* 26(3–4): 114–136 (doi:10.1016/j.apor.2005.01.002).

Wong H, Hu BQ, Ip WC, and Xia J (2006) Change-point analysis of hydrological time series using grey relational method. *Journal of Hydrology* 324: 323–338.

Xiong L and Guo S (2004) Trend test and change-point detection for the annual discharge series of the Yangtze River at the Yichang hydrological station. *Hydrological Sciences Journal* 49(1): 99–112.

Yue S (2001) The bivariate lognormal distribution to model a multivariate flood episode. *Hydrological Processes* 14: 2575–2588 (doi:10.1002/hyp.259).

Yue S, Pilon P, and Cavadias G (2002) Power of the Mann-Kendall and Spearman's rho tests for detecting monotonic trends in hydrological series. *Journal of Hydrology* 259: 254–271.

Zhang L and Singh VP (2006) Bivariate flood frequency analysis using the copula method. *Journal of Hydrologics Engineering* 11(2): 150–164 (doi:10.1061/(ASCE)1084-0699(2006) 11:2 (150)).

Zhang L and Singh VP (2007a) Bivariate rainfall frequency distributions using Archimedean copulas. *Journal of Hydrology* 332(1–2): 93–109 (doi:10.1016/j.jhydrol.2006.06.033).

Zhang L and Singh VP (2007b) Gumbel-Hougaard copula for trivariate rainfall frequency analysis. *Journal of Hydrologics Engineering* 12(4): 409–419 (doi:10.1061/(ASCE)1084-0699 (2007)).

Zhang L and Singh VP (2007c) Trivariate flood frequency analysis using the Gumbel-Hougaard copula. *Journal of Hydrologics Engineering* 12(4): 431–439 (doi:10.1061/(ASCE)1084-0699 (2007)).

Zrinji Z and Burn DH (1996) Regional flood frequency with hierarchical region of influence. *Journal of Water Resources Planning and Management* 122(4): 245–252.

Relevant Websites

http://www.stahy.org
  STAHY- WG, Statistics in Hydrology Working Group.
http://www.iash.info
  What you need, when you need it.

2.19 Scaling and Regionalization in Hydrology

G Blöschl, Vienna University of Technology, Vienna, Austria

© 2011 Elsevier B.V. All rights reserved.

2.19.1	**Introduction**	519
2.19.2	**The Linear Statistical Approach**	520
2.19.2.1	Geostatistics	520
2.19.2.2	Scaling: Variance Reduction of Aggregation	522
2.19.2.3	Regionalization: The Top-Kriging Method	523
2.19.3	**Scaling in Hydrology**	524
2.19.3.1	Self-Similarity and Fractals	524
2.19.3.2	Effective Parameters	524
2.19.3.3	Groundwater Models	525
2.19.3.4	Runoff Models	526
2.19.3.5	Global Circulation Models	527
2.19.4	**Regionalization in Hydrology**	528
2.19.4.1	Similarity Measures	528
2.19.4.2	Floods	529
2.19.4.3	Low Flows	529
2.19.4.4	Runoff Model Parameters	530
2.19.5	**Concluding Remarks**	533
Acknowledgment		533
References		533

2.19.1 Introduction

Hydrological processes exhibit an astounding variability in both space and time. The purpose of the hydrological sciences is to understand this variability, understand where and when the water flows, how much and in what quality. In fact, there is astounding variability at all scales, in both space and time. At the smallest scales of interest in hydrology, water fluxes and composition may vary between individual pores of the soil, and climate and hydrological processes vary over continental scales as well. Infiltration may vary over seconds and groundwater tables may vary over decades and more. Within these limits, variability abounds (Sivapalan, 2003a; Blöschl and Zehe, 2005). Virtually, any quantitative approach to this problem requires the selection of a limited set of spatial and temporal scales. Any particular choice of time and space scales has a major influence on which aspects of this hydrological variability are perceived. Measurements of hydrological quantities are rarely available at the right scales. For example, soil samples, typically, are less than $1\,dm^3$ is size and yet, in modeling the rainfall–runoff processes of catchments, one attempts to represent water flow in the soils at the scale of square kilometers. One is therefore left with the problem of relating the small-scale measurements to larger-scale model descriptions. The process of doing this is usually termed scaling, either upscaling when going from small to large scales or downscaling, when going from large to small scales. Similarly, runoff measurements are never available at all locations of a stream. When estimating runoff characteristics at locations without measurements, procedures usually referred to as regionalization are used, which is a summary term for spatial analysis and estimation in hydrology.

Various aspects of scale issues in hydrology have been reviewed by Blöschl and Sivapalan (1995) and Blöschl (1999, 2005a). A recent concerted effort has been devoted to improving regionalization methods of predicting hydrological variables in ungauged catchments, which has been singled out as one of the major research issues in the hydrological sciences (Sivapalan, 2003b). The International Association of Hydrological Sciences has announced a decade of research on the so-called prediction in ungauged basin (PUB) problem, which is often addressed by regionalization methods (Sivapalan et al., 2003). The purpose of this chapter is to summarize the main concepts of scaling and regionalization in hydrology and assess their applicability to real-world problems.

Before dealing with the individual methods in detail, it is useful to understand the concept of scale as used in hydrology. Natural variability can be characterized by characteristic lengths (Skøien et al., 2003) such as the average distance over which, say, soil properties are correlated. A sampling exercise will rarely reveal the underlying natural variability in full detail because of instrument error and because the spatial and temporal dimensions of the instruments or measurement setup will always be finite. Blöschl and Sivapalan (1995) proposed the concept of a sampling scale triplet to represent the spatial dimensions of measurements. The scale triplet consists of the spacing, extent, and support of the data (**Figure 1**). In dedicated studies, soil hydraulic conductivity measurements, for example, may have spacings of decimeters, while rain gauges in a region are typically spaced at tens of kilometers. The extent, which is the overall size of the domain sampled, may range from meters to hundreds of kilometers in hydrological applications. The support is the integration volume or area of the samples ranging from, say, $1\,dm^2$ in the

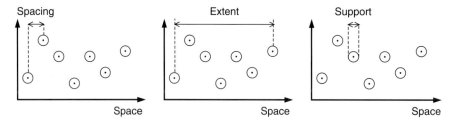

Figure 1 Scale triplet to represent the spatial dimensions of measurements or a model. From Blöschl G and Sivapalan M (1995) Scale issues in hydrological modelling – a review. *Hydrological Processes* 9: 251–290.

case of time domain reflectometry *in situ* probes, to hectares in the case of groundwater pumping tests (Anderson, 1997) or micrometeorological studies of the atmosphere (Schmid, 2002), and square kilometers in the case of remotely sensed data (Western *et al.*, 2002). In hydrology, the catchment area can also be thought of as a support scale (Bierkens *et al.*, 2000). For the case of models, the notion of a scale triplet is similar. For example, for a spatially distributed hydrologic model, the scale triplet may have typical values of, say, 25-m spacing (i.e., the grid spacing), 1-km extent (i.e., the size of the catchment or aquifer to be modeled), and 25-m support (the cell size). Analog scales apply to the temporal domain.

The important point in the context of upscaling and downscaling is that the sampling scale triplet will have some bearing on the data and the modeling scale triplet will have some bearing on the predictions. Generally, if the spacing of the data is large, the small-scale components of the natural variability will not be captured by the measurements. If the extent of the data is small, the large-scale variability will not be captured and will translate into a trend in the data. If the support is large, most of the variability will be smoothed out and the data will appear very smooth. These sampling scale effects can be thought of as a kind of filtering in that the true patterns are filtered by the properties of the measurements (Skøien and Blöschl, 2006c; Blöschl and Grayson, 2000). In the case of models, the scale effects are similar.

The different types of upscaling and downscaling are illustrated in **Figure 2**. Downscaling in terms of spacing (i.e., decreasing the spacing) is usually referred to as interpolation; the opposite is singling out. Upscaling in terms of extent (i.e., increasing the extent) is usually referred to as extrapolation, the opposite is, again, singling out. Upscaling and downscaling in terms of the support is referred to as aggregation and disaggregation, respectively, particularly if the spacing is changing at the same time as the support. The scheme in **Figure 2** can relate to both the sampling step (upscaling/downscaling from the underlying distribution to the data) and the modeling step (upscaling/downscaling from the data to the model predictions). These two steps are conceptually similar.

2.19.2 The Linear Statistical Approach

2.19.2.1 Geostatistics

When one attempts to represent spatial hydrological variability (both in the context of scaling and regionalization), one can choose a geostatistical approach with a number of simplifying assumptions:

- The hydrological quantity can be represented by a spatial random variable.

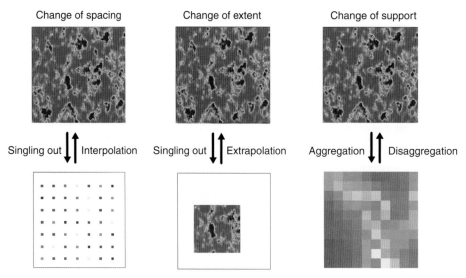

Figure 2 Schematic of upscaling and downscaling by changing the scale triplet. The top row represents the underlying natural variability. The bottom row shows the actual information reflected in the samples (or the model).

- The distribution function and the spatial correlation (represented by the variogram) fully represent the variability of that quantity, and the variogram does not change with location (something termed second-order stationarity).

This then allows us to:

- characterize the spatial structure of hydrological quantities by the variogram.

For example, the variogram is used to characterize rainfall fields and the heterogeneity of aquifers (Skøien *et al.*, 2003; Blöschl, 2005a). The variogram is defined as half the spatial variance of the variable Z at two points with a spacing of h, plotted against that spacing, that is,

$$\gamma(h) = \tfrac{1}{2}E\{[Z(\mathbf{x}+h) - Z(\mathbf{x})]^2\} \quad (1)$$

where $\mathbf{x}$ is the location and E denotes the expected value. The variogram contains equivalent information to the spatial correlation function. There are a number of parametric variogram models that are fitted to the data such as an exponential variogram:

$$\gamma(h) = \sigma^2 \cdot (1 - \exp(-h/\lambda)) \quad (2)$$

where λ is the correlation length, σ^2 is the variance, and h is the distance between two points in the random field as mentioned above. This type of variogram implies that $\gamma(0) = 0$, that is, it is assumed that, for very small distances, the variable Z does not vary because there is no microscale variability and no measurement error.

The variogram embodies the type of correlation, that is, how similar adjacent measurements are. If λ is large, the spatial field will appear smooth. If σ^2 is large, there are big differences between the values in the field. Figure 3 illustrates the relationship between the variogram and the underlying pattern of the variable of interest.

There are additional useful assumptions for an estimation method: the estimation method

- is linear,
- is unbiased (i.e., the expected values of the estimator and the underlying random variable are identical), and
- minimizes the expected estimation error.

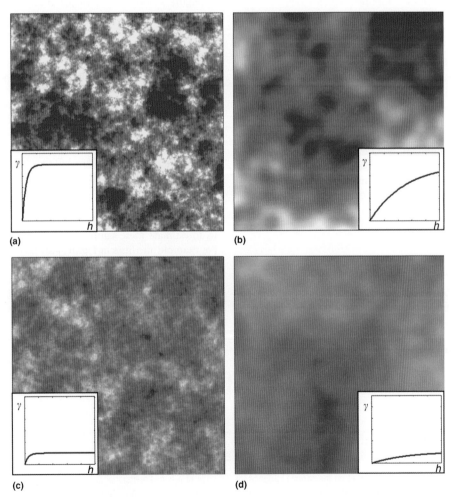

Figure 3 Spatial fields of a random variable (such as rainfall, soil moisture, or hydraulic conductivity) with different variances and correlation lengths and associated variograms (blue lines): (a) large variance, small correlation length; (b) large variance and large correlation length; (c) small variance, small correlation length; and (d) small variance and large correlation length. γ is half the spatial variance of a variable at two points with a spacing of h.

This then allows us to:

- estimate the quantity of interest at locations without measurements, using geostatistical regionalization methods such as kriging.

The linearity assumption above implies that the unknown value $\hat{z}(\mathbf{x}_0)$ of the hydrological variable at position $\mathbf{x}_0$ can be estimated as a weighted average of the variable measured in the neighborhood:

$$\hat{z}(\mathbf{x}_0) = \sum_{i=1}^{n} \lambda_i z(\mathbf{x}_i) \qquad (3)$$

where λ_i is the interpolation weight of the measurement at position $\mathbf{x}_i$ and n is the number of neighboring measurements used for interpolation. The weights λ_i should not be confused with the correlation lengths λ in Equation (2). The similar notation is coincidence and has been chosen here because of their usage in the literature. The weights λ_i can be found by solving the kriging system (which derives from the other two assumptions):

$$\begin{array}{c}\sum_{j=1}^{n}\lambda_{j}\gamma_{ij}-\lambda_{j}\sigma_{i}^{2}+\mu=\gamma_{0i},\quad i=1,\dots,n\\ \sum_{i=1}^{n}\lambda_{i}=0\end{array} \qquad (4)$$

The γ_{ij}'s are the expected semivariance between two measurements i and j, as found from a semivariogram model such as Equation (2). μ is the Lagrange parameter. σ_i^2 represents the measurement error or uncertainty of measurement i. The use of measurement errors in the kriging equations is termed kriging with uncertain data (KUD) (de Marsily, 1986, p. 300; Merz and Blöschl, 2005).

There is an abundant literature on geostatistics (e.g., Journel and Huijbregts, 1978; Isaaks and Srivastava, 1989; Webster and Oliver, 2001) and numerous software packages exist that facilitate application of the method (e.g., GSLIB (Deutsch and Journel, 1997), SURFER (Golden Software, 2009), and the R software environment for statistical computing (R-software, 2009)).

2.19.2.2 Scaling: Variance Reduction of Aggregation

In order to use geostatistical theory for the scaling problem, an additional assumption is needed of how the variable at one support scale is related to the same variable at a different support scale. The most obvious assumption is that:

- the hydrological variable aggregates linearly or, in other words, simple arithmetic averaging applies. For example, the arithmetic average of soil moisture samples in a catchment will give the total moisture volume in that catchment.

This then allows us to:

- scale the distributions for that variable from small to large scales or conversely (upscaling and downscaling).

Note that linear aggregation does not always apply. In Darcy's law, for example, the average hydraulic conductivity over an area does not give the average flux over the same area. Similarly, the average hydraulic roughness of a surface does not give the average flow on that surface. We will now illustrate the effects of the scale triplet (as in **Figures 1** and **2**) on samples and model results for the simplest case of linear upscaling and downscaling based on the geostatistical assumptions above. We will also assume that the spatial correlation can be represented by the exponential variogram of Equation (2).

Figure 4 shows the results from a Monte Carlo analysis where hypothetical samples were drawn from two-dimensional random fields. From the samples, the sample variance was calculated. The results indicate (and this is, of course, consistent with theory) that large spacings (relative to the correlation length of the underlying variability) do not bias the sample variance, but small extents do and will lead to an underestimation of the variance. The latter is because not the entire variability is sampled (see **Figure 2**). Large supports will reduce the variance and this is related to the smoothing effect of the support mentioned above. It is clear that spacing, extent, and support are all scales but they have a different role in upscaling and downscaling. For example, the variance of, say, precipitation tends to increase with scale (if scale is defined as extent) but decreases with scale (if scale is defined as support) because of the filtering involved. Geostatistical methods allow us to estimate these sampling biases in a consistent manner. For example, the lines in **Figure 4** are taken from Western and Blöschl (1999) and there is a rich geostatistical literature on methods for estimating these scale effects, in particular those on the support (e.g., Journel and Huijbregts, 1978; Isaaks and Srivastava, 1989; Goovaerts, 1997; Chilès and Delfiner, 1999).

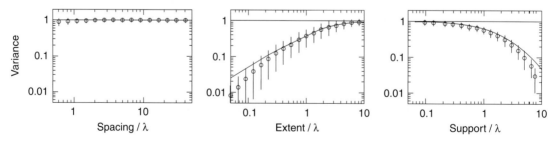

Figure 4 Effect of the sampling scale triplet on the sample variance for the case of a two-dimensional stationary random field. λ is the correlation length of the random field (see **Figure 3**). The circles show the ensemble mean and the error bars represent the standard deviation around the ensemble mean for 100 gridded samples from a Monte Carlo study. The solid lines show the predictions for the ensemble mean. From Skøien JO and Blöschl G (2006a) Sampling scale effects in random fields and implications for environmental monitoring. *Environmental Monitoring and Assessment* 114(1–3): 521–552.

This literature also gives methods for estimating the effects of the scales on the sample variogram. Monte Carlo simulations of these scale effects can be used to assess the uncertainty in the variogram that is introduced by the sampling scales (Skøien and Blöschl, 2006a, 2006b).

The reduction in variance as a result of increasing support is widely used for linear aggregation methods and may also give some guidance for nonlinear cases such as extreme value analyses of rainfall and model parameter aggregation (see, e.g., Sivapalan and Blöschl, 1998).

2.19.2.3 Regionalization: The Top-Kriging Method

In the regionalization problem, we are interested in estimating streamflow-related variables at locations where no measurements are available. The main advantage of using geostatistical methods for this purpose is that they are best linear unbiased estimators as noted above and they can provide estimates of the uncertainty as well. Geostatistical methods have evolved in the mining industry where the main problem consisted of estimating the expected ore grade (and its uncertainty) of a block using point samples of the ore grade in the area (Journel and Huijbregts, 1978). However, the problem in catchment hydrology is quite different in that catchments are organized into nested subcatchments. Water follows a stream network. It is therefore clear that upstream and downstream catchments would have to be treated differently from neighboring catchments that do not share a subcatchment. In order to take the stream network topology into account, a method termed 'top-kriging' has been developed (Skøien *et al.*, 2006).

The main idea behind the method is to combine two groups of hydrological variability. The first group consists of variables that are continuous in space such as rainfall, evapotranspiration, and soil characteristics, which are related to local runoff generation. In top-kriging, the variability of these continuous processes in space is represented by the variogram. The second group of variables, such as runoff, is related to routing in the stream network. These variables are only defined for points on the stream network. In top-kriging, the aggregation effects that lead to these variables are represented by the catchment boundaries associated with each point on the stream network. In the method, the distance between two catchments is measured by the geostatistical distance of the two catchment areas rather than by the Euclidian distance of the stream gauges. The geostatistical distance is the average distance of all pairs of points in the two catchment areas, thus taking the catchment shapes and the pattern of the relative locations of the catchments into account. The Euclidian distance can either be chosen as the distance between the stream gauges or the distance between the centers of gravity of the catchment areas and, hence, contains much less information than the geostatistical distance.

The variogram is then aggregated over the catchment areas (i.e., the support as in **Figures** 1 and 4) and kriging is performed to estimate the variables at the ungauged locations. Variables that can be estimated include the mean annual runoff, flood characteristics, low-flow characteristics, concentrations, turbidity, and stream temperature.

To illustrate the approach, **Figure 5** shows an example of using top-kriging to regionalize the specific 100-year flood in southeastern Austria. Some of the stream gauges are situated directly at the main stream, others at the tributaries. The top-kriging estimates at the main stream are similar to the measurements at the main stream and they do not change much

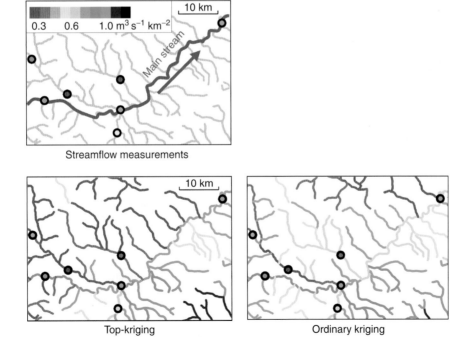

Figure 5 Estimates of the normalized specific 100-year flood Q_{100N} from top-kriging and ordinary kriging color-coded on the stream network of the Mur region. The measurements (i.e., values at the stream gauges) are shown as circles. Units are in m^3 s^{-1} km^{-2}. From Skøien J, Merz R, and Blöschl G (2006) Top-kriging – geostatistics on stream networks. *Hydrology and Earth System Sciences* 10: 277–287.

along the reach as would be expected. The estimates at the northern tributaries are much smaller than those at the main stream, which is consistent with the measurements at the same tributaries. For comparison, the results of using 'ordinary kriging' are also shown. Although the gauge in the center of the panel has measurements of 0.4 (red color), most of the ordinary kriging estimates along this tributary are around 0.6 (yellow to green colors). This is because the estimates along this tributary are too much influenced by the measurements along the main river while they should be mainly influenced by the downstream gauge as is the case in top-kriging. On the other hand, the estimates at the main stream are somewhat underestimated by ordinary kriging as they are too much affected by the measurements at the tributaries. Top-kriging takes both the area and the nested nature of catchments into account while ordinary kriging cannot separate the main stream and the tributaries since Euclidian distances are used.

Because of the minimum number of assumptions, top-kriging is the most natural method of estimating streamflow-related variables on stream networks. It is currently being used for spatially interpolating a range of variables, including run-off time series, flood characteristics, low flow characteristics, and stream temperatures (Skøien and Blöschl, 2007; Merz et al., 2008; Laaha, 2008).

2.19.3 Scaling in Hydrology
2.19.3.1 Self-Similarity and Fractals

The concept of self-similarity and fractals revolves around the idea that the small-scale properties of a variable are similar to the large-scale properties of the same variable:

$$\varphi(\tilde{x}) = v^{-\beta} \cdot \varphi(v \cdot \tilde{x}) \qquad (5)$$

where φ is a property of the variable; $\tilde{x}$ is space or time scale; v is the ratio of the large scale $v \cdot \tilde{x}$ to the small scale $\tilde{x}$; and β is the scaling exponent. The space or time scale $\tilde{x}$ is usually taken as the spacing between two points (in space or time) but, alternatively, it can be taken as the support or the extent (Figure 1).

The idea of fractals was first conceived by Richardson (1961) in the context of estimating the length of the borderline of states to understand whether it is related to the likelihood of armed conflicts. In hydrology, this self-similarity concept is mainly used because it is often able to represent the variability of hydrological variables over many orders of magnitudes of scales – from millimeter to kilometer and from minutes to centuries. Depending on what property, φ, is considered, different types of scaling relationships are used in hydrology:

- Perimeter–area relationships: These are, for example, relationships between the perimeter and the area of clouds (Lovejoy, 1982) or similar relationships for catchments (Hack, 1957).
- Probability density functions (pdf): In this type of fractals, the property, φ, that is self-similar is the pdf and, in essence, suggests that the pdf is a power law. Examples include the distribution of contributing area and the product of area and slope in river basins (Rodríguez-Iturbe et al., 1992).

- Variogram: Here, the variogram is a power law, which is different from the exponential variogram in Equation (2).

$$\gamma(h) = \alpha \cdot h^\beta \qquad (6)$$

where β is the scaling exponent and α is a constant. A power law variogram suggests that there is variability at all scales, that is, there is no scale where the spatial field becomes stationary. This type of variogram is typical of many hydrologic processes where it is possible to fit a straight line to the variogram in a log–log plot. However, whether this type of variogram should be preferred over a nested variogram is not always clear (see Blöschl, 1999).

All the fractal relationships listed above are represented by power laws. This is because a power law is consistent with self-similarity as indicated by Equation (5). From a practical hydrological perspective, the fractal variograms are probably the most relevant scaling relationships. They are scaling relationships as they relate the variance at the small scale (small h) to the variance at the large scale (large h). Fractal variograms can be used to represent the spatial variability of many variables in hydrology such as rainfall, soil characteristics, and aquifer hydraulic conductivity (Skøien et al., 2003). Based on this characterization, one can, for example, derive aggregate parameters for groundwater models (see Section 2.19.3.2), and interpolate variables such as soil parameters with different sampling supports. One can also generate two- or three-dimensional fields similar to those in **Figure 3**, which can then be used as inputs (e.g., rainfall) or parameters (e.g., hydraulic conductivity) of models.

In the time domain, the self-similar temporal characteristics of stream flow have been strongly disputed in hydrology under the topic 'Hurst phenomenon' (Hurst, 1951). In essence, the idea is that variability occurs at all scales – from hours to centuries, which implies long-term persistence. The latter has been usually interpreted as implying the existence of an infinite memory of the hydrologic system, which is difficult to understand from a process perspective as the storage capacity of catchments is always finite.

Klemeš (1974), however, indicated that long-term persistence is not necessarily a consequence of infinite memory and may be related to nonstationarities in the data. There are also issues with the way the data are usually analyzed (Kirchner, 1993). Notwithstanding this discussion, the presence of hydrological variability at all time scales is widely accepted in the literature (see, e.g., Koutsoyiannis, 2002).

There are more complicated types of fractals than those discussed here and a rich literature is available (e.g., Feder, 1988). Multifractals (Sreenivasan, 1991) are one type of more complicated fractals that are used in hydrology. They allow for multiple scaling exponents and are able to represent the intermittency of rainfall well. They are hence mostly used in rainfall models in hydrology (Gupta and Waymire, 1993; Menabde and Sivapalan, 2001; Sivapalan et al., 2005).

2.19.3.2 Effective Parameters

As mentioned above, the sample size in hydrology is, often, much smaller than the model element size. Typical examples are groundwater models. Samples of hydraulic conductivities taken from boreholes have support scales of a decimeter, while the size of the elements of groundwater models ranges from tens of meters to kilometers, depending on the application. Similarly, model equations such as Darcy's law apply at the sample scale of a decimeter as this is the scale at which laboratory tests are made. If the subsurface were homogeneous, these samples could be directly used to specify the parameters (in this case, hydraulic conductivity) of the model and the equations could be directly used as well. However, aquifers and catchments are usually very heterogeneous; therefore, one needs an upscaling procedure to obtain the parameters and the underlying equations at the right scale.

One way of addressing this issue is to assume that the parameters are uniform within each (large scale) model element and that the small-scale equations apply to the whole element. The small-scale parameters are then replaced by effective parameters that are applicable at the (large) scale of the model elements. This gives rise to two questions: (a) Can the small-scale equations be used to describe processes at the large scale (or, in other words, do effective parameters exist)? (b) If so, what is the aggregation rule (or scaling rule) to obtain the large-scale model parameters from the detailed pattern or the distribution of the small-scale parameters. Methods that address (a) and (b) make use of the definition of effective parameters by matching the outputs of the uniform and the heterogeneous systems. If an adequate match can be obtained, an effective parameter exists. The aggregation rule is then derived by relating the effective parameter to the underlying heterogeneous distribution. This is shown schematically in **Figure 6**. The answer to (a) obviously depends on the type of problem. For groundwater flow, effective parameters exist for many cases (see Section 2.19.3.3). For unsaturated flow in porous media, generally, there exists no effective conductivity that is a property of the medium only (Russo, 1992) and approximate effective parameters may exhibit hysteresis as a result of the soil spatial variability (Mantoglou and Gelhar, 1987). For land–atmosphere interactions, again, the existence of effective parameters depends on the degree of nonlinearity (Raupach and Finnigan, 1995; Lhomme et al., 1996).

The upscaling methods for obtaining effective parameters work well if the media is disordered, that is, conductivities look like the patterns in **Figure 3**. In reality, the media are often spatially organized and preferential flow paths may exist. In this case, most upscaling methods based on stochastic theory do not work well as the main assumptions are violated (e.g., the assumption that conductivity is a random variable and can be represented by the variance as in Section 2.19.2.2). The variogram simply cannot capture preferential flow and alternative concepts are needed to represent the connectivity of the flow paths at a range of scales (see, e.g., Western et al., 1998, 2001; Trinchero et al., 2008, Schaap et al., 2008).

Effective parameters are often used inadvertently (and without scaling rules) when assuming uniform model parameters within an element and calibrating the model output to field data. While, from a practical perspective, this is a useful approach for both groundwater and rainfall–runoff modeling, it is important to realize that effective parameters will depend not only on the media characteristics (such as the hydraulic conductivity) but also on the flow pattern. If hydraulic conductivity is anisotropic (depends on direction), changed flow patterns will result in a changed effective conductivity; therefore, the calibrated parameters do not necessarily apply to scenarios with changed flow patterns.

2.19.3.3 Groundwater Models

As mentioned above, scaling in relation to groundwater models is a well-studied research issue (see, e.g., Dagan, 1989; Gelhar, 1993; Wen and Gómez-Hernández, 1996; Paleologos et al., 1996; Rubin, 2003). Here, we will only give a very brief summary of effective parameters that are of interest from a practical perspective.

Consider uniform (parallel flow lines) two-dimensional steady saturated flow through a block of porous medium made up of smaller blocks of different conductivities. For an arrangement of blocks in series, the effective conductivity equals the harmonic mean of the block values while, for an arrangement of blocks in parallel, the effective conductivity equals the arithmetic mean. This scaling rule derives directly from Darcy's law and illustrates the dependence of effective parameters on flow patterns mentioned above. If one assumes that the porous medium is a random field, there are comprehensive theories of how to estimate the effective parameters. Whatever the spatial correlation and distribution function of conductivity and whatever the number of dimensions in space, the average conductivity always ranges between the harmonic mean and the arithmetic mean of the local conductivities (Matheron, 1967; de Marsily, 1986). This is a useful result if the variability of conductivities in the domain is small; but if the variability is large, this provides a very wide margin that can span various orders of magnitude.

For more restrictive assumptions on the nature of variability, more precise results have been found in the literature based on a number of methods (see, e.g., Gelhar, 1993; Wen and Gómez-Hernández, 1996). If the probability density

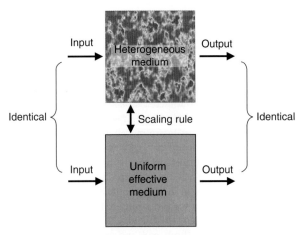

Figure 6 Schematic definition of effective parameters by matching outputs of the uniform and the heterogeneous systems. The aggregation rule (or scaling rule) as indicated by the double arrow is derived by relating the effective parameter to the heterogeneous distribution. From Blöschl G (1996) Scale and scaling in hydrology (Habilitationsschrift). Wiener Mitteilungen, Wasser-Abwasser-Gewässer, Band 132, Institut für Hydraulik, TU Wien, 346pp.

function of the conductivity is lognormal (and for any isotropic spatial correlation) Matheron (1967) and Gelhar (1986) showed that for the two-dimensional case, the effective conductivity equals the geometric mean of the local conductivities. Using the geometric mean is a scaling rule in the sense of **Figure 6**. For the three-dimensional case, the scaling relationships are more complicated and involve the variance and the spatial correlation of the media (e.g., Dykaar and Kitanidis, 1992). For transient conditions, in general, no effective conductivities independent of time may be defined (El-Kadi and Brutsaert, 1985) and for flow systems involving well discharges, the effective conductivity is dependent on pumping rates and boundary conditions (Gómez-Hernández and Gorelick, 1989; Ababou and Wood, 1990; Neuman and Orr, 1993).

It has been observed by a number of authors that effective hydraulic conductivities tend to increase with scale (see, e.g., Ababou and Gelhar, 1990; Clauser, 1992). Sánchez-Vila *et al.* (1996) suggested that this effect may be related to the increasing importance of channelized flow with increasing scale. Blöschl (1996) noted that this scale effect may be related to the dimensionality of the problem. The higher the dimension, the more degrees of freedom are available from which flow can choose the path of lowest resistance (which is consistent with Sánchez-Vila *et al.* (1996)). Hence, for a higher dimension, flow is more likely to encounter a low-resistance path, which tends to increase the effective conductivity. If the dimensionality of the flow problem increases with scale (from essentially one-dimensional flow in the laboratory experiment to a more-dimensional field case), the dimensionality effect translates into increasing effective conductivities with increasing scale (Ababou and Wood, 1990; Indelman *et al.*, 1996).

An alternative to using effective model parameters is to generate two- or three-dimensional fields of the media characteristics (e.g., conductivities) and use them for high-resolution groundwater modeling. Numerous methods for media generation exist, some of which assume fractal media characteristics (e.g., Bellin and Rubin, 1996). Koltermann and Gorelick (1996) and Anderson (1997) provide overviews of media-generation methods.

2.19.3.4 Runoff Models

There are four main issues related to scale and scaling in rainfall–runoff modeling: (a) what is an ideal model grid scale; (b) how do the model parameters change with catchment scale; (c) how does model performance change with catchment scale for a given model structure; and (d) how to best address the scale mismatch between measurements and model elements.

Issue (a) can be addressed by examining how the average values of some property over an area change when increasing the size of that area. This idea was first conceived by Hubbert (1956) in the context of discussing the continuum assumption in groundwater flow theory, which lead to the notion of a representative elementary volume (REV) as the order of magnitude where 'f (porosity) approaches smoothly a limiting value' (i.e., varies only smoothly with changing volume). In analogy, Wood *et al.* (1988) introduced the representative elementary area (REA) in catchment hydrology. The analysis method is to plot peak flow of an event (or many events) versus subcatchment size for a set of nested catchments, and where the variability levels out, one finds the REA. The concept has been re-examined a number of times (e.g., Famiglietti, 1992; Blöschl *et al.*, 1995; Fan and Bras, 1995) suggesting that the REA so obtained very much depends on the catchment scales considered. Blöschl *et al.* (1995) hence suggested that an arbitrary elementary area (AEA) of any size can be used, so choice of element size should depend on the computational resources and the amount of information that is available. More information would justify finer elements. This is also the concept used in sister disciplines such as hydrodynamics and atmospheric sciences.

For addressing issue (b) and (c), results from a recent study are presented that is typical of modeling studies with large data sets. Merz *et al.* (2009) simulated the water balance dynamics of 269 catchments in Austria ranging in size from 10 to 130 000 km^2 using a semi-distributed conceptual model with 11 parameters based on a daily time step. They found that both calibration and verification model efficiencies increase over the scale range of 10 and 10 000 km^2. The authors explained this by the larger number of rainfall stations in each catchment. Indeed, in the very small catchments, it is likely that no raingauge exists, while in the large catchments, there are always a number of raingauges providing for more reliable inputs to the model. The study also showed that the scatter of the model performances decreases with catchment scale, particularly the volume errors as illustrated in **Figure 7**. This result implies that the model simulates the long-term water balance more reliably as one goes up in scale, which is again related to the number of raingauges. Merz *et al.* (2009) also examined how the calibrated model parameters change with catchment scale and found a trend with catchment area of the upper and lower envelope curves of some parameters. Although Merz *et al.* (2009) carefully checked the robustness of the estimated parameters, there may still be issues with parameter uncertainty (Savenije, 2001; Montanari, 2007). From a practical perspective, it seems to be clear that for the case of runoff modeling, data availability should be the main criterion for selecting model grid size and it is also the main driver of scale dependencies in model performance.

The scale mismatch between measurements and model elements (d) has also attracted considerable interest in hydrology (e.g., Grayson *et al.*, 1993). It is increasingly becoming clear that, in catchment hydrological modeling, finer is not necessarily better (Schoups *et al.*, 2008; Savenije, 2009). Stephenson and Freeze (1974) were one of the first to recognize this fact and there is a long track record of studies demonstrating the difficulties in model identification and calibration once the model becomes too complex (Naef, 1981; Loague and Freeze, 1985; Beven, 1989, 2001). This is mainly because the media properties (both soil and vegetation) are highly heterogeneous and essentially always unknown or at least poorly known (Zehe and Blöschl, 2004; Zehe *et al.*, 2007). There will, hence, always be some degree of calibration needed for any model to accurately represent the hydrological processes in a particular case (Blöschl and Grayson, 2002). This also suggests that model element size should mainly depend

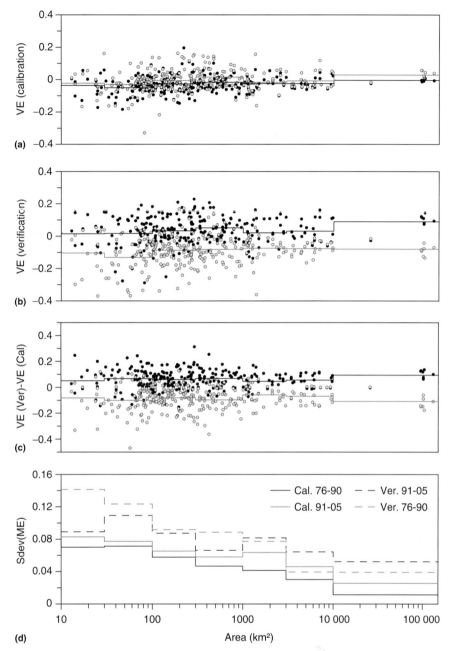

Figure 7 Volume errors, VE, plotted vs. catchment area: (a) calibration; (b) verification; (c) differences of verification and calibration efficiencies; and (d) standard deviation of volume errors for catchment scale classes. Solid lines in ((a)–(c)) show mean efficiencies for a scale range of 3–30 km², 30–300 km², etc. Black and gray dots represent the model performances of catchments calibrated to 1976–90 and 1991–2005, respectively. From Merz R, Parajka J, and Blöschl G (2009) Scale effects in conceptual hydrological modelling. *Water Resources Research* 45: W09405 (doi:10.1029/2009WR007872).

on the amount of relevant hydrological information available, as mentioned above. Grayson *et al.* (2002) pointed out that the scale mismatch between small-scale samples and larger-scale model elements can be effectively addressed by using spatial patterns of hydrological response variables. Such patterns include snow-cover patterns to assess the performance of spatially distributed energy- and water-balance models (Blöschl *et al.*, 1991; Wigmosta *et al.*, 1994), patterns of dominant hydrological processes (Peschke *et al.*, 1999), and soil moisture patterns (Western and Grayson, 2000).

Summaries of several recent case studies of how patterns can be used to address the scale mismatch are reported in Grayson and Blöschl (2000).

2.19.3.5 Global Circulation Models

Outputs of climate simulations from general circulations models (GCMs) cannot be directly used for hydrological impact studies of climate change because of the scale mismatch.

The spatial grid resolution (i.e., support) of GCMs in use today is on the order of several tens of thousands of square kilometers. The useful grid box size is even larger as GCMs are inaccurate at the scale of a single grid box. In contrast, the spatial scale at which inputs to hydrologic impact models are needed is on the order of tens or hundreds of square kilometers. Because of the scale mismatch, the statistical characteristics of the GCM output may be vastly different from those of the local (surface) variable that shares the same name. For example, the maximum rainfall intensities simulated by GCMs tend to be much smaller than those at the point scale. There will also be local effects induced by topography, land cover, etc., which are not captured in the GCM. Downscaling methods can be used to transfer the large-scale GCM output to small-scale variables and to account for local effects.

There are two approaches to downscaling GCM output (IPCC, 2007; Yarnal et al., 2001). The first is dynamic downscaling where deterministic regional climate models are nested into global circulation models. This means the initial conditions and boundary conditions to drive the regional climate model are taken from the GCMs. Dynamic downscaling is discussed, for example, in von Storch (2005).

The second approach is empirical or statistical downscaling, which will be briefly reviewed here in a hydrological context. In empirical or statistical downscaling, explicit relationships between the large-scale GCM output and the observed small-scale or local station data, such as precipitation, are used. Unlike dynamic downscaling, statistical downscaling methods are computationally inexpensive. They can thus be used to generate a large number of realizations to assess the uncertainty of predictions. They can also use climate data from individual stations directly, so that local information can be accounted for in an efficient way. The method, however, hinges on the assumption that the statistical relationships developed for the present-day climate will also hold under the different forcing conditions of possible future climates. This assumption is essentially unverifiable. The relationships can indeed be unstable for many reasons as short-term relationships can be conditional on longer term variations in the climate system (Charles et al., 2004). The application of the method consists of four steps (Yarnal et al., 2001; Blöschl, 2005a):

- Selection of the local variable such as precipitation. The corresponding data are usually collected at individual climate stations in the area.
- Selection of the large-scale GCM-derived variable or variables (termed predictors) such as sea-level pressure or geopotential heights (Wilby and Wigley, 2000; Wilby et al., 2002).
- Deriving relationships between the observed small-scale or local station data and the large-scale GCM-derived variable: the relationships (i.e., the downscaling models) are regression techniques, stochastic models, or the analog method. The latter consists of identifying, from a pool of historical circulation patterns, the one that is most similar to the circulation pattern on the day of interest (Zorita and von Storch, 1999).
- As a final step in the downscaling procedure, the relationships derived above are applied to the GCM output for changed climate scenarios to estimate the local variables for a changed climate. These local variables (mainly precipitation and air temperature) can then be used to drive hydrological models in an impact assessment (IPCC, 2007).

The downscaled data (e.g., precipitation) typically exhibit larger variability than the GCM-derived predictor variables. This is consistent with the effect of support scale on the variance shown in **Figure 4**. A realistic variability of precipitation is extremely important when using precipitation as an input to rainfall–runoff models, as an underestimate in precipitation variability will translate into significant underestimates in runoff because of the nonlinearity of the rainfall–runoff relationship (Komma et al., 2007).

2.19.4 Regionalization in Hydrology

2.19.4.1 Similarity Measures

While the problem of scaling consists of relating variables at different scales, the problem of regionalization consists of relating variables at different locations. A concept that is essential in regionalization is hydrologic similarity (Blöschl, 2001; Wagener et al., 2007; Harman and Sivapalan, 2009). Similarity of hydrological processes can be defined in various ways. Ideally, one would like to relate catchments that are similar in terms of their driving processes. Dunne (1978) suggested that runoff processes are mainly controlled by physioclimatic controls and identified three main types: (1) infiltration of excess runoff, which is generated from partial areas where surface hydraulic conductivities are low; (2) saturation excess runoff, which is generated in areas with shallow water tables or near-channel wetlands; and (3) subsurface storm flow, which is likely to be active and dominant on steep, humid forested hill slopes with very permeable surface soils. In two similar catchments, the relative role of each of these processes would be similar. The characteristics of these processes in natural catchments are never known in full detail so a number of similarity concepts have been proposed in the literature that attempt to represent these processes to various degrees.

Spatial proximity. In the first concept, catchments that are close to each other are assumed to behave hydrologically similarly. The rationale of this concept is that the controls on the rainfall–runoff relationship are likely to vary smoothly in space, or are uniform in predefined regions; therefore, one can expect spatial proximity to be a good indicator of the similarity of catchment response. However, if adjacent catchments are very hydrologically different (see, e.g., the example in Blöschl, 2005b, **Figure 5**), proximity is not a good similarity measure. Geostatistical methods and various methods that are based on homogeneous regions are based on this similarity measure.

Similar catchment attributes. The second similarity concept consists of using measurable catchment attributes as indicators of hydrological similarity such as catchment size, mean areal rainfall, and geological characteristics of the catchments. There are a range of methods that differ in the way the attributes are used in the spatial transfer of information including regression, cluster analysis, classification trees, and the region of influence approach.

Similarity indices. Similarity indices are based on some understanding of the structure of hydrological processes and are usually defined as a dimensionless number. For example, similarity in climate can be quantified by the aridity index of Budyko (1974), which is the ratio of long-term potential evaporation to precipitation. Similarity in runoff generation can be quantified by the topographic wetness index of Beven and Kirkby (1979), which is a function of the area drained per unit contour length at a given point and the local slope gradient (Chirico et al., 2005).

These similarity concepts are the basis of regionalizing floods, low flows, and runoff model parameters in order to obtain estimates for ungauged catchments. The regionalization methods are discussed below. In all cases, the need for regionalization stems from the fact that no local runoff data are available from which floods, low flows, or runoff model parameters could be estimated; therefore, they need to be obtained by regionalization.

2.19.4.2 Floods

A range of methods are available for estimating flood peaks in ungauged catchments (Cunnane, 1988; Bobée and Rasmussen, 1995; Hosking and Wallis, 1997). In the index flood approach (Dalrymple, 1960), the domain is subdivided into regions. Within each region, the flood frequency response is assumed to be similar apart from a scaling factor. The scaling factor, termed index flood, can be the mean annual flood or the median annual flood. This means that the flood frequency curve scaled by the index flood (termed the growth curve) is identical in each region. For an ungauged site, the flood of a given return period is then estimated as the product of a regional growth curve and the local index flood estimated from catchment attributes. The regions are spatially contiguous. They are usually delineated by expert judgment but there exist objective methods such as cluster analyses and methods based on the seasonality of floods (Piock-Ellena et al., 2000). The latter approach is appealing as it allows some process interpretation for each region, such as convective rainfall and glacier runoff as likely flood causes. However, seasonality is only one of the important fingerprints of flood mechanisms. Merz and Blöschl (2003) extended the concept to analyze more complex flood processes at the regional scale.

The region of influence (ROI) approach can be thought of as an extension to the index flood approach. It assumes that every catchment for which flood probabilities are to be estimated is associated with a different homogeneous ROI (Burn, 1990; IH, 2000). Similarity measures are usually based on catchment attributes and seasonality measures. This is the most flexible approach, but it is not straightforward to specify appropriate similarity measures.

An alternative method that uses catchment attributes is the quantile regression method. The term 'quantile' relates to the flood peak discharge of a given return period (e.g., the 100-year flood). The quantile regression approach assumes that the spatial variability is mainly related to catchment attributes such as catchment area and mean annual rainfall. Cunnane (1988), however, noted that the quantile regression method has a number of disadvantages over other methods. Most importantly, methods should be preferred where the first statistical moment (the mean or the median) of the flood distribution is estimated from catchment attributes while the second and the third moments (or, equivalently the shape of the flood frequency curve) should be estimated by pooling regional flood data. This is because of the larger uncertainty of the higher moments.

The geostatistical approach (i.e., some variants of kriging) assumes that the spatial variability is random and only depends on spatial proximity (see Section 2.19.2.1). The method proposed by Merz and Blöschl (2005) regionalizes the three flood moments (mean, coefficient of variation, and skewness) separately. The main advantage of the method is that it can take the different uncertainties of the three moments into account, in line with the note of Cunnane (1988). There are also combinations of the geostatistical approach with regressions, for example, using mean annual precipitation as a catchment attribute (Merz and Blöschl, 2005). This combined approach is termed georegression.

Merz and Blöschl (2005) compared the various types of approaches based on a jack-knifing comparison of locally estimated and regionalized flood quantiles for 575 Austrian catchments. **Figure** 8 shows an example of their results. The biases of the methods (in terms of the normalized mean errors of specific discharges) range between 3% and 12%, depending on the method. The random errors (in terms of the normalized standard deviations of the errors) range from 30% and 50% depending on both the method and the return period of the flood peaks to be estimated. Georegression yields the best predictive performance. The methods that only use catchment attributes (the ROI approach and multiple regressions) perform more poorly than the methods based on spatial proximity (kriging and georegression).

In engineering practice, flood frequency regionalization is often supported by expert judgment (e.g., IH, 1999), which was not represented in the analyses of Merz and Blöschl (2005) in order to obtain an objective comparison. Merz et al. (2008) did take local expert judgment into account to obtain regionalized floods in Austria.

2.19.4.3 Low Flows

Similar to the case of floods, a range of methods are available for estimating low flows in ungauged catchments (Smakhtin, 2001). The methods are usually based on some sort of regression between the low-flow characteristic and catchment attributes. A common low-flow characteristic is the Q_{95} low flow, that is, the discharge that is exceeded on 95% of all days. The catchment attributes include rainfall and geological attributes. If the study domain is large or very heterogeneous in terms of the low-flow processes, it is useful to split the domain into regions and apply a regression relationship to each of the regions independently. This is termed the regional regression approach (Gustard and Irving, 1994). Finding suitable regions is the most critical step in low flow regionalization. The following methods (termed 'grouping methods') exist:

- The residual pattern approach where residuals from an initial, global regression model between flow characteristics

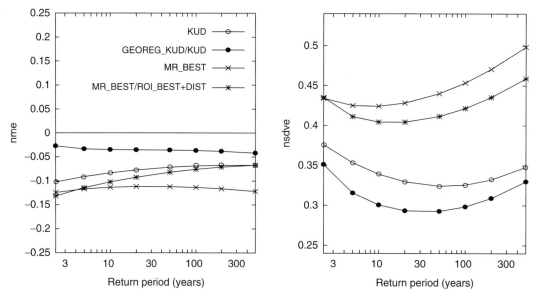

Figure 8 Comparison of bias (left) and random error (right) of flood peak regionalization. Open circles (KUD): variant of kriging; full circles (GEOREG_KUD/KUD): variant of georegression; crosses (MR_BEST): variant of multiple regression; asterisks (MR_BEST/ROI_BEST + DIST): combination of multiple regression and region of influence approach, the latter using catchment attributes and geographical distance. The GEV distribution and product moments are used. From Merz R and Blöschl G (2005) Flood frequency regionalisation – spatial proximity vs. catchment attributes. *Journal of Hydrology* 302(1–4): 283–306.

and catchment attributes are plotted, from which geographically contiguous regions are obtained by manual generalization on a map (e.g., Hayes, 1992; Aschwanden and Kan, 1999).
- Cluster analyses where, usually, both low-flow data and catchment attributes are used in a statistical analysis that obtains regions (or clusters) by maximizing the homogeneity within each region and maximizing the heterogeneity between the regions (Nathan and McMahon, 1990).
- Classification and regression tree (CART) models (Breiman et al., 1984; Laaha and Blöschl, 2006a), where the independent variables in the regression trees are the catchment attributes and the dependent variables are the low flows.
- Seasonality of low flows where the rationale is that differences in the occurrence of low flows within a year are a reflection of differences in the hydrologic processes, such as winter low flows due to snow and freezing processes, and summer low flows due to evaporation (Laaha and Blöschl, 2006b).

Laaha and Blöschl (2006a) performed an analysis similar to that of Merz and Blöschl (2005) referred to in the previous section. They compared four catchment grouping methods in terms of their performance in predicting q_{95} specific low-flow discharges. The grouping methods were the residual pattern approach, weighted cluster analysis, regression trees, and an approach based on seasonality regions. For each group, a regression model between catchment attributes and q_{95} was fitted independently. The performance of the methods was assessed by a jack-knifing comparison of locally estimated and regionalized flood quantiles for 325 catchments in Austria. Their results indicate that the grouping based on seasonality regions performs best and explains 70% of the spatial variance of q_{95}. The favorable performance of this grouping method is likely related to the striking differences in seasonal low-flow processes in the study domain. Winter low flows are associated with the retention of solid precipitation in the seasonal snow pack while summer low flows are related to the relatively large moisture deficits in the lowland catchments during summer. The regression tree grouping performs second best (explained variance of 64%) and the performance of the residual pattern approach is similar (explained variance of 63%). The weighted cluster analysis only explains 59% of the spatial variance of q_{95}, which is only a minor improvement over the global regression model, that is, without using any grouping (explained variance of 57%). Laaha and Blöschl (2007) then applied the seasonality method to obtain regionalized low flows in Austria. Figure 9 shows the results of the regionalization for illustration. The main spatial patterns are the large low flow values at the northern rim of the Alps (West–East band in the center of the country), which are a result of the above average precipitation in the area. The finer scale patterns are mainly due to geological heterogeneities.

2.19.4.4 Runoff Model Parameters

Again similar to the floods and low-flow cases, a range of methods are available for estimating runoff model parameters in ungauged catchments (Blöschl, 2005b). There are a number of engineering procedures for estimating event-scale runoff model parameters. Lag time parameters have been listed as tabulated functions of catchment attributes such as topographic slope, stream slope, and flow length (e.g., USACE, 1994). Similarly, loss parameters (what percentage of rainfall infiltrates) have been listed as tabulated functions of land cover, soil type, and antecedent soil moisture (such as in the

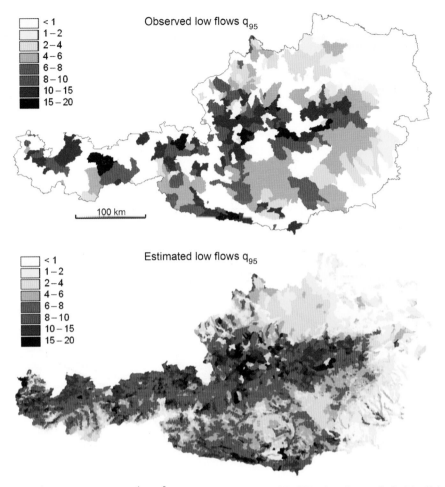

Figure 9 Top: Specific low-flow discharge q_{95} (ls^{-1} km^{-2}) from runoff data observed in 325 subcatchments in Austria. Alpine catchments show higher values and a larger variability. Bottom: Regionalized q_{95} based on the seasonality method. From Laaha G and Blöschl G (2007) A national low flow estimation procedure for Austria. *Hydrological Sciences Journal* 52(4): 625–644.

Curve Number method of the US Soil conservation service, SCS, 1973; Mishra and Singh, 2003).

From the perspective of water science, soil moisture accounting schemes or continuous rainfall–runoff models are of more interest. The most common methods of estimating the parameters of these models are based on relating the calibrated model parameters to catchment attributes. This involves the following main steps:

- Estimating the model parameters for one or more gauged catchments, termed donor catchments, by manual or automatic calibration of the runoff model on observed runoff data.
- Relating each rainfall–runoff model parameter to a set of catchment attributes, for example, by multiple linear regressions, possibly within homogeneous subregions of the entire domain. Identifying homogeneous regions is similar to the low-flow case in Section 2.19.4.3.
- Estimating each parameter of the rainfall–runoff model for the ungauged catchment from the regression model.
- Simulating runoff for the ungauged catchment of interest by applying the runoff model using the regionally transposed model parameters.

A number of authors have tested the relationships between model parameters and catchment attributes and found the correlations to range between 0.5 and 0.8 (Sefton and Howarth, 1998; Seibert, 1999; Beldring *et al.*, 2002). In a recent study, Parajka *et al.* (2005) examined the relative performance of a range of methods for transposing catchment model parameters to ungauged catchments. They calibrated 11 parameters of a semi-distributed conceptual rainfall–runoff model to daily runoff and snow cover data of 320 Austrian catchments and then evaluated the predictive accuracy of the regionalization methods by jack-knife cross-validation against daily runoff data. As an example of the results, **Figure 10** shows the cumulative distribution functions of the model efficiencies of daily runoff (ME, left) and volume errors of runoff (VE, right). As would be expected, the at-site model calibrations (blue lines) show the best model performances (large ME, and VE closest to zero). The regionalization methods show lower performances. The differences between the methods are not large but there is a trend that two methods perform better than the others to a certain extent. The first is a kriging approach (green line) where the model parameters are regionalized independently from each other, based on their spatial correlation. The second is a similarity

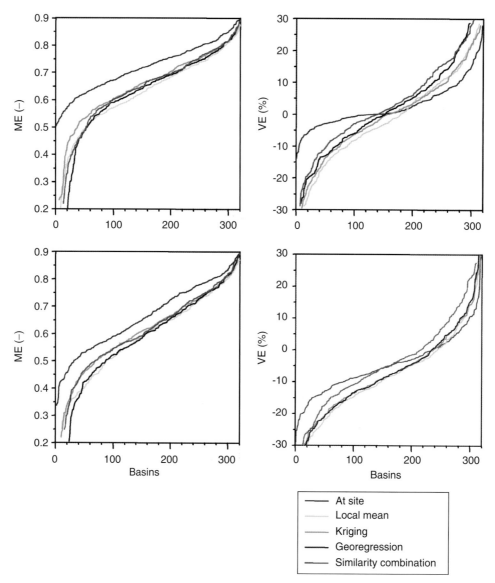

Figure 10 Runoff simulation performance in ungauged catchments for various parameter regionalization methods. Cumulative distribution functions of the model efficiencies of daily runoff (ME, left) and volume errors of runoff (VE, right) are shown. Three-hundred and twenty basins, calibration (top) and verification (bottom) periods. From Parajka J, Merz R, and Blöschl G (2005) A comparison of regionalisation methods for catchment model parameters. *Hydrology and Earth Systems Sciences* 9: 157–171.

approach (red line) where the complete set of model parameters is transposed from a donor catchment that is most similar in terms of its catchment attributes (mean catchment elevation, stream network density, lake index, areal proportion of porous aquifers, land use, soils, and geology).

One of the difficulties with this type of parameter regionalization is that it hinges on a robust calibration of the model parameters. This is similar to the problems with scaling of runoff models discussed in Section 2.19.3.4. As the calibration period increases in length, the parameters can be estimated more robustly; but there is a limit to the degree to which the parameters can be identified from the runoff data (see Figure 10 in Merz *et al.*, 2009). One possibility of obtaining more robust parameters is multiple objective calibration against, say, runoff and snow data (Parajka *et al.*, 2007b). An alternative is regional calibration where the model parameters of a number of catchments are calibrated simultaneously (Parajka *et al.*, 2007a). Still another alternative is based on so-called signatures of runoff such as the mean annual runoff (Zhang *et al.*, 2008). In this approach, the mean annual runoff is regionalized to the ungauged catchment in a first step. In a second step, the model parameters for the ungauged catchment are jointly estimated from the catchment attributes and the mean annual runoff. Another option are methods that optimize a transfer function relating model parameters to catchment characteristics (Hundecha and Bárdossy, 2004). The transfer function assures that similar catchments are mapped to similar parameter sets.

There are also difficulties with finding suitable catchment characteristics that are hydrologically relevant (see, e.g., Merz and Blöschl, 2009); and the spatial arrangements of soils and land use in the catchment may be important but are not captured by the methods mentioned above. A useful strategy hence is to infer the model parameters from field observations in the ungauged catchment. Although the scale difference between point measurements and model elements may be a problem (Blöschl, 2005b), there are frameworks for using local field information to assist in parameter estimation (e.g., Blöschl et al., 2008).

2.19.5 Concluding Remarks

This review has summarized current methods of scaling and regionalizing hydrological variables. What is common to scaling and regionalization is that there is no single best method that should be used in all cases. Rather, the method of choice should depend on the purpose of the study, data availability, and the nature of the underlying hydrological variability. The predictive accuracy of these methods will depend on the same factors. The accuracy will also depend on the degree of nonlinearity present in the system as pointed out by Blöschl and Zehe (2005) and Blöschl (2006). As the degree of nonlinearity increases, predictability tends to decrease. One of the exciting research fields in hydrological scaling and regionalization in the next years will hence be to learn how to separate the predictable and the unpredictable. This will help better understand just how well one can estimate hydrological quantities across space and scale.

Acknowledgment

The author would like to thank FWF project P18993-N10 for financial support.

References

Ababou R and Gelhar LW (1990) Self-similar randomness and spectral conditioning: Analysis of scale effects in subsurface hydrology. In: Cushman JH (ed.) *Dynamics of Fluids in Hierarchical Porous Media*, pp. 393–428. London: Academic Press.

Ababou R and Wood EF (1990) Comment on "Effective groundwater model parameter values: Influence of spatial variability of hydraulic conductivity, leakance, and recharge" by J.J. Gómez-Hernández and S.M. Gorelick. *Water Resources Research* 26: 1843–1846.

Anderson MP (1997) Characterization of geological heterogeneity. In: Dagan G and Neuman SP (eds.) *Subsurface Flow and Transport: A Stochastic Approach*, International Hydrology Series, pp. 23–43. Cambridge: Cambridge University Press.

Aschwanden H and Kan C (1999) Le débit d'étiage Q347 – Etat de la question. *Communications hydrologiques*, vol. 27, Service hydrol. et géol. National, Berne.

Beldring S, Roald LA, and Voksø A (2002) Avrenningskart for Norge (Runoff map for Norway, in Norwegian). *Norwegian Water Resources and Energy Directorate Report*, No. 2, Oslo, Norway.

Bellin A and Rubin Y (1996) Hydrogen: A spatially distributed random field generator for correlated properties. *Stochastic Hydrology and Hydraulics* 10(4): 253–278.

Beven K (1989) Changing ideas in hydrology – the case of physically based models. *Journal of Hydrology* 105: 157–172.

Beven K (2001) How far can we go in distributed hydrological modelling? *Hydrology and Earth System Sciences* 5: 1–12.

Beven KJ and Kirkby MJ (1979) A physically-based, variable contributing area model of basin hydrology. *Hydrological Science Bulletin* 24: 43–69.

Bierkens MFP, Finke PA, and de Willigen P (2000) *Upscaling and Downscaling Methods for Environmental Research*, 190pp. Dordrecht: Kluwer.

Blöschl G (1996) Scale and scaling in hydrology (Habilitationsschrift). Wiener Mitteilungen, Wasser-Abwasser-Gewässer, Band 132, Institut für Hydraulik, TU Wien, 346pp.

Blöschl G (1999) Scaling issues in snow hydrology. *Hydrological Processes* 13: 2149–2175.

Blöschl G (2001) Scaling in hydrology. Invited commentary. *Hydrological Processes* 15: 709–711.

Blöschl G (2005a) Statistical upscaling and downscaling in hydrology. In: Anderson MG (ed.) *Encyclopedia of Hydrological Sciences*, pp. 135–154. Chichester: Wiley.

Blöschl G (2005b) Rainfall-runoff modelling of ungauged catchments. In: Anderson MG (ed.) *Encyclopedia of Hydrological Sciences*, pp. 2061–2080. Chichester: Wiley.

Blöschl G (2006) Hydrologic synthesis – across processes, places and scales. Special section on the vision of the CUAHSI National Center for Hydrologic Synthesis (NCHS). *Water Resources Research* 42: W03S02.

Blöschl G and Grayson R (2000) Spatial observations and interpolation. In: Grayson R and Blöschl G (eds.) *Spatial Patterns in Catchment Hydrology: Observations and Modelling*, ch. 2, pp. 17–50. Cambridge: Cambridge University Press.

Blöschl G and Grayson R (2002) Advances in distributed hydrological modelling – towards a new paradigm. In: *Proceedings of the Third International Conference on Water Resources and Environment Research (ICWRER)*, vol. I, pp. 17–25. Dresden University of Technology, Dresden, Germany, 22–26 July 2002.

Blöschl G and Sivapalan M (1995) Scale issues in hydrological modelling – a review. *Hydrological Processes* 9: 251–290.

Blöschl G and Zehe E (2005) On hydrological predictability. Invited commentary. *Hydrological Processes* 19(19): 3923–3929.

Blöschl G, Grayson RB, and Sivapalan M (1995) On the representative elementary area (REA) concept and its utility for distributed rainfall-runoff modelling. *Hydrological Processes* 9: 313–330.

Blöschl G, Kirnbauer R, and Gutknecht D (1991) Distributed snowmelt simulations in an Alpine catchment. 1. Model evaluation on the basis of snow cover patterns. *Water Resources Research* 27: 3171–3179.

Blöschl G, Reszler C, and Komma J (2008) A spatially distributed flash flood forecasting model. *Environmental Modelling and Software* 23(4): 464–478.

Bobée B and Rasmussen PF (1995) Recent advances in flood frequency analysis. *Reviews of Geophysics* 33(S1): 1111–1116.

Breiman L, Friedman JH, Olshen R, and Stone CJ (1984) *Classification and Regression Trees*. Belmont, CA: Wadsworth International Group.

Budyko MI (1974) *Climate and Life*. 508pp. Orlando, FL: Academic Press.

Burn DH (1990) Evaluation of regional flood frequency analysis with a region of influence approach. *Water Resources Research* 26(19): 2257–2265.

Charles SP, Bates BC, Smith IN, and Hughes JP (2004) Statistical downscaling of daily precipitation from observed and modelled atmospheric fields. *Hydrological Processes* 18(8): 1373–1394.

Chilès J-P and Delfiner P (1999) *Geostatistics: Modeling Spatial Uncertainty*. Wiley Series in Applied Probability and Statistics, 695pp. New York: Wiley.

Chirico GB, Western AW, Grayson RB, and Blöschl G (2005) On the definition of the flow width for calculating specific catchment area patterns from gridded elevation data. *Hydrological Processes* 19(13): 2539–2556.

Clauser C (1992) Permeability of crystalline rocks. *EOS, Transactions of the American Geophysical Union* 73: 233–238.

Cunnane C (1988) Methods and merits of regional flood frequency analysis. *Journal of Hydrology* 100: 269–290.

Dagan G (1989) *Flow and Transport in Porous Formations*, 465pp. Berlin: Springer.

Dalrymple T (1960) *Flood Frequency Methods*, US Geological Survey Water-Supply Paper 1543-A, pp. 11–51. Washington, DC: US Government Printing Office.

de Marsily G (1986) *Quantitative Hydrogeology*, 440pp. San Diego, CA: Academic Press.

Deutsch CV and Journel AG (1997) *GSLIB, Geostatistical Software Library and User's Guide*, 340pp. New York, NY: Oxford University Press.

Dunne T (1978) Field studies of hillslope flow processes. In: Kirkby MJ (ed.) *Hillslope Hydrology*, pp. 227–293. Chichester: Wiley.

Dykaar BB and Kitanidis PK (1992) Determination of the effective hydraulic conductivity for heterogeneous porous media using a numerical spectral approach. 2. Results. *Water Resources Research* 28: 1167–1178.

El-Kadi A and Brutsaert W (1985) Applicability of effective parameters for unsteady flow in nonuniform aquifers. *Water Resources Research* 21: 183–198.

Famiglietti JS (1992) Aggregation and Scaling of Spatially-Variable Hydrological Processes: Local, Catchment-Scale and Macroscale Models of Water and Energy Balance, 207pp. PhD Dissertation, Princeton University.

Fan Y and Bras RL (1995) On the concept of a representative elementary area in catchment runoff. *Hydrological Processes* 9: 821–832.

Feder J (1988) *Fractals*, 283pp. New York, NY: Plenum.

Gelhar LW (1986) Stochastic subsurface hydrology from theory to applications. *Water Resources Research* 22: 135S–145S.

Gelhar LW (1993) *Stochastic Subsurface Hydrology*, 390pp. Englewood Cliffs, NJ: Prentice Hall.

Golden Software (2009) http://www.goldensoftware.com (accessed February 2010).

Gómez-Hernández JJ and Gorelick SM (1989) Effective groundwater model parameter values: Influence of spatial variability of hydraulic conductivity, leakance, and recharge. *Water Resources Research* 25: 405–419.

Goovaerts P (1997) *Geostatistics for Natural Resources Evaluation*. Applied Geostatistics Series, 483pp. New York, NY: Oxford University Press.

Grayson R and Blöschl G (eds.) *Spatial Patterns in Catchment Hydrology: Observations and Modelling* 404pp. Cambridge University Press.

Grayson R, Blöschl G, Western A, and McMahon T (2002) Advances in the use of observed spatial patterns of catchment hydrological response. *Advances in Water Resources* 25: 1313–1334.

Grayson RB, Blöschl G, Barling RD, and Moore ID (1993) Process, scale and constraints to hydrological modelling in GIS. In: Kovar K and Nachtnebel HP (eds.) *Applications of Geographic Information Systems in Hydrology and Water Resources Management*, Proceedings of the Vienna Symposium, April 1993, IAHS Publication No. 211, pp. 83–92. Paris: IAHS.

Gupta VK and Waymire EC (1993) A statistical analysis of mesoscale rainfall as a random cascade. *Journal of Applied Meteorology* 32: 251–267.

Gustard A and Irving KM (1994) Classification of the low flow response of European soils. In: *FRIEND: Flow Regimes from International Experimental and Network Data*, IAHS Publication No. 221, pp. 113–117. Paris: IAHS.

Harman CJ and Sivapalan M (2009) Similarity framework to assess controls on subsurface flow dynamics in hillslopes. *Water Resources Research* 45: W01417 (doi:10.1029/2008WR007067).

Hayes DC (1992) *Low Flow Characteristics of Streams in Virginia*, US Geological Survey, Water Supply Paper 2374, US Department of the Interior, US Geological Survey.

Hosking JRM and Wallis JR (1997) *Regional Frequency Analysis*, 224pp. Cambridge: Cambridge University Press.

Hubbert MK (1956) Darcy's law and the field equations of the flow of underground fluids. *Transactions of the American Institute of Mining Metallurgical and Petroleum Engineering* 207: 222–239.

Hundecha Y and Bárdossy A (2004) Modeling of the effect of land use changes on the runoff generation of a river basin through parameter regionalization of a watershed model. *Journal of Hydrology* 292: 281–295.

Hurst HE (1951) Long-term storage capacity of reservoirs. *Transactions of the American Society of Civil Engineers* 116: 770–808.

Indelman P, Fiori A, and Dagan G (1996) Steady flow toward wells in heterogeneous formations: Mean head and equivalent conductivity. *Water Resources Research* 32: 1975–1983.

IH (Institute of Hydrology) (2000) *Flood Estimation Handbook*. Wallingford: Institute of Hydrology.

IPCC (2007) *Climate Change 2007 – The Physical Science Basis*, 1009pp. Cambridge: Cambridge University Press.

Isaaks EH and Srivastava RM (1989) *An Introduction to Applied Geostatistics*. New York, NY: Oxford University Press.

Journel AG and Huijbregts CJ (1978) *Mining Geostatistics*. London: Academic Press.

Kirchner JW (1993) Statistical inevitability of Horton's laws and the apparent randomness of stream channel networks. *Geology* 21: 591–594.

Klemeš V (1974) The Hurst phenomenon: A puzzle? *Water Resources Research* 10: 675–688.

Koltermann CE and Gorelick SM (1996) Heterogeneity in sedimentary deposits: A review of structure-imitating, process-imitating, and descriptive approaches. *Water Resources Research* 32: 2617–2658.

Komma J, Reszler C, Blöschl G, and Haiden T (2007) Ensemble prediction of floods – catchment non-linearity and forecast probabilities. *Natural Hazards and Earth System Sciences* 7: 431–444.

Koutsoyiannis D (2002) The Hurst phenomenon and fractional Gaussian noise made easy. *Hydrological Sciences Journal* 47(4): 573–595.

Laaha G (2008) Aspekte der statistischen Modellierung raumbezogener Umweltdaten am Beispiel von Abflussdaten (Facets of statistically modelling spatial environmental data illustrated by discharge data). Habilitation Thesis, Universität für Bodenkultur, Vienna, 42pp plus annexes.

Laaha G and Blöschl G (2006a) A comparison of low flow regionalisation methods – catchment grouping. *Journal of Hydrology* 323: 193–214.

Laaha G and Blöschl G (2006b) Seasonality indices for regionalizing low flows. *Hydrological Processes* 20: 3851–3878.

Laaha G and Blöschl G (2007) A national low flow estimation procedure for Austria. *Hydrological Sciences Journal* 52(4): 625–644.

Lhomme J-P, Chehbouni A, and Monteny B (1996) Canopy to region scale translation of surface fluxes. In: Stewart JB, Engman ET, Feddes RA, and Kerr Y (eds.) *Scaling Up in Hydrology Using Remote Sensing*, pp. 161–170. Chichester: Wiley.

Loague KM and Freeze RA (1985) A comparison of rainfall-runoff modeling techniques on small upland catchments. *Water Resources Research* 21: 229–248.

Lovejoy S (1982) Area–perimeter relation for rain and cloud areas. *Science* 216: 185–187.

Hack JT (1957) *Studies of Longitudinal Stream Profiles in Virginia and Maryland*. US Geological Survey Professional Paper 294-B, 97pp. Washington, DC: US Government Printing Office.

Mantoglou A and Gelhar LW (1987) Capillary tension head variance, mean soil moisture content, and effective specific soil moisture capacity of transient unsaturated flow in stratified soils. *Water Resources Research* 23: 47–56.

Matheron G (1967) *Eléments pour une théorie des milieux poreux*. Paris: Masson (cited in de Marsily (1986)).

Menabde M and Sivapalan M (2001) Linking space–time variability of rainfall and runoff fields on a river network: A dynamic approach. *Advances in Water Resources* 24(9–10): 1001–1014.

Merz R and Blöschl G (2003) A process typology of regional floods. *Water Resources Research* 39(12): 1340.

Merz R and Blöschl G (2005) Flood frequency regionalisation – spatial proximity vs. catchment attributes. *Journal of Hydrology* 302(1–4): 283–306.

Merz R and Blöschl G (2009) A regional analysis of event runoff coefficients with respect to climate and catchment characteristics in Austria. *Water Resources Research* 45: W01405 (doi:10.1029/2008WR007163).

Merz R, Blöschl G, and Humer G (2008) National flood discharge mapping in Austria. *Natural Hazards* 46(1): 53–72.

Merz R, Parajka J, and Blöschl G (2009) Scale effects in conceptual hydrological modelling. *Water Resources Research* 45: W09405 (doi:10.1029/2009WR007872).

Mishra SK and Singh VP (2003) *Soil Conservation Service Curve Number (SCS-CN) Methodology*, Water Science and Technology Library, vol. 42. Dordrecht: Kluwer.

Montanari A (2007) What do we mean by 'uncertainty'? The need for a consistent wording about uncertainty assessment in hydrology. *Hydrological Processes* 21(6): 841–845 (doi:10.1002/hyp.6623).

Naef F (1981) Can we model the rainfall-runoff process today? *Hydrological Science Bulletin* 26: 281–289.

Nathan RJ and McMahon TA (1990) Identification of homogeneous regions for the purpose of regionalization. *Journal of Hydrology* 121: 217–238.

Neuman SP and Orr S (1993) Prediction of steady state flow in nonuniform geologic media by conditional moments: Exact nonlocal formalism, effective conductivities, and weak approximations. *Water Resources Research* 29: 341–364.

Paleologos EK, Neuman SP, and Tartakovsky D (1996) Effective hydraulic conductivity of bounded, strongly heterogeneous porous media. *Water Resources Research* 32: 1333–1341.

Parajka J, Blöschl G, and Merz R (2007a) Regional calibration of catchment models: Potential for ungauged catchments. *Water Resources Research* 43: W06406.

Parajka J, Merz R, and Blöschl G (2005) A comparison of regionalisation methods for catchment model parameters. *Hydrology and Earth Systems Sciences* 9: 157–171.

Parajka J, Merz R, and Blöschl G (2007b) Uncertainty and multiple objective calibration in regional water balance modeling – case study in 320 Austrian catchments. *Hydrological Processes* 21: 435–446.

Peschke G, Etzenberg C, Töpfer J, Zimmermann S, and Müller G (1999) Runoff generation regionalization: Analysis and a possible approach to a solution. In: Diekkrüger B, Kirkby MJ, and Schröder U (eds.) *Regionalization in Hydrology*, IAHS Publication No. 254, pp. 147–156. Wallingford: IAHS Press.

Piock-Ellena U, Pfaundler M, Blöschl G, Burlando P, and Merz R (2000) Saisonalitätsanalyse als Basis für die Regionalisierung von Hochwässern (Seasonality analyses as the basis for regionalising floods). *Wasser, Energie, Luft* 92: 13–21.

Raupach MR and Finnigan JJ (1995) Scale issues in boundary-layer meteorology: Surface energy balances in heterogeneous terrain. *Hydrological Processes* 9: 589–612.

Richardson LF (1961) The problem of contiguity. In: *General Systems Yearbook*, vol. 6, pp. 139–187. Washington, DC: Society for General Systems Research.

Rodríguez-Iturbe I, Ijjász-Vásquez EJ, Bras RL, and Tarboton DG (1992) Power law distributions of discharge mass and energy in river basins. *Water Resources Research* 28: 1089–1093.

R-software (2009) http://www.r-project.org (accessed February 2010).

Rubin Y (2003) *Applied Stochastic Hydrogeology*. 416pp. New York, NY: Oxford University Press.

Russo D (1992) Upscaling of hydraulic conductivity in partially saturated heterogeneous porous formation. *Water Resources Research* 28: 397–409.

Sánchez-Vila X, Carrera J, and Girardi JP (1996) Scale effects in transmissivity. *Journal of Hydrology* 183: 1–22.
Savenije HHG (2001) Equifinality, a blessing in disguise? Invited commentary. *Hydrological Processes* 15: 2835–2838.
Savenije HHG (2009) The art of hydrology. *Hydrology and Earth System Sciences* 13: 157–161.
Schaap JD, Lehmann P, Kaestner A, et al. (2008) Measuring the effect of structural connectivity on the water dynamics in heterogeneous porous media using speedy neutron tomography. *Advances in Water Resources* 31: 1233–1241.
Schmid HP (2002) Footprint modeling for vegetation atmosphere exchange studies: A review and perspective. *Agricultural and Forest Meteorology* 113: 159–183.
Schoups G, van de Giesen NC, and Savenije HHG (2008) Model complexity control for hydrologic prediction. *Water Resources Research* 44: 1–14 (W00B03) (doi:10.1029/2008WR006836).
SCS – Soil Conservation Service (1973) *A Method for Estimating Volume and Rate of Runoff in Small Watersheds*. Technical Paper 149. Washington, DC: US Department of Agriculture.
Sefton CEM and Howarth SM (1998) Relationships between dynamic response characteristics and physical descriptors of catchments in England and Wales. *Journal of Hydrology* 211: 1–16.
Seibert J (1999) Regionalisation of parameters for a conceptual rainfall-runoff model. *Agricultural and Forest Meteorology* 98–99: 279–293.
Sivapalan M (2003a) Process complexity at hillslope scale, process simplicity at the watershed scale: Is there a connection? *Hydrology Processes* 17: 1037–1041.
Sivapalan M (2003b) Prediction of ungauged basins: A grand challenge for theoretical hydrology. *Hydrological Processes* 17(15): 3163–3170.
Sivapalan M and Blöschl G (1998) Transformation of point rainfall to areal rainfall: Intensity–duration–frequency curves. *Journal of Hydrology* 204: 150–167.
Sivapalan M, Blöschl G, Merz R, and Gutknecht D (2005) Linking flood frequency to long-term water balance: Incorporating effects of seasonality. *Water Resources Research* 41: W06012.
Sivapalan M, Takeuchi K, Franks SW, et al. (2003) IAHS decade on predictions in ungauged basins (PUB), 2003–2012: Shaping an exciting future for the hydrological sciences. *Hydrological Sciences Journal* 48: 857–880.
Skøien JO and Blöschl G (2006a) Sampling scale effects in random fields and implications for environmental monitoring. *Environmental Monitoring and Assessment* 114(1–3): 521–552.
Skøien JO and Blöschl G (2006b) Scale effects in estimating the variogram and implications for soil hydrology. *Vadose Zone Journal* 5: 153–167.
Skøien J and Blöschl G (2006c) Catchments as space–time filters – a joint spatio-temporal geostatistical analysis of runoff and precipitation. *Hydrology and Earth System Sciences* 10: 645–662.
Skøien JO and Blöschl G (2007) Spatiotemporal topological kriging of runoff time series. *Water Resources Research* 43: W09419.
Skøien JO, Blöschl G, and Western AW (2003) Characteristic space scales and timescales in hydrology. *Water Resources Research* 39(10): 1304.
Skøien J, Merz R, and Blöschl G (2006) Top-kriging – geostatistics on stream networks. *Hydrology and Earth System Sciences* 10: 277–287.
Smakhtin VU (2001) Low flow hydrology: A review. *Journal of Hydrology* 240: 147–186.
Sreenivasan KR (1991) Fractals and multifractals in fluid turbulence. *Annual Reviews Fluid Mechanics* 23: 539–600.
Stephenson GR and Freeze RA (1974) Mathematical simulation of subsurface flow contributions to snowmelt runoff, Reynolds Creek Watershed, Idaho. *Water Resources Research* 10: 284–294.
Trinchero P, Sanchez-Vila X, and Fernandez-Garcia D (2008) Point-to-point connectivity, an abstract concept or a key issue for risk assessment studies? *Advances in Water Resources* 31: 1742–1753.
USACE (1994) *Engineering and Design – Flood-Runoff Analysis*. Publication No. EM 1110-2-1417. Washington, DC: US Army Corps of Engineers.
von Storch H (2005) Models of global and regional climate. In: Anderson MG (ed.) *Encyclopedia of Hydrological Sciences*, pp. 477–490. Chichester: Wiley.
Wagener T, Sivapalan M, Troch PA, and Woods RA (2007) Catchment classification and hydrologic similarity. *Geography Compass* 1/4: 901–931.
Webster R and Oliver MA (2001) *Geostatistics for Environmental Scientists*. 271pp. New York, NY: Wiley.
Wen X-H and Gómez-Hernández JJ (1996) Upscaling hydraulic conductivities in heterogeneous media: An overview. *Journal of Hydrology* 183: ix–xxxii.
Western A and Grayson R (2000) Soil moisture and runoff processes at Tarrawarra. In: Grayson R and Blöschl G (eds.) *Spatial Patterns in Catchment Hydrology: Observations and Modelling*, pp. 209–246. Cambridge: Cambridge University Press.
Western A, Grayson R, and Blöschl G (2002) Scaling of soil moisture: A hydrologic perspective. *Annual Review of Earth and Planetary Sciences* 30: 149–180.
Western AW and Blöschl G (1999) On the spatial scaling of soil moisture. *Journal of Hydrology* 217: 203–224.
Western AW, Blöschl G, and Grayson RB (1998) How well do indicator variograms capture the spatial connectivity of soil moisture? *Hydrological Processes* 12: 1851–1868.
Western AW, Blöschl G, and Grayson RB (2001) Towards capturing hydrologically significant connectivity in spatial patterns. *Water Resources Research* 37(1): 83–97.
Wigmosta MS, Vail LW, and Lettenmaier DP (1994) A distributed hydrology-vegetation model for complex terrain. *Water Resources Research* 30: 1665–1679.
Wilby RL and Wigley TML (2000) Precipitation predictors for downscaling: Observed and general circulation model relationships. *International Journal of Climatology* 20: 641–661.
Wilby RL, Conway D, and Jones PD (2002) Prospects for downscaling seasonal precipitation variability using conditioned weather generator parameters. *Hydrological Processes* 16: 1215–1234.
Wood EF, Sivapalan M, Beven K, and Band L (1988) Effects of spatial variability and scale with implications to hydrologic modeling. *Journal of Hydrology* 102: 29–47.
Yarnal B, Comrie AC, Frakes B, and Brown DP (2001) Developments and prospects in synoptic climatology. *International Journal of Climatology* 21: 1923–1950.
Zehe E and Blöschl G (2004) Predictability of hydrologic response at the plot and catchment scales: Role of initial conditions. *Water Resources Research* 40: W10202 (21pp).
Zehe E, Elsenbeer H, Lindenmaier F, Schulz K, and Blöschl G (2007) Patterns of predictability in hydrological threshold systems. *Water Resources Research* 43: W07434.
Zhang Z, Wagener T, Reed P, and Bhushan R (2008) Reducing uncertainty in predictions in ungauged basins by combining hydrologic indices regionalization and multiobjective optimization. *Water Resources Research* 44: W00B04 (doi:10.1029/2008WR006833).
Zorita E and von Storch H (1999) The analog method as a simple statistical downscaling technique: Comparison with more complicated methods. *Journal of Climate* 12: 2474–2489.

2.20 Stream–Groundwater Interactions

KE Bencala, US Geological Survey, Menlo Park, CA, USA

Published by Elsevier B.V.

2.20.1	Introduction	537
2.20.1.1	The Stream Connected to Its Surroundings – The Stream Is Not a Pipe	537
2.20.1.2	Unidirectional Dominance, but Not Preclusion	537
2.20.2	Hydrology – Range of Interactions	537
2.20.2.1	Perspective	537
2.20.2.2	Hyporheic	538
2.20.2.3	Stream–Catchment	539
2.20.2.4	River–Aquifer	540
2.20.3	Chemical and Ecological Significance	540
2.20.3.1	Solute Chemistry	540
2.20.3.1.1	Non-nutrients	540
2.20.3.1.2	Nutrients	541
2.20.3.2	Ecosystem	542
2.20.3.2.1	Temperature	542
2.20.3.2.2	Biota	542
2.20.4	Field Study Methods and Models	543
2.20.5	Summary and Future Challenges	543
References		544

2.20.1 Introduction

2.20.1.1 The Stream Connected to Its Surroundings – The Stream Is Not a Pipe

Streams exist in connection to their surroundings, specifically to their catchment and the subsurface flows of the catchment. This statement appears to be obvious, particularly if we are envisioning the typical gaining stream increasing in water flow and acquiring solute load as the stream channel proceeds down-valley. A dated view of a stream can lead to an (un-stated) assumption that a stream acts predominantly as a pipe, in effect, open only in certain areas to receive water and draining the catchment at the stream's base. A stream is, however, not a pipe (Bencala, 1993). It is now well recognized that surface waters are connected to groundwater systems in a variety of hydrologic settings and at a variety of scales (Winter et al., 1998; Figure 1). This chapter identifies the diversity and dynamic nature of these connections.

2.20.1.2 Unidirectional Dominance, but Not Preclusion

As represented by two recent journal issues that focused on stream–groundwater interactions (Krause et al., 2009; Borchardt and Pusch, 2009), the topic is of active current hydrologic research and environmental interest. In working to understand the functional significance of stream–groundwater interactions, it is important to recognize, and accept, that these interactions are typically not the dominant process in the stream. Unidirectional, down-stream, down-valley surface and subsurface water flow constitute the context in which stream–groundwater interactions occur. However, the predominant direction of flow in a catchment does not preclude the importance of other flow paths including those of water into (and then back out of) the streambed and into subsurface riparian areas. Further, while the breadth of the interactions can be discussed in a primarily stream-transport-centric viewpoint (e.g., Bencala, 2006; Figure 2), a larger physical scale view explores the interactions from the hydrogeologic perspective of the groundwater system (Woessner, 2000; Sophocleous, 2002; Figure 3).

At the hydrologic scales of extensive aquifers and rivers, issues of water volumes transferred between the subsurface and surface systems seasonally to inter-annually are of importance (e.g., Konikow and Bredehoeft, 1974; Alley et al., 2002). With this chapter focusing on streams and catchment groundwater, the importance of solute exchanges (Runkel et al., 2003), nutrient cycling (Jones and Mulholland, 2000), and functions in aquatic ecosystems such as connectivity and maintenance of habitat structure (Stanford and Ward, 1988) are accentuated.

2.20.2 Hydrology – Range of Interactions

2.20.2.1 Perspective

A variety of perspectives on surface-water–groundwater interactions can be taken, even when focusing on the issues of water and solute exchanges in streams. Conceptually, different possible perspectives can be arrived at according to where you stand (Packman and Bencala, 2000). Are the interactions viewed from the stream, from the streambed interface, or from the subsurface? From each of these perspectives, interactions occur conceptually within a boundary layer (Triska et al., 1989) or across an ecotone (Boulton et al., 1998). Additionally, there are surface-water–groundwater interactions that can be viewed (1) from the interface being essentially a reactive membrane which water and solutes traverse (Schindler and Krabbenhoft, 1998) or (2) from the hydraulic connection

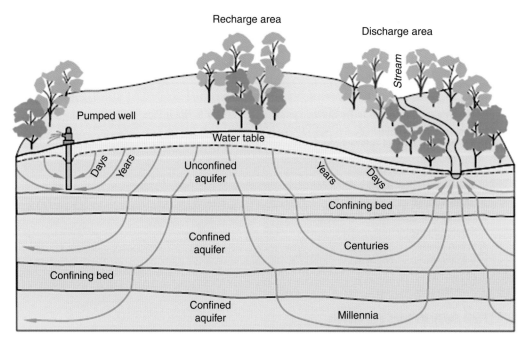

Figure 1 Groundwater flow paths greatly vary in length, depth, and travel time from points of recharge to points of discharge in the groundwater system. Reproduced from Winter TC, Harvey JW, Franke OL, and Alley WM (1998) *Ground Water and Surface Water: A Single Resource*, US Geological Survey Circular 1139. Denver, CO: US Geological Survey.

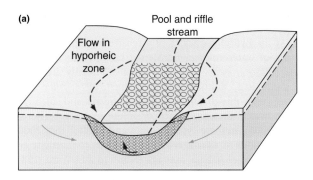

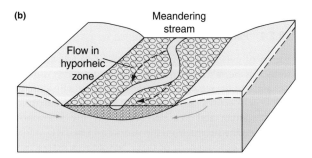

Figure 2 Surface-water exchange with groundwater in the hyporheic zone is associated with abrupt changes in streambed slope (a) and with stream meanders (b). Reproduced from Winter TC, Harvey JW, Franke OL, and Alley WM (1998) *Ground Water and Surface Water: A Single Resource*, US Geological Survey Circular 1139. Denver, CO: US Geological Survey.

causing water to flow from the stream into bank storage (Sharp, 1977) as stream stage rises and from the banks and streambed into the stream as the stage falls. Conceptually, these interactions are quite specific to a given stream. Wörman *et al.* (2006, 2007) have used spectral analyses to significantly generalize solute retention times and effluxes from the channel form to the continents in terms of power law scaling that reflects the fractal nature of surface topography across these scales.

The hydrology of surface- and groundwater interactions is summarized at three scales in Sections 2.20.2.2–2.20.2.4 (hyporheic, stream–catchment, and river–aquifer). Section 2.20.3 addresses the central topic of chemical and ecological significance of surface- and groundwater interactions.

2.20.2.2 Hyporheic

In the hyporheic zone, stream flow infiltrates into the shallow subsurface material forming the channel bed and banks, flows following the general down-valley gradient, and then returns to stream. Although there are no rigorous criteria, hyporheic flow paths are commonly thought of as being tens of meters in length with residence times on the order of hours to days. Variation in stream and catchment characteristics such as hydraulic conductivity, alluvial volume, streambed slope, and turbulence help drive hyporheic flows (Tonina and Buffington, 2009).

Hyporheic flow is routinely observed in various stream settings including gravel-bed steep mountain streams (Bencala, 1984) and naturally ephemeral, wastewater-treatment-dominated, sand-bed streams (Cox *et al.*, 2003). At

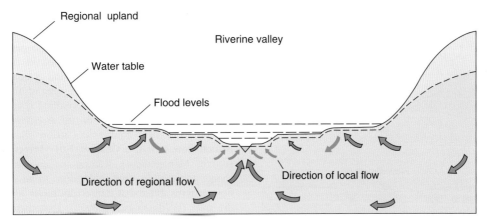

Figure 3 In broad river valleys, small local groundwater flow systems associated with terraces overlie more regional groundwater flow systems. Recharge from floodwaters superimposed on these groundwater flow systems further complicates the hydrology of river. Reproduced from Winter TC, Harvey JW, Franke OL, and Alley WM (1998) *Ground Water and Surface Water: A Single Resource*, US Geological Survey Circular 1139. Denver, CO: US Geological Survey.

the stream channel scale, stepped-channel morphology (Harvey and Bencala, 1993) can set up local hydraulic gradients that define hyporheic flow paths originating at the downstream end of stream pools and returning water to the stream at the base of subsequent steep riffle sections. Larger-scale features of the catchment such as geologic setting and alluvial characteristics control substantial variability observed in hyporheic zone extent and functioning (Morrice et al., 1997). Typically, the water in a hyporheic flow path is a mixture of stream water and local groundwater. Hyporheic flows exist in both surface-water and groundwater-dominated sites (Malcolm et al., 2009). Variation in stream discharge also influences mixture, pathways, and flux of hyporheic flow (Arntzen et al., 2006; Wondzell and Swanson, 1996, 1999).

2.20.2.3 Stream–Catchment

At larger scales, the hyporheic zone represents the boundary across which streams and their catchment exchange water and solutes. Typically, streams are envisioned gaining water from catchment runoff. However, a stream may be both gaining and losing water (**Figure 4**). The relationship between stream discharge and groundwater flow may shift across the mountain to alluvial valley transition (Anderson et al., 1992; Covino and McGlynn, 2007). Fluvial processes and geomorphic features also result in episodic to seasonal modification (Malard et al., 2002). The stream–catchment system may be visualized as settings in a three-dimensional mosaic of multiple scale patches. These patches exhibit cycles of expansion and contraction resulting in differences among patches in their hydrologic exchange rate with the stream and thus the supply of organic matter.

Localized interactions are influenced by (1) groundwater-dominated hydrology, (2) surface-water-dominated hydrology, and (3) transient water tables (Malcolm et al., 2005). These interactions are within the context of valley geomorphology (constrictions and channel confinement), which influences the spatial variation of groundwater movement to the stream. Geomorphic structure affects the bed sediments and flow pathways that govern exchange and mixing with

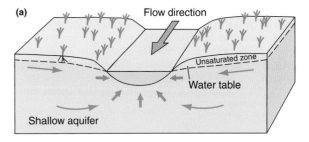

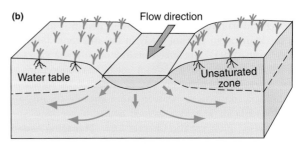

Figure 4 (a) Gaining streams receive water from the groundwater system. (b) Losing streams lose water to the groundwater system. Reproduced from Winter TC, Harvey JW, Franke OL, and Alley WM (1998) *Ground Water and Surface Water: A Single Resource*, US Geological Survey Circular 1139. Denver, CO: US Geological Survey.

floodplains and riparian areas (Huggenberger et al., 1998). Connectivity with the main channel water source varies within alluvial flood plains (Ward et al., 1999) and in turn influences the exchange processes in a specific channel (**Figure 5**).

Stream–groundwater connections do occur in systems losing water and in which flow is not maintained throughout the year. In arid regions, stream–catchment connections may center on losing streams with water spreading into the parafluvial area and later returning to the stream (Dent et al., 2007). Similar connections of streams and groundwater are induced in irrigated areas of the western United States (Fernald and Guldan, 2006; Konrad, 2006). After irrigation flooding, seepage water in the shallow subsurface may return to the river.

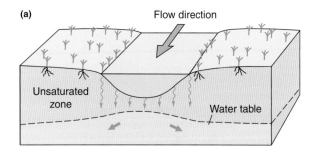

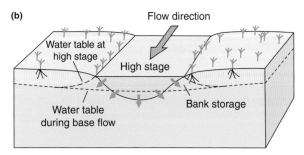

Figure 5 (a) Disconnected streams are separated from the groundwater system by an unsaturated zone. (b) If stream levels rise higher than adjacent groundwater levels, stream water moves into the stream banks as bank storage. Reproduced from Winter TC, Harvey JW, Franke OL, and Alley WM (1998) *Ground Water and Surface Water: A Single Resource*, US Geological Survey Circular 1139. Denver, CO: US Geological Survey.

Losing and gaining reaches have been identified in streams in the Midwestern US, not solely in the arid zones (Silliman and Booth, 1993). Streambed infiltration at a site varies diurnally (Constantz, 1998). Stream–catchment connections are not necessarily permanent. The permanence of subsurface hydrologic connections between non-navigable streams and nearby wetlands is a legal as well as hydrologic issue (Leibowitz *et al.*, 2008; see also **Chapter 1.03 Managing Aquatic Ecosystems**).

2.20.2.4 River–Aquifer

A common assumption of river–aquifer interaction is of subsurface flow to the river. In some cases, down-valley flow, or underflow may dominate. Geomorphic characteristics of channel slope, river sinuosity, incision through the alluvium, width-to-depth ratio, and the fluvial depositional system determine settings in which these flows may be important (Larkin and Sharp, 1992). Aquifer heterogeneity at the scale of 100 m influences spatial distribution of seepage between the river and the aquifer. The arrangement of hydrofacies impacts the connectivity between the river and the aquifer (Fleckenstein *et al.*, 2006). Secondary river channels create zones of hyporheic interaction on the scale of kilometers. Water exchanges in backwaters are related to local variations in streambed porosity. In the Upper Rhone, Germany, groundwater dominates within the exchange flows (Dole-Olivier, 1998).

In rivers of the Northwestern United States, exchanges with aquifers can be a significant aspect of the water budget (Konrad, 2006). The gaining and losing zones along the river are concentrated in a small fraction of the river's length, associated with specific geomorphic and geologic forms (lithologic contacts, tributaries, and thinning of alluvial deposits). Groundwater flow was often longitudinal along the valley axis, rather than directly lateral into the river channel. Irrigation can influence river–aquifer interaction with alterations to natural flow (see also **Chapter 1.01 Integrated Water Resources Management**), leading to both groundwater discharge and surface-water infiltration. In an irrigated system, flow patterns differed at transect locations 100 m apart along the river (Wildman *et al.*, 2009). Hypothetical simulations demonstrated variation in river depletion between settings with confined versus leaky aquifers, which can result from groundwater pumping (Butler *et al.*, 2007). In large-scale interactions between surface water and groundwater in the Everglades, Florida, USA, the observed fluxes of water occur on a decadal timescale (Harvey *et al.*, 2006); water may be resident in the subsurface for years prior to reemergence to the surface.

2.20.3 Chemical and Ecological Significance

2.20.3.1 Solute Chemistry

2.20.3.1.1 Non-nutrients

In natural systems, hyporheic exchange can modulate stream major ion chemistry. In an extreme example of a minimally impacted environment, an experiment using the injection of tracer cations and anions into an Antarctica Dry Valley stream demonstrated the role of hydrologic exchange in increasing ion sorption from the stream water (Gooseff *et al.*, 2004a). Stream–groundwater interactions also regulate water quality in ways that are significant for human impacts on hydrologic systems and human uses of water. Natural attenuation processes in hyporheic flows may limit the movement and availability of mining-derived pollutants at the stream–groundwater interface (Gandy *et al.*, 2007). In Silver Bow Creek, Montana, depletion of mine drainage constituents indicates precipitation, or adsorption occurs along hyporheic flow paths (Benner *et al.*, 1995). Reactive uptake of several dissolved metals by manganese oxide was observed in Pinal Creek, Arizona (Fuller and Harvey, 2000). In Pinal Creek, reactive uptake resulting from hyporheic exchange accounted for removal of approximately 20% of the dissolved manganese flowing out of the drainage basin (Harvey and Fuller, 1998). The role of the hyporheic zone in the transport of arsenic varies spatially in a Virginia mine-influenced stream–aquifer system. Arsenic is retained in hyporheic sediments along segments of the stream, while in other segments arsenic was delivered to the stream by hyporheic flow through mine tailings (Brown *et al.*, 2007).

Bank filtration is a widely utilized scheme in European water supply systems (see also **Chapter 3.04 Emerging Contaminants**). Knowledge of the processes determining geochemical transport is important for human health but remains largely site specific (Hiscock and Grischek, 2002). Infiltration of river water to groundwater is an important water supply component in the River Glatt, Switzerland. With river water transporting anthropogenic ligands, the remobilization of trace metals along the subsurface flow path from the river is shown to be possible (Nowack et al., 1997). Seasonal cycles in the mobilization and precipitation of manganese occurred as river water entered the aquifer adjacent to the River Glatt, Switzerland (Von Gunten et al., 1991). In a study reach of the Cedar River, Iowa, although tributaries aggregate most of the agricultural tile-drain flows, the alluvial aquifer, through bank storage retention and release, contributes a significant portion of the pesticides atrazine and deethylatrazine to the river (Squillace et al., 1993).

2.20.3.1.2 Nutrients

In their textbook, *Stream Ecology: Structure and Function of Running Waters*, Allan and Castillo (2007) introduced the coupling between nutrient dynamics and the physical movement of water (**Figure 6**). The transport of water through a stream–catchment system influences the residence time of conservative solutes and nutrients alike. Dahm et al. (1998), in their review, further discussed the stream–catchment geomorphology establishing the framework in which the surface-water–groundwater interface influences the transport of nutrients. Study of nutrient dynamics at scales from a few meters to several kilometers demonstrates both the significance and the inherent patch-nature of subsurface process influence on stream nutrient dynamics (Dent et al., 2001).

Chloride tracer and nitrate injected into the stream, but observed in hyporheic flow paths, demonstrate that hyporheic flows are an integral component of fluvial structure and function (Triska et al., 1989). The lithology of the material comprising the subsurface hydrologic system as expressed in variation in hydraulic conductivity influences the retention and transformation of nutrients. This coupling of nutrient dynamics to physical processes illustrates the importance of understanding streams as parts of their catchment system (Valett et al., 1996). The geology of a stream's catchment establishes alluvial hydrologic properties that determine the nature of surface-water–groundwater interactions influencing nutrient retention (Valett et al., 1997).

Hydrologic processes contribute to stream–catchment nutrient dynamics. Denitrification observed in shallow hyporheic sediments is controlled by hydrologic exchange (Duff and Triska, 1990). Dissolved oxygen supplied by hyporheic exchange with the stream is reduced through respiration of groundwater-derived dissolved organic carbon (DOC), enabling redox conditions conducive for denitrification. In extreme stream environments in the McMurdo Dry Valleys, Antarctica, nitrate is removed through exchange with both microbial mats as well as denitrification in the hyporheic zones (Gooseff et al., 2004b). Short-term hydrologic events that drive stream flow into the hyporheic zone, for example, the passage of a flood wave, may alter the flow dynamics of the surface-water–groundwater interface creating the opportunity for denitrification to occur in the sediments (Gu et al., 2008).

Multiple flow paths may exist within the stream. In a given groundwater–riparian–stream system, for some constituents, concentrations entering the stream may be controlled by the groundwater flowing through the riparian area. For other constituents, transformation may occur with the near-stream zone (Hill, 1990). Hyporheic zone biogeochemistry varies among stream ecosystems. In streams with high nitrate concentrations, nutrient removal may occur at the stream–streambed interface, leaving an unused potential for nitrogen

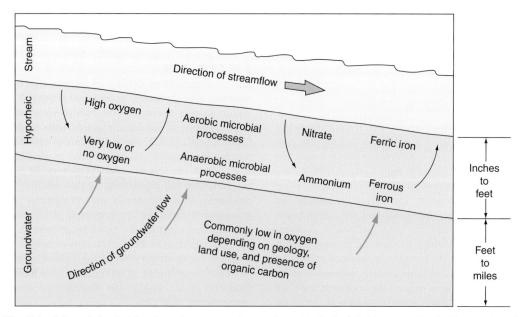

Figure 6 Microbial activity and chemical transformations commonly are enhanced in the hyporheic zone compared with those that take place in groundwater and surface water. This diagram illustrates some of the processes and chemical transformations that may take place in the hyporheic zone. Actual chemical interactions depend on numerous factors including aquifer mineralogy, shape of the aquifer, types of organic matter in surface water and groundwater, and nearby land use. Reproduced from Winter TC, Harvey JW, Franke OL, and Alley WM (1998) *Ground Water and Surface Water: A Single Resource*, US Geological Survey Circular 1139. Denver, CO: US Geological Survey.

depletion in the hyporheic zone at depth (Hill et al., 1998). In addition to the coupling of hydrology and nitrate supply, nitrate removal in stream ecosystems may be dependent on the supply of organic carbon (Hill et al., 2000).

In the South Platte River alluvial aquifer, Colorado, denitrification and mixing of river and groundwaters may substantially decrease the nitrate load transported in the river (McMahon and Bohlke, 1996). Phosphorus uptake in forested streams may also be increased with the occurrence of hyporheic zones (Mulholland et al., 1997). The importance of hydrologic mixing in the subsurface must be accounted for in the nutrient budget of the hydrologic system (Pinay et al., 1998). In floodplain of the River Garonne, France, mixing of river water and groundwater leads to subsurface nitrate concentrations lower than in the groundwater discharging to the river. Also in the floodplain on the River Garonne, France, denitrification rates were higher during snowmelt when subsurface mixing with river water was greatest than during low-flow periods (Baker and Vervier, 2004). Both catchment flow path and in-stream processes work to control stream water nutrient and DOC concentrations. The dominance of each set of processes may shift over the annual hydrologic cycle (Mulholland and Hill, 1997). DOC transported by groundwater can be immobilized in the streambed. This DOC is then available to the stream trophic system (Fiebig and Lock, 1991).

Fluctuations in river stage, sufficient to alter the direction of groundwater–river-water exchange, can establish conditions in which the residence time of solutes in the streambed is increased (McMahon et al., 1995). This leads to an enhanced uptake of oxygen from the river water. In low-gradient streams, nutrient-rich stream water may be retained in the streambed, possibly for days (Puckett et al., 2008). High-flow events can alter the groundwater–stream water exchange and increase the movement of stream water into the streambed. Although a stream may, over its length, be predominately losing water to the subsurface, exchange between the stream and streambed can be an ongoing process. The amount of denitrification in the stream is strongly influenced by the physical transport exchange (Ruehl et al., 2007).

2.20.3.2 Ecosystem

The location of and fluxes across the river–groundwater interface are variable in space and dynamic in time. The geomorphic components can be considered as being linked in a hydrologic continuum (Brunke and Gonser, 1997). The interactions that establish ecological conditions (e.g., nutrient availability, water temperature, and dissolved oxygen) are functionally part of both the fluvial and the groundwater systems. The interface and the physicochemical gradients within it are the result of characteristics and processes occurring across spatial scales. The interface can function as a source and/or a sink for solutes. Hyporheic zone processes and the resulting ecosystem functions occur at the catchment scale, the reach scale, and at the sediment scale (Boulton et al., 1998; see also **Chapter 2.10 Hydrology and Ecology of River Systems**). The continuing challenge is to link the extensive knowledge at each scale to the whole.

Groundwater–surface water connections have a role in the restoration of stream ecosystems. Adding small dams in restoration projects may promote the formation of hot spots of biogeochemical activity in the streambed (Lautz and Fanelli, 2008). The groundwater–surface-water interface may be significantly degraded as a consequence of human activities. To restore streams to their natural ecosystem function, the interface needs to be considered in addition to the surface-water channel form (Hester and Gooseff, 2010). Wood in the stream channel has a role in stream ecosystem function and consequently in restoration. The dynamics of the hyporheic zone are sensitive to the wood load in the channel. Initially, following wood removal, the streambed may be scoured, thus diminishing hyporheic exchange. As the stream adjusts to a reset wood load, sediment accretion may establish new hyporheic zones (Wondzell et al., 2009b).

2.20.3.2.1 Temperature

Due to groundwater–surface water connections, the temperature patterns at the scale of riffles may vary seasonally. Successive riffles along a stream may exhibit different patterns between them. The influence of both the immediate riffle structure and the larger-scale groundwater flows can be observed in temperature patterns (Hannah et al., 2009). Hyporheic-zone temperature patterns may be buffered and lagged relative to the diel cycle of the main channel water (Arrigoni et al., 2008). On a seasonal basis, water returning to the main channel may be either warmer or cooler than the upstream water originally in the channel. The variety of flow-path lengths and residence times point to the need to study streams in the context of their catchments. In the Clackamas River, Oregon, the influence of hyporheic flow on the temperature of the main river channel is small; however, the existence of hyporheic exchange contributes locally to the presence of patches of cooler water (Burkholder et al., 2008). Such patches of cool water may be ecologically important in providing thermal refugia for fish and other aquatic organisms (Torgersen et al., 1999).

2.20.3.2.2 Biota

Aquatic communities develop in response to chemical, geomorphic, and hydraulic conditions (Power et al., 1999). Within a river's catchment connectivity, variation contributes to the creation of the habitat mosaic sustaining biodiversity (Brunke et al., 2003). The incubation of salmonid embryos depends on a supply of oxygen within riverbed gravels (Greig et al., 2007). The flux of oxygen is determined by several factors including bed topography, bed permeability, and surface roughness. Thus, the details and complexities of river- and groundwater exchange are important for establishing this habitat. The variation in groundwater–surface-water interactions contributes to the survival rates of salmonids. In a river system in Scotland, ova survival was reduced in stream reaches in which the influence of groundwater was most substantial (Malcolm et al., 2003). The direction of dominant flow (groundwater discharge vs. stream-water infiltration) was variable with season as well as during hydrologic events.

Hyporheic zones may provide advantages of enriched oxygenated water and moderated temperature for salmon redd sites and spawning habitats. In identifying river characteristics favorable for fish habitat, gravel-bed vertical hydraulic gradient may be a significant factor (Geist and Dauble, 1998). Groundwater–stream-water interactions are significant for the quality of fish habitat at catchment, valley segment, reach, and pool-riffle scales. The presence of groundwater inflows appears significant at the larger scales, while strong intergravel flows and zones of downwelling are characteristic of the location of redd sites (Baxter and Hauer, 2000).

2.20.4 Field Study Methods and Models

Standard methods for the study of groundwater flow can be applied to the range of scales of stream–groundwater interactions. Dahm et al. (2007) review these applications for measuring vertical hydraulic gradients, and hydraulic conductivity. Modern geophysical methods, for example, ground-penetrating radar (Brosten et al., 2009), are increasingly deployed to provide the detailed three-dimensional geometry of the subsurface zone of interaction (see also **Chapter 2.13 Field-Based Observation of Hydrological Processes** and **Chapter 2.15 Hydrogeophysics**).

At the scale of streams and hyporheic exchange, injected tracers (Zellweger et al., 1989), combined with standard physical flow metering, are used to estimate the flow of water in gravel zones adjacent to the open channel. Hyporheic scale exchange can occur along stream reaches which overall are either gaining or losing water. In a stream system with significant, sustained flow loss, physical differential gauging is used to estimate net flow loss. Combined with tracer injections (Ruehl et al., 2006), exchange can also be estimated separately from the loss. Schemes of multiple injections have been devised (Zellweger, 1994; Payn et al., 2009) for estimating the loss. Tracer injections (Castro and Hornberger, 1991) are used to demonstrate the exchange of water and storage solutes from the stream with both short-term areas in the gravel beds and longer-term areas in deeper alluvium. Naturally occurring radon activities are used to estimate groundwater discharge to surface waters. In a river with hyporheic exchanges, the use of injected volatile tracer was used (Cook et al., 2006) to make the corrections for the apparent losses of radon into the hyporheic zone. For identifying connections at the river scale to the regional groundwater, isotopic analyses (Hinkle et al., 2001) are employed (see also **Chapter 2.09 Tracer Hydrology**). Often tracers of flow path are specifically sought for their conservative, minimally reactive properties. An innovative extension is the use of smart tracers (Haggerty et al., 2008) which convert to an alternative chemical form when flow is into microbiological active zones under mildly reducing conditions.

In a manner analogous to the use of chemical tracers, heat is becoming widely used as a tracer in the determination of stream–groundwater exchanges. Constantz (2008) reviews many of these techniques, for example, the use of temperature time-series phase and amplitude analysis (Hatch et al., 2006) to determine seepage, augmentation of temperature data with regulator-mandated water-quality data (Cox et al., 2007), and mapping by Essaid et al. (2008) of spatial temperature profiles to infer near-stream stratigraphic layering. Modern geophysical methods are also deployed to explicitly use information in the temperature signal. Westhoff et al. (2007) describe one of several recent uses of a distributed temperature-sensing fiber-optic cable system to identify sources of subsurface lateral inflow to a stream. Remotely sensed thermal spatial distributions in a stream can be determined using hand-held infrared (IR) imaging cameras (Cardenas et al., 2008a).

As with the standard physical methods for the study of groundwater flow, the standard modeling tools are also applied to stream–groundwater interactions. Applications of modular finite-difference flow (MODFLOW) model (see, e.g., Harbaugh et al., 2000) have been at the hyporheic scale (Lautz and Siegel, 2006) as well as for complex hydraulic analysis of intermittent and ephemeral streams (Niswonger et al., 2008). Wondzell et al. (2009a) examined the issue of validation of groundwater flow models of hyporheic interactions.

The transient storage model (TSM) is a tool used primarily at the hyporheic zone scale of stream–groundwater interactions in the analysis of solute dynamics. The concepts of the model, along with presentation of methods for application with field tracer studies, are given in Webster and Valett (2007). A widely used version of the TSM is available as the one-dimensional transport with inflow and storage (OTIS) code (Runkel, 1998). Runkel (2002) developed a metric based on the parameters of the TSM that provides an indication of the significance of exchange processes to solute dynamics of a stream. Harvey et al. (1996) and Harvey and Wagner (2000) provide quantitative guidance for interpreting the degree of confidence with which the TSM parameters can be determined. The importance of estimation of the lateral inflows in the interpretations of TSM parametrizations was demonstrated by Scott et al. (2003). Although the TSM is most widely used as a tool for parametrization of stream–hyporheic exchange with conservative tracer studies, it has been extended for use with Michaelis–Menten kinetics (Kim et al., 1992; Claessens and Tague, 2009) and geochemical equilibrium (Runkel, 2010).

2.20.5 Summary and Future Challenges

Stream–groundwater interactions are now recognized as significant components of stream aquatic ecosystems. In simple terms, the hydrology underlying this significance may be summarized as "The stream is not a pipe." and "Ground water and surface water: A single resource." This recognition has been built upon decades of field studies demonstrating the ongoing connections between streams and their catchments. Considerable opportunities remain for new and challenging hydrologic work. Flume studies of environmental fluid mechanics (Marion et al., 2008) with sophisticated simulations (Cardenas et al., 2008b) have demonstrated that streambed forms, substrate distributions, and flow conditions can be set up to drive exchange of water and solutes between free-flowing surface and subsurface zones. An actual stream, however, is highly variable in these characteristics. It is the very heterogeneity of the subsurface media and the patchy biogeochemical conditions that create the environments in which the aquatic ecosystem influences are manifest. Stream–groundwater interactions are, in general, secondary processes compared with the bulk water flow or solute transport. Thus, a sequence of challenges is present in first applying mechanistic understanding to measurement of characteristics and processes that vary at relatively fine scales in the field and then in this information to interpret the effects at the scale of the stream–catchment system. A recent series of review articles makes clear that the study of stream–groundwater interactions is truly a multidisciplinary challenge requiring an understanding of the hydrogeomorphology (Poole, 2010), solute dynamics (Mulholland and Webster, 2010), and implications for stream ecology and management (Boulton et al., 2010).

References

Allan JD and Castillo MM (2007) *Stream Ecology: Structure and Function of Running Waters*, 2nd edn. Dordrecht: Springer.

Alley WM, Healy RW, LaBaugh JW, and Reilly TE (2002) Flow and storage in groundwater systems. *Science* 296: 1985–1990.

Anderson TW, Freethey GW, and Tucci P (1992) *Geohydrology and Water Resources of Alluvial Basins in South-Central Arizona and Parts of Adjacent States*, US Geological Survey Professional Paper 1406-B, 67p. Denver, CO: US Geological Survey.

Arntzen EV, Geist DR, and Dresel PE (2006) Effects of fluctuating river flow on groundwater/surface water mixing in the hyporheic zone of a regulated, large cobble bed river. *River Research and Applications* 22: 937–946.

Arrigoni AS, Poole GC, Mertes LAK, O'Daniel SJ, Woessner WW, and Thomas SA (2008) Buffered, lagged, or cooled? Disentangling hyporheic influences on temperature cycles in stream channels. *Water Resources Research* 44: W09418 (doi:10.1029/2007WR006480).

Baker MA and Vervier P (2004) Hydrological variability, organic matter supply and denitrification in the Garonne River ecosystem. *Freshwater Biology* 49: 181–190.

Baxter CV and Hauer FR (2000) Geomorphology, hyporheic exchange, and selection of spawning habitat by bull trout (*Salvelinus confluentus*). *Canadian Journal of Fisheries and Aquatic Sciences* 57: 1470–1481.

Bencala KE (1984) Interactions of solutes and streambed sediment. 2. A dynamic analysis of coupled hydrologic and chemical processes that determine solute transport. *Water Resources Research* 20: 1804–1814.

Bencala KE (1993) A perspective on stream–catchment connections. *Journal of the North American Benthological Society* 12: 44–47.

Bencala KE (2006) Hyporheic exchange flows. In: Anderson MG (ed.) *Encyclopedia of Hydrological Sciences*, part 10, ch. 113 (doi: 10.1002/0470848944.hsa126). Wiley InterScience.

Benner SG, Smart EW, and Moore JN (1995) Metal behavior during surface–groundwater interaction, Silver Bow Creek, Montana. *Environmental Science and Technology* 29: 1789–1795.

Borchardt D and Pusch M (2009) An integrative, interdisciplinary research approach for the identification of patterns, processes and bottleneck functions of the hyporheic zone of running waters. *Advances in Limnology* 61: 1–7.

Boulton AJ, Datry T, Kasshara T, Mutz M, and Stanford JA (2010) Ecology and management of the hyporheic zone: Stream–groundwater interactions of running waters and their floodplains. *Journal of the North American Benthological Society* 29: 25–40.

Boulton AJ, Findlay S, Marmonier P, Stanley EH, and Valett HM (1998) The functional significance of the hyporheic zone in streams and rivers. *Annual Review of Ecology and Systematics* 29: 59–81.

Brosten TR, Bradford J, McNamara JP, et al. (2009) Estimating 3D variation in active-layer thickness beneath arctic streams using ground-penetrating radar. *Journal of Hydrology* 373: 479–486.

Brown BV, Valett HM, and Schreiber ME (2007) Arsenic transport in groundwater, surface water, and the hyporheic zone of a mine-influenced stream–aquifer system. *Water Resources Research* 43: W11404.

Brunke M and Gonser T (1997) The ecological significance of exchange processes between rivers and groundwater. *Freshwater Biology* 37: 1–33.

Brunke M, Hoehn E, and Gonser T (2003) Patchiness of river–groundwater interactions within two floodplain landscapes and diversity of aquatic invertebrate communities. *Ecosystems* 6: 707–722.

Burkholder BK, Grant GE, Haggerty R, Khangaonkar T, and Wampler P (2008) Influence of hyporheic flow and geomorphology on temperature of a large, gravel-bed river, Clackamas River, Oregon, USA. *Hydrological Processes* 22: 941–953.

Butler JJ, Jr., Zhan X, and Zlotnik VA (2007) Pumping-induced drawdown and stream depletion in a leaky aquifer system. *Ground Water* 45: 178–186.

Cardenas MB, Harvey JW, Packman AI, and Scott DT (2008a) Ground-based thermography of fluvial systems at low and high discharge reveals potential complex thermal heterogeneity driven by flow variation and bioroughness. *Hydrological Processes* 22: 980–986.

Cardenas MB, Wilson JL, and Haggerty R (2008b) Residence time of bedform-driven hyporheic exchange. *Advances in Water Resources* 31: 1382–1386.

Castro NM and Hornberger GM (1991) Surface–subsurface water interactions in an alluviated mountain stream channel. *Water Resources Research* 27: 1613–1621.

Claessens L and Tague CL (2009) Transport-based method for estimating in-stream nitrogen uptake at ambient concentration from nutrient addition experiments. *Limnology and Oceanography Methods* 7: 811–822.

Constantz J (1998) Interaction between stream temperature, streamflow, and groundwater exchanges in alpine streams. *Water Resources Research* 34: 1609–1615.

Constantz J (2008) Heat as a tracer to determine streambed water exchanges. *Water Resources Research* 44: W00D10 (doi:10.1029/2008WR006996).

Cook PG, Lamontagne S, Berhane D, and Clark JF (2006) Quantifying groundwater discharge to Cockburn River, southeastern Australia, using dissolved gas tracers 222Rn and SF6. *Water Resources Research* 42: W10411.

Covino TP and McGlynn BL (2007) Stream gains and losses across a mountain-to-valley transition: Impacts on watershed hydrology and stream water chemistry. *Water Resources Research* 43: W10431.

Cox MH, Mendez GO, Kratzer CR, and Reichard EG (2003) Evaluation of tracer tests on the Upper Santa Clara River, Los Angeles and Ventura Counties, California, during October 1999 and May 2000. *US Geological Survey Water-Resources Investigation Report 03-4277*. http://pubs.er.usgs.gov/usgspubs/wri/wri034277 (accessed April 2010).

Cox MH, Su GW, and Constantz J (2007) Heat, chloride, and specific conductance as ground water tracers near streams. *Ground Water* 45: 187–195.

Dahm CN, Grimm NB, Marmonier P, Valett HM, and Vervier P (1998) Nutrient dynamics at the interface between surface waters and groundwaters. *Freshwater Biology* (3): 427–451.

Dahm CN, Valett H, Baxter CV, and Woessner WW (2007) Hyporheic zones. In: Hauer FR and Lamberti GA (eds.) *Methods in Stream Ecology*, 2nd edn., pp. 119–142. London: Elsevier.

Dent CL, Grimm NB, and Fisher SG (2001) Multiscale effects of surface–subsurface exchange on stream water nutrient concentrations. *Journal of the North American Benthological Society* 20: 162–181.

Dent CL, Grimm NB, Martí E, Edmonds JW, Henry JC, and Welter JR (2007) Variability in surface–subsurface hydrologic interactions and implications for nutrient retention in an arid-land stream. *Journal of Geophysical Research G: Biogeosciences* 112: G04004.

Dole-Olivier MJ (1998) Surface water–groundwater exchanges in three dimensions on a backwater of the Rhone River. *Freshwater Biology* 40: 93–109.

Duff JH and Triska FJ (1990) Denitrification in sediments from the hyporheic zone adjacent to a small forested stream. *Canadian Journal of Fisheries and Aquatic Sciences* 47: 1140–1147.

Essaid H, Zamora C, McCarthyu KA, Vogel JR, and Wilson JT (2008) Using heat to characterize streambed water flux variability in four stream reaches. *Journal of Environmental Quality* 37: 1010–1023.

Fernald AG and Guldan SJ (2006) Surface water–groundwater interactions between irrigation ditches, alluvial aquifers, and streams. *Reviews in Fisheries Science* 14: 79–89.

Fiebig DM and Lock MA (1991) Immobilization of dissolved organic matter from groundwater discharging through the stream bed. *Freshwater Biology* 26: 45–55.

Fleckenstein JH, Niswonger RG, and Fogg GE (2006) River–aquifer interactions, geologic heterogeneity, and low-flow management. *Ground Water* 44: 837–852.

Fuller CC and Harvey JW (2000) Reactive uptake of trace metals in the hyporheic zone of a mining-contaminated stream, Pinal Creek, Arizona. *Environmental Science and Technology* 34: 1150–1155.

Gandy CJ, Smith JWN, and Jarvis AP (2007) Attenuation of mining-derived pollutants in the hyporheic zone: A review. *Science of the Total Environment* 373: 435–446.

Geist DR and Dauble DD (1998) Redd site selection and spawning habitat use by fall Chinook Salmon: The importance of geomorphic features in large rivers. *Environmental Management* 40: 655–669.

Gooseff MN, McKnight DM, and Runkel RL (2004a) Reach-scale cation exchange controls on major ion chemistry of an Antarctic glacial meltwater stream. *Aquatic Geochemistry* 10: 221–238.

Gooseff MN, McKnight DM, Runkel RL, and Duff JH (2004b) Denitrification and hydrologic transient storage in a glacial meltwater stream, McMurdo Dry Valleys, Antarctica. *Limnology and Oceanography* 49: 1884–1895.

Greig SM, Sear DA, and Carling PA (2007) A review of factors influencing the availability of dissolved oxygen to incubating salmonid embryos. *Hydrological Processes* 21: 323–334.

Gu C, Hornberger GM, Herman JS, and Mills AL (2008) Effect of freshets on the flux of groundwater nitrate through streambed sediments. *Water Resources Research* 44: W05415.

Haggerty R, Argerich A, and Martí E (2008) Development of a "smart" tracer for the assessment of microbiological activity and sediment–water interaction in natural waters: The resazurin–resorufin system. *Water Resources Research* 44: W00D01 (doi:10.1029/2007WR006670).

Hannah DM, Malcolm IA, and Bradley C (2009) Seasonal hyporheic temperature dynamics over riffle bedforms. *Hydrological Processes* 23: 2178–2194.

Harbaugh AW, Banta ER, Hill MC, and McDonald MG (2000) MODFLOW-2000, the U.S. Geological Survey modular ground-water model – user guide to modularization concepts and the ground-water flow process. *US Geological Survey Open-File Report 00-92*. Report: http://pubs.er.usgs.gov/usgspubs/ofr/ofr200092; software: http://water.usgs.gov/nrp/gwsoftware/modflow.html (accessed April 2010).

Harvey JW and Bencala KE (1993) The effect of streambed topography on surface–subsurface water exchange in mountain catchments. *Water Resources Research* 29: 89–98.

Harvey JW and Fuller CC (1998) Effect of enhanced manganese oxidation in the hyporheic zone on basin-scale geochemical mass balance. *Water Resources Research* 34: 623–636.

Harvey JW, Newlin JT, and Krupa SL (2006) Modeling decadal timescale interactions between surface water and ground water in the central Everglades, Florida, USA. *Journal of Hydrology* 320: 400–420.

Harvey JW and Wagner BJ (2000) Quantifying hydrologic interactions between streams and their subsurface hyporheic zones. In: Jones JB and Mulholland PJ (eds.) *Streams and Ground Waters*, pp. 3–44. London: Elsevier.

Harvey JW, Wagner BJ, and Bencala KE (1996) Evaluating the reliability of the stream tracer approach to characterize stream–subsurface water exchange. *Water Resources Research* 32: 2441–2451.

Hatch CE, Fisher AT, Revenaugh JS, Constantz J, and Ruehl C (2006) Quantifying surface water–groundwater interactions using time series analysis of streambed thermal records: Method development. *Water Resources Research* 42: W10410 (doi:10.1029/2005WR004787).

Hester ET and Gooseff MN (2010) Moving beyond the banks: Hyporheic restoration is fundamental to restoring ecological services and functions of streams. *Environmental Science and Technology* 44: 1521–1525 (doi:10.1021/es902988n).

Hill AR (1990) Ground water flow paths in relation to nitrogen chemistry in the near-stream zone. *Hydrobiologia* 206: 39–52.

Hill AR, Devito KJ, Campagnolo S, and Sanmugadas K (2000) Subsurface denitrification in a forest riparian zone: Interactions between hydrology and supplies of nitrate and organic carbon. *Biogeochemistry* 51: 193–223.

Hill AR, Labadia CF, and Sanmugadas K (1998) Hyporheic zone hydrology and nitrogen dynamics in relation to the streambed topography of a N-rich stream. *Biogeochemistry* 42: 285–310.

Hinkle SR, Duff JH, Triska FJ, et al. (2001) Linking hyporheic flow and nitrogen cycling near the Willamette River – a large river in Oregon, USA *Journal of Hydrology* 244: 157–180.

Hiscock KM and Grischek T (2002) Attenuation of groundwater pollution by bank filtration. *Journal of Hydrology* 266: 139–144.

Huggenberger P, Hoehn E, Beschta R, and Woessner W (1998) Abiotic aspects of channels and floodplains in riparian ecology. *Freshwater Biology* 40: 407–425.

Jones JB and Mulholland PJ (eds.) (2000) *Streams and Ground Waters*. London: Elsevier.

Kim BKA, Jackman AP, and Triska FJ (1992) Modeling biotic uptake by periphyton and transient hyporheic storage of nitrate in a natural stream. *Water Resources Research* 28: 2743–2752.

Konikow LF and Bredehoeft JD (1974) Modeling flow and chemical quality changes in an irrigated stream–aquifer system. *Water Resources Research* 10: 546–562.

Konrad CP (2006) Location and timing of river–aquifer exchanges in six tributaries to the Columbia River in the Pacific Northwest of the United States. *Journal of Hydrology* 329: 444–470.

Krause S, Hannah DM, and Fleckenstein JH (2009) Hyporheic hydrology: Interactions at the groundwater–surface water interface. *Hydrological Processes* 23: 2103–2107 (doi:10.1002/hyp.7366).

Larkin RG and Sharp JM, Jr. (1992) On the relationship between river-basin geomorphology, aquifer hydraulics, and ground-water flow direction in alluvial aquifers. *Geological Society of America Bulletin* 104: 1608–1620.

Lautz LK and Fanelli RM (2008) Seasonal biogeochemical hotspots in the streambed around restoration structures. *Biogeochemistry* 91: 85–104.

Lautz LK and Siegel DI (2006) Modeling surface and ground water mixing in the hyporheic zone using MODFLOW and MT3D. *Advances in Water Resources* 29: 1618–1633.

Leibowitz SG, Wigington PJ, Jr., Rains MC, and Downing DM (2008) Non-navigable streams and adjacent wetlands: Addressing science needs following the Supreme Court's Rapanos decision. *Frontiers in Ecology and the Environment* 6: 364–371.

Malard F, Tockner K, Dole-Olivier M-J, and Ward JV (2002) A landscape perspective of surface–subsurface hydrological exchanges in river corridors. *Freshwater Biology* 47: 621–640.

Malcolm IA, Soulsby C, Youngson AF, and Hannah DM (2005) Catchment-scale controls on groundwater–surface water interactions in the hyporheic zone: Implications for salmon embryo survival. *River Research and Applications* 21: 977–989.

Malcolm IA, Soulsby C, Youngson AF, and Petry J (2003) Heterogeneity in ground water–surface water interactions in the hyporheic zone of a salmonid spawning stream. *Hydrological Processes* 17: 601–617.

Malcolm IA, Soulsby C, Youngson AF, and Tetzlaff D (2009) Fine scale variability of hyporheic hydrochemistry in salmon spawning gravels with contrasting groundwater–surface water interactions. *Hydrogeology Journal* 17: 161–174.

Marion A, Packman AI, Zaramell M, and Bottacin-Busolin A (2008) Hyporheic flows in stratified beds. *Water Resources Research* 44: W09433.

McMahon PB and Bohlke JK (1996) Denitrification and mixing in a stream–aquifer system: Effects on nitrate loading to surface water. *Journal of Hydrology* 186: 105–128.

McMahon PB, Tindall JA, Collins JA, Lull KJ, and Nuttle JR (1995) Hydrologic and geochemical effects on oxygen uptake in bottom sediments of an effluent-dominated river. *Water Resources Research* 31: 2561–2569.

Morrice JA, Valett HM, Dahm CN, and Campana ME (1997) Alluvial characteristics, groundwater–surface water exchange and hydrological retention in headwater streams. *Hydrological Processes* 11: 253–267.

Mulholland PJ and Hill WR (1997) Seasonal patterns in streamwater nutrient and dissolved organic carbon concentrations: Separating catchment flow path and in-stream effects. *Water Resources Research* 33: 1297–1306.

Mulholland PJ, Marzolf ER, Webster JR, Hart DR, and Hendricks SP (1997) Evidence that hyporheic zones increase heterotrophic metabolism and phosphorus uptake in forest streams. *Limnology and Oceanography* 42: 443–451.

Mulholland PJ and Webster JR (2010) Nutrient dynamics in streams and the role of J-NABS. *Journal of the North American Benthological Society* 29: 100–117.

Niswonger RG, Prudic DE, Fogg GE, Stonestrom DA, and Buckland EM (2008) Method for estimating spatially variable seepage loss and hydraulic conductivity in intermittent and ephemeral streams. *Water Resources Research* 44: W05418 (doi:10.1029/2007WR006626).

Nowack B, Xue H, and Sigg L (1997) Influence of natural and anthropogenic ligands on metal transport during infiltration of river water to groundwater. *Environmental Science and Technology* 31: 866–872.

Packman AI and Bencala KE (2000) Modeling surface–subsurface hydrological interactions. In: Jones JB and Mulholland PJ (eds.) *Streams and Ground Waters*, pp. 45–80. London: Elsevier.

Payn RA, Gooseff MN, McGlynn BL, Bencala KE, and Wondzell SM (2009) Channel water balance and exchange with subsurface flow along a mountain headwater stream in Montana, United States. *Water Resources Research* 45: W11427 (doi:10.1029/2008WR007644).

Pinay G, Ruffinoni C, Wondzell S, and Gazelle F (1998) Change in groundwater nitrate concentration in a large river floodplain: Denitrification, uptake, or mixing? *Journal of the North American Benthological Society* 17: 179–189.

Poole GC (2010) Stream hydrogeomorphology as a physical science basis for advances in stream ecology. *Journal of the North American Benthological Society* 29: 12–25.

Power G, Brown RS, and Imhof JG (1999) Groundwater and fish: Insights from northern North America. *Hydrologic Processes* 13: 401–422.

Puckett LJ, Zamora C, Essaid H, et al. (2008) Transport and fate of nitrate at the ground-water/surface-water interface. *Journal of Environmental Quality* 37: 1034–1050.

Runkel RL (1998) One-dimensional transport with inflow and storage (OTIS): A solute transport model for streams and rivers. *US Geological Survey Water-Resources Investigation Report 98–4018*. Report: http://pubs.er.usgs.gov/usgspubs/wri/wri984018; software: http://co.water.usgs.gov/otis (accessed April 2010).

Runkel RL (2002) A new metric for determining the importance of transient storage. *Journal of the North American Benthological Society* 21: 529–543.

Runkel RL (2010) One-dimensional transport with equilibrium chemistry (OTEQ): A reactive transport model for streams and rivers. *US Geological Survey Techniques and Methods Book 6, Chapter B6*. Report: http://pubs.er.usgs.gov/usgspubs/tm/tm6B6; software: http://water.usgs.gov/software/OTEQ (accessed April 2010).

Ruehl C, Fisher AT, Hatch C, Los Huertos M, Stemler G, and Shennan C (2006) Differential gauging and tracer tests resolve seepage fluxes in a strongly-losing stream. *Journal of the North American Benthological Society* 26: 191–206.

Ruehl CR, Fisher AT, Los Huertos M, et al. (2007) Nitrate dynamics within the Pajaro River, a nutrient-rich, losing stream. *Journal of the North American Benthological Society* 26: 191–206.

Runkel RL, McKnight DM, and Rajaram H (2003) Modeling hyporheic zone processes. *Advances in Water Resources* 26: 901–905.

Schindler JE and Krabbenhoft DP (1998) The hyporheic zone as a source of dissolved organic carbon and carbon gases to a temperate forested stream. *Biogeochemistry* 43: 157–174.

Scott DT, Gooseff MN, Bencala KE, and Runkel RL (2003) Automated calibration of a stream solute transport model: Implications for interpretation of biogeochemical parameters. *Journal of the North American Benthological Society* 22: 492–510.

Sharp JM, Jr. (1977) Limitations of bank-storage model assumptions. *Journal of Hydrology* 35: 31–47.

Silliman SE and Booth DF (1993) Analysis of time-series measurements of sediment temperature for identification of gaining vs. losing portions of Juday Creek, Indiana. *Journal of Hydrology* 146: 131–148.

Sophocleous M (2002) Interactions between groundwater and surface water: The state of the science. *Hydrogeology Journal* 10: 52–67.

Squillace PJ, Thurman EM, and Furlong ET (1993) Groundwater as a nonpoint source of atrazine and deethylatrazine in a river during base flow conditions. *Water Resources Research* 29: 1719–1729.

Stanford JA and Ward JV (1988) The hyporheic habitat of river ecosystems. *Nature* 335: 64–66.

Tonina D and Buffington JM (2009) Hyporheic exchange in mountain rivers. I: Mechanics and environmental effects. *Geography Compass* 3: 1063–1086.

Torgersen CE, Price DM, Li HW, and McIntosh BA (1999) Multiscale thermal refugia and stream habitat associations of Chinook salmon in northeastern Oregon. *Ecological Applications* 9: 301–319.

Triska FJ, Kennedy VC, Avanzino RJ, Zellweger GW, and Bencala KB (1989) Retention and transport of nutrients in a third-order stream in northwestern California: Hyporheic processes. *Ecology* 70: 1893–1905.

Valett HM, Dahm CN, Campana ME, Morrice JA, Baker MA, and Fellows CS (1997) Hydrologic influences on groundwater–surface water ecotones: Heterogeneity in nutrient composition and retention. *Journal of the North American Benthological Society* 16: 239–247.

Valett HM, Morrice JA, Dahm CN, and Campana ME (1996) Parent lithology, surface–groundwater exchange, and nitrate retention in headwater streams. *Limnology and Oceanography* 41: 333–345.

Von Gunten HR, Karametaxas G, Krähenbühl U, et al. (1991) Seasonal biogeochemical cycles in riverborne groundwater. *Geochimica et Cosmochimica Acta* 55: 3597–3609.

Ward JV, Malard F, Tockner K, and Uehlinger U (1999) Influence of ground water on surface water conditions in a glacial flood plain of the Swiss Alps. *Hydrological Processes* 13(3): 277–293.

Webster JR and Valett HM (2007) Solute dynamics. In: Hauer FR and Lamberti GA (eds.) *Methods in Stream Ecology*, 2nd edn., pp. 169–185. London: Elsevier.

Westhoff MC, Savenije HHG, Luxemburg WMJ, et al. (2007) A distributed stream temperature model using high resolution temperature observations. *Hydrology and Earth System Sciences* 11: 1469–1480.

Wildman RA, Jr., Domagalski JL, and Hering JG (2009) Hydrologic and biogeochemical controls of river subsurface solutes under agriculturally enhanced ground water flow. *Journal of Environmental Quality* 38: 1830–1840.

Winter TC, Harvey JW, Franke OL, and Alley WM (1998) *Ground Water and Surface Water: A Single Resource*, US Geological Survey Circular 1139. Denver, CO: US Geological Survey.

Woessner WW (2000) Stream and fluvial plain ground water interactions: Rescaling hydrogeologic thought. *Ground Water* 38: 423–429.

Wondzell SM, LaNier J, and Haggerty R (2009a) Evaluation of alternative groundwater flow models for simulating hyporheic exchange in a small mountain stream. *Journal of Hydrology* 364: 142–151.

Wondzell SM, LaNier J, Haggerty R, Woodsmith RD, and Edwards RT (2009b) Changes in hyporheic exchange flow following experimental wood removal in a small, low-gradient stream. *Water Resources Research* 45: W05406.

Wondzell SM and Swanson FJ (1996) Seasonal and storm dynamics of the hyporheic zone of a 4th-order mountain stream. I: Hydrologic processes. *Journal of the North American Benthological Society* 15: 3–19.

Wondzell SM and Swanson FJ (1999) Floods, channel change, and the hyporheic zone. *Water Resources Research* 35: 555–567.

Wörman A, Packman AI, Marklund L, Harvey JW, and Stone SH (2006) Exact three-dimensional spectral solution to surface–groundwater interactions with arbitrary surface topography. *Geophysical Research Letters* 33: L07402 (doi:10.1029/2006GL025747).

Wörman A, Packman AI, Marklund L, Harvey JW, and Stone SH (2007) Fractal topography and subsurface water flows from fluvial bedforms to the continental shield. *Geophysical Research Letters* 34: L07402 (doi:10.1029/2007GL029426).

Zellweger GW (1994) Testing and comparison of four ionic tracers to measure stream flow loss by multiple tracer injection. *Hydrological Processes* 8: 155–165.

Zellweger GW, Avanzino RJ, and Bencala KE (1989) Comparison of tracer dilution and current-meter discharge measurements in a small gravel-bed streams, Little Lost Man Creek, California. *US Geological Survey Water-Resources Investigation Report 89-4150*. http://pubs.er.usgs.gov/usgspubs/wri/wri894150 (accessed April 2010).

Relevant Websites

http://water.usgs.gov
   MODFLOW and Related Programs, US Geological Survey (USGS).

http://co.water.usgs.gov
   OTIS: One-Dimensional Transport with Inflow and Storage, US Geological Survey (USGS).

http://water.usgs.gov
   USGS Water Resources Applications Software: OTEQ.